Handbook of Mitochondrial Dysfunction

Handbook of Mitochondrial Dysfunction

Edited by

Shamim I. Ahmad

CRC Press
Taylor & Francis Group
Boca Raton London New York

CRC Press is an imprint of the
Taylor & Francis Group, an **informa** business

MATLAB® is a trademark of The MathWorks, Inc. and is used with permission. The MathWorks does not warrant the accuracy of the text or exercises in this book. This book's use or discussion of MATLAB® software or related products does not constitute endorsement or sponsorship by The MathWorks of a particular pedagogical approach or particular use of the MATLAB® software.

CRC Press
Taylor & Francis Group
6000 Broken Sound Parkway NW, Suite 300
Boca Raton, FL 33487-2742

First issued in paperback 2021

© 2019 by Taylor & Francis Group, LLC
CRC Press is an imprint of Taylor & Francis Group, an Informa business

No claim to original U.S. Government works

ISBN-13: 978-1-138-33608-7 (hbk)
ISBN-13: 978-1-03-217800-4 (pbk)
DOI: 10.1201/9780429443336

Visit the Taylor & Francis Web site at
http://www.taylorandfrancis.com

and the CRC Press Web site at
http://www.crcpress.com

Dedication

The editor dedicates this book to his late father and mother, Abdul Nasir and Anjuman Ara, who played the important roles to bring him to this stage of academic achievements with their esteemed love, sound care, and sacrifice. Dedication also goes to his wife, Riasat Jan, for her patience and persistent encouragement to produce this book, as well as to his children, Farhin, Mahrin, Tamsin, Alisha, and Arsalan, especially the latter two for providing great pleasure with their innocent interruptions, leading to his energy revitalization. Next, his best wishes go to the ailing patients, suffering from the diseases induced by mitochondrial dysfunction, for their remaining life to run smoothly. Finally his dedication goes to his loving and caring professor Abe Eisenstark, who died in the last week of July 2018 at the age of 99.

Contents

SECTION I Introduction: Structure and Function of Mitochondria

SECTION II Mitochondrial Dysfunction and Diseases

SECTION III Factors Affecting Mitochondrial Function

SECTION IV   Immunity and Toxicity

Preface

Mitochondria are important cellular organelles present ubiquitously in every cell of all eukaryotes with the exception of its absence in mammalian erythrocytes. They also vary in shape, ranging from spherical to elongated and even cup-shaped. They act as a powerhouse, generating energy for the cells from food molecules converted into adenosine triphosphate (ATP) that powers most cell functions such as cell division, synthesis of new cell constituents, osmotic work, nervous conduction, muscle contraction and locomotion, reabsorption in the kidneys, and digestion in the gut; hence an extremely essential component of the body playing roles in health, diseases, and aging. Tricarboxylic acid (TCA) cycle plays most important role in driving these cycles.

Any genetic defect impairing their function, known as Mitochondrial Dysfunction (MD), directly or indirectly has been associated with no less than 80 diseases and syndromes. A number of neurological disorders associated with MD are inflicted by reactive oxygen species (ROS) acting as oxidants, responsible to react rapidly and violently with various cellular components including DNA, proteins, and lipids. Other biological processes of mitochondria include transportation of ATP from inside to outside of the organelles, calcium storage and signalling through mitochondrial ROS, regulations of membrane potentials, participation in apoptosis process, regulation of cellular metabolism, steroid synthesis, and hormonal signalling.

Mitochondrial dysfunctions are induced via two pathways: mutation in mitochondrial DNA (about 15%) and the remaining 85% via mutation in chromosomal DNA leading to their gene products, known as mitochondrial proteins, transported inside mitochondria. Any disruption in any gene, especially by mutation can lead to mitochondrial diseases or syndromes. Examples of diseases linked to mitochondrial gene mutations include Kearns-Sayre syndrome, Pearson syndrome, progressive external ophthalmoplegia, MELAS syndrome, Leber's hereditary optic neuropathy, myoclonic epilepsy with ragged red fibers, and retinitis pigmentosa as well Complex 1 to Complex 4 deficiency. The second pahway leading human diseases is due to mutation in nuclear genes leading to dysfunction of mitochondrial proteins. Diseases identified are Alzheimer's disease, amyotrophic lateral sclerosis, autism, Brugada syndrome, endocrinopathy, cardiomyopathy, epilepsy, sensory axonal neuropathy, pyruvate dehydrogenase deficiency, schizophrenia, dementia, mitochondrial myopathies, and Parkinson's disease. Others include Friedreich's ataxia, hereditary spastic paraplegia, kidney disease, Wilson's disease, Barth syndrome, bipolar disorder, stroke, cardiovascular disease, chronic fatigue syndrome, migraine headaches, mitochondrial cytopathy, neurological myopathy, cholangitis, coronary artery disease, fibromyalgia, Leigh syndrome, diabetes mellitus, and deafness. In addition, the Mitochondrial Theory of Aging points to MD.

MD covers a vast volume of information and it is almost impossible to cover them all in this book; hence, a selection of important diseases as well other features associated with them has been addressed. These have been divided in four sections. **Section 1** includes chapters on the structure and function of mitochondria; **Section 2** covers the selected diseases and syndromes affected by MD; **Section3** includes factors affecting mitochondrial function; and **Section 4** concludes with immunity and toxicity.

In **Section 1**, three chapters are presented by M. R. de Oliveira, S. Roy, and their associates; they intricately describe in detail the mitochondrial structure and function (**Chapter 1**) as well as fundamental aspects of structure and functional features of mitochondria alongside the updated information on their evolutionary facets (**Chapter 2**). Recent findings of the unique features of mitochondrial genome, its damage, and various mechanisms involved in maintenance and repair of this molecule can be studied in **Chapter3**.

Section 2 is the major part of the book with 20 chapters. In **Chapter 4**, K. Tonska intricately covers the human diseases and syndrome induced by nuclear genes through MD. In **Chapter 5**, J. Finsterer presents his esteemed knowledge and skills to describe Kearns-Sayre syndrome (KSS). DNA Deletion of 1000 to 10,000 nucleotides of mitochondrial DNA seems to be the most common reason leading to a number of phenotypic alterations in patients. In **Chapter 6**, S. Claypool et al. have intricately addressed another complex disease, Barth syndrome, which is caused by mutation(s) in the TAZ (tafazzin) gene on human chromosome Xq28. A multisystem disorder characterized by cardiac and skeletal myopathy and neutropenia. No cure is available and morbidity and myopathy in patients is high. Another common human disease, Alzheimer, is addressed by I. Onyango in **Chapter 7**. Brain mitochondria being especially venerable to ROS-induced mitochondrial DNA damage and repair enzymes, also damage to respiratory chain proteins and signaling pathways are the main cause of most neurodegenerative diseases including Alzheimer. In **Chapter 8** E. Miller, using her comprehensive knowledge and polished skill, presents the importance of MD in multiple sclerosis. In **Chapter 9**, A. Erin and her colleagues critically argue the role of MD in kidney function and discussed the distinct form of mitochondrial injury in chronic kidney disease (CKD). In **Chapter 10**, C. Panis and her team highlight various aspects of human breast cancer including physiological aspects, mitochondrial gene's role in breast cancer and mitophagy. In **Chapter 11**, G. Lund presents association between MD and atherosclerosis. This subject is further explored in **Chapter 12** by D. Fu-Dai, including cardiac aging, ischemia reperfusion injury, hypertensive cardiomyopathy, atherosclerosis, and heart failure. In **Chapter 13**, J. Jan discusses a vital role of mitochondria, that is, the decision-making process to determine whether a cell will die or live. In **Chapter 14**, N. Bhatraju and A. Agrawal link obesity and asthma with MD. Hearing loss is another feature of MD, and this is highlighted in **Chapter 15** by M. de la Luz Sordo. In **Chapter 16**, J. Azadmanesh and G. Borgstahl have addressed the importance of oxidants, ROS, and nitrogen species with special emphasis on superoxide and the antioxidants, manganese, and copper/zinc super oxide dismutases in protecting the cells and mitochondria, being damaged, which can lead to several kinds of neurodegenerative diseases. In **Chapter 17**, Vismaya et al. describe the role of post-translational modifications in mitochondrial complex-1 and implication of MD in human brain pathologies. How hemeoxygenase-1 (HO-1) performs its function and can serve as a novel therapeutic is discussed in **Chapter 18** by Lin-Pei Hui and colleagues. In **Chapter 19**, D. Blitzer and his group have shown the importance of mitochondrial transplantation in protecting the patients from ischemic reperfusion injury. In **Chapter 20**, E. Johnson addresses the role of MD in diabetes complications focusing on mitochondrial function in their reduction. **Chapter 21** examines clinical manifestation of MD, specifically in children. Epilepsy is another commonly prevailing multisystem disease and the role of mitochondria in this disease is critically addressed in **Chapter 22** by P. S. Bindu. In **Chapter 23**, K. Khanna and A. Agrawal show that, alongside genetic components, environmental factors (allergens, irritants, and infectious agents) can modulate the allergic phenotype and that mitochondria plays a connecting link between these factors and its pathology.

Section 3 covers the factors affecting mitochondrial functions. In **Chapter 24**, M. R. de Oliveira explains the value of carnosic acid, which can be obtained from the rosemary plant, can prevent mitochondrial impairment in brain cells both *in vivo* and *in vitro*. In **Chapter 25**, T. Seene debates about mitochondrial function in muscle fibers with different oxidative capacity and its interaction with contractile apparatus. Mitochondria play a key role in bioenergetics and redox status maintenance in brain cells, as described by M. R. de Oliveira in **Chapter 26**. Heavy exercise can induce MD and appears as subclinical syndromes such as chronic fatigue syndrome, heart abnormalities, immune disorder, muscle injuries, and brain dysfunction; this is highlighted in **Chapter 27** by

S. M. Ostojic. In **Chapter 28**, C. Ortez and A. Naschave have shown that due to loss of energy production in MD, skeletal muscle and peripheral nerves can lead to exercise intolerance, cramps, recurrent myoglobinuria, or fixed weakness and occasional effect on extra ocular muscles with droopy eyelids (ptosis), progressive external ophthalmoplegia, peripheral ataxia and peripheral polyneuropathy. It can also be induced by a variety of stressors such as DNA damage, oxidative stress, and mitochondrial stress and the ROS is primarily and specifically involved in the signaling pathway; this is addressed in **Chapter 29** by M. Velarde and I. N. Tan. The cellular hallmarks of obesity such as inflammation and endoplasmic reticulum stress can promote MD in the central nervous system and in peripheral organs. Thus, in **Chapter 30**, it is claimed by D. Albuquerque and colleagues that MD in obesity is, on one hand a hallmark of the pathophysiology of obesity, and on other a trigger *per se* of obese phenotype. W. Manucha and his group in **Chapter 31** have identified the chronoregulatory hormone, melatonin, which plays important role in mitochondrial signaling pathways. J. Perry and his colleagues in **Chapter 32** have attempted to highlight the key roles of mitochondria in diabetes complications with emphasis on neuropathy. **Chapter 33** is the last of this section, in which S. Qiazhu and H. Wang have presented the recent breakthrough on mitochondrial pathophysiological functions of the crucial role of mitochondrial crosstalk in the control of metabolic homeostasis and their link with metabolic diseases.

Section 4 deals with the immunity and toxicity linked with mitochondria. In **Chapter 34**, M. Mohsin and his co-contributors have addressed that certain viruses have the ability to modulate the host cell signaling. In a recent study, a new function of mitochondria was discovered; this is discussed by M. Boyer-Guittaut and colleagues in **Chapter 35**. In **Chapter 36**, organophosphate, which is commonly used in agriculture and domestic applications, is shown by A. Hargreaves to have ability to induce acute and delayed neurotoxicity. In **Chapter 37**, Paola Giunti and her co-contributors have elegantly addressed a rare form of human genetic disease – Friedreich's Ataxia, which is associated with mitochondrial dysfunction.

It is anticipated that the materials presented in this book will not only capture the eyes of many thousands of scientists working in this field but also medics, teachers, nurses, libraries, and, more importantly, drug companies striving to find treatments and drugs that will slow down the pathophysiology of diseases or altogether cure the patients from these yet remaining incurable diseases.

Shamim I Ahmad, BSc, MSc, PhD

Acknowledgment

The editor cordially acknowledges the authors of this book for their contributions and in-depth knowledge, skill, and professional presentation. Without their input, it would not have been possible to produce this book on this highly topical and important subject. The editor would also like to acknowledge the staff, especially Jennifer Blaise and Chuck Crumly of CRC Press, for their hard work, friendly approach, and patience, and also CRC Press/Taylor & Francis Group for their efficient and highly professional handling of this project. Finally, he wishes to acknowledge his university for providing this platform and especially the IT service staff members, Iftikhar Mehdi and Ryan Flynn. Finally, acknowledgment is presented to Ananth Ganesan and the staff of the Production Division for their most professional input.

About the Editor

 **Shamim I. Ahmad** obtained his Master's degree in Botany from Patna University, Bihar, India, and his PhD in Molecular Genetics from Leicester University, England. He then joined Nottingham Polytechnic as a grade 1 lecturer and subsequently was promoted to the post of Senior Lecturer. Ahmad now spends much of his time producing/writing medical books. For more than three decades he researched different areas of molecular biology/genetics including thymineless death in bacteria, genetic control of nucleotide catabolism, development of anti-AIDs drug, control of microbial infection of burns, phages of thermophilic bacteria, and microbial flora of Chernobyl after the nuclear power station accident. But his main interest is DNA damage and repair specifically by near-ultraviolet light mostly through the photolysis of biological compounds, production of reactive oxygen species, and their implications on human health including skin cancer. He also investigates near-ultraviolet photolysis of non-biological compounds such as 8-metoxypsoralen, mitomycin C, and their importance in psoriasis treatment and in Fanconi anemia. He has collaborated with the University of Osaka, Japan, and he and his colleagues were able to show that a number of naturally occurring enzymes were able to scavenge the reactive oxygen species.

In 2003 he received a prestigious Asian Jewel Award in Britain for Excellence in Education. His long time ambition to produce medical books started in 2007 and since then he has published *Molecular mechanisms of Fanconi Anemia, Molecular Mechanisms of Xeroderma Pigmentosum; Molecular Mechanisms of Cockayne Syndrome; Molecular Mechanisms of Ataxia Telangiectasia; Diseases of DNA repair; Neurodegenerative diseases, Diabetes: an Old Disease a New Insight; Obesity: a Practical Guide; Thyroid: Basic Science and Clinical Practice, and Ultraviolet light in Human Health, diseases and environment;* also as a co-author the book on Diabetes: *A comprehensive Treatise for patients and Caregivers.* His most recent books are *Reactive Oxygen Species in Biology and Human Health* and *Aging: Exploring a Complex Phenomenon.*

Contributors

Rosella Abeti
Ataxia Centre
Department of Clinical and Movement Neurosciences
University College London
Queen Square Institute of Neurology
London, United Kingdom

Puja Agarwal
Department of Botany
UGC Center for Advanced Studies
The University of Burdwan
West Bengal, India

Anurag Agrawal
Molecular Immunogenetics Laboratory
CSIR Institute of Genomics and Integrative Biology
Delhi, India

Academy of Scientific and Innovative Research
Delhi, India

Baylor College of Medicine
Houston, Texas, USA

David Albuquerque
Research Center for Anthropology and Health (CIAS)
University of Coimbra
Department of Life Sciences
Coimbra, Portugal

Centre for Biomedical Research (CBMR)
University of Algarve
Faro, Portugal

Daniel Álvarez-Vaca
Hospital Universitario Doctor Peset
Valencia, Spain

Célia Aveleira
Center for Neurosciences and Cell Biology (CNC)
University of Coimbra
Coimbra, Portugal

Arianna F. Anzmann
McKusick-Nathans Institute of Genetic Medicine
Johns Hopkins School of Medicine
Baltimore, Maryland, USA

Jahaun Azadmanesh
Department of Biochemistry and Molecular Biology
Nebraska Medical Center
Omaha, Nebraska, USA

Annalisa Baccaro
Ataxia Centre
Department of Clinical and Movement Neurosciences
University College London
Queen Square Institute of Neurology
London, United Kingdom

Muchukunte Mukunda Srinivas Bharath
Department of Neurochemistry
National Institute of Mental Health and
    Neurosciences
Bangalore, Karnataka, India

Neurotoxicology Laboratory
Neurobiology Research Centre
Bangalore, Karnataka, India

Department of Biosciences
Hemagangothri P.G Centre
University of Mysore
Hassan, Karnataka, India

Naveen K. Bhatraju
Molecular Immunogenetics Laboratory
CSIR Institute of Genomics and Integrative Biology
Delhi, India

Michał Bijak
Faculty of Biology and Environmental Protection
Department of General Biochemistry
University of Lodz
Lodz, Poland

David Blitzer
Department of Cardiac Surgery
Harvard Medical School
Boston Children's Hospital
Boston, Massachusetts, USA

Gloria E. O. Borgstahl
Department of Biochemistry and
    Molecular Biology
Eppley Institute for Cancer and Allied Diseases
Nebraska Medical Center
Omaha, Nebraska, USA

Michaël Boyer-Guittaut
Université Bourgogne Franche-Comté
INSERM, EFS BFC, UMR1098
Interactions Hôte-Greffon-Tumeur/Ingénierie
    Cellulaire et Génique
Besançon, France

DImaCell Platform
Université Bourgogne Franche-Comté
Besançon, France

Sara Carmo-Silva
Center for Neurosciences and Cell
    Biology (CNC)
University of Coimbra
Coimbra, Portugal

Yogachar Chithra
Neurotoxicology Laboratory
Neurobiology Research Center
Bangalore, Karnataka, India

Steven M. Claypool
Department of Physiology
Johns Hopkins School of Medicine
Baltimore, Maryland, USA

Thayse Fachin Cormanique
Laboratory of Tumor Biology
State University of West Paraná
Francisco Beltrão, Brazil

Dao-Fu Dai, MD, PhD
University of Iowa
Iowa City, Iowa, USA

Sayanti De
Department of Botany
UGC Center for Advanced Studies
The University of Burdwan
West Bengal, India

Taylor N. Dennis
Department of Biochemistry
University of California
Riverside, California, USA

Emiliano Diez
Instituto de Medicina y Biología Experimental
    de Cuyo (IMBECU)
Consejo Nacional de Investigaciones Científicas
    y Tecnológicas (CONICET)
Mendoza, Argentina

Instituto de Fisiología, Departamento de
    Morfofisiología
Facultad de Ciencias Médicas
Universidad Nacional de Cuyo
Mendoza, Argentina

Pu Duann
Research and Development
Salem Veteran Affairs Medical Center
Salem, Virginia, USA

Angela Dziedzic
Faculty of Biology and Environmental Protection
Department of General Biochemistry
University of Lodz
Lodz, Poland

Alfonso Eirin
Division of Nephrology and Hypertension
Mayo Clinic
Rochester, Minnesota, USA

Salem Veteran Affairs Medical Center
Salem, Virginia, USA

Josef Finsterer
Krankenanstalt Rudolfstiftung
Messerli Institute
Veterinary University of Vienna
Vienna, Austria

Paola Giunti
Ataxia Centre
Department of Clinical and Movement Neurosciences
University College London
Queen Square Institute of Neurology
London, United Kingdom

Alvise Guariento
Department of Cardiac Surgery
Harvard Medical School
Boston Children's Hospital
Boston, Massachusetts, USA

Alan J. Hargreaves
Interdisciplinary Biomedical Research Centre
School of Science and Technology
Nottingham Trent University
Nottingham, UK

Eric Hervouet
Université Bourgogne Franche-Comté
INSERM, EFS BFC, UMR1098
Interactions Hôte-Greffon-Tumeur/Ingénierie Cellulaire
    at Génique
Besançon, France

DImaCell Platform
Université Bourgogne Franche-Comté
Besançon, France

EPIGENEXP Platform
Université Bourgogne Franche-Comté
Besançon, France

Hasan Imam
Division of Infectious Diseases
School of Medicine

University of California
San Diego, California, USA

Maria V. Irazabal, MD
Division of Nephrology and Hypertension
Mayo Clinic
Rochester, Minnesota, USA

Elise Jacquin
INSERM UMR1231 Lipides, Nutrition, Cancer
Université Bourgogne Franche-Comté
Digon, France

Jan Ježek
Department of Molecular Biology
Rowan University
School of Osteopathic Medicine
Stratford, New Jersey, USA

Eric L. Johnson, M.D.
Department of Family and Community Medicine
School of Medicine and Health Sciences
University of North Dakota
Grand Forks, North Dakota, USA

Jennifer Jossart
Department of Biochemistry
University of California
Riverside, California, USA

Rodrigo Kern
Laboratory of Tumor Biology
State University of West Paraná
Francisco Beltrão, Brazil

Mohsin Khan
Division of Infectious Diseases
School of Medicine
University of California
San Diego, California, USA

Kritika Khanna
Molecular Immunogenetics Laboratory
CSIR Institute of Genomics and Integrative Biology
Delhi, India

Elias A. Lianos
Division of Nephrology
Department of Medicine
Virginia Tech Carilion School of Medicine and
    Research Institute
Roanoke, Virginia, USA

Pei-Hui Lin
Davis Heart and Lung Research Institute and
    Department of Surgery
The Ohio State University

Columbus, Ohio, USA

Gertrud Lund
Department of Genetic Engineering
CINVESTAV Irapuato Unit
Irapuato, Gto., Mexico

María de la Luz Arenas Sordo MD, PhD
Instituto Nacional de Rehabilitación LGII
Universidad Nacional Autónoma de México
México City, México

Martin Magner
Department of Paediatrics and Adolescent Medicine
General University Hospital and First Faculty of Medicine
Charles University in Prague
Prague, Czech Republic

Kalyan Mahapatra
Department of Botany
UGC Center for Advanced Studies
The University of Burdwan
West Bengal, India

Walter Manucha
Instituto de Medicina y Biología Experimental de Cuyo
    (IMBECU)
Consejo Nacional de Investigaciones Científicasy
    Tecnológicas (CONICET)
Mendoza, Argentina

Instituto de Fisiología, Departamento
    de Morfofisiología
Facultad de Ciencias Médicas
Universidad Nacional de Cuyo
Mendoza, Argentina

Department of Cellular and Structural Biology
University of Texas Health Science at
    San Antonio
San Antonio, Texas, USA

Área de Farmacología. Departamento de Patología
Facultad de Ciencias Médicas
Universidad Nacional de Cuyo
Mendoza, Argentina

James D. McCully
Department of Cardiac Surgery
Harvard Medical School
Boston Children's Hospital
Boston, Massachusetts, USA

Elzbieta Miller
Department of Physical Medicine
Medical University of Lodz
Lodz, Poland

Neurorehabilitation Ward
General Hospital no III
Lodz, Poland

Saiful Anam Mir
Division of Infectious Diseases
School of Medicine
University of California
San Diego, California, USA

Mehali Mitra
Department of Botany
UGC Center for Advanced Studies
The University of Burdwan
West Bengal, India

Feres José Mocayar Marón
Área de Química Biológica
Departamento de Morfofisiología
Facultad de Ciencias Médicas
Universidad Nacional de Cuyo
Mendoza, Argentina

Andrés Nascimento
Neuromuscular Pathology Unit
Paediatric Neurology Department
Hospital Sant Joan de Déu
Center for Biomedical Research on Rare Diseases
Instituto de Salud Carlos III
Barcelona, Spain

Pedro J. del Nido
Department of Cardiac Surgery
Harvard Medical School
Boston Children's Hospital
Boston, Massachusetts, USA

Clévio Nóbrega
Centre for Biomedical Research (CBMR)
University of Algarve
Faro, Portugal

Center for Neurosciences and Cell Biology (CNC)
University of Coimbra
Coimbra, Portugal

Department of Biomedical Science and Medicine (DCBM)
University of Algarve
Portugal

Algarve Biomedical Center (ABC)
University of Algarve and University Hospital of Algarve
Portugal

Marcos Roberto de Oliveira
Grupo de Estudos em Neuroquímica e Neurobiologia de Moléculas Bioativas
Universidade Federal de Mato Grosso (UFMT
Cuiaba, MT, Brazil

Isaac G. Onyango, DVM, PhD
Gencia Biotechnology
Charlottesville, Virginia, USA

Carlos Ortez
Neuromuscular Pathology Unit
Paediatric Neurology Department
Hospital Sant Joan de Déu
Center for Biomedical Research on Rare Diseases
Instituto de Salud Carlos III
Barcelona, Spain

Sergej M. Ostojic
Faculty of Sport and Physical Education
University of Novi Sad
Novi Sad, Serbia
and University of Belgrade School of Medicine
Belgrade, Serbia

Letícia Madureira Pacholak
Laboratory of Tumor Biology
State University of West Paraná
Francisco Beltrão, Brazil

Carolina Panis
Laboratory of Tumor Biology
State University of West Paraná
Francisco Beltrão, Brazil

J. Jefferson P. Perry
Department of Biochemistry
University of California
Riverside, California, USA

School of Biotechnology
Amrita University
Kollam, Kerala, India

Russel J. Reiter
Department of Cellular and Structural Biology
University of Texas Health Science at San Antonio
San Antonio, Texas, USA

Sujit Roy
Department of Botany
UGC Center for Advanced Studies
The University of Burdwan
West Bengal, India

Joanna Saluk-Bijak
Faculty of Biology and Environmental Protection
Department of General Biochemistry
University of Lodz
Lodz, Poland

Bindu Parayil Sankaran
Department of Neurology
NIMHANS
Bangalore, India

Thalita Basso Scandolara
Laboratory of Tumor Biology
State University of West Paraná
Francisco Beltrão, Brazil

Teet Seene
Institute of Sport Sciences and Physiotherapy
University of Tartu
Tartu, Estonia

Borami Shin
Department of Cardiac Surgery
Harvard Medical School
Boston Children's Hospital
Boston, Massachusetts, USA

Qiaozhu Su
Institute for Global Food Security
School of Biological Sciences
Queen's University Belfast
Belfast, UK

Arun B. Taly
Department of Neurology
NIMHANS
Bangalore, India

Irene L. Tan
University of the Philippines Diliman
Institute of Biology
Quezon City, Philippines

Katarzyna Tońska
Faculty of Biology
Institute of Genetics and Biotechnology
University of Warsaw
Warszawa, Poland

Michael C. Velarde
University of the Philippines Diliman
Institute of Biology
Quezon City, Philippines

Hilary Vernon
McKusick-Nathans Institute of Genetic Medicine
Johns Hopkins School of Medicine
Baltimore, Maryland, USA

Vismaya
Department of Neurochemistry
National Institute of Mental Health and Neurosciences
Bangalore, Karnataka, India

Hao Wang
Institute for Global Food Security
School of Biological Sciences
Queen's University Belfast
Belfast, UK

Silvio Zaina
Department of Medical Sciences
Division of Health Sciences
University of Guanajuato
León, Mexico

Section I

Introduction

Structure and Function of Mitochondria

1 Mitochondrial Structure and Function

Marcos Roberto de Oliveira

Grupo de Estudos em Neuroquímica e Neurobiologia de Moléculas Bioativas, Universidade Federal de Mato Grosso (UFMT), Av. Fernando Corrêa da Costa, Cuiaba, MT, Brazil

CONTENTS

1 INTRODUCTION

The mitochondria exhibit a double-membrane structure that is crucial to the role these organelles present in the fungi, plant, and animal cells [1]. The outer and the inner mitochondrial membranes (OMM and IMM, respectively) limit the intermembrane space (IMS), whose function will be discussed here [1] (Figure 1). The IMM presents several cristae, which are folds of the IMM extended into the mitochondrial matrix, a compartment in which metabolic enzymes and signaling agents are found [2]. Moreover, mitochondria contain DNA in the mitochondrial matrix [3]. The mitochondrial DNA is usually a circular molecule (16,569-bp) and multiple copies are present in each organelle [4]. Both mitochondrial and nuclear genomes are necessary in order to maintain the mitochondrial function [5]. Actually, the mitochondrial genome may encode more than 13 proteins, which are associated with mitochondrial function in humans [6]. Furthermore, the human mitochondrial genome encodes 24 structural RNAs, which are necessary in the translation process in the mitochondria [6]. Nonetheless, the mitochondria depend to a great extent on the nuclear genome in order to maintain the complete expression of their proteins (it has been estimated that more than 1,000 mitochondria-located proteins are encoded by the nuclear genome) [3–5].

The major function of mitochondria is to produce adenosine triphosphate (ATP) from adenosine diphosphate (ADP) and inorganic phosphate (Pi) in the oxidative phosphorylation (OXPHOS) system [7] (Figure 2). Besides, mitochondria play a crucial role in the calcium (Ca^{2+}) ions homeostasis and in the heme biosynthesis [1,7]. The electron transfer chain (ETC), which is found in the IMM, is part of the OXPHOS system and comprises the complexes I (NADH dehydrogenase), II (succinate dehydrogenase), III (coenzyme Q – cytochrome c reductase), and IV (cytochrome c oxidase) [7]. In addition, there are mobile components responsible for the transfer of electrons between the complexes, namely coenzyme Q (so called ubiquinone) and cytochrome c [8]. Coenzyme Q carries electrons from the complexes I and II, as well as from other proteins, to the complex III, from which cytochrome c transfers one electron at a time to the complex IV [9]. The transfer of electrons between the complexes is associated with the pumping of protons (H^+) from the mitochondrial matrix to the IMS by the complexes I, III, and IV [7]. These protein complexes function as H^+ pumps and use the energy released from the flux of electrons in the ETC to generate an electrochemical gradient across the IMM [7,9]. This is the mitochondrial membrane potential (MMP) and it depends on the function of the complexes and on the integrity of the mitochondrial membranes in order to be generated. The accumulation of H^+ in the IMS is necessary to the complex V synthesizes ATP from ADP and Pi [7,9]. Therefore, loss of MMP represents a decrease ability in the mitochondria to produce ATP and also a signal of mitochondrial dysfunction due to several types of insults [7,9].

The major sources of electrons to the ETC include the reduced forms of nicotinamide adenine dinucleotide (NADH) and flavin adenine dinucleotide ($FADH_2$), which contains electrons obtained from several metabolic substrates that are oxidized in certain metabolic pathways, such as the tricarboxylic acid cycle (TCA, the so-called Krebs cycle), the oxidation of fatty acids and ketone bodies, and the oxidation of some α-ketoacids obtained from the degradation of amino acids, depending on the cell type [7]. The glycolysis is also an indirect source of electrons to the ETC by the electron shuttles (namely malate-aspartate shuttle and glycerol-3-phosphate shuttle), since there is not a specific transport to NADH from the cytoplasm to the mitochondrial matrix [10,11]. In the complex IV, these electrons bind to oxygen (O_2), the final electron acceptor, and water is produced [7]. Therefore, O_2 is necessary to mammalian cells in order to amplify the production of ATP in the mitochondria. Without O_2, it is impossible to mitochondria continue

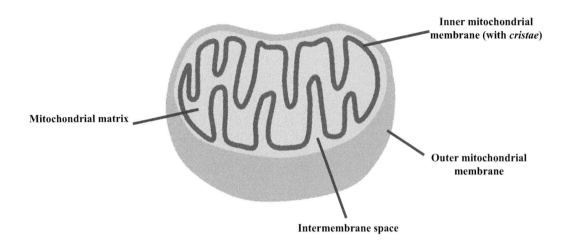

Figure 1 The mitochondrial morphology: The mitochondria contain the outer and the inner mitochondrial membranes (OMM and IMM, respectively) and the intermembrane space (IMS) surrounded by those membranes. The IMM contains the *cristae* that extend into the mitochondrial matrix. The mitochondrion image was obtained from Pixabay website, which gently distributes free images to be used commercially.

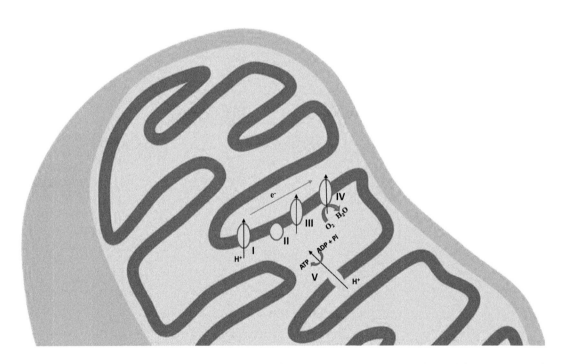

Figure 2 An overview of the mitochondrial oxidative phosphorylation (OXPHOS) system. The OXPHOS system comprises the electron transfer chain (ETC) and the complex V (ATP synthase/ATPase). The mitochondrion image was obtained from Pixabay website, which gently distributes free images to be used commercially.

oxidizing the metabolic substrates and the production of ATP declines. In this context, electron leakage from the ETC leads to the production of ROS, mainly the superoxide anion radical ($O_2^{-\bullet}$) [12,13].

Actually, the mitochondria are the main source of ROS in the mammalian cells [13]. The manganese-dependent superoxide dismutase (Mn-SOD) enzyme (which is located in the mitochondria) converts $O_2^{-\bullet}$ into hydrogen peroxide (H_2O_2), a non-radical species [13]. H_2O_2 then reacts with glutathione peroxidase (GPx), giving rise to water [14]. Other enzymes can metabolize H_2O_2, decreasing the chance this molecule would react with iron and cupper ions (in the Fenton chemistry reaction) or with $O_2^{-\bullet}$ (in the Haber-Weiss reaction) [13–15]. Mitochondrial impairment in the brain is cause of concern due to the increase in the production of ROS it promotes in a very unstable environment [16]. Brain consumes O_2 at very high rates and presents high concentrations of transition metals (mainly iron and cupper), as well as contains lipids in elevated concentrations [17]. Despite this, the brain contains low concentrations of antioxidants [17]. Therefore, the brain is very sensitive to redox disturbances originated or not from the mitochondria. The strategies aiming to promote mitochondrial protection involves upregulation of antioxidant enzymes located in the organelles and induction of the synthesis of reduced glutathione (GSH), the major non-enzymatic antioxidant in mammalian cells, which is found at high concentrations inside the mitochondria [18,19]. Recently, several works published by our research group have demonstrated a crucial role for the transcription factor nuclear factor erythroid 2-related factor 2 (Nrf2) in modulating mitochondrial protection [20–27]. Nrf2 is the master regulator of the redox environment in mammalian cells and presents a role in the maintenance of mitochondrial function and dynamics [28–30]. Nrf2 modulates the expression of the antioxidant enzymes glutathione peroxidase (GPx), glutathione reductase (GR), γ-glutamylcysteine ligase (γ-GCL), and heme oxygenase-1 (HO-1), to cite a few [29]. Moreover, Nrf2 regulates the expression of phases II (conjugation) and III (excretion) detoxification enzymes, including glutathione-S-transferase (GST) and multidrug resistance-associated proteins (MRP), among others [31].

Mitochondrial biogenesis, so called mitogenesis, is the biological event in which new mitochondria are generated in mammalian cells [32]. Mitochondrial biogenesis involves the activation of a signaling pathway modulated by the adenosine monophosphate-activated kinase (AMPK), sirtuin 1 (SIRT1, NAD^+-dependent deacetylase), and peroxisome proliferator-activated receptor-γ coactivator-1α (PGC-1α), as previously reviewed [33]. The activation of PGC-1α causes the upregulation of the transcription factors nuclear respiratory factor-1 (NRF-1) and nuclear respiratory factor-2 (NRF-2) and of the estrogen-related receptor α (ERRα) [34–36]. These proteins mediate the transcription of nuclear genes encoding mitochondria-related proteins [37]. For example, the expression of the mitochondrial transcription factor A (TFAM), B1 (TFB1M), and B2 (TFB2M) is enhanced and causes modifications necessary in the mitochondrial DNA in order to the synthesis of mitochondrial RNA begins. The activation of AMPK (that occurs by an increase in the levels of AMP) leads to specific modulation in the amounts of NAD^+, which upregulates SIRT1 [38]. Then, SIRT1 deacetylates PGC-1α, activating it and triggering mitochondrial biogenesis, as well as increasing the expression of SOD enzymes, among other effects [38]. Therefore, the AMPK/SIRT1/PGC-1α axis is an important target of studies aiming to discover new drugs able to induce mitochondrial biogenesis mainly in mitochondria-related diseases [39–42].

Contrastingly, mitophagy is the destruction of mitochondria presenting damage and is responsible for maintaining the number of mitochondria during bioenergetics and redox alterations in mammalian cells [43,44]. Mitophagy may be triggered by loss of MMP or by mitochondrial unfolded protein response, which activates PTEN-induced kinase 1 (PINK1) inside mitochondria [45]. PINK1, by phosphorylating its targets in the damaged mitochondria, attracts the Parkin (E3 ubiquitin ligase) protein to the organelles [46]. The ubiquitination of proteins located in the OMM leads to the engulfment of the mitochondria by the autophagosome [45]. Thus, mitophagy maintains a control of quality of mitochondria, removing from the cells dysfunctional organelles [47]. The homeostasis between the synthesis and degradation of mitochondria is of crucial importance to maintain both redox and bioenergetics status in mammalian cells.

Moreover, mitochondrial number may vary due to fusion and fission of the organelles as previously reviewed elsewhere [48]. Mitochondrial fusion and fission also lead to alterations in mitochondrial size and morphology according to the circumstance the cells are facing [48]. The dynamin-related protein Drp1, a GTPase, is a major regulator of mitochondrial fission, whereas mitofusin-1 (Mfn1), mitofusin-2 (Mfn2), and optic atrophy 1 (Opa1) proteins play a central role in modulating mitochondrial fusion [49,50].

2   CONCLUSION

The function of mitochondria is complex and is closely related to the architecture of the organelles. Mitochondrial dysfunction may result from alterations in the structure of the organelles, as well as from disruption in the dynamics of mitochondria. Moreover, specific impairments may occur in the components of the OXPHOS system (mutation, inhibition, others), leading to loss of function and a decline in the bioenergetics status. The mitochondria of brain cells are focus of intense research due to the consequences that mitochondrial impairment causes to neurons and glia. Actually, mitochondrial dysfunction has been observed in several neurodegenerative diseases, such as Alzheimer's disease and Parkinson's disease. In this regard, there is increasing interest in discovering molecules able to promote mitochondrial protection in the diverse deleterious circumstances these organelles face in human cells.

REFERENCES

1. Friedman JR, Nunnari J (2014) Mitochondrial form and function. Nature 505:335–343. doi: 10.1038/nature12985
2. Cogliati S, Enriquez JA, Scorrano L (2016) Mitochondrial cristae: where beauty meets functionality. Trends Biochem Sci 41:261–273. doi: 10.1016/j.tibs.2016.01.001
3. D'Souza AR, Minczuk M (2018) Mitochondrial transcription and translation: overview. Essays Biochem 62:309–320. doi: 10.1042/EBC20170102
4. Copeland WC, Longley MJ (2014) Mitochondrial genome maintenance in health and disease. DNA Repair (Amst) 19:190–198. doi: 10.1016/j.dnarep.2014.03.010

5. Kaniak-Golik A, Skoneczna A (2015) Mitochondria-nucleus network for genome stability. Free Radic Biol Med 82:73–104. doi: 10.1016/j.freeradbiomed.2015.01.013

6. Capt C, Passamonti M, Breton S (2016) The human mitochondrial genome may code for more than 13 proteins. Mitochondrial DNA A DNA Mapp Seq Anal 27:3098–3101. doi: 10.3109/19401736.2014.1003924

7. Papa S, Martino PL, Capitanio G, Gaballo A, De Rasmo D, Signorile A, Petruzzella V (2012) The oxidative phosphorylation system in mammalian mitochondria. Adv Exp Med Biol 942:3–37. doi: 10.1007/978-94-007-2869-1_1

8. Chaban Y, Boekema EJ, Dudkina NV (2014) Structures of mitochondrial oxidative phosphorylation supercomplexes and mechanisms for their stabilisation. Biochim Biophys Acta 1837:418–426. doi: 10.1016/j.bbabio.2013.10.004

9. Nath S, Villadsen J (2015) Oxidative phosphorylation revisited. Biotechnol Bioeng 112:429–437. doi: 10.1002/bit.25492

10. Mráček T, Drahota Z, Houštěk J (2013) The function and the role of the mitochondrial glycerol-3-phosphate dehydrogenase in mammalian tissues. Biochim Biophys Acta 1827:401–410. doi: 10.1016/j.bbabio.2012.11.014

11. McDonald AE, Pichaud N, Darveau CA (2017) "Alternative" fuels contributing to mitochondrial electron transport: importance of non-classical pathways in the diversity of animal metabolism. Comp Biochem Physiol B Biochem Mol Biol IN PRESS doi: 10.1016/j.cbpb.2017.11.006

12. Kalyanaraman B (2013) Teaching the basics of redox biology to medical and graduate students: oxidants, antioxidants and disease mechanisms. Redox Biol 1:244–257. doi: 10.1016/j.redox.2013.01.014

13. Sies H, Berndt C, Jones DP (2017) Oxidative stress. Annu Rev Biochem 86:715–748. doi: 10.1146/annurev-biochem-061516-045037

14. Sies H (2015) Oxidative stress: a concept in redox biology and medicine. Redox Biol 4:180–183. doi: 10.1016/j.redox.2015.01.002

15. Lushchak VI (2014) Free radicals, reactive oxygen species, oxidative stress and its classification. Chem Biol Interact 224:164–175. doi: 10.1016/j.cbi.2014.10.016

16. Angelova PR, Abramov AY (2018) Role of mitochondrial ROS in the brain: from physiology to neurodegeneration. FEBS Lett 592:692–702. doi: 10.1002/1873-3468.12964

17. Cobley JN, Fiorello ML, Bailey DM (2018) 13 reasons why the brain is susceptible to oxidative stress. Redox Biol 15:490–503. doi: 10.1016/j.redox.2018.01.008

18. Lu SC (2013) Glutathione synthesis. Biochim Biophys Acta 1830:3143–3153. doi: 10.1016/j.bbagen.2012.09.008

19. Morris G, Anderson G, Dean O, Berk M, Galecki P, Martin-Subero M, Maes M (2014) The glutathione system: a new drug target in neuroimmune disorders. Mol Neurobiol 50:1059–1084. doi: 10.1007/s12035-014-8705-x

20. de Oliveira MR, Ferreira GC, Schuck PF, Dal Bosco SM (2015) Role for the PI3K/Akt/Nrf2 signaling pathway in the protective effects of carnosic acid against methylglyoxal-induced neurotoxicity in SH-SY5Y neuroblastoma cells. Chem Biol Interact 242:396–406. doi: 10.1016/j.cbi.2015.11.003

21. de Oliveira MR, Peres A, Ferreira GC, Schuck PF, Bosco SM (2016) Carnosic acid affords mitochondrial protection in chlorpyrifos-treated Sh-Sy5y Cells. Neurotox Res 30:367–379. doi: 10.1007/s12640-016-9620-x

22. de Oliveira MR, Peres A, Ferreira GC, Schuck PF, Gama CS, Bosco SMD (2017) Carnosic acid protects mitochondria of human neuroblastoma SH-SY5Y cells exposed to paraquat through activation of the Nrf2/HO-1Axis. Mol Neurobiol 54:5961–5972. doi: 10.1007/s12035-016-0100-3

23. de Oliveira MR, Peres A, Gama CS, Bosco SMD (2017) Pinocembrin provides mitochondrial protection by the activation of the Erk1/2-Nrf2 signaling pathway in SH-SY5Y neuroblastoma cells exposed to paraquat. Mol Neurobiol 54:6018–6031. doi: 10.1007/s12035-016-0135-5

24. de Oliveira MR, Fürstenau CR, de Souza ICC, Da Costa Ferreira G (2017) Tanshinone I attenuates the effects of a challenge with H2O2 on the functions of tricarboxylic acid cycle and respiratory chain in SH-SY5Y cells. Mol Neurobiol 54:7858–7868. doi: 10.1007/s12035-016-0267-7

25. de Oliveira MR, Brasil FB, Andrade CMB (2017) Naringenin attenuates H2O2-induced mitochondrial dysfunction by an Nrf2-dependent mechanism in SH-SY5Y cells. Neurochem Res 2017 Nov;42(11):3341–3350. doi: 10.1007/s11064-017-2376-8

26. de Oliveira MR, de Bittencourt Brasil F, Fürstenau CR (2018) Sulforaphane promotes mitochondrial protection in SH-SY5Y cells exposed to hydrogen peroxide by an Nrf2-dependent mechanism. Mol Neurobiol 55:4777–4787. doi: 10.1007/s12035-017-0684-2

27. de Oliveira MR, Brasil FB, Fürstenau CR (2018) Evaluation of the mitochondria-related redox and bioenergetics effects of gastrodin in SH-SY5Y cells exposed to hydrogen peroxide. J Mol Neurosci 2018 Feb;64(2):242–251. doi: 10.1007/s12031-018-1027-0

28. Hayes JD, Dinkova-Kostova AT (2014) The Nrf2 regulatory network provides an interface between redox and intermediary metabolism. Trends Biochem Sci 39:199–218. doi: 10.1016/j.tibs.2014.02.002

29. Dinkova-Kostova AT, Abramov AY (2015) The emerging role of Nrf2 in mitochondrial function. Free Radic Biol Med 88:179–188. doi: 10.1016/j.freeradbiomed.2015.04.036

30. Esteras N, Dinkova-Kostova AT, Abramov AY (2016) Nrf2 activation in the treatment of neurodegenerative diseases: a focus on its role in mitochondrial bioenergetics and function. Biol Chem 397:383–400. doi: 10.1515/hsz-2015-0295

31. Keum YS (2012) Regulation of Nrf2-mediated phase II detoxification and anti-oxidant genes. Biomol Ther (Seoul) 20:144–151. doi: 10.4062/biomolther.2012.20.2.144

32. Whitaker RM, Corum D, Beeson CC, Schnellmann RG (2016) Mitochondrial biogenesis as a pharmacological target: a new approach to acute and chronic diseases. Annu Rev Pharmacol Toxicol 56:229–249. doi: 10.1146/annurev-pharmtox-010715-103155

33. Scarpulla RC (2011) Metabolic control of mitochondrial biogenesis through the PGC-1 family regulatory network. Biochim Biophys Acta 1813:1269–1278. doi: 10.1016/j.bbamcr.2010.09.019

34. Virbasius CA, Virbasius JV, Scarpulla RC (1993) NRF-1, an activator involved in nuclear-mitochondrial interactions, utilizes a new DNA-binding domain conserved in a family of developmental regulators. Genes Dev 7:2431–2445.

35. Virbasius JV, Virbasius CA, Scarpulla RC (1993) Identity of GABP with NRF-2, a multisubunit activator of cytochrome oxidase expression, reveals a cellular role for an ETS domain activator of viral promoters. Genes Dev 7:380–392.

36. Schreiber SN, Knutti D, Brogli K, Uhlmann T, Kralli A (2003) The transcriptional coactivator PGC-1 regulates the expression and activity of the orphan nuclear receptor estrogen-related receptor alpha (ERRalpha). J Biol Chem 278:9013–9018.

37. Anderson R, Prolla T (2009) PGC-1alpha in aging and anti-aging interventions. Biochim Biophys Acta 1790:1059–1066. doi: 10.1016/j.bbagen.2009.04.005

38. Cantó C, Gerhart-Hines Z, Feige JN, Lagouge M, Noriega L, Milne JC, Elliott PJ, Puigserver P, Auwerx J (2009) AMPK regulates energy expenditure by modulating NAD+ metabolism and SIRT1 activity. Nature 458:1056–1060. doi: 10.1038/nature07813

39. Dorn GW 2nd, Vega RB, Kelly DP (2015) Mitochondrial biogenesis and dynamics in the developing and diseased heart. Genes Dev 29:1981–1991. doi: 10.1101/gad.269894.115

40. Guedes-Dias P, Pinho BR, Soares TR, de Proença J, Duchen MR, Oliveira JM (2016) Mitochondrial dynamics and quality control in Huntington's disease. Neurobiol Dis 90:51–57. doi: 10.1016/j.nbd.2015.09.008

41. Deas E, Wood NW, Plun-Favreau H (2011) Mitophagy and Parkinson's disease: the PINK1-parkin link. Biochim Biophys Acta 1813:623–633. doi: 10.1016/j.bbamcr.2010.08.007

42. Picone P, Nuzzo D, Caruana L, Scafidi V, Di Carlo M (2014) Mitochondrial dysfunction: different routes to Alzheimer's disease therapy. Oxid Med Cell Longev 2014:780179. doi: 10.1155/2014/780179

43. Youle RJ, Narendra DP (2011) Mechanisms of mitophagy. Nat Rev Mol Cell Biol 12:9–14. doi: 10.1038/nrm3028

44. Hamacher-Brady A, Brady NR (2016) Mitophagy programs: mechanisms and physiological implications of mitochondrial targeting by autophagy. Cell Mol Life Sci 73:775–795. doi: 10.1007/s00018-015-2087-8

45. Durcan TM, Fon EA (2015) The three 'P's of mitophagy: PARKIN, PINK1, and post-translational modifications. Genes Dev 29:989–999. doi: 10.1101/gad.262758.115

46. Song M, Mihara K, Chen Y, Scorrano L, Dorn GW 2nd (2015) Mitochondrial fission and fusion factors reciprocally orchestrate mitophagic culling in mouse hearts and cultured fibroblasts. Cell Metab 21:273–286. doi: 10.1016/j.cmet.2014.12.011

47. Twig G, Hyde B, Shirihai OS (2008) Mitochondrial fusion, fission and autophagy as a quality control axis: the bioenergetic view. Biochim Biophys Acta 1777:1092–1097. doi: 10.1016/j.bbabio.2008.05.001

48. Lee H, Yoon Y (2016) Mitochondrial fission and fusion. Biochem Soc Trans 44:1725–1735. doi: 10.1042/BST20160129

49. Westermann B (2012) Bioenergetic role of mitochondrial fusion and fission. Biochim Biophys Acta 1817:1833–1838. doi: 10.1016/j.bbabio.2012.02.033

50. Otera H, Ishihara N, Mihara K (2013) New insights into the function and regulation of mitochondrial fission. Biochim Biophys Acta 1833:1256–1268. doi: 10.1016/j.bbamcr.2013.02.002

2 Structure, Function and Evolutionary Aspects of Mitochondria

Puja Agarwal, Mehali Mitra, and Sujit Roy
Department of Botany, UGC Center for Advanced Studies, The University of Burdwan, Golapbag Campus, Burdwan, West Bengal, India

CONTENTS

1 THE STRUCTURE – FUNCTION FEATURES OF MITOCHONDRIA

The structure of mitochondria varies depending on the cell type. Mitochondria are highly dynamic organelle which can change size, shape and position in a short period of time. Mitochondria can fuse with each other or can split into two according to the energy requirement of the cell. These are called mitochondrial fission and fusion respectively which are highly co-ordinated processes (Westermann, 2010). The balance among the continuous mitochondrial fission and fusion maintains the overall morphology of the cell. Mitochondria are large enough to be seen under light microscope. They are bean shaped, independently floating organelles containing an outer membrane, an inner membrane, cristae and mitochondrial matrix, and mitochondrial genome. Mitochondria contain more than thousand proteins and this organelle is one of the most important parts of the cell as well as for our life as they generate ATP which is required to carry out almost all type of cellular functions. In plant cells mitochondria are predominantly present to carry out photosynthesis (Kageyama et al., 2011).

1.1 MITOCHONDRIAL MEMBRANES

Mitochondria contain two membranes – one inner and the other mitochondrial membranes. The outer mitochondrial membrane acts as an outer boundary of the mitochondria whereas the inner mitochondrial membrane has some inward projections (Fig. 1) which are known as Cristae (Hoppins et al., 2011). The outer part of inner mitochondrial membrane is known as inner boundary membrane. This inner boundary membrane and cristae are joined together by narrow and tubular connections which are known as cristae junctions (Douce and Neuberger, 1989).

The membranes of mitochondria divide the organelle in two aqueous parts; one is the matrix which is present inside the inner membrane and another is intermembrane space

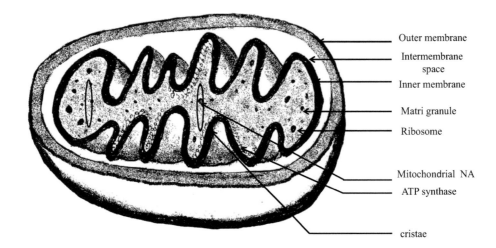

Outer membrane

Intermembrane space

Inner membrane

Matri granule

Ribosome

Mitochondrial NA

ATP synthase

cristae

FIGURE 1   Diagram of mammalian mitochondria. The cristae are folds of inner mitochondrial membrane and on their matrix side they have F_1 particles, ribosome and mitochondrial DNA.

which is present between the outer and inner membrane of mitochondria. The matrix is like a gel due to the presence of thousands of water soluble proteins inside the structure. Half of the outer membrane of mitochondria is composed of lipids by weight and also a mixture of enzyme that have different type of activities. The outer membrane contains porins having large internal channels. These porin channels are permeable to the molecules like Co-A, NAD, ATP when the channels are in wide open configuration. The inner mitochondrial membrane contains 100 non-identical polypeptides and a protein/lipid ratio of 3:1. The inner membrane contains cardiolipin, a phospholipid which have similar characteristics like bacterial plasma membrane. The inner membrane is impermeable to most of the molecules. To enter through inner membrane the molecules need to have a special type of membrane transporter.

1.2   Matrix

Mitochondrial matrix contains mitochondrial DNA, ribosomes, ATP synthase enzyme, soluble enzymes, small organic molecules, nucleotides and their co-factors and inorganic molecules. Also it is more viscous than the cytoplasm of the cell. The ribosomes present in matrix are considerably smaller in size than those present in the cytosol (Sharma et al., 2003). Mitochondria have their own machinery to synthesize their genetic materials. The enzymes in matrix are usually responsible for the reactions that produce ATP and its production takes place through a number of processes like citric acid cycle, oxidation of pyruvate, beta oxidation of fatty acids and oxidative phosphorylation. Several anabolic and catabolic biochemical reactions also take place in the matrix depending on the structures and contents of mitochondria which produce a favourable environment for those reactions (Iborra et al., 2004). In 1963, Nash and Margit

discovered mitochondrial DNA which exists from one to many double stranded DNA molecules and are circular like in most bacterial species. Mitochondrial DNA is rich in Guanine and Cytosine bases and contributes 1% to the total cellular DNA. The matrix has a pH of about 7–8 and the protein content of matrix is 0.8μL/mg.

1.3   Metabolites and Enzymes in Matrix

Mitochondrial matrix contains a wide variety of metabolites. Pyruvate, acetyl CoA, acyl CoA, α-ketoglutarate, isocitrate, succinyl CoA, succinate, malate, fumarate and oxaloacetate are involved in citric acid cycle. L-citrulline, L-ornithine and carbamoyl phosphate are used in urea cycle. Mitochondria DNA, RNA and transfer RNA are used for protein synthesis. Another ions like Ca^{+2}, K^+, Mg^{+2} are also present in the matrix. Moreover CO_2, H_2O, O_2, ATP, ADP and inorganic phosphate are present as metabolites. Enzymes like citrate synthase, Pyruvate dehydrogenase, isocitrate dehydrogenase, aconitase, α-ketoglutarate dehydrogenase, succinyl CoA synthetase, fumerase and malate dehydrogenase facilitates the TCA cycle. Transaminase facilitates amino acid production. β-oxidation uses pyruvate carboxylase, acyl CoA dehydrogenase and β-ketothiolase. The urea cycle is facilitated by carbamoyl phosphate synthetase I and ornithine transcarboxylase.

2   MITOCHONDRIAL FUNCTION

In mitochondria aerobic organisms produce large amount of energy in the form of ATP in presence of O_2 from the two glycolysis products, pyruvate and NADH. Each pyruvate molecule is transported across the inner mitochondrial membrane and then into mitochondrial matrix, where it generates a two carbon acetyl group – CH_3COO^-. The acetyl group along with Co-enzyme A produces acetyl-CoA with simultaneous reduction of NAD^+.

$$Pyruvate + HS\text{-}CoA + NAD^+ \longrightarrow$$

$$Acetyl\ CoA + CO_2 + NADH + H^+$$

The whole process is catalyzed by the multienzyme complex pyruvate dehydrogenase. The acetyl-CoA produced then enters in the citric acid cycle or TCA cycle (Fig. 2). All the enzymes of TCA cycle resides in the mitochondrial matrix except succinate dehydrogenase, which enters through inner mitochondrial membrane (Friedman and Nunnari, 2014; Nunnari and Suomalainen, 2012). In the first step of TCA cycle the acetyl CoA condenses with a 4 carbon molecules oxaloacetate and gives rise to citrate, which is a six carbon molecule (Fermie et al., 2004). This step is catalyzed by citrate synthase. In every cycle this citrate molecule ultimately decreases at chain length giving rise to 4 carbon oxaloacetate molecule so that it can condense with another acetyl-CoA molecule and keeps the cycle going (Rowland and Voeltz, 2012). The two carbons that are released during TCA cycle are completely oxidized to CO_2. The citrate molecule is oxidized to isocitrate by aconitase in the next step. This isocitrate is then converted to a 5-Carbon molecule α-Ketoglutarate along with reduction of a NAD^+ to NADH. In this step one carbon molecule released converts to CO_2. In the whole cycle four

reactions are there in which pair of electrons are transferred to electron accepting co-enzyme. In three of these reactions NAD^+ get reduced to NADH and in another reaction FAD gets reduced to $FADH_2$. The α-Ketoglutarate then gives rise to a 4 carbon molecule succinyl-CoA by α-Ketoglutarate dehydrogenase. This step involves a HS-CoA and another molecule of CO_2 releases from this step. This succinate is ultimately converted to oxaloacetate via fumarate and malate. The rest of the enzymes involved are succinyl Co-A synthase, Succinate dehydrogenase, fumerase and malate dehydrogenase. The overall equation for the TCA Cycle can be summarized in the following two step reactions:

$$Acetyl\ CoA + 2H_2O + FAD + 3NAD^+ + GDP + P_i$$

$$\downarrow$$

$$2CO_2 + FADH_2 + 3NADH + 3H^+ + GTP + HS\text{-}CoA$$

It is evident that every macromolecules of the cell which generally provide energy are broken down in mitochondrial matrix as metabolites of TCA cycle. The acetyl-CoA is the end product of fatty acid cycle. The amino acid catabolism also gives rise to acetyl CoA and glutamate which are major components of TCA cycle. NADH and $FADH_2$ are primary products of TCA cycle (Reyes et al., 2016). Mitochondria are unable to import

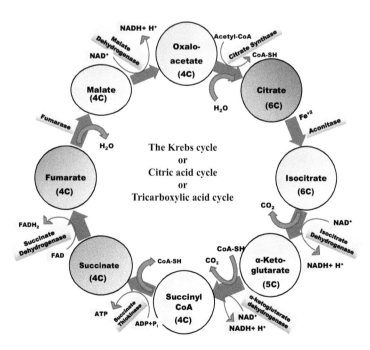

FIGURE 2   **Schematic representation of the tricarboxylic acid cycle (TCA cycle).** It is a series of enzyme-catalysed chemical reactions that form a key part of aerobic respiration in cells. The citric acid cycle – also known as the tricarboxylic acid cycle or the Krebs cycle – is a series of chemical reactions used by all aerobic organisms to release stored energy through the oxidation of acetyl-CoA derived from carbohydrates, fats, and proteins into adenosine triphosphate and carbon dioxide.

cytosolic NADH which is produced from glycolysis. So, the electrons of cytosolic NADH enter the mitochondrial matrix to reduce NAD$^+$ to NADH and FAD TO FADH$_2$. These reduced co-enzymes from TCA cycle are actually responsible for ATP generation in Mitochondria. The reducing equivalents such as NADH and FADH$_2$ which is generated in the TCA cycle are used by mitochondrial electron transport chain to synthesize ATP. By-products from other cycles such as amino acid oxidation and fatty acid oxidation are also used in electron transport chain for energy generation.

$$2H^+ + 2e + \frac{1}{2} O_2 \rightarrow H_2O + Energy$$

A cluster of proteins comprising the ETC [(which is an aggregate of four protein complexes associated with mobile electron carriers (Birsoy et al., 2015)]. Its exact physiological functions for these four complexes are still under debate (Fig. 3). ETC, present in the inner mitochondrial membrane name Cristae, transfer electrons through the membrane creating a gradient of protons that generates energy in the form of ATP. The amount of protons that is pumped from the mitochondrial matrix into the inter-membrane space of the mitochondrion is directly proportional to the ATP production. In ETC, redox reactions take place in which electrons are transferred from the most reduced donor components to an acceptor molecule which creates proton gradient, thereby converting mechanical energy into chemical energy (Junge and Muller, 2011). The proton gradient generates due to "proton motive force' i.e. a chemiosmotic mechanism which drives ATP synthesis.

2.1 COMPLEX-I

The complex-I of ETC is comprised of flavin mono nucleotide derived from vitamin B$_2$, is also known as Riboflavin. Riboflavin is a non-protein molecule required for protein activity. The complex I also contains an iron-sulphur containing protein. Complex I is a large protein complex and is made up of 45 amino acid chain (Efremov and Sazanov, 2011). It can pump 4 hydrogen ions from the matrix into the inter-membrane space across the membrane thereby creating Hydrogen ion gradient between these two mitochondrial compartments separated by inner mitochondrial membrane (Cannino et al., 2012). According to recent research there are four types of NADH dehydrogenase in which two are located on the external face (one NADPH and one oxidizing NADH) and the other two on the inner mitochondrial membrane (Moller, 2002). But the roles of these dehydrogenases have not been confirmed.

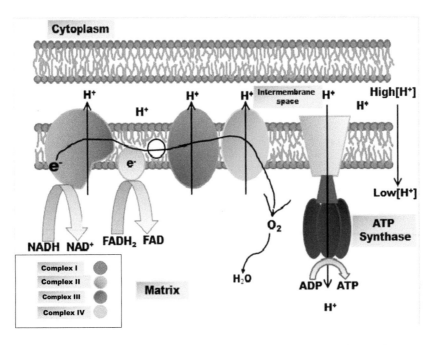

FIGURE 3 **An electron transport chain is a series of complexes that transfer electrons from electron donors to electron acceptors** via redox reactions, and couples this electron transfer with the transfer of protons across a membrane. This diagram shows the flow of electrons and protons (H$^+$) through the four enzyme complexes of the transport chain. Electrons reach quinine (Q) through complex I and II. Q serves as a mobile carrier of electrons and passes them to complex III, which then passes them to cytochrome C. Complex IV then transfers electrons from reduced cytochrome C to O$_2$. Electron flow through complexes I, III and IV is accompanied by proton flow from the mitochondrial matrix to the intermembrane space (cytosolic side).

2.2 COMPLEX-II

Complex II i.e. the enzyme succinate dehydrogenase does not receive electrons from complex I rather it directly receives from $FADH_2$, another by product of TCA cycle (Figueroa et al., 2002). Ubiquinon (Q) connects the first and the second complex to the third one. Ubiquinone can freely move through the hydrophobic core of the membrane as it is lipid soluble (Eubel et al., 2003). The electrons are delivered by reduced ubiquinone (QH_2) to the next complex in the ETC. Electron derived from NADH in complex I and from $FADH_2$ in complex II reduce ubiquinone. The $FADH_2$ bypasses complex I due to which number hydrogen ion pumped from the matrix to the inner mitochondrial membrane is decreased and thereby lowering the number of ATP molecule produced from $FADH_2$ electrons.

2.3 COMPLEX-III

Complex III namely cytochrome oxidoreductase is composed of a group of proteins forming a complex including Cytochrome b, Cytochrome c, a Fe-S protein and a Rieske-centre (2 Fe-S centre).Cytochrome protein contains hemi, which is a prosthetic group and is similar to hemi of haemoglobin (Chan, 2010). This hemi group has iron at its centre which accepts and donates electrons to get reduced and oxidized respectively, obtaining two different oxidation states: Fe^{+2} and Fe^{+3}. Due to different proteins binding the hemi molecule it contributes by slightly different characteristics of each molecule. As electrons are passed from complex III to Cytochrome c, protons are pumped from complex III through the membrane.

2.4 COMPLEX-IV

Cytochrome c, a and a_3 combined is the fourth complex and a and a_3 contains one hemi group each. Cytochrome a_3 contains two copper ions Cu_A and Cu_B. These Cytochromes each are attached with an oxygen molecule between their iron and copper ions until the oxygen is completely reduced by picking up two hydrogen ions available in the surrounding medium to produce water (Bartoli et al., 2000). The formation of water contributes to decrease in hydrogen ion from the system creating an ion gradient used in the process of Chemiosmosis.

3 ATP YIELD

Much of our varied understanding for the ATP yield comes from the catabolism of glucose. The variation could be due to variation between species and the number of hydrogen ions pumped through the membrane in ETC (Ohnishi, 2010). Another reason for this variation may be the transfer of electrons across the membranes of mitochondria which gives rise to ATP yield (Moller and Rasmusson, 1998). The cytosolic NADH generated from glycolysis is not able to enter the mitochondria, which is why the NAD^+/FAD^+ picks up electrons inside mitochondria. These NAD^+ functions as the electron transporter in the liver whereas FAD^+, which generates fewer ATP molecules functions are in the brain.

3.1 ATP SYNTHASE

ATP synthase is composed of proton channel and catalytic unit (Fig. 4). It consists of two components F_0 & F_1. F_0 is a proton channel and F_1 has a catalytic property (von Ballmoos et al., 2008). ATP synthase is a large membrane embedded multi-subunit protein complex, which appears as a balloon and stick (Ferguson, 2010). F_1 subunit consists of five types of polypeptide chain: α-3 chain, β-3 chain, γ- chain, δ-chain & ε- chain (Davies et al., 2012).

α and β subunits are arranged in a hexameric ring of F_1 subunit. β subunits participate directly in catalytic process related to generation of ATP (Fung et al., 2013). Central stock consists of two proteins γ & ε. Each of the β subunits interact with the different phases of α. F_0 is the proton channel situated in the inner mitochondrial membrane; consist of a ring composed of 10–14 C subunits. "A" subunit binds to the outside of this ring. Proton channel depends on both the "A" subunits and C ring for maintaining the structure-function integrity. "A" subunit of F_0 component contains two proton half channels (von Ballmoos et al., 2009.)

3.2 THE BINDING CHAIN MECHANISM

According to this mechanism the role of proton motive force is not to form ADP but to release it from the ATP synthase. ATP does not leave the catalytic site unless proton flows through the enzyme. Hutton and Boyer (1979) proposed a binding chain mechanism, according to which changes in the properties of 3 β subunits allow sequential ADP-P_i binding, ATP synthesis and ATP release. Interaction with the different phases of γ-subunit makes 3 β subunits distinguishable. One β subunit has T or tight conformation which has a high affinity for ATP and it converts bound ADP and P_i to ATP. But this conformation cannot release ATP. Second β subunit is in the L or Loose conformation, binds ADP and P_i but cannot synthesize ATP. The final subunit is in O or open conformation. This conformation releases bound nucleotide. This is associated with the rotation of γ-subunit by 120° which interconvert the 3 β subunits.

4 HEAT PRODUCTION

Under certain conditions protons are diffused into the matrix without contributing two ATP synthesis, in which potential energy of the proton electro chemical gradient is

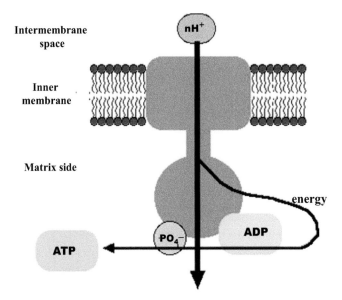

FIGURE 4   ATP synthase is that enzyme which generates adenosine triphosphate which is the most commonly used "energy currency' of cells for all organisms. It is formed from adenosine diphosphate and inorganic phosphate. There are three functional units of ATP synthase; the rotor turns 1200 for every H^+ that crosses the membrane using the molecular "carousel' called the c ring. The catalytic head piece contains the enzyme active site in each of the three β subunits. The stator consists of the α subunit imbedded in the membrane which contains two half channels for protons to enter and exit F_0 component, and a stabilizing arm.

unutilized, releasing heat energy. Thermogenin (33 kDa protein) carry out this process and is commonly found in brown fat or brown adipose tissue responsible for non-shivering thermogenesis (Yun and Finkel, 2014).

5   STORAGE OF CALCIUM IONS

In the cell, concentration of calcium regulates different array of signalling pathways and mitochondria, by storing calcium, contributes to cell-calcium homeostasis. Mitochondria act as group "cytosolic buffer" for calcium as it can rapidly take in and release calcium. Endoplasmic reticulum is also an important site for storing calcium and this element interacts with mitochondria during calcium storage (Christian and Spremulli, 2012). Mitochondrial calcium unit porter (MCU), stored in the inner mitochondrial membrane serves as a passage for movement of calcium across the mitochondrial membrane into the matrix. Movement of calcium is driven by membrane potential of mitochondria. Calcium acts as a secondary messenger to regulate several physiological processes such as muscle contraction, neuronal transmission, movement of flagella and sillia, fertilization, cell proliferation, saliva secretion and apoptosis. Higher level of calcium in mitochondrial matrix also activates a key enzyme of Kreb's cycle, isocitrate dehydrogenase. Mitochondria also play some other important roles such as certain porphyrin heme synthesis, steroid synthesis, hormonal signalling by action of mitochondrial oestrogen receptors in brain and heart, and detoxification of ammonia in liver.

6   UNCOUPLERS AND INHIBITORS

Much studies about electron transport chain and respiratory ATP synthesis shows that there are a few chemical compounds which interfere with one or both of these processes. Early workers described two classes of chemicals that affect electron transport and ATP generation and these were termed as uncouplers and Inhibitors.

6.1   UNCOUPLERS

Uncouplers such as DNP (2,4-dinitrophenol), paratrifluro-methoxy hydrazone etc prevent synthesis of ATP without interfering in electron transport to proceed smoothly (Formentini et al., 2012). but this flow cannot be coupled to ATP synthesis. DNP and para-trifluromethoxy hydrazone are lipophilic in nature and can easily pass through the cytoplasmic membrane (Garlid et al., 2000). They are also acidic in nature and can bind with proton to carry inside from outside of the membrane; thus they prevent generation of the proton gradient.

6.2   INHIBITORS

Inhibitors are chemical compound that blocks electron transport chain and ATP generation including cyanide, carbon monoxide and azide. The inhibitors can bind two iron centre of the Cytochrome and block the electron transport. Without electron transport there will be no proton gradient, so no ATP can be synthesized.

7   MITOCHONDRIAL EVOLUTION

Few years back it was the notion that mitochondria were the sites of aerobic respiration only, a complex biochemical process in which oxidation of pyruvate to CO_2 takes place generating reduced cofactors which drive electron transport chain using chemiosmotic mechanism to synthesize ATP. However, research carried out in the past half a century has shown that mitochondria can perform an array of different functions and which vary among different eukaryotic lineages. Despite this diversity, it is hypothesized that every mitochondrion originated from a single common ancestral organelle that was formed by integration of an endosymbiotic alpha proteobacteria into a host cell related to Archaea. Also the transition of the endosymbiotic bacterium to become permanent organelle took a very large number of evolutionary changes resulting in hundreds of genes and a protein import system being originated, insertion of membrane transporters, integration of reproduction and metabolism, genome reduction, endosymbiotic and lateral gene transfer and the retargeting of proteins. Although research has allowed us to gain many insights into this transition, still controversy remains regarding the original endosymbiont and its nature, its initial interactions with the host and integration into the host based on structural, physiological and genetic backgrounds and also the timing of its integration relative to the origin of other features of eukaryotic cells. In the process of establishment of the organelle many genes and proteins may have been lost, gained and retargeted as mitochondria have diverged in form and function along diverse branches of eukaryotic tree of life.

8   THE ENDOSYMBIOTIC THEORY

The process of integration of a bacterium as an endosymbiont into a eukaryotic host and its transformation into an organelle is explained by two different themes as the "archezoan scenario" and the "symbiogenesis scenario" (Koonin, 2010). In the classical archezoan scenario a eukaryote amitochindriate archezoan served as a host in the proto-mitochondrial endosymbiosis event. This hypothesis is based on small-subunit rRNA sequences (Doolittle, 1980; Margulis, 1970), whereas in the symbiogenesis scenario, alpha-proteobacteriumis was engulfed by an archea, which led to the generation of protomitochondria subsequently followed by compartmentalization of eukaryotic cell (Martin and Muller, 1998).

The transition phase of an autonomous endosymbiotic alphaproteobacteria to a mitochondrial cenancestor (Fig. 5) entailed several significant evolutionary changes including: (1) small molecule transporter machineries were inserted into the endosymbiont inner membrane, (2) evolution and modification of the protein-import machinery, (3) reduction of genome through the loss of irrelevant genes, (4) endosymbiotic gene transfer (EGT) to the nucleus, (5) endosymbiont cell envelope was modified, (6) biochemical pathways and systems between host and symbiont became integrated, (7) origin of an organelle division mechanism to ensure segregation of mitochondrial compartments into daughter cells during cell division that was coordinated with the host-derived cell cycle, (8) specialization of cristae, (9) evolution of contact sites between the endomembrane system and the proto-mitochondria, (10) proteins having diverse origins and localizations were retargeted, and (11) origin of anchors between mitochondria and the cytoskeleton (Roger et al., 2017). These complex alterations that took place in the transformation of pre-mitochondrial alphaproteobacterium were necessarily incremental for the establishment of mitochondria.

Evolutionarily, the origin of the mitochondrial protein import machinery and the mitochondrial envelope are most important changes in organellogenesis. One scenario proposes that first the evolution of mitochondrial protein import machinery happened to insert solute carriers into the proto-mitochondrial inner membrane (Smith, 2006, 2009). These carriers took part in the movement of small molecules such as nucleotides (ATP, ADP), anionic metabolites, amino acids and inorganic ions across the mitochondrial inner membrane. Many inner membrane transporters belongs to eukaryotic "mitochondrial carrier family' (MCFs) which may have originated from a single ancestral carrier that was inserted into the inner membrane of the protomitochondria (Amiri et al., 2003; Andersson et al., 2003; Smith, 2006). Another general view instead suggests that the protein transport machinery evolution took place to insert host proteins into the outer membrane of the proto-mitochondria (Gross and Bhattacharya, 2011). If the latter theory was true, the evolution of the protein import machinery remained advantageous and had happened to gain control over the biogenesis of the proto-mitochondrial endosymbiont envelope.

Fundamentally the biogenesis of mitochondria depends on import and incorporation of protein and lipids into the mitochondrial membranes. Nucleus-encoded mitochondrial proteins in modern mitochondria are delivered into the mitochondria after they are translated by ribosomes in the cytosol (Hewitt et al., 2011; Wiedemann and Pfanner, 2017). Most of the proteins in the mitochondria have 10–100 amino acids in their amino-terminal which are positively charged amphipathic alpha-helical presequences this functions as mitochondrial targeting sequences and are essential for their import (Wiedemann and Pfanner, 2017). Most mitochondrial proteins are imported into the intermembrane space through the Translocase of the Outer mitochondrial Membrane (TOM) complex. Proteins may also be inserted into the outer membrane by the Sorting and Assembly Machinery (SAM) complex, and then they are

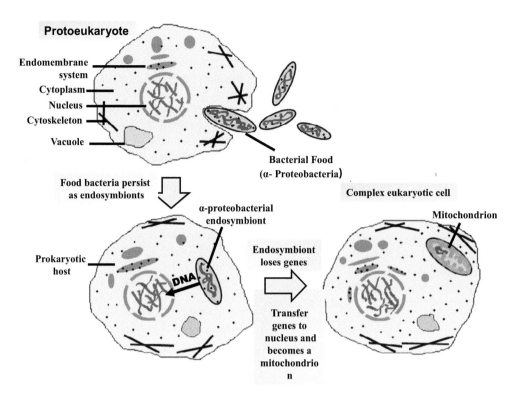

FIGURE 5   Endosymbiotic origin of mitochondria. Mitochondria arose once in evolution, and their origin entailed an endosymbiosis accompanied by gene transfers from the endosymbiont to the host. The endosymbiotic hypothesis for the origin of mitochondria suggests that mitochondria are descended from specialized bacteria (probably purple nonsulfur bacteria) that somehow survived endocytosis by another species of prokaryote or some other cell type, and became incorporated into the cytoplasm. The ability of symbiont bacteria to conduct cellular respiration in host cells that relied on glycolysis and fermentation would have provided a considerable evolutionary advantage. The number of environments in which the cells could survive, has been illustrated.

folded and oxidized by the Mitochondrial Import and Assembly (MIA) machinery to remain in the intermembrane space. They can be further translocated across the mitochondrial inner membrane into the mitochondrial matrix through TIM23–PAM complex, or directly inserted into the inner membrane by the TIM22 insertase (Wiedemann and Pfanner, 2017). Proteins in the matrix can be inserted into the inner membrane by the OXA complex. Proteins in the inner mitochondrial membrane generated by the mitochondrial genome are also inserted by the Cytochrome Oxidase Assembly (OXA) complex (Wiedemann and Pfanner, 2017). Mitochondrial target presequences in the amino-terminus are cleaved off in the imported proteins by a mitochondrial processing peptidase (MPP).

The presence of contact sites between the endomembrane system – mainly the endoplasmic reticulum (ER) – and mitochondria is necessary for the import of lipids into mitochondrial membranes and also the assembly of outer membrane β-barrel proteins (Dimmer and Rapaport, 2017; Ellenrieder et al., 2017). This contact sites are made by the four-subunit ERMES complex, which functions both as a lipid transfer complex and a tether

(Dimmer and Rapaport, 2017; Ellenrieder et al., 2017). ERMES (Endoplasmic reticulum mitochondria endomembrane system) subunits share no known homology with the prokaryotic system, but it was likely present in Last Eukaryotic Common Ancestor (LECA), although it has been lost in major eukaryotic lineages like metazoans and green plants (Wideman and Gomez, 2016). But the loss of ERMES is compensated by the presence of an alternative tethers between the endomembrane system and mitochondria like EMC or vCLAMPs (Dimmer and Rapaport, 2017; Wideman and Gomez, 2016).

Modification of the existing alpha proteobacterial protein export and membrane protein insertion systems gave rise to the mitochondrial protein import machinery. The SAM, TIM23 – PAM, OXA and MPP complexes (or subunits of them) have alphaproteobacterial homologs (Hewitt et al., 2011, 2014). During protomitochondrial evolution these complexes gathered additional subunits and also acquired specific roles. Besides biogenesis of mitochondria, the proto-eukaryote had to consider about the segregation of the mitochondrial compartments into daughter cells during cell division, and to regulate mitochondrial distribution throughout the cell. The division of pre-mitochondrial alphaproteo-bacterium takes place by using a contractile Z ring comprised of polymerized FtsZ protein. Mutually antagonistic system of Min proteins and other bacterial division systems (an array of proteins anchoring the ring to the cell membrane), determined the location of the ring. Among all these components, the FtsZ protein was retained by the mitochondrial cenancestor which underwent duplication prior to LECA (Leger et al., 2015). These proteins are encoded in the nucleus in the extant eukaryotes. Additionally, it had also gained an external dynamin ring which can help in the constriction of the mitochondria at the mid-point (Purkanti and Thattai, 2015). Eukaryotic dynamins also help in vesicle fission. They may have diversified from an ancestral bacterial gene by either being inherited from the pre-mitochondrial alphaproteobacterium and present in the proto-eukaryote host (pre-endosymbiosis), or acquired by Lateral Gene Transfer during proto-eukaryotic evolution.

Mitochondrial cristae that are invaginations in the inner membrane specialized for aerobic respiration also evolved during organellogenesis. Cristae improve the efficiency of aerobic respiration by increasing the surface area for housing large numbers of respiratory complexes. Two main factors are responsible for the development of cristae. First, morphology of cristae is due to the presence of multimers of ATP synthase complexes arranged along cristae membranes which bends it. This capability of the ATP synthase to dimerize and bend cristae is absent in alphaproteobacteria. It appears to have occurred in the proto-mitochondria during the phase of evolution (Kuhlbrandt, 2015). The second factor is the Mitochondrial Contact Site and Cristae Organizing System (MICOS). A small "necks" (cristae junctions) is created by MICOS (multiprotein complex) that compartmentalize cristae and anchors them to the mitochondrial envelope. The core Mic60 subunit of MICOS has an alphaproteobacterial homolog, and the entire complex evolved during the later phase of endosymbiosis (Gomez et al., 2015; Huynen et al., 2016). The core components of MICOS present before endosymbiosis suggests that the respiratory cristae of mitochondria could have evolved from the bioenergetic membrane invaginations known amongst alphaproteobacteria (Gomez et al., 2017).

In modern eukaryotes, membrane transporters and redox/metabolite shuttles coordinated the metabolic pathways and biosynthetic systems in mitochondria and the cytosol. The iron-sulphur (Fe/S) cluster biogenesis machinery is a good example of the pro-eukaryote nucleocytoplasm becoming obligatory dependent on the symbiont-derived system. Almost all modern eukaryotes share a conserved "iron-sulphur cluster" (ISC) system, majority of which have alphaproteobacterial origins (Freibert et al., 2017), present in the mitochondrial matrix comprising 18 proteins (Braymer and Lill, 2017).

Besides synthesizing Fe/S cluster and attaching them to mitochondrial apoproteins, the ISC system is also essential for the synthesis of Fe/S clusters in cytosolic and nuclear Fe/S proteins involved in key pathways (e.g. ribosome assembly and function, nuclear DNA replication and repair) (Braymer and Lill, 2017).

When a free-living bacterium initiates a symbiotic relationship with another cell, it is bathed in the metabolic intermediates of its host. And as long as the symbiont shares the host's metabolites some of its genes are reduced and neutralized by the host's biochemical activities. Subsequently endosymbiotic lifestyle lead to reduction in genome size of proeubacteria compared to alphaproteobacteria (Andersson, 1998; Fraser et al., 1995; 1997, 1998; Hemmelreich, 1997; Stephens et al., 1998). Genes whose functions are dispensable or can be performed by the host nuclear genome are more likely to be deleted from the endosymbiont genome. These genes are involved in functions like amino acid biosynthesis, nucleoside biosynthesis, anaerobic glycolysis and its regulation etc (Andersson, 1998). By comparing genomes of *Rickettsia prowazekki* and *Rickettsia americana* some additional genes were also found to have been lost during the process of evolution of mitochondrial genome. These genes are named as operational genes and are involved in cofactor biosynthesis, fatty acid and phospholipid metabolism, energy and intermediary metabolism, cell envelope synthesis and cell division (Gabaldon and Huynen, 2003). By sequencing and analyzing a complete mitochondrial genome it is quite clear that elimination of DNA from mitochondrial genome is an ongoing process and certain genes have been lost in more than one episode.

9 MITOCHONDRIAL ORIGIN: GENOME AND PHYLOGENETIC ANALYSIS

It is hypothesised that at least one thousand million years ago, the ancestor of modern bacteria collided with an ancestral eukaryote and evolved to form mitochondria. Initially the alphaproteobacteria contained all the genes necessary for its survival. But gradually with the establishment of the organelle genome modification of the endosymbiont took place. According to sequence homology, mitochondria and chloroplast evolved from free living eubacteria via symbiosis but by different pathways (Margulis, 1970). Mitochondria share its origin with purple bacteria while chloroplast shares it with cyanobacteria. Studies of mitochondrial DNA sequences and its expression confirm that mitochondria are direct descendants of alpha-division of the proteobacterial ancestors. Members of Rickettsial subdivision, a group of obligate parasites that includes the genera *Anaplasma, Ehrlichia* and *Rickettsia* are closest eubacterial relative of mitochondria. Phylogenetic reconstructions and distance measurements carried out on the basis of sequences of

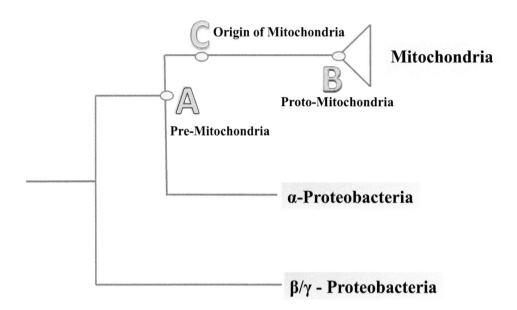

FIGURE 6   Key time points in mitochondrial evolution. Time point A represents pre-mitochondria, the last common ancestor of mitochondria and alpha-proteobacteria. Time point B represents proto-mitochondria, the last common ancestor of all contemporary mitochondria. Time point C represents the origin of mitochondria.

cytochrome c oxidase and cytochrome b posits that mitochondria diverged from bacteria between 1.5 and 2.0 billion years ago (Sicheritz-Ponten et al., 1998). Accordingly, this supports the idea that oxidative respiratory system that was introduced into the eukaryotic lineages was an endosymbiotic alphaproteobacterium (Gray and Doolittle, 1982; Gray et al., 1989, Gray, 1999; Andersson et al., 1998; Sicheritz-Ponten et al., 1998).

Since 1970s and 80s technological progress in the field of DNA cloning and sequencing has been giving a detailed knowledge about mitochondrial genomics and about its origin from eubacteria (Fig. 6). Complete sequencing of ~16kb mitochondrial genome (Anderson et al., 1981, 1982; Bibb et al., 1981) from several mammalian species established that mtDNA encodes protein subunits of electron transport chain, ATP synthase and also small and large subunits of rRNA and tRNA components for translation purpose within the mitochondria. Sequencing of mtDNA from non-mammalian species showed difference in size, physical form, coding capacity, organizational patterns and modes of expression across the eukaryotic domain (Gray et al., 1998). Notably some additional respiratory and ribosomal proteins are encoded by mtDNA of some non-animal taxa. On the other hand, mitochondrial genome of some other taxa is reduced in size loosing many genes for proteins and tRNA. Phylogenetic reconstructions were done based on rRNA gene sequences to trace the evolutionary origin of mitochondria. Although cytosolic and prokaryotic rRNAs vary in size and secondary structure they resemble to some extent in primary

sequence and they can be incorporated into the aligned sequence databases for phylogenetic reconstruction purpose (Cedergren et al., 1988; Gray et al., 1989). Initial phylogenetic tree showed that plant mitochondrial rRNA genes emanated from the alphaproteobacteria and not from Archaebacteria or Eukarya (Schnare and Gray, 1982; Spencer et al., 1984). Subsequently phylogenetic reconstruction done on the basis of mitochondrial and bacterial rRNA sequences have pinpointed Rickettsiales order within alphaproteobacteria (Williams et al., 2007) as a bacterial lineage from which mitochondria evolved (Gray, 1998; Gray and Spencer 1996). Although phylogenetic analysis have confirmed the mitochondria-Rickettsiales connection (Gupta, 1995; Lang et al., 1999; Sicheritz-Ponten et al., 1998; Viale and Arakaki, 1994) it is still questionable whether the two are sister groups or whether mitochondria really branch within the order Rickettsiales (Williams et al., 2007). Still other opinions are that other alphaproteobacterial orders rather than Rickettsiales should also be considered as possible source for mitochondrial origin (Esser et al., 2004). The specific connection between mitochondria and Rickettsiales in phylogenetic tree is still not clear as the inferred relationship may be due to phylogenetic artefacts caused by high rate of sequence divergence and elevated A+T content present in the genome of Rickettsiales taxa and mitochondria, which can also be called LBA (long branch attraction) artefact. Genomic and phylogenetic evidences support monophyletic assemblage of mitochondrial genome. Firstly, as it is considered that mitochondrial genome is a highly reduced version of a much bigger alphaproteobacterial progenitor

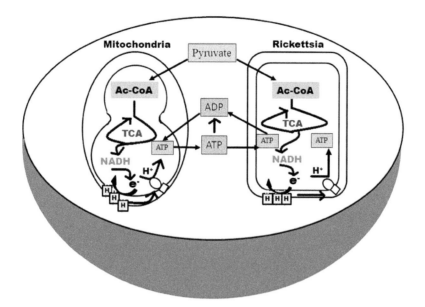

FIGURE 7 Schematic illustration of bio-energetic machineries in mitochondria and *Rickettsia*. ATP production in eukaryotes from glucose and oxygen normally consists of two catabolic processes: the conversion of glucose to pyruvate via a glycolytic pathway, and the oxidative conversion of pyruvate to H2O and CO2. Both Rickettsia and mitochondria rely for pyruvate on the glycolytic systems of their host cells. More than 100 mitochondrial enzymes are involved in the oxidation of pyruvate. The ATP produced by this system is exported into the cytoplasm by the ATP/ADP translocases, which transport 1 molecule of ATP in exchange for 1 molecule of ADP. The Rickettsia genome contains a similar cohort of proteins involved in the pyruvate dehydrogenase complex, the TCA cycle, the respiratory chain complex, and the ATP synthase complex. There are in addition five genes coding for ATP/ADP translocases, but these are required for the import of ATP from the host cell cytosol. Thus, Rickettsia and mitochondria have functionally related systems for ATP production.

genome, but it is not acceptable that independent bacterial symbiont would individually converge on the same small repertoire of ribosomal- protein and respiratory-chain genes. Secondly in many plants the mitochondrial DNA and ribosomal protein genes are clustered together in same transcriptional order as they present in corresponding alphabacterial operon. However in this bacterial operon some additional genes are present which are missing in the mitochondrial gene clusters. These genes were either deleted from the mitochondrial gene clusters or were shifted to the nucleus as a result of endosymbiotic gene transfer. Same deleted version of mitochondrial gene clusters occurs throughout the eukaryotes establishing that these deletions were already present in the common ancestor of mitochondrial genome (Gray et al., 1999).

Amongst eukaryotic groups, mitochondrial proteomes considerably vary in content (Gabaldon and Huynen, 2003; Gabaldon and Pittis, 2015; Szklarczyk and Huynen, 2010). Latest estimation projects that only 10–20% mitochondrial proteome have alphaproteobacterial connection, and an additional 20–30% proteomes of mitochondria have proteobacterial affinity (Gabaldon and Huynen, 2007; Szklarczyk and Huynen, 2010) among which some might have lost their alphaproteobacterial signature and others may be introduced by lateral gene transfer. Approximately 40% mitochondrial proteome have no prokaryotic homologs

(Szklarczyk and Huynen, 2010) and many of the genes encoding these proteins originated before the establishment of cenancestral mitochondrion and divergence of modern eukaryote, in the proto-mitochondrial phase (Gray, 2015).

10 CONCLUSION

Until past few years, very sparse information was available about complete mtDNA sequences. And phylogenetic analysis with this limited information was not fruitful. From this skewed data, tracing the ancestral mitochondrial genome was not easy. Recent database gathered through comprehensive mitochondrial genome sequencing programs has expanded the knowledge about mitochondrial genome evolution. And it is now evident that mtDNA is of two basic types, ancestral and derived. The promitochondrial genome resembles eubacterial ancestry, example 69,034 bp mtDNA of *Rickettsia americana*. In the ancestral mtDNA many additional genes are present including NADH: ubiquinone oxidoreductase (compexI), succinate: ubiquinone oxidoreductase (complex II), ATP synthase (compex V), small and large ribosomal protein genes, and almost complete set of tRNA genes. As well its genome, containing mainly of coding sequences with none or few introns, is tightly packaged gene clusters being arranged like eubacteria. Derived mitochondrial DNA have diverged from the ancestral pattern, with little or no characteristics of ancestral type accompanied by reduction in genome size due to extensive gene loss both in tRNA and protein coding genes. Some other changes in the derived mitochondrial DNA involves, (i) extensive rate of divergence in the sequence and secondary structure of rRNA and protein coding genes, (ii) usage of highly biased codons in protein genes, also elimination of certain codons, (iii) adoption of certain nonstandard codons (Nedelcu and Lee, 1998). Indeed the distinction between the ancestral and derived mitochondrial genome is clear but still some land plants mtDNA share similarity with the ancestral pattern genome content by enlarging the unidentified open reading frames, noncoding sequences and introns (Unseld et al., 1997). Exploring additional sequences of mitochondrial DNA and comparing the ancient mitochondrial genome of early diverged plants with the derived genome of recently diverged plants, it is clear that with time mitochondrial DNA has become recombinationally active and progressively diverged from the ancestral type. The process of divergence is also supported by RNA editing, fragmentation and dispersion of genes, gene transfer to nucleus, gene loss, and incorporation of genes from chloroplast and nucleus.

Phylogenetic studies proved a robust connection between mitochondria and Rickettsiales. Both Rickettsia and mitochondria depends on pyruvate for ATP production and both contains a similar cohort of proteins involved in the pyruvate dehydrogenase complex, the TCA cycle, the respiratory chain complex, and the ATP synthase complex. However, ATP transport is somewhat different in these two systems (Fig. 7). Definitive genome sequencing and data defining mitochondrial proteome of yeast identified 50 mitochondrial proteins that are closely related to alphaproteobacteria. This supports the identification of alphaproteobacteria as ancestor of mitochondria (Degli Esposti et al., 2016). Functional profiling of mitochondrial protein in alphaproteobacteria and eukaryotic homologue explains that bacterial homologues seem to be mainly involved in energy metabolism and translation. In contrast, the eukaryotic proteins participate in transport and regulatory functions. The proteins contributed by the alpha-proteobacteria to the eukaryotes are encoded in both the nuclear and mitochondrial genomes of serving direct and indirect roles in aerobic respiration (Gabaldon and Huynen, 2007; Wang and Wu, 2014).

The phylogenetic analysis of the enzymes of the TCA cycle reveals a complex evolutionary network. The complexities include some genes adopted from alphaproteobacterial ancestor, into the eukaryote nucleus which is retargeted to the mitochondrion or other subcellular compartments, and still other genes seem to have been recruited horizontally from different bacterial or eukaryotic ancestors (Kurland and Andersson, 2000). Overall, extensive phylogenetic studies on mitochondrial origin and evolutionary history have provided meaningful insights for exploration of evolutionary aspects other sub-cellular organelles and the interrelationships with primitive living system or units and their probable existence in other places in the universe apart from earth.

ACKNOWLEDGMENT

The authors gratefully acknowledge The Council of Scientific and Industrial Research, Govt. of India, (Ref. No. 38(1417)/16/EMR-II, dated: 17/05/2016 to SR) and UGC, Govt. of India (Start-Up research grant No.F.30-141/2015 (BSR) for providing financial supports.

REFERENCES

Amiri H, Karlberg O, Andersson SGE (2003) Deep origin of plastid/parasite ATP/ADP translocases. J Mol Evol 56: 137–150.

Anderson S, Bankier AT, Barrell BG, de Bruijn MHL, Coulson AR, Drouin J, Eperon IC, Nierlich DP, Roe BA, Sanger F et al. (1981) Sequence and organization of the human mitochondrial genome. Nature 290: 457–465.

Anderson S, de Bruijn MHL, Coulson AR, Eperon IC, Sanger F, Young IG (1982) Complete sequence of bovine mitochondrial DNA conserved features of the mammalian mitochondrial genome. J Mol Biol 156: 683–717.

Andersson SGE, Karlberg O, Canback B, Kurland CG (2003) On the origin of mitochondria: a genomics perspective. Philos Trans R Soc Lond B Biol Sci 358: 165–179.

Andersson SGE, Zomorodipour A, Andersson JO, Sicheritz-Ponten T, Alsmark UCM, Podowski RM, Na Slund AK, Eriksson AS, Winkler HH, Kurland CG (1998) The genome sequence of *Rickettsia prowazekii* and the origin of mitochondria. Nature 396: 133–140.

Bartoli CG, Pastori GM, Foyer CH (2000) Ascorbate biosynthesis in mitochondria is linked to the electron transport chain between complexes III and IV. Plant Physiol 23: 335–343.

Bibb MJ, Van Etten RA, Wright CT, Walberg MW, Clayton DA (1981) Sequence and gene organization of mouse mitochondrial DNA. Cell 26: 167–180.

Birsoy K, Wang T, Chen WW, Freinkman E, Abu-Remaileh M, Sabatini DM (2015) An essential role of the mitochondrial electron transport chain in cell proliferation is to enable aspartate synthesis. Cell 162: 540–551.

Braymer JJ, Lill R (2017) Iron-sulfur cluster biogenesis and trafficking in mitochondria. J Biol Chem 292: 12754–12763.

Cannino G, El-Khoury R, Pirinen M, Hutz B, Rustin P, Jacobs HT, Dufour E. (2012) Glucose modulates respiratory complex I activity in response to acute mitochondrial dysfunction. J Biol Chem 287: 38729–38740.

Cedergren R, Gray MW, Abel Y, Sankoff D (1988) The evolutionary relationships among known life forms. J Mol Evol 28: 98–112.

Chan SI (2010) Proton pumping in cytochrome c oxidase: the coupling between proton an electron gating. PNAS 107: 8505–8506.

Christian BE, Spremulli LL (2012) Mechanism of protein biosynthesis in mammalian mitochondria. Biochim Biophys Acta 1819: 1035–1054.

Davies KM, Anselmi C, Wittig I, Faraldo-Gomez JD, Kuhlbrandt W (2012) Structure of the yeast F1Fo-ATP synthase dimer and its role in shaping the mitochondrial cristae. Proc Natl Acad Sci USA 109: 13602–13607.

Degli Esposti M, Cortez D, Lozano L, Rasmussen S, Nielsen HB, Martinez RE (2016) Alpha proteobacterial ancestry of the [Fe-Fe]-hydrogenases in anaerobic eukaryotes. Biol Direct 11: 34.

Dimmer KS, Rapaport D (2017) Mitochondrial contact sites as platforms for phospholipid exchange. Biochim Biophys Acta 1862: 69–80.

Doolittle WF (1980) Revolutionary concepts in evolutionary biology. Trends Biochem Sci 5: 146–149.

Douce R, Neuberger M (1989) The uniqueness of plant mitochondria. Annu Rev Plant Physiol Plant Mol Biol 40: 371–414.

Efremov RG, Sazanov LA (2011) Respiratory comple I: "steam engine" of the cell. Curr Opin Struct Biol 21: 532–540.

Ellenrieder L, Rampelt H, Becker T (2017) Connection of protein transport and organelle contact sites in mitochondria. J Mol Biol 429: 2148–2160.

Esser C, Ahmadinejad N, Wiegand C, Rotte C, Sebastiani F, Gelius DG, Henze K, Kretschmann E, Richly E, Leister D et al. (2004) A genome phylogeny for mitochondria among ⍺-Proteobacteria and a predominantly eubacterial ancestry of yeast nuclear genes. Mol Biol Evol 21: 1643–1660.

Eubel H, Jansch L, Braun HP (2003) New insights into the respiratory chain of plant mitochondria. Supercomplexes and a unique composition of complex II. Plant Physiol 133: 274–286.

Ferguson SJ (2010) ATP synthase: from sequence to ring size to the P:O ratio. PNAS 1087: 16755–16756.

Fermie AR, Carrari F, Sweetlove LJ (2004) Respiratory metabolism: glycolysis,the TCA cycle an mitochondrial electron transport. Curr Opin Plant Biol 7: 254–261.

Figueroa P, Leon G, Elorza A, Holuigue L, Araya A, Jordana X (2002) The four subunits of mitochondrial respiratory complex II are encoded by multiple nuclear genes and targeted to mitochondria in Arabidopsis thaliana. Plant Mol Biol 50: 725–734.

Fischer WW (2008) Life before the rise of oygen. Nature 455: 1051–1052.

Formentini L, Sanchez-Arago M, Sanchez-Cenizo L, Cuezva JM (2012) The mitochondrial ATPase inhibitory factor 1 triggers a ROS-mediated retrograde prosurvival and proliferative response. Mol cell 45: 731–742.

Fraser CM, Casjens S, Huang WM, Sutton GG, Clayton R et al. (1997) Genomic sequence of a Lyme disease spirochaete, *Borrelia burgdorferi*. Nature 390: 580–586.

Fraser CM, Gocayne JD, White O, Adams MD, Clayton RA et al. (1995) The minimal gene complement of Mycoplasma genitalium. Science 270: 397–403.

Fraser CM, Norris SJ, Weinstock GM, White O, Sutton GG et al. (1998) Complete genome sequence of Treponema pallidum, the syphilis spirochete. Science 281: 375–388.

Freibert SA, Goldberg AV, Hacker C, Molik S, Dean P et al. (2017) Evolutionary conservation and in vitro reconstitution of microsporidian iron-sulfur cluster biosynthesis. Nat Commun 8: 13932.

Friedman JR, Nunnari J (2014) Mitochondrial form and function. Nature 505: 335–343.

Fung S, Nishimura T, Sasarman F, Shoubridge EA (2013) The conserved interaction of C7orf30 with MRPL14 promotes biogenesis of the mitochondrial large ribosomal subunit and mitochondrial translation. Mol Biol Cell 24: 184–193.

Gabaldon T, Huynen MA (2003) Reconstruction of the protomitochondrial metabolism. Science 301: 609.

Gabaldon T, Pittis AA (2015) Origin and evolution of metabolic sub-cellular compartmentalization in eukaryotes. Biochimie 119: 262–268.

Garlid KD, Jaburek M, Jezek P, Varecha M (2000) How do uncoupling proteins uncouple? Biochim Biophys Acta 1459: 383–389.

Gomez MSA, Slamovits CH, Dacks JB, Baier KA, Spencer KD et al. (2015) Ancient homology of the mitochondrial contact site and cristae organizing system points to an endosymbiotic origin of mitochondrial cristae. Curr Biol 25: 1489–1495.

Gomez MSA, Wideman JG, Roger AJ, Slamovits CH (2017) The origin of mitochondrial cristae from alphaproteobacteria. Mol Biol Evol 34: 943–956.

Gray MW (1998) Rickettsia, typhus and the mitochondrial connection. Nature 396: 109–110.

Gray MW (1999) Evolution of organellar genomes. Curr Opin Genet Dev 9: 678–687.

Gray MW (2015) Mosaic nature of the mitochondrial proteome: implications for the origin and evolution of mitochondria. Proc Natl Acad Sci USA 112: 10133–10138.

Gray MW, Burger G, Lang BF (1999) Mitochondrial evolution. Nature 283: 1476–1481.

Gray MW, Cedergren R, Abel Y, Sankoff D (1989) On the evolutionary origin of the plant mitochondrion and its genome. Proc Natl Acad Sci 86: 2267–2271.

Gray MW, Doolittle WF (1982) Has the endosymbiont hypothesis been proven? Microbiol Rev 46: 1–42.

Gray MW, Spencer DF (1996) Organellar evolution. In Roberts D et al. (eds) Evolution of microbial life (pp. 109–126). Cambridge University Press, Cambridge.

Gross J, Bhattacharya D (2011) Endosymbiont or host: who drove mitochondrial and plastid evolution? Biol Direct 6: 12.

Gupta RS (1995) Evolution of the chaperonin families (Hsp60, Hsp10 and Tcp-1) of proteins and the origin of eukaryotic cells. Mol Microbiol 15: 1–11.

Hewitt V, Alcock F, Lithgow T (2011) Minor modifications and major adaptations: the evolution of molecular machines driving mitochondrial protein import. Biochim Biophys Acta 1808: 947–954.

Hewitt V, Lithgow T, Waller RF (2014) Modifications and innovations in the evolution of mitochondrial protein import pathways. In Loffelhardt W (ed) Endosymbiosis (pp. 19–35), Springer, Vienna.

Hoppins S, Collins SR, Cassidy-Stone A, Hummel E, Devay RM et al. (2011) A mitochondrial-focused genetic interaction map reveals a scaffold-like complex required for inner membrane organization in mitochondria. J Cell Biol 195: 323–340.

Hutton RL, Boyer PD (1979) Subunit interaction during catalysis. Alternating site cooperativity of mitochondrial adenosine triphosphatase. J Biol Chem 254: 9990–9993.

Huynen MA, Muhlmeister M, Gotthardt K, Guerrero CS, Brandt U (2016) Evolution and structural organization of the mitochondrial contact site (MICOS) complex and the mitochondrial intermembrane space bridging (MIB) complex. Biochim Biophys Acta 1863: 91–101.

Iborra FJ, Kimura H, Cook PR (2004) The functional organization of mitochondrial genomes in human cells. BMC Biol 2: 9

Junge W, Muller J (2011) Seeing a molecular motor at work. Science 333: 704–705.

Kageyama Y, Zhang Z, Sesaki H. (2011) Mitochondrial division: molecular machinery an physiological functions. Curr Opin Struct Biol 23: 427–434.

Koonin EV (2010) The origin and early evolution of eukaryotes in the light of phylogenomics. Genome Biol 11: 209.

Kuhlbrandt W (2015) Structure and function of mitochondrial membrane protein complexes. BMC Biol 13: 89.

Kurland CG, Andersson SGE (2000) Origin and evolution of the mitochondrial proteome. Microbiol Mol Biol Rev 64: 786–820.

Lang BF, Gray MW, Burger G (1999) Mitochondrial genome evolution and the origin of eukaryotes. Annu Rev Genet 33: 351–397.

Leger MM, Petru M, Zarsky V, Eme L, Vlcek C et al. (2015) An ancestral bacterial division system is widespread in eukaryotic mitochondria. Proc Natl Acad Sci 112: 10239–10246.

Margulis L (1970) Origin of eukaryotic cells. Yale University Press, New Haven.

Martin W, Muller M (1998) The hydrogen hypothesis for the first eukaryote. Nature 392: 37–41.

Moller IM (2002) A new dawn for plant mitochondrial NAD(P)H dehydrogenases. Trends Plant Sci 7: 235–237.

Moller IM, Rasmusson AG (1998) The role of NADP in the mitochondrial matrix. Trends Plant Sci 3: 21–27.

Nedelcu AM, Lee RW (1998) The molecular biology of chlamydomonas: chloroplast and mitochondria. In Rochaix JD (ed) Kluwer Academic Publishers, Dordrecht, The Netherlands.

Nunnari J, Suomalainen A (2012) Mitochondria: in sickness an in health. Cell 148: 1145–1159.

Ohnishi T (2010) Piston rives a proton pump. Nature 465: 428–429.

Purkanti R, Thattai M (2015) Ancient dynamin segments capture early stages of host-mitochondrial integration. Proc Natl Acad Sci 112: 2800–2805.

Reyes IM, Diebold LP, Kong H, Schieber M, Huang H et al. (2016) TCA cycle an mitochondrial membrane potential are necessary for diverse biological functions. Mol Cell 61: 199–209.

Roger AJ, Munoz GSA, Kamikawa R (2017) The origin and diversification of mitochondria. Curr Biol 27: R1177–R1192.

Rowland AA, Voeltz GK (2012) Endoplasmic reticulum-mitochondria contacts: function of the junction. Nature Rev Mol Cell Biol 13: 607–625.

Schnare MN, Gray MW (1982) 30-Terminal sequence of wheat mitochondrial 18S ribosomal RNA: further evidence of a eubacterial evolutionary origin. Nucleic Acids Res 10: 3921–3932.

Sharma MR, Koc EC, Datta PP, Booth TM, Spremulli LL et al. (2003) Structure of the mammalian mitochondrial ribosome reveals an expanded functional role for its component proteins. Cell 115: 97–108.

Sicheritz-Ponten ST, Kurland CG, Andersson SGE (1998) A phylogenetic analysis of the cytochrome b and cytochrome c oxidase I genes supports an origin of mitochondria from within the Rickettsiaceae. Biochim Biophys Acta 1365: 545–551.

Smith CT (2006) Origin of mitochondria by intracellular enslavement of a photosynthetic purple bacterium. Proc R Soc Lond B Biol Sci 273: 1943–1952.

Smith CT (2009) Predation and eukaryote cell origins: a co-evolutionary perspective. Int J Biochem Cell Biol 41: 307–322.

Spencer DF, Schnare MN, Gray MW (1984) Pronounced structural similarities between the small subunit ribosomal RNA genes of wheat mitochondria and *Escherichia coli*. Proc Natl Acad Sci 81: 493–497.

Stephens RS, Kalman S, Lammel C, Fan J, Marathe R et al. (1998) Genome sequence of an obligate intracellular pathogen of humans: *Chlamydia trachomatis*. Science 282: 754–759.

Szklarczyk R, Huynen MA (2010) Mosaic origin of the mitochondrial proteome. Proteomics 10: 4012–4024.

Unseld M, Marienfeld R, Brandt P, Brennicke A (1997) The mitochondrial genome of Arabidopsis *thaliana* contains 57 genes in 366,924 nucleotides. Nat Genet 15: 57.

Viale AM, Arakaki AK (1994) The chaperone connection to the origins of the eukaryotic organelles. FEBS Lett 341: 146–151.

von Ballmoos C, Cook GM, Dimroth P (2008) Unique rotary ATP synthase and its biological diversity. Annu Rev Biophys 37: 3–64.

von Ballmoos C, Wiedenmann A, Dimroth P (2009) Essentials for ATP synthsis by F_1-F_0 ATP synthases. Ann Rev Biochem 78: 649–672.

Wang Z, Wu M (2014) Phylogenomic reconstruction indicates mitochondrial ancestor was an energy parasite. PLoS One 9: e110685.

Westermann B (2010) Mitochondrial fusion an fission in cell life an earth. Nat Revs Mol Cell Biol 11: 872–884.

Wideman JG, Gomez MSA (2016) The evolution of ERMIONE in mitochondrial biogenesis and lipid homeostasis: an evolutionary view from comparative cell biology. Biochim Biophys Acta 1861: 900–912.

Wiedemann N, Pfanner N (2017) Mitochondrial machineries for protein import and assembly. Annu Rev Biochem 86: 685–714.

Williams KP, Sobral BW, Dickerman AW (2007) A robust species tree for the Alphaproteobacteria. J Bacteriol 189: 4578–4586.

Yun J, Finkel T (2014) Mitohormesis. Cell Metabolism 19: 757–766.

3 Mitochondrial Genome Damage, Dysfunction and Repair

Kalyan Mahapatra, Sayanti De, and Sujit Roy
Department of Botany, UGC Center for Advanced Studies, The University of Burdwan, Golapbag Campus, Burdwan, West Bengal, India

CONTENTS

1 INTRODUCTION

Mitochondria, morphologically distinctive double membrane bound organelles, hypothesized to have evolved from endosymbiotic α-proteobacteria, are indispensable for the vitality of eukaryotic life. They are ubiquitously found throughout the eukaryotic systems. They use electron transport coupled with oxidative phosphorylation to generate ATP (Saraste, 1999). For decades, mitochondria were considered only as cell's power house. But recent findings are that these organelles play vital roles also in apoptosis and survival, the ageing process, and in several diseases, biosynthesis of amino acids and steroids, β-oxidation of fatty acids, FeS metabolism and physiological adaptations to environmental stress (Birsoy et al., 2015). Due to their endosymbiotic origin mitochondria are the only organelles in animal cells, equivalent to plastid in plant cells which have their own genome (Vothknecht and Soll, 2007). The genome of mitochondria contains a minute subset of genes (Anderson et al., 1981) inherited and retained from their α-proteobacterial ancestors, with other genes of prokaryotic endosymbiotic origin either having been lost or else retained by nuclear chromosomes of the cell (Esser et al., 2004). Genes in mitochondria encode mitochondrial proteins and RNAs (Pesole et al., 2012).

2 MITOCHONDRIAL GENOME ARCHITECTURE

Mitochondrial genome of many eukaryotes had been considered to exist in the form of super-coiled circular DNA. However, there is now strong evidence that many of these circular mapped mtDNA consists primarily of linear, multimeric head-to-tail concatemers (Bendich, 1993, 1996). These, for example, have been found in few unrelated organisms such as ciliates, Apicomplexa (Plasmodium and its relatives), fungi, Chlorophycean green algae (*Chlamydomonas* and relatives) and several Cnidarian animals. These linear molecules contain various specialized end structures, such as covalently closed single-stranded DNA termini and terminally attached proteins. They also tend to have telomere-like repeats of differing lengths. In fact, the difference in size of mitochondrial genomes is mostly caused by the variations in the length and organization of intergenic regions which, in some cases are consisting of extensive tandem-repeats or stem-loop motifs.

Mitochondrial DNA is compact and less than 20 kilo bases in length (Wolstenholme, 1992) encoding protein subunits of the mitochondrial respiratory chain (Okimoto et al., 1992). There also exist structures called nucleoids, which are mainly composed of 2-8mtDNAcopies (Legros et al., 2004). Nucleoids are associated with the inner mitochondrial membrane and distributed throughout the mitochondrial network at regular spatial intervals (Prachar, 2010). Generally, mitochondrial DNA is multi copied with a 16,569-bp double-stranded circular

molecule located within the matrix of the mitochondrion, inherited from the maternal oocyte and is self-replicative.

Mammalian mt-DNA (Fig. 1A) contains 37 genes, out of which 13 encode mitochondrial oxidative phos-

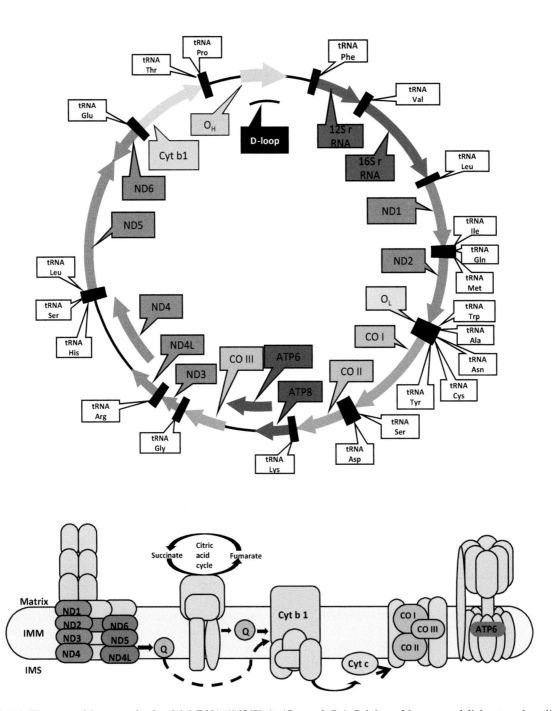

FIGURE 1 (A) The map of human mitochondrial DNA (16569bp). (O_H and O_L) Origins of heavy- and light-strand replication, respectively; (ND1-ND6) subunits of NADH dehydrogenase (ETS complex I) subunits 1–6; (COX1-COX3) subunits of cytochrome oxidase (ETC complex IV); (ATP6 and ATP8) subunit 6 and 8 of mitochondrial ATPase (complex V); (Cyt*b*) cytochrome *b* (complex III). **(B) Schematic representation of mitochondrial electron transport chain**. The mitochondrial DNA encoded subunits are shown in their respective colors as shown in mitochondrial DNA map (Fig1A). Nuclear DNA encoded subunits are shown in light green color.

phorylation complex proteins, 22 encode transfer RNA (tRNA) and others encode ribosomal RNA. In contrast to nuclear genome, mitochondria generally encode two kinds of rRNAs, 16S rRNA and 12S rRNA which participate in the synthesis of 13 mtDNA-encoded proteins. The mRNAs are specific for integral membrane components of the mitochondrial electron transport chain, the rRNAs are involved to produce subunits of mitochondrial ribosome, and the tRNAs mediate in the translation process of the 13 mRNAs by these dedicated mitochondrial ribosomes. Genes are employed using both strands of mtDNA; although they have been classified as heavy (H) and light (L) strands based on their relative buoyant densities in denaturing CsCl gradients.

Transcription initiates at the promoter site using H-strands (called HSP1 andHSP2) and a single L-strand (the LSP). The LSP and HSP1 are located in the major non-coding region of the DNA called the displacement loop (D-loop) regulatory region, whereas HSP2 is located downstream of HSP1 within the tRNA Phe gene. Transcripts derived from HSP2 and LSP are mostly as long as the whole genome and forms polycistronic products, whereas those from HSP1 are terminated at a specific site (in the tRNALeu gene) downstream of the 16S rRNA and produce mainly the two rRNAs (12S and 16S). Because of such unique gene arrangement in mt-DNA i.e., the rRNAs and most mRNAs are immediately flanked by tRNAs, tRNA processing is believed to be the major mechanism that liberates the majority of the 37 mature RNA molecules from the polycistronic primary transcripts.

In mammals, the mature mRNAs lack significant 5′ untranslated sequences and thus the mechanism of ribosome binding and mitochondrial translation initiation remains obscure. MtDNA encodes proteins that are structural subunits of complexes I, III, IV and V of the respiratory chain (RC), (Fig. 1B) as detailed below (only complex II is nuclear encoded).

Seven components of RC complex I (CI, or NADH dehydrogenase complex), ND1, ND2, ND3, ND4, ND4L, ND5 and ND6 constitute a subset of the 14 catalytic proteins. These proteins are involved in the electron transfer and proton pumping activity of the CI (Guzy et al., 2006) cytochrome b (a component of RC complex III (CIII). Cytochrome b transfers the electrons from the ubisemiquinone at the Qo site of the complex) (outer surface of the inner membrane) to the ubisemiquinone located at the Qi site (inner surface of the inner membrane) (Fontanesi et al., 2008). Catalytic subunits 1, 2and 3 of RC complex IV (CIV) along with 10 additional proteins allow the electron transport process (Hong, 2004).Subunits of complex V (CV, or ATPase complex) include ATPase6 and A6L.The ATPase6 (or "a") subunit is a trans-membrane protein of the F_o portion of the complex, involved in the passage of protons from the intermembrane space to the matrix. The function of subunit A6L is still unclear. Recent studies by

Wittig et al. have indicated that both proteins ATPase6 and A6L play important role in the stabilization of ATP synthase dimers/oligomers.

3 MITOCHONDRIAL DNA DAMAGE AND REPAIR

As the histone proteins has the ability to some extent protect the nuclear genome from various kinds of DNA damaging agents but mitochondrial genome lacks them, hence, mitochondrial DNA is more prone to damage than nuclear DNA. The mitochondrial DNA is constantly exposed to various kinds of exogenous and endogenous DNA damaging agents, some of which can lead to various kinds of diseases including of neurodegenerative form, cancer, cardiomyopathy, diabetes and several aging-related disorders.

4 ENDOGENOUS DNA DAMAGE

One of the major products of mitochondrial electron transport chain is reactive oxygen species (ROS) which frequently induces DNA damage (Fig. 2). Mitochondria accounts for about 90% oxygen consumption, about 1–5% of which is converted to superoxide anion (Papa, 1996) which is the main precursor of other ROS including the most toxic superoxide radicalO_2^-• . The reduction potential for the conversion of O_2 to O_2^-• is about –0.160V and the mitochondrial electron transport chain has several redox centres with standard reduction potential ranging from -0.32V to +0.39V giving mitochondria a fairly reductive environment (Wood, 1987). For this reductive environment various respiratory component such as flavoproteins and Fe-S centres are thermodynamically capable monovalent reduction of molecular oxygen giving rise to the superoxide radicals. Seven potential sites for the generation of O_2^-• exist in the mitochondrial matrix with complex I and Complex III showing highest rate of O_2^-• production. These superoxide radicals cannot diffuse through the inner mitochondrial membrane and converted to H_2O_2 by the action of manganese-superoxide dismutase (Mn-SOD) but if this H_2O_2 remains within the matrix it will undergo Fenton chemistry with Fe(II) to form hydroxyl radicals (HO•), which show strong reactivity towards DNA and polyunsaturated fatty acids (PUFA).

The sophisticated antioxidant system of mitochondrial matrix restricts the steady-state concentration of O_2^-• within 10^{-10} M (Cadenas and Davies, 2000). The free radicals escaping from this detoxification process can cause oxidative damage to various biological components. But because mitochondrial DNA is in close proximity to the mitochondrial respiratory chain, it is very much susceptible to oxidative damage by various free radicals. The high rate of oxidative stress operating within the mitochondrial matrix causes a broad spectrum of mitochondrial DNA damage

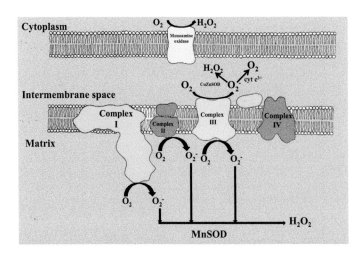

FIGURE 2 Sites of superoxide free radical formation in the mitochondrial electron transport chain. These superoxide anions may be converted to hydrogen peroxide by the action of MnSOD in both Intermembrane space and matrix or may reduce cytochrome c to form oxygen.

including modifications to bases such as 8-oxo-2'-deoxyguanosine (8-oxodG), 8,5'-cyclo-2'deoxynucleosides, thymine glycol, 5,6-dihydroxycytosine, 2,6-diamino-4-hydroxy-5-formamidopyrimidine. Along with this sugar break down products (such as 2-deoxypentose-4-ulose, 2-deoxypentonic acid lactones and erythrose), base free sites or abasic sites, strand breaks and chemical adducts of bases such as aldehyde modifications are also the consequences of the oxidative stress.

Among all these above modifications, oxidatively modified guanine or 8-oxodG is the most extensively studied and it is one of the biomarkers of oxidative stress. 8-oxodG is more important site in DNA than normal deoxyguanosine (dG) to induce a variety of mutations, if not repaired under normal physiological condition and the production increases with aging. It is also found that 8-oxodG accumulates more rapidly in mitochondrial genome than in the nuclear genome.

The metabolic capacity and membrane composition of mitochondria make it a suitable place for generation of reactive aldehydes which are potent to form various DNA adducts. Among various aldehydes, formaldehyde produces N^2-hydroxymethyl-dG, the α, β-unsaturated fatty acids resulting from the oxidation of PUFA and DNA inset a five membered or six-membered ring to the DNA base. LPO also induces several ethno adducts (N^6-ethanodeoxyadenosine and N^4-ethenodeoxycitidine) and MDA creates six membered, exocyclic adducts (M_1 dG, M_1dC and M_1dA). M_1dG is present in higher amount in the mitochondrial genome than in nuclear genome (Jeong et al., 2005). It is because the inner mitochondrial membrane is enriched in PUFA, α, β-unsaturated aldehydes are the major threat to the mitochondrial genome stability and its expression during the oxidative stress.

5   EXOGENOUS DNA DAMAGING AGENTS

Along with the endogenously produced DNA damaging agents, exogenous agents such as various chemicals, metabolites in the dietary components, UV radiation, ozone, pesticides, pharmaceuticals etc. can also cause damage to the mitochondrial DNA.

Also exogenous agents such as the acrolein components of the cigarette, fungal toxin aflatoxin B1, platinum based chemotherapeutic agent like cisplatin, ultraviolet radiation, ozone induce adducts on the DNA and interfere with the mitochondrial gene expression.

Aldehydes are produced in increasing amount during alcohol metabolism and they can also take entry in the body from external environment (Swenberg et al., 1999, 2011). Acetaldehyde is produced in large amount after the ingestion of alcohol which is converted to acetate in mitochondria by the action of aldehydes dehydrogenase 2 or ALDH2, failure of which can form several adducts including N^2-ethylidene-dG, 1, N^2-propano-2'-dG and their derivatives (Wang et al., 2000; Inagaki et al., 2004; Matsuda et al., 2006). Industrial pollutants contain formaldehyde, which together with several alkylating reagents can combine to form N^2-hydroxymethyl-d G (Lu et al., 2011).

Several anticancer drugs are able to bind to DNA causing damage leading to inhibition of tumour growth and also promote apoptosis. These anticancer drugs may have some effects on the mitochondrial genome. Various anticancer drugs such as bleomycin and neocarzinostatin can cause oxidative DNA damage through ROS and reactive aldehydes generation (Dedon et al., 1992, 1998). Another important group of anticancer drugs is platinum (Pt) based chemotherapeutic agents such as cisplatin, oxaliplatin, carboplatin etc which binds to DNA forming single base adducts and intra and inter-strand cross links (Wang et al., 2005). Cisplatin causes intra-strand cross links known as l,2-d(GpG) cross links after drug exposure and this even at low concentrations can cause inhibition to transcription and translation as they prevent strand separation (Ayala-Torres et al., 2000; Todd et al., 2009).

6 MITOCHONDRIAL DNA DAMAGE, REPAIR AND MAINTENANCE

Mitochondria, a semiautonomous organelle probably evolved by endosymbiotically insertion of an unknown whole primitive organism, having their own genome. The mitochondrial genome is the second important element in eukaryotic cells after nuclear genome where most genetic information is being stored. Unlike nuclear DNA, mitochondrial DNA inheritance is almost exclusively maternal and the mitochondrial DNA is redundant with many copies are present within cells.

On the other hand, mitochondrial DNA is continuously replicated even in the terminally differentiated cells such as neurons. Hence somatic mitochondrial DNA damage has more adverse effect in comparison to somatic nuclear DNA damage. And it is not surprising that mutation in mitochondrial genome often results in alterations in mitochondrial functions and ultimately leads to several mitochondrial diseases (Arpa et al., 2003). Mitochondrial diseases can be severe and/or fatal. Along with the mitochondrial diseases, mitochondrial DNA damage is linked to some common diseases including age-related disorders, cancer and diabetes, neurodegenerative disorders and several other symptoms (Coskun et al., 2004; Howell et al., 2005; Taylor and Turnbull, 2005; Wallace, 2005; Chatterjee et al., 2006).

Thus maintaining the integrity of the mitochondrial genome is crucial for the survival of an organism. Mitochondria had long been considered to lack any kind of sophisticated DNA damage repair system like that of nuclear genome because of the discovery that mitochondria is unable to repair UV-induced pyrimidine dimers (Clayton et al., 1974, 1975) and some kinds of alkylation damage (Miyaki et al., 1977). Additionally another idea that mitochondrial fusion and fission events and mitochondrial DNA degradation may allow the organelle to minimize the toxic effect of DNA damage left the mitochondrial DNA damage repair pathways unresolved (Shokolenko et al., 2009). Over the PAST two to three decades the notion of mitochondrial DNA damage repair systems has evolved, and now it has been documented that almost every DNA damage repair system is present in mitochondria like that of in the nucleus (Gredilla, 2011). Important to be noted that for the mitochondria DNA repair this organelle has to depend on the repair enzymes synthesized by the nuclear genome.

Presence of novel DNA damage repair pathway is confirmed by analysing the different components of DNA damage repair pathways in the mitochondrial sub cellular fractionation. Other techniques employed were sub cellular localization of fluorescent fusion protein, immune gold labelling combined with electron microscopy and silico techniques. Overall the understanding of DNA damage repair systems in mitochondria emerged through successive stages. From the preliminary concept of no repair system to recent discoveries of different robust DNA damage repair pathways such as Base excision repair (BER), Nucleotide excision repair (NER), mismatch repair (MMR) and double strand break repair systems. Along with all these repair systems, the elimination of pre-mutagenic damaged dNTPs and selective degradation of severely damaged DNA complement the repair system to maintain the mitochondrial genome integrity.

7 DIRECT REVERSAL REPAIR PATHWAY

Direct reversal repair represents the simplest pathway and also an alternative to photoreactivation of cyclobutane pyrimidine dimers or CPD by photolyase in *E. coli* without cleaving the phosphodiester chain (Sancar, 2008). Although mammalian mitochondria contain homologue of photolyase, transcriptional repressors, cryptochrome1 and 2 (CRY1 and CRY2) (Kobayashi et al., 1998), they are mainly involved in regulating the circadian rhythm. In contrast to yeast photolyase activity, which revert the UV induced damage to DNA (Pasupathy et al., 1992; Yasui et al., 1992), no direct evidence for reversal of UV induced damage could be detected in mitochondria. Along with the yeast, photolyase activity is evidenced in plant mitochondria (Takahashi et al., 2011) and *Xenopus* mitochondria (Ryoji et al., 1996). It is generally seen that higher eukaryote mitochondria do not possess the photolyase activity.

The O^6-methylguanine-DNA methyltransferase (MGMT) is an important enzyme found in mammalian nucleus which is involved in the reversal of alkylated damage to DNA such as O^6-alkylguanines. Because of the presence of a MGMT variant in mitochondria, it is able to repair O^6-methyl-2'-deoxyguanosine and O^6-ethyl-2'-deoxyguanosine (Myers et al., 1988; Satoh et al., 1988). However, the mitochondrial localization of mammalian MGMT is not confirmed yet by Western blotting (Cai et al., 2005). Thus it is likely that this kind of damage in mitochondria is repaired by protein synthesized by nuclear genome.

8 BASE EXCISION REPAIR

As mitochondria are the major sources of $O_2^-\bullet$, the mitochondrial genome remains under continuous oxidative threat. ROS induced oxidative DNA damage, such as 8-doxo-dG, is preferentially repaired by a robust Base Excision Repair (BER) system (Fig. 3). Existence of mitochondrial BER was first discovered after the identification of mitochondrial uracil-DNA glycosylase

(Anderson and Friedberg 1980). The mitochondrial BER pathway that utilizes both mitochondrial and nuclear proteins is even more efficient than nuclear BER pathway because mitochondria remain as the major source of oxidative threat with 8-oxodG is the common DNA lesion (Thorslund et al., 2002; Murphy, 2009).This repair pathway has now been reported in yeast, animal cells and in plant mitochondria also(Mecocci et al., 1993; Higuchi et al., 1995).

Mitochondrial BER pathway follows similar three-step mechanism like that of seen in nuclear BER pathway, i.e., recognition and elimination of damaged base, gap tailoring and DNA synthesis/ligation. Depending on the DNA polymerase mediated insertion of either single nucleotide or short sequence, BER pathway is of two types, i.e., short-patch BER and long-patch BER. Although animal mitochondria show both of them, plant mitochondria are restricted to short-patch BER only (Mecocci et al., 1993).The first step in Short patch BER is recognition of the damaged base and its elimination. The damaged base is recognized by two types of DNA glycosylase i.e., monofunctional and bifunctional.

9   MONOFUNCTIONAL AND BIFUNCTIONAL DNA GLYCOSYLASE

After recognizing the DNA lesion, the monofunctional DNA glycosylase hydrolyzes the N-glycosidic bond at the oxidized bases forming an apurinic/apyrimidinic (AP) site. Two monofunctional glycosylase have been reported in mitochondria till date i.e., uracil-N-glycosylase1 (UNG1) (Anderson and Friedberg 1980) and MutY homolog glycosylase, MUTYH (Ohtsubo et al., 2000). After the action of glycosylase, AP endonuclease (APE1) cleaves the immediate 5′ side of the AP site, leaving a 3′-hydroxyl and 5′-deoxyribose-5-phosphate (5′-dRP) residue. The 5′-deoxyribose-5-phosphate residue is then removed by the 5′DRP lyase activity of DNA polymerase γ for the subsequent gap-filling polymerization reaction.

Besides two monofunctional glycosylases, four bifunctional glycosylases have also been reported present in mitochondria. These are 8-oxoguanine DNA glycosylase (OGG1) (Takao et al., 1998; de Souza-Pinto et al., 2009), nth (homologous to *E. coli* endonuclease III)-like 1 (NTHL1, or NTH1) (Karahalil et al., 2003) and Nei

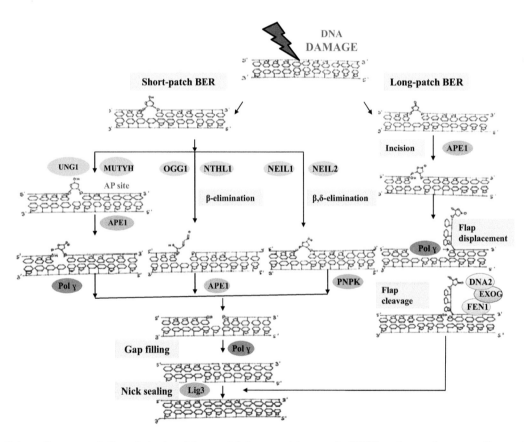

FIGURE 3 Schematic representation of short and long-patch base excision repair (BER) pathways in mammalian system. All the important steps and factors are highlighted. In addition to FEN1, DNA2 is required for efficient flap processing during long-patch BER in mitochondria. Three sub-pathways of short-patch repair are converged at the gap filling step. Short-patch BER (left) and long-patch BER (right) pathways meet at the nick sealing step.

(*E. coli* endonuclease VIII)-like-1 and -2 (NEIL1 and NEIL2) (Hu et al., 2005; Mandal et al., 2012). Unlike monofunctional glycosylase, bifunctional glycosylases have additional AP-lyase activity which incise the apyrimidinic and apurinic (AP) sites by means of β- elimination (OGG1, NTLH1) or β,δ- elimination (NEIL1, NEIL2) (Dodson and Lloyd 2002,).β- elimination reaction results in formation of AP site with 3′ phospho-α,β-unsaturated aldehydes (3′PUA) and 5′P and β,δ- elimination reaction results in single nucleotide gap with terminal 5′P and 3′P. OGG1, NEIL1 and NEIL2 enzymes exhibit overlapping specificity for 8-oxoG but differs in their recognition of other DNA lesions caused by oxidative stress. Though N-methylpurine DNA glycosylase (MPG) is still not reported but mammalian mitochondria able to repair the lesions that are regular substrates of MPG (Chakravarti et al., 1991; Hang et al., 1996).

Removal of blocking groups at 5′ and 3′ end of the products of the first step of BER pathway is known as gap tailoring. The 5′-dRP groups created by the action of UNG1, MUTYH1 and APE1 at the damaged base are excised by the dRPase activity of the Polymerase γ which also has polymerase and 5′-3′ exonuclease activity. As a result, a single strand gap with 5′P and 3′-OH is generated. On the other hand AP site with 3′PUA end generated by the action of OGG1 and NTLH1 is removed by the phosphodiesterase activity of APE1 and the 3′P generated at the site of DNA damage by the action of NEIL1 and NEIL2 and converted to 3′-OH by the phosphatising activity of polynucleotide kinase 3′-phosphatase (Mandal et al., 2012; Tahbaz et al., 2012). The product of the gap tailoring phase of short-patch BER pathway is always a single nucleotide ends with 5′-P and 3′-OH groups. This gap is filled by the activity of polymerase γ and the remaining nick is sealed by the action of DNA ligase IIIα. Importantly mitochondria do not contain XRCC1 which acts as a scaffold protein for the assembly of the nuclear BER components and also brings in ligase III to complete the BER pathway repair (Simsek et al., 2011).

10   LONG-PATCH REPAIR

The dRP-lyase activity of polymerase γ cannot efficiently remove certain 5′ blocking groups e.g. 5′ deoxyribonolactone (5′dL) which is about 72% of the total oxidative sugar damage in DNA (Roginskaya et al., 2005). Removal of this kind of lesions in nuclear DNA takes place through long-patch BER pathway and the mitochondrial long-patch BER pathway has been reported few years back (Akbari et al., 2008; Copeland and Longley 2008). In long-patch BER system, DNA polymerase γ displaces the 5′ blocking group extending from 3′-OH forming a "flap" of about 6–9 nucleotides. This "flap" is subsequently removed by the joint action of two endonucleases, i.e., FEN1 and DNA2 (Copeland and Longley

2008; Zheng et al., 2008; Duxin et al., 2009; Kalifa et al., 2009) forming a nicked DNA duplex. The nick is then sealed by the activity of DNA Lig3 to recover the damaged part of the DNA double helix.

11   NUCLEOTIDE EXCISION REPAIR

Nucleotide excision repair (NER) pathway involves the elimination of and re-synthesis of short DNA fragment on the damaged strand. However, there is no evidence of any typical NER pathway in mitochondria like that of nuclear NER pathway. Though several studies have confirmed that mitochondria do not have any NER mechanism for the removal of UV-induced pyrimidine dimers (Clayton et al., 1974; LeDoux et al. 1992; Pascucci et al., 1997), showed that in yeasts certain NER substrates are repaired by an alternative mechanism; for instance, in *Schizosaccharomyces pombe* mitochondrial cyclobutane pyrimidine dimers (CPDs) and (6–4) pyrimidine-pyrimidone photoproducts [(6–4) PP] are repaired by UV damaged DNA endonuclease-dependent excision repair (UVER) pathway.

Mitochondria also exhibit poor repair of cisplatin induced intra-strand cross links but efficiently repair the inter-strand cross links, which are repaired by Recombination repair system (RRS). Since, both types of cisplatin induced adducts can efficiently be removed by nuclear repair system, mitochondrial damage remains one of the major issues of cisplatin-induced toxicity. Because mitochondria is deficient of a proper NER system, 8,5′-cyclo -2′-deoxypurines (cPu), M_1dG and BaPDE adducts, this causes problems in mitochondria to suffer in many tissues of the body (Wei et al., 1995; Hess et al., 1997; Johnson et al., 1997; Reardon et al., 1999; Brooks et al., 2000; Kuraoka et al., 2000; Lloyd and Hanawalt, 2000). The only evidence of NER-like mechanism in mitochondria is in Cockayne syndrome proteins, CSA and CSB, which are required for transcription-coupled NER (TC-NER) repair system (Aamann et al., 2010; Kamenisch et al., 2010; Lagerwerf et al., 2011).

12   MISMATCH REPAIR

The occurrence of mismatch repair system in mammalian mitochondria was established in-vitro by M-13 based assay in which the rat liver mitochondrial extracts showed small but significant mismatch repair (MMR) activity (Mason et al., 2003). Mismatch repair activity in mammalian mitochondria is efficient in cleaving the G:T and G: G mismatches, without showing any activity for discrimination of the parental strand from the sister strand, thus creating the situation more susceptible to several kind of mutations (Mason et al., 2003). Mammalian mitochondrial mismatch repair system is very different from that of nuclear mismatch repair and is independent of MSH2, one of the master regulators of the nuclear mismatch repair system (de Souza-Pinto et al., 2009). Confocal microscopy also confirmed the absence of other classical MMR members viz. MSH3, MSH6 and MLH1 in the mammalian mitochondria (de Souza-Pinto et al., 2009). In yeast and in *E. coli* mismatch repair component MutS homologue MSH1 is present which repairs the G: A mismatches generated by replication past 8-oxoG (Dzierzbicki et al., 2004). The Y-box binding protein YB-1 protein (NSEP1 and YBX1), previously known to play a major role in nuclear base excision repair pathway and also in repair of crosslinked DNA, is a key component of the mammalian MMR and is mainly involved in mitochondrial mismatch binding activity. RNA interference study also supported the role of YB-1 protein in mitochondrial MMR and YB-1 depletion showed increase in mitochondrial DNA mutagenesis. All these studies strongly suggest the presence of a mismatch repair system in certain mitochondria which is distinct from that of the nuclear mismatch repair, and this pathway is needed to be further characterized.

13   ELIMINATION OF PREMUTAGENIC DNTPS

Not only DNA is subjected to damage by different chemical agents but the free deoxynucleotide triphosphate (dNTPs) pool can be damaged too and may be severely, mainly by ROS oxidation. These damaged (oxidized) dNTPs are the major sources of mismatches during the replication of mitochondrial DNA. On the other hand the fidelity of mitochondrial DNA polymerase ɣ (Pol ɣ) is reduced when the enzyme finds oxidized dGTP (8-oxodGTP), 0.06–0.6% of the total dGTP pool (Pursell et al., 2008). This information indicates that oxidation of dNTPs pool is a greater threat to the stability of mitochondrial DNA. Mitochondria have developed specialized sanitizing enzymes to eliminate the oxidized dNTPsand replace it by dUTPs or 8-oxodGTPS; these are incorrectly incorporated into the mitochondrial DNA during replication in that 8-oxodGTPs is placed opposite to the adenine in the template DNA forming 8-oxodG:dA base pairs which escapes the proof reading activity of Pol ɣ leading to a AT to CG transversions (Hanes et al., 2006).This can lead to mutated mitochondrion and can become dysfunctional

The major mechanism in mitochondria to counter against premutagenic8-oxodGTPs includes the removal of 8-oxodGTPsfrom the dNTPs pool by the activity of mitochondrial MTH1. MTH1 is a mammalian homologue of *E. Coli* MutT protein (Kang et al., 1995; Nakabeppu, 2001). MTH1 actually hydrolyzes the 8-oxo-2′-deoxyguanosine triphosphate to dGMP, the corresponding monophosphate form thus minimizing the chances of incorporation of 8-oxodGTPsby Pol ɣ during mitochondrial DNA replication.MTH1 also protects cell from cytotoxicity of sodium nitroprusside by preventing the build-up of 8-oxoG

(Ichikawa et al., 2008). Although sanitation of oxidized dNTPs is not bonafide DNA damage repair mechanism but elimination of damaged dNTPs reduces the chances of mismatches in the mitochondrial DNA which helps in mitochondrial genome maintenance.

14   DOUBLE STRAND BREAK REPAIR

A well-developed double-strand break repair system is also present in mitochondria as reported in both, yeast and *Drosophila melanogaster* (Contamine and Picard 2000; Morel et al., 2008). In-vitro studies demonstrated the presence of both homologous recombination (HR) and non-homologous end joining (NHEJ) pathways in mammalian mitochondria; also it has been reported that in mitochondria, recombination repair is carried out in both homology dependent and independent manner (Thyagarajan et al., 1996; Coffey et al., 1999; Lakshmipathy et al., 1999; Bacman et al., 2009; Fukui et al., 2009). Presence of both intra-molecular and inter-molecular recombination products in heteroplasmic mice after generating double strand breaks with the help of Pst1 and Sca1 confirmed the presence of both the HR and NHEJ pathways in mammalian mitochondria though the frequency of NHEJ products is rare . The key regulator of nuclear HR pathway in mammal, i.e., Rad51 is also present in mitochondria and thought to be involved in the mitochondrial double strand break repair (Sage et al., 2010). Homologous recombination is also reported in plant mitochondria and later the presence of Rec A homologue in the moss *Physcomitrella patens* is also reported (Manchekar et al., 2006; Odahara et al., 2007). In higher plants the Rec A homologues are mainly involved in the surveillance of recombination pathways.

15   MT-DNA DEGRADATION

Despite all these DNA repair systems, degradation of damaged part of DNA plays an important role in maintaining the genome integrity in mitochondria to avert mitochondrial related diseases. Unlike nuclear genome, mitochondrial genome is present in high copy number which keeps them out of the "repair or die" constraint. Though certain yeast mitochondria have a photolyase that may repair pyrimidine dimers (Yasuhira and Yasui 2000), but mammalian mitochondria cannot repair UV-induced pyrimidine dimers which spurs the concept of mitochondrial DNA degradation in response to DNA damage (Clayton et al., 1974). However, Mita et al. (1988) showed that mitochondrial genome of HeLa cells accumulate very few number of mutations when exposed to various carcinogens such as ethyl methane sulphonate (EMS) suggested that mitochondrial genome is not replicated when large amounts of damage or irreparable damage is accumulated (Mita et al., 1988). Some other studies also showed that many cell lines show resistance to the loss of mitochondrial DNA and can survive gradual loss of mitochondrial DNA after treatment with some intercalating agents such as ethidium bromide (King and Attardi 1989). If the mitochondrial DNA is induced to extensive or persistent double strand breaks by means of site-specific restriction endonuclease enzymes, mitochondrial DNA shows depletion (Alexeyev et al., 2008; Kukat et al., 2008) with a very low amount of DSB leading to recombination (Bacman et al., 2009). This is contrasting feature of the nuclear genome in which persistent DSB can leads to apoptosis. Given the high copy number of mitochondrial DNA, the hypothesis that degradation of the damaged mitochondrial DNA can be a mechanism adopted by mitochondria to cope up with extensive DNA damage. Lack of direct experimental evidence and also discovery of mitochondrial BER pathway (Pettepher et al., 1991) puts doubt on this hypothesis.

The oxidative induced DSB acts as a signal inducing mammalian mitochondrial DNA degradation, presumably generated by stalled DNA or RNA polymerases at the damaged DNA strand. As this kind of DNA breaks and AP sites lack any genetic information, complete elimination of such sites can prevent the mitochondrial mutagenesis and will also protect the integrity of the mitochondrial genome. In these cases, stalled DNA or RNA polymerase may act as signal for mitochondrial DNA degradation.

>A nuclear genome encoded evolutionarily conserved mitochondrial protein Endonuclease G (Endo G) is most abundant and active nuclease in mitochondria and is considered to be primarily involved in generating RNA primer for the mitochondrial DNA replication (Cote and Ruiz-Carrillo, 1993). This Endo G preferentially cleaves the oxidatively damaged mitochondrial DNA (Ikeda and Ozaki, 1997); it is thought to act as an agent for selectively degrading the mitochondrial genome. But in contrary the Endo G null mice or the cells with lower level of Endo G do not show any kinds of defect in mitochondrial genome, its copy number and mutation rate (Zhang et al., 2003; Irvine et al., 2005; David et al., 2006; Huang et al., 2006) making it difficult to understand how Endo G is involved in maintaining the integrity and stability of the mitochondrial genome. Disruption of yeast homologue of EndoG, *NUC1* also showed marginal effects on the integrity of mitochondrial DNA (Zassenhaus et al., 1988), which puts a question mark on the role of Endo G and more studies are required to understand the function of Endo G in mitochondrial DNA degradation.

16 CONCLUSION

Besides certain distinct functional differences between mitochondrial and nuclear genome, they both are composed of DNA double helix with coding sequences. One

healthy living eukaryotic cell contains several mitochondria with each mitochondrion comprising of multiple copies of mitochondrial genome. Recent studies have established that mitochondrial genome has a considerably higher rate of mutation than the nuclear genome, thus generating heterogeneous population mitochondrial genome within the same cell and sometimes even within the same mitochondrion, thus conferring heteroplasmic condition. Although mitochondria are transmitted to the daughter cells during cell division, the segregation occurs in a random manner and irregular fashion as compared to nuclear genome segregation, thus generating additional instability at the cellular level. One relevant question that has been addressed in recent studies includes the high mutation rate in mitochondrial genome. The nuclear gene encoded protein DNA polymerase γ(*POLG*) is responsible for replicating the mitochondrial genome. The catalytic subunit of POLG contains the polymerase activity, while an exonuclease domain recognizes and removes the DNA base-pair mismatches which originate during DNA replication. More detailed studies have indicated a nucleotide imbalance in mitochondria decreases the fidelity of POLG, thus causing higher mutational rates in mitochondrial genome.

Mitochondrial mutations are known to be responsible for a number of common clinical diseases and symptoms. Some examples include diabetes, predisposition to Alzheimer's and Parkinson's disease. Another important mitochondria-associated disorder includes the Leber hereditary optic neuropathy (LHON), which causes loss of vision in both eyes and has been shown to be linked with homoplasmic mitochondrial DNA mutation. However, because of maternal influence in inheritance, males with a mitochondrial disease are generally not considered to be the potential carrier for transmitting the disorder to the next generation. Apart from this, since several mitochondria are present in a cell, carrying potential mutations in the mitochondrial DNA, understanding mitochondrial genome stability in a global scale is important. Since mitochondria function as the power house of the cell, tissues which have high energy demand, such as brain, renal, skeletal and the cardiac muscle tissues are often associated with distinct phenotypes due to mitochondrial mutations. Extensive research has shown that mitochondria affect the stability of the nuclear genome and vice versa. As mentioned before, although mitochondrial genome replicates independently, all classical DNA damage repair pathways participate in repairing mitochondrial genome damage. However, in contrast to nuclear genome stability mechanisms, information is limited about the underlying mechanisms which regulate mitochondrial genome stability. In addition, much less is known about the two-way communication and mutual influence of the nuclear and mitochondrial genomes. Therefore, further future research will provide more insights into the genome stability mechanisms in mitochondria for explanation of various genome instability phenotype associated with the failure of various mitochondrial functions.

ACKNOWLEDGMENTS

The authors gratefully acknowledge The Council of Scientific and Industrial Research, Govt. of India, (Ref. No. 38(1417)/16/EMR-II, dated:17/05/2016 to SR), UGC, Govt. of India Start-Up research grant No.F.30-141/2015 (BSR) and SERB, DST, Govt of India (Ref. No. ECR/2016/000539 to SR) for providing financial supports.

REFERENCES

Aamann MD, Sorensen MM, Hvitby C, Berquist BR, Muftuoglu M, Tian J, de Souza-Pinto NC, Scheibye-Knudsen M, Wilson DM 3rd, Stevnsner T, Bohr VA. Cockayne syndrome group B protein promotes mitochondrial DNA stability by supporting the DNA repair association with the mitochondrial membrane. FASEB J 24 (2010): 2334–2346. [Pubmed: 20181933]

Akbari M, Visnes T, Krokan HE, Otterlei M. Mitochondrial base excision repair of uracil and AP sites takesplace by single-nucleotide insertion and long-patch DNAsynthesis. DNA Repair 7 (2008): 605–616.

Alexeyev MF, Venediktova N, Pastukh V, Shokolenko I, Bonilla G, Wilson GL. Selective elimination of mutant mitochondrial genomes as therapeutic strategy for the treatment of NARP and MILS syndromes. Gene Ther 15 (2008): 516–523.

Anderson CT, Friedberg EC. The presence of nuclear and mitochondrial uracil-DNA glycosylase in extracts of human KB cells. Nucleic Acids Res 8 (1980): 875–888.

Anderson, S, Bankier, AT, Barrell, BG, de Bruijn, MH, Coulson, AR, Drouin, J, Eperon, IC, Nierlich, DP, Roe, BA, Sanger, F., et al. Sequence and organization of the human mitochondrial genome. Nature 290 (1981): 457–465.

Arpa J, Cruz-Martinez A, Campos Y, Gutierrez-Molina M, Garcia-Rio F, Perez-Conde C, Martin MA, Rubio JC, Del Hoyo P, Arpa-Fernandez A, Arenas J. Prevalence and progression of mitochondrial diseases: a study of 50 patients. Muscle Nerve 28 (2003): 690–695.

Ayala-Torres S, Chen Y, Svoboda T, Rosenblatt J, Van Houten B. Analysis of gene-specific DNA damage and repair using quantitative polymerase chain reaction. Methods 22 (2000): 135–147. [Pub Med: 11020328]

Bacman SR, Williams SL, Moraes CT. Intra- and inter-molecular recombination of mitochondrial DNA after in vivo induction of multiple double-strand breaks. Nucleic Acids Res 37 (2009): 4218–4226.

Bendich AJ Reaching for the ring: the study of mitochondrial genome structure. Curr Genet 24 (1993): 279–290.

Bendich AJ. Structural analysis of mitochondrial DNA molecules from fungi and plants using moving pictures and pulse field gel electrophoresis. J Mol Biol 255 (1996): 564–588.

Birsoy K, Wang T, Chen WW, Freinkman E, Abu-Remaileh M, Sabatini DM. An essential role of the mitochondrial electron transport chain in cell proliferation is to enable aspartate synthesis. Cell 162 (2015): 540–551.

Boesch P, Ibrahim N, Paulus F, Cosset A, Tarasenko V, Dietrich A. Plant mitochondria possess a short-patch base excision DNA repair pathway. Nucleic Acids Res 37 (2009): 5690–5700.

Boore JL. Animal mitochondrial genomes. Nucleic Acids Res 27 (1999): 1767–1780).

Brooks PJ, Wise DS, Berry DA, Kosmoski JV, Smerdon MJ, Somers RL, Mackie H, Spoonde AY, Ackerman EJ, Coleman K, Tarone RE, Robbins JH. The oxidative DNA lesion 8,5⊠-(S)- cyclo-2⊠-deoxyadenosine is repaired by the nucleotide excision repair pathway and blocks gene expression in mammalian cells. J Biol Chem 275 (2000): 22355–22362. [PubMed: 10801836]

Buller Well CE, Forget L, Lang BF. Evolution of monoblepharidalean fungi based on complete mitochondrial genome sequences. Nucleic Acids Res 31 (2003): 1614–1623.

Cadenas E, Davies KJ. Mitochondrial free radical generation oxidative stress and aging. Free Radic Biol Med 29 (2000): 222–230.

Cai S, Xu Y, Cooper RJ, Ferkowicz MJ, Hartwell JR, Pollok KE, Kelley MR. Mitochondrial targeting of human O6-methylguanine DNA methyltransferase protects against cell killing by chemotherapeutic alkylating agents. Cancer Res 65 (2005): 3319–3327.

Capaldi RA, Aggeler R, Gilkerson R, Hanson G, Knowles M, Marcus A, Margineantu D, Marusich M, Murray J, Oglesbee D, Remington SJ, Rossignol R. A replicating module as the unit of mitochondrial structure and functioning. Biochim Biophys Acta 1555 (2002): 192–195.

Chakravarti D, Ibeanu GC, Tano K, Mitra S. Cloningand expression in Escherichia coli of a human cDNA encoding the DNA repair protein N-methylpurine-DNAglycosylase. J Biol Chem 266 (1991): 15710–15715.

Chandel NS. Mitochondria as signalling organelles. BMC Biol 12 (2014): 34.

Chatterjee A, Mambo E, Sidransky D. Mitochondrial DNA mutations in human cancer. Oncogene 25 (2006): 4663–4674.

Clayton DA, Doda JN, Friedberg EC. The absence of a pyrimidine dimer repair mechanism in mammalian mitochondria. Proc Natl AcadSci 71 (1974): 2777–2781.

Clayton DA, Doda JN, Friedberg EC. Absence of a pyrimidine dimer repair mechanism for mitochondrial DNA in mouse and human cells. Basic Life Sci 5B (1975): 589–591.

Coffey G, Lakshmipathy U, Campbell C. Mammalian mitochondrial extracts possess DNA end-binding activity. Nucleic Acids Res 27 (1999): 3348–3354.

Contamine V, Picard M. Maintenance and integrity of the mitochondrial genome: a plethora of nuclear genes in the budding yeast. Microbiol Mol Biol Rev 64 (2000): 281–315.

Copeland WC, Longley MJ. DNA2 resolves expandingflap in mitochondrial base excision repair. Mol Cell 32 (2008): 457–458.

Coskun PE, Beal MF, Wallace DC. Alzheimer's brains harbour somatic mtDNA control-region mutations that suppress mitochondrial transcription and replication. Proc Natl Acad Sci U S A 101 (2004): 10726–10731.

Cote J, Ruiz-Carrillo A. Primers for mitochondrial DNA replication generated by endonuclease G. Science 261 (1993): 765–769.

David KK, Sasaki M, Yu SW, Dawson TM, Dawson VL. Endo G is dispensable in embryogenesis and apoptosis. Cell Death Differ 13 (2006): 1147–1155.

de Souza-Pinto NC, Mason PA, Hashiguchi K, Weissman L, Tian J, Guay D, Lebel M, Stevnsner TV, Rasmussen LJ, Bohr VA. Novel DNA mismatch-repair activity involving YB-1 in human mitochondria. DNA Repair (Amst.) 8 (2009): 704–719.

Dedon PC, Goldberg IH. Free-radical mechanisms involved in the formation of sequence dependent bi-stranded DNA lesions by the antitumor antibiotics bleomycin, neocarzinostatin, and calicheamicin. Chem Res Toxicol 5 (1992): 311–332. [PubMed: 1380322]

Dedon PC, Plastaras JP, Rouzer CA, Marnett LJ. Indirect mutagenesis by oxidative DNA damage: formation of the pyrimidopurinone adduct of deoxy guanosine by base propenyl. Proc Natl AcadSci U S A 95 (1998): 11113–11116. [PubMed: 9736698]

Dodson ML, Lloyd RS. Mechanistic comparisonsamong base excision repair glycosylases. Free Radic Biol Med 32 (2002): 678–682.

Duxin JP, Dao B, Martinsson P, Rajala N, Guittat L, Campbell JL, Spelbrink JN, Stewart SA. HumanDna2 is a nuclear and mitochondrial DNA maintenanceprotein. Mol Cell Biol 29 (2009): 4274–4282.

Dzierzbicki P, Koprowski P, Fikus MU, Malc E, Ciesla Z. Repair of oxidative damage in mitochondrial DNA of *Saccharomyces cerevisiae*: involvement of the MSH1-dependent pathway. DNA Repair (Amst.) 3 (2004): 403–411.

Esser C, Ahmadinejad N, Wiegand C, Rotte C, Sebastiani F, Gelius-Dietrich G, Henze K, Kretschmann E, Richly E, Leister D, A genome phylogeny for mitochondria among α-proteobacteria and a predominantly eubacterial ancestry of yeast nuclear genes. Mol Biol Evol 21 (2004): 1643–1660.

Fontanesi F, Soto IC, Barrientos A. Cytochrome-c oxidase biogenesis: new levels of regulation. IUBMB Life 60 (2008): 557–568.

Fukui H, Moraes CT. Mechanisms of formation and accumulation of mitochondrial DNA deletions in aging neurons. Hum Mol Genet 18 (2009): 1028–1036.

Gray MW, Lang BF, Cedergren R, Golding GB, Lemieux C, Sankoff D, Turmel M, Brossard N, Delage E, Littlejohn TG,et al. Genome structure and gene content in protist mitochondrial DNAs. Nucleic Acids Res 26 (1998): 865–878.

Gredilla, R., DNA damage and base excision repair in mitochondria and their role in aging. J Aging Res 2011, Article ID 257093, 2011: 9 pages, https://doi.org/10.4061/2011/257093.

Gutman BL, Niyogi KK. Evidence for base excision repair of oxidative DNA damage in chloroplasts of Arabidopsis thaliana. J Biol Chem 284 (2009): 17006–17012.

Guzy RD, Schumacker Pt. Oxygen sensing by mitochondria at complex III: the paradox of increased reactive oxygen species during hypoxia. Exp Physiol 91 (2006): 807–819.

Hanes JW, Thal DM, Johnson KA. Incorporation and replication of 8-oxo-deoxyguanosine by the human mitochondrial DNA polymerase. J Biol Chem 281 (2006): 36241–36248.

Hang B, Chenna A, Rao S, Singer B. 1, N6-ethenoadenine and 3, N4-ethenocytosine are excised by separate human DNA glycosylases. Carcinogenesis 17 (1996): 155–157. [PubMed: 8565126]

Hess MT, Gunz D, Luneva N, Geacintov NE, Naegeli H. Base pair conformation-dependent excision of benzo[a]pyrene diol epoxide-guanine adducts by human nucleotide excision repair enzymes. Mol Cell Biol 17 (1997): 7069–7076. [PubMed: 9372938]

Higuchi Y, Linn S. Purification of all forms of HeLa cell mitochondrial DNA and assessment ofdamage to it caused by hydrogen peroxide treatment of mitochondria or cells. J Biol Chem 270 (1995): 7950–7956. [PubMed: 7713892]

Hong S, Pedersen PL. Mitochondrial ATP synthase: a bioinformatics approach reveals new insights about the roles of supernumerary subunits g and A6L. J Bioenerg Biomembr 36 (2004): 515–523.

Howell N, Elson JL, Chinnery PF, Turnbull DM. mtDNA mutations and common neurodegenerative disorders. Trends Genet 21 (2005): 583–586.

Hu J, de Souza-Pinto NC, Haraguchi K, Hogue BA, Jaruga P, Greenberg MM, Dizdaroglu M, Bohr VA. Repair offormamidopyrimidines in DNA involves different glycosylases: Role of the OGG1, NTH1, and NEIL1 enzymes. J Biol Chem 280 (2005): 40544–40551.

Huang KJ, Ku CC, Lehman IR. Endonuclease G. A role for the enzyme in recombination and cellular proliferation. Proc Natl AcadSci U S A 103 (2006): 8995–9000.

Ichikawa J, Tsuchimoto D, Oka S, Ohno M, Furuichi M, Sakumi K, Nakabeppu Y. Oxidation of mitochondrial deoxynucleotide pools by exposure to sodium nitroprusside induces cell death. DNA Repair (Amst) 7 (2008): 418–430.

Ikeda S, Ozaki K. Action of mitochondrial endonuclease G on DNA damaged by L-ascorbic acid, peplomycin, and cis-diamminedichloroplatinum (II). Biochem Biophys Res Commun 235 (1997): 291–294.

Inagaki S, Esaka Y, Goto M, Deyashiki Y, Sako M. LC-MS study on the formation of cyclic 1, N2-propano guanine adduct in the reactions of DNA with acetaldehyde in the presence of histone. Biol Pharm Bull 27 (2004): 273–276. [PubMed: 14993787]

Irvine RA, Adachi N, Shibata DK, Cassell GD, Yu K, Karanjawala ZE, Hsieh CL, Lieber MR. Generation and characterization of endonuclease G null mice. Mol Cell Biol 25 (2005): 294–302.

Jeong YC, Nakamura J, Upton PB, Swenberg JA. Pyrimido [1,2-a]-purin-10(3H)-one, M1G, is less prone to artefact than base oxidation. Nucleic Acids Res 33 (2005): 6426–6434. [PubMed:16282591]

Johnson KA, Fink SP, Marnett LJ. Repair of propanodeoxyguanosine by nucleotide excision repair in vivo and in vitro. J Biol Chem 272 (1997): 11434–11438. [PubMed: 9111054]

Kalifa L, Beutner G, Phadnis N, Sheu SS, Sia EA. Evidence for a role of FEN1 in maintaining mitochondrial DNA integrity. DNA Repair 8 (2009): 1242–1249.

Kamenisch Y, Fousteri M, Knoch J, von Thaler AK, Fehrenbacher B, Kato H, Becker T, Dolle ME, Kuiper R, Majora M, et al. Proteins of nucleotide and base excision repair pathways interact in mitochondria to protect from loss of subcutaneous fat, a hallmark of aging. J Exp Med 207 (2010): 379–390. [PubMed: 20100872]

Kang D, Nishida J, Iyama A, Nakabeppu Y, Furuichi M, Fujiwara T, Sekiguchi M, Takeshige K. Intracellular localization of 8-oxo-dGTPase in human cells, with special reference to the role of the enzyme in mitochondria. J Biol Chem 270 (1995): 14659–14665.

Karahalil B, de Souza-Pinto NC, Parsons JL, Elder RH, Bohr VA. Compromised incision of oxidized pyrimidines in liver mitochondria of mice deficient in NTH1and OGG1 glycosylases. J Biol Chem 278 (2003): 33701–33707.

Kennell, JC, Cohen, SM In Handbook of Fungal Biotechnology (Arora, D., ed.), Marcel Dekker King MP, Attardi G. Human cells lacking mtDNA: Repopulation with exogenous mitochondria by complementation. Science 246 (1989): 500–503.

King, MP, Attardi, G. Human cells lacking mtDNA: repopulation with exogenous mitochondria by complementation. Science 246 (1989): 500-503.

Kobayashi K, Kanno S, Smit B, van der Horst GT, Takao M, Yasui A. Characterization of photolyase/blue-light receptor homologs in mouse and human cells. Nucleic Acids Res 26 (1998): 5086–5092.

Kukat A, Kukat C, Brocher J, Schafer I, Krohne G, Trounce IA, Villani G, Seibel P. Generation of rho0 cells utilizing a mitochondrially targeted restriction endonuclease and comparative analyses. Nucleic Acids Res 36 (2008): e44.

Kuraoka I, Bender C, Romieu A, Cadet J, Wood RD, Lindahl T. Removal of oxygen free-radical induced 5′,8-purine cyclodeoxy nucleosides from DNA by the nucleotide excision-repair pathway in human cells. Proc Natl Acad Sci U S A 97 (2000): 3832–3837. [PubMed: 10759556]

Lagerwerf S, Vrouwe MG, Overmeer RM, Fousteri MI, Mullenders LH. DNA damage response and transcription. DNA Repair (Amst) 10 (2011): 743–750. [PubMed: 21622031]

Lakshmipathy U, Campbell C. Double strand break re-joining by mammalian mitochondrial extracts. Nucleic Acids Res 27 (1999): 1198–1204.

LeDoux SP, Wilson GL, Beecham EJ, Stevnsner T, Wassermann K, Bohr VA. Repair of mitochondrial DNA after various types of DNA damage in Chinese hamster ovary cells. Carcinogenesis 13 (1992): 1967–1973. [PubMed: 1423864]

Legros F, Malka F, Frachon P, Lombes A, Rojo M. Organization and dynamics of human mitochondrial DNA. J Cell Sci 117 (2004): 2653–2662.

Lloyd DR, Hanawalt PC. p53-dependent global genomic repair of benzo[a]pyrene-7,8-diol-9,10-epoxide adducts in human cells. Cancer Res 60 (2000): 517–521. [PubMed: 10676627]

Lu K, Moeller B, Doyle-Eisele M, McDonald J, Swenberg JA. Molecular dosimetry of N2- hydroxymethyl-d G DNA adducts in rats exposed to formaldehyde. Chem Res Toxicol 24 (2011): 159–161. [PubMed: 21155545]

Manchekar M, Scissum-Gunn K, Song D, Khazi F, McLean SL, Nielsen, BL. DNA recombination activity in soybean mitochondria. J Mol Biol 356 (2006): 288–299.

Mandal SM, Hegde ML, Chatterjee A, Hegde PM, Szczesny B, Banerjee D, Boldogh I, Gao R, Falkenberg M, Gustafsson CM, et al. Role of human DNA glycosylase Nei-like 2 (NEIL2) and single strand break repair protein polynucleotide kinase 30-phosphatase in maintenance of mitochondrial genome. J Biol Chem 287 (2012): 2819–2829.

Mason PA, Lightowlers RN. why do mammalian mitochondria possess a mismatch repair activity? FEBS Lett 554 (2003): 6–9.

Mason PA, Matheson EC, Hall AG, Lightowlers RN. Mismatch repair activity in mammalian mitochondria. Nucleic Acids Res 31 (2003): 1052–1058.

Matsuda T, Yabushita H, Kanaly RA, Shibutani S, Yokoyama A. Increased DNA damage inALDH2-deficient alcoholics. Chem Res Toxicol 19 (2006): 1374–1378. [PubMed: 17040107]

Mcbride H, Neuspiel M, Wasiak S. Mitochondria: more than just a powerhouse. Curr Biol 16 (2006): R551–R560.

Mecocci P, MacGarvey U, Kaufman AE, Koontz D, Shoffner JM, Wallace DC, Beal MF Oxidative damage to mitochondrial DNA shows marked age-dependent increases in human brain. Ann Neurol 34 (1993): 609–616. [PubMed: 8215249]

Mita S, Monnat RJ, Jr, Loeb LA. Resistance of HeLa cell mitochondrial DNA to mutagenesis by chemical carcinogens. Cancer Res 48 (1988): 4578–4583.

Miyaki M, Yatagai K, Ono T, Strand breaks of mammalian mitochondrial DNA induced by carcinogens. Chem Biol Interact 17 (1977): 321–329.

Morel F, Renoux M, Lachaume P, Alziari S. Bleomycin-induced double-strand breaks in mitochondrial DNA of Drosophila cells are repaired. Mutat Res 637 (2008): 111–117.

Murphy MP. How mitochondria produce reactive oxygen species. Biochem J 417 (2009): 1–13.

Myers KA, Saffhill R, O'Connor PJ. Repair of alkylated purines in the hepatic DNA of mitochondria and nuclei in the rat. Carcinogenesis 9 (1988): 285–292.

Nakabeppu Y. Molecular genetics and structural biology of human MutT homolog. MTH1 Mutat Res 477 (2001): 59–70.

Odahara M, Inouye T, Fujita T, Hasebe M, Sekine Y. Involvement of mitochondrial-targeted Rec-A in the repair of mitochondrial DNA in the moss, *Physcomitrella patens.* Genes Genet Syst 82 (2007): 43–51.

Ogawa S, Yoshino R, Angata K, Iwamoto M, Pi M. The mitochondrial DNA of Dictyostelium discoideum: complete sequence, gene content and genome organization. Mol. Gen. Genet. 263 (2000): 514–519.

Ohtsubo T, Nishioka K, Imaiso Y, Iwai S, Shimokawa H, Oda H, Fujiwara T, Nakabeppu Y. Identificationof human Mut-Y homolog (hMYH) as a repair enzymefor 2-hydroxyadenine in DNA and detection of multiple forms of hMYH located in nuclei and mitochondria. Nucleic Acids Res 28 (2000): 1355–1364.

Okimoto R, Macfarlane JL, Clary DO, Wolstenholme DR The mitochondrial genomes of two nematodes *Caenorhabditis elegans*and *Ascarissuum.* Genetics 130 (1992): 471–498.

Pagliarini D, Rutter J. Hallmarks of a new era in mitochondrial biochemistry. Genes Dev 27 (2013): 2615–2627.

Papa S. Mitochondrial oxidative phosphorylation changes in the life span. Molecular aspects and physio pathological implications. Biochim Biophys Acta 1276 (1996): 87–105.

Pascucci B, Versteegh A, van Hoffen A, van Zeeland AA, Mullenders LH, Dogliotti E. DNA repair of UV photoproducts and mutagenesis in human mitochondrial DNA. J Mol Biol 273 (1997): 417–427. [PubMed: 9344749]

Pasupathy K, Pradhan DS. Evidence for excision repair in pro-mitochondrial DNA of anaerobic cells of *Saccharomyces cerevisiae.* Mutat Res 273 (1992): 281–288.

Pesole G, Allen JF, Lane N, Martin W, Rand DM, Schatz G, Saccone C. The neglected genome. EMBO Rep 13 (2012): 473–474.

Pettepher CC, LeDoux SP, Bohr VA, Wilson GL. Repair of alkali-labile sites within the mitochondrial DNA of RINr 38 cells after exposure to the nitrosoureastreptozotocin. J Biol Chem 266 (1991): 3113–3117.

Prachar J. Mouse and human mitochondrial nucleoid—detailed structure in relation to function. Gen Physiol Biophys 29 (2010): 160–174.

Pursell ZF, McDonald JT, Mathews CK, Kunkel TA. Trace amounts of 8-oxo-dGTP in mitochondrial dNTP pools reduce DNA polymerase g replication fidelity. Nucleic Acids Res 36 (2008): 2174–2181.

Reardon JT, Vaisman A, Chaney SG, Sancar A. Efficient nucleotide excision repair of cisplatin, oxaliplatin, and Bis-aceto-ammine-dichloro-cyclohexylamine-platinum (IV) (JM216) platinum intra strand DNA diadducts. Cancer Res 59 (1999): 3968–3971. [PubMed: 10463593]

Roginskaya M, Bernhard WA, Marion RT, Razskazovskiy Y. The release of 5-methylene-2-furanone from irradiated

DNA catalysed by cationic polyamines and divalent metal cations. Radiat Res 163 (2005): 85–89.

Ryoji M, Katayama H, Fusamae H, Matsuda A, Sakai F, Utano H. Repair of DNA damage in a mitochondrial lysate of *Xenopuslaevis* oocytes. Nucleic Acids Res 24 (1996): 4057–4062.

Sage M, Gildemeister OS, Knight KL. Discovery of a novel function for human Rad51: maintenance of the mitochondrial genome. J Biol Chem 285 (2010): 18984–18990.

Sancar A. Structure and function of photolyase and in vivo enzymology: 50th anniversary. J Biol Chem 283 (2008): 32153–32157.

Saraste M. Oxidative phosphorylation at the fin de siecle. Science 283 (1999): 1488–1492.

Satoh MS, Huh N, Rejewski MF, Kuroki T. Enzymatic removal of O6-ethylguanine from mitochondrial DNA in rat tissues exposed to N-ethyl-N-nitrosourea in vivo. J Biol Chem 263 (1988): 6854–6856.

Schardl CL, Pring DR, Lonsdale DM (1985) Mitochondrial DNA rearrangements associated with fertile revertants of S-type male-sterile maize. Cell 43: 361–368.

Shadel GS. Clayton DA Mitochondrial DNA maintenance in vertebrates. Annu Rev Biochem 66 (1997): 409–435.

Shokolenko I, Venediktova N, Bochkareva A, Wilson GL, Alexeyev MF. Oxidative stress induces degradation of mitochondrial DNA. Nucleic Acids Res 37 (2009): 2539–2548. [PubMed: 19264794]

Simsek D, Furda A, Gao Y, Artus J, Brunet E, Hadjantonakis AK, Van Houten B, Shuman S, McKinnon PJ, Jasin M. Crucial role for DNA ligase III in mitochondria but not in Xrcc1-dependent repair. Nature 471 (2011): 245–248. [PubMed: 21390132]

Stierum RH, Dianov GL, Bohr VA. Single-nucleotide patch base excision repair of uracil in DNA by mitochondrial protein extracts. Nucleic Acids Res 27 (1999): 3712–3719.

Swenberg JA, Bogdanffy MS, Ham A, Holt S, Kim A, Morinello EJ, Ranasinghe A, Scheller N, Upton PB. Formation and repair of DNA adducts in vinyl chloride- and vinyl fluoride-induced carcinogenesis. IARC Sci Publ 150 (1999): 29–43. [PubMed: 10626206]

Swenberg JA, Lu K, Moeller BC, Gao L, Upton PB, Nakamura J, Starr TB. Endogenous versus exogenous DNA adducts: their role in carcinogenesis, epidemiology, and risk assessment. Toxicol Sci 120, Suppl 1 (2011): S130–1145. [PubMed: 21163908]

Tahbaz N, Subedi S, Weinfeld M. Role of polynucleotide kinase/phosphatase in mitochondrial DNA repair. Nucleic Acids Res 40 (2012): 3484–3495.

Takahashi M, Teranishi M, Ishida H, Kawasaki J, Takeuchi A, Yamaya T, Watanabe M, Makino A, Hidema J. Cyclobutane pyrimidine dimer (CPD) photolyase repairs ultraviolet-B-induced CPDs in rice chloroplast and mitochondrial DNA. Plant J 66 (2011): 433–442.

Takao M, Aburatani H, Kobayashi K, Yasui A. Mitochondrial targeting of human DNA glycosylases for repair of oxidative DNA damage. Nucleic Acids Res 26 (1998): 2917–2922.

Taylor RW, Turnbull DM. Mitochondrial DNA mutations in human disease. Nat Rev Genet 6 (2005): 389–402.

Thorslund T, Sunesen M, Bohr VA, Stevnsner T. Repairof 8-oxoG is slower in endogenous nuclear genes than inmitochondrial DNA and is without strand bias. DNA Repair (Amst) 1 (2002): 261–273.

Thyagarajan B, Padua RA, Campbell C. Mammalian mitochondria possess homologous DNA recombination activity. J Biol Chem 271 (1996): 27536–27543.

Todd RC, Lippard SJ. Inhibition of transcription by platinum antitumor compounds. Metallomics 1 (2009): 280–291. [PubMed: 20046924]

Vothknecht UC, Soll J. The endosymbiotic origin of organelles: an ancient process still very much in fashion. Biol Chem 388 (2007): 877.

Wallace DC. A mitochondrial paradigm of metabolic and degenerative diseases, aging, and cancer: A dawn for evolutionary medicine. Annu Rev Genet 39 (2005): 359–407.

Wang D, Lippard SJ. Cellular processing of platinum anticancer drugs. Nat Rev Drug Discov 4 (2005): 307–320. [PubMed: 15789122]

Wang M, Mc. Lntee EJ, Cheng G, Shi Y, Villalta PW, Hecht SS. Identification of DNA adducts of acetaldehyde. Chem Res Toxicol 13 (2000): 1149–1157. [PubMed: 11087437]

Wei D, Maher VM, McCormick JJ. Site-specific rates of excision repair of benzo[a]pyrene diol epoxide adducts in the hypoxanthine phosphoribosyl transferase gene of human fibroblasts: correlation with mutation spectra. Proc Natl AcadSci U S A. 92 (1995): 2204–2208. [PubMed: 7892248]

Wittig I, Meyer B, Heide H, Steger M, Bleier L, Wumaier Z, Karas M, Schagger H. Assembly and oligomerization of human ATP synthase lacking mitochondrial subunits a and A6L. Biochim Biophys Acta 1797 (2010): 1004–1011.

Wolstenholme, DR Animal mitochondrial DNA: structure and evolution. In Mitochondrial genomes. Wolstenholme DR, Jeon KW eds., (San Diego, CA: Academic Press, INC.), (1992) Vol. 141, pp. 173–216.

Wood PM. The two redox potentials for oxygen reduction to superoxide. Trends Biochem Sci 12 (1987): 250–251.

Yasuhira S, Yasui A. Alternative excision repair pathway of UV-damaged DNA in *Schizosaccharomycespombe* operates both in nucleus and in mitochondria. J Biol Chem 275 (2000): 11824–11828.

Yasui A, Yajima H, Kobayashi T, Eker AP, Oikawa A. Mitochondrial DNA repair by photolyase. Mutat Res 273 (1992): 231–236.

Zassenhaus HP, Hofmann TJ, Uthayashanker R, Vincent RD, Zona M. Construction of a yeast mutant lacking the mitochondrial nuclease. Nucleic Acids Res 16 (1988): 3282–3329.

Zhang J, Dong M, Li L, Fan Y, Pathre P, Dong J, Lou D, Wells JM. Olivares-Villagomez D, Van Kaer L, Wang X, Xu M. Endonuclease G is required for early embryogenesis and normal apoptosis in mice. Proc Natl Acad Sci U S A 100 (2003): 15782–15787.

Zheng L, Zhou M, Guo Z, Lu H, Qian L, Dai H, Qiu J, Yakubovskaya E, Bogenhagen DF, Demple B, et al. Human DNA2 is a mitochondrial nuclease/helicase for efficient processing of DNA replication and repair intermediates. Mol Cell 32 (2008): 325–336.

Section II

Mitochondrial Dysfunction and Diseases

4 Instability of Human Mitochondrial DNA, Nuclear Genes and Diseases

Katarzyna Tońska

Faculty of Biology, Institute of Genetics and Biotechnology, University of Warsaw

CONTENTS

1   INTRODUCTION

As the proteins encoded by mitochondrial DNA (mtDNA) are crucial for proper oxidative phosphorylation system function in humans. Copy number and mtDNA integrity are under strict control. While mitochondrial genes encode only 13 proteins (encoding subunits of respiratory chain complexes I, III, IV and ATPase) and RNA apparatus (22 tRNA's and 2 rRNA's) necessary for their translation, all machineries responsible for mtDNA maintenance are encoded in the nucleus (Figure 1). Abnormal function of any element in that machinery may lead to mitochondrial instability presenting as the appearance of multiple large-scale mtDNA deletions (spanning several kilo-bases), mtDNA depletion (decrease in copy-number) or accumulation of point mutations, leading to mitochondrial diseases.

Mitochondrial DNA maintenance defects frequently have the form of multiple large-scale mtDNA deletions but single large mtDNA deletions are also found in patients with sporadic mitochondrial disease. Moreover, single large-scale mtDNA deletions were described in the first molecularly diagnosed cases of mitochondrial disease[1].

Mitochondrial DNA is a multicopy molecule within the cell (hundreds to thousands copies in somatic cells). This means that in the case of mtDNA terms homozygous or heterozygous cannot be used. Homoplasmy is when all the molecules have the same sequence and heteroplasmy are when at least two types of mtDNA molecules are present. For years it was believed that homoplasmy is the natural state in healthy cell. Now in the era of next generation sequencing it turned out that low level heteroplasmy is common, but not visible using routine laboratory techniques. In case of large-scale mtDNA deletions (as well as in case of the most of the pathogenic point mutations) heteroplasmy is clearly visible. This type of mtDNA lesion is never even close to homoplasmy. All mitochondrial encoded genes are essential and a large-scale deletion cuts out at least a few genes, so only an adequate proportion of wild type molecules provides all necessary mtDNA products. In the case of various mtDNA mutations including large-scale deletions the threshold effect can be observed; the respiratory chain defect appears when the percentage of mutated molecules exceeds a certain value. It is considered that for large-scale mtDNA deletions it is around 60% at the cellular level.

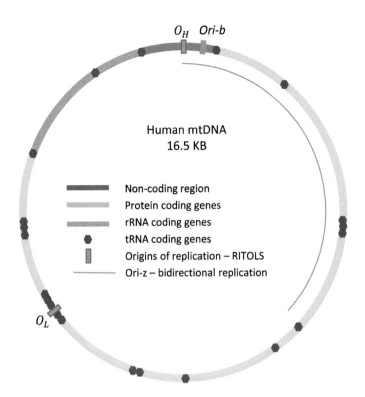

FIGURE 1   The schematic map of human mitochondrial DNA. O_H and O_L are the "classical" origins of replication postulated by Clayton in the strand displacement model. Together with Ori-b they serve as replication origins in RITOLS (RNA incorporation throughout the lagging strand) model. In coupled leading and lagging strand DNA replication model replication may start in any position and it is bidirectional, but most frequently it is initiated in the region marked as Ori-z. The usefulness in certain conditions and truthfulness of these models is still not certain.

mtDNA molecules are organized in nucleoids – nucleoprotein complexes attached to the inner mitochondrial membrane. In one nucleoid one to ten mtDNA molecules are present and they are descendants of one ancestor molecule as the replication process takes place in the nucleoid. This type of organization has an impact on heteroplasmy. When an mtDNA deletion appears and is replicated, all molecules in the nucleoid will have it but it may not affect neighboring nucleoids.

Mitochondrial DNA stability is ensured by the proper action of groups of proteins involved in mitochondrial DNA replication and transcription, nucleotide metabolism but also mitochondrial dynamics. There are even less obvious mechanisms which, when perturbed, indirectly lead to mtDNA instability.

Mitochondrial DNA instability is one of the most frequent causes of mitochondrial diseases with an estimated prevalence of about 4 in 100000[2]. As in other subgroups of mitochondrial diseases, those resulting from mtDNA instability show variable phenotypes with different presentations in patients harboring mutations in the same gene, or even the same mutations. Overlapping phenotypes are also frequent but some of the symptoms may suggest that we are dealing with mitochondrial disease due to mtDNA maintenance defect. Progressive external

ophthalmoplegia (PEO), ptosis, hepatic and gastrointestinal involvement, sensory ataxia and peripheral neuropathy are examples of such symptoms.

The groups of genes encoding proteins responsible for mtDNA stability together with the pathology, responsible for their malfunction, are presented here. Also the possible mechanisms leading to deletion formation and involved in mtDNA deletion behavior in cell tissue and organism are described. At the end the mitochondrial instability in neurodegenerative diseases and ageing is discussed.

2   MITOCHONDRIAL DNA REPLICATION

The first mechanism of mtDNA replication, the Strand Displacement model, was proposed by Clayton[3] in the early 1970s. In subsequent years the model was modified, discussed and negated by scientists. Two other models have been proposed: Bidirectional Coupled Leading and Lagging Strand DNA Replication[4] and RNA Incorporation Throughout the Lagging Strand, RITOLS[5]. Although mitochondrial DNA replication have been studied for many years there is still much to be discovered in this area (Figure 1). In fact it is not yet agreed whether there is only one mtDNA replication mode or more. It is possible that the mtDNA replication depends on the mode of cell, its current bioenergetic state and needs, etc. All the fascinating history of mtDNA discoveries, controversies and mysteries together with the present state of the art has been described[6]. Regardless of the precise model of replication some players are common in every replicative machinery including DNA polymerase, helicase, primase and single strand binding protein.

Polymerase gamma, the only replicative DNA polymerase in human mitochondria, is a heterotrimer composed of one catalytic subunit encoded by the *POLG* gene (15q26.1) and two accessory subunits encoded by the *POLG2* gene (17q23.3). The catalytic subunit is responsible not only for polymerase and exonuclease activities but also have the properties of reverse transcriptase. Accessory subunits increase the processivity of the enzyme.

2.1 POLYMERASE GAMMA AND ASSOCIATED HUMAN DISEASES

In 2001 for the first time *POLG* mutations were shown to induce human disease and dominant and recessive mutations were indicated to be responsible for PEO in three families[7]. Until now hundreds of *POLG* variants have been described, many of them pathogenic and *POLG* is considered one of the most frequently mutated genes in human mitochondrial diseases. *POLG* mutations can lead to the so-called POLG-related disorders including autosomal recessive (more frequent) and dominant (quite rare) PEO and PEO+ (PEO with additional neurological and muscular symptoms), recessive ataxia-neuropathy spectrum (SANDO – sensory ataxic neuropathy, dysarthria/dysphagia and external ophthalmoplegia, MIRAS – mitochondrial recessive ataxia syndrome) and also the recessive Alpers-Huttenlocher syndrome (AHS).

The most devastating is Alpers-Huttenlocher syndrome. It usually appears in young children of 2 to 4 years old but can occur later between 17 to 24 years of age after normal early development. In most patients (over 50%) the disease starts with epilepsy. Liver dysfunction may appear after or before epilepsy but can be accelerated by seizure treatment with valproic acid. The third symptom necessary to diagnose Alpers-Huttenlocher syndrome is developmental regression. The disease is progressive and leads to death within 4 years from its onset[8]. mtDNA depletion can be seen in muscle or liver biopsy of the patients with AHS.

Ataxia neuropathy spectrum onset is in adolescents and young adults. The symptoms include, beside peripheral neuropathy and ataxia, dysarthria, epileptic and myoclonus seizures, also some psychiatric disorders and cognitive impairment. Multiple wide-range mtDNA deletions are found in muscle biopsies of the patients with ataxia neuropathy spectrum.

A phenotype similar to MNGIE (Mitochondrial NeuroGastroIntestinal Encephalomyopathy) is a rare presentation of *POLG* disease. The onset is usually in teenagers. Beside gastrointestinal symptoms leading to cachexia, PEO, ataxia and peripheral neuropathy are also present[7].

PEO and PEO+ are diseases (see below) of adults. The onset of the disease is in over 18, usually before the age of 40 but later cases are also common. mtDNA point mutations and depletion can be seen in muscle biopsies of the patients with PEO/PEO+ and *POLG* mutations but mtDNA wide-range multiple deletions are the most common and easy to analyze.

The spectra of *POLG* mutations are similar in all the above described diseases (except a few dominant pathogenic variants responsible for PEO diseases). Homozygous or compound heterozygous recessive variants can be found in all *POLG*-related diseases but compound heterozygosity with one variant located in the polymerase and the other one in the exonuclease domain seem to lead to more severe AHS phenotype.

The most frequent *POLG* pathogenic variants are recessive p.Trp748Ser, p.Ala467Thr and p.Gly848Ser but the actual spectrum may differ between populations[9]. The number of well documented dominant *POLG* mutations is limited to p.Ala957Ser, p.Tyr955Cys and p.Ser511Asn, although it is sometimes suggested that even recessive mutations like p.Ala467Thr can occasionally lead to a residual phenotype of mitochondrial disease.

While in *POLG* recessive mutations are typical loss of function mutations, the dominance mechanism is unknown. Almost all dominant *POLG* mutations disrupt polymerase function by reducing it up to 1% of the original activity and at the same time significantly lowering its fidelity.

Mutations in the *POLG2* (17q23.3) gene encoding accessory polymerase gamma subunits have also been described[10] and are rare. What is interesting in the biochemical function of *POLG2* is that the two accessory subunits, although they are the same, play different roles during DNA synthesis. One strengthens the interaction of polymerase gamma with DNA and the other increases the speed of polymerization.

POLG2 mutations lead to autosomal dominant PEO with mtDNA multiple wide-range deletions. While the mutations in this gene are rare and confirmed only in a few cases as a cause of the disease, it is still difficult to describe the full phenotype of *POLG2* patients. Beside PEO fatigue, exercise intolerance, hypotonia, gastric problems, seizures and liver dysfunction have also been described.

The autosomal dominant character of *POLG2* mutations is not obvious. The idea that haploinsufficiency may be the case turned out to be false after the mice lacking one *POLG2* allele were found to show no phenotype[11]. It seems that at least some missense pathogenic variants are dominant negative[12] and the effect of the combination of wild type with the mutant one shows five times lower affinity for the catalytic subunit and leads to lower processivity.

2.2 Diseases Resulting from the Other Replication Disorders

The third crucial enzyme in mtDNA replication is Twinkle helicase encoded by the *TWNK* gene (formerly called c10orf2 or PEO1 10q24.31). It was described for the first time in 2001[13] and it turned out that the mutations in *TWNK* lead to autosomal dominant PEO with mtDNA large-scale deletions. Subsequently it turned out that although the main presentation of Twinkle pathology is autosomal dominant PEO recessive *TWNK* mutations are also found (homozygous and compound heterozygous) and lead to more severe phenotypes like mitochondrial depletion syndrome with hepatocerebral symptoms and early onset (9 to 18 months)[14]. The last presentation of TWNK disease caused by recessive mutations is Perrault syndrome. The phenotype of the patients is less severe than in the case of the mitochondrial depletion syndrome. The onset of the disease is in teens with amenorrhea and later sensory axonal neuropathy, ataxia and sensorineural hearing loss. The protein is similar to phage T7 primase/helicase. Taking into account that polymerase gamma belongs to the same class A of polymerases as T7 polymerase[15] the evolutionary past of the mitochondrial replicative machinery seems to be very interesting[16]. While the helicase domain sequence is evolutionary conserved in Twinkle, no significant conservation is observed within the region playing the primase role in T7 helicase-primase. Later studies show that indeed Twinkle maintained only the helicase activity in the system of mammalian mtDNA replication[17].

In 2013 a new enzyme PRIMPOL was described and precharacterized as a protein exhibiting two activities – primase and polymerase with unusual properties: it can by itself start DNA synthesis and is extremely resistant to DNA damage[18]. PRIMPOL acts both in the nucleus and mitochondrion and seems to play similar role in both compartments – it rescues stalled replication forks by adding dNTP at the site of lesion[19]. It does not play a role in the initiation of replication. Only one variant in *PRIMPOL* (4q35.1) (p.Tyr89Asp) was suggested to play role in autosomal dominant high myopia[20]. Although it was proven that this variant negatively influences processivity of the enzyme[21] it is difficult to say whether the proposed phenotype results from alteration of nuclear, mitochondrial or both functions. Moreover, the involvement of PRIMPOL mutation in high myopia was later questioned[22]. The RNA polymerase involved in transcription, playing the role of primase in mtDNA replication will be presented later.

The last element to discuss is mtSSB – single strand binding protein. It functions as a tetramer and stimulates polymerase gamma activity. It was shown that the amount of mtSSB directly correlates with mtDNA copy number[23]. Moreover, lack of mtSSB is lethal in a *Drosophila melanogaster* model[24]. The protein is encoded by the *SSBP* gene (7q34) but until now no pathogenic variants in it have been described.

Mitochondrial DNA replication was thought to be independent on the cell cycle. Recent studies show that although replication takes place during the whole cell cycle it is more intensive during the S phase[25].

3 MITOCHONDRIAL TRANSCRIPTION

Although contribution of the replicative machinery to the stability of mitochondrial DNA seems straightforward it is not so in case of transcription. There are several features that may explain the involvement of transcription machinery in mtDNA maintenance. First, mtDNAs are not lonely molecules floating in the mitochondrial matrix. In fact mitochondrial DNA molecules are organized in so-called nucleoids in which the protein DNA complexes remain attached to the mitochondrial inner membrane. The proteins forming nucleoids are responsible for mtDNA replication, repair and transcription. This means that transcription and replication machinery are acting in the same place. They also, to some extent, act at the same time as RNA polymerase (POLRMT) is involved in priming replication.

mtDNA transcription shows certain similarities to the bacterial system in that the initial transcription products are polycistronic. The main protein elements involved in this process are, besides POLRMT (encoded by *POLRMT*, 19p13.3), TFAM (encoded by *TFAM*, 10q21.1), TFBM2 (encoded by *TFB2M*, 1q44) and TEFM (encoded by *TEFM*, 17q11.2). TFAM is a transcription factor A binding the upstream transcription start and responsible for recruiting POLRMT. TFBM2 is also a transcription factor and is necessary to form a fully active transcription complex. TEFM is a mitochondrial transcription elongation factor responsible for the complex processivity. It also makes a link between transcription and replication. TEFM binding to the transcription complex promotes transcription but when it is absent, the primer for replication is produced[26]. The last core elements, mTERFs – mitochondrial termination factors (1–4) are responsible *inter alia* for transcription termination. mTERF1 terminates the short transcript containing rRNAs while the other ones are involved in termination of the other transcripts and also in mtDNA translation[27].

There are limited numbers of cases with defects of mtDNA transcription. Only mutations in the gene coding TFAM have recently been described[28] and they lead to mitochondrial depletion syndrome with liver failure but with normal function of the nervous system. In the case of the other elements of the system, studies on mice show that their removal affect the mtDNA level[29,30]; this allows the speculation that in humans mutations in these genes may also influence replication and lead to mtDNA depletion or deletions.

4 MTDNA: RECOMBINATION AND REPAIR

In the beginning of human mtDNA studies this molecule was thought not to undergo recombination and repair. Now it is obvious that both processes take place in mitochondria although recombination events are rare and the repair differs from the nuclear one (for example there is no nucleotide excision repair or NER). Mechanisms of mtDNA repair are still not fully understood. The most efficient is base excision repair or BER. It is in two forms: short patch BER (SP-BER) and long patch BER (LP-BER). In mitochondria LP-BER seems to be the most pronounced type of DNA repair[31].

DNA2, MGME1, FEN1 and RNASEH1 are involved in mitochondrial LP-BER. They also form a link between mtDNA repair and replication as they are responsible for the removal of RNA primers necessary to start it. DNA2, FEN1 and RNASEH1 play a role both in the nucleus and mitochondria while MGME1 is only localized in mitochondria.

Recently mutations in *DNA2, MGME1* and *RNASEH1* were described as a cause of human mitochondrial disease[32,33,34]. Mutations in *DNA2* and *RNASEH1* lead to adult PEO with encephalomyopathy and multiple long range deletions, autosomal dominant and recessive, respectively. *MGME1* mutations cause mitochondrial depletion syndrome but in contrast to that described earlier less severe as it starts in teens or in adulthood. It is characterized by PEO with proximal weakness, respiratory failure and sometimes mental retardation, gastrointestinal symptoms and ataxia.

5 NUCLEOTIDE METABOLISM

The next process responsible for mtDNA maintenance is nucleotide metabolism as it delivers dNTPs necessary for mtDNA synthesis. There are two pathways responsible for nucleotide synthesis – *de novo* – producing nucleotides from quite simple molecules like amino acids and carbon dioxide – and the salvage pathway recycling existing nucleosides and deoxynucleosides to produce the final dNTPs and NTPs. There is a huge difference between nucleotide metabolism in replicating cells and postmitotic ones. While proliferating cells use *de novo* nucleotide synthesis in postmitotic cells the salvage pathway plays a crucial role.

The most important protein in the *de novo* dNTP pathway is RNR – ribonucleotide reductase. It is built of two subunits, large R1 and small R2. While there is only one type of R1 subunit, there are two types of R2. One is regular (called simply R2) and is active in proliferating cells. In quiescent cells the expression of R2 is down-regulated and p53R2 takes its place restoring the activity of ribonucleotide reductase.

In the salvage pathway the most important role is played by deoxynucleoside kinases phosphorylating deoxynucleosides to dNMPs. There are four of them: two thymidine kinases TK1 and TK2, deoxycytidine kinase dCK and deoxyguanosine kinase dGK. Regardless of their names they have wider kinase properties. TK1 phosphorylates thymidine (dT) and deoxyuridine (dU), TK2 phosphorylates dT, deoxycytidine (dC), and dU, dCK phosphorylates dC, deoxyadenosine (dA) and deoxyguanosine (dG) and dGK – dG and dA. Because TK1 and dCK are cytosolic while TK2 and dGK are mitochondrial; in both compartments deoxynucleosides can be efficiently recycled. Later, dNMPs are further phosphorylated to dNDPs and dNTPs by nucleoside monophosphate kinases (NMPKs) and nucleoside diphosphate kinases (NDPKs), respectively[35]. Two additional enzymes, both necessary for the proper function of NDPK – SUCL and GABAT are important in the context of nucleotide biosynthesis and human mitochondrial disease. SUCL is mainly a core enzyme of the Krebs cycle catalyzing a reversible reaction in which succinyl-CoA with ATP or GTP is converted to succinate and ADP or GTP, respectively. SUCL acts as a heterodimer built from alpha subunit (encoded by *SUCLG1*) and any of two possible beta subunits responsible for substrate specificity (ATP or GTP). Beta subunits are encoded by *SUCLA2* and *SUCLG2*. GABAT (gamma-aminobutyric acid (GABA) transaminase) is also a bifunctional enzyme involved in GABA catabolism (Table 1).

In mitochondria the salvation pathway is the most important even crucial for mtDNA maintenance and absence of TK2 or dGK leads to severe clinical consequences. Mutations in *TK2* encoding thymidine kinase (2p13.1) and *DGUOK* encoding deoxyguanosine kinase (16q21) are responsible for depletion syndrome. In case of *DGUOK* it is again a hepatocerebral type with very early onset and poor outcome. Mutations in *TK2* lead to the myopathic type of depletion syndrome with muscle weakness and hypotonia and variable age of onset from childhood to adulthood. They can also be responsible for PEO with multiple large-scale mtDNA deletions. All pathogenic variants both in *DGUOK* and *TK2* regardless of the phenotype they manifest are recessive[36,37].

Mutations in *SUCLG1* (2p11.2), *SUCLA2* (13q14.2) (encoding subunits of the SUCL enzyme, as mentioned earlier), are a cause of encephalomyopathic depletion syndrome with or without mild methylmalonic aciduria. The onset of the disease is frequently in neonates with vomiting, hypotonia and respiratory problems, but later can become more variable phenotype[38,39]. Until recently *ABAT* mutations were only shown to lead to GABAT deficiency, psychomotor retardation, seizures, hypotonia and hyperreflexia. In 2015 it was proven that in fact the phenotype of the patients results from dual nature of

TABLE 1

Selected genes encoding proteins involved in nucleotide metabolism important for balanced nucleotide pool in mitochondria.

Gene name	Encoded protein	Protein function	Localization
RRM2B	P52R2-Alternative R2 subunit of the ribonucleotide reductase (RNR)	*De novo* nucleotide biosynthesis. RNR reduces ribonucleotides to deoxyribonucleotides. P52R2 is active regardless of the cell cycle phase, also in quiescent cells. This property makes it important for mitochondrial stability	Cytoplasm
TK1	TK1, thymidine kinase 1	Salvage pathway. Phosphorylates thymidine (dT) and deoxyuridine (dU) to dTMP and dUMP, respectively	Cytoplasm
TK2	TK2, thymidine kinase 2	Salvage pathway. Phosphorylates thymidine (dT), deoxycytidine (dC) and deoxyuridine (dU) to dTMP, dCMP and dUMP, respectively.	Mitochondrion
DCK	dCK, deoxycytidine kinase	Salvage pathway. Phosphorylates deoxycytidine (dC), deoxyguanosine (dG) and deoxyadenosine (dA) to dCMP, dGMP and dAMP, respectively.	Cytoplasm
DGUOK	dGK, deoxyguanosine kinase	Salvage pathway. Phosphorylates deoxyguanosine (dG) and deoxyadenosine (dA) to dCMP, dGMP and dAMP, respectively.	Mitochondrion
SUCLG1 *SUCLA2* *SUCLG2*	Alpha subunit of SUCL Beta subunit of SUCL Beta subunit of SUCL	Salvage pathway. SUCL is necessary for the proper function of the nucleoside monophosphate kinases and nucleoside diphosphate kinases – the enzymes phosphorylating dNMPs to dNDPs and dNDPs to dNTPs, respectively.	Mitochondrion
ABAT	GABAT, gamma-aminobutyric acid transaminase	Salvage pathway. Not well recognized. Physically interacts with SUCLG1, SUCLG2 and SUCLA2	Mitochondrion
TYMP	thymidine phosphorylase	Salvage pathway. Phosphorylates thymidine and deoxyuridine to thymine and uracil	Cytoplasm

GABAT deficiency. GABAT is involved in neurotransmitter metabolism and the nucleotide salvage pathway. *ABAT* mutations lead to mtDNA depletion and respiratory chain deficiency[40].

The absence of p32R2 also leads to autosomal recessive mitochondrial disease with mtDNA depletion expressed as infant, quickly progressing fatal encephalopathy with tubulopathy or adult-onset MNGIE. p53R2, as it was mentioned earlier, is expressed in postmitotic cells and also on a low level in dividing ones, to provide nucleotides necessary in cell cycle phase G1 and G0. Its expression is induced by p53 by its binding to *RRM2B* – a gene encoding p53R2 (8q22.3)[41].

The classical autosomal recessive MNGIE with mitochondrial DNA depletion is caused by mutations in *TYMP*, a gene encoding thymidine phosphorylase (22q13.33). It phosphorylates thymine to thymidine what makes it the basic enzyme necessary for nucleotide salvage biosynthesis pathway. Patients with *TYMP* mutations present with gastrointestinal symptoms, mainly dysmotility, polyneuropathy, PEO with ptosis. They also show the eating disorders, anorexia. The disease leads to cachexia and death. The onset and survival are variable, but the disease does not start in infancy[42].

The last subgroup of mitochondrial diseases with nucleotide pool perturbations is the one resulting from defective transport between mitochondria and the cytosol. Here two main defects can be described – mutations in *SLC25A4* (formerly called *ANT1*) and *MPV* 17.

SLC25A4 (4q35.1) encoding adenosine nucleotides translocator, an inner mitochondrial membrane protein responsible for the transport of ATP, produced in the process of oxidative respiration, from mitochondrial matrix to intermembrane space and ADP, necessary as a substrate for ATP production, from intermembrane space to mitochondrial matrix. Mutations in *SLC25A4* were in fact the first described defect of mtDNA maintenance presenting as autosomal dominant PEO[43]. The dominance of *SLC25A4* mutations can be easily explained by the structure of the adenine nucleotide translocator. The protein forms a homodimer. When proteins produced from one of the alleles are defective it can be predicted that up to 3/4 of translocators can have improper structure and function. Recently the third player in this group has appeared: *AGK* (7q34) gene encoding acylglicerol synthase. It is necessary for membrane lipids synthesis and therefore influences mitochondrial membrane composition. It seems that mutations in *AGK* lead to improper ANT1 assembly, as the nucleotide translocator is bound to the inner membrane phosphatidic acid and cardiolipin[44].

Beside autosomal dominant PEO *SLC25A4* mutations are responsible for autosomal dominant and recessive depletion syndrome with cardiomyopathy being the most pronounced and distinctive feature. Cardiac involvement is the result of specific adenine nucleotide translocator tissue expression – very high in muscle and heart but almost undetectable in other tissues usually affected by mitochondrial diseases like liver or brain (other paralogous translocators function in these tissues). The mechanism leading to multiple large-scale mtDNA deletions in the case of SLC25A4 has not been explained yet.

The case of *MPV17* is even more intriguing. For the first time recessive mutations in *MPV17* were described in 2006[45] and MPV17 protein was localized in the inner mitochondrial membrane. For many years new cases of mitochondrial DNA depletion syndrome with liver failure had been described, but the function of MPV17 was unknown. In 2015 we got closer to its function because it was proved that indeed MPV17 is an inner membrane protein acting as a non-selective channel with gating properties responding to different cellular signals like redox state, pH or membrane potential but it was still not known what type of molecules it transfers through the membrane[46].

A year later it was shown that a Mpv17 defect in mice indeed affects the nucleotide pool by lowering the level of dGTP and dTTP and hence impairment in mtDNA replication[47]. Of course there is still much to be done to find the exact mechanism of *MPV17* mutation action. *AGK* mutations lead to Sengers syndrome, a myopathic mtDNA depletion syndrome with cardiomyopathy and cataracts[48].

6 MITOCHONDRIAL DYNAMICS: FUSION AND FISSION

As it is now commonly known that mitochondria are not just small bean-like organelles freely floating in the cytosol but they are dynamic structures with variable shapes depending on the type of the cell and their metabolic state, capability for active movement based on cellular cytoskeleton, fusion and fission and interacting with other organelles like the endoplasmic reticulum. Hence it should be appropriate to talk about a mitochondrial network rather than about single mitochondrion.

Mitochondrial ability to move, fuse, divide, and change their shape is very important because it enables them to play their roles appropriately as are needed and ensures quality control.

Quality control is one of the main roles of fusion and fission processes. As mitochondria are exposed to reactive oxygen species produced by the respiratory chain damaging mitochondrial lipids, proteins and mtDNA gradually impairing the organelle function. One way of avoiding the collapse of a part of the mitochondrial network is to fuse with a healthy mitochondrion. This enables mtDNA complementation and mitochondrial survival. When, on the other hand, mitochondrial damage goes too far, mitochondrial fission takes place. Mitochondrion undergoes remodelling and "bad" elements are moved to one end where they are cut out by mitochondrial division and directed to mitophagy.

The apparatus necessary for proper fusion and fission is known to a certain extent. Fusion starts with the mitochondrial outer membrane and is performed by mitofusin: mitofusin-1 (MFN1) and mitofusin-2 (MFN2). They act as homo and heterodimers. After outer membrane fusion OPA1, GTPase residing in the inner membrane, also responsible for mitochondrial cristae formation, leads to inner membrane fusion. FIS1, MFF, GDAP1 and DNM1L (dynamin-like protein 1, long called DRP1) are the most important factors responsible for mitochondrial division FIS1 (fission1), MFF (mitochondrial fission factor) MID49 and MID51 (mitochondrial dynamics proteins), outer membrane proteins together with GDAP1 (ganglioside-induced differentiation-associated protein 1) recruit DNM1L forming a kind of a tightening ring leading to mitochondrial fission.

The ability to move is particularly important in neurons as they are the longest cells in the organism. Axonal mitochondrial transport goes along microtubules. From cell body to synapses (anterograde transport) it is conducted via kinesins and is mediated by MFN2 retrograde transport (from synapses to cell body) mitochondria move by dynein and dynactin. Both processes require energy.

Defects of mitochondrial dynamics resulting from mutations of genes encoding basic proteins involved in these processes have pleiotropic effects and some of them have not been recognized as mitochondrial diseases *sensu stricto*. The best example is Charcot-Marie-Tooth disease. Charcot-Marie-Tooth is one of the most common neurologic diseases. It affects motor and sensory neurons. Legs are the most frequently affected with muscle weakness leading to foot drop. The disease is progressive and later may affect other muscles including hands. The onset of the disease is in adolescence or early adulthood. Charcot-Marie-Tooth is characterized by phenotypic but also genetic variability with over 50 genes assigned[49]. Two of these genes are closely related to the topic of this chapter. Mutations in *MFN2* (1p36.2) encoding mitofusin 2, and in *GDAP1* (8q21.11) encoding ganglioside-induced differentiation-associated protein 1 lead to Charcot-Marie-Tooth type 2 disease. In the case of *MNF2* autosomal dominant inheritance is observed while *GDAP1* patients show mostly autosomal recessive pattern but a dominant mutation has been recently described[50]. Until now instability of mitochondrial DNA has not been described in patients, cell culture or model animals. On the contrary, mutations in *MFN2* lead to mtDNA depletion[51], cause mtDNA instability manifesting by an elevated rate of point mutations, multiple large-scale deletions and depletion in a mouse model[52] and beside Charcot-Marie-Tooth type 2 disease cause the autosomal dominant optic atrophy (ADOA) plus phenotype[53] frequent in mitochondrial dynamics disorders.

Autosomal dominant optic atrophy (ADOA, also called Kjer disease) may be seen as a sister disease to Leber hereditary optic neuropathy (LHON) – a classical mitochondrial disease caused by mtDNA point mutations – as it affects the same type of cells which are retinal ganglion cells, leading to their loss and in consequence blindness. In contrast to LHON, ADOA starts early in childhood and is slowly progressive. The most frequent cause of ADOA are *OPA1* (3q28-q29, encoding OPA1 protein) large deletions and point mutations. Beside the classical ADOA *OPA1* mutations sometimes cause the ADOA plus phenotype with PEO, sensorineural hearing loss, ataxia and peripheral neuropathy being typical phenotypic expressions of mtDNA instability and indeed multiple mtDNA deletions have been found in patients muscles[54,55]. Recently the case of two siblings with severe encephalopathy with cardiomyopathy and optic atrophy due to a homozygous mutation in *OPA1* was described[56]. No multiple mtDNA deletions have been found in patients' muscle biopsies but profound mtDNA depletion was present. There was no evidence of any neurological or ophthalmological symptoms in parents of the patients. The fact that OPA1 function as a dimer and the dimerization is necessary for its GTPase activity may explain why *OPA1* mutations are dominant[57].

Mutations in *OPA3* (19q13.32) also lead to ADOA with mitochondrial fragmentation suggesting its involvement in mitochondrial fusion[58]. The function of OPA3 in mitochondrial fusion is not known yet. Until now it is apparent that it is engaged in lipid metabolism and thermogenesis. Mitochondrial DNA instability has not been tested in *OPA3* patients.

The fission process is disturbed in patients harboring the mutations in *DMN1L* (12p11.21) and *MFF* (2q36.3). Autosomal dominant mutations in *DMN1L* lead to dominant optic atrophy while recessive ones cause a lethal phenotype of encephalomyopathy with defect in both, mitochondrial and peroxisomal fission. Dmn1l (Drp1) deficient mice are embryonic lethal[59]. The expression of mutant Dmn1l variants in mouse embryonic fibroblasts suggests that *DMN1L* mutations may have dominant-negative character. Mutations of *MFF* are recessive and also lead to encephalomyopathy with defect in both mitochondrial and peroxisomal fission. MRI in one of the patient was similar to one in patients with Leigh syndrome (basal ganglia involvement). Mitochondrial DNA stability has not been tested in *MFF* and *DMN1L* patients.

Recently[60] a homozygous nonsense mutation in *MIEF2* (17p11.2; encoding MID49 – one of the proteins recruiting DMN1L) was described. It leads to mitochondrial myopathy, disturbed mitochondrial dynamics and a respiratory chain defect. What is interesting, no mtDNA multiple deletions were found in a muscle biopsy but the copy number was elevated.

Until now pathogenic variants in *FIS1* have not been described, but the absence of *Saccharomyces cerevisiae* homologue – *FIS1* leads to mtDNA depletion and a switch from heteroplasmy to homoplasmy in yeast[61].

Although the relationship between impaired mitochondrial dynamics and mtDNA instability is now well established and documented, exact mechanism underlying is still unknown.

7 MISCELLANEOUS

For some of the genes it is still difficult to find a mechanism or common feature explaining their involvement in mtDNA instability. Two such cases are presented here.

One of them is the *FBXL4* gene (6q16.1-q16.2) encoding a protein residing in the inner mitochondrial membrane whose function is unknown. Mutations in this gene lead to encephalomyopathic depletion syndrome with feeding difficulties, microcephaly and growth retardation together with cardiomyopathy, face dysmorphia cerebral atrophy and seizures and are autosomal recessive[62].

Mutations in *SPG7* gene (16q24.3) are known to lead to spastic paraplegia. In 2014 a group of patients suffering from PEO, ptosis, ataxia, spasticity, dysphagia (and additional symptoms depending on the patient), mitochondrial DNA deletions and autosomal recessive mode of inheritance was described[63]. Paraplegin, encoded by *SPG7* shares homology with *S. cerevisiae* inner membrane ATPases AFG3, RCA1, and YME1; exact function is, again, unknown (Figure 2).

8 MECHANISM OF MTDNA DELETION FORMATION

While an elevated level of point mutations or mtDNA depletion are quite easy to explain by a replication defect caused either by a defect in replication machinery (low replication rate, lower polymerase fidelity) or a deficit in nucleotide biosynthesis (shortage of deoxynucleotides or imbalanced nucleotide pools) the mechanism leading to multiple or even single long-range mtDNA deletions is less straightforward and poorly understood.

8.1 Types of mtDNA Large-Scale Deletions

From the genetic perspective mtDNA deletions may be divided in to single and multiple. Single deletions are usually not inherited and treated as sporadic (according to Chinnery[64] recurrence risk is 4–11%). They lead to PEO but also Kearns -Sayre syndrome (PEO, ptosis, retinitis pigmentosa and other) and Pearson syndrome (sideroblastic anemia with pancreatic exocrine dysfunction). It is still not clear whether the deletion is already present in the oocyte at the time of fertilization or appears during embryonic development; maybe both theories are true depending on the case.

Multiple deletions appear as a consequence of the mtDNA maintenance defect caused by mutations in nuclear genes and are inherited in a Mendelian way. Deletions can also accumulate in different tissues with age and are found in brains of patients with neurodegenerative diseases like Parkinson or Alzheimer disease. These can be treated as somatic mutations. The laboratory diagnostic insight into the types of mtDNA deletions is shown on Figure 3.

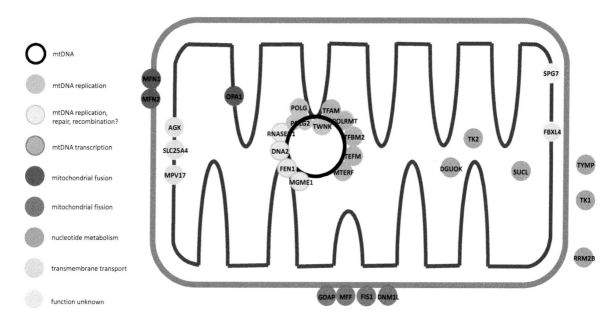

FIGURE 2 The schematic localization of the proteins engaged in mtDNA maintenance. Color code was used to show the processes in which they are involved. Not surprisingly, most of them play their role in mitochondria – in the matrix like the proteins involved in nucleotide salvage pathway or the ones forming nucleoid together with mtDNA like proteins from replication and transcription apparatus. In the inner membrane transport proteins and the proteins responsible for the inner membrane fusion are located. In the outer membrane fusion proteins are present, while the ones responsible for mitochondrial fission are as well recruited to the outer membrane during this process. There are also cytoplasmic proteins involved in nucleotide biosynthesis enabling nucleotide supply in not dividing cells.

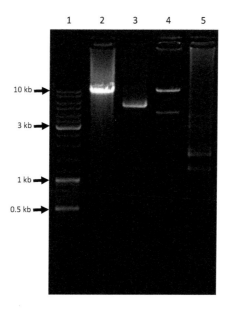

FIGURE 3 The results of the screening for mtDNA deletions with long-range PCR. The wild-type PCR product covers mtDNA major arc and is about 10 kilobases long. DNA ladder is loaded in lane 1. In lane 2 – PCR product obtained on DNA sample from healthy control. In lane 3 – PCR product obtained on DNA sample from the person with single large-scale deletion. Heteroplasmy is easy to observed – both wild-type and single class of deleted molecules are visible. In lanes 4 and 5 – PCR products obtained on DNA sample with multiple mtDNA deletions. Heteroplasmy is visible with numerous classes of shortened molecules. Wild-type molecules are difficult to see in lane 5 probably due to the preferential amplification of shorter molecules during PCR.

Another way of looking at mtDNA deletions is to determine the presence and type of border repeats flanking the deletion. In most deletions, border repeats are present and they can be exact – when the sequence on both sides of the deletion is identical or indirect – when two flanking sequences are similar but not identical. The length of the repeat is from four to fourteen nucleotides according to MITOMAP. It is worth mentioning that the controversy exists whether, in the case of short (a few nucleotide long), direct repeat we are dealing just with short direct repeat or whether there is long (up to 50 nucleotides), indirect repeat consisting of a series of short ones divided by nonhomologous regions[65].

The localization of the deletions within mtDNA molecule can also be taken into account. Most of them lie within the major arc – 2/3 of the length of mtDNA molecule between origins of replication O_H and O_L (Figure 1). It was thought that this is the result of the replication based mechanism of deletion formation. However after taking into account all possible short homology regions within the mtDNA molecule, which may form border repeats (excluding pairs flanking the deletions and removing O_H, O_L or both) over 80% of possible deletions will be localized within the major arc which is consistent with experimental data[66].

Around 10% of large-scale mtDNA deletions remove the replication origin of the light chain, which is necessary for mtDNA replication. Increasing evidence shows that such deletions are accompanied by duplications, probably containing the missing O_L[67].

8.2   The Molecular Mechanism of Deletion Formation

The mechanism of deletion formation is still unknown. Three processes are considered as being responsible for the appearance of mtDNA deletions: replication, recombination and repair and maybe a combination of them. Despite the actual model, in all three cases 3' and 5' homologous flanking regions are actively involved.

In the case of the asynchronous replication model, the single stranded light strand 3' repeat can hybridize with the 5' repeat on the heavy strand forming the single-stranded loop which can be cut out. As a result the two molecules of different lengths (full length and with a deletion) are produced. In fact this model is not strongly supported by the existing data.

The next model involves double-strand break repair. Double-strand breaks are sensitive to exonuclease activity giving long regions of single-stranded DNA. If such activity would expose a possible 5' flanking repeat on one of them and 3' on the other one hybridization and repair following it produces a mtDNA molecule with the deletion[66].

Another, recent idea is a combination of double-strand break repair and replication. Nicks left when the replication gets to its end, when unligated, may lead to a double-strand break in one of the derivative molecules. Replication stalling e.g. as the effect of the mutation affecting polymerase activity, may have a similar effect[68].

8.3   The Fate of Deletions in the Cell, Tissue, and Organism

As it is mentioned above, the diseases caused by single large-scale mtDNA deletions are generally not inherited but the deletion may already be present in the oocyte before fertilization[69,70]. The final level of deletion heteroplasmy in the cell or tissue may depend on multiple factors and result from molecular mechanism existing in a cell and pure chance.

The first thing which may influence the deletion level is replicative advantage of shorter mtDNA molecules. As large mtDNA deletions are several kilobases long they lead to substantially shorter mtDNA molecules. This means that the time needed to replicate them is also shorter what may result in producing more deleted than wild type mtDNA molecules leading to higher heteroplasmy[71,72]. Relaxed replication rate may weaken or even eliminate the advantage[73]. When there is no pressure for quick mtDNA synthesis, similar copy numbers of long and short mtDNA molecules may be produced.

The next factor is the selective pressure in which mtDNA molecules with a deletion do not produce a full set of mitochondrial respiratory chain subunits. In consequence the cell with a high heteroplasmy level is respiratory deficient, shows lower fitness and will not divide. In that model, the cells with high level of mtDNAs with deletion will be eliminated. This, however may only work in dividing cells. This is thought to be a cause of common phenomenon observed in tissues of aging people and patients with mitochondrial as well as neurodegenerative disease. mtDNA deletions can be observed in post mitotic tissues like muscle and brain and are rarely observed in blood and if at all, at a much lower level (the exception is the rare Pearson syndrome where blood is in fact the most affected tissue).

Propagation of mutated mtDNA may also be subject to genetic drift. During cell division mitochondria are randomly distributed to daughter cells (however the mitochondrial network segregation is fully synchronized with the cell division) to assure sufficient ATP supply for this process and equivalent amount of mitochondrial mass in both resulting cells[74]. This may lead to differences in the heteroplasmy level between cells descended from the same ancestor. Probably, in the case of large-scale mtDNA deletions, all the above-described processes cooperate to some extent.

9 MITOCHONDRIAL DELETIONS AND AGEING

The mitochondrial theory of ageing is one of the most popular ones. It proposes that respiratory chain produces free radicals damaging mitochondrial proteins, lipids, and nucleic acids. This leads to respiratory chain dysfunction, higher free radicals production and so-called "vicious circle". Although popular, this theory has not been fully proven. On the other hand no one doubts that mitochondria take part in the process of ageing but in a more complex way. Here we will concentrate on changes involving mtDNA at aging.

The first well documented relationship between ageing and mtDNA appeared in 1990 when the so-called "common deletion" described for the first time in 1988 by Holt was found in post mortem studies in healthy brain and heart tissue of adults but not in neonates or spontaneous abortions[75]. Further studies confirmed that large-scale deletions of different sizes and localizations within the mtDNA molecule indeed accumulate with age in brain, heart and muscles on a low heteroplasmy level, around 0.1%[76]. It was also shown that the deletions accumulate gradually with age[77].

The question is what is the role of mtDNA deletions in ageing? Are they one of the causes of ageing actively contributing to the vicious circle or they are just a consequence of aging with no further impact? Is the accumulation of mtDNA deletions high enough to lead to mitochondrial dysfunction? Single cell studies performed on heart and brain tissue have shown that although the overall heteroplasmy level of mtDNA deletions in aging tissue is low, on a cellular level it might be high enough to lead to dysfunction of this particular cell which in the case of muscle, heart or brain may affect the whole organ[78].

In 2004 the mitochondrial theory of ageing acquired a strong supporter. The mouse model of POLG-related diseases was published[79]. The mouse had a homozygous missense substitution in the exonuclease domain of POLG – an aspartate residue critical for the function of the enzyme was substituted by alanine. Postmortem analysis has shown that, as expected, mtDNA deletions accumulate in different tissues of the animal, the amount of full-length molecules is smaller than in control animals and besides deletions, point mutation accumulation was also observed. A real surprise was the phenotype of so-called mutator mice. They did not present expected mitochondrial disease, rather resembled premature ageing with shorter life span, weight loss, reduced subcutaneous fat, hair loss, kyphosis and osteoporosis.

A year later very similar (with the same homozygous amino acid substitution) model mouse was published[80] and confirmed major findings in respect to the ageing phenotype. H_2O_2 production was measured to address the question of ROS (reactive oxygen species) involvement in this ageing model and no significant differences between mutator and normal mice were observed.

Increased oxidative stress was not seen in these mice. Lipid peroxidation and oxidative damage to nucleic acids was also tested giving no significant differences and leading to the conclusion that mtDNA mutations in the mutator mouse do not cause elevated ROS production and oxidative stress that might be a reason of premature ageing.

As both mice showed accumulation of point mutations together with large-scale deletions the question arises whether mtDNA deletions or point mutations are responsible for this phenomenon to the greatest extent. This question may be addressed by observing another model animal – the so-called deletor mouse – a mouse with mutations in Twinkle. Although it accumulated large-scale mtDNA deletions but did not present an ageing phenotype; however late-onset mitochondrial disease similar to PEO-related disorders in human occurred[81]. The conclusion is that premature ageing of the mutator mouse was rather a consequence of point mtDNA mutation accumulation. The study of the somatic stem cells of the mutator mouse also showed reduced self-renewal[82].

This story also shows again that animal models of human diseases may present dramatically different phenotypes, as the patients with late-onset POLG-dependent disease do not show premature ageing.

10 INSTABILITY OF MTDNA IN NEURODEGENERATIVE DISEASES

Mitochondrial DNA deletions have been described not only in brains of ageing people but also observed, in multiple studies, in specific parts of the brain in patients suffering from specific neurodegenerative diseases, at higher level than in controls. Although early work's sample sizes and methods used did not allow to draw far-reaching conclusions, the last few years brought some new evidence not only about mtDNA instability in neurodegeneration but also on its impact on cell metabolism[83].

Two multifactorial age-related diseases, Parkinson's disease and Alzheimer disease are not only the most common, but also serve as the best examples of the problems with studying mitochondrial involvement in human disease. The evidence for mtDNA deletions in both diseases appeared quite early although the results were inconclusive. New reports show several interesting features of mitochondrial instability in neurodegeneration. First, the defect may be seen not in the whole brain but in specific part like substantia nigra in the case of patients with Parkinson's disease. The same type of analysis in Alzheimer disease showed the presence of mtDNA deletion in three types of neurons coming from the putamen, the frontal cortex, and the substantia nigra; the highest in substantia nigra, but not statistically different from the control group. Second, the mitochondrial dysfunction, confirmed as the appearance of COX negative neurons (histochemical staining for cytochrome c oxidase, complex IV of the respiratory chain, is the standard test used to show a respiratory defect in situ), more profound in hippocampal neurons of patients with Alzheimer disease, indeed correlates with a high level of mtDNA deletions[83,84,85]. Third, not only mtDNA deletions but also depletion may be seen in COX-deficient substantia nigra neurons. The analysis of off target mitochondrial reads obtained from whole exome sequencing of DNA isolated from brain tissue obtained from the patients with neurodegenerative diseases and controls also indicated mtDNA depletion in the case of Alzheimer disease and Creutzfeldt Jacob disease[86]. To sum up, mitochondrial dysfunction, presenting in the form of COX negative fibers accompanied by mtDNA deletions and/or depletion is present in affected parts of the brain in patients with neurodegenerative diseases. However, the actual impact of respiratory chain deficiency on the onset and course of the disease is still unknown.

11   MITOCHONDRIAL DYSFUNCTION DUE TO THE USAGE OF NUCLEOSIDE ANALOGUES

A nucleoside analogue such as zidovudine (azidothymidine, AZT) was the first effective drug tried against AIDS. AZT is unable to form 3' phosphodiester bonds with DNA during its synthesis. It however blocks reverse transcriptase – the HIV enzyme responsible for DNA synthesis based on viral RNA. It was approved by the FDA in 1987. In 1990 it was shown that long term use of this drug leads to myopathy with ragged red fibres and cytochrome c oxidase deficiency – a common sign of mitochondrial disease, also mtDNA depletion[87]. mtDNA depletion in that case is a result of polymerase gamma sensitivity to nucleoside analogues such as AZT. It was also suggested that AZT may also bind to other enzymes like TK2, involved in the nucleotide salvage pathway what can lead to reduced enzymatic activity. The same mechanism underlies mitochondrial dysfunction caused by the use of other nucleoside analogues such as didanosine and zalcitabine.

As Nucleoside/Nucleotide Reverse Transcriptase Inhibitors (NRTIs) are widely used in AIDS therapy, also for treating hepatitis B and C, the patients should be aware of probable mitochondrial dysfunction and mitochondrial function should be regularly monitored. Several other risk factors increase the possibility of mitochondrial involvement in NRTI. These are geneder-women are more sensitive to NRTIs, other drugs like interferon-alpha or protease inhibitors if combined may have synergistic effects. It was also found that *POLG* mutation pArg964Cys sensitizes to NRTI induced mitochondriopathy.

12   IS THERE A CURE FOR PATIENTS WITH MITOCHONDRIAL INSTABILITY?

Mitochondrial diseases are perceived, not without a reason, as ones with no good therapeutic options. In most cases supportive treatment is employed. Medications which may worsen the clinical condition of the patient include classical valproic acid in the case of patients with POLG-related disorder. Dietary supplements are widely used as they may act as antioxidants like CoQ10, idebenone (CoQ10 – derivative), α-lipoic acid, vitamin C and E, mitochondrial substrates as L-carnitine or influenced by the respiratory chain flux, riboflavin. The above-mentioned supplements are used on assumption that they may benefit, but there are no standardized clinical trials showing any beneficial role in the treatment of mitochondrial disease for most of them. Idebenone is an exception drug as it was recently approved for patients with Leber hereditary optic neuropathy.

Although there are several innovative concepts aiming to cure mitochondrial diseases most of them are in an early phase of development. In the case of mitochondrial DNA instability, only one is actually worth consideration. This is nucleoside bypass therapy in the case of TK2 deficit. Deoxynucleosides more easily than deoxynucleotides pass through the digestive system, were first tested in mice with a Tk2 deficit[88]. Now they are tried with at least 18 patients with mutations in TK2 on a "compassionate use" basis giving the opportunity to treat patients with the disease for which there is no satisfactory way of treatment with products under development. According to press releases the therapy works at least with some patients but neither scientific reports nor the clinical trials have confirmed it.

13   CONCLUSIONS

Mitochondrial instability should be treated as a common phenomenon heaving different aspects and nuances. It may appear as single or multiple large-scale mtDNA deletions, mtDNA depletion or accumulation of point mutations. It may be the cause of human genetic diseases: mostly sporadic, being a consequence of single large-scale mtDNA deletions or inherited in a Mendelian fashion related to multiple large-scale mtDNA deletions and mtDNA depletion. Mitochondrial diseases resulting from mitochondrial maintenance defect predominantly present as progressive external ophthalmoplegia (PEO) and PEO+ in the case of adults, ataxia-neuropathy spectrum in the case of adolescents and young adults and severe encephalomyopathy, frequently with liver involvement in case of small children or even neonates. It should, however, be taken into account that this simplified division is far from being precise and complete (Figure 4).

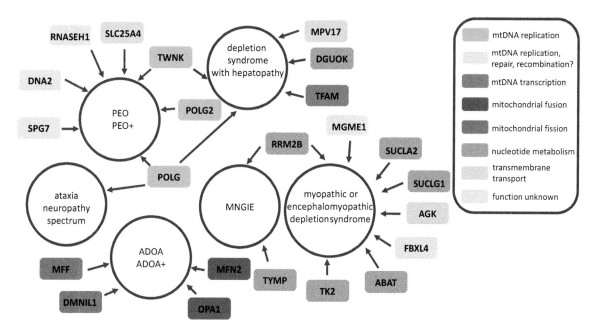

FIGURE 4 Nuclear genes involved in mtDNA stability known to, when mutated, cause mitochondrial disease. Simplified phenotypic expression of their mutations is shown. Color code was used to show the processes in which they are involved. In some cases (*POLG* is the best example) mutations in the same gene may lead to different diseases. As may be noted, mutations in genes involved in nucleotide metabolism tend to cause depletion syndrome without liver failure while mutations in genes encoding proteins responsible for replication lead to PEO (progressive external ophthalmoplegia) and PEO+ but also depletion syndrome with liver dysfunction (Alpers-Huttenlocher syndrome in case of *POLG* mutations). DOA is the main presentation of mitochondrial disease in case of genes responsible for mitochondrial fusion and fission. PEO – progressive external ophthalmoplegia, ADOA – autosomal dominant optic atrophy, MNGIE – mitochondrial neurogastrointestinal encephalopathy.

Large-scale mtDNA deletions accumulate in ageing tissues and together with somatic mtDNA point mutations may influence the process of ageing while large-scale mtDNA deletions and depletion can be seen in respiratory deficient neurons from brains of the patients with neurodegenerative diseases. Both in the case of ageing and neurodegeneration the question of exact contribution of mtDNA deficiency in these processes remains open. Still unanswered are questions of the basic mechanisms underlying mtDNA deletion formation: mtDNA replication and transcription, repair and recombination.

ACKNOWLEDGMENTS

I thank Prof. Ewa Bartnik for the correction of the manuscript and all the helpful comments. This work was supported by Polish National Science Center grant 2014/15/B/NZ5/00434.

INTERNET RESOURCES

ClinVar aggregates information about genomic variation and its relationship to human health. www.ncbi.nlm.nih.gov/clinvar/

Human DNA Polymerase Gamma Mutation Database – the up-to-date list of POLG mutations is present here: https://tools.niehs.nih.gov/polg/

MITOMAP A human mitochondrial genome database https://mitomap.org//MITOMAP

MitoBreak The mitochondrial DNA breakpoints database http://mitobreak.portugene.com/cgi-bin/Mitobreak_home.cgi

REFERENCES

1. Holt, I. J., Harding, A. E. & Morgan-Hughes, J. A. Deletions of muscle mitochondrial DNA in patients with mitochondrial myopathies. *Nature* **331**, 717–719 (1988).
2. Gorman, G. S. *et al.* Mitochondrial diseases. *Nat. Rev. Dis. Primer* **2**, 16080 (2016).
3. Robberson, D. L. & Clayton, D. A. Replication of mitochondrial DNA in mouse L cells and their thymidine kinase – derivatives: displacement replication on a covalently-closed circular template. *Proc. Natl. Acad. Sci. U.S.A.* **69**, 3810–3814 (1972).
4. Holt, I. J., Lorimer, H. E. & Jacobs, H. T. Coupled leading- and lagging-strand synthesis of mammalian mitochondrial DNA. *Cell* **100**, 515–524 (2000).
5. Yasukawa, T. *et al.* Replication of vertebrate mitochondrial DNA entails transient ribonucleotide incorporation throughout the lagging strand. *EMBO J.* **25**, 5358–5371 (2006).

6. Holt, I. J. & Reyes, A. Human mitochondrial DNA replication. *Cold Spring Harb. Perspect. Biol.* **4**, a012971 (2012).

7. Van Goethem, G. *et al.* Novel POLG mutations in progressive external ophthalmoplegia mimicking mitochondrial neurogastrointestinal encephalomyopathy. *Eur. J. Hum. Genet. EJHG* **11**, 547–549 (2003).

8. Saneto, R. P., Cohen, B. H., Copeland, W. C. & Naviaux, R. K. Alpers-Huttenlocher syndrome. *Pediatr. Neurol.* **48**, 167–178 (2013).

9. Piekutowska-Abramczuk D, Kaliszewska M, Sułek A, Jurkowska N, Ołtarzewski M, et al. The frequency of mitochondrial polymerase gamma related disorders in a large Polish population cohort. *Mitochondrion.* 2018 Nov 10. pii: S1567-7249(18)30184-3. doi: 10.1016/j.mito.2018.11.004.

10. Longley, M. J. *et al.* Mutant POLG2 disrupts DNA polymerase gamma subunits and causes progressive external ophthalmoplegia. *Am. J. Hum. Genet.* **78**, 1026–1034 (2006).

11. Humble, M. M. *et al.* Polg2 is essential for mammalian embryogenesis and is required for mtDNA maintenance. *Hum. Mol. Genet.* **22**, 1017–1025 (2013).

12. Young, M. J., Humble, M. M., DeBalsi, K. L., Sun, K. Y. & Copeland, W. C. POLG2 disease variants: analyses reveal a dominant negative heterodimer, altered mitochondrial localization and impaired respiratory capacity. *Hum. Mol. Genet.* **24**, 5184–5197 (2015).

13. Spelbrink, J. N. *et al.* Human mitochondrial DNA deletions associated with mutations in the gene encoding Twinkle, a phage T7 gene 4-like protein localized in mitochondria. *Nat. Genet.* **28**, 223–231 (2001).

14. Nikali, K. *et al.* Infantile onset spinocerebellar ataxia is caused by recessive mutations in mitochondrial proteins Twinkle and Twinky. *Hum. Mol. Genet.* **14**, 2981–2990 (2005).

15. Graziewicz, M. A., Bienstock, R. J. & Copeland, W. C. The DNA polymerase gamma Y955C disease variant associated with PEO and parkinsonism mediates the incorporation and translesion synthesis opposite 7,8-dihydro-8-oxo-2′-deoxyguanosine. *Hum. Mol. Genet.* **16**, 2729–2739 (2007).

16. Shutt, T. E. & Gray, M. W. Bacteriophage origins of mitochondrial replication and transcription proteins. *Trends Genet. TIG* **22**, 90–95 (2006).

17. Korhonen, J. A., Gaspari, M. & Falkenberg, M. TWINKLE Has 5′ -> 3′ DNA helicase activity and is specifically stimulated by mitochondrial single-stranded DNA-binding protein. *J. Biol. Chem.* **278**, 48627–48632 (2003).

18. García-Gómez, S. *et al.* PrimPol, an archaic primase/polymerase operating in human cells. *Mol. Cell* **52**, 541–553 (2013).

19. Krasich, R. & Copeland, W. C. DNA polymerases in the mitochondria: a critical review of the evidence. *Front. Biosci. Landmark Ed.* **22**, 692–709 (2017).

20. Zhao, F. *et al.* Exome sequencing reveals CCDC111 mutation associated with high myopia. *Hum. Genet.* **132**, 913–921 (2013).

21. Keen, B. A., Bailey, L. J., Jozwiakowski, S. K. & Doherty, A. J. Human PrimPol mutation associated with high myopia has a DNA replication defect. *Nucleic Acids Res.* **42**, 12102–12111 (2014).

22. Li, J. & Zhang, Q. PRIMPOL mutation: functional study does not always reveal the truth. *Invest. Ophthalmol. Vis. Sci.* **56**, 1181–1182 (2015).

23. Farr, C. L., Matsushima, Y., Lagina, A. T., Luo, N. & Kaguni, L. S. Physiological and biochemical defects in functional interactions of mitochondrial DNA polymerase and DNA-binding mutants of single-stranded DNA-binding protein. *J. Biol. Chem.* **279**, 17047–17053 (2004).

24. Maier, D. *et al.* Mitochondrial single-stranded DNA-binding protein is required for mitochondrial DNA replication and development in Drosophila melanogaster. *Mol. Biol. Cell* **12**, 821–830 (2001).

25. Sasaki, T., Sato, Y., Higashiyama, T. & Sasaki, N. Live imaging reveals the dynamics and regulation of mitochondrial nucleoids during the cell cycle in Fucci2-HeLa cells. *Sci. Rep.* **7**, 11257 (2017).

26. Hillen, H. S. *et al.* Mechanism of transcription anti-termination in human mitochondria. *Cell* **171**, 1082–1093.e13 (2017).

27. Barshad, G., Marom, S., Cohen, T. & Mishmar, D. Mitochondrial DNA transcription and its regulation: an evolutionary perspective. *Trends Genet. TIG* (2018). doi:10.1016/j.tig.2018.05.009.

28. Stiles, A. R. *et al.* Mutations in TFAM, encoding mitochondrial transcription factor A, cause neonatal liver failure associated with mtDNA depletion. *Mol. Genet. Metab.* **119**, 91–99 (2016).

29. Hyvärinen, A. K., Pohjoismäki, J. L. O., Holt, I. J. & Jacobs, H. T. Overexpression of MTERFD1 or MTERFD3 impairs the completion of mitochondrial DNA replication. *Mol. Biol. Rep.* **38**, 1321–1328 (2011).

30. Kühl, I. *et al.* POLRMT regulates the switch between replication primer formation and gene expression of mammalian mtDNA. *Sci. Adv.* **2**, e1600963 (2016).

31. Zhang, L., Reyes, A. & Wang, X. The role of DNA repair in maintaining mitochondrial DNA stability. *Adv. Exp. Med. Biol.* **1038**, 85–105 (2017).

32. Kornblum, C. *et al.* Loss-of-function mutations in MGME1 impair mtDNA replication and cause multisystemic mitochondrial disease. *Nat. Genet.* **45**, 214–219 (2013).

33. Ronchi, D. *et al.* Mutations in DNA2 link progressive myopathy to mitochondrial DNA instability. *Am. J. Hum. Genet.* **92**, 293–300 (2013).

34. Reyes, A. *et al.* RNASEH1 mutations impair mtDNA replication and cause adult-onset mitochondrial encephalomyopathy. *Am. J. Hum. Genet.* **97**, 186–193 (2015).

35. Wang, L. Mitochondrial purine and pyrimidine metabolism and beyond. *Nucleosides Nucleotides Nucleic Acids* **35**, 578–594 (2016).

36. Mandel, H. *et al.* The deoxyguanosine kinase gene is mutated in individuals with depleted hepatocerebral mitochondrial DNA. *Nat. Genet.* **29**, 337–341 (2001).

37. Saada, A. *et al.* Mutant mitochondrial thymidine kinase in mitochondrial DNA depletion myopathy. *Nat. Genet.* **29**, 342–344 (2001).

38. Ostergaard, E. *et al.* Mitochondrial encephalomyopathy with elevated methylmalonic acid is caused by SUCLA2 mutations. *Brain J. Neurol.* **130**, 853–861 (2007).

39. Ostergaard, E. *et al.* Deficiency of the alpha subunit of succinate-coenzyme A ligase causes fatal infantile lactic acidosis with mitochondrial DNA depletion. *Am. J. Hum. Genet.* **81**, 383–387 (2007).

40. Besse, A. *et al.* The GABA transaminase, ABAT, is essential for mitochondrial nucleoside metabolism. *Cell Metab.* **21**, 417–427 (2015).

41. Bourdon, A. *et al.* Mutation of RRM2B, encoding p53-controlled ribonucleotide reductase (p53R2), causes severe mitochondrial DNA depletion. *Nat. Genet.* **39**, 776–780 (2007).

42. Suomalainen, A. & Isohanni, P. Mitochondrial DNA depletion syndromes–many genes, common mechanisms. *Neuromuscul. Disord. NMD* **20**, 429–437 (2010).

43. Kaukonen, J. *et al.* Role of adenine nucleotide translocator 1 in mtDNA maintenance. *Science* **289**, 782–785 (2000).

44. El-Hattab, A. W., Craigen, W. J. & Scaglia, F. Mitochondrial DNA maintenance defects. *Biochim. Biophys. Acta* **1863**, 1539–1555 (2017).

45. Spinazzola, A. *et al.* MPV17 encodes an inner mitochondrial membrane protein and is mutated in infantile hepatic mitochondrial DNA depletion. *Nat. Genet.* **38**, 570–575 (2006).

46. Antonenkov, V. D. *et al.* The human mitochondrial DNA depletion syndrome gene MPV17 encodes a non-selective channel that modulates membrane potential. *J. Biol. Chem.* **290**, 13840–13861 (2015).

47. Dalla Rosa, I. *et al.* MPV17 loss causes deoxynucleotide insufficiency and slow DNA replication in mitochondria. *PLoS Genet.* **12**, e1005779 (2016).

48. Calvo, S. E. *et al.* Molecular diagnosis of infantile mitochondrial disease with targeted next-generation sequencing. *Sci. Transl. Med.* **4**, 118ra10 (2012).

49. Hoyle, J. C., Isfort, M. C., Roggenbuck, J. & Arnold, W. D. The genetics of Charcot-Marie-Tooth disease: current trends and future implications for diagnosis and management. *Appl. Clin. Genet.* **8**, 235–243 (2015).

50. García-Sobrino, T. *et al.* Phenotypical features of a new dominant GDAP1 pathogenic variant (p.R226del) in axonal Charcot-Marie-Tooth disease. *Neuromuscul. Disord. NMD* **27**, 667–672 (2017).

51. Vielhaber, S. *et al.* Mitofusin 2 mutations affect mitochondrial function by mitochondrial DNA depletion. *Acta Neuropathol. (Berl.)* **125**, 245–256 (2013).

52. Chen, H. *et al.* Mitochondrial fusion is required for mtDNA stability in skeletal muscle and tolerance of mtDNA mutations. *Cell* **141**, 280–289 (2010).

53. Rouzier, C. *et al.* The MFN2 gene is responsible for mitochondrial DNA instability and optic atrophy 'plus' phenotype. *Brain J. Neurol.* **135**, 23–34 (2012).

54. Stewart, J. D. *et al.* OPA1 in multiple mitochondrial DNA deletion disorders. *Neurology* **71**, 1829–1831 (2008).

55. Yu-Wai-Man, P. *et al.* OPA1 mutations cause cytochrome c oxidase deficiency due to loss of wild-type mtDNA molecules. *Hum. Mol. Genet.* **19**, 3043–3052 (2010).

56. Spiegel, R. *et al.* Fatal infantile mitochondrial encephalomyopathy, hypertrophic cardiomyopathy and optic atrophy associated with a homozygous OPA1 mutation. *J. Med. Genet.* **53**, 127–131 (2016).

57. Namba, K. *et al.* Molecular impairment mechanisms of novel OPA1 mutations predicted by molecular modeling in patients with autosomal dominant optic atrophy and auditory neuropathy spectrum disorder. *Otol. Neurotol. Off. Publ. Am. Otol. Soc. Am. Neurotol. Soc. Eur. Acad. Otol. Neurotol.* **37**, 394–402 (2016).

58. Ryu, S.-W., Jeong, H. J., Choi, M., Karbowski, M. & Choi, C. Optic atrophy 3 as a protein of the mitochondrial outer membrane induces mitochondrial fragmentation. *Cell. Mol. Life Sci. CMLS* **67**, 2839–2850 (2010).

59. Ishihara, N. *et al.* Mitochondrial fission factor Drp1 is essential for embryonic development and synapse formation in mice. *Nat. Cell Biol.* **11**, 958–966 (2009).

60. Bartsakoulia, M. *et al.* A novel mechanism causing imbalance of mitochondrial fusion and fission in human myopathies. *Hum. Mol. Genet.* **27**, 1186–1195 (2018).

61. Bradshaw, E., Yoshida, M. & Ling, F. Mitochondrial fission proteins Fis1 and Mdv1, but not Dnm1, play a role in maintenance of heteroplasmy in budding yeast. *FEBS Lett.* **586**, 1245–1251 (2012).

62. El-Hattab, A. W. *et al.* Molecular and clinical spectra of FBXL4 deficiency. *Hum. Mutat.* **38**, 1649–1659 (2017).

63. Pfeffer, G. *et al.* Mutations in the SPG7 gene cause chronic progressive external ophthalmoplegia through disordered mitochondrial DNA maintenance. *Brain J. Neurol.* **137**, 1323–1336 (2014).

64. Chinnery, P. F. *et al.* Risk of developing a mitochondrial DNA deletion disorder. *Lancet Lond. Engl.* **364**, 592–596 (2004).

65. Guo, X. *et al.* Repeats, longevity and the sources of mtDNA deletions: evidence from 'deletional spectra'. *Trends Genet. TIG* **26**, 340–343 (2010).

66. Krishnan, K. J. *et al.* What causes mitochondrial DNA deletions in human cells? *Nat. Genet.* **40**, 275–279 (2008).

67. Rygiel, K. A. *et al.* Complex mitochondrial DNA rearrangements in individual cells from patients with sporadic inclusion body myositis. *Nucleic Acids Res.* **44**, 5313–5329 (2016).

68. Pohjoismäki, J. L. O., Forslund, J. M. E., Goffart, S., Torregrosa-Muñumer, R. & Wanrooij, S. Known unknowns of mammalian mitochondrial DNA maintenance. *BioEssays News Rev. Mol. Cell. Dev. Biol.* e1800102 (2018). doi:10.1002/bies.201800102.

69. Blakely, E. L. *et al.* Mitochondrial DNA deletion in 'identical' twin brothers. *J. Med. Genet.* **41**, e19 (2004).

70. Elson, J. L., Apabhai, S., Gorman, G., Whittaker, R. G. & Krishnan, K. J. Older mothers are not at risk of having grandchildren with sporadic mtDNA deletions. *Genet. Med. Off. J. Am. Coll. Med. Genet.* **12**, 313–314 (2010).

71. Wallace, D. C. Mitochondrial DNA mutations and neuromuscular disease. *Trends Genet. TIG* **5**, 9–13 (1989).

72. Fukui, H. & Moraes, C. T. Mechanisms of formation and accumulation of mitochondrial DNA deletions in aging neurons. *Hum. Mol. Genet.* **18**, 1028–1036 (2009).

73. Campbell, G., Krishnan, K. J., Deschauer, M., Taylor, R. W. & Turnbull, D. M. Dissecting the mechanisms underlying the accumulation of mitochondrial DNA deletions in human skeletal muscle. *Hum. Mol. Genet.* **23**, 4612–4620 (2014).

74. Kanfer, G. & Kornmann, B. Dynamics of the mitochondrial network during mitosis. *Biochem. Soc. Trans.* **44**, 510–516 (2016).

75. Cortopassi, G. A. & Arnheim, N. Detection of a specific mitochondrial DNA deletion in tissues of older humans. *Nucleic Acids Res.* **18**, 6927–6933 (1990).

76. Simonetti, S., Chen, X., DiMauro, S. & Schon, E. A. Accumulation of deletions in human mitochondrial DNA during normal aging: analysis by quantitative PCR. *Biochim. Biophys. Acta* **1180**, 113–122 (1992).

77. Melov, S., Schneider, J. A., Coskun, P. E., Bennett, D. A. & Wallace, D. C. Mitochondrial DNA rearrangements in aging human brain and in situ PCR of mtDNA. *Neurobiol. Aging* **20**, 565–571 (1999).

78. Wiesner, R. J., Zsurka, G. & Kunz, W. S. Mitochondrial DNA damage and the aging process: facts and imaginations. *Free Radic. Res.* **40**, 1284–1294 (2006).

79. Trifunovic, A. *et al.* Premature ageing in mice expressing defective mitochondrial DNA polymerase. *Nature* **429**, 417–423 (2004).

80. Kujoth, G. C. *et al.* Mitochondrial DNA mutations, oxidative stress, and apoptosis in mammalian aging. *Science* **309**, 481–484 (2005).

81. Tyynismaa, H. *et al.* Mutant mitochondrial helicase Twinkle causes multiple mtDNA deletions and a late-onset mitochondrial disease in mice. *Proc. Natl. Acad. Sci. U.S.A.* **102**, 17687–17692 (2005).

82. Ahlqvist, K. J. *et al.* Somatic progenitor cell vulnerability to mitochondrial DNA mutagenesis underlies progeroid phenotypes in Polg mutator mice. *Cell Metab.* **15**, 100–109 (2012).

83. Phillips, N. R., Simpkins, J. W. & Roby, R. K. Mitochondrial DNA deletions in Alzheimer's brains: a review. *Alzheimers Dement. J. Alzheimers Assoc.* **10**, 393–400 (2014).

84. Krishnan, K. J., Ratnaike, T. E., De Gruyter, H. L. M., Jaros, E. & Turnbull, D. M. Mitochondrial DNA deletions cause the biochemical defect observed in Alzheimer's disease. *Neurobiol. Aging* **33**, 2210–2214 (2012).

85. Grünewald, A. *et al.* Mitochondrial DNA depletion in respiratory chain–deficient Parkinson disease neurons. *Ann. Neurol.* **79**, 366–378 (2016).

86. Wei, W. *et al.* Mitochondrial DNA point mutations and relative copy number in 1363 disease and control human brains. *Acta Neuropathol. Commun.* **5**, 13 (2017).

87. Benbrik, E. *et al.* Cellular and mitochondrial toxicity of zidovudine (AZT), didanosine (ddI) and zalcitabine (ddC) on cultured human muscle cells. *J. Neurol. Sci.* **149**, 19–25 (1997).

88. Lopez-Gomez, C. *et al.* Deoxycytidine and deoxythymidine treatment for thymidine kinase 2 deficiency. *Ann. Neurol.* **81**, 641–652 (2017).

5 Kearns-Sayre Syndrome

Josef Finsterer

Krankenanstalt Rudolfstiftung, Messerli Institute, Veterinary University of Vienna, Vienna, Austria

CONTENTS

1 INTRODUCTION

Kearns-Sayre syndrome (KSS) (OMIM 530000) is a genetic, specific mitochondrial multiorgan disorder syndrome (MIMODS), first described by Kearns and Sayre in 1958 (Kearns and Sayre, 1958). The key phenotypic features include ptosis, progressive external ophthalmoplegia (PEO), pigmentary retinopathy, cardiac conduction defects, ataxia, elevated cerebrospinal fluid (CSF) protein, and onset before age 20y (Krishna, 2017). In most of the cases, KSS is due to mitochondrial dysfunction secondary to a single deletion of the mitochondrial DNA (mtDNA) spanning 1000 to 10000 nucleotides (Krishna, 2017; Ortiz et al., 2017). Rarely, mtDNA point mutations or mtDNA duplications, which can be maternally transmitted, manifest phenotypically as KSS (Ortiz et al., 2017; Zhang et al., 2005). Single mtDNA deletions cause cytochrome-c-oxidase (COX) deficiency. Impaired mitochondria are abnormally large and abundant in tissues with high energy demand (Ortiz et al., 2017). This chapter summarises and discusses previous and recent findings concerning the clinical presentation, genetic background, diagnosis, treatment, and prognosis of KSS.

2 CLINICAL PRESENTATION

KSS is a MIMODS affecting the skeletal muscle, eyes, ears, brain, heart, and the endocrine organs. Neonates and children with KSS usually appear normal (Park et al., 2004).

2.1 SKELETAL MUSCLE

The first symptom and sign in KSS is usually ptosis, recognized when children use their brow muscle to elevate the eyelids (Park et al., 2004). Later in the disease course, PEO and other phenotypic features may develop (Park et al., 2004). PEO usually starts after age 5y and typically after onset of ptosis (Park et al., 2004). PEO usually affects horizontal and upward gaze but frequently spares downward gaze (Simaan et al., 1999). Diplopia is unusually due to the slow progression of PEO (Park et al., 2004). Only occasionally, easy fatigability and myopathy of the limb muscles may be seen (Park et al., 2004; van Beynum et al., 2012). Some patients with failure to thrive or dilated cardiomyopathy as initial manifestations have been reported (van Beynum et al., 2012). In a study of 19 Chinese patients with KSS, ophthalmoparesis was the presenting feature at onset in 16 patients. In three patients, short stature was the

55

initial manifestation (Yu et al., 2016). Mean age at onset in this cohort was 9.6y (Yu et al., 2016).

2.2 EYES

The dominant ocular manifestation in KSS is pigmentary retinopathy ("salt and pepper retinopathy"), which usually develops after ptosis and PEO. The retinal pigment epithelium (RPE) is initially affected followed by affection of photoreceptors and choriocapillaries (Ortiz et al., 2017). Pigmentary retinopathy can be distinguished from retinitis pigmentosa since bone spicule formation is uncommon and since it is not confined to the posterior pole (Park et al., 2004). Initial manifestations may be RPE atrophy together with degeneration of photoreceptors (Park et al., 2004). Compromise of rod and cones shows up as an abnormal scotopic response and with a borderline photopic response on electrorentinography (ERG) (Ortiz et al., 2017). Degeneration of photoreceptors is followed by phagocytosis of the photoreceptor debris by macrophages within the affected RPE. The peripheral photoreceptors are relatively spared (Park et al., 2004). The degree of visual impairment depends on the degree of retinal compromise (Ortiz et al., 2017). Other ophthalmologic manifestations of KSS include cataract, polymegathism, subretinal fibrosis, macular hole, and pleomorphism (Kozak et al., 2016).

2.3 BRAIN

In the cerebrum, KSS may manifest as cerebellar atrophy with cerebellar ataxia and nystagmus (Müller et al., 2003), calcifications, elevated CSF protein, respiratory insufficiency, pyramidal signs, seizures, mental retardation, or dementia (Park et al., 2004). In some patients, episodes of aseptic meningitis or encephalitis in childhood have been reported (Ashizawa and Subramony, 2001; Berenbaum et al., 1990; Ishikawa et al., 2000; Kearns and Sayre, 1958). In a study of 32 patients with mtDNA deletions, cognitive decline was reported in 31% of them (Moraes et al., 1989). Only in a few patients stroke-like episodes have been reported (Furuya et al., 1997). Histological work up of brain tissue may reveal spongiform degeneration of the cortex, the basal ganglia, or the brainstem (Park et al., 2004). If these lesions affect the oculomotor nuclei, this may be an additional cause of PEO (Park et al., 2004). If the respiratory center in the medulla is affected respiratory distress and episodic coma may ensue (Chabrol and Paquis, 1997). CSF protein values may occasionally exceed 200mg/dl (Tanji et al., 1999). Cerebral MRI may show cortical atrophy, cerebellar atrophy, white matter lesions, or T2-hyperintensities of the brainstem or the cerebellum (Kang et al., 2017; van Beynum et al., 2012; Yu et al., 2016). Some patients may develop features of Leigh syndrome (van Beynum et al., 2012). In a study of 19 KSS patients, MRI delineated patients with subcortical white matter lesions (n = 6) and patients with predominantly periventricular white matter lesions (n = 9) (Yu et al., 2016).

Hyperintensities of the white matter and the basal ganglia on DWI were found in 6 patients (Yu et al., 2016). MR-spectroscopy may be normal in KSS (Sijens et al., 2008). Rarely, KSS patients may experience ischemic stroke, as documented on autopsy (Müller et al., 2003), most likely due to intracardiac thrombus formation (Finsterer and Zarrouk-Mahjoub, 2016a).

2.4 HEART

Cardiac involvement occurs in about 50% of the cases and is associated with sudden cardiac death (SCD) in about 20% of the cases (Chawla et al., 2008). The heart is particularly vulnerable since COX-deficiency results in NADH accumulation, which inhibits the beta-oxidation (Park et al., 2004). Since the heart mainly relies on fatty acids inhibition of the beta-oxidation may predominantly affect the myocardium. In the conduction system KSS manifests as left anterior hemiblock (van Beynum et al., 2012), right bundle branch block (van Beynum et al., 2012), complete atrio-ventricular block (AV-block III), atrioventricular block I, atrioventricular block II (van Beynum et al., 2012), QT-prolongation (Krishna, 2017; van Beynum et al., 2012), torsades des pointes (van Beynum et al., 2012), monomorphic ventricular tachycardia (Krishna, 2017), or bradycardia-related polymorphic and ventricular tachycardia (PMVT) (Kabunga et al., 2015; Krishna, 2017). PMVT may go along with or without QT-prolongation (Oginosawa et al., 2003). Conduction defects frequently start with left anterior hemiblock, followed by right bundle branch block, which then progress to AV-block III (Park et al., 2004). His-bundle recordings may reveal trifascicular block. ST-depression on ECG is usually without overt concomitant coronary artery disease (Park et al., 2004). Rarely, sinus arrhythmia has been reported (Park et al., 2004). Conduction defects may manifest as palpitations, vertigo, syncopes, cardiac arrest, or SCD. Histopathological investigations of the cardiac conduction system may show fatty infiltration and fibrosis of the sino-atrial node, the AV-node, and the bundle branches (Gallastegui et al., 1987). SCD may even be the initial manifestation of KSS (van Beynum et al., 2012). In the myocardium KSS manifests as arterial hypertension (Ortiz et al., 2017), hypertrophic cardiomyopathy (Kupari, 1984), dilated cardiomyopathy (van Beynum et al., 2012), or right ventricular hypertrophy (Ortiz et al., 2017). Cardiac MRI (cMRI) and autopsy may show myocardial scars (Kabunga et al., 2015; Krishna, 2017). Cardiac involvement in KSS usually becomes apparent after onset of ptosis and PEO (van Beynum et al., 2012). However, in some patients cardiac abnormalities, such as dilated cardiomyopathy, may develop before other phenotypic features arise (Sehgal et al., 2016).

2.5 ENDOCRINE ORGANS

Endocrine abnormalities in KSS include delayed sexual maturation (delayed puberty), short stature, diabetes, thyroid dysfunction, parathyroid dysfunction, hypocorticism, and hyperaldosteronism (Ashizawa and Subramony, 2001; Finsterer and Zarrouk-Mahjoub, 2017; Kearns and Sayre, 1958). Diabetes occurs in about 20% of the cases (Finsterer and Frank, 2015). Only in some patients has hypoparathyroidism been reported (Abramowicz et al., 1996). Dysplasia of the pituitary gland may occur in single patients (Kang et al., 2017).

2.6 EARS

Otologic manifestations in KSS include hearing impairment or hearing loss (Park et al., 2004).

2.7 OTHERS

Rarely, patients may develop Fanconi syndrome (Ho et al., 2014), recurrent episodes of vomiting (Ho et al., 2014), pernicious anemia (Abramowicz et al., 1996), or dysmorphism (facial, pectus carinatum, scoliosis) (Berio and Piazzi, 2007; Kang et al., 2017).

3 EPIDEMIOLOGY

The exact prevalence of KSS is unknown but in a recent study a prevalence of 1.6 cases per 100,000 in the Finnish population has been reported (Remes et al., 2005). Generally, KSS is a rare, orphan disease. In a study of 19 patients with KSS the female to male ratio was 1:1.38 (Yu et al., 2016).

4 GENETIC BACKGROUND

In about 90% of the cases, KSS is due to single mitochondrial DNA (mtDNA) deletions. Single mtDNA deletions span 1.3–8.0kb of mtDNA and usually occur sporadically in oocytes or the zygote (Park et al., 2004). There may be somatic mutations or selective elimination of mutant mitochondria in certain cell lines (Park et al., 2004). Only in single cases is KSS due to mtDNA point mutations, which are usually missed on investigations with restriction fragment length polymorphism (RFLP) (Park et al., 2004). These point mutations include the variants m.3243A>G in the tRNA(Leu) gene (Wilichowski et al., 1998), m.3249G>A in the tRNA(Leu) gene (Seneca et al., 2001), and m.3255G>A in the tRNA(Leu) gene (Finsterer and Zarrouk-Mahjoub, 2016c). Rarely, mtDNA duplications may phenotypically manifest as KSS (Abramowicz et al., 1996). Only in 4% of the cases maternal inheritance of the nutation can be documented (Poulton et al., 2017). This is why it is nonetheless important to take a thorough family history (Finsterer and Zarrouk-Mahjoub, 2016b).

5 DIAGNOSIS

The diagnosis of KSS is made clinically if three criteria are accomplished. These include onset at <20y of age, PEO, and

pigmentary retinopathy (Park et al., 2004). Additionally, at least one of a secondary triad feature should be present, including cardiac conduction defect, CSF protein >100mg/dl, or cerebellar disturbance (Park et al., 2004; Rowland et al., 1991). The diagnosis may be difficult to establish in the early stages of the disease, as some patients may initially present with nonspecific features, with Pearson syndrome (Rahman and Leonard, 2000), or with Leigh syndrome (van Beynum et al., 2012). The main differential diagnosis of KSS is chronic progressive external ophthalmoplegia.

5.1 BLOOD CHEMISTRY

Blood tests usually show elevation of serum lactate and pyruvate (van Beynum et al., 2012). In some patients aminoacids, such as alanine, may be elevated (van Beynum et al., 2012). Occasionally, fasting blood glucose, oral glucose tolerance test, or the HbA1c values may be elevated. Serum aldosterone levels may be elevated in case of hyperaldosteronism (Park et al., 2004). In case of hypothyroidism, thyroid function parameters may be abnormal. In some patients deficiency of ubidecarenone or coenzyme-Q in the serum can be found (Park et al., 2004).

5.2 MUSCLE BIOPSY

Muscle biopsy in KSS patients may show ragged-red fibers, COX-negative fibers, and abnormal mitochondria with both unusual cristae and paracrystalline inclusions (Finsterer and Zarrouk-Mahjoub, 2016b; Marin-Garcia et al., 2000). The number of mitochondria and the number of ragged-red fibers on Gomori tri-chrome staining may be increased. Additionally, there may be aggregation of abnormal mitochondria (Park et al., 2004). Biochemical investigations of the muscle homogenate may show reduced activity of complex-I, complex-II, complex-III, or complex-IV, alone or in combination (Degoul et al., 1991; Finsterer and Zarrouk-Mahjoub, 2016b; Marin-Garcia et al., 2000).

5.3 ELECTRORETINOGRAM

The ERG is usually normal or shows mildly attenuated a-wave and b-wave amplitudes (Park et al., 2004).

5.4 GENETIC WORK-UP

Pathogenic mtDNA deletions can be detected by Southern blot analysis or RFLP. Point mutations can be detected by mtDNA sequencing. Frequently, mtDNA deletions can be detected only in muscle but not in blood.

6 TREATMENT

Patients with complete heart block require pacemaker implantation. Despite implantation of a pacemaker, KSS patients may experience ventricular tachyarrhythmias

(Kabunga et al., 2015; Rashid and Kim, 2002; Skinner et al., 2007; Subbiah et al., 2007). In case of PMVT or other severe ventricular arrhythmias, implantation of an implantable cardioverter defibrillator (ICD) should be considered. Growth hormone substitution seems to be effective depending on the severity of the underlying mitochondrial dysfunction (Quintos et al., 2016). In some patients a significant increase in height can be achieved by growth-hormone substitution (Quintos et al., 2016). Hormonal replacement therapy is also indicated in other endocrinopathies in KSS. Application of coenzyme-Q can be beneficial in single cases, improving cardiac function, exercise tolerance, ataxia, pyruvate metabolism, and CSF protein (Park et al., 2004). However, coenzyme-Q does not seem to have an effect on muscle weakness, hearing impairment, or retinopathy. Ptosis can be corrected by surgery and hearing impairment responds to hearing devices or cochlea implants. In single patients, muscle weakness may deteriorate upon administration of local anesthetics (Finsterer et al., 2005). Tube feeding may be necessary in case of weight loss or dysphagia to guarantee sufficient supply with calories. In some cases, progression or intractable heart failure may necessitate heart transplantation (Subbiah et al., 2007).

7 OUTCOME AND PROGNOSIS

The most important prognostic factor for life expectancy in KSS is cardiac disease (van Beynum et al., 2012). There may be patients with KSS in whom SCD is attributable to complete heart block (Chawla et al., 2008). It is speculated that ventricular arrhythmias are the actual cause of death in KSS since QT-prolongation can be progressive towards torsades des pointes (Subbiah et al., 2007).

8 CONCLUSIONS

KSS is characterized by ptosis, ophthalmoparesis, pigmentary retinopathy, cerebellar ataxia, cardiac conduction defects, short stature, and other, rarer features. In most of the cases, KSS is caused by single, heteroplasmic mtDNA deletions, which can be detected in the skeletal muscle. Since cardiac disease occurs in 50% of the patients and is the most important prognostic factor, it is crucial to monitor cardiac involvement and to select patients requiring a pacemaker or an ICD. Appropriate antiarrhythmic therapy is crucial for the survival of these patients.

ACKNOWLEDGMENTS

COI

There are no conflicts of interest.

Funding

No funding was received.

REFERENCES

Abramowicz MJ, Cochaux P, Cohen LH, Vamos E. Pernicious anaemia and hypoparathyroidism in a patient with Kearns-Sayre syndrome with mitochondrial DNA duplication. J Inherit Metab Dis 1996;19:109–111.

Ashizawa T, Subramony SH. What is Kearns-Sayre syndrome after all? Arch Neurol 2001;58:1053–1054.

Berenbaum F, Cote D, Pradat P, Rancurel G. Kearns-Sayre syndrome. Neurology 1990;40:193–194.

Berio A, Piazzi A. Craniofacial abnormalities in a patient with cytochrome-c-oxidase deficiency subsequently developing Kearns-Sayre syndrome. Panminerva Med 2007;49:97–98.

Chabrol B, Paquis V. Cerebral infarction associated with Kearns-Sayre syndrome. Neurology 1997;49:308.

Chawla S, Coku J, Forbes T, Kannan S. Kearns-Sayre syndrome presenting as complete heart block. Pediatr Cardiol 2008;29:659–662.

Degoul F, Nelson I, Lestienne P, Francois D, Romero N. et al. Deletions of mitochondrial DNA in Kearns-Sayre syndrome and ocular myopathies: genetic, biochemical and morphological studies. J Neurol Sci 1991;101:168–177.

Finsterer J, Frank M. Diabetes in Kearns-Sayre syndrome: more common than anticipated. Can J Diabetes 2015;39:253.

Finsterer J, Haberler C, Schmiedel J. Deterioration of Kearns-Sayre syndrome following articaine administration for local anesthesia. Clin Neuropharmacol 2005;28:148–149.

Finsterer J, Zarrouk-Mahjoub S. Diagnose Kearns-Sayre syndrome genetically and investigate the phenotype comprehensively. Oxf Med Case Reports 2016a;2016:omw059. doi:10.1093/omcr/omw059.

Finsterer J, Zarrouk-Mahjoub S. Diagnosing Kearns-Sayre syndrome requires genetic confirmation. Chin Med J (Engl) 2016b;129:2267–2268.

Finsterer J, Zarrouk-Mahjoub S. Diagnosis of Kearns-Sayre syndrome requires comprehensive work-up. Chin Med J (Engl) 2016c;129:2518–2519.

Finsterer J, Zarrouk-Mahjoub S. Kearns-Sayre syndrome in the absence of a mtDNA deletion? Andrologia 2017;49:10.

Furuya H, Sugimura T, Yamada T, Hayashi K, Kobayashi T. A case of incomplete Kearns-Sayre syndrome with a stroke like episode. Rinsho Shinkeigaku 1997;37:680–684.

Gallastegui J, Hariman RJ, Handler B, Lev M, Bharati S. Cardiac involvement in the Kearns-Sayre syndrome. Am J Cardiol 1987;60:385–388.

Ho J, Pacaud D, Rakic M, Khan A. Diabetes in pediatric patients with Kearns-Sayre syndrome: clinical presentation of 2 cases and a review of pathophysiology. Can J Diabetes 2014;38:225–228.

Ishikawa Y, Goto Y, Ishikawa Y, Minami R. Progression in a case of Kearns-Sayre syndrome. J Child Neurol 2000;15:750–755.

Kabunga P, Lau AK, Phan K, Puranik R, Liang C. et al. Systematic review of cardiac electrical disease in Kearns-Sayre syndrome and mitochondrial cytopathy. Int J Cardiol 2015;181:303–310.

Kang YX, Wang YJ, Zhang Q, Pang XH, Gu W. A case of hypopituitarism accompanying Kearns-Sayre syndrome treated with human chorionic gonadotropin: A case report and literature review. Andrologia 2017 October;49(8). doi:10.1111/and.12711.

Kearns TP, Sayre GP. Retinitis pigmentosa, external ophthalmophegia, and complete heart block: unusual syndrome with histologic study in one of two cases. AMA Arch Ophthalmol 1958;60:280–289.

Kozak I, Oystreck DT, Abu-Amero KK, Nowilaty SR, Alkhalidi H, et al. New observations regarding the retinopathy of genetically confirmed Kearns-Sayre syndrome. Retin Cases Brief Rep 2016 December 19. doi:10.1097/ICB.0000000000000503.

Krishna MR. Kearns Sayre syndrome: looking beyond A-V conduction. Indian Pacing Electrophysiol J 2017;17:78–80.

Marin-Garcia J, Goldenthal MJ, Sarnat HB. Kearns-Sayre syndrome with a novel mitochondrial DNA deletion. J Child Neurol 2000;15:555–558.

Moraes CT, DiMauro S, Zeviani M, Lombes A, Shanske S, et al. Mitochondrial DNA deletions in progressive external ophthalmoplegia and Kearns-Sayre syndrome. N Engl J Med 1989;320:1293–1299.

Müller W, Mennel HD, Bewermeyer K, Bewermeyer H. Is there a final common pathway in mitochondrial encephalomyopathies? Considerations based on an autopsy case of Kearns-Sayre syndrome. Clin Neuropathol 2003;22:240–245.

Oginosawa Y, Abe H, Nagatomo T, Mizuki T, Nakashima Y. Sustained polymorphic ventricular tachycardia unassociated with QT prolongation or bradycardia in the Kearns-Sayre syndrome. Pacing Clin Electrophysiol 2003;26:1911–1912.

Ortiz A, Arias J, Cárdenas P, Villamil J, Peralta M, et al. Macular findings in spectral domain optical coherence tomography and oct angiography in a patient with Kearns-Sayre syndrome. Int J Retina Vitreous 2017 July 10;3:24. doi:10.1186/s40942-017-0077-8.

Park SB, Ma KT, Kook KH, Lee SY. Kearns-Sayre syndrome -3 case reports and review of clinical feature. Yonsei Med J 2004;45:727–735.

Poulton J, Finsterer J, Yu-Wai-Man P. Genetic counselling for maternally inherited mitochondrial disorders. Mol Diagn Ther 2017;21:419–429.

Quintos JB, Hodax JK, Gonzales-Ellis BA, Phornphutkul C, Wajnrajch MP, et al. Efficacy of growth hormone therapy in Kearns-Sayre syndrome: the KIGS experience. J Pediatr Endocrinol Metab 2016;29:1319–1324.

Rahman S, Leonard JV. Early onset of complete heart block in Pearson syndrome. J Inherit Metab Dis 2000;23:753–754.

Rashid A, Kim MH. Kearns-Sayre syndrome: association with long QT syndrome? J Cardiovasc Electrophysiol 2002;13:184e5.

Remes AM, Majamaa-Voltti K, Kärppä M, Moilanen JS, Uimonen S, et al. Prevalence of large-scale mitochondrial DNA deletions in an adult finnish population. Neurology 2005;64:976–981.

Rowland LP, Blake DM, Hirano M, Di Mauro S, Schon EA, et al. Clinical syndromes associated with ragged red fibers. Rev Neurol (Paris) 1991;147:467–473.

Sehgal S, Choudhry S, Debelenko L, L'Ecuyer T. Dilated cardiomyopathy with cardiogenic shock in a child with Kearns-Sayre syndrome. BMJ Case Rep 2016 February 16;2016:bcr2015213813. doi:10.1136/bcr-2015-213813.

Seneca S, Verhelst H, De Meirleir L, Meire F, Ceuterick-De Groote C, et al. A new mitochondrial point mutation in the transfer RNA(Leu) gene in a patient with a clinical phenotype resembling Kearns-Sayre syndrome. Arch Neurol 2001;58:1113–1118.

Sijens PE, Smit GP, Rödiger LA, van Spronsen FJ, Oudkerk M, et al. MR spectroscopy of the brain in Leigh syndrome. Brain Dev 2008;30:579–583.

Simaan EM, Mikati MA, Touma EH, Rötig A. Unusual presentation of Kearns-Sayre syndrome in early childhood. Pediatr Neurol 1999;21:830–831.

Skinner JR, Yang T, Purvis D, Chung SK, Roden DM, et al. Coinheritance of long QT syndrome and Kearns-Sayre syndrome. Heart Rhythm 2007;4:1568e72.

Subbiah RN, Kuchar D, Baron D. Torsades de pointes in a patient with Kearns-Sayre syndrome: a fortunate finding. Pacing Clin Electrophysiol 2007;30:137e9.

Tanji K, Vu TH, Schon EA, DiMauro S, Bonilla E. Kearns-Sayre syndrome: unusual pattern of expression of subunits of the respiratory chain in the cerebellar system. Ann Neurol 1999;45:377–383.

van Beynum I, Morava E, Taher M, Rodenburg RJ, Karteszi J, et al. Cardiac arrest in Kearns-Sayre syndrome. JIMD Rep 2012;2:7–10.

Wilichowski E, Korenke GC, Ruitenbeek W, De Meirleir L, Hagendorff A, et al. Pyruvate dehydrogenase complex deficiency and altered respiratory chain function in a patient with Kearns-Sayre/MELAS overlap syndrome and A3243G mtDNA mutation. J Neurol Sci 1998;157:206–213.

Yu M, Yu L, Wang ZX. Diagnosis and Management of Kearns-Sayre Syndrome Rely on Comprehensive Clinical Evaluation. Chin Med J (Engl) 2016;129:2519–2520.

Zhang Y, Wang ZX, Niu SL, Xu YF, Pei P, et al. Phenotype heterogeneity associated with mitochondrial DNA A3243G mutation. Zhongguo Yi Xue Ke Xue Yuan Xue Bao 2005;27:77–80.

6 Mitochondrial Dysfunction and Barth Syndrome

Arianna F. Anzmann
McKusick-Nathans Institute of Genetic Medicine, Johns Hopkins School of Medicine, Baltimore, Maryland, USA

Steven M. Claypool
Department of Physiology, Johns Hopkins School of Medicine, Baltimore, Maryland, USA

Hilary Vernon
McKusick-Nathans Institute of Genetic Medicine, Johns Hopkins School of Medicine, Baltimore, Maryland, USA

CONTENTS

1 BARTH SYNDROME

In 1980, Barth et al. described a novel X-linked mitochondrial disease affecting cardiac muscle, skeletal muscle and neutrophil leukocytes, at a Neuromuscular Disease Symposium held at Erasmus University[1]. Three years after this preliminary communication, Barth et al. reported a large Dutch family with cardiomyopathy, skeletal myopathy, neutropenia, and high infant mortality due to infection or cardiac failure[2]. These clinical characteristics have since become the cardinal features of what is now known as Barth syndrome (BTHS, MIM#302060) (Figure 1)[3-6]. The incidence of BTHS is estimated to be about 1/300,000–400,000 live births, with fewer than 500 individuals worldwide included in the BTHS Registry & Repository. There is no known racial or ethnic predilection[7,8].

The majority of individuals with BTHS present with cardiomyopathy within their first two years of life[7]. Classically, the cardiomyopathy presents as dilated cardiomyopathy with a component of left ventricular noncompaction, however hypertrophic cardiomyopathy has also been described. Heart function in BTHS patients can follow a waxing and waning pattern, sometimes with a period of relative stability after the toddler years and before puberty[4]. Cardiac arrhythmia and Long QT syndrome have also been described in a number of affected individuals[9].

Skeletal myopathy and skeletal muscle fatigue are also important clinical features of BTHS. Affected individuals have decreased exercise endurance, weakness of proximal leg muscles, and overall lower activity levels. The skeletal muscle weakness and fatigue is independent of the cardiac dysfunction[10].

Neutropenia is also seen in most affected individuals and can present as severe chronic neutropenia, cyclic neutropenia, or intermittent/non-cyclical neutropenia. Low neutrophil counts put affected individuals at risk for severe bacterial infections, ranging from mouth ulcers and gingival inflammation, to sepsis and multi-organ system failure[7,11].

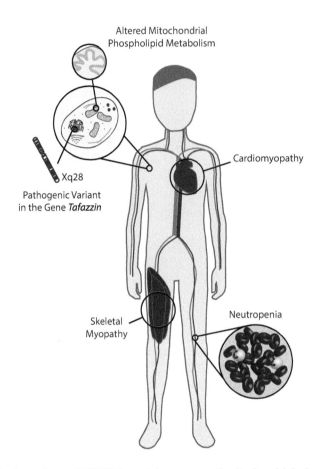

FIGURE 1 Barth syndrome. Barth syndrome (BTHS) is an inborn error of mitochondrial phospholipid metabolism, caused by pathogenic variants in the gene *tafazzin (TAZ)* located in the gene rich region of distal Xq28. *TAZ* encodes for a transacylase involved in the final remodeling step of cardiolipin (CL), which is essential for CL maturation. TAZ deficiency results in altered CL metabolism, with an increased monolyso-CL:CL and an abnormal acyl chain pattern in the remaining CL. Clinically, BTHS is characterized by cardiomyopathy, skeletal myopathy, and neutropenia.

Other clinical features seen in association with BTHS include failure to thrive, pre-pubertal growth delay, gastrointestinal complaints, hypoglycemia, and scoliosis[4,7,12]. There is no specific therapy for BTHS, and treatments are directed toward individual symptoms, including granulocyte-colony stimulating factor (G-CSF) therapy for neutropenia and medical treatment for cardiac failure. In cases of intractable heart failure, cardiac transplantation is performed[5,13].

2 TAFAZZIN

In 1991, almost 10 years after BTHS was first described, and 10 years before the Human Genome Project was completed, Bolhuis et al. analyzed the linkage between the disease locus of BTHS and X-chromosomal markers[14]. Multipoint linkage analysis of the first BTHS family reported by Barth et al., a multi-generational family with seven carriers, four patients, and eight unaffected sons of carriers, mapped the disease locus to the gene rich Xq28 region[14]. Mapping to the distal Xq28 region was further

confirmed in 1993 and 1995 by Ades et al. and Gedeon et al., respectively[3,15]. Then in 1996, Bione et al. reported the identification of the gene, termed *G4.5*, as responsible for BTHS[16]. The open reading frame of *G4.5* suggested several putative proteins ranging from 129 to 229 amino acids in length. Since these putative proteins did not have any homology to any previously identified nucleotide or protein sequence in GenBank at that time, they proposed that the novel proteins be named "tafazzins"[16]. This unique name pays homage to a comic from an Italian sports television show, Tafazzi, who repeatedly hits himself with an empty plastic bottle. 22 years later, the tafazzin gene/protein (*TAZ*/TAZ) maintains its reputation as a difficult subject for masochistic scientists.

TAZ is conserved form yeast to humans and has 11 exons with multiple alternative splicing isoforms[17,18]. Of the identified isoforms there are four major isoforms; the full length transcript (NM_000116), a transcript without exon 5 (Δ5, NM_181311), a transcript without exon 7 (Δ7, NM_181312), and a transcript without both exon 5 and 7 (Δ5Δ7, NM_181313)[17]. Of the four isoforms, the mRNA isoform Δ5 is the predominant isoform, followed by Δ5Δ7, full-length, and Δ7[17,19]. Gene expression studies in patient lymphoblast lines and a TAZ deficient yeast model indicate that of the different isoforms, only two, the Δ5 and full-length isoforms, encode functional TAZ with transacylase (CL remodeling) activity[17,20–22]. Further, only a single polypeptide that co-migrates with the Δ5 isoform is expressed in human skin fibroblasts, human embryonic kidney 293 (HEK293) cells, and murine heart and liver mitochondria[23]. It is interesting to note that the inclusion of exon 5 arose in the hominoid primate lineage after it branched from the Old World monkey lineage, 15–20 million years ago, due to a variant in the splice acceptor consensus sequence (GG>AG)[17]. This primate-specific exon is predicted to correspond to an intrinsically unstructured region of TAZ and has been speculated to contribute to primate-specific molecular interactions[24].

As mentioned, when *TAZ* was first identified it had no sequence homology with any known nucleotide sequence. To date, the closest related protein with a known 3D structure is plant glycerol 3-phosphate acyltransferase (G3PAT), which has 20% sequence identity to human *TAZ*[24]. The mystery of TAZ function was cracked in 1997 when Neuwald reported, based on an alignment of human TAZ with two hypothetical proteins from yeast and worm, that TAZ belongs to a superfamily of acyltransferases with predicted roles in phospholipid biosynthesis[25]. Neuwald further suggested that the mitochondrial dysfunction associated with BTHS may be due to alterations in the mitochondrial membrane phospholipids a prediction that was confirmed by Vreken et al. in 2000[26]. Indeed, *TAZ* is a ubiquitously expressed gene that encodes for a transacylase involved in the final remodeling step of cardiolipin (CL), a major phospholipid of the inner mitochondrial membrane (IMM)[27].

3 CARDIOLIPIN

Mitochondria are unique organelles with two specialized membranes defining separate compartments within the mitochondrion, each with a distinct role and composition: the outer mitochondrial membrane (OMM), the intermembrane space (IMS), the IMM, and the matrix. The OMM, which is more typical of other membranes in terms of its protein to lipid ratio, is believed to be permeable to most metabolites due the presence of pore-forming porins in its bilayer. In contrast, the highly folded IMM is a tight barrier to all ions and molecules, with an atypically high protein to phospholipid ratio[28,29]. The lipid composition of both membranes is characterized by high phosphatidylcholine (PC) and phosphatidylethanolamine (PE) content and low sterol and sphingolipid count[28]. Additionally, the IMM, and

mitochondrial contact sites where the OMM and IMM are in close proximity, display an enrichment of CL, which accounts for up to 20% of total mitochondrial lipids[28,30,31].

CL was actually first purified in 1942 in an attempt to isolate a substance with serological utility in syphilis testing. Investigators termed the new non-nitrogenous phospholipid "cardiolipin" as it was isolated from bovine heart[32]. In 1964, its structure was first described as a phospholipid consisting of two phosphatidyl moieties, each with two acyl chains, that are connected by a glycerol bridge[33]. The four acyl chains allow for an array of CL species that enable the observed tissue specific CL composition; CL in the brain is characterized by a diversified array of acyl chains including polyunsaturated chains, whereas in other tissues CL is predominantly characterized by the tetralinoleoyl form, $(18:2)_4$ CL[34–36]. The enrichment of tetralinoleoyl-CL is most obvious in human heart and skeletal muscle where it accounts for up to 80% of the CL species present[34].

3.1 DE NOVO SYNTHESIS AND REMODELING

In eukaryotes, most phospholipids are synthesized in the endoplasmic reticulum (ER); however, the biosynthetic pathway of CL occurs exclusively in the mitochondrion[37,38]. This multi-step process, elucidated over the past 40 years, begins with the generation of phosphatidic acid (PA), which can be potentially sourced from several pathways[6]. PA, synthesized in the ER, on the outer surface of the OMM, or in the IMS, is then transported to the matrix side of the IMM where a condensation reaction between PA and cytidine triphosphate (CTP), catalyzed by Tam41p (Human Gene Organization (HUGO) Gene Nomenclature Committee (HGNC): *TAMM41*, Saccharomyces Genome Database (SGD): *TAM41*), forms cytidine diphosphate diacylglycerol (CDP-DAG)[6,39,40]. CDP-DAG and glycerol-3-phosphate (G3P) are then converted to phosphatidylglycerol phosphate (PG-P) by PG-P synthase (HGNC: *PGS1*, SGD: *PGS1*)[41,42]. PG-P is rapidly dephosphorylated to PG by PG-P phosphatase (HGNC: *PTPMT1*, SGD: *GEP4*)[43,44]. PG is then bound to another molecule of CDP-DAG by CL synthase (HGNC: *CRLS1*, SGD: *CRD1*), an integral IMM protein whose active site faces the mitochondrial matrix, through a condensation reaction that forms nascent CL (Figure 2)[45–48].

De novo synthesis of nascent CL, characterized by saturated acyl chains of variable length and asymmetry with respect to the two chiral centers due to the limited acyl substrate specificity of CRLS1, is followed by CL remodeling which is essential for CL maturation[48,49]. Remodelling is initiated with the removal of an acyl chain, which generates monolyso-CL (MLCL). In yeast, this is executed by Cld1p; however, there are no known Cld1p orthologs in higher eukaryotes[50,51]. In mammals, several calcium-independent phospholipase A_2 (iPLA$_2$) members have been

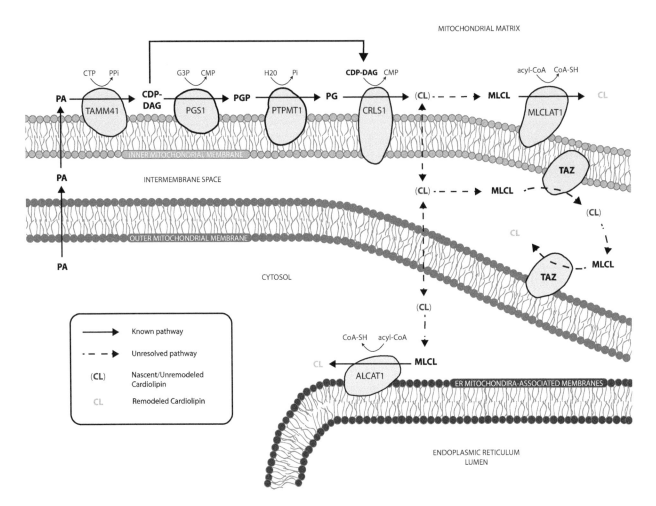

FIGURE 2 Cardiolipin biosynthesis and remodelling. Phosphatidic acid (PA), sourced from several pathways, is transported to the matrix side of the inner mitochondrial membrane (IMM) where a condensation reaction between PA and cytidine triphosphate (CTP), catalyzed by TAMM41, forms cytidine diphosphate diacylglycerol (CDP-DAG). CDP-DAG and glycerol-3-phosphate (G3P) are then converted to phosphatidylglycerol phosphate (PG-P) by PG-P synthase (PGP1). PG-P is rapidly dephosphorylated to PG by PG-P phosphatase (PTPMT1). PG is then bound to another molecule of CDP-DAG by CL synthase (CRLS1) through a condensation reaction that forms nascent CL. *De novo* synthesis of nascent CL is followed by CL remodeling, initiated with the removal of an acyl chain, which generates monolyso-CL (MLCL). In mammals, the enzyme(s) that de-acylates nascent CL and generates MLCL is unresolved. Following de-acylation, MLCL is then re-acylated by one of three enzymes: TAZ, MLCL acyltransferase 1 (MLCLAT1), or acyl-CoA:lysocardiolipin acyltransferase-1 (ALCAT1).

implicated, as these enzymes catalyze the hydrolysis of membrane glycerophospholipids into free fatty acids and lyso-lipids[52,53]. Nonetheless, the question as to which enzyme(s) de-acylates nascent CL and generates MLCL in mammals remains unanswered.

Following de-acylation, MLCL is then re-acylated by one of three enzymes: TAZ, MLCL acyltransferase 1 (MLCLAT1), or acyl-CoA: lysocardiolipin acyltransferase-1 (ALCAT1) (Figure 2)[54–58]. TAZ, the assumed predominant enzyme in CL remodeling, catalyzes a reversible transacylation that acts on both *sn*-1 and *sn*-2 positions, using any number of phospholipids (PLs) and lyso-PLs as acyl chain donors and acceptors, respectively[27,56]. Deficiency of TAZ results in an increase of the intermediate MLCL, a

decrease in mature CL, and an abnormal acyl chain pattern in the remaining CL, the pathognomonic biochemical defect in BTHS[19,26,59–62]. Though the primary biochemical defect in BTHS has been known for several decades, the exact mechanism(s) by which the specific acyl-chain composition of mature CL is generated is still unresolved, and has become the subject of much debate.

3.2 Determinants of the Acyl Chain Composition of Remodeled CL

When considering the acyl specificity of TAZ the debate has centered on two points; whether TAZ specificity is attributed to the physical properties of lipids and the mitochondrial membrane or whether specificity is attributed to the selectivity of TAZ itself. Schlame et al. first proposed that in the presence of TAZ, transacylations will reshuffle the acyl residues to optimize lipid packing in curved membranes[63]. In the mitochondrion, bilayer-prone PLs, such as PC, form a stable arrangement of tightly packed lipid molecules. Bending of the bilayer membrane disturbs the packing order of the bilayer PLs, which is thermodynamically unfavorable, but allows for the intermixing of PLs with non-bilayer characteristics such as CL. CL, which has a relatively small polar head group and four acyl chains, that in its mature form are primarily unsaturated, has a cone like structure and preferentially localizes in regions of high membrane curvature[64]. Therefore, acyl specificity is driven by the packing properties of mitochondrial membranes in the final equilibrium state. This is known as the thermodynamic remodeling hypothesis.

Others have proposed that the thermodynamic remodeling hypothesis does not account for all details of TAZ-mediated CL remodeling. Specifically, the model, which suggests that <1% of endogenous mitochondrial PLs participate in transacylation, does not account for the extensive degree of CL remodeling observed in the mammalian heart and skeletal muscle, where tetralinoleoyl-CL accounts for up to 80% of all CL. Secondly, the model suggests that transacylation only occurs in non-bilayer membranes. However, it has not been established whether the curvature of the mitochondrion, which forms the mitochondrial cristae, can be categorized as non-bilayer domains[65]. Third, since CL preferentially localizes in regions of high membrane curvature, it is difficult to distinguish whether the thermodynamic remodeling hypothesis is the key determinant of TAZ acyl chain specificity. Therefore, in 2016, Abe et al, based on their investigation of purified TAZ in liposomes composed of various sets of acyl donors and acceptors, suggested that TAZ does have acyl chain selectivity, and it may be one of the factors in determining the acyl composition of mature CL[66].

The debate between these two hypotheses has been ongoing. Schlame et al., after repeating and expanding upon the experiments performed by Abe et al., still maintained that acyl specificity of TAZ is driven by the packing properties of non-bilayered mitochondrial membranes in the final equilibrium state. Incubation of yeast TAZ with PC $(18:2)_2$ and MLCL $(18:2)_3$ produced CL $(18:2)_4$ as expected, but incubation of yeast TAZ with PC $(18:1)_2$ and MLCL $(18:2)_3$ produced four different species. Schlame et al. suggest that the incorporation of multiple 18:1 residues in CL is only possible if TAZ transfers not only 18:1 from PC to MLCL, but also transfers 18:2 from CL to lyso-PC, and that the bidirectional exchange of acyl groups cannot be under kinetic control. Abe et al. do agree that there are multiple exchanges of acyl chains that occur in TAZ-mediated transacylation, but they continue to argue that transacylation can occur in rigid lipid bilayers, such as liposomes, and that TAZ does have a degree of acyl chain specificity. Using a series of liposomes composed of different derivatives of dipalmitoleoyl $(16:1)_2$ PC they investigated the TAZ-mediated transacylation reaction between different PCs and MLCL, and found that TAZ strictly discriminates the molecular configuration of the PC acyl chains. Further, they suggest that the final acyl composition may result from a coordinated effort between a deacylase, Cld1p in yeast and yet to be identified in humans, and TAZ

Recently, several groups have demonstrated that Cld1p, the deacylase upstream of Taz1p that was identified in yeast by Beranek et al., has selectivity as to which mitochondrial CLs it deacylates[50,67,68]. Specifically, biochemical studies and computer modeling confirmed the selectivity of Cld1p toward $(16:0)_4$, $(16:1)(16:0)_3$, and $(18:0)_4$-CL substrates and the preservation of $(18:1)_4$, and $(18:2)_4$-CL species. Therefore, the specificity of Cld1p controls the MLCL available for Taz1p to reacylate[51,68]. Even though the deacylase acting upstream of human TAZ has yet to be identified, we must still consider the potential contribution of a deacylase in dictating the final acyl chain composition of CL. Additionally, another determinant of the acyl chain composition of CL could be the availability of acyl chain donors, as dietary supplementation can shift the acyl chain pattern of PLs[69,70]. As the debate continues, perhaps the explanation with regards to the acyl specificity of TAZ will be a combination of each component: the packing properties of the mitochondrial membrane, the acyl specificity of TAZ itself, the role of a deacylase acting upstream of TAZ, and the availability of acyl chain donors.

3.3 GENOTYPE-PHENOTYPE RELATIONSHIP: THE ROLE OF MLCL: CL

BTHS is a result of pathogenic variants within *TAZ*, which can be identified by molecular genetic testing. To date there are over 120 causative variants that have been reported in every exon except for exon 5, including missense, nonsense, splice variants, small insertions/deletions, and large deletions[17]. In addition to molecular genetic testing, a BTHS diagnosis can be established by an increased MLCL:CL ratio, which is detectable in multiple biological specimens including bloodspots, lymphocytes, and muscle tissue. The measurement of MLCL:CL has a diagnostic sensitivity and specificity of 100%.

Investigations of different yeast lines, representing 21 distinct and conserved human *TAZ* pathogenic variants, have established seven functional classes of *TAZ* genetic variation. The first and the largest class, includes variants that result in the production of little to no functional TAZ. The remaining classes include; (2) Mitochondrial mislocalization and aggregate prone; (3) Impaired macromolecular assembly; (4) Catalytically null; (5) Hypomorphs with residual transacylase activity; (6) Degradation due to impaired folding and assembly; and (7) Temperature sensitive[71,72]. Distinct loss of function (LOF) classes with potential for residual TAZ function suggest a possible genotype-phenotype correlation; however, prior clinical phenotyping studies have not identified such a relationship[71,72].

Recently, a possible "CL metabotype"-phenotype relationship was established. In 2015, Bowron et al. reported 7 BTHS individuals with a modified clinical phenotype, increased MLCL, and an increased MLCL: CL ratio, but a normal level of CL[73]. The *TAZ* variants identified in these individuals were: p.Arg57Leu, p. Asn40Asp, and p.Met185Val. The substitution, p. Met185Val (c.553A>G), generates a new splice donor site within exon 7 resulting in the retention of intron 6 and a partial deletion of exon 7, confirming pathogenicity[74]. The variants, p.Arg57Leu (R57L) and p. Asn40Asp (N40D) have both been modeled in yeast; R57L/K65L is a temperature sensitive allele with reduced transacylase activity and N40D/N48D has activity identical to WT[72]. Therefore, the pathogenicity of N40D remains unproven.

The presentation of a modified clinical phenotype associated with a modified metabotype (elevated MLCL and normal CL) was again identified in 2016 in a multidisciplinary investigation of 42 individuals with BTHS. In this study, researchers identified several individuals with a relatively milder skeletal muscle phenotype and the modified metabotype. These modified metabotype-phenotypes were associated with the genotypes c.583 +5G>A and c.873_874dupCCTGG (p.Arg292LeufsX49). They further found that MLCL:CL was inversely correlated to skeletal muscle endurance and directly correlated to left ventricular size, offering further evidence for a genotype-metabotype-phenotype relationship. As a rare disorder with isolated familial variants, continued efforts in establishing a genotype-metabotype-phenotype relationship will not only inform patient care, but help elucidate the disease mechanism of TAZ deficiency and abnormal CL content.

3.4 CARDIOLIPIN FUNCTION: EVERYTHING BUT THE KITCHEN SINK

Since its initial isolation in 1942, and throughout almost 2000 publications, CL has a been implicated in a laundry list of functions within the mitochondrion (Figure 3). Many proteins are known to interact with CL, in both structural and functional capacities, which establishes the foundation for CL's multifaceted role within the mitochondrion[75]. The essentiality of CL is highlighted by the observation that *ptpmt1*[-/-] mice, an enzyme essential for CL synthesis, die *in utero* prior to E8.5[44].

CL composes up to 20% of the phospholipid mass of the IMM and is critical for mitochondrial cristae formation, based on observations of abnormal IMM ultrastructure in CL-deficient models[76–78]. The mitochondrial cristae house the oxidative phosphorylation (OXPHOS) system. Therefore, it is not surprising that CL also plays a critical role in the efficiency and adaptability of the OXPHOS machinery. First, CL stabilizes the higher order assemblies of the individual respiratory complexes into respiratory supercomplexes (SCs), which are combinations of respiratory complexes I, III, and IV[79–87]. SC assembly is thought to increase the efficiency of electron transfer in OXPHOS, maximize OXPHOS function, minimize reactive oxygen species (ROS) production, and thereby reduce oxidative damage. Second, CL is proposed to act as a proton trap, restricting the diffusion of protons and funneling them toward ATP synthase to generate ATP[88]. Lastly, once ATP is formed by OXPHOS, it then passes across the IMM to the IMS via the ADP/ATP carrier. CL is needed for not only the stabilization of the carrier, but is also required for the association of the carrier with the OXPHOS SCs[89,90]. Besides the ADP/ATP carrier, CL also plays a plays a protective and stabilizing role for various mitochondrial carriers (Figure 3)[91].

CL is critical for mitochondrial biogenesis, which requires the import of a large number of proteins from the cytosol. Reduced functionality of OXPHOS in the absence of CL decreases the mitochondrial membrane potential, which negatively affects protein import into the IMM and mitochondrial matrix by IMM translocases, TIM22 and TIM23[92]. Separately, CL is also important for the assembly and function of OM translocases, TOM and SAM[93]. CL is also intimately involved in mitochondrial fusion and fission whose combined actions promote mitochondrial fitness and biogenesis (Figure 3)[94,95]. There are several superb reviews which thoroughly discuss the aforementioned CL-supported functions: Chicco and Sparangna 2007, Houtkooper and Vaz 2008, Lewis and McElhaney 2009, Klingenberg 2009, Claypool and Koehler 2012, Paradies et al. 2014, Li et al. 2015, Musatov and Sedlák 2017, Ikon and Ryan 2017[14,91,96–103].

Recently, there has been an emphasis on CL as a signaling molecule. In healthy cells, CL is localized in the IMM, whereas in damaged cells, CL externalizes to the OMM as an elimination signal. Once externalized, CL can bind and activate LC3-II, which initiates autophagosome formation resulting in mitophagy, the selective degradation of mitochondria. Further, under oxidative stress, the production of ROS results in the peroxidation of CL by Cytochrome c (Cyt c) and its subsequent externalization to the OMM as a damage signal for apoptosis, programmed cell death (Figure 3)[104–106].

Extracellular externalized CL (e.g. outside of plasma membrane) is thought to have an additional role in some immune responses, such as the attraction of phagocytes through the cell surface receptor CD36, resulting in rapid degradation of the cell with extracellular

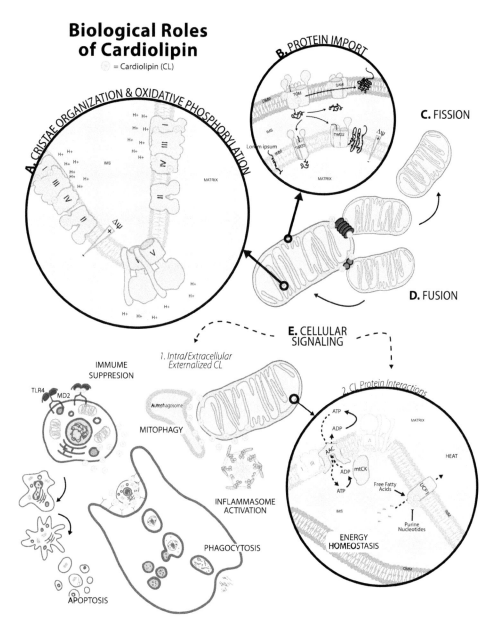

FIGURE 3   Biological roles of cardiolipin. (A) Cardiolipin (CL) is critical for mitochondrial cristae formation[76–78]. Within the cristae, CL plays a critical role in the efficiency and adaptability of the oxidative phosphorylation (OXPHOS) machinery, by stabilizing the assembly of respiratory supercomplexes (SCs). SC assembly is thought to increase the efficiency of electron transfer in OXPHOS, maximize OXPHOS function, minimize reactive oxygen species (ROS) production, and thereby reduce oxidative damage[79–87]. (B) Reduced functionality of OXPHOS decreases the mitochondrial membrane potential, which negatively affects protein import into the IMM and mitochondrial matrix by IMM translocases, TIM22 and TIM23[92]. Separately, CL is also important for the assembly and function of OM translocases, TOM and SAM[93]. Together, CL is critical for mitochondrial biogenesis, which requires the import of a large number of proteins from the cytosol. (C) CL is also crucial for mitochondrial fission and (D) fusion, where cells reorganize the highly dynamic mitochondrial organelles by altering their shape, localization, and cellular number[94,95]. (E1) Intra/extracellular externalized CL is a critical signaling component in mitophagy, apoptosis, and multiple distinct immune responses, such as inflammasome activation and phagocytosis[107,112–114]. Extracellular externalized CL can be bound by the phagocyte Toll-like receptor 4 (TLR4)/Md2, which acts as a structural homologue of antigenically inactive immature Lipid A and suppresses TLR4/Md2 driven cytokine production, thus acting as an immune suppressor[107]. (E2) Many proteins are known to interact with CL, and recently a role for CL in the activation of thermogenic adipocytes was defined. CL may be involved in the direct activation of the thermogenic effector, UCP1, and/or decrease UCP1's binding affinity for inhibitory purine nucleotides[117]. Additionally, CL also physically interacts with creatine kinase (mtCK), an enzyme that drives thermogenesis in beige fat[118].

externalized CL[107]. In addition to signaling the rapid removal of mitochondria and bacteria by phagocytosis, externalized CLs may also play a role in regulating the inflammatory response elicited by bacteria. CL can be bound by the phagocyte Toll-like receptor 4 (TLR4)/Md2, which acts as a structural homologue of antigenically inactive immature Lipid A and suppresses TLR4/Md2 driven cytokine production, thus acting as an immune suppressor[107]. If the innate immune system fails, or if there is a constant release of mitochondria/mitochondrial fragments with externalized CL due to cell death, the continuous presence of CL can activate antigen-presenting cells (APCs). This then stimulates an adaptive immune response resulting in the production of anti-CL antibodies and potentially autoimmune disease[108,109]. In addition to extracellular externalized CL, intracellular externalized CL (e.g. exposed to cytosol) has also been suggested to activate inflammasomes, such as NLRP3[110,111]. To summarize, CL and its oxidized products are critical signaling components in mitophagy, cell death, and multiple distinct immune responses (Figure 3)[112–114]. As an already diverse signaling molecule, there are likely further undiscovered roles for CL in this regard.

Recently, researchers have defined a role for CL in the activation of thermogenic adipocytes. Brown adipose tissue (BAT) converts carbohydrate and lipid substrates to thermal energy in response to cold environments. Lynes et al. measured cold-activated lipid landscapes in blood and adipose tissue by MS/MS^{ALL}, and among the 1,600 unique lipid species profiled, they identified the CL biosynthetic pathway as coordinately activated in brown and beige fat by cold[115]. Further, they found *CRLS1*, the gene encoding CL synthase, to be significantly enriched in cold-induced interscapular BAT as compared with other metabolic tissues, and that loss of CRLS1, and therefore CL, abolished the thermogenic capacity of the adipose tissue[116]. This suggests that CL is likely to have brown and beige fat-specific functions in thermogenesis. Previous studies have demonstrated that CL binds to the uncoupling protein, UCP1, and assists its proper folding. Therefore, CL may function through the direct activation of the thermogenic effector UCP1[117]. Additionally, CL also physically interacts with creatine kinase, an enzyme that drives thermogenesis in beige fat[118]. Together, this suggests that CL is critical for thermogenic mechanisms, adding to the ever growing list of CL functions (Figure 3).

3.5   Cardiolipin Metabolism and Mitochondrial Dysfunction in Human Health: Not Just Barth Syndrome

BTHS, a severe multi-system disorder caused by abnormal CL remodeling, highlights the importance of CL in human health (Figure 1). In addition to the cardinal clinical phenotypes of BTHS (cardiomyopathy, neutropenia, etc.), metabolite profiles of a BTHS cohort compared to age-matched controls revealed broad metabolic dysregulation with vast cellular implications that include: insulin regulation of fatty acid metabolism, lipid metabolism, biogenic amine metabolism, amino acid metabolism, endothelial nitric oxide synthase signaling, and tRNA biosynthesis[119].

As a prominent player in both mitochondrial and cellular activities, it is not surprising that altered CL content and/or quantity is associated with mitochondrial dysfunction in an array of metabolic and aging-related pathological conditions[96,100,101,120]. Mitochondria, and specifically CL, are thought to be intimately involved in the aging process, characterized by the gradual deterioration of various physiologic and metabolic processes, with increased oxidative stress and an attenuated ability to respond to other stresses[121]. While aging itself may not be considered pathologic, with aging there is an increased risk of acquiring age-associated disorders, especially with regards to brain and heart function[122–125]. In Alzheimer's (AD) and Parkinson's (PD), two common age-related neurological disorders, CL is suggested to have a role in disease pathogenesis[126–128]. In an AD transgenic mouse model, lipidomics revealed not only an overall reduction in CL amount, but an altered CL species profile, specifically reduced levels of unsaturated acyl chains[126]. Studies in induced pluripotent stem cells (iPSC) and embryonic stem cells (ESC) expressing variants in *SNCA* (encoding α-synuclein (α-syn)), the causal gene in familial PD, revealed that CL translocates to the OMM and binds to α-syn where it facilitates the folding of α-syn to its α-helical, ordered state[127,128]. Therefore, altered CL content and/or composition may result in misfolding and aggregation of α-syn, a key molecular event in the pathogenesis of PD. Additionally, with age, an individual is also at risk for ischemia/reperfusion (I/R) injury, which contributes to pathology in a range of conditions, such as myocardial infarction and ischemic stroke. In multiple organs, such as the brain, heart, and liver, I/R injury significantly changes the distribution of CL molecular species with an accumulation of oxidized CL[129–131].

Diabetes and obesity, with systemic oxidative stress and mitochondrial dysfunction, is also associated with altered CL content[70,120,132–135]. In a type I diabetic mouse model, there is depletion of tetralinoleoyl-CL as well as a diversification in its molecular speciation in heart tissue, with a significant increase in CL species containing fatty acyl chains longer than 18 carbons[132,133]. Similar to the type I model, a type II diabetic mouse model also has an altered CL profile[132,133]. Cardiac tissue from obese mice, due to a high fat diet, display a significantly altered CL species profile, specifically an increase in n-6 and n-3 polysaturation[70].

Abnormalities in CL, including CL peroxidation, have also been associated with nonalcoholic fatty liver disease (NAFLD), one of the most common forms of liver disease, which has the potential to advance to nonalcoholic steatohepatitis (NASH), liver fibrosis, and liver failure[136,137]. As further evidence for the role of CL in NASH, in a choline-deficient rat model of induced fatty liver disease total CL content is significantly reduced, and peroxidized CL is increased[138,139]. It is currently unclear if the abnormal CL is a product of increased oxidative stress and IMM dysfunction in this disease, or if abnormalities in CL are a primary factor with downstream effects on OXPHOS and augmented production of ROS. Regardless, CL represents a target of interest in unraveling the pathophysiology of NAFLD and NASH.

Abnormalities in CL also have been shown in several models of malignancy. TAZ expression gradually increases from normal cervical tissue to squamous cervical carcinoma, suggesting that TAZ and CL remodeling may contribute to the progression of cervical cancer[140]. Further, shotgun lipidomics from subcutaneously grown brain tumors, including astrocytomas, stem cell tumors, and microgliomas, revealed major abnormalities in CL content/composition[141]. Together, these findings suggest that altered CL is associated with malignant cell metabolism, possibly via inhibition of apoptosis or as a driver of adaptations in global cellular bioenergetics (i.e. the Warburg theory)[140,141].

Abnormalities in CL are the primary cellular defect underlying the Mendelian disease Barth Syndrome, as well as part of the pathologic process in multiple common, multifactorial diseases. As such, models of BTHS have the potential to not only elucidate the underlying disease mechanism of TAZ deficiency in BTHS, but to inform the pathophysiology of common and complex diseases associated with altered CL metabolism.

4 BARTH SYNDROME MODELS

Since the identification of the gene responsible for BTHS, several models of TAZ deficiency have been generated, which all recapitulate the primary BTHS biochemical abnormality; an increased MLCL:CL ratio and an abnormal acyl chain composition of the remaining CL (Table 1). The first model generated was the *S. cerevisiae* yeast model, *taz1Δ*, which contains a null variant in the homologue of human *TAZ*[142,143] Studies in yeast, which represent a proven model for mitochondrial disease, encompass the majority of our understanding of the role TAZ deficiency and abnormal CL metabolism in mitochondrial dysfunction[144,145] Characterization of the yeast model, *taz1Δ*, has connected altered CL metabolism with OXPHOS dysfunction with or without aberrant SC assembly, IMM uncoupling, increased oxidative stress, and impaired iron homeostasis (Table 1)[142,143,146–149].

Three animal models of TAZ deficiency have been developed: a *TAZ* knockout (*TAZKO*) fruit fly, a morpholino induced *TAZ* knockdown (*TAZKD*) zebrafish,

and a doxycycline short hairpin RNA-induced *TAZKD* mouse model[77,78,150,151]. *TAZKO* fruit flies exhibit motor weakness, reduced flying and climbing abilities, and reduced endurance, which parallels the exercise intolerance of many BTHS individuals[150,152]. *TAZKD* zebrafish have severe developmental and growth retardation with the cardiac phenotypes of bradycardia, pericardial effusions, and generalized edema[151]. *TAZKD* mice recapitulate many of the cardiac phenotypes in BTHS, including left ventricular dilation, reduced left ventricular mass, and reduced ejection (Table 1)[77,78].

There are also numerous *in vitro* cellular models of both primary and immortalized cells; BTHS patient neutrophils, BTHS patient-derived fibroblasts and lymphoblast cell lines, *TAZKO* and *TAZKD* mouse embryonic stem cell (ES) lines, *TAZKD* Human pro-myelocytic HL-60 progenitor (HL60) cell lines, *TAZKO* and *TAZ* knock-in (*TAZKI*) HEK293 cell lines, patient derived and engineered *TAZKO* induced pluripotent stem cell (iPSC) lines, *TAZKO* C2C12 myoblast cell lines, and *TAZKO* H9c2 cell lines (Table 1)[21,119,153–161]. Collectively, these cellular models display low basal respiration, low membrane potential, diminished OXPHOS, and destabilization OXPHOS SCs. Two models of note, the *TAZKO* and *TAZKI* HEK293s and patient derived and engineered *TAZKO* iPSCs, allow for human-based studies in a variant and tissue specific context.

To establish a novel mammalian BTHS model, Lu et al. utilized TALEN-mediated genome editing in Flp-In 293 cells, modified HEK293s which contain a single stably integrated FRT site, to generate a *TAZKO* line, *taz*[TALEN23]. A detailed characterization of endogenous TAZ in the *taz*[TALEN] line revealed that mammalian TAZ, as previously determined in yeast, is localized to the mitochondrion, where it associates non-integrally with IMS-facing membranes and assembles in a range of complexes. Further, using the integrated FRT site to express two *TAZ* variants of interest, R57L and H69Q, Lu et al. confirmed LOF mechanisms previously modeled in yeast. This human cellular model substantiated the power of modeling BTHS in yeast, as well as provided a framework with which to investigate the consequences of individual *TAZ* variants, which could prove invaluable in therapeutic discovery and validation.

Like the *TAZKO* and *TAZKI* HEK293 cell lines, the recent development of iPSCs allows for the investigation of TAZ deficiency and abnormal CL metabolism in a *TAZ*-variant specific context. Additionally, unlike the HEK293s, the use of iPSCs allow for investigations in a disease relevant cell type. This is of particular interest in BTHS, which clinically affects specific tissue types and metabolically affects CL composition, which is inherently tissue specific. Wang et al. established two BTHS patient derived iPSC lines, one with a deletion resulting in a frameshift (c.517delG, BTHH) and one with a missense variant (c.328C>T, BTHC), as well as two engineered lines using CRISPR-Cas9 genome editing, one with the

TABLE 1

Barth Syndrome model systems.

Model	Gene	Gene Location	Protein	System	Phenotypes	References
Yeast	*TAZ*1	ChrXVI: 814,349–815, 536	taz1	(KO) *taz1*Δ	• OXPHOS dysfunction with or without aberrant SC assembly • IMM uncoupling • Increased oxidative stress • Impaired iron homeostasis	142, 143, 146–149
Fruit Fly	*Taz*	Chr2R: 12,753,026–12,756,414 [-]	Taz	(KO) *TAZ -/-*	• Motor weakness • Reduced flying and climbing abilities • Reduced endurance	150, 152
Zebra Fish	*Taz*	Chr23: 4,915,595–4,925,724	Taz	(KD) asMO-mediated	• Severe developmental and growth retardation • Cardiac phenotypes: bradycardia, pericardial effusions, and generalized edema	151
Mouse	*Taz*	ChrX: 74,281,912–74,290,151	TAZ	(KD) Dox induction	• Cardiac phenotypes: left ventricular dilation, reduced left ventricular mass, and reduced ejection	77,78
Human	*TAZ*	ChrX: 154,411,518–154,421,726	TAZ	• BTHS Patient neutrophils • BTHS patient – derived fibroblasts • BTHS patient – derived lymphoblasts • BTHS patient – derived iPSCs • KO/KI engineered iPSCs • KD Human pro – myelocytic (HL60) cells • KO/KI Human embryonic kidney cells (HEKs) • KO C2C12 myoblast cells • KO H9c2 cells	• Low basal respiration • OXPHOS dysfunction • Aberrant SC assembly • Smaller and fragmented mitochondria	21, 119, 153–161

same deletion (PGP1-TAZ$^{c.517delG}$) and one with a 14 base pair insertion (PGP1-TAZ$^{c.517ins}$)[156]. Extensive characterization following differentiation into cardiomyocytes (iPSC-CMs), which recapitulated the characteristic BTHS cardiolipin findings, revealed smaller and fragmented mitochondria, reduced peak OXPHOS function, and decreased ATP production in the absence of TAZ. When seeded on micropatterned fibronectin rectangles designed to mimic adult cardiomyocytes, BTHS iPSC-CMs assembled into sparse and irregular sarcomeres, and when seeded onto muscular thin film (MTF) chips the irregular sarcomeres displayed weak contraction.

Each unique model has provided mechanistic insight into the pathophysiology of BTHS (Table 1), and holds potential for the ascertainment of new therapeutic targets and strategies.

5 BARTH SYNDROME THERAPIES

Currently, there are no disease specific treatments for BTHS. Affected individuals are treated symptomatically with vigilant surveillance and management by medical specialists in metabolism, cardiology, hematology, genetic counseling, physical therapy, and nutrition. Therapeutic development has focused on four main therapeutic strategies; lipid replacement, enzyme replacement, gene replacement, and drug therapy (Figure 4).

The pathognomonic metabolic defect in BTHS is an increase in MLCL and a decrease in mature CL, specifically tetralinoleoyl-CL, resulting in an overall increased MLCL: CL. In order to compensate for TAZ activity, the supplementation of CL, or lipid replacement, may be a reasonable approach (Figure 4). This approach was previously tested *in*

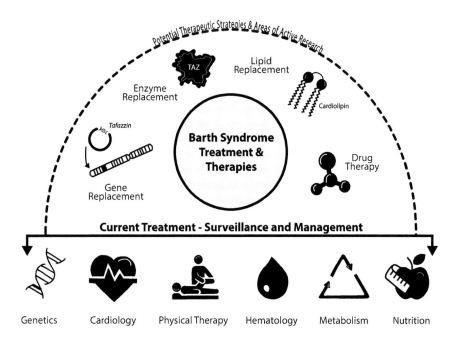

FIGURE 4 Barth syndrome therapeutic strategies. BTHS patients are treated symptomatically with vigilant surveillance and management by clinical teams composed of medical specialists in genetics, cardiology, physical therapy, hematology, metabolism, and nutrition. Therapeutic research has focused on four main areas of therapeutic strategies; gene, replacement, enzyme replacement, lipid replacement, and drug therapy.

vitro, where BTHS fibroblasts, supplemented with linoleic acid, showed a time and dose dependent increase in tetralinoleoyl-CL[162]. Additionally, *TAZ*KD HL-60 cells incubated with CL nanodisks (ND), which delivered tetralinoleoyl-CL to mitochondria, displayed an attenuated apoptotic response[163]. Based on this research, Ikon et al. sought to determine the feasibility of this approach in an *in vivo* model. Unfortunately, lipidomics revealed that the CL-ND failed to alter the CL profile of WT or doxycycline short hairpin RNA-induced *TAZ*KD mice[164]. Though CL-ND are not a viable option, this research highlights the importance of using both *in vitro* and *in vivo* modeling in therapeutic discovery and testing.

In another approach to lipid replacement, Ball et al. suggested supplementing ethanolamine (Etn) in order to increase cellular PE[165]. PE, which can function in an overlapping manner with CL, could then potentially rescue the respiratory chain defects of cells with altered CL metabolism. Interestingly, using the yeast models, *taz1Δ* and *crd1Δ*, Ball et al. showed that Etn ameliorates mitochondrial function, however, without increasing PE levels. Instead, Etn restored OXPHOS SC formation through increased expression of complex III and IV subunits. As an area of ongoing research it will be interesting to see the viability of this approach in mammalian models (*in vitro* and *in vivo*), especially since previous research has demonstrated the difficulties in translating successful treatments between *in vivo* and *in vitro* models.

Direct replacement of the TAZ enzyme is also a therapeutic approach under investigation (Figure 4). Working

toward this goal, Dinca et al. have identified the mitochondrial localization signals in TAZ, TAZ (84–95) and TAZ (185–220). The TAZ (84–95) sequence, located in exon 3, targets TAZ exclusively to the mitochondrion and is sufficient to direct eGFP to mitochondria in TAZ deficient cells[161]. The identification of this 12 amino acid localization sequence may be the first step toward engineering a TAZ protein sequence that can be used for enzyme replacement therapy (ERT). Research into the *in vivo* applicability of TAZ ERT is currently underway.

Replacement of the defective *TAZ* gene, or gene replacement therapy, is another therapeutic approach (Figure 4). As with enzyme replacement, delivery is the biggest hurdle in gene replacement. Recombinant adeno-associated virus (rAAV) vectors are often used for gene delivery because these vectors are derived from a non-pathogenic virus that elicits a minimal immune response and do not carry a risk of insertional mutagenesis. In order to optimize an AAV vector for *TAZ* delivery, Suzuki-Hatano et al. compared three AAV2/9-*TAZ* vectors with different promoters following their intravenous administration into doxycycline short hairpin RNA-induced *TAZ*KD mice[166]. Of the three promoters, the desmin (Des) promoter was found to provide significant improvement in various assessments as compared to *TAZ*KD controls. Of the 28 listed measurements, the Des-*TAZ* treated mice showed significant improvement in 6 measurements: *TAZ* transcription, *TAZ* translation, force mechanics, mitochondrial area, mitochondrial width, and mitochondrial oxygen consumption. It is not yet known how the effects of this gene replacement strategy will translate to patients.

In terms of small molecule/drug therapies for BTHS (Figure 4), there is an ongoing clinical trial testing the safety, tolerability, and efficacy of Elamipretide in Barth Syndrome sponsored by Stealth BioTherapeutics Inc. (ClinicalTrials.gov Identifier: NCT03098797)[167]. Elamipretide, also known as SS-31 (D-Arg-dimethylTyr-Lys-Phe-NH2), was first developed in 2004 by Zhao et al. as part of a series of peptide antioxidants, Szeto-Schiller peptides (SS peptides), that are targeted to the mitochondrion and concentrated in the IMM[168]. Since its development, it has been reported that SS-31 binds with high affinity to CL and that the SS-31/CL complex inhibits CL peroxidation by Cyt. C and optimizes OXPHOS[169–172]. In addition to its therapeutic potential, a greater understanding SS-31's mechanism of action in improving mitochondrial function may provide insight into the pathophysiology of BTHS.

6 CONCLUSION

Even with the significant advancements described in this review, our understanding of the pathophysiology of BTHS remains relatively limited. Specifically, the role of TAZ in determining of the acyl chain composition of remodeled CL (Figure 3), and how TAZ deficiency results in the cardinal phenotypes of BTHS (Figure 1) have not been resolved. In addition to BTHS, altered CL metabolism is also observed in various common and complex diseases are associated with altered CL metabolism. Therefore, additional studies with a variety of model systems (Table 1), including human-based models that are variant and tissue specific, will not only provide answers relevant to BTHS, but address the role of CL metabolism in other common and complex diseases.

ACKNOWLEDGMENTS

This work was supported by the National Institutes of Health grant (R01HL108882) to S.M.C. This work was sponsored by the Maryland Stem Cell Research Fund (Discovery Grant MSCRFD-4237) to H.V.

CONFLICT OF INTEREST

H.V. is the principal investigator on a clinical trial testing the safety, tolerability, and efficacy of Elamipretide in Barth Syndrome sponsored by Stealth BioTherapeutics Inc. in Newton, MA.

REFERENCES

1. Barth P., Van'T Veer-Korthof ET, Van Delden L, Van Damn K, Van der Harten JJ, Kuipers JRG. An X-linked mitochondrial disease affecting cardiac muscle, skeletal muscle and neutrophil leucocytes — Preliminary communication. In: Busch HFM, Jennekens FG., Scholte HR, eds. *Mitochondria and Muscular Disorders: The Proceedings of a Symposium Held at the Erasmus University, Rotterdam, 11th-12th December 1980.* Beetsterzwaag, The Netherlands: Mefar B.V.; 1981:161–164.
2. Barth PG, Scholte HR, Berden JA, et al. An X-linked mitochondrial disease affecting cardiac muscle, skeletal muscle and neutrophil leucocytes. *J Neurol Sci.* 1983;62(1–3):327–355. doi:10.1016/0022-510X(83)90209-5
3. Ades L., Gedeon AK, Wilson MJ, et al. Barth syndrome: clinical features and confirmation of gene localization to distal Xq28. *Am J Med Genet.* 1993;45(3):327–334.
4. Ferreira C, Thompson R, Vernon H. Barth syndrome. In: Adam MP, Ardinger HH, Pagon RA, Wallace SE, eds. *GeneReviews®[Internet].* 1993rd–2018th ed. Seattle, WA: University of Washington; 2014. doi:NBK247162 [bookaccession]
5. Roberts AE, Nixon C, Steward CG, et al. The Barth syndrome registry: distinguishing disease characteristics and growth data from a longitudinal study. *Am J Med Genet Part A.* 2012;158(11):2726–2732. doi:10.1002/ajmg.a.35609
6. Lu YW, Claypool SM. Disorders of phospholipid metabolism: An emerging class of mitochondrial disease due to defects in nuclear genes. *Front Genet.* 2015. doi:10.3389/fgene.2015.00003
7. Clarke SLN, Bowron A, Gonzalez IL, et al. Barth syndrome. *Orphanet J Rare Dis.* 2013;8(23). doi:10.1186/1750-1172-8-23
8. Barth Syndrome Registry and Repository. Toth, MJ. https://barthsyndromeregistry.patientcrossroads.org/
9. Spencer CT, Byrne BJ, Gewitz MH, et al. Ventricular arrhythmia in the X-linked cardiomyopathy Barth syndrome. *Pediatr Cardiol.* 2005;26(5):632–637. doi:10.1007/s00246-005-0873-z
10. Thompson WR, Decroes B, Mcclellan R, et al. New targets for monitoring and therapy in Barth syndrome. *Genet Med.* 2016;18(10):1001–1010. doi:10.1038/gim.2015.204
11. Kelley RI, Cheatham JP, Clark BJ, et al. X-linked dilated cardiomyopathy with neutropenia, growth retardation, and 3-methylglutaconic aciduria. *J Pediatr.* 1991;119(5):738–747. doi:10.1016/S0022-3476(05)80289-6
12. Rigaud C, Lebre A-S, Touraine R, et al. Natural history of Barth syndrome: a national cohort study of 22 patients. *Orphanet J Rare Dis.* 2013;8(70). doi:10.1186/1750-1172-8-70
13. Mangat J, Lunnon-Wood T, Rees P, Elliott M, Burch M. Successful cardiac transplantation in Barth syndrome – Single-centre experience of four patients. *Pediatr Transplant.* 2007;11:327–331. doi:10.1111/j.1399-3046.2006.00629.x
14. Bolhuis PA, Hensels GW, Hulsebos TJ, Baas F, Barth PG. Mapping of the locus for X-linked cardioskeletal myopathy with neutropenia and abnormal mitochondria (Barth syndrome) to Xq28. *Am J Hum Genet.* 1991;48(3):481–485.
15. Gedeon AK, Wilson MJ, Colley AC, Sillence DO, Mulley JC. X linked fatal infantile cardiomyopathy maps to Xq28 and is possibly allelic to Barth syndrome. *J Med Genet.* 1995;32(5):383–388. doi:10.1136/jmg.32.5.383
16. Bione S, D'Adamo P, Maestrini E, Gedeon AK, Bolhuis PA, Toniolo D. A novel X-linked gene, G4.5. is responsible for Barth syndrome. *Nat Genet.* 1996;12(4):385–389. doi:10.1038/ng0496-385
17. Gonzalez IL. Barth syndrome: TAZ gene mutations, mRNAs, and evolution. *Am J Med Genet Part A.* 2005;134(4):409–414. doi:10.1002/ajmg.a.30661
18. Kirwin SM, Manolakos A, Barnett SS, Gonzalez IL. Tafazzin splice variants and mutations in Barth syndrome.

Mol Genet Metab. 2014;111(1):26–32. doi:10.1016/j.ymgme.2013.11.006

19. Houtkooper RH, Turkenburg M, Poll-The BT, et al. The enigmatic role of tafazzin in cardiolipin metabolism. *Biochim Biophys Acta – Biomembr.* 2009;1788(10):2003–2014. doi:10.1016/j.bbamem.2009.07.009

20. Vaz FM, Houtkooper RH, Valianpour F, Barth PG, Wanders RJA. Only one splice variant of the human TAZ gene encodes a functional protein with a role in cardiolipin metabolism. *J Biol Chem.* 2003;278(44):43089–43094. doi:10.1074/jbc.M305956200

21. Xu Y, Sutachan JJ, Plesken H, Kelley RI, Schlame M. Characterization of lymphoblast mitochondria from patients with Barth syndrome. *Lab Investig.* 2005;85(6):823–830. doi:10.1038/labinvest.3700274

22. Xu Y, Zhang S, Malhotra A, et al. Characterization of tafazzin splice variants from humans and fruit flies. *J Biol Chem.* 2009;284(42):29230–29239. doi:10.1074/jbc.M109.016642

23. Lu YW, Galbraith L, Herndon JD, et al. Defining functional classes of Barth syndrome mutation in humans. *Hum Mol Genet.* 2016;25(9):1754–1770. doi:10.1093/hmg/ddw046

24. Hijikata A, Yura K, Ohara O, Go M. Structural and functional analyses of Barth syndrome-causing mutations and alternative splicing in the tafazzin acyltransferase domain. *Meta Gene.* 2015;4:92–106. doi:10.1016/j.mgene.2015.04.001

25. Neuwald AF. Barth syndrome may be due to an acyltransferase deficiency. *Nucleic Acids Res.* 1966;7(8):465–466.

26. Vreken P, Valianpour F, Nijtmans LG, et al. Defective remodeling of cardiolipin and phosphatidylglycerol in Barth syndrome. *Biochem Biophys Res Commun.* 2000;279(2):378–382. doi:10.1006/bbrc.2000.3952

27. Xu Y, Malhotra A, Ren M, Schlame M. The enzymatic function of tafazzin. *J Biol Chem.* 2006;281(51):39217–39224. doi:10.1074/jbc.M606100200

28. Horvath SE, Daum G. Lipids of mitochondria. *Prog Lipid Res.* 2013;52(4):590–614. doi:10.1016/j.plipres.2013.07.002

29. Kühlbrandt W. Structure and function of mitochondrial membrane protein complexes. *BMC Biol.* 2015;13(1):1–11. doi:10.1186/s12915-015-0201-x

30. Ardail D, Privat JP, Egret-Charlier M, Levrat C, Lerme F, Louisot P. Mitochondrial contact sites. Lipid composition and dynamics. *J Biol Chem.* 1990;265(31):18797–18802.

31. Hovius R, Thijssen J, Vanderlinden P, Nicolay K, Dekruijff B. Phospholipid asymmetry of the outer nembrane of rat liver mitochondria: Evidence for the presence of cardiolipin on the outside of the outer membrane. *FEBS Lett.* 1993;330(1):71–76.

32. Pangborn MC. Isolation and purification of a serologically active phospholipid from beef heart. *J Biol Chem.* 1942;143:247–256. doi:10.3181/00379727-48-13365P

33. LeCocq J, Ballou CE. On the structure of cardiolipin. *Biochemistry.* 1964;3(7):976–980. doi:10.1021/bi00895a023

34. Schlame M, Ren M, Xu Y, Greenberg ML, Haller I. Molecular symmetry in mitochondrial cardiolipins. *Chem Phys Lipids.* 2005;138(1-2):38–49. doi:10.1016/j.chemphyslip.2005.08.002

35. Cheng H, Mancuso DJ, Jiang X, et al. Shotgun lipidomics reveals the temporally dependent, highly diversified cardiolipin profile in the mammalian brain: Temporally coordinated postnatal diversification of cardiolipin molecular species with neuronal remodeling. *Biochemistry.* 2008;47(21):5869–5880. doi:10.1021/bi7023282

36. Amoscato AA, Sparvero LJ, He RR, Watkins S, Bayir H, Kagan VE. Imaging mass spectrometry of diversified cardiolipin molecular species in the brain. *Anal Chem.* 2014;86(13):6587–6595. doi:10.1021/ac5011876

37. Hostetler KY, Van Den Bosch H, Van Deenen LLM. Biosynthesis of cardiolipin in liver mitochondria. *Biochim Biophys Acta.* 1971;239(1):113–119. doi:10.1016/0005-2760(71)90201-3

38. Mejia EM, Nguyen H, Hatch GM. Mammalian cardiolipin biosynthesis. *Chem Phys Lipids.* 2014;179:11–16. doi:10.1016/j.chemphyslip.2013.10.001

39. Tamura Y, Harada Y, Yamano K, et al. Identification of Tam41 maintaining integrity of the TIM23 protein translocator complex in mitochondria. *J Cell Biol.* 2006;174(5):631–637. doi:10.1083/jcb.200603087

40. Tamura Y, Endo T. Unveiling the last missing link of the cardiolipin synthetic pathway in mitochondria. *Aging (Albany NY).* 2013;5(6):392–393. doi:10.18632/aging.100572

41. Chang SC, Heacock PN, Clancey CJ, Dowhan W. The PEL1 gene (renamed PGS1) encodes the phosphatidylglycerophosphate synthase of Saccharomyces cerevisiae. *J Biol Chem.* 1998;273(16):9829–9836. doi:10.1074/jbc.273.16.9829

42. Kawasaki K, Kuge O, Chang SC, et al. Isolation of a Chinese hamster ovary (CHO) cDNA encoding phosphatidylglycerophosphate (PGP) synthase, expression of which corrects the mitochondrial abnormalities of a PGP synthase-defective mutant of CHO-K1 cells. *J Biol Chem.* 1999;274(3):1828–1834. doi:10.1074/jbc.274.3.1828

43. Osman C, Haag M, Wieland FT, Brügger B, Langer T. A mitochondrial phosphatase required for cardiolipin biosynthesis: The PGP phosphatase Gep4. *EMBO J.* 2010;29(12):1976–1987. doi:10.1038/emboj.2010.98

44. Zhang J, Guan Z, Murphy AN, et al. Mitochondrial phosphatase PTPMT1 is essential for cardiolipin biosynthesis. *Cell Metab.* 2011;13(6):690–700. doi:10.1016/j.cmet.2011.04.007

45. Schlame M, Haldar D. Cardiolipin is synthesized on the matrix side of the inner membrane in rat liver mitochondria. *J Biol Chem.* 1993;268(1):74–79.

46. Schlame M, Hostetler KY. Cardiolipin synthase from mammalian mitochondria. *Biochim Biophys Acta – Lipids Lipid Metab.* 1997;1348(1–2):207–213. doi:10.1016/S0005-2760(97)00119-7

47. Chang S, Heacock PN, Mileykovskaya E, Voelker DR, Dowhan W. Isolation and characterization of the gene (CLS1) encoding cardiolipin synthase in saccharomyces cerevisiae * have cloned the gene (CLS1) encoding CL synthase in. *J Biol Chem.* 1998;273(24):14933–14941. doi:10.1074/jbc.273.24.14933

48. Houtkooper RH, Akbari H, van Lenthe H, et al. Identification and characterization of human cardiolipin synthase. *FEBS Lett.* 2006;580(13):3059–3064. doi:10.1016/j.febslet.2006.04.054

49. Hostetler KY, Galesloot JM, Boer P, Van Den Bosch H. Further studies on the formation of cardiolipin and phosphatidylglycerol in rat liver mitochondria. Effect of divalent cations and the fatty acid composition of CDP-diglyceride. *Biochim Biophys Acta.* 1975;380(3):382–389.

50. Beranek A, Rechberger G, Knauer H, Wolinski H, Kohlwein SD, Leber R. Identification of a cardiolipin-specific phospholipase encoded by the gene CLD1 (YGR110W) in yeast. *J Biol Chem.* 2009;284(17):11572–11578. doi:10.1074/jbc.M805511200

51. Baile MG, Whited K, Claypool SM. Deacylation on the matrix side of the mitochondrial inner membrane regulates cardiolipin remodeling. *Mol Biol Cell.* 2013;24(12):2008–2020. doi:10.1091/mbc.E13-03-0121

52. Malhotra A, Edelman-Novemsky I, Xu Y, et al. Role of calcium-independent phospholipase A2 in the pathogenesis of Barth syndrome. *Proc Natl Acad Sci.* 2009;106 (7):2337–2341. doi:10.1073/pnas.0811224106

53. Hsu YH, Dumlao DS, Cao J, Dennis EA. Assessing phospholipase A2 activity toward cardiolipin by mass spectrometry. *PLoS One.* 2013;8:3. doi:10.1371/journal. pone.0059267

54. Taylor WA, Hatch GM. Purification and characterization of monolysocardiolipin acyltransferase from pig liver mitochondria. *J Biol Chem.* 2003;278(15):12716–12721. doi:10.1074/jbc.M210329200

55. Cao J, Liu Y, Lockwood J, Burn P, Shi Y. A novel cardiolipin-remodeling pathway revealed by a gene encoding an endoplasmic reticulum-associated acyl-CoA: lysocardiolipinacyltransferase (ALCAT1) in mouse. *J Biol Chem.* 2004;279(30):31727–31734. doi:10.1074/jbc.M402930200

56. Malhotra A, Xu Y, Ren M, Schlame M. Formation of molecular species of mitochondrial cardiolipin. 1. A novel transacylation mechanism to shuttle fatty acids between sn-1 and sn-2 positions of multiple phospholipid species. *Biochim Biophys Acta.* 2009;1791(4):314–320. doi:10.1021/ nn300902w.Release

57. Taylor WA, Hatch GM. Identification of the human mitochondrial linoleoyl-coenzyme a monolysocardiolipin acyltransferase (MLCL AT-1). *J Biol Chem.* 2009;284 (44):30360–30371. doi:10.1074/jbc.M109.048322

58. Li J, Romestaing C, Han X, et al. Cardiolipin remodeling by ALCAT1 links oxidative stress and mitochondrial dysfunction to obesity. *Cell Metab.* 2011;12(2):154–165. doi:10.1016/j.cmet.2010.07.003.Cardiolipin

59. Schlame M, Towbin JA, Heerdt PM, Jehle R, DiMauro S, Blanck TJJ. Deficiency of tetralinoleoylcardiolipin in Barth syndrome. *Ann Neurol.* 2002;51(5):634–637. doi:10.1002/ana.10179

60. Valianpour F, Wanders RJA, Barth PG, Overmars H, Van Gennip AH. Quantitative and compositional study of cardiolipin in platelets by electrospray ionization mass spectrometry: Application for the identification of Barth syndrome patients. *Clin Chem.* 2002;48 (9):1390–1397.

61. Hsu FF, Turk J, Rhoades ER, Russell DG, Shi Y, Groisman EA. Structural characterization of cardiolipin by tandem quadrupole and multiple-stage quadrupole iontrap mass spectrometry with electrospray ionization. *J Am Soc Mass Spectrom.* 2005;16(4):491–504. doi:10.1016/j. jasms.2004.12.015

62. Han X, Yang K, Yang J, Cheng H, Gross RW. Shotgun lipidomics of cardiolipin molecular species in lipid extracts of biological samples. *J Lipid Res.* 2006;47(4):864–879. doi:10.1088/1367-2630/15/1/015008.Fluid

63. Schlame M, Acehan D, Berno B, et al. The physical state of lipid substrates provides transacylation specificity for tafazzin. *Nat Chem Biol.* 2012;8(10):862–869. doi:10.1038/ nchembio.1064

64. Renner LD, Weibel DB. Cardiolipin microdomains localize to negatively curved regions of Escherichia coli membranes. *Proc Natl Acad Sci.* 2011;108(15):6264–6269. doi:10.1073/pnas.1015757108

65. Epand RM, D'Souza K, Berno B, Schlame M. Membrane curvature modulation of protein activity determined by NMR. *Biochim Biophys Acta – Biomembr.* 2015;1848 (1):220–228. doi:10.1016/j.bbamem.2014.05.004

66. Abe M, Hasegawa Y, Oku M, et al. Mechanism for remodeling of the acyl chain composition of cardiolipin catalyzed by

67. Baile MG, Sathappa M, Lu YW, et al. Unremodeled and remodeled cardiolipin are functionally indistinguishable in yeast. *J Biol Chem.* 2014;289(3):1768–1778. doi:10.1074/ jbc.M113.525733

68. Tyurina YY, Lou W, Qu F, et al. Lipidomics characterization of biosynthetic and remodeling pathways of cardiolipins in genetically and nutritionally manipulated yeast cells. *ACS Chem Biol.* 2017;12:265–281. doi:10.1021/ acschembio.6b00995

69. Aoun M, Fouret G, Michel F, et al. Dietary fatty acids modulate liver mitochondrial cardiolipin content and its fatty acid composition in rats with non alcoholic fatty liver disease. *J Bioenerg Biomembr.* 2012;44(4):439–452. doi:10.1007/s10863-012-9448-x

70. Sullivan EM, Fix A, Crouch MJ, et al. Murine diet-induced obesity remodels cardiac and liver mitochondrial phospholipid acyl chains with differential effects on respiratory enzyme activity. *J Nutr Biochem.* 2017;45:94–103. doi:10.1016/j.jnutbio.2017.04.004

71. Claypool SM, Whited K, Srijumnong S, Han X, Koehler CM. Barth syndrome mutations that cause tafazzin complex lability. *J Cell Biol.* 2011;192(3):447–462. doi:10.1083/ jcb.201008177

72. Whited K, Baile MG, Currier P, Claypool SM. Seven functional classes of Barth syndrome mutation. *Hum Mol Genet.* 2013;22(3):483–492. doi:10.1093/hmg/dds447

73. Bowron A, Honeychurch J, Williams M, et al. Barth syndrome without tetralinoleoyl cardiolipin deficiency: a possible ameliorated phenotype. *J Inherit Metab Dis.* 2015;38 (2):279–286. doi:10.1007/s10545-014-9747-y

74. Fan Y, Steller J, Gonzalez IL, et al. A novel exonic splicing mutation in the TAZ (G4.5) gene in a case with atypical Barth syndrome. *JIMD Rep.* 2013;11:99–106. doi:10.1007/ 8904

75. Planas-Iglesias J, Dwarakanath H, Mohammadyani D, Yanamala N, Kagan VE, Klein-Seetharaman J. Cardiolipin interactions with proteins. *Biophys J.* 2015;109(6):1282–1294. doi:10.1016/j.bpj.2015.07.034

76. Acehan D, Xu Y, Stokes DL, Schlame M. Comparison of lymphoblast mitochondria from normal subjects and patients with Barth syndrome using electron microscopic tomography. *Lab Investig.* 2007;87(1):40–48. doi:10.1021/ nn300902w.Release

77. Acehan D, Vaz F, Houtkooper RH, et al. Cardiac and skeletal muscle defects in a mouse model of human Barth syndrome. *J Biol Chem.* 2011;286(2):899–908. doi:10.1074/ jbc.M110.171439

78. Phoon CKL, Acehan D, Schlame M, et al. Tafazzin Knockdown in Mice Leads to a Developmental Cardiomyopathy With Early Diastolic Dysfunction Preceding Myocardial Noncompaction. *J Am Heart Assoc.* 2012;1(2):jah3-e000455-jah3-e000455. doi:10.1161/JAHA.111.000455

79. Fry M, Green DE. Cardiolipin Requirement for Electron Transfer in Complex I and ITl of the Mitochondrial Respiratory Chain*. *J Biol Chem Val.* 1981;256 (4):1874–1880.

80. Paradies G, Petrosillo G, Pistolese M, Ruggiero FM. Reactive oxygen species affect mitochondrial electron transport complex I activity through oxidative cardiolipin damage. *Gene.* 2002;286(1):135–141. doi:10.1016/S0378111901008149 [pii]

81. Gomez B, Robinson NC. Phospholipase digestion of bound cardiolipin reversibly inactivates bovine

cytochrome bc1. *Biochemistry*. 1999;38(28):9031–9038. doi:10.1021/bi990603r

82. Wenz T, Hielscher R, Hellwig P, Schägger H, Richers S, Hunte C. Role of phospholipids in respiratory cytochrome bc1 complex catalysis and supercomplex formation. *Biochim Biophys Acta – Bioenerg*. 2009;1787(6):609–616. doi:10.1016/j.bbabio.2009.02.012

83. Sedlák E, Varhač R, Musatov A, Robinson NC. The kinetic stability of cytochrome c oxidase: effect of bound phospholipid and dimerization. *Biophys J*. 2014;107 (12):2941–2949. doi:10.1016/j.bpj.2014.10.055

84. Sedlák E, Robinson NC. Destabilization of the quaternary structure of bovine heart cytochrome c oxidase upon removal of tightly bound cardiolipin. *Biochemistry*. 2015;54(36):5569–5577. doi:10.1021/acs.biochem.5b00540

85. Pfeiffer K, Gohil V, Stuart RA, et al. Cardiolipin stabilizes respiratory chain supercomplexes. *J Biol Chem*. 2003;278 (52):52873–52880. doi:10.1074/jbc.M308366200

86. Zhang M, Mileykovskaya E, Dowhan W. Gluing the respiratory chain together: cardiolipin is required for supercomplex formation in the inner mitochondrial membrane. *J Biol Chem*. 2002;277(46):43553–43556. doi:10.1074/jbc.C200551200

87. Schwall CT, Greenwood VL, Alder NN. The stability and activity of respiratory Complex II is cardiolipin-dependent. *Biochim Biophys Acta – Bioenerg*. 2012;1817(9):1588–1596. doi:10.1016/j.bbabio.2012.04.015

88. Haines TH, Dencher NA. Cardiolipin: a proton trap for oxidative phosphorylation. *FEBS Lett*. 2002;528(1–3):35–39. doi:10.1016/S0014-5793(02)03292-1

89. Claypool SM, Boontheung P, McCaffery JM, Loo JA, Koehler CM. The cardiolipin transacylase, tafazzin, associates with two distinct respiratory components providing insight into Barth syndrome. *Mol Biol Cell*. 2008;19:5143–5155. doi:10.1091/mbc.E08

90. Claypool SM. Cardiolipin, a critical determinant of mitochondrial carrier protein assembly and function. *Biochim Biophys Acta*. 2009;1788(10):2059–2068. doi:10.1016/j.bbamem.2009.04.020.Cardiolipin

91. Musatov A, Sedlák E. Role of cardiolipin in stability of integral membrane proteins. *Biochimie*. 2017;142:102–111. doi:10.1016/j.biochi.2017.08.013

92. Jiang F, Ryan MT, Schlame M, et al. Absence of cardiolipin in the crd1 null mutant results in decreased mitochondrial membrane potential and reduced mitochondrial function. *J Biol Chem*. 2000;275(29):22387–22394. doi:10.1074/jbc.M909868199

93. Gebert N, Joshi AS, Kutik S, et al. Mitochondrial cardiolipin involved in outer membrane protein biogenesis: implications for Barth syndrome. *Curr Biol*. 2009;19(24):2133–2139. doi:10.1016/j.cub.2009.10.074.Mitochondrial

94. Ban T, Kohno H, Ishihara T, Ishihara N. Relationship between OPA1 and cardiolipin in mitochondrial inner-membrane fusion. *Biochim Biophys Acta – Bioenerg*. 2018;1859(9):951–957. doi:10.1016/j.bbabio.2018.05.016

95. Bustillo-Zabalbeitia I, Montessuit S, Raemy E, Basañez G, Terrones O, Martinou JC. Specific interaction with cardiolipin triggers functional activation of dynamin-related protein 1. *PLoS One*. 2014;9:7. doi:10.1371/journal.pone.0102738

96. Chicco AJ, Sparagna GC. Role of cardiolipin alterations in mitochondrial dysfunction and disease. *AJP Cell Physiol*. 2006;292(1):C33–C44. doi:10.1152/ajpcell.00243.2006

97. Houtkooper RH, Vaz FM. Cardiolipin, the heart of mitochondrial metabolism. *Cell Mol Life Sci*. 2008;65 (16):2493–2506. doi:10.1007/s00018-008-8030-5

98. Lewis RNAH, McElhaney RN. The physicochemical properties of cardiolipin bilayers and cardiolipin-containing lipid membranes. *Biochim Biophys Acta – Biomembr*. 2009;1788 (10):2069–2079. doi:10.1016/j.bbamem.2009.03.014

99. Klingenberg M. Cardiolipin and mitochondrial carriers. *Biochim Biophys Acta – Biomembr*. 2009;1788(10):2048–2058. doi:10.1016/j.bbamem.2009.06.007

100. Claypool SM, Koehler CM. The complexity of cardiolipin in health and disease. *Trends Biochem Sci*. 2012;37(1):32–41. doi:10.1016/j.tibs.2011.09.003

101. Paradies G, Paradies V, Ruggiero FM, Petrosillo G. Cardiolipin and mitochondrial function in health and disease. *Antioxid Redox Signal*. 2014;20(12):1925–1953. doi:10.1089/ars.2013.5280

102. Li XX, Tsoi B, Li YF, Kurihara H, He RR. Cardiolipin and its different properties in mitophagy and apoptosis. *J Histochem Cytochem*. 2015;63(5):301–311. doi:10.1369/0022155415574818

103. Ikon N, Ryan RO. Cardiolipin and mitochondrial cristae organization. *Biochim Biophys Acta*. 2017;1859(6):1156–1163. doi:10.1016/j.clinbiochem.2015.06.023.Gut-Liver

104. Chu CT, Ji J, Dagda RK, et al. Cardiolipin externalization to the outer mitochondrial membrane acts as an elimination signal for mitophagy in neuronal cells. *Nat Cell Biol*. 2013;15(10):1197–1205. doi:10.1038/ncb2837.Cardiolipin

105. Gonzalvez F, Schug ZT, Houtkooper RH, et al. Cardiolipin provides an essential activating platform for caspase-8 on mitochondria. *J Cell Biol*. 2008;183(4):681–696. doi:10.1083/jcb.200803129

106. Kagan VE, Tyurina YY, Tyurin VA, et al. Cardiolipin signaling mechanisms: collapse of asymmetry and oxidation. *Antioxid Redox Signal*. 2015;22(18):1667–1680. doi:10.1089/ars.2014.6219

107. Balasubramanian K, Maeda A, Lee JS, et al. Dichotomous roles for externalized cardiolipin in extracellular signaling: promotion of phagocytosis and attenuation of innate immunity. *Sci Signal*. 2015;8(395). doi:10.1126/scisignal.aaa6179.Dichotomous

108. Broder A, Chan JJ, Putterman C. Dendritic cells: An important link between antiphospholipid antibodies, endothelial dysfunction, and atherosclerosis in autoimmune and non-autoimmune diseases. *Clin Immunol*. 2013;146(3):197–206. doi:10.1088/1367-2630/15/1/015008.Fluid

109. Alessandri C, Conti F, Pendolino M, Mancini R, Valesini G. New autoantigens in the antiphospholipid syndrome. *Autoimmun Rev*. 2011;10(10):609–616. doi:10.1016/j.autrev.2011.04.011

110. Zhou R, Yazdi AS, Menu P, Tschopp J. A role for mitochondria in NLRP3 inflammasome activation. *Nature*. 2011;469(7329):221–226. doi:10.1038/nature09663

111. Iyer SS, He Q, Janczy JR, et al. Mitochondrial cardiolipin is required for Nlrp3 inflammasome activation. *Immunity*. 2013;39(2):311–323. doi:10.1088/1367-2630/15/1/015008.Fluid

112. Maguire JJ, Tyurina YY, Mohammadyani D, et al. Known unknowns of cardiolipin signaling: the best is yet to come. *Biochim Biophys Acta*. 2017;1862(1):8–24. doi:10.1038/nchembio.1527.A

113. Kagan VE, Bayir H, Tyurina YY, et al. Elimination of the unnecessary: intra- and extracellular signaling by anionic phospholipids. *Biochem Biophys Res Commun*. 2017;482 (3):482–490. doi:10.1038/nchembio.1527.A

114. Dudek J. Role of cardiolipin in mitochondrial signaling pathways. *Front Cell Dev Biol.* 2017;5(September):1–17. doi:10.3389/fcell.2017.00090

115. Lynes MD, Shamsi F, Sustarsic EG, et al. Cold-activated lipid dynamics in adipose tissue highlights a role for cardiolipin in thermogenic metabolism. *Cell Rep.* 2018;24 (3):781–790. doi:10.1016/j.celrep.2018.06.073

116. Sustarsic EG, Ma T, Lynes MD, et al. Cardiolipin synthesis in brown and beige fat mitochondria is essential for systemic energy homeostasis. *Cell Metab.* 2018;28(1):159–174.e11. doi:10.1016/j.cmet.2018.05.003

117. Hoang T, Smith MD, Jelokhani-Niaraki M. Expression, folding, and proton transport activity of human uncoupling protein-1 (ucp1) in lipid membranes. *J Biol Chem.* 2013;288(51):36244–36258. doi:10.1074/jbc. M113.509935

118. Paradies G, Paradies V, De Benedictis V, Ruggiero FM, Petrosillo G. Functional role of cardiolipin in mitochondrial bioenergetics. *Biochim Biophys Acta – Bioenerg.* 2014;1837(4):408–417. doi:10.1016/j.bbabio.2013.10.006

119. Sandlers Y, Mercier K, Pathmasiri W, et al. Metabolomics reveals new mechanisms for pathogenesis in Barth syndrome and introduces novel roles for cardiolipin in cellular function. *PLoS One.* 2016;11(3). doi:10.1371/journal. pone.0151802

120. Shi Y. Emerging roles of cardiolipin remodeling in mitochondrial dysfunction associated with diabetes, obesity, and cardiovascular diseases. *J Biomed Res.* 2010;24(1):6–15. doi:10.1016/S1674-8301(10)60003-6

121. Paradies G, Petrosillo G, Paradies V, Ruggiero FM. Oxidative stress, mitochondrial bioenergetics, and cardiolipin in aging. *Free Radic Biol Med.* 2010;48(10):1286–1295. doi:10.1016/j.freeradbiomed.2010.02.020

122. Saini-Chohan HK, Holmes MG, Chicco AJ, et al. Cardiolipin biosynthesis and remodeling enzymes are altered during development of heart failure. *J Lipid Res.* 2009;50:1600–1608. doi:10.1194/jlr.M800561-JLR200

123. Lesnefsky EJ, Minkler P, Hoppel CL. Enhanced modification of cardiolipin during ischemia in the aged heart. *J Mol Cell Cardiol.* 2009;46(6):1008–1015. doi:10.1016/j. yjmcc.2009.03.007

124. Paradies G, Petrosillo G, Paradies V, Ruggiero FM. Mitochondrial dysfunction in brain aging: role of oxidative stress and cardiolipin. *Neurochem Int.* 2011;58(4):447–457. doi:10.1016/j.neuint.2010.12.016

125. Lesnefsky EJ, Chen Q, Hoppel CL. Mitochondrial metabolism in aging heart. *Circ Res.* 2016;118(10):1593–1611. doi:10.1161/CIRCRESAHA.116.307505

126. Monteiro-Cardoso VF, Oliveira MM, Melo T, et al. Cardiolipin profile changes are associated to the early synaptic mitochondrial dysfunction in Alzheimer's disease. *J Alzheimer's Dis.* 2015;43:1375–1392. doi:10.3233/ JAD-141002

127. Ghio S, Kamp F, Cauchi R, Giese A, Vassallo N. Interaction of α-synuclein with biomembranes in Parkinson's disease – Role of cardiolipin. *Prog Lipid Res.* 2016;61:73–82. doi:10.1016/j.plipres.2015.10.005

128. Ryan T, Bamm V V., Stykel MG, et al. Cardiolipin exposure on the outer mitochondrial membrane modulates ⍺-synuclein. *Nat Commun.* 2018;9:817. doi:10.1038/s41467-018-03241-9

129. Martens JC, Keilhoff G, Halangk W, et al. Lipidomic analysis of molecular cardiolipin species in livers exposed to ischemia/reperfusion. *Mol Cell Biochem.* 2014;400(1–2):253–263. doi:10.1007/s11010-014-2282-1

130. Ji J, Baart S, Vikulina AS, et al. Deciphering of mitochondrial cardiolipin oxidative signaling in cerebral ischemia-reperfusion. *J Cereb Blood Flow Metab.* 2015;35(2):319–328. doi:10.1038/jcbfm.2014.204

131. Paradies G, Paradies V, Ruggiero FM, Petrosillo G. Mitochondrial bioenergetics and cardiolipin alterations in myocardial ischemia/reperfusion injury. Implications for pharmacological cardioprotection. *Am J Physiol Circ Physiol.* 2018. doi:10.1152/ajpheart.00028.2018

132. He Q, Han X. Cardiolipin remodeling in diabetic heart. *Chem Phys Lipids.* 2014;179:75–81. doi:10.1016/j. chemphyslip.2013.10.007

133. Han X, Yang J, Yang K, Zhongdan Z, Abendschein DR, Gross RW. Alterations in myocardial cardiolipin content and composition occur at the very earliest stages of diabetes: a shotgun lipidomics study. *Biochemistry.* 2007;46 (21):6417–6428. doi:10.1021/bi7004015

134. Croston TL, Shepherd DL, Thapa D, et al. Evaluation of the cardiolipin biosynthetic pathway and its interactions in the diabetic heart. *Life Sci.* 2013;93(8):313–322. doi:10.1016/j.lfs.2013.07.005

135. Cole LK, Mejia EM, Vandel M, et al. Impaired cardiolipin biosynthesis prevents hepatic steatosis and diet-induced obesity. *Diabetes.* 2016;65(11):3289–3300. doi:10.2337/db16-0114

136. Paradies G, Paradies V, Ruggiero FM, Petrosillo G. Oxidative stress, cardiolipin and mitochondrial dysfunction in nonalcoholic fatty liver disease. *World J Gastroenterol.* 2014;20(39):14205–14218. doi:10.3748/wjg.v20.i39.14205

137. Younossi Z, Anstee QM, Marietti M, et al. Global burden of NAFLD and NASH: trends, predictions, risk factors and prevention. *Nat Rev Gastroenterol Hepatol.* 2018;15:11–20. doi:10.1038/nrgastro.2017.109

138. Petrosillo G, Portincasa P, Grattagliano I, et al. Mitochondrial dysfunction in rat with nonalcoholic fatty liver. Involvement of complex I, reactive oxygen species and cardiolipin. *Biochim Biophys Acta – Bioenerg.* 2007;1767 (10):1260–1267. doi:10.1016/j.bbabio.2007.07.011

139. Hayasaka T, Fuda H, Hui S-P, Chiba H. Imaging mass spectrometry reveals a decrease of cardiolipin in the kidney of NASH model mice. *Anal Sci.* 2016;32(4):473–476. doi:10.2116/analsci.32.473

140. Chen M, Zhang Y, Zheng PS. Tafazzin (TAZ) promotes the tumorigenicity of cervical cancer cells and inhibits apoptosis. *PLoS One.* 2017;12(5):1–13. doi:10.1371/journal.pone.0177171

141. Kiebish MA, Han X, Cheng H, Chuang JH, Seyfried TN. Cardiolipin and electron transport chain abnormalities in mouse brain tumor mitochondria: lipidomic evidence supporting the Warburg theory of cancer. *J Lipid Res.* 2008;49:2545–2556. doi:10.1194/jlr.M800319-JLR200

142. Gu Z, Valianpour F, Chen S, et al. Aberrant cardiolipin metabolism in the yeast taz1 mutant: a model for Barth syndrome. *Mol Microbiol.* 2004;51(1):149–158. doi:10.1046/j.1365-2958.2003.03802.x

143. Ma L, Vaz FM, Gu Z, Wanders RJA, Greenberg ML. The human TAZ gene complements mitochondrial dysfunction in the yeast taz1Δ mutant: Implications for Barth syndrome. *J Biol Chem.* 2004;279(43):44394–44399. doi:10.1074/jbc. M405479200

144. Baile MG, Claypool SM. The power of yeast to model diseases of the powerhouse of the cell. *Front Biosci (Landmark Ed).* 2013;18:241–278. doi:10.2741/4098

145. Saric A, Andreau K, Armand AS, Møller IM, Petit PX. Barth syndrome: from mitochondrial dysfunctions

associated with aberrant production of reactive oxygen species to pluripotent stem cell studies. *Front Genet.* 2016;6:359. doi:10.3389/fgene.2015.00359

146. Brandner K, Mick DU, Frazier AE, Taylor RD, Meisinger C, Rehling P. Taz1, an outer mitochondrial membrane protein, affects stability and assembly of inner membrane protein complexes: implications for Barth syndrome. *Mol Biol Cell.* 2005;16(11):5202–5214. doi:10.1091/mbc.E05-03-0256

147. Chen S, He Q, Greenberg ML. Loss of tafazzin in yeast leads to increased oxidative stress during respiratory growth. *Mol Microbiol.* 2008;68(4):1061–1072. doi:10.1111/j.1365-2958.2008.06216.x

148. Joshi AS, Zhou J, Gohil VM, Chen S, Greenberg ML. Cellular functions of cardiolipin in yeast. *Biochim Biophys Acta – Mol Cell Res.* 2009;1793(1):212–218. doi:10.1016/j.bbamcr.2008.07.024

149. Gaspard GJ, McMaster CR. The mitochondrial quality control protein Yme1 is necessary to prevent defective mitophagy in a yeast model of Barth syndrome. *J Biol Chem.* 2015;290(14):9284–9298. doi:10.1074/jbc.M115.641878

150. Xu Y, Condell M, Plesken H, et al. A drosophila model of Barth syndrome. *Proc Natl Acad Sci USA.* 2006;103 (31):11584–11588. doi:10.1073/pnas.0603242103

151. Khuchua Z, Yue Z, Batts L, Strauss AW. A zebrafish model of human barth syndrome reveals the essential role of tafazzin in cardiac development and function. *Circ Res.* 2006;99 (2):201–208. doi:10.1161/01.RES.0000233378.95325.ce

152. Damschroder D, Reynolds C, Wessells R. Drosophila tafazzin mutants have impaired exercise capacity. *Physiol Rep.* 2018;6(3):e13604. doi:10.14814/phy2.13604

153. Barth PG, Van Den Bogert C, Bolhuis PA, et al. X-linked cardioskeletal myopathy and neutropenia (Barth syndrome): respiratory-chain abnormalities in cultured fibroblasts. *J Inherit Metab Dis.* 1996;19(2):157–160. doi:10.1007/BF01799418

154. Kuijpers TW, Maianski NA, Tool ATJ, et al. Neutrophils in Barth syndrome (BTHS) avidly bind annexin-V in the absence of apoptosis. *Blood.* 2004;103(10):3915–3923. doi:10.1182/blood-2003-11-3940

155. Dudek J, Cheng IF, Balleininger M, et al. Cardiolipin deficiency affects respiratory chain function and organization in an induced pluripotent stem cell model of Barth syndrome. *Stem Cell Res.* 2013;11(2):806–819. doi:10.1016/j.scr.2013.05.005

156. Wang G, McCain ML, Yang L, et al. Modeling the mitochondrial cardiomyopathy of Barth syndrome with induced pluripotent stem cell and heart-on-chip technologies. *Nat Med.* 2014;20(6):616–623. doi:10.1038/nm.3545

157. Lou W, Reynolds CA, Li Y, et al. Loss of tafazzin results in decreased myoblast differentiation in C2C12 cells: a myoblast model of Barth syndrome and cardiolipin deficiency. *Biochim Biophys Acta – Mol Cell Biol Lipids.* 2018;1863(8):857–865. doi:10.1016/j.bbalip.2018.04.015

158. Makaryan V, Kulik W, Vaz FM, et al. The cellular and molecular mechanisms for neutropenia in Barth syndrome. *Eur J Haematol.* 2012;88(3):195–209. doi:10.1111/j.1600-0609.2011.01725.x

159. Acehan D, Khuchua Z, Houtkooper RH, et al. Distinct effects of tafazzin deletion in differentiated and undifferentiated mitochondria. *Mitochondrion.* 2009;9 (2):86–95. doi:10.1016/j.mito.2008.12.001

160. He Q, Harris N, Ren J, Han X. Mitochondria-targeted antioxidant prevents cardiac dysfunction induced by tafazzin gene knockdown in cardiac myocytes. *Oxid Med Cell Longev.* 2014:654198. doi:10.1155/2014/654198

161. Dinca AA, Chien WM, Chin MT. Identification of novel mitochondrial localization signals in human Tafazzin, the cause of the inherited cardiomyopathic disorder Barth syndrome. *J Mol Cell Cardiol.* 2018;114(August 2017):83–92. doi:10.1016/j.yjmcc.2017.11.005

162. Valianpour F, Wanders RJA, Overmars H, Vaz FM, Barth PG, van Gennip AH. Linoleic acid supplementation of Barth syndrome fibroblasts restores cardiolipin levels. *J Lipid Res.* 2003;44(3):560–566. doi:10.1194/jlr.M200217-JLR200

163. Ikon N, Su B, Hsu FF, Forte TM, Ryan RO. Exogenous cardiolipin localizes to mitochondria and prevents TAZ knockdown-induced apoptosis in myeloid progenitor cells. *Biochem Biophys Res Commun.* 2015;464(2):580–585. doi:10.1016/j.bbrc.2015.07.012

164. Ikon N, Hsu F, Shearer J, Forte TM, Ryan RO. Evaluation of cardiolipin nanodisks as lipid replacement therapy for Barth syndrome. *J Biomed Res.* 2018;32(November 2017):107–112.

165. Ball WB, Baker CD, Neff JK, et al. Ethanolamine ameliorates mitochondrial dysfunction in cardiolipin-deficient yeast cells. *J Biol Chem.* 2018;293(28):10870–10883. doi:10.1074/jbc.RA118.004014

166. Suzuki-Hatano S, Saha M, Rizzo SA, et al. AAV-mediated TAZ gene replacement restores mitochondrial and cardioskeletal function in Barth syndrome. *Hum Gene Ther.* 2018. doi:10.1089/hum.2018.020

167. A Trial to Evaluate Safety, Tolerability and Efficacy of Elamipretide in Subjects With Barth Syndrome (TAZPOWER). https://clinicaltrials.gov/ct2/show/NCT 03098797.

168. Zhao K, Zhao GM, Wu D, et al. Cell-permeable peptide antioxidants targeted to inner mitochondrial membrane inhibit mitochondrial swelling, oxidative cell death, and reperfusion injury. *J Biol Chem.* 2004;279:34682–34690. doi:10.1074/jbc.M402999200

169. Birk A V., Liu S, Soong Y, et al. The mitochondrial-targeted compound SS-31 re-energizes ischemic mitochondria by interacting with cardiolipin. *J Am Soc Nephrol.* 2013;24 (8):1250–1261. doi:10.1681/ASN.2012121216

170. Liu S, Soong Y, Seshan S V., Szeto HH. Novel cardiolipin therapeutic protects endothelial mitochondria during renal ischemia and mitigates microvascular rarefaction, inflammation, and fibrosis. *AJP Ren Physiol.* 2014;306(9): F970–9780. doi:10.1152/ajprenal.00697.2013

171. Birk A V., Chao WM, Bracken C, Warren JD, Szeto HH. Targeting mitochondrial cardiolipin and the cytochrome c/ cardiolipin complex to promote electron transport and optimize mitochondrial ATP synthesis. *Br J Pharmacol.* 2014;171(8):2017–2028. doi:10.1111/bph.12468

172. Szeto HH. First-in-class cardiolipin-protective compound as a therapeutic agent to restore mitochondrial bioenergetics. *Br J Pharmacol.* 2014;171(8):2029–2050. doi:10.1111/bph.12461

7 Mitochondrial Dysfunction in the Pathophysiology of Alzheimer's Disease

Isaac G. Onyango, DVM, PhD

Gencia Biotechnology, 706 B Forest Street, Charlottesville, USA

CONTENTS

1 INTRODUCTION

Alzheimer disease (AD) is a devastating neurodegenerative disease and the predominant form of dementia. In 2015, 44 million people worldwide were estimated to have AD or a related dementia. Each year, 4.6 million new cases of dementia are predicted with numbers expected to almost double by 2030 [1].

AD is categorized into two major forms: sporadic AD (sAD) and familial AD (fAD) with < 10% of AD cases being familial [2] and showing autosomal-dominant transmission within affected families. Age is the most prominent biological risk factor for sAD [3] although it has a heterogeneous etiology and heritability of 70 to 80% [4,5] with apolipoprotein E (APOE) gene being an additional risk factor [6]. Female gender is an important contributor partially explainable by aging and lengthened lifespan. A role for dysfunctional mitochondria in AD pathogenesis has been postulated [7]. Cumulative evidence reveal that the regulation of mitochondrial turnover and function becomes impaired as a function of age in the brain and may contribute to neurodegeneration in AD [8]. Cerebral hypometabolism is evident in affected brain regions [9,10] where mitochondrial

structure is altered [11,12]. The expression and activity of mitochondrial enzymes important for metabolism, including cytochrome c oxidase (COX), α-ketoglutarate dehydrogenase complex, and pyruvate dehydrogenase complex is reduced [13–15]. Mitochondrial DNA (mtDNA) haplogroups influence the risk of AD with the demented parent of an AD patient usually being the mother [9,16]. Transferring mtDNA from AD patients into cell lines devoid of mtDNA (rho0 cells) has been shown to induce respiratory enzyme deficiency similar to that seen in AD tissues. This suggests that the deficit is carried at least in part by mtDNA abnormalities [7]. AD brain mitochondria have reduced membrane potential, increased permeability, and produce excess reactive oxygen species (ROS) which damages proteins, lipids, and nucleic acids, and are believed to contribute to the pathogenesis of neurodegeneration. Elevated amyloid-β (Aβ) levels contribute to the mitochondrial abnormalities and although the mechanism is not clearly established, both amyloid precursor protein (APP) and Aβ are found in mitochondrial membranes and interact with mitochondrial proteins. Overproduction of the APP and Aβ may affect dynamics of mitochondrial fusion/fission [17–19], impair mitochondrial transport, disrupt the electron transfer chain, increase ROS production [19–21], and impair mitochondrial function [22–24]. In light of these finding, many researchers and clinicians are making great efforts to better understand the pathophysiology of mitochondrial dysfunction in AD in order to explore novel and effective therapy that target mitochondrial function to slow or stop the neurodegenerative process in AD [25–27].

2   OVERVIEW OF MITOCHONDRIA

Mitochondria play an important in normal cellular function as they are responsible for energy production, synthesis of phospholipids and heme, calcium (Ca^{2+})-homeostasis, apoptotic activation and ROS production, apoptosis, and senescence [28–30]. Each mitochondrion contains 2–10 copies of mtDNA [24]. The human mtDNA consists of a 16.5 kb, double-stranded, circular DNA molecule [31], which encodes 13 of the ~92 polypeptides of the oxidative phosphorylation (OxPhos) system. The remaining structural polypeptides and assembly factors are encoded by nuclear DNA [31,32] and are synthesized in the cytoplasm and are subsequently transported into mitochondria. Nuclear mitochondrial proteins mtDNA are inherited exclusively from the mother and is present in thousands of copies per cell.

3   mtDNA MUTATIONS IN AD

mtDNA in AD brain are more oxidative damaged than can be attributed to aging [33–37]. Alterations in mtDNA are found in AD cells and tissues [38]. These include both qualitative changes, such as the association of specific mtDNA haplogroups with the risk of developing AD [39,40] and the presence of mtDNA deletions [41] and point mutations [42] in patient cells and tissues, and quantitative changes, such as reduced mtDNA levels in AD cerebrospinal fluid [43]. At the gene expression level, the transcription of a number of nuclear-encoded OxPhos subunits was reduced in AD blood, whereas transcription of most mtDNA-encoded subunits was elevated in AD blood [44].

While mitochondria have many antioxidant and DNA repair enzymes including 8-oxoguanine glycosylase 1 (OGG1) and the base excision repair protein MUTYH [34,45,46], the proximity of the mitochondrial genome to the inner mitochondrial membrane where ROS are routinely generated, and lack of protective histone molecules, make them particularly vulnerable to mutations compared to nuclear DNA [47]. mtDNA quality control is crucial for communication with the nucleus. ROS-mediated gene expression elicited by oxidative phosphorylation dysfunction lead to mitochondrial retrograde signaling that stimulates an adaptive nuclear response to mtDNA impairment. Mitochondrial genetic alterations affect the expression of more than 40 nuclear genes [48,49]. Conversely, mtDNA dysfunction can be induced by many signaling molecules that are regulated by nuclear genes, and factors related to mitochondrial metabolism [50–53].

4   MITOCHONDRIA AND γ-SECRETASE IN AD

An intriguing connection between mitochondria and the γ-secretase complex that supports the mitochondrial cascade hypothesis of AD [7] has been established. First, APP and/or Aβ44 as well as components of the γ-secretase complex, have been reported to be at or in mitochondria. Further, there are mitochondria-mediated alterations in APP processing in AD cells and tissues and presenilin 1 (PS1) has been shown to enhance the expression of PGC-1α, the master regulator of mitochondrial biogenesis. This effect is reduced in PS1-mutated cells. Finally, incubation of cultured cells and/or isolated mitochondria with Aβ has deleterious effects on mitochondrial functions, including: respiration, protein import, organellar transport, organellar localization, and organellar dynamics (e.g., mitochondrial fission and fusion) [54].

5   MITOCHONDRIAL BIOGENESIS IN AD

Mitochondrial biogenesis is essential for maintaining an adequate functional neuronal mitochondrial pool by replacing damaged mitochondria that have been eliminated. It is highly regulated and requires coordination and crosstalk between the nuclear and mitochondrial genomes [23]. While mitochondrial biogenesis occurs on a regular basis in healthy cells where mitochondria constantly divide and fuse with each other [24–26]; it is upregulated in response to oxidative stress, increased energy demand, exercise training and certain diseases.

Mitochondrial biogenesis is regulated by the "master regulator" peroxisome proliferator-activated receptor γ coactivator-1α (PGC-1α) which in turn activates different transcription factors, including nuclear respiratory factors 1 and 2 proteins (NRF-1 and NRF-2), estrogen-related receptor alpha (ERR-α) and mitochondrial transcription factor A (TFAM) [8,55,56]. NRF-1 and NRF-2 regulate transcription of nuclear and mitochondrial genes involved in OxPhos, electron transport (complex I–V), mtDNA transcription/replication, heme biosynthesis, protein import/assembly, ion channels, shuttles, and translation [57].

Mitochondrial biogenesis is impaired in AD and the levels of NRF 1, NRF 2, and TFAM along with nuclear levels of PGC-1α are reduced in hippocampal tissues from AD brain compared to age matched control brain [11,58,59]. Overexpression of PGC-1α can reduce the mitochondrial damage and improve biogenesis [59].

6   MITOPHAGY IN AD

Mitophagy is the process through which severely damaged and/or dysfunctional mitochondria are identified and selectively degraded and recycled [60]. It is crucial for cellular homeostasis and it is impaired in AD [61–63] where mitochondria have been shown to be key targets of increased autophagic degradation [12]. Excess ROS can trigger autophagy [64] and dysfunctional mitochondria that overproduce ROS, are indeed selectively targeted for mitophagy [65]. Central to mitochondrial and cellular homeostasis, mitophagy is modulated by the PTEN-induced putative kinase 1 (PINK1)/Parkin pathway [66] which primarily targets mitochondria devoid of membrane potential ($\Delta\Psi$m). Damaged mitochondria can also, independently of Parkin, increase FUNDC1 and Nix expression to recruit autophagosomes to mitochondria via direct interaction with LC3 [67,68]. Ubiquitin ligases, like Smurf1, target depolarized mitochondria for mitophagy [69–71].

7 MITOCHONDRIAL FISSION IN AD

Mitochondrial fission usually occurs during mitochondrial biogenesis when intramitochondrial components are sorted and split into daughter mitochondria [72,73] but also occurs preceding mitophagy or cellular apoptosis [74–77]. Dynamin-related protein 1 (Drp1) is the major protein involved in the division of mitochondrial membranes [78] and while fission occurs regardless of mitochondrial membrane potential, it is upregulated following mitochondrial depolarization, oxidation or nitrosylation, and ETC inhibition which trigger posttranslational modifications, including phophorylation, S-nitrosylation, ubiquitylation, and sumoylation on Drp1 [79,80] resulting in mitochondrial fragmentation [81,82]. In AD, Aβ overproduction is associated with increased number of fragmented mitochondria, increased oxidative stress and loss of $\Delta\psi_m$ and ATP production that is associated with increased expression of Drp1 [83].

Accumulated Aβ enhances Drp1 activity in neurons by increasing Drp1 S-nitrosylation at Cys644. Likewise, AD patients are characterized by having Aβ-Drp1 mediated mitochondrial fragmentation, mtDNA mutations [84] and decrease in OxPhos [85].

8 MITOCHONDRIAL MEMBRANE POTENTIAL IN AD

Several AD animal models and AD patient brains and human cortical neurons ex vivo have decreased mitochondrial membrane potential ($\Delta\psi_m$) [86–88]. The $\Delta\psi_m$ is created when protons are pumped from the mitochondrial matrix to the intermembrane space as electrons pass through the ETC and as a prerequisite for oxidative phosphorylation. However, the higher (more polarized) $\Delta\psi_m$, the more mtROS is generated presumably due to the slowed electron transport [89]. Indeed, ROS generation is decreased when $\Delta\psi_m$ is dissipated by either expressing mitochondrial uncoupling proteins (UCPs) [90] or using chemical uncouplers (Reynolds and Hastings, 1995), such as carbonyl cyanide p-(tri-fluromethoxy)phenyl-hydrazone (FCCP) [91,92]. Small decreases in membrane potential (mild uncoupling) can reduce ROS formation by limiting the life span of reduced electron transport chain (ETC) intermediates capable of generating ROS,

in addition to decreasing local oxygen tensions [93–95] without seriously compromising cellular energetics [94,96].

9 MITOCHONDRIAL OXIDATIVE STRESS IN AD

Nitrogen oxides (Nox)-dependent oxidative stress induce neurodegeneration through the oxidation of DNA, proteins, lipids, amino acids and metals, as well as the activation of redox-sensitive signaling pathways [97]. AD brain's activated Nox are thought to contribute to AD neuropathology [98,99]. ROS can induce permeability transition pore (PTP) opening resulting in mitochondrial swelling, rupture, release of cytochrome c, and neuronal death in the progression of AD.

10 MITOCHONDRIAL STRESS RESPONSE SIGNALING IN AD

Mitochondria are the major source of cellular ROS and subsequent stress signaling that induces cellular senescence and apoptosis]. One of the major consequences of increased ROS and altered cellular redox state is the oxidation of thiol groups in cysteine residues in relevant proteins [100]. Mitochondrial ATM kinase is one of the key modulators of the cellular stress response [101–105]. ATM kinase, which is partly located at the mitochondria, is activated upon mitochondrial uncoupling [106] and while its mitochondrial substrates are not known, loss of ATM in genetically engineered mouse models leads to mitochondrial dysfunction. ATM signaling is reduced in the neurons in vulnerable regions of the AD brain [107].

ATM deficient neurons reenter the cell cycle and die [108,109], suggesting that ATM may protect neuron by stopping cells reentering the cell cycle and lessening DNA damage. ATM impairment in glial cells may also trigger innate immune responses leading to cause neurodegeneration [110]. The histology of microglial cell in ATM knock out (ATM KO) mice was abnormal, and astrocytes from ATM KO mice showed significant expressions of oxidative and endoplasmic reticulum stress and a senescence-like reaction [111,112]. ATM deficiency may disturb DNA repair, trigger apoptosis, and accelerate aging and neuroinflammation.

11 MITOCHONDRIA AND ALTERED CALCIUM HOMEOSTASIS IN AD

Ca^{2+}-homeostasis is perturbed in AD [113–117] and deregulation of Ca^{2+}-homeostasis, arising from mitochondrial dysfunction, is linked to neurotoxicity [113,118–120]. Mitochondria buffer cytosolic Ca^{2+}-by internalizing it mainly through uniporter and releasing it by Na^+-/Ca^{2+}-or H^+-/Ca^{2+}-exchangers [118]. Cytosolic Ca^{2+}-levels is crucial for normal neurotransmission, long and short term plasticity and regulation of gene transcription in the CNS [121–123] and these levels are carefully buffered by mitochondria. The mitochondrial Ca^{2+}- buffering capacity of the CNS declines with age likely due to cumulative oxidative damage to mitochondria [124].

Exposure of phosphatidylserine (PtdS) on the cell surface, a sign of cellular energy deficiency, enhances the ability of Aβ to associate with the membrane [125]. Neurons with reduced cytosolic ATP levels and elevated surface PtdS levels are particularly vulnerable to Aβ toxicity [126,127] and in AD, Aβ oligomers form Ca^{2+}-permeable channels in membranes [128] and calpain is activated in animal models of sporadic AD [129].

12 MITOCHONDRIAL HORMESIS (MITOHORMESIS) IN AD

Mitohormesis occurs when low level stress originating from the mitochondrial respiratory chain induces an adaptive response promote longevity and metabolic health [130]. While ROS can generate detrimental oxidative damage, they also play a crucial role in numerous signaling and stress responses [100,131]. Mitohormesis can be induced by either caloric restriction, exercise [132], or other stimuli trigger an adaptive response that improves overall stress resistance. This is likely through increased endogenous antioxidant defense, which eventually reduces chronic oxidative damage [133] and extends lifespan. Inhibition of glycolysis, impairment of insulin-like signaling and certain mutations in mitochondrial ETC components, are also conditions that may promote longevity via ROS-dependent mitohormesis [134]. Mild mitochondrial insults may also communicate a stress response to induce the expression of mitochondrial chaperones such as HSP-6 and HSP-60. This mitochondrial unfolded protein response (UPR^{mt}) is thought to extend the life span of *C. elegans* by inhibiting the ETC [135,136]. Mild inhibition of mitochondrial respiration extends the lifespan of organisms as diverse as yeast, worms, flies and mice [137] presumably through ROS stimulated HIF-1 activation of gene expression that promote longevity [138]. The activation of the mitohormetic response increases lifespan in different animal models, from worms to mammals and also enhances health span, particularly improving metabolism and immune system [139]. The translational implications of mitohormesis should be considered as an ideal antioxidant therapy that prevents oxidative damage induced AD without interfering with ROS needed for hormesis and cellular signaling.

13 MITOCHONDRIA AND INFLAMMATION IN AD

Immune function is in AD [140]. Mitochondria regulate the pro-inflammatory response of the cell by activating the inflammasome. The activation of the NLR family, pyrin domain containing 3 (NLRP3) inflammasome has been proposed to be associated with mitochondrial dysfunction including: mitochondrial ROS [141], mitochondrion-derived damage associated molecular patterns (DAMPs), such as oxidized mitochondrial DNA [142,143], and translocation of cardiolipin from the inner to the outer mitochondrial membrane [144] and seems crucial in the pathogenesis of AD [145]. Additionally, extracellular ATP at various concentrations can activate microglia and induce neuroprotective or neurotoxic effects by expressing pro- or anti-inflammatory cytokines [146,147]. Several studies in cell lines, genetic rodent models, and humans indicate that redox control might serve as a bidirectional link between energy metabolism, redox control and neuroinflammatory responses in the brain that might serve as an integrated mechanism for AD etiology [148–150].

14 MITOCHONDRIA AND PROTEOSTASIS IN AD

Mitochondrial dysfunction and impairment of the ubiquitin proteasome system (UPS) are two hallmarks of aging and both are implicated in AD [151,152]. Mitochondrial proteins are a substantial part of proteins that are tagged with ubiquitin. Initial studies of ubiquitin-conjugated proteomes identified several mitochondrial proteins in yeast and mammals as ubiquitination substrates and constituting up to 38% of all cellular ubiquitin conjugates [153,154]. Aβ accumulation induces both the UPR^{mt} and mitophagy in a strikingly conserved manner from *C. elegans* to humans. It has been shown that boosting mitochondrial function and proteostasis may decrease the formation of detrimental protein aggregates suggesting that enhancing mitochondrial proteostasis may abrogate the Aβ proteotoxicity in AD [155].

15 CONCLUSION AND PERSPECTIVES

This chapter provides an overview of recent advances in understanding the role of mitochondrial function in the pathogenesis of AD. The present therapeutics for this disease are at best symptomatic and not neuroprotective or neurorestorative. Due to the complex pathophysiology, including a cascade of neurotoxic molecular events involving energy provision, redox and Ca^{2+} homeostasis, cellular and intra-inter-organellar quality control, regulation of cell death/survival pathways resulting in neurodegeneration in AD, significant research effort is still required to elucidate the complexity of the network of multileveled, cross-talk that regulate mitochondrial homeostasis and identify potential multifunctional therapeutic targets that will improve mitochondrial function, attenuate oxidative stress, and optimize mitochondrial quality control (QC) in neurons and slow or halt progressive course of these neurological disorders. This knowledge will likely provide novel and highly effective treatment to preserve neuronal viability and function and delay or reverse the neurodegenerative process in AD.

REFERENCES

1. Van Cauwenberghe C, Van Broeckhoven C, and Sleegers K, *The genetic landscape of Alzheimer disease: clinical implications and perspectives.* Genet Med, 2016. **18**: p. 421–430.

2. Thinakaran G, *The role of presenilins in Alzheimer's disease.* J Clin Invest, 1999. **104**: p. 1321–1327.

3. Carr DB, et al., *Current concepts in the pathogenesis of Alzheimer's disease.* Am J Med, 1997. **103**: p. 3S–10S.

4. Wingo TS, et al., *Autosomal recessive causes likely in early-onset Alzheimer disease.* Arch Neurol, 2012. **69**: p. 59–64.

5. Gatz M, et al., *Role of genes and environments for explaining Alzheimer disease.* Arch Gen Psychiatry, 2006. **63**: p. 168–174.

6. Dorszewska J, et al., *Molecular basis of familial and sporadic Alzheimer's Disease.* Curr Alzheimer Res, 2016. **13**: p. 952–963.

7. Swerdlow RH, Burns JM, and Khan SM, *The Alzheimer's disease mitochondrial cascade hypothesis: progress and perspectives.* Biochim Biophys Acta, 2013. **1842**(8): p. 1219–1231.

8. Scarpulla RC, *Transcriptional paradigms in mammalian mitochondrial biogenesis and function.* Physiol Rev, 2008. **88**: p. 611–638.

9. Mosconi L, et al., *Maternal family history of Alzheimer's disease predisposes to reduced brain glucose metabolism.* Proc Natl Acad Sci U S A, 2007. **104**: p. 19067–19072.

10. Murray J, et al., *FDG and amyloid PET in cognitively normal individuals at risk for late-onset Alzheimer's disease.* Adv J Mol Imaging, 2014. **4**: p. 15–26.

11. Hirai K, et al., *Mitochondrial abnormalities in Alzheimer's disease.* J Neurosci, 2001. **21**: p. 3017–3023.

12. Moreira PI, et al., *Increased autophagic degradation of mitochondria in Alzheimer disease.* Autophagy, 2007. **3**: p. 614–615.

13. Parker WD Jr and Parks JK, *Cytochrome c oxidase in Alzheimer's disease brain: purification and characterization.* Neurology, 1995. **45**: p. 482–486.

14. Gibson GE, Sheu KF, and Blass JP, *Abnormalities of mitochondrial enzymes in Alzheimer disease.* J Neural Transm, 1998. **105**: p. 855–870.

15. Maurer I, Zierz S, and Moller HJ, *A selective defect of cytochrome c oxidase is present in brain of Alzheimer disease patients.* Neurobiol Aging, 2000. **21**: p. 455–462.

16. Silva DF, et al., *Mitochondrial abnormalities in Alzheimer's disease: possible targets for therapeutic intervention.* Adv Pharmacol, 2012. **64**: p. 83–126.

17. Calkins MJ, et al., *Impaired mitochondrial biogenesis, defective axonal transport of mitochondria, abnormal mitochondrial dynamics and synaptic degeneration in a mouse model of Alzheimer's disease.* Hum Mol Genet, 2011. **20**: p. 4515–4529.

18. Manczak M, Calkins MJ, and Reddy PH, *Impaired mitochondrial dynamics and abnormal interaction of amyloid beta with mitochondrial protein Drp1 in neurons from patients with Alzheimer's disease: implications for neuronal damage.* Hum Mol Genet, 2011. **20**: p. 2495–2509.

19. Keller JN, et al., *Impairment of glucose and glutamate transport and induction of mitochondrial oxidative stress and dysfunction in synaptosomes by amyloid beta-peptide: role of the lipid peroxidation product 4-hydroxynonenal.* J Neurochem, 1997. **69**: p. 273–284.

20. Abramov AY, Canevari L, and Duchen MR, *Beta-amyloid peptides induce mitochondrial dysfunction and oxidative stress in astrocytes and death of neurons through activation of NADPH oxidase.* J Neurosci, 2004. **24**: p. 565–575.

21. Manczak M, et al., *Mitochondria are a direct site of A beta accumulation in Alzheimer's disease neurons: implications for free radical generation and oxidative damage in disease progression.* Hum Mol Genet, 2006. **15**: p. 1437–1449.

22. Mattson MP, Partin J, and Begley JG, *Amyloid beta-peptide induces apoptosis-related events in synapses and dendrites.* Brain Res, 1998. **807**: p. 167–176.

23. Lustbader JW, et al., *ABAD directly links Abeta to mitochondrial toxicity in Alzheimer's disease.* Science, 2004. **304**: p. 448–452.

24. Reddy PH, *Mitochondrial medicine for aging and neurodegenerative diseases.* Neuromolecular Med, 2008. **10**: p. 291–315.

25. Calkins MJ, Manczak M, and Reddy PH, *Mitochondria-targeted antioxidant SS31 prevents amyloid beta-induced mitochondrial abnormalities and synaptic degeneration in Alzheimer's disease.* Pharmaceuticals, 2012. **5**: p. 1103–1119.

26. Eckert GP, et al., *Mitochondrial dysfunction a pharmacological target in Alzheimer's disease.* Mol Neurobiol, 2012. **46**: p. 136–150.

27. Reddy PH, et al., *Abnormal mitochondrial dynamics and synaptic degeneration as early events in Alzheimer's disease: implications to mitochondria-targeted antioxidant therapeutics.* Biochim Biophys Acta, 2012. **1822**: p. 639–649.

28. Chen H, Chomyn A, and Chan DC, *Disruption of fusion results in mitochondrial heterogeneity and dysfunction.* J Biol Chem, 2005. **280**: p. 26185–26192.

29. McBride HM, Neuspiel M, and Wasiak S, *Mitochondria: more than just a powerhouse.* Curr Biol, 2006. **16**: p. R551–R560.

30. Yu T, Robotham JL, and Yoon Y, *Increased production of reactive oxygen species in hyperglycemic conditions requires dynamic change of mitochondrial morphology.* Proc Natl Acad Sci U S A, 2006. **103**: p. 2653–2658.

31. Anderson S, et al., *Sequence and organization of the human mitochondrial genome.* Nature, 1981. **290**: p. 457–465.

32. Spelbrink JN, *Functional organization of mammalian mitochondrial DNA in nucleoids: history, recent developments, and future challenges.* IUBMB Life, 2010. **62**: p. 19–32.

33. Onyango I, et al., *Mitochondrial genomic contribution to mitochondrial dysfunction in Alzheimer's disease.* J Alzheimers Dis, 2006. **9**: p. 183–193.

34. Reeve AK, Krishnan KJ, and Turnbull D, *Mitochondrial DNA mutations in disease, aging, and neurodegeneration.* Ann NY Acad Sci, 2008. **1147**: p. 21–29.

35. Smigrodzki R, Parks J, and Parker WD Jr, *High frequency of mitochondrial complex I mutations in Parkinson's disease and aging.* Neurobiol Aging, 2004. **25**: p. 1273–1281.

36. Parker WD Jr and Parks JK, *Mitochondrial ND5 mutations in idiopathic Parkinson's disease.* Biochem Biophys Res Commun, 2005. **326**: p. 667–669.

37. De Coo IF, et al., *A 4-base pair deletion in the mitochondrial cytochrome b gene associated with parkinsonism/MELAS overlap syndrome.* Ann Neurol, 1999. **45**: p. 130–133.

38. Mancuso M, et al., *Is there a primary role of the mitochondrial genome in Alzheimer's disease?* J Bioenerg Biomembr, 2009. **41**: p. 411–416.

39. Ridge PG, et al., *Mitochondrial haplotypes associated with biomarkers for Alzheimer's disease.* PLoS ONE, 2013. **8**: p. e74158.

40. Maruszak A, et al., *The impact of mitochondrial and nuclear DNA variants on late-onset Alzheimer's disease risk.* J Alzheimers Dis, 2011. **27**: p. 197–210.

41. Krishnan KJ, et al., *Mitochondrial DNA deletions cause the biochemical defect observed in Alzheimer's disease.* Neurobiol Aging, 2012. 33: p. 2210–2214.

42. Coskun PE, Beal MF, and Wallace DC, *Alzheimer's brains harbor somatic mtDNA control-region mutations that suppress mitochondrial transcription and replication.* Proc Natl Acad Sci U S A, 2004. **101**: p. 10726–10731.

43. Podlesniy P, et al., *Low cerebrospinal fluid concentration of mitochondrial DNA in preclinical Alzheimer disease.* Ann Neurol, 2013. **74**: p. 655–668.

44. Lunnon K, et al., *Mitochondrial genes are altered in blood early in Alzheimer's disease.* Neurobiol Aging, 2017. **53**: p. 36–47.

45. Liu VWS, Zhang C, and Nagley P, *Mutations in mitochondrial DNA accumulate differentially in three different human tissues during ageing.* Nucleic Acids Res, 1998. **26**: p. 1268–1275.

46. Gu G, et al., *Mitochondrial DNA deletions/rearrangements in parkinson disease and related neurodegenerative disorders.* J Neuropathol Exp Neurol, 2002. **61**: p. 634–639.

47. Tuppen HA, et al., *Mitochondrial DNA mutations and human disease.* Biochim Biophys Acta, 2010. **1797**: p. 113–128.

48. Epstein CB, et al., *Genome-wide responses to mitochondrial dysfunction.* Mol Biol Cell, 2001. **12**: p. 297–308.

49. Yun J and Finkel T, *Mitohormesis.* Cell Metab, 2014. **19**: p. 757–766.

50. Finley LWS and Haigis MC, *The coordination of nuclear and mitochondrial communication during aging and calorie restriction.* Ageing Res Rev, 2009. **8**: p. 173–188.

51. Mootha VK, et al., *PGC-1alpha-responsive genes involved in oxidative phosphorylation are coordinately downregulated in human diabetes.* Nat Genet, 2003. **34**: p. 267–273.

52. Gomes AP, et al., *Declining NAD+ induces a pseudohypoxic state disrupting nuclear-mitochondrial communication during aging.* Cell, 2013. **155**: p. 1624–1638.

53. Zhuang J, et al., *Mitochondrial disulfide relay mediates translocation of p53 and partitions its subcellular activity.* Proc Natl Acad Sci U S A, 2013. **2**: p. 2–7.

54. Area-Gomez E, et al., *A key role for MAM in mediating mitochondrial dysfunction in Alzheimer disease.* Cell Death Dis, 2018. **9**: p. 335.

55. Yin W, et al., *Rapidly increased neuronal mitochondrial biogenesis after hypoxic-ischemic brain injury.* Stroke, 2008. **39**: p. 3057–3063.

56. Medeiros DM, *Assessing mitochondria biogenesis.* Methods, 2008. **46**: p. 288–294.

57. Kelly DP and Scarpulla RC, *Transcriptional regulatory circuits controlling mitochondrial biogenesis and function.* Genes Dev, 2004. **18**: p. 357–368.

58. Qin W, et al., *PGC-1α expression decreases in the Alzheimer disease brain as a function of dementia.* Arch Neurol, 2009. **66**: p. 352–361.

59. Sheng B, et al., *Impaired mitochondrial biogenesis contributes to mitochondrial dysfunction in Alzheimer's disease.* J Neurochem, 2012. **120**: p. 419–429.

60. Kim I, Rodriguez-Enriquez S, and Lemasters JJ, *Selective degradation of mitochondria by mitophagy.* Arch Biochem Biophys, 2007. **462**: p. 245–253.

61. Chen H and Chan DC, *Mitochondrial dynamics- fusion, fission, movement, andmitophagy-in neurodegenerative diseases.* Hum Mol Genet, 2009. **18**: p. R169–R176.

62. Burchell VS, et al., *Targeting mitochondrial dysfunction in neurodegenerative disease part II.* Expert Opin Ther Targets, 2010. **14**: p. 497–511.

63. Burchell VS, et al., *Mitochondrial dysfunction in neurodegenerative disease: part I.* Expert Opin Ther Targets, 2010. **14**: p. 369–385.

64. Kurz T, Terman A, and Brunk UT, *Autophagy, ageing and apoptosis: the role of oxidative stress and lysosomal iron.* Arch Biochem Biophys, 2007. **462**: p. 220–230.

65. Lemasters JJ, *Selective mitochondrial autophagy, or mitophagy, as a targeted defense against oxidative stress, mitochondrial dysfunction, and aging.* Rejuvenation Res, 2005. **8**: p. 3–5.

66. Youle RJ and Narendra DP, *Mechanisms of mitophagy.* Nat Rev Mol Biol, 2011. **12**: p. 9–14.

67. Liu L, et al., *Mitochondrial outer-membrane protein FUNDC1 mediates hypoxia-induced mitophagy in mammalian cells.* Nat Cell Biol, 2012. **14**: p. 177–185.

68. Novak I, et al., *Nix is a selective autophagy receptor for mitochondrial clearance.* EMBO Rep, 2010. **11**: p. 45–51.

69. Ding WX and Yin XM, *Mitophagy, mechanisms, pathophysiological roles, and analysis.* Biol Chem, 2012. **393**: p. 547–564.

70. Fu M, et al., *Regulation of mitophagy by the Gp78 E3 ubiquitin ligase.* Mol Biol Cell, 2013. **24**: p. 1153–1162.

71. Lokireddy S, et al., *The ubiquitin ligase Mul1 induces mitophagy in skeletal muscle in response to muscle-wasting stimuli.* Cell Metab, 2012. **16**: p. 613–624.

72. Horn SR, et al., *Regulation of mitochondrial morphology by APC/CCdh1-mediated control of Drp1 stability.* Mol Biol Cell, 2011. **22**: p. 1207–1216.

73. Kashatus DF, et al., *RALA and RALBP1 regulate mitochondrial fission at mitosis.* Nat Cell Biol, 2011. **13**: p. 1108–1115.

74. Twig G, Hyde B, and Shirihai OS, *Mitochondrial fusion, fission and autophagy as a quality control axis: the bioenergetic view.* Biochim Biophys Acta, 2008. **1777**: p. 1092–1097.

75. Martinou JC and Youle RJ, *Mitochondria in apoptosis: Bcl-2family members and mitochondrial dynamics.* Dev Cell, 2011. **21**: p. 92–101.

76. Autret A and Martin SJ, *Emerging role for members of the Bl-2 family in mitochondrial morphogenesis.* Mol Cell, 2009. **36**: p. 355–363.

77. Youle RJ and van der Bliek AM, *Mitochondrial fission, fusion and stress.* Science, 2012. **337**: p. 1062–1065.

78. Palmer CS, et al., *MiD49 and MiD51, new components of the mitochondrial fission machinery.* EMBO Rep, 2011. **12**: p. 565–573.

79. Cho DH, et al., *S-nitrosylation of Drp1 mediates beta-amyloid-related mitochondrial fission and neuronal injury.* Science, 2009. **324**: p. 102–105.

80. Bossy B, et al., *S-Nitrosylation of DRP1 does not affect enzymatic activity and is not specific to Alzheimer's disease.* J Alzheimers Dis, 2010. **20**(Suppl): p. S513–526.

81. Nakamura T and Lipton SA, *Redox modulation by S-nitrosylation contributes to protein misfolding, mitochondrial dynamics, and neuronal synaptic damage in neurodegenerative diseases.* Cell Death Differ, 2011. **18**: p. 1478–1486.

82. Sauvanet C, et al., *Energetic requirements and bioenergetic modulation of mitochondrial morphology and dynamics.* Semin Cell Dev Biol, 2010. **21**: p. 558–565.

83. Wang H, et al., *Effects of overexpression of Huntingtin proteins on mitochondrial integrity.* Hum Mol Genet, 2008. **18**: p. 737–752.

84. Hamblet NS, et al., *Mutations in mitochondrial-encoded cytochrome c oxidase subunits I, II, and III genes detected in Alzheimer's disease using single-strand conformation polymorphism.* Electrophoresis, 2006. **27**: p. 398–408.

85. Chandrasekaran K, et al., *Evidence for physiological down-regulation of brain oxidative phosphorylation in Alzheimer's disease.* Exp Neurol, 1996. **142**: p. 80–88.

86. Reddy PH, *Amyloid beta, mitochondrial structural and functional dynamics in Alzheimer's disease.* Exp Neurol, 2009. **218**: p. 286–292.

87. Deshpande A, et al., *Different conformations of amyloid beta induce neurotoxicity by distinct mechanisms in human cortical neurons.* J Neurosci, 2006. **26**: p. 6011–6018.

88. Cabezas-Opazo FA, et al., *Mitochondrial dysfunction contributes to the pathogenesis of Alzheimer's disease.* Oxid Med Cell Longev, 2015. **2015**: p. 509–654.

89. Madamanchi NR and Runge MS, *Mitochondrial dysfunction in atherosclerosis.* Circ Res, 2007. **100**: p. 460–473.

90. Ramsden DB, et al., *Human neuronal uncoupling proteins 4 and 5 (UCP4 and UCP5): structural properties, regulation, and physiological role in protection against oxidative stress and mitochondrial dysfunction.* Brain Behav, 2012. **2**: p. 468–478.

91. Votyakova TV and Reynolds IJ, *Δψm dependent and independent production of reactive oxygen species by rat brain mitochondria.* J Neurochem, 2001. **79**: p. 266–277.

92. Brennan JP, et al., *Mitochondrial uncoupling, with low concentration FCCP, induces ROS-dependent cardioprotection independent of KATP channel activation.* Cardiovasc Res, 2006. **72**: p. 313–321.

93. Skulachev VP, *Uncoupling: new approaches to an old problem of bioenergetics.* Biochim Biophys Acta, 1998. **1363**: p. 100–124.

94. Korshunov SS, Skulachev VP, and Starkov AA, *High protonic potential actuates a mechanism of production of reactive oxygen species in mitochondria.* FEBS Lett, 1997. **416**: p. 15–18.

95. Brookes PS, *Mitochondrial H(+) leak and ROS generation: an odd couple.* Free Radic Biol Med, 2005. **38**: p. 12–23.

96. Brand MD, Buckingham JA, and Esteves TC, *Mitochondrial superoxide and aging: uncoupling-protein activity and superoxide production.* Biochem Soc Symp, 2004. 203–213.

97. Hernandes MS and Britto LR, *NADPH oxidase and neurodegeneration.* Curr Neuropharmacol, 2012. **10**: p. 321–327.

98. Shimohama S, et al., *Activation of NADPH oxidase in Alzheimer's disease brains.* Biochem Biophys Res Commun, 2000. **273**: p. 5–9.

99. de la Monte SM and Wands JR, *Molecular indices of oxidative stress and mitochondrial dysfunction occur early and often progress with severity of Alzheimer's disease.* J Alzheimers Dis, 2006. **9**: p. 167–181.

100. Hamanaka RB and Chandel NS, *Mitochondrial reactive oxygen species regulate cellular signaling and dictate biological outcomes.* Trends Biochem Sci, 2010. **35**: p. 505–513.

101. Bera AK, Ghosh S, and Das S, *Mitochondrial VDAC can be phosphorylated by cyclic AMP-dependent protein kinase.* Biochem Biophys Res Commun, 1995. **209**: p. 213–217.

102. Mihaylova MM and Shaw RJ, *The AMPK signaling pathway coordinates cell growth, autophagy and metabolism.* Nat Rev Cell Biol, 2011. **13**: p. 1016–1023.

103. Alto NM, Soderling J, and Scott JD, *Rab32 is an A-kinase anchoring protein and participates in mitochondrial dynamics.* J Cell Sci, 2002. **158**: p. 659–6810.

104. Das S, et al., *Glycogen synthase kinase 3 inhibition slows mitochondrial adenine nucleotide transport and regulates*

105. Robey RB and Hay N, *Is Akt the "Warburg kinase"? – Akt energy metabolism interactions and oncogenesis.* Semin Cancer Biol, 2009. **19**: p. 25–31.

106. Valentin-Vega YA, et al., *Mitochondrial dysfunction in ataxia-telangiectasia.* Blood, 2012. **119**: p. 1490–5001.

107. Shen X, et al., *Neurons in vulnerable regions of the Alzheimer's disease brain display reduced ATM signaling.* eNeuro, 2016. **3**. doi:10.1523/ENEURO.0124-15.2016.

108. Rimkus SA, et al., *Mutations in String/CDC25 inhibit cell cycle re-entry and neurodegeneration in a Drosophila model of Ataxia telangiectasia.* Genes Dev, 2008. **22**: p. 1205–1220.

109. Yang Y and Herrup K, *Loss of neuronal cell cycle control in Ataxia-telangiectasia: A unified disease mechanism.* J Neurosci, 2005. **25**: p. 2522–2529.

110. Petersen AJ, Rimkus SA, and Wassarman DA, *ATM kinase inhibition in glial cells activates the innate immune response and causes neurodegeneration in Drosophila.* Proc Natl Acad Sci U S A, 2012. **109**: p. E656–E664.

111. Kuljis RO, et al., *Degeneration of neurons, synapses, and neuropil and glial activation in a murine ATM knockout model of Ataxia-telangiectasia.* Proc Natl Acad Sci U S A, 1997. **94**: p. 12688–12693.

112. Liu N, et al., *ATM deficiency induces oxidative stress and endoplasmic reticulum stress in astrocytes.* Lab Investig, 2005. **85**: p. 1471–1480.

113. Bezprozvanny I and Mattson MP, *Neuronal calcium mishandling and the pathogenesis of Alzheimer's disease.* Trends Neurosci, 2008. **31**: p. 454–463.

114. Green KN and LaFerla FM, *Linking calcium to Abeta and Alzheimer's disease.* Neuron, 2008. **59**: p. 190–194.

115. Mattson MP, *Pathways towards and away from Alzheimer's disease.* Nature, 2004. **430**: p. 631–639.

116. Mattson MP, *Calcium and neurodegeneration.* Aging Cell, 2007. **6**: p. 337–350.

117. Selkoe DJ, *Alzheimer's disease: genes, proteins, and therapy.* Physiol Rev, 2001. **81**: p. 741–766.

118. Wojda U, Salinska E, and Kuznicki J, *Calcium ions in neuronal degeneration.* IUBMB Life, 2008. **60**: p. 575–590.

119. Mattson MP, et al., *beta-Amyloid peptides destabilize calcium homeostasis and render human cortical neurons vulnerable to excitotoxicity.* J Neurosci, 1992. **12**: p. 376–389.

120. Canevari L, Abramov AY, and Duchen MR, *Toxicity of amyloid beta peptide: tales of calcium, mitochondria, and oxidative stress.* Neurochem Res, 2004. **29**: p. 637–650.

121. Duchen MR, *Mitochondria, calcium-dependent neuronal death and neurodegenerative disease.* Pflugers Arch, 2012. **464**: p. 111–121.

122. Nicholls DG, *Mitochondrial calcium function and dysfunction in the central nervous system.* Biochim Biophys Acta, 2009. **1787**: p. 1416–1424.

123. Pivovarova NB and Andrews SB, *Calcium-dependent mitochondrial function and dysfunction in neurons.* FEBS J, 2010. **227**: p. 3622–3636.

124. Toescu EC and Verkhratsky A, *The importance of being subtle: small changes in calcium homeostasis control cognitive decline in normal aging.* Aging Cell, 2007. **6**: p. 267–273.

125. Lee G, Pollard HB, and Arispe N, *Annexin 5 and apolipoprotein E2 protect against Alzheimer's amyloid-β-peptide cytotoxicity by competitive inhibition at a common phosphatidylserine interaction site.* Peptides, 2002. **23**: p. 1249–1263.

126. Simakova O and Arispe NJ, *The cell-selective neurotoxicity of the Alzheimer's Abeta peptide is determined by surface phosphatidylserine and cytosolic ATP levels. Membrane*

binding is required for Abeta toxicity. J Neurosci, 2007. **27**: p. 13719–13729.

127. Kuchibhotla KV, et al., *Abeta plaques lead to aberrant regulation of calcium homeostasis in vivo resulting in structural and functional disruption of neuronal networks.* Neuron, 2008. **59**: p. 214–225.

128. De Felice FG, et al., *Abeta oligomers induce neuronal oxidative stress through an N-methyl-D-aspartate receptor-dependent mechanism that is blocked by the Alzheimer drug memantine.* J Biol Chem, 2007. **282**: p. 11590–11601.

129. Vosler PS, Brennan CS, and Chen J, *Calpain-mediated signaling mechanisms in neuronal injury and neurodegeneration.* Mol Neurobiol, 2008. **38**: p. 78–100.

130. Calabrese EJ and Baldwin LA, *Defining hormesis.* Hum Exp Toxicol, 2002. **21**: p. 91–97.

131. Mammucari C and Rizzuto R, *Signaling pathways in mitochondrial dysfunction and aging.* Mech Ageing Dev, 2010. **131**: p. 536–543.

132. Ristow M, et al., *Antioxidants prevent health-promoting effects of physical exercise in humans.* Proc Natl Acad Sci U S A, 2009. **106**: p. 8665–8670.

133. Ristow M and Zarse K, *How increased oxidative stress promotes longevity and metabolic health: the concept of mitochondrial hormesis (mitohormesis).* Exp Gerontol, 2010. **45**: p. 410–418.

134. Kenyon CJ, *The genetics of ageing.* Nature, 2010. **464**: p. 504–512.

135. Bennett CF and Kaeberlein M, *The mitochondrial unfolded protein response and increased longevity: cause, consequence, or correlation?* Exp Gerontol, 2014. **56**: p. 42–46.

136. Bennett CF, et al., *Activation of the mitochondrial unfolded protein response does not predict longevity in Caenorhabditis elegans.* Nat Commun, 2014. **5**: p. 3483.

137. Feng J, Bussière F, and Hekimi S, *Mitochondrial electron transport is a key determinant of life span in Caenorhabditis elegans.* Dev Cell, 2001. **1**: p. 633–644.

138. Durieux J, Wolff S, and Dillin A, *The cell-non-autonomous nature of electron transport chain-mediated longevity.* Cell, 2011. **144**: p. 79–91.

139. Barcena C, Mayoral P, and Quiro PM, *Mitohormesis, an Antiaging Paradigm.* Int Rev Cell Mol Bio, 2018. **340**: p. 35–77.

140. Heneka MT, et al., *Neuroinflammation in Alzheimer's disease.* Lancet Neurol, 2015. **14**: p. 388–405.

141. Zhou R, et al., *A role for mitochondria in NLRP3 inflammasome activation.* Nature, 2011. **469**: p. 221–225.

142. Wilkins HM and et Al., *Mitochondrial lysates induce inflammation and Alzheimer's disease-relevant changes in microglial and neuronal cells.* J Alzheimers Dis, 2015. **45**: p. 305–318.

143. Shimada K, et al., *Oxidized mitochondrial DNA activates the NLRP3 inflammasome during apoptosis.* Immunity, 2012. **36**: p. 401–414.

144. Iyer SS, et al., *Mitochondrial cardiolipin is required for Nlrp3 inflammasome activation.* Immunity, 2013. **39**: p. 311–323.

145. Walsh JG, Muruve Da, et al., *Inflammasomes in the CNS.* Nat Rev Neurosci, 2014. **15**: p. 84–97.

146. Davalos D, et al., *ATP mediates rapid microglial response to local brain injury in vivo.* Nat Neurosci, 2005. **8**: p. 752–758.

147. Inoue K, *Microglial activation by purines and pyrimidines.* Glia, 2002. **40**: p. 156–163.

148. Zong H, et al., *AMP kinase is required for mitochondrial biogenesis in skeletal muscle in response to chronic energy deprivation.* Proc Natl Acad Sci U S A, 2002. **99**: p. 15983–15987.

149. Yin J, et al., *NLRP3 inflammasome inhibitor ameliorates amyloid pathology in a mouse model of Alzheimer's disease.* Mol Neurobiol, 2018. **55**: p. 1977–1987.

150. Dempsey C, et al., *Inhibiting the NLRP3 inflammasome with MCC950 promotes non-phlogistic clearance of amyloid-β and cognitive function in APP/PS1 mice.* Brain Behav Immun, 2017. **61**: p. 306–316.

151. Ross JM, Olson L, and Coppotelli G, *Mitochondrial and ubiquitin proteasome system dysfunction in ageing and disease: two sides of the same coin?* Int J Mol Sci, 2015. **16**: p. 19458–19476.

152. Riederer BM, et al., *The role of the ubiquitin proteasome system in Alzheimer's disease.* Exp Biol Med (Maywood), 2011. **236**: p. 268–276.

153. Peng J, et al., *A proteomics approach to understanding protein ubiquitination.* Nat Biotechnol, 2003. **21**: p. 921–926.

154. Matsumoto M, et al., *Large-scale analysis of the human ubiquitin-related proteome.* Proteomics, 2005. **5**: p. 4145–4151.

155. Sorrentino V, et al., *Enhancing mitochondrial proteostasis reduces amyloid-β proteotoxicity.* Nature, 2017. **552**: p. 187–193.

8 Mitochondrial Dysfunction in Multiple Sclerosis

Angela Dziedzic

Faculty of Biology and Environmental Protection, Department of General Biochemistry, University of Lodz, Pomorska 141/143, 90–236 Lodz, Poland

Elzbieta Miller

Department of Physical Medicine, Medical University of Lodz, Pl. Hallera 1, Poland
Neurorehabilitation Ward, General Hospital no III, Lodz, Milionowa 14, Poland

Joanna Saluk-Bijak and Michał Bijak

Faculty of Biology and Environmental Protection, Department of General Biochemistry, University of Lodz, Pomorska 141/143, 90–236 Lodz, Poland

CONTENTS

1 INTRODUCTION

Mitochondria are one of the double-membrane organelles in eukaryotic cells. One of the most important roles of mitochondria is oxidative energy metabolism. Mitochondria contain the respiratory chain where energy is most efficiently produced in the form of adenosine triphosphate (ATP) (Chan 2006). The mitochondrial respiratory chain is located in the inner mitochondrial membrane and consists of five complexes (complexes I–V), whereof, the complex V is directly involved in ATP synthesis (Friedman and Nunnari 2014). Mitochondria are a significant source of reactive oxygen species (ROS) within most mammalian cells. Therefore, any damage that impairs the function of the respiratory chain might also have an impact on cell survival. To protect cells from oxidative burst, mitochondria contain an intricate defense system to exonerate from ROS and repair ROS-induced damages (Nita and Grzybowski 2016).

Disturbances in mitochondrial function are manifested through decrease in ATP levels, and gradually enhance neuronal dysfunction or neurodegeneration. Recent research indicates that mitochondria have been increasingly linked to the pathogenesis of many neurological disorders, including multiple sclerosis (MS) (Patergnani et al. 2017).

Neurodegenerative diseases, due to presence of the chronic oxidative stress and dysregulation of the inflammatory response, represent a major threat to human health. There are a diverse group of disorders with progressive degeneration of the structure and abnormal organic function of the central nervous system (CNS) or peripheral nervous system. The chronic course of neurodegenerative diseases makes their treatment an increasing medical, social and economic problem worldwide. Annually, an estimated around 7 million people die as a result of neurological disorders. In Europe, the economic cost of neurological diseases was assessed at about 139 billion euros in 2004 (Andlin-Sobocki et al. 2005).

2 PATHOGENESIS OF MS

MS is responsible for the majority of neurological disability of young adults – under the age of 30, and is presently incurable. Data from the World Health Organization (WHO) indicate that its median prevalence is of 80 per 1,00,000 people in Europe. MS is a chronic immune-mediated disease of CNSduring which demyelination, inflammation, and axonal damage occurs. MS is considered as a heterogeneous neurological disease with variable clinical manifestations. In the majority of cases, the disease is typified initially by episodes of reversible neurological deficits, which is often followed by progressive neurological deterioration over time (Miller 2012). More than 30% of MS patients have intensified spasticity (typically in the legs). Primary clinical findings in MS patients are often sensory disturbances, such as paresthesias (numbness and tingling), dysesthesias (burning and "pins and needles"), diplopia, ataxia, vertigo and bladder (urinary sphincter) disturbances. Another common presenting sign of MS is optic neuritis, highlighted by complete or partial loss of vision. As a result of complications from ongoing neurological disorders there is a shortening of the natural life time. Since the onset of illness, the average lifespan is estimated at 25–35 years (Hauser and Goodwin 2008).

Depending on progression of clinical symptoms, four distinct MS subtypes can be distinguished: relapsing-remitting (RRMS), secondary-progressive (SPMS), primary-progressive (PPMS) and progressive-relapsing (PRMS). PPMS occur only slightly in about 15% of patients and mainly affect patients in whom the onset of the disease appeared at a later age. The most common is RRMS subtype, affecting about 85% of patients with MS. After about 10–20 years of the disease in some patients RRMS is transformed into SPMS, characterized by a slow and irreversible deterioration of neurological changes. Increasing constant neurological changes lead to progressive disability, which is accompanied by exacerbation of the disease with no remission (Fig 8.1) (Miller 2012).

Traditionally, MS has been considered an autoimmune disorder in which T cells, macrophages, soluble mediators of inflammation and autoantibodies contribute to multifocal demyelination. It is generally known that immune cells, such as T cells, B cells, and macrophages, play essential roles in both mediating and regulating MS pathophysiology. Peripherally activated T cells

cross the blood-brain barrier (BBB) *via* interaction of adhesion molecules, such as vascular cell adhesion molecule-1 (VCAM-1) into CNS, where they are reactivated and secrete proinflammatory cytokines (Palmer 2013). Due to crucial immune cell infiltration and their cytokines, the inflammation of the white and gray matter tissues in the CNS spread is the primary cause of damage in MS. Numerous studies have suggested that T helper (Th) cell (also known as CD4+ T cells) intervention and adaptive immune responses, which initiated by interaction between antigen presenting cells (APCs) with T lymphocytes, play a significant role in the initiation and progression of MS (Gandhi et al. 2010). T cells may be divided into two groups on the basis of their expression of either the CD4+ or CD8+ surface molecules. Differentiation of such CD4+ T cells results in Th1 and Th17 cells, all of which produce proinflammatory cytokines. Th1 and Th17 cells infiltrate into the CNS through the BBB, where they are reactivated and initiate the destruction of myelin sheath (demyelination) and axonal/neuronal degeneration in MS. Th1 cells produce their lineage-defining cytokines, interferon-γ (INF-γ), as well as tumor necrosis factor-α (TNF-α), while Th17 cells secrete their defining interleukin-17 (IL-17), as well as IL-21 and IL-22, which contributes to their pathogenicity. Cytokines IL-17, TNF-α and IL-6 have both overlapping and distinct roles in neuroinflammation. IL-17 enhances the production of proinflammatory cytokines and chemokines by other cells *via* activating the nuclear factor kappa B (NFκB) pathway. IL-17 appears to play pivotal role in the pathogenesis of chronic inflammatory disorders and in many autoimmune diseases, including MS. What is the most important; this cytokine secretion leads to the enhanced production of ROS. These tiered mechanisms may be crucial in causing of mitochondrial dysfunction, neuronal demyelination- and irreversible

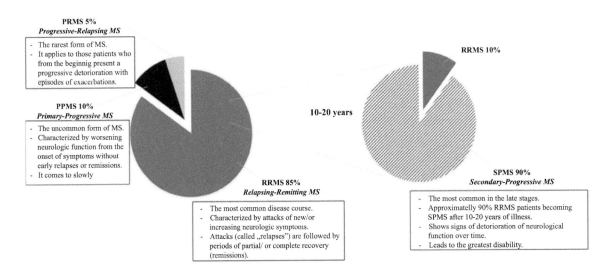

FIGURE 1 Classification of MS types with characteristic clinical symptoms (Hurwitz 2009; Miller 2012; Lublin 2014)

tissue damage distinguished by axonal loss and gliosis (Li 2013; Michel et al. 2015; Golubovskaya and Wu 2016; Solleiro-Villavicencio and Rivas-Arancibia 2018).

Cellular degeneration at all MS stages entailed by oxidative injury is manifested as the accumulation of oxidized proteins, lipids, and DNA. This can be recognized as the presence of active and slowly-expanding lesions in both the white and grey matter in CNS. Due to the progressive stage, aging processes such as chronic mitochondrial injury is enhanced and can trigger the formation of demyelinating lesions (Mahad et al. 2015).

3 MITOCHONDRIAL DYSFUNCTION IN MS

Mitochondrial dysfunction manifests by carrying out incorrectly processes, such as ATP synthesis, Ca^{2+} homeostasis, central metabolic pathways or ROS production. Nonetheless, the mechanisms and regulation of mitochondrial dysfunction remain poorly described. Given the central role of the mitochondria in many important cellular functions including energy production, it is reasonable that its dysfunction is the significant contributor to neurodegenerative process (Ivannikov 2013; Pathak et al. 2013).

Neurons are extremely dependent on oxidative energy metabolism. Axons consume major amounts of ATP, which is used primarily to fuel the sodium/potassium ATPase, or sodium pump that functions to remove the sodium ions that enter the axon during impulse activity. The brain and the CNS are particularly injured by alterations of both mitochondrial shape and function (McFarland et al. 2010).

In human neurons, the largest number of mitochondria is located at synaptic sites, where energy demand is especially high. Although postsynaptic dendritic spines lack mitochondria, many presynaptic boutons contain them and promote neurotransmission. Interestingly, neurons are the most sensitive to bioenergetic fluctuations, which indicates that mitochondria regulate fundamental aspects of brain function (Hara et al. 2014).

Variations in mitochondrial activity, which could be induced by genetic and external factors such as nutrients, hormones, inhibitors of respiratory chain (like rotenenon) and hypoxia, may elicit detrimental mitochondrial stress and generate increased levels of ROS (Liu and Hajnóczky 2011). Mitochondrial structural changes, different mitochondrial gene expression and enzyme activities, increased free radical production and oxidative damage have all been reported in patients with MS (Mao and Reddy 2010). Particularly, electron microscopic studies position abnormal mitochondrial shape as an emerging disease biomarker and potential cause of neurodegenerative disorders (Burté et al. 2015), laterally with oxidative stress and neuroinflammation (Lin and Beal 2006).

One of the mechanisms explaining the diffuse neurodegeneration found in MS patients involves mitochondrial dysfunction. Looking for evidence, linking mitochondrial dysfunction to neurodegeneration has been shown by postmortem analysis of MS patient brain tissue, including lesioned areas, normal appearing white matter (NAWM) and non-lesional cortex. N-acetyl-aspartate (NAA) is detected in the adult brain in neurons, oligodendrocytes and myelin and is produced in the mitochondria from the aspartic acid and acetyl-coenzyme A. NAA may function as a neurotransmitter in the brain by acting on metabotropic glutamate receptors (Nordengen et al. 2015). NAA is a marker of axonal degeneration neuronal integrity that gives off the largest signal in the technique of magnetic resonance spectroscopy (MRS) of the human brain. Decrease in NAA has been correlated with relapses and neurological decline in MS patients. NAA is almost exclusively found in active mitochondria of neurons and axons and is thus a marker of axonal function and integrity. Reduced levels of NAA are suggestive for axonal degeneration and/or axonal loss. That decrease may correlate to histopathological alterations and suggested histotoxic mechanisms in MS (Aboul-Enein et al. 2010). Additionally, NAA levels are decreased in MS white matter lesions and NAWM as well as gray matter, suggesting that neurodegenerative processes are widespread throughout the CNS (Lu et al. 2004; Moffett et al. 2007; Benarroch 2008). White matter atrophy rates were constant across the MS stages (3-fold increase). The fact is surprising that gray matter atrophy rates have been correlated with getting worse disability in atrophy from clinically isolated syndromes to RRMS (3.4-fold increase) and in atrophy from RRMS to SPMS (14-fold increase) (Fisher et al. 2008). NAA decreases in white matter lesions and NAWM also may suggest mitochondrial dysfunction within neurons as well as neuronal damage. Application of the electron microscopy of demyelinated spinal cord lesions demonstrated dramatically decreased numbers of microtubules and mitochondria (Bjartmar et al. 2000; Dutta et al. 2006).

Neuronal mitochondria have a high capacity to store Ca^{2+} and can therefore protect neurons against transient elevations in intracellular calcium concentrations during neuronal hyperactivity. Glutamate is a major excitatory neurotransmitter in the CNS. Excitotoxicity of glutamate causes neuronal dysfunction and has also been linked to chronic degeneration (Lau and Tymianski 2010). It is well documented that glutamate is a formidable neurotoxin, capable of killing neurons in the CNS when its extracellular concentration is elevated. Glutamate exposure results in a massive accumulation of Ca^{2+} by neurons (Manev et al. 1989). Mitochondria are implicated at multiple stages of glutamate excitotoxicity, that triggers the ischemic release of the amino acid, the sequestration of Ca^{2+} entering via the N-Methyl-D-aspartic acid (NMDA) receptor after reperfusion, the generation of ROS and triggering of an apoptotic cascade (Nicholls et al. 2000; Hajnóczky et al. 2006).

According to Su et al. (2013), mitochondrial dysfunction in MS may be mediated by the pathological opening of the mitochondrial permeability transition pore (PTP).

The PTP is a transient pore located in the inner mitochondrial membrane. The pore transiently opens to certain stimuli, including calcium and ROS. Mitochondrial dysfunction, like loss of the membrane potential and equilibrium of ionic gradients, which can prevent ATP synthesis, mitochondrial swelling and eventual cell death are caused by pathological pore opening, triggered by the influx of solutes in the mitochondrial matrix. Continued study of mitochondrial dysfunction in the context of the PTP has led to the investigation of crucial promoters of PTP opening, such as ROS (especially, hydrogen peroxide, superoxide an ion and hydroxyl radicals). There are numerous stressors which can dramatically increase ROS levels leading to significant intracellular damage. This applies in particular to mitochondria, which are main sites of ROS production due to aerobic respiration and oxidative phosphorylation (Bernardi et al. 2006).

Oxidative stress is a main factor of cell injury in aging CNS and chronic inflammation, as well as in vascular and neurodegenerative diseases, including every subtype of MS. Very high production of both ROS and RNS lead to axonal injury, leukocyte migration and oligodendrocyte damage. Subsequently, oxidative stress is involved in the functioning of different types of CNS cells, and affects some cellular components which exacerbate inflammation and neurodegenerative processes. ROS are created at a high rate in the CNS of MS patients, mainly by microglia and activated macrophages. (Gilgun-Sherki et al. 2004; Lee et al. 2012)

Numerous studies demonstrate that over-production of nitric oxide (NO) plays a significant role in damage of oligodendrocytes in MS. Increased expression of inducible nitric oxide synthase (iNOS) has been detected in MS lesions (Liu et al. 2001; Boullerne and Benjamins 2006; Witherick et al. 2011). What more, both endogenous NO, released by glial cells and NO generated from exogenous NO-donors, are known to induce death of oligodendrocytes (Jana and Pahan 2013). Generation of mitochondrial ROS mainly takes place at the respiratory chain located on the inner mitochondrial membrane during the process of oxidative phosphorylation. ROS production by mitochondria can lead to oxidative damage to mitochondrial proteins, membranes and DNA, impairing the ability of mitochondria to synthesized ATP (Murphy 2009).

Mitochondria contain their own genetic material – the mitochondrial DNA (mtDNA), which encodes essential molecular elements required for electron transport by the respiratory chain where oxygen is consumed. mtDNA alterations, impaired activity of mitochondrial enzyme complexes in damaged tissues (Shokolenko et al. 2009) and reduced mitochondrial gene expression, specific to neurons in non-damaged tissues, have been observed in MS (Bonawitz et al. 2006; Krüger et al. 2010). Furthermore, numerous deletions of mtDNA in cells of gray matter are allegedly important contributors to neurodegeneration in MS. Oxidative damage to mtDNA in active MS lesions leads to mitochondrial dysfunction and axonal degeneration (Lu et al. 2000; Tranah et al. 2015). mtDNA encodes thirteen polypeptides, twenty-two transfer RNAs (tRNAs) and two ribosomal RNAs (rRNAs), all of which are essential for electron transport and ATP generation by oxidative phosphorylation. This requires the assembly of the protein products of both the mitochondrial and nuclear genomes into functional respiratory complexes (Anderson et al. 1981). Even milder mtDNA sequence variants, or single nucleotide polymorphisms (SNPs) can confer risk of different diseases belonging to neurodegenerative disorders (van der Walt et al. 2004). All potential factors affecting mitochondrial dysfunction in MS described in this work are presented in Table 1.

TABLE 1
Potential mitochondrial damage factors and their effects on neural cells in MS

Potential mitochondrial damage factors	Effect on the neural cells	References
Reduced level of NAA in active mitochondria	-Axonal and neuronal degeneration -Axonal loss -Reduction of neuronal integrity -Increasing the number of relapses and neurological decline in MS patients	Fisher et al. (2008) Aboul-Enein et al. (2010)
Elevated extracellular concentration of glutamate	- The massive sequestration of the Ca^{2+} - Influxe of ROS leading to intracellular damage - Apoptotic cascade activation	Manev et al. (1989) Hajnóczky et al. (2006)
Increased permeability of PTP	- Loss of the membrane potential and equilibrium of ionic gradients - Prevention of ATP synthesis	Bernardi et al. (2006) Su et al. (2013)
Increased expression of iNOS	- Inducing death of oligodendrocytes	Liu et al. (2001) Jana and Pahan (2013)
Oxidative damage to mtDNA	- Impaired activity of mitochondrial enzyme complexes - Decreased ATP synthesis - Neuronal death or degeneration	Lu et al. (2000) Krüger et al. (2010)

4 CONCLUSION

Degeneration of chronically demyelinated axons is a major cause of the continuous, irreversible neurological disability that occurs in the chronic stages of MS. In this context, it becomes important to fully understand the exact function of the surviving axons as well as the contribution of mitochondria to the axonal dysfunction and degeneration in MS patients. MS is not a mitochondrial disease, such as certain inherited mitochondrial diseases but we believe mitochondrial dysfunction is a critical component of axonal injury within the acute focal inflammatory lesions and in the progressive neurodegenerative phase of the illness. Mitochondria are potential therapeutic targets in MS, particularly with respect to prevention of disease progression. Complex mitochondria-targeted neuroprotective therapies may be in the future part will become a standard treatment routine for not only MS but other neurodegenerative diseases as well.

REFERENCES

Aboul-Enein F, Krsššák M, Höftberger R, Prayer D, Kristoferitsch W. Reduced NAA-levels in the NAWM of patients with MS is a feature of progression. A study with quantitative magnetic resonance spectroscopy at 3 Tesla. Plos One. 2010;5: e11625.

Anderson S, Bankier AT, Barrell BG, de Bruijn MHL, Coulson AR, Drain J, Eperon IC, Nierlich DP, Roe BA, Sanger F, Schreier PH, Smith AJH, Staden R, Young IG. Sequence and organization of the human mitochondrial genome. Nature. 1981;290:457–465.

Andlin-Sobocki P, Jönsson B, Wittchen HU. Costs of disorder of the brain in Europe. Eur J Neurol. 2005;12:1–27.

Benarroch EE. N-acetylaspartate and N-acetylaspartylglutamate: neurobiology and clinical significance. Neurology. 2008;70:1353–1357.

Bernardi P, Krauskopf A, Basso E, Petronilli V, Blachly-Dyson E, Di Lisa F, Forte MA. The mitochondrial permeability transition from in vitro artifact to disease target. FEBS J. 2006;273:2077–2099.

Bjartmar C, Kidd G, Mörk S, Rudick R, Trapp BD. Neurological disability correlates with spinal cord axonal loss and reduced N-acetyl aspartate in chronic multiple sclerosis patients. Ann Neurol. 2000;48:893–901.

Bonawitz ND, Rodeheffer MS, Shadel GS. Defective mitochondrial gene expression results in reactive oxygen species-mediated inhibition of respiration and reduction of yeast life span. Mol Cell Biol. 2006;26:4818–4829.

Boullerne AI, Benjamins JA. Nitric oxide synthase expression and nitric oxide toxicity in oligodendrocytes. Antioxid Redox Sign. 2006;8:967–980.

Burté F, Carelli V, Chinnery PF, Yu-Wai-Man P. Disturbed mitochondrial dynamics and neurodegenerative diseases. Nat Rev Neurol. 2015;11:11–24.

Chan DC. Mitochondria: dynamic organelles in disease, aging, and development. Cell. 2006;125:1241–1252.

Dutta R, McDonough J, Yin X, Peterson J, Chang A, Torres T, Gudz T, Macklin WB, Lewis DA, Fox RJ, Rudick R, Mirnics K, Trapp BD. Mitochondrial dysfunction as a cause of axonal degeneration in multiple sclerosis patients. Ann Neurol. 2006;59:478–489.

Fisher E, Lee JC, Nakamura K, Rudick RA. Gray matter atrophy in multiple sclerosis: a longitudinal study. Ann Neurol. 2008;64:255–265.

Friedman JR, Nunnari J. Mitochondrial form and function. Nature. 2014;505:335–343.

Gandhi R, Laroni A, Weiner HL. Role of the innate immune system in the pathogenesis of multiple sclerosis. J Neuroimmunol. 2010;221:7–14.

Gilgun-Sherki Y, Melamed E, Offen D. The role of oxidative stress in the pathogenesis of multiple sclerosis: the need for effective antioxidant therapy. J Neurol. 2004;251:261–268.

Golubovskaya V, Wu L. Different subsets of T cells, memory, effector, functions, and CAR-T immunotherapy. Cancers (Basel). 2016;8:36.

Hajnóczky G, Csordás G, Das S, Garcia-Perez C, Saotome M, Roy SS, Yi M. Mitochondrial calcium signalling and cell death: approaches for assessing the role of mitochondrial Ca2+ uptake in apoptosis. Cell Calcium. 2006;40:553–560.

Hara Y, Yuk F, Puri R, Janssen WG, Rapp PR, Morrison JH. Presynaptic mitochondrial morphology in monkey prefrontal cortex correlates with working memory and is improved with estrogen treatment. Proc Natl Acad Sci U S A. 2014; 111:486–491.

Hauser SL, Goodwin DS. Multiple sclerosis and other demyelinating diseases. In Harrison's Principles of Internal Medicine. Editors: Fauci AS, Braunwald E, Kasper DL, Hauser SL. 17th ed. II. Publishing house: New York: McGraw-Hill Medical 2008. pp. 2611–2621.

Hurwitz BJ. The diagnosis of multiple sclerosis and the clinical subtypes. Ann Indian Acad Neurol. 2009;12:226–230.

Ivannikov MV, Sugimori M, Llinás RR. Synaptic vesicle exocytosis in hippocampal synaptosomes correlates directly with total mitochondrial volume. J Mol Neurosci. 2013;49:223–230.

Jana M, Pahan K. Down-regulation of myelin gene expression in human oligodendrocytes by nitric oxide: implications for demyelination in multiple sclerosis. J Clin Cell Immunol. 2013;4. doi:10.4172/2155-9899.1000157.

Krüger J, Hinttala R, Majamaa K, Remes AM. Mitochondrial DNA haplogroups in early-onset Alzheimer's disease and frontotemporal lobar degeneration. Mol Neurodegener. 2010;5:8.

Lau A, Tymianski M. Glutamate receptors, neurotoxicity and neurodegeneration. Pflugers Arch. 2010;460:525–542.

Lee DH, Gold R, Linker RA. Mechanisms of oxidative damage in multiple sclerosis and neurodegenerative diseases: therapeuthic modulation via fumaric acid esters. Int J Mol Sci. 2012;13:11783–11803.

Li N. CD4+ T cells in atherosclerosis: regulation by platelets. Thromb Haemost. 2013;109:980–990.

Lin MT, Beal MF. Mitochondrial dysfunction and oxidative stress in neurodegenerative diseases. Nature. 2006;442:787–795.

Liu JS, Zhao ML, Brosnan CF, Lee SC. Expression of inducible nitric oxide synthase and nitrotyrosine in multiple sclerosis lesions. Am J Pathol. 2001;158:2057–2066.

Liu X, Hajnóczky G. Altered fusion dynamics underlie unique morphological changes in mitochondria during hypoxia-reoxygenation stress. Cell Death Differ. 2011;18:1561–1572.

Lu F, Selak M, O'Connor J, Croul S, Lorenzana C, Butunoi C, Kalman B. Oxidative damage to mitochondrial DNA and activity of mitochondrial enzymes in chronic active lesions of multiple sclerosis. J Neurol Sci. 2000;177:95–103.

Lu ZH, Chakraborty G, Ledeen RW, Yahya D, Wu G. N-Acetylaspartate synthase is bimodally expressed in microsomes and mitochondria of brain. Brain Res Mol Brain Res. 2004;122:71–78.

Lublin FD. New multiple sclerosis phenotypic classification. Eur Neurol. 2014;72:1–5.

Mahad DH, Trapp BD, Lassmann H. Pathological mechanisms inprogressive multiple sclerosis. Lancet Neurol. 2015;14:183–193.

Manev H, Favaron M, Guidotti A, Costa E. Delayed increase of Ca2+ influx elicited by glutamate: role in neuronal death. Mol Pharmacol. 1989;36:106–112.

Mao P, Reddy PH. Is multiple sclerosis a mitochondrial disease? Biochim Biophys Acta. 2010;1802:66–79.

McFarland R, Tyler RW, Turnbull DM. A neurological perspective on mitochondrial disease. Lancet Neurol. 2010;9:829–840.

Michel L, Touil H, Pikor NB, Gommerman JL, Prat A, Bar-Or A. B cells in the multiple sclerosis central nervous system: trafficking and contribution to CNS -compartmentialized inflammation. Front Immunol. 2015;6:636.

Miller E. Multiple sclerosis. Adv Exp Med Biol. 2012;724:222–238.

Moffett JR, Ross B, Arun P, Madhavarao CN, Namboodiri AM. N-acetylaspartate in the CNS: from neurodiagnostics to neurobiology. Prog Neurobiol. 2007;81:89–131.

Murphy MP. How mitochondria produce reactive oxygen species. Biochem J. 2009;417:1–13.

Nicholls DG, Budd N, Budd SL. Mitochondria and neuronal survival. Physiol Rev. 2000;80:315–360.

Nita M, Grzybowski A. The role of the reactive oxygen species and oxidative stress in the pathomechanism of the age-related ocular diseases and other pathologies of the anterior and posterior eye segments in adults. Oxid Med Cell Longev. 2016;2016:3164734.

Nordengen K, Heuser C, Rinholm JE, Matalon R, Gundersen V. Localisation of N-acetylaspartate in oligodendrocytes/myelin. Brain Struct Funct. 2015;220:899–917.

Palmer AM. Multiple sclerosis and the blood-central narvous system barrier. Cardiovasc Psychiatry Neuro. 2013. doi:10.1155/2013/530356.

Patergnani S, Fossati V, Bonora M, Giorgi C, Marchi S, Missiroli S, Rusielewicz T, Wieckowski MR, Pinton P. Mitochondria in multiple sclerosis: molecular mechanisms of pathogenesis. Int Rev Cell Mol Biol. 2017:49–103. doi:10.1016/bs.ircmb.2016.08.003.

Pathak D, Berthet A, Nakamura K. Energy failure: does It contribute to neurodegeneration? Ann Neurol. 2013;74:506–516.

Shokolenko I, Venediktova N, Bochkareva WGL, Alexeyev MF. Oxidative stress induces degradation of mitochondrial DNA. Nucleic Acids Res. 2009;37:2539–2548.

Solleiro-Villavicencio H, Rivas-Arancibia S. Effect of chronic oxidative stress on neuroinflammatory response mediated by CD4+ T cells in neurodegenerative diseases. Front Cell Neurosci. 2018;12:114.

Su K, Bourdette D, Forte M. Mitochondrial dysfunction and neurodegeneration in multiple sclerosis. Front Physiol. 2013;4:169.

Tranah GJ, Santaniello A, Caillier SJ, D'Alfonso S, Martinelli Boneschi F, Hauser SL, Oksenberg JR. Mitochondrial DNA sequence variation in multiple sclerosis. Neurology. 2015;85:325–330.

van der Walt JM, Dementieva YA, Martin ER, Scott WK, Nicodemus KK, Kroner CC, Welsh-Bohmer KA, Saunders AM, Roses AD, Small GW, Schmechel DE, Murali Doraiswamy P, Gilbert JR, Haines JL, Vance JM, Pericak-Vance MA. Analysis of European mitochondrial haplogroups with Alzheimer disease risk. Neurosci Lett. 2004;365:28–32.

Witherick J, Wilkins A, Scolding N, Kemp K. Mechanisms of oxidative damage in multiple sclerosis and cel therapy approach to tratment. Autoimmune Dis. 2011;2011:164608.

9 Mitochondrial Dysfunction in Chronic Kidney Disease

Maria V. Irazabal, MD and Alfonso Eirin, MD

Division of Nephrology and Hypertension, Mayo Clinic, Rochester, MN, USA

CONTENTS

1 INTRODUCTION

Chronic kidney disease (CKD) remains a growing health concern, defined as the gradual decline in glomerular filtration rate to less than $60 mL/min/1.73m^2$ or the presence of kidney damage for at least 3 months (Levey et al. 2003). CKD affects almost 15% of the U.S. population (Ojo 2014), and its prevalence is expected to increase in the next decade (Hoerger et al. 2015), imparting a significant economic burden (Honeycutt et al. 2013; Levey et al. 2007). The leading causes of CKD are hypertension and diabetes, which are responsible for almost 60% of the cases (Hoerger et al. 2015); CKD may also arise from several disease conditions, including inherited kidney disorders, glomerulonephritis, infections, or obstructive uropathy.

Patients with CKD gradually progress to end-stage renal disease (ESRD), requiring renal replacement therapy with dialysis or kidney transplantation. Progression of CKD is associated with several complications, including increased incidence of cardiovascular disease, anemia, and metabolic bone disease (Thomas et al. 2008). Eventually, all these conditions can progress toward ESRD, increasing overall mortality rates. CKD contributed to more than 10% of all deaths in 2015, but these numbers are projected to increase within the next 10 years (Hoerger et al. 2015). Elucidating the mechanisms responsible for CKD progression may assist in developing adequate interventions to reduce morbidity and mortality in these patients.

The kidney is a highly energetic organ that regulates the body fluid composition through filtering and reabsorbing materials. Reabsorption activity within the proximal and distal tubules is mainly due to adenosine triphosphate (ATP)-dependent active transport. To accomplish this, renal tubular cells are equipped with a significant number of mitochondria (Pfaller and Rittinger 1980), which also modulate other important cellular pathways, including proliferation, apoptosis, oxidative stress, and calcium signaling (McFarland et al. 2007). Therefore, mitochondrial injury can have an important impact on overall renal function.

Over the past few years, experimental studies have shown that mitochondrial structural abnormalities and dysfunction coexist with many forms of CKD (Che et al. 2014). Importantly, results from these studies prompted development of novel mitochondria-targeted interventions to ameliorate renal injury and dysfunction, some of which are currently tested in clinical trials in patients with CKD (Eirin et al. 2017a; Tabara et al. 2014). The purpose of this chapter is to discuss distinct forms of mitochondrial injury in the context of CKD. Specifically, we will focus on hypertension, which occurs in 70–90% of patients with CKD, and polycystic kidney disease (PKD), an inherited disorder that represents the fourth leading cause of kidney failure worldwide. We will further discuss the primary mechanisms of action and potential renoprotective effects of emerging therapies targeting mitochondria in CKD.

2 MECHANISMS OF MITOCHONDRIAL INJURY IN HYPERTENSIVE RENAL DISEASE

Hypertension is an important risk factor for CKD that affects almost 80 million adults in the United States, and its prevalence is expected to increase to 7.2% by 2030 (Dorans et al. 2018). Patients with hypertension have an increased risk for cardiovascular disease and premature mortality (Collaborators 2016; Danaei et al. 2009; Mills et al. 2016; Murray et al. 2013), associated with a significant economic burden, reflected in annual estimated direct costs of around $50 billion dollars (Merai et al. 2016). Approximately 70–90% of individuals with hypertension have concomitant CKD, and its prevalence increases as renal function declines (Tedla et al. 2011), currently representing the second cause to ESRD in the United States (Saran et al. 2018). Therefore, efforts to develop novel therapeutic interventions to prevent hypertension-induced renal damage are urgently needed.

Renal mitochondrial damage is a common finding in several murine models of hypertension (Table 1). Renal tubular epithelial and glomerular cells from salt-sensitive and spontaneously hypertensive rats exhibit mitochondrial injury, reflected in decreased membrane potential and increased production of reactive oxygen species (ROS) (de Cavanagh et al. 2006; Zheleznova et al. 2012). Similarly, hypertension impairs mitochondrial membrane potential, decreases complex-I activity and gluthatione activity, and increases production of ROS in rat tubular cells (Marquez-Ramirez et al. 2018). Furthermore, transcriptomic analysis of the kidneys of hypertensive rats reveals decreased expression of several genes encoding for mitochondrial proteins involved in ATP production (Evans et al. 2018), linking mitochondrial injury to renal damage in hypertension.

Mitochondrial dysfunction has been also reported in models of obesity-induced hypertension. Obesity is an epidemic condition that increases the risk of developing hypertension by several mechanisms, including sodium retention and activation of the sympathetic nervous system or the renin-angiotensin-aldosterone system (RAAS) (Re 2009). Obesity-induced hypertension in swine induces mitochondrial structural abnormalities in the renal medulla, associated with mitochondria-dependent apoptosis and oxidative stress, and decreased ATP synthesis (Eirin et al. 2017b). Importantly, obesity-induced mitochondrial damage in swine contributed to renal microvascular loss, endothelial dysfunction, glomerular hyperfiltration, and medullary hypoxia (Eirin et al. 2017b, 2018), implicating mitochondrial damage in renal injury in experimental hypertension.

Renovascular disease (RVD), an important cause of renal dysfunction in the aging population (Eirin and Lerman 2013), may also induce significant mitochondrial injury. Studies in RVD pigs have shown that the post-stenotic kidney mitochondria exhibit important structural abnormalities (Figure 1), associated with impaired energy production (Eirin et al. 2014a). In addition, RVD decreases mitochondrial biogenesis in the post-stenotic pig kidney (Eirin et al. 2012) and upregulates mitophagy in the clipped rat kidney (Fedorova et al. 2013), indicating hypertension-induced changes in mitochondrial homeostasis. Importantly, mitochondrial protection ameliorated renal injury and dysfunction in the ischemic kidney (Eirin et al. 2012, 2014a), positioning mitochondria as an important therapeutic target in RVD-induced renal damage.

Mechanisms implicated in hypertension-induced renal damage include activation of the RAAS, increased production of ROS, extracellular matrix turnover, and fibrosis (Eirin et al. 2015). Angiotensin-II type-2 receptors are located in the inner mitochondrial membrane in renal tubular cells, and its activation modulates mitochondrial respiration, membrane potential, and ROS generation (Abadir et al. 2011). Furthermore, activation of these receptors stimulates production of mitochondrial ROS (Dikalov et al. 2014; Griendling et al. 1994), and induces apoptosis in renal tubular cells (Bhaskaran et al. 2003).

TABLE 1
Evidence of renal mitochondrial injury and dysfunction in hypertension

Reference	Species/model	Approach	Findings
de Cavanagh et al. (2006)	Spontaneously hypertensive rats	Fluorimetric methods/western blotting	Decreased bioenergetics and antioxidant defences
Zheleznova et al. (2012)	Dahl salt-sensitive rats	Proteomic analysis	Decreased bioenergetics
Eirin et al. (2017b)	Swine obesity-induced hypertension	Electron microscopy, colorimetric/fluorimetric methods/immunofluorescent staining/western blotting	Structural damage/Increased apoptosis and oxidative stress/Decreased ATP synthesis and antioxidant defenses
Eirin, et. al. (2014a)	Swine RVD	Immunofluorescent staining/western blotting	Structural damage/Decreased energy production
Eirin et al. (2012)	Swine RVD	Immunofluorescent staining/western blotting	Decreased mitochondrial biogenesis/Increased mitophagy
Fedorova et al. (2013)	Goldblatt's 2 kidney 1 clip rat	Immunofluorescent staining/western blotting	Decreased mitochondrial mass, protein, and DNA level
Marquez-Ramirez et al. (2018)	Hypertensive rats induced with L-NAME	Colorimetric/fluorimetric methods	Decreased membrane potential and glutathione activity/Increased ROS levels
Evans et al. (2018)	Dahl salt-sensitive rats	RNA sequencing analysis	Decreased expression of genes encoding for mitochondrial respiratory proteins

ATP: Adenosine triphosphate, SOD: Superoxide dismutase, ROS: Reactive oxygen species, RVD: Renovascular disease. L-NAME: Nω-nitro-l-arginine methyl ester.

Studies in hypertensive salt-sensitive rats have shown that renal tubular cells display mitochondria-dependent apoptosis and severe renal injury, establishing a link between mitochondrial injury and renal damage in hypertensive nephrosclerosis (Ying and Sanders 2001).

Angiotensin-II may also increase cytosolic levels of ROS, which may in turn oxidize mitochondrial proteins and lipids (Andrukhiv et al. 2006), and induce mutations in mitochondrial-DNA (Navarro 2004), preventing transcription of mitochondrial proteins implicated in ATP synthesis (Paradies et al. 2002). In addition, ROS can cause peroxidation of the inner mitochondrial membrane phospholipid cardiolipin, favoring formation of the mitochondrial permeability transition pore, and the subsequent release of mitochondrial ROS and cytochrome-c to the cytosol (Baines 2009; Szeto 2014). Cytosolic cytochrome-c may activate the caspase-3 and -9 apoptotic pathway and contribute to cellular oxidative stress (Niimi et al. 2012). Concomitantly, ROS-induced disruption of mitochondrial membranes may interfere with mitochondrial antioxidant defenses, exacerbating cellular oxidative stress.

Renal fibrosis may also contribute to mitochondrial injury in hypertension-induced renal damage. Accumulating evidence indicates that matrix metalloproteinases can directly damage mitochondrial-DNA and promote apoptosis (Santos et al. 2013). Notably, mitoprotection attenuates renal fibrosis in experimental hypertension (Eirin et al. 2012, 2014b; Singh et al. 2005), implying that mitochondrial damage secondary to hypertension may be implicated in the pathogenesis of renal fibrosis. Taken together, these mechanisms link renal mitochondrial injury and dysfunction to experimental hypertension, but a cause-and-effect relationship warrants further investigation.

3 MITOCHONDRIAL DYSFUNCTION IN THE PATHOPHYSIOLOGY OF PKD

PKD is a group of monogenic disorders characterized by progressive development of tubular cysts, associated with enlargement of the kidneys and destruction of the renal parenchyma (Irazabal and Torres 2013). The most common forms of PKD constitute autosomal dominant PKD (ADPKD) and autosomal recessive PKD (ARPKD). ADPKD typically presents in adults and is caused most commonly by mutations in either *PKD1* or *PKD2* genes, whereas ARPKD primarily affects children and results from mutations in the polycystic kidney and hepatic disease 1 (*PKHD1*) gene (Irazabal and Torres 2013). Both ADPKD and ARPKD progress toward ESRD, requiring renal replacement therapy. Furthermore, these patients present with a myriad of extra-renal manifestations including cysts in other organs (liver, pancreas), congenital hepatic fibrosis (mainly ARPKD), intracranial aneurysms and abnormalities in the cardiovascular, gastrointestinal, and genitourinary systems.

Increasing evidence supports the notion that PKD is associated with structural mitochondrial alterations (Table 2). Human epithelial cells obtained from patients with PKD and ESRD have fewer mitochondria and decreased cytochrome oxidase activity compared to normal tubular cells (Wilson et al. 1986). Ultrastructural analysis of cystic epithelial cells from mice with PKD revealed that in the early stages of cyst expansion mitochondrial number is well preserved (Thomson et al. 2003), but as cyst expansion proceeds, they become much less prominent, and are almost absent in the final stages of cyst expansion. Studies using comparative proteomics analysis identified a large number of mitochondrial proteins involved in electron transport, energy production, and antioxidant defenses downregulated in ADPKD and ARPKD cystic kidney tissues compared to normal kidney tissues (Li et al. 2012; Liu et al. 2008). Human ADPKD cells treated with ouabain, a hormone that enhances several characteristics of the ADPKD cystic phenotype, show increased release of cytochrome-c from the mitochondria to the cytoplasm, and activation of the caspase-3 pathway, suggesting mitochondrial-dependent apoptosis (Venugopal and Blanco 2016). In agreement, treatment with peptides that mimics the mitochondria-derived activator of caspase, a mitochondrial protein that binds to and antagonizes inhibitors of apoptosis (Du et al. 2000), reduced cyst growth in PKD mice (Fan et al. 2013). Therefore, these observations implicate mitochondrial structural abnormalities and dysfunction in the pathogenesis of PKD.

Importantly, mitochondria regulate several cellular processes implicated in cyst formation and progression in PKD, including calcium signaling (Mangolini et al. 2016; Torres and Harris 2014b), cell proliferation, apoptosis (Goilav 2011), energy metabolism (Rowe et al. 2013), oxidative stress (Maser et al. 2002), and endothelial function (Klawitter et al. 2014) (Figure 2).

The polycystins, PC1 (Hughes et al. 1995) and PC2 (Hayashi et al. 1997), protein products of *PKD1* and *PKD2* respectively (Kottgen 2007), localize principally to the plasma membranes of the primary cilia, but also in a number of other locations such as the endoplasmic reticulum (ER) (likely a major site of PC2). PC1 participates in cell-cell and cell-matrix interactions, whereas PC2 is a nonselective cation channel with great permeability to calcium (Cai et al. 1999). PC1 and PC2 interact with several proteins involved in calcium signaling, including inositol 1,4,5-trisphosphate receptor (Li et al. 2005, 2009; Santoso et al. 2011), ryanodine-2 (Anyatonwu et al. 2007), and others (Kottgen et al. 2008; Tsiokas et al. 1999), modulating calcium release from intracellular deposits (Nauli et al. 2003, 2006). PC1 and PC2 dysfunction activates calcium-sensitive adenylyl cyclases, increasing cyclic adenosine monophosphate (cAMP) (Torres and Harris 2014b). This in turn activates several pathways that modulate cell migration, differentiation, proliferation, and apoptosis, including the mammalian target of rapamycin (mTOR) and extracellular-signal-regulated kinases (ERK).

TABLE 2

Evidence of renal mitochondrial injury and dysfunction in PKD

Reference	Species/model	Approach	Findings
Wilson et al. (1986)	Human PKD tubular epithelial cells	Electron microscopy and cytochrome oxidase staining	Decreased mitochondrial density and cytochrome oxidase activity
Thomson et al. (2003)	$Pkd2^{WS25/-}$ cystic epithelial cells	Electron microscopy	Decreased mitochondrial density with cyst size
Liu et al. (2008)	Human cystic kidney tissue ADPKD	Comparative proteomics	A large number of mitochondrial enzymes were downregulated
Li et al. (2012)	Human cystic kidney tissue ARPKD	Comparative proteomics	Most of the identified proteins were associated with mitochondria and/or localized in mitochondria
Fan et al. (2013)	Pkd1 null MEK cells, human cystic epithelial cells, $Pkd1^{flox/flox}$:Ksp-Cre mice	Western blot analysis	Smac-mimetic-induced TNFⓍ-dependent cystic apoptosis
Venugopal and Blanco (2016)	Human PKD tubular epithelial cells	Immunoblot and immunocytochemistry analysis	Ouabain-induced mitochondrial cytochrome-c release and apoptosis
Padovano et al. (2016)	Pkd1$^{-/-}$ and Pkd1$^{flox/-}$ mouse renal proximal tubule cells	Seahorse Analyzer	Pkd1$^{-/-}$ cells have reduced OCR and mitochondrial Ca^{2+} uptake

PKD: Polycystic kidney disease, ARPKD: Autosomal recessive polycystic kidney disease, ADPKD: Autosomal dominant polycystic kidney disease, cAMP: cyclic adenosine monophosphate, TNF: Tumor necrosis factor, OCR: Oxygen consumption rate, MEK: Mouse embryonic kidney.

Recent studies suggest that deficiency of PC1 is associated with impaired mitochondrial calcium uptake (Padovano et al. 2016) due to alterations in the expression of mitochondria associated membranes, which regulate physiological functions between ER and mitochondria (Patergnani et al. 2011). Interestingly, the cAMP-protein kinase-A (PKA) complex phosphorylates the translocase of the outer membrane complex (Schmidt et al. 2011), impairing the import of metabolite carriers. However, whether the increase in cAMP-PKA signaling observed PKD is associated defective protein import remains to be elucidated.

Mitochondria are also important regulators of cell proliferation and apoptosis. One of the major phenotypic changes associated with PKD is the excessive cell proliferation, which partly results from activation of proliferative signaling pathways, including mTOR and ERK (Distefano et al. 2009; Spirli et al. 2010; Yamaguchi et al. 2000). ERK1/2 are trafficked into the mitochondrion and subsequently activated by cysteine oxidations (Antico Arciuch et al. 2012). ERK signaling has been shown to activate mTOR (Distefano et al. 2009; Spirli et al. 2010), and mTOR complex-1 phosphorylation regulates mitochondrial biogenesis by upregulating the expression of peroxisome proliferator-activated receptor gamma coactivator (PGC)-1α (Peterson et al. 2011; Wullschleger et al. 2006). Therefore, mTOR and ERK may represent potential mechanisms by which mitochondria modulate cell proliferation and survival in PKD. Yet, upregulation of other pro-proliferative transcription factors, including cAMP response element-binding protein (Aguiari et al. 2012), paired box gene-2 (Qin et al. 2012), and signal transducer and activator of transcription-3 (Talbot et al. 2014; Torres and Harris 2014a) have been linked to cAMP-PKA-mediated cell proliferation in PKD.

Tubular cell apoptosis is an important pathological feature of PKD and has been reported both in experimental models and humans (Ali et al. 2000). B-cell lymphoma-2 deficiency in mice is associated with increased apoptosis and severe PKD-like changes in the proximal and distal tubular segments (Nakayama et al. 1994; Veis et al. 1993). Contrarily, renal expression of Bcl-2-associated X protein remains unchanged in PKD rat kidneys (Ecder et al. 2002), suggesting that an imbalance between pro- and anti-apoptotic factors may contribute to programed cell death in experimental PKD.

PKD cells have been reported to have higher content of ATP compared to wildtype controls, which may contribute to fluid accumulation and cyst enlargement through autocrine and paracrine ATP signaling stimulation. ADPKD cells release significant amounts of ATP into the cyst fluid (Wilson et al. 1999) and have slower rates of ATP degradation due to decreased activity of ATPases and other hydrolyzing enzymes (Solini et al. 2015). Interestingly, mitochondrial ATP synthase inhibitors have minor effects in PKD cells, implying that glucose dependent alternative pathway may be responsible for ATP production (Rowe et al. 2013). This condition, known as "Warburg effect" is less effective than mitochondrial respiration in generating ATP (Vander Heiden et al. 2009), and may reflect mitochondrial dysfunction (Senyilmaz and Teleman 2015). In line with this contention, mutations in several genes encoding for proteins involved in the mitochondrial respiratory chain have been previously reported in patients with PKD (Bohm et al. 1982; Wilson et al. 1989). Nevertheless, additional mitochondrial functional and structural studies are needed to define the role of mitochondrial energy metabolism in PKD.

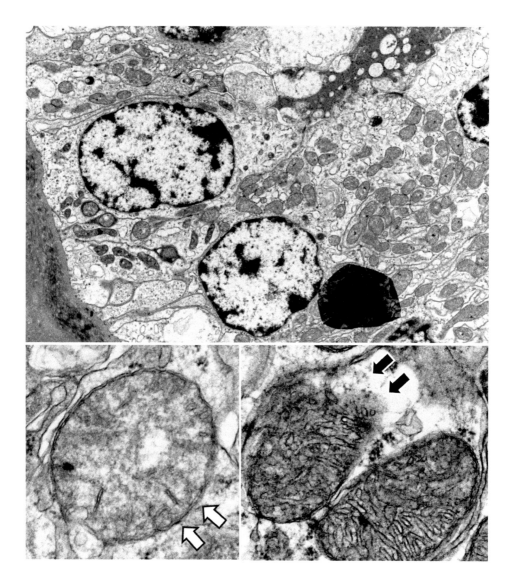

FIGURE 1 Mitochondrial structural abnormalities in swine renovascular disease. Representative transmission electron microscopy images showing renal tubular cell structural mitochondrial abnormalities in the post-stenotic kidney of pigs with renovascular disease. Renal tubular cell mitochondria are swollen with loss of cristae membranes and matrix density (white arrows), and disruption of the outer mitochondrial membrane with release of matrix content to the cytosol (black arrows).

Experimental and clinical data suggest that oxidative injury plays an important role in the pathogenesis of PKD (Cowley et al. 1996; Hochman et al. 2000; Torres et al. 1994). Circulating isoprostanes are elevated in patients with ADPKD (Menon et al. 2011), and antioxidant therapy in PKD mice prevents progressive renal dysfunction (Nagao et al. 2000). Likewise, expression of the mitochondrial superoxide dismutase (SOD)-2 is downregulated in PKD mice (Maser et al. 2002) and patients with ADPKD (Menon et al. 2011). Furthermore, glutathione peroxidase activity, which neutralizes mitochondrial hydrogen peroxide, is blunted in cystic kidneys of PKD mice and rats (Maser et al. 2002), indicating impaired mitochondrial antioxidant defenses.

Renal vascular abnormalities are commonly seen in experimental and clinical PKD, and partly result from stretching and compression of normal renal tissue by cyst expansion. Renal vascular damage in PKD is associated with defective angiogenesis, microvascular loss (Bello-Reuss et al. 2001; Franchi et al. 2015; Wei et al. 2006; Xu et al. 2013), and endothelial dysfunction (Peterson et al. 2013; Wang et al. 2000, 2003). Endothelial cell mitochondria only comprise less than 5% of total cellular volume and have a small contribution to energy production, yet these organelles are important regulators of endothelial function (Szewczyk et al. 2015). Mechanisms triggering endothelial cell mitochondrial injury in PKD may include activation of the nicotinamide adenine dinucleotide phosphate oxidase-4 (Graham et al. 2010; Van Buul et al. 2005), which oxidizes nitric oxide (NO) and decreases its availability (Forstermann

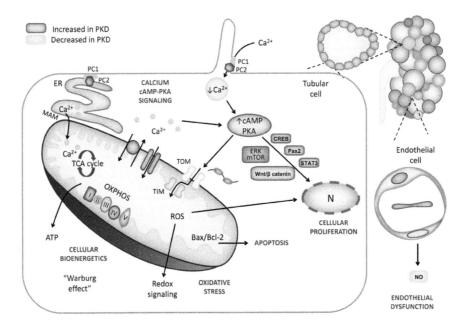

FIGURE 2 Schematic of proposed mitochondria-mediated mechanisms of disease progression in PKD. Mitochondria are primarily responsible for cellular respiration and ATP production, but also modulate several functions of the cell including calcium signaling, cell proliferation, survival, apoptosis, production of oxidative stress, and endothelial function, all of which have been implicated in the pathogenesis of PKD. PC: polycystin, cAMP: cyclic adenosine monophosphate, mTOR: mammalian target of rapamycin, ERK: extracellular-signal-regulated kinases (ERK), MAM: mitochondria associated membranes, ER: endoplasmic reticulum, TOM: translocase of the outer membrane complex, CREB: cAMP response element-binding protein, Pax2: paired box gene-2, STAT3:signal transducer and activator of transcription-3, Bcl2: B-cell lymphoma-2, Bax: Bcl-2-associated X protein, NO: nitric oxide.

and Munzel 2006), and impaired mitochondrial antioxidant defenses (Maser et al. 2002; Menon et al. 2011; Wu et al. 2016). In addition, activation of angiotensin-II-type-II receptors in the inner mitochondrial membrane (Tkachenko et al. 2013), which are coupled to mitochondrial NO production, can increase generation of mitochondrial ROS (Abadir et al. 2011), creating a vicious cycle of oxidative stress and endothelial dysfunction. Collectively, these observations suggest that mitochondria may represent an attractive target in PKD. Nevertheless, a cause-effect relationship and the clinical importance of mitochondrial injury in the development and progression of PKD remain to be confirmed.

4 MITOCHONDRIA-TARGETED THERAPIES FOR CKD

Given the implication of mitochondria in renal physiology and in the pathogenesis of leading causes of CKD, it is reasonable to postulate that targeted interventions that preserve these organelles may alleviate progressive renal injury and dysfunction. Over the last couple of years, novel therapeutic approaches to enhance mitochondrial function have shown promising results in both experimental and clinical settings (Eirin et al. 2017a). This includes mitochondrial antioxidants, biogenesis activators, cardiolipin-protective drugs, and modulators of fatty acid oxidation.

Delivery of antioxidants agents that selectively accumulate in mitochondria may be beneficial in CKD. Previous studies have shown that administration of antioxidant SOD-2 mimetics attenuates oxidative stress and tissue injury in rats with hypertensive-induced renal damage (Peixoto et al. 2009), imitating the effects of antioxidants vitamins in rats with hypertension-induced renal dysfunction (Seifi et al. 2009). Pharmaceutical activation of mitochondrial biogenesis may also preserve renal structure and function in experimental CKD. β2-adrenergic receptor agonists promote mitochondrial biogenesis in rat tubular cells by upregulating PGC-1α (Wills et al. 2012). Activation of PGC-1 α increases renal mitochondrial density and ameliorates necrosis in rats with chronic ischemic injury (Fedorova et al. 2013), suggesting that biogenesis activators might preserve renal structure in experimental CKD.

Strategies aimed to preserve the integrity of cardiolipin have shown important reno-protective properties in experimental CKD. Cardiolipin content is significantly reduced in the kidneys of pigs with RVD (Eirin et al. 2014a, 2016a) and obesity-induced renal injury (Eirin et al. 2017b, 2018), but its restoration attenuates renal structural and functional damage. Furthermore, cardiolipin protection before and during renal angioplasty preserves renal function both in pigs (Eirin et al. 2012) and humans (Saad et al. 2017), underscoring the potential of this intervention to improve kidney outcomes after revascularization in RVD. Importantly, all patients tolerated this therapy and no adverse clinical effects have been identified, supporting the safety and efficacy of cardiolipin protection in RVD.

Finally, therapies that modulate fatty acid oxidation may exert beneficial effects in CKD. Treatment with peroxisome proliferator-activated receptor-α agonists, which regulate the expression of the rate-limiting enzyme in fatty acid oxidation carnitine palmitoyltransferase-1 (Szeto 2017), attenuates hypertension-induced renal damage in rats (Hou et al. 2010). However, carefully designed prospective experimental and clinical studies are needed to determine the optimal utilization of this intervention to ultimately reduce complications and slow progression of CKD.

5   CONCLUSION

CKD remains a major challenge in nephrology and a growing health problem in the U.S. and worldwide. Emerging experimental evidence suggests that mitochondrial abnormalities and dysfunction may be implicated in the pathogenesis of common etiologies of CKD like hypertension and PKD. Mitochondria-targeting is also gaining momentum as a novel therapy for several forms of renal disease, and is currently being tested in clinical trials. Although preliminary results from these studies implicate mitochondrial dysfunction in renal disease progression, the extent to which actual mitochondrial damage modulates renal injury remains to be fully elucidated. Therefore, further experimental and clinical studies are needed to explore the mechanisms regulating mitochondrial function in CKD and confirm the safety and efficacy of mitoprotection to prevent disease progression in these patients.

ACKNOWLEDGMENTS

This work was partly supported by the National Institutes of Health (DK106427 and DK118391).

REFERENCES

Abadir PM, Foster DB, Crow M, Cooke CA, Rucker JJ, Jain A, Smith BJ, Burks TN, Cohn RD, Fedarko NS et al. (2011) Identification and characterization of a functional mitochondrial angiotensin system. Proc Natl Acad Sci U S A 108: 14849–54.

Aguiari G, Bizzarri F, Bonon A, Mangolini A, Magri E, Pedriali M, Querzoli P, Somlo S, Harris PC, Catizone L et al. (2012) Polycystin-1 regulates amphiregulin expression through CREB and AP1 signalling: Implications in ADPKD cell proliferation. J Mol Med (Berl) 90: 1267–82.

Ali SM, Wong VY, Kikly K, Fredrickson TA, Keller PM, DeWolf WE Jr., Lee D, Brooks DP (2000) Apoptosis in polycystic kidney disease: Involvement of caspases. Am J Physiol Regul Integr Comp Physiol 278: R763–9.

Andrukhiv A, Costa AD, West IC, Garlid KD (2006) Opening mitoKATP increases superoxide generation from complex I of the electron transport chain. Am J Physiol Heart Circ Physiol 291: H2067–74.

Antico Arciuch VG, Elguero ME, Poderoso JJ, Carreras MC (2012) Mitochondrial regulation of cell cycle and proliferation. Antioxid Redox Signal 16: 1150–80.

Anyatonwu GI, Estrada M, Tian X, Somlo S, Ehrlich BE (2007) Regulation of ryanodine receptor-dependent calcium signaling by polycystin-2. Proc Natl Acad Sci U S A 104: 6454–59.

Baines CP (2009) The molecular composition of the mitochondrial permeability transition pore. J Mol Cell Cardiol 46: 850–7.

Bello-Reuss E, Holubec K, Rajaraman S (2001) Angiogenesis in autosomal-dominant polycystic kidney disease. Kidney Int 60: 37–45.

Bhaskaran M, Reddy K, Radhakrishanan N, Franki N, Ding G, Singhal PC (2003) Angiotensin II induces apoptosis in renal proximal tubular cells. Am J Physiol Renal Physiol 284: F955–65.

Bohm N, Uy J, Kiessling M, Lehnert W (1982) Multiple acyl-CoA dehydrogenation deficiency (glutaric aciduria type II), congenital polycystic kidneys, and symmetric warty dysplasia of the cerebral cortex in two newborn brothers. II. Morphology and pathogenesis. Eur J Pediatr 139: 60–5.

Cai Y, Maeda Y, Cedzich A, Torres VE, Wu G, Hayashi T, Mochizuki T, Park JH, Witzgall R, Somlo S (1999) Identification and characterization of polycystin-2, the PKD2 gene product. J Biol Chem 274: 28557–65.

Che R, Yuan Y, Huang S, Zhang A (2014) Mitochondrial dysfunction in the pathophysiology of renal diseases. Am J Physiol Renal Physiol 306: F367–78.

Collaborators GBDRF (2016) Global, regional, and national comparative risk assessment of 79 behavioural, environmental and occupational, and metabolic risks or clusters of risks, 1990–2015: A systematic analysis for the global burden of disease study 2015. Lancet 388: 1659–724.

Cowley BD Jr., Grantham JJ, Muessel MJ, Kraybill AL, Gattone VH 2nd (1996) Modification of disease progression in rats with inherited polycystic kidney disease. Am J Kidney Dis 27: 865–79.

Danaei G, Ding EL, Mozaffarian D, Taylor B, Rehm J, Murray CJ, Ezzati M (2009) The preventable causes of death in the United States: Comparative risk assessment of dietary, lifestyle, and metabolic risk factors. PLoS Med 6: e1000058.

de Cavanagh EM, Toblli JE, Ferder L, Piotrkowski B, Stella I, Inserra F (2006) Renal mitochondrial dysfunction in spontaneously hypertensive rats is attenuated by losartan but not by amlodipine. Am J Physiol Regul Integr Comp Physiol 290: R1616–25.

Dikalov SI, Nazarewicz RR, Bikineyeva A, Hilenski L, Lassegue B, Griendling KK, Harrison DG, Dikalova AE (2014) Nox2-induced production of mitochondrial superoxide in angiotensin II-mediated endothelial oxidative stress and hypertension. Antioxid Redox Signal 20: 281–94.

Distefano G, Boca M, Rowe I, Wodarczyk C, Ma L, Piontek KB, Germino GG, Pandolfi PP, Boletta A (2009) Polycystin-1 regulates extracellular signal-regulated kinase-dependent phosphorylation of tuberin to control cell size through mTOR and its downstream effectors S6K and 4EBP1. Mol Cell Biol 29: 2359–71.

Dorans KS, Mills KT, Liu Y, He J (2018) Trends in prevalence and control of hypertension according to the 2017 American College of Cardiology/American Heart Association (ACC/AHA) guideline. J Am Heart Assoc 7.

Du C, Fang M, Li Y, Li L, Wang X (2000) Smac, a mitochondrial protein that promotes cytochrome c-dependent caspase activation by eliminating IAP inhibition. Cell 102: 33–42.

Ecder T, Melnikov VY, Stanley M, Korular D, Lucia MS, Schrier RW, Edelstein CL (2002) Caspases, Bcl-2 proteins and apoptosis in autosomal-dominant polycystic kidney disease. Kidney Int 61: 1220–30.

Eirin A, Ebrahimi B, Kwon SH, Fiala JA, Williams BJ, Woollard JR, He Q, Gupta RC, Sabbah HN, Prakash YS et al. (2016a) Restoration of mitochondrial cardiolipin attenuates cardiac damage in swine renovascular hypertension. J Am Heart Assoc 5.

Eirin A, Ebrahimi B, Zhang X, Zhu XY, Woollard JR, He Q, Textor SC, Lerman A, Lerman LO (2014a) Mitochondrial protection restores renal function in swine atherosclerotic renovascular disease. Cardiovasc Res 103: 461–72.

Eirin A, Hedayat AF, Ferguson CM, Textor SC, Lerman A, Lerman LO (2018) Mitoprotection preserves the renal vasculature in porcine metabolic syndrome. Exp Physiol 103: 1020–9.

Eirin A, Lerman A, Lerman LO (2015) Mitochondria: A pathogenic paradigm in hypertensive renal disease. Hypertension 65: 264–70.

Eirin A, Lerman A, Lerman LO (2016b) The emerging role of mitochondrial targeting in kidney disease. Handb Exp Pharmacol.

Eirin A, Lerman A, Lerman LO (2017) The emerging role of mitochondrial targeting in kidney disease. Handb Exp Pharmacol 240: 229–50.

Eirin A, Lerman LO (2013) Darkness at the end of the tunnel: Poststenotic kidney injury. Physiology (Bethesda) 28: 245–53.

Eirin A, Li Z, Zhang X, Krier JD, Woollard JR, Zhu XY, Tang H, Herrmann SM, Lerman A, Textor SC et al. (2012) A mitochondrial permeability transition pore inhibitor improves renal outcomes after revascularization in experimental atherosclerotic renal artery stenosis. Hypertension 60: 1242–9.

Eirin A, Williams BJ, Ebrahimi B, Zhang X, Crane JA, Lerman A, Textor SC, Lerman LO (2014b) Mitochondrial targeted peptides attenuate residual myocardial damage after reversal of experimental renovascular hypertension. J Hypertens 32: 154–165.

Eirin A, Woollard JR, Ferguson CM, Jordan KL, Tang H, Textor SC, Lerman A, Lerman LO (2017b) The metabolic syndrome induces early changes in the swine renal medullary mitochondria. Transl Res 184: 45–56 e9.

Evans LC, Dayton A, Yang C, Liu P, Kurth T, Ahn KW, Komas S, Stingo FC, Laud PW, Vannucci M et al. (2018) Transcriptomic analysis reveals inflammatory and metabolic pathways that are regulated by renal perfusion pressure in the outer medulla of Dahl-S rats. Physiol Genomics 50: 440–7.

Fan LX, Zhou X, Sweeney WE Jr., Wallace DP, Avner ED, Grantham JJ, Li X (2013) Smac-mimetic-induced epithelial cell death reduces the growth of renal cysts. J Am Soc Nephrol 24: 2010–22.

Fedorova LV, Sodhi K, Gatto-Weis C, Puri N, Hinds TD Jr., Shapiro JI, Malhotra D (2013) Peroxisome proliferator-activated receptor delta agonist, HPP593, prevents renal necrosis under chronic ischemia. PLoS One 8: e64436.

Forstermann U, Munzel T (2006) Endothelial nitric oxide synthase in vascular disease: From marvel to menace. Circulation 113: 1708–14.

Franchi F, Peterson KM, Xu R, Miller B, Psaltis PJ, Harris PC, Lerman LO, Rodriguez-Porcel M (2015) Mesenchymal stromal cells improve renovascular function in polycystic kidney disease. Cell Transplant 24: 1687–98.

Goilav B (2011) Apoptosis in polycystic kidney disease. Biochim Biophys Acta 1812: 1272–80.

Graham KA, Kulawiec M, Owens KM, Li X, Desouki MM, Chandra D, Singh KK (2010) NADPH oxidase 4 is an oncoprotein localized to mitochondria. Cancer Biol Ther 10: 223–31.

Griendling KK, Minieri CA, Ollerenshaw JD, Alexander RW (1994) Angiotensin II stimulates NADH and NADPH oxidase activity in cultured vascular smooth muscle cells. Circ Res 74: 1141–8.

Hayashi T, Mochizuki T, Reynolds DM, Wu G, Cai Y, Somlo S (1997) Characterization of the exon structure of the polycystic kidney disease 2 gene (PKD2). Genomics 44: 131–6.

Hochman A, Liang H, Offen D, Melamed E, Sternin H (2000) Developmental changes in antioxidant enzymes and oxidative damage in kidneys, liver and brain of bcl-2 knockout mice. Cell Mol Biol (Noisy-le-grand) 46: 41–52.

Hoerger TJ, Simpson SA, Yarnoff BO, Pavkov ME, Rios Burrows N, Saydah SH, Williams DE, Zhuo X (2015) The future burden of CKD in the United States: A simulation model for the CDC CKD initiative. Am J Kidney Dis 65: 403–11.

Honeycutt AA, Segel JE, Zhuo X, Hoerger TJ, Imai K, Williams D (2013) Medical costs of CKD in the medicare population. J Am Soc Nephrol 24: 1478–83.

Hou X, Shen YH, Li C, Wang F, Zhang C, Bu P, Zhang Y (2010) PPARalpha agonist fenofibrate protects the kidney from hypertensive injury in spontaneously hypertensive rats via inhibition of oxidative stress and MAPK activity. Biochem Biophys Res Commun 394: 653–9.

Hughes J, Ward CJ, Peral B, Aspinwall R, Clark K, San Millan JL, Gamble V, Harris PC (1995) The polycystic kidney disease 1 (PKD1) gene encodes a novel protein with multiple cell recognition domains. Nat Genet 10: 151–60.

Irazabal MV, Torres VE (2013) Experimental therapies and ongoing clinical trials to slow down progression of ADPKD. Curr Hypertens Rev 9: 44–59.

Klawitter J, Reed-Gitomer BY, McFann K, Pennington A, Klawitter J, Abebe KZ, Klepacki J, Cadnapaphornchai MA, Brosnahan G, Chonchol M et al. (2014) Endothelial dysfunction and oxidative stress in polycystic kidney disease. Am J Physiol Renal Physiol 307: F1198–206.

Kottgen M (2007) TRPP2 and autosomal dominant polycystic kidney disease. Biochim Biophys Acta 1772: 836–50.

Kottgen M, Buchholz B, Garcia-Gonzalez MA, Kotsis F, Fu X, Doerken M, Boehlke C, Steffl D, Tauber R, Wegierski T et al. (2008) TRPP2 and TRPV4 form a polymodal sensory channel complex. J Cell Biol 182: 437–47.

Levey AS, Atkins R, Coresh J, Cohen EP, Collins AJ, Eckardt KU, Nahas ME, Jaber BL, Jadoul M, Levin A et al. (2007) Chronic kidney disease as a global public health problem: Approaches and initiatives - a position statement from kidney disease improving global outcomes. Kidney Int 72: 247–59.

Levey AS, Coresh J, Balk E, Kausz AT, Levin A, Steffes MW, Hogg RJ, Perrone RD, Lau J, Eknoyan G et al. (2003) National Kidney Foundation practice guidelines for chronic kidney disease: Evaluation, classification, and stratification. Ann Intern Med 139: 137–47.

Li QW, Lv XY, You Y, Sun H, Liu XY, Ai JZ, Tan RZ, Chen TL, Liu LH, Duan WW et al. (2012) Comparative proteomic analysis suggests that mitochondria involve in autosomal recessive polycystic kidney disease. Proteomics.

Li Y, Santoso NG, Yu S, Woodward OM, Qian F, Guggino WB (2009) Polycystin-1 interacts with inositol 1,4,5-

trisphosphate receptor to modulate intracellular Ca2+ signaling with implications for polycystic kidney disease. J Biol Chem 284: 36431–41.

Li Y, Wright JM, Qian F, Germino GG, Guggino WB (2005) Polycystin 2 interacts with type I inositol 1,4,5-trisphosphate receptor to modulate intracellular Ca2+ signaling. J Biol Chem 280: 41298–306.

Liu Y, Dai B, Mei C, Zhang Y, Xiong X, Sandford R (2008) Identification of phosphoproteins in kidney tissues from patients with autosomal dominant polycystic kidney disease. Proteomics Clin Appl 2: 1153–66.

Mangolini A, de Stephanis L, Aguiari G (2016) Role of calcium in polycystic kidney disease: From signaling to pathology. World J Nephrol 5: 76–83.

Marquez-Ramirez CA, Hernandez de la Paz JL, Ortiz-Avila O, Raya-Farias A, Gonzalez-Hernandez JC, Rodriguez-Orozco AR, Salgado-Garciglia R, Saavedra-Molina A, Godinez-Hernandez D, Cortes-Rojo C (2018) Comparative effects of avocado oil and losartan on blood pressure, renal vascular function, and mitochondrial oxidative stress in hypertensive rats. Nutrition 54: 60–7.

Maser RL, Vassmer D, Magenheimer BS, Calvet JP (2002) Oxidant stress and reduced antioxidant enzyme protection in polycystic kidney disease. J Am Soc Nephrol 13: 991–9.

McFarland R, Taylor RW, Turnbull DM (2007) Mitochondrial disease–its impact, etiology, and pathology. Curr Top Dev Biol 77: 113–55.

Menon V, Rudym D, Chandra P, Miskulin D, Perrone R, Sarnak M (2011) Inflammation, oxidative stress, and insulin resistance in polycystic kidney disease. Clin J Am Soc Nephrol 6: 7–13.

Merai R, Siegel C, Rakotz M, Basch P, Wright J, Wong B, Dhsc TP (2016) CDC grand rounds: A public health approach to detect and control hypertension. MMWR Morb Mortal Wkly Rep 65: 1261–4.

Mills KT, Bundy JD, Kelly TN, Reed JE, Kearney PM, Reynolds K, Chen J, He J (2016) Global disparities of hypertension prevalence and control: A systematic analysis of population-based studies from 90 countries. Circulation 134: 441–50.

Murray CJ, Atkinson C, Bhalla K, Birbeck G, Burstein R, Chou D, Dellavalle R, Danaei G, Ezzati M, Fahimi A, et al. (2013) The state of US health, 1990–2010: Burden of diseases, injuries, and risk factors. JAMA 310: 591–608.

Nagao S, Yamaguchi T, Kasahara M, Kusaka M, Matsuda J, Ogiso N, Takahashi H, Grantham JJ (2000) Effect of probucol in a murine model of slowly progressive polycystic kidney disease. Am J Kidney Dis 35: 221–6.

Nakayama K, Nakayama K, Negishi I, Kuida K, Sawa H, Loh DY (1994) Targeted disruption of Bcl-2 alpha beta in mice: Occurrence of gray hair, polycystic kidney disease, and lymphocytopenia. Proc Natl Acad Sci U S A 91: 3700–04.

Nauli SM, Alenghat FJ, Luo Y, Williams E, Vassilev P, Li X, Elia AE, Lu W, Brown EM, Quinn SJ et al. (2003) Polycystins 1 and 2 mediate mechanosensation in the primary cilium of kidney cells. Nat Genet 33: 129–37.

Nauli SM, Rossetti S, Kolb RJ, Alenghat FJ, Consugar MB, Harris PC, Ingber DE, Loghman-Adham M, Zhou J (2006) Loss of polycystin-1 in human cyst-lining epithelia leads to ciliary dysfunction. J Am Soc Nephrol 17: 1015–25.

Navarro A (2004) Mitochondrial enzyme activities as biochemical markers of aging. Mol Aspects Med 25: 37–48.

Niimi K, Yasui T, Hirose M, Hamamoto S, Itoh Y, Okada A, Kubota Y, Kojima Y, Tozawa K, Sasaki S et al. (2012) Mitochondrial permeability transition pore opening induces the initial process of renal calcium crystallization. Free Radic Biol Med 52: 1207–17.

Ojo A (2014) Addressing the global burden of chronic kidney disease through clinical and translational research. Trans Am Clin Climatol Assoc 125: 229–43; discussion 243–6.

Padovano V, Kuo IY, Stavola LK, Aerni HR, Flaherty BJ, Chapin HC, Ma M, Somlo S, Boletta A, Ehrlich BE et al. (2016) The polycystins are modulated by cellular oxygen sensing pathways and regulate mitochondrial function. Mol Biol Cell.

Paradies G, Petrosillo G, Pistolese M, Ruggiero FM (2002) Reactive oxygen species affect mitochondrial electron transport complex I activity through oxidative cardiolipin damage. Gene 286: 135–41.

Patergnani S, Suski JM, Agnoletto C, Bononi A, Bonora M, De Marchi E, Giorgi C, Marchi S, Missiroli S, Poletti F et al. (2011) Calcium signaling around Mitochondria Associated Membranes (MAMs). Cell Commun Signal 9: 19.

Peixoto EB, Pessoa BS, Biswas SK, Lopes de Faria JB (2009) Antioxidant SOD mimetic prevents NADPH oxidase-induced oxidative stress and renal damage in the early stage of experimental diabetes and hypertension. Am J Nephrol 29: 309–18.

Peterson KM, Franchi F, Loeffler DL, Psaltis PJ, Harris PC, Lerman LO, Lerman A, Rodriguez-Porcel M (2013) Endothelial dysfunction occurs prior to clinical evidence of polycystic kidney disease. Am J Nephrol 38: 233–40.

Peterson TR, Sengupta SS, Harris TE, Carmack AE, Kang SA, Balderas E, Guertin DA, Madden KL, Carpenter AE, Finck BN et al. (2011) mTOR complex 1 regulates lipin 1 localization to control the SREBP pathway. Cell 146: 408–20.

Pfaller W, Rittinger M (1980) Quantitative morphology of the rat kidney. Int J Biochem 12: 17–22.

Qin S, Taglienti M, Cai L, Zhou J, Kreidberg JA (2012) c-Met and NF-kappaB-dependent overexpression of Wnt7a and -7b and Pax2 promotes cystogenesis in polycystic kidney disease. J Am Soc Nephrol 23: 1309–18.

Re RN (2009) Obesity-related hypertension. Ochsner J 9: 133–6.

Rowe I, Chiaravalli M, Mannella V, Ulisse V, Quilici G, Pema M, Song XW, Xu H, Mari S, Qian F et al. (2013) Defective glucose metabolism in polycystic kidney disease identifies a new therapeutic strategy. Nat Med 19: 488–93.

Saad A, Herrmann SMS, Eirin A, Ferguson CM, Glockner JF, Bjarnason H, McKusick MA, Misra S, Lerman LO, Textor SC (2017a) Phase 2a clinical trial of mitochondrial protection (elamipretide) during stent revascularization in patients with atherosclerotic renal artery stenosis. Circ Cardiovasc Interv 10.

Santos JM, Tewari S, Lin JY, Kowluru RA (2013) Interrelationship between activation of matrix metalloproteinases and mitochondrial dysfunction in the development of diabetic retinopathy. Biochem Biophys Res Commun 438: 760–4.

Santoso NG, Cebotaru L, Guggino WB (2011) Polycystin-1, 2, and STIM1 interact with IP(3)R to modulate ER Ca release through the PI3K/Akt pathway. Cell Physiol Biochem 27: 715–26.

Saran R, Robinson B, Abbott KC, Agodoa LYC, Bhave N, Bragg-Gresham J, Balkrishnan R, Dietrich X, Eckard A, Eggers PW et al. (2018) US renal data system 2017 annual data report: Epidemiology of kidney disease in the United States. Am J Kidney Dis 71: A7.

Schmidt O, Harbauer AB, Rao S, Eyrich B, Zahedi RP, Stojanovski D, Schonfisch B, Guiard B, Sickmann A, Pfanner

N et al. (2011) Regulation of mitochondrial protein import by cytosolic kinases. Cell 144: 227–39.

Seifi B, Kadkhodaee M, Karimian SM, Zahmatkesh M, Shams S, Bakhshi E (2009) Reduction of kidney damage by supplementation of vitamins C and E in rats with deoxycorticosterone-salt-induced hypertension. Iran J Kidney Dis 3: 197–202.

Senyilmaz D, Teleman AA (2015) Chicken or the egg: Warburg effect and mitochondrial dysfunction. F1000Prime Rep 7: 41.

Singh D, Chander V, Chopra K (2005) Cyclosporine protects against ischemia/reperfusion injury in rat kidneys. Toxicology 207: 339–347.

Solini A, Usuelli V, Fiorina P (2015) The dark side of extracellular ATP in kidney diseases. J Am Soc Nephrol 26: 1007–16.

Spirli C, Okolicsanyi S, Fiorotto R, Fabris L, Cadamuro M, Lecchi S, Tian X, Somlo S, Strazzabosco M (2010) ERK1/2-dependent vascular endothelial growth factor signaling sustains cyst growth in polycystin-2 defective mice. Gastroenterology 138: 360–71 e7.

Szeto HH (2014) First-in-class cardiolipin-protective compound as a therapeutic agent to restore mitochondrial bioenergetics. Br J Pharmacol 171: 2029–50.

Szeto HH (2017) Pharmacologic approaches to improve mitochondrial function in AKI and CKD. J Am Soc Nephrol 28: 2856–65.

Szewczyk A, Jarmuszkiewicz W, Koziel A, Sobieraj I, Nobik W, Lukasiak A, Skup A, Bednarczyk P, Drabarek B, Dymkowska D et al. (2015) Mitochondrial mechanisms of endothelial dysfunction. Pharmacol Rep 67: 704–10.

Tabara LC, Poveda J, Martin-Cleary C, Selgas R, Ortiz A, Sanchez-Nino MD (2014) Mitochondria-targeted therapies for acute kidney injury. Expert Rev Mol Med 16: e13.

Talbot JJ, Song X, Wang X, Rinschen MM, Doerr N, Lariviere WB, Schermer B, Pei YP, Torres VE, Weimbs T (2014) The cleaved cytoplasmic tail of polycystin-1 regulates Src-dependent STAT3 activation. J Am Soc Nephrol.

Tedla FM, Brar A, Browne R, Brown C (2011) Hypertension in chronic kidney disease: Navigating the evidence. Int J Hypertens 2011: 132405.

Thomas R, Kanso A, Sedor JR (2008) Chronic kidney disease and its complications. Prim Care 35: 329–44, vii.

Thomson RB, Mentone S, Kim R, Earle K, Delpire E, Somlo S, Aronson PS (2003) Histopathological analysis of renal cystic epithelia in the Pkd2WS25/- mouse model of ADPKD. Am J Physiol Renal Physiol 285: F870–80.

Tkachenko O, Helal I, Shchekochikhin D, Schrier RW (2013) Renin-Angiotensin-aldosterone system in autosomal dominant polycystic kidney disease. Curr Hypertens Rev 9: 12–20.

Torres VE, Harris PC (2014a) Strategies targeting cAMP signaling in the treatment of polycystic kidney disease. J Am Soc Nephrol 25.

Torres VE, Harris PC (2014b) Strategies targeting cAMP signaling in the treatment of polycystic kidney disease. J Am Soc Nephrol 25: 18–32.

Torres VE, Mujwid DK, Wilson DM, Holley KH (1994) Renal cystic disease and ammoniagenesis in Han: SPRDrats. J Am Soc Nephrol 5: 1193–200.

Tsiokas L, Arnould T, Zhu C, Kim E, Walz G, Sukhatme VP (1999) Specific association of the gene product of PKD2 with the TRPC1 channel. Proc Natl Acad Sci U S A 96: 3934–9.

Van Buul JD, Fernandez-Borja M, Anthony EC, Hordijk PL (2005) Expression and localization of NOX2 and NOX4 in primary human endothelial cells. Antioxid Redox Signal 7: 308–17.

Vander Heiden MG, Cantley LC, Thompson CB (2009) Understanding the Warburg effect: The metabolic requirements of cell proliferation. Science 324: 1029–33.

Veis DJ, Sorenson CM, Shutter JR, Korsmeyer SJ (1993) Bcl-2-deficient mice demonstrate fulminant lymphoid apoptosis, polycystic kidneys, and hypopigmented hair. Cell 75: 229–40.

Venugopal J, Blanco G (2016) Ouabain enhances ADPKD cell apoptosis via the intrinsic pathway. Front Physiol 7: 107.

Wang D, Iversen J, Strandgaard S (2000) Endothelium-dependent relaxation of small resistance vessels is impaired in patients with autosomal dominant polycystic kidney disease. J Am Soc Nephrol 11: 1371–6.

Wang D, Iversen J, Wilcox CS, Strandgaard S (2003) Endothelial dysfunction and reduced nitric oxide in resistance arteries in autosomal-dominant polycystic kidney disease. Kidney Int 64: 1381–8.

Wei W, Popov V, Walocha JA, Wen J, Bello-Reuss E (2006) Evidence of angiogenesis and microvascular regression in autosomal-dominant polycystic kidney disease kidneys: A corrosion cast study. Kidney Int 70: 1261–68.

Wills LP, Trager RE, Beeson GC, Lindsey CC, Peterson YK, Beeson CC, Schnellmann RG (2012) The beta2-adrenoceptor agonist formoterol stimulates mitochondrial biogenesis. J Pharmacol Exp Ther 342: 106–18.

Wilson GN, de Chadarevian JP, Kaplan P, Loehr JP, Frerman FE, Goodman SI (1989) Glutaric aciduria type II: Review of the phenotype and report of an unusual glomerulopathy. Am J Med Genet 32: 395–401.

Wilson PD, Hovater JS, Casey CC, Fortenberry JA, Schwiebert EM (1999) ATP release mechanisms in primary cultures of epithelia derived from the cysts of polycystic kidneys. J Am Soc Nephrol 10: 218–29.

Wilson PD, Schrier RW, Breckon RD, Gabow PA (1986) A new method for studying human polycystic kidney disease epithelia in culture. Kidney Int 30: 371–8.

Wu M, Gu J, Mei S, Xu D, Jing Y, Yao Q, Chen M, Yang M, Chen S, Yang B et al. (2016) Resveratrol delays polycystic kidney disease progression through attenuation of nuclear factor kappaB-induced inflammation. Nephrol Dial Transplant 31: 1826–34.

Wullschleger S, Loewith R, Hall MN (2006) TOR signaling in growth and metabolism. Cell 124: 471–84.

Xu R, Franchi F, Miller B, Crane JA, Peterson KM, Psaltis PJ, Harris PC, Lerman LO, Rodriguez-Porcel M (2013) Polycystic kidneys have decreased vascular density: A micro-CT study. Microcirculation 20: 183–9.

Yamaguchi T, Pelling JC, Ramaswamy NT, Eppler JW, Wallace DP, Nagao S, Rome LA, Sullivan LP, Grantham JJ (2000) cAMP stimulates the in vitro proliferation of renal cyst epithelial cells by activating the extracellular signal-regulated kinase pathway. Kidney Int 57: 1460–71.

Ying WZ, Sanders PW (2001) Cytochrome c mediates apoptosis in hypertensive nephrosclerosis in Dahl/Rapp rats. Kidney Int 59: 662–72.

Zheleznova NN, Yang C, Ryan RP, Halligan BD, Liang M, Greene AS, Cowley AW Jr. (2012) Mitochondrial proteomic analysis reveals deficiencies in oxygen utilization in medullary thick ascending limb of Henle in the Dahl salt-sensitive rat. Physiol Genomics 44: 829–42.

10 Mitochondrial Dysfunction in Breast Cancer

A Potential Target from the Bench to Clinical Therapeutics

Thalita Basso Scandolara, Letícia Madureira Pacholak, Thayse Fachin Cormanique, Rodrigo Kern, and Carolina Panis
Laboratory of Tumor Biology, State University of West Paraná, Unioeste, Francisco Beltrão, Brazil.

CONTENTS

1 INTRODUCTION

Mitochondria are one of the main sources of oxidative stress in cells. Under physiological conditions, the leakage of reactive oxygen species (ROS) during the respiratory chain immediately activates the antioxidant system, which neutralizes most of these excessive ROS. When pathological conditions occurs, this system is frequently deregulated (Sinha et al. 2013; Ng Kee Kwong et al. 2017; Zhang et al. 2017). Mitochondrial dysfunction has been reported in several chronic disorders, described as both cause and consequence of oxidative stress generation, which affects directly cell fate by ROS-driven destructive mechanisms (Georgieva, Ivanova, and Zhelev 2017). As consequence, DNA, lipids and proteins are negatively modified, allowing the occurrence of cellular abnormalities which causes diseases sustained by redox imbalance (Pagano et al. 2014).

Mitochondrial dysfunction is implicated in cancer induction, by altering cell responsiveness to apoptotic signals and modifying its capability to deal with the energetic metabolism (Księżakowska-Łakoma, Zyła, and Wilczyński 2014). This event is a hallmark of cancer, and in association with sustained chronic inflammation and oxidative stress generation, contributes to fuel the cellular growth and proliferation (Samadi et al. 2015). Mitochondrial reprogramming energy metabolism emerged as a new hallmark of cancer few years ago, although its basis have had described decades ago (Hanahan and Weinberg 2011). Named as "Warburg effect" [reporting to Otto Warburg, more than 50 years ago first described this metabolic switch in cancer cells Warburg 1931, 1956a, 1956b]; this phenomenon occurs in cancer cells that, in spite of with oxygen availability, prefer the aerobic glycolysis route for getting energy.

Beyond this, cancer is a ROS-related disease in which sustained oxidative stress consists in a disease signature. Several studies have demonstrated the occurrence of systemic oxidative stress in cancer patients (Panis et al. 2015), and experimental evidence has implicated mitochondrial failure in this context (Banerjee et al. 2017; Ng Kee Kwong et al. 2017; Zhang et al. 2017). Interestingly, at the same time that mitochondria ensure the fuel for cancer cells survival, it can be attacked by chemotherapy, leading to either cell death or chemoresistance. Undoubtedly, cancer cells are tightly adapted to mitochondrial-induced oxidative changes and handle very well in this situation.

Considering the number of studies focusing on breast cancer, this chapter presents the main events that correlate with mitochondrial role in different aspects of breast cancer biology.

2 MITOCHONDRIA IN NORMAL BREAST TISSUES AND MITOCHONDRIAL GENES FOR BREAST CANCER

Breast cancer is a leading cause of cancer death among women in the world. There are many factors involved in the development and progression of cancer, such as family history, obesity and genetic predisposition (American Cancer Society 2017). It demonstrates a wide heterogeneity in morphological features, a diversity in immunohistochemical subtypes that have different clinical outcomes (Barnard, Boeke and Tamimi 2015; Makki 2015). Because of a huge variety of subtypes with different biological behaviors and clinicopathological features that can lead to different prognosis, therapeutic protocols are not solely dependent on traditional clinicopathological features, such as age, menopausal status, tumor size, lymph node involvement, distant metastasis, histological grade, and type (Prat et al. 2015). Although there are established protocols that leads the treatment against different subtypes, there is still a need to identify more targets to improve treatment efficacy for this disease (Shao, Sun, and Deng 2017).

The interest in the mitochondrial role in carcinogenesis was initiated due to findings of respiratory deficits in dividing cells, especially the one's which had intensive proliferation rates (Grzybowska-Szatkowska et al. 2014). In this context, mitochondria, involved in many important cell mechanisms such as energy regulation, production of ROS, modulation of cell redox status and inhibition of apoptosis, could be intrinsically involved in cancer development and progression (Wallace 2012).

To exercise its functions, mitochondria has a number of genes, that when translated, produce enzymes that are indispensable for maintenance of cellular structure (Ahn and Metallo 2015). One of these enzymes is human pyrroline-5-carboxylate reductase (PYCR or P5CR, EC1.5.1.2), a cytoplasmic house-keeping enzyme that catalyzes the last step in proline biosynthesis (Δ1- pyrroline-5-carboxylate (P5C) to proline), with concomitant oxidation of NAD(P)H to NAD(P)$^{+}$, which demonstrates an ability to protect cells from mitochondrial fragmentation upon oxidative stress (Ahn and Metallo 2015; Ding et al. 2017).

Studies have found that PYCR1 significantly enhanced the aggressiveness of breast cancer and is significantly associated with poor survival of breast cancer patients. Ding et al. (2017) discovered that inhibition of PYCR1 could significantly enhance cytotoxicity of doxorubicin on breast cancer cells and mRNA expression levels of PYCR1 were significantly higher in aggressive molecular subtypes, including Luminal B, HER-2 positive and basal-like breast cancer. Therefore, mRNA levels of PYCR1 were significantly associated with poor differentiation and aggressive phenotype of breast cancer (Ding et al. 2017).

Another protein that is important to maintain mitochondria activities is dynamin-related protein 1 (Drp1), a major mitochondrial fission-related protein, which is mainly located in cytoplasm and translocated into mitochondria during its fission. Mitochondrial morphology is dynamic and involves continuous fission and fusion. Fusion occurs to recover the activities of damaged mitochondria, while fission is necessary for redistribution of mitochondrial DNA during cell division and for transporting mitochondria to the daughter cells in mitosis and meiosis. Drp1 regulates the process of mitochondrial fission through its post-translational modifications, as phosphorylation ubiquination. Phosphorylation at SER616 induces Drp1 activity, and this upregulation of Drp1 is associated with breast cancer progression (Hu, Huang, and Li 2017; Liu and Chan 2015).

Studies had shown that estrogen regulates cellular functions by modifying mitochondrial morphology. Estrogen has direct and indirect effects in mitochondrial structure and biogenesis, which are intermediated by the genomic and membrane-initiated activities of receptors. It was also found that 17-β-estradiol was reported to change the profile of breast cancer cells, turning them into secretory cells containing large, clear mitochondria with well-defined cristae formation, which has happened only in Estrogen Receptor (ER) positive cell lines, suggesting that these effects are mediated through ER. Some reports even suggest that estrogen regulates cellular functions by modifying mitochondrial morphology; however, this mechanism remains unclear (Rangel, Villegas, and Rondón-Lagos 2017; Oo et al. 2018).

Estrogen is an important regulator for breast cancer progression due to its ability to preserve mitochondrial structure and biogenesis in various tissues (Liao et al. 2015). However, Oo et al. (2018) found that mitochondrial morphology had changed from a tubular pattern to a short and small mitochondrial in MCF7 cells, induced by Drp1 phosphorylation at SER616, which occurred by estrogen binding in ERα, indicating that estrogen and ERα are important for regulating mitochondrial morphology through Drp1 phosphorylation in MCF7 breast cancer cells.

Cancer cells have the characteristics of continuous proliferation and apoptosis resistance. Therefore, it is possible to observe that abnormal function of mitochondria contribute to these characteristics (Wallace 2012). In this context, *mitochondrial fusion* can limit damage and autophagy activation, inducing mitophagy (detail later) (Senft and Jolla 2016).

Mitochondrial biogenesis is a process involving replication of the mitochondrial genome and coordinated expression of both nuclear and mitochondrial encoded proteins and assembly of the oxidative phosphorylation complexes (Zong, Rabinowitz, and White 2016). Studies have found that many proteins, including mitofusin-2 (MFN2) (Xu et al. 2017), PTEN-induced putative kinase 1 (PINK1) (O'Flanagan and O'Neill 2014), peroxisome proliferator-activated receptor gamma coactivator 1-alpha (PGC-1α) (Jornayvaz and Shulman 2010) and mitochondrial transcription factor A (MT-TFA) (Kang, Kim, and Hamasaki 2007) plays critical roles in regulating mitochondrial fusion, biogenesis and maintaining its integrity. Despite that, the role of mitochondrial fusion and biogenesis in breast cancer development and progression remains largely unknown.

Researchers have found that a vesicular endocytosis-associated protein, called SH3GL2 might act as a potential breast cancer suppressor in breast cancer cells and is associated with enhanced expression of mitochondrial fusion and biogenesis proteins such as MFN2, PGC-1α and PINK1. The overexpression of SH3GL2 have reduced the ability of breast cancer cells to metastasize to the lung and liver and when this cell loses SH3GL2 and MFN2 in primary tumor, progression of breast cancer may ensue (Kannan et al. 2016).

Mitochondria are the important metabolic component of cancer cells as well one of the most important pathways of cancer involving mutations in the TP53 gene (Kandoth et al. 2014). This gene encodes a protein, called "Guardian of the Genome" (Lane 1992). p53, the most extensively studied and most multifaceted tumor suppressor protein identified yet is a product of TP53. In majority of cancer cases, mutations in TP53 commonly occur and contribute to tumor progression (White 2016). However, this protein is not just a tumor suppressor, it is also an important regulator of metabolic homeostasis, pivotal in most major cellular process (Park et al. 2016). Studies found that p53 is associated with modulation of key proteins in mitochondrial metabolism, involved in many metabolic pathways such as glycolysis, regulation of glutaminase-2 and regulation of cytochrome C oxidase (COX) complex (Moulder et al. 2018).

Mitochondria are not just the powerhouses of human cancer cells, as we discussed before, but also their *Achilles' heel* (Hockenbery 2002). It is because cancer cells can amplify their capacity from mitochondrial oxidative metabolism and obtain high-energy mitochondrial fuels from other cells in adjacent stromal tissue, which are undergoing aerobic glycolysis, the "Warburg effect" (Martinez-Outschoorn et al. 2011).

This research has identified a series of new mitochondrial biomarkers capable to predict tumor recurrence, lymph node metastasis, tamoxifen resistance and poor clinical outcomes in human breast cancer (Sotgia, Fiorillo, and Lisanti 2017). In addition, 15 marker proteins, associated with mitochondrial biogenesis, have been found to be highly expressed in human breast cancer cells and largely absent in adjacent stromal cells; this demonstrates "the two-compartment tumor metabolism," which postulates that epithelial cancer cells amplify oxidative mitochondrial metabolism, while cancer-associated fibroblasts undergo predominantly glycolysis, suffering from a mitochondrial deficiency or dysfunction (Sotgia et al. 2012).

Mitochondria have two markers associated with poor clinical outcomes, TIMM17A and TOMM34, which can be predictive of higher tumor grade and size, lymph-vascular invasion and lymph node metastasis (Aleskandarany et al. 2012; Salhab et al. 2012). Another marker GOLPH3, associated with increased mitochondrial lipid

synthesis and mitochondrial function, promotes tumor growth without significant increase in tumor angiogenesis in triple-negative breast cancer cells (MDA-MB-231) (Sotgia et al. 2012).

Studies show that MYC, a proto-oncogene, activates a stereotypic transcriptional program of genes involved in cell growth mammary epithelial cells. The activity of MYC oncoprotein is mediated by a phospholipase of the mitochondrial outer membrane, PDL6, that alters mitochondrial fusion and fission dynamics, which activates AMPK, inhibiting YAP/TAZ co-activators, that maintain the clonogenic potential of these cell (von Eyss et al. 2015).

Another gene, miR-4485, translocates to mitochondria and binds in mitochondrial 16S rRNA, modulating mitochondrial I complex activity and production of ATP, ROS, caspase 3/7 and apoptosis. As MYC oncogene, miR-4485 reduces clonogenic ability of breast cancer cells this may act as a tumor suppressor by negatively regulating mitochondrial RNA processing and mitochondrial functions (Sripada et al. 2017).

Mitochondrial ribosomal protein L12 (MRPL12), mitochondrial DNA polymerase gamma (POLG), and mitochondrial ribonuclease H1 (RNASEH1) have also been found to strongly up-regulated in cancer cells, indicating that mitochondria tend to proliferate in breast cancer cells due to increased expression of genes involved in mitochondrial genetic replication (Zhang et al. 2016).

3 MITOPHAGY IN BREAST CANCER

Mitochondrial homeostasis is crucial to efficient cell function. When damaged, caused either by mutation or reactive oxygen species (ROS), sophisticated mechanisms activate to repair or remove the damaged organelles. If can't be repaired, they are eliminated by an specialized form of autophagy, a lysosome-dependent degradation, called *mitophagy* (Zou et al. 2016; Gkikas, Palikaras, and Tavernarakis 2018; Ježek, Cooper, and Strich 2018). This selective mechanism then maintain the proper balance of responses and prevent accumulation of dysfunctional mitochondria, associated with several diseases (Rodger, Mcwilliams, and Ganley 2017; Biel and Rao 2018; Gkikas, Palikaras, and Tavernarakis 2018).

Defective mitochondria results in bioenergetics change (the Warburg effect) and an increased mitochondrial ROS (mtROS), which could impel an inflammation response, both situations connected with cancer and its progression (Zou et al. 2016; Rodger, Mcwilliams, and Ganley 2017). The damaged and compromised mitochondria, through mitophagy, may reduce ROS produced and diminish tumorigenesis caused by it (Jin 2006). However, the role of mitophagy in cancer is controversial and direct evidence for it is still lacking (Zong, Rabinowitz, and White 2016). Inside the tumor microenvironment, autophagy allows tumor cells to survive through a hypoxia and nutrient deprivation, situation that increases mitochondria damage and aggravate ROS production (Dower et al. 2017). When under this microenvironment for extensive period, mitophagy can provide cells with energy to maintain cellular homeostasis and promotes tumor-survival.

Several pathways have been identified to be involved in mitophagy, such an example is BNIP3 gene with tumor suppression properties and that's known to be induced by hypoxia, FOXO3, NFKB, TP53 and others tumor-related transcription factors (Chourasia and Macleod 2015; Chourasia et al. 2015; Qiao et al. 2016). Loss of BNIP3-dependent mitophagy has been shown to predict increased ROS production and progression to metastasis due to augmented expression of Hypoxia-inducible factor 1α (HIF-1α) (Chourasia and Macleod 2015; Chourasia et al. 2015). This fact is also associated with glycolysis and angiogenesis, specially in the triple-negative subtype when compared to other subtypes of breast cancer (Chourasia and Macleod 2015). Furthermore, in mouse models, loss of BNIP3 showed an increased in mitochondrial mass and a function reduction of mitochondria both *in vivo* and *in vitro* (Chourasia et al. 2015).

The protein SIRT3, a mitochondrial sirtuin deacetylase, promotes mitophagy through reduction of the interaction between the protein VDAC1 and hexokinase II, an enzyme that participates in glycolysis and is upregulated in many cancer types, acting as an inhibitor of apoptosis and helping the evasion of tumor cell death (Qiao et al. 2016; Campello 2017). SIRT3 enhances the association of VDAC1 with Parkin, an E3 ubiquitin ligase that plays a major role in Parkinson's disease and PINK1/Parkin-mediated mitophagy. Besides, silencing of SIRT3, autophagy is also decreased in breast cancer cells under hypoxia, demonstrating its role as a positive regulator of mitophagy and as a tumor suppressor protein (Geisler et al. 2010; Qiao et al. 2016; Campello 2017).

There is evidence that the adaptor protein, MITA, inhibits the turnover of mitochondria through mitophagy and decreases the mitochondrial biogenesis by downregulating PGC1α, leading to an accumulation of damaged mitochondria and acting as a tumor suppressor (Bhatelia et al. 2017). Another potential tumor suppressor and possibly a mitophagy regulator in breast cancer is FBP1, an enzyme involved in gluconeogenesis. It has been suggested that higher expression of FBP1 is capable of promoting apoptosis in breast cancer cell lines MCF-7 and MDA-MB-231; furthermore, that both groups had elevated levels of pro-apoptotic proteins. Besides, FBP1 acts as a mitophagy suppressor by blocking HIF1a/BNIP3 pathway and enhancing the association between Beclin 1 and Bcl-2, causing inhibition of mitophagy, causing an accumulation of ROS levels and consequently an intensification of apoptosis rate in tumor cells (Liu et al. 2017a).

There are still a lot of questions remain unanswered, as how cancer cells adapt rapidly to the amount of ROS accumulated in the microenvironment. Is it by blocking

mitophagy and if so how much time mitochondria require becoming dysfunctional once mitophagy is inhibited? Understanding these changes in metabolic responses is essential to determine if mitophagy is an adaptative response of cancer, protecting cells from metabolic stress like hypoxia and starvation, or is inhibiting cancer-associated metabolic reprogramming; protecting cells from ROS induced damage and mutagenesis.

4 MITOCHONDRIAL DYSFUNCTION, MOLECULAR SUBTYPES OF BREAST CANCER AND ITS ROLE IN CHEMORESISTANCE

Mitochondria are organelles that have an important function in maintaining cellular homeostasis and play a major role in metabolic reprogramming during carcinogenesis, since most signaling pathways are associated with these organelles (Guaragnella, Giannattasio, and Moro 2014; Guerra, Arbini, and Moro 2017). Therefore, mitochondria have the capacity to sense cellular stress and provides cell adaptation to aggressive and harmful microenvironments, contributing to cancer cell survival and progression (Sciacovelli et al. 2014).

The molecular subtypes of breast cancer are actually considered as one of the main parameters for defining patients' treatment and prognosis. In this context, breast cancer have been divide into tumors of good prognosis (hormone dependents, named as Luminal) and those with poorly responsive (hormone-independents as the triple negative ones or hormone dependents that are highly proliferative as the Luminal B subtype). Thus, researches have demonstrated that patients with luminal breast cancer subtype have a better prognosis and a positive response to hormone therapy However, patients who exhibit HER2 overexpression (receptor for the human epidermal growth factor 2), although its aggressiveness has a good prognosis because of its treatment with monoclonal antibodies; while patients with the triple negative subtype do not have any specific antitumor therapy and are categorized as of the poor prognosis subsets (Goldhirsch et al. 2011; Liu, Zhang, and Zhang 2014).

Studies have shown that there are differences in subtypes of breast cancer depending upon mitochondria, suggesting that the glycolytic inhibition could be a potential antitumor activity, mainly for triple negative breast cancer. Due to low mitochondrial respiration and increased glycolysis in triple negative breast cancer when compared with estrogen positive receptor, for example, possibly is the cause of the activation of AMPK and decrease of p70S6K expression (Pelicano et al. 2014).

It is known that PI3K pathway can be altered in some cases of breast cancer. A study demonstrated that PI3K activation was associated with hormonal positive status, estrogen and progesterone, also showed that the alterations were associated with death from breast cancer. As most patients analyzed had an alteration in this pathway, suggests that PI3K could be a potential target for a novel antitumor therapy, mainly in basal-like subtype (López-Knowles et al. 2010).

In order to determine the dependency, capacity and flexibility of triple negative breast cancer cells, researchers showed that these cells have different fuel needs in mitochondria and glycolytic process could be a target for novel antineoplastic drugs (Lopez et al. 2017). After analyzing cancerous tissues and cancer cells, it was demonstrated that there was an increased defect in mitochondria especially in triple negative breast cancer subtypes. Also, it was demonstrated that genes responsible for structuring mitochondria and determining their function were different between triple and non-triple negative breast cancer (Guha et al. 2018).

Chemotherapy resistance often occurs due to a failure to induce cancer cell death, resulted in drug-resistance acquired by tumor cells (Su et al. 2014). A better understanding of how cancer cells modify the tumor microenvironment (TME) and adapt its energy metabolism might explain some of the chemoresistance mechanisms and help to overcome drug resistance in cancer therapy (Guaragnella, Giannattasio, and Moro 2014).

Doxorubicin (DOX) is an anthracycline antibiotic widely used as an anti-cancer cytotoxic drug, inducing DNA damage, increasing mitochondrial membrane permeability and leading to apoptosis. However, the occurrence of drug resistance by the cancer cells occur with a high frequency (Chamberlain, Tulumello, and Kelley 2013; Liu et al. 2017b). An accumulation of DOX in mitochondria might be a way to counteract DOX resistance. Mitochondrial accumulation of DOX is decreased in the presence of ATP and increased when ATP-dependent efflux pumps, such as breast cancer resistance protein (BCRP) and multidrug resistance-associated protein (MRP), specifically MRP1, were inhibited (Dartier et al. 2017).

MicroRNAs (miRNAs) are small non-coding RNAs that are involved in many biological process, such as proliferation, differentiation and apoptosis (Catalanotto, Cogoni, and Zardo 2016). Many miRNAs have been shown to sensitize tumor cells to chemotherapy, as miRNA133a in DOX-resistant breast cancer cells, by decreasing the expression of UCP-2 in both RNA and protein levels, plays an essential role of miRNA133a in this chemotherapy development (Yuan et al. 2015). Down-regulation of miRNA489, through LAPTM4B-autophagy pathway, contributes to increased autophagy flux and chemotherapy resistance. Inhibition of autophagy by miRNA489 in triple negative breast cancer (TNBC) MDA-MB-231 had significant cytotoxic effect compared to other cell lines and its restoration-increased sensitization to DOX chemotherapy. On the other hand, patients with higher miRNA489 expression show significantly better survival response than patients with lower expression of miRNA489 (Soni et al. 2018). Furthermore, miRNA125b, a possible tumor suppressor in breast cancer, is able to

return sensitivity to DOX resistance cancer cells by acting through down-regulation of Mcl-1, a member of Bcl-2 family, which is up-regulated in doxorubicin-resistant cells (Xie et al. 2015).

Hematopoietic cell-specific protein 1-associated protein X-1(HAX-1), a mitochondria protein and an important regulator of apoptosis, has a higher expression in breast cancer cells and is up-regulated in DOX-resistant cells as in cisplatin (CIS) resistant ones (Fadeel and Grzybowska 2009; Yang et al. 2017). Besides, knockdown of HAX-1 increased apoptosis induction of CIS/DOX in resistant MDA-MB-231 cells and enhanced proliferation inhibition induced by these anti-cancer drugs (Yang et al. 2017).

An adjuvant hormonal therapy usually utilized in estrogen positive breast cancer, Tamoxifen (TAM), also has its potential benefit limited by chemoresistance, as most of patients treated with TAM has been shown to develop resistance (Qadir et al. 2008). Mechanisms behind this were already explained, as an increased expression of genes involved in mitochondrial functions. Sirtuin-3 (SIRT3), localized in mitochondrial matrix and a member of sirtuin family, is associated with the regulation of aging and disease (Van de Ven, Santos, and Haigis 2017) and plays a key role in regulating both cell death and survival (Chen et al. 2014). SIRT3 participates in modulation of TAM-resistant breast cancer, as augmented expression of SIRT3 decrease cellular sensitivity, protecting cells from damage and maintaining mitochondrial functions and cellular integrity while suppressing TAM-induced apoptosis(Alhazzazi et al. 2011; Chen et al. 2014).

Higher expression of NRF-1 has been observed in TAM-resistant breast cancer cells. NRF-1 is a major regulator of mitochondrial gene transcription involved in regulating the mitochondrial unfolded protein response (UPR), which have cytoprotective activity in breast cancer cells and probably acts decreasing TAM-induced apoptosis (Luo and Lee 2012; Radde et al. 2016). Suppression of BIK, a pro-apoptotic protein, inducible by estrogen starvation and important for TAM response, is probably associated with the resistance to apoptosis due to induction of proteins like Bcl-2 and Mcl-1, inhibitors of cytochrome C release, preventing mitochondrial pore formation and thus corroborating for tumor cell survival (Viedma-Rodriguez et al. 2013).

Autophagy has pro-survival properties and has been shown to be responsible for acquired resistance against TAM, Paclitaxel (PAX) and Epirubicin (EPB)-induced cell death, and its inhibition has re-sensitized tumor cells to chemotherapy (Qadir et al. 2008; Samaddar et al. 2008; Vazquez-Martin, Oliveras-Ferraros, and Menendez 2009; Chittaranjan et al. 2014; Wen et al. 2015). Paclitaxel, a microtubule-stabilizing agent, is a drug widely used to treat TNBC patients (Maddalena et al. 2013). Basal autophagy in MDA-MB-231 cell lines is enhanced by PAX pulse treatment, to mimic clinic treatment, and this upregulation plays a cytoprotective function under PAX administration, assisting to resistance to its cytotoxic effect. Additionally, MDA-MB-231 PAX-resistant cells show resistance to another chemotherapy drugs, cisplatin and Epirubicin, suggesting that autophagy is correlated with tumor cells chemoresistance and its inhibition could be an important target to overcome this function (Wen et al. 2015).

Although many mechanisms explained in this chapter have been associated with chemoresistance, it is still an unsolved problem in clinical oncology. Recent studies suggest that cancer stem cells (CSCs), which might share stem cell-like properties such as self-renew, could be responsible for multidrug resistance, leading to tumor recurrence and metastasis (Mani et al. 2008; Farnie, Sotgia, and Lisanti 2015; Lyakhovich and Lleonart 2016; Lee et al. 2017; Wang et al. 2018). Cells with high mitochondrial mass were linked to resistance to PAX treatment, with less DNA damage induced by chemotherapy, and although the authors did not know the mechanism behind it, high levels of mTORC1, PGC-1a and expression of MYC, mediators of mitochondrial biogenesis, were assumed (Farnie, Sotgia, and Lisanti 2015). Expression of MYC and MCL1 was found increased in TNBC, both enhancing mitochondrial oxidative phosphorylation (mtOXPHOS) and also ROS production, enriching CSCs and thus maintaining tumor growth and its progression (Lee et al. 2017).

5   ARE MITOCHONDRIA AS A POTENTIAL TARGET FOR BREAST CANCER THERAPY?

Apoptosis performs an important role in tissue homeostasis and it is a well-established pathway involving caspase, a cysteine protease, which regulates this process (Kim and Han 2001). In addition, there are other pathways regulating this programmed cell death, such as the mechanism that involves mitochondria, which is regulated by pro and anti-apoptotic factors, Bax and Bcl-2 family, respectively. Also, these proteins are able to act in the mitochondrial outer membrane permeabilization by regulating Mfn1(Lin et al. 2014).

Surveys reveal that studying the function of mitochondria in relation to breast cancer may contribute to find novel strategies to treat the disease, such as drugs that are able to target mitochondria (Frantz and Wipf 2010; Carden et al. 2017). Synthetic α-methylene-δ-lactones that target the mitochondria through induced apoptosis and causing alterations in the polarization of this organelle has been evaluated in breast cancer. It shows suppression of migration and invasion of cancer cells and authors concluded that this drug may be a potential novel anticancer drug against this type of neoplasia (Wyrębska et al. 2013).

Delocalized lipophilic cations (DLCs) are used as biomarkers to cancer cells, due to its specificity for cells of tumorigenic origin (Modica-Napolitano and Aprille 2001). Studies show that DLCs can accumulate in cancer and normal cells in different concentrations because of the differences of the higher mitochondrial membrane potential

(Murphy and Smith 2007; Neuzil et al. 2013). Therefore, a study with DLC F16, covalently linked with a DNA-damaging agent, chlorambucil (CBL), showed that this combination can be employed in as an effective antitumor therapy that target mitochondria. This combinational-mitochondrial anticancer agent may act selectively killing cancer cells and these authors also concluded that this drug can overcome the common dosage resistance caused by classical chemotherapeutic drugs (Peng et al. 2017).

In a review, authors demonstrated that dichloroacetate may contribute to suppress tumor growth, by contributing to the mitochondrial apoptosis in cancer cells. This drug acts through inhibiting the pyruvate dehydrogenase kinase and, consequently, increases the concentration of pyruvate in the mitochondria and improving the glycolysis (Michelakis, Webster, and Mackey 2008).

Mitophagy inhibition has also been studied as a novel therapy and researchers observed that liensinine, an alkaloid, is able to inhibit this mechanism, preventing the autophagosome-lysosome fusion. They also analyzed the use of liensinine at the same time with other antineoplastic drugs and observed that this co-treatment significantly increased the breast cancer cells death (Zhou et al. 2015).

6   CONCLUSION

Further extensive studies into mitochondria machinery and their roles in breast cancer can allow understanding that this organelle is fundamental to tumor cell removal by a wide range of mechanisms. It is clear that most cancer cells change the energetic metabolism to allow them to grow and survive under the most challenging environments, such as hypoxia and nutrient deprivation, although the mechanisms are not the same between different cancer types, proving to be an obstacle to drug development and therapy. Mitochondrial dysfunction has been associated with tumor progression, metastasis, and poor prognosis and with chemoresistance, thus is an important subject of research. Better understanding of such plasticity mechanisms which tumor cells use to evade apoptosis, activate autophagy and how tumor cell microenvironment influence this metabolic alteration will have a key role to overcome chemotherapy resistance, associated with combined therapy strategies that might be more efficient to induce cell-death. Studies are reaching a more comprehensive stage of mitochondria in cancer, and this advancement may constitute a future key for developing new strategies for treating breast cancer and the comprehension of tumor biology.

REFERENCES

Ahn, Christopher S., and Christian M. Metallo. 2015. "Mitochondria as Biosynthetic Factories for Cancer Proliferation." *Cancer and Metabolism* 3 (1): 1–10. https://doi.org/10.1186/s40170-015-0128-2.

Aleskandarany, Mohammed A., Ola H. Negm, Emad A. Rakha, Mohamed A. H. Ahmed, Christopher C. Nolan, Graham

R. Ball, Carlos Caldas, Andrew R. Green, Patrick J. Tighe, and Ian O. Ellis. 2012. "TOMM34 Expression in Early Invasive Breast Cancer: A Biomarker Associated with Poor Outcome." *Breast Cancer Research and Treatment* 136 (2): 419–427. https://doi.org/10.1007/s10549-012-2249-4.

Alhazzazi, Turki Y., Pachiyappan Kamarajan, Eric Verdin, and Yvonne L. Kapila. 2011. "SIRT3 and Cancer: Tumor Promoter or Suppressor?" *Biochimica et Biophysica Acta – Reviews on Cancer* 1816 (1): 80–88. https://doi.org/10.1016/j.bbcan.2011.04.004.

American Cancer Society. 2017. "Breast Cancer Facts & Figures 2017–2018." *American Cancer Society.* https://doi.org/10.1007/s10549-012-2018-4.Mesothelin.

Banerjee, Sumita, Saikat Mukherjee, Sanjib Mitra, and Pallav Singhal. 2017. "Altered Expression of Mitochondrial Antioxidants in Oral Squamous Cell Carcinoma." *Journal of Oral Science* 59 (3): 439–446.

Barnard, Mollie E., Caroline E. Boeke, and Rulla M. Tamimi. 2015. "Established Breast Cancer Risk Factors and Risk of Intrinsic Tumor Subtypes." *Biochimica et Biophysica Acta – Reviews on Cancer* 1856 (1): 73–85. https://doi.org/10.1016/j.bbcan.2015.06.002.

Bhatelia, Khyati, Kritarth Singh, Paresh Prajapati, Lakshmi Sripada, Milton Roy, and Rajesh Singh. 2017. "MITA Modulated Autophagy Flux Promotes Cell Death in Breast Cancer Cells." *Cellular Signalling* 35 (March): 73–83. https://doi.org/10.1016/j.cellsig.2017.03.024.

Biel, Thomas G., and V. Ashutosh Rao. 2018. "Mitochondrial Dysfunction Activates Lysosomal-Dependent Mitophagy Selectively in Cancer Cells." *Oncotarget* 9 (1): 995–1011. https://doi.org/10.18632/oncotarget.23171.

Campello, Silvia. 2017. "The Close Interconnection between Mitochondrial Dynamics and Mitophagy in Cancer." *Frontiers in Oncology* 7 (81): 1–9. https://doi.org/10.3389/fonc.2017.00081.

Carden, Trevor, Bhupendra Singh, Ved Mooga, Prachi Bajpai, and Keshav K. Singh. 2017. "Epigenetic Modification of MiR-663 Controls Mitochondriatonucleus Retrograde Signaling and Tumor Progression." *Journal of Biological Chemistry* 292 (50): 20694–20706. https://doi.org/10.1074/jbc.M117.797001.

Catalanotto, Caterina, Carlo Cogoni, and Giuseppe Zardo. 2016. "MicroRNA in Control of Gene Expression: An Overview of Nuclear Functions." *International Journal of Molecular Sciences* 17 (10). https://doi.org/10.3390/ijms17101712.

Chamberlain, Graham R., David V. Tulumello, and Shana O. Kelley. 2013. "Targeted Delivery of Doxorubicin to Mitochondria." *ACS Chemical Biology* 8 (7): 1389–1395. https://doi.org/10.1021/cb400095v.

Chen, Yu-long, L L Fu, Xiaowei Wen, Xiang Yuan Wang, Jianping Liu, Yan Cheng, and Jian Huang. 2014. "Sirtuin-3 (SIRT3), a Therapeutic Target with Oncogenic and Tumor-Suppressive Function in Cancer." *Cell Death and Disease* 5: 1047. https://doi.org/10.1038/cddis.2014.14.

Chittaranjan, Suganthi, Svetlana Bortnik, Wieslawa H. Dragowska, Jing Xu, Namal Abeysundara, Amy Leung, Nancy E. Go, et al. 2014. "Autophagy Inhibition Augments the Anticancer Effects of Epirubicin Treatment in Anthracycline-Sensitive and -Resistant Triple-Negative Breast Cancer." *Clinical Cancer Research* 3159–3174. https://doi.org/10.1158/1078-0432.CCR-13-2060.

Chourasia, Aparajita H., and Kay F. Macleod. 2015. "Tumor Suppressor Functions of BNIP3 and Mitophagy." *Autophagy* 11 (10): 1937–1938. https://doi.org/10.1080/15548627.2015.1085136.

Chourasia, Aparajita H., Kristin Tracy, Casey Frankenberger, Michelle L. Boland, Marina N. Sharifi, Lauren E. Drake, Joseph R. Sachleben, et al. 2015. "Mitophagy Defects Arising from BNip 3 Loss Promote Mammary Tumor Progression to Metastasis." *EMBO Reports* 16 (9): 1–19. https://doi.org/10.15252/embr.201540759.

Dartier, Julie, Elsa Lemaitre, Igor Chourpa, Caroline Goupille, Stéphane Servais, Stéphan Chevalier, Karine Mahéo, and Jean François Dumas. 2017. "ATP-Dependent Activity and Mitochondrial Localization of Drug Efflux Pumps in Doxorubicin-Resistant Breast Cancer Cells." *Biochimica et Biophysica Acta – General Subjects* 1861 (5): 1075–1084. https://doi.org/10.1016/j.bbagen.2017.02.019.

Ding, Jiefeng, Mei Ling Kuo, Leila Su, Lijun Xue, Frank Luh, Hang Zhang, Jianghai Wang, et al. 2017. "Human Mitochondrial Pyrroline-5-Carboxylate Reductase 1 Promotes Invasiveness and Impacts Survival in Breast Cancers." *Carcinogenesis* 38 (5): 519–531. https://doi.org/10.1093/carcin/bgx022.

Dower, Christopher M., Neema Bhat, Edward W. Wang, and Hong-gang Wang. 2017. "Selective Reversible Inhibition of Autophagy in Hypoxic Breast Cancer Cells Promotes Pulmonary Metastasis." *Cancer Research* 77 (3): 646–657. https://doi.org/10.1158/0008-5472.CAN-15-3458.Selective.

Fadeel, Bengt, and Ewa Grzybowska. 2009. "HAX-1: A Multifunctional Protein with Emerging Roles in Human Disease." *Biochimica et Biophysica Acta – General Subjects* 1790 (10): 1139–1148. https://doi.org/10.1016/j.bbagen.2009.06.004.

Farnie, Gillian, Federica Sotgia, and Michael P. Lisanti. 2015. "High Mitochondrial Mass Identifies a Sub-Population of Stem-like Cancer Cells That Are Chemo-Resistant." *Oncotarget* 6 (31): 30472–30486. https://doi.org/10.18632/oncotarget.5401.

Frantz, Marie Céline, and Peter Wipf. 2010. "Mitochondria as a Target in Treatment." *Environmental and Molecular Mutagenesis* 51 (5): 462–475. https://doi.org/10.1002/em.20554.

Geisler, Sven, Kira M. Holmström, Diana Skujat, Fabienne C. Fiesel, Oliver C. Rothfuss, and Philipp J. Kahle. 2010. "PINK1/Parkin-Mediated Mitophagy Is Dependent on VDAC1 and P62/SQSTM1." *Nature Cell Biology* 12 (2): 119–131. https://doi.org/10.1038/ncb2012.

Georgieva, Ekaterina, Donika Ivanova, and Zhivko Zhelev. 2017. "Mitochondrial Dysfunction and Redox Imbalance as a Diagnostic Marker of 'Free Radical Diseases.'." *Anticancer Research* 37 (10): 5373–5381. https://doi.org/10.21873/anticanres.11963.

Gkikas, Ilias, Konstantinos Palikaras, and Nektarios Tavernarakis. 2018. "The Role of Mitophagy in Innate Immunity." *Frontiers in Immunology* 9 (1283): 1–15. https://doi.org/10.3389/fimmu.2018.01283.

Goldhirsch, Aron, William C. Wood, Alan S. Coates, Richard D. Gelber, Beat Thürlimann, and H. J. Senn. 2011. "Strategies for Subtypes-Dealing with the Diversity of Breast Cancer: Highlights of the St Gallen International Expert Consensus on the Primary Therapy of Early Breast Cancer 2011." *Annals of Oncology* 22 (8): 1736–1747. https://doi.org/10.1093/annonc/mdr304.

Grzybowska-Szatkowska, Ludmiła, Brygida Ślaska, Jolanta Rzymowska, Anna Brzozowska, and Floriańczyk Bolesław. 2014. "Novel Mitochondrial Mutations in the ATP6 and ATP8 Genes in Patients with Breast Cancer." *Molecular*

Medicine Reports 10 (4): 1772–1778. https://doi.org/10.3892/mmr.2014.2471.

Guaragnella, Nicoletta, Sergio Giannattasio, and Loredana Moro. 2014. "Mitochondrial Dysfunction in Cancer Chemoresistance." *Biochemical Pharmacology* 92 (1): 62–72. https://doi.org/10.1016/j.bcp.2014.07.027.

Guerra, Flora, Arnaldo A. Arbini, and Loredana Moro. 2017. "Mitochondria and Cancer Chemoresistance." *Biochimica et Biophysica Acta – Bioenergetics* 1858 (8): 686–699. https://doi.org/10.1016/j.bbabio.2017.01.012.

Guha, Manti, Satish Srinivasan, Pichai Raman, Yuefu Jiang, Brett A. Kaufman, Deanne Taylor, Dawei Dong, et al. 2018. "Aggressive Triple Negative Breast Cancers Have Unique Molecular Signature on the Basis of Mitochondrial Genetic and Functional Defects." *Biochimica et Biophysica Acta – Molecular Basis of Disease* 1864 (4): 1060–1071. https://doi.org/10.1016/j.bbadis.2018.01.002.

Hanahan, Douglas, and Robert A. Weinberg. 2011. "Hallmarks of Cancer: The Next Generation Douglas." *Cell* 144: 646–674. https://doi.org/10.1016/j.cell.2011.02.013.

Hockenbery, David M. 2002. "A Mitochondrial Achilles' Heel in Cancer?" *Cancer Cell* 2 (1): 1–2. https://doi.org/10.1016/S1535-6108(02)00087-9.

Hu, Chenxia, Yong Huang, and Lanjuan Li. 2017. "Drp1-Dependent Mitochondrial Fission Plays Critical Roles in Physiological and Pathological Progresses in Mammals." *International Journal of Molecular Sciences* 18: 1. https://doi.org/10.3390/ijms18010144.

Ježek, Jan, Katrina Cooper, and Randy Strich. 2018. "Reactive Oxygen Species and Mitochondrial Dynamics: The Yin and Yang of Mitochondrial Dysfunction and Cancer Progression." *Antioxidants* 7 (1): 13. https://doi.org/10.3390/antiox7010013.

Jin, Shengkan. 2006. "Autophagy, Mitochondrial Quality Control, and Oncogenesis." *Autophagy* 2 (2): 80–84. https://doi.org/10.4161/auto.2.2.2460.

Jornayvaz, François R., and Gerald I. Shulman. 2010. "Regulation of Mitochondrial Biogenesis." *Essays in Biochemistry* 47: 69–84. https://doi.org/10.1042/bse0470069.Regulation.

Kandoth, Cyriac, Michael D. McLellan, Mingchao Xie, Qunyuan Zhang, Joshua F. Mcmichael, Matthew A. Wyczalkowski, Michael C. Wendl, Timothy J. Ley, Richard K. Wilson, and Benjamin J. Raphael. 2014. "Mutational Landscape and Significance across 12 Major Cancer Types." *Nature* 502 (7471): 333–339. https://doi.org/10.1038/nature12634.Mutational.

Kang, Dongchon, Sang Ho Kim, and Naotaka Hamasaki. 2007. "Mitochondrial Transcription Factor A (TFAM): Roles in Maintenance of MtDNA and Cellular Functions." *Mitochondrion* 7 (1–2): 39–44. https://doi.org/10.1016/j.mito.2006.11.017.

Kannan, Anbarasu, Robert B. Wells, Subramaniam Sivakumar, Satoshi Komatsu, Karan P. Singh, Buka Samten, Julie V. Philley, et al. 2016. "Mitochondrial Reprogramming Regulates Breast Cancer Progression." *Clinical Cancer Research* 22 (13): 3348–3360. https://doi.org/10.1158/1078-0432.CCR-15-2456.

Kim, Sung Ouk, and Jiahuai Han. 2001. "Pan-Caspase Inhibitor ZVAD Enhances Cell Death in RAW246.7 Macrophages." *Journal of Endotoxin Research* 7 (4): 292–296. https://doi.org/10.1179/096805101101532873.

Ksiⓧazakowska-Łakoma, Kinga, Monika Zyła, and Jacek R. Wilczyⓧski. 2014. "Mitochondrial Dysfunction in Cancer." *Przeglad Menopauzalny* 18 (2): 136–144. https://doi.org/10.5114/pm.2014.42717.

Lane, David P. 1992. "Cancer. P53, Guardian of the Genome." *Nature.* https://doi.org/10.1038/358015a0.

Lee, Kyung min, Jennifer M. Giltnane, Justin M. Balko, Luis J. Schwarz, Angel L. Guerrero-Zotano, Katherine E. Hutchinson, Mellissa J. Nixon, et al. 2017. "MYC and MCL1 Cooperatively Promote Chemotherapy-Resistant Breast Cancer Stem Cells via Regulation of Mitochondrial Oxidative Phosphorylation." *Cell Metabolism* 26 (4): 633–647. e7. https://doi.org/10.1016/j.cmet.2017.09.009.

Liao, Tien Ling, Chii Ruey Tzeng, Chao Lan Yu, Yi Pei Wang, and Shu Huei Kao. 2015. "Estrogen Receptor-ⓧ in Mitochondria: Implications for Mitochondrial Bioenergetics and Tumorigenesis." *Annals of the New York Academy of Sciences* 1350 (1): 52–60. https://doi.org/10.1111/nyas.12872.

Lin, Meng Liang, Shih Shun Chen, Ren Yu Huang, Yao Cheng Lu, Yu Ren Liao, Mopuru Vijaya Bhaskar Reddy, Chuan Chun Lee, and Tian Shung Wu. 2014. "Suppression of PI3K/Akt Signaling by Synthetic Bichalcone Analog TSWU-CD4 Induces ER Stress- and Bax/Bak-Mediated Apoptosis of Cancer Cells." *Apoptosis* 19 (11): 1637–1653. https://doi.org/10.1007/s10495-014-1031-y.

Liu, Raymond, and David C. Chan. 2015. "The Mitochondrial Fission Receptor Mff Selectively Recruits Oligomerized Drp1." *Molecular Biology of the Cell* 26 (24): 4466–4477. https://doi.org/10.1091/mbc.E15-08-0591.

Liu, Yifeng, Yulin Jiang, Nian Wang, Qianni Jin, Feihu Ji, Changli Zhong, Zhiqiang Zhang, Junhong Yang, Xiangsen Ye, and Tingmei Chen. 2017a. "Invalidation of Mitophagy by FBP1-Mediated Repression Promotes Apoptosis in Breast Cancer." *Tumor Biology* 39: 6. https://doi.org/10.1177/1010428317708779.

Liu, Zhaoqi, Xiang Sun Zhang, and Shihua Zhang. 2014. "Breast Tumor Subgroups Reveal Diverse Clinical Prognostic Power." *Scientific Reports* 4: 1–9. https://doi.org/10.1038/srep04002.

Liu, Zhenyu, Aiping Shi, Dong Song, Bing Han, Zhiru Zhang, Le Ma, Dongxu Liu, and Zhimin Fan. 2017b. "Resistin Confers Resistance to Doxorubicin-Induced Apoptosis in Human Breast Cancer Cells through Autophagy Induction." *American Journal of Cancer Research* 7 (3): 574–583.

Lopez, Marcos, Micaël Hardy, Olivier Ouari, and Sandra Milena Sanabria-Barrera. 2017. "Mitochondrial Fuel Metabolic Differences in Triple Negative Breast Cancer." *Free Radical Biology and Medicine* 112: 171. https://doi.org/10.1016/j.freeradbiomed.2017.10.266.

López-Knowles, Elena, Sandra A. O'Toole, Catriona M. McNeil, Ewan K.A. Millar, Min R. Qiu, Paul Crea, Roger J. Daly, Elizabeth A. Musgrove, and Robert L. Sutherland. 2010. "PI3K Pathway Activation in Breast Cancer Is Associated with the Basal-like Phenotype and Cancer-Specific Mortality." *International Journal of Cancer* 126 (5): 1121–1131. https://doi.org/10.1002/ijc.24831.

Luo, Bihui, and Amy S. Lee. 2012. "The Critical Roles of Endoplasmic Reticulum Chaperones and Unfolded Protein Response in Tumorigenesis and Anticancer Therapies." *Oncogene* February: 1–14. https://doi.org/10.1038/onc.2012.130.

Lyakhovich, Alex, and Matilde E. Lleonart. 2016. "Bypassing Mechanisms of Mitochondria-Mediated Cancer Stem Cells Resistance to Chemo- and Radiotherapy." *Oxidative Medicine and Cellular Longevity* 2016. https://doi.org/10.1155/2016/1716341.

Maddalena, Francesca, Lorenza Sisinni, Giacomo Lettini, Valentina Condelli, Danilo Swann Matassa, Annamaria Piscazzi,

Maria Rosaria Amoroso, Giuseppe La Torre, Franca Esposito, and Matteo Landriscina. 2013. "Resistance to Paclitxel in Breast Carcinoma Cells Requires a Quality Control of Mitochondrial Antiapoptotic Proteins by TRAP1." *Molecular Oncology* 7 (5): 895–906. https://doi.org/10.1016/j.molonc.2013.04.009.

Makki, Jaafar. 2015. "Diversity of Breast Carcinoma: Histological Subtypes and Clinical Relevance." *Clinical Medicine Insights: Pathology* 8 (1): 23–31. https://doi.org/10.4137/CPath.s31563.

Mani, Sendurai A., Wenjun Guo, Mai Jing Liao, Elinor Ng Eaton, Ayyakkannu Ayyanan, Alicia Y. Zhou, Mary Brooks, et al. 2008. "The Epithelial-Mesenchymal Transition Generates Cells with Properties of Stem Cells." *Cell* 133 (4): 704–715. https://doi.org/10.1016/j.cell.2008.03.027.

Martinez-Outschoorn, Ubaldo E., Richard G. Pestell, Anthony Howell, Mark L. Tykocinski, Fnu Nagajyothi, Fabiana S. Machado, Herbert B. Tanowitz, Federica Sotgia, and Michael P. Lisanti. 2011. "Energy Transfer in 'Parasitic' Cancer Metabolism: Mitochondria Are the Powerhouse and Achilles' Heel of Tumor Cells." *Cell Cycle* 10 (24): 4208–4216. https://doi.org/10.4161/cc.10.24.18487.

Michelakis, E. D., L. Webster, and J. R. Mackey. 2008. "Dichloroacetate (DCA) as a Potential Metabolic-Targeting Therapy for Cancer." *British Journal of Cancer* 99 (7): 989–994. https://doi.org/10.1038/sj.bjc.6604554.

Modica-Napolitano, Josephine S., and June R. Aprille. 2001. "Delocalized Lipophilic Cations Selectively Target the Mitochondria of Carcinoma Cells." *Advanced Drug Delivery Reviews* 49 (1–2): 63–70. https://doi.org/10.1016/S0169-409X(01)00125-9.

Moulder, David E., Diana Hatoum, Enoch Tay, Yiguang Lin, and Eileen M. McGowan. 2018. "The Roles of P53 in Mitochondrial Dynamics and Cancer Metabolism: The Pendulum between Survival and Death in Breast Cancer?" *Cancers* 10: 6. https://doi.org/10.3390/cancers10060189.

Murphy, Michael P., and Robin A.J. Smith. 2007. "Targeting Antioxidants to Mitochondria by Conjugation to Lipophilic Cations." *Annual Review of Pharmacology and Toxicology* 47: 629–656. https://doi.org/10.1146/annurev.pharmtox.47.120505.105110.

Neuzil, Jiri, Lan Feng Dong, Jakub Rohlena, Jaroslav Truksa, and Stephen J. Ralph. 2013. "Classification of Mitocans, Anti-Cancer Drugs Acting on Mitochondria." *Mitochondrion* 13 (3): 199–208. https://doi.org/10.1016/j.mito.2012.07.112.

Ng Kee Kwong, Francois, Andrew G. Nicholson, Celeste L. Harrison, Philip M. Hansbro, Ian M. Adcock, and Kian Fan Chung. 2017. "Is Mitochondrial Dysfunction a Driving Mechanism Linking COPD to Nonsmall Cell Lung Carcinoma?" *European Respiratory Review* 26: 146. https://doi.org/10.1183/16000617.0040-2017.

O'Flanagan, Ciara H., and Cora O'Neill. 2014. "PINK1 Signalling in Cancer Biology." *Biochimica et Biophysica Acta – Reviews on Cancer* 1846 (2): 590–598. https://doi.org/10.1016/j.bbcan.2014.10.006.

Oo, Phyu Synn, Yuya Yamaguchi, Akira Sawaguchi, Myat Tin Htwe Kyaw, Narantsog Choijookhuu, Mohmand Noor Ali, Naparee Srisowanna, Shin-ichiro Hino, and Yoshitaka Hishikawa. 2018. "Estrogen Regulates Mitochondrial Morphology through Phosphorylation of Dynamin-Related Protein 1 in MCF7 Human Breast Cancer Cells." *Acta Histochemica et Cytochemica* 51 (1): 21–31. https://doi.org/10.1267/ahc.17034.

Pagano, Giovanni, Annarita Aiello Talamanca, Giuseppe Castello, Mario D. Cordero, Marco D'Ischia, Maria Nicola Gadaleta, Federico V. Pallardó, Sandra Petrović, Luca Tiano, and Adriana Zatterale. 2014. "Oxidative Stress and Mitochondrial Dysfunction across Broad-Ranging Pathologies: Toward Mitochondria-Targeted Clinical Strategies." *Oxidative Medicine and Cellular Longevity* 2014. https://doi.org/10.1155/2014/541230.

Panis, Carolina, Vanessa Jacob Victorino, Ana Cristina da Silva do Amaral Herrera, Alessandra Lourenço Cecchini, Andréa Name Colado Simão, Luciana Yuki Tomita, and Rubens Cecchini. 2015. "Can Breast Tumors Affect the Oxidative Status of the Surrounding Environment? A Comparative Analysis among Cancerous Breast, Mammary Adjacent Tissue, and Plasma." *Oxidative Medicine and Cellular Longevity* 2015: 6429812. https://doi.org/10.1155/2016/6429812.

Park, Ji-Hoon, Jie Zhuang, Jie Li, and Paul M. Hwang. 2016. "P53 as Guardian of the Mitochondrial Genome." *FEBS Letters* 590 (7): 924–934. https://doi.org/10.1002/1873-3468.12061.

Pelicano, Hélène, Wan Zhang, Jinyun Liu, Naima Hammoudi, Jiale Dai, Rui-Hua Xu, Lajos Pusztai, and Peng Huang. 2014. "Mitochondrial Dysfunction in Some Triple-Negative Breast Cancer Cell Lines: Role of MTOR Pathway and Therapeutic Potential." *Breast Cancer Research: BCR* 16 (5): 434. https://doi.org/10.1186/s13058-014-0434-6.

Peng, Yong Bo, Zi Long Zhao, Teng Liu, Guo Jian Xie, Cheng Jin, Tang Gang Deng, Yang Sun, et al. 2017. "A Multi-Mitochondrial Anticancer Agent that Selectively Kills Cancer Cells and Overcomes Drug Resistance." *ChemMedChem* 12 (3): 250–256. https://doi.org/10.1002/cmdc.201600538.

Prat, Aleix, Estela Pineda, Barbara Adamo, Patricia Galván, Aranzazu Fernández, Lydia Gaba, Marc Díez, Margarita Viladot, Ana Arance, and Montserrat Muñoz. 2015. "Clinical Implications of the Intrinsic Molecular Subtypes of Breast Cancer." *The Breast* 24: S26–S35. https://doi.org/10.1016/j.breast.2015.07.008.

Qadir, Mohammed A., Benjamin Kwok, Wieslawa H. Dragowska, Kam H. To, David Le, Marcel Bally, and Sharon M. Gorski. 2008. "Macroautophagy Inhibition Sensitizes Tamoxifen-Resistant Breast Cancer Cells and Enhances Mitochondrial Depolarization." *Breast Cancer Research and Treatment* 112: 389–403. https://doi.org/10.1007/s10549-007-9873-4.

Qiao, Aimin, Kuansong Wang, Yunsheng Yuan, Yidi Guan, Xingcong Ren, Li Lanya, Xisha Chen, et al. 2016. "Sirt3-Mediated Mitophagy Protects Tumor Cells against Apoptosis under Hypoxia." *Oncotarget* 7 (28): 43390–43400. https://doi.org/10.18632/oncotarget.9717.

Radde, Brandie N., Margarita M. Ivanova, Huy Xuan Mai, Negin Alizadeh-Rad, Kellianne Piell, Patrick Van Hoose, Marsha P. Cole, et al. 2016. "Nuclear Respiratory Factor-1 and Bioenergetics in Tamoxifen- Resistant Breast Cancer Cells." *Experimental Cell Research* 347 (1): 222–231. https://doi.org/10.1016/j.yexcr.2016.08.006.

Rangel, Nelson, Victoria E. Villegas, and Milena Rondón-Lagos. 2017. "Profiling of Gene Expression Regulated by 17☒-Estradiol and Tamoxifen in Estrogen Receptor-Positive and Estrogen Receptor-Negative Human Breast Cancer Cell Lines." *Breast Cancer: Targets and Therapy* 9: 537–550. https://doi.org/10.2147/BCTT.S146247.

Rodger, Catherine E., Thomas G. Mcwilliams, and Ian G. Ganley. 2017. "Mammalian Mitophagy – from in Vitro Molecules to in Vivo Models." *FEBS Journal* 285: 1185–1202. https://doi.org/10.1111/febs.14336.

Salhab, Mohamed, Neill Patani, Wen Jiang, and Kefah Mokbel. 2012. "High TIMM17A Expression Is Associated with Adverse Pathological and Clinical Outcomes in Human Breast Cancer." *Breast Cancer* 19 (2): 153–160. https://doi.org/10.1007/s12282-010-0228-3.

Samaddar, Julia S., Virgil T. Gaddy, Jennifer Duplantier, Sudharsan Periyasamy Thandavan, Manish Shah, Marlena J. Smith, Darren Browning, et al. 2008. "A Role for Macroautophagy in Protection against 4-Hydroxytamoxifen – Induced Cell Death and the Development of Antiestrogen Resistance." *Molecular Cancer Therapeutics* 7 (September): 2977–2988. https://doi.org/10.1158/1535-7163.MCT-08-0447.

Samadi, Abbas K., Alan Bilsland, Alexandros G. Georgakilas, Amedeo Amedei, Amr Amin, Anupam Bishayee, Asfar S. Azmi, et al. 2015. "A Multi-Targeted Approach to Suppress Tumor-Promoting Inflammation." *Seminars in Cancer Biology* 35: 151–184. https://doi.org/10.1016/j.semcancer.2015.04.010.Targeting.

Sciacovelli, Marco, Edoardo Gaude, Mika Hilvo, and Christian Frezza. 2014. *The Metabolic Alterations of Cancer Cells. Methods in Enzymology.* 1st ed. Vol. 542. Elsevier Inc. https://doi.org/10.1016/B978-0-12-416618-9.00001-7.

Senft, Daniela, and La Jolla. 2016. "Regulators of Mitochondrial Dynamics in Cancer." *Current Opinion in Cell Biology* 39: 43–52. https://doi.org/10.1016/j.ceb.2016.02.001. Regulators.

Shao, Fangyuan, Heng Sun, and Chu-Xia Deng. 2017. "Potential Therapeutic Targets of Triple-Negative Breast Cancer Based on Its Intrinsic Subtype." *Oncotarget* 8 (42): 73329–73344. https://doi.org/10.18632/oncotarget.20274.

Sinha, Krishnendu, Joydeep Das, Pabitra Bikash Pal, and Parames C. Sil. 2013. "Oxidative Stress: The Mitochondria-Dependent and Mitochondria-Independent Pathways of Apoptosis." *Archives of Toxicology* 87 (7): 1157–1180. https://doi.org/10.1007/s00204-013-1034-4.

Soni, Mithil, Yogin Patel, Eleni Markoutsa, Chunfa Jie, Shou Liu, Peisheng Xu, and Hexin Chen. 2018. "Autophagy, Cell Viability and Chemo-Resistance Are Regulated by MiR-489 in Breast Cancer." *Molecular Cancer Research*, molcanres.0634.2017. https://doi.org/10.1158/1541-7786.MCR-17-0634.

Sotgia, Federica, Marco Fiorillo, and Michael P. Lisanti. 2017. "Mitochondrial Markers Predict Recurrence, Metastasis and Tamoxifen-Resistance in Breast Cancer Patients: Early Detection of Treatment Failure with Companion Diagnostics." *Oncotarget* 8 (40): 68730–68745. https://doi.org/10.18632/oncotarget.19612.

Sotgia, Federica, Diana Whitaker-Menezes, Ubaldo E. Martinez-Outschoorn, Ahmed F. Salem, Aristotelis Tsirigos, Rebecca Lamb, Sharon Sneddon, James Hulit, Anthony Howell, and Michael P. Lisanti. 2012. "Mitochondria 'Fuel' Breast Cancer Metabolism: Fifteen Markers of Mitochondrial Biogenesis Label Epithelial Cancer Cells, but Are Excluded from Adjacent Stromal Cells." *Cell Cycle* 11 (October 2017): 4390–4401. https://doi.org/10.4161/cc.22777.

Sripada, Lakshmi, Kritarth Singh, Anastasiya V. Lipatova, Aru Singh, Paresh Prajapati, Dhanendra Tomar, Khyati Bhatelia, et al. 2017. "Hsa-MiR-4485 Regulates Mitochondrial Functions and Inhibits the Tumorigenicity of Breast Cancer Cells." *Journal of Molecular Medicine* 95 (6): 641–651. https://doi.org/10.1007/s00109-017-1517-5.

Su, Jing, Lei Zhou, Mei-hui Xia, Ye Xu, Xi-yan Xiang, and Liankun Sun. 2014. "Bcl-2 Family Proteins Are Involved in the Signal Crosstalk between Endoplasmic Reticulum Stress and Mitochondrial Dysfunction in Tumor Chemotherapy

Resistance." *BioMed Research International* 2014: 234370. https://doi.org/10.1155/2014/234370.

Van de Ven, Robert A. H., Daniel Santos, and Marcia C. Haigis. 2017. "Mitochondrial Sirtuins and Molecular Mechanisms of Aging." *Trends in Molecular Medicine* 23 (4): 320–331. https://doi.org/10.1016/j.cogdev.2010.08.003.Personal.

Vazquez-Martin, Alejandro, Cristina Oliveras-Ferraros, and Javier A. Menendez. 2009. "Autophagy Facilitates the Development of Breast Cancer Resistance to the Anti-HER2 Monoclonal Antibody Trastuzumab." *PLoS One* 4 (7). https://doi.org/10.1371/journal.pone.0006251.

Viedma-Rodriguez, Rubí, Luis Arturo Baiza-Gutman, Alejandro García-Carran̄a, Leticia Moreno-Fierros, Fabio Salamanca-Gómez, and Diego Arenas-Aranda. 2013. "Suppression of the Death Gene BIK Is a Critical Factor for Resistance to Tamoxifen in MCF-7 Breast Cancer Cells." *International Journal of Oncology* 43 (6): 1777–1786. https://doi.org/10.3892/ijo.2013.2127.

von Eyss, Björn, Laura A. Jaenicke, Roderik M. Kortlever, Nadine Royla, Katrin E. Wiese, Sebastian Letschert, Leigh Anne McDuffus, et al. 2015. "A MYC-Driven Change in Mitochondrial Dynamics Limits YAP/TAZ Function in Mammary Epithelial Cells and Breast Cancer." *Cancer Cell* 28 (6): 743–757. https://doi.org/10.1016/j.ccell.2015.10.013.

Wallace, Douglas C. 2012. "Mitochondria and Cancer." *Nature Reviews Cancer* 12 (10): 685–698. https://doi.org/10.1038/nrc3365.Mitochondria.

Wang, Tianyi, Johannes Francois Fahrmann, Heehyoung Lee, Yi Jia Li, Satyendra C. Tripathi, Chanyu Yue, Chunyan Zhang, et al. 2018. "JAK/STAT3-Regulated Fatty Acid β-Oxidation Is Critical for Breast Cancer Stem Cell Self-Renewal and Chemoresistance." *Cell Metabolism* 27: 1–15. https://doi.org/10.1016/j.cmet.2017.11.001.

Warburg, Otto. 1931. "The Metabolism of Tumours: Investigations from the Kaiser Wilhelm Institute for Biology, Berlin-Dahlem." *BJS*.

——— 1956a. "On Respiratory Impairment in Cancer Cells." *Science* 124 (3215): 269–270.

——— 1956b. "On the Origin of Cancer Cells." *Science* 123 (3191): 309–314.

Wen, Jian, Syn Yeo, Chenran Wang, Song Chen, Shaogang Sun, Micheal A. Haas, Wei Tu, Feng Jin, and Jun-Lin Guan. 2015. "Autophagy Inhibition Re-Sensitizes Pulse Stimulation-Selected Paclitaxel-Resistant Triple Negative Breast Cancer Cells to Chemotherapy-Induced Apoptosis." *Breast Cancer Research and Treatment* 149 (3): 619–629. https://doi.org/10.1016/j.cogdev.2010.08.003.Personal.

White, Eileen. 2016. "Autophagy and P53." *Cold Spring Harbor Perspectives in Medicine* 6 (4): 1–9. https://doi.org/10.1101/cshperspect.a026120.

Wyr̄bska, Anna, Katarzyna Gach, Urszula Lewandowska, Karolina Szewczyk, El̄bieta Hrabec, Jakub Modranka, Rafał Jakubowski, Tomasz Janecki, Jacek Szymāski, and Anna Janecka. 2013. "Anticancer Activity of New Synthetic α-Methylene-δ-Lactones on Two Breast Cancer Cell Lines." *Basic & Clinical Pharmacology & Toxicology* 113 (6): 391–400. https://doi.org/10.1111/bcpt.12120.

Xie, Xiaohong, Yuanyuan Hu, Leilai Xu, Yongqing Fu, Jue Tu, Hong Zhao, Shuo Zhang, Ri Hong, and Xidong Gu. 2015. "The Role of MiR-125b-Mitochondria-Caspase-3 Pathway in Doxorubicin Resistance and Therapy in Human Breast Cancer." *Tumor Biology* 36 (9): 7185–7194. https://doi.org/10.1007/s13277-015-3438-7.

Xu, Ke, Guo Chen, Xiaobo Li, Xiaoqin Wu, Zhijie Chang, Jianhua Xu, Yu Zhu, Peihao Yin, Xin Liang, and Lei Dong. 2017.

"MFN2 Suppresses Cancer Progression through Inhibition of MTORC2/Akt Signaling." *Scientific Reports* 7 (September 2016): 1–13. https://doi.org/10.1038/srep41718.

Yang, Ji, Yue Wu, Xiao Wang, Liqian Xu, Xiaohong Zhao, and Yunmei Yang. 2017. "Chemoresistance Is Associated with Overexpression of HAX-1, Inhibition of Which Resensitizes Drug-Resistant Breast Cancer Cells to Chemotherapy." *Tumor Biology* 39 (3): 1–10. https://doi.org/10.1177/1010428317692228.

Yuan, Yuan, Yu Feng Yao, Sai Nan Hu, Jin Gao, and Li Li Zhang. 2015. "MiR-133a Is Functionally Involved in Doxorubicin-Resistance in Breast Cancer Cells MCF-7 via Its Regulation of the Expression of Uncoupling Protein 2." *PLoS One* 10 (6): 1–11. https://doi.org/10.1371/journal.pone.0129843.

Zhang, Qinglin, Zhi Liang, Yongxiang Gao, Maikun Teng, and Liwen Niu. 2016. "Differentially Expressed Mitochondrial Genes in Breast Cancer Cells: Potential New Targets for Anti-Cancer Therapies." *Gene.* https://doi.org/10.1016/j.gene.2016.10.005.

Zhang, Xiaolong, Linhua Lan, Lili Niu, Juping Lu, Changxi Li, Miaomiao Guo, Shouyong Mo, Jing Lu, Yongzhang Liu, and Bin Lu. 2017. "Oxidative Stress Regulates Cellular Bioenergetics in Esophageal Squamous Cell Carcinoma Cell." *Bioscience Reports* 37 (6): BSR20171006. https://doi.org/10.1042/BSR20171006.

Zhou, Jing, Guobing Li, Yi Zheng, Han-ming Shen, Xiaoye Hu, Qian-liang Ming, Cheng Huang, Peng Li, and Ning Gao. 2015. "A Novel Autophagy/Mitophagy Inhibitor Liensinine Sensitizes Breast Cancer Cells to Chemotherapy through DNM1L-Mediated Mitochondrial Fi Ssion." *Autophagy* 11 (September 2016): 1259–1279. https://doi.org/10.1080/15548627.2015.1056970.

Zong, Wei-Xing, Joshua D. Rabinowitz, and Eileen White. 2016. "Mitochondria and Cancer." *Molecular Cell* 61 (5): 667–676. https://doi.org/10.1016/j.cogdev.2010.08.003.Personal.

Zou, Peng, Longhua Liu, Louise D. Zheng, Kyle K. Payne, Masoud H. Manjili, Michael O. Idowu, Jinfeng Zhang, Eva M. Schmelz, and Zhiyong Cheng. 2016. "Coordinated Upregulation of Mitochondrial Biogenesis and Autophagy in Breast Cancer Cells: The Role of Dynamin Related Protein-1 and Implication for Breast Cancer Treatment." *Oxidative Medicine and Cellular Longevity* 2016. https://doi.org/10.1155/2016/4085727.

11 Mitochondrial Dysfunction and DNA Methylation in Atherosclerosis

Silvio Zaina

Department of Medical Sciences, Division of Health Sciences, León Campus, University of Guanajuato, León, Gto., Mexico.

Gertrud Lund

Department of Genetic Engineering, CINVESTAV Irapuato Unit, Irapuato, Gto., Mexico.

CONTENTS

1   INTRODUCTION

1.1   ATHEROSCLEROSIS: A BRIEF OVERVIEW

Atherosclerosis is a disease characterized by the development of lesions (referred to as atheromas or plaques) in the vascular wall of medium size and large arteries. In the initial phase of atherosclerosis, excess circulating apolipoprotein B (ApoB)-containing lipoproteins (very low, intermediate and low density lipoproteins, VLDL, IDL and LDL, respectively) causes endothelial dysfunction, which results in loss of endothelial barrier function (Hoff et al. 1975; Feng et al. 2018). Consequently, ApoB lipoproteins infiltrate into the underlying vascular wall, where they undergo a range of chemical modifications including oxidation and proteolysis. Modified lipoproteins evoke an inflammatory response marked by the accumulation of macrophages and other immune cells within the vascular tissue. These processes are accompanied by the accumulation of extracellular matrix (ECM) and a phenotypic switch of vascular smooth muscle cells (VSMCs) from contractile to synthetic. While contractile VSMCs are terminally differentiated cells specialized in regulating blood pressure, the synthetic counterparts are capable of migration and proliferation. Sustained cellular proliferation, ECM deposition and chronic inflammation lead to the development of a fibrocellular lesion. In humans, this process is clinically silent for decades, yet the accumulation of apoptotic/necrotic cell debris, erosion of the ECM by inflammatory cell-derived proteolytic enzymes and calcification can promote plaque weakening and rupture (Panh et al. 2017; Tabas 2017). The resulting thrombosis is the underlying cause of the clinical complications of atherosclerosis, depending on the anatomical location of the ruptured atheroma: stroke, myocardial infarction and peripheral vascular disease. Epidemiologically, despite evidence of a decreasing trend, atherosclerosis-related cardiovascular disease (CVD) ranks among the leading cause of mortality worldwide and is expected to hold that grim record for the next two decades (Mathers and Loncar 2006; Hackam and Spence 2018).

2   MITOCHONDRIAL FUNCTIONALITY IN ATHEROSCLEROSIS

The notion that mitochondria play a fundamental role in cell homeostasis and metabolism grounds the hypothesis that mitochondrial dysfunction is an important under-lying cause of atherosclerosis. The field has been intensively studied for decades and continues to awake a considerable interest. One of the earliest studies of that kind, established a link between atherogenic diet, induced magnesium deficiency and defective cardiac oxidative phosphorylation (OXPHOS) in a rodent model (Vitale et al. 1957). In a later example, following a logic that is echoed by recent studies of CVD and environmental epigenetics (Baccarelli et al. 2009; Fiorito et al. 2017), decreased mitochondrial metabolism resulting in lipid accumulation was proposed to explain the association between CVD risk and exposure to ambient carbon monoxide (Whereat 1970). Despite subsequent results showing no impairment of aortic mitochondrial activity in two animal models of atherosclerosis (Morrison et al. 1973), a flurry of studies followed. In the next paragraphs, we will present the major directions taken by research in the field.

2.1 Mitochondrial Loss

Epidemiologically, the observation that cardiovascular complications are a major cause of death in patients with mitochondrial diseases, a heterogeneous set of disorders, supports a causal role of overall loss of mitochondrial activity in atherosclerosis. Mitochondrial function impairment may be due to decreased enzymatic activity, loss of mitochondrial mass, loss of mitochondrial DNA (mtDNA) or a combination thereof (Barends et al. 2016).

2.2 Mitochondrial DNA Abundance

Perhaps the clearest mechanistic evidence that mitochondrial DNA (mtDNA) abundance affects the atherosclerosis burden, is that expression of a defective mtDNA polymerase in a apolipoprotein E (ApoE)-null background in mice, a model of hyperlipidaemia-induced atherosclerosis, increases lesion size (Zhang et al. 1992; Yu et al. 2013). mtDNA copy number is a practical and useful proxy of mitochondrial activity in atherosclerosis. Leukocyte mtDNA content was shown to be inversely correlated with ischemic stroke incidence (Lien et al. 2017). Also, the study revealed a synergy between low mtDNA copy number and an array of CVD risk factors including smoke and diabetes mellitus. Interestingly, mtDNA copy number correlated inversely although weakly with a marker of mitochondrial biogenesis (Peroxisome proliferator-activated receptor (PPAR)γ coactivator 1α) and a stimulator of thermogenesis (Uncoupling protein 2). The latter results reveal a compensatory response to low mtDNA aimed at restoring mitochondrial functionality, yet at the same time illustrate one *caveat* in interpreting mitochondria-related gene expression as proxy of mitochondrial functionality. Another

study in peripheral blood leukocytes of >20,000 participants from three independent cohorts reached similar conclusions, with a particularly strong independent association with coronary heart disease (Ashar et al. 2017). Importantly from a clinical viewpoint, mtDNA copy number improved the identification of patients to be recommended for preventive treatment with statin. A more recent study took one step further, analysing a common deletion in mtDNA that affects a number of respiratory chain components, in addition to surveying mtDNA abundance *per se* (Vecoli et al. 2018). That prospective study clearly identified mtDNA integrity as a predictor of cardiac events.

Vascular tissue mtDNA copy number studies are uncommon. One such study addresses the causal role of mtDNA abundance in human and murine atheromas (Yu et al. 2017). First, the authors confirmed previous findings that mtDNA copy number and mitochondrial respiration are reduced in human plaques and by exploiting a cultured VSMC model, assigned a mtDNA-depleting activity to oxidized LDL. Subsequently, the study addressed causality in ApoE-null mice. Aortic root lesions in that model mirrored human counterparts with respect to mtDNA abundance and mitochondrial respiration rate. Crucially, overexpression of Twinkle from a transgene in the ApoE-null background improved mtDNA copy number in macrophages although not significantly in whole aortic plaques or isolated VSMCs. Twinkle is a helicase that participates in a protein complex necessary for mtDNA replication. Although Twinkle exerted complex, cell-specific effects on ApoE-null mtDNA, it markedly improved markers of plaque stability, i.e. increased fibrous cap thickness and decreased necrotic core mass. Overall, this important work established causality and might therefore open novel therapeutic avenues. Another noticeable study addresses mtDNA integrity in atherosclerosis (Docherty et al. 2018). The authors performed a detailed analysis of protein expression, mtDNA integrity and cellular metabolism in carotid plaques obtained by endarterectomy, therefore mainly advanced lesions. VSMCs of the lesion cap - the roughly central part of the lesion, on the luminal side of the lipid core - showed diminished mtDNA integrity and increased Pink1 kinase expression. Pink1 anchors to the surface of mitochondria with defective OXPHOS and tags them for mitophagy. The finding may be clinically relevant as cap VSMCs are implicated in plaque stability.

2.3 Loss of Mitochondrial Enzyme Activity

An early study addressed the activity of succinic dehydrogenase, one of the enzymes of the tricarboxylic acid cycle (TCAC) and electron transport chain, in the myocardium of a rabbit model of atherosclerosis (Rosnowski and Kujawa 1977). Here, a biphasic trend was observed, with an increase in young animals and a decrease in advanced atherosclerosis. To our knowledge, this study is unique in addressing mitochondrial function at different CVD stages. We present further implications of these findings in the paragraph dedicated to the atheroma DNA methylome. In a follow-up study of monocyte cytochrome oxidase expression, a predictive value for coronary events was assigned to low expressing subjects, thus clearly making the case for a potential clinical application (Holvoet et al. 2016).

2.4 REACTIVE OXYGEN SPECIES

A major lead in the field is the link between cellular oxidative stress and mitochondrial function. The concept stems from the consensus that oxidative stress contributes to endothelial dysfunction and consequently to atherosclerosis by creating reactive oxygen species (ROS). ROS are two-edged swords that are physiologically produced as by-products of mitochondrial respiration but can favour mitochondrial dysfunction by targeting mitochondrial DNA (mtDNA), lipids and metabolic enzymes. Additionally, oxidized LDL can exacerbate ROS-induced mitochondrial damage. The readers are referred to an article by Puddu et al. that discusses the basic concepts and reviews the extensive literature on ROS (Puddu et al. 2005) and to a recent thorough review more focused on atherosclerosis (Yang et al. 2017). Here, we provide an outline of the main findings on ROS, mitochondria, and atherosclerosis. Remarkably, despite the large body of general literature on ROS and atherosclerosis, studies on the subjects continue to emerge, a clear indication that the topic is complex and not completely understood. Studies in LDL receptor (LDLR)-null mice, a widely used mouse model of atherosclerosis, offer correlative evidence pointing to ROS as major culprits in vascular damage. Multiple LDLR-null mouse tissues showed ROS increase in comparison with wild-type controls, pointing to the interesting and little explored concept that altered lipoprotein metabolism has widespread consequences besides the arteries and the liver (Oliveira et al. 2005). Another signature distinguishing mutant from wild-type mitochondria was a loss of matrix NAPDH, a major source of reducing capacity. These results were echoed by a functional study that demonstrated a decrease in proinflammatory markers subsequent to loss of NADPH oxidase 4 (NOX4) activity by RNA interference in cultured mouse VSMCs (Lozhkin et al. 2017). As supporting evidence, the same study showed that NOX4 is increased in ApoE-null mouse lesions. Furthermore, a survey of ageing LDLR-null mice revealed a strong correlation between aortic lesion size and mitochondrial ROS (Dorighello et al. 2018). Besides NOX4, cytochrome oxidases are additional players in ROS formation and specifically the mitochondria-encoded 1 enzyme is decreased in vascular lesions of

mouse and pigs, coinciding with low expression of a range of guardians of mitochondrial homeostasis such as PPARs (Holvoet et al. 2017). Taking a step further, Wang et al. aimed at buffering the effects of ROS by expressing a catalase transgene, an enzyme responsible for hydrogen peroxide degradation, in the LDLR-null background (Wang et al. 2017). Catalase expression strongly decreased lesion size, despite no reduction in cholesterol. Taken together, the data underline the importance and clarify molecular mechanisms of ROS accumulation to explain ageing as a cardiovascular risk factor.

The importance of ROS-related genes in atherosclerosis has been recently confirmed by an association study between SNPs in antioxidant genes and atherosclerosis in a Chinese cohort (Yeh et al. 2018). The Val16Ala polymorphism in the manganese superoxide dismutase (MnSOD) was proposed as an independent for coronary artery disease.

Finally, on the topic of mitophagy and ROS, the reader is referred to a recent review that addresses the molecular mechanisms involved and the contradictory data in the literature (Wible and Bratton 2018).

2.5 ROLE OF LIPIDS

Lipids play fundamental roles both as cardiovascular risk factors and molecules participating in mitochondrial damage. An early fundamental study established more than four decades ago that free fatty acids (FFAs) induce mitochondrial damage (Acosta and Wenzel 1974). A nearly contemporary work uncovered a link between hypercholesterolemia and loss of VSMC mitochondrial function (Cheney et al. 1977). The issue is still subject to intense research, and lipoproteins are an obvious point of interest. Mechanistically, oxidized LDL (oxLDL) exposure results in cytochrome C leaking, damaged mitochondria in endothelial cells, a response that accompanies a range of adverse cellular phenomena that ultimately lead to apoptosis (Lin et al. 2018). The classical functional distinction between LDL and VLDL (risk enhancers) and high-density lipoprotein (HDL; protective) is generally maintained when ROS generation and mitochondrial function are considered (White et al. 2017). Yet, as the latter review points out, HDL-operated reverse cholesterol transport is thought to be responsible for maintaining low ROS, but selected HDL components such as apolipoprotein O (APOO), have been shown to increase ROS (Turkieh et al. 2014). Interestingly, APOO is mitochondrial, thus it provides an interesting example of an apolipoprotein with cellular functions other than scaffolds for or homeostasis regulators of lipoproteins. As for fatty acids (FAs), supplementation of chow diet with a set of polyunsaturated FAs (PUFAs)-rich oils improved mitochondrial respiration and decreased ROS in mouse peritoneal macrophages, although one of the oils used (echium oil) unexpectedly increased ROS (Shen et al. 2017). The latter observation is a reminder of the underlying complexity of oil-based dietary supplementation. On the other hand, a relatively low dose (100 μM) of palmitic acid (PA), a proinflammatory FA, increases ROS and, counter intuitively, also increases mitochondrial mass in endothelial cells, thus adding novel conflicting evidence that is bound to improve our understanding of the cellular effects of FAs (Dymkowska et al. 2017). A master regulator of mitochondrial activity and FAs is PPARα, a transcription factor that stimulates the production of fatty acid oxidation (FAO) enzymes (Gulick et al. 1994). It follows that low PPARα activity would promote FA accumulation and increased atherosclerosis. Yet, PPARα deficiency decreases atherosclerosis in ApoE-null mice, despite a concomitant hypertriglyceridemia that was patent after 3 weeks of a high-fat diet (Tordjman et al. 2001). Furthermore, a high-fat diet increases hepatic PPARα and its target gene expression (Zou et al. 2009). PPARα is expressed in macrophages and in the atheroma, and is believed to participate in the documented beneficial effects of fibrates (Chinetti et al. 1998; Fruchart et al. 1999). Clearly, further research is needed to understand the intricate interrelationship between mitochondria, PPARα-mediated FA metabolism and atherosclerosis (see later).

2.6 INFLAMMATION

Some of the coincidences between inflammation and ROS generation have been outlined above. An additional important player is the NLRP3 inflammasome, a complex that is assembled in response to immunity-related stress, including atherosclerosis (reviewed in Yu and Lee 2016). Mitochondrial dysfunction is intimately related to NLRP3 inflammasome activation, as the latter lies downstream to ROS generation. ROS act by increasing the availability of at least two inflammasome assemblers, namely thioredoxin-interacting protein (TXNIP) and oxidised mtDNA, a by-product of elevated cellular oxidation potential (Zhou et al. 2010; Shimada et al. 2012).

A further link between mitochondrial metabolism and inflammation is the observation that mitochondrial activity is influenced by macrophage phenotype. Traditionally, macrophage phenotypes are classified into M1 and M2, pro-inflammatory and anti-inflammatory, respectively. M2 macrophages are believed to promote inflammation resolution in the atheroma (Tabas and Bornfeldt 2016). Interestingly, M2 macrophages rely on FAO, whereas M1 counterparts produce cellular energy through aerobic glycolysis, suggesting a fundamentally different mitochondrial metabolic profile in resolution-prone, in principle stable, plaque (Huang et al. 2014; Ouimet et al. 2015). This difference also accounts for the general coincidence between inflammation and defective FA catabolism.

3 PERIVASCULAR ADIPOCYTE MITOCHONDRIA

An emerging topic in atherosclerosis is the role of signals generated by the perivascular adipose tissue (PVAT). Although traditionally regarded as an inert depot of energy for the adjacent vasculature, more complex functions for PVAT are increasingly uncovered. Adipose tissue type-specific deletion experiments in mice showed that PVAT inhibits atherosclerosis, similar to the recently documented effects of brown fat depot deletion (Chang et al. 2012; Xiong et al. 2018). Efforts to uncover the mechanisms of the protective effects of PVAT have focused on factors related to tissue fat browning, in particular DGSH iron sulphur domain 1 (MitoNEET). MitoNEET is an outer mitochondrial membrane protein that catalyses NADPH reduction to control oxygen levels (Wang et al. 2017). MitoNEET is downregulated in thermogenesis-impaired PVAT and, crucially, expression of a MitoNEET transgene in ApoE-null mice decreases atherosclerosis (Xiong et al. 2017). These results are difficult to translate into cardiovascular therapies though, since MitoNEET overexpression induces lipid accumulation and glucose intolerance, and is indeed a target of diabetes treatment drugs (Kusminski et al. 2016). The relevance of the PVAT mitochondria-brown fat-atherosclerosis axis was further supported by a proteomics survey of pericardial fat (Tang et al. 2018). The study determined that ribosomal protein S3A (RPS3A) downregulation in the periaortic fat exacerbates atherosclerosis in ApoE-null mice. The authors propose that the observed effects are explained by the ability of RPS3A to migrate to mitochondria to preserve brown adipocyte functions.

4 RECENT "OMICS"-GENERATED EVIDENCE

The relevance of mitochondrial function in atherosclerosis has been confirmed by two large-scale expression studies. First, proteomics analysis of different portions of early human aortic atherosclerosis revealed that differences in mitochondrial protein abundance is a predominant signature of affected aorta, although mitochondrial protein abundance greatly varies between aortic locations (Herrington et al. 2018). Another study exploited the Reversa model mouse, in which the hepatic microsomal triglyceride transfer protein gene can be silenced by Cre-lox recombination following interferon administration, thus blocking VLDL assembly and reversing hyperlipidaemia in a Ldlr-null /ApoB100 knock-in background (Lieu et al. 2003). A transcriptome analysis of the Reversa mouse revealed a causal relationship between mitochondrial protein abundance and hyperlipidaemia-induced atherosclerosis (Vilne et al. 2017).

5 THE LINK BETWEEN MITOCHONDRIAL FUNCTION AND EPIGENETIC MARKS

Transcription of virtually any gene is affected directly or indirectly by epigenetic marks – DNA methylation, histone post-translational modifications – present at regulatory loci. It is therefore not surprising that alterations in the epigenome would impact mitochondrial function. In a remarkable example, the inhibitor of DNA methylation azacytidine lowers mitochondria-derived ROS and improves stemness in adipose tissue-derived stem cells (Kornicka et al. 2017). Although understanding how mitochondrial function can be improved by tuning the epigenome is of great importance, it will be achieved by dissecting relatively well-understood mechanisms of transcriptional regulation. Rather, in this chapter we will focus on the mitochondria-to-nucleus retrograde signalling, i.e.,

how the epigenome is affected by factors produced by the mitochondria. The field that covers such phenomena is epigenetics and metabolism, a relatively new and fast-growing focus of interest that has been covered by excellent reviews (Keating and El-Osta 2015; Tzika et al. 2018). Figure 1 resumes the different interactions mentioned in this paragraph.

A central component of mitochondria-to-nucleus signalling is the TCAC. A number of metabolites participate in the TCAC, which regulate critical DNA- and histone-modifying enzymes. First, acetyl-CoA is the source of acetyl groups used by histone acetyltransferases (HATs) to establish transcription-permissive, acetylated histone-rich chromatin (Marushige 1976). On the other hand, the conversion of pyruvate produced by glycolysis to acetyl-CoA reduces NAD^+, a cofactor for the NAD^+-dependent deacetylases Sirtuins. Sirtuins

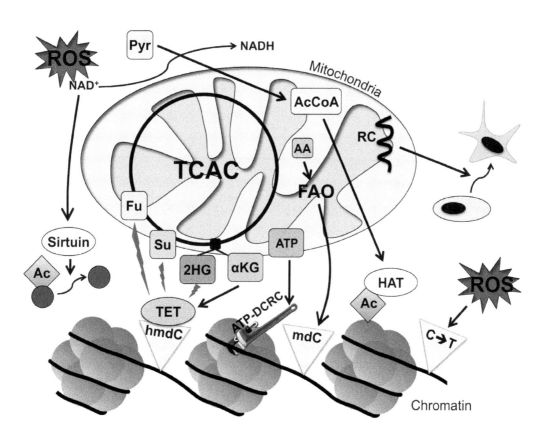

FIGURE 1 Signalling from the mitochondria to the nucleus. Chromatin is represented as DNA (black line) wrapped around histone octamers (green shaded spheres). Stimulation (black arrows) and inhibition (red lightnings) of TET, HATs, Sirtuins. α-ketoglutarate and 2-hydroxyglutarate are both produced by isocitrate dehydrogenase. Fatty acid oxidation mediates arachidonic acid-induced DNA methylation. The respiratory chain supports cellular differentiation (mid right). ROS indirectly promotes Sirtuin activity by increasing NAD^+ (upper left). Also, ROS create oxidized DNA bases that are prone to deamination (C-to-T transition; lower right). See the text for details. 2HG, 2-hydroxyglutarate; AA, arachidonic acid; Ac, acetyl group; AcCoA, acetyl-CoA; αKG, α-ketoglutarate; ATP-DCRC, ATP-dependent chromatin remodelling complex; FAO, fatty acid oxidation; Fu, fumarate; HAT, histone acetyltransferase; hmdC, 5-hydroxymethylcytosine; mdC, 5-methylcytosine; Pyr, pyruvate; RC, respiratory chain; ROS, reactive oxygen species; Su, succinate; TCAC, tricaboxylic acid cycle; TET, Ten-eleven translocation. Cellular structures are not drawn to scale.

deacetylate a variety of different proteins including histones and transcription coactivators involved in metabolism, such as PPARγ coactivator-1α (Houtkooper et al. 2012). Thus, acetylation reactions are highly sensitive to the relative rates of FAO, glycolysis and TCAC progression. Indeed, pyruvate is a major source of acetate production for acetylation reactions and is increased by hyperactive glycolysis (Liu et al. 2018).

As for DNA methylation, α-ketoglutarate (or 2-oxoglutarate) is a cofactor of the Ten-eleven translocation (TET) enzymes, a group (3 in humans) of dioxygenase that initiate the active DNA demethylation cascade by converting 5-methyldeoxycytosine to 5-hydroxymethyldeoxycytosine (Tahiliani et al. 2009). It has been proposed that ROS participate in TET regulation by increasing NAD^+ and consequently Sirt activity, which can increase isocitrate dehydrogenase activity by deacetylation, resulting in augmented α-ketoglutarate (Chia et al. 2011). In addition, succinate and fumarate have inhibitor activity by competing with α-ketoglutarate for the active site of TETs (Xiao et al. 2012). Another TET inhibitor worth mentioning is 2-hydroxyglutarate (2-HG), an oncometabolite produced by mutated isocitrate dehydrogenase, the same TCAC enzyme that in physiological conditions produces α-ketoglutarate (Xu et al. 2011).

ROS can also act directly on DNA by oxidizing cytosine residues, which become prone to deamination and C-to-T transition, the predominant source of mutations in mammals (Kreutzer and Essigmann 1998). The latter phenomenon may account for a significant portion of single nucleotide polymorphisms and somatic mutations.

Additionally, ATP is a cofactor for ATP-dependent chromatin remodelling complexes, a class of proteins that are responsible for the dynamic changes in chromatin structure during transcription regulation (Tsukiyama and Wu 1995). For a review on these factors, see (Swygert and Peterson 2014).

Reverse mitochondria-to-nucleus signalling does not merely reflect unspecific consequences of impaired cellular metabolism, as illustrated by the fact that mitochondrial respiratory chain impacts selected cellular functions: it is necessary to contain both 2-HG and DNA methylation within physiological range to sustain cellular differentiation, but is not necessary for proliferation (Ansó et al. 2017).

6 THE CROSS-TALK BETWEEN THE EPIGENOME AND MITOCHONDRIA IN ATHEROSCLEROSIS

Since the first report of aberrant DNA methylation profiles in atherosclerosis, a number of gene-specific and epigenomic-based studies have followed (Laukkanen et al. 1999; Ordovás and Smith 2010; Zaina 2014). In particular, descriptive and functional evidence points to a causal role for widespread artery DNA hypermethylation in atherosclerosis. During early phases of the disease and in fibrocellular stable plaques (AHA grade III-VII),

DNA hypermethylation is detected both at loci that remain hypermethylated during lesion progression or are further hypermethylated as histological lesion complexity increases (Zaina et al. 2014; Valencia-Morales et al. 2015; Yamada et al. 2018). Genes involved in VSMC and macrophage functions are enriched among hypermethylated loci. The causal importance of the early DNA hypermethylation has been confirmed by the anti-atherogenic effects of either biochemical DNA methylation inhibition or TET2 expression manipulation in mouse models (Cao et al. 2014; Dunn et al. 2014; Peng et al. 2016). Consistently, TET2 expression induces a contractile VSMC phenotype, indicating a coincidence between DNA hypermethylation and proatherogenic VSMC phenotype (Liu et al. 2013). Following rupture, the plaque shows a post-symptomatic event time-dependent demethylation that targets anti-inflammatory and cellular oxidation-counteracting genes (Zaina et al. 2015).

Subsequent work suggested that the mitochondria could participate in shaping the DNA methylome of the atherosclerotic artery. Arachidonic acid (AA) induces DNA hypermethylation in cultured THP-1 monocytes and HEK-293 cells (Silva-Martínez et al. 2016). The response to AA is greatly reduced by etomoxir, an inhibitor of FA import into mitochondria, and by PPARα inhibitors. Furthermore, a comparison of AA-induced DNA methylation profiles with published data in a variety of diseases and normal tissues, revealed significant clustering with atherosclerosis-specific and the proinflammatory PA-induced profiles. Thus, the data suggest that increased mitochondrial FAO results in DNA hypermethylation, possibly by generating incomplete oxidation products, some of which can act as PPAR ligands (Figure 1) (Forman et al. 1997). Accordingly, PA increases mitochondrial mass as outlined above (Dymkowska et al. 2017). Also, enhanced long-chain FA oxidation has been reported in atherosclerosis (Gillies and Bell 1979). Interestingly, the dynamic DNA hypermethylation followed by demethylation observed in the previously discussed human study, mirrors a corresponding biphasic increase/decrease trend of mitochondrial abundance in a rabbit model of atherosclerosis (Rosnowski and Kujawa 1977).

Nonetheless, data are present in the literature, apparently contradicting the logic outlined above, reminding of the intrinsic complexity of the phenomena outlined here. Etomoxir reduced tubular necrosis and maintained baseline PPARα expression in a rat model of acute kidney failure, although no direct data on PPARα expression or activation was reported (Portilla et al. 2000). Another example is a functional study that addresses mitochondrial signalling to the epigenome (Lozoya et al. 2018). The authors exploited a model of inducible mitochondria depletion consequent to expression of a dominant negative DNA polymerase gamma (POLG), which is essential for mitochondrial proliferation. During mitochondrial depletion, the authors report an increase in

DNA methylation, coinciding with a spike in S-adenosyl-methionine (SAM), a universal donor of methyl groups for DNA methylation. A *caveat* to this interesting study is that it describes short-term responses (≤ 9 days) and indeed suggests a biphasic trend, with a tendency to DNA demethylation and a patent SAM decline at the latest time point, which does not contradict the direct correlation between mitochondrial number and DNA methylation outlined above. Also, the data reveal a sizeable variability of DNA methylation profiles. For a further discussion of the complexity in the interplay between atherogenesis, DNA methylation and mitochondrial function, the reader is referred to a published review (Zaina and Lund 2017).

7 DIFFERENTIAL METHYLATION OF CPT1A IS A RECURRENT FINDING IN CVD-RELATED METHYLOME STUDIES

Carnitine Palmitoyltransferase 1A (CPT1A) is a critical protein for FA import into mitochondria (Fritz and Yue 1963). Consistent with the importance of the mitochondrial machinery in atherosclerosis, differential methylation in the *CPT1A* locus is consistently detected in epigenome-wide association studies (EWAS) of atherosclerosis-related metabolic conditions. In particular, significant associations with VLDL cholesterol, metabolic syndrome, plasma triglyceride, adiponectin, body-mass index (BMI) and smoking have been reported (Irvin et al. 2014; Aslibekyan et al. 2015; Das et al. 2016; Aslibekyan et al. 2017; Braun et al. 2017; Fernández-Sanlés et al. 2018). Associations of a 5′UTR CpG site (Illumina's Infinium HumanMethylation450 BeadChip ID: cg00574958) have been reported by all cited studies and an inverse correlation between methylation and expression has been generally observed. Interestingly, the reported inverse correlations of cg00574958 methylation with BMI and triglycerides, and a direct one with adiponectin, are consistent with increased FAO in conditions predisposing to atherosclerosis (see above). Recently, causality was determined for triglycerides on *CPT1A* cg00574958 methylation, an important result that echoes previous conclusions (Dekkers et al. 2016; Sayols-Baixeras et al. 2018). Adding to that wealth of data, methylation of an intragenic region in *CPT1A* is modified by prenatal malnutrition in the Dutch Hunger Winter cohort, where it significantly correlates with LDL cholesterol (Tobi et al. 2014). Not surprisingly, CPT1A was proposed as a priority focus point to design therapies for chronic degenerative diseases (Irvin et al. 2014).

8 CONCLUSION

The link between mitochondrial function and cardiovascular disease is well established. Yet, knowledge of the interplay between the mitochondrial metabolic machinery and the epigenome is still relatively limited. Future exciting work will aim at understanding how mitochondria-generated factors modulate epigenetic marks and whether that information can be used to design new treatments.

REFERENCES

Acosta, Daniel, and Duane G. Wenzel. 1974. "Injury Produced by Free Fatty Acids to Lysosomes and Mitochondria in Cultured Heart Muscle and Endothelial Cells." *Atherosclerosis* 20 (3): 417–426. doi:10.1016/0021-9150(74)90023-9.

Ansó, Elena, Samuel E. Weinberg, Lauren P. Diebold, Benjamin J. Thompson, Sébastien Malinge, Paul T. Schumacker, Xin Liu, et al. 2017. "The Mitochondrial Respiratory Chain Is Essential for Haematopoietic Stem Cell Function." *Nature Cell Biology* 19 (6): 614–625. doi:10.1038/ncb3529.

Ashar, Foram N., Yiyi Zhang, Ryan J. Longchamps, John Lane, Anna Moes, Megan L. Grove, Josyf C. Mychaleckyj, et al. 2017. "Association of Mitochondrial DNA Copy Number With Cardiovascular Disease." *JAMA Cardiology* 2 (11): 1247. doi:10.1001/jamacardio.2017.3683.

Aslibekyan, S., A.N. Do, H. Xu, S. Li, M.R. Irvin, D. Zhi, H. K. Tiwari, et al. 2017. "CPT1A Methylation Is Associated with Plasma Adiponectin." *Nutrition, Metabolism and Cardiovascular Diseases* 27 (3): 225–233. doi:10.1016/j.numecd.2016.11.004.

Aslibekyan, Stella, Ellen W. Demerath, Michael Mendelson, Degui Zhi, Weihua Guan, Liming Liang, Jin Sha, et al. 2015. "Epigenome-Wide Study Identifies Novel Methylation Loci Associated with Body Mass Index and Waist Circumference." *Obesity (Silver Spring, MD)* 23 (7): 1493–1501. doi:10.1002/oby.21111.

Baccarelli, Andrea, Robert O. Wright, Valentina Bollati, Letizia Tarantini, Augusto A. Litonjua, Helen H. Suh, Antonella Zanobetti, David Sparrow, Pantel S. Vokonas, and Joel Schwartz. 2009. "Rapid DNA Methylation Changes after Exposure to Traffic Particles." *American Journal of Respiratory and Critical Care Medicine* 179 (7): 572–578. doi:10.1164/rccm.200807-1097OC.

Barends, Marlieke, Lotte Verschuren, Eva Morava, Victoria Nesbitt, Doug Turnbull, and Robert Mcfarland. 2016. "Causes of Death in Adults with Mitochondrial Disease." *Journal of Inherited Metabolic Disorders Reports* 26: 103–113. doi:10.1007/8904_2015_449.

Braun, Kim V.E., Klodian Dhana, Paul S. de Vries, Trudy Voortman, Joyce B. J. van Meurs, Andre G. Uitterlinden, Albert Hofman, et al. 2017. "Epigenome-Wide Association Study (EWAS) on Lipids: The Rotterdam Study." *Clinical Epigenetics* 9 (1): 15. doi:10.1186/s13148-016-0304-4.

Cao, Qiang, Xianfeng Wang, Lin Jia, Ashis K. Mondal, Abdoulaye Diallo, Gregory A. Hawkins, Swapan K. Das, et al. 2014. "Inhibiting DNA Methylation by 5-Aza-2′-Deoxycytidine Ameliorates Atherosclerosis through Suppressing Macrophage Inflammation." *Endocrinology* 155 (12): 4925–4938. doi:10.1210/en.2014-1595.

Chang, Lin, Luis Villacorta, Rongxia Li, Milton Hamblin, Wei Xu, Chunyan Dou, Jifeng Zhang, Jiarui Wu, Rong Zeng, and Y. Eugene Chen. 2012. "Loss of Perivascular Adipose Tissue on Peroxisome Proliferator-Activated Receptor-γ Deletion in Smooth Muscle Cells Impairs Intravascular Thermoregulation and Enhances Atherosclerosis." *Circulation* 126 (9): 1067–1078. doi:10.1161/CIRCULATIONAHA.112.104489.

Cheney, C.P., R.F. Scott, J. Frick, and E.S. Morrison. 1977. "Mitochondrial Ca2+ Metabolism in Aortic Smooth Muscle of Swine. Effect of Mild Hypercholesterolemia." *Atherosclerosis* 28 (2): 141–154. doi:10.1016/0021-9150(77)90151-4.

Chia, Nancy, Luan Wang, Xiangyi Lu, Mary-Claude Senut, Carol A. Brenner, and Douglas M. Ruden. 2011. "Hypothesis: Environmental Regulation of 5-Hydroxymethylcytosine by Oxidative Stress." *Epigenetics* 6 (7): 853–856. doi:10.4161/epi.6.7.16461.

Chinetti, G., S. Griglio, M. Antonucci, I.P. Torra, P. Delerive, Z. Majd, J.C. Fruchart, J. Chapman, J. Najib, and B. Staels. 1998. "Activation of Proliferator-Activated Receptors Alpha and Gamma Induces Apoptosis of Human Monocyte-Derived Macrophages." *The Journal of Biological Chemistry* 273 (40): 25573–25580.

Das, Mithun, Jin Sha, Bertha Hidalgo, Stella Aslibekyan, Anh N. Do, Degui Zhi, Dianjianyi Sun, et al. 2016. "Association of DNA Methylation at CPT1A Locus with Metabolic Syndrome in the Genetics of Lipid Lowering Drugs and Diet Network (GOLDN) Study." Edited by Tom R. Gaunt. *PloS One* 11 (1): e0145789. doi:10.1371/journal.pone.0145789.

Dekkers, Koen F., Maarten van Iterson, Roderick C. Slieker, Matthijs H. Moed, Marc Jan Bonder, Michiel van Galen, Hailiang Mei, et al. 2016. "Blood Lipids Influence DNA Methylation in Circulating Cells." *Genome Biology* 17 (1): 138. doi:10.1186/s13059-016-1000-6.

Docherty, Craig K., Andy Carswell, Elaine Friel, and John R. Mercer. 2018. "Impaired Mitochondrial Respiration in Human Carotid Plaque Atherosclerosis: A Potential Role for Pink1 in Vascular Smooth Muscle Cell Energetics." *Atherosclerosis* 268 (January): 1–11. doi:10.1016/J.ATHEROSCLEROSIS.2017.11.009.

Dorighello, Gabriel G., Bruno A. Paim, R. Leite Ana Catarina, Anibal E. Vercesi, and Helena C.F. Oliveira. 2018. "Spontaneous Experimental Atherosclerosis in Hypercholesterolemic Mice Advances with Ageing and Correlates with Mitochondrial Reactive Oxygen Species." *Experimental Gerontology* 109 (August): 47–50. doi:10.1016/j.exger.2017.02.010.

Dunn, Jessilyn, Haiwei Qiu, Soyeon Kim, Daudi Jjingo, Ryan Hoffman, Chan Woo Kim, Inhwan Jang, et al. 2014. "Flow-Dependent Epigenetic DNA Methylation Regulates Endothelial Gene Expression and Atherosclerosis." *Journal of Clinical Investigation* 124 (7): 3187–3199. doi:10.1172/JCI74792.

Dymkowska, Dorota, Maria Kawalec, Tomasz Wyszomirski, and Krzysztof Zabłocki. 2017. "Mild Palmitate Treatment Increases Mitochondrial Mass but Does Not Affect EA. Hy926 Endothelial Cells Viability." *Archives of Biochemistry and Biophysics* 634 (November): 88–95. doi:10.1016/j.abb.2017.10.006.

Feng, Ma, Fabiana Rached, Anatol Kontush, and M. John Chapman. 2018. "Impact of Lipoproteins on Atherobiology." *Cardiology Clinics* 36 (2): 193–201. doi:10.1016/j.ccl.2017.10.001.

Fernández-Sanlés, Alba, Sergi Sayols-Baixeras, Santiago Curcio, Isaac Subirana, Jaume Marrugat, and Roberto Elosua. 2018. "DNA Methylation and Age-Independent Cardiovascular Risk, an Epigenome-Wide Approach." *Arteriosclerosis, Thrombosis, and Vascular Biology*, January. doi:10.1161/ATVBAHA.117.310340.

Fiorito, Giovanni, Jelle Vlaanderen, Silvia Polidoro, John Gulliver, Claudia Galassi, Andrea Ranzi, Vittorio Krogh, et al. 2017. "Oxidative Stress and Inflammation Mediate the Effect of Air Pollution on Cardio- and Cerebrovascular Disease: A Prospective Study in Nonsmokers." *Environmental and Molecular Mutagenesis*, November. doi:10.1002/em.22153.

Forman, B.M., J. Chen, and R.M. Evans. 1997. "Hypolipidemic Drugs, Polyunsaturated Fatty Acids, and Eicosanoids Are Ligands for Peroxisome Proliferator-Activated Receptors Alpha and Delta." *Proceedings of the National Academy of Sciences of the United States of America* 94 (9): 4312–4317.

Fritz, I.B., and K.T. Yue. 1963. "Long-Chain Carnitine Acyltransferase and the Role of Acylcarnitine Derivatives in the Catalytic Increase of Fatty Acid Oxidation Induced by Carnitine." *Journal of Lipid Research* 4 (July): 279–288.

Fruchart, J.C., P. Duriez, and B. Staels. 1999. "Peroxisome Proliferator-Activated Receptor-Alpha Activators Regulate Genes Governing Lipoprotein Metabolism, Vascular Inflammation and Atherosclerosis." *Current Opinion in Lipidology* 10 (3): 245–257.

Gillies, Peter J., and Frank P. Bell. 1979. "Carnitine Palmitoyltransferase Activity in Mitochondrial Fractions Isolated from Aortas of Rabbits Fed Cholesterol-Supplemented Diets." *Atherosclerosis* 34 (1): 25–34. doi:10.1016/0021-9150(79)90102-3.

Gulick, T., S. Cresci, T. Caira, D.D. Moore, and D.P. Kelly. 1994. "The Peroxisome Proliferator-Activated Receptor Regulates Mitochondrial Fatty Acid Oxidative Enzyme Gene Expression." *Proceedings of the National Academy of Sciences of the United States of America* 91 (23): 11012–11016.

Hackam, Daniel G., and J. David Spence. 2018. "Decline in the Severity of Carotid Atherosclerosis and Associated Risk Factors From 2002 to 2014." *Stroke* 49 (11): 2786–2788. doi:10.1161/STROKEAHA.118.021445.

Herrington, David M., Chunhong Mao, Sarah J. Parker, Zongming Fu, Guoqiang Yu, Lulu Chen, Vidya Venkatraman, et al. 2018. "Proteomic Architecture of Human Coronary and Aortic Atherosclerosis." *Circulation* 137 (25): 2741–2756. doi:10.1161/CIRCULATIONAHA.118.034365.

Hoff, H.F., C.L. Heideman, G.P. Noon, and J.S. Meyer. 1975. "Localization of Apo-Lipoproteins in Human Carotid Artery Plaques." *Stroke* 6 (5): 531–534.

Holvoet, Paul, Maarten Vanhaverbeke, Katarzyna Bloch, Pieter Baatsen, Peter Sinnaeve, and Stefan Janssens. 2016. "Low MT-CO1 in Monocytes and Microvesicles Is Associated With Outcome in Patients With Coronary Artery Disease." *Journal of the American Heart Association* 5 (12). doi:10.1161/JAHA.116.004207.

Holvoet, Paul, Maarten Vanhaverbeke, Benjamine Geeraert, Dieuwke De Keyzer, Maarten Hulsmans, and Stefan Janssens. 2017. "Low Cytochrome Oxidase 1 Links Mitochondrial Dysfunction to Atherosclerosis in Mice and Pigs." *PloS One* 12 (1): e0170307. doi:10.1371/journal.pone.0170307.

Houtkooper, Riekelt H., Eija Pirinen, and Johan Auwerx. 2012. "Sirtuins as Regulators of Metabolism and Healthspan." *Nature Reviews. Molecular Cell Biology* 13 (4): 225–238. doi:10.1038/nrm3293.

Huang, Stanley Ching-Cheng, Bart Everts, Yulia Ivanova, David O'Sullivan, Marcia Nascimento, Amber M. Smith, Wandy Beatty, et al. 2014. "Cell-Intrinsic Lysosomal Lipolysis Is Essential for Alternative Activation of Macrophages." *Nature Immunology* 15 (9): 846–855. doi:10.1038/ni.2956.

Irvin, Marguerite R., Stella Aslibekyan, Bertha Hidalgo, and Donna K. Arnett. 2014. "CPT1A: The Future of Heart Disease Detection and Personalized Medicine?" *Clinical Lipidology* 9 (1): 9–12. doi:10.2217/clp.13.75.

Irvin, Marguerite R., Degui Zhi, Roby Joehanes, Michael Mendelson, Stella Aslibekyan, Steven A. Claas, Krista S. Thibeault, et al. 2014. "Epigenome-Wide Association Study of Fasting Blood Lipids in the Genetics of Lipid-Lowering Drugs and Diet Network Study." *Circulation* 130 (7): 565–572. doi:10.1161/CIRCULATIONAHA. 114.009158.

Keating, S.T., and A. El-Osta. 2015. "Epigenetics and Metabolism." *Circulation Research* 116 (4): 715–736. doi:10.1161/CIRCRESAHA.116.303936.

Kornicka, Katarzyna, Krzysztof Marycz, Monika Marędziak, Krzysztof A. Tomaszewski, and Jakub Nicpoń. 2017. "The Effects of the DNA Methyltranfserases Inhibitor 5-Azacitidine on Ageing, Oxidative Stress and DNA Methylation of Adipose Derived Stem Cells." *Journal of Cellular and Molecular Medicine* 21 (2): 387–401. doi:10.1111/jcmm.12972.

Kreutzer, D.A., and J.M. Essigmann. 1998. "Oxidized, Deaminated Cytosines Are a Source of C → T Transitions *in vivo*." *Proceedings of the National Academy of Sciences of the United States of America* 95 (7): 3578–3582. doi:10.1073/ PNAS.95.7.3578.

Kusminski, Christine M., Shiuhwei Chen, Risheng Ye, Kai Sun, Qiong A. Wang, Stephen B. Spurgin, Phillip E. Sanders, et al. 2016. "MitoNEET-Parkin Effects in Pancreatic α- and β-Cells, Cellular Survival, and Intrainsular Cross Talk." *Diabetes* 65 (6): 1534–1555. doi:10.2337/db15-1323.

Laukkanen, M.O., S. Mannermaa, M.O. Hiltunen, S. Aittomäki, K. Airenne, J. Jänne, and S. Ylä-Herttuala. 1999. "Local Hypomethylation in Atherosclerosis Found in Rabbit Ec-Sod Gene." *Arteriosclerosis, Thrombosis, and Vascular Biology* 19 (9): 2171–2178.

Lien, Li-Ming, Hung-Yi Chiou, Hsu-Ling Yeh, Shang-Yen Chiu, Jiann-Shing Jeng, Huey-Juan Lin, Chaur-Jong Hu, Fang-I. Hsieh, and Yau-Huei Wei. 2017. "Significant Association between Low Mitochondrial DNA Content in Peripheral Blood Leukocytes and Ischemic Stroke." *Journal of the American Heart Association* 6 (11). doi:10.1161/ JAHA.117.006157.

Lieu, Hsiao D., Shannon K. Withycombe, Quinn Walker, James X. Rong, Rosemary L. Walzem, Jinny S. Wong, Robert L. Hamilton, Edward A. Fisher, and Stephen G. Young. 2003. "Eliminating Atherogenesis in Mice by Switching off Hepatic Lipoprotein Secretion." *Circulation* 107 (9): 1315– 1321.

Lin, Fei, Likai Pei, Qingbin Zhang, Weizhong Han, Shiliang Jiang, Yanliang Lin, Bo Dong, Lianqun Cui, and Min Li. 2018. "Ox-LDL Induces Endothelial Cell Apoptosis and Macrophage Migration by Regulating Caveolin-1 Phosphorylation." *Journal of Cellular Physiology* 233 (10): 6683–6692. doi:10.1002/jcp.26468.

Liu, Renjing, Yu Jin, Wai Ho Tang, Lingfeng Qin, Xinbo Zhang, George Tellides, John Hwa, Jun Yu, and Kathleen A. Martin. 2013. "Ten-Eleven Translocation-2 (TET2) Is a Master Regulator of Smooth Muscle Cell Plasticity." *Circulation* 128 (18): 2047–2057. doi:10.1161/ CIRCULATIONAHA.113.002887.

Liu, Xiaojing, Daniel E. Cooper, Ahmad A. Cluntun, Marc O. Warmoes, Steven Zhao, Michael A. Reid, Juan Liu, et al. 2018. "Acetate Production from Glucose and Coupling to Mitochondrial Metabolism in Mammals." *Cell* 175 (2): 502–513. doi:10.1016/j.cell.2018.08.040.

Lozhkin, Andrey, Aleksandr E. Vendrov, Hua Pan, Samuel A. Wickline, Nageswara R. Madamanchi, and Marschall S. Runge. 2017. "NADPH Oxidase 4 Regulates Vascular Inflammation in Aging and Atherosclerosis." *Journal of Molecular and Cellular Cardiology* 102: 10–21. doi:10.1016/j.yjmcc.2016.12.004.

Lozoya, Oswaldo A., Inmaculada Martinez-Reyes, Tianyuan Wang, Dagoberto Grenet, Pierre Bushel, Jianying Li, Navdeep Chandel, Richard P. Woychik, and Janine H. Santos. 2018. "Mitochondrial Nicotinamide Adenine Dinucleotide Reduced (NADH) Oxidation Links the Tricarboxylic Acid (TCA) Cycle with Methionine Metabolism and Nuclear DNA Methylation." *PLoS Biology* 16 (4): e2005707. doi:10.1371/journal.pbio.2005707.

Marushige, K. 1976. "Activation of Chromatin by Acetylation of Histone Side Chains." *Proceedings of the National Academy of Sciences of the United States of America* 73 (11): 3937–3941.

Mathers, Colin D., and Dejan Loncar. 2006. "Projections of Global Mortality and Burden of Disease from 2002 to 2030." *PLoS Medicine* 3 (11): e442. doi:10.1371/journal. pmed.0030442.

Morrison, E.S., R.F. Scott, M. Kroms, and J. Frick. 1973. "Aortic Mitochondrial Function in Experimentally Induced Atherosclerosis in Swine and Rabbits." *Biochemical Medicine* 7 (2): 308–315.

Oliveira, Helena C.F., Ricardo G. Cosso, Luciane C. Alberici, Evelise N. Maciel, Alessandro G. Salerno, Gabriel G. Dorighello, Jesus A. Velho, Eliana C. de Faria, and Aníbal E. Vercesi. 2005. "Oxidative Stress in Atherosclerosis-Prone Mouse Is Due to Low Antioxidant Capacity of Mitochondria." *The FASEB Journal* 19 (2): 278–280. doi:10.1096/fj.04-2095fje.

Ordovás, José, and Caren Smith. 2010. "Epigenetics and Cardiovascular Disease." *Nature Reviews. Cardiology* 7 (9): 510– 519. doi:10.1038/nrcardio.2010.104.

Ouimet, Mireille, Hasini N. Ediriweera, U. Mahesh Gundra, Frederick J. Sheedy, Bhama Ramkhelawon, Susan B. Hutchison, Kaitlyn Rinehold, et al. 2015. "MicroRNA-33–Dependent Regulation of Macrophage Metabolism Directs Immune Cell Polarization in Atherosclerosis." *The Journal of Clinical Investigation* 125 (12): 4334–4348. doi:10.1172/JCI81676.

Panh, Loïc, Olivier Lairez, Jean-Bernard Ruidavets, Michel Galinier, Didier Carrié, and Jean Ferrières. 2017. "Coronary Artery Calcification: From Crystal to Plaque Rupture." *Archives of Cardiovascular Diseases* 110 (10): 550–561. doi:10.1016/j.acvd.2017.04.003.

Peng, Juan, Qin Yang, A-Fang Li, Rong-Qing Li, Zuo Wang, Lu-Shan Liu, Zhong Ren, et al. 2016. "Tet Methylcytosine Dioxygenase 2 Inhibits Atherosclerosis via Upregulation of Autophagy in ApoE−/− Mice." *Oncotarget* 7 (47): 76423–76436. doi:10.18632/oncotarget.13121.

Portilla, Didier, Gonghe Dai, Jeffrey M. Peters, Frank J. Gonzalez, Mark D. Crew, and Alan D. Proia. 2000. "Etomoxir-Induced PPARⓍ-Modulated Enzymes Protect during Acute Renal Failure." *American Journal of Physiology-Renal Physiology* 278 (4): F667–F675. doi:10.1152/ ajprenal.2000.278.4.F667.

Puddu, Paolo, Giovanni M. Puddu, Livia Galletti, Eleonora Cravero, and Antonio Muscari. 2005. "Mitochondrial Dysfunction as an Initiating Event in

Atherogenesis: A Plausible Hypothesis." *Cardiology* 103 (3): 137–141. doi:10.1159/000083440.

Rosnowski, Andrzej, and Marek Kujawa. 1977. "Succinic Dehydrogenase Activity of the Myocardium in Experimental Atherosclerosis." *Journal of Molecular and Cellular Cardiology* 9 (1): 81–85. doi:10.1016/0022-2828(77) 90027-X.

Sayols-Baixeras, Sergi, Hemant K. Tiwari, and Stella W. Aslibekyan. 2018. "Disentangling Associations between DNA Methylation and Blood Lipids: A Mendelian Randomization Approach." *BMC Proceedings* 12 (S9): 23. doi:10.1186/s12919-018-0119-8.

Shen, Lulu, Yan Yang, Tiantong Ou, Chia-Chi C. Key, Sarah H. Tong, Russel C. Sequeira, Jonathan M. Nelson, et al. 2017. "Dietary PUFAs Attenuate NLRP3 Inflammasome Activation via Enhancing Macrophage Autophagy." *Journal of Lipid Research* 58 (9): 1808–1821. doi:10.1194/jlr. M075879.

Shimada, Kenichi, Timothy R. Crother, Justin Karlin, Jargalsaikhan Dagvadorj, Norika Chiba, Shuang Chen, V. Krishnan Ramanujan, et al. 2012. "Oxidized Mitochondrial DNA Activates the NLRP3 Inflammasome during Apoptosis." *Immunity* 36 (3): 401–414. doi:10.1016/j. immuni.2012.01.009.

Silva-Martínez, Guillermo A., Dalia Rodríguez-Ríos, Yolanda Alvarado-Caudillo, Alejandro Vaquero, Manel Esteller, F. Javier Carmona, Sebastian Moran, et al. 2016. "Arachidonic and Oleic Acid Exert Distinct Effects on the DNA Methylome." *Epigenetics* 11 (5): 321–334. doi:10.1080/15592294.2016.1161873.

Swygert, Sarah G., and Craig L. Peterson. 2014. "Chromatin Dynamics: Interplay between Remodeling Enzymes and Histone Modifications." *Biochimica et Biophysica Acta (BBA) - Gene Regulatory Mechanisms* 1839 (8): 728–736. doi:10.1016/j.bbagrm.2014.02.013.

Tabas, Ira. 2017. "2016 Russell Ross Memorial Lecture in Vascular Biology." *Arteriosclerosis, Thrombosis, and Vascular Biology* 37 (2): 183–189.

Tabas, Ira, and Karin E. Bornfeldt. 2016. "Macrophage Phenotype and Function in Different Stages of Atherosclerosis." *Circulation Research* 118 (4): 653–667.

Tahiliani, Mamta, Kian P. Koh, Yinghua Shen, William A. Pastor, Hozefa Bandukwala, Yevgeny Brudno, Suneet Agarwal, et al. 2009. "Conversion of 5-Methylcytosine to 5-Hydroxymethylcytosine in Mammalian DNA by MLL Partner TET1." *Science (New York, NY)* 324 (5929): 930–935. doi:10.1126/science.1170116.

Tang, Yan, Yi He, Chen Li, Wenjuan Mu, Ying Zou, Conghui Liu, Shuwen Qian, et al. 2018. "RPS3A Positively Regulates the Mitochondrial Function of Human Periaortic Adipose Tissue and Is Associated with Coronary Artery Diseases." *Cell Discovery* 4: 52. doi:10.1038/s41421-018-0041-2.

Tobi, Elmar W., Jelle J. Goeman, Ramin Monajemi, Hongcang Gu, Hein Putter, Yanju Zhang, Roderick C. Slieker, et al. 2014. "DNA Methylation Signatures Link Prenatal Famine Exposure to Growth and Metabolism." *Nature Communications* 5 (November): 5592. doi:10.1038/ncomms6592.

Tordjman, K., C. Bernal-Mizrachi, L. Zemany, S. Weng, C. Feng, F. Zhang, T.C. Leone, T. Coleman, D.P. Kelly, and C.F. Semenkovich. 2001. "PPARalpha Deficiency Reduces Insulin Resistance and Atherosclerosis in ApoE-Null Mice." *The Journal of Clinical Investigation* 107 (8): 1025–1034. doi:10.1172/JCI11497.

Tsukiyama, T., and C. Wu. 1995. "Purification and Properties of an ATP-Dependent Nucleosome Remodeling Factor." *Cell* 83 (6): 1011–1020.

Turkieh, Annie, Céline Caubère, Manon Barutaut, Franck Desmoulin, Romain Harmancey, Michel Galinier, Matthieu Berry, et al. 2014. "Apolipoprotein O Is Mitochondrial and Promotes Lipotoxicity in Heart." *The Journal of Clinical Investigation* 124 (5): 2277–2286. doi:10.1172/JCI74668.

Tzika, Evangelia, Tobias Dreker, and Axel Imhof. 2018. "Epigenetics and Metabolism in Health and Disease." *Frontiers in Genetics* 9 (September): 361. doi:10.3389/fgene.2018.00361.

Valencia-Morales, María del Pilar, Silvio Zaina, Holger Heyn, F. Javier Carmona, Nuray Varol, Sergi Sayols, Enric Condom, et al. 2015. "The DNA Methylation Drift of the Atherosclerotic Aorta Increases with Lesion Progression." *BMC Medical Genomics* 8 (1): 7. doi:10.1186/s12920-015-0085-1.

Vecoli, Cecilia, Andrea Borghini, Silvia Pulignani, Antonella Mercuri, Stefano Turchi, Clara Carpeggiani, Eugenio Picano, and Maria Grazia Andreassi. 2018. "Prognostic Value of Mitochondrial DNA4977 Deletion and Mitochondrial DNA Copy Number in Patients with Stable Coronary Artery Disease." *Atherosclerosis* 276 (September): 91–97. doi:10.1016/j.atherosclerosis.2018. 07.015.

Vilne, Baiba, Josefin Skogsberg, Hassan Foroughi Asl, Husain Ahammad Talukdar, Thorsten Kessler, Johan L. M. Björkegren, and Heribert Schunkert. 2017. "Network Analysis Reveals a Causal Role of Mitochondrial Gene Activity in Atherosclerotic Lesion Formation." *Atherosclerosis* 267 (December): 39–48. doi:10.1016/J. ATHEROSCLEROSIS.2017.10.019.

Vitale, J.J., P.L. White, M. Nakamura, D.M. Hegsted, N. Zamcheck, and E.E. Hellerstein. 1957. "Interrelationships between Experimental Hypercholesteremia, Magnesium Requirement, and Experimental Atherosclerosis." *The Journal of Experimental Medicine* 106 (5): 757–766.

Wang, Yiming, Aaron P. Landry, and Huangen Ding. 2017. "The Mitochondrial Outer Membrane Protein MitoNEET Is a Redox Enzyme Catalyzing Electron Transfer from FMNH2 to Oxygen or Ubiquinone." *The Journal of Biological Chemistry* 292 (24): 10061–10067. doi:10.1074/jbc. M117.789800.

Wang, Ying, Wei Wang, Nan Wang, Alan R. Tall, and Ira Tabas. 2017. "Mitochondrial Oxidative Stress Promotes Atherosclerosis and Neutrophil Extracellular Traps in Aged Mice." *Arteriosclerosis, Thrombosis, and Vascular Biology* 37 (8): e99–e107. doi:10.1161/ATVBAHA.117. 309580.

Whereat, A.F. 1970. "Is Atherosclerosis a Disorder of Intramitochondrial Respiration?" *Annals of Internal Medicine* 73 (1): 125. doi:10.7326/0003-4819-73-1-125.

White, C. Roger, Geeta Datta, and Samantha Giordano. 2017. "High-Density Lipoprotein Regulation of Mitochondrial Function." *Advances in Experimental Medicine and Biology* 982: 407–429. doi:10.1007/978-3-319-55330-6_22.

Wible, Daric J., and Shawn B. Bratton. 2018. "Reciprocity in ROS and Autophagic Signaling." *Current Opinion in Toxicology* 7 (February): 28–36. doi:10.1016/j.cotox.2017. 10.006.

Xiao, M., H. Yang, W. Xu, S. Ma, H. Lin, H. Zhu, L. Liu, et al. 2012. "Inhibition of -KG-Dependent Histone and DNA Demethylases by Fumarate and Succinate That Are

Accumulated in Mutations of FH and SDH Tumor Suppressors." *Genes & Development* 26 (12): 1326–1338. doi:10.1101/gad.191056.112.

Xiong, Wenhao, Xiangjie Zhao, Minerva T. Garcia-Barrio, Jifeng Zhang, Jiandie Lin, Y. Eugene Chen, Zhisheng Jiang, and Lin Chang. 2017. "MitoNEET in Perivascular Adipose Tissue Blunts Atherosclerosis under Mild Cold Condition in Mice." *Frontiers in Physiology* 8: 1032. doi:10.3389/fphys.2017.01032.

Xiong, Wenhao, Xiangjie Zhao, Luis Villacorta, Oren Rom, Minerva T. Garcia-Barrio, Yanhong Guo, Yanbo Fan, et al. 2018. "Brown Adipocyte-Specific PPARγ (Peroxisome Proliferator-Activated Receptor γ) Deletion Impairs Perivascular Adipose Tissue Development and Enhances Atherosclerosis in Mice." *Arteriosclerosis, Thrombosis, and Vascular Biology*, June. doi:10.1161/ATVBAHA.118. 311367.

Xu, Wei, Hui Yang, Ying Liu, Ying Yang, Ping Wang, Se-Hee Kim, Shinsuke Ito, et al. 2011. "Oncometabolite 2-Hydroxyglutarate Is a Competitive Inhibitor of α-Ketoglutarate-Dependent Dioxygenases." *Cancer Cell* 19 (1): 17–30. doi:10.1016/j.ccr.2010.12.014.

Yamada, Yoshiji, Hideki Horibe, Mitsutoshi Oguri, Jun Sakuma, Ichiro Takeuchi, Yoshiki Yasukochi, Kimihiko Kato, and Motoji Sawabe. 2018. "Identification of Novel Hyper-ï¿½or Hypomethylated CpG Sites and Genes Associated with Atherosclerotic Plaque Using an Epigenome-Wide Association Study." *International Journal of Molecular Medicine*, February. doi:10.3892/ijmm.2018.3453.

Yang, Xinyu, Yang Li, Yanda Li, Xiaomeng Ren, Xiaoyu Zhang, Dan Hu, Yonghong Gao, Yanwei Xing, and Hongcai Shang. 2017. "Oxidative Stress-Mediated Atherosclerosis: Mechanisms and Therapies." *Frontiers in Physiology* 8: 600. doi:10.3389/fphys.2017.00600.

Yeh, Hseng-Long, Li-Tang Kuo, Fung-Chang Sung, and Chih-Ching Yeh. 2018. "Association between Polymorphisms of Antioxidant Gene (MnSOD, CAT, and GPx1) and Risk of Coronary Artery Disease." *BioMed Research International* 2018: 5086869. doi:10.1155/2018/5086869.

Yu, Emma, Patrick A. Calvert, John R. Mercer, James Harrison, Lauren Baker, Nichola L. Figg, Sheetal Kumar, et al. 2013. "Mitochondrial DNA Damage Can Promote Atherosclerosis Independently of Reactive Oxygen Species Through Effects on Smooth Muscle Cells and Monocytes and Correlates With Higher-Risk Plaques in Humans." *Circulation* 128 (7): 702–712. doi:10.1161/ CIRCULATIONAHA.113.002271.

Yu, Emma P.K., Johannes Reinhold, Haixiang Yu, Lakshi Starks, Anna K. Uryga, Kirsty Foote, Alison Finigan, et al. 2017. "Mitochondrial Respiration Is Reduced in Atherosclerosis, Promoting Necrotic Core Formation and Reducing Relative Fibrous Cap Thickness." *Arteriosclerosis, Thrombosis, and Vascular Biology* 37 (12): 2322–2332. doi:10.1161/ATVBAHA.117.310042.

Yu, Je-Wook, and Myung-Shik Lee. 2016. "Mitochondria and the NLRP3 Inflammasome: Physiological and Pathological Relevance." *Archives of Pharmacal Research* 39 (11): 1503–1518. doi:10.1007/s12272-016-0827-4.

Zaina, S., H. Heyn, F. Javier Carmona, N. Varol, S. Sayols, E. Condom, J. Ramirez-Ruz, et al. 2014. "DNA Methylation Map of Human Atherosclerosis." *Circulation: Cardiovascular Genetics* 7 (5): 692–700. doi:10.1161/ CIRCGENETICS.113.000441.

Zaina, Silvio. 2014. "Unraveling the DNA Methylome of Atherosclerosis." *Current Opinion in Lipidology* 25 (2): 148–153. doi:10.1097/MOL.0000000000000059.

Zaina, Silvio, Isabel Gonçalves, F. Javier Carmona, Antonio Gomez, Holger Heyn, Inês G. Mollet, Sebastian Moran, Nuray Varol, and Manel Esteller. 2015. "DNA Methylation Dynamics in Human Carotid Plaques after Cerebrovascular Events." *Arteriosclerosis, Thrombosis, and Vascular Biology* 35 (8): 1835–1842. doi:10.1161/ atvbaha.115.305630.

Zaina, Silvio, and Gertrud Lund. 2017. "Connecting the Dots Between Fatty Acids, Mitochondrial Function, and DNA Methylation in Atherosclerosis." *Current Atherosclerosis Reports* 19 (9): 36. doi:10.1007/s11883-017-0673-y.

Zhang, S.H., R.L. Reddick, J.A. Piedrahita, and N. Maeda. 1992. "Spontaneous Hypercholesterolemia and Arterial Lesions in Mice Lacking Apolipoprotein E." *Science (New York, NY)* 258 (5081): 468–471.

Zhou, Rongbin, Aubry Tardivel, Bernard Thorens, Inpyo Choi, and Jürg Tschopp. 2010. "Thioredoxin-Interacting Protein Links Oxidative Stress to Inflammasome Activation." *Nature Immunology* 11 (2): 136–140. doi:10.1038/ni.1831.

Zou, Yanyan, Hui Du, Miao Yin, Liang Zhang, Liufeng Mao, Ning Xiao, Guocheng Ren, Cong Zhang, and Jie Pan. 2009. "Effects of High Dietary Fat and Cholesterol on Expression of PPARα, LXRα, and Their Responsive Genes in the Liver of ApoE and LDLR Double Deficient Mice." *Molecular and Cellular Biochemistry* 323 (1–2): 195–205. doi:10.1007/s11010-008-9982-3.

12 Mitochondrial Dysfunction and Heart Diseases

Dao-Fu Dai MD, PhD

University of Iowa, Iowa city, IA 52242, USA

CONTENTS

1 INTRODUCTION

1.1 MITOCHONDRIA: POWERHOUSE, ROS PRODUCTION, AND SIGNALING ORGANELLE

As powerhouse of cells, mitochondria generate most of the cell's supply of energy source in the form of adenosine triphosphate (ATP). The ATP is generated by oxidizing pyruvate and acetyl CoA, the major metabolic products of glycolysis and fatty acid oxidation through complex process called oxidative phosphorylation. During oxidative phosphorylation, the redox energy from NADH and $FADH_2$ is transferred to oxygen via the electron transport chain through mitochondrial respiratory complex I-IV located in the mitochondrial membranes, followed by ATP synthesis by complex V (ATP synthase). Since formation of ATP from ADP and Pi is energetically unfavorable, Complex V obtains the energy required for ATP synthesis from the electrochemical gradient created by the difference in proton concentration across the mitochondrial membrane. This electrochemical proton gradient is generated when a high energy electron is passed along the electron transport chains in the inner mitochondrial membrane. During this process, a fraction of electrons may prematurely reduce oxygen, forming reactive oxygen species (ROS), such as superoxide radicals. This can cause oxidative stress in the mitochondria and may contribute to the decline in mitochondrial function. In addition to generating ATP, mitochondria are involved in several other essential tasks, such as biosynthetic pathways, redox regulation, cellular signaling and Ca^{2+} homeostasis. The cytoplasmic Ca^{2+} concentrations are critical to regulate a vast array of reactions essential for cellular signal transduction. Mitochondria can transiently store Ca^{2+} and regulate Ca^{2+} homeostasis. Indeed their ability to rapidly take up and later release Ca^{2+} makes them very good buffers for cytosolic Ca^{2+}, secondary to the endoplasmic reticulum (ER) as the most important storage site of Ca^{2+}.[1] Mitochondria closely interact with ER to regulate calcium signaling,[2] which will be discussed in Section 4. This chapter will focus on mitochondrial ROS and calcium; while briefly discuss the role of autophagy and mitochondrial dynamics in cardiac aging, hypertensive cardiomyopathy, heart failure and cardiac arrhythmia.

2 MITOCHONDRIAL DYSFUNCTION IN CARDIAC AGING

2.1 THE PHENOTYPES OF CARDIAC AGING

Aging is a major risk factor for cardiovascular diseases, as shown by the exponential increase in the prevalence of cardiovascular diseases in the elderly population. According to the Framingham Heart Study and the Baltimore

Longitudinal Study on Aging, there is an age-dependent increase in left ventricular hypertrophy and diastolic dysfunction, as measured by echocardiography. In these apparently healthy elderly individuals without hypertension,[3] the significant increase in the prevalence of LV hypertrophy and diastolic dysfunction likely result from intrinsic cardiac aging. Diastolic dysfunction predispose to the development of heart failure with preserved ejection fraction (HFpEF), which also becomes increasingly prevalent with advanced age, especially in women. The resting systolic function is relatively preserved with age, however, exercise capacity and cardiovascular reserve after prolonged exercise significantly decline with age.[4] The decline in exercise capacity in the elderly is likely attributed to a modest decrease in systolic function and a more prominent decline in maximal heart rate during peak exercise. In addition, the increased contribution of LV filling from atrial contraction because of increased LV stiffness and/or impaired LV relaxation also increases atrial pressure. The latter adversely results in atrial hypertrophy and dilatation, which subsequently lead to atrial fibrillation in the elderly.[5-7] Atrial fibrillation increases the risk of stroke, compromises exercise capacity, and predisposes to the development of HFpEF. Indeed, HFpEF accounts for more than half of all heart failure cases in patients older than 75 years, especially in those without structural or ischemic heart diseases. There is also age-dependent endocardial fibrosis and valvular calcification, which contribute to valvular stenosis, especially of the aortic valve.[8] Histological examinations of normal aged human myocardium demonstrate cardiomyocyte hypertrophy with enlarged nuclei, increases in interstitial accumulation of fat, collagen, elastin and lipofuscin.[9]

2.2 MITOCHONDRIAL OXIDATIVE STRESS IN CARDIAC AGING

The intracellular reactive oxygen species (ROS) is a major determinant of lifespan, as proposed by Denham Harman.[10] According to the free radical theory of aging, the decline in cellular and organ functions in old age could be attributed to the deleterious effects of ROS. In living cells, ROS are generated in multiple compartments by multiple enzymes, including NADPH oxidase at the plasma membrane, lipid oxidation within peroxisomes, oxidative phosphorylation within mitochondria, and xanthine oxidase and various cyclooxygenases in the cytoplasm. Although all of these sources contribute to the overall oxidative burden, the majority of ROS are produced as byproducts of oxidative phosphorylation during ATP generation within mitochondria.[11] This has led to the mitochondrial variant of free radical theory,[12] which implicates that mitochondrial are the main sites for age-related ROS production and ROS induced damage. Mitochondrial ROS might attack various mitochondrial constituents, causing mitochondrial DNA mutations and respiratory enzymes

oxidative damage. A defect in mitochondrial respiratory enzymes would increase mitochondrial production of ROS, causing further mitochondrial damage and dysfunction, then leading to further functional declines of cellular and organ function that can eventually progress to death.[11]

Previous publications have shown evidence both supporting and arguing against the free radical theory of aging. Evidence arguing against free radical theory has been reported in several knockout and transgenic mouse models for cellular antioxidants.[13] For instance, mice with deletion of major cellular antioxidant enzymes show a relatively mild phenotype and rarely demonstrate a lifespan decrease despite significant increases in ROS. Conversely, overexpression of antioxidant enzymes has generally failed to extend mouse lifespan. Consistent with this, oral antioxidant supplementation in humans with good nutritional status has generally failed to demonstrate any beneficial effects. However, as mitochondrial variant of the free radical theory of aging postulates that free radical generated by mitochondria is the main determinant of lifespan. The lack of significant impact of the reported knockout and transgenic mice on lifespan might be explained by lack of appropriate targeting of ROS-induced mitochondrial damage.

As a metabolically active organ relying on ATP generated by mitochondria, the heart is exposed to ROS as byproducts of mitochondrial oxidative phosphorylation, and hence cardiac mitochondria are more susceptible to oxidative damage. Direct evidence of the role of mitochondrial ROS in aging was shown by mice overexpressing catalase targeted to the mitochondria, mCAT, which had 18% prolongation of lifespan when compared with WT littermates, whereas mice overexpressing catalase targeted to peroxisomes (pCAT, the natural site of catalase) or the nucleus had little or no lifespan extension.[14] In mouse hearts, mCAT overexpression attenuates the phenotypes of cardiac aging,[15] including age-dependent left ventricular hypertrophy indicated by increase in left ventricular mass index (LVMI, $P<0.01$ for age-dependent change), a modest decline in systolic function with age (FS%, $P =0.03$) and a significant decline in diastolic function measured by Ea/Aa ($P<0.01$) as well as impairment of myocardial performance with age, indicated by increase in myocardial performance index (indicated a greater fraction of systole was spent during isovolemic contraction and relaxation, $P<0.01$)[15] (Figure 1). These cardiac aging phenotypes in mice closely recapitulate those found in human cardiac aging discussed earlier. In addition, there is left atrial dilatation in old age. All of the above phenotypes of cardiac aging were significantly attenuated in mCAT mice (p<0.01 for old WT vs. old mCAT). There was an increase in mitochondrial protein carbonyls and increase in mitochondrial DNA deletion frequencies compared with young wild type hearts, both of which indicate increased mitochondrial oxidative damage in the aged hearts (Figure 2). These oxidative damage markers were attenuated by mCAT.[15]

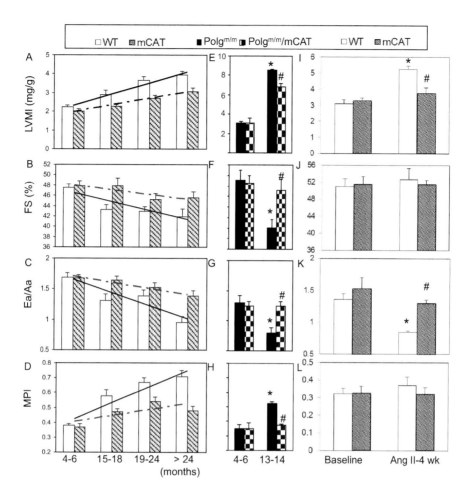

Figure 1   Echocardiography of mouse cardiac aging and hypertensive cardiomyopathy; modified from Dai, et al. Echocardiography in WT and mCAT mice (A–D), Polg [m/m] mice and 4-week Angiotensin II treated mice (I-L) in the presence or absence of mCAT. (A, E, I) Left ventricular mass index (LVMI) defined as heart weight in mg divided by body weight in gram, (B, F, J) percentage of fractional shortening (FS), indicating systolic function, (C,G, K) Ea/Aa by tissue Doppler imaging indicating diastolic function, (D, H,L). the myocardial performance index (MPI) calculated by the sum of isovolemic contraction and relaxation time divided by ejection time (higher is worse). The increased linear trends across ages in WT mice were significant for all parameters ($p < 0.05$ for all, left panels). The beneficial effect of mCAT versus WT was analyzed by the interaction between genotype and the linear age trend, and was significant in all cases ($p < 0.01$ for all except fractional shortening, $p = 0.03$). Middle panels $*p < 0.05$ Polg [m/m] at 13–14 months old vs WT littermate, $\#p < 0.05$ Polg [m/m] at 13–14 months old versus Polg [m/m]/mCAT at the same age. Right panels:. $*p < 0.05$ WT-Ang II-4 wk vs. WT at baseline before Ang II, $\#p < 0.05$ mCAT-Ang II-4 wk vs. WT-Ang II-4 wk.; mCAT, catalase targeted to mitochondria. Modified from Dai et al.[15,20,32]

Additional implication for the role of mitochondrial dysfunction in aging was shown in mice with homozygous mutation of mitochondrial polymerase gamma at D257A (Polga$^{D257A/D257A}$ designated as Polg$^{m/m}$), which have impaired proofreading capacity of mtDNA polymerase leading to time-dependent accumulation of mtDNA point mutations and deletions.[16,17] These mice have shortened lifespan (~15 months) and several "accelerated" aging phenotypes, including kyphosis, graying and loss of hair, anemia, osteoporosis, age-related loss of muscle mass and hearing loss.[18] The accumulation of mitochondrial DNA mutation with age has been shown to increase apoptosis[19] and mitochondrial oxidative damage in the mouse heart,

leading to cardiomyopathy around middle age (13–14 months) in Polg$^{m/m}$ mice, to a degree that is even more severe than the normal cardiac aging in WT mice.[20] Cardiac hypertrophy is much more prominent in middle age Polg$^{m/m}$ mice, as are impairment of both systolic and diastolic function, when compared with normal cardiac aging (Figure 1). The observations that mCAT partially rescued the cardiomyopathy phenotypes in Polg$^{m/m}$ mice suggests that mitochondrial ROS and mitochondrial DNA damage are part of a vicious cycle of ROS-induced ROS releases.[20] Interestingly, endurance exercise has been reported to attenuate skeletal muscle and cardiac phenotypes in Polg$^{m/m}$ mice,[21] through upregulation of mitochondrial biogenesis

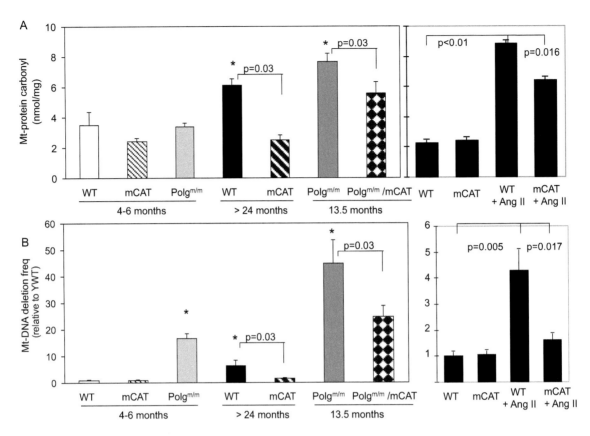

Figure 2 Mitochondrial oxidative damage and mtDNA deletions in cardiac aging and hypertensive cardiomyopathy. (A). Mitochondrial protein carbonyl (nmol/mg) significantly increased in old WT (OWT, > 24 months) and even more in middle-aged Polg (13.5 months) mouse hearts when compared with young WT mouse hearts. mCAT significantly reduced the age-dependent mitochondrial protein carbonylation. Right panel shows significant increase in cardiac mitochondrial protein carbonyl after 4-week of Ang II in WT mice, which is significantly attenuated by mCAT mice (B). Mitochondrial DNA deletion frequency significantly increased in old WT (> 24 months) and young Polg (4 months) when compared with young WT, and this is dramatically increased in middle-aged Polg (13.5 months). mCAT overexpression significantly reduced the deletion frequency for both. *p < 0.05 compared with YWT. Right panel shows significant increase in mtDNA deletion frequency after 4 week of Ang II in WT mice, which is significantly ameliorated by mCAT. Modified from Dai et al.[15,20,32]

seen with exercise in these mice. As exercise is known to induce low levels of ROS, this observation also supports the concept of beneficial effect of mitohormesis. Mitohormesis is a mitochondrial stress response signaling to activate endogenous protective mechanisms in response to low level of stresses such as ROS, decline in energetic substrates, misfolded/damaged proteins or mitochondrial toxins. Mitohormesis due to low level of ROS has been shown to upregulate endogenous antioxidants, which reduce chronic oxidative damage and subsequently improve healthspan.[19]

Another evidence of mitochondrial ROS in cardiac aging was shown by mice with a targeted mutation of the p66[Shc] gene. These mice display prolonged lifespan, reduced production of ROS, and increased resistance to ROS-mediated apoptosis.[22]. The p66[Shc] localizes to the mitochondrial intermembranous space and has been shown to be a mitochondrial redox enzyme, forming

ROS by using electrons from the respiratory chain to produce H_2O_2.[23] The p66[Shc] was phosphorylated by PKC-beta together with prolyl isomerase Pin-1, then the phosphorylated p66[Shc] accumulated within mitochondria to activate mitochondrial Ca^{2+} response, and subsequently induce apoptosis.[24] Disruption of p66[Shc] prevents Angiotensin-II induced LV hypertrophy and cardiomyocytes apoptosis as well as reducing oxidative damage in endothelial cells in a diabetic mouse model.[25]

2.3 NEUROHORMONAL REGULATION OF CARDIAC AGING

2.3.1 Renin-Angiotensin-Aldosterone System

The renin-angiotensin-aldosterone system (RAAS) is the key player implicated in a broad spectrum of cardiovascular diseases, including hypertension, coronary heart disease, congestive heart failure and atrial fibrillation. The prevalence of these diseases is known to increase with age as reported by the American Heart Association Heart Disease and Stroke Statistics.[26] Both renin and angiotensin II (Ang II) have been shown to induce cardiac hypertrophy and cardiomyocytes apoptosis, increases cardiac fibrosis and impairs cardiomyocyte relaxation.[27] These changes are similar to cardiac aging changes. The fact that there is an increase in the tissue level of Ang II in the aged rodent hearts suggests that Ang II is one of the mediators of cardiac aging changes.[15,28] Increased tissue Ang II is presumably related to increase in local angiotensin II converting enzyme (ACE) level in cardiac and vascular tissues.[29] Although the mechanism of increased ACE in aging is not well understood, long-term inhibition with angiotensin receptor blockers or disruption of angiotensin receptor type I have been shown to reduce age-dependent cardiac pathology and prolong rat[30] and mouse[31] survival. Thus, the activation of RAAS might play a critical role in cardiac aging and age-associated cardiovascular diseases. Ang II has been shown to induce mitochondrial and total cellular ROS and scavenging of mitochondrial ROS by mitochondrial targeted catalase attenuate cardiac hypertrophy and diastolic dysfunction[32] (see Section 3). Together with the finding of increased tissue Ang II in the aged hearts, these finding suggests that Angiotensin II induced mitochondrial ROS is likely one of the mechanisms of cardiac aging changes.

2.3.2 Adrenergic Signaling

Chronic adrenergic stimulation by catecholamine is hazardous to the heart, as it increased heart rate, contractility, blood pressure (afterload) and cardiac wall stress, leading to increased cardiac metabolic demand. Adenylate cyclase is a key enzyme producing c-AMP as a secondary messenger downstream to β-adrenergic signaling. In the heart, adenylate cyclase type 5 (AC5) is the major form. Disruption of AC5 has been shown to protect against chronic pressure overload-induced cardiac hypertrophy, apoptosis and failure by chronic catecholamine stimulation or aortic banding.[33,34] Mice with AC5 disruption had prolonged lifespan, the mechanisms of which might be mediated through upregulation of the Raf-1/pMEK/pERK pathway, which confers protection against various stresses[35.] These mice were also shown to attenuate aging changes in the heart, including attenuation of cardiac hypertrophy, systolic dysfunction, apoptosis and fibrosis.[36] Chronic β-adrenergic stimulation caused mitochondrial membrane depolarization and apoptosis in adult rat cardiomyocytes which was inhibited by Mn-SOD/catalase mimetics and by overexpression of catalase,[37] suggesting the role of mitochondrial ROS. Another study demonstrates that β-adrenergic stimulation increased cardiomyocytes Ca^{2+} transient amplitude that was diminished by N-acetyl cysteine antioxidant as well as the mitochondrial-targeted SS31, an indirect mitochondrial antioxidant.[38] Taken together, current evidences support the roles of mitochondrial ROS in β-adrenergic signaling.

2.3.3 Growth Hormone and IGF1 Signaling

Decreased growth hormone (GH) and its downstream insulin-like growth factor-1 (IGF-1) has been shown to associate with age-related cardiovascular morbidity and mortality in humans as well as animal models.[39] Consistent with this, previous studies showed that the cardiovascular beneficial effects of IGF-1 might be partly related to the mitochondrial protection mechanisms. For instance, IGF-1 treatment decreases mitochondrial superoxide production in cultured endothelial cells and cardiomyocytes.[40] Furthermore, Ames dwarf mice which have low levels of plasma GH and IGF-1 due to defective development of the pituitary gland because of Prop-1 mutation, is associated with increased mitochondrial oxidative stress both in the vasculature and the heart,[40] mimicking the aging phenotype. Interestingly, mitochondrial oxidative stress in the heart of Ames dwarf mice appears to be associated with impaired contractile function.[41] Recent studies show that treatment of aged rodents with IGF-1 confers mitochondrial protection, including an attenuation of mitochondrial ROS generation in the liver.[42] In summary, studies show that treatments that increase circulating IGF-1 levels exert cardiovascular protective effects in aging.[28,43,44]

2.4 MITOCHONDRIAL DYNAMICS AND AUTOPHAGY IN CARDIAC AGING AND HEART FAILURE

Mitochondria are dynamic organelles continuously undergoing fusion and fission to maintain normal shape and function.[45] Previous studies reported significant morphological changes of mitochondria in cardiac aging and cardiomyopathy,[46,47] suggesting that mitochondrial dynamics may play critical roles in mitochondrial maintenance and quality control in aging and during progression of cardiomyopathy. For instance, mitochondrial fission has been shown to generate mitochondrial fragments with lower oxidative phosphorylation activity, resulting in lower membrane potential, and these mitochondrial fragments will be targeted for degradation by ubiquitination through Pink and Parkin-mediated pathway,[48] as part of quality control mechanism.

Mitofusin1 (Mfn1), mitofusins2 (Mfn2) and OPA1 are proteins involved in mitochondrial fusion. Abnormality of mitochondrial fusion can cause abnormal mitochondrial structure and function, resulting in inefficient cellular respiration in many tissues,[49–51] including heart.[52,53] Disruption of both mitochondrial fusion proteins both mitofusin 1 (Mfn1) and 2 (Mfn2) is embryonic lethal. Mfn1 deletion has been shown to cause cardiac hypertrophy and dysfunction due to mitochondrial fragmentation and dysfunction.[53] Cardiac specific deficiency of Mfn2 has been reported to cause cardiac hypertrophy and failure due to reduced oxidative metabolism and altered mitochondrial permeability transition pore (mPTP).[53] Likewise, defective mitochondrial fusion due to partial deletion of both Mfn1 and 2 has been shown to cause mitochondrial fragmentation, disrupted cristae structure,[53,54] impaired mitochondrial respiration, leading to cardiomyopathy.[55] These findings emphasize the importance of mitochondrial fusion to maintain mitochondrial function. Indeed, downregulation of Mfn2 has been reported in several experimental models of heart failure.[56]

Autophagy is a cellular recycling process of cytoplasmic macromolecules and organelles to maintain cellular homeostasis. Cells remove damaged and aggregated proteins through autophagy, replenish damaged organelles such as mitochondria, and break down macromolecules to recycle biomolecules and energy to cells.[57] Mitophagy is a quality control mechanism to degrade damaged mitochondria to prevent hazardous signals (such as ROS) to propagate and cause further damage. One mechanism of mitophagy involves PTEN-induced putative kinase 1 (PINK1)-Parkin-mitofusin 2 (Mfn2) complex, which marks a depolarized mitochondria to be engulfed by autophagosomes through an LC3-receptor – dependent mechanism.[58] PINK1 induces phosphorylation of ubiquitin, which recruits LC3 receptor proteins and other autophagy factors to mitochondria.[59] Deletion of Parkin led to accumulation of disorganized mitochondria in aged mouse cardiomyocytes.[60] Atg5 is a key protein involved in formation of autophagic vacuoles. It has been shown that mice with Atg5 overexpression demonstrate extended longevity, while cardiac-specific mutation of Atg5 develop left ventricular hypertrophy and heart failure in mice.[61] These mice also demonstrate accumulation of ubiquitinated proteins and p62, suggesting proteotoxic effect due to impairment of autophagy.[61,62] In contrast, induction of autophagy in cardiomyocytes has been shown to ameliorate ROS-induced protein damage, aggregation and protein ubiquitination, resulting in improved mitochondrial function and reduced cell death. Taken together, these studies suggest critical role of autophagy in maintaining mitochondrial quality,[63] and the impairment of autophagy is likely a downstream mechanism of cardiac aging[64] and heart failure.

3   MITOCHONDRIAL DYSFUNCTION IN HYPERTENSION AND HEART FAILURES

The prevalence of hypertension exponentially increases in old age, with up to 70% of the population older than 70 years old having hypertensive cardiovascular disease. Hypertension is the most common cause of LV hypertrophy, which predisposes to LV dilatation, heart failure and sudden cardiac death. Angiotensin II as the key molecule of the Renin-Angiotensin Aldosterone System (RAAS) is a key mediator molecule in hypertension.

Ang II increases blood pressure and is well known to cause left ventricular hypertrophy and fibrosis.[65] Whether the effects of Ang II on LV remodeling are dependent on blood pressure or direct effects is still controversial.[66]

Our previous studies demonstrate that 4-week continuous pressor dose of Ang II by subcutaneous minipump increased blood pressure, induced cardiac hypertrophy, cardiac fibrosis and diastolic dysfunction[67] (Figure 1, right panels, I–L). This cardiac hypertrophy is associated with increased cardiac mitochondrial protein carbonyl content and the frequency of mitochondrial DNA deletions, indicating oxidative damage to mitochondria (Figure 2, right panels), leading to mitochondrial dysfunction as shown by the decline in mitochondrial respiration.[32] Furthermore, the accumulation of mitochondrial oxidative damage activated mitophagy, which in turn increased signaling for mitochondrial biogenesis through activation of peroxisome proliferator-activated receptor gamma coactivator-1 alpha (PGC-1α) and its target genes. This is consistent with the study showing that PGC-1α is transcriptionally upregulated by ROS.[68] The fact that mice overexpressing mitochondrial catalase (mCAT) but not mice overexpressing the natural peroxisomal catalase, were resistant to cardiac hypertrophy, fibrosis and diastolic dysfunction emphasizes the critical role of ROS within mitochondria in Ang II-induced cardiomyopathy.[32]

Mechanistically, Ang II binds to ATR1, a Gαq coupled-receptor, and then activates NADPH oxidase-2 (NOX2) on the plasma membrane through a PKC-dependent manner to produce ROS.[69] In addition, NOX4 isoform in the mitochondrial membrane might also increase mitochondrial ROS production, as previously shown in endothelial and vascular smooth muscle cells.[70,71] Additional studies show that disruption of p66[Shc] prevents Ang II-induced LV hypertrophy and cardiomyocyte apoptosis as well as reducing oxidative damage in cardiac progenitor cells, cardiomyocytes and endothelial cells in a diabetic mouse model.[25,72,73] Likewise, mice deficient in mitochondrial deacetylase SIRT3 demonstrate LV hypertrophy and fibrosis, associated with mitochondrial swelling due to increased mPTP opening, and increased mortality in response to pressure overload.[74]

As mentioned above and shown in Figures 1–2, Polg[m/m] mice have increased mitochondrial DNA mutations and develop heart failure at middle age, which is ameliorated by mCAT. Treatment with 4-week pressor dose of Ang II in Polg[m/m] mice resulted in cardiomyopathy and heart failure, which is also attenuated by mCAT.[15,32] This suggests that primary damage to mitochondrial DNA contributes directly to the phenotype of heart failure with reduced ejection fraction, through increased mt ROS. Taken together, the protective effects of mCAT in Ang-induced cardiac hypertrophy and Gαq-induced heart failure provide direct evidence that amplification of ROS (vicious cycle) within mitochondria is a key mediator in these murine models.[32] Using the transverse-aortic constriction (TAC) pressure overload model, we further demonstrate that TAC-induced heart failure is associated with remodeling of the mitochondrial proteome, including decreased abundance of proteins involved in fatty acid metabolism and increased abundance of proteins in glycolysis, apoptosis, mitochondrial unfolded protein response and proteolysis. Overexpression of mCAT mitigates the phenotype of heart failure, better preserves proteins involved in fatty acid metabolism and attenuates the increases in apoptotic and proteolytic enzymes.[75] Consistent with this, we also demonstrated that the mitochondrial protective peptide SS31 attenuates cardiac hypertrophy and diastolic dysfunction induced by chronic Angiotensin II, and the heart failure phenotypes induced by overexpression of Gαq or transverse aortic constriction. Furthermore, SS-31 has also been shown to prevent hypoxia-reoxygenation induced apoptosis in renal tubular epithelial cell by down-regulation of p66[Shc].[76] Thus, breaking the ROS vicious cycle within mitochondria by mCAT is effective in attenuating both cardiac hypertrophy and failure (Figure 3).

Myocardial infarction is a major cause of heart failure. Ischemic-reperfusion (IR) injury often occurs during acute myocardial infarction, either due to spontaneous recanalization of the occluded artery or as a result of a reperfusion therapy. ROS are well known to be primary mediators in IR injury. ROS accumulate during ischemia,[77] causing mitochondrial respiratory complex dysfunction, which leads to a burst of ROS during reperfusion. Further, post-ischemic reperfusion is known to cause ROS accumulation, acidic pH and a rise in $[Ca^{2+}]$ (overload). These conditions predispose to mitochondrial ROS-induced ROS release.[78] The aged myocardium has less tolerance to ischemia and hemodynamic stress than the young myocardium.[79] Aged cardiomyocytes have a lower threshold for ROS induced ROS release and increased susceptibility to mPTP opening.[80] Ischemic

preconditioning, which is beneficial in reducing damage during I/R, is impaired in the aged myocardium.[80] This loss of endogenous protective mechanisms of ischemic preconditioning in the aged heart might be due to a decrease in mitochondrial heat shock protein-70,[81] reduced nitric oxide bioavailability,[82] damaged mitochondria that are vulnerable to stress, and diminished PKC translocation into mitochondria, all of which are required for the protective effect of ischemic preconditioning.[83,84]

Consistent with the findings in experimental animals, mutations of genes encoding mitochondrial proteins can cause various forms of idiopathic hypertrophic and dilated cardiomyopathies, as seen in human mitochondrial diseases.[85] Furthermore, mitochondrial dysfunction is well documented in human cardiac explants from patients with end-stage heart failure, due to both ischemic and non-ischemic dilated cardiomyopathies.[86,87] In these failing hearts, there is evidence of mitochondrial DNA depletion (reduced mt DNA copy number), increased mt DNA deletion frequency, decline in protein levels and activities of several mitochondrial respiratory enzymes, including complex I, II and IV, as well as disruption of mitochondrial ultrastructures.[86,87] Possible molecular mechanisms may include mitochondrial biogenesis that does not keep up with the increasing demand of the heart during hypertrophy,[88–90] mitochondrial uncoupling and decreased substrate availability,[64] and increased mitochondrial DNA deletions[32] which can produce defective respiratory enzymes. This deficiency of mitochondrial number/function is likely to cause energetic failure. Studies using[31] P NMR spectroscopy have demonstrated that the ATP content of human failing hearts is generally 20–30% lower than that of normal young hearts.[91] Furthermore, it has been shown that phosphocreatine significantly declines (by up to 60%) in elderly heart failure patients.[92] Phosphocreatine is an important short-term energy reserve that maintains

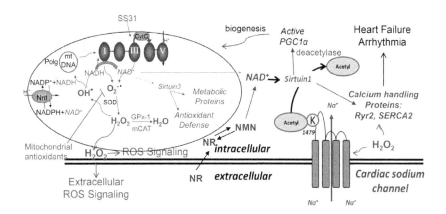

Figure 3   Proposed signaling mechanisms of mitochondrial ROS and sirtuins in cardiac aging, heart failure and arrhythmia. NR/NMN=Nicotinamide Riboside/Mononucleotide.

a high phosphorylation potential to cope with acute increases in energy demand such as during stress or exercise. The magnitude of the decline in ATP and phosphocreatine is related to the severity of heart failure[93] and predicts mortality rate in patients with dilated cardiomyopathy.[94] In summary, mitochondrial dysfunction is evident in end-stage heart failure due to various etiologies, and mitochondrial protective strategies might be a beneficial method of treatment and prevention of heart failure.

4   MITOCHONDRIAL DYSFUNCTION IN ARRHYTHMIA

The normal function of the heart requires coordinated and synchronized rhythmic electrical activity that is coupled with contractile action. Emerging evidence indicates that mitochondrial dysfunction can adversely affect cardiac electrical activity by impairing intracellular ion homeostasis and membrane excitability leading to increased susceptibility to cardiac arrhythmias. Potential mechanisms may include reduced ATP production, as many ion channels are ATP-dependent pumps, excessive reactive oxygen species (ROS) that impair the function of ion channels and calcium handling proteins.

Mitochondrial dysfunction impairs ATP production, increases ROS production and leads to the accumulation of some metabolites, such as NADH, ADP and lactate, which could be deleterious. Mitochondrial dysfunction may also impair endogenous antioxidant defense system and deplete intra- mitochondrial reduced glutathione (GSH), which is required for redox regulation within mitochondria. This will further increase oxidative stress and aggravate cellular damage. Increased ROS has been shown to oxidize cellular lipids and proteins, resulting in cardiomyocyte damage and dysfunction. Excessive amount of mitochondrial ROS may lead to opening of mitochondrial permeability transition pore (mPTP), causing cytochrome c leakage which further activate apoptotic cell death pathway.[95]

Increased mitochondrial superoxide can activate mitochondrial uncoupling proteins, causing increased proton leak and uncoupling of mitochondrial ATP production from oxygen consumption,[96] leading to reduced ATP production and further increased in mitochondrial ROS. Mitochondrial ROS can cause oxidative damage to components of the electron transport chain, including complexes I,[97] III and IV,[98] ATP synthase,[97] as well as cytochrome c oxidase,[99] leading to impaired ATP production and increased ETC electron leak that favors ROS generation. These phenomena are known as ROS induced ROS release.[100] Moreover, increased mitochondrial ROS can impair cellular electrical function by ROS-mediated signaling and oxidative damage of ion channels (as discussed later), as well as by reducing ATP production that is essential for ion channel/transporters

to function. Other post-translational modification, such as acetylation, has recently been shown to affect ion channel function, as summarized in Figure 3.

4.1   CARDIAC SODIUM CHANNELS AND MITOCHONDRIAL DYSFUNCTION

Cardiac voltage-gated sodium channels (Nav) consist of heteromeric assembly of a pore-forming α subunit and auxiliary β subunits. Nav 1.5 (SCN5A) is the major Nav α subunit expressed in myocardium.[101] Nav channels play a critical role by generating the rapid upstroke (phase 0) of the action potential to excite cardiomyocytes. Nav channels also regulate the conduction of electrical impulse within the myocardium, together with cardiac gap junctions. Abnormal cardiac Nav channels has been described in several hereditary cardiac diseases, in myocardial ischemia[102] and heart failure.[103] Gain of function (GOF) mutation of Nav 1.5 can cause long QT syndrome type III and rare familial atrial fibrillation,[104] while loss of function (LOF) mutation have been reported to cause Brugada syndrome,[105] progressive cardiac conduction defect, sinus node dysfunction, atrial arrhythmias, and poor pacemaker capture.[106] Mutation of SCN5A has also been reported in multifocal ectopic premature Purkinje-related complexes (GOF mutations), isolated cardiac conduction defect (LOF mutations), sick sinus syndrome (LOF mutations), atrial fibrillation (LOF or GOF mutations) [see review[107]]. In addition, both loss- and gain-of-function mutations may cause dilated cardiomyopathy, which increase the risk of arrhythmia.

Mitochondria play critical roles in modulating cardiac Nav channel function through various mechanisms. Increased ROS increased the slowly inactivating component of sodium current (late INa), leading to prolongation of action potential duration (APD), early after depolarizations (EADs), increased Na^+/Ca^{2+} exchange, and subsequent cellular Ca^{2+} overload, which increased susceptibility of arrhythmia.[108] It has been shown that elevated NADH level leads to decreased peak INa via a PKC-dependent increase in mitochondrial ROS, and this was attenuated by mito-TEMPO, a mitochondrial antioxidant, but not by inhibitors of NADPH oxidase, xanthine oxidase, or NO synthase.[109] The fact that only inhibition of mitochondrial ROS but not ROS produced by other oxidant pathways reduced peak INa and increased the risk of arrhythmia emphasize the critical role of mitochondrial ROS in modulating cardiac Nav function.[109]

Recent report using mice with cardiomyocyte specific knock-out of sirtuins 1 demonstrated that hyperacetylation of Nav1.5 due to sirt1 deficiency impaired the trafficking of the sodium channel to plasma membranes, thereby reducing INa and this has been shown to cause cardiac conduction defect.[110] In nonischemic cardiomyopathy, increased in both cytosolic NADH and mitochondrial ROS have been reported to decrease *INa* without changing the membrane expression of Nav channel protein. The fact that treatment with NAD+ or mito-TEMPO can partially restore *INa* suggests that hyperacetylation and/or oxidative modification may impact the function of Nav channels and affect cardiac conduction system[111] (Figure 3). Consistently, NAD+ treatment has been shown to ameliorate impaired cardiac conduction in human heart failure.[111] Furthermore, the A280V mutation in glycerol-3-phosphate dehydrogenase 1-like (GPD1-L) protein that cause Brugada syndrome, demonstrated increased NADH levels and mitochondrial ROS.[111,112] Treatment with NAD+ or mitoTEMPO ameliorates A280V GPD1-L the decline in *INa* caused by A280V GPD1-L mutation.[111] These observations highlight the critical role of mitochondrial ROS, altered cardiac metabolism reflected by NAD+/NADH ratio in affecting cardiac sodium channel function in ischemia or heart failure models, part of which through modification by sirtuins deacetylases (Figure 3).

4.2 CARDIAC CA^{2+} HOMEOSTASIS, ROS AND ARRHYTHMIA

One of the hallmarks of cardiomyocytes is highly fluctuated calcium transients, which are defined as cyclic and transient oscillations of the intracytoplasmic Ca^{2+} at high concentration during systole and at low concentration during diastole. The Ca^{2+} transients underlie the rhythmic beating of cardiomyocytes through tightly regulated excitation–contraction (EC) coupling mechanisms. In addition to the critical roles in EC coupling, localized Ca^{2+} regulates activities of numerous enzymes and transcription factors. Impairment of calcium handling has been linked to mechanical dysfunction and arrhythmia in human and experimental models of cardiac hypertrophy,[113] heart failure[114] and myocardial ischemia.[115]

Mechanistically, cyclic Ca^{2+} transients are initiated by Ca^{2+} entry through voltage-gated Ca^{2+} channels, which trigger sarcoplasmic reticulum (SR) Ca^{2+} release by activating/opening ryanodine receptor (RyR2), resulting in high intracytoplasmic [Ca^{2+}] that activate EC coupling leading to sarcomere contraction (systolic phase). Diastolic phase (relaxation) is initiated predominantly by active reuptake of [Ca^{2+}] into SR by Sarcoplasmic reticulum Ca- ATPase-2 (SERCA2) and exit of Ca^{2+} via sodium calcium exchanger NCX.

There are two types of voltage-gated calcium channels: the L- (low threshold) and T- (transient) type channels. The L-type Ca^{2+} channel (LTCC) is predominantly in ventricular cardiomyocytes, while the T-type Ca^{2+} channel is mainly in pacemaker and conduction systems.[116] ROS has been shown to have various effects on LTCC, although whether ROS increase or decrease L-type Ca^{2+} channel current (*ICaL*) in cardiomyocytes remain controversial, depending on the experimental models.

ROS also affect the function of various Ca^{2+}-handling proteins. Increased ROS enhances the open probability of RyR2 to increase cytoplasmic Ca^{2+} from the SR. More specifically, increased mitochondrial ROS production triggers localized calcium release from RyR2,[117] termed Ca^{2+} sparks. In contrast to ROS-induced activation (opening) of RyR2, SERCA2 ATPase reuptake of Ca^{2+} in cardiomyocytes are indeed inhibited by ROS,[118] either due to decreased ATP supply by mitochondria or oxidative modification[119] secondary to mitochondrial dysfunction. In both scenarios of activation of Ryr2 and inhibition of SERCA2 by ROS, the outcome is overload of [Ca^{2+}], which may activate several downstream pathways. The effect of ROS on calcium handling proteins may also involve activation of CaMKII, a Ca^{2+}/calmodulin-dependent protein kinase that are activated by increases in the intracellular [Ca^{2+}].[120] The other calcium handling protein NCX, an antiporter that removes one Ca^{2+} ion from the cytosol in exchange for the import of three Na+ ions, has been shown to be activated by ROS.[121] Taken together, the net effects of ROS on these Ca^{2+}-handling proteins result in increased cytosolic Ca^{2+} (overload) and depletion of the SR Ca^{2+} store, leading to activation of several downstream pathways, such as CaMKII, calcineurin, and NFAT.[122] Intracytoplasmic Ca^{2+} overload has also been reported to cause a prolonged action potential, delayed after-depolarizations (DADs), contractile dysfunction, and cardiac arrhythmia.[123,124] The fact that most of these Ca^{2+} handling proteins contain thiol groups or methionine residues suggest their susceptibility to direct regulation by ROS by oxidative modification.

In addition to the roles in regulation of ROS and acetylation (Figure 3), mitochondria are capable of directly and rapidly taking up calcium, affecting both matrix energetics within mitochondria or releasing Ca^{2+} into cytosol. Ca^{2+} import across the outer mitochondrial membrane occurs via the voltage-dependent anion channels (VDAC).[125] VDAC is as a large voltage-gated channel, fully opened with high-conductance and low anion-selectivity at low transmembrane potentials (less than 20–30 mV), but switching to high cation selectivity and lower conductance at higher potentials.[126–128] Ca^{2+} transport across the inner mitochondrial membrane occurs through two possible mechanisms: First, through mitochondrial calcium uniporter (MCU), which transfers Ca^{2+} down the potential gradient generated during oxidative phosphorylation. Pharmacological inhibition of MCU with Ru360 has been reported to reduce the incidence of ventricular arrhythmias induced by ischemia–reperfusion in the rat heart.[129] Second, through a Na+/Ca^{2+} exchanger and a H+/Ca^{2+} exchanger. During pathological conditions a large increase in mitochondrial calcium levels is thought to activate the mitochondrial permeability transition pore, resulting in cell death.

The capacity of mitochondria to uptake and release Ca^{2+} can have substantial impact on the spatial and temporal dynamics of Ca^{2+} signaling in cardiomyocytes.

Impaired mitochondrial function has been reported to cause alterations of calcium transient by affecting the capacity of the mitochondrial Ca^{2+} handling.[130] This change increases susceptibility to cardiac arrhythmias. Increased ROS has also been shown to alter mitochondrial Ca^{2+} concentration[131] by modulating mitochondrial NCX function or cytosolic Ca^{2+} levels, contributing to the perturbation of intracellular Ca^{2+}.[115,132] Ischemia-induced ROS can lead to collapse of mitochondrial membrane potential and release Ca^{2+} into via a reverse transport mode through mitochondrial NCX,[133] resulting in $[Ca^{2+}]$ overload.[134] Conversely, Ca^{2+} overload can induced mitochondrial ROS production. Mitochondrial Ca^{2+} is known to stimulate oxidative phosphorylation, which also increase the chance of respiratory chain electron leakage leading to ROS production.[135] Ca^{2+} is also known to activate nitric oxide synthase to generate NO, which may inhibit complex IV and enhance ROS production.[136] Increased Ca^{2+} can affect cytochrome c dislocation by competing for cardiolipin binding sites, which may cause complex III inhibition and increased ROS production.[137]

In summary, current evidence suggests that mitochondrial Ca^{2+} can be part of feed-forward cycle of ROS amplification: increased mitochondrial Ca^{2+} can induce ROS and mitochondrial ROS can induce Ca^{2+} overload. In normal physiological function, there is tight regulation of mitochondrial Ca^{2+} and ROS by controlling Ca-handling proteins and endogenous antioxidant systems. During myocardial ischemia and in response to pressure overload, multiple mechanisms can lead to increased mitochondrial ROS production, which will increase mitochondrial Ca^{2+}. These vicious cycles consist of ROS-induced ROS release, ROS-induced Ca^{2+} overload and Ca^{2+}-induced ROS production, leading to ROS and/or Ca^{2+} overload, resulting in mPTP opening, cellular damage, and death.

4.3 OTHER MECHANISMS CONTRIBUTING TO ARRHYTHMIA

Various voltage-gated K^+ channels (K_v) contribute to repolarization of action potential in the myocardium. Increased ROS have been shown to reduce the expression or function of K_v channels, resulting in declined K currents in ventricular cardiomyocytes,[138–140] which can further lead to delayed repolarization, prolonged APD, and increased susceptibility to ventricular arrhythmias.[101] Many of these can be reversed by enhancing mitochondrial antioxidants, such as reduced glutathione (GS).[141,142]

Gap junctions mediate intercellular communication in cardiomyocytes, which can freely transfer small metabolites and ions between cardiomyocytes, through which electrical impulse can be rapidly and synchronously conducted to the whole ventricles.[143] There are three main connexin isoforms expressed in the heart: connexin (Cx)

40 predominantly in the atria and conduction system, Cx43 in both the atrial and the ventricular cardiomyocytes and Cx45 mostly in the sinoatrial node, atrioventricular node, and His bundle.[144] Increased ROS by activation of renin-angiotensin-aldosterone system have been shown to decrease ventricular Cx43[145,146] and increased the risk of conduction block and ventricular arrhythmia.

5 CONCLUSION

Heart diseases are among the leading causes of death worldwide. Hypertension and ageing are not only the most common causes of left ventricular hypertrophy and diastolic dysfunction, but also impose significant risks of coronary heart disease, congestive heart failure, stroke and sudden death (arrhythmia). Understanding the role of mitochondrial dysfunction and signaling in aging and hypertensive cardiomyopathy, pressure-overload induced heart failure and arrhythmia will assist the development of strategies to prevent or ameliorate cardiac hypertrophy and failure, or even to delay cardiac aging changes. Since mitochondrial dysfunction is one major mechanism leading to heart diseases, the small molecule drugs that can protect mitochondria may become attractive potential novel therapeutics for patients with heart diseases. Several drugs have been developed, including mitochondrial targeted antioxidants, such as triphenylalkylphosphonium (TPP+) conjugated coenzyme Q (MitoQ) and plastoquinone (SkQ1)[147,148] and Elamipretide (SS31) that targeted to mitochondrial inner membranes enriched with cardiolipin. These small molecules have shown promising results in various preclinical models of heart diseases. Several phase II and phase III clinical trials are being performed to study the effect of Elamipretide in heart failure and mitochondrial myopathies.[149,150]. The roles of these mitochondrial protective interventions in cardiac aging and arrhythmia also warrant further clinical investigations.

REFERENCES

1. Santulli G, Marks AR. Essential roles of intracellular calcium release channels in muscle, brain, metabolism, and aging. *Curr Mol Med.* 2015;8:206–222.
2. Pizzo P, Pozzan T. Mitochondria-endoplasmic reticulum choreography: Structure and signaling dynamics. *Trends Cell Biol.* 2007;17:511–517.
3. Bursi F, Weston SA, Redfield MM, Jacobsen SJ, Pakhomov S, Nkomo VT, Meverden RA, Roger VL. Systolic and diastolic heart failure in the community. *JAMA.* 2006;296:2209–2216.
4. Correia LC, Lakatta EG, O'Connor FC, Becker LC, Clulow J, Townsend S, Gerstenblith G, Fleg JL. Attenuated cardiovascular reserve during prolonged submaximal cycle exercise in healthy older subjects. *J Am Coll Cardiol.* 2002;40:1290–1297.
5. Lakatta EG. Arterial and cardiac aging: Major shareholders in cardiovascular disease enterprises: Part III:

Cellular and molecular clues to heart and arterial aging. *Circulation.* 2003;107:490–497.

6. Lakatta EG, Levy D. Arterial and cardiac aging: Major shareholders in cardiovascular disease enterprises: Part II: The aging heart in health: Links to heart disease. *Circulation.* 2003;107:346–354.

7. Lakatta EG, Levy D. Arterial and cardiac aging: Major shareholders in cardiovascular disease enterprises: Part I: Aging arteries: A "set up" for vascular disease. *Circulation.* 2003;107:139–146.

8. Olsen MH, Wachtell K, Bella JN, Gerdts E, Palmieri V, Nieminen MS, Smith G, Ibsen H, Devereux RB. Aortic valve sclerosis relates to cardiovascular events in patients with hypertension (a life substudy). *Am J Cardiol.* 2005;95:132–136.

9. Brunk UT, Terman A. Lipofuscin: Mechanisms of age-related accumulation and influence on cell function. *Free Radic Biol Med.* 2002;33:611–619.

10. Harman D. Aging: A theory based on free radical and radiation chemistry. *J Gerontol.* 1956;11:298–300.

11. Balaban RS, Nemoto S, Finkel T. Mitochondria, oxidants, and aging. *Cell.* 2005;120:483–495.

12. Harman D. The biologic clock: The mitochondria? *J Am Geriatr Soc.* 1972;20:145–147.

13. Perez VI, Bokov A, Van Remmen H, Mele J, Ran Q, Ikeno Y, Richardson A. Is the oxidative stress theory of aging dead? *Biochim Biophys Acta.* 2009;1790:1005–1014.

14. Schriner SE, Linford NJ, Martin GM, Treuting P, Ogburn CE, Emond M, Coskun PE, Ladiges W, Wolf N, Van Remmen H, Wallace DC, Rabinovitch PS. Extension of murine life span by overexpression of catalase targeted to mitochondria. *Science.* 2005;308:1909–1911.

15. Dai DF, Santana LF, Vermulst M, Tomazela DM, Emond MJ, MacCoss MJ, Gollahon K, Martin GM, Loeb LA, Ladiges WC, Rabinovitch PS. Overexpression of catalase targeted to mitochondria attenuates murine cardiac aging. *Circulation.* 2009;119:2789–2797.

16. Vermulst M, Bielas JH, Kujoth GC, Ladiges WC, Rabinovitch PS, Prolla TA, Loeb LA. Mitochondrial point mutations do not limit the natural lifespan of mice. *Nat Genet.* 2007;39:540–543.

17. Vermulst M, Wanagat J, Kujoth GC, Bielas JH, Rabinovitch PS, Prolla TA, Loeb LA. DNA deletions and clonal mutations drive premature aging in mitochondrial Mutator mice. *Nat Genet.* 2008;40:392–394.

18. Trifunovic A, Wredenberg A, Falkenberg M, Spelbrink JN, Rovio AT, Bruder CE, Bohlooly YM, Gidlof S, Oldfors A, Wibom R, Tornell J, Jacobs HT, Larsson NG. Premature ageing in mice expressing defective mitochondrial DNA polymerase. *Nature.* 2004;429:417–423.

19. Kujoth GC, Hiona A, Pugh TD, Someya S, Panzer K, Wohlgemuth SE, Hofer T, Seo AY, Sullivan R, Jobling WA, Morrow JD, Van Remmen H, Sedivy JM, Yamasoba T, Tanokura M, Weindruch R, Leeuwenburgh C, Prolla TA. Mitochondrial DNA mutations, oxidative stress, and apoptosis in mammalian aging. *Science.* 2005;309:481–484.

20. Dai DF, Chen T, Wanagat J, Laflamme M, Marcinek DJ, Emond MJ, Ngo CP, Prolla TA, Rabinovitch PS. Age-dependent cardiomyopathy in mitochondrial Mutator mice is attenuated by overexpression of catalase targeted to mitochondria. *Aging Cell.* 2010;9:536–544.

21. Safdar A, Bourgeois JM, Ogborn DI, Little JP, Hettinga BP, Akhtar M, Thompson JE, Melov S, Mocellin NJ, Kujoth GC, Prolla TA, Tarnopolsky MA. Endurance exercise rescues progeroid aging and induces systemic mitochondrial rejuvenation in mtDNA Mutator mice. *Proc Natl Acad Sci U S A.* 2011;108:4135–4140.

22. Migliaccio E, Giorgio M, Mele S, Pelicci G, Reboldi P, Pandolfi PP, Lanfrancone L, Pelicci PG. The p66shc adaptor protein controls oxidative stress response and life span in mammals. *Nature.* 1999;402:309–313.

23. Giorgio M, Migliaccio E, Orsini F, Paolucci D, Moroni M, Contursi C, Pelliccia G, Luzi L, Minucci S, Marcaccio M, Pinton P, Rizzuto R, Bernardi P, Paolucci F, Pelicci PG. Electron transfer between cytochrome c and p66shc generates reactive oxygen species that trigger mitochondrial apoptosis. *Cell.* 2005;122:221–233.

24. Pinton P, Rimessi A, Marchi S, Orsini F, Migliaccio E, Giorgio M, Contursi C, Minucci S, Mantovani F, Wieckowski MR, Del Sal G, Pelicci PG, Rizzuto R. Protein kinase c beta and prolyl isomerase 1 regulate mitochondrial effects of the life-span determinant p66shc. *Science.* 2007;315:659–663.

25. Camici GG, Schiavoni M, Francia P, Bachschmid M, Martin-Padura I, Hersberger M, Tanner FC, Pelicci P, Volpe M, Anversa P, Luscher TF, Cosentino F. Genetic deletion of p66(shc) adaptor protein prevents hyperglycemia-induced endothelial dysfunction and oxidative stress. *Proc Natl Acad Sci U S A.* 2007;104:5217–5222.

26. Roger VL, Go AS, Lloyd-Jones DM, Adams RJ, Berry JD, Brown TM, Carnethon MR, Dai S, de Simone G, Ford ES, Fox CS, Fullerton HJ, Gillespie C, Greenlund KJ, Hailpern SM, Heit JA, Michael Ho P, Howard VJ, Kissela BM, Kittner SJ, Lackland DT, Lichtman JH, Lisabeth LD, Makuc DM, Marcus GM, Marelli A, Matchar DB, McDermott MM, Meigs JB, Moy CS, Mozaffarian D, Mussolino ME, Nichol G, Paynter NP, Rosamond WD, Sorlie PD, Stafford RS, Turan TN, Turner MB, Wong ND, Wylie-Rosett J. Executive summary: Heart disease and stroke statistics–2011 update: A report from the American heart association. *Circulation.* 2011;123:459–463.

27. Domenighetti AA, Wang Q, Egger M, Richards SM, Pedrazzini T, Delbridge LM. Angiotensin II-mediated phenotypic cardiomyocyte remodeling leads to age-dependent cardiac dysfunction and failure. *Hypertension.* 2005;46:426–432.

28. Groban L, Pailes NA, Bennett CD, Carter CS, Chappell MC, Kitzman DW, Sonntag WE. Growth hormone replacement attenuates diastolic dysfunction and cardiac angiotensin ii expression in senescent rats. *J Gerontol A Biol Sci Med Sci.* 2006;61:28–35.

29. Wang M, Takagi G, Asai K, Resuello RG, Natividad FF, Vatner DE, Vatner SF, Lakatta EG. Aging increases aortic mmp-2 activity and angiotensin ii in nonhuman primates. *Hypertension.* 2003;41:1308–1316.

30. Basso N, Cini R, Pietrelli A, Ferder L, Terragno NA, Inserra F. Protective effect of long-term angiotensin II inhibition. *Am J Physiol Heart Circ Physiol.* 2007;293: H1351–11358.

31. Benigni A, Corna D, Zoja C, Sonzogni A, Latini R, Salio M, Conti S, Rottoli D, Longaretti L, Cassis P, Morigi M, Coffman TM, Remuzzi G. Disruption of the ANG II type 1 receptor promotes longevity in mice. *J Clin Invest.* 2009;119:524–530.

32. Dai DF, Johnson SC, Villarin JJ, Chin MT, Nieves-Cintron M, Chen T, Marcinek DJ, Dorn GW, 2nd, Kang YJ, Prolla TA, Santana LF, Rabinovitch PS.

Mitochondrial oxidative stress mediates angiotensin II-induced cardiac hypertrophy and galphaq overexpression-induced heart failure. *Circ Res.* 2011;108:837–846.

33. Okumura S, Takagi G, Kawabe J, Yang G, Lee MC, Hong C, Liu J, Vatner DE, Sadoshima J, Vatner SF, Ishikawa Y. Disruption of type 5 adenylyl cyclase gene preserves cardiac function against pressure overload. *Proc Natl Acad Sci U S A.* 2003;100:9986–9990.

34. Okumura S, Vatner DE, Kurotani R, Bai Y, Gao S, Yuan Z, Iwatsubo K, Ulucan C, Kawabe J, Ghosh K, Vatner SF, Ishikawa Y. Disruption of type 5 adenylyl cyclase enhances desensitization of cyclic adenosine monophosphate signal and increases Akt signal with chronic catecholamine stress. *Circulation.* 2007;116:1776–1783.

35. Dai DF, Rabinovitch PS. Cardiac aging in mice and humans: The role of mitochondrial oxidative stress. *Trends Cardiovasc Med.* 2009;19:213–220.

36. Yan L, Vatner DE, O'Connor JP, Ivessa A, Ge H, Chen W, Hirotani S, Ishikawa Y, Sadoshima J, Vatner SF. Type 5 adenylyl cyclase disruption increases longevity and protects against stress. *Cell.* 2007;130:247–258.

37. Remondino A, Kwon SH, Communal C, Pimentel DR, Sawyer DB, Singh K, Colucci WS. Beta-adrenergic receptor-stimulated apoptosis in cardiac myocytes is mediated by reactive oxygen species/c-Jun nh2-terminal kinase-dependent activation of the mitochondrial pathway. *Circ Res.* 2003;92:136–138.

38. Andersson DC, Fauconnier J, Yamada T, Lacampagne A, Zhang SJ, Katz A, Westerblad H. Mitochondrial production of reactive oxygen species contributes to the beta-adrenergic stimulation of mouse cardiomycytes. *J Physiol.* 2011;589:1791–1801.

39. Khan AS, Sane DC, Wannenburg T, Sonntag WE. Growth hormone, insulin-like growth factor-1 and the aging cardiovascular system. *Cardiovasc Res.* 2002;54:25–35.

40. Csiszar A, Labinskyy N, Perez V, Recchia FA, Podlutsky A, Mukhopadhyay P, Losonczy G, Pacher P, Austad SN, Bartke A, Ungvari Z. Endothelial function and vascular oxidative stress in long-lived GH/IGF-deficient Ames dwarf mice. *Am J Physiol Heart Circ Physiol.* 2008;295:H1882-11894.

41. Ren J, Brown-Borg HM. Impaired cardiac excitation-contraction coupling in ventricular myocytes from Ames dwarf mice with IGF-I deficiency. *Growth Horm IGF Res.* 2002;12:99–105.

42. Puche JE, Garcia-Fernandez M, Muntane J, Rioja J, Gonzalez-Baron S, Castilla Cortazar I. Low doses of insulin-like growth factor-i induce mitochondrial protection in aging rats. *Endocrinology.* 2008;149:2620–2627.

43. Rivera EJ, Goldin A, Fulmer N, Tavares R, Wands JR. de la Monte SM. Insulin and insulin-like growth factor expression and function deteriorate with progression of Alzheimer's disease: Link to brain reductions in acetylcholine. *J Alzheimers Dis.* 2005;8:247–268.

44. Lopez-Lopez C, Dietrich MO, Metzger F, Loetscher H, Torres-Aleman I. Disturbed cross talk between insulin-like growth factor I and amp-activated protein kinase as a possible cause of vascular dysfunction in the amyloid precursor protein/presenilin 2 mouse model of Alzheimer's disease. *J Neurosci.* 2007;27:824–831.

45. Bereiter-Hahn J, Voth M. Dynamics of mitochondria in living cells: Shape changes, dislocations, fusion, and fission of mitochondria. *Microsc Res Tech.* 1994;27:198–219.

46. Hom J, Sheu SS. Morphological dynamics of mitochondria–a special emphasis on cardiac muscle cells. *J Mol Cell Cardiol.* 2009;46:811–820.

47. Ong SB, Hausenloy DJ. Mitochondrial morphology and cardiovascular disease. *Cardiovasc Res.* 2010;88:16–29.

48. Matsuda N, Sato S, Shiba K, Okatsu K, Saisho K, Gautier CA, Sou YS, Saiki S, Kawajiri S, Sato F, Kimura M, Komatsu M, Hattori N, Tanaka K. PINK1 stabilized by mitochondrial depolarization recruits Parkin to damaged mitochondria and activates latent Parkin for mitophagy. *J Cell Biol.* 2010;189:211–221.

49. Chen H, Chomyn A, Chan DC. Disruption of fusion results in mitochondrial heterogeneity and dysfunction. *J Biol Chem.* 2005;280:26185–26192.

50. Szabadkai G, Simoni AM, Chami M, Wieckowski MR, Youle RJ, Rizzuto R. Drp-1-dependent division of the mitochondrial network blocks intraorganellar ca2+ waves and protects against ca2+-mediated apoptosis. *Mol cell.* 2004;16:59–68.

51. Westermann B. Mitochondrial fusion and fission in cell life and death. *Nat Rev Mol Cell Biol.* 2010;11:872–884.

52. Bossy-Wetzel E, Barsoum MJ, Godzik A, Schwarzenbacher R, Lipton SA. Mitochondrial fission in apoptosis, neurodegeneration and aging. *Curr Opin Cell Biol.* 2003;15:706–716.

53. Papanicolaou KN, Khairallah RJ, Ngoh GA, Chikando A, Luptak I, O'Shea KM, Riley DD, Lugus JJ, Colucci WS, Lederer WJ, Stanley WC, Walsh K. Mitofusin-2 maintains mitochondrial structure and contributes to stress-induced permeability transition in cardiac myocytes. *Mol Cell Biol.* 2011;31:1309–1328.

54. Piquereau J, Caffin F, Novotova M, Prola A, Garnier A, Mateo P, Fortin D, Huynh le H, Nicolas V, Alavi MV, Brenner C, Ventura-Clapier R, Veksler V, Joubert F. Down-regulation of opa1 alters mouse mitochondrial morphology, PTP function, and cardiac adaptation to pressure overload. *Cardiovasc Res.* 2012;94:408–417.

55. Chen Y, Liu Y, Dorn GW, 2nd. Mitochondrial fusion is essential for organelle function and cardiac homeostasis. *Circ Res.* 2011;109:1327–1331.

56. Fang L, Moore XL, Gao XM, Dart AM, Lim YL, Du XJ. Down-regulation of mitofusin-2 expression in cardiac hypertrophy in vitro and in vivo. *Life Sci.* 2007;80:2154–2160.

57. Kroemer G, Marino G, Levine B. Autophagy and the integrated stress response. *Mol cell.* 2010;40:280–293.

58. Chen Y, Dorn GW, 2nd. Pink1-phosphorylated mitofusin 2 is a Parkin receptor for culling damaged mitochondria. *Science.* 2013;340:471–475.

59. Lazarou M, Sliter DA, Kane LA, Sarraf SA, Wang C, Burman JL, Sideris DP, Fogel AI, Youle RJ. The ubiquitin kinase PINK1 recruits autophagy receptors to induce mitophagy. *Nature.* 2015;524:309–314.

60. Kubli DA, Quinsay MN, Gustafsson AB. Parkin deficiency results in accumulation of abnormal mitochondria in aging myocytes. *Commun Integr Biol.* 2013;6:e24511.

61. Taneike M, Yamaguchi O, Nakai A, Hikoso S, Takeda T, Mizote I, Oka T, Tamai T, Oyabu J, Murakawa T, Nishida K, Shimizu T, Hori M, Komuro I, Takuji Shirasawa TS, Mizushima N, Otsu K. Inhibition of autophagy in the heart induces age-related cardiomyopathy. *Autophagy.* 2010;6:600–606.

62. Wohlgemuth SE, Calvani R, Marzetti E. The interplay between autophagy and mitochondrial dysfunction in

oxidative stress-induced cardiac aging and pathology. *J Mol Cell Cardiol.* 2014;71:62–70.

63. Dutta D, Xu J, Kim JS, Dunn WA, Leeuwenburgh C. Upregulated autophagy protects cardiomyocytes from oxidative stress-induced toxicity. *Autophagy.* 2013;9:328–344.

64. Leon LJ, Gustafsson AB. Staying young at heart: Autophagy and adaptation to cardiac aging. *J Mol Cell Cardiol.* 2016;95:78–85.

65. Lifton RP, Gharavi AG, Geller DS. Molecular mechanisms of human hypertension. *Cell.* 2001;104:545–556.

66. Kurdi M, Booz GW. New take on the role of angiotensin ii in cardiac hypertrophy and fibrosis. *Hypertension.* 2011;57:1034–1038.

67. Dai DF, Chen T, Szeto H, Nieves-Cintron M, Kutyavin V, Santana LF, Rabinovitch PS. Mitochondrial targeted antioxidant peptide ameliorates hypertensive cardiomyopathy. *J Am Coll Cardiol.* 2011;58:73–82.

68. St-Pierre J, Drori S, Uldry M, Silvaggi JM, Rhee J, Jager S, Handschin C, Zheng K, Lin J, Yang W, Simon DK, Bachoo R, Spiegelman BM. Suppression of reactive oxygen species and neurodegeneration by the PGC-1 transcriptional coactivators. *Cell.* 2006;127:397–408.

69. Mollnau H, Wendt M, Szocs K, Lassegue B, Schulz E, Oelze M, Li H, Bodenschatz M, August M, Kleschyov AL, Tsilimingas N, Walter U, Forstermann U, Meinertz T, Griendling K, Munzel T. Effects of angiotensin ii infusion on the expression and function of NAD(P)H oxidase and components of nitric oxide/cGMP signaling. *Circ Res.* 2002;90:E58–565.

70. Doughan AK, Harrison DG, Dikalov SI. Molecular mechanisms of angiotensin ii-mediated mitochondrial dysfunction: Linking mitochondrial oxidative damage and vascular endothelial dysfunction. *Circ Res.* 2008;102:488–496.

71. Kimura S, Zhang GX, Nishiyama A, Shokoji T, Yao L, Fan YY, Rahman M, Abe Y. Mitochondria-derived reactive oxygen species and vascular map kinases: Comparison of angiotensin ii and diazoxide. *Hypertension.* 2005;45:438–444.

72. Rota M, LeCapitaine N, Hosoda T, Boni A, De Angelis A, Padin-Iruegas ME, Esposito G, Vitale S, Urbanek K, Casarsa C, Giorgio M, Luscher TF, Pelicci PG, Anversa P, Leri A, Kajstura J. Diabetes promotes cardiac stem cell aging and heart failure, which are prevented by deletion of the p66shc gene. *Circ Res.* 2006;99:42–52.

73. Francia P, Cosentino F, Schiavoni M, Huang Y, Perna E, Camici GG, Luscher TF, Volpe M. P66(shc) protein, oxidative stress, and cardiovascular complications of diabetes: The missing link. *J Mol Med (Berl).* 2009;87:885–891.

74. Hafner AV, Dai J, Gomes AP, Xiao CY, Palmeira CM, Rosenzweig A, Sinclair DA. Regulation of the mPTP by SIRT3-mediated deacetylation of CypD at lysine 166 suppresses age-related cardiac hypertrophy. *Aging (Albany NY).* 2010;2:914–923.

75. Dai DF, Hsieh EJ, Liu Y, Chen T, Beyer RP, Chin MT, MacCoss MJ, Rabinovitch PS. Mitochondrial proteome remodelling in pressure overload-induced heart failure: The role of mitochondrial oxidative stress. *Cardiovasc Res.* 2012;93:79–88.

76. Zhao WY, Han S, Zhang L, Zhu YH, Wang LM, Zeng L. Mitochondria-targeted antioxidant peptide ss31 prevents hypoxia/reoxygenation-induced apoptosis by down-regulating p66shc in renal tubular epithelial cells. *Cell Physiol Biochem.* 2013;32:591–600.

77. Becker LB, Vanden Hoek TL, Shao ZH, Li CQ, Schumacker PT. Generation of superoxide in cardiomyocytes during ischemia before reperfusion. *Am J Physiol.* 1999;277:H2240–22246.

78. Zorov DB, Juhaszova M, Sollott SJ. Mitochondrial ROS-induced ROS release: An update and review. *Biochim Biophy Acta.* 2006;1757:509–517.

79. Isoyama S, Nitta-Komatsubara Y. Acute and chronic adaptation to hemodynamic overload and ischemia in the aged heart. *Heart Fail Rev.* 2002;7:63–69.

80. Juhaszova M, Rabuel C, Zorov DB, Lakatta EG, Sollott SJ. Protection in the aged heart: Preventing the heart-break of old age? *Cardiovasc Res.* 2005;66:233–244.

81. Nitta Y, Abe K, Aoki M, Ohno I, Isoyama S. Diminished heat shock protein 70 mRNA induction in aged rat hearts after ischemia. *Am J Physiol.* 1994;267:H1795–11803.

82. Chou TC, Yen MH, Li CY, Ding YA. Alterations of nitric oxide synthase expression with aging and hypertension in rats. *Hypertension.* 1998;31:643–648.

83. Korzick DH, Holiman DA, Boluyt MO, Laughlin MH, Lakatta EG. Diminished alpha1-adrenergic-mediated contraction and translocation of PKC in senescent rat heart. *Am J Physiol Heart Circ Physiol.* 2001;281:H581–5589.

84. Tani M, Honma Y, Hasegawa H, Tamaki K. Direct activation of mitochondrial K(ATP) channels mimics preconditioning but protein kinase c activation is less effective in middle-aged rat hearts. *Cardiovasc Res.* 2001;49:56–68.

85. DiMauro S, Schon EA. Mitochondrial respiratory-chain diseases. *N Engl J Med.* 2003;348:2656–2668.

86. Karamanlidis G, Nascimben L, Couper GS, Shekar PS, del Monte F, Tian R. Defective DNA replication impairs mitochondrial biogenesis in human failing hearts. *Circ Res.* 2010;106:1541–1548.

87. Ahuja P, Wanagat J, Wang Z, Wang Y, Liem DA, Ping P, Antoshechkin IA, Margulies KB, Maclellan WR. Divergent mitochondrial biogenesis responses in human cardiomyopathy. *Circulation.* 2013;127:1957–1967.

88. Rabinowitz M, Zak R. Mitochondria and cardiac hypertrophy. *Circ Res.* 1975;36:367–376.

89. Goffart S, von Kleist-Retzow J-C, Wiesner RJ. Regulation of mitochondrial proliferation in the heart: Power-plant failure contributes to cardiac failure in hypertrophy. *Cardiovasc Res.* 2004;64:198–207.

90. Murray AJ, Anderson RE, Watson GC, Radda GK, Clarke K. Uncoupling proteins in human heart. *Lancet.* 2004;364:1786–1788.

91. Beer M, Seyfarth T, Sandstede J, Landschutz W, Lipke C, Kostler H, von Kienlin M, Harre K, Hahn D, Neubauer S. Absolute concentrations of high-energy phosphate metabolites in normal, hypertrophied, and failing human myocardium measured noninvasively with (31) P-SLOOP magnetic resonance spectroscopy. *J Am Coll Cardiol.* 2002;40:1267–1274.

92. Weiss RG, Gerstenblith G, Bottomley PA. ATP flux through creatine kinase in the normal, stressed, and failing human heart. *Proc Natl Acad Sci U S A.* 2005;102:808–813.

93. Nakae I, Mitsunami K, Omura T, Yabe T, Tsutamoto T, Matsuo S, Takahashi M, Morikawa S, Inubushi T, Nakamura Y, Kinoshita M, Horie M. Proton magnetic resonance spectroscopy can detect creatine depletion associated with the progression of heart failure in cardiomyopathy. *J Am Coll Cardiol.* 2003;42:1587–1593.

94. Neubauer S, Horn M, Cramer M, Harre K, Newell JB, Peters W, Pabst T, Ertl G, Hahn D, Ingwall JS,

Kochsiek K. Myocardial phosphocreatine-to-ATP ratio is a predictor of mortality in patients with dilated cardiomyopathy. *Circulation*. 1997;96:2190–2196.

95. Halestrap AP, Richardson AP. The mitochondrial permeability transition: A current perspective on its identity and role in ischaemia/reperfusion injury. *J Mol Cell Cardiol*. 2015;78:129–141.

96. Echtay KS, Roussel D, St-Pierre J, Jekabsons MB, Cadenas S, Stuart JA, Harper JA, Roebuck SJ, Morrison A, Pickering S, Clapham JC, Brand MD. Superoxide activates mitochondrial uncoupling proteins. *Nature*. 2002;415:96–99.

97. Murray J, Taylor SW, Zhang B, Ghosh SS, Capaldi RA. Oxidative damage to mitochondrial complex I due to peroxynitrite: Identification of reactive tyrosines by mass spectrometry. *J Biol Chem*. 2003;278:37223–37230.

98. Tatarkova Z, Kuka S, Racay P, Lehotsky J, Dobrota D, Mistuna D, Kaplan P. Effects of aging on activities of mitochondrial electron transport chain complexes and oxidative damage in rat heart. *Physiological Res*. 2011;60:281–289.

99. Paradies G, Petrosillo G, Pistolese M, Ruggiero FM. The effect of reactive oxygen species generated from the mitochondrial electron transport chain on the cytochrome c oxidase activity and on the cardiolipin content in bovine heart submitochondrial particles. *FEBS Lett*. 2000;466:323–326.

100. Zorov DB, Juhaszova M, Sollott SJ. Mitochondrial reactive oxygen species (ROS) and ROS-induced ROS release. *Physiol Rev*. 2014;94:909–950.

101. Nerbonne JM, Kass RS. Molecular physiology of cardiac repolarization. *Physiol Rev*. 2005;85:1205–1253.

102. Janse MJ, Wit AL. Electrophysiological mechanisms of ventricular arrhythmias resulting from myocardial ischemia and infarction. *Physiol Rev*. 1989;69:1049–1169.

103. Valdivia CR, Chu WW, Pu J, Foell JD, Haworth RA, Wolff MR, Kamp TJ, Makielski JC. Increased late sodium current in myocytes from a canine heart failure model and from failing human heart. *J Mol Cell Cardiol*. 2005;38:475–483.

104. Makiyama T, Akao M, Shizuta S, Doi T, Nishiyama K, Oka Y, Ohno S, Nishio Y, Tsuji K, Itoh H, Kimura T, Kita T, Horie M. A novel scn5a gain-of-function mutation m1875t associated with familial atrial fibrillation. *J Am Coll Cardiol*. 2008;52:1326–1334.

105. Schott JJ, Alshinawi C, Kyndt F, Probst V, Hoorntje TM, Hulsbeek M, Wilde AA, Escande D, Mannens MM, Le Marec H. Cardiac conduction defects associate with mutations in scn5a. *Nat Genet*. 1999;23:20–21.

106. Chiang DY, Kim JJ, Valdes SO, de la Uz C, Fan Y, Orcutt J, Domino M, Smith M, Wehrens XH, Miyake CY. Loss-of-function scn5a mutations associated with sinus node dysfunction, atrial arrhythmias, and poor pacemaker capture. *Circ Arrhythm Electrophysiol*. 2015;8:1105–1112.

107. Wilde AAM, Amin AS. Clinical spectrum of SCN5A mutations: Long QT syndrome, Brugada syndrome, and cardiomyopathy. *JACC Clin Electrophysiol*. 2018;4:569–579.

108. Song Y, Shryock JC, Wagner S, Maier LS, Belardinelli L. Blocking late sodium current reduces hydrogen peroxide-induced arrhythmogenic activity and contractile dysfunction. *J Pharmacol Exp Ther*. 2006;318:214–222.

109. Liu M, Liu H, Dudley SC, Jr. Reactive oxygen species originating from mitochondria regulate the cardiac sodium channel. *Circ Res*. 2010;107:967–974.

110. Vikram A, Lewarchik CM, Yoon JY, Naqvi A, Kumar S, Morgan GM, Jacobs JS, Li Q, Kim YR, Kassan M, Liu J, Gabani M, Kumar A, Mehdi H, Zhu X, Guan X, Kutschke W, Zhang X, Boudreau RL, Dai S, Matasic DS, Jung SB, Margulies KB, Kumar V, Bachschmid MM, London B, Irani K. Sirtuin 1 regulates cardiac electrical activity by deacetylating the cardiac sodium channel. *Nat Med*. 2017;23:361–367.

111. Liu M, Gu L, Sulkin MS, Liu H, Jeong EM, Greener I, Xie A, Efimov IR, Dudley SC, Jr. Mitochondrial dysfunction causing cardiac sodium channel downregulation in cardiomyopathy. *J Mol Cell Cardiol*. 2013;54:25–34.

112. Liu M, Sanyal S, Gao G, Gurung IS, Zhu X, Gaconnet G, Kerchner LJ, Shang LL, Huang CL, Grace A, London B, Dudley SC, Jr. Cardiac Na+ current regulation by pyridine nucleotides. *Circ Res*. 2009;105:737–745.

113. Bentivegna LA, Ablin LW, Kihara Y, Morgan JP. Altered calcium handling in left ventricular pressure-overload hypertrophy as detected with aequorin in the isolated, perfused ferret heart. *Circ Res*. 1991;69:1538–1545.

114. Winslow RL, Rice J, Jafri S, Marban E, O'Rourke B. Mechanisms of altered excitation-contraction coupling in canine tachycardia-induced heart failure, II: Model studies. *Circ Res*. 1999;84:571–586.

115. Varadarajan SG, An J, Novalija E, Smart SC, Stowe DF. Changes in [Na(+)](i), compartmental [Ca(2+)], and NADH with dysfunction after global ischemia in intact hearts. *Am J Physiol Heart Circ Physiol*. 2001;280:H280–H293.

116. Grant AO. Cardiac ion channels. *Circ Arrhythm Electrophysiol*. 2009;2:185–194.

117. Kawakami M, Okabe E. Superoxide anion radical-triggered Ca2+ release from cardiac sarcoplasmic reticulum through ryanodine receptor Ca2+ channel. *Mol Pharmacol*. 1998;53:497–503.

118. Kukreja RC, Kearns AA, Zweier JL, Kuppusamy P, Hess ML. Singlet oxygen interaction with Ca(2+)-ATPase of cardiac sarcoplasmic reticulum. *Circ Res*. 1991;69:1003–1014.

119. Xu KY, Zweier JL, Becker LC. Hydroxyl radical inhibits sarcoplasmic reticulum Ca(2+)-ATPase function by direct attack on the ATP binding site. *Circ Res*. 1997;80:76–81.

120. Rokita AG, Anderson ME. New therapeutic targets in cardiology: Arrhythmias and ca2+/calmodulin-dependent kinase II (CaMKII). *Circulation*. 2012;126:2125–2139.

121. Eigel BN, Gursahani H, Hadley RW. ROS are required for rapid reactivation of Na+/Ca2+ exchanger in hypoxic reoxygenated guinea pig ventricular myocytes. *Am J Physiol Heart Circ Physiol*. 2004;286:H955–9963.

122. Choudhary G, Dudley SC, Jr. Heart failure, oxidative stress, and ion channel modulation. *Congest Heart Fail (Greenwich, Conn)*. 2002;8:148–155.

123. Splawski I, Timothy KW, Sharpe LM, Decher N, Kumar P, Bloise R, Napolitano C, Schwartz PJ, Joseph RM, Condouris K, Tager-Flusberg H, Priori SG, Sanguinetti MC, Keating MT. Ca(v)1.2 calcium channel dysfunction causes a multisystem disorder including arrhythmia and autism. *Cell*. 2004;119:19–31.

124. Pereira L, Cheng H, Lao DH, Na L, van Oort RJ, Brown JH, Wehrens XH, Chen J, Bers DM. Epac2 mediates cardiac beta1-adrenergic-dependent sarcoplasmic reticulum Ca2+ leak and arrhythmia. *Circulation*. 2013;127:913–922.

125. Simamura E, Shimada H, Hatta T, Hirai K. Mitochondrial voltage-dependent anion channels (VDACS) as novel pharmacological targets for anti-cancer agents. *J Bioenerg Biomembr*. 2008;40:213–217.

126. Colombini M. The published 3d structure of the VDAC channel: Native or not? *Trends Biochem Sci.* 2009;34:382–389.

127. De Pinto V, Reina S, Guarino F, Messina A. Structure of the voltage dependent anion channel: State of the art. *J Bioenerg Biomembr.* 2008;40:139–147.

128. Shoshan-Barmatz V, De Pinto V, Zweckstetter M, Raviv Z, Keinan N, Arbel N. VDAC, a multi-functional mitochondrial protein regulating cell life and death. *Mol Aspects Med.* 2010;31:227–285.

129. Garcia-Rivas Gde J, Carvajal K, Correa F, Zazueta C. Ru360, a specific mitochondrial calcium uptake inhibitor, improves cardiac post-ischaemic functional recovery in rats in vivo. *Br J Pharmacol.* 2006;149:829–837.

130. Williams GS, Boyman L, Chikando AC, Khairallah RJ, Lederer WJ. Mitochondrial calcium uptake. *Proc Natl Acad Sci U S A.* 2013;110:10479–10486.

131. Yan Y, Liu J, Wei C, Li K, Xie W, Wang Y, Cheng H. Bidirectional regulation of ca2+ sparks by mitochondria-derived reactive oxygen species in cardiac myocytes. *Cardiovasc Res.* 2008;77:432–441.

132. McElnea EM, Quill B, Docherty NG, Irnaten M, Siah WF, Clark AF, O'Brien CJ, Wallace DM. Oxidative stress, mitochondrial dysfunction and calcium overload in human lamina cribrosa cells from glaucoma donors. *Mol Vis.* 2011;17:1182–1191.

133. Boyman L, Williams GS, Khananshvili D, Sekler I, Lederer WJ. NCLX: The mitochondrial sodium calcium exchanger. *J Mol Cell Cardiol.* 2013;59:205–213.

134. Wagner S, Rokita AG, Anderson ME, Maier LS. Redox regulation of sodium and calcium handling. *Antioxid Redox Signal.* 2013;18:1063–1077.

135. Przygodzki T, Sokal A, Bryszewska M. Calcium ionophore A23187 action on cardiac myocytes is accompanied by enhanced production of reactive oxygen species. *Biochim Biophys Acta.* 2005;1740:481–488.

136. Cleeter MW, Cooper JM, Darley-Usmar VM, Moncada S, Schapira AH. Reversible inhibition of cytochrome c oxidase, the terminal enzyme of the mitochondrial respiratory chain, by nitric oxide. Implications for neurodegenerative diseases. *FEBS Lett.* 1994;345:50–54.

137. Ott M, Robertson JD, Gogvadze V, Zhivotovsky B, Orrenius S. Cytochrome c release from mitochondria proceeds by a two-step process. *Proc Natl Acad Sci U S A.* 2002;99:1259–1263.

138. Cerbai E, Ambrosio G, Porciatti F, Chiariello M, Giotti A, Mugelli A. Cellular electrophysiological basis for oxygen radical-induced arrhythmias. A patch-clamp study in guinea pig ventricular myocytes. *Circulation.* 1991;84:1773–1782.

139. Liang H, Li X, Li S, Zheng MQ, Rozanski GJ. Oxidoreductase regulation of KV currents in rat ventricle. *J Mol Cell Cardiol.* 2008;44:1062–1071.

140. Li X, Tang K, Xie B, Li S, Rozanski GJ. Regulation of kv4 channel expression in failing rat heart by the thioredoxin system. *Am J Physiol Heart Circ Physiol.* 2008;295: H416–4424.

141. Li S, Li X, Li YL, Shao CH, Bidasee KR, Rozanski GJ. Insulin regulation of glutathione and contractile phenotype in diabetic rat ventricular myocytes. *Am J Physiol Heart Circ Physiol.* 2007;292:H1619–H1629.

142. Xu Z, Patel KP, Lou MF, Rozanski GJ. Up-regulation of K(+) channels in diabetic rat ventricular myocytes by insulin and glutathione. *Cardiovasc Res.* 2002;53:80–88.

143. Lo CW. Role of gap junctions in cardiac conduction and development: Insights from the connexin knockout mice. *Circ Res.* 2000;87:346–348.

144. Litchenberg WH, Norman LW, Holwell AK, Martin KL, Hewett KW, Gourdie RG. The rate and anisotropy of impulse propagation in the postnatal terminal crest are correlated with remodeling of Cx43 gap junction pattern. *Cardiovasc Res.* 2000;45:379–387.

145. De Mello WC. Is an intracellular renin-angiotensin system involved in control of cell communication in heart? *J Cardiovasc Pharmacol.* 1994;23:640–646.

146. Xiao HD, Fuchs S, Campbell DJ, Lewis W, Dudley SC, Jr., Kasi VS, Hoit BD, Keshelava G, Zhao H, Capecchi MR, Bernstein KE. Mice with cardiac-restricted angiotensin-converting enzyme (ace) have atrial enlargement, cardiac arrhythmia, and sudden death. *Am J Pathol.* 2004;165:1019–1032.

147. Skulachev VP, Anisimov VN, Antonenko YN, Bakeeva LE, Chernyak BV, Erichev VP, Filenko OF, Kalinina NI, Kapelko VI, Kolosova NG, Kopnin BP, Korshunova GA, Lichinitser MR, Obukhova LA, Pasyukova EG, Pisarenko OI, Roginsky VA, Ruuge EK, Senin, II, Severina, II, Skulachev MV, Spivak IM, Tashlitsky VN, Tkachuk VA, Vyssokikh MY, Yaguzhinsky LS, Zorov DB. An attempt to prevent senescence: A mitochondrial approach. *Biochim Biophys Acta.* 2009;1787:437–461.

148. Smith RA, Hartley RC, Cocheme HM, Murphy MP. Mitochondrial pharmacology. *Trends Pharmacol Sci.* 2012;33:341–352.

149. Daubert MA, Yow E, Dunn G, Marchev S, Barnhart H, Douglas PS, O'Connor C, Goldstein S, Udelson JE, Sabbah HN. Novel mitochondria-targeting peptide in heart failure treatment: A randomized, placebo-controlled trial of **Elamipretide**. Circ Heart Fail. 2017 Dec;10(12):pii: e004389. doi: 10.1161/CIRCHEARTFAILURE.117. 004389. PMID:29217757.

150. El-Hattab AW, Zarante AM, Almannai M, Scaglia F. Therapies for mitochondrial diseases and current **clinical trials**. Mol Genet Metab. 2017 Nov; 122(3):1-9. doi: 10.1016/j.ymgme.2017.09.009. Epub 2017 Sep 18. Review.

13 The Role of Mitochondrial Dysfunction in Mitophagy and Apoptosis and Its Relevance to Cancer

Jan Ježek

Department of Molecular Biology, Rowan University, School of Osteopathic Medicine, Stratford, NJ, 08084 USA

CONTENTS

1 INTRODUCTION

1.1 MITOCHONDRIAL POWERHOUSES

As the terminal site for oxygen utilization, mitochondria are essential organelles for maintaining the bioenergetic and redox balance in healthy cells. From this point of view, the primary function of mitochondria is to generate energy via oxidative phosphorylation (OXPHOS) to drive cellular processes and sustain life. The catabolic pathways of carbohydrate, lipid, and amino acid oxidation generate reduced nicotinamide (NADH) and flavin adenine dinucleotide (FADH$_2$) cofactors. NADH and FADH$_2$ serve as substrates for the mitochondrial respiratory chain Complex I (NADH-ubiquinone dehydrogenase) and II (succinate dehydrogenase), respectively. These and other dehydrogenases feed electrons derived from the oxidation of reducing equivalents into the respiratory chain [1]. In Complex I, III (cytochrome bc_1), and IV (cytochrome c oxidase), the transfer of electrons along the linear chain of redox sites toward oxygen is coupled to the pumping of protons from the mitochondrial matrix into the inter-membrane space (IMS) [2]. This pumping of protons in actively respiring mitochondria generates a proton gradient across the inner mitochondrial membrane (IMM), referred to as the proton motive force when expressed in millivolts [3]:

$$\Delta p = -2.303 \frac{RT}{F} \Delta pH + \Delta \psi_m = -Z \Delta pH + \Delta \psi_m$$

Δp – proton motive force
R – universal gas constant (8.314 J.K^{-1}.mol^{-1})
T – absolute temperature (in K)
F – Faraday's constant (96,485 J.V^{-1}.mol^{-1})
ΔpH – mitochondrial pH gradient ($pH_{in} - pH_{out}$)
$\Delta \psi_m$ – mitochondiral membrane potential ($\psi_{in} - \psi_{out}$)
Z – conversion factor (0.059V at 298K)

According to the chemiosmotic theory, proton motive force drives both ATP synthesis and metabolite transport across IMM upon the translocation of protons back into the mitochondrial matrix via Complex V (ATP synthase) and the mitochondrial carrier system, respectively [4].

1.2 To be or Not to be, a Decision Left for Mitochondria

Apart from being involved in the production of ATP, mitochondria lie at the center of a key stress signaling junction where decisions whether to endure and survive or undergo intrinsic apoptosis, a form of programmed cell death (PCD), are being made [5]. One of the early manifestations of mitochondrial dysfunction in response to stress is mitochondrial fragmentation. The continuous balance between mitochondrial fission and fusion events dictates the overall mitochondrial network morphology, which shows predominantly elongated and reticular form in unstressed, healthy cells [6]. Important exception to this pattern is dividing cells, which display an overall fragmented mitochondrial phenotype before undergoing cytokinesis. This stems from the fact that mitochondrial fission facilitates equal distribution of mitochondria between daughter cells formed during mitosis [7].

In response to endogenous stress such as hypoxia or exogenous stress such as oxidative stress, UV and ionizing radiation, and drug-induced chemotherapy, mitochondrial ultrastructure rapidly shifts from elongated to fragmented morphology [6]. Depending on the severity and persistence of stress, cells either execute pro-survival or pro-death pathways, in both of which mitochondrial fission serves a pivotal role. Upon mild stress, mitochondrial fragmentation acts as the first-line defense mechanism that facilitates the elimination of dysfunctional mitochondria via mitophagy, a selective form of the lysosomal degradation pathway [8]. On the contrary, when exposed to overwhelming or chronic stress, cells are destined to undergo apoptotic or necrotic cell death. However, unlike necrosis, apoptosis is considered to be immunologically silent, wherein dead apoptotic cells are cleared from tissues by phagocytes without inducing an inflammatory response [9].

2 MITOCHONDRIAL FRAGMENTATION AS AN EARLY INDICATOR OF MITOCHONDRIAL DYSFUNCTION

2.1 Inner Workings of the Mitochondrial Dynamics Machinery

The core component of the mitochondrial fission machinery is the large guanosine 5′-triphosphatase (GTPase) dynamin-related protein 1 (Drp1), which, upon stimulation, translocates from the cytosol to circumscribe the outer mitochondrial membrane (OMM) and physically mediate the simultaneous constriction of the OMM and inner mitochondrial membrane (IMM) leading to the division of the organelle [10]. Apart from Drp1, several accessory proteins participate in the orchestration of mitochondrial fragmentation. Mitochondrial fission factor (Mff), mitochondrial fission 1 (Fis1), mitochondrial dynamic proteins 49 and 51 (MiD49 and MiD51) [11,12],

and sorting assembly machinery 50 kDa subunit (Sam50) [13] are adaptor proteins that recruit Drp1 to OMM. Furthermore, recent research has identified another mechanoenzyme to be actively involved in mitochondrial dynamics. During the late stages of mitochondrial fission, the classic GTPase dynamin 2 (Dnm2) is recruited to OMM to complete scission of both mitochondrial membranes [14].

Large GTPases are also involved in the process of mitochondrial fusion. Namely, mitofusins 1 and 2 (Mfn1 and Mfn2) and optic atrophy 1 (Opa1) are integral membrane proteins that mediate OMM and IMM fusion, respectively [15]. Mitofusins form homo- and heterotypic dimeric complexes, which serve as molecular tethers between two opposing mitochondria being fused [16]. Following intermitochondrial complex formation, GTP hydrolysis induces conformational changes in each of the mitofusin to drive OMM fusion [17]. Similarly, IMM fusion is mediated by the GTP-dependent activity of the long form of Opa1 (L-Opa1). Unlike the short form (S-Opa1), L-Opa1 is not proteolytically cleaved and hence is still associated with the IMM [18].

2.2 Mitochondrial Dynamics and Signal Transduction

The function of Drp1 is regulated by post-translational modifications including phosphorylation, ubiquitination, sumoylation, S-nitrosylation, and O-GlcNAcylation [19]. For example, phosphorylation of Drp1 at Ser616 and Ser637 promotes and inhibits mitochondrial fission, respectively [14]. As an example of signaling pathways that converge on Drp1 in a physiological setting, Taguchi et al. have shown in thymidine-synchronized HeLa cells transfected with rat Drp1 that the cyclin-dependent kinase 1 (Cdk1)/cyclin B mitotic kinase complex is required for Drp1 phosphorylation on Ser585 and consequent mitochondrial fission during cell cycle progression from prophase to anaphase [20]. Consistently, siRNA-mediated knockdown of Drp1 prevented mitochondrial fragmentation in dividing cells. Via this pathway, cell cycle-specific activation of Drp1 ensures equal segregation of the nascent mitochondrial fragments between daughter cells during mitotic cell division. In addition, Cdk1/cyclin B-dependent and Drp1-mediated mitochondrial fission was shown to rely on the activity of another mitotic kinase Aurora A, which activates the small GTPase RalA and its effector RalA binding protein 1 (RALBP1) [21].

2.3 Mitochondrial Fission as a Therapeutic Target

Aberrant activation of mitochondrial fragmentation via Ser616 phosphorylation of Drp1 has been associated with the pathology of cancer [22], diabetes [23], cardiovascular [24], and neurodegenerative diseases such as Parkinson's [25] and Alzheimer's [26] diseases, suggesting that mitochondrial dynamics plays a cardinal role in maintaining normal bioenergetics and cellular homeostasis. Although it is currently unclear whether excessive mitochondrial fission is a cause or consequence of these disease states, Drp1 as well as other functional components of the mitochondrial fission machinery gain increased recognition as prospective therapeutic targets in the etiology of cancer and other diseases [27].

During tumorigenesis, changes in mitochondrial shape have been linked to oncogenic signaling networks such as Ras/mitogen-activated protein kinase (MAPK), phosphoinositide 3-kinase (PI3K)/Akt, or Myc pathways [28]. Whereas Ras/MAPK- and PI3K/Akt-dependent transformation is associated with Drp1 activation and gives rise to a fragmented mitochondrial network morphology, Myc-mediated signaling stimulates the expression of Mfn2 and Opa1 resulting in an interconnected mitochondrial phenotype and increased mitochondrial biogenesis mediated by peroxisome proliferator-activated receptor-gamma coactivator-1β (PGC-1β) [29]. One of the mechanisms by which mitochondrial fission confers growth advantage to cancer cells is by switching the mode of metabolism from OXPHOS to glycolysis while simultaneously increasing the production of ROS. Indeed, metabolic reprogramming toward aerobic glycolysis, collectively known as the Warburg's effect, was widely observed in conjunction with augmented production of ROS in tumors and can therefore be considered as one of the most prominent hallmarks of cancer [30]. Developing therapies targeting mitochondrial dynamics, such as using the small molecule Drp1 inhibitor Mdivi-1, may therefore proof effective in combating at least some forms of the malignant disease [31].

3 THE ROLE OF MITOCHONDRIAL DYSFUNCTION IN MITOPHAGY

3.1 When Mitochondria Fuse with Lysosomes

Mitophagy is a lysosome-dependent quality control pathway that selectively targets and removes dysfunctional mitochondria [32]. Similar to autophagy, basic molecular metabolites that are produced during this process can be either reused for biosynthesis of new cellular material or oxidized for energy production. Mitophagy is initiated by the generation of a double-membrane structure called phagophore, which is recruited to mitochondria in a receptor-specific manner [33]. Elongated phagophores eventually engulf damaged mitochondria to form mitophagosomes [34]. These vesicular organelles were named in analogy to autophagosomes that are assembled by a similar mechanism during the onset of general autophagy. Subsequent fusion between mitophagosomes and lysosomes leads to the formation of so-called mitophagolysosomes, in which the engulfed mitochondrial cargo is proteolytically degraded by lysosomal hydrolases.

3.2 Parking Mitochondria for Mitophagy

Mechanisms by which phagophores recognize dysfunctional mitochondria can be divided into ubiquitin-dependent and ubiquitin-independent. Whereas ubiquitin-independent recognition systems rely on mitophagy-specific receptors located in the OMM such as Bcl2/E1B 19 kDa-interacting protein 3-like protein (BNIP3) and NIX (also known as BNIP3L) as direct binding partners for microtubule-associated protein light 1 chain 3 (LC3B), which is anchored to the phagophore, dysfunctional mitochondria can also be primed for lysosomal degradation by adapter proteins that selectively recognize polyubiquitinated substrates on the OMM [32]. Among the latter belong for example p62 (also known as sequestosome 1), optineurin, neighbor of BRCA1 gene 1 (NBR1), nuclear dot protein 52 (NDP52), and Tax1 binding protein 1 (TAX1BP1) [35].

One of the key ubiquitin ligases that polyubiquitinate mitochondrial substrates on dysfunctional mitochondria during mitophagy is the E3 ubiquitin ligase Parkin. This cytosolic protein relies on the mitochondrial serine/threonine kinase PTEN-induced putative kinase 1 (PINK1) for its recruitment to damaged mitochondria. Parkin-mediated mitophagy proceeds by an intricate mechanism, in which PINK1 fulfills the role of an endogenous molecular tag to ensure that only dysfunctional organelles are polyubiquitinated by Parkin for lysosomal degradation [36]. In healthy mitochondria, PINK1 is constitutively engaged by the mitochondrial protein import machinery and subsequently degraded in the proteasome. This process involves partial import of the nascent polypeptide chain of PINK1 by translocases of the outer (TOM) and inner (TIM) mitochondrial membrane, which is followed by its concomitant cleavage inside the mitochondrial matrix mediated by the mitochondrial processing peptidase (MPP) and presenilin-associated rhomboid-like protease (PARL) [37]. Proteolytically-truncated form of PINK1 is consequently retro-translocated back into the cytosol and degraded in the proteasome. The targeting signal for the import of most proteins into mitochondria, including that of PINK1, is a positively-charged pre-sequence located at the N-terminal end of the translocated protein precursor. The mitochondrial targeting sequence is recognized by receptor subunits of the TOM complex and then electrostatically attracted by the negative charge formed inside the mitochondrial matrix of energized mitochondria, which drives the translocation of the protein through the TIM complex into the mitochondrial matrix [38]. The fact that the translocation of mitochondrial proteins, and particularly that of PINK1, into the mitochondrial matrix depends on mitochondrial membrane potential ensures that only dysfunctional mitochondria are turned over. As a result, PINK1 specifically accumulates on mitochondria with compromised mitochondrial membrane potential to selectively recruit Parkin to the OMM during the onset of mitophagy.

3.3 On the Relationship between Mitophagy and Cancer

Although mutations in *PARK2*, the gene encoding Parkin, were originally identified in patients with Parkinson's disease, Parkin was also found to play an important tumor suppressive role in the context of various types of cancer [39]. For example, study by Poulogiannis et al. identified heterozygous mutations in *PARK2* in primary colorectal patient tumor samples [40]. Loss-of-function mutations or deletions in *PARK2* were also observed in breast, lung, and ovarian cancers [41]. On the other hand, activation of mitochondrial quality control pathways in response to exogenous stress was linked to increased ROS detoxification and enhanced cancer cell survival [34]. The role of mitophagy in tumor progression can also be documented by the self-renewal and proliferative capacity of cancer stem cells. In an elegant set of experiments, Katajisto et al. have discovered the propensity of human mammary stem-like cells (SLCs) to undergo asymmetric division, during which mitochondria were unequally apportioned between the two daughter cells based on their age [42]. The authors went on to show that such mechanism increased the mammosphere-forming ability, a marker of stemness, in SLCs in a Parkin-dependent manner, suggesting the involvement of mitophagy. Taken together, mitophagy plays both growth-promoting and growth-suppressive roles in cancer.

4 THE ROLE OF MITOCHONDRIAL DYSFUNCTION IN INTRINSIC APOPTOSIS

4.1 Mitochondria as the Tool Box for Programmed Cell Death

Intrinsic apoptosis is a highly conserved stress response pathway, whereby an elaborate network of signaling mechanisms impacts on mitochondria to trigger a sequence of events, which lead to highly controlled and energy-dependent self-decomposition of the cell [43]. Among the most prominent proteins involved in apoptotic signaling are members of the B-cell lymphoma 2 (Bcl-2) family, which share a common pattern of evolutionarily conserved sequence motifs known as the Bcl-2 homology (BH) domains [44]. Despite having a high degree of structural similarity, the Bcl-2 family proteins perform diametrically opposed functions, acting as either pro-apoptotic or pro-survival regulators [45]. The core pro-apoptotic effectors, Bcl-2-associated X (Bax), Bcl-2 homologous antagonist/killer (Bak), and Bcl-2 related ovarian killer (Bok) contain three BH domains (BH1–3) [46]. Upon activation, these proteins oligomerize into high-molecular weight complexes capable of eliciting mitochondrial outer membrane permeabilization (MOMP). In this process, apoptogenic factors such as cytochrome *c*, apoptosis-inducing factor

(AIF), second mitochondria-derived activator of caspase/direct inhibitor of apoptosis-binding protein with low pI (Smac/DIABLO), Omi/high temperature requirement protein A2 (Omi/HtrA2), and endonuclease G (endoG) are released from IMS into the cytosol [47]. Mitochondrial release of pro-apoptotic second messengers leads to the activation of a cascade of caspases, which are aspartate-specific cysteine proteases that ultimately drive cellular disassembly. For example, mitochondrial release of cytochrome *c* is a common step in intrinsic apoptosis, which serves as a common indicator of MOMP [43]. Cytosolic cytochrome *c* binds to apoptotic protease activating factor-1 (Apaf-1) to form a multimeric protein complex called apoptosome [48]. Mature apoptosome recruits and activates initiator caspase-9, which subsequently cleaves and activates executioner caspase-3 and caspase-7 [49]. Executioner caspases, in their activated dimeric form, further cleave a plethora of cellular substrates such as poly(ADP-ribose) polymerase (PARP) and inhibitor of caspase-activated DNase (iCAD) [50]. The proteolytic cleavage of iCAD by caspase-3 leads to the dimerization and activation of caspase-activated DNase (CAD), which generates double-stranded DNA breaks [51]. PARP cleavage [52] and DNA fragmentation, giving rise to a characteristic "DNA laddering" pattern [53], serve as valid markers for caspase-mediated cell death. Other key features of apoptosis include externalization of phosphatidylserine on the outer leaflet of the plasma membrane, chromatin condensation (pyknosis), nuclear fragmentation (karyolysis), plasma membrane blebbing, cell shrinkage and detachment, and the ensuing phagocytosis of dead cells [54]. The ability of Bax, Bak, and Bok to permeabilize membranes stems from their structural resemblance to bacterial pore-forming proteins such as colicins [55] and diphtheria toxin [56]. Although numerous models have been proposed for the activation of these apoptotic effectors, the exact mechanism by which Bax or Bak, and Bok mediate MOMP is not completely understood and it has been only recently that Bax-mediated OMM pores were visualized in mammalian cells using nanoscopy techniques [57]. The main difference between the mechanism of activation of Bax and Bak stems from their differential compartmentalization within a healthy cell. Whereas Bak is thought to be constitutively associated with the OMM, the steady-state levels of mitochondrial Bax are governed by the rates of its retro-translocation between mitochondria and cytosol [46]. This dynamic shuttling is mediated by the pro-survival members of the Bcl-2 family, namely Bcl-2, B-cell lymphoma-extra large (Bcl-xL), and myeloid cell leukemia 1 (Mcl-1) proteins, which contain all four BH domains (BH1–4) [58]. Similarly to Bcl-2, Bcl-xL, and Mcl-1, other anti-apoptotic proteins such as Bcl-w, Bcl-2-related protein A1 (also known as Bfl-1), and Bcl-B participate in the inhibitory sequestration of Bax and Bak [59].

In addition to direct MOMP effectors, another class of pro-apoptotic factors is comprised of proteins, that harbor only one BH domain (BH3) in their primary sequence, known as the BH3-only protein sub-family

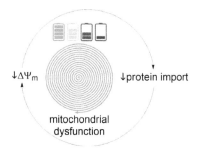

FIGURE 1 The vicious cycle between decreasing mitochondrial membrane potential and protein import leading to mitochondrial dysfunction. The fact that mitochondrial respiration and OXPHOS depend on the nuclear genome for the delivery of vast functional proteins creates a selection pressure against mitochondrial populations that are unable to maintain sufficiently high levels of mitochondrial membrane potential such as during the progression of apoptosis. Since mitochondrial import of most proteins strictly depends on mitochondrial fitness, these mitochondria are more susceptible to further degeneration due their decreasing ability to transport nuclearly encoded proteins from the cytosol and are therefore likely to be degraded via mitophagy.

[60]. BH3-only proteins such as Bcl-2-associated death promoter (Bad), Bcl-2 interacting killer (Bik), Bcl-2 modifying factor (Bmf), harakiri (Hrk), and Noxa sensitize cells to apoptosis by sequestering the pro-survival proteins from interaction with Bax or Bak. Hence, BH3-only proteins are in general referred to as apoptotic sensitizers. In addition, a sub-group of so-called direct activators has been identified among the BH3-only proteins. Apart from acting as sensitizers, proteins such as truncated BH3-interacting domain death agonist (tBid), Bcl-2-like protein 11 (Bim), p53 upregulated modulator of apoptosis (Puma), and Noxa, can directly bind to and activate the MOMP effectors Bax and Bak.

4.2 The Death Spiral of Mitochondrial Dysfunction

Given that MOMP is the earliest irreversible step in apoptosis, this event is regarded as the point of no return in the execution of apoptosis. Since cytochrome c is an indispensable mobile carrier of electrons between Complexes III and IV during mitochondrial respiration, its mitochondrial release during MOMP triggers a sequence of events that eventually lead to the loss of mitochondrial function [61]. Although the initial decline in the levels of mitochondrial membrane potential induced by MOMP can be rescued by a compensatory mechanism, in which Complex V operates in reverse mode acting as a proton pump to re-establish the proton gradient at the expense of ATP, long-term activity of MOMP inevitably results in the deterioration of mitochondrial health. Under such circumstances, mitochondria are exposed to an inherent selection pressure imposed by the dependence of mitochondrial protein transport on membrane potential [62]. In this scenario, defective mitochondria are less likely to import protein constituents encoded by the nuclear genome. Since many of these proteins play essential roles in the maintenance of mitochondrial health, such mechanism may further

contribute to the deterioration of mitochondrial respiration as well as OXPHOS resulting in the loss of mitochondrial membrane potential in a cyclic fashion (Figure 1). Moreover, given that mitophagy preferentially targets dysfunctional mitochondria, this paradigm could also explain the phenomenon of simultaneous activation of apoptosis and mitophagy by the same stress stimuli [63].

5 CONCLUSIONS

Intensity and duration of stress dictates whether cells engage mitophagy or trigger intrinsic apoptosis to effect cell survival or demise, respectively. Mitochondrial dynamics represents a common signaling step in the induction of both mitochondrial quality control and cell death pathways. This stress-sensing mechanism may involve redox pathways converging upon the mitochondrial fission and fusion machineries [6]. The fact that mitophagy and apoptosis are often simultaneously activated by identical cues can be explained by the concept of a self-perpetuating cycle that progressively drives mitochondrial dysfunction. This vicious cycle involves a reciprocal decay of mitochondrial membrane potential and import of proteins required to sustain the rates of mitochondrial respiration and ATP synthesis.

REFERENCES

1. Wong, H. S., Dighe, P. A., Mezera, V., Monternier, P. A., and Brand, M. D. 2017. Production of superoxide and hydrogen peroxide from specific mitochondrial sites under different bioenergetic conditions. *J Biol Chem* 292:16804–9.
2. Sun, F., Zhou, Q., Pang, X., Xu, Y., and Rao, Z. 2013. Revealing various coupling of electron transfer and proton pumping in mitochondrial respiratory chain. *Curr Opin Struct Biol* 23:526–38.
3. Berry, B. J., Trewin, A. J., Amitrano, A. M., Kim, M., and Wojtovich, A. P. 2018. Use the protonmotive force: mitochondrial uncoupling and reactive oxygen species. *J Mol Biol* 430:3873–91.

4. Kocherginsky, N. 2009. Acidic lipids, H(+)-ATPases, and mechanism of oxidative phosphorylation. Physico-chemical ideas 30 years after P. Mitchell's Nobel Prize award. *Prog Biophys Mol Biol* 99:20–41.

5. Cooper, K. F. 2018. Till death do us part: the marriage of autophagy and apoptosis. *Oxid Med Cell Longev* 2018:4701275.

6. Jezek, J., Cooper, K. F., and Strich, R. 2018. Reactive oxygen species and mitochondrial dynamics: the Yin and Yang of mitochondrial dysfunction and cancer progression. *Antioxidants (Basel)* 7 :E13.

7. Yamano, K., and Youle, R. J. 2011. Coupling mitochondrial and cell division. *Nat Cell Biol* 13:1026–27.

8. Kulikov, A. V., Luchkina, E. A., Gogvadze, V., and Zhivotovsky, B. 2017. Mitophagy: link to cancer development and therapy. *Biochem Biophys Res Commun* 482:432–9.

9. Arandjelovic, S., and Ravichandran, K. S. 2015. Phagocytosis of apoptotic cells in homeostasis. *Nat Immunol* 16:907–17.

10. Kraus, F., and Ryan, M. T. 2017. The constriction and scission machineries involved in mitochondrial fission. *J Cell Sci* 130:2953–60.

11. Kalia, R., Wang, R. Y., Yusuf, A., et al. 2018. Structural basis of mitochondrial receptor binding and constriction by DRP1. *Nature* 558:401–5.

12. Atkins, K., Dasgupta, A., Chen, K. H., Mewburn, J., and Archer, S. L. 2016. The role of Drp1 adaptor proteins MiD49 and MiD51 in mitochondrial fission: implications for human disease. *Clin Sci (Lond)* 130:1861–74.

13. Liu, S., Gao, Y., Zhang, C., et al. 2016. SAMM50 affects mitochondrial morphology through the association of Drp1 in mammalian cells. *FEBS Lett* 590:1313–23.

14. Tilokani, L., Nagashima, S., Paupe, V., and Prudent, J. 2018. Mitochondrial dynamics: overview of molecular mechanisms. *Essays Biochem* 62:341–60.

15. Silva Ramos, E., Larsson, N. G., and Mourier, A. 2016. Bioenergetic roles of mitochondrial fusion. *Biochim Biophys Acta* 1857:1277–83.

16. Schrepfer, E., and Scorrano, L. 2016. Mitofusins, from mitochondria to metabolism. *Mol Cell* 61:683–94.

17. Mishra, P. 2016. Interfaces between mitochondrial dynamics and disease. *Cell Calcium* 60:190–198.

18. MacVicar, T., and Langer, T. 2016. OPA1 processing in cell death and disease – the long and short of it. *J Cell Sci* 129:2297–306.

19. Pagliuso, A., Cossart, P., and Stavru, F. 2017. The ever-growing complexity of the mitochondrial fission machinery. *Cell Mol Life Sci* 75:355–74.

20. Taguchi, N., Ishihara, N., Jofuku, A., Oka, T., and Mihara, K. 2007. Mitotic phosphorylation of dynamin-related GTPase Drp1 participates in mitochondrial fission. *J Biol Chem* 282:11521–9.

21. Kashatus, D. F., and Counter, C. M. 2011. Breaking up is hard to do: RalA, mitochondrial fission and cancer. *Small GTPases* 2:329–33.

22. Oo, P. S., Yamaguchi, Y., Sawaguchi, A., et al. 2018. Estrogen regulates mitochondrial morphology through phosphorylation of dynamin-related protein 1 in MCF7 human breast cancer cells. *Acta Histochem Cytochem* 51:21–31.

23. Shi, Y., Fan, S., Wang, D., et al. 2018. FOXO1 inhibition potentiates endothelial angiogenic functions in diabetes via suppression of ROCK1/Drp1-mediated mitochondrial fission. *Biochim Biophys Acta* 1864:2481–94.

24. Xu, W., Yu, H., Ma, R., et al. 2017. Apelin protects against myocardial ischemic injury by inhibiting dynamin-related protein 1. *Oncotarget* 8:100034–44.

25. Zhang, Z., Liu, L., Jiang, X., Zhai, S., and Xing, D. 2016. The essential role of Drp1 and its regulation by S-Nitrosylation of Parkin in dopaminergic neurodegeneration: implications for Parkinson's disease. *Antioxid Redox Signal* 25:609–22.

26. He, Y., Jia, K., Li, L., et al. 2018. Salvianolic acid B attenuates mitochondrial stress against Abeta toxicity in primary cultured mouse neurons. *Biochem Biophys Res Commun* 498:1066–72.

27. Serasinghe, M. N., and Chipuk, J. E. 2017. Mitochondrial fission in human diseases. *Handb Exp Pharmacol* 240:159–88.

28. Nagdas, S., and Kashatus, D. F. 2017. The interplay between oncogenic signaling networks and mitochondrial dynamics. *Antioxidants (Basel)* 6 E33.

29. Trotta, A. P., and Chipuk, J. E. 2017. Mitochondrial dynamics as regulators of cancer biology. *Cell Mol Life Sci* 74:1999–2017.

30. Gwangwa, M. V., Joubert, A. M., and Visagie, M. H. 2018. Crosstalk between the Warburg effect, redox regulation and autophagy induction in tumourigenesis. *Cell Mol Biol Lett* 23:20.

31. Rosdah, A. A., Holien, J. K., Delbridge, L. M., Dusting, G. J., and Lim, S. Y. 2016. Mitochondrial fission – a drug target for cytoprotection or cytodestruction? *Pharmacol Res Perspect* 4:e00235.

32. Pickles, S., Vigie, P., and Youle, R. J. 2018. Mitophagy and quality control mechanisms in mitochondrial maintenance. *Curr Biol* 28:R170–85.

33. Hamacher-Brady, A., and Brady, N. R. 2016. Mitophagy programs: mechanisms and physiological implications of mitochondrial targeting by autophagy. *Cell Mol Life Sci* 73:775–95.

34. Yan, C., and Li, T. S. 2018. Dual role of mitophagy in cancer drug resistance. *Anticancer Res* 38:617–21.

35. Wu, H., Wei, H., Sehgal, S. A., Liu, L., and Chen, Q. 2016. Mitophagy receptors sense stress signals and couple mitochondrial dynamic machinery for mitochondrial quality control. *Free Radic Biol Med* 100:199–209.

36. Panicker, N., Dawson, V. L., and Dawson, T. M. 2017. Activation mechanisms of the E3 ubiquitin ligase Parkin. *Biochem J* 474:3075–86.

37. Yamano, K., and Youle, R. J. 2013. PINK1 is degraded through the N-end rule pathway. *Autophagy* 9:1758–69.

38. Straub, S. P., Stiller, S. B., Wiedemann, N., and Pfanner, N. 2016. Dynamic organization of the mitochondrial protein import machinery. *Biol Chem* 397:1097–1114.

39. Liu, J., Zhang, C., Hu, W., and Feng, Z. 2018. Parkinson's disease-associated protein Parkin: an unusual player in cancer. *Cancer Commun (Lond)* 38:40.

40. Poulogiannis, G., McIntyre, R. E., Dimitriadi, M., et al. 2010. PARK2 deletions occur frequently in sporadic colorectal cancer and accelerate adenoma development in Apc mutant mice. *Proc Natl Acad Sci U S A* 107:15145–50.

41. Bernardini, J. P., Lazarou, M., and Dewson, G. 2017. Parkin and mitophagy in cancer. *Oncogene* 36:1315–27.

42. Katajisto, P., Dohla, J., Chaffer, C. L., et al. 2015. Stem cells. Asymmetric apportioning of aged mitochondria between daughter cells is required for stemness. *Science* 348:340–43.

43. Solano-Galvez, S. G., Abadi-Chiriti, J., Gutierrez-Velez, L., et al. 2018. Apoptosis: activation and inhibition in health and disease. *Med Sci (Basel)* 6:E54.

44. Cui, J., and Placzek, W. J. 2018. Post-transcriptional regulation of anti-apoptotic BCL2 family members. *Int J Mol Sci* 19:E308.

45. Popgeorgiev, N., Jabbour, L., and Gillet, G. 2018. Subcellular localization and dynamics of the Bcl-2 family of proteins. *Front Cell Dev Biol* 6:13.
46. Pena-Blanco, A., and Garcia-Saez, A. J. 2017. Bax, Bak and beyond – mitochondrial performance in apoptosis. *FEBS J*:416–31.
47. Kilbride, S. M., and Prehn, J. H. 2013. Central roles of apoptotic proteins in mitochondrial function. *Oncogene* 32:2703–2711.
48. Dorstyn, L., Akey, C. W., and Kumar, S. 2018. New insights into apoptosome structure and function. *Cell Death Differ* 25:1194–1208.
49. Garner, T. P., Lopez, A., Reyna, D. E., Spitz, A. Z., and Gavathiotis, E. 2017. Progress in targeting the BCL-2 family of proteins. *Curr Opin Chem Biol* 39:133–42.
50. McArthur, K., and Kile, B. T. 2018. Apoptotic caspases: multiple or mistaken identities? *Trends Cell Biol* 28:475–493.
51. Larsen, B. D., and Sorensen, C. S. 2017. The caspase-activated DNase: apoptosis and beyond. *FEBS J* 284:1160–70.
52. Aredia, F., and Scovassi, A. I. 2014. Poly(ADP-ribose): a signaling molecule in different paradigms of cell death. *Biochem Pharmacol* 92:157–163.
53. Majtnerova, P., and Rousar, T. 2018. An overview of apoptosis assays detecting DNA fragmentation. *Mol Biol Rep*:1469–78.
54. Rybczynska, A. A., Boersma, H. H., de Jong, S., et al. 2018. Avenues to molecular imaging of dying cells: focus on cancer. *Med Res Rev*:1713–68.
55. Gilbert, R. J. 2016. Protein-lipid interactions and non-lamellar lipidic structures in membrane pore formation and membrane fusion. *Biochim Biophys Acta* 1858:487–99.
56. Antignani, A., and Youle, R. J. 2006. How do Bax and Bak lead to permeabilization of the outer mitochondrial membrane? *Curr Opin Cell Biol* 18:685–9.
57. Sahl, S. J., Hell, S. W., and Jakobs, S. 2017. Fluorescence nanoscopy in cell biology. *Nat Rev Mol Cell Biol* 18:685–701.
58. Hantusch, A., Das, K. K., Garcia-Saez, A. J., Brunner, T., and Rehm, M. 2018. Bax retrotranslocation potentiates Bcl-xL's antiapoptotic activity and is essential for switch-like transitions between MOMP competency and resistance. *Cell Death Dis* 9:430.
59. Zheng, J. H., Viacava Follis, A., Kriwacki, R. W., and Moldoveanu, T. 2016. Discoveries and controversies in BCL-2 protein-mediated apoptosis. *FEBS J* 283:2690–700.
60. Hantusch, A., Rehm, M., and Brunner, T. 2018. Counting on death – Quantitative aspects of Bcl-2 family regulation. *FEBS J*:4124–38.
61. Tait, S. W., Ichim, G., and Green, D. R. 2014. Die another way–non-apoptotic mechanisms of cell death. *J Cell Sci* 127:2135–44.
62. Stewart, J. B., and Chinnery, P. F. 2015. The dynamics of mitochondrial DNA heteroplasmy: implications for human health and disease. *Nat Rev Genet* 16:530–42.
63. Kagan, V. E., Chu, C. T., Tyurina, Y. Y., Cheikhi, A., and Bayir, H. 2014. Cardiolipin asymmetry, oxidation and signaling. *Chem Phys Lipids* 179:64–9.

14 Mitochondrial Dysfunction Linking Obesity and Asthma

Naveen K. Bhatraju

Molecular Immunogenetics Laboratory, CSIR Institute of Genomics and Integrative Biology, Delhi, India

Anurag Agrawal

Molecular Immunogenetics Laboratory, CSIR Institute of Genomics and Integrative Biology, Delhi, India
Academy of Scientific and Innovative Research, Delhi, India
Baylor College of Medicine, Houston, Texas

CONTENTS

Abbreviations

12/15 LOX	12/15- Lipoxygenase
12-S-HETE	12-S-hydroxyeicosatetraenoic acid
13-S-HODE	13-S-hydroxyoctadecadienoic acid
ADMA	Asymmetric dimethyl arginine
ADP	Adenosine diphosphate
ATP	Adenosine triphosphate
BMI	Body mass index
Drp1	Dynamin related protein 1
ER	Endoplasmic reticulum
ETC	Electron transport chain
$FADH_2$	Dihydro flavin adenine dinucleotide
Fis1	Fission protein 1
IL-4	Interleukin 4
MAMs	Mitochondria-associated ER membranes
MetS	Metabolic syndrome
Mff	Mitochondrial fission factor
Mfn1/2	Mitofusin 1/2
mtDNA	Mitochondrial deoxy ribonucleic acid
NAD^+	Nicotinamide adenine dinucleotide (oxidized)
NADH	Nicotinamide adenine dinucleotide (reduced)
NO	Nitric oxide
NRF	Nuclear respiratory factor
Opa1	Optic atrophy-1
OXPHOS	Oxidative phosphorylation
PGC-1α	Peroxisome proliferator-activated receptor-gamma coactivator (PGC)-1α
ROS	Reactive oxygen species
SIRT	Silent mating type information regulation 2 homolog
TCA	Tricarboxylic acid
TFAM	Mitochondrial transcription factor A
T_H2	Type 2 helper T cells

1 INTRODUCTION

Obesity dramatically alters normal lung homeostasis and thus increases the risk for a number of chronic lung diseases including asthma. A strong epidemiological basis currently exists for a bidirectional association between obesity and asthma (Dixon et al. 2010; Gonzalez-Muniesa et al. 2017; Peters et al. 2018a). While, there is limited evidence pointing toward the predisposition of asthmatics to obesity, plenty of evidence from observational and experimental studies suggests that obesity precedes asthma incidence, and is associated with increased asthma severity (Dixon & Poynter 2016; Peters et al. 2018a, 2018b). In fact, a meta-analysis involving ~108,000 mother-child pairs has suggested that obesity-associated asthma risk can start *in utero*. In this study, they have found maternal obesity during pregnancy and gestation weight gain to be associated with an approximate 15–30%

increased risk of asthma in offspring (Forno et al. 2014) and more importantly, child's own obesity does not mediate this risk (Harskamp-van Ginkel et al. 2015). Similarly, a dose-response relationship between obesity and incident asthma has been reported in another meta-analysis of several prospective studies involving more than 300,000 adults (Beuther & Sutherland 2007).

Obese-asthma, both in children and adults, is characterized by increased asthma severity, poor asthma control, increased risk of hospitalization, reduced response to conventional asthma medications and decreased quality of life. The clinical presentation of Obese-asthma is affected by several factors including sex, age at disease onset, atopy status, and airway inflammation status (Dixon & Poynter 2016; Peters et al. 2018a). Based on these variations, two major obese-asthma phenotypes have been described in epidemiological studies: The first cluster is characterized by late-onset asthma, lower markers of allergic inflammation (low levels of exhaled nitric oxide (NO) or sputum eosinophils) and airway reactivity that improves with weight loss, and the characteristics of the second cluster include high T_H2 inflammation (allergic inflammation), early-onset disease and has the most severe disease among the obese-asthmatics (Dixon et al. 2011; Holguin et al. 2011, 2013; Sutherland et al. 2012). Simple excess abdominal and thoracic mass loading mediated mechanical effects fail to explain the striking clinical differences between the phenotypes. This becomes more intuitive considering the following lines of evidences: (1) not all obese people develop airway disease, (2) improvements in airway reactivity with weight loss (observed in the first cluster) cannot be directly related to changes in lung volumes, and (3) Metabolic abnormalities associated with obesity, but not body mass, are more frequently observed in asthmatics (Cottrell et al. 2011;Dixon & Poynter 2016;Perez & Piedimonte 2014; Peters et al. 2018a). Altered airway structure and function as well as immune dysfunction, consequent to obesity-associated changes in diet and metabolic abnormalities, could either lead to the development of asthma or complicate the pre-existing disease (Bates & Dixon 2015).

Experimental models of obesity and asthma support these epidemiological observations and further broaden the definition of obesity, beyond excess mass loading, to include metabolic and functional anomalies associated with obesity. In mice with diet-induced metabolic syndrome, predisposition to asthma-like airway hyperresponsiveness and remodelling was found to be mainly dependent on the metabolic abnormalities, but not weight gain (Singh et al. 2015). Further, the characteristics such as minimal airway inflammation and low levels of exhaled NO observed in these mice resemble the late-onset asthma cluster described above. Given a role for NO in bronchodilation and lung homeostasis, decreased NO in the context of obesity may have pathological consequences (Haldar et al. 2008; Moore et al. 2010). Decreased bioavailability of arginine, a substrate for nitric oxide synthase, due

to competition from increased levels of asymmetric dimethyl arginine (ADMA) could be a reason for the observed low levels of NO (Holguin et al. 2013). Interestingly, such mice upon allergen challenge and sensitization develop enhanced allergic inflammation and airway remodelling (Leishangthem et al. 2013). Such exaggerated allergic response in the setting of obesity could be explained by shared pathological processes between these two seemingly separate diseases. It is now well established that mitochondrial dysfunction is a critical component of both obesity-associated metabolic abnormalities and asthma pathophysiology (Bhatti et al. 2017; de Mello et al. 2018; Hemachandra Reddy 2011; Mabalirajan & Ghosh 2013). In the current chapter, we examine the evidence supporting the contribution of mitochondrial dysfunction to either of the diseases and then look into the unifying mitochondria-centric pathological mechanisms connecting obesity and asthma. Finally we present the mitochondrial health-centric strategies that may be actively considered for the prevention and treatment of both obesity and asthma.

2 MULTIFACETED ROLES OF MITOCHONDRIA IN CELLULAR AND METABOLIC CONTEXT

Mammalian mitochondria are considered to be the descendants of ancient alphaproteobacterium engulfed by a eukaryotic progenitor which remained as endosymbionts since then. Mammalian mitochondria contain their own genome (~16.5 kb) that remains sequestered from the cytoplasm by two functionally distinct membranes. While the outer membrane is semipermeable, which is reflected in the similarity of the composition of intermembrane space and cytoplasm, inner membrane is largely impermeable and is inwardly curved into the matrix to form cristae. Considering the mismatch between the mitochondrial proteome (~1500 proteins) and the contribution of the mitochondrial DNA (mtDNA) toward the same (13 proteins), it is evident that a complex coordination between nucleus and mitochondria is essential to maintain mitochondrial integrity and function (Nunnari & Suomalainen 2012).

Mitochondrion is the primary site for the cellular energy production by oxidative phosphorylation (OXPHOS). OXPHOS involves transfer of electrons from NADH and FADH2, reducing equivalents generated during tricarboxylic acid (TCA) cycle in the mitochondrial matrix, through a series of complexes (I-IV) constituting the electron transport chain (ETC) finally to oxygen. The proton gradient generated by complexes I, III, and IV is utilized to drive the phosphorylation of ADP to ATP by ATP synthase complex or complex V. OXPHOS is an imperfect process, at times causing premature transfer of electrons to oxygen at complex I and III, leading to incomplete reduction of oxygen thus generating reactive oxygen species (ROS) such as superoxide radical, hydroxyl radical and hydrogen peroxide. ROS can cause oxidative damage of proteins, lipids and nucleic acids leading to oxidative stress and mitochondrial dysfunction observed in a wide range of diseases and aging. However, at low levels, mitochondrial ROS has been shown to serve as a secondary messenger that helps in coordinating complex responses involved in the regulation of cell proliferation and differentiation, and adaptive stress signalling pathways such as hypoxia (Chandel et al. 2014; Nunnari & Suomalainen 2012). It is important to note that oxidative damage of mitochondria does occur even under physiological conditions. Nevertheless, continued mitochondrial activity and homeostasis is ensured by activation of an intricate homeostatic system that involves balanced mitochondrial biogenesis, mitophagy, mitochondria-specific protein turnover pathways and mitochondrial fusion and fission (Whitaker et al. 2016). Disruption of one or more of these mitochondrial quality control mechanisms result in loss of mitochondrial homeostasis and mitochondrial dysfunction which has been implicated in a number of diseases including asthma (Piantadosi & Suliman 2017; Whitaker et al. 2016).

Inside a cell, mitochondria are highly dynamic, motile and exhibit variable morphologies coupled to their functional state. A balance between the fission and fusion processes regulates the spatiotemporal organization of mitochondria into either reticular networks of elongated mitochondria or short fragments dispersed throughout the cytoplasm. Mitochondrial fusion/fission is a tightly regulated process involving a coordinated interplay of a number of proteins. The main proteins involved in the regulation of mammalian mitochondrial fusion are mitofusin 1 and 2 (Mfn1 and 2) and autosomal dominant optic atrophy-1 (Opa1). Mitofusins and Opa1 are respectively localized in the outer and the inner mitochondrial membrane controlling the outer and inner membrane fusions. Additional roles for both Mfn2 and Opa1 have also been reported. While Opa1 plays a role in cristae remodelling, Mfn2 has been shown to be implicated in the structural and functional connection between mitochondria and endoplasmic reticulum (ER) (Theurey & Rieusset 2017). On the other hand, mitochondrial fission is regulated by dynamin related protein 1 (Drp1) and fission protein 1 (Fis1). Fis1 is an outer mitochondrial membrane protein that helps in the recruitment of Drp1, a cytosolic protein, during mitochondrial fragmentation (Nunnari & Suomalainen 2012). Mitochondrial fission factor (Mff) is another protein localized in the outer mitochondrial membrane that has been shown to have similar function to that of Fis1. The reasons for such high dynamicity of mitochondrial structure are currently unclear. Perhaps, reducing mtDNA heterogeneity through matrix content mixing via mitochondrial fusion may connect the mitochondrial dynamics to the maintenance of mitochondrial health (Chen et al. 2010). Further, mitochondrial fission has been shown to be critical to cellular adaptation to stress and thus cell survival. The strategic positioning of mitochondria in the metabolic flux allows it to sense and adapt to alterations in nutrient availability (Naviaux 2014). Mitochondrial capacity to switch between the fatty acid and glucose oxidation during fasting

and fed states forms the core of the metabolic flexibility concept (Smith et al. 2018). Metabolic flexibility is defined as the remarkable ability of a cell/organ/organism to tune metabolism to maintain the energy homeostasis depending on the substrate availability and environmental demand. Mitochondrial response to altered energy demands involve both acute changes such as changes in mitochondrial dynamics and post-translational modifications associated with prompt activity modulation, and long term transcriptional responses such as changes in mitochondrial biogenesis and mitophagy that regulate mitochondrial volume and density (Gao et al. 2014; Smith et al. 2018). However, dysregulation of either of these adaptive responses has been shown to result in metabolic inflexibility, a key pathology associated with a number of metabolic disorders such as insulin resistance, type-II diabetes and metabolic syndrome.

3 MITOCHONDRIAL DYSFUNCTION IN OBESITY AND METABOLIC SYNDROME

Coordinated regulation of food intake, energy expenditure and thermogenesis, and fat storage, by neural and endocrine systems, is critical to the stable maintenance of body weight and composition. An alteration in one or more of these processes has devastating effects on the metabolic and systemic health of the individual. Obesity is an important outcome of a chronic imbalance in the energy intake and expenditure. Obesity is defined as a disproportionate body weight for height with an excessive accumulation of adipose tissue that is usually accompanied by mild, chronic, systemic inflammation. Sedentary lifestyle translating to reduced energy expenditure associated with an increase in the consumption of the high calorie diet is thought to be the primary cause of excessive adipose tissue accumulation in obesity. However, in reality, obesity is much more complex and has multifactorial aetiology involving interplay among several behavioral, environmental, physiological, genetic, epigenetic and socio-economic factors. The complex interplay among these aetiological factors affects food intake, nutrient utilization and thermogenesis, fatty acid oxidation, and fat storage distribution between adipose and non-adipose tissues (Gonzalez-Muniesa et al. 2017). An unprecedented rise in obesity has been noted, in the last three decades, globally.

It would not be an overstatement to say that obesity is the modern world's pandemic. According to the world health organization (WHO) estimates, approximately one third of the current global population are either overweight or obese. By 2030, it has been forecast that an estimated 57.8% of the world's adult population would be overweight or obese (Finkelstein et al. 2012; Gonzalez-Muniesa et al. 2017). It is worth noting here that obesity prevalence has tripled since 1975 and is on the rise in both developed and developing nations. What is most alarming is that the prevalence of obesity, in children and adolescents, has also been found to be doubled since 1980 (Ng et al. 2014). Childhood obesity has been shown to be a potential risk factor for metabolic complications and chronic diseases in adulthood (Koletzko et al. 2011). A number of methods have been described in literature to determine whether a person is obese and if so, to assess the extent of obesity and fat distribution (Box 1). Of all the methods available, body mass index (BMI) is the most commonly used indicator to define obesity.

BMI is expressed as weight in kilograms (Kg) divided by square of height in meters (m^2). WHO classifies adult people (> 19 years age) with BMI $\geq$ 30 Kg/m^2 as obese and a value of BMI between 25 and 29.9 Kg/m^2 is defined as overweight. However, these threshold values do not hold true for children. In children aged between 5 and 19 years, BMI-for-age greater than one and two standard deviations, above WHO growth reference median, are referred to as overweight and obesity respectively. In case of children less than 5 years, overweight and obesity are defined by the weight-for-height greater than two and three standard deviations above the WHO child growth standards median respectively. BMI, being an indirect measure of adiposity, fails to capture the information regarding the fat distribution among the regional fat depots. The heterogeneity observed in the physiology and metabolic health of obese individuals, with similar BMI values, could at least in part be explained by their fat distribution profiles (Tchernof & Despres 2013). Subcutaneous obesity and visceral obesity are the two distinct types of obesity that form the end points of the fat distribution continuum. These two

BOX 1 ANTHROPOMETRIC MEASURES USED FOR ASSESSING FAT DISTRIBUTION PROFILES IN OBESE PEOPLE

Waist circumference, Umbilical circumference, Thigh circumference, Hip circumference, Waist-to-height ratio, Waist-to-hip ratio, Waist-to-thigh ratio, Coronal diameter, Sagittal abdominal diameter, Central and peripheral skin fold ratio and Conicity index (Pereira et al. 2015).

Apart from these measurements, several other *in vivo* methods for assessment of body composition and fat distribution are currently available: Densitometry, bioimpedance analysis, dual-energy X-ray absorptiometry (DXA), computed tomography (CT), magnetic resonance imaging (MRI), nuclear magnetic resonance spectroscopy (MRS) and ultrasound (US). However, the use of either one or a combination of these techniques in research settings is dependent mainly on the research question being addressed (Fosbol & Zerahn 2015; Seabolt et al. 2015).

obesity phenotypes substantially vary in their cardiometabolic disease risk. Table1 illustrates the major differences between the subcutaneous and visceral obesities. While subcutaneous adiposity, especially in glutealfemoral fat depot, is considered healthy, visceral adiposity is a potential risk factor for cardiorespiratory and metabolic disorders, and some forms of cancer. Waist circumference could be used in combination with BMI to determine visceral obesity.

Visceral adiposity directly alters the mechanical properties of the lung. Accumulation of fat in the thoracic and abdominal cavities restricts the downward movement of the diaphragm which in turn causes an increase in pleural pressure and functional residual capacity (FRC) to decrease (Peters et al. 2018b). These fat dependent mechanical effects on the lungs and chest wall are substantial, with a 10%, 22% and 33% FRC decrease observed in overweight, mildly obese and severely obese non-asthmatic subjects respectively. In addition visceral adiposity also causes decrement in pulmonary compliance which likely contributes to the respiratory symptoms observed in obese people. Furthermore, Forno et al., have reported dysanapsis (dissociation between airway growth and lung size) as a causal factor for the development of lung disease in obese children (Forno et al. 2017). However, the fact that not all obese people develop lung diseases suggests that obesity is beyond simple excess mass loading. This further extends the obesity-asthma link to metabolic and functional changes associated with obesity (Agrawal et al. 2011; Dixon & Poynter 2016; Peters et al. 2018b; Singh et al. 2013).

Visceral obesity is associated with a number of metabolic abnormalities collectively termed as metabolic syndrome (MetS). Metabolic syndrome is defined as the co-existence of at least three of the five following abnormalities of metabolic origin: visceral obesity, hypertriglyceridaemia, decreased high density lipoprotein (HDL) cholesterol, hypertension and insulin resistance (Gonzalez-Muniesa et al. 2017). It is now well established that metabolic syndrome is a manifestation of metabolic inflexibility (Smith et al. 2018). Given the central role of mitochondrial function in the control of metabolic flexibility, mitochondrial dysfunction may play a significant role in obesity related diseases. Indeed, mitochondrial dysfunction has been demonstrated in all the main metabolic tissues, such as liver, muscle and adipose tissues, associated with obesity/MetS (de Mello et al. 2018; Gao et al. 2014; Putti et al. 2015). Also, compromised mitochondrial function accompanied by altered mitochondrial dynamics (promoting mitochondrial fission) have been observed in liver and skeletal muscle tissues of genetic/diet-induced obesity mice models (Lionetti et al. 2014; Putti et al. 2015). Similar observations were made in obese Zucker rats with insulin resistance (Bach et al. 2003). In fact, a reduction in Mfn2 expression was also observed in the skeletal muscle of obese type-II diabetic subjects (Bach et al. 2003; Putti et al. 2015). In addition, Mfn2 deficiency in mice has been

TABLE 1

Major difference between Visceral and Subcutaneous obesities

Feature	Visceral/Android obesity	Subcutaneous/Gynoid obesity
Fat distribution	Mainly concentrated in abdominal region (mesenteric adipose tissue)	Mainly concentrated in hip and thigh regions
Body shape	Apple-like shape	Pear-like shape
Gender predominance	More common in men	more common in women
Anthropometric measurements	High waist-to -Hip ratio	Low waist-to-Hip ratio
Ectopic fat accumulation	Associated with accumulation of fat in liver, heart, pancreas and muscle	Not associated
Metabolic status	Metabolically infavorable and associated with a number of metabolic abnormalities and increased cardiovascular risk	Metabolically favorable and considered protective

shown to induce mitochondrial dysfunction and endoplasmic reticulum stress parallel to impaired insulin signalling in liver and muscle (Sebastian et al. 2012; Schrepfer & Scorrano 2016). A positive correlation identified between Mfn2 levels and insulin sensitivity in morbidly obese people, who underwent bariatric surgery, further strengthens the implication of mitochondrial dynamics in insulin resistance and obesity (Bach et al. 2005). Other than its role in mitochondrial fusion, Mfn2 is also localized to ER-mitochondria contact sites called mitochondria-associated ER membranes (MAMs) (Theurey & Rieusset 2017). These sites are hot spots for calcium, lipid and metabolite exchange between the two organelles and are thought to play a key role in nutrient sensing and hepatic insulin signalling. Arruda and colleagues have shown that compromised mitochondrial function and oxidative stress observed in liver, during obesity, could be a result of chronic enrichment of MAMs leading to mitochondrial calcium overload (Arruda et al. 2014). In line with this, several other studies have highlighted the implications of MAMs in metabolic homeostasis (Rieusset 2018). Apart from this, variations in the dietary fat source have also been found to have differential effects on the mitochondrial function and dynamics (Lionetti et al. 2014). High fat diet rich in saturated fatty acids (HL) was found to promote hepatic fat accumulation and insulin resistance in parallel to mitochondrial dysfunction, smaller mitochondria and increased mitochondrial fission machinery *in vivo*. Similar structural and functional alterations in mitochondria were found in immortalized mouse myoblast C2C12 cells upon treatment with saturated fatty acids (Jheng et al. 2012). In accordance, mitochondria from obese people were found to be smaller with reduced bioenergetic capacity compared to those from lean individuals (Kelley et al. 2002). On the contrary, omega 3 polyunsaturated fatty acids were found to improve mitochondrial function by shifting mitochondrial dynamics toward fusion and reducing ROS generation both *in vitro* and *in vivo* (Lionetti et al. 2014;

Zhang et al. 2011). Similarly, in obese mice, inhibition of mitochondrial fission has been shown to improve muscle insulin signalling and systemic insulin sensitivity (Civitarese & Ravussin 2008). Although, increased mitochondrial fragmentation during conditions of mitochondrial overload makes a case for an adaptive response via increasing the total surface area for metabolic substrate accessibility to carrier proteins and thus increased dietary fatty acid oxidation, it results in mitochondrial ROS production surpassing the antioxidant capacity causing oxidative stress (Lionetti et al. 2014; Putti et al. 2015).

Low ATP utilization combined with high membrane potential, a condition such as nutrient excess, potentially increases mitochondrial ROS production causing oxidative damage affecting the activity of several redox sensitive metabolic enzymes (Smith et al. 2018). An increased $NADH/NAD^+$ ratio, indicative of a low electron flux through ETC, reduces the carbon influx into TCA cycle via redox inhibition of TCA cycle enzymes leading to accumulation of acetyl-CoA and other acyl-CoA molecules. This in turn increases the protein acetylation of several mitochondrial proteins resulting in impaired bioenergetic capacity (Gao et al. 2014; Smith et al. 2018). Further, incomplete fatty acid oxidation and accumulation of fatty-acyl-CoA leads to impaired insulin signalling via increased insulin receptor phosphorylation and impaired activation of Akt/PKB (Brons & Grunnet 2017). Several of these changes such as increased mitochondrial ROS, dysfunctional oxidative respiration have been observed in both genetically obese mice as well as mice with diet-induced obesity (Anderson et al. 2009; Xu et al. 2016b). Similarly, increased oxidative stress and abnormal adipokine production, associated with reduced lipid metabolism and lower mitochondrial oxidative capacities, were noticed in obese individuals (Montgomery & Turner 2015). In contrast, conditions such as caloric restriction and starvation were found to be associated with reduced oxidative damage vis-à-vis changes in mitochondrial dynamics (Montgomery & Turner 2015; Putti et al. 2015).

Mitochondrial biogenesis has also been found to be affected in obesity. A decrease in peroxisome proliferator-activated receptor-gamma coactivator (PGC)-1α, the master regulator of mitochondrial biogenesis, has been observed in obese individuals and mouse models of obesity (Crunkhorn et al. 2007; Semple et al. 2004). In addition, a marked decrease in two other transcription factors associated with mitochondrial biogenesis, mitochondrial transcription factor A (TFAM) and nuclear respiratory factor-1 (NRF-1) has been observed in the liver tissue of high fat diet fed mice (Chen et al. 2016). Hyperacetylation of mitochondrial proteins is an acute response to caloric excess that is associated with reduction in mitochondrial OXPHOS, fatty acid oxidation and ROS metabolism (Gao et al. 2014; Smith et al. 2018). The balance between acetylation and deacetylation is regulated by the activity of SIRT3, a mitochondrial NAD^+-dependent protein deacetylase (Hirschey et al. 2010). A marked decrease in SIRT3 activity has been observed in the liver of high fat fed mice (Kendrick et al. 2011). The role of SIRT3 in the pathophysiology of MetS is further substantiated by the reversal of mitochondrial dysfunction and MetS features in a SIRT3 activity dependent manner (Mabalirajan & Ghosh 2013). Taken together, these studies indicate that mitochondrial dysfunction is a key pathological mechanism implicated in obesity related diseases. Further, accumulating evidence suggests that obesity and/or MetS associated local disturbances in the metabolic tissues leads to global metabolic inflexibility increasing the risk of several chronic diseases such as asthma.

4 MITOCHONDRIAL DYSFUNCTION IN ASTHMA

4.1 Asthma Pathophysiology and Heterogeneity

Asthma is the most common chronic inflammatory disease of the lung affecting >300 million people worldwide. Although, the prevalence of asthma is more in developed countries, increasing Westernization of lifestyle, in the recent decades, has led to the rise in asthma prevalence in developing countries as well (Global Initiative for Asthma 2018; Holgate et al. 2015). Further, asthma is estimated to cause 346,000 global deaths annually (Lozano et al. 2012). Despite the large socioeconomic burden associated with asthma, it remains poorly defined. Global Strategy for Asthma Management and Prevention by the Global Initiative for Asthma (GINA) defined asthma as a,

> heterogeneous disease, usually characterized by chronic airway inflammation and variable remodelling that results in a range of clinical presentations, treatment responses defined by history of respiratory symptoms such as wheeze, shortness of breath, chest tightness and cough that vary

over time and in intensity, together with variable airflow limitation.

(GINA 2018)

Asthma is a complex and heterogeneous syndrome with several recognized endotypes [grouping of patients based on established pathophysiologic and mechanistic parameters] (Martin et al. 2014; Wenzel 2012). Recent efforts to understand the asthma pathophysiology have revealed the critical role of mitochondrial dysfunction and oxidative stress in the development and progression of asthma (Hemachandra Reddy 2011).

Description of mitochondrial structural abnormalities, in the airways of asthmatic patients, traces back to 1985 (Konradova et al. 1985). In this study, Konradova et al., have reported ultrastructural changes such as swollen mitochondria, and dilated endoplasmic reticulum and golgi complex in the bronchial epithelial cells of three asthmatic girls. Later, Mabalirajan and colleagues have made similar observations in mice models of allergic asthma. They have found that the dysmorphic mitochondria are associated with T_H2 inflammation and thus are an integral part of asthma pathophysiology (Mabalirajan et al. 2008). Subsequently, Aguilera-Aguirre and colleagues reported that allergen sensitization and challenge in mice with genetic deletion of mitochondrial ubiquinol-cytochrome C-reductase core II protein in airway epithelium increases the severity of asthma (Aguilera-Aguirre et al. 2009). Taken together, these studies suggest a causal role for mitochondrial dysfunction in asthma. Recently, work from the author's group presented evidence for an intriguing cross-talk between IL-4 signalling and arginine metabolism in airway epithelial dysfunction (Pattnaik et al. 2016). According to this study, IL-4 promotes the intracellular accumulation of ADMA, by altering both its synthesis and degrading enzymes, contributing to the high ADMA levels observed in allergically inflamed lungs. Further, increased oxo-nitrative stress and hypoxic response induced mitochondrial loss has been observed in airway epithelial cells upon sequential treatment with IL-4 and ADMA. In another study from author's group, it has been shown that excessive induction of hypoxic response results in the development of severe asthma phenotype in mice where allergen challenge can even lead to fatal bronchoconstriction (Ahmad et al. 2012). Interestingly, restoration of mitochondrial homeostasis either through increased mitochondrial biogenesis or through high dose of arginine supplementation have been shown to attenuate asthma features in these models (Mabalirajan et al. 2010a, 2010b).

Although, dysmorphic mitochondria, increased mitochondrial ROS and reduction in mitochondrial content in airway epithelial cells of several asthmatic mouse models has been widely reported, preserved mitochondrial respiration associated with increased mitochondrial number and volume has been reported in the airway epithelial cells and smooth muscle cells of non-severe asthmatics patients (Girodet et al. 2016; Leishangthem et al. 2013;

Xu et al. 2016a). These observations were attributed to the presence of adaptive mechanisms such as increased *de novo* arginine synthesis, and arginine metabolism by arginase 2 (ARG2) that help in maintenance of cellular bioenergetics through an increase in the TCA cycle flux. Further, *arg2* knockout mice have been reported to exhibit metabolic dysfunction, hypertension and increased susceptibility to asthma (Xu et al. 2016a). Moreover, *Arg2* genomic variants have been found to be associated with risk and severity of asthma in genome-wide association studies. Taken together, the data suggests that arginine metabolism is a strong determinant of mitochondrial health connecting asthma and MetS.

5 UNIFYING MITOCHONDRIA-CENTRIC MECHANISMS CONNECTING OBESITY, METABOLIC SYNDROME AND ASTHMA

Mitochondrial dysfunction is a central component of a wide range of chronic pathologies including neurodegenerative (e.g., Parkinson's disease, Alzheimer's disease), metabolic (e.g., type-2 diabetes, MetS), cardiovascular (e.g., coronary artery disease, atherosclerosis), renal (e.g., chronic kidney disease) and respiratory diseases (e.g., asthma, idiopathic pulmonary fibrosis) (Whitaker et al. 2016). While in most of these pathologies mitochondrial dysfunction is localized to a specific tissue/organ, mitochondrial dysfunction in case of obesity and metabolic syndrome results in systemic alteration in metabolism. This explains the increased risk, associated with obesity and MetS, for the development of other diseases including respiratory diseases such as asthma. However, as discussed in other sections of this chapter, obesity is also a potent disease modifier of asthma. This could be explained by the existence of overlapping pathophysiological mechanisms causally associated with mitochondrial dysfunction in both of these diseases. Some of the known overlapping mechanisms are discussed here.

Increased levels of 12/15 lipoxygenase (12/15 LOX), in adipocytes, has been associated with macrophage infiltration and adipose tissue inflammation (Nunemaker et al. 2008). 12/15 LOX is a non-heme iron dioxygenase that catalyzes the hydroperoxidation of polyunsaturated fatty acids such as arachidonic acid and linoleic acid. This enzymatic oxidation reaction results in the production of active lipid metabolites such as 12-S-hydroxyeicosatetraenoic acid (12-S-HETE) and 13-S-hydroxyoctadecadienoic acid (13-S-HODE) (Dobrian et al. 2011). The induction of 12/15 LOX is essential for early onset adipose tissue inflammation and insulin resistance in high-fat fed mice. This was corroborated by studies showing the protective effects of fat-specific deletion of this enzyme against obesity-associated complications wherein 12/15 LOX deficiency improved insulin sensitivity and reduced both local and systemic inflammation (Cole et al. 2012). Interestingly, genetic ablation of 12/15 LOX has been shown to alleviate

asthmatic features (Andersson et al. 2008). 12/15 LOX has also been shown to cause mitochondrial dysfunction either by direct oxidation of mitochondrial membranes or through its metabolites, 12-S-HETE and 13-S-HODE (Mabalirajan et al. 2013a,2013b).

Derailed nitric oxide metabolism is another intersecting pathophysiological mechanism observed both in obesity and asthma (Mabalirajan et al. 2010a). In airway epithelium, NO is implicated in several cellular processes such as ciliary beating, maintenance of bronchodilation, bactericidal functions, and mitochondrial biogenesis (Holguin 2013). Therefore, alterations in NO bioavailability may have pathogenic consequences. Reduced bioavailability of NO has been shown to be a composite effect of decreased bioavailability of L-arginine (NOS substrate) and ADMA-mediated competitive inhibition of endogenous NOS. ADMA binding has been shown to uncouple NOS, and promote the production of ROS without generating NO. Increased ADMA levels have been reported in asthma and obesity, with L-arginine supplementation being sufficient to rescue eNOS function and attenuate asthma features in mice (Mabalirajan & Ghosh 2013). Recently, IL-4, a key pro-inflammatory cytokine, has been shown to promote intracellular ADMA accumulation in airways leading to mitochondrial loss, in an oxidative stress and hypoxia dependent manner (Pattnaik et al. 2016). This provides a novel understanding of molecular underpinnings of the bidirectional associations between obesity and asthma. Taken together, these studies suggest that mitochondrial dysfunction connects both asthma and obesity, and mitochondria health-centric strategies may be actively considered for the management of obese-asthma.

6 THERAPEUTIC STRATEGIES TARGETING MITOCHONDRIA IN ASTHMA AND OBESITY

The refractory nature of obese-asthma to conventional anti-inflammatory asthma medications demands for novel strategies toward better clinical management of this epidemic. Given the central role of mitochondrial dysfunction in the pathophysiology of metabolic syndrome and asthma, mitochondria-targeted therapeutics could be an attractive option. In this section, we will highlight the different types of mitochondria-targeted interventions with potential benefit in metabolic syndrome and asthma.

Restoration of mitochondrial function and thus metabolic flexibility is pivotal to gaining optimal control over this growing epidemic. This could be achieved through the use of scavengers to neutralize harmful ROS, simple lifestyle modifications or by drugs/small molecules that mimic these modifications, through mitochondrial donation. These methods based on the fundamental principles of repair, reprogram or replace (referred to a "3R model") could be tailored to be used individually or in combination to develop personalized treatment strategies. Saying this, weight loss through caloric restriction

(CR) and exercise can be considered as first line of recommendations, in obese-asthmatics, for their established beneficial roles both in obesity and metabolic syndrome (improves insulin sensitization), and asthma control (Mendes et al. 2010; Sideleva et al. 2013). The contextual and general health benefits of exercise and CR are thought to be coordinated through their ability to reset mitochondrial function by epigenetic reprogramming (Cheng & Almeida 2014). However, the contextual utilization of this strategy is restricted in some patients due to impaired exercise capacity and/or lack of compliance to energy-restricted diets over long time. In such conditions, chemical molecules that could mimic CR could be of great application. Metformin falls into this category and has been shown to improve insulin sensitivity and induce weight loss possibly via effects on gut microbiome (Hur & Lee 2015; Park et al. 2012). Metformin has also been shown to attenuate asthma features in obese mice (Calixto et al. 2013). This pinpoints to an old evolutionary microbe-mitochondria link that could be harnessed for efficient mitochondrial reprogramming. It may be noted that gut microbiome has been associated with both asthma and obesity (Cho and Blaser 2012; Fujimura & Lynch 2015; Riiser 2015). In this context, microbiome-derived metabolites, such as pyrroloquinoline quinone (PQQ), which is mitochondria-targeted antioxidant and potent inducer of mitochondrial biogenesis warrants further investigation.

Oxidative stress related tissue damage in the pathophysiology of asthma and metabolic syndrome is well known. However, supplementation of general antioxidants such as α-tocopherol (vitamin E), vitamin C did not show any beneficial effects in humans despite some effects in mouse models (Hemachandra Reddy 2011). Considering the fact that mitochondria are the major source of ROS, the failure of these antioxidants could be attributed to the non-targeted nature of their activity. This has been overcome by the mitochondria-targeted antioxidants, which alone or in combination with other general antioxidants, have been shown to be beneficial both in metabolic syndrome and asthma. For instance, an 8 week supplementation of 100mg coenzyme Q10 (CoQ10), a redox component of ETC, has been shown to be effective in improving insulin resistance of MetS patients (Raygan et al. 2016). In another open-label study involving steroid-dependent asthmatics, reduction in steroid usage has been observed after supplementation with a daily antioxidant cocktail containing 120 mg CoQ10, 400 mg Vitamin E and 250 mg vitamin C (Gazdik et al. 2002; Gvozdjáková et al. 2005). Mitochondria-targeted antioxidants such as mitoquinone mesylate (mitoQ), mitochondria-targeted vitamin E (mito-vit E), contain covalently linked alkyl triphenyl phosphonium that allows them to accumulate in the mitochondrial matrix at high concentrations. This helps in scavenging ROS at the source and thus could be more effective. However, human data for their beneficial effects is awaited.

Replacement is the most recent advancement in the mitochondrial therapeutics. The author's group and others have shown that mesenchymal stem cells (MSCs) mediated mitochondrial donation to damaged lung epithelial cells leads to restoration of cellular homeostasis (Ahmad et al. 2014; Islam et al. 2012). In asthma, MSC-mediated mitochondrial transfer has been shown to be critical to therapeutic response. This was substantiated by alterations in rescue efficacy by over-expression or knockdown of protein mitochondrial Rho GTPase, a guanosine triphosphatase that regulates mitochondria trafficking (Ahmad et al. 2014). Considering the properties of MSCs such as homing behavior, the authors are of the opinion that MSC-mediated mitochondrial donation could help improve features of metabolic syndrome. Although, MSC based therapies are in clinical trials for some of the diseases, the interindividual variability in mitochondrial donation capacity limits the immediate application of this method. Understanding the mechanistic underpinnings of such variability could not only help in developing strategies to improve the same, but also, could economise the MSC therapeutics.

7 CONCLUSION

The intertwined epidemiology among obesity, metabolic syndrome and asthma could at least in part be explained by shared pathophysiological mechanisms. Now, there is ample evidence indicating the key role of mitochondrial dysfunction in each of these conditions highlighting it as a unifying mechanism linking these complex diseases. Although mitochondria-targeted therapeutics has been shown to be beneficial in various experimental models, supporting clinical data is limited. Nevertheless, mitochondria-targeted therapeutics seems to be a viable strategy for the clinical management of obese-asthmatics, who are refractory to conventional asthma medications.

REFERENCES

Agrawal A, Mabalirajan U and Ahmad T, et al. Emerging interface between metabolic syndrome and asthma. Am J Respir Cell Mol Biol 2011; 44: 270–275.

Aguilera-Aguirre L, Bacsi A, Saavedra-Molina A, et al. Mitochondrial dysfunction increases allergic airway inflammation. J Immunol (Baltimore, MD: 1950)2009; 183: 5379–5387.

Ahmad T, Kumar M, Mabalirajan U, et al. Hypoxia response in asthma: differential modulation on inflammation and epithelial injury. Am J Respir Cell Mol Biol 2012; 47: 1–10.

Ahmad T, Mukherjee S, Pattnaik B, et al. Miro1 regulates intercellular mitochondrial transport and enhances mesenchymal stem cell rescue efficacy. EMBO J 2014; 33: 994–1010.

Anderson EJ, Lustig ME, Boyle KE, et al. Mitochondrial H2O2 emission and cellular redox state link excess fat intake to insulin resistance in both rodents and humans. J Clin Invest 2009; 119: 573–581.

Andersson CK, Claesson HE, Rydell-Tormanen K, et al. Mice lacking 12/15-lipoxygenase have attenuated airway allergic inflammation and remodeling. Am J Respir Cell Mol Biol 2008; 39: 648–656.

Arruda AP, Pers BM, Parlakgul G, et al. Chronic enrichment of hepatic endoplasmic reticulum-mitochondria contact leads to mitochondrial dysfunction in obesity. Nat Med 2014; 20: 1427–1435.

Bach D, Naon D, Pich S, et al. Expression of Mfn2, the Charcot-Marie-Tooth neuropathy type 2A gene, in human skeletal muscle: effects of type 2 diabetes, obesity, weight loss, and the regulatory role of tumor necrosis factor alpha and interleukin-6. Diabetes 2005; 54: 2685–2693.

Bach D, Pich S, Soriano FX, et al. Mitofusin-2 determines mitochondrial network architecture and mitochondrial metabolism. A novel regulatory mechanism altered in obesity. J Biol Chem 2003; 278: 17190–17197.

Bates JH and Dixon AE. Potential role of the airway wall in the asthma of obesity. J Appl Physiol (1985) 2015;118: 36–41.

Beuther DA and Sutherland ER. Overweight, obesity, and incident asthma: a meta-analysis of prospective epidemiologic studies. Am J Respir Crit Care Med 2007; 175: 661–666.

Bhatti JS. Bhatti GK and Reddy PH. Mitochondrial dysfunction and oxidative stress in metabolic disorders – A step towards mitochondria based therapeutic strategies. Biochim Biophys Acta Mol Basis Dis 2017; 1863: 1066–1077.

Brons C and Grunnet LG. Mechanisms in endocrinology: skeletal muscle lipotoxicity in insulin resistance and type 2 diabetes: a causal mechanism or an innocent bystander?Eur J Endocrinol 2017; 176: R67-R78.

Calixto MC, Lintomen L, André DM, et al. Metformin attenuates the exacerbation of the allergic eosinophilic inflammation in high fat-diet-induced obesity in mice. PLoS One 2013; 8: e76786.

Chandel NSN, Liu X, Kim CN, et al. Mitochondria as signaling organelles. BMC Biol 2014; 12: 34.

Chen CC, Lee TY, Kwok CF, et al. Cannabinoid receptor type 1 mediates high-fat diet-induced insulin resistance by increasing forkhead box O1 activity in a mouse model of obesity. Int J Mol Med 2016; 37: 743–754.

Chen H, Vermulst M, Wang YE, et al. Mitochondrial fusion is required for mtDNA stability in skeletal muscle and tolerance of mtDNA mutations. Cell 2010; 141: 280–289.

Cheng Z and Almeida FA. Mitochondrial alteration in type 2 diabetes and obesity: an epigenetic link. Cell Cycle 2014; 13: 890–897.

Cho I and Blaser MJJ. The human microbiome: at the interface of health and disease. Nat Rev Genet 2012; 13: 260–270.

Civitarese AE and Ravussin E. Mitochondrial energetics and insulin resistance. Endocrinology 2008; 149: 950–954.

Cole BK, Morris MA, Grzesik WJ, et al. Adipose tissue-specific deletion of 12/15-lipoxygenase protects mice from the consequences of a high-fat diet. Mediators Inflamm 2012; 2012: 851798.

Cottrell L, Neal WA, Ice C, et al. Metabolic abnormalities in children with asthma. Am J Respir Crit Care Med 2011; 183: 441–448.

Crunkhorn S, Dearie F, Mantzoros C, et al. Peroxisome proliferator activator receptor gamma coactivator-1 expression is reduced in obesity: potential pathogenic role of saturated fatty acids and p38 mitogen-activated protein kinase activation. J Biol Chem 2007; 282: 15439–15450.

de Mello AH, Costa AB, Engel JDG, et al. Mitochondrial dysfunction in obesity. Life Sci 2018; 192: 26–32.

Dixon AE, Holguin F, Sood A, et al. An official american thoracic society workshop report: obesity and asthma. Proc Am Thorac Soc 2010; 7: 325–335.

Dixon AE and Poynter ME. Mechanisms of asthma in obesity. Pleiotropic aspects of obesity produce distinct asthma phenotypes. Am J Respir Cell Mol Biol 2016; 54: 601–608.

Dixon AE, Pratley RE, Forgione PM, et al. Effects of obesity and bariatric surgery on airway hyperresponsiveness, asthma control, and inflammation. J Allergy Clin Immunol 2011; 128: 508-515.e501–502.

Dobrian AD, Lieb DC, Cole BK, et al. Functional and pathological roles of the 12- and 15-lipoxygenases. Prog Lipid Res 2011; 50: 115–131.

Finkelstein EA, Khavjou OA, Thompson H, et al. Obesity and severe obesity forecasts through 2030. Am J Prev Med 2012; 42: 563–570.

Forno E, Weiner DJ, Mullen J, et al. Obesity and airway dysanapsis in children with and without asthma. Am J Respir Crit Care Med 2017; 195: 314–323.

Forno E, Young OM, Kumar R, et al. Maternal obesity in pregnancy, gestational weight gain, and risk of childhood asthma. Pediatrics 2014; 134: e535-5546.

Fosbol MO and Zerahn B. Contemporary methods of body composition measurement. Clin Physiol Funct Imaging 2015; 35: 81–97.

Fujimura KE and Lynch SV. Microbiota in allergy and asthma and the emerging relationship with the gut microbiome. Cell Host and Microbe 2015; 17: 592–602.

Gao AW,Canto C and Houtkooper RH. Mitochondrial response to nutrient availability and its role in metabolic disease. EMBO Mol Med 2014; 6: 580–589.

Gazdik F, Gvozdjakova A, Horvathova M, et al. Levels of coenzyme Q10 in asthmatics. Bratisl Lek Listy 2002; 103: 353–356.

Girodet PO, Allard B, Thumerel M, et al. Bronchial smooth muscle remodeling in nonsevere asthma. Am J Respir Crit Care Med 2016; 193: 627–633.

Global Initiative for Asthma. Global Strategy for Asthma Management and Prevention. 2018. Available from: www.ginasthma.org

Gonzalez-Muniesa P, Martinez-Gonzalez MA, Hu FB, et al. Obesity. Nat Rev Dis Primers 2017; 3: 17034.

Gvozdjáková A, Kucharská J, Bartkovjaková M, et al. Coenzyme Q10 supplementation reduces corticosteroids dosage in patients with bronchial asthma. BioFactors(Oxford, England)2005; 25: 235–240.

Haldar P, Pavord ID, Shaw DE, et al. Cluster analysis and clinical asthma phenotypes. Am J Respir Crit Care Med 2008; 178: 218–224.

Harskamp-van Ginkel MW, London SJ, Magnus MC, et al. A study on mediation by offspring BMI in the association between maternal obesity and child respiratory outcomes in the amsterdam born and their development study cohort. PLoS One 2015; 10: e0140641.

Hemachandra Reddy P. Mitochondrial dysfunction and oxidative stress in asthma: implications for mitochondria-targeted antioxidant therapeutics. Pharmaceuticals 2011; 4: 429–456.

Hirschey MD, Shimazu T, Goetzman E, et al. SIRT3 regulates mitochondrial fatty-acid oxidation by reversible enzyme deacetylation. Nature 2010; 464: 121–125.

Holgate ST, Wenzel S, Postma DS, et al. Asthma. Nat Rev Dis Primers 2015; 1: 15025.

Holguin F. Arginine and nitric oxide pathways in obesity-associated asthma. J Allergy (Cairo) 2013; 2013: 714595.

Holguin F, Bleecker ER, Busse WW, et al. Obesity and asthma: an association modified by age of asthma onset. J Allergy Clin Immunol 2011; 127: 1486–1493 e1482.

Holguin F, Comhair SAA, Hazen SL, et al. An association between L-arginine/asymmetric dimethyl arginine balance, obesity, and the age of asthma onset phenotype. Am J Respir Crit Care Med 2013; 187: 153–159.

Hur KY and Lee MS. New mechanisms of metformin action: focusing on mitochondria and the gut. J Diabetes Investig 2015; 6: 600–609.

Islam MN, Das SR, Emin MT, et al. Mitochondrial transfer from bone-marrow-derived stromal cells to pulmonary alveoli protects against acute lung injury. Nat Med 2012; 18: 759–765.

Jheng HF, Tsai PJ, Guo SM, et al. Mitochondrial fission contributes to mitochondrial dysfunction and insulin resistance in skeletal muscle. Mol Cell Biol 2012; 32: 309–319.

Kelley DE, He J, Menshikova EV, et al. Dysfunction of mitochondria in human skeletal muscle in type 2 diabetes. Diabetes 2002; 51: 2944–2950.

Kendrick AA, Choudhury M, Rahman SM, et al. Fatty liver is associated with reduced SIRT3 activity and mitochondrial protein hyperacetylation. Biochem J 2011; 433: 505–514.

Koletzko B, Symonds ME and Olsen SF. Programming research: where are we and where do we go from here?Am J Clin Nutr 2011; 94: 2036S–2043S.

Konradova V, Copova C, Sukova B, et al. Ultrastructure of the bronchial epithelium in three children with asthma. Pediatr Pulmonol 1985; 1: 182–187.

Leishangthem GD, Mabalirajan U, Singh VP, et al. Ultrastructural changes of airway in murine models of allergy and diet-induced metabolic syndrome. ISRN Allergy 2013; 2013: 261297.

Lionetti L, Mollica MP, Donizzetti I, et al. High-lard and high-fish-oil diets differ in their effects on function and dynamic behaviour of rat hepatic mitochondria. PLoS One 2014; 9: e92753.

Lozano R, Naghavi M, Foreman K, et al. Global and regional mortality from 235 causes of death for 20 age groups in 1990 and 2010: a systematic analysis for the Global Burden of Disease Study 2010. Lancet 2012; 380: 2095–2128.

Mabalirajan U, Ahmad T, Leishangthem GD, et al. L-Arginine reduces mitochondrial dysfunction and airway injury in murine allergic airway inflammation. Int Immunopharmacol 2010a; 10: 1514–1519.

Mabalirajan U, Ahmad T, Leishangthem GD, et al. Beneficial effects of high dose of L-arginine on airway hyperresponsiveness and airway inflammation in a murine model of asthma. J Allergy Clin Immunol 2010b; 125: 626–635.

Mabalirajan U, Dinda AK, Kumar S, et al. Mitochondrial structural changes and dysfunction are associated with experimental allergic asthma. J Immunol 2008; 181: 3540–3548.

Mabalirajan U and Ghosh B. Mitochondrial dysfunction in metabolic syndrome and asthma. J Allergy (Cairo) 2013; 2013: 340476.

Mabalirajan U, Rehman R, Ahmad T, et al. 12/15-lipoxygenase expressed in non-epithelial cells causes airway epithelial injury in asthma. Sci Rep 2013a; 3: 1540.

Mabalirajan U, Rehman R, Ahmad T, et al. Linoleic acid metabolite drives severe asthma by causing airway epithelial injury. Sci Rep 2013b; 3: 1349.

Martin RA, Hodgkins SR, Dixon AE, et al. Aligning mouse models of asthma to human endotypes of disease. Respirology 2014; 19: 823–833.

Mendes FAR, Gonçalves RC, Nunes MPT, et al. Effects of aerobic training on psychosocial morbidity and symptoms in patients with asthma: a randomized clinical trial. Chest 2010; 138: 331–337.

Montgomery MK and Turner N. Mitochondrial dysfunction and insulin resistance: an update. Endocr Connect 2015; 4: R1-R15.

Moore WC, Meyers DA, Wenzel SE, et al. Identification of asthma phenotypes using cluster analysis in the severe asthma research program. Am J Respir Crit Care Med 2010; 181: 315–323.

Naviaux RK. Metabolic features of the cell danger response. Mitochondrion 2014; 16: 7–17.

Ng M, Fleming T, Robinson M, et al. Global, regional, and national prevalence of overweight and obesity in children and adults during 1980–2013: a systematic analysis for the Global Burden of Disease Study 2013. Lancet 2014; 384: 766–781.

Nunemaker CS, Chen M, Pei H, et al. 12-Lipoxygenase-knockout mice are resistant to inflammatory effects of obesity induced by western diet. Am J Physiol Endocrinol Metab 2008; 295: E1065-11075.

Nunnari J and Suomalainen A. Mitochondria: in sickness and in health. Cell 2012; 148: 1145–1159.

Park CS, Bang BR, Kwon HS, et al. Metformin reduces airway inflammation and remodeling via activation of AMP-activated protein kinase. Biochem Pharmacol 2012; 84: 1660–1670.

Pattnaik B, Bodas M, Bhatraju NK, et al. IL-4 promotes asymmetric dimethylarginine accumulation, oxo-nitrative stress, and hypoxic response-induced mitochondrial loss in airway epithelial cells. J Allergy Clin Immunol 2016; 138: 130–141.e139.

Pereira PF, Serrano HM, Carvalho GQ, et al. [Measurements of location of body fat distribution: an assessment of colinearity with body mass, adiposity and stature in female adolescents]. Rev Paul Pediatr 2015; 33: 63–71.

Perez MK and Piedimonte G. Metabolic asthma: is there a link between obesity, diabetes, and asthma?Immunol Allergy Clin North Am 2014; 34: 777–784.

Peters U, Dixon AE and Forno E. Obesity and asthma. J Allergy Clin Immunol 2018a; 141: 1169–1179.

Peters U, Suratt BT, Bates JHT, et al. Beyond BMI: obesity and lung disease. Chest 2018b; 153: 702–709.

Piantadosi CA and Suliman HB. Mitochondrial dysfunction in lung pathogenesis. Annu Rev Physiol 2017; 79: 495–515.

Putti R, Sica R, Migliaccio V, et al. Diet impact on mitochondrial bioenergetics and dynamics. Front Physiol 2015; 6: 109.

Raygan F, Rezavandi Z, Dadkhah Tehrani S, et al. The effects of coenzyme Q10 administration on glucose homeostasis parameters, lipid profiles, biomarkers of inflammation and oxidative stress in patients with metabolic syndrome. Eur J Nutr 2016; 55: 2357–2364.

Rieusset J. Mitochondria-associated membranes (MAMs): an emerging platform connecting energy and immune sensing to metabolic flexibility. Biochem Biophys Res Commun 2018; 500: 35–44.

Riiser A. The human microbiome, asthma, and allergy. Allergy Asthma Clin Immunol 2015; 11: 1–7.

Schrepfer E and Scorrano L. Mitofusins, from mitochondria to metabolism. Mol Cell 2016; 61: 683–694.

Seabolt LA, Welch EB and Silver HJ. Imaging methods for analyzing body composition in human obesity and cardiometabolic disease. Ann N Y Acad Sci 2015; 1353: 41–59.

Sebastian D, Hernandez-Alvarez MI, Segales J, et al. Mitofusin 2 (Mfn2) links mitochondrial and endoplasmic reticulum function with insulin signaling and is essential for normal glucose homeostasis. Proc Natl Acad Sci USA 2012; 109: 5523–5528.

Semple RK, Crowley VC, Sewter CP, et al. Expression of the thermogenic nuclear hormone receptor coactivator PGC-1alpha is reduced in the adipose tissue of morbidly obese subjects. Int J Obes Relat Metab Disord 2004; 28: 176–179.

Sideleva O, Black K and Dixon AE. Effects of obesity and weight loss on airway physiology and inflammation in asthma. Pulm Pharmacol Ther 2013; 26: 455–458.

Singh S, Prakash YS, Linneberg A, et al. Insulin and the lung: connecting asthma and metabolic syndrome. J Allergy (Cairo) 2013; 2013: 627384.

Singh VP, Aggarwal R, Singh S, et al. Metabolic syndrome is associated with increased oxo-nitrative stress and asthma-like changes in lungs. PLoS One 2015; 10: e0129850.

Smith RL, Soeters MR, Wust RCI, et al. Metabolic flexibility as an adaptation to energy resources and requirements in health and disease. Endocr Rev 2018; 39: 489–517.

Sutherland ER, Goleva E, King TS, et al. Cluster analysis of obesity and asthma phenotypes. PLoS One 2012; 7: e36631.

Tchernof A and Despres JP. Pathophysiology of human visceral obesity: an update. Physiol Rev 2013; 93: 359–404.

Theurey P and Rieusset J. Mitochondria-associated membranes response to nutrient availability and role in metabolic diseases. Trends Endocrinol Metab 2017; 28: 32–45.

Wenzel SE. Asthma phenotypes: the evolution from clinical to molecular approaches. Nat Med 2012; 18: 716–725.

Whitaker RM, Corum D, Beeson CC, et al. Mitochondrial biogenesis as a pharmacological target: a new approach to acute and chronic diseases. Annu Rev Pharmacol Toxicol 2016; 56: 229–249.

Xu W, Ghosh S, Comhair SAA, et al. Increased mitochondrial arginine metabolism supports bioenergetics in asthma. J Clin Invest 2016a; 126: 2465–2481.

Xu XJ, Babo E, Qin F, et al. Short-term caloric restriction in db/db mice improves myocardial function and increases high molecular weight (HMW) adiponectin. IJC Metab Endocr 2016b; 13: 28–34.

Zhang Y, Jiang L, Hu W, et al. Mitochondrial dysfunction during in vitro hepatocyte steatosis is reversed by omega-3 fatty acid-induced up-regulation of mitofusin 2. Metabolism 2011; 60: 767–775.

15 Mitochondrial Dysfunction and Hearing Loss

María de la Luz Arenas Sordo MD, PhD
Instituto Nacional de Rehabilitación LGII. Secretaría de Salud, Universidad Nacional Autónoma de México, México City, México

CONTENTS

1 INTRODUCTION

The mitochondrion is one of the most important organelles of the cell. It plays an important role in cell function providing energy and also in cell death (apoptosis). Although the mitochondria have characteristics from their ancestors, having the double membrane, it has changed in its form and composition. With time the mitochondrial genome changed, losing and acquiring material, because it has to do many new functions and stop doing others.[1]

We have to remember the inner ear has nervous tissue with high metabolism rate, especially in the vascularis stria; for this reason, it is an organ with high sensitivity to damage because lower energy. However, we know now that there are many diseases for mitochondrial dysfunction with no relationship to the respiratory chain.

Mutations in genes encoding factors such as TFAM, DNA polymerase γ, and others proteins such as the replication helicase twinkle, cause a spectrum of human mitochondrial diseases.[1,2] As we know, the number of mitochondria is related to the metabolic activity of the cells and in the cochlea in the basal turn, the density of them is higher, also in the infra-nuclear area of the hair cells.[3]

In the case of hearing problems, the *mitochondrial disease score* has a section for multisystemic disease (max 3 points), that helps to diagnose mitochondrial disease, but this is only important in cases when many other organs are involved, not only the ear.[4]

It is important to note that homoplasmic and heteroplasmic conditions in the mutations or variants of the mitochondrial genes are related to different diseases and expressions (clinical characteristics).[5]

2   BRIEF EMBRYOLOGY

The ear starts its developmental during the 3 weeks after the fertilization; then during the 23 days period the otic placode is evident. As in any organ, the development of the ear has specific temporal and spatial changes, with many genes involved. At this time, there is a trilaminar embryo (ectoderm, mesoderm and endoderm). In the median ectoderm, the subjacent cells constitute the prechordal plate and notochordal plate. In the prechordal plate, the stomodeum forms. The mesoderm located lateral to the midline (paraxial mesoderm) is segmented in somitomeres. There are seven somitomeres from prosencephalon to the level of the otic placode. When the anterior neural tube closes, the otic placode and neural epithelium are separated for thin ectodermic, then the placode epithelium thickens.[6] It is important to point out, that all the skeleton and connective tissue of the face are from neural crest origin.[7]

2.1   EXTERNAL EAR

This part of the ear derives from the first and second branchial arches. During the second month on both sides of the pharyngeal cleft, three nodular masses of mesenchyme begin their development. Finally, they come together and create a structure that can be identified as an external ear (see Figs. 1 and 2). During their formation, the ears move from the neck base to their final normal location.[7]

2.2   MIDDLE EAR

This part is also associated with first and second branchial arches, the cavity and auditive tube have their origin in an expansion of the pharyngeal pouches; for this reason these areas have epithelium from endodermic origin; otherwise the tympanic membrane has tissues from the 3 germinal layers. Around the sixth week there is a mesenchymal condensation whose origin was the neural crest; this condensation will start the ossicles of the ear. It is until the last stages of gestation that the unnecessary mesenchyme completely disappears, due to the process of reabsorption and programmed cell death and the ossicles stay suspended inside the cavity. The somitomeres develops the muscle cells related to the first and second arches, that include the tensor tympani and stapedius muscles.[6,7]

2.3   INNER EAR

The epithelium of the inner ear is derived from otic placodes. Recent studies show that FGF-19 (fibroblast growth factor 19) causes the expression of Wnt-8c (Wingless/Integrated 8c) in the rhombencephalon neuroepithelium, which stimulates to form FGF-3 (fibroblast growth factor 3). The FGF family members have many activities, especially around mitogenic and cell survival functions that involve processes that include embryonic development, for example

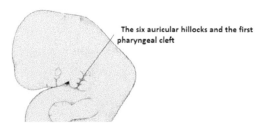

The six auricular hillocks and the first pharyngeal cleft

FIGURE 1   The six auricular hillocks.

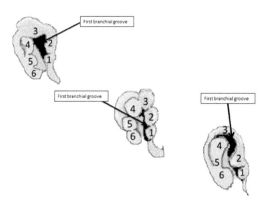

First branchial groove

First branchial groove

First branchial groove

FIGURE 2   Origin of external ear.

the initiation of inner ear when the FGF-19 chick homolog and Wnt-8c interact.[7–9] The FGF-3 gene will act at the end of the 4th week in the embryogenic period, in the transformation of the otic placode first into the otic cup, after that, this one closes off creating the otic vesicle. (see Fig. 3). The Pax-2 homeobox is involved in the early stages of the development of the otic vesicle. If this gene fails, the formation of the cochlea or the spiral ganglion is not possible. For the Semicircular ducts formation, the action of Otx-1 is necessary for the lateral one, and for the rest of them, anterior and posterior, the homeobox Dlx-5.[7,9,10]

3 EAR FUNCTION

The ear has 3 areas that perform different functions with a common final end to send the sound to the central pathways of hearing. In this section is presented about diseases of the inner ear that involve mitochondria in different ways.

3.1 OUTER EAR

The outer ear receives the sound waves and changes them to mechanical system, when they produce vibrations in the tympanum. The form of this part of the ear, acts like a trumpet, has the purpose of capturing and conducting the sound waves into the ear.[11]

3.2 MIDDLE EAR

called resonance chamber, is a cavity in the temporal bone where the ossicles move with the vibration changes from the tympanum. For a good function of this chamber the same pressure is necessary in both sides of the tympanum, which is possible because of the auditive tube. There are also two muscles, the tensor tympani and the stapedius which are responsible to protect the inner ear from the loud sounds modifying the transmission. Their maximal effect is about 20 dB.[11]

3.3 INNER EAR, COCHLEAR FUNCTION

The function of the cochlea is to convert the sound waves into an electrochemical stimulus that can be transmitted to the Central Nervous System (CNS). The receptor of this sensory function is the inner hair cells (IHC) which stereocilia movement produces ion channels opening to allow entry of K+ and Ca++. This result in a transduction current that generates the activation the calcium channels which are in the wall and the base of the IHC. The glutamate neurotransmitter is released and it binds afferent nerve terminals surrounding the HC, then an action potential is propagated for the afferent nerve. The outer hair cells (OHC) provide the amplification of the signal, this happen because of the elongation and contraction of these cells that cause depolarization and hyperpolarization of the cell augmenting the displacement of the basilar membrane.[12,13]

Mitochondria are especially important in the inner ear because of their neurological functions the cells need a large amount of energy that is provided by these organelles. It has been shown that there are different mitochondrial subpopulations in the different areas of ears and that the striated organelle is the one that spends more energy. Inside this striated organelle the mitochondria are of type 1 hair cells, two times larger in surface area and four times larger in volume when they are compared with those in type 2 cells.[14]

The first two areas are related to the conduction of sound, and the third with the sensory function, the creation of the electrical stimuli and transmission to the central nervous system.

4 TYPES OF HEARING LOSS (HL) SECONDARY TO MITOCHONDRIAL DYSFUNCTION

4.1 Non-syndromic HL
4.2 Syndromic HL

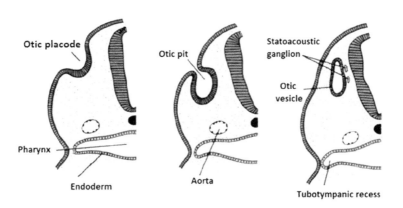

FIGURE 3 Otic placode, otic pit and otic vesicle.

4.1 Non-Syndromic HL

The first time it was identified as non-syndromic HL from mitochondrial origin, was due to the hereditary transmission pattern, only through the mother; it is also important to say that it was noted that patients were affected with different degrees of severity of the HL (moderate to deep) and that the onset could also be from birth to adulthood. Table 1.

4.1.1 Sensorineural Deafness with or without Aminoglycosides Exposure

In the cochlea, the ribosomes and the mitochondria seem to be the cause of HL with the exposure to aminoglycoside, because the natural target of these drugs is the evolutionarily bacterial ribosome. There is a homoplasmic mutation (patogenic variant) A1555G in the mitochondrial 12S ribosomal RNA gene (*MT-RNR1*), that can produce HL with or without exposure to aminoglycoside. In early stage of studies, the possibility of suffering HL was thought to be only through the exposure to the drug; but now we know that this is not true.[15]

This cause of HL is very frequent in Spanish people, but not in Mexican patients. This is due to the lack of cells with mitochondria from Spanish origin, in most of the Mexican people; because the great majority of the Spanish people who came to Mexico were males and the admixture (miscegenation) were with native Mexican females.[16–18]

In the Spanish people it is very probable that the 1555A-G mitochondrial mutation is the cause of a large proportion of families with late-onset sensorineural deafness. It seems there is an age-dependent penetrance for HL and enhanced for aminoglycosides exposure.[16]

There are other mitochondrial mutations, with similar phenotype, the *MT-TS1* 7476insC, and 7445A-G, which changes the stop codon AGA of the heavy-strand-encoded mRNA for subunit COI of cytochrome c oxidase to an equivalent AGG stop codon and changes a U to C transition in the-light strand-encoded tRNA-ser precursor.[17] In these cases, the HL can be the only feature, but it can be accompanied by neurological features.[2]

4.1.2 Noise Induced Hearing Loss (NIHL)

Noise Induced Hearing Loss (NIHL) is a big public health problem. We know now, there are a lot of nuclear genes that predispose us to suffer NIHL, but also the mitochondrial dysfunction without nuclear gene mutations could play a role. In these cases apparently the glutamate toxicity, increased levels of reactive oxygen species, glutathione depletion and oxidative stress could be involved.[2]

4.1.3 Presbycusis

Presbycusis also called hearing loss associated with aging, may lead to mitochondrial dysfunction through both, the accumulation of mitochondrial DNA mutation and variations in control region as Falah et al found (16,223 C>T, 16,311 T>C, 16,249 T>C, and 15,954 A>C) comparing patients and controls, and others factors as noise and vascular and metabolic diseases that produce bioenergetically

---

TABLE 1
Mitochondrial mutations and hearing loss. Non-syndromic

Syndrome	Mutations	Clinical findings
SND wiht or without aminoglycosides exposure	A1555G (12S ribosomal RNA gene (*MT-RNR1*). Homoplasmic	Sensorineural hearing loss with early or late onset, enhanced for aminoglycosides exposure
NIHL	Mitochondrial dysfunction without mutations. ROS and others.	Hearing loss especially of the high frequencies after different grades and time of noise exposure
Presbycusis	Mitochondrial dysfunction: accumulation of mitochondrial DNA mutation and others factors.	Hearing loss in adulthood. Usually between moderate and severe. More affected high frequencies.
Nonsyndromic Sensorineural Deafness. NSHL	*MT-TS1* gene Variants: • T-to-C transition at nucleotide 7510. Heteroplasmic • 7511T-C transition. Homoplasmic • 7455A-C transversion • 7505T-C transition. Homoplasmic	Progressive, postlingual, sensorineural hearing loss involving high frequencies
Late-onset nonsyndromic HL	3243A>G mutation in tRNA Leu (UUR)	Late-onset HL. Progressive High frequencies

SND = Sensorineural deafness; NIHL = Noise induced hearing loss; ROS = reactive oxigen species; NSHL = Nonsyndromic Sensorineural Hearing Loss); HL = hearing loss.

deficiencies in the cell.[19] For all these reasons it is proposed to give patients' treatment with antioxidants and some dietary restrictions. With those recommendations it is supposed the mitochondria function will be better and several age-related hearing loss processes could be reversed.

4.1.4 Nonsyndromic Sensorineural Deafness (NSHL)

The MT-TS1 gene, which causes some syndromic HL, has several allelic variants that also give rise only HL: Heteroplasmic T-to-C transition at nucleotide 7510, homoplasmic 7511T-C transition, 7445A-C transversion, homoplasmic 7505T-C transition.[2] Other involved genes are: MT-RNR1, MT-CO1, MT-TH, MT-ND1 and MT-T1.

4.1.5 Late-Onset Nonsyndromic Hearing Loss

The m.3243A>G mutation in tRNA Leu (UUR) is associated with this type of HL.[20]

4.2 SYNDROMIC HL

Systemic neuromuscular syndromes due to heteroplasmic mitochondrial DNA mutations, such as in Kearns–Sayre syndrome, mitochondrial encephalomyopathy, lactic acidosis and stroke-like episodes (MELAS), mitochondrial encephalomyopathy with ragged red fibers, diabetes and deafness, frequently have hearing loss as one of their signs and symptoms. Table 2.

4.2.1 Kearns-Sayre Syndrome

Synonyms are: Ophthalmoplegia, pigmentary degeneration of retina, and cardiomyopathy; Oculocraniosomatic syndrome; Ophthalmoplegia-plus syndrome; mitochondrial cytopathy; Ophthalmoplegia progressive external with ragged-red fibers; chronic progressive external ophthalmoplegia with myopathy; CPEO with myopathy; CPEO with ragged-red fibers. This syndrome is caused by various mitochondrial deletions and the onset is before age 20 and most of the cases are sporadic. Its clinical findings include: short stature, craniofacial features: microcephaly, sensorineural hearing loss, progressive external ophthalmoplegia, pigmentary retinopathy and ptosis.[21]

Muscular system: includes weakness of facial, pharyngeal, trunk and extremity muscles and ragged-red fibers on muscle biopsy.

Cardiovascular system: includes complete heart block, cardiomyopathy, different cardiac conductions defects.

Neurologic system: includes sensory-motor neuropathy, cerebellar ataxia, calcification of the basal ganglia, dementia, diffuse abnormality of the white matter, seizures.

Genitourinary: includes renal tubular acidosis and Fanconi syndrome. Endocrine: Diabetes mellitus, hypoparathyroidism, Addison disease. Other abnormalities included are Lactic acidosis, sideroblastic anemia,

increased cerebrospinal fluid protein (>100 mg/dl), decreased serum and muscle coenzyme Q.[22]

Aetiology: Point mutations in the mitochondrial tRNA (leucine)-1 gene (MT-TL1) (A3243G, A8344G) and deletion of multiple genes in the mitochondrial DNA.[22,23]

4.2.2 Mitochondrial Encephalomyopathy, Lactic Acidosis and Stroke-Like Episodes (MELAS)

This disease is very heterogenous with variable clinical phenotype. The age of onset is also variable. The patients have a number of clinical manifestations: Sensorineural bilateral and progressive HL, cataracts and other ophthalmological features as ophthalmoplegia, hemianopsia and cortical blindness, cardiovascular abnormalities: left ventricular hypertrophy with dysfunction, heart failure, Wolf-Parkinson-White syndrome and hypertension. The patients have also episodic vomiting, myopathy, ragged-red fibers, migraine, epilepsy, hemiparesis, dementia, encephalopathy, stroke-like episodes. The metabolic and endocrine features are lactic acidosis, diabetes mellitus and elevated serum lactate.[24,25]

Aetiology: The MTTL1 c.3243A-G transition is the most common mutation but it was found the 3243A>T mutation also. Other genes involved are MTTL1, MTTQ, MTTH, MITK, MTTC, MTTS1, MTTS2, MTDN1, MTDN6. In some patients who have MELAS syndrome with phenotype overlapping with other mitochondrial diseases or with different phenotype, heteroplasmic variants in the MTND5 gene have been identified: 12770A-G transition, 13045A-C transversion, 13513G-A transition and 13084A-T transversion.[26]

4.2.3 Myoclonic Epilepsy Associated with Ragged Red Fibers (MERRF)

The main clinical findings are muscle weakness, myoclonus epilepsy, ataxia, spasticity, Sensorineural HL and ragged-red muscle fibers. The patients present elevated serum pyruvate or pyruvate and lactate. Also defect in translation of all mtDNA-encoded genes. The difference between various patients could be explained by the heteroplasmic percentage of the mitochondria or even being homoplasmic. This syndrome can be produced by many mitochondrial genes: MT-Tk (mitochondrially encoded tRNA lysine), MT-TL1 (mitochondrially encoded tRNA leucine 1 (UUA/G)), MT-TH (mitochondrially encoded tRNA histidine), MT-TS1 (mitochondrially encoded tRNA serine 1), MT-TS2 (mitochondrially encoded tRNA serine 2 (AGU/C)), MT-TF (mitochondrially encoded tRNA phenylalanine), MT-NDA5 (Mitochondrially encoded NADH: ubiquinone oxidoreductase core subunit 5).[27,28]

MT-Tk: nucleotides 8295–8364
MT-TL1: nucleotides 3230–3304.
MT-TH: nucleotides 12138–12206
MT-TS1: nucleotides 7445–7516.

TABLE 2
Mitochondrial mutations and hearing loss. Syndromic

Syndrome	Mutations	Clinical findings
Kearns-Sayre	Point mutations tRNA (leucine)-1 gene *MT-TL1* (A3243G, A8344G) and deletions in other genes	SHL, progressive external ophthalmoplegia, pigmentary retinopathy and ptosis. Muscular system: weakness of facial, pharyngeal, trunk and extremity muscles and ragged-red fibers on muscle biopsy. Cardiovascular and neurological features
MELAS	*MTTL1* c.3243A-G transition (most common mutation) and 3243A>T Other genes involved are *MTTL1, MTTQ, MTTH, MITK, MTTC, MTTS1, MTTS2, MTDN1, MTDN6*	Age of onset variable. Encephalomyopathy, lactic acidosis, diabetes, elevated serum lactate and stroke-like episodes. HL sensorineural, bilateral and progressive, cortical blindness, cardiovascular problems, other manifestations of CNS and ragged-red fibers.
MERRF	Many genes mutations are involved: *MT-Tk:* 8295–8364 *MT-TL1* 3230–3304. *MT-TH:* 12138–12206 *MT-TS1:* 7445–7516. *MT-TS2:* 12207–12265 *MT-NDA5:* 12337–14148	Muscle weakness, myoclonus epilepsy, ataxia, spasticity, Sensorineural HL and ragged-red muscle fibers.
Keratoderma palmo-plantar with deafness	*MT-TS1*, 7445A-G	HL: postlingual, progressive and high frequencies and palmoplantar keratoderma
Mitochondrial Cytochrome c Oxidase Deficiency	*MT-TS1* m.7471_7472insC	HL progressive and neurologic features: ataxia, myoclonic epilepsy, dysarthria and mental retardation.
MERRF/MELAS overlap syndrome	*MT-TS1* 7512T-C transition Heteroplasmic	HL, ataxia, myoclonic epilepsy and mental retardation
MIDD (Maternal inherited diabetes and deafness)	*MT-TL1* m.3243A>G	HL of high frequencies. Affectations in muscle, gastrointestinal tract, nephropaty, cardiomyopathy, neuropsychiatric symptoms, macular dystrophy.
Mitochondrial encephalomyopathy	*VDAC1*(nuclear gene) The protein is a transporter for metabolites directly involved in oxidative phosphorylation. *SLC25A4*, PEOA2 and SLC25A20 mitochondrial membrane	Global developmental delay, failure to thrive, hypotonia, seizures, cortical dysplasia, metabolic alterations, some craniofacial dysmorphic features and HL.
Cerebellar ataxia, cataract and diabetes mellitus	*MT-TS2* 12258C-A heteroplasmic	HL, cataracts, cerebellar ataxia and diabetes in adult life.

SHL = Sensorineural Hearing Loss; CNS = Central Nervous System; HL.

MT-TS2: nucleotides 12207–12265
MT-NDA5: nucleotides 12337–14148

4.2.3.1 Keratoderma Palmoplantar with Deafness

The first time this disease was described it was known as Deafness *nonsyndromic sensorineural*. Sevior et al. in 1998 found that many relatives had palmoplantar keratoderma in addition to deafness. This HL was associated to mutations in *MT-TS1*, 7445A-G genes.[27] The HL clinical characteristics are: postlingual, progressive and high frequencies are involved. These findings are very similar to some autosomal dominant HL.[27] The 7445A-G mutation changes the stop codon AGA of the heavy-strand (H-strand)-encoded mRNA for subunit COI of cytochrome c oxidase to an AGG codon, and at the same time, changes a U-to-C

transition in the light-strand (L-strand)-encoded tRNA-ser (UCN) precursor.[28,29]

4.2.3.2 MITOCHONDRIAL CYTOCHROME C OXIDASE DEFICIENCY

In this allelic variant the patients have: Sensorineural HL and neurologic features: ataxia, myoclonic epilepsy, dysarthria and mental retardation. The HL is progressive, and it is a predominant feature. Tiranti el al in 1995 found the cause, the insertion of a cytosine nucleotide position 7472 of the *MT-TS1* gene.[30]

4.2.3.3 MERRF/MELAS OVERLAP SYNDROME

Nakamura et al. (1995) identified a heteroplasmic 7512T-C transition in *MT-TS1* gene in a family with this syndrome, otherwise Jaksch reported the same mutation in 2 patients with mitochondrial encephalopathy and Cytochrome c Oxidase deficiency associated with sensorineural HL, ataxia, myoclonic epilepsy, and mental retardation.[31,32]

4.2.4 Maternal Inherited Diabetes and Deafness (MIDD)

This type of diabetes is not common when we think of diabetes in a global way, but it is the most common endocrinopathy related to mitochondrial diseases. The most common mutation is the m.3243A>G in *MT-TL1* gene, and it accounts for 0.2–3% of all the cases of diabetes. The HL is especially of high tones. The symptoms in general, can occur at any age. The patients can develop affectations in all tissues/organs: muscle, gastrointestinal tract, nephropathy, cardiomyopathy, neuropsychiatric symptoms, and specific macular pattern dystrophy lesions which are asymptomatic are present in most cases.[33]

Aaetiology: in addition to the *MT-TL1* gene, other rare genes are involved such as *MT-TE* and *MT-TK.* Mitochondrial diabetes is more frequent when the m3243A>G mutation is present in the *MT-TL1* gene.

4.2.5 Mitochondrial Encephalomyopathy

In this disease, **HL** is not the main problem, because the patients suffer from neurological signs and symptoms and many others such as especially metabolic. The patients have: failure to thrive, in terms of difficulty to grow and gain weight, prominent forehead, poor visual fixation, depressed nasal bridge, high-arched palate, umbilical hernia, hypoplastic nails, hypotonia, global developmental delay, seizures, dilated lateral ventricles, cortical dysplasia, polymicrogyria, lactic acidosis, bone marrow failure, aminoaciduria, ketonuria, electrolyturia, and abnormal transaminases.[34]

Aetiology: There is a deficiency of VDAC1 (voltage-dependent anion channel-1) in skeletal muscle and fibroblasts. The VDAC1 protein is a transporter for metabolites directly involved in oxidative phosphorylation. In other cases other mitochondrial membrane carriers have been reported, ATN (SLC25A4) causing PEOA2 and defects in SLC25A20 producing metabolic problems.[35,36]

4.2.6 Cerebellar Ataxia, Cataract, and Diabetes Mellitus

Although in the name of this mitochondrial disease **HL** does not appear, it is one of the cardinal clinical features. A case of mother and daughter who developed cataracts, deafness, cerebellar ataxia and diabetes, all of them in adult life but not at the same time, was described, and a 12258C-A heteroplasmic patogenic variant of the *MT-TS2* gene was found. In other cases reported with the same patogenic variant, the characteristics were different, the patients have progressive sensorineural hearing loss and retinitis pigmentosa (*Retinitis pigmentosa-deafness syndrome*), these symptoms were very similar to those in the USHER syndrome type III. Patients begin with hearing problems first, shortly after

in their 20s they note the ophthalmologic symptoms: impairment of night vision and loss of peripheral visual fields. It is important to note that the electroretinography responses are abnormal before the symptoms begin.[36,37]

We must point out that in diseases caused by mitochondrial dysfunction, pathogenic variants (mutations) are expressed in a ubiquitous and specific tissue way and, certainly, patients have very variable clinical manifestations and almost any organ can be involved. It is also true that nuclear-encoded genes play important roles in modifying the expression of mitochondrial genes, since different diseases caused by identical heteroplasmic mtDNA mutations have been found. All these contribute to the difficulty to define mitochondrial diseases with precision. However, we believe that in near future we could have more knowledge to look for the possible pharmacological targets for the treatment of these patients.[38]

5 CONCLUSIONS

Despite HL being very common in the general population, the roles of mitochondrial mutations is only partially understood. There are many reports of families studied in which known or new mutations are found. It is possible that the cause of HL secondary to mitochondrial dysfunction could be more frequent than we initially thought. It is necessary to think about mitochondrial dysfunction in those patients with severe neurological symptoms and HL problems, or in those with non-syndromic HL without nuclear genes mutations. We have to remember that there are also differences between populations; each one has to analyze their own and know the frequent variants to be more assertive in the diagnosis, in genetic counselling, and also in possible treatment.

REFERENCES

1. Friedman JR, Jodi Nunnar J. Mitochondrial form and function. Nature. 2014 January 16; 505(7483): 335–343. doi:10.1038/nature12985.
2. Fischel-Ghodsian N, Kopkeb RD, Ge X. Mitochondrial dysfunction in hearing loss. Mitochondrion. 2004; 4: 675–694.
3. Beurg M, Nam JH, Chen Q, Fettiplace R. Calcium balance and mechanotransduction in rat cochlear hair cells. J Neurophysiol. 2010 July; 104(1): 18–34. doi:10.1152/jn.00019.2010.
4. Morava E, van den Heuvel L, Hol F, de Vries MC, Hogeveen M, Rodenburg RJ, Smeitink JAM. Mitochondrial disease criteria. Diagnostic applications in children. Neurol. 2006; 67: 1823–1826.
5. Moggio M, Colombo I, Peverelli L, Villa L, Xhani R, Testolin S, Di Mauro S, Sciacco M. Mitochondrial disease heterogeneity: a prognostic challenge. Acta Myologica. 2014; XXXIII: 86–93.
6. Sulik K, Cotanche D. Embriology of the ear. In: Toriello, H.; Reardon, W.; Gorlin, R. (eds.): Hereditary hearing loss and its syndromes. 4th Chapter. Second Edition. Oxford University Press 2004. pp. 17–36.
7. Gascón-Rubio MC, Díaz de Cerio Canduela P, Lacosta Nicolás JL. Embriología del oído. En: Libro Virtual de formación en ORL. Capítulo 5. Sociedad Española de

Otorrinolaringología y Patología Cervico facial. España 2015. https://booksmedicos.org/libro-virtual-de-formacion-en-otorrinolaringologia-seorl/.

8. Gene Data Base: www.ncbi.nlm.nih.gov/gene/?term=gene+FGF19

9. García-Peláez I, Arteaga-Martínez SM, Flores-Peña L. Desarrollo del ojo y el oído. In: (ed.): Embriología Humana y Biología del Desarrollo. 25th Chapter. 2th edition. México: Panamericana 2017. pp. 459–478.

10. Gene Data Base: www.ncbi.nlm.nih.gov/gene/?term=gene+FGF3

11. Tortora G, Derrickson B. Principles of anatomy and physiology. 15th edition. John Willey and sons, Ltd. USA. 17th Chapter. pp. 601–616.

12. Brown M.C. Auditory system: efferent systems to auditory periphery in module in neuroscience and biobehavioral psychology. In: Encyclopedia of Neuroscience. USA, 2009. pp. 753–757.

13. Fettiplace R. Hair cell transduction, tuning and synaptic transmission in the mammalian cochlea. Compr Physiol. 2017; 7(4): 1197–1227. doi:10.1002/cphy.c160049.

14. Vranceanu F, Perkins GA, Terada M, Chidavaenzi RL, Ellisman MH, Lysakowski A. Striated organelle, a cytoskeletal structure positioned to modulate hair-cell transduction. PNAS. 2012; 109(12): 4473–4478.

15. Braverman I, Jaber L, Levi H, Adelman C, Arnos KS, Fischel-Ghodsian N, Shohat M, Elidan J. Audiovestibular findings in patients with deafness caused by a mitochondrial susceptibility mutation and precipitated by an inherited nuclear mutation or aminoglycosides. Arch Otolaryngol Head Neck Surg. 1996; 122: 1001–1004.

16. Estivill X, Govea N, Barcelo A, Perello E, Badenas C, Romero E, Moral L, Scozzari R, D'Urbano L, Zeviani M, Torroni A. Familial progressive sensorineural deafness is mainly due to the mtDNA A1555G mutation and is enhanced by treatment with aminoglycosides. Am J Hum Genet. 1998; 62: 27–35.

17. Bykhovskaya Y, Estivill X, Taylor K, Hang T, Hamon M, Casano RAMS, Yang H, Rotter JI, Shohat M, Fischel-Ghodsian N. Candidate locus for a nuclear modifier gene for maternally inherited deafness. Am J Hum Genet. 2000; 66: 1905–1910.

18. de la Luz Arenas-Sordo M, Menendez I, Hernandez E, Simarci A, Gutierrez D, McGetrick M, Murphy P, Leyva X, Huesca F, Dominguez-Aburtio J, Tekin M. Unique spectrum of GJB2 mutations in Mexico. Int J Pediatr Otorhinolaryngol. 2012; 76(11): 1678–1680.

19. Seidman MD, Khan MJ, Bai U, Shirwany N, Quirk WS. Biologic activity of mitochondrial metabolites on aging and age-related hearing loss. Am J Otol. 2000; 21: 161–167.

20. Deschauer M, Muller T, Wieser T, Schulte-Mattler W, Kornhuber M, Zierz S. Hearing impairment is common in various phenotypes of the mitochondrial DNA A3243G mutation. Arch Neurol. 2001; 58: 1885–1888.

21. Yu M, Yu L, Wang, ZX. Diagnosis and management of Kearns-Sayre syndrome rely on comprehensive clinical evaluation. Chin Med J (Engl). 2016; 129(20): 2519–2520. doi:10.4103/0366-6999.191855.

22. https://omim.org/entry/530000

23. https://omim.org/entry/540000

24. Ikeda T, Osaka H, Shimbo H, Tajika M, Yamazaki M, Ueda A, Murayama K, Yamagata T. Mitochondrial DNA 3243A>T mutation in a patient with MELAS syndrome. Hum Genome Var. 2018 September 4; 5: 25. doi:10.1038/s41439-018-0026-6.

25. Anderson S, Bankier AT, Barrell BG, de Bruijn MHL, Coulson AR, Drouin J, Eperon IC, Nierlich DP, Roe BA, Sanger F, Schreier PH, Smith AJH, Staden R, Young IG. Sequence and organization of the human mitochondrial genome. Nature. 1981; 290: 457–465.

26. Wallace DC, Lott MT, Torroni A, Brown MD, Shoffner JM. Report of the committee on human mitochondrial DNA. In: Cuticchia A. J.; Pearson P. L. (eds.): Human gene mapping, 1993: a compendium. Baltimore: Johns Hopkins Univ. Press 1994. pp. 813–845.

27. Sevior KB, Hatamochi A, Stewart IA, Bykhovskaya Y, Allen-Powell DR, Fischel-Ghodsian N, Maw MA. Mitochondrial A7445G mutation in two pedigrees with palmoplantar keratoderma and deafness. Am J Med Genet. 1998; 75: 179–185.

28. Hutchin TP, Lench NJ, Arbuzova S, Markham AF, Mueller RF. Maternally inherited hearing impairment in a family with the mitochondrial DNA A7445G mutation. Europ J Hum Genet. 2001; 9: 56–58.

29. Martin L, Toutain A, Guillen C, Haftek M, Machet MC, Toledano C, Arbeille B, Lorette G, Rötig A, Vaillant L. Inherited palmoplantar keratoderma and sensorineural deafness associated with A7445G point mutation in the mitochondrial genome. Br J Dermatol. 2000; 143(4): 876–883.

30. Tiranti V, Chariot, P, Carella F, Toscano A, Soliveri P, Girlanda P, Carrara F, Fratta GM, Reid FM, Mariotti C, Zeviani M. Maternally inherited hearing loss, ataxia and myoclonus associated with a novel point mutation in mitochondrial tRNA-ser(UCN) gene. Hum Molec Genet. 1995; 4: 1421–1427.

31. Nakamura M, Nakano S, Gato Y-i, Ozawa M, Nagahama Y, Fukuyama H, Akiguch I, Kaji R, Kimura J. A novel point mutation in the mitochondrial tRNA (ser(UCN)) gene detected in a family with MERRF/MELAS overlap syndrome. Biochem Biophys Res Commun. 1995; 214: 86–93.

32. Jaksch M, Hofmann S, Kleinle S, Liechti-Gallati S, Pongratz DE, Muller-Hocker J, Jedele KB, Meitinger TD. A systematic mutation screen of 10 nuclear and 25 mitochondrial candidate genes in 21 patients with cytochrome c oxidase (COX) deficiency shows tRNA-ser(UCN) mutations in a subgroup with syndromal encephalopathy. J Med Genet. 1998; 35: 895–900.

33. Yee ML, Wong R, Datta M, Fazlo TN, Ebrahim MM, Mcnamara EC, De Jong G, Gilfillan C. Mitochondrial disease: an uncommon but important cause of diabetes mellitus. Endocrinol Diabetes Metab Case Rep. 2018 September 25. doi:10.1530/EDM-18-0091.

34. De Pinto V, Messina A, Schmid A, Simonetti S, Carnevale F, Benz R. Characterization of channel-forming activity in muscle biopsy from a porin-deficient human patient. J Bioenerg Biomembr. 2000; 32: 585–593.

35. Huizing M, Ruitenbeek W, Thinnes FP, De Pinto V, Wendel U, Trijbels FJM, Smit LME, Ter Laak HJ, Van Den Heuvel LP. Deficiency of the voltage-dependent anion channel: a novel cause of mitochondriopathy. Pediat Res. 1996; 39: 760–765.

36. https://omim.org/entry/590085.

37. Mansergh FC, Millington-Ward S, Kennan A, Kiang AS, Humphries M, Farrar GJ, Humphries P, Kenna PF. Retinitis pigmentosa and progressive sensorineural hearing loss caused by a C12258A mutation in the mitochondrial MTTS2 gene. Am J Hum Genet. 1999; 64: 971–985.

38. Naviaux RK. Developing a systematic approach to the diagnosis and classification of mitochondrial disease. Mitochondrion. 2004; 4: 351–361.

16 Superoxide Dismutase, Mitochondrial Dysfunction, and Neurodegenerative Diseases

Jahaun Azadmanesh
Department of Biochemistry and Molecular Biology, 985870 Nebraska Medical Center, Omaha, USA

Gloria E. O. Borgstahl
Department of Biochemistry and Molecular Biology, 985870 Nebraska Medical Center, Omaha, USA
Eppley Institute for Cancer and Allied Diseases, 986805 Nebraska Medical Center, Omaha, USA

CONTENTS

1 INTRODUCTION

Excessive amounts of ROS and RNS correlate with the presence of several pathologies including cancers, cardiovascular disorders, and neurodegenerative diseases [1–4]. Mitochondrion produce up to 90% of cellular ROS/RNS, and if their levels are not regulated, irreversible damage to biological macromolecules, occurs through oxidative damage [5–8].

Imbalance in ROS/RNS levels is associated with mitochondrial dysfunction, which is considered to be the main cause of neurodegenerative diseases, such as Alzheimer's disease (AD), amyotrophic lateral sclerosis (ALS), Parkinson's disease (PD) and Huntington's disease (HD) [9,10].

The mitochondrial electron transfer complex (ETC), which is essential in supplying cells with energy, is responsible for the generation of the majority of endogenous ROS/RNS [8]. Sequentially, $O_2^{\bullet-}$ is the first reactive species produced by the ETC. Electrons being shuttled among the complexes of the ETC have the potential to leak off onto O_2 to form $O_2^{\bullet-}$. Whereas $O_2^{\bullet-}$ is poorly reactive toward biological macromolecules, it is a precursor for the formation of highly reactive oxygen and nitrogen species that damage the DNA, proteins, and lipids of cellular systems including mitochondria [5,6]. If left unmanaged, excessive amounts of $O_2^{\bullet-}$ lead to mitochondrial degeneration and/or apoptosis that are the basis of several diseases [11].

SODs are antioxidant metalloproteins that are 'the first line of defense' against excessive ROS/RNS by lowering the levels of $O_2^{\bullet-}$. These metalloprotein enzymes convert $O_2^{\bullet-}$ into either O_2 or H_2O_2 depending on the redox state of the active-site metal. For all SODs excluding the CuZn isoform, a trivalent metal oxidizes $O_2^{\bullet-}$ to form O_2 while a divalent metal reduces $O_2^{\bullet-}$ with the coupling of two proton transfers to form H_2O_2 (**Scheme 1**) [12]. The Cu of CuZnSOD instead shuffles between divalent and monovalent states while the divalent Zn plays only a structural role (**Scheme 2**). The catalytic activity of SODs can thereby consist of cyclic redox reactions of the active site metal to convert $O_2^{\bullet-}$ into its products.

$$\text{Scheme 1}: M^{3+} + O_2^{\bullet-} \leftrightarrow M^{2+} + O_2$$
$$M^{2+} + O_2^{\bullet-} + 2H^+ \leftrightarrow M^{3+} + H_2O_2$$
$$M = Mn, Fe, \text{ or } Ni$$

$$\text{Scheme 2}: Cu^{2+} + O_2^{\bullet-} \leftrightarrow Cu^+ + O_2$$
$$Cu^+ + O_2^{\bullet-} + 2H^+ \leftrightarrow Cu^{2+} + H_2O_2$$

The different isoforms of SODs can be defined by their core metal(s) coordinated at the active site: Ni, Fe, Cu/Zn or Mn. NiSOD is exclusive to prokaryotes while the other three are found in both prokaryotes and eukaryotes. In eukaryotes, FeSOD is found within chloroplasts, and CuZnSOD resides within the cytoplasm, MIMS, and extracellular environment. MnSOD dwells within the mitochondrial matrix. The CuZn and Mn SODs are the only isoforms found in humans. The capacity for SODs to decrease $O_2^{\bullet-}$ levels within cells is associated with longevity and the presence or absence of disease states. Each human isoform seems to play a preventive role in specific pathologies that reflects their location [1].

The two human isoforms of CuZnSOD, intracellular-cytoplasmic/MIMS and extracellular forms (hereafter referred to as IC-CuZnSOD and EC-CuZnSOD, respectively), are 60% homologous with similar active site architecture and are likely to follow the same catalytic mechanism [13,14]. EC-CuZnSOD residing within the extracellular environment provides cells a defense against $O_2^{\bullet-}$ generated from exogenous or environmental stress. EC-CuZnSOD does not seem to play a direct role in

shielding mitochondria against ROS/RNS [15]. It is nonetheless vital in preventing oxidative effects outside cells, such as fibrosis in chronic obstructive pulmonary disease [13]. In regards to preserving mitochondrial function, the residence of IC-CuZnSOD within the MIMS is of major importance in conjunction with the mitochondrial MnSOD.

For humans, IC-CuZnSOD compromises 90% of total SOD within a cell [16]. While MnSOD constitutes only a small fraction of total SOD, its specific concentration within mitochondria is significant due to the high levels of ROS/RNS production by the organelle [8]. IC-CuZnSOD and MnSOD work in tandem to avert mitochondrial degeneration, as observed in knockout mice [15]. Mice lacking IC-CuZnSOD have ~70% of the lifespan compared to their wildtype counterparts [17]. These knockout mice appear normal during weaning but develop adult-onset neuropathies as a consequence of mitochondrial dysfunction, which includes motor neuron degeneration and axonal damage [18]. Knockout mice of MnSOD die within the first ten days of life due to impairment of mitochondrial function in neurons and cardiac myocytes [19,20]. Together, IC-CuZnSOD and MnSOD are vital in preserving the mitochondria and thereby cellular function.

2 SUPEROXIDE PRODUCTION BY THE MITOCHONDRIA AND OTHER SOURCES

The primary function of the mitochondria is to generate adenosine triphosphate (ATP), the 'energy currency' that powers processes within the cell. The ETC forms the proton gradient that drives the production of ATP within the mitochondria and is the primary source of intracellular $O_2^{\bullet-}$. The $O_2^{\bullet-}$ generated during the shuttling of electrons across the complexes of the ETC has the potential to lead to oxidative stress if ample amounts of antioxidant systems are not present [8]. The trade-off for meeting the energetic demands of cellular processes is the production of $O_2^{\bullet-}$ and downstream ROS and RNS.

Among the mitochondrial ETC members involved in shuttling electrons, complexes I and III are the dominant sites of electron leakage for $O_2^{\bullet-}$ formation (Fig. 1). Complex I leaks toward the mitochondrial matrix whereas complex III leaks toward the intermembrane space [21]. Due to the permeability of the outer mitochondrial membrane, ROS/RNS may freely diffuse to the cytoplasm after passing through the intermembrane space. Conversely, $O_2^{\bullet-}$ that is generated because of complex I electron leakage is confined within the matrix. This compartmentalization of $O_2^{\bullet-}$ generation requires similar assortment of SODs to counteract the deleterious effects of the ROS/RNS.

IC-CuZnSOD cannot compensate for lack of MnSOD and vice versa which reflects their non-redundant and strategic locations [15]. The presence of IC-CuZnSOD in

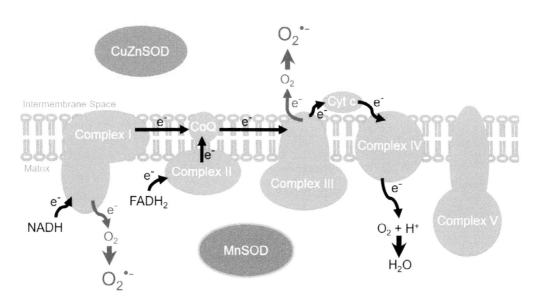

FIGURE 1 $O_2^{\cdot-}$ production by the ETC. Complexes I-V are depicted on the inner mitochondrial membrane along with coenzyme Q (CoQ) and cytochrome c (Cyt c). Black arrows show the electron path normally used for ATP generation. Red arrows display sites of electron leakage from the ETC leading to $O_2^{\cdot-}$ formation. Intracellular CuZnSOD dismutes $O_2^{\cdot-}$ in the intermembrane space and cytoplasm while MnSOD protects against $O_2^{\cdot-}$ produced in the matrix.

the intermembrane space and cytoplasm safeguards against $O_2^{\cdot-}$ produced from complex III while the residence of MnSOD in the matrix dismutes $O_2^{\cdot-}$ formed by complex I. However, $O_2^{\cdot-}$ within the matrix is more detrimental because of its confinement by the inner mitochondrial membrane, as ROS/RNS produced in the intermembrane space can diffuse to the cytoplasm. This explains why mice lacking MnSOD have a significantly short lifespan (< 10 days) compared to mice without IC-CuZnSOD (120 weeks) [15]. Of note, knockout of EC-CuZnSOD did not affect lifespan in these studies. Nevertheless, the mitochondria require SODs on both sides of the inner membrane to protect against $O_2^{\cdot-}$ made from the ETC.

Other sources of endogenous $O_2^{\cdot-}$ are xanthine oxidases (XOs), cytosolic enzymes used for purine metabolism that are widely distributed in tissues, and NADPH oxidases (NOXs), extracellular facing membrane-bound protein complexes found in the plasma membrane of phagosomes and brain cells [22]. $O_2^{\cdot-}$ is a byproduct of purine metabolism by XOs whereas phagocytic NOXs produces $O_2^{\cdot-}$ extracellularly to kill bacteria and fungi within the body. NOX residing on the membrane of intracellular compartments is capable of generating $O_2^{\cdot-}$ within cells. Though, the physiological significance of intracellular NOX in brain cells such as astrocytes, microglia, and neurons is not well understood [23–25]. During neuroinflammation, XOs and NOXs increase $O_2^{\cdot-}$ levels up to 1000-fold [26], placing a heavy burden of ROS/RNS throughout the cell and extracellular space. Therefore, XOs and NOXs must be considered in

pathogenesis as they are substantial contributors to excesses in $O_2^{\cdot-}$ and subsequent oxidative stress.

3 EFFECTS OF EXCESSIVE SUPEROXIDE AND ITS DERIVATIVES ON MITOCHONDRIA

Declines in the mitochondrial population eventually can lead to cell death [27]. Excessive ROS and RNS levels are the cause of a significant number of diseases that manifest due to mitochondrial-related cell death [28]. The presence of these damaging amounts of ROS require copious amounts of $O_2^{\cdot-}$. While $O_2^{\cdot-}$ itself reacts poorly with biological macromolecules, it reacts with other endogenous small molecules, such as nitric oxide ($^{\cdot}$NO), to create products that are capable of harming mitochondria [5,6]. Within this section, the mechanism in which $O_2^{\cdot-}$ and its derivatives harm mitochondria is briefly discussed.

The basis for the bio-protective significance of SODs is the removal of the $O_2^{\cdot-}$ precursor that contributes to the formation of detrimental ROS and RNS. $O_2^{\cdot-}$ is capable of reacting with several molecules to produce injurious reactive species, such as $ONOO^-$, $^{\cdot}OH$, and nitrogen dioxide ($^{\cdot}NO_2$) (Fig. 2). All these molecules are highly reactive and capable of damaging the cellular macromolecules such as DNA, proteins, and lipids [7,8].

3.1 PEROXYNITRITE

$ONOO^-$ is a product of a reaction between $O_2^{\cdot-}$ and $^{\cdot}NO$. The source of endogenous $^{\cdot}NO$ are nitric oxide synthases

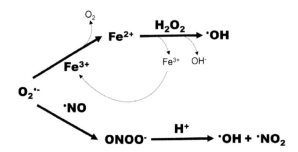

FIGURE 2 ROS and RNS are formed from $O_2^{\cdot-}$. $O_2^{\cdot-}$ may act as a reducing agent for free or enzyme-bound Fe^{3+} atoms to produce Fe^{2+}. Subsequent Fenton reaction of Fe^{2+} with H_2O_2 yields highly reactive $\cdot OH$. Fe^{3+} is reformed as a by-product and can be reused. Alternatively, $O_2^{\cdot-}$ reacts with $\cdot NO$ at diffusion-limited rates to form $ONOO^-$. $ONOO^-$ directly damages macromolecules of the cell but can also decompose to $\cdot OH$ and $\cdot NO_2$ upon its protonation.

(NOSs), which carry out physiological signaling in processes such as vasodilation and inflammation [26]. There are three isoforms of NOS: NOS1, NOS2, and NOS3, each being cell/tissue type specific [29]. Depending on the cell type, more than one NOS isoform may be expressed with subcellular-specific localization, such as the Golgi apparatus or mitochondria [30]. $\cdot NO$ is a potent signaling molecule due to its ability to diffuse across membranes freely. Produced in the mitochondria it encounters the $O_2^{\cdot-}$ confined in the mitochondria may encounter •NO produced within or out the organelle, formation of harmful $ONOO^-$ within the mitochondria [31]. The likelihood of $ONOO^-$ formation both within and outside the mitochondria is high given that the reaction between $O_2^{\cdot-}$ and $\cdot NO$ is diffusion-limited, meaning that every time a pair of the two molecules collide, $ONOO^-$ is formed [26]. Another contributing factor to the likelihood of $ONOO^-$ production is the concentration of $\cdot NO$. At basal physiological conditions, the concentrations of $\cdot NO$ are similar to the concentrations of endogenous SODs [26], meaning there is a delicate balance among the concentrations of $O_2^{\cdot-}$, SODs, and $\cdot NO$ that minimize $ONOO^-$ formation in physiologically normal (i.e., under low oxidative stress) cells. A disruption in this balance increases oxidative stress, such as neuroinflammation that increases the levels of $O_2^{\cdot-}$ and $\cdot NO$ production, raising the amount of $ONOO^-$ proportionally.

There are several ways that $ONOO^-$ contributes to mitochondrial dysfunction. It can react with Fe-S clusters found in several crucial mitochondrial proteins, removing a Fe^{2+} atom, and inactivating them. Free Fe^{2+} can then lead to $\cdot OH$ formation via the Fenton reaction (see Section 3.2) [32]. One example of vital mitochondrial protein inactivation is aconitase, which isomerizes citrate to isocitrate in the citric acid cycle. Aconitase Fe-S cluster inactivation contributes to mitochondrial dysfunction, cell death, and neurodegeneration as a result of the release of free Fe^{2+} [33]. Similarly, $ONOO^-$ can react with Fe-S clusters found in complexes of the ETC, irreversibly inactivating them and disrupting ATP synthesis [32]. However, it

is debated whether the Fe-S clusters of the ETC are somehow resistant to oxidative damage [34]. $ONOO^-$ contributes to a cascade of reactions that promote $\cdot OH$ formation and ATP synthesis inactivation.

$ONOO^-$ can also inactivate proteins through reaction with amino acids. For example, $ONOO^-$ can oxidize critical cysteine active site residues of enzymes and inactivate catalysis. Inactivation of tyrosine phosphatases by $ONOO^-$ oxidation is especially noteworthy [26], as this foregoes loss of anti-apoptotic signaling cascades in brain cells, leading to programmed cell death [35]. $ONOO^-$ can also oxidize methionine to form methionine sulfoxide, which can modulate enzyme activities. Excessive methionine sulfoxide formation correlates with development of Alzheimer's disease in brain tissues [26]. Both IC- and EC-CuZnSODs can be inactivated by $ONOO^-$ by reacting with the active site metal and active site histidine to form a histidinyl radical, decreasing functional amounts of antioxidant systems [26]. $ONOO^-$ oxidizes and inactivates complexes of the ETC and antioxidant proteins that contribute to mitochondrial dysfunction.

A large part of mitochondrial toxicity is lipid peroxidation initiated by the decomposition products of $ONOO^-$, $\cdot OH$ and $\cdot NO_2$ (Figure 2). These decomposition products are discussed in Sections 3.2 and 3.3, respectively. Peroxidation of phospholipids is a prominent marker of mitochondrial dysfunction and neurodegeneration [36,37]. The direct effects of lipid peroxidation by $\cdot OH$ and $\cdot NO_2$ are also explored in Sections 3.2 and 3.3 in this chapter, respectively.

$ONOO^-$ also damages mitochondrial DNA and culminates in cell death by dysregulation of mitochondrial processes [11,26]. In particular, guanine is the most susceptible to oxidation by $ONOO^-$, which results in its fragmentation. This process induces mutagenesis and double-strand breaks of mitochondrial DNA. $ONOO^-$ can also attack the sugar backbone of DNA to cause single-strand breaks. These damaging processes of $ONOO^-$ to mitochondrial DNA (and nuclear DNA) are strongly related to cell death and inflammation, which ultimately generate additional ROS/RNS [26].

3.2 HYDROXYL RADICAL

$^\bullet$OH is highly reactive and can damage neighboring molecules at near diffusion-limited rates [38–40]. Considered as the most biologically active and damaging free radical, $^\bullet$OH is a strong contributor to mitochondrial dysfunction that can lead to neurodegenerative pathologies [41,42]. Generation of $^\bullet$OH can occur through two reactions within mitochondria (Figure 2). First, a redox reaction of H_2O_2 with free Fe^{2+} leads to the formation of $^\bullet$OH with Fe^{3+} and hydroxide ion (OH^-) as byproducts [41]. This reaction is also known as the Fenton reaction. In the presence of $O_2^{\bullet-}$, pools of the reactants for this pathway of $^\bullet$OH generation (H_2O_2 and Fe^{3+}) are increased. Concentration of H_2O_2 is increased during an increase of $O_2^{\bullet-}$ concentration due to catalysis by SODs (Scheme 1). Free Fe^{2+} becomes available as a result of $O_2^{\bullet-}$ donating an electron to Fe^{3+} found in Fe-S cluster proteins of the mitochondria to generate Fe^{2+} and oxygen (Figure 2; top pathway). Fe in the divalent form favors disassociation from these clusters. Of note, these Fe-S clusters are abundant among proteins. Free Fe^{3+}, a product of Fenton chemistry, can also be used to produce Fe^{2+}. Second, at physiological conditions, $ONOO^-$ is protonated and degrades to $^\bullet$OH and $^\bullet NO_2$ (Figure 2; bottom pathway) [43]. At 37 °C, the pK_a of $ONOO^-$ is 7.5. The mitochondrial matrix has a pH of 7 while the cytosol has a pH of 7.4, meaning that protonation of $ONOO^-$ is likely, with protonation more common in the mitochondrial matrix due to its lower pH value [44].

$^\bullet$OH plays prominent roles during oxidative stress, specifically in the presence of $O_2^{\bullet-}$. Furthermore, neurodegenerative diseases correlate with the prevalence of $^\bullet$OH formation [45]. $O_2^{\bullet-}$ is a precursor of $^\bullet$OH, meaning the concentrations of $O_2^{\bullet-}$ and the activity of SODs influence the amount of $^\bullet$OH formation within the mitochondria and cytosol. This highly biologically reactive free radical interacts with mitochondrial proteins, lipids, and DNA, leading to the deterioration of the mitochondria.

Macromolecular proteins are highly vulnerable to denaturation by $^\bullet$OH because tyrosine and tryptophan amino acid residues are susceptible to oxidation. Upon exposure to $^\bullet$OH, tyrosine residues bind other tyrosines covalently, and insoluble protein aggregates form due to this covalent modification [46]. Tryptophan is also rapidly degraded by $^\bullet$OH [46]. Given the abundance of these two residues in proteins, the presence of $^\bullet$OH leads to aggregation, direct fragmentation, and intracellular proteolysis of critical proteins central for mitochondrial function.

$^\bullet$OH initiates the peroxidation of phospholipids and is a prominent marker of mitochondrial dysfunction and neurodegeneration [36,37]. Phospholipid peroxidation increases membrane permeability, which is catastrophic to the mitochondrial inner membrane because a large part of its biochemical function relies on a gradient of molecules between the mitochondrial matrix and MIMS [47]. A notable example of a process affected by increased membrane permeability is the generation of ATP by oxidative phosphorylation. The proton gradient that drives phosphorylation of ADP to ATP is disrupted by membrane permeability, leading to attenuated ATP synthesis. Various transporters and respiratory enzymes needed to maintain mitochondrial function and ATP synthesis are also affected by MIMS permeability. Peroxidation can also exacerbate neurological dysfunction during neuroinflammation. Lipids of myelin sheaths can undergo peroxidation by $^\bullet$OH leading to demyelination, causing impaired conductance of nerve signals [48]. Hence, peroxidation of lipids by $^\bullet$OH poses deleterious consequences to both mitochondrial function and neurological function.

One of the principal products of DNA oxidation is 8-oxo-2′-deoxyguanosine (8-oxo-dG), which is generated by the addition of $^\bullet$OH [49,50]. Mitochondrial DNA is especially susceptible to oxidation compared to nuclear DNA given that $O_2^{\bullet-}$ generation is more prominent within the mitochondrial matrix. The DNA adduct, 8-oxo-dG can lead to mutagenesis [51]. Mutations of the mitochondrial DNA may lead to aberrant protein function, including those of the ETC, leading to further electron leakage, subsequent $O_2^{\bullet-}$ generation, and exacerbating oxidative insults to the mitochondria [49,50]. Also, $^\bullet$OH can attack the sugar moieties of mitochondrial DNA that result in DNA strand breaks. Thus, $^\bullet$OH contributes to mitochondrial dysfunction by damaging the protein-encoding genes of mitochondrial DNA.

3.3 NITROGEN DIOXIDE

Similarly to other reactive species, $^\bullet NO_2$ also contributes to mitochondrial dysfunction by damaging molecules that maintain the vitality of this organelle. For example, $^\bullet NO_2$ reacts with tyrosine to form 3-nitrotyrosine, adding a bulky element and lowering the pK_a of the hydroxyl group 2–3 units [52]. When the nitro group causes steric restrictions for catalysis, or if a protonated hydroxyl group is required for hydrogen bonding critical to activity, then 3-nitrotyrosine can inactivate the enzyme. MnSOD is especially susceptible to inactivation by nitration of Tyr34, located at its active site [53]. The nitro group sterically impedes the substrate access funnel that $O_2^{\bullet-}$ can enter and lead to nearly complete inhibition of MnSOD activity. This inactivation of MnSOD promotes further oxidative stress by compromising the central mitochondrial anti-oxidant system. Other critical mitochondrial proteins that are vulnerable to inactivation by nitration include cytochrome c, voltage-dependent anion channel, and enzymes of the citric acid cycle [54,55]. Excessive nitration ultimately leads to the formation of the permeability transition pore that leads to apoptosis [52].

$^\bullet$NO$_2$ initiates lipid peroxidation [54] and causes mitochondria DNA damage [56]. As a radical, $^\bullet$NO$_2$ can abstract hydrogen atoms from either lipids or the sugar backbone of DNA to form lipid or sugar radicals, respectively. Lipid radicals either propagate formation of other lipid radicals by further abstracting hydrogen atoms from neighboring lipids or end up as peroxyl radical groups from the addition of O$_2$ at the radical site. Membrane-bound proteins, such as those of the ETC, require particular lipid environments for their actions and are hampered by adjacent products of lipid peroxidation [57]. Sugar radicals lead to strand breaks of the mitochondrial DNA. Studies suggest that mitochondrial DNA damage plays a causative role to neurodegeneration [58,59]. Lipid peroxidation and mitochondria DNA damage are hallmarks of mitochondrial dysfunction and neurodegenerative diseases.

3.4   Superoxide and Hydrogen Peroxide

While the majority of O$_2^{\bullet-}$ mediated damage is a consequence of it acting as a precursor to form reactive species, there exists a direct mechanism of O$_2^{\bullet-}$ damage to the mitochondria. Like ONOO$^-$, O$_2^{\bullet-}$ liberates Fe atoms from Fe-S clusters that normally act as needed cofactors for the enzymatic function of several mitochondrial proteins [60]. In particular, several enzymes of the citric acid cycle and amino acid biosynthesis pathway contain these clusters and are vulnerable to O$_2^{\bullet-}$ mediated inactivation [60,61]. Consequently, excessive O$_2^{\bullet-}$ levels in the mitochondria cause an elevation of free-iron levels. High free-iron levels coupled with large amounts of H$_2$O$_2$ (such as from increased SOD activity due to high amounts of O$_2^{\bullet-}$) instigates Fenton reaction to produce copious sums of $^\bullet$OH (Figure 2) [41]. The damage to Fe-S by O$_2^{\bullet-}$ and increased Fe levels underlie mitochondrial dysfunction and neurodegenerative disease progression [62,63].

High levels of H$_2$O$_2$ are cytotoxic as it is a ROS and a precursor to the highly reactive $^\bullet$OH [64]. As SOD activity forms H$_2$O$_2$, it seems cells favor H$_2$O$_2$ production over the presence of O$_2^{\bullet-}$. Several molecular attributions explain this phenomenon; first, for every two O$_2^{\bullet-}$ molecules removed by SODs one molecule of H$_2$O$_2$ is produced, which effectively reduces the concentration of ROS within a cell. Second, H$_2$O$_2$ is capable of crossing membranes while O$_2^{\bullet-}$ is not [65]. For mitochondria, this is important because the inner mitochondrial membrane traps O$_2^{\bullet-}$ within the organelle while H$_2$O$_2$ can diffuse out. Third, H$_2$O$_2$ is poorly reactive against biological molecules [64]. Its harmful effects come from reacting with Fe^{2+} to produce $^\bullet$OH, meaning the cytotoxicity of H$_2$O$_2$ is dependent on free iron concentrations [66]. Finally, multiple enzyme systems within cells eliminate H$_2$O$_2$, including catalases, peroxidases, and thioredoxin-interacting proteins [67]. Cells tolerate the production of H$_2$O$_2$ by SODs as a trade-off for eliminating O$_2^{\bullet-}$.

4   MNSOD AND IC-CUZNSOD CATALYSIS

A crucial component of the antioxidant power of SODs is their enzyme kinetics. SODs have one of the fastest and most efficient rates of all enzymes. MnSOD is "diffusion-limited" (k_{cat}/K_m near 10^9 M^{-1} s^{-1}), meaning the enzyme is so efficient that the rate-limiting steps for enzymatic activity are the diffusion of substrate and products into and out of the active site, respectively [14,68]. Catalysis by IC-CuZnSOD surpasses rates that are diffusion limited (k_{cat}/K_m of 2 × 10^9 M^{-1} s^{-1}) by being inherently catalytically efficient and using charged electrostatic potentials on the enzyme surface to guide substrate to the active site and products out [14,69,70]. MnSOD also enhances diffusion with its electrostatic surface. However, unlike IC-CuZnSOD, MnSOD is product-inhibited by peroxide and thus does not have the catalytic efficiency of IC-CuZnSOD [71]. The high catalytic efficiency of SODs is paramount in maintaining mitochondrial and cell vitality. As seen later in this chapter, disruption of this catalytic efficiency through electrostatic-altering mutations or protein aggregation can yield deleterious consequences.

In addition to electrostatic diffusion, a second integral component for the high catalytic efficiency of SODs is a rapid proton transfer relay [14,70,72]. The electron transfers that are responsible for converting O$_2^{\bullet-}$ to its products are coupled to proton transfers (i.e., changes in charge) to drive energetically favorable redox potentials for catalysis by the active site metal ion [70]. Since the active site metal shuffles between oxidation states (e.g., Mn^{3+} and Mn^{2+} for MnSOD; Cu^{2+} and Cu$^+$ for CuZnSOD), the consequent changes in charge by the metal ion are counter balanced via H$^+$ transfers to allow the active site to retain the net charge needed for electron transfers to occur. Given the high catalytic speed and efficiency of SODs, a systematic relay of proton transfers along ionizable amino acids is thought to deliver protons to molecules that are coordinated to the active site metal [72]. While the exact path of the relay has yet to be determined, mutagenesis studies of the active site perturb catalysis significantly [68,69,71,73–75]. From these studies, it is difficult to determine which amino acids are responsible for the transfer of protons because they can fulfill more than one role for catalysis, such as stabilizing hydrogen bonds for substrate or maintenance of electrostatic charge. Nonetheless, a proton shuttling network contributes to the catalytic efficiency of SODs.

While MnSOD and CuZnSOD both use electrostatic guidance and a proton transfer relay to facilitate exceptional catalytic rates, the two are structurally different, and the active site configurations are dissimilar [69,70]. The following sections detail O$_2^{\bullet-}$ entry and binding to the active site as well as the current theories of the proton-based enzymatic mechanism.

5 MNSOD

5.1 ACTIVE SITE STRUCTURE OF MNSOD

Human MnSOD consists of a homotetramer, with each of the four subunits containing a manganese ion at its catalytic site (Figure 3a). At each site, the manganese ion is coordinated and surrounded by evolutionarily conserved amino acids that fine-tune its redox potential (probability of an electron transfer) to facilitate catalysis [76]. The manganese covalently binds the "inner sphere" residues (His26, His74, His163, Asp159) and a single-oxygen containing molecule thought to be either OH^- or H_2O (denoted WAT1) (Figure 3b) [77]. The 'outer sphere' consists of a second layer of amino acids that are integral to catalytic function but not bound to the manganese. These are His30, Tyr34, Phe77, Trp78, Trp123, Gln143, and Trp161. The contribution of all of these residues to efficient catalysis has been verified by mutagenesis studies except Trp78 [71,75,78–81]. Of note, Glu162 from the adjacent subunit (magenta residue in Figure 3b) forms a hydrogen bond with His163 to stabilize oligomerization [81]. Our studies speculate that Glu162 also participates in the electrostatic guidance of $O_2^{\bullet-}$ to the active site [82]. The role of the inner sphere amino acids is to hold the manganese in place while the outer sphere amino acids facilitate catalysis.

The outer sphere residues are composed of a cluster of hydrogen-bonding residues and a cluster of hydrophobic residues. Residues His30, Tyr34, Gln143, and a water molecule (denoted WAT2) form a hydrogen bond network to the manganese-ligated WAT1. This network is thought to be involved in shuttling protons near the manganese ion to facilitate proton-coupled electron transfer; whether the protons actually shuttle along this network, is not clear due to experimental limitations [75,78–80,83]. A ~5 Å gap formed between His30 and Tyr34 is the only solvent accessible area to the active site and $O_2^{\bullet-}$ likely interacts with the manganese at this site in a position opposite Asp159 [82]. Residues Phe77, Trp78, Trp123, and Trp161 form a hydrophobic cage around a portion of the active site to exclude water and promote $O_2^{\bullet-}$ interaction with the manganese once $O_2^{\bullet-}$ enters through His30 and Tyr34. These bulky hydrophobic residues also constrain the nearby residues, as the active site residues are evolutionarily fine-tuned to precise positions to allow for an optimized redox potential. Any relaxation of the hydrophobic packing (i.e., a W161F mutant) significantly impedes catalysis [71]. These evolutionarily conserved residues are integral to catalytic function.

5.2 ELECTROSTATIC GUIDANCE OF SUPEROXIDE BY MNSOD

A contributing factor to the fast catalysis of MnSOD is the electrostatic guidance of its substrate $O_2^{\bullet-}$. The positive surfaces (i.e., clusters of positively-charged residues) near the active site cavity along with the manganese

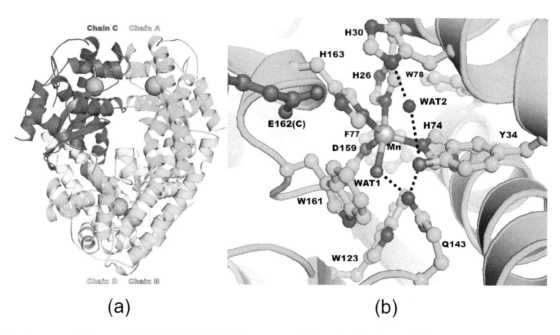

(a)   (b)

FIGURE 3 Structure of human MnSOD. (a) The MnSOD tetramer with each subunit containing a manganese ion (pink sphere) at the catalytic center. (b) Structure of the active site. Green, red and blue spheres denote carbon, oxygen, and nitrogen atoms, respectively. Magenta denotes carbon atoms from the adjacent chain C. Glu162 hydrogen bonds with His163 from across the dimer interface. The dashed lines represent a hydrogen bond network thought to be involved in relaying protons near the manganese for proton-coupled electron transfer. Adapted from Azadmanesh et al., 2017 PDB entry 5VF9 [82].

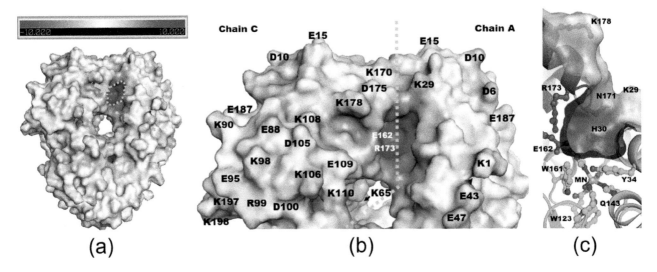

FIGURE 4 Charged surfaces of human MnSOD colored in kiloteslas. (a) Electrostatic surfaces of the tetramer. Only one of the four active sites is seen in this view, indicated by the orange dashed circle and arrow. (b) A view of two subunits that contribute to forming an active site cavity. The orange, dashed lines indicate the dimer interface between the two subunits. The amino acids labeled in white indicate their location on the concave surface of the active site cavity. (c) Cross section of the active site cavity rotated 90° along the horizontal axis. Chain A is green, and chain C is magenta. Adapted from Azadmanesh et al., 2017, PDB entry 5VF9 [82]. Partial charges of the manganese and bound molecules were derived from molecular dynamics simulations performed by Neves and colleagues [88].

cation guide the negatively-charged $O_2^{\bullet-}$ to the active site (Figure 4a) [82]. Guided diffusion appears to be a long-range process, meaning that charged residues far from the active site generate a net field vector that guides the substrate near the active site [84]. Neutralization of charged amino acids, such as acetylation of lysine residues, decreases the enzymatic activity of the enzyme, likely due to perturbation of the net field vectors [85–87]. Dysregulation of mitochondrial acetylation systems and thereby a decrease in MnSOD activity is a contributing factor to mitochondrial dysfunction [85]. Consequently, the electrostatic surfaces of MnSOD are necessary to preserve the mitochondria.

Clusters of negatively charged amino acids on the surface also contribute to the guided diffusion of $O_2^{\bullet-}$ to the active site [82]. At the outer ridge of the site, these clusters repel $O_2^{\bullet-}$ toward the positive surfaces that subsequently direct the substrate toward the location of catalysis (Figure 4b). These negative residues far from the active site also contribute to this electrostatic guidance, as they influence the direction and magnitude of the electric potential in tandem with the positive residues [82]. Although negative potentials on the surface of MnSOD repel $O_2^{\bullet-}$, the contributing negatively-charged residues appear to be just as important as the positively-charged for guiding the substrate to the active site.

At the active site pit, there is a cluster of three positively charged residues that are integral to the electropositive potential, Lys29 of chain A and Arg173 and Lys178 of chain C. Near the manganese ion, there are

two charged amino acids, Arg173 and Glu162, which are especially critical in maneuvering $O_2^{\bullet-}$ to the correct position for catalysis (Figure 4c) [82]. As Arg173 is the closest positive residue to the manganese ion and the only positive residue of the cavity, it is responsible for drawing $O_2^{\bullet-}$ toward the positively charged metal. Previous studies demonstrate that Arg173 is integral for enzyme activity [89,90]. Upon entering the cavity, the negative Glu162 repels the substrate into the active site. This negative residue is necessary for promoting association of $O_2^{\bullet-}$ with the manganese cation rather than Arg173. A pair of oppositely charged amino acids are found in all SODs and are necessary for the guidance of $O_2^{\bullet-}$ to the catalytic site [82,84].

5.3 MANNER OF SUPEROXIDE BINDING TO THE ACTIVE SITE

The orientation of $O_2^{\bullet-}$ binding to the active site remains unclear owing to its short half-life and high reactivity. This information has been sought after because the enzymatic mechanism also remains unclear, and researchers have pursued the knowledge of substrate binding as a means to decipher the mechanism. Several groups have attempted to infer the manner of $O_2^{\bullet-}$ binding using a variety of approaches, but these have yielded several competing theories [82,91–93]. These theories can be roughly divided into two categories, inner-sphere binding, and outer-sphere binding.

For a substrate to bind in an inner-sphere manner, $O_2^{\bullet-}$ binds directly to the manganese ion. The location of binding is likely opposite the manganese-bound aspartate residue (Asp159) as it is the only location that would sterically allow $O_2^{\bullet-}$ to interact closely with the catalytic metal [82]. Upon direct binding to the manganese, it remains unclear whether a manganese-interacting molecule dissociates. The bound oxygen-containing molecule or the aspartate residue would be the most likely candidates if substrate interaction were to displace a metal-bound molecule [92]. Inner-sphere $O_2^{\bullet-}$ binding garners support from its simplicity.

Another group of studies supports $O_2^{\bullet-}$ binding at the active site in an outer-sphere manner, described as substrate binding and catalysis without coordinating to the manganese ion [93]. Instead, $O_2^{\bullet-}$ is closer to the amino acids that form the gap into the active site, Tyr34, and His30, and its orientation stabilized by hydrogen bonding from other molecules of the hydrogen bond network such as Gln143, WAT1, and WAT2 [72]. This prospective method of substrate binding assumes that the electron transfers occurring between $O_2^{\bullet-}$ and the manganese can occur when these two molecules are not directly interacting. Alternatively, outer-sphere substrate binding may occur for only some parts of catalysis, depending on the oxidation state of the manganese metal [93].

Computational chemistry supports a compromise between the two binding modes. These studies suggest $O_2^{\bullet-}$ binds directly to Mn(III) but not to Mn(II) [82,93]. Said another way, the substrate at the active site alternates between inner and outer-sphere binding depending on which half-reaction of the catalytic cycle MnSOD is using.

5.4 THE PROTON-BASED CATALYTIC MECHANISM

A systematic proton relay is thought to drive catalysis by MnSOD [69,70,72]. Not only are the protons used during reduction of $O_2^{\bullet-}$ into H_2O_2, but they also influence the charge at the active site to increase the likelihood of electron transfers. The notion that an orderly proton shuttling mechanism exists comes from the extreme speed MnSOD harbors at physiological pH as well as the number of protons required within a - given second (80,000/s per catalytic site) [68]. Many studies have attempted to decipher the exact path of the proton relay, but it remains elusive owing to the experimental difficulty in direct detection of the protons.

The proton transfers required for MnSOD catalysis are thought to occur along the hydrogen bond network of the active site at least partially (dashed lines, Figure 3b) [69,71,74,75,78,79,81,94]. In particular, ionizable amino acids and water molecules are the most likely to be involved in proton shuttling. Direct evidence for this relay has yet to be observed as visualization of protonation states is difficult through conventional structural

biology techniques, such as X-ray diffraction (XRD) and nuclear magnetic resonance (NMR). This is a consequence of XRD and NMR harboring limited sensitivity to hydrogen atoms. Instead, studies of MnSOD with neutron diffraction have recently been explored to probe the protonation states at the active site [95]. Neutrons are much more sensitive to the presence of hydrogen and show promise in elucidating the proton-based enzymatic mechanism of MnSOD.

5.5 PRODUCT INHIBITION

The enzymatic mechanism of MnSOD is unique to others SODs because it is capable of being reversibly inhibited by its product, H_2O_2 [96]. MnSOD product-inhibition is thought to prevent the formation of overwhelming amounts of H_2O_2 at once and allowing the systems that clear H_2O_2, such as catalases and peroxidases, to 'catch up' [70,93,96]. This "self-regulation" is thought to use an alternative proton relay. Kinetics suggest this alternative pathway occurs half the time when the enzyme is in steady-state conditions and may be a result of copious amounts of $O_2^{\bullet-}$ overwhelming the rate at which protons can be delivered [69,71].

A cryogenic X-ray crystal structure of *Escherichia coli* soaked with peroxide displays a putative product-inhibited complex (PDB entry 3K9S) [97]. A molecule consisting of two oxygen atoms is observed bound to the manganese ion in two differing orientations, both in a side-on manner. Since excessive amounts of peroxide are known to push the backward reaction of MnSOD to produce $O_2^{\bullet-}$, it is unclear whether this molecule is actually peroxide or $O_2^{\bullet-}$ [96]. The resolution of the structure impeded the determination of the protonation states and the actual bond length between the two oxygen atoms. As noted in Section 5.4, neutron diffraction would be helpful in determining the protonation state of the dual-oxygen molecule and in deciphering a product-inhibition mechanism.

5.6 DYSFUNCTION OF MnSOD

Given the need for adequate function of MnSOD for cellular life [19,98], dysfunction of its catalysis increases the susceptibility of an individual to mitochondrial degeneration and related pathologies [99]. The most prominent causes of dysfunction are polymorphisms (i.e., mutations) and aberrant regulation of post-translation modifications [85,99]. Three polymorphic mutants and dysregulated acetylation, which neutralizes charged lysine residues that leads $O_2^{\bullet-}$ to the active site, are responsible for inadequate clearance of oxidative stress in mitochondrial degeneration.

Polymorphisms that affect MnSOD activity appear to cause late-onset diseases through various mechanisms. Some individuals with PD symptoms have an Ala16 to

Val mutation in MnSOD [100], yet the same mutation has been reported in healthy individuals of Asian descent indicating that the genetic background for polymorphisms must be considered [101]. Residue Ala16 is in the mitochondrial signaling sequence, and this mutation impairs translocation of MnSOD to the mitochondria. MnSOD is a nuclear-encoded gene that is translated with a 24-amino acid N-terminal signaling sequence that mediates translocation across the mitochondrial membranes to the matrix. After translocation, the signaling peptide is cleaved off. Individuals with the Ala16Val variant have 40% lower MnSOD activity. While this mutation does not directly affect catalysis, it is a marker for several cancers [101], and it demonstrates the importance of localization of MnSOD within the mitochondria to prevent degeneration. Individuals with PD and breast cancer can also have an Ile58 to Thr polymorphism [101]. The Ile58 to Thr polymorphic variant causes two packing defects in each of the two four-helix bundles of the tetrameric interface. These cavities in the interface substantially decrease the stability of the enzyme, causes the tetramer to dissociate into dimers, and causes the enzyme to heat-inactivate at normal body temperatures with a resultant decrease in cells by one-third [102,103]. Also, a Leu60 to Phe variant found on the opposite side of the α-helix from I58 is prevalent among those with T-cell leukemia [104]. It appears that MnSOD polymorphic variants with translocation defects, decreased stability, and resultantly decreased activity appear to cause pathologies involving mitochondrial dysfunction.

Post-translational modifications regulate the activity of MnSOD probably by altering its electrostatic surface [82]. Acetylation of lysine residues neutralizes positive charges and alters the net electrostatic vectors that draw negatively charged $O_2^{\bullet-}$ to the active site. For example, conserved residues, Lys29, Lys65, and Lys98 are not in the active site (see Figure 4b), but their acetylation has a substantial negative effect on the activity of MnSOD [85], which highlights the influence of long-range electrostatic effects on $O_2^{\bullet-}$ diffusion. The enzyme responsible for deacetylating MnSOD and promoting its activity is Sirtuin 3 (SIRT3). SIRT3 is mostly localized to the mitochondria and has been shown to interact with MnSOD directly [92,93]. Mitochondrial-related pathologies show decreased SIRT3 levels. Prominent examples are neurodegenerative diseases such as AD, PD, ALS and HD [105]. In these diseases, SIRT3 mRNA and protein levels are decreased and correlated with hyperacetylation and decreased activity of MnSOD [106]. Consequently, oxidant levels overwhelm the proper function of the mitochondria partially due to a lack of $O_2^{\bullet-}$ clearance [86,106].

Another means of hampered MnSOD catalysis is overwhelming amounts of ROS/RNS. The conserved Tyr34 residue that is necessary for efficient enzymatic activity is susceptible to nitration by ONOO⁻ *in vitro* and *in vivo* [107]. Nitration of this residue impedes the substrate access funnel which abolishes MnSOD activity

completely [53]. A characteristic of several cancers is the upregulation of MnSOD with the absence of its activity as a result of ONOO⁻ mediated inactivation.

5.7 ROLE OF MnSOD IN MITOCHONDRIAL DYSFUNCTION AND NEUROLOGICAL DISEASE

The preservation of mitochondrial integrity and homeostasis relies on the efficient clearance of ROS and RNS [108]. This clearance is heavily dependent on the removal of $O_2^{\bullet-}$ residing in the mitochondrial matrix. However, other studies seeking to define the role of MnSOD in mitochondrial dysfunction leading to neurodegeneration suggest the relationship is correlative rather than causative (i.e., the Ile58Thr mutant found in individuals with PD does not necessarily *cause* the disease) [108,109]. In other words, does an increase in oxidative stress cause neurodegeneration or is it a symptom of neurodegeneration?

In the early stages of AD, neurons upregulate MnSOD [110]. However, such a characteristic is likely a means of compensation from the elevated levels of ROS/RNS that are a characteristic of individuals of older age. There is evidence that increased incidences of oxidative stress in such individuals are a consequence of amyloid plaques hampering mitochondrial bioenergy production, not dysregulation of MnSOD [111]. While MnSOD does not *cause* AD, it seems that there is a relationship between the ability of MnSOD to clear ROS/RNS from the mitochondria and disease progression [112]. The extent of oxidative stress from the reactive species in the mitochondria is directly related to the extent of plaque deposition and neuron loss. Said another way, any defect in MnSOD function will exacerbate the progression of AD.

A molecular marker that distinguishes PD is the loss of complex I activity within the mitochondria of neurons [113]. Complex I is susceptible to direct damage by $O_2^{\bullet-}$ and H_2O_2 due to its use of Fe-S clusters [60,61], which suggests a link between MnSOD clearance of $O_2^{\bullet-}$ and PD pathogenesis. However, whether there is direct involvement of MnSOD in the progression of the disease is still debated. Studies which measured the MnSOD expression levels and activities in disease progression, have yielded conflicting results due to the complexity of the disease. Many molecular changes occur, and PD is more than just a loss of mitochondria-generated energy [114,115]. Regardless, MnSOD polymorphic variants found in individuals with PD exhibit significantly decreased catalytic activity [100,101]. The effect of these mutants on disease progression is unclear.

The majority of ALS cases (90–95%) are idiopathic, meaning cases manifest sporadically without known cause [116]. The remainder of cases (5–10%) are believed to be caused by familial mutations (over 100) in the gene encoding IC-CuZnSOD, as mice with the mutations manifest ALS (discussed in detail in the subsequent IC-CuZnSOD section) [117]. Mitochondrial dysfunction is a paramount symptom of ALS, with motor neuron degeneration linked to impaired energy production [118]. This link poses the question of how MnSOD is involved in ALS. A subset of ALS mutant mouse models with a copy of the SOD2 gene knocked out exhibit increased disease progression while others do not [119]. These experiments suggest mitochondrial oxidative stress playing a role in the pathogenesis of ALS in only some of the familial mutations, suggesting multiple mechanisms for the manifestation of the disease.

Like other neurodegenerative pathologies, HD progression includes the presence of oxidative stress. The levels and activity of MnSOD are significantly upregulated in post-mortem HD patients while the activities of the ETC complexes are remarkably reduced [120]. Fe-S clusters found within the ETC complexes are susceptible to inactivation by both $O_2^{\bullet-}$ and H_2O_2, meaning both the substrate and product of MnSOD can contribute to the impairment of oxidative metabolism [60]. Of note, enzymes that remove H_2O_2, catalase and glutathione peroxidases, are also elevated [120]. These post-mortem samples do not capture the role of antioxidant enzymes during disease progression. In an *in vivo* model of HD progression, MnSOD upregulation is found to be neuroprotective [121].

A unifying theme of MnSOD and neurological pathologies is that MnSOD appears to slow down neurodegeneration as a result of mitochondrial dysfunction [109]. However, the elevation of MnSOD activities, either exogenously or endogenously, does not act as a 'magic bullet' to treat these diseases. It is still unclear whether oxidative stress is a cause or symptom of neurodegeneration, but neuron death appears to be a multifaceted event that includes more than just oxidative stress.

6 IC-CUZNSOD

6.1 ACTIVE SITE STRUCTURE OF IC-CuZnSOD

IC-CuZnSOD is composed of a homodimer with each subunit containing a Cu and Zn ion at the active site (Figure 5a). The Cu ion is the site of electron transfers with $O_2^{\bullet-}$ and shuffles between an oxidation state of divalent and monovalent during catalysis while the Zn ion remains divalent [14]. The Zn ion contributes to the positive charge at the active site, attracting $O_2^{\bullet-}$ and maintaining a favorable redox potential for catalysis [70]. Zn also contributes to the structural integrity of the active site needed for catalysis, as the bridging histidine between the two metal ions, His63, is oriented by Zn in such a way to fine-tune the redox potential at the Cu ion for favorable catalysis (Figure 5b) [122]. Even though the Zn ion contributes to efficient enzymatic activity, Cu is responsible for the redox reactions, where Zn influences the redox potential of catalysis by Cu through its charge [123].

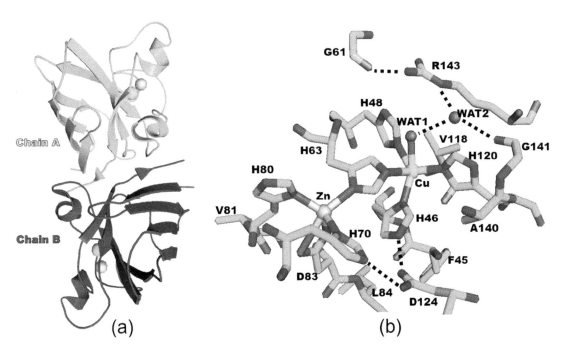

FIGURE 5 Structure of Human IC-CuZnSOD (SOD1). (a) The IC-CuZnSOD dimer with each subunit containing a Cu ion (cyan sphere) and Zinc ion (gray sphere) at the catalytic center. (b) Structure of the active site with green, red and blue denoting carbon, oxygen, and nitrogen atoms, respectively. The dashed lines represent the hydrogen bonds. The cluster of hydrogen bonds involving Gly61, Arg143, WAT1, WAT2, and Gly141 are thought to be involved in relaying protons for proton-coupled electron transfer. There is no direct crystallographic evidence for this proton relay. Figures were made using PDB entry 1PU0.

The 'inner-sphere' residues at the active site consist of the amino acids directly ligated to the ions. Residues His63, His70, His80, and Asp83, are covalently bound to the Zn ion whereas His46, His48, His63, His120, and a water molecule (denoted WAT1) are bound to the Cu ion (Figure 5b) [122]. Of note, His63 and WAT1 bind the Cu ion only in the resting, divalent state. When the Cu ion is in the monovalent state, the bond between it and His63 is broken, and WAT1 is not present. $O_2^{\cdot-}$ is thought to occupy the position of WAT1 during catalysis [122]. The 'outer-sphere' residues, the residues necessary to catalysis but are not interacting directly with the metal ions, consist of a hydrogen bonding portion and a hydrophobic portion. Asp124 hydrogen bonds to His46 and His70, ligands of Cu and Zn, respectively, to maintain an orientation that fine-tunes the redox potential (slight movement of residues at the active site greatly affect redox potential) [70]. Gly61, Gly141, Arg143, and WAT2, form a hydrogen bond network in conjunction with the Cu-ligated WAT1 molecule (Figure 5b). This network is believed to be involved in shuttling protons for proton-coupled electron transfer [70,122]. Arg143 also is necessary for the efficient electrostatic guidance of $O_2^{\cdot-}$ to the active site [70]. The hydrophobic portion of the outer-sphere residues consists of Phe45, Val81, Val118, and Ala140. These residues provide rigidity to the active site, maintaining the redox potential and limiting access of water molecules to the Cu ion. There is only enough room for one substrate or a water molecule to bind directly to the Cu ion at a time as a consequence of the small, ~5 Å wide opening. The hydrophobic residues are also thought to help 'push' the uncharged products out of the active site, whereas the charged $O_2^{\cdot-}$ substrate can overcome the hydrophobicity to bind to the Cu ion [122]. Like MnSOD, IC-CuZnSOD has a hydrogen bond network and a bottom hydrophobic 'cage' to facilitate catalysis.

6.2 ELECTROSTATIC GUIDANCE OF SUPEROXIDE BY IC-CuZnSOD

Like MnSOD, a contributing factor to the fast catalysis of IC-CuZnSOD is the electrostatic guidance of the $O_2^{\cdot-}$ substrate to the active site. The Cu and Zn ions contribute significantly to the strong positive electrostatic potential found at the active site (Figure 6a). The back face of the protein opposite the active site consists of primarily negatively-charged residues, meaning the electrostatics of the protein facilitate repulsion of $O_2^{\cdot-}$ from the back-side of the subunit to the active site (see Chain B, Figure 6a). Charged residues far from the active site also contribute to the net electrostatic potentials that lead $O_2^{\cdot-}$ to the active site, fine-tuning the angle and magnitude of the electrostatic field [84]. Both MnSOD and IC-CuZnSOD use electrostatics to accelerate substrate diffusion.

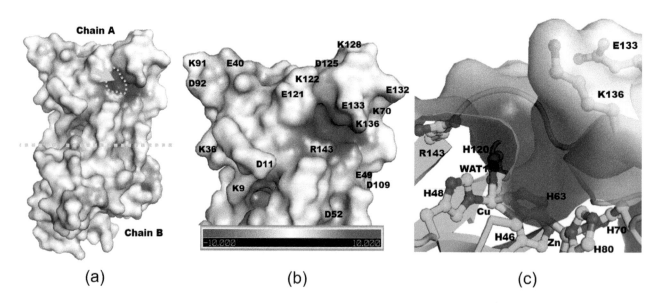

(a)   (b)   (c)

FIGURE 6 Charged surfaces of human IC-CuZnSOD colored in kiloteslas. (a) Electrostatic surfaces of the dimer. Only one of the two active sites is seen in this view, indicated by the orange arrow in Chain A. The horizontal dotted orange line displays the dimer interface. Chain B displays the back face of the subunit, opposite the face exposing the active site. (b) A zoom-in of the Chain A subunit with charged amino acids observable in this orientation labeled. (c) Cross section of the active site cavity rotated 90° along the horizontal axis. PROPKA 3.1 was used at pH 7.0 to generate partial charges. Then the partial charges were assigned on a per-atom basis in PQR file format using PDB2PQR within the same automated pipeline [125]. Electrostatic surfaces were generated with APBS [126] using PDB entry 1PU0.

The surrounding charged amino acids surrounding the active site "cavity" have been shown to be integral to efficient diffusion of $O_2^{\cdot-}$ to the active site, most notably Glu133, Lys136, and Arg143 (Figures 6b and 6c) [70,84]. Mutation of any of these residues to a non-charged counterpart changes the angular direction of the electrostatic field by at least 20° or greater while also significantly hampering catalysis [70,84]. Glu133 and Lys136 are the pair of oppositely charged amino acids found in other SODs that function to 'lure' and "push" substrate into the active site. Lys136 and its positive charge are well exposed at the protein surface, which acts to draw the negatively charged $O_2^{\cdot-}$ in (Figure 6c). Upon attraction by Lys136, $O_2^{\cdot-}$ is repelled by Glu133 toward Arg143 and the Cu ion. Arg143 acts to draw $O_2^{\cdot-}$ toward the Cu ion. Once in close enough proximity, the greater charge of the Cu ion dominates to pull in $O_2^{\cdot-}$ for catalysis.

Unlike the Mn ion of MnSOD, the Cu ion of IC-CuZnSOD is surface exposed, but the active site cavity is only accessible by molecules less than ~5 Å (Figure 6c) [14]. While the Zn ion appears to be surface exposed as well, it is actually occluded from the solvent by the above alpha helix where Lys136 resides. However, the occluded Zn contributes significantly to the electrostatic field that draws $O_2^{\cdot-}$ into the active site and is structurally integral to the catalytic mechanism [122].

Lys122 has been shown to be post-translationally modified by acetylation *in vivo*, a modification that neutralizes its positive charge [124]. Acetylation of this residue is thought to reduce the activity of IC-CuZnSOD, changing the angle of the electrostatic field vectors by 22° [84]. This may serve as an endogenous regulator of activity but is also linked to sporadic ALS [124]. Acetyl-Lys123 IC-CuZnSOD is found in prominent density in the neurons of the central nervous system, forming a peculiar link between electrostatic guidance and ALS pathogenesis.

6.3 MANNER OF SUPEROXIDE BINDING TO THE ACTIVE SITE

Directly determining the manner of $O_2^{\cdot-}$ binding at the active site for catalysis has yet to yield successful results as a consequence of the short half-life and high reactivity of the anion. Alternative methods have yielded some information, most notably by using substrate analogs such as azide and H_2O_2. Studies with these analogs have yielded two competing theories, inner-sphere and outer-sphere binding [122,127]. Inner-sphere binding suggests that $O_2^{\cdot-}$ directly binds the catalytic Cu ion, replacing the position of the water molecule (WAT1, Figure 5b), opposite His46. This position is large enough to accommodate only one atom, meaning that $O_2^{\cdot-}$ binds to the Cu in an end-on manner, with one end bound to the Cu ion and the other hydrogen-bonded to Arg143 and a water molecule (denoted WAT2, Figure 5b)

[122]. Outer-sphere binding proposes that a redox reaction can occur without $O_2^{\cdot-}$ directly interacting with the catalytic metal [127]. Instead, $O_2^{\cdot-}$ hydrogen bonds to nearby residues and/or water molecules within the active site. Hart and colleagues constructed a model based on crystallographic studies where $O_2^{\cdot-}$ binds in an outer-sphere manner only when the Cu ion is in the monovalent state [127]. In this model, $O_2^{\cdot-}$ hydrogen bonds to His63 (His63 disassociates when Cu is monovalent, exposing a nitrogen atom to be used as a hydrogen bond donor) and Arg143. However, this type of binding brings about the issue of how a systematic electron transfer can occur over a distance greater than 3 Å.

6.4 THE PROTON-BASED CATALYTIC MECHANISM

Like all SODs, a systematic proton relay is thought to exist at the active site for IC-CuZnSOD. Of note, IC-CuZnSOD and MnSOD are believed to have evolved independently, representing examples of convergent evolution where both enzymes adopted different but highly efficient proton shuttles [70]. The principal use of the proton transfers is to drive electron transfers (i.e., redox reactions) by changing the charge at the active site in addition to protonating $O_2^{\cdot-}$ to H_2O_2 [69,70,72]. The efficiency of the proton transfers and electron transfers for IC-CuZnSOD is represented by the rate of 2×10^9 M^{-1} s^{-1}. This rate exceeds "diffusion controlled" rates at physiological pH, in part due to electrostatic guidance of the substrate to the active site [69,70]. An orderly series of proton transfers contribute to the extreme catalytic efficiency of IC-CuZnSOD, though the series of transfers is unclear due to difficulty in probing protons.

Like other SODs, the proton relay for IC-CuZnSOD has yet to be directly observed due to experimental limitations in XRD and NMR. Though, it appears SODs share similar active site attributes. Similar to MnSOD, IC-CuZnSOD has a hydrogen bond network around the catalytic metal, which includes ordered water molecules and ionizable amino acids (Figure 5b). Neutron diffraction provides an avenue to decipher the path of the proton relay due to its sensitivity to hydrogen atoms. The recent surge of neutron diffraction instrumentation is making visualization of proton shuttling feasible [128,129].

6.5 DYSFUNCTION OF IC-CuZnSOD

Mitochondrial degeneration and other non-neurological age-related phenotypes (e.g., muscle atrophy and hearing loss) correlate with deficient IC-CuZnSOD activity [18,70]. In mouse models, a total absence of IC-CuZnSOD activity does not yield a neurodegenerative phenotype, but "accelerated aging" instead. However, aberrations of the IC-CuZnSOD structure that lead to its aggregation underlie neurodegeneration. In fact, IC-CuZnSOD fibrils probably cause motor neuron death in ALS [116]. There are two categories of dysfunction for IC-CuZnSOD: a low activity that leads to increased ROS/RNS, and fibril formation that leads to motor neuron deterioration. Moreover, there is overlap between the two categories. Extensive reviews of the relationship between IC-CuZnSOD and ALS are found in the following references [130,131].

The over 160 ALS-associated IC-CuZnSOD variants so far identified include missense mutations, indels, and truncations [132]. The resultant enzymatic activity ranges from complete abolishment to increases up to 150%. Mutations that abolish activity are those that affect the binding of Cu or Zn to the active site, as both of these are necessary for catalytic activity [116]. Other mutants that do not affect metal binding and exhibit normal or increased enzymatic activity have decreased protein or mRNA stability. As an example, the Gly37Arg mutant has 150% intrinsic activity compared to the wild-type, but the overall activity is 40% of the wild-type due to factors involving the turnover of the protein [133]. The increased activity is presumed to be due to the increased positive charge the mutation provides. While the activity of IC-CuZnSOD is impaired in ALS mutations, there is no correlation between IC-CuZnSOD activity and ALS disease severity (i.e., IC-CuZnSOD deficits do not cause motor neuron death), rather IC-CuZnSOD activity relates to the extent motor neurons are vulnerable to non-lethal dysfunction from oxidative injury [70,116].

ALS prognosis correlates with the extent of IC-CuZnSOD aggregation into fibrils, much like amyloid-β fibrils in Alzheimer's disease [70]. Fibrils form by IC-CuZnSOD misfolding and self-association into high-order, low energy structures. Examples are metal-deficient variants that destabilize the tertiary structure of the protein, and non-native interactions are energetically favored [116]. Other variants, affecting monomer structure and stability of the dimer interface, are thought to aggregate into fibrils as a result of lower thermostability and increased disorder [70,116]. Protein fold stability is further compromised through a lack of disulfide bond formation. Despite the cytosol being a strong reducing environment, IC-CuZnSOD harbors an oxidized disulfide bond that is buried near the dimer interface in WT. Variants that decrease dimer stability and expose the bond are more susceptible to disulfide reduction by glutathione/glutaredoxin systems [134]. A lack of Cu or Zn metallation also decreases dimer stability, meaning variants that compromise metal binding also expose the disulfide bond for reduction. The propensity for IC-CuZnSOD to form fibrils is related to factors compromising its stability, such as mutations and reduced disulfide bonds.

6.6   ROLE OF IC-CuZnSOD IN MITOCHONDRIAL DYSFUNCTION AND NEUROLOGICAL DISEASE

IC-CuZnSOD is peculiar in that its absence yields only mild phenotypes in mice with no neurodegeneration and yet variants result in ALS, a mitochondrial-related pathology that involves motor neuron death [18,70]. Apparently, IC-CuZnSOD variants cause disease through gain-of-function toxicity rather than by creating a toxic cellular environment through a deficiency in clearing $O_2^{\bullet-}$. In familial ALS, this is especially apparent in aberrations of the IC-CuZnSOD structure (either through mutation or lack of post-translational modifications) leading to heavy aggregation [116]. IC-CuZnSOD aggregates exhibit amyloid fibril characteristics, where IC-CuZnSOD aggregates can transmit their dysfunction to other cells, much like prion proteins [70]. IC-CuZnSOD is believed to underlie the disease etiology of mitochondrial abnormalities in ALS [116].

ALS is thought to be a consequence of an interplay between mitochondrial pathology and dysfunctional IC-CuZnSOD [135]. Mutant IC-CuZnSOD aggregates accumulate in the mitochondrial matrix, MIMS, and the outer membrane of the mitochondria [136]. It is not clear how these aggregates become localized within the matrix of the mitochondria, with speculation pointing to misfolded IC-CuZnSOD somehow affecting mitochondrial import and/or retention. Association of aggregated IC-CuZnSOD with the mitochondria is related to abnormal clustering of the organelle along with extensive fragmentation [137]. Damage to the inner mitochondrial membrane by the aggregates also causes a loss in the mitochondrial membrane potential that leads to perturbed ATP production [138]. Altogether, IC-CuZnSOD fibrils directly damage mitochondria through physical association and instigate mitochondrial-mediated apoptosis.

An indirect mitochondrial effect of aggregated IC-CuZnSOD is disruption of mitochondrial transport machinery, where aggregates physically impair axonal transport of mitochondria and other material (proteins and other organelles) for neurological processes in motor neurons [138]. The ability of a motor neuron to transport such cellular material from the cell body to its far-reaching axons is critical for the vitality of the neuron. Under normal physiological conditions, mitochondria are abundant in neurons and are transported to the axons, where intense ATP demands are present [138]. An inability to meet the ATP demands at these distal neuromuscular junctions has been shown to lead to motor axon degeneration and eventual death of the motor neuron, the defining characteristic of ALS [139].

The difference in the phenotypes between loss of IC-CuZnSOD activity and IC-CuZnSOD aggregation is a consequence of the former being a loss-of-function phenotype while the latter is a gain-of-function. The pathologies that manifest when IC-CuZnSOD is not present are a result of oxidative insults (e.g., muscle atrophy) while aggregate-related pathologies (e.g., ALS) are a result of impaired energy production in neurons [18,139]. However, these two phenotypes are not mutually exclusive, as IC-CuZnSOD mutants may both exhibit decreased activity and aggregative tendencies. IC-CuZnSOD has not been shown to be involved in other neurodegenerative diseases such as AD, PD, or HD.

7 CONCLUSION

SODs are crucial redox enzymes that keep cells alive by acting as biological antioxidants. MnSOD and IC-CuZnSOD contribute to cell vitality by preserving the mitochondria, the cell's source of energy, through clearance of $O_2^{\bullet-}$. Excessive levels of $O_2^{\bullet-}$ inhibit the energy-producing components of the mitochondria and is a precursor to ROS and RNS that damage DNA, proteins, and lipids. Clearance of $O_2^{\bullet-}$ through the redox activities of MnSOD and IC-CuZnSOD involves extraordinary efficiency through electrostatic guidance to the active site and systematic proton shuttling near a catalytic metal ion. Any impairment of this process through mutations or aberrant post-translational modifications elevates the oxidative stress at the mitochondria and underlies neurodegenerative pathologies. Dysfunction of MnSOD is associated with AD, PD, ALS, and HD while dysfunction of IC-CuZnSOD is thought to be the cause of ALS. Altogether, SODs are one of the most important class of redox enzymes for mitochondrial function and cell vitality.

ACKNOWLEDGMENTS

We would like to thank William Lutz for useful discussion, technical assistance, and expertise. This work was supported by NASA EPSCoR funding (44-0307-1021-201) and the Fred & Pamela Buffett Cancer Center Support Grant (P30CA036727). Jahaun Azadmanesh also acknowledges University of Nebraska Medical Center and NASA Nebraska Space Grant for fellowships.

REFERENCES

1. Landis, G.N.; Tower, J. Superoxide Dismutase Evolution and Life Span Regulation. *Mech. Ageing Dev.* 2005, *126*, 365–379.
2. Warner, H.R. Superoxide Dismutase, Aging, and Degenerative Disease. *Free Radic. Biol. Med.* 1994, *17*, 249–258.
3. Fukai, T.; Ushio-Fukai, M. Superoxide Dismutases: Role in Redox Signaling, Vascular Function, and Diseases. *Antioxid. Redox Signal.* 2011, *15*, 1583–1606.
4. Miao, L.; St Clair, D.K. Regulation of Superoxide Dismutase Genes: Implications in Disease. *Free Radic. Biol. Med.* 2009, *47*, 344–356.
5. Nita, M.; Grzybowski, A. The Role of the Reactive Oxygen Species and Oxidative Stress in the Pathomechanism of the Age-Related Ocular Diseases and Other Pathologies of the Anterior and Posterior Eye Segments in Adults. *Oxid. Med. Cell. Longev.* 2016, *2016*, 3164734.
6. Muller, F.L.; Lustgarten, M.S.; Jang, Y.; Richardson, A.; Van Remmen, H. Trends in Oxidative Aging Theories. *Free Radic. Biol. Med.* 2007, *43*, 477–503.
7. Gao, L.; Laude, K.; Cai, H. Mitochondrial Pathophysiology, Reactive Oxygen Species, and Cardiovascular Diseases. *Vet. Clin. North Am. Small Anim. Pract.* 2008, *38*, 137–155, vi.
8. Jastroch, M.; Divakaruni, A.S.; Mookerjee, S.; Treberg, J.R.; Brand, M.D. Mitochondrial Proton and Electron Leaks. *Essays Biochem.* 2010, *47*, 53–67.
9. Bhat, A.H.; Dar, K.B.; Anees, S.; Zargar, M.A.; Masood, A.; Sofi, M.A.; Ganie, S.A. Oxidative Stress, Mitochondrial Dysfunction and Neurodegenerative Diseases; A Mechanistic Insight. *Biomed. Pharmacother.* 2015, *74*, 101–110.
10. Federico, A.; Cardaioli, E.; Da Pozzo, P.; Formichi, P.; Gallus, G.N.; Radi, E. Mitochondria, Oxidative Stress and Neurodegeneration. *J. Neurol. Sci.* 2012, *322*, 254–262.
11. Van Houten, B.; Woshner, V.; Santos, J.H. Role of Mitochondrial DNA in Toxic Responses to Oxidative Stress. *DNA Repair (Amst)* 2006, *5*, 145–152.
12. McCord, J.M.; Fridovich, I. Superoxide Dismutase: The First Twenty Years (1968–1988). *Free Radic. Biol. Med.* 1988, *5*, 363–369.
13. van der Vliet, A. Antioxidant Defense in the Lung. In *Comparative Biology of the Normal Lung* (Second Edition), Elsevier Inc., London, UK: 2015, pp. 489–507.
14. Perry, J.J.; Shin, D.S.; Getzoff, E.D.; Tainer, J.A. The Structural Biochemistry of the Superoxide Dismutases. *Biochim. Biophys. Acta* 2010, *1804*, 245–262.
15. Sentman, M.L.; Granstrom, M.; Jakobson, H.; Reaume, A.; Basu, S.; Marklund, S.L. Phenotypes of Mice Lacking Extracellular Superoxide Dismutase and Copper- and Zinc-Containing Superoxide Dismutase. *J. Biol. Chem.* 2006, *281*, 6904–6909.
16. Noor, R.; Mittal, S.; Iqbal, J. Superoxide Dismutase – Applications and Relevance to Human Diseases. *Med. Sci. Monit* 2002, *8*, 210–215.
17. Elchuri, S.; Oberley, T.D.; Qi, W.; Eisenstein, R.S.; Jackson Roberts, L.; Van Remmen, H.; Epstein, C.J.; Huang, T.T. Cuznsod Deficiency Leads to Persistent and Widespread Oxidative Damage and Hepatocarcinogenesis Later in Life. *Oncogene* 2005, *24*, 367–380.
18. Saccon, R.A.; Bunton-Stasyshyn, R.K.; Fisher, E.M.; Fratta, P. Is Sod1 Loss of Function Involved in Amyotrophic Lateral Sclerosis? *Brain* 2013, *136*, 2342–2358.
19. Li, Y.; Huang, T.T.; Carlson, E.J.; Melov, S.; Ursell, P.C.; Olson, J.L.; Noble, L.J.; Yoshimura, M.P.; Berger, C.; Chan, P.H.; Wallace, D.C.; Epstein, C.J. Dilated Cardiomyopathy and Neonatal Lethality in Mutant Mice Lacking Manganese Superoxide Dismutase. *Nat. Genet.* 1995, *11*, 376–381.
20. Lebovitz, R.M.; Zhang, H.; Vogel, H.; Cartwright, J., Jr.; Dionne, L.; Lu, N.; Huang, S.; Matzuk, M.M. Neurodegeneration, Myocardial Injury, and Perinatal Death in Mitochondrial Superoxide Dismutase-Deficient Mice. *Proc. Natl. Acad. Sci. U. S. A.* 1996, *93*, 9782–9787.
21. Drose, S.; Brandt, U. The Mechanism of Mitochondrial Superoxide Production by the Cytochrome bc1 Complex. *J. Biol. Chem.* 2008, *283*, 21649–21654.
22. Gandhi, S.; Abramov, A.Y. Mechanism of Oxidative Stress in Neurodegeneration. *Oxid. Med. Cell. Longev.* 2012, *2012*, 428010.
23. Buetler, T.M.; Krauskopf, A.; Ruegg, U.T. Role of Superoxide as a Signaling Molecule. *Physiology* 2004, *19*, 120–123.

24. Block, M.L. Nadph Oxidase as a Therapeutic Target in Alzheimer's Disease. *BMC Neurosci.* 2008, *9, Suppl 2*, S8.

25. Bylund, J.; Brown, K.L.; Movitz, C.; Dahlgren, C.; Karlsson, A. Intracellular Generation of Superoxide by the Phagocyte Nadph Oxidase: How, Where, and What for? *Free Radic.Biol. Med.* 2010, *49*, 1834–1845.

26. Pacher, P.; Beckman, J.S.; Liaudet, L. Nitric Oxide and Peroxynitrite in Health and Disease. *Physiol. Rev.* 2007, *87*, 315–424.

27. Melser, S.; Lavie, J.; Benard, G. Mitochondrial Degradation and Energy Metabolism. *Biochim. Biophys. Acta* 2015, *1853*, 2812–2821.

28. Jezek, J.; Cooper, K.F.; Strich, R. Reactive Oxygen Species and Mitochondrial Dynamics: The Yin and Yang of Mitochondrial Dysfunction and Cancer Progression. *Antioxidants (Basel)* 2018, *7*.

29. Villanueva, C.; Giulivi, C. Subcellular and Cellular Locations of Nitric Oxide Synthase Isoforms as Determinants of Health and Disease. *Free Radic. Biol. Med.* 2010, *49*, 307–316.

30. Giulivi, C.; Kato, K.; Cooper, C.E. Nitric Oxide Regulation of Mitochondrial Oxygen Consumption I: Cellular Physiology. *Am. J. Physiol. Cell Physiol.* 2006, *291*, C1225–C1231.

31. Lacza, Z.; Pankotai, E.; Busija, D.W. Mitochondrial Nitric Oxide Synthase: Current Concepts and Controversies. *Front Biosci. (Landmark Ed)* 2009, *14*, 4436–4443.

32. Brown, G.C.; Borutaite, V. Inhibition of Mitochondrial Respiratory Complex I by Nitric Oxide, Peroxynitrite and S-Nitrosothiols. *Biochim. Biophys. Acta* 2004, *1658*, 44–49.

33. Cantu, D.; Schaack, J.; Patel, M. Oxidative Inactivation of Mitochondrial Aconitase Results in Iron and H2o2-Mediated Neurotoxicity in Rat Primary Mesencephalic Cultures. *PLoS One* 2009, *4*, e7095.

34. Pearce, L.L.; Martinez-Bosch, S.; Manzano, E.L.; Winnica, D.E.; Epperly, M.W.; Peterson, J. The Resistance of Electron-Transport Chain Fe-S Clusters to Oxidative Damage During the Reaction of Peroxynitrite with Mitochondrial Complex Ii and Rat-Heart Pericardium. *Nitric Oxide* 2009, *20*, 135–142.

35. Lopez, C.J.; Qayyum, I.; Mishra, O.P.; Delivoria-Papadopoulos, M. Effect of Nitration on Protein Tyrosine Phosphatase and Protein Phosphatase Activity in Neuronal Cell Membranes of Newborn Piglets. *Neurosci. Lett.* 2005, *386*, 78–81.

36. Sultana, R.; Perluigi, M.; Allan Butterfield, D. Lipid Peroxidation Triggers Neurodegeneration: A Redox Proteomics View into the Alzheimer Disease Brain. *Free Radic. Biol. Med.* 2013, *62*, 157–169.

37. Fernandez-Moriano, C.; Gonzalez-Burgos, E.; Gomez-Serranilos, M.P. Lipid Peroxidation and Mitochondrial Dysfunction in Alzheimer's and Parkinson's Diseases: Role of Natural Products as Cytoprotective Agents. In *Neuroprotective Natural Products: Clinical Aspects and Mode of Action*, Brahmachari, G., Ed., Wiley-VCH, Weinheim, Germany: 2017, pp. 107–135.

38. Riley, P.A. Free Radicals in Biology: Oxidative Stress and the Effects of Ionizing Radiation. *Int. J. Radiat. Biol.* 1994, *65*, 27–33.

39. Sun, Y.; Yin, Y.; Zhang, J.; Yu, H.; Wang, X.; Wu, J.; Xue, Y. Hydroxyl Radical Generation and Oxidative Stress in Carassius Auratus Liver, Exposed to Pyrene. *Ecotoxicol. Environ. Saf.* 2008, *71*, 446–453.

40. Lipinski, B. Hydroxyl Radical and Its Scavengers in Health and Disease. *Oxid. Med. Cell. Longev.* 2011, *2011*, 809696.

41. Thomas, C.; Mackey, M.M.; Diaz, A.A.; Cox, D.P. Hydroxyl Radical Is Produced Via the Fenton Reaction in Submitochondrial Particles under Oxidative Stress: Implications for Diseases Associated with Iron Accumulation. *Redox Rep* 2013, *14*, 102–108.

42. Cadenas, E.; Davies, K.J. Mitochondrial Free Radical Generation, Oxidative Stress, and Aging. *Free Radic. Biol. Med.* 2000, *29*, 222–230.

43. Goldstein, S.; Merenyi, G. The Chemistry of Peroxynitrite: Implications for Biological Activity. *Methods Enzymol.* 2008, *436*, 49–61.

44. Beckman, J.S.; Beckman, T.W.; Chen, J.; Marshall, P.A.; Freeman, B.A. Apparent Hydroxyl Radical Production by Peroxynitrite: Implications for Endothelial Injury from Nitric Oxide and Superoxide. *Proc. Natl. Acad. Sci. U. S. A.* 1990, *87*, 1620–1624.

45. Kell, D.B. Toward a Unifying, Systems Biology Understanding of Large-Scale Cellular Death and Destruction Caused by Poorly Liganded Iron: Parkinson's, Huntington's, Alzheimer's, Prions, Bactericides, Chemical Toxicology and Others as Examples. *Arch. Toxicol.* 2010, *84*, 825–889.

46. Davies, K.J. Protein Damage and Degradation by Oxygen Radicals. I. General Aspects. *J. Biol. Chem.* 1987, *262*, 9895–9901.

47. Guo, C.; Sun, L.; Chen, X.; Zhang, D. Oxidative Stress, Mitochondrial Damage and Neurodegenerative Diseases. *Neural Regen Res* 2013, *8*, 2003–2014.

48. van der Veen, R.C.; Roberts, L.J. Contrasting Roles for Nitric Oxide and Peroxynitrite in the Peroxidation of Myelin Lipids. *J. Neuroimmunol.* 1999, *95*, 1–7.

49. Giulivi, C.; Boveris, A.; Cadenas, E. Hydroxyl Radical Generation During Mitochondrial Electron Transfer and the Formation of 8-Hydroxydesoxyguanosine in Mitochondrial DNA. *Arch. Biochem. Biophys.* 1995, *316*, 909–916.

50. Agarwal, A.; Tvrda, E.; Mulgund, A. Oxidative Stress in Preeclampsia. In *Handbook of Fertility*, Watson, R.R., Ed., Elsevier Inc., London, UK: 2015, pp. 283–290.

51. Yasui, M.; Kanemaru, Y.; Kamoshita, N.; Suzuki, T.; Arakawa, T.; Honma, M. Tracing the Fates of Site-Specifically Introduced DNA Adducts in the Human Genome. *DNA Repair (Amst)* 2014, *15*, 11–20.

52. Radi, R. Nitric Oxide, Oxidants, and Protein Tyrosine Nitration. *Proc. Natl. Acad. Sci. U. S. A.* 2004, *101*, 4003–4008.

53. Quint, P.; Reutzel, R.; Mikulski, R.; McKenna, R.; Silverman, D.N. Crystal Structure of Nitrated Human Manganese Superoxide Dismutase: Mechanism of Inactivation. *Free Radic. Biol. Med.* 2006, *40*, 453–458.

54. Radi, R.; Cassina, A.; Hodara, R.; Quijano, C.; Castro, L. Peroxynitrite Reactions and Formation in Mitochondria. *Free Radic. Biol. Med.* 2002, *33*, 1451–1464.

55. Turko, I.V.; Li, L.; Aulak, K.S.; Stuehr, D.J.; Chang, J.Y.; Murad, F. Protein Tyrosine Nitration in the Mitochondria from Diabetic Mouse Heart. Implications to Dysfunctional Mitochondria in Diabetes. *J. Biol. Chem.* 2003, *278*, 33972–33977.

56. Kilinc, K.; Kilinc, A. Mutagenic Actions of Nitrogen Oxides. *Indoor Built Environ.* 2005, *14*, 503–512.

57. Bindoli, A. Lipid Peroxidation in Mitochondria. *Free Radic. Biol. Med.* 1988, *5*, 247–261.

58. Yang, J.L.; Weissman, L.; Bohr, V.A.; Mattson, M.P. Mitochondrial DNA Damage and Repair in Neurodegenerative Disorders. *DNA Repair (Amst)* 2008, *7*, 1110–1120.

59. Nissanka, N.; Moraes, C.T. Mitochondrial DNA Damage and Reactive Oxygen Species in Neurodegenerative Disease. *FEBS Lett.* 2018, *592*, 728–742.

60. Imlay, J.A. Iron-Sulphur Clusters and the Problem with Oxygen. *Mol. Microbiol.* 2006, *59*, 1073–1082.

61. Wallace, M.A.; Liou, L.L.; Martins, J.; Clement, M.H.; Bailey, S.; Longo, V.D.; Valentine, J.S.; Gralla, E.B. Superoxide Inhibits 4fe-4s Cluster Enzymes Involved in Amino Acid Biosynthesis. Cross-Compartment Protection by Cuzn-Superoxide Dismutase. *J. Biol. Chem.* 2004, *279*, 32055–32062.

62. Urrutia, P.J.; Mena, N.P.; Nunez, M.T. The Interplay between Iron Accumulation, Mitochondrial Dysfunction, and Inflammation During the Execution Step of Neurodegenerative Disorders. *Front. Pharmacol.* 2014, *5*, 38.

63. Xu, W.; Barrientos, T.; Andrews, N.C. Iron and Copper in Mitochondrial Diseases. *Cell Metab.* 2013, *17*, 319–328.

64. Halliwell, B.; Clement, M.V.; Long, L.H. Hydrogen Peroxide in the Human Body. *FEBS Lett.* 2000, *486*, 10–13.

65. Bienert, G.P.; Schjoerring, J.K.; Jahn, T.P. Membrane Transport of Hydrogen Peroxide. *Biochim. Biophys. Acta* 2006, *1758*, 994–1003.

66. Liochev, S.L. The Role of Iron-Sulfur Clusters in in Vivo Hydroxyl Radical Production. *Free Radic. Res.* 1996, *25*, 369–384.

67. Netto, L.E.; Antunes, F. The Roles of Peroxiredoxin and Thioredoxin in Hydrogen Peroxide Sensing and in Signal Transduction. *Mol. Cells* 2016, *39*, 65–71.

68. Guan, Y.; Hickey, M.J.; Borgstahl, G.E.O.; Hallewell, R. A.; Lepock, J.R.; O'Connor, D.; Hsieh, Y.; Nick, H.S.; Silverman, D.N.; Tainer, J.A. Crystal Structure of Y34f Mutant Human Mitochondrial Manganese Superoxide Dismutase and the Functional Role of Tyrosine 34. *Biochemistry* 1998, *37*, 4722–4730.

69. Abreu, I.A.; Cabelli, D.E. Superoxide Dismutases-a Review of the Metal-Associated Mechanistic Variations. *Biochim. Biophys. Acta* 2010, *1804*, 263–274.

70. Sheng, Y.; Abreu, I.A.; Cabelli, D.E.; Maroney, M.J.; Miller, A.F.; Teixeira, M.; Valentine, J.S. Superoxide Dismutases and Superoxide Reductases. *Chem. Rev.* 2014, *114*, 3854–3918.

71. Hearn, A.S.; Stroupe, M.E.; Cabelli, D.E.; Lepock, J.R.; Tainer, J.A.; Nick, H.S.; Silverman, D.N. Kinetic Analysis of Product Inhibition in Human Manganese Superoxide Dismutase. *Biochemistry* 2001, *40*, 12051–12058.

72. Azadmanesh, J.; Borgstahl, G.E.O. A Review of the Catalytic Mechanism of Human Manganese Superoxide Dismutase. *Antioxidants (Basel)* 2018, *7*.

73. Hearn, A.S.; Fan, L.; Lepock, J.R.; Luba, J.P.; Greenleaf, W.B.; Cabelli, D.E.; Tainer, J.A.; Nick, H.S.; Silverman, D.N. Amino Acid Substitution at the Dimeric Interface of Human Manganese Superoxide Dismutase. *J. Biol. Chem.* 2004, *279*, 5861–5866.

74. Hearn, A.S.; Stroupe, M.E.; Cabelli, D.E.; Ramilo, C. A.; Luba, J.P.; Tainer, J.A.; Nick, H.S.; Silverman, D. N. Catalytic and Structural Effects of Amino Acid Substitution at Histidine 30 in Human Manganese Superoxide Dismutase: Insertion of Valine C Gamma into the Substrate Access Channel. *Biochemistry* 2003, *42*, 2781–2789.

75. Perry, J.J.; Hearn, A.S.; Cabelli, D.E.; Nick, H.S.; Tainer, J. A.; Silverman, D.N. Contribution of Human Manganese Superoxide Dismutase Tyrosine 34 to Structure and Catalysis. *Biochemistry* 2009, *48*, 3417–3424.

76. Bannister, J.V.; Bannister, W.H.; Rotilio, G. Aspects of the Structure, Function, and Applications of Superoxide Dismutase. *CRC Crit. Rev. Biochem.* 1987, *22*, 111–180.

77. Borgstahl, G.E.; Parge, H.E.; Hickey, M.J.; Beyer, W.F., Jr.; Hallewell, R.A.; Tainer, J.A. The Structure of Human Mitochondrial Manganese Superoxide Dismutase Reveals a Novel Tetrameric Interface of Two 4-Helix Bundles. *Cell* 1992, *71*, 107–118.

78. Edwards, R.A.; Whittaker, M.M.; Whittaker, J.W.; Baker, E.N.; Jameson, G.B. Outer Sphere Mutations Perturb Metal Reactivity in Manganese Superoxide Dismutase. *Biochemistry* 2001, *40*, 15–27.

79. Leveque, V.J.; Stroupe, M.E.; Lepock, J.R.; Cabelli, D.E.; Tainer, J.A.; Nick, H.S.; Silverman, D.N. Multiple Replacements of Glutamine 143 in Human Manganese Superoxide Dismutase: Effects on Structure, Stability, and Catalysis. *Biochemistry* 2000, *39*, 7131–7137.

80. Ramilo, C.A.; Leveque, V.; Guan, Y.; Lepock, J.R.; Tainer, J.A.; Nick, H.S.; Silverman, D.N. Interrupting the Hydrogen Bond Network at the Active Site of Human Manganese Superoxide Dismutase. *J. Biol. Chem.* 1999, *274*, 27711–27716.

81. Quint, P.S.; Domsic, J.F.; Cabelli, D.E.; McKenna, R.; Silverman, D.N. Role of a Glutamate Bridge Spanning the Dimeric Interface of Human Manganese Superoxide Dismutase. *Biochemistry* 2008, *47*, 4621–4628.

82. Azadmanesh, J.; Trickel, S.R.; Borgstahl, G.E.O. Substrate-Analog Binding and Electrostatic Surfaces of Human Manganese Superoxide Dismutase. *J. Struct. Biol.* 2017, *199*, 68–75.

83. Abreu, I.A.; Rodriguez, J.A.; Cabelli, D.E. Theoretical Studies of Manganese and Iron Superoxide Dismutases: Superoxide Binding and Superoxide Oxidation. *J. Phys. Chem. B* 2005, *109*, 24502–24509.

84. Getzoff, E.D.; Tainer, J.A.; Weiner, P.K.; Kollman, P.A.; Richardson, J.S.; Richardson, D.C. Electrostatic Recognition between Superoxide and Copper, Zinc Superoxide Dismutase. *Nature* 1983, *306*, 287–290.

85. Ozden, O.; Park, S.H.; Kim, H.S.; Jiang, H.; Coleman, M. C.; Spitz, D.R.; Gius, D. Acetylation of Mnsod Directs Enzymatic Activity Responding to Cellular Nutrient Status or Oxidative Stress. *Aging (Albany NY)* 2011, *3*, 102–107.

86. Tao, R.; Coleman, M.C.; Pennington, J.D.; Ozden, O.; Park, S.H.; Jiang, H.; Kim, H.S.; Flynn, C.R.; Hill, S.; Hayes McDonald, W.; Olivier, A.K.; Spitz, D.R.; Gius, D. Sirt3-Mediated Deacetylation of Evolutionarily Conserved Lysine 122 Regulates Mnsod Activity in Response to Stress. *Mol. Cell* 2010, *40*, 893–904.

87. Qiu, X.; Brown, K.; Hirschey, M.D.; Verdin, E.; Chen, D. Calorie Restriction Reduces Oxidative Stress by Sirt3-Mediated Sod2 Activation. *Cell Metab.* 2010, *12*, 662–667.

88. Neves, R.P.; Sousa, S.F.; Fernandes, P.A.; Ramos, M.J. Parameters for Molecular Dynamics Simulations of Manganese-Containing Metalloproteins. *J. Chem. Theory Comput.* 2013, *9*, 2718–2732.

89. Borders, C.L., Jr.; Chain, V.W.; Bjerrum, M.J. The Positive Charge at Position 189 Is Essential for the Catalytic Activity of Iron- and Manganese-Containing Superoxide Dismutases. *Free Radic. Res. Commun.* 1991, *12–13 Pt, 1*, 279–285.

90. Chan, V.W.; Bjerrum, M.J.; Borders, C.L., Jr. Evidence That Chemical Modification of a Positively Charged

Residue at Position 189 Causes the Loss of Catalytic Activity of Iron-Containing and Manganese-Containing Superoxide Dismutases. *Arch. Biochem. Biophys.* 1990, *279*, 195–201.

91. Lah, M.S.; Dixon, M.M.; Pattridge, K.A.; Stallings, W.C.; Fee, J.A.; Ludwig, M.L. Structure-Function in Escherichia Coli Iron Superoxide Dismutase: Comparisons with the Manganese Enzyme from Thermus Thermophilus. *Biochemistry* 1995, *34*, 1646–1660.

92. Whittaker, M.M.; Whittaker, J.W. Low-Temperature Thermochromism Marks a Change in Coordination for the Metal Ion in Manganese Superoxide Dismutase. *Biochemistry* 1996, *35*, 6762–6770.

93. Srnec, M.; Aquilante, F.; Ryde, U.; Rulisek, L. Reaction Mechanism of Manganese Superoxide Dismutase Studied by Combined Quantum and Molecular Mechanical Calculations and Multiconfigurational Methods. *J. Phys. Chem. B* 2009, *113*, 6074–6086.

94. Edwards, R.A.; Whittaker, M.M.; Whittaker, J.W.; Baker, E.N.; Jameson, G.B. Removing a Hydrogen Bond in the Dimer Interface of Escherichia Coli Manganese Superoxide Dismutase Alters Structure and Reactivity. *Biochemistry* 2001, *40*, 4622–4632.

95. Azadmanesh, J.; Trickel, S.R.; Weiss, K.L.; Coates, L.; Borgstahl, G.E. Preliminary Neutron Diffraction Analysis of Challenging Human Manganese Superoxide Dismutase Crystals. *Acta Cryst. F* 2017, *73*, 235–240.

96. Hearn, A.S.; Tu, C.; Nick, H.S.; Silverman, D.N. Characterization of the Product-Inhibited Complex in Catalysis by Human Manganese Superoxide Dismutase. *J. Biol. Chem.* 1999, *274*, 24457–24460.

97. Porta, J.; Vahedi-Faridi, A.; Borgstahl, G.E.O. Structural Analysis of Peroxide-Soaked Mnsod Crystals Reveals Side-on Binding of Peroxide to Active-Site Manganese. *J. Mol. Biol.* 2010, *399*, 377–384.

98. Sun, J.; Folk, D.; Bradley, T.J.; Tower, J. Induced Overexpression of Mitochondrial Mn-Superoxide Dismutase Extends the Life Span of Adult Drosophila Melanogaster. *Genetics* 2002, *161*, 661–672.

99. Gamarra, D.; Elcoroaristizabal, X.; Fernandez-Martinez, M.; de Pancorbo, M.M. Association of the C47t Polymorphism in Sod2 with Amnestic Mild Cognitive Impairment and Alzheimer's Disease in Carriers of the Apoeepsilon4 Allele. *Dis. Markers* 2015, *2015*, 746329.

100. Shimoda-Matsubayashi, S.; Matsumine, H.; Kobayashi, T.; Nakagawa-Hattori, Y.; Shimizu, Y.; Mizuno, Y. Structural Dimorphism in the Mitochondrial Targeting Sequence in the Human Manganese Superoxide Dismutase Gene. A Predictive Evidence for Conformational Change to Influence Mitochondrial Transport and a Study of Allelic Association in Parkinson's Disease. *Biochem. Biophys. Res. Commun* 1996, *226*, 561–565.

101. Kim, A. Modulation of Mnsod in Cancer: Epidemiologicaland Experimental Evidence. *Toxicol Res.* 2010, *26*, 83–93.

102. Borgstahl, G.E.; Parge, H.E.; Hickey, M.J.; Johnson, M.J.; Boissinot, M.; Hallewell, R.A.; Lepock, J.R.; Cabelli, D.E.; Tainer, J.A. Human Mitochondrial Manganese Superoxide Dismutase Polymorphic Variant Ile58thr Reduces Activity by Destabilizing the Tetrameric Interface. *Biochemistry* 1996, *35*, 4287–4297.

103. Zhang, H.J.; Yan, T.; Oberley, T.D.; Oberley, L.W. Comparison of Effects of Two Polymorphic Variants of Manganese Superoxide Dismutase on Human Breast Mcf-7 Cancer Cell Phenotype. *Cancer Res.* 1999, *59*, 6276–6283.

104. Hernandez-Saavedra, D.; McCord, J.M. Paradoxical Effects of Thiol Reagents on Jurkat Cells and a New Thiol-Sensitive Mutant Form of Human Mitochondrial Superoxide Dismutase. *Cancer Res.* 2003, *63*, 159–163.

105. Salvatori, I.; Valle, C.; Ferri, A.; Carri, M.T. Sirt3 and Mitochondrial Metabolism in Neurodegenerative Diseases. *Neurochem. Int.* 2017, *109*, 184–192.

106. Kim, H.S.; Patel, K.; Muldoon-Jacobs, K.; Bisht, K.S.; Aykin-Burns, N.; Pennington, J.D.; van der Meer, R.; Nguyen, P.; Savage, J.; Owens, K.M.; Vassilopoulos, A.; Ozden, O.; Park, S.H.; Singh, K.K.; Abdulkadir, S.A.; Spitz, D.R.; Deng, C.X.; Gius, D. Sirt3 is a Mitochondria-Localized Tumor Suppressor Required for Maintenance of Mitochondrial Integrity and Metabolism During Stress. *Cancer Cell* 2010, *17*, 41–52.

107. Candas, D.; Li, J.J. Mnsod in Oxidative Stress Response-Potential Regulation Via Mitochondrial Protein Influx. *Antioxid. Redox Signal.* 2014, *20*, 1599–1617.

108. Franco-Iborra, S.; Vila, M.; Perier, C. Mitochondrial Quality Control in Neurodegenerative Diseases: Focus on Parkinson's Disease and Huntington's Disease. *Front. Neurosci.* 2018, *12*, 342.

109. Flynn, J.M.; Melov, S. Sod2 in Mitochondrial Dysfunction and Neurodegeneration. *Free Radic. Biol. Med.* 2013, *62*, 4–12.

110. De Leo, M.E.; Borrello, S.; Passantino, M.; Palazzotti, B.; Mordente, A.; Daniele, A.; Filippini, V.; Galeotti, T.; Masullo, C. Oxidative Stress and Overexpression of Manganese Superoxide Dismutase in Patients with Alzheimer's Disease. *Neurosci. Lett.* 1998, *250*, 173–176.

111. Casley, C.S.; Canevari, L.; Land, J.M.; Clark, J.B.; Sharpe, M.A. Beta-Amyloid Inhibits Integrated Mitochondrial Respiration and Key Enzyme Activities. *J. Neurochem.* 2002, *80*, 91–100.

112. Melov, S.; Adlard, P.A.; Morten, K.; Johnson, F.; Golden, T.R.; Hinerfeld, D.; Schilling, B.; Mavros, C.; Masters, C.L.; Volitakis, I.; Li, Q.X.; Laughton, K.; Hubbard, A.; Cherny, R.A.; Gibson, B.; Bush, A.I. Mitochondrial Oxidative Stress Causes Hyperphosphorylation of Tau. *PLoS One* 2007, *2*, e536.

113. Puschmann, A. Monogenic Parkinson's Disease and Parkinsonism: Clinical Phenotypes and Frequencies of Known Mutations. *Parkinsonism Relat. Disord.* 2013, *19*, 407–415.

114. Ihara, Y.; Chuda, M.; Kuroda, S.; Hayabara, T. Hydroxyl Radical and Superoxide Dismutase in Blood of Patients with Parkinson's Disease: Relationship to Clinical Data. *J. Neurol. Sci.* 1999, *170*, 90–95.

115. Marttila, R.J.; Lorentz, H.; Rinne, U.K. Oxygen Toxicity Protecting Enzymes in Parkinson's Disease. Increase of Superoxide Dismutase-Like Activity in the Substantia Nigra and Basal Nucleus. *J. Neurol. Sci.* 1988, *86*, 321–331.

116. Valentine, J.S.; Hart, P.J. Misfolded Cuznsod and Amyotrophic Lateral Sclerosis. *Proc. Natl. Acad. Sci. U. S. A.* 2003, *100*, 3617–3622.

117. Gurney, M.E.; Pu, H.; Chiu, A.Y.; Dal Canto, M.C.; Polchow, C.Y.; Alexander, D.D.; Caliendo, J.; Hentati, A.; Kwon, Y.W.; Deng, H.X. et al. Motor Neuron Degeneration in Mice that Express a Human Cu,Zn Superoxide Dismutase Mutation. *Science* 1994, *264*, 1772-1775.

118. Muyderman, H.; Chen, T. Mitochondrial Dysfunction in Amyotrophic Lateral Sclerosis – A Valid Pharmacological Target? *Br. J. Pharmacol.* 2014, *171*, 2191–2205.

119. Muller, F.L.; Liu, Y.; Jernigan, A.; Borchelt, D.; Richardson, A.; Van Remmen, H. Mnsod Deficiency Has

a Differential Effect on Disease Progression in Two Different Als Mutant Mouse Models. *Muscle Nerve* 2008, 38, 1173–1183.

120. Kumar, A.; Ratan, R.R. Oxidative Stress and Huntington's Disease: The Good, the Bad, and the Ugly. *J. Huntingtons Dis.* 2016, 5, 217–237.

121. Madhavan, L.; Ourednik, V.; Ourednik, J. Neural Stem/Progenitor Cells Initiate the Formation of Cellular Networks that Provide Neuroprotection by Growth Factor-Modulated Antioxidant Expression. *Stem Cells* 2008, 26, 254–265.

122. Shin, D.S.; Didonato, M.; Barondeau, D.P.; Hura, G.L.; Hitomi, C.; Berglund, J.A.; Getzoff, E.D.; Cary, S.C.; Tainer, J.A. Superoxide Dismutase from the Eukaryotic Thermophile Alvinella Pompejana: Structures, Stability, Mechanism, and Insights into Amyotrophic Lateral Sclerosis. *J. Mol. Biol.* 2009, 385, 1534–1555.

123. Goto, J.J.; Zhu, H.; Sanchez, R.J.; Nersissian, A.; Gralla, E. B.; Valentine, J.S.; Cabelli, D.E. Loss of in Vitro Metal Ion Binding Specificity in Mutant Copper-Zinc Superoxide Dismutases Associated with Familial Amyotrophic Lateral Sclerosis. *J. Biol. Chem.* 2000, 275, 1007–1014.

124. Kaliszewski, M.; Kennedy, A.K.; Blaes, S.L.; Shaffer, R.S.; Knott, A.B.; Song, W.; Hauser, H.A.; Bossy, B.; Huang, T. T.; Bossy-Wetzel, E. Sod1 Lysine 123 Acetylation in the Adult Central Nervous System. *Front. Cell. Neurosci.* 2016, 10, 287.

125. Dolinsky, T.J.; Czodrowski, P.; Li, H.; Nielsen, J.E.; Jensen, J.H.; Klebe, G.; Baker, N.A. PDB2PQR: Expanding and Upgrading Automated Preparation of Biomolecular Structures for Molecular Simulations. *Nucleic Acids Res.* 2007, 35, W522–W525.

126. Baker, N.A.; Sept, D.; Joseph, S.; Holst, M.J.; McCammon, J.A. Electrostatics of Nanosystems: Application to Microtubules and the Ribosome. *Proc. Natl. Acad. Sci. U. S. A.* 2001, 98, 10037–10041.

127. Hart, P.J.; Balbirnie, M.M.; Ogihara, N.L.; Nersissian, A.M.; Weiss, M.S.; Valentine, J.S.; Eisenberg, D. A Structure-Based Mechanism for Copper-Zinc Superoxide Dismutase. *Biochemistry* 1999, 38, 2167–2178.

128. O'Dell, W.B.; Bodenheimer, A.M.; Meilleur, F. Neutron Protein Crystallography: A Complementary Tool for Locating Hydrogens in Proteins. *Arch. Biochem. Biophys.* 2016, 602, 48–60.

129. Coates, L.; Cao, H.B.; Chakoumakos, B.C.; Frontzek, M. D.; Hoffmann, C.; Kovalevsky, A.Y.; Liu, Y.; Meilleur, F.;

Dos Santos, A.M.; Myles, D.A.A.; Wang, X.P.; Ye, F. A. Suite-Level Review of the Neutron Single-Crystal Diffraction Instruments at Oak Ridge National Laboratory. *Rev. Sci. Instrum* 2018, 89.

130. Chattopadhyay, M.; Valentine, J.S. Aggregation of Copper-Zinc Superoxide Dismutase in Familial and Sporadic Als. *Antioxid. Redox Signal.* 2009, 11, 1603–1614.

131. Zarei, S.; Carr, K.; Reiley, L.; Diaz, K.; Guerra, O.; Altamirano, P.F.; Pagani, W.; Lodin, D.; Orozco, G.; Chinea, A. A Comprehensive Review of Amyotrophic Lateral Sclerosis. *Surg. Neurol. Int.* 2015, 6, 171.

132. Guegan, C.; Przedborski, S. Programmed Cell Death in Amyotrophic Lateral Sclerosis. *J. Clin. Invest.* 2003, 111, 153–161.

133. Borchelt, D.R.; Lee, M.K.; Slunt, H.S.; Guarnieri, M.; Xu, Z.S.; Wong, P.C.; Brown, R.H., Jr.; Price, D.L.; Sisodia, S.S.; Cleveland, D.W. Superoxide Dismutase 1 with Mutations Linked to Familial Amyotrophic Lateral Sclerosis Possesses Significant Activity. *Proc. Natl. Acad. Sci. U. S. A.* 1994, 91, 8292–8296.

134. Bouldin, S.D.; Darch, M.A.; Hart, P.J.; Outten, C.E. Redox Properties of the Disulfide Bond of Human Cu,Zn Superoxide Dismutase and the Effects of Human Glutaredoxin 1. *Biochem. J.* 2012, 446, 59–67.

135. Bruijn, L.I.; Miller, T.M.; Cleveland, D.W. Unraveling the Mechanisms Involved in Motor Neuron Degeneration in Als. *Annu. Rev. Neurosci.* 2004, 27, 723–749.

136. Kawamata, H.; Manfredi, G. Different Regulation of Wild-Type and Mutant Cu, Zn Superoxide Dismutase Localization in Mammalian Mitochondria. *Hum. Mol. Genet.* 2008, 17, 3303–3317.

137. Shi, P.; Gal, J.; Kwinter, D.M.; Liu, X.; Zhu, H. Mitochondrial Dysfunction in Amyotrophic Lateral Sclerosis. *Biochim. Biophys. Acta* 2010, 1802, 45–51.

138. Borthwick, G.M.; Johnson, M.A.; Ince, P.G.; Shaw, P.J.; Turnbull, D.M. Mitochondrial Enzyme Activity in Amyotrophic Lateral Sclerosis: Implications for the Role of Mitochondria in Neuronal Cell Death. *Ann. Neurol.* 1999, 46, 787–790.

139. Dupuis, L.; Gonzalez de Aguilar, J.L.; Echaniz-Laguna, A.; Eschbach, J.; Rene, F.; Oudart, H.; Halter, B.; Huze, C.; Schaeffer, L.; Bouillaud, F.; Loeffler, J.P. Muscle Mitochondrial Uncoupling Dismantles Neuromuscular Junction and Triggers Distal Degeneration of Motor Neurons. *PLoS One* 2009, 4, e5390.

17 Role of Post-Translational Modifications of Mitochondrial Complex I in Mitochondrial Dysfunction and Human Brain Pathologies

Vismaya
Department of Neurochemistry, National Institute of Mental Health and Neurosciences (NIMHANS), Karnataka, India

Yogachar Chithra
Department of Biosciences, Hemagangothri P.G Centre, University of Mysore, Karnataka, India

Muchukunte Mukunda Srinivas Bharath
Department of Neurochemistry, National Institute of Mental Health and Neurosciences (NIMHANS), Karnataka, India
Neurotoxicology Laboratory, Neurobiology Research Centre, NIMHANS, Karnataka, India

CONTENTS

1 MAIN CHAPTER

1.1 STRUCTURE OF COMPLEX I

Mitochondrial oxidative phosphorylation (OXPHOS) involves two processes: transfer of electrons across a series of oxidation and reduction reactions and ATP production (1). Among the five OXPHOS protein complexes, NADH: Ubiquinone Oxidoreductase or Mitochondrial Complex I (CI) predominantly mediates the entry of electrons into the electron transport chain (ETC) that are transferred via Coenzyme Q (CoQ) or Ubiquinone, onto subsequent complexes (2). Concomitantly, CI facilitates proton-dependent transmembrane electrochemical membrane potential (1–3), which ultimately results in ATP production (4,5).

CI is the biggest and the most important enzyme in the eukaryotic ETC (4). Mammalian CI is a 1 MDa complex consisting of ~45 different subunits, encoded both by the nuclear and mitochondrial genome (6,7). Among these, 7 subunits [NADH Dehydrogenase 1–6 (ND1-6) and ND4L] are encoded by the mitochondrial DNA (mtDNA) (6), and implicated in CoQ binding and proton pumping functions (1,8). The other 38 subunits are encoded by the nuclear genome (6,9,10). These include 7 core subunits (NDUFV1-2, NDUFS1-3, NDUFS7 and NDUFS8; NDUF=NADH Dehydrogenase Flavoprotein) (1,11) and 31 supernumerary/

accessory subunits, whose function is not completely known (6). CI has L-shaped structure with peripheral and hydrophobic arms (12–14), that is conserved from *E.coli* to Human (4). The Peripheral arm has 2 functional modules: (i) NADH (N) (electron input module) containing the NADH oxidation site and (ii) Quinone (Q) (electron output module) with CoQ reduction site (15). The hydrophobic arm has 2 functional modules (i) Proximal Pump (PP) and (ii) Distal Pump (DP) involved in proton translocation (16) (Figure 1). The details of the structural organization of CI are beyond the scope of the current chapter.

1.2 COMPLEX I DYSFUNCTION IN HUMAN BRAIN PATHOLOGIES

Since CI is critical for mitochondrial metabolism, it has emerged as a central regulator of bioenergetics in the Central Nervous System (CNS). Consequently, minor changes in CI function either due to mutations or inhibitors can severely affect energy production, potentially leading to mitochondrial damage and neuronal dysfunction (17). In addition, CI is a major source of reactive oxygen species (ROS) in mitochondria, which induces oxidative damage that ultimately contributes to aging and neurodegeneration (18–22). Genetic mutations and post-translational modifications (PTMs) in critical subunits of CI (23) predominantly contribute to its dysfunction and ensuing mitochondrial damage during brain pathologies.

Deficiency of CI is the most frequently encountered single enzyme deficiency in patient with mitochondrial

disorders (24). Many neurological diseases linked with OXPHOS involve CI dysfunction, which could be due to genetic mutations in its subunits. Among OXPHOS complexes, mutated CI genes are extensively studied by candidate gene analysis, genetic linkage and homozygosity mapping, and exome sequencing (25). Approximately 33% of CI deficiency is due to mutations in CI subunits, while others are linked with dysfunctional CI assembly factors (26) or other factors that induce mutation independent effects (27).

Mutations in CI subunits are reported in many mitochondrial disorders such as mitochondrial encephalomyopathy, lactic acidosis and Stroke-like episodes (MELAS), myoclonic epilepsy with ragged-red fibers (MERRF), progressive epilepsy (28) and Leigh's syndrome (LS) (29). CI mutations are also associated with neurodegenerative diseases such as Parkinson's disease (PD) (30) and Alzheimer's disease (AD) (31), Down's syndrome (32), progressive encephalopathy (33), Hypertrophic Cardiomyopathy (HCM), fatal infantile lactic acidosis (FILA) or severe infantile onset encephalopathies, and other human diseases (34,35).

Loss of CI activity is evident in neurodegenerative diseases. Decreased CI activity in Substantia Nigra *pars compacta* (SNpc) region of the ventral mid-brain and cortex has been reported in PD patients (36). Apart from mtDNA defects (37) and somatic mutation in CI subunits, CI is also the main target for oxidative PTM in PD brain since these subunits are susceptible to oxidative damage, resulting in CI disassembly and dysfunction (38). Mitochondrial dysfunction could contribute to early AD pathogenesis (39). Aggregation of Amyloid

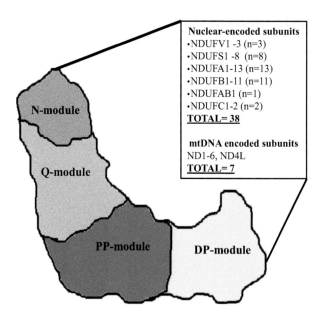

FIGURE 1   Schematic representation of the structure of mammalian Complex I highlighting the different functional modules.

beta (Aβ) protein is linked with mitochondrial dysfunction (40,41). Impaired OXPHOS, especially CI activity has be reported in AD patients and animal models (32,42,43). Proteomic studies have correlated accumulation of Aβ and tau proteins with CI defects and mitochondrial dysfunction (44,45). CI mediated ROS production is involved in the amyloidogenic Amyloid precursor proteins (APP) processing (46,47).

1.3   Role of PTMs in Brain Diseases

PTMs have emerged as an important mechanism contributing to protein function during homeostatic and disease conditions. This also bears importance to mitochondrial dysfunction and altered CI activity in the brain. Since proteins containing oxidative PTMs accumulate during aging and in some pathological conditions (48), it is pertinent to understand the underlying chemistry of PTMs and their role in loss of protein function.

Although the amino acid sequence of a protein determines its folding, conformation, activity and stability, at any instant, the proteome of a cell/tissue would be two to three times more complex than the prediction of the encoding genome (49). Such expansion of the proteomic complexity occurs predominantly due to PTMs among others. Interestingly, in eukaryotes, around 5% of the genome codes for enzymes that contribute to PTMs of cellular proteins (49). PTMs by enzymatic and non-enzymatic mechanisms influence protein structure and function (50). Some PTMs are amino-acid specific. For instance, phosphorylation affects Ser, Thr, and Tyr residues; acetylation and ubiquitination occur on Lys and methylation on Lys and Arg (51). PTMs of cellular proteins directly contribute to the pathogenesis of several human disorders including cardiovascular and chronic kidney disorder (CKD) (52), aging and age related disorders (50), transmissible spongiform encephalopathies (TSEs) or prion disorders (53) and predominantly in neurodegenerative disorders.

In PD for instance, phosphorylation of α-synuclein (α-syn) is correlated with disease pathogenis (54). Phosphorylation at Ser129 (pS129) of α-syn is considered to be disease-associated in human post-mortem samples and animal models. During dopaminergic neurodegeneration, this PTM represents >90% of the total α-syn found in Lewy bodies, compared with ~4% under normal physiological conditions (54,55). In transgenic models over-expressing α-syn (56–58), the levels of pS129 α-syn positively correlated with neurodegenerative pathology.

In AD, hyper-phosphorylation of Tau protein in the brain is considered to be the major PTM associated with neurodegeneration (59). In addition, 3 Ser and 13 Thr residues in Tau protein are potentially modified by O-linked-*N*-acetyl glucosaminylation (OGlcNAcylation or glycation) (60–64). Acetylation of Tau protein has been reported in the microtubule binding domain, which potentially disrupts microtubule assembly (65). Further, elevated phosphorylation of Aβ at Ser8 and Ser26 has been reported in AD brains (66,67).

In Huntington's disease (HD), most PTMs are reported in the huntingtin (htt) protein (68). Htt protein is modified by phosphorylation (69,70), SUMOylation (SUMO=small ubiquitin-like modifier) and Ubiquitination (71). Acetylation at Lys444 facilitates the autophagic clearance of mutated htt (72). The htt protein is palmitoylated at Cys214 (73,74) and protein palmitoylation-resistant mutation increases inclusion formation, and neurotoxicity (74).

1.4   PTMs in CI

PTMs could have a physiological role including cell signaling (75) and regulation of kinase pathways (76) with implications for mitochondrial function. PTMs, both oxidative and non-oxidative could contribute to altered CI organization and function. Although oxidative PTMs of CI are linked to diseases (23), resulting in disassembly and loss of activity (38), they could have physiological role. Taylor et al. (77) reported extensive oxidative modification of Trp residues in CI and CV subunits, which could contribute to their dysfunction under physiological conditions. Similarly, phosphorylation of CI could have a potential role during physiological conditions and brain disease (78). However, the functional outcome and regulatory role of PTMs and structural implications on specific subunit/sub-complex/entire CI have not been completely defined.

1.4.1 Oxidative PTMs in CI

Oxidative stress causes deleterious effects via oxidative PTMs of neuronal proteins including mitochondrial proteins with implications for human diseases (79). Oxidation of proteins by reactive species include hydroperoxide (ROOH) formation, hydroxylation of aromatic and aliphatic amino acids, nitrosation of sulfhydryl (SH) group, oxidation of SH groups [-SH could form disulfide bond with Sulfinic acid (–SOH), Sulfenic acid (-SO$_2$H) or Sulfonic acid (-SO$_3$H) derivatives], chlorination of aromatic groups and primary amino groups, and carbonylation of amino acid chains (80). Cys, Met and Thr residues in proteins are relatively more susceptible to oxidation (reversible and irreversible) leading to intra and inter-protein cross-linking, disruption of polypeptide chain and loss of protein function (80,81). Table 1 lists the amino acids in proteins with respective oxidized products.

CI is very sensitive to redox damage and its subunits could undergo reversible/irreversible oxidative PTMs (23), which could inhibit its activity leading to electron

TABLE 1

Oxidation of amino acids in proteins, with different oxidized products (HNE = 4-hydroxy nonenal, PHA = p-hydroxyphenyl acetaldehyde)

Amino acids	Oxidized products	Reference
Lys	Aminadipic semialdehyde (metal catalyzed), HNE-Lys, Acrolein-Lys, Carboxymethy-Lys, PHA-Lys	(81,82)
His	2-Oxo-histidine (metal catalyzed), Asn, Asp, HNE-His	(81,82)
Phe	2,3-Dihydroxy Phe, 2,-3, & 4 hydroxy Phe	(81)
Tyr	Di-Tyr, 3-nitro Tyr, Chlor-Tyr	
Trp	Hydroxy-& nitro Trp, Oxindolylalanine, N-formyl Kynurenine, Kynurenine	(81,83)
Met	Met sulfoxide, Met sulfone	(81)
Cys	Disulfide, Mixed disulfide (e.g., glutathionylation), S-Nitosation, HNE-Cys, Cys acid, sulfinic, sulfenic and sulfonic acid	(84,85)
Val, Leu	Hydro peroxides	(81)
Glu	Oxalic acid, Pyruvic acid	(81)
Arg	Glutamic semialdehyde, Chloramines	(81)
Thr	2-Amino 3- ketobutyric acid	(81)

leakage and exacerbated oxidative damage (86). PTMs of CI could entail irreversible modification of Cys of Fe-S centers or release of Fe from Fe-S centre or irreversible oxidation of Cys, Trp and nitration of Tyr in different subunits.

1.4.1.1 PTM OF CYS RESIDUES

Oxidation of thiol group of Cys in proteins include formation of intra-molecular disulfide bond with adjacent thiols, S-nitrosation and formation of higher oxidation states: -SOH, -SO$_2$H and SO$_3$H (23), with the last two being irreversible PTMs. Although studies on Cys oxidation of CI subunits are available, systematic mapping and quantification of all cys PTMs is incomplete due to technical limitations. Lin et al. (87) developed a novel method to study the mitochondrial proteome and quantify oxidized Cys by labeling with 4-iodobutyl triphenyl phosphonium (IBTP). IBTP localizes within the mitochondria and reacts with thiols to generate thioether adducts, thereby indicating the redox status of the cell.

The redox status of neurons and the constituent protein thiols could be affected by depletion of GSH. SNpc region of the human brain displays depletion of glutathione-reduced (GSH) compared to other brain regions (88), making it more susceptible to oxidative damage of Cys thiols with implications for PD (88). Danielson et al. (89) developed a method to detect oxidative PTM of thiols in a mouse model of GSH-depletion with implications for PD, and mapped the reversible oxidation status of ~26% (34 out of 130) of total Cys in CI. Differential alkylation and mass spectrometry of isolated CI detected increased oxidation of 6 Cys residues in 3 subunits in the PD mouse model, compared to controls. Of these, three residues were present within Fe-S clusters (89).

1.4.1.1.1 S-Glutathionylation Oxidative stress is usually associated with decreased ratio of GSH to oxidized glutathione (GSSG) (84). Increased GSSG could inactivate critical enzymes via glutathionylation, a PTM where thiol-disulfide exchange takes place between protein thiols and GSSG to form adducts called protein GSH disulfides (Pr-SSG) (84). This is a spontaneous and enzyme-independent reaction that varies with GSH/GSSG ratio and is reversible by glutaredoxin (Grx) (84). Glutathionylated protein either maintains Pr-SSG levels or may displace GSH by an adjacent thiol to form an intra disulfide bridge (90).

Glutathionylation of Cys thiols could potentially alter CI function. CI is glutathionylated at NDUFS1 (75kDa) and NDUFV1 (51 kDa) subunits (83), inhibiting its activity and enhancing the production of superoxide (83). This can be reversed by Grx2 or thioredoxin2 (Trx2), which restores the superoxide level (90). Both Grx2 and Trx2 maintain thiol homeostasis in brain cells (91). Grx2 mutant can neither reverse CI glutathionylation nor restore CI activity or elevated superoxide indicating that Grx2 is necessary to maintain the redox status and activity of CI (23). Not only increased GSSG, but reactive thiyl radical could also cause CI S-glutathionylation (92).

NDUFS1, a critical subunit of CI containing 16 Cys residues (10 involved in Fe-S clusters and 6 in other subunits) is reported to undergo glutathionylation (93). Among these, Cys531 and Cys704 are glutathionylated as these two residues are present on the surface of mammalian CI and exposed to the glutathione pool. While glutathionylation of Cys531 and 704 on NDUFS1 inhibited CI activity but did not lead to superoxide production, its reversal did not restore the CI activity (93). Apart from these residues, Cys226, Cys367 and Cys727 are also glutathionylated in NDUFS1 (92,94), apart from S-sulfonation of Cys554 and Cys727 (92).

Another subunit NDUFV1 displayed glutathionylation at Cys206 and Cys187 (94). In addition, Cys425, which acts as a ligand residue of 4Fe-4S N3 centre was also glutathionylated, indicating that damage of Fe-S centers could affect CI function (94). Zhang et al. (95) detected intra-polypeptide disulfide bridges in NDUFV1 at Cys125/Cys142, Cys187/Cys206, and Cys142/Cys206 in response to oxidative stress.

1.4.1.1.2 S-Nitrosation Reactive nitrogen species (RNS) such as Nitric oxide (NO) influences cellular redox status, mitochondrial function and affects CI activity by modifying Cys and Tyr residues thereby altering the structure of the complex (23). NO mediates PTM of CI via formation of S-nitrosothiols (RSNO) at Cys residues leading to CI inhibition (54). Exposure to NO or RSNO [S-nitrosoglutathione (SNOG) and S-nitrosoacetyl penicillamine (SNAP)] inhibited CI activity (96), and the same could be reversed by thiol reducing agents (97). S-nitrosation mediated inhibition of CI activity increases superoxide levels (85). Dopaminergic neurons are more sensitive to NO mediated CI inhibition indicating the role of S-nitrosylation in PD pathogenesis (98). Increased mitochondrial NO levels and decreased GSH could cause reversible CI inhibition in dopaminergic neurons (54,99).

Bioinformatics data speculated 43 Cys residues in 22 subunits of rodent CI and 35 Cys residues in 16 subunits of human CI as potential targets of S-nitrosation (100). Cys residues of subunits NDUFA9, NDUFS2 were more susceptible for S-nitrosation. Cys137 in NDUFS1 subunit was reported to be targeted for S-nitrosation (99). Continuous exposure of NO causes irreversible damage by destruction of assembly of Fe-S centers (101).

1.4.1.2 NITRATION

Mitochondrial metabolism is known to generate peroxynitrite (PN), a RNS that induces nitration of Tyr residues in proteins to generate 3-nitrotyrosine (3-NT) (102). Increased protein nitration has been noted in the human brain during physiological aging (especially in hippocampus and frontal cortex) and PD (103,104). Exposure to NO inhibited CI activity, which was partially reversed by Superoxide Dismutase (SOD) or PN blockers, but not by thiol antioxidants (105), indicating that NO mediates 3-NT modification in addition to S-nitrosation. Curcumin, a dietary polyphenol and potential PN scavenger reversed CI inhibition mediated by direct exposure to PN (106). GSH depletion exacerbated PN-dependent CI inhibition, which was reversed by curcumin (107). Proteomics data demonstrated that PN-dependent 3-NT modification occurs in five CI subunits: NDUFS2, NDUFS8, B17.2, NDUFB4, and NDUFA6 (108). However, the structural basis of 3-NT modification at specific sites, on CI dysfunction is not completely understood.

1.4.1.3 CARBONYLATION AND TRP OXIDATION

Protein carbonylation is another important PTM which occurs in response to oxidative stress during aging and other pathological conditions. The side chain of amino acids such as Lys, Arg, Pro and Thr are susceptible to carbonylation which might alter mitochondrial function (48). Our previous studies demonstrated elevated protein carbonylation during brain aging and PD (103,104). Keeney et al. (38) demonstrated ~47% increase in protein carbonyls in the catalytic subunits of CI in PD patients. Carbonylation of mtDNA encoded subunits ND4 and ND5 and six nuclear DNA-encoded subunits NDUFS1, NDUFS2, NDUFV1, NDUFB5, NDUFB6 and NDUFB7 induced disassembly of CI structure and inhibition of catalytic activity. Proteomic analysis in a mouse model of epilepsy identified carbonylation of Arg76 on NDUFS1 subunit of CI, which caused substantial structural changes in the whole complex (109).

In addition to carbonylation, Trp moieties in proteins can also undergo oxidation which results in increase in molecular weight by +16, +32 and +4 Da corresponding to the three oxidation states i.e., Oxindolylalanine, N-formyl Kynurenine, and Kynurenine respectively (23). Our previous study in a mouse model of myodegeneration and muscle biopsies from subjects with muscle pathologies demonstrated Trp oxidation in mitochondrial proteins with structural implications (110). Taylor et al. (83) identified *N*-formyl Kynurenine in multiple CI subunits (NDUFA9, NDUFV1, NDUFS3, NDUFS1, NDUFS7, NDUFS8, NDUFS4 and NDUFA5). However, the structural effects of these modifications in CI are currently unknown.

1.4.2 Non-Oxidative PTMs in CI

1.4.2.1 PHOSPHORYLATION

Among the non-oxidative PTMs, phosphorylation is the mostcommon modification that influences the function of mitochondrial proteins (111) including CI (112,113). Phosphorylation of specific subunits of CI (114) could be mediated by specific kinases thereby altering CI activity (115) and assembly (116). Phosphorylation also influences the higher order assembly of OXPHOS complexes (117).

Mitochondrial function can be regulated by Tyr phosphorylation of mitochondrial proteins, which in turn is regulated by different Kinases including Tyr Src Kinases (118). Tyr phosphorylation contributes to cell survival, proliferation and metastasis (119). During proliferative phase, Src Kinases are activated, leading to Tyr phosphorylation of mitochondrial proteins including CI subunits (118). Tyr Src Kinases-mediated phosphorylation of NDUFB10 subunit of CI preserved its activity in cancer cells (120), which could control mitochondrial redox potential and $NAD^+/NADH$ ratio. Increased NADH consumption ensures sufficient flux through

TCA cycle, thus producing TCA cycle intermediates. Preservation of membrane potential by importing mitochondrial proteins such as TCA cycle enzyme into mitochondria enables biosynthesis and proliferation. Since NDUFB10 is involved in Fe-S cluster assembly, its phosphorylation could affect electron transfer within CI and increase its activity (119).

Elevated intracellular cAMP promotes phosphorylation of NDUFS4 subunit of CI by cAMP dependent protein kinase (PKA), which stimulates its activity and NAD-linked mitochondrial respiration (121). Alteration of this phosphorylation site abolished cAMP-mediated activation of CI and impaired its normal assembly indicating that cAMP cascade could regulate bioenergetics of mammalian tissue (121). Loss of the phosphorylation site in NDUFS4 is linked with neurological disorders (122). In addition to these phosphorylation events, Ser59 and Ser231 of NDUFA10 in bovine heart and Thr96 of NDUFA13 and phosphorylation in NDUFB11 also altered CI enzyme activity (78).

Analysis of bovine heart CI (123) revealed site-specific phosphorylation in NDUFA10 subunit (124,125). PINK1, a mitochondrial kinase phosphorylates NDUFA10 at Ser250 with implications for PD (126). Loss of PINK1 function lowered CI activity indicating that phosphorylation at a single site of an accessory subunit could regulate CI activity. These data highlight the significance of phosphorylation in the assembly, activity and regulation of CI. To understand their role in CI dynamics, chemical modifications mimicking PTMs are employed (127).

Studies attempting to systematically analyze the total protein phosphorylation profile in the subunits of CI and correlate their role are non-existent. Covian and Balaban (128) reported 69 unique phosphorylation sites in 26 subunits of CI, mostly from total phosphoproteome analysis of mitochondria isolated from different tissues. Among these, 18 sites are in the peripheral arm of CI, while one is in the hydrophobic core. The other 50 sites are spread among 18 accessory subunits. Phosphorylation could affect the orientation of co-factors/Fe-S clusters, structure of subunits and their interaction with adjacent subunits. For instance, phosphorylation at Tyr141 in the subunit NDUFS2 (129,130) has structural importance since it is positioned at the interface with NDUFS7, a subunit that contains the last Fe-S cluster (N2). Since this interface is important for CoQ reduction (123), it is crucial for phosphorylation- mediated regulation.

The accessory subunits in the membrane domain of CI including NDUFA3, NDUFA4, NDUFB4, NDUFB10, and NDUFB7 harbor several phosphorylation sites (128). Phosphorylation in the accessory subunits could potentially affect CI assembly (123) but not the proton pumping activity. Phosphorylation at Ser59 and Ser231 of NDUFA10 (131) and Ser55 of NDUFA1 in bovine heart CI (112) affect

the assembly of the complex (132). NDUFA13, a subunit vital for CI assembly is phosphorylated at Thr96 (133). On the other hand, phosphorylation of NDUFB11 altered the enzyme activity of CI (132).

In a recent study, we carried out bioinformatics analysis to predict the potential phosphorylation sites in different subunits of CI and their structural consequence (78). Our study predicted phosphorylation to be highest among the core subunits of CI, with NDUFS1 having significantly higher number and % phosphorylation sites compared to others. Structural modelling of subunits harboring Fe-S cluster, NADH and FMN binding sites and CoQ binding sites highlighted the close proximity of phosphorylation sites to the binding sites of these cofactors with potential structural implications. However, mitochondrial phosphoproteomics carried out in rat and human muscle to validate the bioinformatics study identified very few phosphorylation sites in CI subunits. This could be due to tissue specificity, transient nature of phosphorylation events, sensitivity of the experimental methods, loss of peptides during proteomics and limitations of phosphopeptide enrichment. Similar to other non-oxidative PTMs, phosphorylation status depends on age, physiological stimuli and kinase profile. Some of the phosphorylated residues detected in the proteomics experiment may not match the bioinformatics data indicating the limitation of the theoretical prediction. We propose that protein phosphorylation is a transient and regulatory event that potentially influences CI structure and function, affecting the bioenergetics and thereby contributing to aging and brain diseases. However, the limitations of predictive studies and experimental approaches need to be considered while assigning phosphorylation events in CI.

1.4.2.2 ACETYLATION

Apart from phosphorylation, N-α-acetylation is one of the most common PTMs in eukaryotic proteins. In general up to 50% cellular proteins and under specific conditions, up to 80–90% mammalian proteins are subjected to acetylation. Many times, acetylation could occur co-translationally after the synthesis of 20–30 amino acids of the nascent polypeptide or by removal of the initiator Met followed by N-acetylation of the second residue (134). Around 78% of the modified subunits of CI are acetylated by this pathway (134). In general, Gly, Ala, Ser, Met, and Asp are the predominant N-terminal residues of N-acetylated proteins (135).

Proteomic analysis of bovine CI demonstrated that, subunits NDUFB6, NDUFB9, NDUFA13, NDUFB4, NDUFA7, NDUFC2, NDUFA6, NDUFB5, NDUFA2, and NDUFA3 are N-acetylated at the second residue, while NDUFA11 and NDUFA12 are acetylated at the initiator residue (134).

1.4.2.3 METHYLATION

Methylation is a reversible non-oxidative PTM, involving side chains of Arg, Lys and to a limited extent, His. In CI isolated from bovine mitochondria, N-acetylated NDUFB3 undergoes methylation, with 1, 2 or 3 methyl groups attached to His at position 4, 6 and 8 in various combinations (134). However, its biological significance and structural implications are currently unclear, since the orientation of the modified residue with respect to the inner mitochondrial membrane has not been ascertained (134).

NDUFAF7 is a CI assembly factor present in the mitochondrial matrix. It is a protein methylase which methylates the NDUFS2 subunit of human CI at Arg85 at ω-N^G,$N^{G'}$ nitrogen atoms of its guanidine group using S-adenosyl Met as the methyl donor and protein Arg methyltransferases (PRMTs) type II (136,137). This is physiologically significant since it acts early in CI assembly, thereby stabilizing the 400 kDa sub-complex that includes NDUFS2, NDUFS7, ND1, NDUFS3 and NDUFS8. These subunits form the initial part of peripheral arm and later form the junction to the membrane and peripheral arm (137). This PTM may also influence the enzyme activity of CI. NDUFS2, NDUFS7 and ND1 form the CoQ binding pocket, while NDUFS7 also harbors the N2 Fe-S cluster. Methylation of Arg85 could increase its hydrophobicity and decrease its potential to form hydrogen bonds thereby influencing the redox potential of cluster N2 (137).

1.4.2.4 MYRISTOYLATION

Protein N-myristoylation is the attachment of Myristic acid, a 14-carbon saturated fatty acid, to the N-terminal Gly of proteins (138,139). It is estimated that approximately 0.5–1.5% of eukaryotic proteins are N-myristoylated. In general, myristic acid is co-translationally attached to the N-terminal Gly residue after removal of the initiating Met. In addition, post-translational N-myristoylation can also occur in many caspase-cleavage products in apoptotic cells (140,141). Both co-translational and post-translational N-myristoylation are catalyzed by N-myristoyltransferase (NMT), a member of the GCN5-related N-acetyltransferase superfamily of proteins (142).

The biological significance of myristoylation among mitochondrial proteins is obscure. Its role in CI function has not been extensively studied. Bovine heart CI is N-α-myristoylated on Gly 2 of the nuclear encoded subunit NDUFB7 (143). Since this subunit is a component of sub-complex I that represents a part of the membrane arm, the myristoyl group may be responsible for binding of the subunit to the inner mitochondrial membrane (134).

1.4.2.5 OTHER NON-OXIDATIVE PTMS

In bovine CI, certain PTMs in mtDNA encoded subunits are transient including formylation where all subunits retain N α-formyl groups on their translational initiator Met residues (136). On the other hand, stable PTMs include loss of mitochondrial import sequence in 18 subunits of bovine CI and the introduction of Fe-S clusters into five subunits NDUFS1, NDUFV2, NDUFV1, NDUFS8 and NDUFS7 (136).

Although other non-oxidative PTMs like SUMOylation, Palmitoylation, Ubiquitination, Prenylation, Glycosylation (both N and O linked) are present, they are not reported in human, especially in CI. Since PTMs have emerged as an important mechanism contributing to

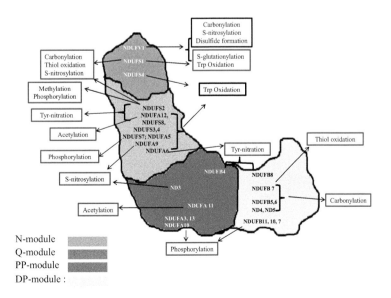

FIGURE 2 Schematic illustration of some of the reported PTMs in different CI subunits.

altered CI activity in various neurodegenerative disorders, assessment of these PTMs warrants further investigation.

2 CONCLUSION

CI plays a crucial role in OXPHOS, mitochondrial bioenergetics and generation of reactive species with implications for many pathophysiological processes. Many human diseases are associated with altered CI activity and subsequent mitochondrial dysfunction. Research evidences highlight the influence of PTMs on CI function (Summarized in Figure 2). PTMs could contribute either to CI damage or in cytoprotection, depending on the site of modification with implications for CNS diseases. In terms of structural biology, analysis of PTMs could assist in understanding the structure-function relationship of the complex. Studies on PTMs in CI with disease implications have limitations. Firstly, most of the studies are in experimental models, while their status in human tissues is relatively unknown. Secondly, the structural implications of PTMs are currently unclear. Thirdly, studies addressing CI PTMs concentrate on a particular domain and/or subunit, such as NDUFS1 (75 kD) and NDUFV1 (51 kD) subunits, and not the entire complex. Fourthly, since PTMs could be transient and influenced by tissue, age, physiological and pathological stimuli among others, an exhaustive list of all PTMs is not available. Hence, Proteomics-based comprehensive analysis of PTMs throughout the complex under different physiological and pathological conditions is required to get a complete picture of their role in any mitochondrial disorder.

ACKNOWLEDGMENTS

This work is supported by the Science and Engineering Research Board-National Postdoctoral Fellowship (SERB-NPDF).

REFERENCES

1. Janssen RJRJ, Nijtmans LG, van den Heuvel LP, Smeitink JAM. Mitochondrial complex I: Structure, function and pathology. Journal of Inherited Metabolic Disease. 2006 Aug 01;29(4):499–515.
2. Zickermann V, Kerscher S, Zwicker K, Tocilescu MA, Radermacher M, Brandt U. Architecture of complex I and its implications for electron transfer and proton pumping. Biochimica et biophysica acta. 2009 Jun;1787(6):574–83. PubMed PMID: 19366614. Pubmed Central PMCID: PMC2699368. Epub 2009/04/16.eng.
3. Yano T. The energy-transducing NADH: quinone oxidoreductase, complex I. Molecular Aspects of Medicine. 2002 Oct 1;23(5):345–68.
4. Schultz BE, Chan SI. Structures and proton-pumping strategies of mitochondrial respiratory enzymes. Annual Review of Biophysics and Biomolecular Structure. 2001;30(1):23–65. PubMed PMID: 11340051.
5. Mimaki M, Wang X, McKenzie M, Thorburn DR, Ryan MT. Understanding mitochondrial complex I assembly in health and disease. Biochimica et Biophysica Acta (BBA) – Bioenergetics. 2012 Jun 1;1817(6):851–62.
6. Carroll J, Fearnley IM, Shannon RJ, Hirst J, Walker JE. Analysis of the subunit composition of complex I from bovine heart mitochondria. Molecular & Cellular Proteomics: MCP. 2003 Feb;2(2):117–126. PubMed PMID: 12644575. Epub 2003/03/20.eng.
7. Carroll J, Fearnley IM, Skehel JM, Shannon RJ, Hirst J, Walker JE. Bovine complex I Is a complex of 45 different subunits. Journal of Biological Chemistry. 2006 Oct 27;281(43):32724–7.
8. Friedrich T, Böttcher B. The gross structure of the respiratory complex I: a Lego System. Biochimica et Biophysica Acta (BBA) – Bioenergetics. 2004 Jan 30;1608(1):1–9.
9. Stojanovski D, Johnston AJ, Streimann I, Hoogenraad NJ, Ryan MT. Import of nuclear-encoded proteins into mitochondria. Experimental Physiology. 2003;88(1):57–64.
10. Hoogenraad NJ, Ward LA, Ryan MT. Import and assembly of proteins into mitochondria of mammalian cells. Biochimica et Biophysica Acta (BBA) – Molecular Cell Research. 2002 Sep 2;1592(1):97–105.
11. Lazarou M, Thorburn DR, Ryan MT, McKenzie M. Assembly of mitochondrial complex I and defects in disease. Biochimica et Biophysica Acta (BBA) – Molecular Cell Research. 2009 Jan 1;1793(1):78–88.
12. Hofhaus G, Weiss H, Leonard K. Electron microscopic analysis of the peripheral and membrane parts of mitochondrial NADH dehydrogenase (Complex I). Journal of Molecular Biology. 1991 Oct 5;221(3):1027–43.
13. Vinothkumar KR, Zhu J, Hirst J. Architecture of mammalian respiratory complex I. Nature. 2014 Sep 7 online;515:80.
14. Zickermann V, Wirth C, Nasiri H, Siegmund K, Schwalbe H, Hunte C, et al. Mechanistic insight from the crystal structure of mitochondrial complex I. Science. 2015;347(6217):44–9.
15. Sazanov LA, Hinchliffe P. Structure of the hydrophilic domain of respiratory complex I from Thermus thermophilus. Science. 2006;311(5766):1430–1436.
16. Mathiesen C, Hägerhäll C. Transmembrane topology of the NuoL, M and N subunits of NADH: quinoneoxidoreductase and their homologues among membrane-bound hydrogenases and bona fide antiporters. Biochimica et Biophysica Acta (BBA) – Bioenergetics. 2002 Dec 2;1556(2):121–32.
17. Schapira AH. Human complex I defects in neurodegenerative diseases. Biochimica et biophysica acta. 1998 May 6;1364(2):261–270. PubMed PMID: 9593927. Epub 1998/06/19. eng.
18. Balaban RS, Nemoto S, Finkel T. Mitochondria, oxidants, and aging. Cell. 2005 Feb 25;120(4):483–95.
19. Dawson TM, Dawson VL. Molecular pathways of neurodegeneration in Parkinson's disease. Science. 2003;302(5646):819–22.
20. Giachin G, Bouverot R, Acajjaoui S, Pantalone S, Soler-Lopez M. Dynamics of human mitochondrial complex I assembly: implications for neurodegenerative diseases. Frontiers in Molecular Biosciences. 2016;3:43. PubMed PMID: 27597947. Pubmed Central PMCID: PMC4992684. Epub 2016/ 09/07.eng.
21. Hur JH, Stork DA, Walker DW. Complex-I-ty in aging. Journal of Bioenergetics and Biomembranes. 2014 Aug 1;46(4):329–35.

22. Scheffler IE. Mitochondrial disease associated with complex I (NADH-CoQ oxidoreductase) deficiency. Journal of Inherited Metabolic Disease. 2015 May 1;38(3):405–15.

23. Srinivas Bharath MM. Post-translational oxidative modifications of mitochondrial complex I (NADH: Ubiquinone Oxidoreductase): implications for pathogenesis and therapeutics in human diseases. Journal of Alzheimer's Disease: JAD. 2017;60(s1):S69–686. PubMed PMID: 28582861. Epub 2017/ 06/07.eng.

24. Rodenburg RJT. Biochemical diagnosis of mitochondrial disorders. Journal of Inherited Metabolic Disease. 201105/0412/21/received03/16/revised03/17/accepted;34(2):283–92. PubMed PMID: PMC3063578.

25. Fassone E, Rahman S. Complex I deficiency: clinical features, biochemistry and molecular genetics. Journal of Medical Genetics. 2012;49(9):578.

26. Calvo SE, Tucker EJ, Compton AG, Kirby DM, Crawford G, Burtt NP, et al. High-throughput, pooled sequencing identifies mutations in NUBPL and FOXRED1 in human complex I deficiency. Nature Genetics. 2010 Sep 5 online;42:851.

27. Nouws J, Nijtmans LGJ, Smeitink JA, Vogel RO. Assembly factors as a new class of disease genes for mitochondrial complex I deficiency: cause, pathology and treatment options. Brain. 2012;135(1):12–22.

28. Yagi T, Seo BB, Bernardo SD, Nakamaru-Ogiso E, Kao M-C, Matsuno-Yagi A. NADH dehydrogenases: from basic science to biomedicine. Journal of Bioenergetics and Biomembranes. 2001 Jun 1;33(3):233–42.

29. Rodenburg RJ. Mitochondrial complex I-linked disease. Biochimica et Biophysica Acta (BBA) – Bioenergetics. 2016 Jul 1;1857(7):938–45.

30. Parker WD, Parks JK. Mitochondrial ND5 mutations in idiopathic Parkinson's disease. Biochemical and Biophysical Research Communications. 2005 Jan 21;326 (3):667–9.

31. Imanishi H, Yokota M, Mori M, Shimizu A, Nakada K, Hayashi J-I. Nuclear but not mitochondrial DNA involvement in respiratory complex I defects found in senescence-accelerated mouse strain, SAMP8. Experimental Animals. 2011;60(4):397–404.

32. Kim SH, Vlkolinsky R, Cairns N, Fountoulakis M, Lubec G. The reduction of NADH ubiquinone oxidoreductase 24- and 75-kDa subunits in brains of patients with Down syndrome and Alzheimer's disease. Life Sciences. 2001 May 4;68(24):2741–2750. PubMed PMID: 11400916. Epub 2001/06/13.eng.

33. Ogilvie I, Kennaway NG, Shoubridge EA. A molecular chaperone for mitochondrial complex I assembly is mutated in a progressive encephalopathy. Journal of Clinical Investigation. 200506/20/received07/26/accepted;115 (10):2784–92. PubMed PMID: PMC1236688.

34. Pagliarini DJ, Calvo SE, Chang B, Sheth SA, Vafai SB, Ong SE, et al. A mitochondrial protein compendium elucidates complex I disease biology. Cell. 2008 Jul 11;134 (1):112–123. PubMed PMID: 18614015. Pubmed Central PMCID: PMC2778844. Epub 2008/07/11.eng.

35. Carilla-Latorre S, Gallardo ME, Annesley SJ, Calvo-Garrido J, Grana O, Accari SL, et al. MidA is a putative methyltransferase that is required for mitochondrial complex I function. Journal of Cell Science. 2010 May 15;123 (Pt 10):1674–1683. PubMed PMID: 20406883. Epub 2010/ 04/22.eng.

36. Haelterman NA, Yoon WH, Sandoval H, Jaiswal M, Shulman JM, Bellen HJ. A mitocentric view of Parkinson's

disease. Annual Review of Neuroscience. 2014;37:137–59. PubMed PMID: 24821430. Pubmed Central PMCID: PMC4659514. Epub 2014/05/14.eng.

37. Gu M, Cooper JM, Taanman JW, Schapira AH. Mitochondrial DNA transmission of the mitochondrial defect in Parkinson's disease. Annals of Neurology. 1998 Aug;44 (2):177–186. PubMed PMID: 9708539. Epub 1998/ 08/26. eng.

38. Keeney PM, Xie J, Capaldi RA, Bennett JP, Jr. Parkinson's disease brain mitochondrial complex I has oxidatively damaged subunits and is functionally impaired and misassembled. The Journal of Neuroscience: The Official Journal of the Society for Neuroscience. 2006 May 10;26 (19):5256–5264. PubMed PMID: 16687518. Epub 2006/ 05/12.eng.

39. Valla J, Berndt JD, Gonzalez-Lima F. Energy hypometabolism in posterior cingulate cortex of Alzheimer's patients: superficial laminar cytochrome oxidase associated with disease duration. The Journal of Neuroscience: The Official Journal of the Society for Neuroscience. 2001 Jul 1;21(13):4923–4930. PubMed PMID: 11425920. Epub 2001/06/27.eng.

40. Tillement L, Lecanu L, Papadopoulos V. Alzheimer's disease: effects of beta-amyloid on mitochondria. Mitochondrion. 2011 Jan;11(1):13–21. PubMed PMID: 20817045. Epub 2010/09/08.eng.

41. Cha MY, Han SH, Son SM, Hong HS, Choi YJ, Byun J, et al. Mitochondria-specific accumulation of amyloid beta induces mitochondrial dysfunction leading to apoptotic cell death. PLoS One. 2012;7(4):e34929. PubMed PMID: 22514691. Pubmed Central PMCID: PMC3325919. Epub 2012/04/20.eng.

42. Hroudová J, Singh N, Fišar Z. mitochondrial dysfunctions in neurodegenerative diseases: relevance to Alzheimer's disease. BioMed Research International. 2014;2014:9.

43. Kim SH, Vlkolinsky R, Cairns N, Lubec G. Decreased levels of complex III core protein 1 and complex V beta chain in brains from patients with Alzheimer's disease and Down syndrome. Cellular and Molecular Life Sciences: CMLS. 2000 Nov;57(12):1810–16. PubMed PMID: 11130185. Epub 2000/12/29.eng.

44. Chou JL, Shenoy DV, Thomas N, Choudhary PK, Laferla FM, Goodman SR, et al. Early dysregulation of the mitochondrial proteome in a mouse model of Alzheimer's disease. Journal of Proteomics. 2011 Apr 1;74 (4):466–479. PubMed PMID: 21237293. Epub 2011/01/18. eng.

45. Zhang L, Zhang S, Maezawa I, Trushin S, Minhas P, Pinto M, et al. Modulation of mitochondrial complex I activity averts cognitive decline in multiple animal models of familial Alzheimer's disease. EBioMedicine. 2015 Apr 1;2(4):294–305. PubMed PMID: 26086035. Pubmed Central PMCID: PMC4465115. Epub 2015/06/19.eng.

46. Tamagno E, Guglielmotto M, Monteleone D, Tabaton M. Amyloid-β production: major link between oxidative stress and BACE1. Neurotoxicity Research. 2012 Oct 1;22 (3):208–19.

47. Bobba A, Amadoro G, Valenti D, Corsetti V, Lassandro R, Atlante A. Mitochondrial respiratory chain complexes I and IV are impaired by β-amyloid via direct interaction and through complex I-dependent ROS production, respectively. Mitochondrion. 2013 Jul 1;13(4):298–311.

48. Berlett BS, Stadtman ER. Protein oxidation in aging, disease, and oxidative stress. Journal of Biological Chemistry. 1997 Aug 15;272(33):20313–20316.

49. Walsh CT, Garneau-Tsodikova S, Gatto GJ, Jr. Protein posttranslational modifications: the chemistry of proteome diversifications. Angewandte Chemie (International ed in English). 2005 Dec 1;44(45):7342–72. PubMed PMID: 16267872. Epub 2005/11/04.eng.

50. Santos AL, Lindner AB. Protein posttranslational modifications: roles in aging and age-related disease. Oxidative Medicine and Cellular Longevity. 2017;2017:5716409. PubMed PMID: 28894508. Pubmed Central PMCID: PMC5574318. Epub 2017/09/13.eng.

51. Reimand J, Wagih O, Bader GD. Evolutionary constraint and disease associations of post-translational modification sites in human genomes. PLoS Genetics. 2015 Jan;11(1): e1004919. PubMed PMID: 25611800. Pubmed Central PMCID: PMC4303425. Epub 2015/01/23.eng.

52. Gajjala PR, Fliser D, Speer T, Jankowski V, Jankowski J. Emerging role of post-translational modifications in chronic kidney disease and cardiovascular disease. Nephrology, Dialysis, Transplantation: Official Publication of the European Dialysis and Transplant Association – European Renal Association. 2015 Nov;30(11):1814–24. PubMed PMID: 25862763. Epub 2015/04/12.eng.

53. Prusiner SB. Prions. Proceedings of the National Academy of Sciences of the United States of America. 1998 Nov 10;95(23):13363–13383. PubMed PMID: 9811807. Pubmed Central PMCID: PMC33918. Epub 1998/11/13. eng.

54. Anderson JP, Walker DE, Goldstein JM, de Laat R, Banducci K, Caccavello RJ, et al. Phosphorylation of Ser-129 is the dominant pathological modification of alpha-synuclein in familial and sporadic Lewy body disease. Journal of Biological Chemistry. 2006 Oct 6;281 (40):29739–29752. PubMed PMID: 16847063. Epub 2006/07/19.eng.

55. Fujiwara H, Hasegawa M, Dohmae N, Kawashima A, Masliah E, Goldberg MS, et al. alpha-Synuclein is phosphorylated in synucleinopathy lesions. Nature Cell Biology. 2002 Feb;4(2):160–164. PubMed PMID: 11813001. Epub 2002/01/29.eng.

56. Kahle PJ, Neumann M, Ozmen L, Müller V, Jacobsen H, Spooren W, et al. Hyperphosphorylation and insolubility of α-synuclein in transgenic mouse oligodendrocytes. EMBO Reports. 200202/15/received03/22/revised03/28/ accepted;3(6):583–588. PubMed PMID: PMC1084143.

57. Yamada M, Iwatsubo T, Mizuno Y, Mochizuki H. Overexpression of alpha-synuclein in rat substantia nigra results in loss of dopaminergic neurons, phosphorylation of alpha-synuclein and activation of caspase-9: resemblance to pathogenetic changes in Parkinson's disease. Journal of Neurochemistry. 2004 Oct;91(2):451–461. PubMed PMID: 15447678. Epub 2004/09/28.eng.

58. Chen L, Feany MB. Alpha-synuclein phosphorylation controls neurotoxicity and inclusion formation in a Drosophila model of Parkinson disease. Nature Neuroscience. 2005 May;8(5):657–663. PubMed PMID: 15834418. Epub 2005/04/19.eng.

59. Rissman RA, Poon WW, Blurton-Jones M, Oddo S, Torp R, Vitek MP, et al. Caspase-cleavage of tau is an early event in Alzheimer disease tangle pathology. Journal of Clinical Investigation. 200411/25/received05/07/ accepted;114(1):121–130. PubMed PMID: PMC437967.

60. Arnold CS, Johnson GV, Cole RN, Dong DL, Lee M, Hart GW. The microtubule-associated protein tau is extensively modified with O-linked N-acetylglucosamine. Journal of Biological Chemistry. 1996 Nov 15;271 (46):28741–28744. PubMed PMID: 8910513. Epub 1996/11/15.eng.

61. Wang Z, Udeshi ND, O'Malley M, Shabanowitz J, Hunt DF, Hart GW. Enrichment and site mapping of O-linked N-acetylglucosamine by a combination of chemical/enzymatic tagging, photochemical cleavage, and electron transfer dissociation mass spectrometry. Molecular & Cellular Proteomics: MCP. 2010 Jan;9(1):153–160. PubMed PMID: 19692427. Pubmed Central PMCID: PMC2808261. Epub 2009/08/21.eng.

62. Smet-Nocca C, Broncel M, Wieruszeski JM, Tokarski C, Hanoulle X, Leroy A, et al. Identification of O-GlcNAc sites within peptides of the Tau protein and their impact on phosphorylation. Molecular BioSystems. 2011 May;7 (5):1420–29. PubMed PMID: 21327254. Epub 2011/02/18. eng.

63. Ledesma MD, Bonay P, Avila J. Tau protein from Alzheimer's disease patients is glycated at its tubulin-binding domain. Journal of Neurochemistry. 1995 Oct;65(4):1658–1664. PubMed PMID: 7561862. Epub 1995/10/01.eng.

64. Nacharaju P, Ko L, Yen SH. Characterization of in vitro glycation sites of tau. Journal of Neurochemistry. 1997 Oct;69(4):1709–1719. PubMed PMID: 9326300. Epub 1997/11/05.eng.

65. Cohen TJ, Guo JL, Hurtado DE, Kwong LK, Mills IP, Trojanowski JQ, et al. The acetylation of tau inhibits its function and promotes pathological tau aggregation. Nature Communications. 2011;2:252. PubMed PMID: 21427723. Pubmed Central PMCID: PMC3120096. Epub 2011/03/24.eng.

66. Kumar S, Walter J. Phosphorylation of amyloid beta (Abeta) peptides – a trigger for formation of toxic aggregates in Alzheimer's disease. Aging. 2011 Aug;3(8):803–12. PubMed PMID: 21869458. Pubmed Central PMCID: PMC3184981. Epub 2011/ 08/27.eng.

67. Milton NG. Phosphorylation of amyloid-beta at the serine 26 residue by human cdc2 kinase. Neuroreport. 2001 Dec 4;12(17):3839–3844. PubMed PMID: 11726805. Epub 2001/12/01.eng.

68. Ehrnhoefer DE, Sutton L, Hayden MR. Small changes, big impact: posttranslational modifications and function of huntingtin in Huntington disease. The Neuroscientist: A Review Journal Bringing Neurobiology, Neurology and Psychiatry. 2011 Oct;17(5):475–92. PubMed PMID: 21311053. Pubmed Central PMCID: PMC3200085. Epub 2011/ 02/12.eng.

69. Colin E, Zala D, Liot G, Rangone H, Borrell-Pagès M, Li X-J, et al. Huntingtin phosphorylation acts as a molecular switch for anterograde/retrograde transport in neurons. The EMBO Journal. 200807/1002/21/received06/ 17/accepted;27(15):2124–34. PubMed PMID: PMC2516882.

70. Gauthier LR, Charrin BC, Borrell-Pages M, Dompierre JP, Rangone H, Cordelieres FP, et al. Huntingtin controls neurotrophic support and survival of neurons by enhancing BDNF vesicular transport along microtubules. Cell. 2004 Jul 9;118(1):127–38. PubMed PMID: 15242649. Epub 2004/07/10.eng.

71. Steffan JS, Agrawal N, Pallos J, Rockabrand E, Trotman LC, Slepko N, et al. SUMO modification of Huntingtin and Huntington's disease pathology. Science. 2004 Apr 2;304(5667):100–4. PubMed PMID: 15064418. Epub 2004/04/06.eng.

72. Jeong H, Then F, Melia TJ, Jr., Mazzulli JR, Cui L, Savas JN, et al. Acetylation targets mutant huntingtin to autophagosomes for degradation. Cell. 2009 Apr 3;137 (1):60–72. PubMed PMID: 19345187. Pubmed Central PMCID: PMC2940108. Epub 2009/04/07.eng.

73. Huang K, Sanders S, Singaraja R, Orban P, Cijsouw T, Arstikaitis P, et al. Neuronal palmitoyl acyl transferases exhibit distinct substrate specificity. FASEB Journal: Official Publication of the Federation of American Societies for Experimental Biology. 2009 Aug;23(8):2605–2615. PubMed PMID: 19299482. Pubmed Central PMCID: PMC2717768. Epub 2009/03/21.eng.

74. Yanai A, Huang K, Kang R, Singaraja RR, Arstikaitis P, Gan L, et al. Palmitoylation of huntingtin by HIP14 is essential for its trafficking and function. Nature Neuroscience. 2006 Jun;9(6):824–831. PubMed PMID: 16699508. Pubmed Central PMCID: PMC2279235. Epub 2006/05/16.eng.

75. Burgoyne JR, Oka S-I, Ale-Agha N, Eaton P. Hydrogen peroxide sensing and signaling by protein kinases in the cardiovascular system. Antioxidants & Redox Signaling. 201307/16/received8/07/accepted;18(9):1042–52. PubMed PMID: PMC3567777.

76. Burgoyne JR, Eaton P. Oxidant sensing by protein kinases A and G enables integration of cell redox state with phosphoregulation. Sensors (Basel, Switzerland). 201003/2601/11/received03/19/revised03/22/accepted;10(4):2731–51. PubMed PMID: PMC3274199.

77. Taylor RW, Giordano C, Davidson MM, d'Amati G, Bain H, Hayes CM, et al. A homoplasmic mitochondrial transfer Ribonucleic Acid mutation as a cause of maternally inherited hypertrophic cardiomyopathy. Journal of the American College of Cardiology. 2003 May 21;41 (10):1786–96.

78. Gowthami N, Sunitha B, Kumar M, Keshava Prasad TS, Gayathri N, Padmanabhan B, et al. Mapping the protein phosphorylation sites in human mitochondrial complex I (NADH: Ubiquinone oxidoreductase): A bioinformatics study with implications for brain aging and neurodegeneration. Journal of Chemical Neuroanatomy. 2019 Jan;95:13–28.

79. Chen X, Guo C, Kong J. Oxidative stress in neurodegenerative diseases. Neural Regeneration Research. 201209/24/received11/22/accepted;7(5):376–385. PubMed PMID: PMC4350122.

80. Ponczek M, Wachowicz B. Interaction of reactive oxygen and nitrogen species with proteins. Postepy Biochem. 2005;51(2):140–5.

81. Ahmad S, Khan H, Shahab U, Rehman S, Rafi Z, Khan MY, et al. Protein oxidation: an overview of metabolism of sulphur containing amino acid, cysteine. Frontiers in Bioscience (Scholar edition). 2017 Jan 1;9:71–87. PubMed PMID: 27814576. Epub 2016/ 11/05.eng.

82. Uchida K. Histidine and lysine as targets of oxidative modification. Amino Acids. 2003 Dec; 25(3–4): 249–57. PubMed PMID: 14661088. Epub 2003/ 12/09.eng.

83. Taylor ER, Hurrell F, Shannon RJ, Lin T-K, Hirst J, Murphy MP. Reversible glutathionylation of complex I increases mitochondrial superoxide formation. Journal of Biological Chemistry. 2003 May 30;278(22):19603–10.

84. Fratelli M, Demol H, Puype M, Casagrande S, Eberini I, Salmona M, et al. Identification by redox proteomics of glutathionylated proteins in oxidatively stressed human T lymphocytes. Proceedings of the National Academy of Sciences. 2002;99(6):3505–10.

85. Clementi E, Brown GC, Feelisch M, Moncada S. Persistent inhibition of cell respiration by nitric oxide: Crucial role of S-nitrosylation of mitochondrial complex I and protective action of glutathione. Proceedings of the National Academy of Sciences. 1998;95(13):7631–6.

86. Tsang AHK, Chung KKK. Oxidative and nitrosative stress in Parkinson's disease. Biochimica et Biophysica Acta (BBA) – Molecular Basis of Disease. 2009 Jul 1;1792 (7):643–50.

87. Lin T-K, Hughes G, Muratovska A, Blaikie FH, Brookes PS, Darley-Usmar V, et al. Specific modification of mitochondrial protein thiols in response to oxidative stress: A proteomics approach. Journal of Biological Chemistry. 2002 May 10;277(19):17048–56.

88. Perry TL, Godin DV, Hansen S. Parkinson's disease: A disorder due to nigral glutathione deficiency?Neuroscience Letters. 1982 Dec 13;33(3):305–10.

89. Danielson SR, Held JM, Oo M, Riley R, Gibson BW, Andersen JK. Quantitative mapping of reversible mitochondrial complex I cysteine oxidation in a Parkinson disease mouse model. Journal of Biological Chemistry. 201101/0109/29/received12/15/revised;286(9):7601–8. PubMed PMID: PMC3045014.

90. Beer SM, Taylor ER, Brown SE, Dahm CC, Costa NJ, Runswick MJ, et al. Glutaredoxin 2 catalyzes the reversible oxidation and glutathionylation of mitochondrial membrane thiol proteins: implications for mitochondrial redox regulation and antioxidant defense. Journal of Biological Chemistry. 2004 Nov 12;279(46):47939–51.

91. Balijepalli S, Boyd MR, Ravindranath V. Human brain thioltransferase: constitutive expression and localization by fluorescence in situ hybridization. Molecular Brain Research. 2000 Dec 28;85(1):123–32.

92. Kang PT, Zhang L, Chen C-L, Chen J, Green KB, Chen Y-R. Protein thiyl radical mediates S-glutathionylation of complex I. Free Radical Biology and Medicine. 2012 Aug 15;53(4):962–73.

93. Hurd TR, Requejo R, Filipovska A, Brown S, Prime TA, Robinson AJ, et al. Complex I within oxidatively stressed bovine heart mitochondria is glutathionylated on Cys-531 and Cys-704 of the 75-kDa subunit: potential role of cys residues in decreasing oxidative damage. Journal of Biological Chemistry. 200805/06/received06/11/revised;283 (36):24801–24815. PubMed PMID: PMC2529008.

94. Chen C-L, Zhang L, Yeh A, Chen C-A, Green-Church KB, Zweier JL, et al. Site-specific S-glutathiolation of mitochondrial NADH ubiquinone reductase. Biochemistry. 2007 Apr 20;46(19):5754–65. PubMed PMID: PMC2527596.

95. Zhang L, Xu H, Chen C-L, Green-Church KB, Freitas MA, Chen Y-R. Mass spectrometry profiles superoxide-induced intra-molecular disulfide in the FMN-binding subunit of mitochondrial complex I. Journal of the American Society for Mass Spectrometry. 2008 Aug 12;19(12):1875–1886. PubMed PMID: PMC2614441.

96. Beltrán B, Orsi A, Clementi E, Moncada S. Oxidative stress and S-nitrosylation of proteins in cells. British Journal of Pharmacology. 200011/02/received12/03/revised12/09/accepted;129(5):953–60. PubMed PMID: PMC1571926.

97. Jekabsone A, Ivanoviene L, Brown GC, Borutaite V. Nitric oxide and calcium together inactivate mitochondrial complex I and induce cytochrome c release. Journal of Molecular and Cellular Cardiology. 2003 Jul 1;35 (7):803–9.

98. Antunes F, Han D, Rettori D, Cadenas E. Mitochondrial damage by nitric oxide is potentiated by dopamine in PC12 cells. Biochimica et biophysica acta. 200212; 1556 (2–3): 233–238. PubMed PMID: 12460681. eng.

99. Hsu M, Srinivas B, Kumar J, Subramanian R, Andersen J. Glutathione depletion resulting in selective mitochondrial complex I inhibition in dopaminergic cells is via an NO-mediated pathway not involving peroxynitrite: implications for Parkinson's disease. Journal of Neurochemistry. 2005;92(5):1091–103.

100. Chinta SJ, Andersen JK. Nitrosylation and nitration of mitochondrial complex I in Parkinson's disease. Free Radical Research. 2011 Jan 1;45(1):53–8.

101. Galkin A, Moncada S. S-nitrosation of mitochondrial complex I depends on its structural conformation. Journal of Biological Chemistry. 2007 Dec 28;282(52):37448–53.

102. Alvarez B, Radi R. Peroxynitrite reactivity with amino acids and proteins. Amino Acids. 2003 Dec 01;25 (3):295–311.

103. Venkateshappa C, Harish G, Mahadevan A, Srinivas Bharath MM, Shankar SK. Elevated oxidative stress and decreased antioxidant function in the human hippocampus and frontal cortex with increasing age: implications for neurodegeneration in Alzheimer's disease. Neurochemical Research. 2012 Aug 1;37(8):1601–14.

104. Mythri RB, Venkateshappa C, Harish G, Mahadevan A, Muthane UB, Yasha TC, et al. Evaluation of markers of oxidative stress, antioxidant function and astrocytic proliferation in the striatum and frontal cortex of Parkinson's disease brains. Neurochemical Research. 2011 Aug;36 (8):1452–1463. PubMed PMID: 21484266. Epub 2011/ 04/ 13.eng.

105. RiobóNa, Clementi E, Melani M, Boveris A, Cadenas E, Moncada S, et al. Nitric oxide inhibits mitochondrial NADH: ubiquinonereductase activity through peroxynitrite formation. Biochemical Journal. 2001;359(1):139–45.

106. Mythri RB, Jagatha B, Pradhan N, Andersen J, Bharath MMS. Mitochondrial complex I inhibition in Parkinson's disease: how can curcumin protect mitochondria?Antioxidants & Redox Signaling. 2007;9(3):399–408. PubMed PMID: 17184173.

107. Jagatha B, Mythri RB, Vali S, Bharath MMS. Curcumin treatment alleviates the effects of glutathione depletion in vitro and in vivo: therapeutic implications for Parkinson's disease explained via in silico studies. Free Radical Biology and Medicine. 2008 Mar 1;44(5):907–17.

108. Murray J, Taylor SW, Zhang B, Ghosh SS, Capaldi RA. Oxidative damage to mitochondrial complex I due to peroxynitrite: identification of reactive tyrosines by mass spectrometry. Journal of Biological Chemistry. 2003 Sep 26;278(39):37223–30.

109. Ryan K, Backos DS, Reigan P, Patel M. Post-translational oxidative modification and inactivation of mitochondrial complex I in epileptogenesis. The Journal of Neuroscience. 2012;32(33):11250–8.

110. Sunitha B, Gayathri N, Kumar M, Keshava Prasad TS, Nalini A, Padmanabhan B, et al. Muscle biopsies from human muscle diseases with myopathic pathology reveal common alterations in mitochondrial function. Journal of Neurochemistry. 2016 Jul;138(1):174–191. PubMed PMID: 27015874. Epub 2016/ 03/27.eng.

111. Pawson T. Dynamic control of signaling by modular adaptor proteins. Current Opinion in Cell Biology. 2007 Apr 1;19(2):112–6.

112. Chen R, Fearnley IM, Peak-Chew SY, Walker JE. The phosphorylation of subunits of complex I from bovine heart mitochondria. Journal of Biological Chemistry. 2004 Jun 18;279(25):26036–45.

113. Papa S, Sardanelli AM, Cocco T, Speranza F, Scacco SC, Technikova-Dobrova Z. The nuclear-encoded 18 kDa (IP) AQDQ subunit of bovine heart complex I is phosphorylated by the mitochondrial cAMP-dependent protein kinase. FEBS Letters. 1996;379(3):299–301.

114. Scheffler IE, Yadava N, Potluri P. Molecular genetics of complex I-deficient Chinese hamster cell lines. Biochimica et Biophysica Acta (BBA) – Bioenergetics. 2004 Dec 6;1659(2):160–71.

115. Papa S, Sardanelli A, Scacco S, Petruzzella V, Technikova-Dobrova Z, Vergari R, et al. The NADH: Ubiquinone oxidoreductase (complex I) of the mammalian respiratory chain and the cAMP cascade. Journal of Bioenergetics and Biomembranes. 2002;34(1):1–10.

116. Petruzzella V, Papa S. Mutations in human nuclear genes encoding for subunits of mitochondrial respiratory complex I: the NDUFS4 gene. Gene. 2002 Mar 6;286 (1):149–54.

117. Lenaz G, Genova ML. Supramolecular organisation of the mitochondrial respiratory chain: a new challenge for the mechanism and control of oxidative phosphorylation. In: Kadenbach B, editor. Mitochondrial oxidative phosphorylation: nuclear-encoded genes, enzyme regulation, and pathophysiology. New York: Springer New York; 2012. pp. 107–44.

118. Tibaldi E, Brunati AM, Massimino ML, Stringaro A, Colone M, Agostinelli E, et al. Src-Tyrosine kinases are major agents in mitochondrial tyrosine phosphorylation. Journal of Cellular Biochemistry. 2008;104(3):840–9.

119. Hebert-Chatelain E, Jose C, Gutierrez Cortes N, Dupuy J-W, Rocher C, Dachary-Prigent J, et al. Preservation of NADH ubiquinone-oxidoreductase activity by Src kinase-mediated phosphorylation of NDUFB10. Biochimica et Biophysica Acta (BBA) – Bioenergetics. 2012 May 1;1817(5):718–25.

120. Fajer M, Meng Y, Roux B. The activation of c-Src tyrosine kinase: conformational transition pathway and free energy landscape. The Journal of Physical Chemistry B. 2017 Apr 20;121(15):3352–63.

121. Papa S. The NDUFS4 nuclear gene of complex I of mitochondria and the cAMP cascade. Biochimica et Biophysica Acta (BBA) – Bioenergetics. 2002 Sep 10;1555(1):147–53.

122. Papa S, Scacco S, Sardanelli AM, Petruzzella V, Vergari R, Signorile A, et al. Complex I and the cAMP cascade in human physiopathology. Bioscience Reports. 2002 Feb 1;22(1):3–16.

123. Brandt U. Energy converting NADH: Quinone oxidoreductase (complex I). Annual Review of Biochemistry. 2006;75(1):69–92. PubMed PMID: 16756485.

124. Schilling B, Aggeler R, Schulenberg B, Murray J, Row RH, Capaldi RA, et al. Mass spectrometric identification of a novel phosphorylation site in subunit NDUFA10 of bovine mitochondrial complex I. FEBS Letters. 2005;579(11):2485–90.

125. Schulenberg B, Aggeler R, Beechem JM, Capaldi RA, Patton WF. Analysis of steady-state protein phosphorylation in mitochondria using a novel fluorescent phosphosensor dye. Journal of Biological Chemistry. 2003 Jul 18;278(29):27251–5.

126. Valente EM, Abou-Sleiman PM, Caputo V, Muqit MMK, Harvey K, Gispert S, et al. Hereditary early-onset Parkinson's disease caused by mutations in PINK1. Science. 2004;304(5674):1158–60.

127. Murai M, Miyoshi H. Chemical modifications of respiratory complex I for structural and functional studies. Journal of Bioenergetics and Biomembranes. 2014 Aug 1;46(4):313–21.

128. Covian R, Balaban RS. Cardiac mitochondrial matrix and respiratory complex protein phosphorylation. American Journal of Physiology – Heart and Circulatory Physiology. 201208/1001/26/received08/01/accepted;303(8):H940–9466. PubMed PMID: PMC3469641.

129. Rikova K, Guo A, Zeng Q, Possemato A, Yu J, Haack H, et al. Global survey of phosphotyrosine signaling identifies oncogenic kinases in lung cancer. Cell. 2007 Dec 14;131 (6):1190–203.

130. Aponte AM, Phillips D, Hopper RK, Johnson DT, Harris RA, Blinova K, et al. Use of (32)P to study dynamics of the mitochondrial phosphoproteome. Journal of Proteome Research. 2009;8(6):2679–95. PubMed PMID: PMC3177856.

131. Deng N, Zhang J, Zong C, Wang Y, Lu H, Yang P, et al. Phosphoproteome analysis reveals regulatory sites in major pathways of cardiac mitochondria. Molecular & Cellular Proteomics: MCP. 201105/2204/29/received;10(2): M110.000117.PubMed PMID: PMC3033665.

132. Yadava N, Potluri P, Scheffler IE. Investigations of the potential effects of phosphorylation of the MWFE and ESSS subunits on complex I activity and assembly. The International Journal of Biochemistry & Cell Biology. 2008 Jan 1;40(3):447–60.

133. Boja ES, Phillips D, French SA, Harris RA, Balaban RS. Quantitative mitochondrial phosphoproteomics using iTRAQ on an LTQ-Orbitrap with high energy collision dissociation. Journal of Proteome Research. 2009 Oct 2;8 (10):4665–75.

134. Carroll J, Fearnley IM, Skehel JM, Runswick MJ, Shannon RJ, Hirst J, et al. the post-translational modifications of the nuclear encoded subunits of complex I from bovine heart mitochondria. Molecular & Cellular Proteomics. 2005 May 1;4(5):693–99.

135. Tsunasawa S, Sakiyama F. [14] Amino-terminal acetylation of proteins: an overview. Methods in Enzymology. Academic Press; 1984;106:165–70.

136. Carroll J, Ding S, Fearnley IM, Walker JE. Posttranslational modifications near the quinone binding site of mammalian complex I. Journal of Biological Chemistry. 2013 Aug 23;288(34):24799–808.

137. Rhein VF, Carroll J, Ding S, Fearnley IM, Walker JE. NDUFAF7 methylates arginine 85 in the NDUFS2 subunit of human complex I. Journal of Biological Chemistry. 201311;288(46):33016–33026. PubMed PMID: 24089531. eng.

138. Farazi TA, Waksman G, Gordon JI. The biology and enzymology of protein N-myristoylation. Journal of Biological Chemistry. 2001 Oct 26;276(43):39501–39504. PubMed PMID: 11527981. Epub 2001/08/31.eng.

139. Resh MD. Fatty acylation of proteins: new insights into membrane targeting of myristoylated and palmitoylated proteins. Biochimica et biophysica acta. 1999 Aug 12;1451(1):1–16. PubMed PMID: 10446384. Epub 1999/ 08/14.eng.

140. Zha J, Weiler S, Oh KJ, Wei MC, Korsmeyer SJ. Posttranslational N-myristoylation of BID as a molecular switch for targeting mitochondria and apoptosis. Science. 2000 Dec 1;290(5497):1761–1765. PubMed PMID: 11099414. Epub 2000/ 12/02.eng.

141. Martin DD, Beauchamp E, Berthiaume LG. Posttranslational myristoylation: fat matters in cellular life and death. Biochimie. 2011 Jan;93(1):18–31. PubMed PMID: 21056615. Epub 2010/ 11/09.eng.

142. Dyda F, Klein DC, Hickman AB. GCN5-related N-acetyltransferases: a structural overview. Annual Review of Biophysics and Biomolecular Structure. 2000;29:81–103. PubMed PMID: 10940244. Pubmed Central PMCID: PMC4782277. Epub 2000/ 08/15.eng.

143. Towler DA, Adams SP, Eubanks SR, Towery DS, Jackson-Machelski E, Glaser L, et al. Myristoyl CoA: proteinN-myristoyltransferase activities from rat liver and yeast possess overlapping yet distinct peptide substrate specificities. Journal of Biological Chemistry. 1988 Feb 5;263(4):1784–90.

18 Heme Oxygenase-1 in Kidney Health and Disease

Perspective of Mitochondrial Homeostasis

Pu Duann
Research and Development, Salem Veteran Affairs Medical Center, Salem, VA, USA

Elias A. Lianos
Salem Veteran Affairs Medical Center, Salem, VA and Virginia Tech Carilion School of Medicine, Department of Basic Science Education, Roanoke, VA, USA

Pei-Hui Lin
Davis Heart and Lung Research Institute and Department of Surgery, The Ohio State University, Columbus, USA

CONTENTS

Abbreviations

AKI	Acute kidney injury
ARE	Antioxidant responsive element
BVR	Biliverdin reductase
CO	Carbon monoxide
CKD	Chronic kidney disease
CPR	Cytochrome P450 reductase
ER	Endoplasmic reticulum
eNOS	endothelial nitric oxide synthase
IRI	Ischemia reperfusion injury
GN	Glomerulonephritis
HO	Heme oxygenase
Mφ	Macrophage
MPS	Mononuclear phagocytic system
NADPH	Nicotinamide adenine dinucleotide phosphate
Nrf2	Nuclear factor erythroid 2-related factor 2
OS	Oxidative stress
OXPHOS	Oxidative phosphorylation
PTC	Proximal tubule cells
PPIX	Protoporphyrin IX
ROS	Reactive oxygen species
SPP	Signal peptide peptidase
SER	Smooth endoplasmic reticulum
UUO	Unilateral ureteral obstruction

1 INTRODUCTION

Heme oxygenase (HO) was initially identified as a liver microsomal protein with activity to degrade heme to bilirubin (1). HO is the rate-limiting enzyme that catalyzes the degradation of heme to equimolar quantities of

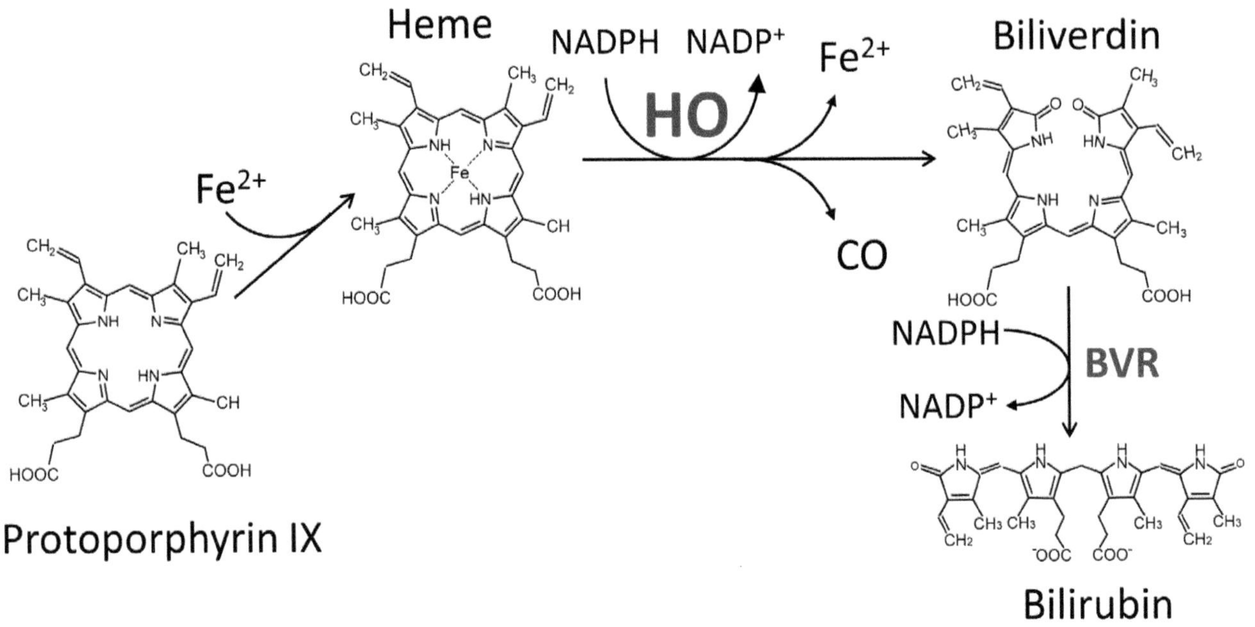

Figure 1   HO enzymatic reaction. Protoporphyrin IX (PPIX) is the final intermediate in heme biosynthesis. The toxic free-heme in cells and tissues is removed by HO degradation pathway. HO catalyzes the breakdown of heme into CO, iron, and biliverdin (BV) which is quickly converted into bilirubin (BR) by biliverdin reductase. CO, iron and BV/BR are bioactive molecules influence important cellular processes including redox homeostasis, tissue health with both beneficial and potential adverse effects.

carbon monoxide (CO), iron and biliverdin (Figure 1). Biliverdin is subsequently reduced to bilirubin by biliverdin reductase (BVR) (2). Heme is pro-oxidant and synthesized in the mitochondria. In physiology, significant amount of heme could arise from the destruction of aged red blood cells, and therefore should be tightly controlled. Two major HO isoforms exist. HO-2 is constitutively active, whereas HO-1 is inducible.

Cellular oxidative damage is one of the major pathogenic determinants in several clinical disorders, including inflammatory diseases, ischemia-reperfusion injury (IRI) and chemical-induced cytotoxicity. Mitochondria are critical cellular organelles best known for their role in providing efficient energetic support through the chemiosmotic process of oxidative phosphorylation (OXPHOS). Mitochondria also perform roles in reactive oxygen species (ROS) free radicals production (3) and transduction of metabolic and stress signals (4,5). Persistent mitochondria damage is a major source of oxidants. As such, mitochondria fitness decides the body's health.

During oxidative stress induced orang injury, HO-1 is often induced. The beneficial effects of HO-1 expression and its by-products provide anti-oxidant, anti-inflammatory, and anti-apoptotic effects. Beyond its enzymatic activity, HO-1 exerts non-canonic signaling cascades through its various binding partners and different post-translational processing that leads to its translocation from endoplasmic reticulum (ER) to different subcellular compartments, such as cytosol, nucleus, mitochondria, and caveolae, and functions.

The kidney is composed of multiple cell populations which are involved in several vital functions that maintain body's homeostasis such as acid-base and electrolytes balance, blood pressure regulation, nutrients reabsorption and hormone secretion (6). Mitochondria are critically important in maintaining the health and function of the metabolically active kidney. HO-1 plays important roles in mitochondria quality control via its action on mitochondria biogenesis, bioenergetics, dynamics, and autophagy as dysfunctional mitochondria is involved in various kidney diseases such as acute kidney injury (AKI), chronic kidney disease (CKD), diabetic nephropathy (DN), and glomerulonephritis (GN). In this chapter, we discuss how HO-1 performs it function and serves as novel therapeutics for organ injury and repair.

2   HEME OXYGENASES (HOs)

Heme is an iron-containing porphyrin complex and constitutes the prosthetic group of several hemoproteins with important biological functions. Heme is synthesized in the mitochondria with protoporphyrins supplied from its precursor succinyl-CoA from mitochondria TCA (Kreb) cycle, which then subsequently exported out via the mitochondrial transporter ATP-binding cassette (ABC) B10 after biosynthesis (7). Hemes are most commonly recognized as components of hemoglobin from red blood cells (erythrocytes). Some other examples of hemoproteins include myoglobin (enriched in muscle), catalases, heme peroxidase, cytochromes, and endothelial nitric oxide synthase (eNOS) (8). The redox-active nature of iron makes heme critically involved in modulation of oxidating-reducing activities of hemoproteins which engaged in oxygen transport (hemoglobin) and storage (myoglobin), mitochondrial electron transfer and energy transformation (cytochromes), hydrogen peroxide activation (heme peroxidase) or inactivation (catalases) and nitro oxide synthesis (eNOS) (9). In physiology, significant level of heme could arise from the destruction of aged red blood cells. Because heme also catalyzes the formation of toxic reactive oxygen species (ROS) and free hydroxyl radicals to induce pro-oxidant and cytotoxic effects, level of "free-heme" must be tightly regulated. Disturbed heme metabolism causes mitochondrial decay, oxidative stress, and iron accumulation has been linked to age-related diseases (10).

Heme oxygenase (HO) is the rate limiting enzyme catalyzing oxidative cleavage, with electrons delivered from NADPH via cytochrome P450 reductase (CPR), of the porphyrin ring of heme into equimolar by-products of ferrous iron (Fe2+), carbon monoxide (CO) and biliverdin (2) (Figure 1). Biliverdin is quickly converted by biliverdin reductase (BVR) into bilirubin which is ultimately expelled from body (11). Three different isoforms of heme oxygenase (HO) have been reported in mammalian, including the inducible HO-1 (encoded by HMOX1 gene) (12), the homeostatic low level and constitutively active HO-2 isoform (encoded by HMOX2 gene) (13), and the non-enzymatic HO-3 originally identified in rat brain (14). Based on phylogenic alignment, HO-3 evolved as a splicing variant from the HO-2 gene which binds heme but is deprived of catalytic activity (15). Both HO-1 and HO-2 are ubiquitously expressed and catalytically active. HO-2 is present in relatively high concentration in the brain, testis, and vascular endothelial cells (16,17), while HO-1 is more widely distributed. HO-1 is abundant in tissues rich with its substrate heme expression, such as muscle, erythroid-phagocytic system (liver and bone marrow) and spleen where abundant heme was released from senescent RBC processing and hemoglobin recycling for new heme synthesis (18,19).

3   STRUCTURE AND THE REDOX FUNCTION OF HOs

The two major HO isoforms contain different primary structures with human HO-1 as a 32 kDa (288 amino acid residues, UniProtKB: P09601; also named as heat shock protein (HSP32)) and HO-2 as a 36 kDa (315 residues, UniProtKB: P30519) and showing only 58% homology in primary structure (20). The catalytic mechanism of the HOs reaction has been determined gradually by studies of its crystal structure and HO mutants in

perspective of heme binding and its catalytic center (21–25) and www.rcsb.org/structure/1N3U). The two isoforms share a region with 100% secondary structure homology at the catalytic site of the proteins (26). Further crystallography study also revealed a novel inducible binding pocket in HO-1 with heme (27).

Heme is sandwiched between a proximal A-helix (Leu13-Glu29) and a distal F-helix (Leu129-Met155) of the HO-1 which consists of eight (namely A–H) α-helices. The His25 serves as the proximal ligand in direct contact with the heme iron center and Gly139 and Gly143 are close to the distal ligand of the heme iron. The conserved Gly143 near the oxygen binding site allows close contact between the helix backbone and heme was shown to provide the flexibility required for substrate binding and product release during HO-1 catalysis (22,23,25). HO contains a C-terminal membrane binding motif responsible for its microsomal ER localization which would not affect its enzymatic activity (detailed in the section of Cellular localization of HO-1 activation).

Along with its catabolic products (biliverdin/bilirubin, CO, Fe2+), HO and its metabolites which often are induced in response to cellular injury and stress and play significant functions (reviewed in refs (28–30). CO is an autocrine and paracrine vasodilator in cerebral and systemic circulations (31,32) and functions as the second gasotransmitter in the GI tract (33,34). Bilirubin is a free radical scavenger and cellular antioxidant (35–37). Iron and the induced ferritin, a highly conserved iron-binding protein, is a key protein in protection from oxidative stress and the maintenance of cellular iron homeostasis (38,39).

4   HO-1 INDUCTION AND ORGAN PROTECTION FROM OXIDATIVE INJURY

HO-1 is induced as physiological response to free heme-induced oxidative stress and non-heme related stimuli from hypoxia, oxidative stress, heavy metal, septic toxins, polyphenols, statins and inflammatory insults of cytokines activation (Reviewed in ref. 40). Many disease models have shown that HO-1 and its products have protective effects, including anti-inflammation, anti-apoptosis, anti-proliferation and anti-oxidation properties. For examples, the cytoprotective HO-1 activation have been demonstrated in several disease or organ stress states such as cardiovascular system (41,42), murine colon colitis model (43), the anti-inflammatory effects in myeloid phagocyte system (44,45), in traumatic brain injury (46), in radiation-induced hippocampal neurons injury (47), in toxin induced acute lung injury (48). Placenta HO-1 protects fetus during pregnancy which was thought in analogy as a specified allograft-protection (49,50).

The cytoprotective roles of HO-1 in kidney are well documented. Kidney nephrons encounter and filter toxic oxidants and chemicals from circulation. Renal HO-1 is induced as physiological response to heme-induced and non-heme induced oxidative stress from hypoxia, heavy metal, septic toxins, and inflammatory cytokines (Reviewed in refs (28,51–54). Studies from animal models of ischemia reperfusion injury (IRI) (55–57), AKI from cisplatin nephrotoxicity (58), rhabdomyolysis-associated AKI (59), or from septic kidney injury (60,61), and CKD such as diabetes (62,63), hypertension (64) and glomerulonephritis (GN) (65,66) have revealed the importance of HO-1 induction in mediating oxidative stress response and iron homeostasis in many biological reactions. Moreover, HO-1 was shown to enhance vascular endothelia resistance and minimize complement deposition injury through glomerular activation of Decay Accelerating Factor (DAF), a suppressor for complement mediated inflammation in GN (67,68). In human AKI patients, elevated plasma and urinary HO-1 was detected and have been proposed as biomarkers of AKI to reflect intrarenal HO-1 gene activity (69).

5 THE NON-CANONICAL FUNCTION OF HO-1

Recently, the "non-heme catalytic" functions have been explored which indicate HO-1 regulates important cellular functions through its non-canonical signaling cascades (70,71). The multiplex HOs functions are achieved through the differential subcellular compartmentation

and function, and also specific interaction partners of HOs.

5.1 THE CELLULAR LOCALIZATION OF HO-1 ACTIVATION

HO-1 forms oligomers for its enhanced stability as a smooth endoplasmic reticulum (SER)/microsomal protein (72). The C-terminal 23 amino acids hydrophobic transmembrane sequence (TMS) tail is responsible for its SER association (73) (Figure 2). However, HO-1 is also processed and translocates to nuclei, cytosol (soluble HO-1 (sHO-1)) and mitochondria under various stress conditions and have different activities, also a part of HO-1's non-canonic functions (Figure 2A).

The signal peptide peptidase (SPP), a member of aspartyl protease family 42 kDa glycoprotein located in the ER membrane, catalyzes proteolysis of many tail-anchored ER proteins including HO-1 (74,75). Under hypoxic condition, HO-1 (but not HO-2) undergoes SPP-mediated intramembrane cleavage which resulted in HO-1 cytosolic and nuclear translocation, a step independent of HO-1 catalytic activity (75,76) (Figure 2A and B). The SPP cleavage site was identified between a conserved F275 and L276 dipeptide within its C-terminal 23 amino acid membrane anchor sequence (75,77) (Figure 2B). Full cleavage by SPP and the subsequent

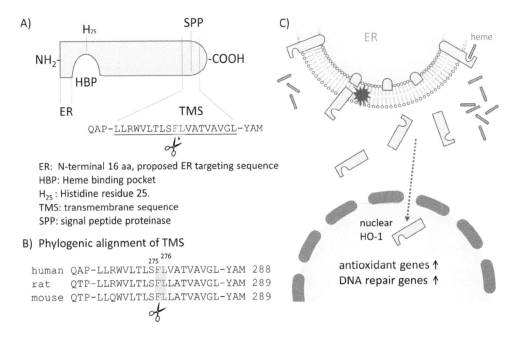

Figure 2 Cellular localization of HO-1 activation. (A) Proposed functional domain structure of HO-1. The proposed endoplasmic reticulum (ER) targeting sequences (N-terminal 16 amino acids), HBP (heme binding pocket), catalytic Histidine 25 (H25), C-terminal hydrophobic transmembrane sequence (TMS) and the signal peptide peptidase (SPP) cleavage site are shown. HO-1 is normally present as a smooth ER protein. (B) Under hypoxic condition, HO-1 undergoes SPP-mediated intramembrane cleavage (between F275 and L276) which results in HO-1 cytosolic and nuclear translocation. (C) The nuclear translocated HO-1 plays important non-canonic (catalytic activity-independent) functions. HO-1 is reported to translocate to mitochondria, but the detailed mechanism is less known.

nuclear translocation also relies on the neighboring degradation signal sequence (PEST domain, amino acids 239–254) and a leucine-rich region of nuclear shuttle sequence (amino acids 207–221) in HO-1 (75).

Both cytosolic and nuclear HO-1, whether enzymatic active or not, protected cells against hydrogen peroxide-mediated injury equally well, indicating a non-canonical HO-1 signaling pathway (78). This truncated and enzymatically inactive nuclear HO-1 (Figure 2C) was demonstrated to alter binding of transcription factors to alter gene expression. For examples, nuclear factor erythroid 2-related factor 2 (NF-E2-related factor 2, Nrf2), a master transcriptional factor of the cellular redox homeostasis, which binds to a cis-acting element called the antioxidant responsive element (ARE) and regulates numerous antioxidants and cytoprotective proteins, including HO-1 itself. Nrf2 is accumulated in nucleus during OS. Nuclear HO-1 interacts with Nrf2 and stabilizes it from ubiquitin-proteasomal degradation (79). Interestingly, Nrf2 also support mitochondria biogenesis and integrity (80) (Figure 3A). High level of nuclear HO-1 is linked to numerous tumor growth and drug resistance (81). The underlying mechanism has been attributed to its activation of a panel of transcription factors during oxidative stress such as activator protein-1, Nrf2 (70,78). The nuclear HO-1 also regulates its own expression in prostate cancer (82) and has been shown when undergoing acetylation by post-translational modification could promote tumor cells proliferation and invasion (77,83).

Normally, HO-1 is rarely localized in the mitochondria, an organelle for heme biosynthesis. However, under hypoxia stress condition, HO-1 could translocate to mitochondria in activated macrophage. It was proposed that the 16 amino acids of N-terminal are an ER-targeting domain, and deletion which causes HO-1 mitochondria targeting. In chronic alcohol hepatoxicity, this mitochondria-targeting HO-1 fraction increased with increased ROS production and diminished cytochrome C oxidase activity (84). On the contrary, beneficiary cytoprotective and anti-apoptotic effects of mitochondria-targeting HO-1 in gastric mucosal cells against nonsteroid anti-inflammatory drug (NSAID)-induced oxidative injury was reported (85). Additionally, renal epithelial cells overexpressing an engineered mitochondria leader sequence (derived from MnSOD) tagged HO-1 (mito-HO-1, 36 kDa) mitigates hypoxia-mediated mitochondria injury with resistance to heme-mediated cell death. Within mitochondria, this mito-HO-1 was actually processed to a molecule slightly larger than native HO-1. The nature of this processing was not identified yet (86). More work need to be done to entail the function of this mitochondria-targeted HO-1.

Besides intracellular locations, HO-1 could exist in extracellular space such as in plasma and urine of AKI patients which had been suggested to correlates with intrarenal HO-1 gene activity (69), in cerebrospinal fluid from infants and children after severe traumatic brain injury (87) and in human milk which might play a non-

catalytic while immune-regulatory role (88). The mechanism of HO-1 body fluid excretion and whether it could serve as a tissue specific biomarker is still not known.

5.2 THE HOs INTERACTION PARTNERS

The interaction between HO and the electron donor cytochrome P450 reductase (CPR) is essential for HO catalytic activity. However, CPR also prevents nuclear translocation of HO-1 under hypoxic conditions and promotes oligomerization of HO-1 into higher ordered complexes (76,89). HO-1 was first shown to interact with caveolin-1 and -2 in a mouse renal mesangial cell upon cadmium (Cd) treatment (90). The interaction of HO-1 and caveolin-1was identified in caveolin-1scaffolding domain (residues 82–101, namely CAV (82–101)) (91). Disruption of HO-1/caveolin-1 interaction with caveolin-1 scaffolding domain peptides CAV (82–101) enhances the anti-inflammatory effect of HO-1 and modulates HO-1 activity (92,93). Furthermore, through a multi "omics" approach, HO-1 induction was shown to orchestrate multi-molecular network that regulates tumor cytoskeletal cell compartments, cell adhesion and cell–cell communication in prostate cancer cells (94).

6 HO-1 IN KIDNEY HEALTH AND DISEASE-COMBINATION OF CATALYTIC AND NON-CANONIC FUNCTIONS

Kidney is a complex organ with orchestrated different cell types of parenchyma and mononuclear phagocytic system (MPS) of myeloid origin including macrophages (Mφ) and dendritic cells (DC). HO-1 is induced in most kidney substructures including proximal tubules, glomeruli, interstitium and renal resident MPS during injury. Much of HO-1 effects on kidney functions have been attributed to seminal studies from global deletion of HO-1 (HO-1$^{-/-}$) (95), conditional or nephron cell type-specific HO-1 knockout or overexpressing animals (96–98) and HO-1 deficient patients derived from allelic microsatellite polymorphism in HO-1 gene promoter (99). With those valuable transgenic animals and human genetic studies, the cell type-specific HO-1 functions are revealed which demonstrate orchestrated signal crosstalk among different renal compartments in kidney homeostasis. The topics of HO-1 in kidney health and diseases have recently thoroughly reviewed (28,51,54) and here we provide only a glimpsed overview in its anti-inflammatory response.

Macrophage (Mφ) is critical in heme-associated iron homeostasis (100), in modulations of innate and adaptive immune response via plethora inflammatory mediators, in wound healing and tissue remodeling during repair process (101–103). Heme toxicity and elevated iron is a major culprit in AKI animals (104) and a pathogenic factor in cardiac surgery-associated AKI patients (99).

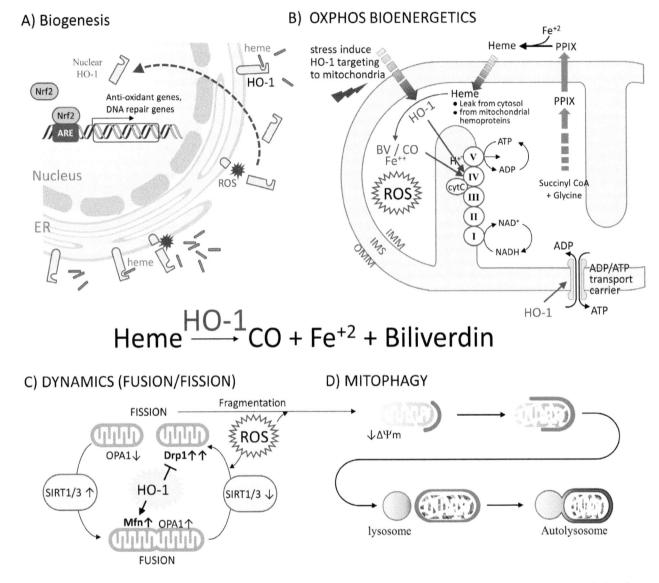

Figure 3 HO-1 in mitochondrial quality control. (A) The nuclear HO-1 interacts and stabilizes Nrf2, a master transcription factor binds to cis-acting ARE, which modulates many anti-oxidant and DNA repair genes and thus plays important roles in redox homeostasis and mitochondria biogenesis. (B) HO-1 regulates OXPHOS bioenergetics. Mitochondria are responsible for ATP synthesis through OXPHOS process by respiratory electron transfer chain (complexes I-V) and cytochrome C (cyt C). This is accompanied with ROS production. Mitochondria are also the organelle for heme biosynthesis. HO-1 increases ADP/ATP transport carrier (AAC) and cyt C oxidase (complex IV) activity in ATP bioenergetics. (C) HO-1 regulates mitochondria dynamics and enhances mitochondria fusion via increasing Mfn2 and decreasing Drp1, two of the important mitochondrial dynamics-related proteins. (D) HO-1 regulates autophagy. In AKI models, HO-1 induction is generally associated with cytoprotective effects and altered autophagy. Nrf2: Nuclear factor erythroid 2-related factor 2; ARE: antioxidant responsive element; IMM: inner mitochondrial membrane; OMM: outer mitochondrial membrane; IMS: mitochondria intermembrane space.

As critical role of HO-1 in managing heme toxicity and iron metabolism, substantial evidences support cross-talk between renal parenchyma and myeloid system to regulate inflammatory response and disease progression (105). For examples, HO-1 alters temporal dynamics of myeloid immune cell trafficking and subsequent inflammatory response and fibrogenesis in IRI kidney and

syngenic renal transplantation experiments (97). Lack of HO-1 promotes TGFβ-mediated epithelial-mesenchymal transition (EMT) and renal fibrosis associated with abnormal Mϕ recruitment in an UUO kidney inflammatory model (106). Upregulated HO-1 by HO inducer like hemin was shown to regulate Mϕ polarization toward anti-inflammation to abate chronic renal inflammatory conditions such as diabetic nephropathy (107). Taken together, the integration effects of HO-1 on anti-oxidant, anti-inflammatory and anti-apoptotic signaling uniquely qualify HO-1 as a key regulator of kidney functions.

7   AUTOPHAGY IN RENAL HOMEOSTASIS

Mitochondria are important in metabolically active organs. In human body, kidney receives roughly 20% of cardiac output and consumes 10 % of the body's oxygen and 7% of body's daily ATP energy expenditure to perform its function (108,109). Renal tubule cells and glomerular podocytes have abundant mitochondria densities for their high energy demands to maintain body homeostasis. Distinct mitochondria distributions and densities were found in various nephron segments with highest density in the S1 proximal tubule segment responsible for reabsorption and secretion against chemical gradients; whereas podocytes provide cytoskeletal structure stability, foot process remodeling and uptake of filtered proteins also heavily rely on normal mitochondrial function. Strong association of mitochondria dysfunction and kidney diseases is well documented (109,110).

Autophagy refers degradation and recycling process of organelles and macromolecules by the cellular lysosomal machinery to tightly maintain metabolism and turnover of cellular components (111). Depending on different ways of transporting intracellular constituents to lysosomes, autophagy is classified into three types: macroautophagy (mediated via autophagosomes), microautophagy (direct lysosomal transporting), and chaperone-mediated autophagy. For simplicity, macroautophagy is named autophagy hereafter. Mitophagy is a mitochondrial quality control, selective degradation process to turnover defective mitochondria by autophagy which occurs following damage or oxidative stress during tissue injury. Recent studies have revealed striking links of autophagy to human health and disease such as aging, cancer, neurodegeneration, infection and many disease (reviewed in (112)). Studies of animal models and clinical relevance in kidney disease have uncovered importance of autophagy in modulating kidney homeostasis (reviewed in (113,114). Protective roles of activated autophagy have been demonstrated to against cisplatin-induced (115) or septic AKI (116,117). Whereas impaired autophagic turnover has been linked to diabetic nephropathy (118,119), glomerular disease (120), fibrotic FSGS (121), in UUO rat model (122) and in prosenescent changes during AKI transition to CKD (123).

8   HO-1 IN MITOCHONDRIA QUALITY CONTROL

The mitochondria quality control includes homeostatic regulation of its biogenesis, oxidative phosphorylation for ATP bioenergetics, dynamics of fusion/fission and autophagy (109) (Figure 3). The HO-1/CO system induces mitochondrial biogenesis in skeletal, cardiac, Mϕ and hepatic systems (124–126) (Figure 3A). The induced mitochondrial biogenesis is also linked to anti-inflammatory cytokines such as IL-10, IL-1Ra expressions in sepsis in mouse liver and human macrophage cells (127). HO-1 enhances renal mitochondrial ADP/ATP transport carriers (AAC) and cytochrome C oxidase (complex IV, COX, also a hemoprotein) activity in experimental diabetes (128) (Figure 3B). Moreover, HO-1 induction protects against intermittent hypoxia-induced cardiac dysfunction and myocardial fibrosis through the inhibition of mitochondrial fission via regulating mitochondrial dynamics-related proteins (increases Mfn2 and decreases Drp1) and cell apoptosis (129) (Figure 3C). All these support the essential roles of HO-1 in mitochondria quality control.

HO-1 was linked to autophagy modulation in neural tissues (130) and cardiomyocytes (131). Strong evidence supports HO-1 in modulating autophagy in various renal compartments from injury (Figure 3D). Autophagy is induced in several AKI models including IRI, nephrotoxin- and sepsis-mediated AKI and in human kidney transplantation. Depending on the degree of injury, autophagy induction may have different outcomes as promoting survival or converging into apoptotic cell death (132,133). Study by Bolisetty et al. demonstrated HO-1$^{-/-}$ proximal tubule cells (PTC) have upregulated basal autophagy and impaired autophagy progression which converges into apoptosis induction; whereas HO-1 overexpression delays autophagy progression and inhibit apoptosis to protect against cisplatin nephrotoxicity (134). This HO-1 cytoprotective mechanism includes mitigation of ferroptosis, an iron-dependent programmed cell death, in PTC (135). Interestingly, in another cisplatin-induced AKI model, the cytoprotective effect is attributed to the cross-talk between HO-1 and Toll-like receptor 2 (TLR2), a class of proteins with important function in innate immune response, in association with upregulated autophagy (136).

Induced HO-1 had been shown to protect mouse podocytes from hyperglycemia induced apoptosis through a Beclin-1 mediate autophagy upregulation, which suggest HO-1 induction and regulation of autophagy are potential therapeutic targets for diabetic nephropathy (137). However, HO-1 induction alleviates H_2O_2 induced oxidative stress and cellular apoptosis in glomerular mesangial cells with concomitant reduce of excessive autophagy (138). Taken together, HO-1 mediated autophagy modulation decides cell fate. More work is needed to understand HO-1's role in autophagy flux regulation for its underlying molecular mechanism.

9   TRANSLATIONAL SIGNIFICANCE OF HO-1 AS THERAPEUTIC TARGET IN DISEASES

As HO system plays versatile roles suggest its potential as novel drug target in alleviating different disease states such as cancers (139), chronic and acute myeloid lymphomas (140), inflammation (141,142), AKI (54), IRI (143), obesity and metabolic syndromes (144). Several clinical trials targeting HO/CO system are in progress. However, conflicting or ambiguous results also exist. Furthermore, it is also a challenge to develop HO-1 selective therapeutic strategies. Detailed structural activity research (145) and the recent development of human genome population screen in cohorts of patients (99) would certainly enhance our knowledge and future studies.

10 CONCLUSION AND FUTURE PERSPECTIVES

During the past several decades, the important roles for HO-1 in the pathophysiology of kidney diseases and other organs support the following notions. (i) HO-1 is induced in various kidney diseases from several animal models and human studies. (ii) HO-1 exerts cytoprotective effects via its anti-oxidant, anti-inflammatory and anti-apoptotic effects. (iii) Both catalytic and non-canonic pathways co-exist in HO-1 induction. (iv) Although HO-1 is mainly an ER localized protein, different processes leads to distinct subcellular localization to nucleus, cytosol, caveolae, mitochondria, and extracellular spaces occur with different cellular functions. (v) Urinary and plasma levels of HO-1 was implicated as biomarker for AKI in both animals models and human. (vi) HO-1 modulates mitochondria quality control by affecting mitochondria biosynthesis, bioenergetics, dynamics and autophagy. (vii) Allelic polymorphisms in human HO-1 gene promoter determine HO-1 expression and their susceptibility to several kidney diseases. (viii) HO-1 system is aimed as a novel therapeutic target for kidney, heart, and liver disease with several ongoing clinical trials. In summary, with better understandings of molecular mechanisms underlying the pathophysiologic roles of HO-1 will lead to significant translational medicines to ameliorate kidney (organs) injuries and diseases in the future.

REFERENCES

1. Tenhunen R., Marver H. S. & Schmid R. The enzymatic conversion of heme to bilirubin by microsomal heme oxygenase. *Proceedings of the National Academy of Sciences of the United States of America.* 1968; 61(2):748–55.
2. Yoshida T. & Migita C. T. Mechanism of heme degradation by heme oxygenase. *Journal of inorganic biochemistry.* 2000; 82(1–4):33–41.
3. Sena L. A. & Chandel N. S. Physiological roles of mitochondrial reactive oxygen species. *Molecular cell.* 2012; 48 (2):158–67.
4. Galluzzi L., Kepp O. & Kroemer G. Mitochondria: master regulators of danger signalling. *Nature reviews molecular cell biology.* 2012; 13(12):780–8.
5. Angelova P. R. & Abramov A. Y. Functional role of mitochondrial reactive oxygen species in physiology. *Free radical biology & medicine.* 2016; 100:81–85.

6. Skorecki K., Glenn M., Chertow G. M., et al. Brenner and Rector's the kidney. 10 ed. Philadelphia: Saunders; 2015. 2748 p.
7. Bayeva M., Khechaduri A., Wu R., et al. ATP-binding cassette B10 regulates early steps of heme synthesis. *Circulation research.* 2013; 113(3):279–87.
8. Paoli M., Marles-Wright J. & Smith A. Structure-function relationships in heme-proteins. *DNA and cell biology.* 2002; 21(4):271–80.
9. Chiabrando D., Vinchi F., Fiorito V., et al. Heme in pathophysiology: a matter of scavenging, metabolism and trafficking across cell membranes. *Frontiers in pharmacology.* 2014; 5:61.
10. Atamna H. Heme, iron, and the mitochondrial decay of ageing. *Ageing research reviews.* 2004; 3(3):303–18.
11. O'Brien L., Hosick P. A., John K., et al. Biliverdin reductase isozymes in metabolism. *Trends in endocrinology and metabolism: TEM.* 2015; 26(4):212–20.
12. Alam J., Cai J. & Smith A. Isolation and characterization of the mouse heme oxygenase-1 gene. Distal 5′ sequences are required for induction by heme or heavy metals. *The journal of biological chemistry.* 1994; 269(2):1001–9.
13. McCoubrey W. K., Jr., Ewing J. F. & Maines M. D. Human heme oxygenase-2: characterization and expression of a full-length cDNA and evidence suggesting that the two HO-2 transcripts may differ by choice of polyadenylation signal. *Archives of biochemistry and biophysics.* 1992; 295(1):13–20.
14. McCoubrey W. K., Jr., Huang T. J. & Maines M. D. Isolation and characterization of a cDNA from the rat brain that encodes hemoprotein heme oxygenase-3. *European journal of biochemistry.* 1997; 247(2):725–32.
15. Hayashi S., Omata Y., Sakamoto H., et al. Characterization of rat heme oxygenase-3 gene. Implication of processed pseudogenes derived from heme oxygenase-2 gene. *Gene.* 2004; 336(2):241–50.
16. Bellner L., Martinelli L., Halilovic A., et al. Heme oxygenase-2 deletion causes endothelial cell activation marked by oxidative stress, inflammation, and angiogenesis. *The journal of pharmacology and experimental therapeutics.* 2009; 331(3):925–32.
17. Chen R. J., Yuan H. H., Zhang T. Y., et al. Heme oxygenase-2 suppress TNF-alpha and IL6 expression via TLR4/MyD88-dependent signaling pathway in mouse cerebral vascular endothelial cells. *Molecular neurobiology.* 2014; 50(3):971–78.
18. Delaby C., Rondeau C., Pouzet C., et al. Subcellular localization of iron and heme metabolism related proteins at early stages of erythrophagocytosis. *PLoS one.* 2012; 7(7): e42199.
19. Maines M. D. Heme oxygenase: function, multiplicity, regulatory mechanisms, and clinical applications. *FASEB journal: official publication of the Federation of American Societies for experimental biology.* 1988; 2(10):2557–68.
20. Lad L., Ortiz de Montellano P. R. & Poulos T. L. Crystal structures of ferrous and ferrous-NO forms of verdoheme in a complex with human heme oxygenase-1: catalytic implications for heme cleavage. *Journal of inorganic biochemistry.* 2004; 98(11):1686–95.
21. Lad L., Friedman J., Li H., et al. Crystal structure of human heme oxygenase-1 in a complex with biliverdin. *Biochemistry.* 2004; 43(13):3793–801.
22. Lad L., Schuller D. J., Shimizu H., et al. Comparison of the heme-free and -bound crystal structures of human

heme oxygenase-1. *The journal of biological chemistry.* 2003; 278(10):7834–43.

23. Schuller D. J., Wilks A., Ortiz de Montellano P. R., et al. Crystal structure of human heme oxygenase-1. *Nature structural biology.* 1999; 6(9):860–67.

24. Schuller D. J., Zhu W., Stojiljkovic I., et al. Crystal structure of heme oxygenase from the gram-negative pathogen Neisseria meningitidis and a comparison with mammalian heme oxygenase-1. *Biochemistry.* 2001; 40(38):11552–8.

25. Sugishima M., Omata Y., Kakuta Y., et al. Crystal structure of rat heme oxygenase-1 in complex with heme. *FEBS letters.* 2000; 471(1):61–66.

26. Sugishima M., Higashimoto Y., Oishi T., et al. X-ray crystallographic and biochemical characterization of the inhibitory action of an imidazole-dioxolane compound on heme oxygenase. *Biochemistry.* 2007; 46(7):1860–67.

27. Rahman M. N., Vlahakis J. Z., Vukomanovic D., et al. X-ray crystal structure of human heme oxygenase-1 with (2R,4S)-2-[2-(4-chlorophenyl)ethyl]-2-[(1H-imidazol-1-yl)methyl]-4 [((5-trifluorom ethylpyridin-2-yl)thio)methyl]-1,3-dioxolane: a novel, inducible binding mode. *Journal of medicinal chemistry.* 2009; 52(15):4946–50.

28. Lever J. M., Boddu R., George J. F., et al. Heme oxygenase-1 in kidney health and disease. *Antioxidants & redox signaling.* 2016; 25(3):165–83.

29. Agarwal A. & Bolisetty S. Adaptive responses to tissue injury: role of heme oxygenase-1. *Transactions of the American Clinical and Climatological Association.* 2013; 124:111–22.

30. Wegiel B., Nemeth Z., Correa-Costa M., et al. Heme oxygenase-1: a metabolic nike. *Antioxidants & redox signaling.* 2014; 20(11):1709–22.

31. Thorup C., Jones C. L., Gross S. S., et al. Carbon monoxide induces vasodilation and nitric oxide release but suppresses endothelial NOS. *The American journal of physiology.* 1999; 277(6 Pt 2):F882–29.

32. Leffler C. W., Parfenova H. & Jaggar J. H. Carbon monoxide as an endogenous vascular modulator. *American journal of physiology heart and circulatory physiology.* 2011; 301(1):H1–11.

33. Wu L. & Wang R. Carbon monoxide: endogenous production, physiological functions, and pharmacological applications. *Pharmacological reviews.* 2005; 57(4):585–630.

34. Kasparek M. S., Linden D. R., Kreis M. E., et al. Gasotransmitters in the gastrointestinal tract. *Surgery.* 2008; 143(4):455–9.

35. Sedlak T. W., Saleh M., Higginson D. S., et al. Bilirubin and glutathione have complementary antioxidant and cytoprotective roles. *Proceedings of the National Academy of Sciences of the United States of America.* 2009; 106(13):5171–6.

36. Zelenka J., Muchova L., Zelenkova M., et al. Intracellular accumulation of bilirubin as a defense mechanism against increased oxidative stress. *Biochimie.* 2012; 94(8):1821–7.

37. Dore S., Takahashi M., Ferris C. D., et al. Bilirubin, formed by activation of heme oxygenase-2, protects neurons against oxidative stress injury. *Proceedings of the National Academy of Sciences of the United States of America.* 1999; 96(5):2445–50.

38. Hatcher H. C., Tesfay L., Torti S. V., et al. Cytoprotective effect of ferritin H in renal ischemia reperfusion injury. *PLoS one.* 2015; 10(9):e0138505.

39. Arosio P., Carmona F., Gozzelino R., et al. The importance of eukaryotic ferritins in iron handling and cytoprotection. *The biochemical journal.* 2015; 472(1):1–15.

40. Waza A. A., Hamid Z., Ali S., et al. A review on heme oxygenase-1 induction: is it a necessary evil. *Inflammation research: official journal of the European Histamine Research Society [et al].* 2018; 67(7):579–88.

41. Issan Y., Kornowski R., Aravot D., et al. Heme oxygenase-1 induction improves cardiac function following myocardial ischemia by reducing oxidative stress. *PLoS one.* 2014; 9(3):e92246.

42. Morita T., Imai T., Sugiyama T., et al. Heme oxygenase-1 in vascular smooth muscle cells counteracts cardiovascular damage induced by angiotensin II. *Current neurovascular research.* 2005; 2(2):113–20.

43. Sheikh S. Z., Hegazi R. A., Kobayashi T., et al. An anti-inflammatory role for carbon monoxide and heme oxygenase-1 in chronic Th2-mediated murine colitis. *Journal of immunology (Baltimore, Md: 1950).* 2011; 186(9):5506–5513.

44. Vijayan V., Baumgart-Vogt E., Naidu S., et al. Bruton's tyrosine kinase is required for TLR-dependent heme oxygenase-1 gene activation via Nrf2 in macrophages. *Journal of immunology (Baltimore, Md: 1950).* 2011; 187(2):817–27.

45. Hsu H. Y., Chu L. C., Hua K. F., et al. Heme oxygenase-1 mediates the anti-inflammatory effect of Curcumin within LPS-stimulated human monocytes. *Journal of cellular physiology.* 2008; 215(3):603–12.

46. Li Z., Wang Y., Zeng G., et al. Increased miR-155 and heme oxygenase-1 expression is involved in the protective effects of formononetin in traumatic brain injury in rats. *American journal of translational research.* 2017; 9 (12):5653–61.

47. Liao G., Li R., Chen X., et al. Sodium valproate prevents radiation-induced injury in hippocampal neurons via activation of the Nrf2/HO-1 pathway. *Neuroscience.* 2016; 331:40–51.

48. Yin H., Li X., Yuan B., et al. Heme oxygenase-1 ameliorates LPS-induced acute lung injury correlated with downregulation of interleukin-33. *International immunopharmacology.* 2011; 11(12):2112–17.

49. Ahmed A., Rahman M., Zhang X., et al. Induction of placental heme oxygenase-1 is protective against TNFalpha-induced cytotoxicity and promotes vessel relaxation. *Molecular medicine (Cambridge, Mass).* 2000; 6(5):391–409.

50. Schumacher A., Wafula P. O., Teles A., et al. Blockage of heme oxygenase-1 abrogates the protective effect of regulatory T cells on murine pregnancy and promotes the maturation of dendritic cells. *PLoS one.* 2012; 7(8):e42301.

51. Abraham N. G., Cao J., Sacerdoti D., et al. Heme oxygenase: the key to renal function regulation. *American journal of physiology renal physiology.* 2009; 297(5):F1137–52.

52. Ferenbach D. A., Kluth D. C. & Hughes J. Hemeoxygenase-1 and renal ischaemia-reperfusion injury. *Nephron experimental nephrology.* 2010; 115(3):e33–337.

53. Nath K. A. Heme oxygenase-1 and acute kidney injury. *Current opinion in nephrology and hypertension.* 2014; 23 (1):17–24.

54. Bolisetty S., Zarjou A. & Agarwal A. Heme oxygenase 1 as a therapeutic target in acute kidney injury. *American journal of kidney diseases: the official journal of the National Kidney Foundation.* 2017; 69(4):531–45.

55. Chok M. K., Ferlicot S., Conti M., et al. Renoprotective potency of heme oxygenase-1 induction in rat renal ischemia-reperfusion. *Inflammation & allergy drug targets.* 2009; 8(4):252–59.

56. Ferenbach D. A., Nkejabega N. C., McKay J., et al. The induction of macrophage hemeoxygenase-1 is protective during acute kidney injury in aging mice. *Kidney international.* 2011; 79(9):966–76.

57. Tracz M. J., Juncos J. P., Croatt A. J., et al. Deficiency of heme oxygenase-1 impairs renal hemodynamics and exaggerates systemic inflammatory responses to renal ischemia. *Kidney international.* 2007; 72(9):1073–80.

58. Shiraishi F., Curtis L. M., Truong L., et al. Heme oxygenase-1 gene ablation or expression modulates cisplatin-induced renal tubular apoptosis. *American journal of physiology renal physiology.* 2000; 278(5):F726–36.

59. Wei Q., Hill W. D., Su Y., et al. Heme oxygenase-1 induction contributes to renoprotection by G-CSF during rhabdomyolysis-associated acute kidney injury. *American journal of physiology renal physiology.* 2011; 301(1):F162–70.

60. Kang K., Nan C., Fei D., et al. Heme oxygenase 1 modulates thrombomodulin and endothelial protein C receptor levels to attenuate septic kidney injury. *Shock (Augusta, Ga).* 2013; 40(2):136–43.

61. Poole B., Wang W., Chen Y. C., et al. Role of heme oxygenase-1 in endotoxemic acute renal failure. *American journal of physiology renal physiology.* 2005; 289(6):F1382–5.

62. Lee S. C., Han S. H., Li J. J., et al. Induction of heme oxygenase-1 protects against podocyte apoptosis under diabetic conditions. *Kidney international.* 2009; 76(8):838–48.

63. Nicolai A., Li M., Kim D. H., et al. Heme oxygenase-1 induction remodels adipose tissue and improves insulin sensitivity in obesity-induced diabetic rats. *Hypertension (Dallas, Tex: 1979).* 2009; 53(3):508–15.

64. Ndisang J. F. & Chibbar R. Heme oxygenase improves renal function by potentiating podocyte-associated proteins in Nomega-Nitro-l-Arginine-Methyl Ester (l-NAME)-induced hypertension. *American journal of hypertension.* 2015; 28(7):930–42.

65. Datta P. K., Duann P. & Lianos E. A. Long-term effect of heme oxygenase (HO)-1 induction in glomerular immune injury. *The Journal of laboratory and clinical medicine.* 2006; 147(3):150–5.

66. Duann P. & Lianos E. A. GEC-targeted HO-1 expression reduces proteinuria in glomerular immune injury. *American journal of physiology renal physiology.* 2009; 297(3): F629–38.

67. Kinderlerer A. R., Pombo Gregoire I., Hamdulay S. S., et al. Heme oxygenase-1 expression enhances vascular endothelial resistance to complement-mediated injury through induction of decay-accelerating factor: a role for increased bilirubin and ferritin. *Blood.* 2009; 113(7):1598–607.

68. Detsika M. G., Duann P., Atsaves V., et al. Heme oxygenase 1 up-regulates glomerular decay accelerating factor expression and minimizes complement deposition and injury. *The American journal of pathology.* 2016; 186 (11):2833–45.

69. Zager R. A., Johnson A. C. & Becker K. Plasma and urinary heme oxygenase-1 in AKI. *Journal of the American Society of Nephrology: JASN.* 2012; 23(6):1048–57.

70. Dulak J. & Jozkowicz A. Novel faces of heme oxygenase-1: mechanisms and therapeutic potentials. *Antioxidants & redox signaling.* 2014; 20(11):1673–6.

71. Vanella L., Barbagallo I., Tibullo D., et al. The non-canonical functions of the heme oxygenases. *Oncotarget.* 2016; 7(42):69075–86.

72. Hwang H. W., Lee J. R., Chou K. Y., et al. Oligomerization is crucial for the stability and function of heme oxygenase-1 in the endoplasmic reticulum. *The journal of biological chemistry.* 2009; 284(34):22672–9.

73. Yoshida T. & Sato M. Posttranslational and direct integration of heme oxygenase into microsomes. *Biochemical and biophysical research communications.* 1989; 163(2):1086–92.

74. Boname J. M., Bloor S., Wandel M. P., et al. Cleavage by signal peptide peptidase is required for the degradation of selected tail-anchored proteins. *The journal of cell biology.* 2014; 205(6):847–62.

75. Schaefer B., Moriishi K. & Behrends S. Insights into the mechanism of isoenzyme-specific signal peptide peptidase-mediated translocation of heme oxygenase. *PLoS one.* 2017; 12(11):e0188344.

76. Linnenbaum M., Busker M., Kraehling J. R., et al. Heme oxygenase isoforms differ in their subcellular trafficking during hypoxia and are differentially modulated by cytochrome P450 reductase. *PLoS one.* 2012; 7(4):e35483.

77. Hsu F. F., Yeh C. T., Sun Y. J., et al. Signal peptide peptidase-mediated nuclear localization of heme oxygenase-1 promotes cancer cell proliferation and invasion independent of its enzymatic activity. *Oncogene.* 2015; 34 (18):2360–70.

78. Lin Q., Weis S., Yang G., et al. Heme oxygenase-1 protein localizes to the nucleus and activates transcription factors important in oxidative stress. *The journal of biological chemistry.* 2007; 282(28):20621–33.

79. Biswas C., Shah N., Muthu M., et al. Nuclear heme oxygenase-1 (HO-1) modulates subcellular distribution and activation of Nrf2, impacting metabolic and anti-oxidant defenses. *The journal of biological chemistry.* 2014; 289 (39):26882–94.

80. Holmstrom K. M., Kostov R. V. & Dinkova-Kostova A. T. The multifaceted role of Nrf2 in mitochondrial function. *Current opinion in toxicology.* 2016; 1:80–91.

81. Nitti M., Piras S., Marinari U. M., et al. HO-1 induction in cancer progression: a matter of cell adaptation. *Antioxidants (Basel, Switzerland).* 2017; 6(2) :pii:E29.

82. Sacca P., Meiss R., Casas G., et al. Nuclear translocation of haeme oxygenase-1 is associated to prostate cancer. *British journal of cancer.* 2007; 97(12):1683–9.

83. Hsu F. F., Chiang M. T., Li F. A., et al. Acetylation is essential for nuclear heme oxygenase-1-enhanced tumor growth and invasiveness. *Oncogene.* 2017; 36(49):6805–14.

84. Bansal S., Biswas G. & Avadhani N. G. Mitochondria-targeted heme oxygenase-1 induces oxidative stress and mitochondrial dysfunction in macrophages, kidney fibroblasts and in chronic alcohol hepatotoxicity. *Redox biology.* 2014; 2:273–83.

85. Bindu S., Pal C., Dey S., et al. Translocation of heme oxygenase-1 to mitochondria is a novel cytoprotective mechanism against non-steroidal anti-inflammatory drug-induced mitochondrial oxidative stress, apoptosis, and gastric mucosal injury. *The journal of biological chemistry.* 2011; 286(45):39387–402.

86. Bolisetty S., Traylor A., Zarjou A., et al. Mitochondria-targeted heme oxygenase-1 decreases oxidative stress in renal epithelial cells. *American journal of physiology renal physiology.* 2013; 305(3):F255–64.

87. Cousar J. L., Lai Y., Marco C. D., et al. Heme oxygenase 1 in cerebrospinal fluid from infants and children after severe traumatic brain injury. *Developmental neuroscience.* 2006; 28(4–5):342–7.

88. Li Volti G., Galvano F., Frigiola A., et al. Potential immunoregulatory role of heme oxygenase-1 in human milk: a

combined biochemical and molecular modeling approach. *The journal of nutritional biochemistry.* 2010; 21(9):865–71.

89. Spencer A. L., Bagai I., Becker D. F., et al. Protein/protein interactions in the mammalian heme degradation pathway: heme oxygenase-2, cytochrome P450 reductase, and biliverdin reductase. *The journal of biological chemistry.* 2014; 289(43):29836–58.

90. Jung N. H., Kim H. P., Kim B. R., et al. Evidence for heme oxygenase-1 association with caveolin-1 and -2 in mouse mesangial cells. *IUBMB life.* 2003; 55(9):525–32.

91. Taira J., Sugishima M., Kida Y., et al. Caveolin-1 is a competitive inhibitor of heme oxygenase-1 (HO-1) with heme: identification of a minimum sequence in caveolin-1 for binding to HO-1. *Biochemistry.* 2011; 50(32):6824–31.

92. Jin L. G., Zeng S., Sun X. Q., et al. Deletion 101 residue at caveolin-1 scaffolding domain peptides impairs the ability of increasing heme oxygenase-1 activity. *International immunopharmacology.* 2018; 63:137–144.

93. Weng P., Zhang X. T., Sheng Q., et al. Caveolin-1 scaffolding domain peptides enhance anti-inflammatory effect of heme oxygenase-1 through interrupting its interact with caveolin-1. *Oncotarget.* 2017; 8(25):40104–14.

94. Paez A. V., Pallavicini C., Schuster F., et al. Heme oxygenase-1 in the forefront of a multi-molecular network that governs cell-cell contacts and filopodia-induced zippering in prostate cancer. *Cell death & disease.* 2016; 7(12):e2570.

95. Poss K. D. & Tonegawa S. Reduced stress defense in heme oxygenase 1-deficient cells. *Proceedings of the National Academy of Sciences of the United States of America.* 1997; 94(20):10925–30.

96. Kim J., Zarjou A., Traylor A. M., et al. In vivo regulation of the heme oxygenase-1 gene in humanized transgenic mice. *Kidney international.* 2012; 82(3):278–91.

97. Hull T. D., Kamal A. I., Boddu R., et al. Heme oxygenase-1 regulates myeloid cell trafficking in AKI. *Journal of the American Society of Nephrology: JASN.* 2015; 26(9):2139–51.

98. Bolisetty S., Traylor A., Joseph R., et al. Proximal tubule-targeted heme oxygenase-1 in cisplatin-induced acute kidney injury. *American journal of physiology renal physiology.* 2016; 310(5):F385–94.

99. Leaf D. E., Body S. C., Muehlschlegel J. D., et al. Length polymorphisms in heme oxygenase-1 and AKI after cardiac surgery. *Journal of the American Society of Nephrology: JASN.* 2016; 27(11):3291–97.

100. Haldar M., Kohyama M., So A. Y., et al. Heme-mediated SPI-C induction promotes monocyte differentiation into iron-recycling macrophages. *Cell.* 2014; 156(6):1223–34.

101. Cao Q., Harris D. C. & Wang Y. Macrophages in kidney injury, inflammation, and fibrosis. *Physiology (Bethesda, Md).* 2015; 30(3):183–94.

102. Huen S. C. & Cantley L. G. Macrophages in renal injury and repair. *Annual review of physiology.* 2017; 79:449–69.

103. Meng X. M., Tang P. M., Li J., et al. Macrophage phenotype in kidney injury and repair. *Kidney diseases (Basel, Switzerland).* 2015; 1(2):138–46.

104. Paller M. S. Hemoglobin- and myoglobin-induced acute renal failure in rats: role of iron in nephrotoxicity. *The American journal of physiology.* 1988; 255(3 Pt 2):F539–44.

105. Hull T. D., Agarwal A. & George J. F. The mononuclear phagocyte system in homeostasis and disease: a role for heme oxygenase-1. *Antioxidants & redox signaling.* 2014; 20(11):1770–88.

106. Kie J. H., Kapturczak M. H., Traylor A., et al. Heme oxygenase-1 deficiency promotes epithelial-mesenchymal transition and renal fibrosis. *Journal of the American Society of Nephrology: JASN.* 2008; 19(9):1681–91.

107. Ndisang J. F., Jadhav A. & Mishra M. The heme oxygenase system suppresses perirenal visceral adiposity, abates renal inflammation and ameliorates diabetic nephropathy in Zucker diabetic fatty rats. *PLoS one.* 2014; 9(1):e87936.

108. Hoenig M. P. & Zeidel M. L. Homeostasis, the milieu interieur, and the wisdom of the nephron. *Clinical Journal of the American Society of Nephrology: CJASN.* 2014; 9 (7):1272–81.

109. Duann P. & Lin P. H. Mitochondria Damage and Kidney Disease. *Advances in experimental medicine and biology.* 2017; 982:529–51.

110. Galvan D. L., Green N. H. & Danesh F. R. The hallmarks of mitochondrial dysfunction in chronic kidney disease. *Kidney international.* 2017; 92(5):1051–57.

111. Klionsky D. J., Abdelmohsen K., Abe A., et al. Guidelines for the use and interpretation of assays for monitoring autophagy (3rd edition). *Autophagy.* 2016; 12(1):1–222.

112. Choi A. M., Ryter S. W. & Levine B. Autophagy in human health and disease. *The New England journal of medicine.* 2013; 368(19):1845–6.

113. Wang Z. & Choi M. E. Autophagy in kidney health and disease. *Antioxidants & redox signaling.* 2014; 20(3):519–37.

114. Lenoir O., Tharaux P. L. & Huber T. B. Autophagy in kidney disease and aging: lessons from rodent models. *Kidney international.* 2016; 90(5):950–64.

115. Takahashi A., Kimura T., Takabatake Y., et al. Autophagy guards against cisplatin-induced acute kidney injury. *The American journal of pathology.* 2012; 180(2):517–25.

116. Mei S., Livingston M., Hao J., et al. Autophagy is activated to protect against endotoxic acute kidney injury. *Scientific reports.* 2016; 6:22171.

117. Leventhal J. S., Ni J., Osmond M., et al. Autophagy Limits Endotoxemic acute kidney injury and alters renal tubular epithelial cell cytokine expression. *PLoS one.* 2016; 11(3):e0150001.

118. Ding Y. & Choi M. E. Autophagy in diabetic nephropathy. *The journal of endocrinology.* 2015; 224(1):R15–30.

119. Kume S., Yamahara K., Yasuda M., et al. Autophagy: emerging therapeutic target for diabetic nephropathy. *Seminars in Nephrology.* 2014; 34(1):9–16.

120. Hartleben B., Wanner N. & Huber T. B. Autophagy in glomerular health and disease. *Seminars in Nephrology.* 2014; 34(1):42–52.

121. Kawakami T., Gomez I. G., Ren S., et al. Deficient autophagy results in mitochondrial dysfunction and FSGS. *Journal of the American Society of Nephrology: JASN.* 2015; 26(5):1040–52.

122. Kim W. Y., Nam S. A., Song H. C., et al. The role of autophagy in unilateral ureteral obstruction rat model. *Nephrology (Carlton, Vic).* 2012; 17(2):148–59.

123. Baisantry A., Bhayana S., Rong S., et al. Autophagy induces prosenescent changes in proximal tubular S3 segments. *Journal of the American Society of Nephrology: JASN.* 2016; 27(6):1609–16.

124. Pecorella S. R., Potter J. V., Cherry A. D., et al. The HO-1/CO system regulates mitochondrial-capillary density relationships in human skeletal muscle. *American journal of physiology lung cellular and molecular physiology.* 2015; 309(8):L857–71.

125. Hull T. D., Boddu R., Guo L., et al. Heme oxygenase-1 regulates mitochondrial quality control in the heart. *JCI insight.* 2016; 1(2):e85817.

126. Yu W., Zhang X., Wu H., et al. HO-1 is essential for tetra-hydroxystilbene glucoside mediated mitochondrial biogenesis and anti-inflammation process in LPS-treated RAW264.7 macrophages. *Oxidative medicine and cellular longevity.* 2017; 2017:1818575.

127. Piantadosi C. A., Withers C. M., Bartz R. R., et al. Heme oxygenase-1 couples activation of mitochondrial biogenesis to anti-inflammatory cytokine expression. *The journal of biological chemistry.* 2011; 286(18):16374–85.

128. Di Noia M. A., Van Driesche S., Palmieri F., et al. Heme oxygenase-1 enhances renal mitochondrial transport carriers and cytochrome C oxidase activity in experimental diabetes. *The journal of biological chemistry.* 2006; 281(23):15687–93.

129. Han Q., Li G., Ip M. S., et al. Haemin attenuates intermittent hypoxia-induced cardiac injury via inhibiting mitochondrial fission. *Journal of cellular and molecular medicine.* 2018; 22(5):2717–26.

130. Zukor H., Song W., Liberman A., et al. HO-1-mediated macroautophagy: a mechanism for unregulated iron deposition in aging and degenerating neural tissues. *Journal of neurochemistry.* 2009; 109(3):776–91.

131. Suliman H. B., Keenan J. E. & Piantadosi C. A. Mitochondrial quality-control dysregulation in conditional HO-1(-/-) mice. *JCI insight.* 2017; 2(3):e89676.

132. Duann P., Lianos E. A., Ma J., et al. Autophagy, innate immunity and tissue repair in acute kidney injury. *International journal of molecular sciences.* 2016; 17(5).

133. Decuypere J. P., Ceulemans L. J., Agostinis P., et al. Autophagy and the kidney: implications for ischemia-reperfusion injury and therapy. *American journal of kidney diseases: the official journal of the National Kidney Foundation.* 2015; 66(4):699–709.

134. Bolisetty S., Traylor A. M., Kim J., et al. Heme oxygenase-1 inhibits renal tubular macroautophagy in acute kidney injury. *Journal of the American Society of Nephrology: JASN.* 2010; 21(10):1702–12.

135. Adedoyin O., Boddu R., Traylor A., et al. Heme oxygenase-1 mitigates ferroptosis in renal proximal tubule cells. *American journal of physiology renal physiology.* 2018; 314(5):F702–7f14.

136. Andrade-Silva M., Cenedeze M. A., Perandini L. A., et al. TLR2 and TLR4 play opposite role in autophagy associated with cisplatin-induced acute kidney injury. *Clinical science (London, England: 1979).* 2018; 132(16):1725–39. doi: 10.1042/CS20170262.

137. Dong C., Zheng H., Huang S., et al. Heme oxygenase-1 enhances autophagy in podocytes as a protective mechanism against high glucose-induced apoptosis. *Experimental cell research.* 2015; 337(2):146–59.

138. Xu J., Li J., Wang J., et al. Heme oxygenase1 protects H2O2insulted glomerular mesangial cells from excessive autophagy. *Molecular medicine reports.* 2016; 13(6):5269–75.

139. Chau L. Y. Heme oxygenase-1: emerging target of cancer therapy. *Journal of biomedical science.* 2015; 22:22.

140. Salerno L., Romeo G., Modica M. N., et al. Heme oxygenase-1: a new druggable target in the management of chronic and acute myeloid leukemia. *European journal of medicinal chemistry.* 2017; 142:163–78.

141. Zhao M., Yang M., Que W., et al. Myeloid heme oxygenase-1: a new therapeutic target in anti-inflammation. *Frontiers in bioscience (Landmark edition).* 2018; 23:2001–15.

142. Ryter S. W. & Choi A. M. Targeting heme oxygenase-1 and carbon monoxide for therapeutic modulation of inflammation. *Translational research: the journal of laboratory and clinical medicine.* 2016; 167(1):7–34.

143. Cheng Y. & Rong J. Therapeutic potential of heme oxygenase-1/carbon monoxide system against ischemia-reperfusion injury. *Current pharmaceutical design.* 2017; 23(26):3884–98.

144. Abraham N. G., Junge J. M. & Drummond G. S. Translational significance of heme oxygenase in obesity and metabolic syndrome. *Trends in pharmacological sciences.* 2016; 37(1):17–36.

145. Rahman M. N., Vukomanovic D., Vlahakis J. Z., et al. Structural insights into human heme oxygenase-1 inhibition by potent and selective azole-based compounds. *Journal of the royal society, interface.* 2013; 10(78):20120697.

19 Mitochondrial Transplantation in Myocardial Ischemia and Reperfusion Injury

David Blitzer, Borami Shin, Alvise Guariento, Pedro J. del Nido, and James D. McCully

Department of Cardiac Surgery, Harvard Medical School, Boston Children's Hospital, Boston, USA

CONTENTS

1 INTRODUCTION

Ischemic heart disease and its sequelae is one of the leading causes of morbidity across the globe and it is the leading cause of mortality [1]. As the leading cause of death worldwide, ischemia-reperfusion injury has been the subject of extensive investigation with the aim of understanding the mechanisms that lead to irreversible myocardial injury and subsequent death [2–5]. These studies revealed the pathways leading to myocardial irreversible injury, which include alterations in cellular energy and homeostatic ions, and accumulation of reactive oxygen species, all of which converge on mitochondria to create dysfunction and initiate the signaling pathways that result in apoptosis and necrosis [6–9]. The study of myocardial injury, which exists along a spectrum from stunning and hibernation to apoptosis and necrosis, has also lead to the discovery of protective mechanisms which include preconditioning and post-conditioning [10–22].

The broader understanding of the pathways leading from the ischemia-reperfusion injury and the discovery of protective pathways has spawned multiple attempts to utilize these pathways via pharmacological, genetic and procedural therapies. These investigative therapies have targeted multiple steps along the myocardial injury pathway, including the depletion of energy substrates and antioxidants, the accumulation of intracellular calcium ($Ca2+$) and sodium, the downstream mitochondrial signaling, and the inflammatory system activation. Despite positive results with these interventions in preclinical studies, subsequent clinical trials with these therapies have demonstrated equivocal or negative results [23–26].

These initial biochemical investigations and the corresponding attempts at therapeutic interventions have consistently supported the prominent role of mitochondria, more specifically mitochondrial dysfunction, as a final common pathway of cardiomyocyte response to ischemia-reperfusion injury. The prominent role of the mitochondrial response to ischemia-reperfusion injury leads to hypotheses that mitochondria may represent a primary end target for myocardial salvage and cardio-protection. As such, many have attempted to target individual targets in the injury or signaling pathways leading from ischemia-reperfusion to mitochondrial dysfunction and cell death; however, an emerging therapeutic modality delivers healthy mitochondria in order to replace damaged cardiomyocyte mitochondria. In the following chapter, we will further explore the underlying physiology of myocardial ischemia-reperfusion injury with an emphasis on mitochondrial metabolism and the mitochondrial response to ischemia-reperfusion injury. With this foundational knowledge, we will then explore the potential mechanisms by which mitochondrial transplantation might afford a protective benefit to the cardiomyocyte and the methodologies by mitochondrial transplantation is being performed.

2   NORMAL PHYSIOLOGY OF MYOCARDIAL MITOCHONDRIA

The heart relies on the continuous production of adenosine tri-phosphate (ATP), by myocardial mitochondria, to maintain cellular electrochemical homeostasis and to power the repeated cycles of contraction and relaxation which comprise the cardiac cycle and to maintain. Myocardial mitochondria produce ATP at rates that approach approximately 15 times the weight of the human heart. The reliance of the myocardium on adequate mitochondrial function is demonstrated by the fact that mitochondria comprise approximately 30% of myocardium by volume [27]. There is, likewise, a tight relationship between effective mitochondrial metabolism, namely ATP production and myocardial contraction.

The general mechanistic pathways through which mitochondria provide cellular energy are preserved throughout the human organism. ATP is produced within the mitochondria, more specifically along the inner mitochondrial membrane wherein a series of highly interconnected multi-subunit complexes, referred to collectively as the Electron Transport Chain (ETC), conduct oxidative phosphorylation. The ETC is comprised of five subunits: Complex I (NADH dehydrogenase), II (succinate ubiquinone oxidoreductase), III (ubiquinol: cytochrome c oxidoreductase), IV (cytochrome c oxidase), V (ATP synthase), and two main electron carriers, ubiquinone and cytochrome *c*. Of the 76 protein that make up the five complexes of the ETC, thirteen are encoded within the mitochondrial DNA and nuclear DNA encodes the remaining 63. In short, the intermediary steps of the catabolic process of ATP production

generate reducing equivalents in the form of NADH and FADH2. Electrons from these reducing equivalents flow down a gradient, by which protons are pumped into the intermembrane space of the mitochondria resulting in an electrochemical gradient that is then used to phosphorylate ADP to ATP. After transportation into the intermembrane space by an inner membrane channel adenine nucleotide translocase (ANT), where mitochondrial creatine kinase (miCK) transfers the high energy phosphate of the ATP to create phosphocreatine which can then be stored in the cytosol as energy reserves. In cardiomyocytes, the energy from ATP is vital for myocardial contraction, most notably as an energy source for Myosin-ATPase. The energy from ATP is also used to maintain cellular homeostasis, as in the case of Ca2+/Mg2+-ATPase which maintains stores of Ca2+ in the sarcoplasmic reticulum and Na+/K+-ATPase which maintains cellular excitability and Ca2+ efflux [28].

Within the myocardium, there are two distinct populations of mitochondria which are distinguished spatially as either subsarcolemmal mitochondria (SSM) or interfibrillar mitochondria (IFM), with multiple investigations revealing further distinguishing characteristics [29–33]. SSM are variable in lengths, measuring between 0.4 and 3.0 µm and they generally possess more lamelliform and densely packed cristae, whereas IFM are more elongated, measuring 1.5 to 2.0 µm in length and have cristae that are more tubular in nature [33,34]. Using three dimensional modeling and specific mitochondrial staining studies have demonstrated that IFM are arranged in a more orderly pattern while SSM are arranged more randomly [35,36]. Other studies have differentiated the two mitochondrial population by substrate utilization, enzyme function rates, and ATP synthesis [29,31,37–40]. The response of these subpopulations to insult or injury, such as ischemia-reperfusion, also appears to be different, however there is conflicting evidence of which subpopulation is more susceptible injury based on the animal model utilized [41–50]. As such, the clinical significance and therapeutic implications of the differing mitochondrial subpopulations remains a topic of investigation.

3   MYOCARDIAL ISCHEMIA

When blow flow to the heart, via the coronary arteries, is attenuated or occluded, oxygen delivery to the myocardium is reduced to the point where there is insufficient oxygen delivery to support cellular function. This is referred to as ischemic injury. Myocardium is particularly susceptible to ischemic injury, as it only has sufficient energy stored to sustain myocardial function for as little as 8 seconds [49]. Furthermore, at baseline the myocardium extracts 75% of the oxygen supplied by the coronary vasculature, as such the only innate response to reduced oxygen delivery is to increase coronary blood flow. Ischemic insults that exceed the energy reserve of the myocardium and limit the efficacy of the coronary vasodilatory response directly lead to diminished myocardial function and ultimately to myocardial infarction [50–55]. Myocardial injury is first evident in the subendocardium and papillary muscle within 20 minutes of ischemia. The injury subsequently extends into the mid-myocardium within 60 to 90 minutes. The ischemic injury and any infarction progress to transmural in 3 to 4 hours [56–61]. Subjected to global ischemia, both ex-vivo and in-vivo blood perfused hearts have demonstrated downregulation of proteins of mitochondrial structure and function by transcriptomic and proteomic analysis. Furthermore, hearts that are subjected to global ischemia and treated with cardioplegia for cardioprotection demonstrate preserved mitochondrial function among other measures, indicating a key role for the mitochondria in the pathophysiology of ischemic injury and the benefit afforded from cardioprotective measures [62].

4 MYOCARDIAL REPERFUSION INJURY

Restoration of blood flow is the best measure to arrest ischemic injury, however reperfusion is accompanied by and activates a cascade of inflammatory pathways that can augment the ischemic injury to heart itself and even extends to distant organs [62–64]. Pathologically, reperfusion can cause hemorrhagic infarct with accompanying contraction band necrosis. Paradoxically, reperfusion can also lead to disruption of the microvasculature, creating a "no-reflow" phenomenon in which blood flow to the ischemic region remains diminished after addressing the arterial occlusion [65–67]. The mechanism by which the reperfusion augments ischemic injury has been the subject of extensive investigations. The investigations have demonstrated that mitochondrial response plays a crucial role in that the rate of oxygen consumption, and thus the synthesis of high energy phosphates, immediately declines with the reduction of blood flow and oxygen supply [55]. When assessed by oximetry in animal models, mitochondrial function is affected by ischemia in a number of ways, notably the decline in state three oxygen consumption and respiratory control index marked by decreases in malate, a substrate of complex I, succinate, a substrate of complex III, and cytochrome c [48,68–76].

5 CALCIUM OVERLOADING

Upon reperfusion, the cell experiences a rapid wash out of accumulated lactic acid via the Na+/H+ exchanger (NHE) and Na+/HCO3− exchanger, resulting in a rapid restoration of physiologic intracellular pH. With the normalization of intracellular pH the mitochondrial membrane potential also normalizes and the mitochondrial Ca2+-uniporter (MCU) then opens. Mitochondria, which already experience Ca2+-overloading as a result of the initial ischemic injury, then experience further Ca2+-overloading after MCU opening [77–79]. Furthermore, ischemia induced damage to the sarcolemmal membrane and dysfunction of the sarcoplasmic

reticulum, are believed to play a role in increasing cytosolic Ca2+ further altering the intracellular homeostasis of the cardiac myocyte.

6 REACTIVE OXYGEN SPECIES

Reactive Oxygen Species (ROS), which include H_2O_2, HO−, HOCl, and O2⁻, have also been implicated as mediators of reperfusion injury, and injured mitochondria are the primary source of these ROS. Quantitative increases in ROS upon reperfusion have been demonstrated using numerous quantitative modalities and across multiple animal models, including humans [80,82]. It has been recently proposed that mitochondrial succinate accumulation plays a central role in the oxidative injury that results from ischemia-reperfusion. During ischemia, succinate, the substrate of complex II of the ETC accumulates within the mitochondria. Succinate is also the only component derived from mitochondria which is a precursor to succinate, xanthine, and hypoxanthine, three metabolites that accumulate with mitochondrial ischemia [80,81]. Upon reperfusion, complex II hydrolyzes the accumulated succinate at a rate that exceeds that of ATP synthesis, leading to an accumulation of electrons that then drives the ETC in reverse, known as Reverse Electron Transport (RET), generating the superoxide [82–84]. ROS accumulation also reduces the autoprotective mechanisms available to the cell, such as that of nitric oxide [85]. Inhibition of mitochondrial complex I, during the early reperfusion, has been demonstrated to reduce ROS accumulation and affords protective benefits [86,87]. The various mechanisms by which ROS lead to cellular damage, including but not limited to mitochondrial and nuclear DNA breaks, and SER dysfunction, converge to activate apoptotic and necrotic signaling pathways.

7 FINAL PATHWAY

The opening of the mitochondrial permeability transition pore (mPTP), a voltage gated pore within the inner mitochondrial membrane, is proposed as the key step in the realization of the apoptotic and necrotic pathways. With the mTPT opened any molecule smaller than 1.5 kDa can freely pass through the mitochondrial membrane. The result is collapse of the mitochondrial membrane potential, uncoupling of oxidative phosphorylation and disruption of the mitochondrial membrane leading to swelling and eventual rupture. Upon rupture of the mitochondrial membrane, cytochrome, a component of the ETC and a potent activator of apoptotic signaling is released, leading to cardiomyocyte death [88,89]. This pathway also releases systemic inflammatory signals leading to leukocyte aggregation and leukocyte-mediated tissue destruction. The same signals also cause endothelial dysfunction leading to platelet aggregation and microvascular injury.

8   ISCHEMIA VERSUS REPERFUSION INJURY

The investigation of reperfusion injury as an independent mediator of cardiomyocyte injury, is distinct from ischemic injury and complicated by difficulties in accurately assessing the progression of necrosis and the amount attributable to ischemic or reperfusion injury [23]. The challenges of elucidating reperfusion injury are also demonstrated by the equivocal results of studies targeting the accepted mediators of reperfusion injury. Several presumed mediators of reperfusion injury, such as the accumulation of Ca2+, ROS, and inflammatory signals or release of ETC components have frustrated attempts at clinical application despite promising preclinical data. Investigators have targeted sarcolemmal Ca2+ ion channels, the mitochondrial Ca2+ uniporter, and the Na+/H+ exchanger in order to mitigate damage due to Ca2+ accumulation, but promising results from animal studies have not been borne out in translation to humans [33,90–97]. The deleterious effects of ROS accumulation have been targeted using antioxidants such as superoxide dismutase, trimetazidine, and vitamins C and E. Once again, despite positive results in animal studies, similar benefit was lacking in human trials [98–100]. Similarly, attempts to mitigate the injurious inflammatory response through leukocyte depleted blood, antibodies against the cell adhesion molecule P selection, or inhibition of the complement system have been promising in animal trials and disappointing in attempts at translation to humans [101–107]. Despite the indisputable role of mitochondrial injury in myocardial dysfunction secondary to ischemia and reperfusion injury, the inability to translate preclinical results into the clinical sphere emphasizes the need for further evaluation of the mechanism by which ischemia-reperfusion injury creates mitochondrial damage and subsequent myocardial dysfunction. Only through a thorough understanding of the underlying pathophysiology can successful interventions be discovered and disseminated for clinical benefit.

9   REVERSIBLE MYOCARDIAL INJURY

9.1   Myocardial Stunning

Myocardial ischemia-reperfusion injury may be reversible, depending on the severity and duration of ischemia, in which case the myocardium is said to experience stunning. This is a myocardial physiologic survival adaptation that results from changes in gene expressions due to ischemia-reperfusion injury that promote survival. Stunned myocardium is characterized by myocytes that have not undergone irreversible damage and yet demonstrate persistent contractile dysfunction despite adequate reperfusion [108]. Myocardial stunning also plays a prominent role in patients after cardiac surgery, where despite continued improvements in cardioprotective strategies, patients often experience prolonged myocardial dysfunction, which manifests as a prolonged requirement for inotropic support, which eventually resolves [109-112].

9.2   Myocardial Hibernation

Hibernation, in regards to myocardium, refers to a chronic state wherein viable myocardium exhibits a reversible reduction in contractility after experiencing a local reduction in perfusion [109,110,113–115]. This is an adaptive mechanism whereby myocardial metabolism and thus contractility is purposefully minimized such that metabolic demands can still be met by the limited perfusion [116]. Several studies have demonstrated that this chronic myocardial dysfunction at a global level in the human heart is typified by myofibrillar loss, increased glucose dependence, and the accumulation of glycogen at a cellular level [117–119]. Such a state cannot be maintained indefinitely and revascularization is necessary to prevent irreversible injury and associated morbidity [120,121]. Stunning and hibernation are now believed to be part of a continuous spectrum of ischemic injury, wherein repetitive episodes of stunning initiate the signaling that leads to myocardial hibernation [122,123].

9.3   Irreversible Myocardial Ischemia

Despite the auto-protective mechanisms, prolonged ischemia is inevitably succeeded by irreversible myocardial injury. Numerous studies have been performed to elucidate the intracellular pathways of this irreversible ischemic injury and have identified a central role for mitochondrial injury in the intracellular cascade leading to myocyte death. With the depletion of ATP, multiple changes occur in the mitochondria at the biochemical and ultrastructural levels. These include cellular acidosis and the accumulation of Ca2+ within the cytoplasm, the mitochondria, and the nucleus. Oxygen depletion within the myocardium signals for a change from fatty acid oxidation to anaerobic glycolysis with a resulting accumulation of lactate and intracellular acidosis, which is readily measured with a pH probe or by tissue PCO2 and 21P NMR spectroscopy [124,125]. With the accumulation of intracellular hydrogen (H+) the sodium-hydrogen exchanger (NHE) opens, with resulting increases in extracellular H+ and intracellular Na+ [126]. The sodium-calcium exchanger (NCX) on the plasma membrane then acts, in reverse, to reduce the intracellular Na+ load. This, in turn causes an increase in cytosolic Ca2+ which is augmented by opening of the L-type Ca2+ channels as a result of the depolarization at the plasma membrane. Furthermore, with the diminished availability of ATP sarcoplasmic reticulum by the sarco/endoplasmic reticulum Ca2+-ATPase (SERCA) stops taking up excess Ca2+ [127]. With the accumulation of cytoplasmic Ca2+ comes activation of Ca2+-dependent phospholipases and proteases leading to injury of the cellular membrane and continued accumulation of intracellular Ca2+.

With the build-up in intracellular Ca2+ the mitochondrial experience a corresponding build-up which has detrimental effects on mitochondrial structure and function [128]. In normal physiology, the inner membrane of the mitochondria pump H+ to the cytosol which creates the voltage gradient that allows for passive influx of Ca2+ into the mitochondrial matrix through the Ca2+-uniporter. With the accumulation of intracellular Ca2+ from ischemia, the mitochondria increases Ca2+ uptake, destabilizing the inner membrane potential. The mitochondria uses ATP to extrude the excess Ca2+ against the electrochemical gradient in a pathway named 'futile Ca2+ cycling' [129,130]. Mitochondrial Ca2+ overloading is followed by passive intra-mitochondrial water accumulation swelling; A process that can be visualized microscopically [74,75,131,132–135]. With the alterations in mitochondrial structure come changes in function. Proposed mechanisms for these functional changes include dissociation of adenine nucleotide translocase (ANT), mitochondrial creatine kinase (miCK), and the voltage-dependent anion carrier (VDAC) complex. Mitochondrial swelling increases the distance between these complexes which compromises their function [136]. Studies which have targeted Ca2+ overloading for therapeutic intervention results of these investigations have informed the development of cardioprotective cardioplegia solutions [55,61,103,135–138]. For example, the addition of diazoxide to cardioplegic solution opens mitochondrial ATP-sensitive K+ channels which reduce Ca2+ accumulation and in turn limit mitochondrial swelling, leading to improved post-ischemia function [136,138–141].

10   MITOCHONDRIAL TRANSPLANTATION

With the demonstration that mitochondrial injury and dysfunction plays a key role in the pathophysiology of ischemia-reperfusion injury, mitochondria have become a target for intervention in the treatment of such injury. Multiple interventions have been proposed to act at points in the pathway delineated above. Rather than target a single step or mediator of this complex pathway, mitochondrial transplantation has been proposed and investigated. Mitochondrial transplantation, using mitochondria isolated from non-ischemic tissue of the same subject, is hypothesized to replace damaged mitochondria with respiration-competent mitochondria as a means to mitigate the numerous pathologies associated with ischemia-reperfusion injury and the subsequent mitochondrial dysfunction.

10.1   MITOCHONDRIAL SOURCE AND VIABILITY

Initial studies in mitochondrial transplantation utilized mitochondria isolated from remote, non-ischemic, tissue of the same organism (autogenic) [142,143] This was initially motivated by a desire to augment clinical applicability and reduce potential immunogenic complications from mitochondria sourced from a different organism (allogeneic). In terms of immunogenic response, recent evidence demonstrates there is no immunogenic or inflammatory response to autogeneic or allogeneic mitochondrial transplantation in mice, as confirmed by multiplex and ELISA assays [144]. Different sources of tissue have been investigated as a source of mitochondria, and as expected different sources have differing concentrations of mitochondria, with liver having the highest, followed by skeletal muscle and then cardiac muscle [143]. Despite differences in mitochondrial concentration, there are no detectable differences in the cardioprotective effect of the differing mitochondrial populations [145,146].

While mitochondrial source makes no detectable cardioprotective difference, mitochondrial viability is a major determinant on cardioprotective effect. Studies have demonstrated that nonviable mitochondria, as confirmed by oxygen consumption testing, provide no cardioprotective effect. Similar lack of benefit is seen when transplanting mitochondrial complex proteins or mitochondrial RNA, DNA, exogenous ATP and ADP, or ATP synthesis promoters [145,147]. This is explained in part, by lability and short half-life of ATP, although there has been some evidence to suggest a modest benefit from the administration of mitochondrial complexes, such as cytochrome c or coenzyme Q10, albeit at very high doses [148,149].

10.2   MITOCHONDRIAL ISOLATION AND PURIFICATION

Since mitochondrial viability has been demonstrated as a key component in efficacious transplantation, a standardized technique of mitochondrial isolation was developed. Such a process also needed to be simple and cost-effective to insure clinical practicality. Furthermore, the process needed to be rapid in order to make mitochondrial transplantation a realistic therapy in the clinical setting. In the case of a myocardial infarction, for example, the current accepted standard is a period of no longer than 90 minutes from the patient's presentation to revascularization via percutaneous coronary intervention (PCI). Previous methodologies for isolation of mitochondria took well over this 90-minute allotment.

With all of this in consideration, a new methodology was developed by Preble et al. to isolate mitochondria rapidly, reliably, and in a cost effective manner [150]. This methodology has since been validated for the purity and viability of the isolate using Clarke-type electrode for oxygen consumption rates, ATP assays, and fluorescent probe analysis. Light microscopy and transmission electron microscopy have also successfully validated this method [142,145,151,152].

11   CARDIOPROTECTION BY MITOCHONDRIAL TRANSPLANTATION

Mitochondrial transplantation, as a therapeutic intervention for ischemic heart disease, was first validated through a Langendorff-perfused rabbit heart [145]. In these initial studies, mitochondria was isolated from the left ventricle of healthy rabbits as donors and injected directly into myocardium after 30 minutes of ischemia. After mitochondrial transplantation, ATP content and myocardial function was significantly enhanced while infarct size was significantly decreased in the hearts receiving mitochondrial transplantation. Confocal microscopy in these hearts demonstrated viable mitochondria present after completing 120 minutes of reperfusion and distributed widely from the epicardium to subendocardium. With these initial results came heightened interest in mitochondrial therapy and the mechanisms of transfer and uptake by which this novel therapy affords a cardioprotective benefit.

11.1   MITOCHONDRIAL DELIVERY, MODE OF UPTAKE, AND DISTRIBUTION

Various modes of delivery have been investigated for mitochondrial transplantation. These include direct injection into myocardium and intracoronary delivery. Initial studies utilized direct injection into multiple sites of an ischemic left ventricular free wall, and these demonstrated significant reductions in infarct size and improvements in myocardial function. More recently, in similar models, intracoronary infusion was validated as a delivery model. Distribution of transplanted mitochondria by intravascular delivery was confirmed using positron emission tomography, microcomputed tomography, and magnetic resonance imaging with 18F-rhodamine 6G and iron oxide labeled mitochondria [143]. While intracoronary delivery resulted in a more widespread distribution, direct injection resulted in higher mitochondrial concentrations in the targeted region. Intracoronary delivery demonstrated equivalent cardioprotection by functional measures and by infarct size [142]. Intracoronary delivery, via a catheter, expands the potential applications of mitochondrial transplantation as it provides an easy avenue for global delivery of mitochondria and one that is readily applicable at the time of other coronary interventions such as PCI for myocardial revascularization.

Histologically, tagged mitochondria are visible in the interstitial space surrounding cardiomyocytes within 10 minutes of injection. Within 1–2 hours of injection, these mitochondria are detectable within the cardiomyocytes themselves, in clusters around damaged endogenous mitochondria in a location adjacent to the sarcolemma between Z-lines of the sarcomeres and also to the nucleus. Using mitochondria isolated from the HeLa human cell line in a rabbit heart model, internalization of transplanted mitochondria was confirmed through immune-gold staining and visualization under electron microscopy [141,144].

There are several proposed mechanisms for cardiomyocyte internalization for transplanted mitochondria, which include actin-meditated endocytosis, Caveolae-dependent-clathrin dependent endocytosis, tunneling nanotubes and macro-pinocytosis [152–154]. The widely accepted endosymbiotic origin theory of mitochondria would support the actin-mediated endocytosis hypothesis, and indeed recent in vitro work demonstrated that only Cyctochalasin D, a specific inhibitor of actin polymerization, decreased cardiomyocyte internalization of mitochondria, and, subsequently, deceased ATP content. This further supports actin-mediated endocytosis as the mechanism by which mitochondria are internalized by cardiomyocytes [155]. Once internalized, there is in-vivo evidence to suggest that mitochondria are transferred between cells and through the syncytium by a process of tunneling nanotubes [156].

12   MECHANISMS OF CARDIOPROTECTION

There are numerous mediators of mitochondrial damage during ischemic injury; as such there are numerous means by which mitochondrial transplantation affords cardioprotection. These pathways include increased myocardial ATP content and metabolic recovery, changes in proteomic and transcriptomics, upregulations of cytokine and chemokines that enhance post-infarct myocardial function and replacement of damaged mitochondrial DNA.

There is consensus within the literature in that mitochondrial transplantation increases cardiomyocytes ATP content and oxygen consumption. In an in-vivo model of regional ischemia-reperfusion injury in rabbit hearts, mitochondrial transplantation significant increased the total tissue ATP content in the area at risk. This increase was noted within 15 minutes of transplantation and was persistent out to 21 days. ATP content alone is not protective, as administration of exogenous ATP or ADP has not proven to restore high-energy phosphate stores nor benefit post-ischemic function [141,146,157]. This is explained, in part, to the short half-life of ATP and the evidence outlined above, demonstrating that transplantation of intact, viable and respiration competent, mitochondria is necessary for a cardioprotective effect.

Proteomic analysis demonstrates differentially expressed myocardial proteins after mitochondrial transplantation. Such analysis in post-ischemic rabbit myocardium tissue revealed significant increases in mitochondrial proteome clusters, metabolic energy precursors, and other protein populations after mitochondrial transplantation [141].

With mitochondrial transplantation there is also evidence of downregulation in inflammatory cytokines and an upregulation of chemokines which mediate pathways of angiogenesis and arteriogenesis. These same chemokines are part of the signaling cascade for progenitor cell migration and protection against apoptosis, all of which enhance cardiac functional recovery. In-vivo rabbit hearts exposed to regional ischemia and treated with autologous mitochondria demonstrated a significant decrease in inflammatory markers such as tumor necrosis factor alpha (TNFα), interleukin-6 and -10 (IL-6, -10), monocyte chemoattractant protein- 1(MCP-1) and high-sensitivity C-reactive protein (hsCRP). Multiplex analysis in such hearts also demonstrated significant upregulation of epidermal growth factor (EGF), growth-related oncogene (GRO), and monocyte chemoattractant protein-3 (MCP-3). EGF has been implicated as a key protective factor after myocardial ischemic injury through the stimulation of cell growth, proliferation and migration [158,159]. GRO has been identified as a chemoattractant promoting neovascularization and inhibiting apoptosis after myocardial ischemic injury [160]. EGF, GRO, and other chemokines act in coordination with MCP-3 to enhance myocardial function and promote beneficial myocardial remodeling after sustaining an ischemic injury.

The effect of mitochondrial transplantation on ROS and the implications on the ischemia-reperfusion injury pathway remain to be fully elucidated. When Langendorff-perfused rabbit hearts are exposed to regional ischemia, ROS populations, as measured by levels of thiobarbituric acid-reactive substances (TBARS), were significantly decreased after mitochondrial transplantation compared to controls. However, administration of 2-Mercaptopropionylglycine (MPG), an ROS scavenger, throughout reperfusion or in conjunction with mitochondrial delivery, failed to inhibit the benefit from mitochondrial transplantation. This suggests that the cardioprotection afforded by mitochondrial transplantation is not acting through ROS modulation [161,162]. Further investigation is warranted to establish a more defined role for ROS in mitochondrial injury and the cardioprotection of mitochondrial transplantation.

Recent investigations have demonstrated that, in conjunction with the mechanism outlined above, mitochondrial transplantation can also replace damaged mitochondrial DNA [163]. In vitro studies using HeLa p0 cells, a special cell line depleted of mitochondrial DNA and thus incapable of oxygen consumption, demonstrated cell rescue when co-incubated with mitochondria containing intact mtDNA. After coincubation, ATP content and oxygen consumption rate were significantly increased for up to 2 weeks. Verification of mtDNA replacement as verified through quantitative real-time RT-PCR analysis. Such an observation warrants further investigation as it may be a major factor in the long term cardioprotective benefit seen with mitochondrial transplantation.

13 MITOCHONDRIAL TRANSPLANTATION IN HUMAN SUBJECTS

In 2016, the first in-human clinical application of mitochondrial transplantation therapy was performed. Five pediatric patients at Boston Children's Hospital who suffered from myocardial ischemia-reperfusion injury and were dependent on extracorporeal membrane oxygenation (ECMO) support as a result of that injury. Mitochondria were isolated from the patients' rectus abdominis muscle and injected at ten hypokinetic regions of the myocardium as noted by echocardiography [164,165]. Myocardial function was significantly improved in all five patients and four patients were successfully weaned off of ECMO support within 48 hours of mitochondrial transplantation. The patient who remained ECMO dependent had suffered severe multi-organ failure and expired despite evidence of myocardial recovery. There was no increase in rate of adverse short-term events including arrhythmia, myocardial hematoma, or scarring. While further investigations will be necessary to evaluate appropriate dosing and modes of delivery for mitochondrial transplantation, these initial results are promising for future applications of this novel technique and the potential benefit to patients across the globe.

14 CONCLUSION

The key feature of myocardial ischemic injury at the cellular level is mitochondrial dysfunction. Such dysfunction begins with the onset of ischemia and progresses with reperfusion leading to persistent myocardial dysfunction. The alterations experienced at the mitochondrial level include changes to structure, ionic homeostasis, enzyme activity, complex function, with resulting alterations in high energy synthesis, mitochondrial DNA structure, transcriptomics, and proteomics. Furthermore, the apoptotic pathways that are initiated by ischemia-reperfusion are mediated by mitochondria. Thus, mitochondrial injury and dysfunction has become a key target for the treatment of ischemia-reperfusion injury. Mitochondrial transplantation, which aims to replace damaged mitochondria with respiration competent mitochondria isolated from the patient's own body. As a novel intervention, mitochondrial transplantation has demonstrated decreased myocardial infarct severity and improvements in cellular viability, and ventricular function. These benefits have been observed in numerous animal models and have also been confirmed in preliminary trials in humans. Mitochondrial transplantation can afford numerous cardioprotective benefits which include increased myocardial ATP content and the upregulation of cytokine and chemokines that enhance post-infarct myocardial function. Further notable changes include alterations in mitochondrial proteomics and transcriptomics and replacement of damaged mitochondrial DNA. Mitochondrial damage underlies the pathophysiology in a number of disorders including ischemia-reperfusion injury, but also Alzheimer's disease, Parkinson's disease and the numerous yet rare mitochondrial genetic disorders. As such, there are numerous possible applications of mitochondrial transplantation. Recently, mitochondrial transplantation has been successfully applied in animals model for Parkinson's disease as well and ischemia-reperfusion in liver, lung, and neuronal cells [166–169]. While further investigation regarding the mechanism by which mitochondrial transplantation can influence on cell survival as exert a role in numerous disease pathologies is needed, current evidence identifies mitochondrial transplantation as a promising strategy for cardioprotection before and after ischemic heart disease. Indeed there is great hope a more thorough understanding of mitochondrial function in normal cellular physiology will bring advancement in therapeutic potential and a boon to patients across the clinical spectrum.

ACKNOWLEDGMENTS

This work was supported by: The Richard A. and Susan F. Smith Foundation, President's Innovation Award, Boston Children's Hospital, Michael B. Klein and Family, the Sidman Family Foundation, the Bulens/Capozzi Foundation, the Boston Children's Hospital Anesthesia Foundation, the Kenneth C. Griffin Charitable Research Fund and the Boston Investment Council.

DISCLOSURES

Dr. McCully, Dr. Cowan and Dr. del Nido have patents pending for the isolation and usage of mitochondria. There are no other conflicts of interest by any of the authors. The authors attest they had full freedom to explore the data, analyze the results independent from any sponsor and that they had sole authority to make the final decision to submit the material for publication.

REFERENCES

1. Benjamin EJ, Blaha MJ, Chiuve SE, Cushman M, Das SR, Deo R, de Ferranti SD, Floyd J, Fornage M, Gillespie C, Isasi CR, Jiménez MC, Jordan LC, Judd SE, Lackland D, Lichtman JH, Lisabeth L, Liu S, Longenecker CT, Mackey RH, Matsushita K, Mozaffarian D, Mussolino ME, Nasir K, Neumar RW, Palaniappan L, Pandey DK, Thiagarajan RR, Reeves MJ, Ritchey M, Rodriguez CJ, Roth GA, Rosamond WD, Sasson C, Towfighi A, Tsao CW, Turner MB, Virani SS, Voeks JH, Willey JZ, Wilkins JT, Wu JH, Alger HM, Wong SS, Muntner P, American Heart Association Statistics Committee and Stroke Statistics Subcommittee. Heart disease and stroke statistics-2017 update: a report from the American Heart Association. Circulation. 2017;135:e146–1603.
2. Jennings RB, Reimer KA. Lethal myocardial ischemic injury. Am J Pathol. 1981;102:241–55.
3. Buja LM, Willerson JT. Abnormalities of volume regulation and membrane integrity in myocardial tissue slides after early ischemic injury in the dog: effects of mannitol, polyethylene glycol, and propranolol. Am J Pathol. 1981;103:79–95.
4. Santulli G, Nakashima R, Yuan Q, Marks AR. Intracellular calcium release channels: an update. J Physiol. 2017; (in press). doi: 10.1113/JP272781.
5. Braunwald E, Kloner RA. Myocardial reperfusion: a double-edge sword? J Clin Invest. 1985;76:1713–19.
6. Kolwicz SC Jr, Purohit S, Tian R. Cardiac metabolism and its interactions with contraction, growth and survival of cardiomyocytes. Circ Res. 2013;113:603–16.
7. Carden DL, Granger DN. Pathophysiology of ischemia-reperfusion injury. J Pathol. 2000;190:255–66.
8. Collard CD, Gelman S. Pathophysiology, clinical manifestations, and prevention of ischemia reperfusion injury. Anesthesiology. 2001;94:1133–38.
9. Turer AT, Hill JA. Pathogenesis of myocardial ischemia-reperfusion injury and rationale for therapy. Am J Cardiol. 2010;106:360–68.
10. Fassina L, Rozzi G, Rossi S, Scacchi S, Galetti M, Lo Muzio FP, Del Bianco F, Colli Franzone P, Petrilli G, Faggian G, Miragoli M. Cardiac kinematic parameters computed from video of in situ beating heart. Sci Rep. 2017;7:46143. doi: 10.1038/srep46143.
11. Kloner RA, Przyklenk K, Patel B. Altered myocardial states: the stunned and hibernating myocardium. Am J Med. 1989;86:14–22.
12. Murry CE, Jennings RB, Reimer KA. Preconditioning with ischemia: a delay of lethal cel injury in ischemic myocardium. Circulation. 1986;74:1124–36.
13. Murry CE, Richard VJ, Reimer KA, Jennings RB. Ischemic preconditioning slows energy metabolism and delays ultrastructural damage during a sustained ischemic episode. Circ Res. 1990;66:913–31.
14. Zhao ZQ, Corvera JS, Halkos ME, Kerendi F, Wang NP, Guyton RA, Vinten-Johansen J. Inhibition of myocardial injury by ischemic postconditioning during reperfusion: comparison with ischemic preconditioning. Am J Physiol Heart Circ Physiol. 2003;285:H579–88.
15. Zhao ZQ, Vinten-Johansen J. Postconditioning: reduction of reperfusion-induced injury. Cardiovasc Res. 2006;70:200–11.
16. Cohen MV, Yang XM, Downey JM. The pH hypothesis of postconditioning: staccato reperfusion reintroduces oxygen and perpetuates myocardial acidosis. Circulation. 2007;115:1895–903.
17. Penna C, Perrelli MG, Raimondo S, Tullio F, Merlino A, Moro F, Geuna S, Mancardi D, Pagliaro P. Postconditioning induces an anti-apoptotic effect and preserves mitochondrial integrity in isolated rat hearts. Biochim Biophys Acta. 2009;1787:794–801, 891,892 128.
18. Penna C, Rastaldo R, Mancardi D, Raimondo S, Cappello S, Gattullo D, Losano G, Pagliaro P. Post-conditioning induced cardioprotection requires signaling through a redox-sensitive mechanism, mitochondrial ATP-sensitive K+ channel and protein kinase C activation. Basic Res Cardiol. 2006;101:180–189.
19. Przyklenk K, Bauer B, Ovize M, Kloner RA, Whittaker P. Regional ischemic 'preconditioning' protects remote virgin myocardium from subsequent sustained coronary occlusion. Circulation. 1993;87:893–9,910.
20. Quarrie R, Lee DS, Steinbaugh G, Cramer B, Erdahl W, Pfeiffer DR, Zweier JL, Crestanello JA. Ischemic preconditioning preserves mitochondrial membrane potential and limits reactive oxygen species production. J Surg Res. 2012 Nov;178(1):8-17.
21. Ruiz-Meana M, Nunez E, Miro-Casas E, Martinez-Acedo P, Barba I, Rodriguez-Sinovas A, Inserte J, Fernandez-Sanz C, Hernando V, Vazquez J, Garcia-Dorado D. Ischemic preconditioning protects cardiomyocyte mitochondria through mechanisms independent of cytosol. J Mol Cell Cardiol. 2014;68:79–88.
22. Argaud L, Gateau-Roesch O, Raisky O, Loufouat J, Robert D, Ovize M. Postconditioning inhibits mitochondrial permeability transition. Circulation. 2005;111:194–7,494.

23. Yellon DM, Hausenloy DJ. Myocardial reperfusion injury. N Engl J Med. 2007;357:1121–35.

24. Cung TT, Morel O, Cayla G, Rioufol G, Garcia-Dorado D, Angoulvant D, Bonnefoy-Cudraz E, Guerin P, Elbaz M, Delarche N, Coste P, Vanzetto G, Metge M, Aupetit JF, Jouve B, Motreff P, Tron C, Labeque JN, Steg PG, Cottin Y, Range G, Clerc J, Claeys MJ, Coussement P, Prunier F, Moulin F, Roth O, Belle L, Dubois P, Barragan P, Gilard M, Piot C, Colin P, De Poli F, Morice MC, Ider O, Dubois-Rande JL, Unterseeh T, Le Breton H, Beard T, Blanchard D, Grollier G, Malquarti V, Staat P, Sudre A, Elmer E, Hansson MJ, Bergerot C, Boussaha I, Jossan C, Derumeaux G, Mewton N, Ovize M. Cyclosporine before PCI in patients with acute myocardial infarction. N Engl J Med. 2015;373:1021–31.

25. Song K, Wang S, Qi D. Effects of cyclosporine on reperfusion injury in patients: a meta-analysis of randomized controlled trials. Oxid Med Cell Longev. 2015;2015:287058.

26. Meerson FZ, Zaletayeva TA, Lagutchev SS, Pshennikova MG. Structure and mass of mitochondria in the process of compensatory hyperfunction and hypertrophy of the heart. Exp Cell Res. 1964;36:568–78.

27. Feinberg H, Levitsky S. Biochemical rationale of cardioplegia. In: Engelman RM, Levitsky S, editors. A textbook of clinical cardioplegia. Mt Kisco, NY: Futura: 1982; pp. 131–39.

28. Palmer JW, Tandler B, Hoppel CL. Biochemical properties of subsarcolemmal and interfibrillar mitochondria isolated from rat cardiac muscle. J Biol Chem. 1977;252:8731–39.

29. Riva A, Tandler B, Loffredo F, Vazquez E, Hoppel C. Structural differences in two biochemically defined populations of cardiac mitochondria. Am J Physiol Heart Circ Physiol. 2005;289:H868–72.

30. Weinstein ES, Benson DW, Fry ED. Subpopulations in human heart mitochondria. J Surg Res. 1986;40:495–498.

31. Dalen H, Odegarden S, Saetersdal T. The application of various electron microscopic techniques for ultrastructural characterization of the human papillary heart muscle cell in biopsy material. Virchows Arch A Pathol Anat Histopathol. 1987;410:265–79.

32. Shimada T, Horita K, Murakami M, Ogura R. Morphological studies of different mitochondrial populations in monkey myocardial cells. Cell Tissue Res. 1984;238:577–82.

33. Lukyanenko V, Chikando A, Lederer WJ. Mitochondria in cardiomyocyte Ca2+ signaling. Int J Biochem Cell Biol. 2009;41:1957–71.

34. Birkedal R, Shiels HA, Vendelin M. Three-dimensional mitochondrial arrangement in ventricular myocytes: from chaos to order. Am J Physiol Cell Physiol. 2006;291:C1148–458.

35. Hollander JM, Thapa D, Shepherd D. Physiological and structural differences in spatially distinct subpopulations of cardiac mitochondria. Am J Physiol Heart Circ Physiol. 2014;307:H1–14.

36. Muller W. Subsarcolemmal mitochondria and capillarization of soleus muscle fibers in young rats subjected to an endurance training. A morphometric study of semithin sections. Cell Tissue Res. 1976;174:367–89.

37. Kalkhoran SB, Munro P, Qiao F, Ong SB, Hall AR, Cabrera-Fuentes H, Chakraborty B, Boisvert WA, Yellon DM, Hausenloy DJ. Unique morphological characteristics of mitochondrial subtypes in the heart: the effect of ischemia and ischemic preconditioning. Discoveries (Craiova). 2017 Jan-Mar;5(1):pii: e71.

38. Rosca MG, Hoppel CL. Mitochondrial dysfunction in heart failure. Heart Fail Rev. 2013;18:607–22.

39. Baseler WA, Thapa D, Jagannathan R, Dabkowski ER, Crsoton TL, Hollander JM. miR-141 as a regulator of the mitochondrial phosphate carrier (Slc25a3) in the type 1 diabetic heart. Am J Physiol Cell Physiol. 2012;303:C1244–51.

40. Croston TL, Shepherd DL, Thapa D, Nichols CE, Lewis SE, Dabkowski ER, Jagannathan R, Baseler WA, Hollander JM. Evaluation of the cardiolipin biosynthetic pathway and its interactions in the diabetic heart. Life Sci. 2013;93:313–22.

41. Chen Q, Moghaddas S, Hoppel CL, Lesnefsky EJ. Ischemic defects in the electron transport chain increase the production of reactive oxygen species from isolated rat heart mitochondria. Am J Physiol Cell Physiol. 2008;294:C460–6.

42. Lesnefsky EJ, Gudz TI, Migita CT, Ikeda-Saito M, Hassan MO, Turkaly PJ, Hoppel CL. Ischemic injury to mitochondrial electron transport in the aging heart: damage to the iron-sulfur protein subunit of electron transport complex III. Arch Biochem Biophys. 2001;385:117–28.

43. Lesnefsky EJ, Chen Q, Moghaddas S, Hassan MO, Tandler B, Hoppel CL. Blockade of electron transport chain during ischemia protects cardiac mitochondria. J Biol Chem. 2004;279:47961–7.

44. Rosca MG, Vazquez EJ, Kerner J, Parland W, Chandler MP, Stanely W, Sabbah HN, Hoppel CL. Cardiac mitochondria in heart failure: decrease in respirasomes and oxidative phosphorylation. Cardiovasc Res. 2008;80:30–9.

45. Dabkowski ER, Williamson CL, Hollander JM. Mitochondria-specific transgenic overexpression of phospholipid hydroperoxide glutathione peroxidase (GPx4) attenuates ischemia/reperfusion-associated cardiac dysfunction. Free Radic Biol Med. 2008;45:855–65.

46. Duan J, Karmazyn M. Relationship between oxidative phosphorylation and adenine nucleotide translocase activity of two populations of cardiac mitochondria and mechanical recovery of ischemic hearts following reperfusion. Can J Physiol Pharmacol. 1989;67:704–9.

47. Lesnefsky EJ, Chen Q, Slabe TJ, Stoll MS, Minkler PE, Hassan MO, Tandler B, Hoppel CL. Ischemia, rather than reperfusion, inhibits respiration through cytochrome oxidase in the isolated, perfused rabbit heart: role of cardiolipin. Am J Physiol Heart Circ Physiol. 2004;287:H258–67.

48. Kubler W, Spieckermann PG. Regulation of glycolysis in the ischemic and anoxic myocardium. J Mol Cell Cardiol. 1970;1:351–77.

49. Boengler K, Heusch G, Schulz R. Nuclear-encoded mitochondrial proteins and their role in cardioprotection. Biochim Biophys Acta. 2011;1813:1286–94.

50. Conway MA, Allis J, Ouwerkerk R, Niioka T, Rajagopalan B, Radda GK. Detection of low phosphocreatine to ATP ratio in failing hypertrophied human myocardium by 31P magnetic resonance spectroscopy. Lancet. 1991;338:973–6.

51. Hardy CJ, Weiss RG, Bottomley PA, Gerstenblith G. Altered myocardial high-energy phosphate metabolites in patients with dilated cardiomyopathy. Am Heart J. 1991;112:795–801.

52. Neubauer S, Krahe T, Schindler R, Horn M, Hillenbrand H, Entzeroth C, Mader H, Kromer EP, Riegger GA, Lackner K. 31P magnetic resonance spectroscopy in dilated cardiomyopathy and coronary artery disease. Altered cardiac high-energy phosphate metabolism in heart failure. Circulation. 1992;86:1810–18.

53. Neubauer S, Horn M, Cramer M, Harre K, Newell JB, Peters W, Pabst T, Ertl G, Hahn D, Ingwall JS, Kochsiek K. Myocardial phosphocreatine-to-ATP ratio is a predictor of mortality in patients with dilated cardiomyopathy. Circulation. 1997;96:2190–6.

54. Tsukube T, McCully JD, Metz KR, Cook CU, Levitsky S. Amelioration of ischemic calcium overload correlates with high-energy phosphates in the senescent myocardium. Am J Phys. 1997;273:H418–25.

55. Reimer KA, Lowe JE, Rasmussen MM, Jennings RB. The wavefront phenomenon of ischemic cell death: 1. Myocardial infarct size vs duration of coronary occlusion in dogs. Circulation. 1977;56:786–94.

56. Luongo TS, Lambert JP, Gross P, Nwokedi M, Lombardi AA, Shanmughapriya S, Carpenter AC, Kolmetzky D, Gao E, van Berlo JH, Tsai EJ, Molkentin JD, Chen X, Madesh M, Houser SR, Elrod JW. The mitochondrial Na(+)/Ca(2+) exchanger is essential for Ca(2+) homeostasis and viability. Nature. 2017;545:93–97.

57. Reimer KA, Jennings RB. The 'wavefront phenomenon' of myocardial ischemic cell death: II. Transmural progression of necrosis within the framework of ischemic bed size (myocardium at risk) and collateral flow. Lab Investig. 1979;40:633–44.

58. Reimer KA, Jennings RB. Myocardial ischemia, hypoxia, and infarction. In: Fozzard HA, Haber E, Jennings RB, Katz AM, Morgan HE, editors. The heart and cardiovascular system: scientific foundations, vol. II. 2nd ed. New York: Raven Press; 1992. pp. 1875–973.

59. Chen YR, Zweier JL. Cardiac mitochondria and reactive oxygen species generation. Circ Res. 2014;114:524–37.

60. McCully JD, Levitsky S. Mitochondrial ATP-sensitive potassium channels in surgical cardioprotection. Arch Biochem Biophys. 2003;420:237–45.

61. McCully JD, Bhasin MK, Daly C, Guerrero MC, Dillon S, Liberman TA, Cowan DB, Mably JD, McGowan FX, Levitsky S. Transcriptomic and proteomic analysis of global ischemia and cardioprotection in the rabbit heart. Physiol Genomics. 2009;38:125–37.

62. Ito H. No-reflow phenomenon and prognosis in patients with acute myocardial infarction. Nat Clin Pract Cardiovasc Med. 2006;499:506.

63. Bolli R, Jeroudi MO, Patel BS, DuBose CM, Lai EK, Roberts R, McCay PB. Direct evidence that oxygen-derived free radicals contribute to postischemic myocardial dysfunction in the intact dog. Proc Natl Acad Sci U S A. 1989;86:4695–9.

64. Jennings RB, Sommers HM, Smyth GA, Flack HA, Linn H. Myocardial necrosis induced by temporary occlusion of a coronary artery in the dog. Arch Pathol. 1960;70:68–78.

65. Barry MC, Kelly C, Burke P, Sheehan S, Redmond HP, Bouchier-Hayes D. Immunological and physiological responses to aortic surgery: effect of reperfusion on neutrophil and monocyte activation and pulmonary function. Br J Surg. 1997;84:513–19.

66. Buja LM. Myocardial ischemia and reperfusion injury. Cardiovasc Pathol. 2005;14:170–75.

67. Chen Q, Moghaddas S, Hoppel CL, Lesnefsky EJ. Reversible blockade of electron transport during ischemia protects mitochondria and decreases myocardial injury following reperfusion. J Pharmacol Exp Ther. 2006;319 (3):1405–12.

68. Kornfeld OS, Hwang S, Disatnik MH, Chen CH, Qvit N, Mochly-Rosen D. Mitochondrial reactive oxygen species at the heart of the matter: new therapeutic approaches for cardiovascular diseases. Circ Res. 2015;116:1783–99.

69. Argaud L, Loufouat J, Gateau-Roesch O, Gomez L, Robert D, Ovize M. Persistent inhibition of mitochondrial permeability transition by preconditioning during the first hours of reperfusion. Shock. 2008;30:552–6.

70. Baines CP, Kaiser RA, Purcell NH, Blair NS, Osinska H, Hambleton MA, Brunskill EW, Sayen MR, Gottlieb RA, Dorn GW, Robbins J, Molkentin JD. Loss of cyclophilin D reveals a critical role for mitochondrial permeability transition in cell death. Nature. 2005;434:658–62.

71. Boengler K, Hilfiker-Kleiner D, Heusch G, Schulz R. Inhibition of permeability transition pore opening by mitochondrial STAT3 and its role in myocardial ischemia/reperfusion. Basic Res Cardiol. 2010;105:771–85.

72. Di Lisa F, Menabo R, Canton M, Barile M, Bernardi P. Opening of the mitochondrial permeability transition pore causes depletion of mitochondrial and cytosolic NAD+ and is a causative event in the death of myocytes in postischemic reperfusion of the heart. J Biol Chem. 2001;276:2571–75.

73. Gross GJ, Auchampach JA. Reperfusion injury: does it exist? J Mol Cell Cardiol. 2007;42:12–18.

74. Rousou AJ, Ericsson M, Federman M, Levitsky S, McCully JD. Opening of mitochondrial KATP channels enhances cardioprotection through the modulation of mitochondrial matrix volume, calcium accumulation and respiration. Am J Physiol Heart Circ Physiol. 2004;287: H1967–76.

75. Alavian KN, Beutner G, Lazrove E, Sacchetti S, Park HA, Licznerski P, Li H, Nabili P, Hockensmith K, Graham M, Porter GA Jr, Jonas EA. An uncoupling channel within the c-subunit ring of the F1FO ATP synthase is the mitochondrial permeability transition pore. Proc Natl Acad Sci U S A. 2014;111:10580–85.

76. McCully JD, Rousou AJ, Parker RA, Levitsky S. Age and gender-related differences in mitochondrial oxygen consumption and calcium with cardioplegia and diazoxide. Ann Thorac Surg. 2007;83:1102–09.

77. Faulk EA, McCully JD, Hadlow NC, Tsukube T, Krukenkamp IB, Federman M, Levitsky S. Magnesium cardioplegia enhances mRNA levels and the maximal velocity of cytochrome oxidase I in the senescent myocardium during global ischemia. Circulation. 1995;92(II):405–12.

78. Pan X, Liu J, Nguyen T, Liu C, Sun J, Teng Y, Fergusson MM, Rovira II, Allen M, Springer DA, Aponte AM, Gucek M, Balaban RS, Murphy E, Finkel T. The physiological role of mitochondrial calcium revealed by mice lacking the mitochondrial calcium uniporter. Nat Cell Biol. 2013;15:1464–72.

79. Arroyo CM, Kramer JH, Dickens BF, Weglicki WB. Identification of free radicals in myocardial ischemia/reperfusion by spin trapping with nitrone DMPO. FEBS Lett. 1987;221:101–4.

80. Ferrari R, Alfieri O, Curello S, Ceconi C, Cargnoni A, Marzollo P, Pardini A, Caradonna E, Visioli O. Occurrence of oxidative stress during reperfusion of the human heart. Circulation. 1990;81:201–11.

81. Pacher P, Nivorozhkin A, Szabó C. Therapeutic effects of xanthine oxidase inhibitors: renaissance half a century

after the discovery of allopurinol. Pharmacol Rev. 2006;58:87–114.

82. Chouchani ET, Pell VR, Guade E, Aksentijevic D, Sundier SY, Robb EL, Logan A, Nadtochiy SM, Ord EN, Smith AC, Eyassu F, Shirley R, Hu CH, Dare AJ, James AM, Rogatti S, Hartley RC, Eaton S, Costa AS, Brookes PS, Davidson SM, Duchen MR, Saeb-Parsy K, Shattock MJ, Robinson AJ, Work LM, Frezza C, Krieg T, Murphy MP. Ischaemic accumulation of succinate controls reperfusion injury through mitochondrial ROS. Nature. 2014;515:431–5.

83. Ashrafian H, Czibik G, Bellahcene M, Aksentijevic D, Smith AC, Mitchell SJ, Dodd MS, Kirwan J, Byrne JJ, Ludwig C, Isackson H, Yavari A, Stottrup NB, Contractor H, Cahill TJ, Sahgal N, Ball DR, Birkler RI, Hargreaves I, Tennant DA, Land J, Lygate CA, Johannsen M, Kharbanda RK, Neubauer S, Redwood C, de Cabo R, Ahmet I, Talan M, Günther UL, Robinson AJ, Viant MR, Pollard PJ, Tyler DJ, Watkins H. Fumarate is cardioprotective via activation of the Nrf2 antioxidant pathway. Cell Metab. 2012;15:361–71.

84. Murphy MP. How mitochondria produce reactive oxygen species. Biochem J. 2009;417:1–3.

85. Zweier JL, Talukder MA. The role of oxidants and free radicals in reperfusion injury. Cardiovasc Res. 2006;70:181–90.

86. Chouchani ET, Methner C, Nadtochiy SM, Logan A, Pell VR, Ding S, James AM, Cochemé HM, Reinhold J, Lilley KS, Partridge L, Fearnley IM, Robinson AJ, Hartley RC, Smith RA, Krieg T, Brookes PS, Murphy MP. Cardioprotection by S-nitrosation of cysteine switch on mitochondrial complex I. Nat Med. 2013;19:753–9.

87. Stewart S, Lesnefsky EJ, Chen Q. Reversible blockade of electron transport chain with amobarbital at the onset of reperfusion attenuates cardiac injury. Transl Res. 2009;153:224–31.

88. Burwell LS, Nadtochiy SM, Tompkins AJ, Young S, Brookes PS. Direct evidence for S-nitrosation of mitochondrial complex I. Biochem J. 2006;394:627–34.

89. Methner C, Chouchani ET, Buonincontri G, Pell VR, Sawiak SJ, Murphy MP, Krieg T. Mitochondria selective S-nitrosation by mitochondria-targeted S-nitrosothiol protects against post-infarct heart failure in mouse hearts. Eur J Heart Fail. 2014;16:712–17.

90. Crompton M. The mitochondrial permeability transition pore and its role in cell death. Biochem J. 1999;341:233–249.

91. Halestrap AP, Clarke SJ, Javadov SA. Mitochondrial permeability transition pore opening during myocardial reperfusion- a target for cardioprotection. Cardiovasc Res. 2004;61:372–85.

92. Klein HH, Pich S, Lindert S, Nebendahl K, Warneke G, Kreuzer H. Treatment of reperfusion injury with intracoronary calcium channel antagonists and reduced coronary free calcium concentration in regionally ischemic, reperfused porcine hearts. J Am Coll Cardiol. 1989;13:1395–401.

93. Carry MM, Mrak RE, Murphy ML, Peng CF, Straub KD, Fody EP. Reperfusion injury in ischemic myocardium: protective effects of ruthenium red and of nitroprusside. Am J Cardiovasc Pathol. 1989;2:335–44.

94. Gumina RJ, Buerger E, Eickmeier C, Moore J, Daemmgen J, Gross GJ. Inhibition of the Na+/H+ exchanger confers greater cardioprotection against 90 minutes of myocardial ischemiathan ischemic preconditioning in dogs. Circulation. 1999;100:2519–26.

95. Boden WE, van Gilst W, Scheldewaert RG, Starkey IR, Carlier MF, Julian DG, Whitehead A, Bertrand ME, Col JJ, Pedersen OL, Lie KI, Santoni JP, Fox KM. Diltiazem in acute myocar- dial infarction treated with thrombolytic agents: a randomized placebo-controlled trial. Incomplete Infarction Trial of European Research Collaborators Evaluating Prognosis post-Thrombolysis (INTERCEPT). Lancet. 2000;355:1751–56.

96. Théroux P, Chaitman BR, Danchin N, Erhardt L, Meinertz T, Schroeder JS, Tognoni G, White HD, Willerson JT, Jessel A. Inhibition of the sodium-hydrogen exchanger with cariporide to prevent myocardial infarction in high-risk ischemic situations. Main results of the GUARDIAN trial. Guard during ischemia against necrosis (GUARDIAN) investigators. Circulation. 2000;102:3032–8.

97. Zeymer U, Suryapranata H, Monassier JP, Opolski G, Davies J, Rasmanis G, Linssen G, Tebbe U, Schröder R, Tiemann R, Machnig T, Neuhaus KL, ESCAMI Investigators. The Na(+)/H(+) exchange inhibitor eniporide as an adjunct to early reperfusion therapy for acute myocardial infarction. Results of the evaluation of the safety and cardioprotective effects of eniporide in acute myocardial infarction (ESCAMI) trial. J Am Coll Cardiol. 2001;38:1644–50.

98. Flaherty JT, Pitt B, Gruber JW, Heuser RR, Rothbaum DA, Burwell LR, George BS, Kereiakes DJ, Deitchman D, Gustafson N. Recombinant human superoxide dismutase (h-SOD) fails to improve recovery of ventricular function in patients undergoing coronary angioplasty for acute myocardial infarction. Circulation. 1994;89:1982–91.

99. Downey JM. Free radicals and their involvement during long-term myocardial ischemia and reperfusion. Annu Rev Physiol. 1990;52:487–504.

100. Sesso HD, Buring JE, Christen WG, Kurth T, Belanger C, MacFadyen J, Manson JE, Glynn RJ, Gaziano JM. Vitamins E and C in the prevention of cardiovascular disease in men: the Physicians' Health Study II randomized controlled trial. JAMA. 2008;300:2123–33.

101. Litt MR, Jeremy RW, Weisman HF, Winkelstein JA, Becker LC. Neutrophil depletion limited to reperfusion reduces myocardial infarct size after 90 minutes of ischemia: evidence for neutrophil-mediated reperfusion injury. Circulation. 1989;80:1816–27.

102. Hayward R, Campbell B, Shin YK, Scalia R, Lefer AM. Recombinant soluble P-selectin glycoprotein ligand-1 protects against myocardial ischemic reperfusion injury in cats. Cardiovasc Res. 1999;41:65–76.

103. Vakeva AP, Agah A, Rollins SA, Matis LA, Li L, Stahl GL. Myocardial infarction and apoptosis after myocardial ischemia and reperfusion: role of the terminal complement components and inhibition by anti-C5 therapy. Circulation. 1998;97:2259–67.

104. Baran KW, Nguyen M, McKendall GR, Lambrew CT, Dykstra G, Palmeri ST, Gibbons RJ, Borzak S, Sobel BE, Gourlay SG, Rundle AC, Gibson CM, Barron HV, Limitation of Myocardial Infarction Following Thrombolysis in Acute Myocardial Infarction (LIMIT AMI) Study Group. Double-blind, randomized trial of anti-CD18 antibody in conjunction with recombinant tissue plasminogen activator for acute myocardial infarction: limitation of myocardial infarction following thrombolysis in acute myocardial infarction (LIMIT AMI) study. Circulation. 2001;104:2778–83.

105. Faxon DP, Gibbons RJ, Chronos NA, Gurbel PA, Sheehan F, HALT-MI Investigators. The effect of blockade of the

CD11/CD18 integrin receptor on infarct size in patients with acute myocardial infarction treated with direct angioplasty: the results of the HALT-MI study. J Am Coll Cardiol. 2002;40:1199–204.

106. Mertens P, Maes A, Nuyts J, Belmans A, Desmet W, Esplugas E, Charlier F, Figueras J, Sambuceti G, Schwaiger M, Mortelmans L, Van de Werf F, PSALM investigators. Recombinant P-selectin glycoprotein ligand-immunoglobulin, a P-selectin antagonist, as an adjunct to thrombolysis in acute myocardial infarction. The P-Selectin Antagonist Limiting Myonecrosis (PSALM) trial. Am Heart J. 2006 Jul;152(1):125.e1-8.

107. Mahaffey KW, Granger CB, Nicolau JC, Ruzyllo W, Weaver WD, Theroux P, Hochman JS, Filloon TG, Mojcik CF, Todaro TG, Armstrong PW, COMPLY Investigators. Effect of pexelizumab, an anti-C5 complement antibody, as adjunctive therapy to fibrinolysis in acute myocardial infarction: the COMPlement inhibition in myocardial infarction treated with thromboLYtics (COMPLY) trial. Circulation. 2003;108:1176–83.

108. Granger CB, Mahaffey KW, Weaver WD, Theroux P, Hochman JS, Filloon TG, Rollins S, Todaro TG, Nicolau JC, Ruzyllo W, Armstrong PW, COMMA Investigators. Pexelizumab, an anti-C5 complement antibody, as adjunctive therapy to primary percutaneous coronary intervention in acute myocardial infarction: the COMplement inhibition in Myocardial infarction treated with Angioplasty (COMMA) trial. Circulation. 2003;108:1184–90.

109. Heyndrickx GR, Millard RW, McRitchie RJ, Maroko PR, Vatner SF. Regional myocardial functional and electrophysiological alterations after brief coronary artery occlusion in conscious dogs. J Clin Invest. 1975;56:978–985.

110. Gray R, Maddahi J, Bernan D, Raymond M, Waxman A, Ganz W, Matloff J, Swan HJ. Scintigraphic and hemodynamic demonstration of transient left ventricular dysfunction immediately after uncomplicated coronary artery bypass grafting. J Thorac Cardiovasc Surg. 1979;77:504–10.

111. Kloner RA, Przyklenk K, Kay GL. Clinical evidence for stunned myocardium after coronary artery bypass surgery. J Card Surg. 1994;9:397–402.

112. Rahimtoola SH. A perspective on the three large multicenter randomized clinical trials of coronary bypass surgery for chronic stable angina. Circulation. 1985;72:V123–1235.

113. Buja LM, Vander Heide RS. Pathobiology of ischemic heart disease: past, present and future. Cardiovasc Pathol. 2016;25:214–20.

114. Chen C, Liu J, Hua D, Ma L, Lai T, Fallon JT, Knibbs D, Gillam L, Mangion J, Knight DR, Waters D. Impact of delayed reperfusion of myocardial hibernation on myocardial ultrastructure and function and their recoveries after reperfusion in a pig model of myocardial hibernation. Cardiovasc Pathol. 2000;9:67–84.

115. Lai T, Fallon JT, Liu J, Mangion J, Gillam L, Waters D, Chen C. Reversibility and pathohistological basis of left ventricular remodeling in hibernating myocardium. Cardiovasc Pathol. 2000;9:323–35.

116. Shah BN, Khattar RS, Senior R. The hibernating myocardium: current concepts, diagnostic dilemmas, and clinical challenges in the post-STICH era. Eur Heart J. 2013;34:1323–36.

117. Borgers M, Ausma J. Structural aspects of the chronic hibernating myocardium in man. Basic Res Cardiol. 1995;90:44–6.

118. Arai AE, Pantely GA, Anselone CG, Bristow J, Bristow JD. Active downregulation of myocardial energy requirements during prolonged moderate ischemia in swine. Circ Res. 1991;69:1458–69.

119. Schulz R, Rose J, Martin C, Brodde OE, Heusch G. Development of short-term myocardial hibernation. Its limitation by the severity of ischemia and inotropic stimulation. Circulation. 1993;88:684–95.

120. Bax JJ, Schinkel AF, Boersma E, Rizzello V, Elhendy A, Maat A, Roelandt JR, van der Wall EE, Poldermans D. Early versus delayed revascularization in patients with ischemic cardiomyopathy and substantial viability: impact on outcome. Circulation. 2003;108:II39–42.

121. Beanlands RS, Hendry PJ, Masters RG, de Kemp RA, Woodend K, Ruddy TD. Delay in revascularization is associated with increased mortality rate in patients with severe left ventricular dysfunction and viable myocardium on fluorine 18-fluorodeoxyglucose positron emission tomography imaging. Circulation. 1998;98:II51–6.

122. Vanoverschelde JL, Wijns W, Depre C, Essamri B, Heyndrickx GR, Borgers M, Bol A, Melin JA. Mechanisms of chronic regional postischemic dysfunction in humans. New insights from the study of non-infarcted collateral-dependent myocardium. Circulation. 1993;87:1513–23.

123. Kim SJ, Peppas A, Hong SK, Yang G, Huang Y, Diaz G, Sadoshima J, Vatner DE, Vatner SF. Persistent stunning induces myocardial hibernation and protection: flow/function and metabolic mechanisms. Circ Res. 2003;92:1233–9.

124. Walker CA, Crawford FA Jr, Spinale FG. Myoctye contractile dysfunction with hypertrophy and failure: relevance to cardiac surgery. J Thorac Cardiovasc Surg. 2000;119:388–400.

125. Khabbaz KR, Zankoul F, Warner KG. Operative metabolic monitoring of the heart II. Online measurement of myocardial tissue pH. Ann Thorac Surg. 2001;72:S2227–33.

126. Karmazyn M. The role of the myocardial sodium-hydrogen exchanger (NHE) and its role in mediating ischemic and reperfusion injury. Keio J Med. 1998;47:65–72.

127. Kaplan P, Hendrikx M, Mattheussen M, Mubagwa K, Flameng W. Effect of ischemia and reperfusion on sarcoplasmic reticulum calcium uptake. Circ Res. 1992;71:1123–30.

128. Santulli G, Xie W, Reiken SR, Marks AR. Mitochondrial calcium overload is a key determinant in heart failure. Proc Natl Acad Sci U S A. 2015;112:11389–94.

129. Frolkis VV, Frolkis RA, Mkhitarian LS, Shevchuk VG, Fraifeld VE, Vakulenko LG, Syrovy I. Contractile function and Ca2+ transport system of myocardium in ageing. Gerontology. 1988;34:64–74.

130. Peng CF, Kane JJ, Murphy ML, Straub KD. Abnormal mitochondrial oxidative phosphorylation of ischemic myocardium by calcium chelating agents. J Mol Cell Cardiol. 1977;9:897–908.

131. Suleiman MS, Halestrap AP, Griffiths EJ. Mitochondria: a target for myocardial protection. Pharmacol Ther. 2001;89:29–46.

132. Ozcan C, Holmuhamedov EL, Jahangir A, Terzic A. Diazoxide protects mitochondria from anoxic injury: implication for myopreservation. J Thorac Cardiovasc Surg. 2001;121:298–306.

133. Steenbergen C, Murphy E, Watts JA, London RE. Correlation between cytosolic free calcium, contracture, ATP, and irreversible ischemic injury in perfused rat heart. Circ Res. 1990;66:135–46.

134. Ataka K, Chen D, Levitsky S, Jimenez E, Feinberg H. Effect of aging on intracellular Ca2+, pHi, and contractility during ischemia and reperfusion. Circulation. 1992;86:II371–6.

135. Tsukube T, McCully JD, Faulk EA, Federman M, LoCicero J 3rd, Krukenkamp IB, Levitsky S. Magnesium cardioplegia reduces cytosolic and nuclear calcium and DNA fragmentation in the senescent myocardium. Ann Thorac Surg. 1994;58:1005–11.

136. Garlid KD. Opening mitochondrial K(ATP) in the heart-what happens, and what does not happen. Basic Res Cardiol. 2000;95:275–9.

137. Hearse DJ, Garlick PB, Humphrey SM. Ischemic contracture of the myocardium: mechanism and prevention. Am J Cardiol. 1977;39:986–93.

138. Hearse DJ, Stewart DA, Braimbridge MV. Myocardial protection during ischemic cardiac arrest: the importance of magnesium in cardioplegic infusates. J Thorac Cardiovasc Surg. 1978;75:877–85.

139. Faulk EA, McCully JD, Tsukube T, Hadlow NC, Krukenkamp IB, Levitsky S. Myocardial mitochondrial calcium accumulation modulates nuclear calcium accumulation and DNA fragmentation. Ann Thorac Surg. 1995;60:338–44.

140. McCully JD, Wakiyama H, Cowan DB, Federman M, Levitsky S. Diazoxide amelioration of myocardial injury and mitochondrial damage during cardiac surgery. Ann Thorac Surg. 2002;74:2138–45.

141. Wakiyama H, Cowan DB, Toyoda Y, Federman M, Levitsky S, McCully JD. Selective opening of mitochondrial ATP-sensitive potassium channels during surgically induced myocardial ischemia increases necrosis and apoptosis. Eur J Cardiothorac Surg. 2002;21:424–33.

142. Halestrap AP, Kerr PM, Javadov S, Suleiman S. The mitochondrial permeability transition: role in ischemia/reperfusion injury. Sepsis. 1998;2:313–25.

143. Ramirez-Barbieri G, Moskowitzova K, Shin B, Blitzer D, Orfany A, Guariento A, Iken K, Friehs I, Zurakowski D, Del Nido PJ, McCully JD. Alloreactivity and allorecognition of syngeneic and allogeneic mitochondria. Mitochondrion. 2018 Mar 26;pii:S1567-7249(18)30029–1. doi: 10.1016/j.mito.2018.03.002.

144. Holmuhamedov EL, Jovanovic S, Dzeja PP, Jovanovic A, Terzic A. Mitochondrial ATP-sensitive K+ channels modulate cardiac mitochondrial function. Am J Phys. 1998;275:H1567–76.

145. Holmuhamedov EL, Wang L, Terzic A. ATP-sensitive K +m channel openers prevent CA2+ overload in rat cardiac mitochondria. J Physiol. 1999;519:347–60.

146. Murata M, Akao M, O'Rourke B, Marban E. Mitochondrial ATP-sensitive potassium channels attenuate matrix CA2+b overload during stimulated ischemia and reperfusion: possible mechanism of cardioprotection. Circ Res. 2001;89:891–8.

147. Masuzawa A, Black KM, Pacak CA, Ericsson M, Barnett RJ, Drumm C, Seth P, Bloch DB, Levitsky S, Cowan DB, McCully JD. Transplantation of autologously-derived mitochondria protects the heart from ischemia-reperfusion injury. Am J Phys Heart Circ Physiol. 2013;304:H966–82.

148. Cowan DB, Yao R, Akurathi V, Snay ER, Thedsanamoorthy JK, Zurakowski D, Ericsson M, Friehs I, Wu Y, Levitsky S, Del Nido PJ, Packard AB, McCully JD. Intracoronary delivery of mitochondria to the ischemic heart for cardioprotection. PLoS One. 2016;11:e0160889.

149. McCully JD, Cowan DB, Pacak CA, Toumpoulis IK, Dayalan H, Levitsky S. Injection of isolated mitochondria during early reperfusion for cardioprotection. Am J Physiol Heart Circ Physiol. 2009;296:H94–105.

150. Bolling SF, Le B, Bove EL. Effect of ATP synthesis promoters on postischemic myocardial recovery. J Surg Res. 1990;49:205–211.

151. Piel DA, Gruber PJ, Weinheimer CJ, Courtois MR, Robertson CM, Coopersmith CM, Deutschman CS, Levy RJ. Mitochondrial resuscitation with exogenous cytochrome c in the septic heart. Crit Care Med. 2007;35:2120–7.

152. Pepe S, Marasco SF, Haaas SJ, Sheeran FL, Krum H, Rosenfeldt FL. Coenzyme Q10 in cardiovascular disease. Mitochondrion. 2007;7(Suppl):S154–67.

153. Preble JM, Pacak CA, Kondo H, MacKay AA, Cowan DB, McCully JD. Rapid isolation and purification of mitochondria for transplantation by tissue dissociation and differential filtra- tion. J Vis Exp. 2014;91:e51682. doi: 10.3791/51682.

154. Preble JM, Kondo H, Levitsky S, McCully JD. Quality control parameters for mitochondria transplant in cardiac tissue. JSM Biochem Mol Biol. 2014;2:1008.

155. Pfeiffer ER, Wright AT, Edwards AG, Stowe JC, McNall K, Tan J, Niesman I, Patel HH, Roth DM, Omens JH, McCulloch AD. Caveolae in ventricular myocytes are required for stretch-dependent conduction slowing. J Mol Cell Cardiol. 2014;76:265–74.

156. Lou E, Fujisawa S, Morozov A, Barlas A, Romin Y, Dogan Y, Gholami S, Moreira AL, Manova-Todorova K, Moore MA. Tunneling nanotubes provide a unique conduit for intercellular transfer of cellular contents in human malignant pleural mesothelioma. PLoS One. 2012;7(3):e33093.

157. Kitani T, Kami D, Matoba S, Gojo S. Internalization of isolated functional mitochondria: involvement of micropinocytosis. J Cell Mol Med. 2014;18:1694–703.

158. Pacak AP, Preble JM, Kondo H, Seibel P, Levitsky S, Del Nido PJ, Cowan DB, McCully JD. Actin-dependent mitochondrial internalization in cardiomyocytes: evidence for rescue of mitochondrial function. Biol Open. 2015;4:622–6.

159. Huang X, Sun L, Ji S, Zhao T, Zhang W, Xu J, Zhang J, Wang Y, Wang X, Franzini-Armstrong C, Zheng M, Cheng H. Kissing and nanotunneling mediate intermitochondrial communication in the heart. Proc Natl Acad Sci USA. 2013;110(8):2846–51.

160. Stark G, Domanowits H, Sterz F, Startk U, Bachernegg M, Kickenweiz E, Decrinis M, Laggner AN, Tritthart HA. Action of ATP on ventricular automaticity. J Cardiovasc Pharmacol. 1994;24:740–4.

161. Lorita J, Soley M, Ramirez I. Epidermal growth factor protects the heart against low-flow ischemia-induced injury. J Phys Biochem. 2012;66:55–62.

162. Roche S, D'Ippolito G, Gomez LA, Bouckenooghe T, Lehmann S, Monteromenei CN, Schiller PC. Comparative analysis of protein expression of three stem cell populations: models of cytokine delivery system in vivo. Int J Pharm. 2013 Jan 2;440(1):72–82.

163. Kocher AA, Schuster MD, Bonaros N, Lietz K, Xiang G, Martens TP, Kurlansky PA, Sondermeijer H, Witkowski P, Boyle A, Homma S, Wang SF, Itescu S. Myocardial homing and neovascularization by human bone marrow blasts is regulated by IL-8/Gro CXC chemochines. J Mol Cell Cardiol. 2006;40:455–64.

164. Emani SM, Piekarski BL, Baird C, Kaza A, Harrild D, del Nido PJ, McCully JD. Autologous mitochondria transplantation by ventricular dysfunction following myocardial ischemia-reperfusion injury. (personal communication). N.d.

165. Emani SM, McCully JD. Mitochondrial transplantation: applications for pediatric patients with congenital heart disease. Transl Pediatr. 2018 Apr;7(2):169–75. doi: 10.21037/tp.2018.02.02.

166. Chang JC, Wu SL, Liu KH, Chen YH, Chuang CS, Cheng FC, Su HL, Wei YH, Kuo SJ, Liu CS. Allogenic/xenogenic transplantation of peptide-labeled mitochondria in Parkinson's dis- ease: restoration of mitochondria functions and attenuation of 6-hydroxydo-pamine-induced neurotoxicity. Transl Res. 2016;170: 40–56.

167. Lin HC, Liu SY, Lai HS, Lai IR. Isolated mitochondria infusion mitigates ischemia-reperfusion injury of the liver in rats. Shock. 2013;39:304–10.

168. Zhu L, Zhang J, Zhou J, Lu Y, Huang S, Xiao R, Yu X, Zeng X, Liu B, Liu F, Sun M, Dai M, Hao Q, Li J, Wang T, Li T, Lu Q. Mitochondrial transplantation attenuates hypoxic pulmonary hypertension. Oncotarget. 2016 Aug 2;7(31):48925-40.

169. Hayakawa K, Esposito E, Wang X, Terasaki Y, Liu Y, Xing C, Ji X, Lo EH. Transfer of mitochondria from astrocytes to neurons after stroke. Nature. 2016;535:551–5.

20 Mitochondrial Dysfunction in Diabetes

Eric L. Johnson M.D

Department of Family and Community Medicine, School of Medicine and Health Sciences, Department of Education Resources,University of North Dakota, USA

CONTENTS

1 INTRODUCTION

Mitochondria (singular: mitochondrion) are "organelles within eukaryotic cells that produce adenosine triphosphate (ATP), the main energy molecule used by the cell. Containing features of their known bacterial evolutionary predecessors, mitochondria have outer and inner membranes that are functionally separate and distinct, as well as intermembrane space and a matrix (3,4). Through the electron transport chain, energy is used to transport proteins across the inner membrane (3). Classically, the mitochondrion is sometimes referred to as "the powerhouse of the cell, and is found in all eukaryotes, producing ATP through cellular respiration as well as oxidative phosphorylation (OXPHOS). Over 90% of the energy required by a cell is produced by OXPHOS. Long chain fatty acids and sugars are broken down within these organelles while ADP is converted back to ATP. As well, steroids and lipids are synthesized along with other important functions (4). Through more recent research, mitochondria are known to have a more dynamic role in cell function and disease (5).

Mitochondria appear to have many dynamic and interconnected functions in response to cellular signals. As a steady state reticulum, these organelles are regulated by fission, fusion, and motility. Fission is dependent upon a dynamin GTPase, DRP-1 (6–8). Several different GTPases are part of fusion: Mitofusin 1 (Mfn 1) and Mitofusin 2 (Mfn 2) are integrated in the outer membrane. Exposure to the cytosol occurs via their GTPase and coiled domains. MFn1 and Mfn 2 can form between adjacent organelles as both heterotopic and homotopic complexes, GTPase in regulation (9–11). Opa-1, a dynamin-like GTPase is associated with the inner membrane with a role in fusion and is genetically linked to Mfn 1 (12). Other proteins along with GTPases are proposed to be involved with both fission and fusion (13). Signaling appears to exist between other cellular components and mitochondria with the involvement of a "GTPase switch," integrating function to cellular demands related to changes in metabolism, development, division, and cell death. A comprehensive inventory of proteins in different eucaryocytes appears to exist through proteomic, genomic, and bioinformatics approaches; over 1,500 mammalian mitochondria are currently known (14–16).

Mitochondrial metabolism is dependent on molecular components that sense energy status, including NAD+/NADH ratio, AMP/ATP ratio, levels of acetyl-CoA, as well as transcription factors, hormones, cofactors, kinases, and nuclear receptors (5)

Mitochondrial dysfunction appears to play a role in many human diseases across many diverse systems, organs, and disease states, including cardiac, metabolic, cancer, obesity, and diabetes. These are genetically heterogenous with mutations in mitochondrial and nuclear DNA implicated (17,18). These can occur at any age, and may be inherited from an X chromosome, an autosome, or maternally. No known cures are known at this time, with treatment directed at symptomatology with other research underway. Alterations of mitochondrial mass and activity are thought to be contributory factors in obesity and metabolic syndrome (19,20).

2 ENERGY BALANCE AND GLUCOSE METABOLISM

A balance between gastrointestinal tract, uptake and utilization in peripheral muscle and adipose tissue, and glucose production by the liver are a summary of multiple tissue specific mechanisms of blood glucose homeostasis. These processes are normally tightly regulated by different hormones, including insulin, glucagon, amylin, and gut incretins. Uptake and regulation of substrates into cells is mediated by insulin for ATP generation. Beta cells in the pancreas respond to an increase in blood glucose by secreting insulin, allowing uptake of glucose into the cell, ultimately leading to utilization of glucose, while decreasing gluconeogenesis and driving glucose storage in the liver. Glucagon is secreted from pancreatic alpha cells responding to lower glucose levels in the blood, which in turn increases glycogen breakdown and gluconeogenesis in the liver. In other insulin sensitive tissues, such as adipose tissue and muscle, decreased glucose uptake occurs in response to glucagon (21–26).

Non-insulin glucose uptake has a less tightly regulated system. The glucose transporter- 1 (GLUT-1) is a non-insulin glucose co-transporter present on the membranes of cells in many tissues, but particularly noted in the blood brain barrier. GLUT-3 works with GLUT-1 as a co-transporter in neurons. GLUT-2 allows passive movement of glucose across cell membranes, and is present on beta cells, renal tubular, gut epithelium, liver, and blood. In contrast, GLUT-4 relies on insulin for facilitation of diffusion. As well, in the renal and gut epithelium, glucose uptake can be independent of insulin via sodium-dependent co-transporters regulated by the kidney proximal tubules (21,27–30).

3 MITOCHONDRIAL ACTIVITY AND PATHOPHYSIOLOGY IN DIABETES

Brownlee (31) proposed the unifying hypothesis of diabetes and vascular injury encompassing many diabetes complication mechanisms. Originally, he proposed that reactive oxygen species (ROS) formed in the mitochondria migrate to the nucleus causing DNA damage activating Poly [ADP-ribose] polymerase 1 (PARP-1). As a result, Glyceraldehyde-3-dehydrogenase (GAPDH) undergoes ADP-ribosylation inactivating glycolytic enzyme in the nucleus. The decline in GAPDH activity disrupts glycolysis, resulting in glycolytic intermediates being diverted to pathways of hyperglycemic injury. This was later updated to include the transfer of superoxide from the mitochondria to the cell nucleus DNA, which uncouples the nitric oxide system and oxidative species generated by diabetic dyslipidemia.

Other pathways thought to be activated by mitochondrial O2- overproduction in endothelial cells include polyol pathway flux, increased formation of AGEs (advanced glycation end products), increased expression of the receptor for AGEs and its activating ligands, activation of protein kinase C isoforms, and overactivity of the hexosamine pathway. These processes also inactivate endothelial nitric oxide synthase and prostacyclin, both of which are antiatherosclerotic enzymes (31,32). Resultant ROS activate a number of proinflammatory pathways. Pathway-selective insulin resistance increases mitochondrial ROS production from free fatty acids and by inactivation of antiatherosclerosis enzymes by ROS also exist. Specifically, a form of ROS known as superoxide (O_2^-) are released via OXPHOS by complexes I and III. In case of diabetes the O_2^- is involved with hyperglycemic tissue injury (33). Teodoro et al. (34) demonstrated that hyperglycemia decreases mitochondrial function. Long term exposure of hepatic G2 cells to high glucose concentrations lead to loss of respiratory capacity and decreased mitochondrial biogenesis by a mechanism of accelerated coenzyme Q reduction and generation of ROS in their study.

3.1 INSULIN RESISTANCE

In terms of insulin resistance, it appears to be established that although impaired mitochondrial function in skeletal muscle occurs in type 2 diabetes, it may not be a central feature. It may be that it is sufficient in those with specific insulin resistance subtypes, as noted in those with mutations that impair mitochondrial DNA. In the case of mitochondrial myopathies, insulin resistance is noted (35–37). Impaired glucose and lipid oxidative capacity in skeletal muscle has also been demonstrated. It's unclear whether this is from reduced mitochondrial density or reduced OXPHOS capacity (38–41). Phelix and colleagues have demonstrated that ADP stimulated respiration, maximal respiratory capacity, and in vivo mitochondrial function are reduced in persons with type 2 diabetes (42). Overall, it appears that these may be due to environmental factors, not genetically acquired (43–45), but are not clear-cut. There is also debate regarding proliferator activated receptor co-activator 1 α expression and mitochondrial phenotype in persons with type 2 diabetes and offspring with respect to the role of physical inactivity (46–51). Studies show alterations in mitochondrial morphology may occur after the onset of insulin resistance (52–55). In addition, differences exist between rodent and human models of mitochondrial phenotypes, as obese rodents have increased mitochondrial gene expression and fatty acid oxidation, but humans are the converse (56).

In comparison to skeletal muscle, data is more limited for the liver. Mitochondrial dysfunction is noted in animal models (57). In humans, hepatic insulin resistance was associated with reduced ATP, ATP turnover, and inorganic phosphate levels (58,59), but were inverse to waist circumference, body mass index (BMI), and fasting plasma glucose, which may indicate a role for whole body homeostasis and insulin resistance in the liver (59). Similarly, in adipose tissue, there is some debate about whole body insulin resistance and mitochondrial dysfunction. This is supported by data using thiazolidinediones in humans and rodents, which showed improvement in expression of mitochondrial genes in subcutaneous adipose tissue (60–63).

Insulin resistance and mitochondrial dysfunction data exists for both type 1 and type 2 subjects. Dysfunction in oxidative metabolism and reduce pyruvate activity and increased fatty acid oxidation with decreased cardiac efficiency have been noted (64–69). Another study was supportive in that the anti-oxidant metallothione improved cardiac function in diabetes (70). However, data is limited to date and mitochondrial role in cardiac insulin resistance is not completely clear.

Potential mechanisms linking impaired mitochondrial dysfunction and insulin resistance include ectopic lipid accumulation (71–77), ROS production (78–82), inflammation (83–87), other mitochondrial retrograde signals (Ca++, Fe2+, nitric oxide, CO), and mitochondrial peptides (88,89). (see Fig 1) Impairments of complexes of the electron transport chain do not appear to cause insulin resistance (90).

3.2 Type 1 Diabetes

Many factors are thought to be associated with the development of type 1 diabetes, most notably autoimmunity antibody to beta cell production, possibly associated with viral exposure (i.e., coxsackie, rubella), nitrates, or cow's milk protein. Genetic susceptibility may be associated with MHC Class II histocompatibility complex mutations in the HLA-DR/DQ and IDDM genes. These autoimmune factors along with environmental factors are proposed mechanisms for beta cell destruction (21).

Type 1 diabetes has been associated with a Mt5178C polymorphism as a genetic mechanism. Conversely, a single nucleotide polymorphism (SNP) for Complex 1 mt-Nd2 may confer resistance to type 1 diabetes by generation of less ROS (91).

In streptozocin treated mice, increased ROS and nitrogen species were noted with increased complex I and IV activity and decreased complex II and II activity in renal and liver tissue (92–94). Likewise, in the retinal tissue of streptozocin treated mice, increased expression of OXPHOS proteins and damaged mitochondrial DNA replication were noted (95,96). Mitochondrial dysfunction is also related to neuropathy with respect to impaired

adenosine monophosphate protein kinase signaling (97). In cardiac tissue in both streptozocin treated mice and human cultures, increased mitochondrial fusion protein (OPA-1) and decreased mitochondrial fission protein (DRP1) were noted, but with decreased state 3 respiration and TCA activity with increased mitochondrial density, which did not confer increased function. In skeletal muscle, alterations of mitochondrial ultrastructure and bioenergenics have been demonstrated (98). Mitochondria appear to play a role in type 1 diabetes and complication development, but it's clear that further investigation is needed.

3.3 Type 2 Diabetes

As noted, type 2 diabetes has multiple mechanisms, most notably insulin resistance and beta cell dysfunction. Mitochondrial metabolism has also been implicated. Mitochondrial DNA is susceptible to oxidative damage. Interestingly, in the genetic diabetes syndrome Maternally-inherited insulin dependent diabetes with deafness (MIDD) is noted in the setting of a 10.4 kb deletion of mitochondrial DNA. This results in decreased OXPHOS activity. A nonobese rodent model of type 2 diabetes has been developed with mitochondria from FHH rats and nuclear DNA from Wistar rats (46,72,78,99–101).

More recently, a proposed mechanism involving defective insulin signaling, inflammation and mitochondrial dysfunction as underlying factors tying type 2 diabetes to Alzheimer's dementia (102–104). Hyperglycemia increases the risk of dementia in diabetes by 40% (103,104). Impaired energy metabolism and oxidative stress are implicated in aging (105,106). High ROS levels in the brain are thought to be related to the molecular etiology of Alzheimer's disease. Oxidized forms of lipids, protein, and DNA are all noted in the Alzheimer's brain (107). Brain insulin signaling and oxidative stress are connected as ABO-induced neuronal oxidative stress processes are blocked by insulin (108,109). Activation of Akt and prevention of abnormal NDMA-R activation appear to be an insulin protective mechanism (110). Recently, mitochondrial dysfunction and subsequent ROS levels were found to cause insulin resistance in skeletal muscle and liver via c-Jun N-terminal kinase (JNK) activation (111). A Mitochondria, insulin, and JNK signaling may play a role in neuronal dysfunction (112).

4 MITOCHONDRIAL DYSFUNCTION AND DIABETES COMPLICATIONS

Clinically, it's apparent that there is variability in diabetic complication manifestation across different patients, not always tightly associated with glycemic control. Onset and severity may be related to mitochondrial

S.D Martin, S L. McGee / Biochimica et Biophysica Acta 1840 (2014) 1303-1312

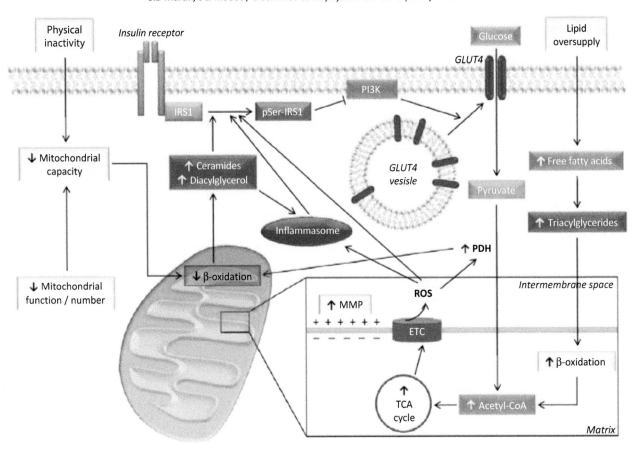

FIGURE 1   Proposed mechanisms by which impaired mitochondrial function inhibits insulin action. Physical inactivity or intrinsic impairments in mitochondrial function and/or number reduce oxidative capacity, leading to reduced β-oxidation and accumulation of lipid species such as diacylglycerides and ceramides, which can ultimately impair the canonical insulin signalling pathway and GLUT4 translocation to the plasma membrane in skeletal muscle, adipose and the heart that results in insulin resistance. Alternatively, lipid oversupply increases β-oxidation that leads to hper-reduction of the electron trasport chain, increased mitochondrial membrane potential (MMP) and electron leak in the form of reactive oxygen species (ROS). Elevated ROS levels can impair insulin signalling either directly through as yet unresolved mechanisms, or through inflammasome activation. Elevated ROS levels also increase PDH activity that will lead to decreased β-oxidation and accumulation of intracellular lipids, which can also activate the inflammasome to induce insulin resistance.

DNA variation and mismatched nuclear and mitochondrial DNA (113–115). Substrate availability is increased with intracellular hyperglycemia. This in turn drives the citric acid cycle and subsequent pyruvate oxidation, leading to increases in NADH and FADH2. Mitochondrial membrane threshold is increased through OXPHOS, blocking the electron transport chain; electrons then reduce molecular oxygen forming O_2^-. This goes on to inhibit GADPH. As well the O_2^- may cause DNA breakage, which can activate PARP-1. This inhibition of GADPH results in glycolytic pathway metabolites, which pass on to polyol, PKC activation, AGE product, and hexosamine pathways (115,116). OXPHOS is complex and unique in that 13 of

approximately 92 structural protein subunits are encoded in mtDNA, some of which are associated with diabetic phenotypes. Some specific haplotypes have been possibly implicated in diabetes complication development to include coronary artery disease, retinopathy, and nephropathy (21). As with type 1 diabetes, mitochondrial dysfunction may play a role in skeletal muscle insulin resistance (117), yet in another study, insulin resistance was the result of oxidative stress from lipid overload (118). However, insulin resistant skeletal muscle appears to be resistant to adipokines, which have been demonstrated to improve mitochondrial dysfunction, and treatment with thyroid hormone has mixed results (119–123).

4.1 DIABETIC NEPHROPATHY

Diabetes globally remains the leading cause of end stage renal disease or nephropathy (124,125). Chronic hyperglycemia contributes to glomerular injury, glomerular hypertrophy, and podocyte depletion (126). Extracellular matrix deposition is triggered by profibrotic genes signaling, also by vascular endothelial growth factor (VEGF), cytokines, and nitric oxide. These also result in vascular permeability, chronic inflammation, and vasodilation (127).

Diabetic kidney disease (DKD) is also associated with atherosclerotic cardiovascular disease (ASCVD). Historical treatments for DKD include angiotensin converting enzyme inhibitors (ACE), angiotensin receptor blockers (ARB), overall blood pressure control, and glycemic control (128). More recently, certain glucagon-peptide-like 1 receptor agonists (GLP-1 RA) and sodium glucose transport inhibitors 2 (SGLT-2) have been shown to slow the progression of DKD (129,130).

As with other diabetes complications, ROS has been implicated in DKD. Increased ROS have been thought to cause inflammation leading to renal fibrosis by promoting lipid peroxidation, DNA damage, and mitochondrial dysfunction (31,131–133). Urinary excretion of citric acid cycle metabolites and genes regulating these metabolites are reduced in persons with DKD, which supports the idea of mitochondrial dysfunction in DKD (134). Excessive levels of mtROS can damage intramitochondrial lipids, proteins and DNA (135). Reduced mitochondrial content, decreased biogenesis, and increased total urinary ROS have been demonstrated, with O_2^- production reduced in kidney tissue in streptozocin type 1 diabetes with no evidence of increased mitochondrial superoxide production, in contrast to the unifying theory of ROS in diabetes complications (136–138). Antioxidant treatments in DKD did not show positive results, which suggests that aggressive "removal" of all ROS may not be beneficial, as some ROS are responsible for vascular homeostasis and immune function (139,140). However, increased mtROS has been demonstrated in kidney tissue in diabetic mouse models, suggesting that location and type of ROS may be important (141). According to Daehn (126), "oxidative damage is context dependent in DKD," and "glomerular endothelium could be the weakest link" suggesting that established podocyte depletion may be contingent on mitochondrial dysfunction.

4.2 DIABETIC RETINOPATHY

With hyperglycemia, a number of pathways are affected in retinopathy, which is still one of the leading causes of blindness (1,2). Insulin-like growth factor and VEGF lead to microvascular abnormalities, ultimately leading to proliferative growth of vessels on the retina. Hyperglycemia-induced mitochondrial overproduction of reactive oxygen species has been shown to induce altered gene transcription by covalent modification of coregulatory proteins (142).

4.3 CARDIOVASCULAR DISEASE

Hyperglycemia and insulin resistance are associated with atherosclerosis, heart failure (both heart failure with preserved ejection fraction and heart failure with reduced ejection fraction), and increased post-MI fatality rates. Hyperglycemia-linked oxidative stress has been shown to contribute to chronic inflammation and endothelial cell abnormalities that lead to atherosclerosis and heart disease (143). Prolonged increases in ROS production in diabetic cardiovascular cells lie at the root of these issues. ROS species hydrogen peroxide (H_2O_2) function as signaling molecules essential for normal cellular homeostasis (144,145). In contrast, ROS production from hyperglycemia leads to impaired cellular function and the stated cardiovascular pathologies by previously noted ROS mechanisms, including shifting the balance from mitochondrial fusion to fission (146). Insulin resistance causes excessive cardiomyocyte ROS production by increasing fatty acid flux and oxidation, in turn stimulating overexpression of the nuclear receptor PPARα and nuclear translocation of forkhead box O 1, leading to cardiomyopathy(147). According to Shah and Brownlee, less is known about higher post-MI fatality rates in persons with diabetes, but the role of mitochondrial ROS effects in gap junction remodeling and fatal arrhythmia generation is evident (147). Autonomic denervation may also play a role in this process (148).

5 DIABETES TREATMENTS AND MITOCHONDRIAL FUNCTION

Current investigations suggest that insulin stimulates mitochondrial protein synthesis, which is conversely decreased in insulin deficiency (149,150). The exact mechanisms are uncertain, but insulin may stimulate OXPHOS in different tissues by increasing protein transcription and mitochondrial function (151,152). Thiazolidinediones may also play a role as noted above (60–63).

Intracellular metformin (given orally) appears to inhibit mitochondrial function through AMP-activated protein kinase as well as other AMPK independent pathways (153–155).

Historically, a number of "mitochondrial cocktails" have been trialed in the past, including many vitamins and supplements. These may include L-carnitine, CoQ-10, riboflavin, nicotinamide, and thiamin, but the results are mixed, showing little if any benefit (156–158).

Current research focusing on treatments based on genome discovery and manipulation of mtDNA including the use of restriction endonucleases (RE), and CRISPR technology. Gene therapy to prevent transmission of mitochondrial diseases involving the removal of mutant mtDNA using genetically engineered nucleases

TABLE 1

Mitochondrial forms of diabetes.

mtDNA mutation	Prevalence of diabetes	Other signs and symptoms	Other abnormal lab testing
m.3243A>G (MELAS, MIDD)	38%	Stroke symptoms, Seizure,Dementia, Hearing loss,Migraine, Cognitive impairment, Myopathy, Progressive external Opthalmoplegia (Kearns Sayre syndrome	Increased CSF and serum lactate Increased serum alanine Decreased arginine
single mtDNA deletions/duplication	11–14%	Ptosis, External ophthalmoplegia, Dysarthria, Retinitis, Conduction block, CMP, Endocrine disease, Pearson syndrome (Sideroblastic anemia,Pancytopenia Exocrine pancreatic insufficiency), Malabsorption, Nephropathy, Hepatopathy	
m.8344A>G Myoclonic Epilepsy with Ragged-Red Fibers (MERRF)	10%	Multiple lipomatosis, RRF in muscle tissue, Myoclonus, Myopathy,	Elevated lactic acid in plasma and/or CSF, elevated protein in CSF, elevated CK, elevated alanine
m.12258C>A	100%	Cerebellar ataxia, cataracts	
POLG related disorders nDNA mutations Alpers–Huttenlocher syndrome	11%	Seizures,muscle regression, liver disease	Possibly elevated LFT, CK
OPA-1		Proximal renal tubulopathy with nephrocalcinosis, Seizures/global developmental delay, Microcephaly, Hearing loss, Bilateral vision loss,	
RRM2B	4.5%	Myopathy, GI dysmotility	
Friedereich's ataxia 32%	32%	Gait ataxia, dysarthria, hearing loss, vision loss, scoliosis, heart disease	

Abbreviations: CK, creatine kinase; CSF, cerebrospinal fluid; LFT, liver function tests; MELAS, mitochondrial encephalomyopathy with lactic acidosis and stroke-like episodes; MIDD, maternally inherited diabetes and deafness; mtRNA, mitochondrial RNA; PEO progressive external ophthalmoplegia; RRF, ragged red fibers; POLG, polymerase gamma.

(159). Other strategies, including mitochondrial replacement are currently being investigated. See the consensus statement of the Mitochondrial Medicine Society for further information in this area (160).

6   MITOCHONDRIAL FORMS OF DIABETES

Although rare (1:5000), there are several mitochondrial disorders with diabetes as a prominent feature, including m.3243A>G mitochondrial encephalomyopathy with lactic acidosis and stroke-like episodes (MELAS), maternally inherited diabetes and deafness (MIDD), single mtDNA, m.8344A>G (MERRF), m.12258C>A deletions/ duplication, nDNA mutations, RRM2B, and OPA1. These

may present with neurologic, metabolic, endocrine, and organ dysfunction. Most of these syndromes present with a diabetes incidence of less than 15%, but is as high as 38% for m.3243A>G (MELAS, MIDD) and 100% for m.12258C>A. These may present from infancy to adulthood (Table 1). For more on diagnosis and treatment, see the complete review by Karaa and Goldsand Goldstein (161).

7   CONCLUSION

Diabetes remains a global disease burden, costing billions of dollars annually in US alone. Current diabetes treatment options are multiple and generally high quality, but an understanding of cellular processes of mitochondrial dysfunction and ROS in disease onset and complication development may lead to entirely new areas of therapeutics in the future. Genetic treatments show promise for mitochondrial disease in general, particularly with research involving targeting and prevention of specific conditions underway.

REFERENCES

1. World Health Organization (WHO). Diabetes fact sheet www.who.int/diabetes/en/. Accessed November 19, 2018.
2. Center for Disease Control and Prevention (CDC) Diabetes data. www.cdc.gov/diabetes/data/. Accessed November 19, 2018.
3. Biology Dictionary. https://biologydictionary.net/mitochondria/. Accessed October 12, 2018
4. McBride HM, Neuspiel M, Wasiak S. Mitochondria: more than just a powerhouse. Curr. Biol. 2006 Jul 25;16 (14):R551–R560.
5. Nunnari J, Suomalainen A. Mitochondria in sickness and in health. Cell. 2012 Mar16;148(6):1145–1159. doi: 10.1016/j.cell.2012.02.035.
6. Pitts, KR, Yoon, Y, Krueger, EW, McNiven, MA. The dynamin-like protein DLP1 Is essential for normal distribution an morphology of the endoplasmic reticulum and mitochondria in mammalian cells. Mol. Biol. Cell. 1999;10:4403–4417.
7. Yoon Y, Pitts KR, and McNiven MA. Mammalian dynamin-like protein dlp1 tubulates membranes. Mol. Biol. Cell. 2001;12:2894–2905.
8. Smirnova, E, Griparic, L, Shurland, DL, van Der Bliek, AM. Dynamin-related protein drp1 is required for mitochondrial division in mammalian cells. Mol. Biol. Cell. 2001;12:2245–2256.
9. Chen H, Detmer SA, Ewald AJ, Griffin EE, Fraser SE, Chan DC. Mitofusins Mfn1 and Mfn2 coordinately regulate mitochondrial fusion and are essential for embryonic development. J. Cell Biol. 2003;160:189–200.
10. Koshiba T, Detmer SA, Kaiser JT, Chen H, McCaffery JM, Chan DC. Structural basis of mitochondrial tethering by mitofusin complexes. Science. 2004;305:858–862.
11. Ishihara N, Eura Y, Mihara K. Mitofusin 1 and 2 play distinct roles inmitochondrial fusion reactions via GTPase activity. J. Cell Sci. 2004;117:6535–6546.
12. Cipolat S, Martins de Brito O, Dal Zilio B, Scorrano L. OPA1 requiresmitofusin 1 to promotemitochondrial fusion. Proc. Natl. Acad. Sci. U. S. A. 2004;101:15927–15932.
13. Chan DC. Mitochondrial fusion and fission in mammals. Annu. Rev. Cell Dev. Biol. 2006 May; 22:79–99.
14. Chen H, Chan DC. Emerging functions of mammalian mitochondrial fusion and fission. Hum. Mol. Genet. 2005;14:R283–R289.
15. Gaston D, Tsaousis AD, Roger AJ. Predicting proteomes of mitochondria and related organelles from genomic and expressed sequence tag data. Methods Enzymol. 2009;457:21–47.
16. Mootha VK, Bunkenborg J, Olsen JV, Hjerrild M, Wisniewski JR, Stahl E, Bolouri MS, Ray HN, Sihag S, Kamal M, et al. Integrated analysis of protein composition, tissue diversity, and gene regulation in mouse mitochondria. Cell. 2003;115:629–640.
17. Pagliarini DJ, Calvo SE, Chang B, Sheth SA, Vafai SB, Ong SE, Walford GA, Sugiana C, Boneh A, Chen WK, et al. A mitochondrial protein compendium elucidates complex I disease biology. Cell. 2008;134:112–123.
18. Sickmann A, Reinders J, Wagner Y, Joppich C, Zahedi R, Meyer HE, Schonfisch B, Perschil I, Chacinska A, Guiard B, et al. The proteome of Saccharomyces cerevisiae mitochondria. Proc. Natl. Acad. Sci. U. S. A.. 2003;100:13207–13212.
19. Ylikallio E, Suomalainen A Mechanisms of mitochondrial diseases. Ann Med. 2012 Feb;44(1):41-59. doi: 10.3109/07853890.2011.598547.
20. Suomalainen A. Semin Fetal Neonatal Med. 2011 Aug;16 (4):236-240. doi: 10.1016/j.siny.2011.05.003
21. Blake R, Trounce IA. Mitochondrial dysfunction and complications associated with diabetes. Biochim. Biophys. Acta. 2014 Apr;1840(4):1404–1412. doi: 10.1016/j.bbagen.2013.11.007.
22. Shaw JE, Sicree RA, Zimmet PZ. Global estimates of the prevalence of diabetes for 2010 and 2030. Diabetes Res. Clin. Pract. 2010;87:4–14.
23. Berne RM, Levy RM, Koeppen BM, Stanton BA. Berne & Levy physiology, Mosby/Elsevier, Philadelphia, PA, 2008.
24. Giacco F, Brownlee M. Oxidative stress and diabetic complications. Circ. Res. 2010;107:1058–1070.
25. Johnston AM, Pirola L, Van Obberghen E. Molecular mechanisms of insulin receptor substrate protein-mediated modulation of insulin signalling FEBS Lett. 2003;546:32–36.
26. Kumar V, Abbas AK, Fausto N, Robbins SL, Cotran RS. Robbins and Cotran pathologic basis of disease, 7th ed., edited by Vinay Kumar, Abul K Abbas, Nelson Fausto with illustrations by James A. Perkins. ed., Elsevier Saunders, Philadelphia, PA; London, 2005.
27. Whitehead JP, Clark SF, Urso B, James DE. Signalling through the insulin receptor Curr. Opin. Cell Biol. 2000;12:222–228.
28. Thorens B. Glucose transporters in the regulation of intestinal, renal, and liver glucose fluxes Am. J. Physiol. 1996;270:G541–G553.
29. Vallon V, Sharma K. Sodium–glucose transport: role in diabetes mellitus and potential clinical implications Curr. Opin. Nephrol. Hypertens. 2010;19:425–431.
30. Seyer P, Vallois D, Poitry-Yamate C, Schutz F, Metref S, Tarussio D, Maechler P, Staels B, Lanz B, Grueter R, et al. Hepatic glucose sensing is required to preserve beta cell glucose competence J. Clin. Invest. 2013;123:1662–1676.
31. Brownlee M. Biochemistry and molecular cell biology of diabetic complications. Nature 2001;414:813–820.
32. Giacco F, Brownlee M. Oxidative stress and diabetic complications. Circ. Res. 2010;107:1058–1070.
33. Ha H, Hwang IA, Park JH, Lee HB. Role of reactive oxygen species in the pathogenesis of diabetic nephropathy. Diabetes Res. Clin. Pract. 2008 Nov 13;82 (Suppl 1):S42–S45.
34. Teodoro JS, Gomes AP, Varela AT, Duarte FV, Rolo AP, Palmeira CM. Uncovering the beginning of diabetes: the cellular redox status and oxidative stress as starting players in hyperglycemic damage. Mol. Cell. Biochem. 2013;376:103–110.
35. Reardon W, Ross RJ, Sweeney MG, Luxon LM, Pembrey ME, Harding AE, Trembath RC. Diabetes

mellitus associated with a pathogenic point mutation in mitochondrial DNA. Lancet. 1992;340:1376–1379.

36. Kadowaki T, Kadowaki H, Mori Y, Tobe K, Sakuta R, Suzuki Y, Tanabe Y, Sakura H, Awata T, Goto Y. A subtype of diabetes mellitus associated with a mutation of mitochondrial DNA. N. Engl. J. Med. 1994;330:962–968.

37. Becker R, Laube H, Linn T, Damian MS. Insulin resistance in patients with themitochondrial tRNA(Leu(UUR)) gene mutation at position 3243. Exp. Clin. Endocrinol. Diabetes. 2002;110:291–297.

38. Kelley DE, He J, Menshikova EV, Ritov VB. Dysfunction of mitochondria in human skeletal muscle in type 2 diabetes. Diabetes. 2002;51:2944–2950.

39. Boushel R, Gnaiger E, Schjerling P, Skovbro R, Kraunsøe R, Dela F. Patients with type 2 diabetes have normal mitochondrial function in skeletal muscle. Diabetologia. 2007;50:790–796.

40. Ritov VB, Menshikova EV, He J, Ferrell RE, Goodpaster BH, Kelley DH. Deficiency of subsarcolemmal mitochondria in obesity and type 2 diabetes. Diabetes. 2005;54:8–14.

41. Mogensen M, Sahlin K, Fernström M, Glintborg D, Vind BF, Beck-Nielsen H, Højlund K. Mitochondrial respiration is decreased in skeletal muscle of patients with type 2 diabetes. Diabetes. 2007;56:1592–1599.

42. Phielix E, Schrauwen-Hinderling VB, Mensink M, Lenaers E, Meex R, Hoeks J, Kooi ME, Moonen-Kornips E, Sels JP, Hesselink MK, et al. Lower intrinsic ADP-stimulated mitochondrial respiration underlies in vivo mitochondrial dysfunction in muscle of male type 2 diabetic patients. Diabetes. 2008;57:2943–2949.

43. Huang X, Eriksson KF, Vaag A, Lehtovirta M, Hansson M, Laurila E, Kanninen T, Olesen BT, Kurucz I, Koranyi L, et al. Insulin-regulated mitochondrial gene expression is associated with glucose flux in human skeletal muscle. Diabetes. 1999;48:1508–1514.

44. Mustelin L, Pietiläinen KH, Rissanen A, Sovijärvi AR, Piirilä P, Naukkarinen J, Peltonen L, Kaprio J, Yki-Järvinen H. Acquired obesity and poor physical fitness impair expression of genes of mitochondrial oxidative phosphorylation in monozygotic twins discordant for obesity. Am. J. Physiol. Endocrinol. Metab. 2008;295: E148–E154.

45. Rönn T, Poulsen P, Tuomi T, Isomaa B, Groop L, Vaag A, Ling C. Genetic variation in ATP5O is associated with skeletal muscle ATP50 mRNA expression and glucose uptake in young twins. PLoS One. 2009;4:e4793. doi: 10.1371/journal.pone.0004793.

46. Patti ME, Butte AJ, Crunkhorn S, Cusi K, Berria R, Kashyap S, Miyazaki Y, Kohane I, Costello M, Saccone R, et al. Coordinated reduction of genes of oxidative metabolism in humans with insulin resistance and diabetes: potential role of PGC1 and NRF1. Proc. Natl. Acad. Sci. U. S. A. 2003 Jul 8;100(14):8466–8471.

47. Mootha VK, Lindgren CM, Eriksson KF, Subramanian A, Sihag S, Lehar J, Puigserver P, Carlsson E, Ridderstråle M, Laurila E, et al. PGC-1alpha-responsive genes involved in oxidative phosphorylation are coordinately downregulated in human diabetes. Nat Genet. 2003;34:267–273.

48. Baar K, Wende AR, Jones TE, Marison M, Nolte LA, Chen M, Kelly DP, Holloszy JO. Adaptations of skeletal muscle to exercise: rapid increase in the transcriptional coactivator PGC-1. FASEB J. 2002;14:1879–1886.

49. Morino K, Petersen KF, Shulman GI. Molecular mechanisms of insulin resistance in humans and their potential links with mitochondrial dysfunction. Diabetes. 2006;55: S9–S15.

50. Karlsson HK, Ahlsén M, Zierath JR, Wallberg-Henriksson H, Koistinen HA. Insulin signaling and glucose transport in skeletal muscle from first-degree relatives of type 2 diabetic patients. Diabetes. 2006;55:1283–1288.

51. Choi CS, Befroy DE, Codella R, Kim S, Reznick RM, Hwang YJ, Liu ZX, Lee HY, Distefano A, Samuel VT, et al. Paradoxical effects of increased expression of PGC-1alpha on muscle mitochondrial function and insulin-stimulated muscle glucose metabolism. Proc. Natl. Acad. Sci. U. S. A. 2008;105:19926–19931.

52. Chomentowski P, Coen PM, Radiková Z, Goodpaster BH, Toledo FG. Skeletal muscle mitochondria in insulin resistance: differences in intermyofibrillar versus subsarcolemmal subpopulations and relationship to metabolic flexibility. J. Clin. Endocrinol. Metab. 2011;96:494–503.

53. Fabris R, Mingrone G, Milan G, Manco M, Granzotto M, Dalla Pozza A, Scarda A, Serra R, Greco AV, Federspil G, Vettor R. Further lowering of muscle lipid oxidative capacity in obese subjects after biliopancreatic diversion. J. Clin. Endocrinol. Metab. 2004;89:1753–1759.

54. Berggren JR, Boyle KE, Chapman WH, Houmard JA. Skeletal muscle lipid oxidation and obesity: influence of weight loss and exercise. Am. J. Physiol. Endocrinol. Metab. 2008;294:E726–E732.

55. Toledo FG, Menshikova EV, Azuma K, Radiková Z, Kelley CA, Ritov VB, Kelley DE. Mitochondrial capacity in skeletal muscle is not stimulated by weight loss despite increases in insulin action and decreases in intramyocellular lipid content. Diabetes. 2008;57:987–994.

56. Simoneau JA, Veerkamp JH, Turcotte LP, Kelley DE. Markers of capacity to utilize fatty acids in human skeletal muscle: relation to insulin resistance and obesity and effects of weight loss. FASEB J. 1999;13:2051–2060.

57. Bouderba S, Sanz MN, Sánchez-Martín C, El-Mir MY, Villanueva GR, Detaille D. Koceïr EA.Hepatic mitochondrial alterations and increased oxidative stress in nutritional diabetes-prone Psammomys obesus model. Exp. Diabetes Res. 2012;2012:430176.

58. Szendroedi J, Chmelik M, Schmid AI, Nowotny P, Brehm A, Krssak M, Moser E, Roden M. Abnormal hepatic energy homeostasis in type 2 diabetes. Hepatology 2009;50:1079–1086.

59. Schmid AI, Szendroedi J, Chmelik M, Krssák M, Moser E, Roden M. Liver ATP synthesis is lower and relates to insulin sensitivity in patients with type 2 diabetes. Diabetes Care. 2011;34:448–453.

60. Wilson-Fritch L, Nicoloro S, Chouinard M, Lazar MA, Chui PC, Leszyk J, Straubhaar J, Czech MP, Corvera S. Mitochondrial remodeling in adipose tissue associated with obesity and treatment with rosiglitazone. J. Clin. Invest. 2004;114:1281–1289.

61. Choo HJ, Kim JH, Kwon OB, Lee CS, Mun JY, Han SS, Yoon YS, Yoon G, Choi KM, Ko YG. Mitochondria are impaired in the adipocytes of type 2 diabetic mice. Diabetologia. 2006;49:784–791.

62. Rong JX, Qiu Y, Hansen MK, Zhu L, Zhang V, Xie M, Okamoto Y, Mattie MD, Higashiyama H, Asano S, et al. Adipose mitochondrial biogenesis is suppressed in db/db

and high-fat diet-fed mice and improved by rosiglitazone. Diabetes. 2007;56:1751–1760.

63. Bogacka I, Xie H, Bray GA, Smith SR. Pioglitazone induces mitochondrial biogenesis in human subcutaneous adipose tissue in vivo. Diabetes. 2005;54:1392–1399.

64. Kuo TH, Giacomelli F, Wiener J. Defective oxidative metabolism of heart mitochondria from genetically diabetic mice. Diabetes. 1983;32:781–787.

65. Kuo TH, Giacomelli F, Wiener J. Lapanowski-Netzel K. Pyruvate dehydrogenase activity in cardiac mitochondria from genetically diabetic mice. Diabetes. 1995;34:1075–1081.

66. Boudina S, Sena S, O'Neill BT, Tathireddy P, Young ME, Abel ED. Reduced mitochondrial oxidative capacity and increased mitochondrial uncoupling impair myocardial energetics in obesity. Circulation. 2005;112:2686–2695.

67. Buchanan J, Mazumder PK, Hu P, Chakrabarti G, Roberts MW, Yun UJ, Cooksey RC, Litwin SE, Abel ED. Reduced cardiac efficiency and altered substrate metabolism precedes the onset of hyperglycemia and contractile dysfunction in two mouse models of insulin resistance and obesity. Endocrinology. 2005;146:5341–5349.

68. How OJ, Aasum E, Severson DL, Chan WY, Essop MF, Larsen TS. Increased myocardial oxygen consumption reduces cardiac efficiency in diabetic mice. Diabetes. 2006;55:466–473.

69. Boudina S, Sena S, Theobald H, Sheng X, Wright JJ, Hu XX, Aziz S, Johnson JI, Bugger H, Zaha VG, et al. Mitochondrial energetics in the heart in obesity related diabetes: direct evidence for increased uncoupled respiration and activation of uncoupling protein. Diabetes. 2007;32:781–787.

70. Ye G, Metreveli NS, Ren J, Epstein PN. Metallothionein prevents diabetes-induced deficits in cardiomyocytes by inhibiting reactive oxygen species production. Diabetes. 2003;52:777–783.

71. Lowell BB, Shulman GI. Mitochondrial dysfunction and type 2 diabetes. Science. 2005;307:384–387.

72. Morino K, Petersen KF, Dufour S, Befroy D, Frattini J, Shatzkes N, Neschen S, White MF, Bilz S, Sono S, et al. Reduced mitochondrial density and increased IRS-1 serine phosphorylation in muscle of insulin-resistant offspring of type 2 diabetic parents. J. Clin. Invest. 2005;115:3587–3593.

73. Bruce CR, Hoy AJ, Turner N, Watt MJ, Allen TL, Carpenter K, Cooney GJ, Febbraio MA, Kraegen EW. Overexpression of carnitine palmitoyltransferase-1 in skeletal muscle is sufficient to enhance fatty acid oxidation and improve high-fat diet-induced insulin resistance. Diabetes. 2009;58:550–558.

74. Gnaiger E. Capacity of oxidative phosphorylation in human skeletal muscle: new perspectives of mitochondrial physiology. Int. J. Biochem. Cell Biol. 2009;41:1837–1845.

75. Koves TR, Ussher JR, Noland RC, Slentz D, Mosedale M, Ilkayeva O, Bain J, Stevens R, Dyck JR, Newgard CB, et al. Mitochondrial overload and incomplete fatty acid oxidation contribute to skeletal muscle insulin resistance. Cell Metab. 2008;7:45–56.

76. Hoehn KL, Hohnen-Behrens C, Cederberg A, Wu LE, Turner N, Yuasa T, Ebina Y, James DE. IRS1-independent defects define major nodes of insulin resistance. Cell Metab. 2008;7:421–433.

77. Anderson EJ, Lustig ME, Boyle KE, Woodlief TL, Kane DA, Lin CT, Price JW 3rd, Kang L, Rabinovitch PS, Szeto HH, et al. Mitochondrial H2O2 emission and cellular redox state link excess fat intake to insulin resistance in both rodents and humans. J. Clin. Invest. 2009;119:573–581.

78. Houstis N, Rosen ED, Lander ES. Reactive oxygen species have a causal role in multiple forms of insulin resistance. Nature. 2006;440:944–948.

79. Fisher-Wellman KH, Neufer PD. Linking mitochondrial bioenergetics to insulin resistance via redox biology. Trends Endocrinol. Metab. 2012;23:142–153.

80. Li Q, Spencer NY, Oakley FD, Buettner GR, Engelhardt JF. Endosomal Nox2 facilitates redox-dependent induction of NF-kappaB by TNF-alpha Antioxid. Redox Signal. 2009;11:1249–1263.

81. Morgan MJ, Liu ZG. Reactive oxygen species in TNF⊠-induced signalling and cell death. Mol. Cells. 2010;30:1–12.

82. James AM, Collins Y, Logan A, Murphy MP. Mitochondrial oxidative stress and the metabolic syndrome. Trends Endocrinol. Metab. 2012;23:429–434.

83. Johnson AM, Olefsky JM. The origins and drivers of insulin resistance. Cell. 2013;152:673–684.

84. Zhou R, Yazdi AS, Menu P, Tschopp J. A role for mitochondria in NLRP3 inflammasome activation. Nature. 2011;469:221–225.

85. Lane T, Flam B, Lockey R, Kolliputi N. TXNIP shuttling: themissing link between oxidative stress and inflammasome activation. Front. Physiol. 2013;4:50.

86. Shimada K, Crother TR, Karlin J, Dagvadorj J, Chiba N, Chen S, Ramanujan VK, Wolf AJ, Vergnes L, Ojcius DM, et al. Oxidized mitochondrial DNA activates the NLRP3 inflammasome during apoptosis. Immunity. 2012;36:401–414.

87. Vandanmagsar B, Youm YH, Ravussin A, Galgani JE, Stadler K, Mynatt RL, Ravussin E, Stephens JM, Dixit VD. The NLRP3 inflammasome instigates obesity induced inflammation and insulin resistance. Nat. Med. 2011;17:179–188.

88. Martin SD, McGee SL. The role of mitochondria in the aetiology of insulin resistance and type 2 diabetes. Biochim Biophys Acta. 2014 Apr;1840(4): 1303–1312.

89. Lee C, Yen K, Cohen P. Humanin: a harbinger of mitochondrial derived peptides? Trends Endocrinol. Metab. 2013;24:222–228.

90. Han DH, Hancock CR, Jung SR, Higashida K, Kim SH, Holloszy JO. Deficiency of the mitochondrial electron transport chain in muscle does not cause insulin resistance. PLoS One. 2011;12:e19739. doi: 10.1371/journal.pone.0019739.

91. Chen J, Gusdon AM, Mathews CE. Role of genetics in resistance to type 1 diabetes. Diabetes Metab. Res. Rev. 2011;27:849–853.

92. Raza H, Prabu SK, John A, Avadhani NG. Impaired mitochondrial respiratory functions and oxidative stress in streptozotocin-induced diabetic rats. Int. J. Mol. Sci. 2011;12:3133–3147.

93. Bugger H, Chen D, Riehle C, Soto J, Theobald HA, Hu XX, Ganesan B, Weimer BC, Abel ED. Tissue-specific remodeling of the mitochondrial proteome in type 1 diabetic Akita mice. Diabetes. 2009;58:1986–1997.

94. Liu HY, Yehuda-Shnaidman E, Hong T, Han J, Pi J, Liu Z, Cao W. Prolonged exposure to insulin suppresses-mitochondrial production in primary hepatocytes. J. Biol. Chem. 2009;284:14087–14095.

95. Adachi-Uehara N, Kato M, Nimura Y, Seki N, Ishihara A, Matsumoto E, Iwase K, Ohtsuka S,

Kodama H, Mizota A, et al. Up-regulation of genes for oxidative phosphorylation and protein turnover in diabetic mouse retina. Exp. Eye Res. 2006;83:849–857.

96. Tewari S, Santos JM, Kowluru RA. Damaged mitochondrial DNA replication system and the development of diabetic retinopathy Antioxid. Redox Signal. 2012;17:492–504.

97. Roy Chowdhury SK, Smith DR, Saleh A, Schapansky J, Marquez A, Gomes S, Akude E, Morrow D, Calcutt NA, Fernyhough P. Impaired adenosine monophosphate-activated protein kinase signalling in dorsal root ganglia neurons is linked to mitochondrial dysfunction and peripheral neuropathy in diabetes. Brain. 2012;135:1751–1766.

98. Monaco CMF, Hughes MC, Ramos SV, Varah NE, Lamberz C, Rahman FA, McGlory C, Tarnopolsky MA, Krause MP, Laham R, et al. Altered mitochondrial bioenergetics and ultrastructure in the skeletal muscle of young adults with type 1 diabetes. Diabetologia. 2018 Jun;61(6):1411–1423.

99. Befroy DE, Petersen KF, Dufour S, Mason GF, de Graaf RA, Rothman DL, Shulman GI. Impaired mitochondrial activity in the insulin-resistant offspring of patients with type 2 diabetes. N. Engl. J. Med. 2004;350:664–671.

100. Morino K, Petersen KF, Sono S, Choi CS, Samuel VT, Lin A, Gallo A, Zhao H, Kashiwagi A, Goldberg IJ, et al. Regulation of mitochondrial biogenesis by lipoprotein lipase in muscle of insulin-resistant offspring of parents with type 2 diabetes. Diabetes. 2012;61:877–887.

101. Ballinger SW, Shoffner JM, Hedaya EV, Trounce I, Polak MA, Koontz DA, Wallace DC. Maternally transmitted diabetes and deafness associated with a 10.4 kb mitochondrial DNA deletion. Nat. Genet. 1992;1:11–15.

102. Ott A, Stolk RP, van Harskamp F, Pols HAP, Hofman A, Breteler MMB. Diabetes mellitus and the risk of dementia: the Rotterdam Study. Neurology. 1999;53:1937–1942.

103. Crane PK, Walker R, Hubbard RA, Li G, Nathan DM, Zheng H, Haneuse S, Craft S, Montine TJ, Kahn SE, et al. Glucose levels and risk of dementia. N. Engl. J. Med. 2013;369:540–548.

104. Ott A, Stolk RP, Hofman A, van Harskamp F, Grobbee DE, Breteler MMB. Association of diabetes mellitus and dementia: the Rotterdam Study. Diabetologia. 1996;39:1392–1397.

105. Petersen KF, Befroy D, Dufour S, Dziura J, Ariyan C, Rothman DL, DiPietro L, Cline GW, Shulman GI. Mitochondrial dysfunction in the elderly: possible role in insulin resistance. Science. 2003;300:1140–1142.

106. Cadenas E, Davies KJ. Mitochondrial free radical generation, oxidative stress, and aging. Free Radic. Biol. Med. 2000;29:222–230.

107. Mattson MP. Pathways towards and away from Alzheimer's disease. Nature. 2004;430:631–639.

108. De Felice FG, Vieira MNN, Bomfim TR, Decker H, Velasco PT, Lambert MP, Viola KL, Zhao WQ, Ferreira ST, Klein WL. Protection of synapses against Alzheimer's-linked toxins: insulin signaling prevents the pathogenic binding of Abeta oligomers. Proc. Natl. Acad. Sci. U. S. A. 2009;106:1971–1976.

109. Picone P, Giacomazza D, Vetri V, Carrotta R, Militello V, San Biagio PL, Di Carlo M. Insulin-activated Akt rescues Ab oxidative stress-induced cell death by orchestrating molecular trafficking. Aging Cell. 2011;10:832–843.

110. Decker H, Jürgensen S, Adrover MF, Brito-Moreira J, Bomfim TR, Klein WL, Epstein AL, De Felice FG, Jerusalinsky D, Ferreira ST. N-methyl-D-aspartate receptors are required for synaptic targeting of Alzheimer's toxic amyloid-b peptide oligomers. J. Neurochem. 2010;115:1520–1529.

111. Sebastián D, Hernández-Alvarez MI, Segalés J, Sorianello E, Muñoz JP, Sala D, Waget A, Liesa M, Paz JC, Gopalacharyulu P, et al. Mitofusin 2 (Mfn2) links mitochondrial and endoplasmic reticulum function with insulin signaling and is essential for normal glucose homeostasis. Proc. Natl. Acad. Sci. U. S. A. 2012;109:5523–5528.

112. Yin F, Jiang T, Cadenas E. Metabolic triad in brain aging: mitochondria, insulin/IGF-1 signalling and JNK signalling. Biochem. Soc. Trans. 2013;41:101–105.

113. McKenzie M, Chiotis M, Pinkert CA, Trounce IA. Functional respiratory chain analyses in murid xenomitochondrial cybrids expose coevolutionary constraints of cytochrome b and nuclear subunits of complex III. Mol. Biol. Evol. 2003;20:1117–1124.

114. McKenzie M, Trounce I. Expression of Rattus norvegicus mtDNA in Mus musculus cells results in multiple respiratory chain defects. J. Biol. Chem. 2000;275:31514–31519.

115. McKenzie M, Trounce IA, Cassar CA, Pinkert CA. Production of homoplasmic xenomitochondrial mice. Proc. Natl. Acad. Sci. U. S. A. 2004;101:1685–1690.

116. Du XL, Edelstein D, Rossetti L, Fantus IG, Goldberg H, Ziyadeh F, Wu J, Brownlee M. Hyperglycemia-induced mitochondrial superoxide overproduction activates the hexosamine pathway and induces plasminogen activator inhibitor-1 expression by increasing Sp1 glycosylation. Proc. Natl. Acad. Sci. U. S. A. 2000;97:12222–12226.

117. Halvatsiotis PG, Turk D, Alzaid A, Dinneen S, Rizza RA, Nair KS. Insulin effect on leucine kinetics in type 2 diabetes mellitus. Diabetes Nutr. Metab. 2002;15:136–142.

118. Bonnard C, Durand A, Peyrol S, Chanseaume E, Chauvin M-A, Morio B, Vidal H, Rieusset J. Mitochondrial dysfunction results from oxidative stress in the skeletal muscle of diet-induced insulin-resistant mice. J. Clin. Invest. 2008;118:789–800.

119. Doar JW, Stamp TC, Wynn V, Audhya TK. Effects of oral and intravenous glucose loading in thyrotoxicosis. Studies of plasma glucose, free fatty acid, plasma insulin and blood pyruvate levels. Diabetes. 1969;18:633–639.

120. Hoenig M, Ferguson DC. Impairment of glucose tolerance in hyperthyroid cats. J. Endocrinol. 1989;121:249–251.

121. Dimitriadis G, Baker B, Marsh H, Mandarino L, Rizza R, Bergman R, Haymond M, Gerich J. Effect of thyroid hormone excess on action, secretion, and metabolism of insulin in humans. Am. J. Physiol. 1985;248:E593–E601.

122. Weinstein SP, Watts J, Haber RS. Thyroid hormone increases muscle/fat glucose transporter gene expression in rat skeletal muscle. Endocrinology. 1991;129:455–464.

123. Shen DC, Davidson MB, Kuo SW, Sheu WH. Peripheral and hepatic insulin antagonism in hyperthyroidism. J. Clin. Endocrinol. Metab. 1988;66:565–569.

124. Saran R, Li Y, Robinson B, Abbott KC, Agodoa LY, Ayanian J, Bragg-Gresham J, Balkrishnan R, Chen JL, Cope E, Eggers PW, et al. US renal data system 2015 annual data report: epidemiology of kidney disease in the United States. Am. J. Kidney Dis. 2016;67(3 Suppl 1):S1–S305. doi: 10.1053/j.ajkd.2015.12.014.

125. Collins AJ, Kasiske B, Herzog C, Chavers B, Foley R, Gilbertson D, Grimm R, Liu J, Louis T, Manning W, et al. Excerpts from the United States renal data system 2004 annual data report: atlas of end-stage renal disease in the United States. Am. J. Kidney Dis. 2005;45(1Suppl 1):A5–A7. doi: 10.1053/j.ajkd.2004.10.009.

126. Daehn IS. Glomerular endothelial cells stress and cross-talk with podocytes in the development of diabetic kidney disease. Front Med. (Lausanne). 2018 Mar;23 (5):76.

127. Costa PZ, Soares R. Neovascularization in diabetes and its complications. Unraveling the angiogenic paradox. Life Sci. 2013;92:1037–1045.

128. American Diabetes Association. Standards of medical care in diabetes-2018. Diabetes Care. 2018 Jan; 41(Supplement 1);S124–1238

129. Wanner C, Inzucchi SE, Lachin JM, Fitchett D, von Eynatten M, Mattheus M, Johansen OE, Woerle HJ, Broedl UC, Zinman B, et al. Empagliflozin and progression of kidney disease in type 2 diabetes. N. Engl. J. Med. 2016;375(4):323–334. doi: 10.1056/NEJMoa1515920.

130. Mann JFE, Orsted DD, Brown-Frandsen K, Marso SP, Poulter NR, Rasmussen S, Tornøe K, Zinman B, Buse JB, LEADER Steering Committee and Investigators. Liraglutide and renal outcomes in type 2 diabetes. N. Engl. J. Med. 2017;377(9):839–848. doi: 10.1056/NEJMoa1616011.

131. Nishikawa T, Edelstein D, Du XL, Yamagishi S, Matsumura T, Kaneda Y, Yorek MA, Beebe D, Oates PJ, Hammes HP, et al. Normalizing mitochondrial superoxide production blocks three pathways of hyperglycaemic damage. Nature. 2000;404(6779):787–790. doi: 10.1038/35008121.

132. Baynes JW. Role of oxidative stress in development of complications in diabetes. Diabetes. 1991;40(4):405–412. doi: 10.2337/diab.40.4.405.

133. Balaban RS, Nemoto S, Finkel T. Mitochondria, oxidants, and aging. Cell. 2005;120(4):483–495. doi: 10.1016/j.cell.2005.02.001.

134. Hallan S, Afkarian M, Zelnick LR, Kestenbaum B, Sharma S, Saito R, Darshi M, Barding G, Raftery D, Ju W, et al. Metabolomics and gene expression analysis reveal down-regulation of the citric acid (TCA) cycle in non-diabetic CKD patients. EBioMedicine. 2017;26:68–77. 10.1016/j.ebiom.2017.10.027.

135. Palmeira CM, Rolo AP, Berthiaume J, Bjork JA, Wallace KB. Hyperglycemia decreases mitochondrial function: the regulatory role of mitochondrial biogenesis. Toxicol Appl. Pharmacol. 2007;225(2):214–220. doi: 10.1016/j.taap.2007.07.015.

136. Wang W, Wang Y, Long J, Wang J, Haudek SB, Overbeek P, Chang BH, Schumacker PT, Danesh FR. Mitochondrial fission triggered by hyperglycemia is mediated by ROCK1 activation in podocytes and endothelial cells. Cell Metab. 2012;15(2):186–200. doi: 10.1016/j.cmet.2012.01.009.

137. Dugan LL, You YH, Ali SS, Diamond-Stanic M, Miyamoto S, DeCleves AE, Andreyev A, Quach T, Ly S, Shekhtman G, et al. AMPK dysregulation promotes diabetes-related reduction of superoxide and mitochondrial function. J. Clin. Invest. 2013;123(11):4888–4899. doi: 10.1172/JCI66218.

138. Sharma K, Karl B, Mathew AV, Gangoiti JA, Wassel CL, Saito R, Pu M, Sharma S, You YH, Wang L, et al. Metabolomics reveals signature of mitochondrial dysfunction in diabetic kidney disease. J. Am. Soc. Nephrol. 2013;24 (11):1901–1912. doi: 10.1681/ASN.2013020126.

139. Lonn E, Yusuf S, Hoogwerf B, Pogue J, Yi Q, Zinman B, Bosch J, Dagenais G, Mann JF, Gerstein HC, et al. Effects of vitamin E on cardiovascular and microvascular outcomes in high-risk patients with diabetes: results of the HOPE study and MICRO-HOPE substudy. Diabetes Care. 2002;25 (11):1919–1927. doi: 10.2337/diacare.25.11.1919.

140. de Zeeuw D, Akizawa T, Audhya P, Bakris GL, Chin M, Christ-Schmidt H, Goldsberry A, Houser M, Krauth M, Lambers Heerspink HJ, et al. Bardoxolone methyl in type 2 diabetes and stage 4 chronic kidney disease. N. Engl. J. Med. 2013;369(26):2492–2503. doi: 10.1056/NEJMoa1306033.

141. Galvan DL, Badal SS, Long J, Chang BH, Schumacker PT, Overbeek PA, Danesh FR. Real-time in vivo mitochondrial redox assessment confirms enhanced mitochondrial reactive oxygen species in diabetic nephropathy. Kidney Int. 2017;92(5):1282–1287. doi: 10.1016/j.kint.2017.05.015.

142. Hammes HP, Feng Y, Pfister F, Brownlee M. Diabetic retinopathy: targeting vasoregression. Diabetes. 2011 Jan;60 (1):9–16.

143. Fowler M. Microvascular and macrovascular complications of diabetes Clin. Diabetes. 2008;26:77–82.

144. Ye G, Metreveli NS, Donthi RV, Xia S, Xu M, Carlson EC, Epstein PN. Catalase protects cardiomyocyte function in models of type 1 and type 2 diabetes. Diabetes. 2004;53:1336–1343.

145. Finkel T. Signal transduction by reactive oxygen species. J. Cell Biol. 2011;194:7–15. doi: 10.1083/jcb.201102095.

146. Montaigne D, Marechal X, Coisne A, Debry N, Modine T, Fayad G, Potelle C, El Arid JM, Mouton S, Sebti Y, et al. Myocardial contractile dysfunction is associated with impaired mitochondrial function and dynamics in type 2 diabetic but not in obese patients. Circulation. 2014;130:554–564.

147. Shah MS, Brownlee M. Molecular and cellular mechanisms of cardiovascular disorders in diabetes. Circ. Res. 2016 May 27;118(11): 1808–1829.

148. Stevens MJ, Dayanikli F, Raffel DM, Allman KC, Sandford T, Feldman EL, Wieland DM, Corbett J, Schwaiger M. Scintigraphic assessment of regionalized defects in myocardial sympathetic innervation and blood flow regulation in diabetic patients with autonomic neuropathy. J. Am. Coll. Cardiol. 1998;31:1575–1584.

149. Stump CS, Short KR, Bigelow ML, Schimke JM, Nair KS. Effect of insulin on human skeletal muscle mitochondrial ATP production, protein synthesis, and mRNA transcripts, Proc. Natl. Acad. Sci. U. S. A. 2003;100:7996–8001.

150. Karakelides H, Asmann YW, Bigelow ML, Short KR, Dhatariya K, Coenen-Schimke J, Kahl J, Mukhopadhyay D, Nair KS. Effect of insulin deprivation on muscle mitochondrial ATP production and gene transcript levels in type 1 diabetic subjects. Diabetes. 2007;56:2683–2689.

151. Boirie Y, Short KR, Ahlman B, Charlton M, Nair KS. Tissue-specific regulation of mitochondrial and cytoplasmic protein synthesis rates by insulin. Diabetes. 2001;50:2652–2658.

152. Nair KS, Bigelow ML, Asmann YW, Chow LS, Coenen-Schimke JM, Klaus KA, Guo ZK, Sreekumar R, Irving BA. Asian Indians have enhanced skeletal muscle mitochondrial capacity to produce ATP in association with severe insulin resistance. Diabetes. 2008;57:1166–1175.

153. Zhou G, Myers R, Li Y, Chen Y, Shen X, Fenyk-Melody J, Wu M, Ventre J, Doebber T, Fujii N, et al. Role of AMP-activated protein kinase in mechanism of metformin action. J. Clin. Invest. 2001;108:1167–1174.

154. Foretz M, Hébrard S, Leclerc J, Zarrinpashneh E, Soty M, Mithieux G, Sakamoto K, Andreelli F, Viollet B. Metformin inhibits hepatic gluconeogenesis in mice independently of the LKB1/AMPK pathway via a decrease in hepatic energy state. J. Clin. Invest. 2010;120:2355–2369.

155. Rena G, Pearson ER, Sakamoto K. Molecular mechanism of action of metformin: old or new insights? Diabetologia. 2013;56:1898–1906.

156. Rodriguez MC, MacDonald JR, Mahoney DJ, Parise G, Beal MF, Tarnopolsky MA. Beneficial effects of creatine, CoQ10, and lipoic acid in mitochondrial disorders. Muscle Nerve. 2007 Feb;35(2): 235–242.

157. Klopstock T, Yu-Wai-Man P, Dimitriadis K, Rouleau J, Heck S, Bailie M, Atawan A, Chattopadhyay S, Schubert M, Garip A, et al. A randomized placebo-controlled trial of idebenone in Leber's hereditary optic neuropathy. Brain. 2011 Sep;134(Pt 9);():2677–2686.

158. Kerr DS. Review of clinical trials for mitochondrial disorders: 1997–2012. Neurotherapeutics. 2013 Apr;10(2):307–319. doi: 10.1007/s13311-013-0176-7.

159. Craven L, Tang MX, Gorman GS, De Sutter P, Heindryckx B. Novel reproductive technologies to prevent mitochondrial disease. Hum. Reprod. Update. 2017 Sep 1;23(5):501–519. doi: 10.1093/humupd/dmx018.

160. Parikh, S, Goldstein A, Karaa A, Koenig MK, Anselm I, Brunel-Guitton C, Christodoulou J, Cohen BH, Dimmock D, Enns GM, etal. Patient care standards for primary mitochondrial disease: a consensus statement from the Mitochondrial Medicine Society. Genet Med.. 2017 Dec;19(12). doi: 10.1038/gim.2017.107. Epub 2017 Jul 27

161. Karaa A, Goldstein A. The spectrum of clinical presentation, diagnosis, and management of mitochondrial forms of diabetes. Pediatr Diabetes. 2015 Feb;16(1):1–9. doi: 10.1111/pedi.12223.

21 Clinical Manifestation of Mitochondrial Disorders in Childhood

Martin Magner

Department of Paediatrics and Adolescent Medicine, General University Hospital and First Faculty of Medicine, Charles University in Prague, Ke Karlovu 2, 128 08, Prague 2, Czech Republic

CONTENTS

1 INTRODUCTION

Mitochondrial disorders is an umbrella term for group of clinically, biochemically and genetically variable metabolic disorders with the common link of impaired mitochondrial function of the oxidative phosphorylation system (OXPHOS) and pyruvate dehydrogenase complex. The clinical presentation and course of patients with mitochondrial syndromes are extremely diverse, even among patients or relatives with identical enzymatic or genetic defects. Unfortunately, factors influencing the development either of full syndromic illness or just a part of the full spectrum of the disease remain still hidden in the shadows. The range of symptoms of mitochondrial disorders is broad and includes almost all CNS functions, vision and hearing, heart and skeletal muscle, gastrointestinal tract, kidneys, endocrine glands and haematological changes. The presentation usually depends on generalized or tissue-specific decrease in ATP production. Some mitochondrial disorders affect a single organ (e.g., the eye in Leber hereditary optic neuropathy), but many involve multiple organ systems. Virtually any organ may be impaired, but the organs with the highest energetic demands are most frequently involved, including brain, muscle, heart and liver. Mitochondrial diseases may manifest at any age from birth until late-adulthood with acute manifestation or as a chronic progressive disease (Böhm et al., 2006).

Molecular genetics as a diagnostic tool for these diseases came into focus in the late 1980s. Two research groups reported mutations in mtDNA and linked them to specific clinical presentation. Anita Harding and coworkers at the Institute of Neurology in London, UK, reported on large numbers of mitochondria with smaller, deleted, non-functional genomes arising spontaneously in patients with eye-muscle paralysis. Wallace and colleagues described a maternally inherited point mutation in a human family with a history of with Leber's hereditary optic neuropathy in the same year (Holt et al., 1988; Wallace et al., 1988a; Schon et al., 1994). Following first pioneering works, researchers began to suppose that many other neurological and muscle diseases could be related to mtDNA deletions, single point mutations or rearrangements of mtDNA sequences. From 1989 to 2001, the rate of newly discovered pathogenic mutations in mtDNA was about 10 mutations per year. Until now, more than 275 disease-causing mtDNA mutations have been reported across the mitochondrial genome (Saneto and Sedensky, 2013; Alston et al., 2017). Moreover, the respiratory chain is under dual genomic control, and one should bear in mind that only about 15% of the mutations related to mitochondrial disorders come under the mitochondrial DNA mutations (DiMauro and Davidzon, 2005). The first nuclear mitochondrial gene mutation was described in 1995 (Bourgeron et al., 1995). Since discovery of this mutation in Succinate dehydrogenase A (SDHA), encoding a structural subunit of complex II, monumental progress in search for mitochondrial disease candidate among more than nuclear genes encoding mitochondrial proteins has been made. As the current number of proteins required for mitochondrial function is estimated stands at 1,158 in MitoCarta2.0 (Burki, 2016), the potential number of various mitochondrial disorder is immense. Generally, mtDNA mutations are more frequently described in adults, whereas mutations in nuclear genes are rather represented in paediatric population (DiMauro and Schon, 2003).

The dual involvement of the mitochondrial and nuclear genomes results in several inheritance models. In fact, all possible inheritance patterns can be found in mitochondrial patients (maternal, X linked, autosomal recessive, autosomal dominant, and *de novo* occurrence) (Craven et al., 2017). Exceptionally, paternal inheritance has also been proven in an adult patient harbouring a 2-bp deletion in the *ND2* gene, which resulted in decrease of complex I activity below 5% of normal in the patient's muscle (Schwartz and Vissing, 2002).

Mitochondrial disorders were previously considered as rare; however, despite the fact that the precise prevalence of mitochondria disorders is difficult to find, current epidemiological studies suggest otherwise. The minimum disease prevalence in adults is established to about 12.5 per 100,000 in adults and 4.7 per 100,000 in children (Skladal et al., 2003; Thorburn, 2004; Schaefer et al., 2008; Bannwarth et al., 2013; Gorman et al., 2015). The research team of Patrick Chinnery in the United Kingdom estimated population prevalence even of one in 200 for mtDNA point mutations (including small indel mutations) (Chinnery et al., 2012). The most frequent primary mtDNA disorders are caused by point mutations and large-scale mtDNA deletions. Whereas point mtDNA mutations are usually maternally inherited, large-scale mtDNA deletions typically arise *de novo* during embryonic development (Alston et al., 2017). Genetic and biochemical causes of mitochondrial disorders summarizes Table 1.

2   CLINICAL SYMPTOMS

As the mitochondria house the machinery necessary for production of ATP as the main energy source for the cell, body tissues requiring the most energy are most commonly affected. The diagnosis suspicion on mitochondrial disorders is thus initially raised by specific clinical manifestations involving multiple organ systems but predominantly presenting with neurologic and myopathic features. It can also cluster with additional diagnostic hallmarks such as lactic acidosis or other evidence of impaired mitochondrial respiratory chain function. Establishing the diagnosis of a specific inherited disorder that affects mitochondrial function should include particularly thorough examination of brain and sensomotory systems, skeletal muscle, heart, liver and kidney, and endocrine glands.

Many patients display an assemblage of clinical features that can be ascribed to a discrete clinical syndrome, such as the Kearns-Sayre syndrome (KSS), chronic progressive external ophthalmoplegia (CPEO) (Moraes et al., 1989), neurogenic weakness with ataxia and retinitis pigmentosa (NARP) (Holt et al., 1990), myoclonic epilepsy with ragged-red fibers (MERRF) (Hammans et al., 1993), or mitochondrial encephalomyopathy with lactic

TABLE 1

Summary of syndromic clinical phenotypes of mitochondrial disorders. CPEO = chronic progressive external ophthalmoplegia, KSS = Kearns-Sayre syndrome, LHON = Leber hereditary optic neuropathy; MELAS = mitochondrial encephalomyopathy with lactic acidosis and stroke-like episodes; MEMSA = myoclonic epilepsy myopathy sensory ataxia; MERRF = myoclonic epilepsy with ragged-red fibers; MIRAS = mitochondrial recessive ataxia syndrome; NARP = neurogenic weakness with ataxia and retinitis pigmentosa; SANDO = sensory ataxia neuropathy, dysarthria ophtalmoparesis; SCAE = spinocerebellar ataxia with epilepsy. Modified according to (Finsterer, 2006a; Saneto and Sedensky, 2013; Chinnery et al., 2012; Liang et al., 2014).

Clinical phenotype	Type of mutation typically associated with phenotype	Age at onset	Common clinical features	Additional Features
MELAS syndrome	tRNA point mutations: m.3243A>G, m.3271T>C, m. 3291T>C, m.13513G>A	Child or adulthood	Short stature Stroke-like episodes at age <40 years with clinical prodrome Seizures and/or dementia Ragged-red fibers and/or lactic acidosis Areas of high signal on T2 weighted MRI that do not conform to vascular territories	Diabetes mellitus Cardiomyopathy (initially hypertrophic; later dilated) Bilateral deafness Pigmentary retinopathy Cerebellar ataxia
MERRF syndrome	tRNA Point mutation: m.8344A>G, m. 8356T>C m.8363G>A	Child or adult	Myoclonus, spasticity Seizures, epilepsy Cerebellar ataxia Myopathy	Multiple lipomata in "collar" distribution Hearing loss Dementia Optic atrophy Peripheral neuropathy
MEMSA (also referred to as MIRAS, SANDO and SCAE; ataxia neuropathy syndromes)	Mutations in the catalytic subunit of the mitochondrial DNA polymerase γ (POLG)	Typical age of onset 13 to 25 years of age	Myopathy Seizures SANDO Sensory axonal neuropathy/ variable sensory & cerebellar ataxia	Dementia Peripheral neuropathy Spasticity
Mitochondrial neurogastrointestinal encephalopathy (MNGIE)	Mutations in *ECGF1* gene	Late childhood/ adolescence	Chronic progressive ophtalmoplegia, ptosis Gastrointestinal dysmotility Leukoencephalopathy	Peripheral neuropathy Myopathy
Pearson syndrome	mitochondrial DNA deletions leading to deficiency in mitochondrial respiratory chain function	Early childhood	Sideroblastic anemia of childhood Pancytopenia Exocrine pancreatic failure	Renal tubular defects
NARP	m.8993T>C/G heteroplasmic *de novo* c.8989G>C mutations in the subunit 6 of mitochondrial H(+)-ATPase gene	Late child or adult	Late-childhood or adult- onset peripheral neuropathy Ataxia Pigmentary retinopathy	Basal ganglia lucencies Abnormal electroretinogram Sensorimotor neuropathy
Kearns-Sayre syndrome/ CPEO	Single deletion Deletion/duplications (rare)	KSS: Later childhood CPEO: onset at age <20 years	Ptosis Pigmentary retinopathy Short stature Ataxia Proximal and bulbar weakness. One of the following: CSF protein >1g/L, cerebellar ataxia, heart block	Bilateral deafness Mild proximal myopathy Dysphagia Diabetes mellitus Hypoparathyroidism Dementia
Alpers-Huttenlocher syndrome	> 60 mutations in polymerase gamma gene The most common: trans compound heterozygote mutation within	Infancy and early childhood	Hypotonia Seizures Liver failure	Renal tubulopathy

(Continued)

TABLE 1 (Cont.)

Clinical phenotype	Type of mutation typically associated with phenotype	Age at onset	Common clinical features	Additional Features
	the linker region and polymerase region, p.A467T/p.W748S			
Leigh's disease (Leigh syndrome)	Point mutation in mtDNA for protein subunits	Infancy and early childhood	Bilateral characteristic basal ganglia lesions Psychomotor retardation, Seizures, movement disorders Lactic acidosis Subacute relapsing encephalopathy Infantile onset	Cerebellar and brain stem signs Maternal history of neurologic disease or Leigh syndrome
Leber's hereditary optic neuropathy (LHON)	Point mutation in protein subunit: m.11778G>A, m.3460G>A, m.14484T>C	Adult (third to fourth decade)	Rapidly progressive, usually sequential (subacute) painless bilateral visual loss commonly in young males Males:females ~4:1 Median age of onset 24 years	Dystonia Cardiac pre-excitation syndromes

acidosis and stroke-like episodes (MELAS) (Hirano et al., 1992). On the other hand, many other mitochondrial disorders do not clearly fit into one particular and recognizable syndrome category, and the spectrum of disease phenotypes can be broadly overlapping among several syndromes.

Despite the diversity of clinical phenotypes, there are specific "red-flags" that can provide clues to the classification of a mitochondrial disease. Independently on involvement of a mitochondrial or nuclear gene, in adults they include ophthalmological abnormalities (pigmentary retinopathy, ptosis, progressive external ophthalmoplegia, and optic atrophy), proximal myopathy and exercise intolerance, cardiomyopathy, sensorineural deafness, and diabetes mellitus (van den Ouweland et al., 1992; Liang et al., 2014). Central nervous system involvement embraces fluctuating encephalopathy, seizures, dementia, migraine, stroke-like episodes, ataxia, and spasticity; chorea can also be a prominent feature (Nelson et al., 1995). Endocrine abnormalities should be always considered in more detail. To explain, mitochondrial disorder associated diabetes is usually seen in younger and non-obese patients during periods of physiological stress (Gebhart et al., 1996; Shigemoto et al., 1998). In women, mid- and late pregnancy loss is a common but often unrecognized event (Tay et al., 2004).

Common clinical presentation of children is often different from adults. Typical manifestation includes failure to thrive, psychomotor regression, metabolic encephalopathy, seizures, ptosis, external ophthalmoplegia and cardiomyopathy (Liang et al., 2014). Some of these common syndromic clinical phenotypes of mitochondrial disease summarizes Table 2.

Findings that can suggest a mitochondrial disorder are summed up in Table 2.

2.1 NERVOUS SYSTEM

Neurons are heavily dependent on mitochondria and are especially vulnerable to deficits in energy generation. Neurological deficits are the most common features occurring in more than 90% of children with mitochondrial disease (Chi, 2015). They can also represent the largest contributor to morbidity and mortality of these patients. The range of neurological manifestations of mitochondrial disorders is extremely variable.

The CNS manifestations of mitochondrial disorders in children typically include seizures, developmental delay, altered level of consciousness, floppiness, spasticity, mental retardation, psychomotor delay or regression, sucking difficulty, involuntary movement, headache, external ocular motility limitation, tremor, apnea and apneustic respiration, dystonia, stroke, sudden infant death syndrome, and/or hypoventilation (Chi, 2015). Cognitive impairment in older children and adults involves nonverbal cognitive impairments and verbal short-term memory deficits, deficits in working memory, with hypometabolism in the temporal lobes on neuroradiological imaging (Turconi et al., 1999). In the older children in whom the psychological assessment is possible, all of abstract reasoning, verbal memory, visual memory, language (naming, fluency), executive or constructive functions, calculation, attention (attention deficit disorder, decreased attention span), or visuo-spatial functions may be impaired (Finsterer, 2006a).

TABLE 2

Genetic and biochemical classification of the mitochondrial diseases. According to (DiMauro, 2005; Liang et al., 2014).

Genome	Gene		Biochemistry	Clinical phenotype
mtDNA		single Δ	↓ protein synthesis	KSS; ocular myopathy; PS
	tRNA^Leu(UUR)	m.3243A>G	↓ protein synthesis	MELAS
		m.3271T>C m. 3291T>C		
		m.13513G>A		
	tRNA^Lys	m.8344A>G	↓ protein synthesis	MERRF
		m. 8356T>C		
		m.8363G>A		
	other tRNAs	position refers to the L-chain	↓ protein synthesis	multiple phenotypes
		m.3243A>G		CPEO
		m.4274T>C		
		m.14709T>C		myopathy
		m.12320A>G		
		m.3243A>G		cardiomyoptahy
		m.4269A>G		
		m.3243A>G		diabetes and deafness
		m.12258C>A		
		m.1606G>A		encephalomyopathy
		m.10010T>C		
		m.7445A>G		nonsyndromic sensorineural deafness
	rRNA	m.1555A>G		aminoglycoside-induced nonsyndromic deafness
	ND6	m.14484T>C	↓ complex I	primary LHON mutations
	ND4	m.11778G>A		
	ND1	m.3460G>A		
	ATPase6	m. 8993T > G	↓ ATP synthesis	NARP/MILS
	ND1, ND4		↓ complex II	myopathy
	Cyt b		↓ complex III	myopathy
	COX II		↓ complex IV	myopathy
nDNA	*NDUFS1 NDUFS4 NDUFS7 NDUFS8 NDUFV1*	genes encoding structural subunits	↓ complex I	Leigh syndrome with complex I deficiency
	SDHA		↓ complex II	Leigh syndrome with complex II deficiency
	SDHAF1		↓ succinate dehydrogenase complex assembly factor 1	leukodystrophy with complex II deficiency
	NDUFS2		↓ complex I	cardiomyopathy and encephalopathy
	UQCRB		↓ complex III	hypokalemia and lactic acidosis
	BCS1L	genes encoding assembly factors	↓ complex III	GRACILE
	SURF1		↓ complex IV	Leigh syndrome
	SCO1		↓ complex IV	hepatoencephalomyopathy
	SCO2		↓ complex IV	cardioencephalomyopathy
	COX10		↓ complex IV	nephroencephalomyopathy
	COX15		↓ complex IV	cardioencephalomyopathy
	ATP12		↓ complex V	fatal infantile multisystemic
	ATPAF2		↓ complex V	encephalopathy
	GFM1	genes encoding translation factors	↓ protein synthesis	Leigh syndrome, liver failure, and lactic acidosis
	MRPS16			lactic acidosis, developmental failure, and dysmorphism
	PUS1			myopathy and sideroblastic anaemia
	TUFM			leukodystrophy and polymicrogyria
	TACO1			Leigh syndrome and optic atrophy with COX deficienc
	TYMP	multiple Δ		MNGIE

(Continued)

TABLE 2 (Cont.)

Genome	Gene		Biochemistry	Clinical phenotype
	SLC25A4	multiple Δ	↓ protein synthesis	progressive external ophthalmoplegia (PEO)
	TWNK	multiple Δ	↓ protein synthesis	PEO, ataxia neuropathy spectrum
	TWNK	Y508C	leads to mtDNA depletion	infantile-onset spinocerebellar ataxia
	POLG	multiple Δ	↓ protein synthesis	autosomal progressive external ophthalmoplegia
	DGUOK	mtDNAdepletion	↓ enzyme activity	encephalomyopathy and liver failure
	TK2	mtDNA depletion	↓ enzyme activity	infantile myopathy/spinal muscular atrophy
	SUCLA2	mtDNA depletion	↓ enzyme activity	hypotonia, movement disorder, and/or Leigh syndrome with methylmalonic aciduria
	RRM2B	mtDNA depletion	impairment of oxidative phosphorylation	hypotonia, encephalopathy, renal tubulopathy, lactic acidosis
	GFER	R194H	↓ enzyme activity ↓ protein folding in the mitochondrial intermembrane space	progressive myopathy and partial combined respiratory-chain deficiency, congenital cataract, sensorineural hearing loss, and developmental delay
	TRMU	Y77H G272D	↓ activity of mitochondrial tRNA-modifying enzyme ↓ uridine thiolation	infantile, transient liver failure, elevated liver enzymes, jaundice, vomiting, coagulopathy, hyperbilirubinemia, increased serum lactate
	SLC25A3	G72E dbSNP:rs104894375	↓ enzyme activity ↓ phosphate transport into the mitochondrial matrix	mitochondrial phosphate carrier deficiency (MPCD) with lactic acidosis, hypertrophic cardiomyopathy and muscular hypotonia
	TMEM70	c.317-2A > G c.470T > A c.628A > C c.118_119insGT or c.251delC	↓ complex V	developmental delay, hypotonia, microcephaly, encephalopathy, cardiomyopathy
	TAZ (G4.5)	R94S C118R G197R G240R	↓ cardiolipin	Barth syndrome (proximal skeletal myopathy, growth retardation, neutropenia, and organic aciduria)
	OPA1	2826delT other deletions	↓ mit. motility	optic atrophy type 1

Abbreviations and symbols: Δ, deletion; KSS, Kearns–Sayre syndrome; PS, Pearson syndrome; MELAS, Mitochondrial encephalomyopathy, lactic acidosis and stroke-like episodes; LS, Leigh syndrome; MILS, maternally inherited LS; GRACILE, growth retardation, aminoaciduria, cholestasis, lactic acidosis, early death; CPEO; chronic progressive external ophthalmoplegia.

Epilepsy with seizures is a common phenotypic feature during the evolution of mitochondrial disorders (Finsterer and Zarrouk Mahjoub, 2012). 30–50% of children with mitochondrial disorders can suffer epileptic seizures (Scaglia et al., 2004; Desguerre et al., 2014). Epilepsy may start at infancy as infantile spasms, West syndrome, myoclonic jerks, astatic seizures, or myoclonic epilepsy. Electro-encephalogram (EEG) may document the generalized slow wave activity or the specific epileptiform discharges.

Seizure may be associated with stroke-like episodes. They can be best defined as an episode of focal cerebral metabolic crisis, increasing the energy requirements of the metabolically already compromised neuronal tissue (Lax et al., 2017). Stroke-like episodes are one of the main syndromic features of patients with MELAS; however, they may occur also in other mitochondrial disorders (Kaufman et al., 2010; Lax et al., 2017). During the development of stroke-like episode, patients might develop positive visual phenomena including phosphenes and moving objects, and the episode is often associated with a throbbing headache. Negative visual phenomena including homonymous hemianopia and cortical blindness can follow (Lax et al., 2017). Other symptoms associated with stroke-like episodes typically include focal weakness, paresthesia, speech disturbance, hemihypesthesia, or seizure. Stroke-like episodes are accompanied by difficultly treatable migrainous phenomena. Stroke-like episodes usually have a similar manifestation under the picture of posterior reversible encephalopathy syndrome (PRES). For PRES, headache, encephalopathy, seizures and visual disturbances with cerebral oedema in the occipital poles are typical (Graham and Pylypchuk, 2014). Using magnetic resonance imaging (MRI), stroke-like episodes are typically detected as T2-weighted hyperintensities in posterior brain regions (Lax et al., 2017).

Mitochondrial movement disorders, dysarthria, pyramidal signs, spasticity or hyperreflexia are often associated with degeneration of particular parts of the motor control pathways. Basal ganglia degeneration and cerebellar ataxia is common in both adult and paediatric patients (Scaglia et al., 2005; Lax et al., 2012; Bargiela et al., 2015). Ataxia is especially prevalent in patients harbouring many different mtDNA mutations; however, cerebellar changes are also reported in patients harbouring nuclear DNA (nDNA) mutations (e.g., POLG syndrome, where ataxia can represent one of the prevailing neurological features). It may also be prevalent in MEMSA (Mignarri et al., 2015), MELAS, MERRF, Leigh syndrome (Rahman et al., 1996), KSS, NARP, and Friedreich ataxia. Cerebellar atrophy or brainstem involvement can also manifest as more rare clinical features of dysphagia and nystagmus (Finsterer, 2006b).

The manifestation of psychiatric disturbances also occurs more frequently in patients with MD than in the general population. For example, in one group of 36 patients with MD and fatty acid oxidation disorders, the lifetime prevalence of psychiatric diagnosis was 69% (Fattal et al., 2007). Psychiatric comorbidities of mitochondrial disorders are highly prevalent in adults; however, children can also manifest a variety of psychiatric features. Psychiatric symptom (especially depression) may be even the first symptom of the mitochondrial disease. All of affective disorders with major depressive disorder and bipolar disorder, cognitive deterioration, psychosis, and anxiety, may be present, especially in MELAS or mitochondrial DNA deletions and other point mutations.

Neuroimaging provides useful information in the diagnosis of mitochondrial syndromes. Neuroimaging should be indicated in children with suspected CNS disease. In general, predominant cerebellar volume loss with progressive cerebellar atrophy, cerebellar hypoplasia, and widespread white matter hyperintensity, can be found. Cerebellar atrophy is a prominent feature in children (Barragan-Campos et al., 2005; Scaglia et al., 2005). The characteristic MRI finding associated with mitochondrial disorders in children is Leigh syndrome. It is characterized by vascular proliferation and demyelination, which lead to necrosis and cavitation in typical locations, including the basal ganglia, midbrain, pons, posterior column of the spinal cord, and characteristic putaminal involvement (Valanne et al., 1998). The finding of Leigh syndrom results from several defects in the mitochondrial oxidative phosphorylation system (OXPHOS), the most common involving cytochrome c oxidase (COX, complex IV) (e.g., SURF1 or SCO2 deficit), nicotinamide adenine dinucleotide (NADH): ubiquinone oxidoreductase (complex I), and the pyruvate dehydrogenase complex (PDHC). Additionaly, brain CT may show basal ganglia calcification.

Peripheral polyneuropathy (typically axonal sensorimotor polyneuropathy) is also often present.

2.2 VISION AND HEARING

The most common eye disorders in children include strabism, bilateral optic atrophy, ptosis, nystagmus and pigmental retinitis (Scaglia et al., 2004). Progressive external ophtalmoplegia (PEO), which is typical mainly for mtDNA depletion syndromes, is not often in children. Symptomatically, an early sign of retinitis pigmentosa leading to degeneration of the photoreceptor cells in the retina can manifest as night blindness. With progression, patients may complain of worsening peripheral vision, further prograding to central blindness (Listernick, 2007). Retinopathy can be attended by blurring, and diplopia occurs in patients with ophthalmoplegia (Richardson et al., 2005). However, blurring and double-vision are more usually transient and occur in association with generalized muscle fatigue.

Sensorineural hearing loss is detected in about 20% of the children with mitochondrial disorders (Scaglia et al., 2004). It is usually symmetrical and may be induced or aggravated by potentially ototoxic antibiotics of aminoglycosides group such as gentamicin, kanamycin, and streptomycin (Nye et al., 2000). Deafness may be present in patients with various syndromes: MELAS, DIDMOAD, MERRF, Mohr-Tranebjaerg or *SUCLA2* gene mutations (Carrozzo et al., 2007).

2.3 HEART

The most frequent cardiac manifestation is cardiomyopathy with minimal occurence of 20–30% of children with mitochondrial disorder (Yaplito-Lee et al., 2007). This frequency varies over age, being more frequent in neonatal and paediatric patients (40% out of 120 neonates in the study of Honzik et al., 2012, 58% in 113 children in the study of Scaglia et al., 2004) than in adults with various mtDNA or nDNA mutations (cardiac involvement in 30% out of 260 patients in the study of Wahbi et al., 2015). The most common is a hypertrophic cardiomyopathy (Lev et al., 2004; Yaplito-Lee et al., 2007; Bates et al., 2012), when up to 60% of patients can manifest with this particular cardiomyopathy form (Scaglia et al., 2004). Other cardiomyopathies include dilated, restrictive, histiocytoid or unclassified cardiomyopathy, or Takotsubo syndrome (Finsterer and Kothari, 2014). The differential diagnosis of cardiomyopathy as a hallmark of the disease is very broad includes TMEM70, Barth syndrome, Sengers syndrome, SCO2, AARS2, AIFM, TK2, DNAJC19, and many others.

The cardiac conductive system, the connective tissue, the valves, the aortic root, the coronary arteris or the pericardium can also be affected. Functional abnormalities include impulse generation or conduction abnormalities, systolic dysfunction leading eventually to heart failure, or affection of the autonomic cardiac nerves (Finsterer and

Kothari, 2014). Frequent electrocardiographic findings are atrioventricular (AV) block (Lee et al., 2001), and QT prolongation, Wolff-Parkinson-White (WPW), bundle branch block, ST and T-wave abnormalities, atrial fibrillation, and episodic ventricular tachyarrhythmias (Finsterer and Kothari, 2014).

2.4 SKELETAL MUSCLE

Skeletal muscle manifestations disorders resulting from primary dysfunction of the mitochondrial respiratory chain are among the most common presentation of mitochondrial disease (van Adel and Tarnopolsky, 2009). Mitochondrial myopathies symptoms can range from relatively non-specific exercise intolerance, resting muscle pain or exercise-induced myalgia to axial and predominantly proximal weakness. CK elevations are not always present (Pfeffer and Chinnery, 2013). All symptoms are exacerbated by exercise or inflammatory stress. The infant-onset mitochondrial myopathies can have a severe clinical presentation with hypotonia. It can mimic facioscapulohumeral muscle dystrophy (Filosto et al., 2008), or spinal muscular atrophy (SMA-like) (Magner et al., 2010).

Majority of paediatric patients with mitochondrial disorders (>90%) carry nDNA mutations in genes causing defective OXPHOS (Taylor et al., 2004). This explains highly prevalent absence of mosaic ragged red fibres (RRF) and/or COX-negative fibres in biopsies of these patients. RRF and/or COX-negative were observed only in 17% of biopsies without detectable mtDNA mutations in a study involving 117 children with mitochondrial diseases (Lamont et al., 1998).

2.5 GASTROINTESTINAL TRACT

Gastrointestinal tract manifestations present with disorders of peristalsis caused by affection of smooth muscle tissue, the autonomic nervous system, and the enteral neural plexus (Blondon et al., 2005; Amiot et al., 2009). Typical gastrointestinal symptoms include delayed gastric emptying with nausea and vomiting, the attacks of abdominal pain and discomfort, dysphagia, constipation, malabsorption, diarrhoea with potential villous atrophy of the small intestines, intestinal pseudoobstruction, hepatopathy with steatosis hepatis and liver insufficiency. The severe gastrointestinal manifestation is present in patients with myoneurogastrointestinal encephalopathy (MNGIE) syndrome. The combination of encephalopathy and hepatopathy is typical for Alpers-Huttenlocher syndrome. The syndromes with mtDNA depletion may present the malabsorption due to exocrine pancreatic insufficiency.

2.6 KIDNEYS

Kidneys as organs with high energetic demands may be also affected in mitochondrial disorders, although the kidney disease are rarely in the limelight of general clinical picture. Kidneys are affected frequently in children than in adults (Niaudet and Rotig, 1997; Emma and Salviati, 2017). Renal manifestations include polycystic kidneys, nonspecific nephritis, focal, segmental glomerulosclerosis, or tubular dysfunction, which may turn into chronic renal failure, event. requiring haemodialysis. Fanconi's syndrome may occur in an isolated form or as part of a multisystem disease. Mitochondrial syndromes associated with renal manifestations include the KSS, Pearson, DIDMOAD, and Leigh syndromes (Finsterer and Scorza, 2017). Occurrence of tubular acidosis, Bartter's syndrome, chronic tubulointerstitial nephritis, or nephritic syndrome is rare (Rotig, 2003).

2.7 ENDOCRINE GLANDS

Children with mitochondrial disorders may present with short stature, hypothyroidism, and hypogonadism. 2–3% of non-insulin dependent diabetes mellitus is given on account of OXPHOS dysfunction. Endocrinological symptoms are in the limelight of maternally inherited diabetes and deafness (MIDD), and diabetes insipidus, diabetes mellitus, optic atrophy, deafness (DIDMOAD or Wolfram) syndrome. Non-sense and missense germline mutations in complex II genes have been found in patients with familiar pheochromocytoma, with or without paraganglioma, and in sporadic pheochromocytoma (Astuti et al., 2004).

2.8 HAEMATOLOGICAL MANIFESTATION

Haematological disorders are long-known dominant or collateral features of mitochondrial disorders (Finsterer, 2007). They result from the involvement of mitochondria in haematopoiesis, particularly in oxidative phosphorylation processes, heme synthesis, iron metabolism, iron-sulfur [Fe-S] cluster biogenesis, and apoptosis (Fontenay et al., 2006). The most frequent haematological abnormalities of mitochondrial disorders include anaemia, thrombocytopenia, and leukopenia (Finsterer and Frank, 2015). Cyclic neutropenia is a typical finding in Barth syndrome. Sideroblastic macrocytic anemia as well as thrombocytopenia may be found in mitochondrial depletion syndrome as KSS, CPEO, and Pearson syndrome. Other mitochondrial disorders with predominant haematological manifestation are MLASA (autosomal recessive mitochondrial myopathy, lactic acidosis, and sideroblastic anaemia), XLSA (X-linked sideroblastic anaemia), XLSA-A (X-linked sideroblastic anaemia with ataxia) PUS1 deficiency, and COX10 deficiency (Finsterer and Frank, 2015; Tesarova, 2018). In single cases either permanent or recurrent eosinophilia can be observed, not attributable to any of the established causes.

3 A SHORT OVERVIEW OF SELECTED MITOCHONDRIAL DISORDERS

3.1 mtDNA Point Mutations

3.1.1 Mitochondrial Encephalomyopathy, Lactic Acidosis, and Stroke-Like Episodes (MELAS)

One of the most frequent maternally inherited mitochondrial disorders is MELAS, a condition resulting from one of more than 30 casual mutations of mtDNA. However, more than 80% of patients with MELAS have m.3243A>G point mutation of *MT-TL1* gene, which codes mitochondrial tRNA-Leu (UUA/UUG). It results in decreased stability of tRNA-Leu, thus leading to decreased mitochondrial proteosynthesis with combined OXPHOS deficiency. It was first delineated in 1984 (Pavlakis et al., 1984). Symptoms usual onset in children or young adults after normal early development. Almost one-half of affected individuals are diagnosed before age of 16 (juvenile form of the disease), neonatal presentation is very rare (Dvorakova et al., 2016).

Original clinical diagnostic criteria for MELAS syndrome were established in 1992. The MELAS clinical diagnosis was based on (1) stroke-like episodes before age 40 years, (2) encephalopathy characterized by seizures and/or dementia, and (3) mitochondrial myopathy evident by lactic acidosis and/or ragged-red fibers (RRFs). The diagnosis was considered confirmed if patient meets at least two of the following criteria: (1) normal early psychomotor development, (2) recurrent headaches, and (3) recurrent vomiting episodes (Hirano et al., 1992). However, since the original reports patients with milder or incomplete manifestations not fullfilling the mentioned diagnostic criteria have been broadly diagnosed. Many carriers of the disease causing mutation also remain asymptomatic. The median survival time based of fully symptomatic patients is significantly lower and death rate more than 17-fold higher if compared to unaffected carriers (Kaufmann et al., 2011).

Clinical symptoms of this disorder vary broadly – sensorineural hearing loss, ptosis, epilepsy, muscle weakness and pain, general myopathy, myalgias following exercise, cardiomyopathy, cerebellar symptomatology, severe headaches resembling migraines, repeated stroke-like episodes, and diabetes mellitus. Early symptoms in children, leading to MELAS suspicion, can include failure to thrive, short stature and psychomotoric development delay, particularly if other symptoms are seen in combination.

Up to one-half of patients with fully-expressed form of the disease present with stroke-like episodes. The episodes are marked by seizures, visual field sensory loss, at least partially reversible aphasia, hemianopsia, and they are preceded with severe headaches resembling migraines in 70% of patients. Transient hemiplegia and hemianopsia can ensue and last for several hours or weeks, with eventual progres and permament results. The MRI neuroimaging shows affected areas that do not correspond to the distribution of classic major vessels and are asymmetric, impair predominantly the cortical areas of temporal, parietal, and occipital lobes, and can involve subcortical white matter (El-Hattab et al., 2015). Cognitive deficit usually features as mild intellectual disability. Patients with chronic progressive external ophthalmoplegia (CPEO) usually present with progressive paralysis of external eye muscles resulting in limited sidewards or upwards gaze. Patients can suffer with skeletal muscle tiredness and weakness with exercise intoleration, and pain. Additional symptoms can also by psychiatric (Fattal et al., 2007).

The administration of L-arginine and/or citrulline during the acute and interictal periods is recommended. It can reduce impairement of vasodilation in intracerebral arteries owing to nitric oxide depletion (Dvorakova et al., 2016).

3.1.2 Myoclonus Epilepsy with Ragged Red Fibres (MERRF or MERF)

The first described human maternally inherited disease with mtDNA mutation was MERRF syndrome (Wallace et al., 1988b). Even more, MERRF was also the first disorder in which an epilepsy syndrome was matched with a molecular defect. The most common cause of MERRF is m.8344A>G mutation in the tRNA^{Lys} gene. More than 80% of the MERRF patients harbour this mutation (DiMauro and Hirano, 2015). A number of other mutations have been found in MERRF patients, mostly in genes encoding for mitochondrial tRNA^{Lys} (m. 8356T>C, m.8363G>A) or the tRNA^{Phe} (m.611363G>A) (Mancuso et al., 2004).

The disease onset is typical for the second or third decade of life (Altmann et al., 2016). The course of MERRF is usually slowly progressive. The major clinical manifestations of this common mitochondrial encephalomyopathy are myoclonus, generalized epilepsy, cerebellar ataxia, and ragged red fibres in muscles. Other presentation may include hearing loss, peripheral neuropathy, dementia, short stature, exercise intolerance, multiple lipomas in a cervical distribution, and optic atrophy (Mancuso et al., 2013; Finsterer et al., 2018). Ataxia and central nervous system involvement may dominate the phenotype in adults; on the other hand, it may manifest only half of the patients of older age. However, children and adolescents may initially manifest with epilepsy and the disease progression may be rapid with a fatal outcome (Catteruccia et al., 2015).

EMG reveals myogenic pattern. Hallmark biopsy finding in fibres stained with the modified Gomori trichrome stain is ragged red fibres (>90% of patients) (DiMauro et al., 2002).

3.1.3 Neuropathy, Ataxia, and Retinitis Pigmentosa (NARP)/Maternally Inherited Leigh Syndrome (MILS)

The NARP and MILS syndromes are part of a continuum of progressive neurodegenerative disorders caused by mutations at nt-8993 of the *ATPase6* gene (Holt et al., 1990). As for other mtDNA mutations, the phenotypic threshold of the m. 8993T>G mutation is relatively high. NARP occurs when mutation load reaches up to 70%, whereas when the degree of heteroplasmy is very high (about 90%), patients manifest with MILS (Tatuch et al., 1992). Thus, mutation is directly related to the clinical phenotype, an excellent example of the pathogenic importance of heteroplasmy level, not so clear in other diseases associated with mtDNA mutations (Uziel et al., 1997).

The clinical signs of NARP include proximal neurogenic muscle weakness with sensory neuropathy, ataxia, and pigmentary retinopathy. The syndrome usually affects young adults; however, symptoms onset is also common in early childhood. Children may manifest particularly with ataxia and learning difficulties (Thorburn et al., 2017). Maternally inherited Leigh syndrome (MILS or subacute necrotizing encephalomyelopathy) manifests as a more severe infantile encephalopathy with characteristic symmetrical lesions in the basal ganglia and the brainstem, and is often linked to a viral infection. Typically, a child decompensate during an intercurrent illness, which is accompanied with elevated lactate levels in blood and/or cerebrospinal fluid. Psychomotor retardation or regression follows, with neurologic features including hypotonia, spasticity, movement disorders (including chorea), cerebellar ataxia, and peripheral neuropathy. About half of patients die by age three years (Thorburn et al., 2017).

3.1.4 Leber's Hereditary Optic Neuropathy (LHON)

LHON is by far the most frequent mitochondrial disease affecting about 1:30 000 of the population (Yu-Wai-Man et al., 2010). Pathogenic homoplasmic point mutations of mitochondrial DNA at positions 11778G>A/*MTND4*, 3460G>A/*MTND1* and 14484T>C/*MTND6* decrease the complex I activity and increase the oxidative stress (Yu-Wai-Man et al., 2010). LHON belongs to a subgroup of optic neuropathise characterized by acute or subacute vision loss. It usually manifests in the second or third decade of life; however, it may occur at any age (Wallace et al., 1988a). Males are more affected than females, in an approximate 4–5:1 ratio (Yu-Wai-Man and Chinnery, 2016).

The disease is characterized by progressive and painless visual deterioration, usually resulting in irreversible blindness. Nonspecific symptoms, such as headache, flashes of light and blurring of vision may preceed (Riordan-Eva et al., 1995). In the vast majority of cases, visual loss becomes binocular, with involvement of the fellow eye either simultaneously or sequentially within 6–8 weeks. Visual function usually stabilizes in the chronic phase after the mean of 3.7 months. Visual acuity (0.1 or less) as well as colour discrimination and contrast sensitivity are severely altered; perimetry typically reveals central or caecocentral scotomas with relative sparing of the nasal quadrants (Riordan-Eva et al., 1995). Minority of patients may present associated features, such as cardiac conduction abnormalities, sensory and motor neuropathies, tremor, ataxia or less commonly basal ganglia lesions (LHON plus).

Therapeutic management of LHON is still limited, but recent studies found benefits from quinone analogs (Idebenone) enabling to bypass the complex I defect and exert the antioxidant effect if used during the early phase (Gueven et al., 2015). Avoidance of agents that may compromise mitochondrial function (smoking, alcohol and certain antibiotics) is highly recommended as is optimization of assistive devices and vision rehabilitation.

3.2 MITOCHONDRIAL DNA DELETIONS

3.2.1 Progressive External Ophtalmoplegia (PEO), Kearns-Sayre Syndrome (KSS) and Pearson Syndrome

Three clinical syndromes associated with single large-scale deletions of mtDNA are progressive external ophtalmoplegia (PEO), Kearns-Sayre syndrome (KSS) and Pearson syndrome. Approximately one-third of patients harbors a "common deletion" of about 5 kilobase (4977 bp), the other have 1.3–10 kb mtDNA deletions (Mancuso et al., 2015). A common deletion affects seven protein–encoding genes (four from complex I) and five tRNA genes are lost and is the most common. Large-scale duplications of mtDNA were also reported in some patients with KSS (Poulton et al., 1989).

Clinical presentation of ever-increasing number of patients points to the fact that these three syndromes actually are a continuum of clinical phenotypes, varying from the most severe manifestation with multisystem involvement (Pearson syndrome) to mild, organ-specific forms (isolated ptosis or PEO) (Mancuso et al., 2015). The overall variety of clinical manifestation is extremely broad, including in addition to myopathy exercise intolerance, hearing loss, muscle wasting, muscle pain, ataxia, retinopathy, failure to thrive, diabetes mellitus, migraine, cardiac conduction defects, cardiomyopathy, and hypothyreoidism. Muscle biopsy can reveal characteristic ragged red fibres and increased SDH staining. The diagnosis is usually carried out by detection of mtDNA deletions in buccal smear or cells of urine sediment. No causative therapy is available, patients with AV block profit from pacemaker implantation.

PEO has traditionally featured with ptosis and paralysis of the extraocular muscles (ophthalmoplegia). Some patients with PEO may have particular manifestations of KSS but do not fulfil all the clinical criteria for the diagnosis. Such a situation is termed "KSS minus" or "PEO plus." Kearns-Sayre syndrome (Kearns and Sayre, 1958) manifests with PEO, typical onset before age 20 years, pigmentary retinopathy, and cardiac conduction block or hyperproteinorhachia or ataxia. Other frequently present clinical features include cerebellar ataxia, cardiac conduction defects with conduction block, raised cerebrospinal fluid protein content, and proximal myopathy. Proximal or distal tubular acidosis may occur. If the patient survives into the fourth decade of life, bilateral sensorineural hearing loss is typical. Patients may more rarely manifest also with deafness, limb-girdle myopathy, multiple endocrinopathies and diabetes mellitus (about 13% of patients) (Harvey and Barnett, 1992), and dementia. Hypoparathyroidism and Addison disease may also be present (Berio and Piazzi, 2013). Funduscopy reveals an atypical "salt and pepper" retinopathy caused by rod-cone dystrophy. Retinal dystrophy is present in ERG, with normal visual fields. More than half of KSS patients have cardiac involvement, including recurrent syncope, bundle branch blocks, fascicular blocks and non-specific intraventricular conduction disturbance. Cardiac causes contribute to about 20% of deaths of KSS patients (Charles et al., 1981). The histopathological diagnosis of KSS is usually based on the presence of ragged red fibres with positive succinate dehydrogenase staining.

Pearson syndrome first descibed in 1979 (Pearson et al., 1979) was defined as paediatric refractory sideroblastic anaemia associated with exocrine pancreas dysfunction. Other features include early poor growth in infancy, and failure to thrive, and lactic acidosis. Children also present with common renal tubular dysfunction, more rarely with progressive external ophthalmoplegia, hepatopathy, insulin-dependent diabetes, myopathy, cardiomyopathy, dementia, seizures, and metabolic strokes (Broomfield et al., 2015). If children survive early childhood, they go on to develop the features of Kearns-Sayre syndrome.

3.3 Mitochondrial Disorders Caused by Nuclear DNA Mutations

Nuclear gene defects can be classified according to their function: defects of genes encoding structural components of the mitochondrial respiratory chain, genes encoding assembly proteins for the mitochondrial respiratory chain, genes involved in mitochondrial proteins, genes affecting the stability or maintenance of mtDNA, and genes indirectly involved in mitochondrial function.

3.3.1 Mitochondrial Neurogastrointestinal Encephalopathy (MNGIE)

MNGIE syndrome is an autosomal recessive progressive multisystem disease due to thymidine phosphorylase deficiency and by homozygous or compound heterozygous pathogenic mutations in *TYMP* gene. An excess of thymidine is damaging to mtDNA and causes mtDNA depletion, deletion, and point mutations (Taanman et al., 2009). Molecular genetic studies reveal multiple deletions of mitochondrial DNA (mtDNA) (Honzík et al., 2006).

Clinical symptoms usually start in adolescence with reccurent attacks of abdominal pain and gastrointestinal dysmotility and pseudoobstruction due to mitochondrial dysfunction of the intestinal smooth muscle. Patients manifest with nausea, dysphagia, vomiting, early satiety, gastroparesis, borborygmi, meteorism, diarrhoea, convulsive abdominal pain, and pseudo-obstruction syndrome resulting in cachexia with mean weight loss of 14 kg between the onset of first clinical symptoms and the time of diagnosis (Hirano, 2016). Gastrointestinal dysmotility is caused primarily by enteric myopathy, progresses slowly over several decades, may affect any part of the gastrointestinal tract (Giordano et al., 2008). Neurological symptoms are often mild in contrast to severe gastrointestinal problems. Most of patients gradually develop periferal neuropathy with paresthesias occur in a stocking-glove distribution, hearing loss, ptosis and external ophtalmoplegia, and glaucoma-like symptoms. Ultimately, almost all patients present with diffuse leukoencephalopathy visible on MRI which is usually asymptomatic (Hirano et al., 1994; Hirano, 2016). Laboratory findings include significantly increased CSF protein, lactic acidemia and common defects are in cytochrome c oxidase (complex IV).

There is no specific treatment; the mean survival is 38 years (26–58 years) (Hirano, 2016). Management for intestinal bacterial overgrowth includes administration of metronidazole and probiotics. Celiac plexus block or lysis was described to reduce abdominal pain and symtoms of pseudo-obstruction. The therapeutic potential of haematopoetic stem cell transplantation is also under consideration. Adequate nutrition is crucial, home parenteral nutrition is often started. MNGIE is a progressive disorder with a poor prognosis. The mean age of death is 35 years (Garone et al., 2011). Late onset of the disease can occur in some patients harbouring less severe TYMP pathogenic variants (Marti et al., 2005).

3.3.2 Leigh Syndrome Associated with Mitochondrial Complex I Deficiency (LS)

Leigh syndrome (Leigh, 1951) itself has two different meanings. The first represents the radiological or pathological findings of focal bilaterally symmetrical lesions, especially in the thalamus and brainstem regions. The other broadens this meaning to the clinical unit also known as subacute necrotizing encephalomyelopathy. Genetically, LS is very heterogenous and should be defined in by specific mutation or protein deficit where possible, as some particular may specifically differ in their clinical manifestation (e.g., SURF1 or pyruvatedehydrogenase complex deficiency). In general, LS may be caused by deficts of respiratory chain complex subunits (complex I, II, IV, and V) and their cofactors (e.g., coenzyme Q10), mutations in nDNA (e.g., SCO2, SURF1), mtDNA encoded tRNA, or the pyruvate dehydrogenase complex (Loeffen et al., 2000; Finsterer, 2008). Mitochondrial respiratory chain complex I (nicotinamide adenine dinucleotide-ubiquinone oxidoreductase) is the largest enzymatic complex of the mitochondrial respiratory chain. Defects in complex I due to nuclear DNA mutations are one of the most frequent casuses of LS. Various mutations in subunits of complex I encoded by nDNA (NDUFV1, NDUFV2, NDUFS1, NDUFS2, NDUFS3, NDUFS4, NDUFS7, and NDUFS8 were reported (Marin et al., 2013).

The hallmarks of LS are psychomotor retardation, central respiratory disturbance, nystagmus, ophthalmoparesis, optic atrophy, ataxia, and dystonia in infancy with typical onset between 3 and 12 months of age (DiMauro and Schon, 2003; Marin et al., 2013; Thorburn et al., 2017). The most common neurological presentations are developmental delay, seizures, and altered level of consciousness (Huntsman et al., 2005). Other clinical signs include abnormalities in tone, muscle weakness, movement disorders, tremor, peripheral neuropathy, bulbar symptoms such as dysarthria or dysphagia, and abnormalities of thermoregulation (Thorburn et al., 2017). Some patients can exhibit first symptoms of LS later in life, even in adulthood (Huntsman et al., 2005). The course of illness includes common decompensation during intercurrent illness with stable periods of incomplete recovery, sustained presentation or even progress (Thorburn et al., 2017). Most patients die within early childhood, about half of them by 3 years of age. The main causes of death are respiratory insufficiency or cardiac failure (Rahman et al., 1996).

3.3.3 Mitochondrial Complex IV Assembly Protein SURF1 Deficiency

SURF1 deficiency due to mutations in *SURF1* gene is an autosomal recessively inherited mitochondrial disorder. SURF1 is an assembly factor of mitochondrial complex IV and is an integral component of the mitochondrial inner membrane (Tiranti et al., 1998; Zhu et al., 1998). More than 100 SURF1 mutations have been reported so far (Li et al., 2018) and linked to Leigh syndrome due to cytochrome c oxidase deficiency

Clinical symptoms in most children with *SURF1* mutations usually start at the end of infancy with attacks of hyperventilation, hypotonia (93% of children), failure to thrive (95%), vomiting, growth retardation and delay of motoric development (88%). Strabism, ophthalmoplegia, hypertrichosis and regression of psychomotor development are usually present in the second year of life, the social communication is usually preserved a little longer (Fig. 1) (Wedatilake et al., 2013). Less frequent symptoms included optic atrophy (23%), seizures or cardiomyopathy (14%). The disease is progressive and eventually leads to respiratory failure. Median survival is about 6 years, only 16% of patients survive 10 years of age. Missense mutations in *SURF1* are rare, but they seem to be associated with *longer survival of patients with LS due to* mitochondrial complex IV deficiency. The disease onset is delayed and the develomepent of symptoms is slower in patients harbouring at least one missense mutation in the *SURF1* gene (Piekutowska-Abramczuk et al., 2009).

Lactate level in cerebrospinal fluid (CSF) is increased in most patients. Lactate level in blood is elevated and the ratio between lactate and pyruvate may be increased, but lactatemia may also be normal. EMG shows peripheral neuropathy in most children, ofter accompanied

FIGURE 1 Two-year-old girl with SURF1 deficiency. She manifested with growth disorder, ataxia, excessive hair growth, and lactic acidosis. Mental development may remain relatively spared for a long time even after the regression of gross motor development. Note the characteristic facial appearance.

with demyelination, less frequently of axonal form. The activity of COX is low in isolated lymphocytes, cultured fibroblasts and muscle biopsy (Piekutowska-Abramczuk et al., 2009), but the diagnosis is based on detection of pathogenic mutations in *SURF1* gene. The prognosis is poor, no casual therapy is available so far. The metabolic acidosis may be corrected with alkalinisation therapy.

3.3.4 Fatal Infantile Cardioencephalomyopathy due to SCO2 *Gene Mutations*

Mutations in the *SCO2* gene are the second most common cause of assembly disorders of COX (respiratory chain complex IV). SCO2 protein is a copper metallochaperone essential for the synthesis and maturation of cytochrome c oxidase subunit II. Mutation in SCO2 gene reduce copper transport or delivery to COX subunits I and II, which results in defective catalytic function of COX (Leary et al., 2004). Interestingly, a relatively high residual COX activity is seen in fibroblasts of the patients (Jaksch et al., 2000).

Clinical symptoms usually start in infancy with hypotonia, encephalopathy, hypertrophic cardiomyopathy and developmental delay. Some children have inspiratory stridor since neonatal period and phenotype resembling infantile spinal muscular atrophy (Magner et al., 2010). The onset of the disease is often precipitated by an acute febrile infection, anaesthesia or surgery in more than one half of patients. The course of the disease is progressive with developmental regress and respiratory failure (Pronicka et al., 2013).

Children with early onset of the disease may develop severe cardiomyopathy within the first month of life, some children with delayed onset and more attenuated

form of the disease can reach the stage of unassisted walking before the onset of regression period. In some children, ptosis, tremor, tongue fasciculation, opistotonus, dystonia, ataxia, abnormal eye movement, nystagmus, or strabismus can develop. Atrophy of optic nerve, hearing disorder and seizures are not present. Persistent fever of possible central origin is present in 25% of patients (Pronicka et al., 2013). Prognosis in children with disease onset in early infancy is poor; the mean age upon death is 11.7±7 months (Pronicka et al., 2013).

MRI of the brain shows progressive cerebral and cerebellar atrophy. Leigh-like changes can be observed in half of patients. Blood and CSF-lactate levels are increased. Muscle biopsy can reveal decreased COX activity, neuropathic changes with histological pattern resembling spinal muscular atrophy and the COX deficient fibres, but the diagnosis is based on detection of pathogenic mutations in *SCO2* gene.

3.3.5 Mitochondrial Complex V (ATP Synthase) Deficiency, Nuclear Type 2 Due to TMEM70 *Mutations*

The disease was first described in the Roma population, with neonatal onset of the disease and poor prognosis associated with homozygous mutation c.317-2A>G in *TMEM70* gene (Fig. 2.) (Cizkova et al., 2008). It represented then very unique finding of nuclear autosomal recessive mutation of OXPHOS, encoding ancillary factor of ATP synthase biogenesis. Patients usually present in utero or in the first days of life; with metabolic crisis with hypertrophic cardiomyopathy, cardiorespiratory failure lactic acidosis, hyperammonemia. Oligohydroamnion, intrauterine growth restriction (IUGR), facial dysmorphism and hypospadias in boys are also

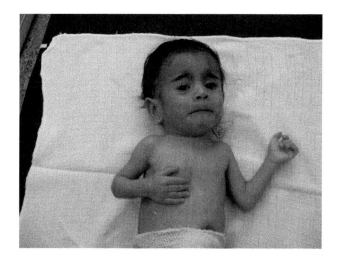

FIGURE 2 The TMEM70 deficiency was first described in the Roma population with prevalent c.317-2A>G mutation. Note the microcephaly, mild facial dysmorphy, low-set ears, and severe dystrophy despite adequate gastric tube feeding in this one-year-old Roma boy. The laboratory signs include lactic acidosis, hyperammonemic crises, and increased excretion of 3-methylglutaconic acid.

common. Laboratory findings reveal lactic acidosis and 3-methylglutaconic aciduria. Neonatal period is extremely critical for metabolic crisis management, when complicated with heart or respiratory failure and pulmonary hypertension. Metabolic failures with hyperammonemia can also result in encephalopathy and developmental regression. Common triggers of metabolic crisis can include surgery, acute febrile illness and acute gastroenteritis. The succesful management of acute metabolic crises is crucial for prognosis, developement and life expectancy of the child. Ten-year survival is 63%, with the most critical neonatal period (Magner et al., 2015). An anaplerotic therapy with supplementation of anaplerotic amino acids, lipids, and symptomatic treatment for affected children during metabolic crises was proposed as a promising treatment approach (Braczynski et al., 2015).

3.3.6 Defects of Mitochondrial DNA Polymerase Gamma Gene (POLG1)

The mitochondrial DNA polymerase gamma (POLG1) comprises of a C-terminal polymerase ("pol") domain and an amino-terminal exonuclease ("exo") domain. The exo domain affects mitochondrial DNA replication by conferring a proofreading activity to the enzyme. POLG1 mutations lead to mtDNA mutations, depletion, and decrease oxidative phosphorylation (Lamantea et al., 2002). Mutations in POLG1 cause a variety of neurological syndromes. These include:

3.3.6.1 MITOCHONDRIAL DNA DEPLETION SYNDROME 4A (ALPERS-HUTTENLOCHER SYNDROME)

This is a severe autosomal recessive hepatocerebral syndrome. Psychomotor development delay with developmental regression, intractable (often myoclonic) epilepsy, and liver failure in infants and young children are the hallmarks of the disease (Alpers, 1931). Combination of severe progressive cerebral atrophy with liver cirrhosis is typical. Patients usually die from hepatic failure or status epilepticus by the age of 3 years (Milone and Massie, 2010).

3.3.6.2 MITOCHONDRIAL DNA DEPLETION SYNDROME 4B (MNGIE TYPE)

It is an autosomal recessive inherited disorder clinically indistinguishable from that known as mitochondrial gastrointestinal encephalopathy (MNGIE) due to mutation in the TYMP gene (Vissing et al., 2002; Van Goethem et al., 2003). As distinct from MNGIE due to mutation in the TYMP gene, mitochondrial DNA depletion syndrome 4B patients show normal brain MRI without signs of leukoencephalopathy (Van Goethem et al., 2003; Giordano et al., 2009).

3.3.6.3 SPINOCEREBELLAR ATAXIA WITH EPILEPSY (SCAE) AND SENSORY ATAXIC NEUROPATHY, DYSARTHRIA, AND OPHTHALMOPARESIS (SANDO)

These are characterized by sensory ataxic neuropathy, dysarthria, and ophthalmoparesis. SANDO manifests mainly in adulthood with ataxic gait, loss of distal proprioception and vibration, areflexia in the lower limbs, positive Romberg sign, and electrophysiologic and pathologic evidence of a peripheral axonal neuropathy (Fadic et al., 1997). SCAE is a similar disorder with earlier onset in childhood or adolescence and higher frequency of migraine headaches and seizures, dysarthria, progressive external ophthalmoplegia, myoclonus, or status epilepticus (Winterthun et al., 2005).

3.3.6.4 AUTOSOMAL RECESSIVE PROGRESSIVE EXTERNAL OPHTHALMOPLEGIA WITH MITOCHONDRIAL DNA DELETIONS

They represents a severe and clinically more heterogeneous (Filosto et al., 2003; Luoma et al., 2004) is characterized by progressive external ophthalmoplegia, generalized muscle weakness, including the neck and facial muscles, increased serum creatine kinase, and areflexia. Patients may also manifest with distal sensory loss, abnormal nerve conduction studies, mitral valve prolapse, cardiomyopathy, and gastrointestinal dysmotility, and psychiatric symptoms.

3.3.7 Pyruvate Dehydrogenase Complex (PDHC) Defects

Pyruvate oxidation defects are represented not only by deficiency of protein subunits of the pyruvate dehydrogenase complex (PDHC) but also by disorders of PDHC regulation. The most common cause of PDHC deficiency are mutations in the PDHA1 gene, which encode the E1-alpha subunit (Cameron et al., 2004). Despite the fact that the PDHA1 gene is located on the X chromosome (Xp22.1), numbers of affected girls and boys are almost equal. This can be explained by different mutation spectrum in boys and girls (mutations occuring in girls can be lethal for boys), and by skewed X-inactivation (Cameron et al., 2004).

Clinical symptoms vary broadly. They represent continuous spectrum covering neonatal onset of the disease with poor prognosis, as well as milder forms with dystonia as a isolated disease manifestation in adulthood (Bachmann-Gagescu et al., 2009). The most common features are developemental delay, epilepsy, neurodevelopement disorders or cerebellar atrophy. Patients can also present with hypotonia, less commonly hypertonia, hearing and visual disorders and failure to thrive. Neurological manifestations include microcephaly, ataxia, neuropathy, ptosis, dystonia, dystonic hemiplegia and choreoathetoid movements. Epilepsy is particulary frequent, manifesting with spasm and focal seizures Respiratory dysfunction (Kussmaul breathing, respiratory failure) may occur, especially in children with neonatal-onset of the disease. Mild craniofacial dysmorphy is more frequent in boys than in girls, with narrow head, head frontal bossing, prominent philtrum, and a depressed flat nasal bridge (Sperl et al., 2015). Most patients have elevated lactate level in blood with decreased ratio between lactate and pyruvate. Leigh syndrome may be sometimes found also in milder forms of the disease (Magner et al., 2011). Diagnosis can be proved by measurement of PDHC activity in muscle tissue or cultivated fibroblasts, but the diagnosis is based upon molecular-genetic testing, next-generation sequencing technologies come into focus.

Ketogenic diet can stimulate fatty acid metabolism, provides an alternative energy source and supplements intermediates missing in PDHC (coenzyme A of Krebs cycle) (Gano et al., 2014). In patients with milder forms of the disease, some treatment effects were observed after fast carbohydrates restriction (improved muscle weakness and dystonic disorder) (Debray et al., 2006). A very small number of about 5% patients with mutations in the *PDHA1* gene are thiamine-responsive, however, initial therapy with thiamine (50 mg/kg/day) may be indicated especially for those presenting with a dystonic disorder. Dichloroacetate has been used but significant side effects, such as peripheral neuropathy, may limit effectiveness. The benefit of treatment combination of high dose of thiamine and biotin has been established in some patients with thiamine transporters deficiency (SLC19A3) (Subramanian et al., 2006). Metabolic acidosis can be partially compensated using alkalinisation therapy.

3.3.8 Barth Syndrome

Barth syndrome (BTHS) is an X-linked recessive disorder that is caused by mutations in Taffazin gene (*TAZ*). The protein defect causes an important reduction in incorporation of linoleic acid into the side chain of cardiolipin. Moreover, there is a decreased pool size of cardiolipin which indicates accelerated cardiolipin degradation leading to severe cardiolipin deficiency. Cardiolipin constitutes a component of the inner mitochondrial membrane, provides stability to the mitochondrial respiratory chain and its loss results in respiratory chain dysfunction. BTHS was first described by Barth et al. in 1983 in an extended pedigree with dilated cardiomyopathy (CMP), skeletal myopathy and neutropenia. The disease course can be variable, even within families. It occurs almost exclusively in males. Clinical features include dilated cardiomyopathy with endocardial fibroelastosis, predominantly proximal skeletal myopathy, growth retardation, neutropenia, and organic aciduria, and particular excess of 3-methylglutaconic acid. Patients may manifest with hypertrophic cardiomyopathy, isolated left ventricular non-compaction, ventricular arrhythmia, motor delay, poor appetite, fatigue and exercise intolerance, hypoglycemia, lactic acidosis, hyperammonemia (Ichida et al., 2001).

3.3.9 Dominant Optic Atrophy (DOA)

Nuclear gene *OPA1* encodes one of four large GTPases involved in mitochondrial membrane dynamics and cell homeostasis. Its mutations result in selective loss of retinal ganglion cells exhibiting the same susceptibility to energetic failure at the level of lamina cribrosa that is believed to occur in LHON. (Yu-Wai-Man et al., 2010). Most cases of DOA type 1 manifest in first two decades of life with painless bilateral visual deterioration with the development of generalized dyschromatopsia and central, caecocentral and paracentral scotomas that may remain small and stable over long periods of time. Compared to LHON, the disease progression in DOA type 1 is usually much slower with milder vision loss. Spontaneous visual recovery has never been described in DOA type 1. Oftentimes, DOA type 1 is discovered incidentally in asymptomatic individuals during routine vision testing or as a part of screening of family members of a proband (penetrance 84–88%). Extra-ocular neurological complications in DOA type 1 (DOA plus) have been described in up to 30% of carriers (Yu-Wai-Man et al., 2010) and they may involve sensorineural deafness, ataxia, myopathy, peripheral neuropathy, external ophthalmoplegia, spastic paraparesis and multiple sclerosis-like illness.

3.3.10 Friedreich's Ataxia

Friedreich's ataxia (FRDA) is autosomal-recessively disease results from mutations in gene encoding for frataxin protein. Deficient expression of frataxin leads to the accumulation of inorganic iron aggregates in the mitochondrial matrix, despairs cellular antioxidant defense and exacerbates oxidative stress (Chiang et al., 2016). Usual onset is between 5 and 15 years of age. It is characterized by progressive ataxia, loss of deep tendon reflexes and of vibration sense in the lower limbs, distal sensory loss, cerebellar dysarthria, and pyramidal signs with muscle weakness, and Babinski sign. Most patients show signs of hypertrophic cardiomyopathy and diabetes is present in about 30% cases. Skeletal deformities include scoliosis and pes cavus and are found in two-thirds of the patients (Calabrese et al., 2005).

3.3.11 Wolfram Disease (WFS)/DIDMOAD Syndrome

Wolfram syndrome is also referred to by the abbreviation DIDMOAD (diabetes insipidus, diabetes mellitus, optic atrophy, and deafness). Typical first manifestaion of this autosomal recessive neurodegenerative disease is onset of non-autoimmune juvenile-onset diabetes mellitus followed by optic atrophy, central diabetes insipidus and progressive high-frequency sensori-neural hearing loss (Barrett et al., 1995; Barrett and Bundey, 1997). Other neurodegenerative symptoms include cerebellar ataxia, bulbar signs, central apnea, autonomic and peripheral neuropathy, anhidrosis or hyperhidrosis, gastroparesis, hypothermia, and hyperpyrexia, hypo/anosmia or hypersomnolence, tremor, and seizures or myoclonus (Barrett et al., 1995; Urano, 2016). Other features may include cardiomyopathy, recurrent urinary tract infections, and hypothyroidism. Patients with WFS2 differ from classical WFS in lack of diabetes insipidus and a defective platelet aggregation pattern (Mozzillo et al., 2014). The prognosis is poor; patients usually die within first three decade of life. The main causes of death include neurological disabilities and central apnea related to brain stem atrophy (Barrett et al., 1995).

3.3.12 Mohr-Tranebjaerg Syndrome (MTS)

Mohr-Tranebjaerg syndrome, also known as deafness-dystonia-optic neuronopathy (DDON) syndrome (Mohr and Mageroy, 1960) is inherited in X-linked recessive pattern, with some female carriers also showing signs of minor neuropathy and mild hearing impairment. DDP (deafness dystonia peptide) protein is involved in the import of nuclear-encoded mitochondrial proteins into the inner mitochondrial membrane (Wallace and Murdock, 1999). Except of deafness, other symptoms may include visual disability leading to cortical blindness, focal, segmental, or multifocal dystonia primarily in the upper body and with progressive generalization, fractures, cortical atrophy predominantly in the parieto-occipital cortex, and mental deficiency (Tranebjaerg et al., 1995). The dystonia with age of onset ranging from the first to the fourth decade tends to be focal, segmental, or multifocal at onset with progressive generalization (Ujike et al., 2001). Another protein linked with MTS is translocase of mitochondrial inner membrane 8A (TIMM8A). Mutation in TIMM8A also underlies a distinct disorder called Jensen syndrome (Tranebjaerg et al., 2001).

4   CONCLUSION

Clinical manifestations of mitochondrial disorder are extremely complex. They affect, in addition to the tissues of the highest energy demand (central and peripheral nervous system, senses, muscle and heart), also visceral organs, such as endocrine organs, gastrointestinal tract, liver, kidneys, or haematopoietic system. This chapter focused on the interplay of the unique pathophysiology and genetic cause of mitochondrial diseases in the current knowledge of genetically governed mitochondrial disease.

ACKNOWLEDGMENT

The work was supported by project of Charles University in Prague UNCE 204064, AZV 17-30965A and RVO-VFN 64165/2012.

REFERENCES

Alpers, B. J. (1931). Diffuse progressive degeneration of gray matter of cerebrum. *Arch Neurol Psychiatry, 25*, 469–505.

Alston, C. L., Rocha, M. C., Lax, N. Z., Turnbull, D. M., & Taylor, R. W. (2017). The genetics and pathology of mitochondrial disease. *J Pathol, 241*(2), 236–250.

Altmann, J., Buchner, B., Nadaj-Pakleza, A., Schafer, J., Jackson, S., Lehmann, D., ... Klopstock, T. (2016). Expanded phenotypic spectrum of the m.8344A>G "MERRF" mutation: data from the German mitoNET registry. *J Neurol, 263*(5), 961–972.

Amiot, A., Tchikviladze, M., Joly, F., Slama, A., Hatem, D. C., Jardel, C., ... Lombes, A. (2009). Frequency of mitochondrial defects in patients with chronic intestinal pseudoobstruction. *Gastroenterology, 137*(1), 101–109.

Astuti, D., Morris, M., Krona, C., Abel, F., Gentle, D., Martinsson, T., ... Maher, E. R. (2004). Investigation of the role of SDHB inactivation in sporadic phaeochromocytoma and neuroblastoma. *Br J Cancer, 91*(10), 1835–1841.

Bachmann-Gagescu, R., Merritt, J. L., 2nd, Hahn, S. H. (2009). A cognitively normal PDH-deficient 18-year-old man carrying the R263G mutation in the PDHA1 gene. *J Inherit Metab Dis*, (Suppl 1), 1101-1104.

Bannwarth, S., Procaccio, V., Lebre, A. S., Jardel, C., Chaussenot, A., Hoarau, C., ... Paquis-Flucklinger, V. (2013). Prevalence of rare mitochondrial DNA mutations in mitochondrial disorders. *J Med Genet, 50*(10), 704–714.

Bargiela, D., Shanmugarajah, P., Lo, C., Blakely, E. L., Taylor, R. W., Horvath, R., ... Hadjivassiliou, M. (2015). Mitochondrial pathology in progressive cerebellar ataxia. *Cerebellum Ataxias, 2*, 16.

Barragan-Campos, H. M., Vallee, J. N., Lo, D., Barrera-Ramirez, C. F., Argote-Greene, M., Sanchez-Guerrero, J., ... Chiras, J. (2005). Brain magnetic resonance imaging findings in patients with mitochondrial cytopathies. *Arch Neurol, 62*(5), 737–742.

Barrett, T. G., & Bundey, S. E. (1997). Wolfram (DIDMOAD) syndrome. *J Med Genet, 34*(10), 838–841.

Barrett, T. G., Bundey, S. E., & Macleod, A. F. (1995). Neurode eneration and diabetes: UK nationwide study of Wolfram (DIDMOAD) syndrome. *Lancet, 346*(8988), 1458–1463.

Barth, P. G., Scholte, H. R., Berden, J. A., Van der Klei-Van Moorsel, J. M., Luyt-Houwen, I. E., ... Sobotka-Plojhar, M. A. (1983). An X-linked mitochondrial disease affecting cardiac muscle, skeletal muscle and neutrophil leucocytes. *J Neurol Sci, 62*, 327–355.

Bates, M. G., Nesbitt, V., Kirk, R., He, L., Blakely, E. L., Alston, C. L., ... McFarland, R. (2012). Mitochondrial respiratory chain disease in children undergoing cardiac transplantation: a prospective study. *Int J Cardiol, 155*(2), 305–306.

Berio, A., & Piazzi, A. (2013). Multiple endocrinopathies (growth hormone deficiency, autoimmune hypothyroidism and diabetes mellitus) in Kearns-Sayre syndrome. *Pediatr Med Chir, 35*(3), 137–140.

Blondon, H., Polivka, M., Joly, F., Flourie, B., Mikol, J., & Messing, B. (2005). Digestive smooth muscle mitochondrial myopathy in patients with mitochondrial-neuro-gastrointestinal encephalomyopathy (MNGIE). *Gastroenterol Clin Biol, 29*(8–9), 773–778.

Böhm, M., Pronicka, E., Karczmarewicz, E., Pronicki, M., Piekutowska-Abramczuk, D., Sykut-Cegielska, J., ... Zeman, J. (2006). Retrospective, multicentric study of 180 children with cytochrome C oxidase deficiency. *Pediatr Res., 59*(1), 21–26.

Bourgeron, T., Rustin, P., Chretien, D., Birch-Machin, M., Bourgeois, M., Viegas-Pequignot, E., … Rotig, A. (1995). Mutation of a nuclear succinate dehydrogenase gene results in mitochondrial respiratory chain deficiency. *Nat Genet, 11*(2), 144–149.

Braczynski, A. K., Vlaho, S., Muller, K., Wittig, I., Blank, A. E., Tews, D. S., … Mittelbronn, M. (2015). ATP synthase deficiency due to TMEM70 mutation leads to ultrastructural mitochondrial degeneration and is amenable to treatment. *Biomed Res Int, 2015*, 462592.

Broomfield, A., Sweeney, M. G., Woodward, C. E., Fratter, C., Morris, A. M., Leonard, J. V., … Rahman, S. (2015). Paediatric single mitochondrial DNA deletion disorders: an overlapping spectrum of disease. *Inherit Metab Dis, 38*(3), 445–457.

Burki, F. (2016). Mitochondrial evolution: going, going, gone. *Curr Biol, 26*(10), R410–R412.

Calabrese, V., Lodi, R., Tonon, C., D'Agata, V., Sapienza, M., Scapagnini, G., … Butterfield, D. A. (2005). Oxidative stress, mitochondrial dysfunction and cellular stress response in Friedreich's ataxia. *J Neurol Sci, 233*(1–2), 145–162.

Cameron, J. M., Levandovskiy, V., Mackay, N., Tein, I., Robinson, B. H. (2004). Deficiency of pyruvate dehydrogenase caused by novel and known mutations in the E1alpha subunit. *Am J Med Genet A, 131*, 59–66.

Carrozzo, R., Dionisi-Vici, C., Steuerwald, U., Lucioli, S., Deodato, F., Di Giandomenico, S., … Wevers, R. A. (2007). SUCLA2 mutations are associated with mild methylmalonic aciduria, Leigh-like encephalomyopathy, dystonia and deafness. *Brain, 130*(Pt 3), 862–874.

Catteruccia, M., Sauchelli, D., Della Marca, G., Primiano, G., Cuccagna, C., Bernardo, D., … Servidei, S. (2015). "Myocardiomyopathy" is commonly associated with the A8344G "MERRF" mutation. *J Neurol, 262*(3), 701–710.

Charles, R., Holt, S., Kay, J. M., Epstein, E. J., & Rees, J. R. (1981). Myocardial ultrastructure and the development of atrioventricular block in Kearns-Sayre syndrome. *Circulation, 63*(1), 214–219.

Chi, C. S. (2015). Diagnostic approach in infants and children with mitochondrial diseases. *Pediatr Neonatol, 56*(1), 7–18.

Chiang, S., Kovacevic, Z., Sahni, S., Lane, D. J., Merlot, A. M., Kalinowski, D. S., … Richardson, D. R. (2016). Frataxin and the molecular mechanism of mitochondrial iron-loading in Friedreich's ataxia. *Clin Sci (Lond), 130*(11), 853–870.

Chinnery, P. F., Elliott, H. R., Hudson, G., Samuels, D. C., & Relton, C. L. (2012). Epigenetics, epidemiology and mitochondrial DNA diseases. *Int J Epidemiol, 41*(1), 177–187.

Cizkova, A., Stranecky, V., Mayr, J. A., Tesarova, M., Havlickova, V., Paul, J., … Kmoch, S. (2008). TMEM70 mutations cause isolated ATP synthase deficiency and neonatal mitochondrial encephalocardiomyopathy. *Nat Genet, 40*(11), 1288–1290.

Craven, L., Alston, C. L., Taylor, R. W., & Turnbull, D. M. (2017). Recent Advances in Mitochondrial Disease. *Annu Rev Genomics Hum Genet, 18*, 257–275.

Debray, F. G., Lambert, M., Vanasse, M., Decarie, J. C., Cameron, J., Levandovskiy, V., … Mitchell, G. A. (2006). Intermittent peripheral weakness as the presenting feature of pyruvate dehydrogenase deficiency. *Eur J Pediatr, 165*, 462–466.

Desguerre, I., Hully, M., Rio, M., & Nabbout, R. (2014). Mitochondrial disorders and epilepsy. *Rev Neurol (Paris), 170*(5), 375–380.

DiMauro, S., & Davidzon, G. (2005). Mitochondrial DNA and disease. *Ann Med, 37*(3), 222–232.

DiMauro, S., & Hirano, M. (2015). MERRF. In M. P. Adam, H. H. Ardinger, R. A. Pagon, S. E. Wallace, L. J. H. Bean, K. Stephens & A. Amemiya (eds.), *GeneReviews((R))*. Seattle, WA.

DiMauro, S., Hirano, M., Kaufmann, P., Tanji, K., Sano, M., Shungu, D. C., … DeVivo, D. C. (2002). Clinical features and genetics of myoclonic epilepsy with ragged red fibers. *Adv Neurol, 89*, 217–229.

DiMauro, S., & Schon, E. A. (2003). Mitochondrial respiratory-chain diseases. *N Engl J Med, 348*(26), 2656–2668.

Dvorakova, V., Kolarova, H., Magner, M., Tesarova, M., Hansikova, H., Zeman, J., & Honzik, T. (2016). The phenotypic spectrum of fifty Czech m.3243A>G carriers. *Mol Genet Metab, 118*(4), 288–295.

El-Hattab, A. W., Adesina, A. M., Jones, J., & Scaglia, F. (2015). MELAS syndrome: clinical manifestations, pathogenesis, and treatment options. *Mol Genet Metab, 116*(1–2), 4–12.

Emma, F., & Salviati, L. (2017). Mitochondrial cytopathies and the kidney. *Nephrol Ther, 13*(Suppl 1), S23–S28.

Fadic, R., Russell, J. A., Vedanarayanan, V. V., Lehar, M., Kuncl, R. W., & Johns, D. R. (1997). Sensory ataxic neuropathy as the presenting feature of a novel mitochondrial disease. *Neurology, 49*(1), 239–245.

Fattal, O., Link, J., Quinn, K., Cohen, B. H., Franco, K. (2007). Psychiatric comorbidity in 36 adults with mitochondrial cytopathies. *CNS Spectr, 12*(6), 429–438.

Filosto, M., Mancuso, M., Nishigaki, Y., Pancrudo, J., Harati, Y., Gooch, C., … DiMauro, S. (2003). Clinical and genetic heterogeneity in progressive external ophthalmoplegia due to mutations in polymerase gamma. *Arch Neurol, 60*(9), 1279–1284.

Filosto, M., Tonin, P., Scarpelli, M., Savio, C., Greco, F., Mancuso, M., … Tomelleri, G. (2008). Novel mitochondrial tRNA Leu(CUN) transition and D4Z4 partial deletion in a patient with a facioscapulohumeral phenotype. *Neuromuscul Disord, 18*(3), 204–209.

Finsterer, J. (2006a). Central nervous system manifestations of mitochondrial disorders. *Acta Neurol Scand, 114*(4), 217–238.

Finsterer, J. (2006b). Overview on visceral manifestations of mitochondrial disorders. *Neth J Med, 64*(3), 61–71.

Finsterer, J. (2007). Hematological manifestations of primary mitochondrial disorders. *Acta Haematol, 118*(2), 88–98.

Finsterer, J. (2008). Leigh and Leigh-like syndrome in children and adults. *Pediatr Neurol, 39*(4), 223–235.

Finsterer, J., & Frank, M. (2015). Haematological abnormalities in mitochondrial disorders. *Singapore Med J, 56*(7), 412–419.

Finsterer, J., & Kothari, S. (2014). Cardiac manifestations of primary mitochondrial disorders. *Int J Cardiol, 177*(3), 754–763.

Finsterer, J., & Scorza, F. A. (2017). Renal manifestations of primary mitochondrial disorders. *Biomed Rep, 6*(5), 487–494.

Finsterer, J., & Zarrouk Mahjoub, S. (2012). Epilepsy in mitochondrial disorders. *Seizure, 21*(5), 316–321.

Finsterer, J., Zarrouk-Mahjoub, S., & Shoffner, J. M. (2018). MERRF classification: implications for diagnosis and clinical trials. *Pediatr Neurol, 80*, 8–23.

Fontenay, M., Cathelin, S., Amiot, M., Gyan, E., & Solary, E. (2006). Mitochondria in hematopoiesis and hematological diseases. *Oncogene, 25*(34), 4757–4767.

Gano, L. B., Patel, M., Rho, J. M. (2014). Ketogenic diets, mitochondria, and neurological diseases. *J Lipid Res, 55*(11), 2211–2228.

Garone, C., Tadesse, S., & Hirano, M. (2011). Clinical and genetic spectrum of mitochondrial neurogastrointestinal encephalomyopathy. *Brain, 134*(Pt 11), 3326–3332.

Gebhart, S. S., Shoffner, J. M., Koontz, D., Kaufman, A., & Wallace, D. (1996). Insulin resistance associated with maternally inherited diabetes and deafness. *Metabolism, 45*(4), 526–531.

Giordano, C., Powell, H., Leopizzi, M., De Curtis, M., Travaglini, C., Sebastiani, M., … d'Amati, G. (2009). Fatal congenital myopathy and gastrointestinal pseudo-obstruction due to POLG1 mutations. *Neurology, 72*(12), 1103–1105.

Giordano, C., Sebastiani, M., De Giorgio, R., Travaglini, C., Tancredi, A., Valentino, M. L., … Carelli, V. (2008). Gastrointestinal dysmotility in mitochondrial neurogastrointestinal encephalomyopathy is caused by mitochondrial DNA depletion. *Am J Pathol, 173*(4), 1120–1128.

Gorman, G. S., Schaefer, A. M., Ng, Y., Gomez, N., Blakely, E. L., Alston, C. L., … McFarland, R. (2015). Prevalence of nuclear and mitochondrial DNA mutations related to adult mitochondrial disease. *Ann Neurol, 77*(5), 753–759.

Graham, B. R., & Pylypchuk, G. B. (2014). Posterior reversible encephalopathy syndrome in an adult patient undergoing peritoneal dialysis: a case report and literature review. *BMC Nephrol, 15*, 10.

Gueven, N., Woolley, K., Smith, J. (2015). Border between natural product and drug: comparison of the related benzoquinones idebenone and coenzyme Q10. *Redox Biol, 4*, 289–295.

Hammans, S. R., Sweeney, M. G., Brockington, M., Lennox, G. G., Lawton, N. F., Kennedy, C. R., … Harding, A. E. (1993). The mitochondrial DNA transfer RNA(Lys)A–>G (8344) mutation and the syndrome of myoclonic epilepsy with ragged red fibres (MERRF). Relationship of clinical phenotype to proportion of mutant mitochondrial DNA. *Brain, 116*(Pt 3), 617–632.

Harvey, J. N., & Barnett, D. (1992). Endocrine dysfunction in Kearns-Sayre syndrome. *Clin Endocrinol (Oxf), 37*(1), 97–103.

Hirano, M. (2016). Mitochondrial Neurogastrointestinal Encephalopathy Disease. In M. P. Adam, H. H. Ardinger, R. A. Pagon, S. E. Wallace, L. J. H. Bean, K. Stephens & A. Amemiya (eds.), *GeneReviews((R))*. Seattle, WA.

Hirano, M., Ricci, E., Koenigsberger, M. R., Defendini, R., Pavlakis, S. G., DeVivo, D. C., … Rowland, L. P. (1992). Melas: an original case and clinical criteria for diagnosis. *Neuromuscul Disord, 2*(2), 125–135.

Hirano, M., Silvestri, G., Blake, D. M., Lombes, A., Minetti, C., Bonilla, E., … DiMauro, S. (1994). Mitochondrial neurogastrointestinal encephalomyopathy (MNGIE): clinical, biochemical, and genetic features of an autosomal recessive mitochondrial disorder. *Neurology, 44*(4), 721–727.

Holt, I. J., Harding, A. E., & Morgan-Hughes, J. A. (1988). Deletions of muscle mitochondrial DNA in patients with mitochondrial myopathies. *Nature, 331*(6158), 717–719.

Holt, I. J., Harding, A. E., Petty, R. K., & Morgan-Hughes, J. A. (1990). A new mitochondrial disease associated with mitochondrial DNA heteroplasmy. *Am J Hum Genet, 46*(3), 428–433.

Honzík, T., Tesarová, M., Hansíková, H., Krijt, J., Benes, P., Zámecník, J., … Zeman, J. (2006). Mitochondrial neurogastrointestinal encephalomyopathy (MNGIE). *Cas Lek Cesk, 145*, 665–670.

Honzik, T., Tesarova, M., Magner, M., Mayr, J., Jesina, P., Vesela, K., … Zeman J. (2012). Neonatal onset of mitochondrial disorders in 129 patients: clinical and laboratory characteristics and a new approach to diagnosis. *J Inherit Metab Dis, 35*(5), 749–759.

Huntsman, R. J., Sinclair, D. B., Bhargava, R., & Chan, A. (2005). Atypical presentations of leigh syndrome: a case series and review. *Pediatr Neurol, 32*(5), 334–340.

Ichida, F., Tsubata, S., Bowles, K. R., Haneda, N., Uese, K., Miyawaki, T., … Towbin, J. A. (2001). Novel gene mutations in patients with left ventricular noncompaction or Barth syndrome. *Circulation, 103*(9), 1256–1263.

Jaksch, M., Ogilvie, I., Yao, J., Kortenhaus, G., Bresser, H. G., Gerbitz, K. D., & Shoubridge, E. A. (2000). Mutations in SCO2 are associated with a distinct form of hypertrophic cardiomyopathy and cytochrome c oxidase deficiency. *Hum Mol Genet, 9*(5), 795–801.

Kaufman, K. R., Zuber, N., Rueda-Lara, M. A., & Tobia, A. (2010). MELAS with recurrent complex partial seizures, nonconvulsive status epilepticus, psychosis, and behavioral disturbances: case analysis with literature review. *Epilepsy Behav, 18*(4), 494–497.

Kaufmann, P., Engelstad, K., Wei, Y., Kulikova, R., Oskoui, M., Sproule, D. M., … De Vivo, D. C. (2011). Natural history of MELAS associated with mitochondrial DNA m.3243A>G genotype. *Neurology, 77*(22), 1965–1971.

Kearns, T. P., & Sayre, G. P. (1958). Retinitis pigmentosa, external ophthalmophegia, and complete heart block: unusual syndrome with histologic study in one of two cases. *AMA Arch Ophthalmol, 60*(2), 280–289.

Kozak, I., Oystreck, D. T., Abu-Amero, K. K., Nowilaty, S. R., Alkhalidi, H., Elkhamary, S. M., … Bosley, T. M. (2018). New observations regarding the retinopathy of genetically confirmed Kearns-Sayre syndrome. *Retin Cases Brief Rep, 12*(4), 349–358.

Lamantea, E., Tiranti, V., Bordoni, A., Toscano, A., Bono, F., Servidei, S., … Zeviani, M. (2002). Mutations of mitochondrial DNA polymerase gammaA are a frequent cause of autosomal dominant or recessive progressive external ophthalmoplegia. *Ann Neurol, 52*(2), 211–219.

Lamont, P. J., Surtees, R., Woodward, C. E., Leonard, J. V., Wood, N. W., & Harding, A. E. (1998). Clinical and laboratory findings in referrals for mitochondrial DNA analysis. *Arch Dis Child, 79*(1), 22–27.

Lax, N. Z., Gorman, G. S., & Turnbull, D. M. (2017). Review: central nervous system involvement in mitochondrial disease. *Neuropathol Appl Neurobiol, 43*(2), 102–118.

Lax, N. Z., Hepplewhite, P. D., Reeve, A. K., Nesbitt, V., McFarland, R., Jaros, E., … Turnbull, D. M. (2012). Cerebellar ataxia in patients with mitochondrial DNA disease: a molecular clinicopathological study. *J Neuropathol Exp Neurol, 71*(2), 148–161.

Leary, S. C., Kaufman, B. A., Pellecchia, G., Guercin, G. H., Mattman, A., Jaksch, M., & Shoubridge, E. A. (2004). Human SCO1 and SCO2 have independent, cooperative functions in copper delivery to cytochrome c oxidase. *Hum Mol Genet, 13*(17), 1839–1848.

Lee, K. T., Lai, W. T., Lu, Y. H., Hwang, C. H., Yen, H. W., Voon, W. C., & Sheu, S. H. (2001). Atrioventricular block in Kearns-Sayre syndrome: a case report. *Kaohsiung J Med Sci, 17*(6), 336–339.

Leigh, D. (1951). Subacute necrotizing encephalomyelopathy in an infant. *J Neurol Neurosurg Psychiatry, 14*(3), 216–221.

Lev, D., Nissenkorn, A., Leshinsky-Silver, E., Sadeh, M., Zeharia, A., Garty, B. Z., … Lerman-Sagie, T. (2004). Clinical presentations of mitochondrial cardiomyopathies. *Pediatr Cardiol, 25*(5), 443–450.

Li, Y., Wen, S., Li, D., Xie, J., Wei, X., Li, X., … Lyu, J. (2018). SURF1 mutations in Chinese patients with Leigh syndrome: novel mutations, mutation spectrum, and the functional consequences. *Gene, 674*, 15-54.

Liang, C., Ahmad, K., & Sue, C. M. (2014). The broadening spectrum of mitochondrial disease: shifts in the diagnostic paradigm. *Biochim Biophys Acta, 1840*(4), 1360–1367.

Listernick, R. (2007). A 9-year-old boy with labored breathing. *Pediatr Ann, 36*(5), 254–257.

Loeffen, J. L., Smeitink, J. A., Trijbels, J. M., Janssen, A. J., Triepels, R. H., Sengers, R. C., & van Den Heuvel, L. P. (2000). Isolated complex I deficiency in children: clinical, biochemical and genetic aspects. *Hum Mutat, 15*(2), 123–134.

Luoma, P., Melberg, A., Rinne, J. O., Kaukonen, J. A., Nupponen, N. N., Chalmers, R. M., … Suomalainen, A. (2004). Parkinsonism, premature menopause, and mitochondrial DNA polymerase gamma mutations: clinical and molecular genetic study. *Lancet, 364*(9437), 875–882.

Magner, M., Dvorakova, V., Tesarova, M., Mazurova, S., Hansikova, H., Zahorec, M., … Honzik, T. (2015). TMEM70 deficiency: long-term outcome of 48 patients. *J Inherit Metab Dis, 38*(3), 417–426.

Magner, M., Vesela, K., Honzik, T., Jesina, P., Vobruba, V., Petrak, B., … Klement, P. (2010). Mitochondriální encefalomyopatie na podkladě deficitu proteinu Sco2 s obrazem SMA-like neurogenní svalové atrofie – kazuistiky. *Cesk Slov Neurol N, 106*(1), 73–75.

Magner, M., Vinšová, K., Tesařová, M., Hájková, Z., Hansíková, H., Wenchich, L., … Honzík, T. (2011). Two patients with clinically distinct manifestation of pyruvate dehydrogenase deficiency due to mutations in PDHA1 gene. *Prague Med Rep, 112*(1): 18-28.

Mancuso, M., Filosto, M., Mootha, V. K., Rocchi, A., Pistolesi, S., Murri, L., … Siciliano, G. (2004). A novel mitochondrial tRNAPhe mutation causes MERRF syndrome. *Neurology, 62*(11), 2119–2121.

Mancuso, M., Orsucci, D., Angelini, C., Bertini, E., Carelli, V., Comi, G. P., Donati, M. A., … Siciliano, G. (2015). Redefining phenotypes associated with mitochondrial DNA single deletion. *J Neurol, 262*, 1301–1309.

Mancuso, M., Orsucci, D., Angelini, C., Bertini, E., Carelli, V., Comi, G. P., … Siciliano, G. (2013). Phenotypic heterogeneity of the 8344A>G mtDNA "MERRF" mutation. *Neurology, 80*(22), 2049–2054.

Marin, S. E., Mesterman, R., Robinson, B., Rodenburg, R. J., Smeitink, J., & Tarnopolsky, M. A. (2013). Leigh syndrome associated with mitochondrial complex I deficiency due to novel mutations In NDUFV1 and NDUFS2. *Gene, 516*(1), 162–167.

Marti, R., Verschuuren, J. J., Buchman, A., Hirano, I., Tadesse, S., van Kuilenburg, A. B., … Hirano, M. (2005). Late-onset MNGIE due to partial loss of thymidine phosphorylase activity. *Ann Neurol, 58*(4), 649–652.

Mignarri, A., Cenciarelli, S., Da Pozzo, P., Cardaioli, E., Malandrini, A., Federico, A., & Dotti, M. T. (2015). Mitochondrial recessive ataxia syndrome: a neurological rarity not to be missed. *J Neurol Sci, 349*(1-2), 254–255.

Milone, M., & Massie, R. (2010). Polymerase gamma 1 mutations: clinical correlations. *Neurologist, 16*(2), 84–91.

Mohr, J., & Mageroy, K. (1960). Sex-linked deafness of a possibly new type. *Acta Genet Stat Med, 10*, 54–62.

Moraes, C. T., DiMauro, S., Zeviani, M., Lombes, A., Shanske, S., Miranda, A. F., … Rowland, L. P. (1989). Mitochondrial DNA deletions in progressive external ophthalmoplegia and Kearns-Sayre syndrome. *N Engl J Med, 320*(20), 1293–1299.

Mozzillo, E., Delvecchio, M., Carella, M., Grandone, E., Palumbo, P., Salina, A., … Franzese, A. (2014). A novel CISD2 intragenic deletion, optic neuropathy and platelet aggregation defect in Wolfram syndrome type 2. *BMC Med Genet, 15*, 88.

Nelson, I., Hanna, M. G., Alsanjari, N., Scaravilli, F., Morgan-Hughes, J. A., & Harding, A. E. (1995). A new mitochondrial DNA mutation associated with progressive dementia and chorea: a clinical, pathological, and molecular genetic study. *Ann Neurol, 37*(3), 400–403.

Niaudet, P., & Rotig, A. (1997). The kidney in mitochondrial cytopathies. *Kidney Int, 51*(4), 1000–1007.

Nye, J. S., Hayes, E. A., Amendola, M., Vaughn, D., Charrow, J., McLone, D. G., … Pandya, A. (2000). Myelocystocele-cloacal exstrophy in a pedigree with a mitochondrial 12S rRNA mutation, aminoglycoside-induced deafness, pigmentary disturbances, and spinal anomalies. *Teratology, 61*(3), 165–171.

Pavlakis, S. G., Phillips, P. C., DiMauro, S., De Vivo, D. C., & Rowland, L. P. (1984). Mitochondrial myopathy, encephalopathy, lactic acidosis, and strokelike episodes: a distinctive clinical syndrome. *Ann Neurol, 16*(4), 481–488.

Pearson, H. A., Lobel, J. S., Kocoshis, S. A., Naiman, J. L., Windmiller, J., Lammi, A. T., … Marsh, J. C. (1979). A new syndrome of refractory sideroblastic anemia with vacuolization of marrow precursors and exocrine pancreatic dysfunction. *J Pediatr, 95*(6), 976–984.

Pfeffer, G., & Chinnery, P. F. (2013). Diagnosis and treatment of mitochondrial myopathies. *Ann Med, 45*(1), 4–16.

Piekutowska-Abramczuk, D., Magner, M., Popowska, E., Pronicki, M., Karczmarewicz, E., Sykut-Cegielska, J., … Pronicka, E. (2009). SURF1 missense mutations promote a mild Leigh phenotype. *Clin Genet, 76*(2), 195–204.

Poulton, J., Deadman, M. E., & Gardiner, R. M. (1989). Duplications of mitochondrial DNA in mitochondrial myopathy. *Lancet, 1*(8632), 236–240.

Pronicka, E., Piekutowska-Abramczuk, D., Szymańska-Dębińska, T., Bielecka, L., Kowalski, P., Luczak, S., … Krajewska-Walasek, M. (2013). The natural history of **SCO2** deficiency in 36 Polish children confirmed the genotype-phenotype correlation. *Mitochondrion, 13*, 810–816.

Rahman, S., Blok, R. B., Dahl, H. H., Danks, D. M., Kirby, D. M., Chow, C. W., … Thorburn, D. R. (1996). Leigh syndrome: clinical features and biochemical and DNA abnormalities. *Ann Neurol, 39*(3), 343–351.

Richardson, C., Smith, T., Schaefer, A., Turnbull, D., & Griffiths, P. (2005). Ocular motility findings in chronic progressive external ophthalmoplegia. *Eye (Lond), 19*(3), 258–263.

Riordan-Eva, P., Sanders, M. D., Govan, G. G., Sweeney, M. G., Da Costa, J., & Harding, A. E. (1995). The clinical features of Leber's hereditary optic neuropathy defined by the presence of a pathogenic mitochondrial DNA mutation. *Brain, 118*(Pt 2), 319–337.

Rotig, A. (2003). Renal disease and mitochondrial genetics. *J Nephrol, 16*(2), 286–292.

Saneto, R. P., & Sedensky, M. M. (2013). Mitochondrial disease in childhood: mtDNA encoded. *Neurotherapeutics, 10*(2), 199–211.

Scaglia, F., Towbin, J. A., Craigen, W. J., Belmont, J. W., Smith, E. O., Neish, S. R., ... Vogel, H. (2004). Clinical spectrum, morbidity, and mortality in 113 pediatric patients with mitochondrial disease. *Pediatrics, 114*(4), 925–931.

Scaglia, F., Wong, L. J., Vladutiu, G. D., & Hunter, J. V. (2005). Predominant cerebellar volume loss as a neuroradiologic feature of pediatric respiratory chain defects. *AJNR Am J Neuroradiol, 26*(7), 1675–1680.

Schaefer, A. M., McFarland, R., Blakely, E. L., He, L., Whittaker, R. G., Taylor, R. W., ... Turnbull, D. M. (2008). Prevalence of mitochondrial DNA disease in adults. *Ann Neurol, 63*(1), 35–39.

Schon, E. A., Hirano, M., & DiMauro, S. (1994). Mitochondrial encephalomyopathies: clinical and molecular analysis. *J Bioenerg Biomembr, 26*(3), 291–299.

Schwartz, M., & Vissing, J. (2002). Paternal inheritance of mitochondrial DNA. *N Engl J Med, 347*(8), 576–580.

Shigemoto, M., Yoshimasa, Y., Yamamoto, Y., Hayashi, T., Suga, J., Inoue, G., ... Nakao, K. (1998). Clinical manifestations due to a point mutation of the mitochondrial tRNAleu(UUR) gene in five families with diabetes mellitus. *Intern Med, 37*(3), 265–272.

Skladal, D., Halliday, J., & Thorburn, D. R. (2003). Minimum birth prevalence of mitochondrial respiratory chain disorders in children. *Brain, 126*(Pt 8), 1905–1912.

Sperl, W., Fleuren, L., Freisinger, P., Haack, T. B., Ribes, A., Feichtinger, R. G., ... Mayr, J. A. (2015). The spectrum of pyruvate oxidation defects in the diagnosis of mitochondrial disorders. *J Inherit Metab Dis, 38*(3), 391–403.

Subramanian, V. S., Marchant, J. S., Said, H. M. (2006). Biotin-responsive basal ganglia disease-linked mutations inhibit thiamine transport via hTHTR2: biotin is not a substrate for hTHTR2. *Am J Physiol Cell Physiol, 291*, C851–C859.

Taanman, J. W., Daras, M., Albrecht, J., Davie, C. A., Mallam, E. A., Muddle, J. R., ... Ginsberg, L. (2009). Characterization of a novel TYMP splice site mutation associated with mitochondrial neurogastrointestinal encephalomyopathy (MNGIE). *Neuromuscul Disord, 19*(2), 151–154.

Tatuch, Y., Christodoulou, J., Feigenbaum, A., Clarke, J. T., Wherret, J., Smith, C., ... Robinson, B. H. (1992). Heteroplasmic mtDNA mutation (T—G) at 8993 can cause Leigh disease when the percentage of abnormal mtDNA is high. *Am J Hum Genet, 50*(4), 852–858.

Tay, S. K., Shanske, S., Kaplan, P., & DiMauro, S. (2004). Association of mutations in SCO2, a cytochrome c oxidase assembly gene, with early fetal lethality. *Arch Neurol, 61*(6), 950–952.

Taylor, R. W., Schaefer, A. M., Barron, M. J., McFarland, R., & Turnbull, D. M. (2004). The diagnosis of mitochondrial muscle disease. *Neuromuscul Disord, 14*(4), 237–245.

Thorburn, D. R. (2004). Mitochondrial disorders: prevalence, myths and advances. *J Inherit Metab Dis, 27*(3), 349–362.

Tesarova, M., Vondrackova, A., Stufkova, H., Veprekova, L., Stranecky, V., Berankova K., ... Zeman, J. (2019) Sideroblastic anemia associated with multisystem mitochondrial disorders. *Pediatr Blood Cancer 66*(4), e27591.

Thorburn, D. R., Rahman, J., & Rahman, S. (2017). Mitochondrial DNA-Associated Leigh Syndrome and NARP. In M. P. Adam, H. H. Ardinger, R. A. Pagon, S. E. Wallace, L. J. H. Bean, K. Stephens & A. Amemiya (eds.), *GeneReviews ((R))*. Seattle, WA.

Tiranti, V., Hoertnagel, K., Carrozzo, R., Galimberti, C., Munaro, M., Granatiero, M., ... Zeviani, M. (1998). Mutations of SURF-1 in Leigh disease associated with cytochrome c oxidase deficiency. *Am J Hum Genet, 63*(6), 1609–1621.

Tranebjaerg, L., Jensen, P. K., Van Ghelue, M., Vnencak-Jones, C. L., Sund, S., Elgjo, K., ... Skullerud, K. (2001). Neuronal cell death in the visual cortex is a prominent feature of the X-linked recessive mitochondrial deafness-dystonia syndrome caused by mutations in the TIMM8a gene. *Ophthalmic Genet, 22*(4), 207–223.

Tranebjaerg, L., Schwartz, C., Eriksen, H., Andreasson, S., Ponjavic, V., Dahl, A., ... Barker, D. (1995). A new X linked recessive deafness syndrome with blindness, dystonia, fractures, and mental deficiency is linked to Xq22. *J Med Genet, 32*(4), 257–263.

Turconi, A. C., Benti, R., Castelli, E., Pochintesta, S., Felisari, G., Comi, G., ... Bresolin, N. (1999). Focal cognitive impairment in mitochondrial encephalomyopathies: a neuropsychological and neuroimaging study. *J Neurol Sci, 170*(1), 57–63.

Ujike, H., Tanabe, Y., Takehisa, Y., Hayabara, T., & Kuroda, S. (2001). A family with X-linked dystonia-deafness syndrome with a novel mutation of the DDP gene. *Arch Neurol, 58*(6), 1004–1007.

Urano, F. (2016). Wolfram syndrome: diagnosis, management, and treatment. *Curr Diab Rep, 16*(1), 6.

Uziel, G., Moroni, I., Lamantea, E., Fratta, G. M., Ciceri, E., Carrara, F., & Zeviani, M. (1997). Mitochondrial disease associated with the T8993G mutation of the mitochondrial ATPase 6 gene: a clinical, biochemical, and molecular study in six families. *J Neurol Neurosurg Psychiatry, 63*(1), 16–22.

Valanne, L., Ketonen, L., Majander, A., Suomalainen, A., & Pihko, H. (1998). Neuroradiologic findings in children with mitochondrial disorders. *AJNR Am J Neuroradiol, 19*(2), 369–377.

van Adel, B. A., & Tarnopolsky, M. A. (2009). Metabolic myopathies: update 2009. *J Clin Neuromuscul Dis, 10*(3), 97–121.

van den Ouweland, J. M., Lemkes, H. H., Ruitenbeek, W., Sandkuijl, L. A., de Vijlder, M. F., Struyvenberg, P. A., ... Maassen, J. A. (1992). Mutation in mitochondrial tRNA (Leu)(UUR) gene in a large pedigree with maternally transmitted type II diabetes mellitus and deafness. *Nat Genet, 1*(5), 368–371.

Van Goethem, G., Schwartz, M., Lofgren, A., Dermaut, B., Van Broeckhoven, C., & Vissing, J. (2003). Novel POLG mutations in progressive external ophthalmoplegia mimicking mitochondrial neurogastrointestinal encephalomyopathy. *Eur J Hum Genet, 11*(7), 547–549.

Vissing, J., Ravn, K., Danielsen, E. R., Duno, M., Wibrand, F., Wevers, R. A., & Schwartz, M. (2002). Multiple mtDNA deletions with features of MNGIE. *Neurology, 59*(6), 926–929.

Wahbi K, Bougouin W, Béhin A, Stojkovic T, Bécane HM, Jardel C, ... Laforêt P. (2015). Long-term cardiac prognosis and risk stratification in 260 adults presenting with mitochondrial diseases. *Eur Heart J, 36*(42), 2886–2893.

Wallace, D. C., & Murdock, D. G. (1999). Mitochondria and dystonia: the movement disorder connection? *Proc Natl Acad Sci U S A, 96*(5), 1817–1819.

Wallace, D. C., Singh, G., Lott, M. T., Hodge, J. A., Schurr, T. G., Lezza, A. M., ... Nikoskelainen, E. K. (1988a).

Mitochondrial DNA mutation associated with Leber's hereditary optic neuropathy. *Science*, *242*(4884), 1427–1430.

Wallace, D. C., Zheng, X. X., Lott, M. T., Shoffner, J. M., Hodge, J. A., Kelley, R. I., ... Hopkins, L. C. (1988b). Familial mitochondrial encephalomyopathy (MERRF): genetic, pathophysiological, and biochemical characterization of a mitochondrial DNA disease. *Cell*, *55*(4), 601–610.

Wedatilake, Y., Brown, R.M., McFarland, R., Yaplito-Lee, J., Morris, A.A., Champion, M. (2013). SURF1 deficiency: a multi-centre natural history study. *Orphanet J Rare Dis*, *8*, 96.

Winterthun, S., Ferrari, G., He, L., Taylor, R. W., Zeviani, M., Turnbull, D. M., ... Bindoff, L. A. (2005). Autosomal recessive mitochondrial ataxic syndrome due to mitochondrial polymerase gamma mutations. *Neurology*, *64*(7), 1204–1208.

Yaplito-Lee, J., Weintraub, R., Jamsen, K., Chow, C. W., Thorburn, D. R., & Boneh, A. (2007). Cardiac manifestations in oxidative phosphorylation disorders of childhood. *J Pediatr*, *150*(4), 407–411.

Yu-Wai-Man, P., & Chinnery, P. F. (2016). Leber Hereditary Optic Neuropathy. In M. P. Adam, H. H. Ardinger, R. A. Pagon, S. E. Wallace, L. J. H. Bean, K. Stephens & A. Amemiya (eds.), *GeneReviews((R))*. Seattle, WA.

Yu-Wai-Man, P., Griffiths, P. G., Burke, A., Sellar, P. W., Clarke, M. P., Gnanaraj, L., ... Chinnery, P. F. (2010). The prevalence and natural history of dominant optic atrophy due to OPA1 mutations. *Ophthalmology*, *117*, 1538–1546.

Zhu, Z., Yao, J., Johns, T., Fu, K., De Bie, I., Macmillan, C., ... Shoubridge, E. A. (1998). SURF1, encoding a factor involved in the biogenesis of cytochrome c oxidase, is mutated in Leigh syndrome. *Nat Genet*, *20*(4), 337–343.

22 Mitochondrial Dysfunction and Epilepsies

Bindu Parayil Sankaran and Arun B. Taly
Department of Neurology, NIMHANS, Bangalore, India

CONTENTS

1 INTRODUCTION

Mitochondria are highly specialized subcellular organelles present in almost all eukaryotic cells whose primary function is generation of adenosine tri phosphate (ATP) through the oxidative phosphorylation [OXPHOS]. (Chinnery and Hudson 2013) Oxidative phosphorylation involves five enzymatic complexes situated in the inner mitochondrial membrane, the mitochondrial respiratory chain (MRC). (Hatefi 1985) The most important and elegant aspect of MRC is that it is the result of genetic complementation between nuclear DNA and mitochondrial DNA. (Chinnery and Hudson 2013) Mitochondrial biogenesis requires the concerted interaction of more than 1500 proteins derived from these two distinct genomes. (Raymond, Horvath, and Chinnery 2018). MtDNA is a circular, double-stranded, 16,569 base-pair molecule of DNA that encodes 37 genes, including 13 polypeptides essential for the formation and function of four of the five MRC complexes, namely complex I, III, IV and V, two ribosomal RNAs (12S and 16S rRNA), and 22 transfer RNAs (tRNA). (Anderson et al. 1981) All other OXPHOS-related proteins, including most of the MRC subunits or factors controlling their expression, assembly function, and turnover is specifically targeted, sorted and imported to their final mitochondrial location (Mokranjac and Neupert 2005). Consequently OXPHOS dysfunction results from mutations in either mitochondrial genes or mito-nuclear genes. (Smeitink, van Den Heuvel, and DiMauro 2001).

Mitochondrial disorders are clinical syndromes resulting from dysfunction of the common final pathway of the mitochondrial energy metabolism, OXPHOS. The implication of mitochondrial dysfunction in health and human diseases are exemplified in the ever increasing genotypes and phenotypes of mitochondrial disorders. (Lightowlers, Taylor, and Turnbull 2015). Tissues with high metabolic demand such as skeletal muscle, brain and heart are usually affected (McFarland, Taylor, and Turnbull 2010). Several factors make the diagnosis of mitochondrial disorders difficult, including oligo symptomatic presentations, poor phenotype genotype correlations and overlapping phenotypes (Raymond, Horvath, and Chinnery 2018).

Brain is the second most commonly affected organ in mitochondrial dysfunction (Finsterer 2006). Mitochondrial diseases are among the most frequently inherited neurological disorders (Gorman et al. 2015). Even though brain accounts for only 2% of the body weight it represents 20% of resting whole body metabolism (Silver and Erecinska 1998). A single cortical neuron uses 4.7 billion ATP molecule/second in a resting human brain, which reflects the high energy demand of the brain (Zhu et al. 2012). Because of this high dependence on ATP, brain is particularly vulnerable to mitochondrial dysfunction. Besides, the glycolytic activity of neurons is limited and therefore mitochondrial oxidative phosphorylation is essential for neuronal ATP supply.

Central nervous system manifestations in mitochondrial disorders are diverse and any part of the neuraxis can get affected. (McFarland, Taylor, and Turnbull 2010) Epilepsy is one of the important central nervous system manifestations and is part of the clinical presentation in a variety of mitochondrial disorders. (El Sabbagh et al. 2010; Finsterer and Zarrouk Mahjoub 2012; Khurana et al. 2008). Conversely, analysis of refractory pediatric epilepsies have shown that one-third of individuals with refractory seizures have biochemical evidence of mitochondrial dysfunction. (Parikh et al. 2008) Studies on temporal lobe epilepsy have shown that mitochondrial and metabolic dysfunction develops during pathological neuronal activity. (Kunz et al. 2000) It is postulated that, studying the overlap between mitochondrial epilepsy and mitochondrial dysfunction in well characterized epilepsy syndromes and their models may help gaining insight into pathogenesis of epileptogenesis and epileptic cell death. (Cock and Schapira 1999).

This chapter describes the possible pathophysiological mechanisms leading to epilepsy and status epilepticus in mitochondrial disorders, the 'mitochondrial epilepsy' and the clinical characteristics. An over view of common clinical phenotypes and genotypes associated with mitochondrial epilepsy and therapeutic considerations are also being discussed.

2 PATHOPHYSIOLOGICAL MECHANISMS IN MITOCHONDRIAL EPILEPSY

The exact pathophysiological processes underpinning mitochondrial epilepsy remain enigmatic. (Rahman 2015) Mitochondria in neurons are highly dynamic, and there exists a fine-tuned coupling between neuronal activity and mitochondrial function. (Kann and Kovacs 2007) Given the critical role of mitochondria in neuronal function, various reviewers have reported many interconnected areas in the pathogenesis of mitochondrial disorders,

epileptogenesis and epileptic cell death. (Cock and Schapira 1999; Kunz 2002) These include the role of reactive oxygen species (ROS) damage, calcium homeostasis and excitotoxic cell death.

The heavy metabolic demand of neurons is due to the fact that they are highly differentiated cells that need a large amount of ATP for maintenance of ionic gradients across the cell membrane and synaptic transmission. (Kann and Kovacs 2007) Neurons primarily depend on the OXPHOs for energy supply (Ames 2000). Conversely, neuronal function and survival are very sensitive to mitochondrial dysfunction (Nicholls and Budd 2000). Mitochondrial dysfunction leads to reduced energy production and compromised neuronal and glial integrity. Besides, mitochondria have diverse roles ranging from neurotransmitter synthesis, calcium homeostasis, redox signalling, production and modulation of reactive oxygen species, and neuronal death (Kann and Kovacs 2007).

2.1 REACTIVE OXYGEN SPECIES DAMAGE

Mitochondria are the primary source of reactive oxygen species. When the mitochondrial respiratory chain (MRC) is inhibited, the reactive oxygen species generation is increased which in turn causes oxidative damage. A vicious circle is formed since reactive oxygen species themselves are potent inhibitors of normal MRC function. Reactive oxygen species are implicated in the pathogenesis of epilepsy both in primary mitochondrial diseases and acquired epilepsy. (Patel 2004) In astrocyte models it was shown that production of superoxide preceded ATP depletion, suggesting an important role of reactive oxygen species generation in epilepsy. (Jacobson et al. 2005) It has been demonstrated that prolonged seizure like activity led to calcium independent free radical production by NADPH and xanthine oxidase in the plasma membrane through N-methyl-D-aspartate (NMDA) receptor activation.

2.2 CALCIUM HOMEOSTASIS

Mitochondria has a role in intracellular calcium sequestration in neurons. (Nicholls 1985) Ca^{2+} ions play vital roles in neuronal excitability and synaptic transmission. (Sanganahalli et al. 2013) Aberrant calcium signalling in astrocytes and neurons are likely to be the factors in epileptogenesis probably by causing excessive synchronization of neurons. Impaired calcium handling in mitochondrial dysfunction can increase seizures. (Kann and Kovacs 2007) Mitochondrial Ca^{2+} transport depends on the balance between Ca^{2+} uptake and extrusion. Recently a calcium uniporter MCU and its regulatory subunits MICU1, MICU2, EMRT have been identified. (Baughman et al. 2011; De Stefani, Patron, and Rizzuto 2015) MICU mutations cause proximal myopathy, learning difficulties and a progressive extrapyramidal disorder. (Musa et al. 2018) Epilepsy occurs rarely in patients with MICU mutations. (Personal communication from Dr. Richard Webster, paediatric neurologist, Sydney, Australia).

It has also been demonstrated that impaired cellular Ca2+ homeostasis due to substantial alterations of mitochondrial Ca2+ handling was the predominant feature of cybrid cells harboring the mitochondrial T8356C mutation, which is associated with Myoclonic epilepsy with ragged red fiber (MERRF) syndrome. (Brini et al. 1999) Abnormal Ca2+ handling has also been shown in m.3243 A>G mutations. (Moudy et al. 1995)

2.3 EXCITOTOXICITY

It has been shown that mitochondria plays a central role in excitotoxic cell death (Ankarcrona et al. 1995) Excitotoxic cell damage/death is implicated in neuronal damage resulting from prolonged status epilepticus. (Wasterlain et al. 1993) Excitotoxicity is also postulated to be the mechanisms in hippocampal sclerosis caused by febrile seizures. Excessive glutamate causes chronic over stimulation of post synaptic glutamate receptors resulting in accumulation of intracellular calcium leading to cell death. On the other hand mitochondrial dysfunction has been implicated in epileptogenesis and chronic epilepsy (Kunz 2002). This is evidenced by the mitochondrial ultrastructural pathology in hilar neurons in Ammon's horn sclerosis. (Blumcke et al. 1999) In addition to the pathological abnormalities, functional deficits of mitochondria have been reported in areas of epileptogenesis. Severe impairment of respiratory chain complex 1 activity was observed in CA3 neurons in hippocampus. (Kunz et al. 2000) It is interesting to note that mitochondrial abnormality have been observed only in epileptic focus and not in surrounding tissue.

3 PREVALENCE

The Prevalence rate of mitochondrial epilepsy differs in various reports. (Bindoff and Engelsen 2012; Rahman 2012) Many studies have focused on the characteristics of epilepsy associated with mitochondrial disorders. (Chevallier, Von Allmen, and Koenig 2014; Khurana et al. 2008; Lee et al. 2008; Saneto 2017; Whittaker et al. 2015) Epilepsy occurs in >20% of adult patients and 40–60% of pediatric cohorts approximately. (Rahman 2018) A large cross-sectional study from the United Kingdom has assessed the prevalence and classification of epilepsy in an adult cohort. (Whittaker et al. 2015) The study also evaluated the epilepsy-related morbidity and mortality. Most of the studies have used respiratory chain enzyme deficiency as the inclusion criteria. As the genetic diagnostic rates are improved, the classification and description of mitochondrial epilepsy are more likely to be genetically based. (Steele and Chinnery 2015)

4 EPILEPTIC PHENOTYPES

The epileptic phenotypes and syndromes described in mitochondrial disorders are diverse. (El Sabbagh et al. 2010; Lee et al. 2008) The most frequently reported phenotypes include focal motor seizures with or without secondary generalization, generalised tonic clonic seizures, myoclonic seizures, and epilepsia partialis continua.

Different types of epilepsy syndromes have also been described in children with respiratory chain defects which include Otohara syndrome, West syndrome, Lennox Gastaut syndrome and Landau Kleffner syndrome. (Lee et al. 2008) Neonatal refractory status and multiorgan failure, neonatal myoclonic epilepsy, infantile spasms, refractory and recurrent status epilepticus, epilepsia partialis continua and myoclonic epilepsy have also been described (El Sabbagh et al. 2010). In one series about 60% had several seizure types emphasizing the complexity of mitochondrial epilepsies. Even though any seizure types or combinations can occur, classical absence epilepsy with three per second spike waves appears to be rare. (Bindoff and Engelsen 2012)

The seizure phenotypes largely depend on the age at presentation. In older children and adults predominant phenotypes include the focal motor seizures, focal seizures progressing to bilateral convulsive seizures, myoclonic seizures and epilepsia partialis continua. Epilepsia partialis continua and status epilepticus have been described mainly in those with m3243 A>G, and POLG1 mutations. As pointed out above the occurrence of the seizures and seizure types are due to the dominant cortical involvement in patients with m3243 A>G mutations. (Cock and Schapira 1999) This may be related to the mutation load in different tissues with a particularly high mutation load in cerebral vascular smooth muscle cells in patients with m.3243 A>G mutation (Betts et al. 2006).

Elucidating the mitochondrial etiology in patients with epilepsy is difficult. Careful analysis of the history and physical signs to look for mitochondrial signatures may be helpful. (Rahman 2012) Meticulous evaluation often reveals clinical, laboratory or imaging evidence of mitochondrial dysfunction in the majority of cases. In the cohort described by El Sabbagh et al. (2010), the first seizures were preceded by failure to thrive, developmental delay, ataxia or multisystem dysfunction in about 82.5% of the children.

5 ELECTROENCEPHALOGRAPHY (EEG) IN MITOCHONDRIAL EPILEPSY

There are no characteristic EEG abnormalities in patients with mitochondrial epilepsy. Nevertheless, there are certain EEG traits which should alert the physician about the mitochondrial etiology. Epileptic abnormalities with a posterior predilection, periodic lateralized epileptiform discharges (PLEDS) and focal high-voltage delta waves with polyspikes (FHDPS) and rhythmic high amplitude delta with superimposed spikes and polyspikes (RHADS) are also described. (Steele and Chinnery 2015).

Recurrent episodes of complex partial status epilepticus and distinctive pattern of periodic lateralized epileptiform discharges have also been described in patients with MELAS. (Bindu et al. 2018; Funakawa et al. 1997; Leff et al. 1998) Fujimoto et al. (1999) in detail have reported the electroencephalographic findings in MELAS according to the time related to the stroke-like episode. These findings included focal high-voltage delta waves with polyspikes (FHDPS), which were recognized as ictal electroencephalography findings. In the subacute and chronic stages, focal spikes or sharp waves and 14- and 6-Hz positive bursts were frequently recorded. The FHDPS seems to be similar to the rhythmic high amplitude delta activity with superimposed spikes (RHADS) noted during status

epilepticus in children with *POLG* related disorders. (Wolf et al. 2009) The typical pattern includes periodic lateralized epileptiform discharges and rhythmic high amplitude delta waves with superimposed spikes during the stroke-like episodes. Interictal findings included focal discharges from the parietooccipital regions as reported by Canafoglia et al. (2001)

[Fig. 1 demonstrates the spectrum of the EEG findings in MELAS.]

The Pattern of focal high amplitude rhythmic delta activity with or without spikes with a posterior dominance has been described in patients with POLG related disorders. (Engelsen et al. 2008)

(Fig. 2 EEG in POLG1-related disorder.)

Focal epileptiform discharges were the dominating findings reported in many studies. (Chevallier, Von Allmen, and Koenig 2014) The most common EEG finding in a large pediatric series was an interictal pattern of multifocal spike discharges [MISD] over the

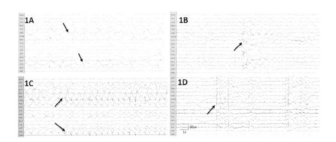

FIGURE 1   1A: EEG during the stroke-like episode in a patient with MELAS syndrome and m.3243 A>G mutation demonstrating periodic lateralized epileptiform discharges(PLEDS Arrows) from right posterior leads. MRI showed right-sided parieto-occipital cortical lesion. 1B: Interictal EEG demonstrating posterior predominant spike and slow waves (arrows) in an 11-year-old girl with MELAS. 1C: EEG demonstrating rhythmic high voltage delta waves with superimposed spikes (arrows) in a 5-year-old girl with MELAS. 1D: Interictal EEG showing generalised spike wave discharges (arrows) in an 11-year-old boy with MELAS. MRI demonstrated multiple small stroke-like lesions.

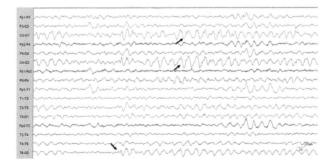

FIGURE 2   Interictal EEG showing posterior predominant rhythmic delta waves (arrows) in a 6-year-old girl with homozygous L304R mutation in *POLG1*. She had photosensitive eyelid myoclonus and generalized tonic clonic seizures.

right and left hemisphere with background slowing present in 58 of the 85 patients presenting with epilepsy (Saneto 2017). This EEG finding has also been described as multiple independent spike foci. This is defined as epileptiform discharges arising from multiple cerebral lobes, including at least one area of both hemispheres. In this cohort, 14 patients had an EEG pattern of hypsarrhythmia which in majority evolved to multifocal spike discharges. Generalized spike-wave pattern was infrequent.

6 MOLECULAR BASIS OF MITOCHONDRIAL EPILEPSY

Epilepsy has been described in mitochondrial disorders caused by both mitochondrial DNA and nuclear mutations.

6.1 MITOCHONDRIAL DNA MUTATIONS WITH EPILEPSY

There are certain mitochondrial DNA genes which are hot spots for mutations causing mitochondrial epilepsy. Mutations in MT-TK &MT-TL1 are considered as prototypes of mitochondrial epilepsies.

6.1.1 MT- TK

MT-TK is a hot spot for mutation causing mitochondrial epilepsy. Mutations in MT-TK results in the classical mitochondrial syndrome- Myoclonic epilepsy with ragged red fibre syndrome denoted by the acronym- MERRF. The most common mutation in MERRF is the 8344A>G substitution in the gene encoding mt tRNA Lys present in more than 90% of the patients reported . (DiMauro and Hirano 1993) The other mutations reported are m.8356T>C, and m. 8363G>A mutations. (Silvestri et al. 1992) The canonical features of MERRF syndrome include myoclonus, generalized epilepsy, ataxia and ragged red fibres [RRF] or identification of a pathogenic variant. (DiMauro and Hirano 1993) MERRF is a multisystemic disorder. Besides

epilepsy, other systemic manifestations are common which include hearing loss, peripheral neuropathy, cognitive decay and eventually dementia, short stature, exercise intolerance, optic atrophy among others. (Altmann et al. 2016; Mancuso et al. 2013) Less common clinical signs (seen in <50% of the patients) include cardiomyopathy, pigmentary retinopathy, pyramidal signs, ophthalmoparesis, and the appearance of multiple lipomas, particularly in the neck and upper trunk. (Lamperti and Zeviani 2016).

Seizures are a frequent problem and have been reported in 46% and 92% of patients in different series. Status epilepticus occurred in 2 of 24 (8%) of cases in an Italian patient registry. (Mancuso et al. 2013) Myoclonus in MERRF may be constant or intermittent photosensitive or intensified by action mainly writing or eating. (Lamperti and Zeviani 2016) EEG findings include spikes or polyspikes. There may be suppression of epileptic activity with eye-opening. Even though myoclonus and epilepsy are tightly associated in the MERRF acronym, it has been demonstrated that myoclonus was more frequently linked to ataxia than to generalized seizures. (Mancuso et al. 2013) This reflects the role of the cerebellum in the production of cortical reflex myoclonus. It has been noted that Purkinje cells, neurons of the dentate nucleus and inferior olivary are especially vulnerable to this mutation. (Lax et al. 2012) Magnetic resonance imaging (MRI) usually demonstrates cerebral and cerebellar atrophy (Altmann et al. 2016; Bindu et al. 2015). (Fig.3 :EEG in MERRF syndrome)

6.1.2 MT-TL1

The m.3243 A>G mutation in *MT-TL1* most commonly cause mitochondrial encephalopathy lactic acidosis and stroke like episodes syndrome. Nevertheless, wide variety of overlapping spectra of clinical features are seen. Seizures are most often associated with the stroke like episodes. The stroke-like episodes each time are heralded by headache, hemianopia, visual hallucinations and focal seizures, which progress to epilepsia partialis continua, and typical stroke-like lesions on magnetic resonance imaging. Seizures

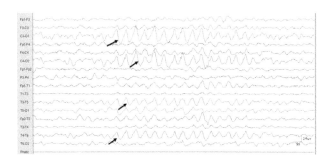

FIGURE 3 EEG demonstrates the occipital predominant high voltage, monomorphic rhythmic delta activity (arrows) in a 10-year-old boy with MERRF syndrome due to m.8334G>G mutation.

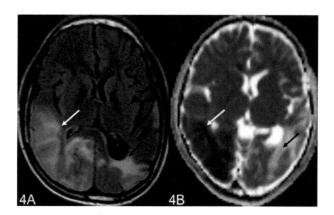

FIGURE 4   4A: MRI brain (Fluid Attenuated Inversion recovery (FLAIR)- axial view) shows the hyper intense signal changes (white arrow) in bilateral parieto occipital lesions in a 17-year-old boy with MELAS syndrome during a stroke-like episode. Apparent Diffusion Coefficient (ADC) image demonstrates restricted diffusion on the right side (white arrow) suggesting the acute nature of the lesion while the lesion on the left side shows free diffusion (black arrow) indicating an old lesion.

occur during the stroke-like episodes and usually correspond to side of the cortical lesion. Those with an earlier age of onset and longer duration of symptoms usually have multiple small cortical lesions, and they had seizures independent of the stroke-like episodes. (Bindu et al. 2015)

Epilepsy is a dominant feature of all patients with MELAS syndrome and 3243A>G mutation. Epilepsy usually occurs during the stroke-like episodes and corresponds to the side of the cortical lesion.

(Fig. 4 MRI in MELAS syndrome).

Electroencephalogram usually shows epileptiform discharges corresponding to the side of stroke-like episodes. In the latter stages, multiple cortical lesions and refractory epilepsy is seen.

In an adult cohort status epilepticus occurred in 7.9% of patients with MELAS (Whittaker et al. 2015). Stauts epilepticus has been an initial presentation in adults SE may present as focal status/non-convulsive status. Aggressive confusional state and occipital status epilepticus have also been described.

6.2   Mito- Nuclear DNA Genes with Epilepsy

6.2.1 POLG1

Mutations in POLG1 encoding the catalytic subunits of polymerase gamma is one of the most common gene defects associated with epilepsy in mitochondrial disorders. (Hikmat et al. 2017) The clinical syndromes caused by POLG1 mutations represent the disease spectra with a wide range of overlapping phenotypes. The most severe of them is represented by an early onset invariably fatal, the Alpers Huttenlocher syndrome. The disease spectrum also includes other entities such as juvenile or adult-onset mitochondrial spinocerebellar ataxia epilepsy (MASCE), myoclonic

epilepsy myopathy sensory ataxia (MEMSA)/spinocerebellar ataxia epilepsy syndromes/sensory ataxia neuropathy dysarthria ophthalmoplegia, chronic progressive ophthalmoplegia and late-onset myopathies. Both autosomal recessive and dominant mutations cause these neurological disorders.

Epilepsy is a common manifestation in POLG related disorder. The estimated frequency varies from 50–80%. (Anagnostou et al. 2016; Engelsen et al. 2008; Hikmat et al. 2017) Alpers's syndrome is characterized by refractory epilepsy, progressive encephalopathy and liver involvement. Majority develops epilepsia partialis continua and generalized status epilepticus. (Nguyen et al. 2005)

Most common seizure type in juvenile and adult-onset patients is focal motor seizures-particularly affecting the upper limbs, neck or proximal muscles or trunk which evolves into bilateral convulsive seizures. There is an occipital predilection in the early stages, characterized by visual disturbances including coloured light scotoma, visual blurring, ictal visual loss and oculo clonus. (Engelsen et al. 2008) Status epilepticus and epilepsia partialis continua are common. The onset is frequently explosive and fulminant. Additional neurological features include migraine neuropathy ataxia and chronic progressive external ophthalmoplegia. The specific EEG findings include rhythmic high amplitude delta with superimposed spikes and polyspikes (RHODES) (Wolf et al. 2009).

6.2.2 Complex 1 Nuclear Mutations

Complex 1 deficiency is by far the most common cause of respiratory chain deficiency. Of the nuclear subunits causing complex 1 deficiency, epilepsy has been associated with mutations in *NDUFV1, NDUFS4, NDUFS8,* and *NDUFA1*. (Bindu et al. 2018; Fernandez-Moreira et al. 2007; Schuelke et al. 1999; van Den Heuvel et al.

1998). Myoclonic epilepsy has also been reported in complex I assembly defects. (Fassone et al. 2010) Myoclonic seizures in nuclear-encoded complex I deficiency were responsive to antiepileptic drugs compared to patients with *POLG* related disorders. (Bindu et al. 2018; Fassone et al. 2010; Fernandez-Moreira et al. 2007) Complex I defects are more potent generators of epilepsy than defects affecting the other complexes (Rahman 2012), perhaps because this complex is working close to its maximum in neurons (Kann and Kovacs 2007).

6.2.3 SLC25A22

The *SLC25A22* encodes mitochondrial carriers, that transport metabolites across the inner mitochondrial membrane (Molinari 2010). Mutations in this gene have been implicated in two types of infantile epilepsy syndrome and has been designated as EIEE-3 (early infantile epileptic encephalopathy- 3) (OMIM 609304) First is the neonatal epileptic encephalopathy with suppression-burst (NEESB). The clinical features in this group of children were characterized by a very early onset or neonatal onset intractable myoclonic epilepsy, progressive microcephaly, periodic EEG with suppression bursts, abnormal VEPs evolving to a severe encephalopathy and spasticity. Characteristically the seizures start in the early hours of life in these patients.

The second phenotypic expression is migrating partial seizures of infancy. (Poduri et al. 2013) This has been described in a consanguineous pedigree with two affected individuals. The seizures appeared at one week of age as hemi convulsions which progressed to the opposite side. The other salient features included hypotonia and brisk DTRs. There was a profound psychomotor delay and subsequent arrest of development. EEG in these patients showed independent focal spikes in multifocal distribution and partial seizures of multifocal origin.

Brain MRI in these patients was characterized by delayed myelination and thinning of corpus callosum and atrophy at later stages. The seizures were refractory to treatment.

6.2.4 Mitochondrial Aminoacyl tRNA Synthetase Genes

Mutations in genes coding mitochondrial aminoacyl-tRNA synthetases cause diseases with diverse neurological phenotypes including epileptic encephalopathies. The mitochondrial aminoacyl synthetases genes implicated in early onset epileptic encephalopathy include *RARS2, FARS2, NARS2, PARS2,* and *VARS2*. The epilepsy phenotypes are characterized by drug-resistant intractable seizures and episodes of status.

6.2.4.1 RARS2

RARS2 encode arginyl-tRNA synthetase synthetase and has been implicated in an early onset mitochondrial encephalopathy. Mutations in this gene have been classically described to be associated with ponto-cerebellar hypoplasia 6. The manifestations are neonatal or infantile onset epileptic encephalopathy, neonatal hypoglycemia, feeding difficulties, abnormal visual behaviour, hypotonia acquired microcephaly and early raised serum and/CSF lacate (Ngoh et al. 2016). In addition to the common seizure types of myoclonic seizures, generalized tonic-clonic seizures and focal clonic seizures, infantile spasms also have been described (Ngoh et al. 2016).

6.2.4.2 QARS

The phenotype associated with mutations in *QARS* encoding glutaminyl-tRNA synthetase, is particularly of note given its resemblance to developmental disorders and characterized by microcephaly at birth with profound developmental delay (Zhang et al. 2014). The seizure onset in reported patients was within the first hour of life. Subsequently, there were multiple seizures and patients also had pharmaco resistant status epilepticus or frequent recurrent long-lasting seizures. Seizures with a migrating pattern also have been described. Also described were clonic seizures with apnoea and cyanosis/clusters of polymorphic seizures/eye deviation chewing and apnoea. In all the patients, the head circumference was less than the 3rd centile for age and there was associated hypotonia. MRI findings include hypo or delayed myelination, thin corpus callosum, reduced white matter, moderately enlarged ventricles, small cerebellar vermis and mild atrophy of the cerebellar hemisphere.

6.2.4.3 FARS2

FARS2 encodes mitochondrial phenylalanine tRNA synthetase. (Raviglione et al. 2016) The most common features are early-onset epileptic encephalopathy, global delay and poor prognosis. Early onset disease is especially associated with epileptic encephalopathy. Liver disease was also a feature in many of them. Seizures were the presenting manifestation in most of them. Many had facial eye twitching and myoclonic jerks of the extremities. Seizures were difficult to control and progressed quickly at an early age to intractable seizures with frequent status epilepticus. EEG showed multifocal spikes. All had a profound delay, axial hypotonia with appendicular spasticity. They developed long tract pyramidal signs over serial examination. The additional features included failure to thrive and microcephaly. None had significant dysmorphic features. MRI findings included marked cerebral atrophy with relative sparing of the basal ganglia and adjacent deep white matter and deep cerebellar white matter hyperintensity. An early onset encephalopathy consistent with fatal infantile Alpers encephalopathy has been described. (Elo et al. 2012) and the child presented with early-onset encephalopathy, developmental delay and arrest, uncontrolled seizures starting early in infancy as well as elevated blood lactate level, and liver dysfunction.

6.2.5 CARS2

homozygous mutations in CARS2 is associated with symptoms resembling MERRF syndrome with severe myoclonic epilepsy as the presenting manifestations (Hallmann et al. 2014). It is also reported in patients with severe epileptic encephalopathy associated with a complex movement disorder (Coughlin et al. 2015).

7 THERAPEUTIC CONSIDERATIONS

Seizures in mitochondrial disorders are triggered, or exacerbated, by metabolic disturbance. (Steele and Chinnery 2015) Appropriate measures should be taken to maintain hydration, normalizing blood glucose, managing acidosis and treating concomitant infections.

Mitochondrial "cocktail" therapy consisting of coenzyme Q10, Vitamin B complex, vitamin C, E and L-carnitine has been used in mitochondrial respiratory chain disorders (Finsterer and Bindu 2015). However, a recent Cochrane review has not found any benefit from these treatments.

7.1 ANTIEPILEPTIC DRUGS

The drugs recommended include levetiracetam, topiramate, zonisamide, and piracetam. Studies have suggested that the anticonvulsants lamotrigine and levetiracetam may be neuroprotective, at least in experimental models of mitochondriopathies (Gibbs, Walker, and Cock 2006; Lee, Shen, and Chang 2000) Even though sodium channel blockers appear effective, there may be worsening of myoclonic seizures by lamotrigine and gabapentin. Other drugs which potentially aggravate myoclonus include phenytoin, carbamazepine, oxcarbazepine, vigabatrine, gabapentin, pregabalin, and tiagabine. Care should be taken to avoid mitochondrial toxins especially sodium valproate, which has been shown to precipitate fatal liver failure in patients with POLG related disorders (Tzoulis et al. 2006).

Aggressive management of status epilepticus is recommended. In status epilepticus, benzodiazepine infusion is a first-line drug followed by phosphenytoin, levetiracetam, or lacosamide.

The ketogenic diet is found to have an increasing evidence base for use in mitochondrial disease. (El Sabbagh et al. 2010; Kang et al. 2007; Lee et al. 2008; Martikainen et al. 2012) This diet has been shown to influence key metabolic pathways in mitochondria. (Bough et al. 2006; Bough and Rho 2007) In a pediatric series, 7/21 patients on the ketogenic diet showed favorable response with more than 75% reduction in seizure frequency. Introduction of ketogenic diet requires expertise from a specialized dietician for close monitoring and adjustments.

8 CONCLUSION

Mitochondrial epilepsies are challenging to diagnose and treat due to their heterogeneous clinical manifestations and are grossly under-recognized. A high index of suspicion and meticulous analysis of clinical, biochemical, and neuroimaging features can help in early diagnosis of mitochondrial epilepsies.

Mitochondrial etiology is suspected when there is epilepsy in association with multisystem disease. The Explosive onset of focal epilepsy/epilepsia partialis continua/status epilepticus with an apparently good inter-ictal period is often a red flag. Recurrent physiological triggers could be identified in many. Worsening following sodium valproate is another clinical pointer. Progressive myoclonic epilepsy syndromes, especially in association with multisystem presentation such as peripheral neuropathy, elevated CPK, deafness, exercise intolerance, should alert the clinician toward a mitochondrial etiology. There are various mechanisms postulated to explain the pathophysiological process in mitochondrial epilepsy. With the application of newer sequencing technologies in the diagnosis of patients with epileptic encephalopathies, more mitochondrial proteins implicated in mitochondrial epilepsies may be identified. Even though there is an advancement in the diagnostic evaluation, the therapeutic options in mitochondrial epilepsy remain limited and await further research.

REFERENCES

Altmann, J., B. Buchner, A. Nadaj-Pakleza, J. Schafer, S. Jackson, D. Lehmann, M. Deschauer, R. Kopajtich, R. Lautenschlager, K. A. Kuhn, K. Karle, L. Schols, J. B. Schulz, J. Weis, H. Prokisch, C. Kornblum, K. G. Claeys, and T. Klopstock. 2016. "Expanded phenotypic spectrum of the m.8344A>G "MERRF" mutation: data from the German mitoNET registry." *J Neurol* 263 (5):961–972. doi:10.1007/s00415-016-8086-3.

Ames, A., 3rd. 2000. "CNS energy metabolism as related to function." *Brain Res Brain Res Rev* 34(1–2):42–68.

Anagnostou, M. E., Y. S. Ng, R. W. Taylor, and R. McFarland. 2016. "Epilepsy due to mutations in the mitochondrial polymerase gamma (POLG) gene: a clinical and molecular genetic review." *Epilepsia* 57(10):1531–1545. doi:10.1111/epi.13508.

Anderson, S., A. T. Bankier, B. G. Barrell, M. H. de Bruijn, A. R. Coulson, J. Drouin, I. C. Eperon, D. P. Nierlich, B. A. Roe, F. Sanger, P. H. Schreier, A. J. Smith, R. Staden, and I. G. Young. 1981. "Sequence and organization of the human mitochondrial genome." *Nature* 290 (5806):457–465.

Ankarcrona, M., J. M. Dypbukt, E. Bonfoco, B. Zhivotovsky, S. Orrenius, S. A. Lipton, and P. Nicotera. 1995. "Glutamate-induced neuronal death: a succession of necrosis or apoptosis depending on mitochondrial function." *Neuron* 15(4):961–973.

Baughman, J. M., F. Perocchi, H. S. Girgis, M. Plovanich, C. A. Belcher-Timme, Y. Sancak, X. R. Bao, L. Strittmatter, O. Goldberger, R. L. Bogorad, V. Koteliansky, and V. K. Mootha. 2011. "Integrative genomics identifies MCU as an essential component of the mitochondrial calcium uniporter." *Nature* 476(7360):341–345. doi:10.1038/nature10234.

Betts, J., E. Jaros, R. H. Perry, A. M. Schaefer, R. W. Taylor, Z. Abdel-All, R. N. Lightowlers, and D. M. Turnbull. 2006. "Molecular neuropathology of MELAS: level of heteroplasmy in individual neurones and evidence of extensive vascular involvement." *Neuropathol Appl Neurobiol* 32 (4):359–373. doi:10.1111/j.1365-2990.2006.00731.x.

Bindoff, L. A., and B. A. Engelsen. 2012. "Mitochondrial diseases and epilepsy." *Epilepsia* 53(Suppl 4):92–97. doi:10.1111/j.1528-1167.2012.03618.x.

Bindu, P. S., H. R. Arvinda, A. B. Taly, C. Govindaraju, K. Sonam, S. Chiplunkur, R. Kumar, N. Gayathri, S. Bharath, M. Nagappa, S. Sinha, N. A. Khan, P. Govindraj, N. Vandana, A. Paramsivam, and K. Thangaraj. 2015. "Magnetic resonance imaging correlates of genetically characterized patients with mitochondrial disorders: a study from south India." *Mitochondrion*. doi:10.1016/j.mito.2015.08.002.

Bindu, P. S., K. Sonam, P. Govindaraj, C. Govindaraju, S. Chiplunkar, M. Nagappa, R. Kumar, C. C. Vekhande, H. R. Arvinda, N. Gayathri, M. M. Srinivas Bharath, J. N. J. Ponmalar, M. Philip, V. P. Vandana, N. A. Khan, V. Nunia, A. Paramasivam, S. Sinha, K. Thangaraj, and A. B. Taly. 2018. "Outcome of epilepsy in patients with mitochondrial disorders: phenotype genotype and magnetic resonance imaging correlations." *Clin Neurol Neurosurg* 164:182–189. doi:10.1016/j.clineuro.2017.12.010.

Blumcke, I., W. Zuschratter, J. C. Schewe, B. Suter, A. A. Lie, B. M. Riederer, B. Meyer, J. Schramm, C. E. Elger, and O. D. Wiestler. 1999. "Cellular pathology of hilar neurons in Ammon's horn sclerosis." *J Comp Neurol* 414(4):437–453.

Bough, K. J., and J. M. Rho. 2007. "Anticonvulsant mechanisms of the ketogenic diet." *Epilepsia* 48(1):43–58. doi:10.1111/j.1528-1167.2007.00915.x.

Bough, K. J., J. Wetherington, B. Hassel, J. F. Pare, J. W. Gawryluk, J. G. Greene, R. Shaw, Y. Smith, J. D. Geiger, and R. J. Dingledine. 2006. "Mitochondrial biogenesis in the anticonvulsant mechanism of the ketogenic diet." *Ann Neurol* 60(2):223–235. doi:10.1002/ana.20899.

Brini, M., P. Pinton, M. P. King, M. Davidson, E. A. Schon, and R. Rizzuto. 1999. "A calcium signaling defect in the pathogenesis of a mitochondrial DNA inherited oxidative phosphorylation deficiency." *Nat Med* 5(8):951–954. doi:10.1038/11396.

Canafoglia L, Franceschetti S, Antozzi C, Carrara F, Farina L, Granata T, Lamantea E, Savoiardo M, Uziel G, Villani F, Zeviani M, Avanzini G. Epileptic phenotypes associated with mitochondrial disorders. Neurology. 2001 May 22;56 (10):1340-6.

Chevallier, J. A., G. K. Von Allmen, and M. K. Koenig. 2014. "Seizure semiology and EEG findings in mitochondrial diseases." *Epilepsia* 55(5):707–712. doi:10.1111/epi.12570.

Chinnery, P. F., and G. Hudson. 2013. "Mitochondrial genetics." *Br Med Bull* 106:135–159. doi:10.1093/bmb/ldt017.

Cock, H., and A. H. Schapira. 1999. "Mitochondrial DNA mutations and mitochondrial dysfunction in epilepsy." *Epilepsia* 40(Suppl 3):33–40.

Coughlin, C. R., 2nd, G. H. Scharer, M. W. Friederich, H. C. Yu, E. A. Geiger, G. Creadon-Swindell, A. E. Collins, A. V. Vanlander, R. V. Coster, C. A. Powell, M. A. Swanson, M. Minczuk, J. L. Van Hove, and T. H. Shaikh. 2015. "Mutations in the mitochondrial cysteinyl-tRNA synthase gene, CARS2, lead to a severe epileptic encephalopathy and

complex movement disorder." *J Med Genet* 52(8):532–540. doi:10.1136/jmedgenet-2015-103049.

De Stefani, D., M. Patron, and R. Rizzuto. 2015. "Structure and function of the mitochondrial calcium uniporter complex." *Biochim Biophys Acta* 1853(9):2006–2011. doi:10.1016/j.bbamcr.2015.04.008.

DiMauro, S., and M. Hirano. 1993. "MERRF." In *GeneReviews((R))*, edited by M. P. Adam, H. H. Ardinger, R. A. Pagon, S. E. Wallace, L. J. H. Bean, K. Stephens, and A. Amemiya. Seattle: University of Washington.

El Sabbagh, S., A. S. Lebre, N. Bahi-Buisson, P. Delonlay, C. Soufflet, N. Boddaert, M. Rio, A. Rotig, O. Dulac, A. Munnich, and I. Desguerre. 2010. "Epileptic phenotypes in children with respiratory chain disorders." *Epilepsia* 51(7):1225–1235. doi:10.1111/j.1528-1167.2009.02504.x.

Elo, J. M., S. S. Yadavalli, L. Euro, P. Isohanni, A. Gotz, C. J. Carroll, L. Valanne, F. S. Alkuraya, J. Uusimaa, A. Paetau, E. M. Caruso, H. Pihko, M. Ibba, H. Tyynismaa, and A. Suomalainen. 2012. "Mitochondrial phenylalanyl-tRNA synthetase mutations underlie fatal infantile Alpers encephalopathy." *Hum Mol Genet* 21 (20):4521–4529. doi:10.1093/hmg/dds294.

Engelsen, B. A., C. Tzoulis, B. Karlsen, A. Lillebo, L. M. Laegreid, J. Aasly, M. Zeviani, and L. A. Bindoff. 2008. "POLG1 mutations cause a syndromic epilepsy with occipital lobe predilection." *Brain* 131(Pt 3):818–828. doi:10.1093/brain/awn007.

Fassone, E., A. J. Duncan, J. W. Taanman, A. T. Pagnamenta, M. I. Sadowski, T. Holand, W. Qasim, P. Rutland, S. E. Calvo, V. K. Mootha, M. Bitner-Glindzicz, and S. Rahman. 2010. "FOXRED1, encoding an FAD-dependent oxidoreductase complex-I-specific molecular chaperone, is mutated in infantile-onset mitochondrial encephalopathy." *Hum Mol Genet* 19(24):4837–4847. doi:10.1093/hmg/ddq414.

Fernandez-Moreira, D., C. Ugalde, R. Smeets, R. J. Rodenburg, E. Lopez-Laso, M. L. Ruiz-Falco, P. Briones, M. A. Martin, J. A. Smeitink, and J. Arenas. 2007. "X-linked NDUFA1 gene mutations associated with mitochondrial encephalomyopathy." *Ann Neurol* 61(1):73–83. doi:10.1002/ana.21036.

Finsterer, J. 2006. "Central nervous system manifestations of mitochondrial disorders." *Acta Neurol Scand* 114(4):217–238. doi:10.1111/j.1600-0404.2006.00671.x.

Finsterer, J., and P. S. Bindu. 2015. "Therapeutic strategies for mitochondrial disorders." *Pediatr Neurol* 52(3):302–313. doi:10.1016/j.pediatrneurol.2014.06.023.

Finsterer, J., and S. Zarrouk Mahjoub. 2012. "Epilepsy in mitochondrial disorders." *Seizure* 21(5):316–321. doi:10.1016/j.seizure.2012.03.003.

Fujimoto S, Mizuno K, Shibata H, Kanayama M, Kobayashi M, Sugiyama N, Ban K, Ishikawa T, Itoh T, Togari H, Wada Y. Serial electroencephalographic findings in patients with MELAS. Pediatr Neurol. 1999 Jan;20(1):43–8.

Funakawa, I., T. Yasuda, and A. Terao. 1997. "Periodic lateralized epileptiform discharges in mitochondrial encephalomyopathy." *Electroencephalogr Clin Neurophysiol* 103(3):370–375.

Gibbs, J. E., M. C. Walker, and H. R. Cock. 2006. "Levetiracetam: antiepileptic properties and protective effects on mitochondrial dysfunction in experimental status epilepticus." *Epilepsia* 47(3):469–478. doi:10.1111/j.1528-1167.2006.00454.x.

Gorman, G. S., A. M. Schaefer, Y. Ng, N. Gomez, E. L. Blakely, C. L. Alston, C. Feeney, R. Horvath, P. Yu-Wai-Man,

P. F. Chinnery, R. W. Taylor, D. M. Turnbull, and R. McFarland. 2015. "Prevalence of nuclear and mitochondrial DNA mutations related to adult mitochondrial disease." *Ann Neurol* 77(5):753–759. doi:10.1002/ana.24362.

Hallmann, K., G. Zsurka, S. Moskau-Hartmann, J. Kirschner, R. Korinthenberg, A. K. Ruppert, O. Ozdemir, Y. Weber, F. Becker, H. Lerche, C. E. Elger, H. Thiele, P. Nurnberg, T. Sander, and W. S. Kunz. 2014. "A homozygous splice-site mutation in CARS2 is associated with progressive myoclonic epilepsy." *Neurology* 83(23):2183–2187. doi:10.1212/wnl.0000000000001055.

Hatefi, Y. 1985. "The mitochondrial electron transport and oxidative phosphorylation system." *Annu Rev Biochem* 54:1015–1069. doi:10.1146/annurev.bi.54.070185.005055.

Hikmat, O., T. Eichele, C. Tzoulis, and L. A. Bindoff. 2017. "Understanding the epilepsy in POLG related disease." *Int J Mol Sci* 18:9. doi:10.3390/ijms18091845.

Hikmat, O., C. Tzoulis, W. K. Chong, L. Chentouf, C. Klingenberg, C. Fratter, L. J. Carr, P. Prabhakar, N. Kumaraguru, P. Gissen, J. H. Cross, T. S. Jacques, J. W. Taanman, L. A. Bindoff, and S. Rahman. 2017. "The clinical spectrum and natural history of early-onset diseases due to DNA polymerase gamma mutations." *Genet Med* 19(11):1217–1225. doi:10.1038/gim.2017.35.

Jacobson, J., M. R. Duchen, J. Hothersall, J. B. Clark, and S. J. Heales. 2005. "Induction of mitochondrial oxidative stress in astrocytes by nitric oxide precedes disruption of energy metabolism." *J Neurochem* 95(2):388–395. doi:10.1111/j.1471-4159.2005.03374.x.

Kang, H. C., Y. M. Lee, H. D. Kim, J. S. Lee, and A. Slama. 2007. "Safe and effective use of the ketogenic diet in children with epilepsy and mitochondrial respiratory chain complex defects." *Epilepsia* 48(1):82–88. doi:10.1111/j.1528-1167.2006.00906.x.

Kann, O., and R. Kovacs. 2007. "Mitochondria and neuronal activity." *Am J Physiol Cell Physiol* 292(2):C641–C657. doi:10.1152/ajpcell.00222.2006.

Khurana, D. S., L. Salganicoff, J. J. Melvin, E. F. Hobdell, I. Valencia, H. H. Hardison, H. G. Marks, W. D. Grover, and A. Legido. 2008. "Epilepsy and respiratory chain defects in children with mitochondrial encephalopathies." *Neuropediatrics* 39(1):8–13. doi:10.1055/s-2008-1076737.

Kunz, W. S. 2002. "The role of mitochondria in epileptogenesis." *Curr Opin Neurol* 15(2):179–184.

Kunz, W. S., A. P. Kudin, S. Vielhaber, I. Blumcke, W. Zuschratter, J. Schramm, H. Beck, and C. E. Elger. 2000. "Mitochondrial complex I deficiency in the epileptic focus of patients with temporal lobe epilepsy." *Ann Neurol* 48(5):766–773.

Lamperti, C., and M. Zeviani. 2016. "Myoclonus epilepsy in mitochondrial disorders." *Epileptic Disord* 18(S2):94–102. doi:10.1684/epd.2016.0846.

Lax, N. Z., P. D. Hepplewhite, A. K. Reeve, V. Nesbitt, R. McFarland, E. Jaros, R. W. Taylor, and D. M. Turnbull. 2012. "Cerebellar ataxia in patients with mitochondrial DNA disease: a molecular clinicopathological study." *J Neuropathol Exp Neurol* 71(2):148–161. doi:10.1097/NEN.0b013e318244477d.

Lee, W. T., Y. Z. Shen, and C. Chang. 2000. "Neuroprotective effect of lamotrigine and MK-801 on rat brain lesions induced by 3-nitropropionic acid: evaluation by magnetic resonance imaging and in vivo proton magnetic resonance spectroscopy." *Neuroscience* 95(1):89–95.

Lee, Y. M., H. C. Kang, J. S. Lee, S. H. Kim, E. Y. Kim, S. K. Lee, A. Slama, and H. D. Kim. 2008. "Mitochondrial respiratory chain defects: underlying etiology in various epileptic conditions." *Epilepsia* 49(4):685–690. doi:10.1111/j.1528-1167.2007.01522.x.

Leff, A. P., A. W. McNabb, M. G. Hanna, C. R. Clarke, and A. J. Larner. 1998. "Complex partial status epilepticus in late-onset MELAS." *Epilepsia* 39(4):438–441.

Lightowlers, R. N., R. W. Taylor, and D. M. Turnbull. 2015. "Mutations causing mitochondrial disease: what is new and what challenges remain?" *Science* 349(6255):1494–1499. doi:10.1126/science.aac7516.

Mancuso, M., D. Orsucci, C. Angelini, E. Bertini, V. Carelli, G. P. Comi, C. Minetti, M. Moggio, T. Mongini, S. Servidei, P. Tonin, A. Toscano, G. Uziel, C. Bruno, E. Caldarazzo Ienco, M. Filosto, C. Lamperti, D. Martinelli, I. Moroni, O. Musumeci, E. Pegoraro, D. Ronchi, F. M. Santorelli, D. Sauchelli, M. Scarpelli, M. Sciacco, M. Spinazzi, M. L. Valentino, L. Vercelli, M. Zeviani, and G. Siciliano. 2013. "Phenotypic heterogeneity of the 8344A>G mtDNA "MERRF" mutation." *Neurology* 80(22):2049–2054. doi:10.1212/WNL.0b013e318294b44c.

Martikainen, M. H., M. Paivarinta, S. Jaaskelainen, and K. Majamaa. 2012. "Successful treatment of POLG-related mitochondrial epilepsy with antiepileptic drugs and low glycaemic index diet." *Epileptic Disord* 14 (4):438–441. doi:10.1684/epd.2012.0543.

McFarland, R., R. W. Taylor, and D. M. Turnbull. 2010. "A neurological perspective on mitochondrial disease." *Lancet Neurol* 9(8):829–840. doi:10.1016/s1474-4422(10)70116-2.

Mokranjac, D., and W. Neupert. 2005. "Protein import into mitochondria." *Biochem Soc Trans* 33(Pt 5):1019–1023. doi:10.1042/bst20051019.

Molinari, F. 2010. "Mitochondria and neonatal epileptic encephalopathies with suppression burst." *J Bioenerg Biomembr* 42(6):467–471. doi:10.1007/s10863-010-9323-6.

Moudy, A. M., S. D. Handran, M. P. Goldberg, N. Ruffin, I. Karl, P. Kranz-Eble, D. C. DeVivo, and S. M. Rothman. 1995. "Abnormal calcium homeostasis and mitochondrial polarization in a human encephalomyopathy." *Proc Natl Acad Sci U S A* 92(3):729–733.

Musa, S., W. Eyaid, K. Kamer, R. Ali, M. Al-Mureikhi, N. Shahbeck, F. Al Mesaifri, N. Makhseed, Z. Mohamed, W. A. AlShehhi, V. K. Mootha, J. Juusola, and T. Ben-Omran. 2018. "A middle eastern founder mutation expands the genotypic and phenotypic spectrum of mitochondrial MICU1 deficiency: a report of 13 patients." *JIMD Rep*. doi:10.1007/8904_2018_107.

Ngoh, A., J. Bras, R. Guerreiro, E. Meyer, A. McTague, E. Dawson, K. Mankad, R. Gunny, P. Clayton, P. B. Mills, R. Thornton, M. Lai, R. Forsyth, and M. A. Kurian. 2016. "RARS2 mutations in a sibship with infantile spasms." *Epilepsia* 57(5):e97–e102. doi:10.1111/epi.13358.

Nguyen, K. V., E. Ostergaard, S. H. Ravn, T. Balslev, E. R. Danielsen, A. Vardag, P. J. McKiernan, G. Gray, and R. K. Naviaux. 2005. "POLG mutations in Alpers syndrome." *Neurology* 65(9):1493–1495. doi:10.1212/01.wnl.0000182814.55361.70.

Nicholls, D. G. 1985. "A role for the mitochondrion in the protection of cells against calcium overload?" *Prog Brain Res* 63:97–106. doi:10.1016/s0079-6123(08)61978-0.

Nicholls, D. G., and S. L. Budd. 2000. "Mitochondria and neuronal survival." *Physiol Rev* 80(1):315–360. doi:10.1152/physrev.2000.80.1.315.

Parikh, S., B. H. Cohen, A. Gupta, D. K. Lachhwani, E. Wyllie, and P. Kotagal. 2008. "Metabolic testing in the pediatric epilepsy unit." *Pediatr Neurol* 38(3):191–195. doi:10.1016/j.pediatrneurol.2007.10.011.

Patel, M. 2004. "Mitochondrial dysfunction and oxidative stress: cause and consequence of epileptic seizures." *Free Radic Biol Med* 37(12):1951–1962. doi:10.1016/j.freeradbiomed.2004.08.021.

Poduri, A., E. L. Heinzen, V. Chitsazzadeh, F. M. Lasorsa, P. C. Elhosary, C. M. LaCoursiere, E. Martin, C. J. Yuskaitis, R. S. Hill, K. D. Atabay, B. Barry, J. N. Partlow, F. A. Bashiri, R. M. Zeidan, S. A. Elmalik, M. M. Kabiraj, S. Kothare, T. Stodberg, A. McTague, M. A. Kurian, I. E. Scheffer, A. J. Barkovich, F. Palmieri, M. A. Salih, and C. A. Walsh. 2013. "SLC25A22 is a novel gene for migrating partial seizures in infancy." *Ann Neurol* 74(6):873–882. doi:10.1002/ana.23998.

Rahman, S. 2012. "Mitochondrial disease and epilepsy." *Dev Med Child Neurol* 54(5):397–406. doi:10.1111/j.1469-8749.2011.04214.x.

Rahman, S. 2015. "Pathophysiology of mitochondrial disease causing epilepsy and status epilepticus." *Epilepsy Behav* 49:71–75. doi:10.1016/j.yebeh.2015.05.003.

Rahman, S. 2018. "Mitochondrial diseases and status epilepticus." *Epilepsia* 59(Suppl 2):70–77. doi:10.1111/epi.14485.

Raviglione, F., G. Conte, D. Ghezzi, C. Parazzini, A. Righini, R. Vergaro, A. Legati, L. Spaccini, S. Gasperini, B. Garavaglia, and M. Mastrangelo. 2016. "Clinical findings in a patient with FARS2 mutations and early-infantile-encephalopathy with epilepsy." *Am J Med Genet A*. doi:10.1002/ajmg.a.37836.

Raymond, F. L., R. Horvath, and P. F. Chinnery. 2018. "First-line genomic diagnosis of mitochondrial disorders." *Nat Rev Genet* 19(7):399–400. doi:10.1038/s41576-018-0022-1.

Saneto, Russell P. 2017. "Epilepsy and mitochondrial dysfunction." *J Inborn Errors Metab Screen* 5. doi:10.1177/2326409817733012.

Sanganahalli, B. G., P. Herman, F. Hyder, and S. S. Kannurpatti. 2013. "Mitochondrial calcium uptake capacity modulates neocortical excitability." *J Cereb Blood Flow Metab* 33(7):1115–1126. doi:10.1038/jcbfm.2013.61.

Schuelke, M., J. Smeitink, E. Mariman, J. Loeffen, B. Plecko, F. Trijbels, S. Stockler-Ipsiroglu, and L. van Den Heuvel. 1999. "Mutant NDUFV1 subunit of mitochondrial complex I causes leukodystrophy and myoclonic epilepsy." *Nat Genet* 21(3):260–261. doi:10.1038/6772.

Silver, I., and M. Erecinska. 1998. "Oxygen and ion concentrations in normoxic and hypoxic brain cells." *Adv Exp Med Biol* 454:7–16.

Silvestri, G., C. T. Moraes, S. Shanske, S. J. Oh, and S. DiMauro. 1992. "A new mtDNA mutation in the tRNA(Lys) gene associated with myoclonic epilepsy and ragged-red fibers (MERRF)." *Am J Hum Genet* 51(6):1213–1217.

Smeitink, J., L. van Den Heuvel, and S. DiMauro. 2001. "The genetics and pathology of oxidative phosphorylation." *Nat Rev Genet* 2(5):342–352. doi:10.1038/35072063.

Steele, H. E., and P. F. Chinnery. 2015. "Mitochondrial causes of epilepsy: evaluation, diagnosis, and treatment." *Semin Neurol* 35(3):300–309. doi:10.1055/s-0035-1552624.

Tzoulis, C., B. A. Engelsen, W. Telstad, J. Aasly, M. Zeviani, S. Winterthun, G. Ferrari, J. H. Aarseth, and L. A. Bindoff. 2006. "The spectrum of clinical disease caused by the A467T and W748S POLG mutations: a study of 26 cases." *Brain* 129(Pt 7):1685–1692. doi:10.1093/brain/awl097.

van Den Heuvel, L., W. Ruitenbeek, R. Smeets, Z. Gelman-Kohan, O. Elpeleg, J. Loeffen, F. Trijbels, E. Mariman, D. de Bruijn, and J. Smeitink. 1998. "Demonstration of a new pathogenic mutation in human complex I deficiency: a 5-bp duplication in the nuclear gene encoding the 18-kD (AQDQ) subunit." *Am J Hum Genet* 62(2):262–268. doi:10.1086/301716.

Wasterlain, C. G., D. G. Fujikawa, L. Penix, and R. Sankar. 1993. "Pathophysiological mechanisms of brain damage from status epilepticus." *Epilepsia* 34(Suppl 1):S37–S53.

Whittaker, R. G., H. E. Devine, G. S. Gorman, A. M. Schaefer, R. Horvath, Y. Ng, V. Nesbitt, N. Z. Lax, R. McFarland, M. O. Cunningham, R. W. Taylor, and D. M. Turnbull. 2015. "Epilepsy in adults with mitochondrial disease: a cohort study." *Ann Neurol* 78:949–957. doi:10.1002/ana.24525.

Wolf, N. I., S. Rahman, B. Schmitt, J. W. Taanman, A. J. Duncan, I. Harting, G. Wohlrab, F. Ebinger, D. Rating, and T. Bast. 2009. "Status epilepticus in children with Alpers' disease caused by POLG1 mutations: EEG and MRI features." *Epilepsia* 50(6):1596–1607. doi:10.1111/j.1528-1167.2008.01877.x.

Zhang, X., J. Ling, G. Barcia, L. Jing, J. Wu, B. J. Barry, G. H. Mochida, R. S. Hill, J. M. Weimer, Q. Stein, A. Poduri, J. N. Partlow, D. Ville, O. Dulac, T. W. Yu, A. T. Lam, S. Servattalab, J. Rodriguez, N. Boddaert, A. Munnich, L. Colleaux, L. I. Zon, D. Soll, C. A. Walsh, and R. Nabbout. 2014. "Mutations in QARS, encoding glutaminyl-tRNA synthetase, cause progressive microcephaly, cerebral-cerebellar atrophy, and intractable seizures." *Am J Hum Genet* 94(4):547–558. doi:10.1016/j.ajhg.2014.03.003.

Zhu, X. H., H. Qiao, F. Du, Q. Xiong, X. Liu, X. Zhang, K. Ugurbil, and W. Chen. 2012. "Quantitative imaging of energy expenditure in human brain." *Neuroimage* 60(4):2107–2117. doi:10.1016/j.neuroimage.2012.02.013.

23 Mitochondrial Dysfunction and Allergic Disease

Kritika Khanna

Molecular Immunogenetics Laboratory, CSIR Institute of Genomics and Integrative Biology, Delhi, India

Anurag Agrawal

Molecular Immunogenetics Laboratory, CSIR Institute of Genomics and Integrative Biology, Delhi, India
Academy of Scientific and Innovative Research, Delhi, India
Baylor College of Medicine, Houston, Texas

CONTENTS

Abbreviations

ROS	Reactive Oxygen Species
ETC	Electron Transport Chain
ATP	Adenosine triphosphate
DAMP	Damage Associated Molecular Pattern
PAMP	Pathogen Associated Molecular Pattern
PRR	Pathogen Recognition Receptor
NLR	Nucleotide binding domain / leucine-rich repeat receptors
TLR	Toll-like Receptor
CLR	C-type lectin receptors
RLR	Retinoic acid-inducible gene (RIG-1)- like receptors
NLRP3	Nod like receptor family pyrin domain containing 3
DC	Dendritic cells
MHC II	Major Histocompatibility Complex Class II
Th2	type 2 T helper cells
IL	Interleukins
IgE	Immunoglobulin E
OVA	Ovalbumin
HDM	House Dust Mite
LPS	Lipopolysaccharide
UQCRC2	Ubiquinol-cytochrome c reductase core protein II

ASC	Apoptosis-associated speck-like protein containing a caspase recruitment domain
siRNA	Small interfering RNA
PKC-δ	Protein kinase C delta type
TNF-α	Tumor necrosis factor alpha
CaMKII	Ca^{2+}/calmodulin-dependent protein kinase II
TCR	T-cell Receptor
MCU	Mitochondrial Calcium Uniporter
PCP	hosphatidylcholine
PBMCs	Peripheral Blood Mononuclear Cells
NADPH	Nicotinamide adenine dinucleotide phosphate, reduced
NADH	Nicotinamide adenine dinucleotide (reduced)
FADH$_2$	Dihydro flavin adenine dinucleotide
PM2.5P	articular Matter below 2.5 μm diameter
MAVS	Mitochondrial Antiviral Signalling protein
MARCH5	Membrane-associated ring-CH-type finger 5
DRP1	Dynamin related protein 1
OPA1	Optic atrophy protein 1
MAPK	Mitogen-activated protein kinases
NF-κB	Nuclear factor kappa-light-chain-enhancer of activated B cells
MMP-8	Matrix Metalloproteinase-8
NFP	N-Formyl Peptides
P2X7	Purinoceptor 7
Treg	Regulatory T cells
MSC	Mesenchymal Stem Cells
PQQ	Pyrroloquinoline quinone

1 INTRODUCTION

The global prevalence of allergic diseases has increased alarmingly in the recent decades. These diseases include allergic asthma, allergic rhinitis, atopic dermatitis, allergic conjunctivitis, urticaria, angioedema, food allergies, drug allergies, and insect allergies. It has been reported that more than 300 million people suffer from asthma and about 400 million from rhinitis. 240–550 million people suffer from food allergies and one tenth of the population suffers from drug related allergies (Pawankar et al. 2013). These epidemiological studies highlight the large socio-economic burden associated with allergic disorders. Of note, children and young adults have been reported to carry the greatest burden of allergies, with a stark decrease in the quality of life. Moreover, many allergic diseases increase the risk of other allergies, leading to concomitant occurrence of multiple allergic diseases in the same subject. Apart from genetic predisposition, the effect of gene-environment interaction is also a major player in pathogenesis of allergic diseases. Environmental factors like climate change and pollution have a significant effect on development of allergies. Due to exposure to outdoor pollutants like particulate matter,

ozone, tobacco smoke and indoor pollutants like biomass fuels, prevalence of allergic diseases is on the rise in developing countries (Pawankar 2014). Taken together, allergens and these pollutants induce aberrant immune responses and lead to inflammation which is a major feature in allergic disorders.

Evidence has now highlighted that mitochondrial dysfunction is also a central component in the pathophysiology of allergic diseases. Allergens can induce mitochondrial dysfunction and inherent defects in mitochondria can also reciprocally aggravate allergic diseases (Iyer, Mishra, and Agrawal 2017). Mitochondria have classically been recognized for their role in cellular bioenergetics. However, recent advances demonstrate that they are also involved in other key processes of the cell like signal transduction, sensing cellular stress and inducing cell death pathways (Chandel 2014). Thus, association of mitochondria with diseases is now not just restricted to the monogenic, inheritable diseases, but has expanded to various neurodegenerative diseases, cardiovascular diseases, metabolic disorders and cancers wherein mitochondrial dysfunction is a critical factor in disease pathophysiology (Nunnari and Suomalainen 2012; López-Armada et al. 2013; Sorrentino, Menzies, and Auwerx 2018). In allergic diseases, mitochondrial dysfunction leads to oxidative stress and reactive oxygen species (ROS) production in the cells, stimulates the innate and adaptive immune responses and ultimately triggers the inflammatory cascade (Iyer, Mishra, and Agrawal 2017).

In this chapter, we discuss the pleiotropic roles of mitochondria in regulating cellular homeostasis, the clinical and immunological aspects of allergic diseases, followed by the experimental links connecting mitochondrial dysfunction with these disorders. We have also examined the role of allergens and environmental pollutants in inducing mitochondrial dysfunction in the cells and how this dysfunction deregulates the immune responses, ultimately causing inflammation and other allergic features. Placed at the centre-stage of pathogenesis of allergic diseases, mitochondria provide a novel therapeutic handle for these pathologies. We, finally, explore the three plausible strategies of mitochondrial medicine – repair, reprogramming and replacement.

2 MITOCHONDRIA AND THEIR PLEIOTROPIC ROLES IN MAINTENANCE OF CELLULAR HOMEOSTASIS

Mitochondria are semi-autonomous cellular organelles that are descendants of alphaproteobacteria and were endosymbiosed into the eukaryotic cell along the evolutionary path (Archibald 2015). Human mitochondria have their own circular double stranded DNA that codes for 13 proteins through their own transcription and translational machinery. These organelles are double membraned structures with each compartment – the inner membrane, intermembrane space and mitochondrial matrix – housing important enzymes, tRNAs and proteins involved in key processes like cellular respiration, signalling and apoptosis (Nunnari and Suomalainen 2012).

Mitochondria are powerhouses of the cell as they carry out the essential function of harnessing energy by oxidation of glucose, fatty acids and amino acid metabolites. Taking place in the mitochondrial matrix, Krebs cycle leads to the generation of electron carriers like NADH and $FADH_2$ that enter into the electron transport chain (ETC) in the inner mitochondrial membrane, eventually leading to the synthesis of ATP (Nunnari and Suomalainen 2012). This movement of high energy electrons through ETC makes the mitochondrial milieu vulnerable to generation of ROS such as superoxide ($O^{-\cdot}$), hydrogen (H_2O_2) peroxide and hydroxyl radicals (.OH), levels of which are kept in check by various antioxidants in the cell. However, during pathological conditions, these highly reactive free radicals are produced in inordinate amounts that can induce oxidative damage of proteins and mitochondrial DNA (mtDNA) and ultimately trigger inflammation. Interestingly, recent evidence has also shown that low levels of ROS can act as signalling molecules and have cytoprotective effects (Hamanaka and Chandel 2010; Chandel 2014). This positive adaptive response, termed as mitochondria hormesis or mitohormesis, illustrates that the levels of ROS need to be finely balanced for maintenance of cellular homeostasis (Yun and Finkel 2014). Along with their roles in cellular bioenergetics and signal transduction, mitochondria are also involved in homeostatic sensing. They act as sensors for cellular stresses like oxygen and nutrient deprivation, as well threats like pathogen invasions (Naviaux 2014). In addition, the role of mitochondrial matrix protein cytochrome c in apoptosis places them as at the hub of cell survival decision making.

With such essential and pleiotropic roles of mitochondria, several sophisticated pathways for mitochondrial quality control are in place. Following cellular stress, damaged mitochondria can be selectively degraded by a specific type of macroautophagy termed as mitophagy (Youle and Narendra 2011).

Mitochondria are highly dynamic in nature and constantly undergo fission and fusion processes to maintain their network. While mitochondrial fusion serves to mitigate stress by fusing together partially defective mitochondria and buffering the damage, mitochondrial fission enables clearing of damaged mitochondria and can facilitate apoptosis (Archer 2013). Another method of quality control of mitochondria is the unfolded protein response, which senses the misfolded proteins in mitochondria and enables restoration of proteostasis (Jovaisaite and Auwerx 2015).

Extreme cellular stress and decline in these quality control pathways leads to mitochondrial dysfunction that changes the metabolic profile of the cell, deregulates the signal transduction cascades and ultimately triggers apoptosis. Mitochondrial stress also leads to release of mitochondrial components like mtDNA, ATP, N-formyl peptides into the cytoplasm or extracellular milieu. These components act as damage associated molecular patterns (DAMPs) and are intercepted by the same receptors of the immune system that recognize pathogen associated molecular patterns and activate the inflammatory cascades (López-Armada et al. 2013; Meyer et al. 2018). Causing deterioration of numerous pathways in the cell, mitochondrial dysfunction has been a signature of multitude of diseases namely inflammatory disorders, autoimmune diseases, chronic diseases, neurodegenerative diseases, cardiovascular diseases, metabolic disorders, and cancers.

3 ALLERGIC DISEASES

Allergic diseases such as allergic asthma, allergic rhinitis, atopic dermatitis, allergic conjunctivitis and food allergies have now reached epidemic proportions, with 20–30% of the population suffering from certain types of allergy (Pawankar et al. 2013). These diseases are manifested when the immune system generates a maladaptive over-reaction toward otherwise innocuous agents like pollens, animal dander, moulds, food components, house dust mites and insects (Kay 2001).

The mechanism of development of allergic diseases involves both arms of the immune system – innate and adaptive immunity. Innate immunity is at the interface of the external environment and the body's internal system. Cells of the innate immune system such as granulocytes, macrophages, dendritic cells, epithelial cells and innate immune lymphocytes harbour membrane bound and intracellular pathogen recognition receptors (PRRs) such as Toll-like receptors (TLRs), nucleotide binding domain/leucine-rich repeat receptors (NLRs) and C-type lectin receptors (CLRs). These PRRs sense pathogen associated molecular patterns (PAMPs) and damage associated molecular patterns (DAMPs). Interaction of PRRs with allergens in healthy individuals leads to immune tolerance. However, in allergic subjects, this interaction initiates an inflammatory cascade and induces the adaptive immune responses (Minnicozzi, Sawyer, and Fenton 2011).

Dendritic cells (DCs) are a type of antigen presenting cells that act as a bridge between the innate and adaptive responses. The external allergen is taken up by DCs and transported to the local lymph node where it is processed and presented to the naïve T cells via the class II Major Histocompatibility Complex (MHC II) molecule. Upon binding of allergens, DCs drive the differentiation of naïve T cells into T helper cell type 2 (Th2) cells that secrete cytokines IL-4 and IL-13 to initiate inflammatory cascade by recruitment of inflammatory cells. These primed Th2 cells also activate B cells and convert them to plasma cells which secrete the allergen specific IgE antibodies (Kay et al. 2008). The IgE antibodies then binds to their receptors present on mast cells, sensitizing them to the allergen. Re-exposure of the body to that allergen mediates crosslinking of these IgE receptors, leading to degranulation of the mast cells and release of histamine and other mediators. Histamine increases smooth muscle contraction, vasodilation and increased capillary permeability. Other mediators like Prostaglandins and leukotrienes induce eosinophil chemotaxis and drive inflammation (Galli and Tsai 2012).

Apart from the classical innate and adaptive immune responses, epithelial cell barrier is also an important factor in the pathogenesis of allergic diseases. Epithelial cells are the first line of defense of the body from the damaging environmental agents. However, the epithelium doesn't merely act as a physical barrier but is also involved in immunomodulation by activation of the dendritic cells and release of inflammatory cytokines that promote Th2 responses. Several allergens are proteases that impair the epithelial barrier function and cause perturbations in the innate immune responses. Other factors like exposure to microbes and cigarette smoke also affect the epithelial functions and subsequent manifestation of allergic features (Mattila et al. 2011). Box 1 highlights the mouse models that are used to understand the allergic responses.

Pathogenesis of allergic diseases is complex and multifactorial. Genetic influences along with environmental factors can lead to loss of tolerance to allergens and mounting of an exaggerated response to these not so harmful agents. While the immunological hypersensitivity and epithelial barrier breach associated with allergies have been widely studied, underlying molecular causes are currently being investigated. Recent studies highlight mitochondrial dysfunction to be lying at the hub of allergy pathogenesis; we examine these associations subsequently.

BOX 1 MURINE MODELS TO STUDY ALLERGIC DISEASES

Mouse models are extensively used to study allergic responses as they can elicit IgE release upon allergen exposure. In a standard murine model of allergic asthma or allergic airway inflammation, mice are sensitized with ovalbumin (OVA) along with alum adjuvant, through intra-peritoneal injections. This is followed by challenges with aerosolized ovalbumin. Allergic features like airway inflammation characterized by eosinophil infiltration, secretion of IgE, release of Th2 cytokines like IL-4 and IL-13, and airway hyper responsiveness are observed. In more clinically relevant models, mice are exposed to allergens like cockroach extract and house dust mite (HDM) (Gershwin 2015).

4 MITOCHONDRIA AND ALLERGY – EXPERIMENTAL LINKS

4.1 GENETIC AND FUNCTIONAL STUDIES

Epidemiological studies have revealed maternal history of asthma and atopy as an important risk factor for development of allergies and asthma in children. Such an epidemiological link along with the exclusive transmission of mitochondria from the maternal lineage indicates the role of mitochondrial genome sequence variations in atopic diseases. Indeed, initial reports of associations between mitochondria and atopy show a genetic link between the two. Common mitochondrial haplogroup U has been shown to be significantly associated with total serum IgE levels. This haplogroup also showed trends of associations with greater skin prick reactivity and higher incidence of atopic dermatitis (Raby et al. 2007). Another study demonstrates haplogroup in D loop region of the mitochondrial DNA to be associated with cow milk allergy expressed as atopic dermatitis and gastrointestinal disease (Muino et al. 2014). Apart from genetic associations, many functional studies have highlighted the role of mitochondria dysfunction in allergic diseases. The bronchial epithelium of the human asthma subjects has been demonstrated to have swollen mitochondrial morphology (Konrádová et al. 1985). Such ultrastructural changes, with respect to loss of cristae and mitochondrial swelling, were also observed in murine model of allergic inflammation. This was associated with reduced expression of cytochrome c oxidase and mitochondrial respiratory complex I in the bronchial epithelium and increased cytochrome c in the lung cytosols (Mabalirajan and Ghosh 2013). In another study, Aguilera et al. demonstrated that pre-existing mitochondrial defects can aggravate allergic airway inflammation. Exposure of airway epithelial cells to ragweed pollen extract was shown to induce oxidative damage to the mitochondrial respiratory chain complex proteins. Among these proteins, destruction of the ubiquinol-cytochrome c reductase core protein II (UQCRC2), which is an important structural protein of the complex III of the mitochondrial respiratory chain, was found to increase mitochondrial ROS generation. Interestingly, siRNA mediated downregulation of the core protein in mice, prior to exposure of ragweed pollen, led to aggravated allergic responses in terms of bronchial hyperresponsiveness, increased eosinophil recruitment and elevated mucin production in the airways. Thus, any such pre-existing defects in the mitochondrial respiratory chain can lead to an enhanced allergic response to ragweed pollen (Aguilera-Aguirre et al. 2009). These observations also have important implications on the environmental factor associated exacerbation of allergies and atopy. Oxidant environmental pollutants like diesel exhaust particles, ozone, tobacco smoke particles and factors like respiratory virus infections, can induce mitochondrial dysfunction which could further aggravate the allergic symptoms in atopic individuals.

4.2 MECHANISTIC CONNECTION VIA IMMUNE RESPONSE MODULATION

Mechanistic connection between mitochondrial dysfunction and allergies has also been shown with respect to regulation of mast cell degranulation. In a model of mucosal mast cells, exposure to ragweed pollen extract was found to increase H_2O_2 production from the mitochondrial respiratory complex III. These mtROS were shown to further trigger secretion of biogenic amines like histamine and serotonin in an IgE independent manner. This mtROS induced histamine release was found to occur in absence of complete exocytosis of the secretory vesicles and was independent of the intracellular free Ca^{2+} levels. It was rather governed by Protein kinase C delta type (PKC-δ) and microtubule dependent migration of the secretory granules. The release of stored biogenic amines from the granules was triggered by inhibition of vacuolar H^+-ATPase activity which consequently increased the intra-vesicular pH. These observations highlight the differences between mtROS induced and IgE mediated mast cell degranulation. Apart from such antigen independent activation of mast cells, mtROS can also synergistically increase the production of IL-4 from the antigen-treated sensitized mast cells. Thus, mitochondrial dysfunction can not only act independent of adaptive immunity, but it can also enhance the Th2 responses (Chodaczek et al. 2009). Mitochondrial dysfunction has also been reported to activate other innate immune responses in allergic inflammation. The total and mitochondrial ROS levels have been reported to be elevated in the airway inflammatory and epithelial cells of OVA, LPS, and HDM induced murine models of allergic inflammation (Kim et al. 2014). Treatment of such mice with a novel mitochondrial ROS inhibitor NecroX-5 has been shown to reduce the mtROS levels and improve allergic features like airway inflammation and hyper-responsiveness. These effects of NecroX-5 seem to be governed by modulation of NLRP3 inflammasome activation. Inflammasomes are multi-protein complexes of the innate immunity that sense the inflammatory stimuli like pathogen associated molecular pattern molecules (PAMPs) and damage associated molecular pattern molecules (DAMPs) and respond by subsequent production of inflammatory cytokines like IL-1β and IL-18. Mitochondrial ROS has been reported to be an important activator of the NLRP3 inflammasome activation cascade. Along with this, damaged and improperly repaired mitochondrial DNA, which is released in the cytoplasm, can also act as a mitochondrial DAMP and be a coactivator of the inflammasome activation. Kim et al. have shown that in addition to elevated mtROS, mitochondrial DNA was

also damaged and reduced in the lungs of OVA and HDM treated mice. Furthermore, there was increased expression of inflammasome proteins like NLRP3 and caspase1, and mature form of IL-1β in the lung tissues of these allergic mice. Treatment with NecroX-5 led to restoration of the levels and integrity of the mtDNA and reduced the expression of NLRP3, caspase 1 and IL-1β in the lung tissues. These observations suggest that mtROS generation and mtDNA damage can induce inflammasome activation in allergic airway inflammation. Additionally, mtROS was also demonstrated to increase the translocation of NF-kB to the nucleus, along with an increase in the various inflammatory cytokines like IL-4, IL-5, IL-13, IL-17 and TNF-α. These effects were reversed by the administration of NecroX-5 (Kim et al. 2014). In line with this report, exposure to ovalbumin or *Aspergillus fumigatus* in a murine model has also been shown to activate oxidant Ca^{2+}/calmodulin-dependent protein kinase II (CaMKII) expressed in the mitochondria and consequently increase mtROS generation. mtROS was further shown to activate of NF-kB and NLRP3 inflammasome, and thereafter induce Th2 cytokines that eventually mediate airway inflammation and airway hyperactivity (Sebag et al. 2017). In a similar mechanism, a study of chromium allergy which is a common occupational skin disease showed that chromium (VI) can induce mtROS generation, subsequently activate NLRP3 inflammasome and cause cytotoxicity (Adam et al. 2017). Along with innate immune responses, mitochondria have also been shown to regulate adaptive immune responses in allergic diseases. Oxidation signals from the mitochondrial respiratory complex I can induce T cell activation mediated expression of IL-2 and IL-4. In T cells isolated from the patients of atopic dermatitis, the inhibition of complex I abrogates the spontaneous expression and TCR-induced expression of IL-4 (Kaminski et al. 2010). Taken together, these studies highlight the role of mitochondrial dysfunction in the pathogenesis of allergic disorders via modulation of the innate and adaptive immune responses.

4.3 MECHANISTIC CONNECTION VIA EPITHELIAL BARRIER DISRUPTION

Apart from regulating immune responses, mitochondria are also important for maintaining epithelial barrier function. Recent advances highlight the active role of epithelial cells in pathogenesis of allergic disorders. Epithelial cells are the first line of defense as they are the primary barrier between the noxious external agents like pollutants, allergens, cigarette smoke and pathogens, and body's internal environment. Disruption of the epithelial barrier has been reported in allergic diseases like atopic dermatitis, peanut allergy, pollen allergies and allergic asthma. Compromised epithelial barrier can not only increase the risk of allergen sensitization but also initiate immune responses and signal transduction cascades, altering tissue homeostasis (Mattila et al. 2011). Recent studies show the role of mitochondrial function in maintaining this epithelial integrity. Sebag et al. demonstrated that mitochondrial calcium uniporter (MCU), which resides in the organelle's inner membrane and regulates Ca^{2+} uptake in mitochondrial matrix, is critical in regulating epithelial barrier function. Mitochondrial calcium overload leads to accumulation of mtROS, which consequently leads to dissipation of mitochondrial membrane potential and swelling of mitochondria. This, in turn, leads to cytochrome c release and induction of cellular apoptosis. In this study, the authors show that inhibition of MCU in primary human airway epithelial cells decreased IL-13 mediated mitochondrial Ca^{2+} uptake, abolished mtROS production and preserved mitochondrial membrane potential. It also protected against cytokine induced cellular apoptosis and decreased impairment of barrier function caused by IL-13. These *in vitro* findings were also confirmed in MCU-/- mice. Upon ovalbumin induced allergic inflammation, MCU deletion decreased apoptosis within large airway epithelial cells and preserved expression of tight junction protein ZO-1, thus maintaining the epithelial barrier integrity (Sebag et al. 2018). Another study highlights the link between mitochondrial homeostasis and epithelial barrier permeability via a phospholipid transfer protein Stard-7. Stard-7 promotes the uptake of phosphatidylcholine (PC) in mitochondria. PC is the major phospholipid species in mitochondria and is important for maintaining mitochondrial function. Knockdown of Stard7 in bronchial epithelial cells and its targeted knockout in lung epithelial cells in mice (Stard7$^{epi\Delta/\Delta}$) led to altered mitochondrial structure, along with perturbed mitochondrial respiration, increased mtROS and mtDNA damage. This was also associated with impaired epithelial barrier permeability, both *in vitro* and *in vivo*. Interestingly, the reduced levels of tight junction proteins as seen in the Stard7 downregulated epithelial cells were restored upon treatment with mitochondrial targeted antioxidant – Mito TEMPO. Concordantly, paracellular barrier leak in the tracheal epithelial cells isolated from the Stard7$^{epi\Delta/\Delta}$ mice was also restored upon MitoTEMPO treatment (Yang et al. 2017).

4.4 Miscellaneous Studies

The effect of allergens on mitochondria dynamics has also been reported. House dust mite allergen modulates mitochondrial dynamics via endoplasmic reticulum-mitochondria interactions. In human bronchial epithelial cells, exposure of house dust mite allergen led to stress induced increase in ER-mitochondria contacts, followed by recruitment of mitochondrial fission protein – Dynamin

related protein, DRP1 and induction of mitochondrial fragmentation. This was shown to consequently increase the release of pro-inflammatory cytokines like IL-8 and IL-1β (Cahoon and Anathy 2015). Apart from atopic dermatitis, allergic asthma and allergic rhinitis, mitochondrial dysfunction has also been demonstrated in food allergies. In a murine model of peanut allergy, hepatic mitochondrial dysfunction was observed in terms of impaired mitochondrial respiration and increased mtROS (Trinchese et al. 2018).

While the above mentioned studies have mechanistically explored the role of mitochondrial dysfunction in airway inflammatory cells and the airway epithelial cells in model systems, there is also evidence that these pathways are operational in humans. Eleven patients with a history of grass pollen allergic rhinitis were recruited and 2 nasal challenges of either grass pollen extract or placebo were carried out in a randomized order, 4 weeks apart. Peripheral blood was collected from the subjects before and 6 and 24 hours post the challenge, followed by isolation of peripheral blood mononuclear cells (PBMCs). Mitochondrial respiratory function of the PBMCs was analyzed by measuring maximal respiration rate and specific mitochondrial respiratory chain complex activities using high resolution respirometry. Exposure to pollen led to significant reduction of the combined activities of mitochondrial complex I, III, and IV, 6 hours after the allergen challenge. This effect was, however, restored 24 hours post the challenge. The nasal allergen challenge had deleterious effect on the mitochondrial respiratory chain complex I activity, while the complex II, III, and IV activities were not significantly impaired. Additionally, there was a significant reduction in the mitochondrial coupling of oxidation and phosphorylation 6 hours after the allergen exposure. This effect of allergen exposure to the mitochondrial function of circulating cells of allergic patients is the first report to highlight the mitochondrial function involvement in a systemic manner in pollen allergy pathophysiology (Qi et al. 2017).

5 ALLERGEN AND PARTICULATE MATTER INDUCED MITOCHONDRIAL DYSFUNCTION

While the above reports highlight the experimental links between mitochondrial dysfunction and allergies, in this section we examine the mechanisms of allergen and environmental pollutant driven mitochondrial dysfunction.

5.1 Allergen Induced Mitochondrial Dysfunction

The mechanisms for allergen driven immune responses can take multiple parallel paths. Pollen allergens have inherent proteases and NADPH oxidases that can trigger innate immune responses. The oxidases can generate reactive oxygen species that oxidize glutathione and cause DNA damage, leading to production of IL-8 and recruitment of neutrophils. These activated inflammatory cells, in turn, generate more ROS (Hosoki, Boldogh, and Sur 2015). In addition, as previously mentioned, pollen allergens can also directly cause mitochondrial dysfunction by affecting respiratory complex III and increasing mtROS production (Aguilera-Aguirre et al. 2009). Moreover, the protease activity in allergens can disrupt the epithelial barrier and increase allergen penetration. Binding of the allergens on the PRRs present on innate immune cells like epithelial cells and mononuclear phagocytes can activate downstream inflammatory cascade. Such activation of the innate response stimulates the activation of adaptive immune response wherein DCs present the allergenic peptides to the T cells and trigger an allergen specific immune response (Lambrecht and Hammad 2014).

5.2 Particulate Matter Induced Mitochondrial Dysfunction

Environmental pollutants have been strongly associated with allergic diseases and asthma. There are several studies demonstrating that air borne particulate matter can trigger mitochondrial dysfunction via multiple mechanisms. Exposure of rat lungs to particular matter below 2.5 μm diameter (PM2.5) disrupts mitochondrial dynamics by affecting the expression of proteins regulating mitochondrial fission (dynamin related protein, DRP1) and mitochondrial fusion (optic atrophy protein, OPA1) (Li et al. 2015). Early life exposure of PM is also associated with increased mitochondrial oxidative DNA damage in both mothers and their newborns (Grevendonk et al. 2016). Pollutants like diesel exhaust particles have also been shown to induce mitochondrial dysfunction via increased ROS generation, disruption of mitochondrial membrane potential, release of cytochrome c and subsequent activation of the apoptotic pathway (Wang et al. 2007). Moreover, diesel exhaust particles also decrease the ubiquitin proteasome system which is important for mitochondrial quality control (Kipen et al. 2011). Studies have shown that exposure to other pollutants like cadmium increases lipid peroxidation which induces mitochondrial membrane permeability and release of mitochondrial calcium (Müller 1986). Overall, these studies highlight the effect of air pollutants on mitochondrial function suggesting mitochondria as the common, connecting link between rise in pollution levels and increased prevalence of allergic diseases.

6 MITOCHONDRIA MEDIATED REGULATION OF THE IMMUNE SYSTEM

Mitochondria are critical to the cellular defense mechanisms as they regulate both innate and adaptive immune responses. The earlier mentioned studies highlight how allergens and air borne particulate matter induce mitochondrial dysfunction. With mitochondria at centre of immune regulation, any alteration in their function is detrimental to the immune balance and can trigger inflammatory responses.

In intact cells, mitochondria act as platforms for pathogen recognition receptor (PRR) responses. Upon stimulation of the PRRs, mitochondria interact with the downstream signalling molecules and consequently activate the signalling cascade. Additionally, in stressed and damaged cells, mitochondrial components can also act as damage associated molecular patterns (DAMPs) and trigger the immune responses. Mitochondrial respiration also governs the polarization of T helper cells when stimulated by the dendritic cells, thus regulating the adaptive immune responses as well (López-Armada et al. 2013; Meyer et al. 2018).

6.1 Mitochondria in Innate Responses

Mitochondria are important contributors in PRR responses. One of the PRRs, Retinoic acid-inducible gene (RIG-1)-like receptors (RLRs), are cytosolic RNA sensors that recognize viral double-stranded RNA (dsRNA). During a viral infection, binding of viral RNA to the RLRs leads to their stimulation and subsequent binding to mitochondrial adaptor protein- mitochondrial antiviral signalling protein (MAVS) (López-Armada et al. 2013). This interaction is critical for activation of the downstream signalling pathway that eventually leads to production of antiviral molecules like type I interferon (Seth et al. 2005). Any deregulations in mitochondrial function that enhance the mtROS generation or change the mitochondrial dynamics with respect to mitochondrial fusion have been shown to modulate the RLR signalling in a MAVS dependent manner. Predominantly, such dysregulations play an important role in inflammatory pathologies such as autoimmune diseases (Meyer et al. 2018). However, a recent study also shows the link between MAVS and allergic diseases through gut dysbiosis. MAVS -/- mice were reported to have a changed gut microbiota which compromised the intestinal barrier function and was associated with increased delayed type hypersensitivity. Thus, MAVS can also act as a negative regulator of allergic dermatitis by modulating gut microbiota (Plantamura et al. 2018).

Other types of PRRs are the transmembrane Toll-like Receptors (TLRs) that recognize PAMPs and DAMPs and induce proinflammatory cytokines through MYD88 and TRIF signalling cascades (Meyer et al. 2018). Mitochondria play an important part in the TLR signal transduction. Upon TLR engagement in macrophages, mitochondria position near the phagosomes and enhance ROS generation which is important for bactericidal activity (West et al. 2011). In another study, mtROS has also been shown to enhance MAPK activation upon TLR stimulation (Kamata et al. 2005). Apart from the involvement of the mitochondrial respiratory chain, a ubiquitin ligase present in the mitochondrial outer membrane – membrane-associated ring-CH-type finger 5 (MARCH5) has been shown to regulate TLR induced NF-kB activation (Shi et al. 2011).

Mitochondria have also been shown to be key players in the NLRP3 inflammasome activation. NLRP3 is a type of nucleotide-binding oligomerization domain (NOD)-like receptor (NLR) that is a cytoplasmic sensor for PAMPs and DAMPs. Upon stimulation, NLRP3 gets oligomerized and recruits an adaptor protein named apoptosis-associated speck-like protein containing a caspase recruitment domain (ASC) and procaspase-1. This multiprotein complex, called inflammasome, then induces maturation of procaspase-1 to caspase-1 which further cleaves pro IL-18 and pro IL-1β into mature forms (Gross et al. 2011). Mitochondria contribute to the inflammasome modulation through multiple mechanisms. Physical interaction between mitochondria and inflammasome is important. Upon activation, inflammasome complex has been shown to translocate to the perinuclear area and colocalize with the mitochondria (Zhou et al. 2011; Misawa et al. 2013). Moreover, like RLR activation, NLRP3 inflammasome activity has also been shown to depend on interaction with mitochondrial MAVS protein (Subramanian et al. 2013). The vicinity between mitochondria and inflammasome enables easy modulation of NLRP3 inflammasome through the mitochondrial components. mtROS has been shown to be essential for inflammasome activation and reducing their generation decreases the inflammasome signalling (Kepp, Galluzzi, and Kroemer 2011; Zhou et al. 2011). Additionally, mitochondrial components like mtDNA, cardiolipin and ATP in stressed or damaged cells can also act as DAMPs and active the NLRP3 inflammasome. As described in the previous section, multiple studies have shown the role of mitochondria dependent regulation of NLRP3 inflammasome in allergic inflammation.

6.2 Mito-DAMPs

Due to their microbial origin, mitochondria share many molecular features with bacteria. For instance, similar to bacteria, mtDNA has conserved unmethylated CpG motifs, mitochondrial peptides have N-formyl methionine residues and cardiolipin, the phospholipid in mitochondrial membrane which is distinct from the eukaryotic membrane phospholipids. These distinct molecular features and the endosymbiotic evolution of mitochondria has enabled them to act as sensors for any stress or damage in the cell. Any cellular injury or stress leads to release of these mitochondrial components either in the cytoplasm or outside of the cells whereby

they act as DAMPs and illicit immune responses to deal with the pathological insult. Elaborated below are these mito-DAMPs and the immune responses generated.

6.2.1 mtDNA

Mitochondrial DNA is intercepted by innate immunity through TLRs or NLRP3 inflammasome. Like bacterial DNA, CpG rich mtDNA, if released in the extracellular milieu upon cell damage, is sensed by TLR9 on neutrophils, monocytes and endothelial cells (Zhang et al. 2010; Hong et al. 2013). The mtDNA driven activation of TLR9 on neutrophils leads to increased release of proinflammatory cytokine – matrix metalloproteinase-8 (MMP-8) which causes severe inflammation in the lungs (Wei et al. 2015). Moreover, the mtDNA-TLR interaction also leads to activation of NF-κB and downstream expression of proinflammatory cytokines such as TNF-α and IL-6. This eventually promotes chemotaxis and recruitment of inflammatory cells to the site of tissue injury (Zhang et al. 2014). Mitochondrial DNA can also be released into the cytoplasm upon cellular stress which induces mitochondrial dysfunction, increased ROS production and opening of mitochondrial transition pore. While mtROS acts as a trigger, cytoplasmic mtDNA acts as a coactivator to amplify the NLRP3 inflammasome signalling leading to release of IL-1β and IL-18 (Nakahira et al. 2011; Shimada et al. 2012). In allergic disease models, studies have shown mtDNA driven NLRP3 inflammasome and NF-κB activation that aggravate the inflammatory conditions (Kim et al. 2014).

6.2.2 ATP

Mitochondria generate high amounts of ATP that are released in the cytoplasm where they serve as the main source of energy for the cell. Under physiological conditions, the ATP secreted out of the cells is maintained in a proper concentration by the action of ecto ATP/ADPases (Lohman, Billaud, and Isakson 2012). However, under pathological conditions, the levels of extracellular ATP increase and serve as "alert signals" for the immune system. Extracellular ATP is recognized by purinergic receptor – P2X7, an ATP gated channel receptor, which is highly expressed on immune cells. Stimulation of these receptors activates the NLRP3 inflammasome signalling and induces inflammatory chemotaxis (Iyer et al. 2009).

Extracellular ATP has been shown to be a critical mediator of asthma. In patients of allergic asthma and in experimental models of allergic inflammation, elevated levels of extracellular ATP have been reported, which correlated with eosinophilic inflammation and Th2 immune responses (Idzko et al. 2007). Another study has reported a crucial role of ATP in promoting Th-17 mediated neutrophilic asthma (Zhang et al. 2017). Extracellular ATP also acts as a danger signal for a common aeroallergen, the fungus *Alternaria alternate*, by inducing IL-33 release followed by innate Th2 type responses (Kouzaki et al. 2011).

6.2.3 Cardiolipin

Cardiolipin is a unique phospholipid present only in the inner mitochondrial membrane in the eukaryotic cell (Daum 1985). Mitochondrial dysfunction can lead to extrusion of cardiolipin outside of the cells where it can act as a mito-DAMP and activate NLRP3 inflammasome (Iyer et al. 2013). Cardiolipin has also been shown to inhibit the anti-inflammatory pathways by blocking the production of IL-10 (Chakraborty et al. 2017).

6.2.4 N-Formyl Peptides

Akin to bacteria, mitochondria initiate protein synthesis with N-formyl methionine instead of methionine residue (Smith and Marcker 1968). Such N-formyl peptides (NFPs) are recognized by formyl peptide receptors that are expressed by immune cells and endothelial cells. Binding of these NFPs activates these cells, subsequently driving inflammatory cytokine secretion, promoting chemotaxis and increasing oxidative burst in the target cells (Panaro et al. 2006; Crouser et al. 2009).

6.3 Mitochondria in Adaptive Immune Responses

Activation of the innate immune pathways upon allergen exposure has been shown to stimulate the adaptive immune responses as well. Dendritic Cells, acting as a link between the innate and adaptive immunity, sample the allergens and present them to the naïve T cells. Metabolic state of the cells at the immune synapse directs the polarization of T lymphocytes into either proinflammatory Th cells or anti-inflammatory T-regulatory cells (Tregs) (Meyer et al. 2018).

Mitochondrial respiration dependent expression of costimulatory molecules on DCs governs the T cell differentiation. For instance, while increased glycolysis in DCs is associated with expression of TLR-induced costimulatory molecules, improved mitochondrial biogenesis is negatively associated with expression of DC costimulatory molecules leading to Th-cell anergy (Klotz et al. 2007; Krawczyk et al. 2010). T-lymphocytes, present at the other end of the immune synapse, also depend on mitochondrial respiration for the subsequent differentiation, as seen by the translocation of mitochondria to the immune synapse and the activation of ROS signalling (Quintana et al. 2007; Sena et al. 2013). Glycolytic respiration is associated with proinflammatory Th17 cells whereas high lipid oxidation rate is associated with Tregs (Michalek et al. 2011).

7 MITOCHONDRIA AS POTENTIAL THERAPEUTIC TARGETS

With such multifaceted roles, any dysfunction in mitochondria affects key cellular processes and positions them at the hub of pathogenesis of allergic diseases. This makes mitochondria a novel, potential target for treatment and clinical management of patients with allergies. Given the multitude of diseases with mitochondrial dysfunction as the central pathology, three prominent strategies have emerged for invigorating the mitochondrial function – Repair, Reprogram and Replace (Agrawal and Mabalirajan 2016).

The repair strategy aims toward scavenging harmful mitochondrial ROS. General antioxidants like alpha tocopherol and vitamin C have been used toward this, with limited therapeutic success (Reddy 2011). This ineffectiveness of general antioxidant therapy could be due to enormous levels of mtROS which could not be attenuated using these antioxidants or that the mtROS mediated effects had already reached damaging proportions and could not be mitigated by simply quenching secondary ROS. Interestingly, researchers are now showing promising results using mitochondrial targeted antioxidants like coenzyme Q10, Mito-Q and Mito-TEMPO, that have been earmarked for mitochondria via lipophilic cations or mitochondria targeted peptides (Sorrentino, Menzies, and Auwerx 2018). In a study of steroid dependent patients of bronchial asthma, a combination of coenzyme Q10, alpha tocopherol and vitamin c improved steroid sensitivity in the patients (Gvozdjáková et al. 2005). As detailed in the preceding sections, in experimental models of OVA and HDM induced allergic inflammation, administration of a mitochondria-targeted antioxidant – NecroX5 diminished the mtROS levels and alleviated airway inflammation and hyperresponsiveness (Kim et al. 2014). Another study illustrates the role of mito-TEMPO in restoring the mitochondrial and epithelial barrier defects both *in vitro* and *in vivo* (Yang et al. 2017). Although these studies highlight the beneficial effects of mitochondria targeted antioxidants, caution should be exercised since the oxidant-antioxidant balance is critical to the maintenance of cellular homeostasis. Moreover, different cell types require different levels of ROS for their optimal functions, making mito targeted antioxidant therapy a cell type dependent approach (Iyer, Mishra and Agrawal 2017).

The other strategy of mitochondrial therapeutics is reprogramming the organelle by modulating its regulatory pathways. Stimulators like resveratrol and L-arginine act through sirtuins and NO pathway respectively, and have been shown to enhance mitochondrial biogenesis and function in models of allergic inflammation (Ahmad et al. 2010; Mabalirajan, et al. 2010a, 2010b;

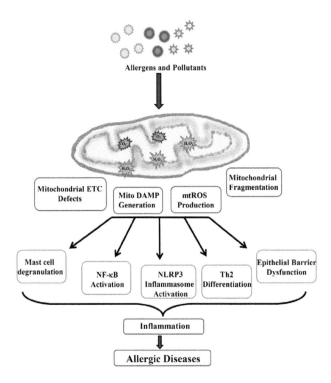

Figure 1 A schematic overview of the allergen and pollutant driven mitochondrial dysfunction in pathogenesis of allergic diseases.

Aich et al. 2012). Studies have also elaborated the role of miRNAs like mitomirs and hypoxamirs in improving the mitochondrial function (Agrawal and Mabalirajan 2016). With recent advances indicating a link between the gut microbiome and allergies, a microbial derivative – Pyrroloquinoline quinone (PQQ) which is an enhancer of mitochondrial biogenesis, also appears to be a strong candidate for mitochondria based therapies (Agrawal and Mabalirajan 2016). Other pharmacological agents that modulate mitochondrial bioenergetics, mitochondrial fission-fusion dynamics and enhance mitophagy are currently under assessment as therapies in various diseases where mitochondrial dysfunction is the central component (Sorrentino, Menzies, and Auwerx 2018). Although, they have not yet been reported in allergic diseases but they could also be explored as attractive therapeutic options.

Third and the most recent arm of mitochondrial medicine is mitochondrial replacement therapy. Defective mitochondria can be replaced by healthy, functional mitochondria through microinjection of isolated mitochondria; peptide tagging of naked mitochondria and through Mesenchymal Stem Cells (MSCs) based transfer of mitochondria (Agrawal and Mabalirajan 2016). MSCs have been shown to be effective in donating their mitochondria to cells that have dysfunctional mitochondria via tunnelling nanotubes or though extracellular vesicles. A recent study highlights the donation of mitochondria from MSCs to injured epithelial cells in murine models of allergic inflammation, leading to alleviation of the allergic features. This restoration was governed by Rho GTPase Miro1 which regulates the trafficking of mitochondria (Ahmad et al. 2014). Although MSC based therapy of donating mitochondria seems promising, it warrants deeper investigation for standardizing dose and source of MSCs and other factors related to MSC donor efficiency before taking to clinical trials.

8 CONCLUSION

Multiple studies highlight a bidirectional connection between allergic diseases and mitochondrial dysfunction. While allergens and air borne particulate matter have been shown to cause mitochondrial dysfunction, allergic features are also aggravated by pre-existing mitochondrial defects. The dysfunctional mitochondria then leads to exaggerated innate and adaptive immune responses, ultimately precipitating into inflammation and other allergic disease manifestations (Figure 1). This positioning of mitochondria at the center of pathogenesis of allergic diseases has opened newer therapeutic avenues like specific targeting of mitochondrial ROS by novel chemical quenchers and MSC based mitochondrial transfer therapies. With increased prevalence of allergies due to persistent air pollution, such strategies offer promise of improved clinical interventions.

REFERENCES

Adam, Christian, Jonas Wohlfarth, Maike Haußmann, Helga Sennefelder, Annette Rodin, Mareike Maler, Stefan F. Martin, et al. 2017. "Allergy-Inducing Chromium Compounds Trigger Potent Innate Immune Stimulation Via ROS-Dependent Inflammasome Activation." *The Journal of Investigative Dermatology* 137 (2): 367–376. doi:10.1016/j.jid.2016.10.003.

Agrawal, Anurag, and Ulaganathan Mabalirajan. 2016. "Rejuvenating Cellular Respiration for Optimizing Respiratory Function: Targeting Mitochondria." *American Journal of Physiology-Lung Cellular and Molecular Physiology* 310 (2): L103–L113. doi:10.1152/ajplung.00320.2015.

Aguilera-Aguirre, Leopoldo, Attila Bacsi, Alfredo Saavedra-Molina, Alexander Kurosky, Sanjiv Sur, and Istvan Boldogh. 2009. "Mitochondrial Dysfunction Increases Allergic Airway Inflammation." *Journal of Immunology (Baltimore, Md.: 1950)* 183 (8): 5379–5387. doi:10.4049/jimmunol.0900228.

Ahmad, Tanveer, Ulaganathan Mabalirajan, Balaram Ghosh, and Anurag Agrawal. 2010. "Altered Asymmetric Dimethyl Arginine Metabolism in Allergically Inflamed Mouse Lungs." *American Journal of Respiratory Cell and Molecular Biology* 42 (1): 3–8. doi:10.1165/rcmb.2009-0137RC.

Ahmad, Tanveer, Shravani Mukherjee, Bijay Pattnaik, Manish Kumar, Suchita Singh, Manish Kumar, Rakhshinda Rehman, et al. 2014. "Miro1 Regulates Intercellular Mitochondrial Transport & Enhances Mesenchymal Stem Cell Rescue Efficacy." *The EMBO Journal* 33 (9): 994–1010. doi:10.1002/embj.201386030.

Aich, Jyotirmoi, Ulaganathan Mabalirajan, Tanveer Ahmad, Kritika Khanna, Rakshinda Rehman, Anurag Agrawal, and Balaram Ghosh. 2012. "Resveratrol Attenuates Experimental Allergic Asthma in Mice by Restoring Inositol Polyphosphate 4 Phosphatase (INPP4A)." *International Immunopharmacology* 14 (4): 438–443. doi:10.1016/j.intimp.2012.08.017.

Archer, Stephen L. 2013. "Mitochondrial Dynamics – Mitochondrial Fission and Fusion in Human Diseases." Edited by Dan L. Longo. *New England Journal of Medicine* 369 (23): 2236–2251. doi:10.1056/NEJMra1215233.

Archibald, John M. 2015. "Endosymbiosis and Eukaryotic Cell Evolution." *Current Biology* 25 (19): R911–R921. doi:10.1016/j.cub.2015.07.055.

Cahoon, Jonathon, and Vikas Anathy. 2015. "Endoplasmic Reticulum – Mitochondrial Interactions in House Dust Mite Induced Inflammation." *UVM College of Arts and Sciences College Honors Theses*, January. https://scholarworks.uvm.edu/castheses/12.

Chakraborty, Krishnendu, Mahesh Raundhal, Bill B. Chen, Christina Morse, Yulia Y. Tyurina, Anupriya Khare, Timothy B. Oriss, et al. 2017. "The Mito-DAMP Cardiolipin Blocks IL-10 Production Causing Persistent Inflammation during Bacterial Pneumonia." *Nature Communications* 8 (January): 13944. doi:10.1038/ncomms13944.

Chandel, Navdeep S. 2014. "Mitochondria as Signaling Organelles." *BMC Biology* 12 (1). BioMed Central: 34. doi:10.1186/1741-7007-12-34.

Chodaczek, Grzegorz, Attila Bacsi, Nilesh Dharajiya, Sanjiv Sur, Tapas K. Hazra, and Istvan Boldogh. 2009. "Ragweed Pollen-Mediated IgE-Independent Release of Biogenic Amines from Mast Cells via Induction of Mitochondrial Dysfunction." *Molecular Immunology* 46 (13): 2505–2514. doi:10.1016/j.molimm.2009.05.023.

Crouser, Elliott D., Guohong Shao, Mark W. Julian, Jennifer E. Macre, Gerald S. Shadel, Susheela Tridandapani, Qin Huang, et al. 2009. "Monocyte Activation by Necrotic Cells Is Promoted by Mitochondrial Proteins and Formyl Peptide Receptors." *Critical Care Medicine* 37 (6): 2000–2009. doi:10.1097/CCM.0b013e3181a001ae.

Daum, Günther. 1985. "Lipids of Mitochondria." *Biochimica et Biophysica Acta* 822 (1): 1–42. www.ncbi.nlm.nih.gov/pubmed/2408671.

Galli, Stephen J., and Mindy Tsai. 2012. "IgE and Mast Cells in Allergic Disease." *Nature Medicine* 18 (5). Nature Publishing Group: 693–704. doi:10.1038/nm.2755.

Gershwin, Laurel J. 2015. "Comparative Immunology of Allergic Responses." *Annual Review of Animal Biosciences* 3 (1). Annual Reviews: 327–346. doi:10.1146/annurev-animal-022114-110930.

Grevendonk, Lotte, Bram G. Janssen, Charlotte Vanpoucke, Wouter Lefebvre, Mirjam Hoxha, Valentina Bollati, and Tim S. Nawrot. 2016. "Mitochondrial Oxidative DNA Damage and Exposure to Particulate Air Pollution in Mother-Newborn Pairs." *Environmental Health: A Global Access Science Source* 15 (1): 10. doi:10.1186/s12940-016-0095-2.

Gross, Olaf, Christina J. Thomas, Greta Guarda, and Jurg Tschopp. 2011. "The Inflammasome: An Integrated View." *Immunological Reviews* 243 (1): 136–151. doi:10.1111/j.1600-065X.2011.01046.x.

Gvozdjáková, Anna, Jarmila Kucharská, Mária Bartkovjaková, Katarína Gazdíková, and Frantis Ek Gazdík. 2005. "Coenzyme Q10 Supplementation Reduces Corticosteroids Dosage in Patients with Bronchial Asthma." *BioFactors (Oxford, England)* 25 (1–4): 235–240. www.ncbi.nlm.nih.gov/pubmed/16873952.

Hamanaka, Robert B., and Navdeep S. Chandel. 2010. "Mitochondrial Reactive Oxygen Species Regulate Cellular Signaling and Dictate Biological Outcomes." *Trends in Biochemical Sciences* 35 (9). Elsevier: 505–513. doi:10.1016/j.tibs.2010.04.002.

Hong, Elizabeth E., Cindy Y. Okitsu, Andrew D. Smith, and Chih-Lin Hsieh. 2013. "Regionally Specific and Genome-Wide Analyses Conclusively Demonstrate the Absence of CpG Methylation in Human Mitochondrial DNA." *Molecular and Cellular Biology* 33 (14): 2683–2690. doi:10.1128/MCB.00220-13.

Hosoki, Koa, Istvan Boldogh, and Sanjiv Sur. 2015. "Innate Responses to Pollen Allergens." *Current Opinion in Allergy and Clinical Immunology* 15 (1). NIH Public Access: 79–88. doi:10.1097/ACI.0000000000000136.

Idzko, Marco, Hamida Hammad, Menno van Nimwegen, Mirjam Kool, Monique A.M. Willart, Femke Muskens, Henk C. Hoogsteden, et al. 2007. "Extracellular ATP Triggers and Maintains Asthmatic Airway Inflammation by Activating Dendritic Cells." *Nature Medicine* 13 (8): 913–919. doi:10.1038/nm1617.

Iyer, Divyaanka, Navya Mishra, and Anurag Agrawal. 2017. "Mitochondrial Function in Allergic Disease." *Current Allergy and Asthma Reports* 17 (5): 29. doi:10.1007/s11882-017-0695-0.

Iyer, Shankar S., Qiong He, John R. Janczy, Eric I. Elliott, Zhenyu Zhong, Alicia K. Olivier, Jeffrey J. Sadler, et al. 2013. "Mitochondrial Cardiolipin Is Required for Nlrp3 Inflammasome Activation." *Immunity* 39 (2): 311–323. doi:10.1016/j.immuni.2013.08.001.

Iyer, Shankar S., Wilco P. Pulskens, Jeffrey J. Sadler, Loes M. Butter, Gwendoline J. Teske, Tyler K. Ulland, Stephanie

C. Eisenbarth, et al. 2009. "Necrotic Cells Trigger a Sterile Inflammatory Response through the Nlrp3 Inflammasome." *Proceedings of the National Academy of Sciences of the United States of America* 106 (48): 20388–20393. doi:10.1073/pnas.0908698106.

Jovaisaite, Virginija, and Johan Auwerx. 2015. "The Mitochondrial Unfolded Protein Response – Synchronizing Genomes." *Current Opinion in Cell Biology* 33 (April): 74–81. doi:10.1016/j.ceb.2014.12.003.

Kamata, Hideaki, Shi-Ichi Honda, Shin Maeda, Lufen Chang, Hajime Hirata, and Michael Karin. 2005. "Reactive Oxygen Species Promote TNFalpha-Induced Death and Sustained JNK Activation by Inhibiting MAP Kinase Phosphatases." *Cell* 120 (5): 649–661. doi:10.1016/j.cell.2004.12.041.

Kaminski, Marcin M., Sven W. Sauer, Claus-Detlev Klemke, Dorothee Süss, Jürgen G. Okun, Peter H. Krammer, and Karsten Gülow. 2010. "Mitochondrial Reactive Oxygen Species Control T Cell Activation by Regulating IL-2 and IL-4 Expression: Mechanism of Ciprofloxacin-Mediated Immunosuppression." *Journal of Immunology (Baltimore, MD: 1950)* 184 (9): 4827–4841. doi:10.4049/jimmunol.0901662.

Kay, A. Barry 2001. "Allergy and Allergic Diseases." Edited by Ian R. Mackay and Fred S. Rosen. *New England Journal of Medicine* 344 (1). Massachusetts Medical Society: 30–37. doi:10.1056/NEJM200101043440106.

Kay, A. Barry, Allen P. Kaplan, Jean Bousquet, and Patrick G. Holt. 2008. *Allergy and Allergic Diseases.* Edited by A. Barry Kay, Allen P. Kaplan, Jean Bousquet, and Patrick G. Holt. Second Edition. Blackwell Publishing Ltd., Chicester, United Kingdom.

Kepp, Oliver, Lorenzo Galluzzi, and Guido Kroemer. 2011. "Mitochondrial Control of the NLRP3 Inflammasome." *Nature Immunology* 12 (3): 199–200. doi:10.1038/ni0311-199.

Kim, So Ri, Dong ImKim, SoonHa Kim, Hern-Ku Lee, Kyung Sun Lee, Seong Ho Cho, and Yong Chul Lee. 2014. "NLRP3 Inflammasome Activation by Mitochondrial ROS in Bronchial Epithelial Cells Is Required for Allergic Inflammation." *Cell Death & Disease* 5 (10): e1498. doi:10.1038/cddis.2014.460.

Kipen, Howard M., Sampada Gandhi, David Q. Rich, Pamela Ohman-Strickland, Robert Laumbach, Zhi-Hua Fan, Li Chen, et al. 2011. "Acute Decreases in Proteasome Pathway Activity after Inhalation of Fresh Diesel Exhaust or Secondary Organic Aerosol." *Environmental Health Perspectives* 119 (5). National Institute of Environmental Health Science: 658–663. doi:10.1289/ehp.1002784.

Klotz, Luisa, Indra Dani, Frank Edenhofer, Lars Nolden, Bernd Evert, Bianca Paul, Waldemar Kolanus, et al. 2007. "Peroxisome Proliferator-Activated Receptor Gamma Control of Dendritic Cell Function Contributes to Development of CD4+ T Cell Anergy." *Journal of Immunology (Baltimore, MD: 1950)* 178 (4): 2122–2131. www.ncbi.nlm.nih.gov/pubmed/17277116.

Konrádová, Václava, C. Copová, Blanka Suková, and Josef Houstĕk. 1985. "Ultrastructure of the Bronchial Epithelium in Three Children with Asthma." *Pediatric Pulmonology* 1 (4): 182–187. Accessed November 3. www.ncbi.nlm.nih.gov/pubmed/4069806.

Kouzaki, Hideaki, Koji Iijima, Takao Kobayashi, Scott M. O'Grady, and Hirohito Kita. 2011. "The Danger Signal, Extracellular ATP, Is a Sensor for an Airborne Allergen and Triggers IL-33 Release and Innate Th2-Type Responses." *Journal of Immunology (Baltimore, MD:*

1950) 186 (7). NIH Public Access: 4375–4387. doi:10.4049/jimmunol.1003020.

Krawczyk, Connie M., Thomas Holowka, Jie Sun, Julianna Blagih, Eyal Amiel, Ralph J. DeBerardinis, Justin R. Cross, et al. 2010. "Toll-Like Receptor-Induced Changes in Glycolytic Metabolism Regulate Dendritic Cell Activation." *Blood* 115 (23): 4742–4749. doi:10.1182/blood-2009-10-249540.

Lambrecht, Bart N., and Hamida Hammad. 2014. "Allergens and the Airway Epithelium Response: Gateway to Allergic Sensitization." *The Journal of Allergy and Clinical Immunology* 134 (3): 499–507. doi:10.1016/j.jaci.2014.06.036.

Li, Ruijin, Xiaojing Kou, Hong Geng, Jingfang Xie, Zhenhua Yang, Yuexia Zhang, Zongwei Cai, et al. 2015. "Effect of Ambient PM(2.5) on Lung Mitochondrial Damage and Fusion/Fission Gene Expression in Rats." *Chemical Research in Toxicology* 28 (3): 408–418. doi:10.1021/tx5003723.

Lohman, Alexander W., Marie Billaud, and Brant E. Isakson. 2012. "Mechanisms of ATP Release and Signalling in the Blood Vessel Wall." *Cardiovascular Research* 95 (3): 269–280. doi:10.1093/cvr/cvs187.

López-Armada, María J., Romina R. Riveiro-Naveira, Carlos Vaamonde-García, and Marta N. Valcárcel-Ares. 2013. "Mitochondrial Dysfunction and the Inflammatory Response." *Mitochondrion* 13 (2): 106–118. doi:10.1016/j.mito.2013.01.003.

Mabalirajan, Ulaganathan, Tanveer Ahmad, Geeta Devi Leishangthem, Amit Kumar Dinda, Anurag Agrawal, and Balaram Ghosh. 2010a. "L-Arginine Reduces Mitochondrial Dysfunction and Airway Injury in Murine Allergic Airway Inflammation." *International Immunopharmacology* 10 (12): 1514–1519. doi:10.1016/j.intimp.2010.08.025.

Mabalirajan, Ulaganathan, Tanveer Ahmad, Geeta Devi Leishangthem, Duraisamy Arul Joseph, Amit Kumar Dinda, Anurag Agrawal, and Balaram Ghosh. 2010b. "Beneficial Effects of High Dose of L-Arginine on Airway Hyperresponsiveness and Airway Inflammation in a Murine Model of Asthma." *The Journal of Allergy and Clinical Immunology* 125 (3). Elsevier: 626–635. doi:10.1016/j.jaci.2009.10.065.

Mabalirajan, Ulaganathan, and Balaram Ghosh. 2013. "Mitochondrial Dysfunction in Metabolic Syndrome and Asthma." *Journal of Allergy* 2013: 1–12. doi:10.1155/2013/340476.

Mattila, Pirkko, Sakari Joenväärä, Jutta Renkonen, Sanna Toppila-Salmi, and Risto Renkonen. 2011. "Allergy as an Epithelial Barrier Disease." *Clinical and Translational Allergy* 1 (1). BioMed Central: 5. doi:10.1186/2045-7022-1-5.

Meyer, Alain, Gilles Laverny, Livio Bernardi, Anne Laure Charles, Ghada Alsaleh, Julien Pottecher, Jean Sibilia, et al. 2018. "Mitochondria: An Organelle of Bacterial Origin Controlling Inflammation." *Frontiers in Immunology* 9. Frontiers Media SA: 536. doi:10.3389/fimmu.2018.00536.

Michalek, Ryan D., Valerie A. Gerriets, Sarah R. Jacobs, Andrew N. Macintyre, Nancie J. MacIver, Emily F. Mason, Sarah A. Sullivan, et al. 2011. "Cutting Edge: Distinct Glycolytic and Lipid Oxidative Metabolic Programs Are Essential for Effector and Regulatory CD4+ T Cell Subsets." *Journal of Immunology (Baltimore, MD: 1950)* 186 (6): 3299–3303. doi:10.4049/jimmunol.1003613.

Minnicozzi, Michael, Richard T. Sawyer, and Matthew J. Fenton. 2011. "Innate Immunity in Allergic Disease." *Immunological Reviews* 242 (1): 106–127. doi:10.1111/j.1600-065X.2011.01025.x.

Misawa, Takuma, Michihiro Takahama, Tatsuya Kozaki, Hanna Lee, Jian Zou, Tatsuya Saitoh, and Shizuo Akira. 2013. "Microtubule-Driven Spatial Arrangement of Mitochondria Promotes Activation of the NLRP3 Inflammasome." *Nature Immunology* 14 (5): 454–460. doi:10.1038/ni.2550.

Muino, Juan Carlos, Raul Boudet, Maria Chaig, Roberto Chaig, Nelida Gerez, and Juan Carlos Copioli. 2014. "The Relationship Between Mitochondrial Haplogroups Variant on Children with Cow Milk Allergy Expressed As Atopic Dermatitis and Gastrointestinal Disease." *Journal of Allergy and Clinical Immunology* 133 (2). Elsevier: AB114. doi:10.1016/j.jaci.2013.12.424.

Müller, Ludwig. 1986. "Consequences of Cadmium Toxicity in Rat Hepatocytes: Mitochondrial Dysfunction and Lipid Peroxidation." *Toxicology* 40 (3): 285–295. www.ncbi.nlm.nih.gov/pubmed/3750329.

Nakahira, Kiichi, Jeffrey Adam Haspel, Vijay A.K. Rathinam, Seon-Jin Lee, Tamas Dolinay, Hilaire C. Lam, Joshua A. Englert, et al. 2011. "Autophagy Proteins Regulate Innate Immune Responses by Inhibiting the Release of Mitochondrial DNA Mediated by the NALP3 Inflammasome." *Nature Immunology* 12 (3): 222–230. doi:10.1038/ni.1980.

Naviaux, Robert K. 2014. "Metabolic Features of the Cell Danger Response." *Mitochondrion* 16 (May): 7–17. doi:10.1016/j.mito.2013.08.006.

Nunnari, Jodi, and Anu Suomalainen. 2012. "Mitochondria: In Sickness and in Health." *Cell* 148 (6). Elsevier: 1145–1159. doi:10.1016/j.cell.2012.02.035.

Panaro, Maria Antonietta, Angela Acquafredda, Margherita Sisto, Sabrina Lisi, Angela Bruna Maffione, and Vincenzo Mitolo. 2006. "Biological Role of the N-Formyl Peptide Receptors." *Immunopharmacology and Immunotoxicology* 28 (1): 103–127. doi:10.1080/08923970600625975.

Pawankar, Ruby. 2014. "Allergic Diseases and Asthma: A Global Public Health Concern and a Call to Action." *World Allergy Organization Journal* 7 (1). BioMed Central: 1–3. doi:10.1186/1939-4551-7-12.

Pawankar, Ruby, Giorgio W. Canonica, Stephen T. Holgate, Richard F. Lockey, and Michael Blaiss. 2013. *WAO White Book on Allergy 2013 Update.* www.worldallergy.org/UserFiles/file/WhiteBook2-2013-v8.pdf.

Plantamura, Emilie, Amiran Dzutsev, Mathias Chamaillard, Sophia Djebali, Lyvia Moudombi, Lilia Boucinha, Morgan Grau, et al. 2018. "MAVS Deficiency Induces Gut Dysbiotic Microbiota Conferring a Proallergic Phenotype." *Proceedings of the National Academy of Sciences of the United States of America* 115 (41). National Academy of Sciences: 10404–10409. doi:10.1073/pnas.1722372115.

Qi, Shanshan, Cindy Barnig, Anne-Laure Charles, Anh Poirot, Alain Meyer, Raphaël Clere-Jehl, Fréderic de Blay, et al. 2017. "Effect of Nasal Allergen Challenge in Allergic Rhinitis on Mitochondrial Function of Peripheral Blood Mononuclear Cells." *Annals of Allergy, Asthma & Immunology: Official Publication of the American College of Allergy, Asthma, & Immunology* 118 (3): 367–369. doi:10.1016/j.anai.2016.11.026.

Quintana, Ariel, Christian Schwindling, Anna S. Wenning, Ute Becherer, Jens Rettig, Eva C. Schwarz, and Markus Hoth. 2007. "T Cell Activation Requires Mitochondrial Translocation to the Immunological Synapse." *Proceedings of the National Academy of Sciences* 104 (36): 14418–14423. doi:10.1073/pnas.0703126104.

Raby, Benjamin A., Barbara Klanderman, Amy Murphy, Salvatore Mazza, Carlos A. Camargo, Edwin K. Silverman, and Scott T. Weiss. 2007. "A Common Mitochondrial Haplogroup Is Associated with Elevated Total Serum IgE Levels." *The Journal of Allergy and Clinical Immunology* 120 (2): 351–358. doi:10.1016/j.jaci.2007.05.029.

Reddy, P. Hemachandra. 2011. "Mitochondrial Dysfunction and Oxidative Stress in Asthma: Implications for Mitochondria-Targeted Antioxidant Therapeutics." *Pharmaceuticals (Basel, Switzerland)* 4 (3): 429–456. doi:10.3390/ph4030429.

Sebag, Sara C., Olha M. Koval, John D. Paschke, Christopher J. Winters, Alejandro P. Comellas, and Isabella M. Grumbach. 2018. "Inhibition of the Mitochondrial Calcium Uniporter Prevents IL-13 and Allergen-Mediated Airway Epithelial Apoptosis and Loss of Barrier Function." *Experimental Cell Research* 362 (2): 400–411. doi:10.1016/j.yexcr.2017.12.003.

Sebag, Sara C., Olha M. Koval, John D. Paschke, Christopher J. Winters, Omar A. Jaffer, Ryszard Dworski, Fayyaz S. Sutterwala, et al. 2017. "Mitochondrial CaMKII Inhibition in Airway Epithelium Protects against Allergic Asthma." *JCI Insight* 2 (3): e88297. doi:10.1172/jci.insight.88297.

Sena, Laura A., Sha Li, Amit Jairaman, Murali Prakriya, Teresa Ezponda, David A. Hildeman, Chyung-Ru Wang, et al. 2013. "Mitochondria Are Required for Antigen-Specific T Cell Activation through Reactive Oxygen Species Signaling." *Immunity* 38 (2): 225–236. doi:10.1016/j.immuni.2012.10.020.

Seth, Rashu B., Lijun Sun, Chee-Kwee Ea, and Zhijian J. Chen. 2005. "Identification and Characterization of MAVS, a Mitochondrial Antiviral Signaling Protein That Activates NF-KappaB and IRF 3." *Cell* 122 (5): 669–682. doi:10.1016/j.cell.2005.08.012.

Shi, He-Xin, Xing Liu, Qiang Wang, Pei-Pei Tang, Xin-Yi Liu, Yu-Fei Shan, and Chen Wang. 2011. "Mitochondrial Ubiquitin Ligase MARCH5 Promotes TLR7 Signaling by Attenuating TANK Action." Edited by Barbara Sherry. *PLoS Pathogens* 7 (5): e1002057. doi:10.1371/journal.ppat.1002057.

Shimada, Kenichi, Timothy R. Crother, Justin Karlin, Jargalsaikhan Dagvadorj, Norika Chiba, Shuang Chen, V. Krishnan Ramanujan, et al. 2012. "Oxidized Mitochondrial DNA Activates the NLRP3 Inflammasome during Apoptosis." *Immunity* 36 (3): 401–414. doi:10.1016/j.immuni.2012.01.009.

Smith, Alan E., and Kjeld A. Marcker. 1968. "N-Formylmethionyl Transfer RNA in Mitochondria from Yeast and Rat Liver." *Journal of Molecular Biology* 38 (2): 241–243. www.ncbi.nlm.nih.gov/pubmed/5760639.

Sorrentino, Vincenzo, Keir J. Menzies, and Johan Auwerx. 2018. "Repairing Mitochondrial Dysfunction in Disease." *Annual Review of Pharmacology and Toxicology* 58 (1): 353–389. doi:10.1146/annurev-pharmtox-010716-104908.

Subramanian, Naeha, Kannan Natarajan, Menna R. Clatworthy, Ze Wang, and Ronald N. Germain. 2013. "The Adaptor MAVS Promotes NLRP3 Mitochondrial Localization and Inflammasome Activation." *Cell* 153 (2). NIH Public Access: 348–361. doi:10.1016/j.cell.2013.02.054.

Trinchese, Giovanna, Lorella Paparo, Rosita Aitoro, Carmela Fierro, Michela Varchetta, Rita Nocerino, Maria Pina Mollica, et al. 2018. "Hepatic Mitochondrial Dysfunction and Immune Response in a Murine Model of Peanut Allergy." *Nutrients* 10 (6): 744. doi:10.3390/nu10060744.

Wang, Xiao-Bo, Hui-Yuan Gao, Bai-Ling Hou, Jian Huang, Rong-Gang Xi, and Li-Jun Wu. 2007. "Nanoparticle Realgar Powders Induce Apoptosis in U937 Cells through

Caspase MAPK and Mitochondrial Pathways." *Archives of Pharmacal Research* 30 (5): 653–658. www.ncbi.nlm.nih.gov/pubmed/17615687.

Wei, Xiawei, Bin Shao, Zhiyao He, Tinghong Ye, Min Luo, Yaxiong Sang, Xiao Liang, et al. 2015. "Cationic Nanocarriers Induce Cell Necrosis through Impairment of Na(+)/K(+)-ATPase and Cause Subsequent Inflammatory Response." *Cell Research* 25 (2): 237–253. doi:10.1038/cr.2015.9.

West, A Phillip, Igor E Brodsky, Christoph Rahner, Dong Kyun Woo, Hediye Erdjument-Bromage, Paul Tempst, Matthew C. Walsh, et al. 2011. "TLR Signalling Augments Macrophage Bactericidal Activity through Mitochondrial ROS." *Nature* 472 (7344): 476–480. doi:10.1038/nature09973.

Yang, Li, Cheng-Lun Na, Shiyu Luo, David Wu, Simon Hogan, Taosheng Huang, and Timothy E. Weaver. 2017. "The Phosphatidylcholine Transfer Protein Stard7 Is Required for Mitochondrial and Epithelial Cell Homeostasis." *Scientific Reports* 7 (1): 46416. doi:10.1038/srep46416.

Youle, Richard J., and Derek P. Narendra. 2011. "Mechanisms of Mitophagy." *Nature Reviews. Molecular Cell Biology* 12 (1): 9–14. doi:10.1038/nrm3028.

Yun, Jeanho, and Toren Finkel. 2014. "Mitohormesis." *Cell Metabolism* 19 (5): 757–766. doi:10.1016/j.cmet.2014.01.011.

Zhang, Fang, Xin Su, Gang Huang, Xiao-Feng Xin, E-Hong Cao, Yi Shi, and Yong Song. 2017. "Adenosine Triphosphate Promotes Allergen-Induced Airway Inflammation and Th17 Cell Polarization in Neutrophilic Asthma." *Journal of Immunology Research* 2017 (May). Hindawi: 1–10. doi:10.1155/2017/5358647.

Zhang, Jian-Zheng, Zhi Liu, Jia Liu, Ji-Xin Ren, and Tian-Sheng Sun. 2014. "Mitochondrial DNA Induces Inflammation and Increases TLR9/NF-⊠B Expression in Lung Tissue." *International Journal of Molecular Medicine* 33 (4): 817–824. doi:10.3892/ijmm.2014.1650.

Zhang, Qin, Mustafa Raoof, Yu Chen, Yuka Sumi, Tolga Sursal, Wolfgang Junger, Karim Brohi, et al. 2010. "Circulating Mitochondrial DAMPs Cause Inflammatory Responses to Injury." *Nature* 464 (7285). NIH Public Access: 104–107. doi:10.1038/nature08780.

Zhou, Rongbin, Amir S. Yazdi, Philippe Menu, and Jürg Tschopp. 2011. "A Role for Mitochondria in NLRP3 Inflammasome Activation." *Nature* 469 (7329): 221–225. doi: 10.1038/nature09663.

Section III

Factors Affecting Mitochondrial Function

24 Role of Carnosic Acid in Protection of Brain Mitochondria

Marcos Roberto de Oliveira

Grupo de Estudos em Neuroquímica e Neurobiologia de Moléculas Bioativas, Universidade Federal de Mato Grosso (UFMT), Av. Fernando Corrêa da Costa, 2367, CEP 78060-900, Cuiaba, MT, Brazil

CONTENTS

1 BRAIN MITOCHONDRIA

The mammalian brain presents specific characteristics that increase its susceptibility to reactive oxygen species (ROS) and reactive nitrogen species (RNS, among others. Firstly, the brain metabolism is highly dependent on oxygen gas (O_2) to maintain the production of adenosine triphosphate (ATP) in the mitochondria [1–3]. It is estimated that the brain consumes about 20% of the O_2 humans breathe [4]. Moreover, O_2 is utilized in other reactions by both neuronal and glial cells in the brain of mammals [5]. O_2 is a source for the generation of other reactive species, such as the radical anion superoxide ($O_2^{\cdot-}$), which is generated by the mitochondria due to the electron leakage from the electron transfer chain (ETC) activity [6–8]. Actually, the mitochondria are the main site of reactive oxygen species (ROS) production in virtually every nucleated mammalian cell, including neurons and glia [9–11]. The high utilization of glucose by neurons and glia also leads to redox consequences, since glucose may undergo auto-oxidation reaction by reacting with transition metals, giving rise to reactive species [12]. Besides, reactive molecules may arise from the glycolysis, such as glyoxal and methylglyoxal [13–16].Glucose also reacts with some proteins generating glycated proteins, which loss their function and may aggregate in cytoplasm and extracellular space [17]. Moreover, the brain is rich in lipids (in both plasma membranes and myelin sheath) that may undergo lipid peroxidation, giving rise to lipid peroxide intermediates able to impair neuronal and glial function, leading to cell death [18]. In this regard, the brain contains high levels of cholesterol, which generates oxysterols after undergoing autoxidation [19,20]. The concentration of transition metals (such as iron and copper ions) is high in the brain cells, favoring the formation of the highly reactive free hydroxyl radical ($^{\cdot}OH$) by the Fenton chemistry reaction [21,22]. Also, neurotransmitters can react with O_2 in auto-oxidation reactions, leading to the formation of semiquinone radicals (as is observed during the auto-oxidation of dopamine), which presents neurotoxic activity, as previously reviewed [23]. The semiquinone radicals may also give rise to $O_2^{\cdot-}$ after reacting with O_2 [24–26]. The degradation of neurotransmitters also generates hydrogen peroxide (H_2O_2), as occurs with dopamine during the reaction mediated by monoamine oxidase (MAO) enzymes [27,28]. MAO is present in mitochondrial membranes and its active site faces the cytoplasm [27,29]. In this regard, the neurotransmitter-induced fluctuations in the Ca^{2+} ions levels in both cytosol and mitochondria are associated with redox shifts that favor pro-oxidant pulses in the brain mitochondria, as observed during the glutamate-elicited excitotoxicity [30–32]. Neuroinflammation, which is mediated by the microglia located in the brain, is also a cause to the impairment of the redox biology in the mammalian brain [33,34]. Finally, the brain exhibits a modest antioxidant defense when compared to other organs, such as liver and heart [21,35,36]. Therefore, redox impairment in brain cells is virtually easy to occur and may induce harmful consequences regarding the cell fate. Actually, disruption in the redox environment is a cause for cell death in the mammalian brain, as observed in neurodegenerative diseases [21].

The mitochondria are double-membrane organelles presenting a mitochondrial matrix in which several metabolic reactions occur. The ETC is located in the inner mitochondrial membrane (IMM) and is comprised by four protein complexes, as follows: complex I (NADH dehydrogenase), complex II (succinate dehydrogenase), complex III (ubiquinol-cytochrome c reductase), and complex IV (cytochrome c oxidase) [1,37]. The ETC is responsible for the transfer of electrons between the complexes and consequent pumping of protons from the mitochondrial matrix to the intermembrane space, in which the protons accumulate and generate the electrochemical gradient, which is utilized by the complex V (ATP synthase/ATPase) to produce ATP from adenosine diphosphate (ADP) and inorganic phosphate (Pi) [1]. Coenzyme Q10 (also known as ubiquinone) and cytochrome c are specific electron transfer agents between the mitochondrial complexes I-III and III-IV, respectively [38,39]. The electron flux releases energy, which is used by the complexes I, III and IV in order to pump the protons to the intermembrane space [40,41]. Therefore, in addition to maintaining physiological levels of O_2 in the brain, it is of crucial importance keeping the electron transfer agents in their right place and in a chemically useful form. The oxidation of coenzyme Q10 or cytochrome c by reactive species may lead to mitochondrial dysfunction and loss of the electrochemical gradient, consequently causing bioenergetics deficits due to a decrease in the synthesis of ATP. Additionally, the redox state of cytochrome c, for example, may affect cell fate when mitochondria-dependent apoptosis is triggered by certain stimuli: oxidation of cytochrome c suppresses its pro-apoptotic function and may favor necrosis in mammalian cells, giving rise to inflammation [42–44]. Thus, it is particularly relevant in the brain cells due to their high demand for ATP and limited tissue regeneration.

2 MITOCHONDRIAL ANTIOXIDANT DEFENSES

Although mitochondria are a major source of ROS, these organelles exhibit both non-enzymatic and enzymatic antioxidant defenses, as will be summarized here. The major non-enzymatic antioxidant defense found in mitochondria is glutathione (GSH) [45,46]. GSH is produced mainly in the cytosol and its transport to the mitochondrial matrix is complex and needs to be well coordinated in order to avoid mitochondria-related redox impairment and carryout the normal function of the organelles [47,48]. Mitochondria also present enzymatic antioxidant defenses, such as Mn-superoxide dismutase (Mn-SOD) and glutathione peroxidase (GPx) [49,50]. Mn-SOD is responsible for the conversion of O_2^- into H_2O_2, which generates water in a GSH-dependent reaction mediated by GPx [51]. In this reaction, GSH is consumed and GS-SG (oxidized GSH; two molecules of GSH are oxidized in the conversion of H_2O_2 into water by GPx) is produced and recycled into GSH by the enzyme glutathione reductase (GR), which depends on NADPH to reduce GS-SG [52,53]. Moreover, brain mitochondria contain a thioredoxin/peroxiredoxin system utilized to scavenge peroxides, including H_2O_2 [54–57]. Besides, there is evidence pointing to the existence of the enzyme catalase (CAT) in the mitochondria of brain cells [57]. Thus, brain mitochondria contain an appreciable antioxidant defence against ROS these organelles produce by performing reactions needed to maintain the bioenergetics status of neurons and glia physiologically.

3 MITOCHONDRIAL DYSFUNCTION IN NEURODEGENERATIVE DISEASES: AN OVERVIEW

In spite of the antioxidant system present in mitochondria, there are several work reporting dysfunction and redox impairment in these organelles during neurodegeneration.

4 PARKINSON'S DISEASE

Parkinson's disease (PD) affects more than 7 million people worldwide and is characterized as a progressive neurodegenerative disease involving severe motor limitation, such as tremor (mainly at rest), slowness of movement (bradykinesia), limb rigidity, and postural impairment [58,59]. Furthermore, there are non-motor symptoms, such as constipation, apathy, decline in cognition, depression, and sleep disturbances [60]. The number of dopaminergic neurons is reduced in the *substantia nigra pars compacta*, causing a decrease in the dopamine levels in the *striatum* [23]. Disruption in the catecholaminergic signaling is not restricted to the nigro-striatal axis, and other catecholaminergic areas are affected in PD [23]. Sporadic (the so called *idiopathic* form) PD has been associated with environmental risk factors (*e.g.*, pollutants, general toxicants, drugs, brain trauma associated or not with sports, redox impairment, mitochondrial dysfunction, to cite a few), but the familial form of PD is linked to mutations in the proteins α-synuclein and parkin [61–65]. At the molecular level, it has been observed accumulation of protein aggregates (containing mainly α-synuclein, subunits of the proteasome, ubiquitin, neurofilaments, heat-shock proteins, among others, forming the Lewy bodies), which take a role in perturbing the cellular homeostasis and trigger cell death by different ways [66–69]. The formation of dopamine-quinone molecules (which are originated from the auto-oxidation of dopamine in alkaline pH) also induces modification of α-synuclein by the formation of adducts with α-synuclein, causing cellular damage [70–72]. There is evidence indicating a vicious cycle involving α-synuclein and the formation of dopamine-quinone molecules, since α-synuclein may bind to dopamine containing vesicles, permeabilizing it and promoting dopamine release in alkaline pH in which dopamine is more reactive and auto-oxidizes [71]. Mitochondrial dysfunction is mainly represented by reduced complex I activity in PD, as will be discussed below.

Parker et al. (1989) reported impaired function of the ETC in platelets obtained from patients with idiopathic PD [73]. Also, Schapira (1990) found decreased activity of the complexes I and III in the *substantia nigra* of patients who died with PD [74]. Parker et al. (2008) demonstrated that the activity of the mitochondrial complex I is reduced in the frontal cortex of patients with PD [75]. Recently, Thomas et al. (2012) have reported that the synthesis of new complex I is impaired in the frontal cortex of patients with PD [76]. On the other hand, Janetzky et al. (1994) have published unaltered aconitase in the *substantia nigra* obtained from patients with Parkinson's disease [77]. In this regard, Penn et al. (1995) described generalized mitochondrial impairment in the muscle of subjects with PD [78]. Actually, experimental models aiming to induce PD-like symptoms in animals utilize chemical stressors (*e.g.*, rotenone, 6-hydroxydopamine, 1-methyl-4-(2′-methylphenyl)-1,2,3,6-tetrahydropyridine) that affect the function of the complex I in the brain mitochondria [79–82]. This strategy is also used in *in vitro* experimental models, in which cell lines such as the SH-SY5Y human neuroblastoma or the rat PC12 cells serve as interesting dopaminergic models in the study of possible mechanisms associated with PD [79].

In addition to the effects observed in the function of mitochondria, the dynamics of these organelles may also be altered in the brain of subjects affected by PD. As is the case of DJ-1, a protein found in both IMS and matrix of the mitochondria, as well as in other parts of the cell [83]. DJ-1 is a redox-sensitive protein that migrates to the mitochondria during redox impairment [84]. Mutation in DJ-1 has been associated with autosomal recessive early-onset Parkinsonism and involves mitochondrial dysfunction [85]. Even though the function of DJ-1 is not completely understood yet, a role for this protein in the maintenance of both mitochondrial function and redox state may not be discarded. Mutations in the PTEN-induced putative kinase 1 (PINK1) gene also lead to impaired mitochondrial function and dynamics, including reduced activities of the complexes I and IV, loss of the mitochondrial membrane potential, impaired ATP production, and increased production of ROS by the

mitochondria [86–88]. Additionally, it has been described decreased ability to maintain mitochondrial import and reduced levels of mitochondrial DNA (mtDNA) in the organisms presenting PINK1 mutations [89,90]. Interestingly, it was reported that mutations in PINK1 increase the susceptibility of neurons to cell death promoted by Ca^{2+} ions [91].

Parkin, an ubiquitin protein ligase, is other protein whose mutation leads to mitochondrial dysfunction. Müftüoglu et al. (2004) observed decreased activity of the complex I in the leukocytes obtained from patients presenting parkin mutations [92]. Similarly, Mortiboys et al. (2008) have found a disruption in the function of mitochondria, as well as abnormalities associated with the architecture of these organelles obtained from the fibroblasts of patients with mutations in the gene of parkin [93]. The authors described decreased complex I activity and reduced synthesis of ATP by the mitochondria obtained from parkin-mutated fibroblasts. Additionally, the duo parkin-PINK1 also presents a potential role in the maintenance of mitochondrial dynamics, *i.e.* the equilibrium between mitochondrial fusion (when two or more mitochondria fuse to generate one mitochondrion) and fission (when a mitochondrion is divided in two or more mitochondria) [94,95]. The link between parkin and PINK1 mutations and PD/parkinsonism may be studied in excellent reviews [96–100]. Other protein whose mutations are associated with PD and mitochondrial dysfunction is α-synuclein, a major component of the Lewy bodies found in both familial and sporadic PD [101]. It was described that the binding of α-synuclein to mitochondria triggers cytochrome c release and increases the levels of Ca^{2+} ions within the mitochondria [102]. Interestingly, the N-terminal 32 amino acids residues of α-synuclein seems to be a signal targeting to mitochondria, in which it binds to and decreases the activity of complex I, as well as promotes an increase in the production of reactive species, as observed in human fetal dopaminergic primary neuronal cultures [103]. Actually, the mitochondria obtained from the *substantia nigra* and *striatum* of patients suffering from PD presented increased amounts of α-synuclein [103]. Nonetheless, α-synuclein knockout mice exhibited reduced amounts of cardiolipin and decreased complexes I-III activities in their brains [104]. Thus, either overexpression or mutations of α-synuclein seems to be related to mitochondrial dysfunction in the human brain.

5 ALZHEIMER'S DISEASE

Mitochondrial dysfunction has been also observed in samples obtained from patients suffering from Alzheimer's disease (AD). AD affects over 46 million people in the world and is characterized by a decline in cognitive function, mainly memory ability and alterations in personality [105]. At the molecular, AD presents extracellular amyloid plaques composed by amyloid-β peptide in an aggregated form and intracellular neurofibrillary tangles, which contain high levels of hyperphosphorylated tau protein [106–108]. At the cellular level, AD is characterized by increased rates of neuronal loss by different types of cell death [109]. The amyloid precursor protein (APP) is the source of amyloid-β peptides through the reaction mediated by β-secretase (BACE-1) and γ-secretase, leading to the formation of amyloid-β peptides containing different number of amino acid residues [110,111]. The most common amyloid-β peptides are 1–40 and 1–42 fragment size [112]. The amyloid-β peptides originated from the action of the secretases are secreted into the extracellular fluid, in which the peptides accumulate, causing several deleterious consequences to brain cells [110].

In that context, impaired redox biology and mitochondrial dysfunction, as well as endoplasmic reticulum stress, have been seen in the brain of patients affected by AD and also in both *in vitro* and *in vivo* experimental models of AD [113–118]. Importantly, there is evidence pointing to a mechanistic link between mitochondrial impairment and cell death by enhanced expression of the voltage-dependent anion channel 1 (VDAC1) in the brain of AD patients [105,119,120]. VDAC1 is a channel protein that mediates the release of pro-apoptotic factors (among other molecules, including metabolites) to the cytosol, triggering cell death by apoptosis [121].

In the mitochondria-related field, Sorbi et al. (1983) have found decreased pyruvate dehydrogenase complex (PDHC) activity in the brain of patients with AD [122]. Sheu et al. (1985) confirmed this information by performing an immunochemical study in the brain of AD patients [123]. Sheu et al. (1994) also observed abnormalities in the α-ketoglutarate dehydrogenase complex (α-KGDHC) in the fibroblasts obtained from subjects affected by familial AD [124]. Alterations in the activities of enzymes of the Krebs cycle and of the respiratory chain (mainly complex IV) were observed by different research groups and clearly indicate mitochondria-related bioenergetics deficits in the brain of those patients [125–128]. PDHC is an important link between the glycolysis and the Krebs cycle by producing acetyl-CoA from pyruvate, and α-KGDHC is part of the Krebs cycle, generating succinyl-CoA from α-ketoglutarate [129]. The α-KGDHC produces NADH + H$^+$ by reducing NAD$^+$, and NADH is a source of electrons to the complex I (NADH dehydrogenase) in the IMM. Therefore, important sources of the electrons that would be utilized to generate an electrochemical gradient across the IMM are affected in the brain of subjects suffering from AD.

In this regard, Parker and Parks (1995) reported reduced complex IV (cytochrome c oxidase) activity in both peripheral and nervous tissues obtained from Alzheimer's disease subjects [130]. Hirai et al. (2001) observed increased amounts of mtDNA and cytochrome oxidase enzyme levels in the neurons presenting the

higher levels of oxidative stress markers in the brain of AD patients [131]. However, such molecules were found in the cytoplasm of the neurons. In this regard, the authors also reported a decrease in the number of mitochondria in the samples. Mecocci et al. (1994) have found increased oxidative damage in the mtDNA isolated from the postmortem brain tissue of AD patients [132]. Similarly, Wang et al. (2005) reported enhanced levels of mtDNA oxidation in the brain (mainly in the temporal lobe) of patients with AD [133].

Devi et al. (2006) have reported that the accumulation of APP in the channels involved in the import of molecules inside mitochondria leads to impairment in the function of the organelles [134]. The blockade of these channels caused a decrease in the activity of the complex IV activity due to a deficit in the import of this protein from the cytoplasm [135]. Moreover, that research group observed increased generation of H$_2$O$_2$ in the mitochondria obtained from AD patients [135]. Studies using mice presenting mutations in APP also demonstrated impaired mitochondrial function and redox biology, involving increased production of O$_2^-$ and H$_2$O$_2$, for example [136]. Actually, redox impairment, and not only genetic abnormalities, has been associated with mitochondria-related bioenergetics disruption in the case of AD [105,137,138]. Indeed, it was previously demonstrated that α-KGDHC is vulnerable to reactive species due to the presence of sulfhydryl groups in the FAD group of the E3 subunit [139,140]. Moreover, it was reported partial or total inactivation of complex I and succinate dehydrogenase (SDH) in Chinese hamster ovary (CHO) cells exposed to hyperoxia (98% O$_2$ for three days) [140]. During hyperoxia, the mitochondria produce higher amounts of reactive species that cause widespread redox impairment [141,142]. Furthermore, hyperoxia favors auto-oxidation reactions, among others, that cooperate in the disruption of the redox homeostasis in mammalian brain cells [143,144].

Importantly, it has been observed altered mitochondrial dynamics, *i.e.* fusion and fission, in patients suffering from AD. Wang et al. (2008) described decreased amounts of dynamin-related protein 1 (DRP-1) in the fibroblasts of patients with sporadic AD [145]. DRP-1 regulates mitochondrial fission and distribution in the cells [146]. The anormal distribution of mitochondria in the cells favors bioenergetics deficits and the maintenance of ATP-dependent reactions [147]. The increased production of amyloid-β in the neurons of AD patients may also be associated with impairment in mitochondrial dynamics [148].

Mitochondrial deficits associated with AD were also observed experimentally. Several research groups have demonstrated loss of the mitochondrial membrane potential (MMP), decreased ATP production, spontaneous generation of reactive species, and release of pro-apoptotic factors (*e.g.*, cytochrome c), among others, in both *in vitro* and *in vivo* experimental models [135,136,149–153].

6 HUNTINGTON'S DISEASE

Mitochondrial dysfunction also appears in some brain areas of Huntington's disease (HD) patients. HD patients exhibit an elongated polyglutamine sequence at the amino terminus of huntingtin (HTT) due to a CAG tri-nucleotide repeat expansion observed in the exon 1 of the gene of htt [154]. This additional polyglutamine sequence in the htt facilitates protein aggregation and altered function of the protein, giving rise to neurotoxicity due to a gain-of-function without beneficial biological effects [155–157]. There is evidence demonstrating a link between mutated HTT and mitochondrial dysfunction [158–160]; however, it remains to be fully understood how mutated htt impairs mitochondrial function and dynamics.

In relation to mitochondrial function, Sorbi et al. (1983) have described reduced activity of PDHC in the caudate and putamen of patients suffering from HD [122]. Interestingly, the authors observed decreases PDHC activity also in the hippocampus of those patients. Moreover, Browne et al. (1997) found decreased complexes II-III in both caudate and putamen of patients affected by HD [161]. On the other hand, the activity of complex IV was observed decreased in the putamen of those patients. The authors also examined the redox state of mtDNA and found enhanced levels of oxidized mtDNA only in the caudate. The activity of the SOD enzyme was found to be reduced in the parietal cortex and cerebellum of the patients. In that context, Polidori et al. (1999) have demonstrated increased oxidative damage in the mtDNA obtained from the parietal cortex of HD patients [162]. On the other hand, evidences of general impaired redox environment were not observed in the samples obtained from the brain of HD patients, as reported by Alam et al. (2000) [163]. The authors did not find any change regarding redox alterations in lipids, proteins, and DNA isolated from the brain of such patients.

Some research groups also found impaired mitochondrial function and trafficking in animal models of HD [159]. Those abnormalities may be due to the interaction of N-terminal mutant HTT fragments with the mitochondria, causing a disruption in the flux of electrons in the respiratory chain and the association of the organelles with the microtubules [159]. Mutated HTT also induces mitochondrial fragmentation (fission) and disrupts cristae architecture, favoring cell death by apoptosis through a Ca^{2+}-dependent mechanism [164]. Indeed, it was demonstrated in both mitochondria obtained by human [165] and experimental animals presenting mutant HTT [166] increased vulnerability to Ca^{2+} loads when compared to control, since mitochondrial depolarization was observed at lower levels of Ca^{2+}. Therefore, the mitochondria-related Ca^{2+} metabolism (including Ca^{2+} capacity) is affected in HD and takes a role in the triggering of apoptosis in the affected brain areas.

7 MITOCHONDRIAL DYSFUNCTION IN OTHER NEUROLOGICAL DISTURBANCES

Mitochondrial dysfunction has been seen also in other neurological disturbances, such as schizophrenia, bipolar disorder, major depression, epilepsy, and multiple sclerosis, as well as in inborn errors of metabolism, among others. Some of the alterations observed in brain mitochondria of the patients suffering from such disturbances are similar with the effects seen in PD, AD, and HD. Nonetheless, the causes of mitochondrial impairment are different. It has been described loss of MMP, increased generation of reactive species, altered activity of enzymes of the Krebs cycle and respiratory chain, and enhanced susceptibility of mitochondria to certain stressors [167–173]. Moreover, increased rates of mitochondria-dependent cell death are observed in such cases in different brain areas [170]. In addition to biochemical alterations regarding mitochondrial function and dynamics, there are genetic abnormalities leading to the disruption of the function of the organelles in these neurological disturbances [174,175]. Depending on the circumstance, it may be suggested using alternative pharmacological therapies based on bioactive molecules during the treatment of these maladies. This alternative strategy would involve, for example, the use of bioactive molecules in combination with classical treatments. However, it depends on new studies in order to evaluate the efficacy and to avoid intoxication of patients.

8 CARNOSIC ACID

Carnosic acid (CA; $C_{20}H_{28}O_4$) is characterized as a phenolic diterpene and is a major component of the *Rosmarinus officinalis* L. (known as rosemary or "alecrim") (Fig. 1) and *Salvia officinalis* L. (salvia) plants [176]. CA is an ortho-dihydroquinone-type molecule that produces an electrophilic compound after some free radical-dependent reactions [177]. The conversion of a pro-electrophilic to an electrophilic quinone leads to the interaction of CA with the transcription factor nuclear factor erythroid 2-related factor 2 (Nrf2), which is considered a master regulator of the redox environment in mammalian cells [177–179]. There is evidence that electrophilic agents reacts with some cysteines in the Keap1 protein, leading to the release of Nrf2 from the Nrf2-Keap1 complex and the consequent migration of Nrf2 to the cell nucleus [180–182]. Nrf2 binds to the ARE region of phase II detoxification enzymes, as well as to the ARE region in the genes of antioxidant enzymes [183,184]. Moreover, CA is also a direct antioxidant molecule due to the presence of two O-phenolic hydroxyl moieties in C11 and C12 [185,186].

Due to its chemical properties, CA exerts potent antioxidant and anti-inflammatory effects in different cells and tissues of mammals [187–190]. Furthermore, CA is an anti-tumor agent, as demonstrated in several

FIGURE 1 The *Rosmarinus officinalis* L. (also known as rosemary or "alecrim") plant (A) and carnosic acid chemical structure (B). The *Rosmarinus officinalis* L. image was obtained from https://pixabay.com/pt/alecrim-flores-azul-violet-1090419/.

experimental models [187,191]. CA is an apolar molecule and easily crosses the blood-brain barrier [177], causing neuroprotection by the modulation of specific signaling pathways associated with neuronal survival [187]. Doolaege et al. (2011) demonstrated that the bioavailability of CA was about 40.1% after 360 min of oral administration of this diterpene at 64.3 ± 5.8 mg/kg [192]. Additionally, Romo Vaquero et al. (2013) found CA in the brain of Zucker female rats after the ingestion of rosemary extract enriched in CA [193]. Importantly, CA administration to experimental animals caused neuroprotection in models of traumatic brain injury [194], retinitis pigmentosa [195], memory impairment [196], and cyanide intoxication [197]. CA also stimulated neurotrophin upregulation in the rat brain after intranasal administration [198,199]. Learning and memory abilities in mice were also improved after the ingestion of CA-enriched extract [200]. Therefore, it is very likely that CA is a neuromodulatory agent presenting benefits in chemically and biologically diverse challenges and also exhibits promising therapeutic utilization, as recently reviewed [187]. Importantly, CA induced cytoprotection in a greater extent when compared with other bioactive molecules in some experimental models [187].

9 THE EFFECTS OF CARNOSIC ACID ON BRAIN MITOCHONDRIA

In that context, it is easy to understand why mitochondria of brain cells face a chemically intense environment. Therefore, it is necessary to think about exogenous agents that may promote mitochondrial protection in brain cells. In this regard, prevention of mitochondrial dysfunction by the ingestion of bioactive molecules is of particular interest due to the cost of the pharmacological treatment. Nonetheless, it remains to be fully understood how bioactive molecules cause mitochondrial protection.

Furthermore, studies aiming to reveal whether bioactive molecules may cause intoxication in humans need to be performed in order to better apply the preventive utilization of these agents in the maintenance of human health. In this regard, CA has attracted attention due to its ability in promoting mitochondrial protection without inducing toxic effects experimentally (Fig. 2). CA affects parameters related with mitochondria redox biology and bioenergetics, as well as modulates signaling pathways associated with mitochondria-dependent apoptotic cell death, as summarized in Table 1.

10 IN VITRO EXPERIMENTAL MODELS

The ability of CA in inducing mitochondrial protection in brain cells has been seen in *in vitro* experimental models using the human neuroblastoma SH-SY5Y cell line.

Meng et al. (2015) demonstrated that a pretreatment with CA at 10 μM for 1 h attenuated the effects of amyloid-β peptides 1–42 and 1–43 on SH-SY5Y by blocking the mitochondria-associated cell death pathway [201]. The authors found that CA prevented the amyloid-β-induced upregulation of caspase-9 and caspase-3, among others, which are activated after the release of cytochrome c from the mitochondria during the triggering of apoptosis [208]. Therefore, it is expected that CA downregulated cell death-related signaling pathways that mediate the activation of the mitochondrial step involved in the onset of apoptosis and/or CA directly blocked the release of pro-apoptotic factors by an unknown mechanism. In this regard, different research groups have reported that CA (at 1 μM) is an activator of the phosphoinositide 3-kinase/Akt (PI3K/Akt) signaling pathway, resulting in an anti-apoptotic action in SH-SY5Y cells exposed to different toxicants [202,203,209]. The inhibition of the PI3K/Akt axis abrogated the anti-apoptotic and mitochondria-related effects induced by

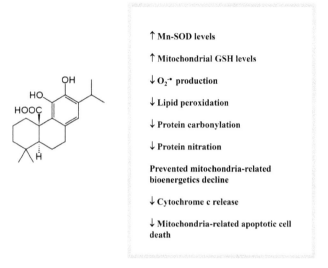

↑ Mn-SOD levels

↑ Mitochondrial GSH levels

↓ O$_2^{•}$ production

↓ Lipid peroxidation

↓ Protein carbonylation

↓ Protein nitration

Prevented mitochondria-related bioenergetics decline

↓ Cytochrome c release

↓ Mitochondria-related apoptotic cell death

FIGURE 2   A summary of the effects induced by carnosic acid on brain mitochondria. As discussed in the text, carnosic acid activates the Nrf2 transcription factor by different mechanisms, leading to mitochondrial protection in different experimental models.

TABLE 1
Summary of the effects induced by CA on mitochondria.

Experimental model	Effects	References
CA at 1 mg/kg administrated via i.p. 15 min after the induction of traumatic brain injury in male CF-1 mice	restored the activity of complex I and complex II, ↓ the levels of 4-HNE and 3-nitrotyrosine in mitochondria of the mice cerebral cortex	[194]
CA at 10 μM for 1 h before exposure of SH-SY5Y cells to amyloid-β peptide	↓ the activation of the pro-apoptotic caspases-9 and -3 ↓ the cleavage of PARP ↓ the amounts of the mature form of AIF ↓ the number of apoptotic cells	[201]
CA at 1 μM for 12 h before exposure of SH-SY5Y cells to different chemical stressors (paraquat, methylglyoxal, chlorpyrifos, H$_2$O$_2$)	↓ the amounts of the pro-apoptotic protein Bax ↓ the levels of markers of mitochondria-related apoptotic cell death (cytochrome c release, caspases-9 and -3 activity, cleavage of PARP, fragmentation of DNA) ↓ the effect of the toxicants on the loss of MMP ↑ the amounts of GSH and Mn-SOD in mitochondria ↓ the amounts of markers of oxidative and nitrosative stress in the membranes of mitochondria Prevented the effects of the chemical stressors on the activity of mitochondria-located enzymes associated with the maintenance of the bioenergetics state (aconitase, α-ketoglutarate dehydrogenase, succinate dehydrogenase, complex I, complex V) Prevented the chemically-induced decrease in the amounts of ATP Inhibition of the HO-1 enzyme attenuated the CA-elicited mitochondrial protection	[202–206]
CA at 1 mg/kg via i.p. 48 h before mitochondrial isolation and chemical challenge with 4-HNE in an *ex vivo* experimental model	Prevented the 4-HNE-induced dysfunction in complexes I and II Prevented the increase in 4-HNE in mitochondrial samples CA at 1 mg/kg induced mitochondrial protection in a very similar way when compared to sulforaphane at 5 mg/kg	[207]

CA in SH-SY5Y cells [203]. The PI3K/Akt signaling pathway upregulation was associated with the activation of the transcription factor Nrf2, a well-known mediator of cytoprotection in mammalian cells [210,211]. This transcription factor modulated the CA-induced expression of Mn-SOD, a mitochondria-located enzyme responsible for the conversion of O_2^- into H_2O_2 in the mitochondrial matrix [212,213]. Indeed, it was observed decreased levels of markers of redox impair in the mitochondrial membranes obtained from SH-SY5Y treated with CA and challenged with certain chemical stressors [202–204]. Thus, CA caused mitochondrial protection by a mechanism associated with the activation of the PI3K/Akt/Nrf2 signaling pathway.

In other works, it was reported that CA also prevented mitochondria-related bioenergetics disturbances caused by exposing SH-SY5Y cells to chemical stressors such as paraquat [205] and H_2O_2 [206]. In this regard, it was recently demonstrated that the enzyme HO-1 may play a pivotal role in mediating the CA-elicited mitochondrial protection in SH-SY5Y cells, since the inhibition of this enzyme by zinc protoporphyrin IX (ZnPP IX) suppressed the effects resulting from the pretreatment with CA in that experimental model [205]. Moreover, the CA-induced expression of HO-1 seems to be dependent on the PI3K/Akt axis, since blocking of this signaling pathway abrogated HO-1 induction in CA-treated SH-SY5Y cells [203]. Upregulation of the Nrf-2/HO-1 signaling pathway has been associated with both antioxidant and anti-inflammatory effects in different experimental models [214]. Therefore, the discovery of new agents able to modulate the Nrf2-/HO-1 axis is pharmacologically interesting.

Even though the activation of the PI3K/Akt/Nrf2/HO-1 signaling pathway appears to be crucial in the CA-mediated mitochondrial protection, as observed in vitro, this is not the only way to decrease the impact of chemical stressors on the organelles. Lin et al. (2016) have found that CA (1 μM) depends on parkin to block 6-hydroxydopamine-induced mitochondria-related apoptosis in SH-SY5Y cells [215]. Parkin is an ubiquitin E3 ligase that mediates the binding of ubiquitin to target proteins that should be degraded by the 26S proteasome [216]. Additionally, parkin directly regulates the activity of the 26S proteasome, activating it [217]. Protein degradation is necessary to remove damaged proteins from mammalian cells, and impairment in the 26S proteasomal activity leads to protein accumulation and aggregation, as observed in different neurodegenerative diseases [218,219]. Similarly, Liu et al. (2016) have reported that the inhibition of autophagy attenuated the mitochondrial protection mediated by CA in SH-SY5Y cells exposed to amyloid-β_{25-35}, as assessed through the measurement of the MMP [220]. Thus, it is very likely that proteasomal activity and/or autophagy are necessary to preserve mitochondrial function in brain cells exposed to stressful conditions. The need for one or another signaling pathway may be dependent on the type of stress to which the cells are submitted.

It is apparent that several signaling pathways may be modulated by the same bioactive molecule; however, the effects seen on mitochondria may be more dependent on one or another. CA prevented mitochondrial impairment in SH-SY5Y cells by different mechanisms, and in vivo experimental models should be performed in order to investigate whether and how CA would benefit mitochondria in different conditions.

11 EX VIVO EXPERIMENTAL MODELS

The role of CA as a mitochondrial protective agent has also been studied in ex vivo experimental models. Miller et al. (2013) have found that mitochondria obtained from cerebral cortex of male CF-1 mice ingesting CA at 1 mg/kg for two days are more resistant to an ex vivo challenge with 4-hydroxynonenal (4-HNE) [207], which is a reactive aldehyde that inhibits mitochondrial respiration [221]. The levels of 4-HNE are high in the brain and cerebrospinal fluid of patients with AD, and this reactive aldehyde may be an important inducer of mitochondrial dysfunction and cell death in such disease [222]. Thus, it is very likely that the ingestion of CA modulated neuroprotective signaling pathways associated with the maintenance of mitochondrial function. This modulation caused a decrease in the vulnerability of the mitochondria during the challenge with 4-HNE. Since 4-HNE is a redox stressor, the CA-induced mitochondrial protection may be associated with the ability CA exhibits in upregulating the synthesis of GSH, which may be transported into mitochondria after its production in the cytosol [223]. Indeed, de Oliveira et al. (2015) demonstrated that CA upregulated the levels of GSH in SH-SY5Y cells by a mechanism dependent on the PI3K/Akt/Nrf-2 axis [202]. Nonetheless, it remains to be clarified whether this is triggered in in vivo experimental models.

12 IN VIVO EXPERIMENTAL MODELS

There is a lack of data regarding the mitochondria-related in vivo effects triggered by CA. Miller et al. (2015) described that the administration of CA at 1 mg/kg to male CF-1 mice 15 min after traumatic brain injury preserved mitochondrial function and reduced the levels of 4-HNE and 3-nitrotyrosine in mitochondria obtained from the cerebral cortex [194]. The exact mechanism by which CA modulated such parameters was not analyzed by the authors; however, the activation of cytoprotective signaling pathways such as PI3K/Akt/Nrf2 may not be discarded. Further research would be necessary in order to elucidate how CA induced mitochondrial protection in in vivo experimental models. Zhang et al. (2015) demonstrated that CA administration at 25 mg/kg.day^{-1} (added to the food) during 29 days protected some brain areas (neocortex, hippocampus, striatum) of male NSA mice exposed to cyanide, a mitochondrial toxicant that inhibits complex IV activity [197]. The authors did not analyze mitochondria-related effects of CA in that experimental model, but it is expected that CA caused mitochondrial protection attenuating the cyanide-induced neuronal death. However, it remains to be investigated in details.

13 CONCLUSION

CA is effective in attenuating mitochondrial impairment in brain cells, as observed in different experimental models using a wide range of chemical stressors. Nonetheless, it remains to be elucidated whether CA would be able to protect mitochondria in other mammalian tissues. Importantly, it is recommended to increase the efforts regarding the investigation of the *in vivo* effects of CA in experimental animals and also in humans. Even though *in vitro* data are crucial to understand the mechanism underlying the mitochondrial protection elicited by CA, *in vivo* experiments are pivotal to reveal whether CA would be able to access a specific brain area under stress, rescuing mitochondria. Regarding CA bioavailability, studies involving nanotechnology-related pharmacological strategies would be welcome, since optimizing CA distribution in the different brain areas would be important to improve the neuroprotective action of this diterpene. In this context, the involvement of other signaling pathways in mediating the CA-induced cytoprotection should be addressed in future research.

REFERENCES

1. Chance B, Williams GR (1995) Respiratory enzymes in oxidative phosphorylation. I. Kinetics of oxygen utilization. J Biol Chem 217:383–393.
2. Magistretti PJ, Allaman I (2015) A cellular perspective on brain energy metabolism and functional imaging. Neuron 86:883–901. doi: 10.1016/j.neuron.2015.03.035.
3. Cobley JN, Fiorello ML, Bailey DM (2018) 13 reasons why the brain is susceptible to oxidative stress. Redox Biol 15:490–503. doi: 10.1016/j.redox.2018.01.008.
4. Mink JW, Blumenschine RJ, Adams DB (1981) Ratio of central nervous system to body metabolism in vertebrates: Its constancy and functional basis. Am J Physiol 241: R203–R212.
5. Fukuto JM, Carrington SJ, Tantillo DJ, Harrison JG, Ignarro LJ, Freeman BA, Chen A, Wink DA (2012) Small molecule signaling agents: The integrated chemistry and biochemistry of nitrogen oxides, oxides of carbon, dioxygen, hydrogen sulfide, and their derived species. Chem Res Toxicol 25:769–793. doi: 10.1021/tx2005234.
6. Kudin AP, Bimpong-Buta NY, Vielhaber S, Elger CE, Kunz WS (2004) Characterization of superoxide-producing sites in isolated brain mitochondria. J Biol Chem 279:4127–4135.
7. Adam-Vizi V (2005) Production of reactive oxygen species in brain mitochondria: Contribution by electron transport chain and non-electron transport chain sources. Antioxid Redox Signal 7:1140–1149.
8. Kudin AP, Debska-Vielhaber G, Kunz WS (2005) Characterization of superoxide production sites in isolated rat brain and skeletal muscle mitochondria. Biomed Pharmacother 59:163–168.
9. Naoi M, Maruyama W, Shamoto-Nagai M, Yi H, Akao Y, Tanaka M (2005) Oxidative stress in mitochondria: Decision to survival and death of neurons in neurodegenerative disorders. Mol Neurobiol 31:81–93.
10. Lin MT, Beal MF (2006) Mitochondrial dysfunction and oxidative stress in neurodegenerative diseases. Nature 443:787–795.
11. Schon EA, Przedborski S (2011) Mitochondria: The next (neurode)generation. Neuron 70:1033–1053. doi: 10.1016/j.neuron.2011.06.003.
12. Prebil M, Jensen J, Zorec R, Kreft M (2011) Astrocytes and energy metabolism. Arch Physiol Biochem 117:64–69. doi: 10.3109/13813455.2010.539616.
13. Richard JP (1991) Kinetic parameters for the elimination reaction catalyzed by triosephosphate isomerase and an estimation of the reaction's physiological significance. Biochemistry 30:4581–4585.
14. Richard JP (1993) Mechanism for the formation of methylglyoxal from triosephosphates. Biochem Soc Trans 21:549–553.
15. Bélanger M, Allaman I, Magistretti PJ (2011) Brain energy metabolism: focus on astrocyte-neuron metabolic cooperation. Cell Metab 14:724–738. doi: 10.1016/j.cmet.2011.08.016.
16. Allaman I, Bélanger M, Magistretti PJ (2015) Methylglyoxal, the dark side of glycolysis. Front Neurosci 9:23. doi: 10.3389/fnins.2015.00023.
17. Abate G, Marziano M, Rungratanawanich W, Memo M, Uberti D (2017) Nutrition and AGE-ing: Focusing on Alzheimer's disease. Oxid Med Cell Longev 2017:7039816. doi: 10.1155/2017/7039816.
18. Bazinet RP, Layé S (2014) Polyunsaturated fatty acids and their metabolites in brain function and disease. Nat Rev Neurosci 2014 Dec;15(12):771–785. doi: 10.1038/nrn3820.
19. Miyoshi N, Iuliano L, Tomono S, Ohshima H (2014) Implications of cholesterol autoxidation products in the pathogenesis of inflammatory diseases. Biochem Biophys Res Commun 446:702–708. doi: 10.1016/j.bbrc.2013.12.107.
20. Zerbinati C, Iuliano L (2017) Cholesterol and related sterols autoxidation. Free Radic Biol Med 111:151–155. doi: 10.1016/j.freeradbiomed.2017.04.013.
21. Halliwell B (2006) Oxidative stress and neurodegeneration: Where are we now? J Neurochem 97:1634–1658.
22. Pavlin M, Repič M, Vianello R, Mavri J (2016) The chemistry of neurodegeneration: Kinetic data and their implications. Mol Neurobiol 53:3400–3415. doi: 10.1007/s12035-015-9284-1.
23. Lotharius J, Brundin P (2002) Pathogenesis of Parkinson's disease: Dopamine, vesicles and alpha-synuclein. Nat Rev Neurosci 3:932–942.
24. Heikkila RE, Cohen G (1973) 6-Hydroxydopamine: Evidence for superoxide radical as an oxidative intermediate. Science 181:456–457.
25. Cohen G, Heikkila RE (1974) The generation of hydrogen peroxide, superoxide radical, and hydroxyl radical by 6-hydroxydopamine, dialuric acid, and related cytotoxic agents. J Biol Chem 249:2447–2452.
26. Miller DM, Buettner GR, Aust SD (1990) Transition metals as catalysts of "autoxidation" reactions. Free Radic Biol Med 8:95–108.

27. Youdim MB, Edmondson D, Tipton KF (2006) The therapeutic potential of monoamine oxidase inhibitors. Nat Rev Neurosci 7:295–309.

28. Mousseau DD, Baker GB (2012) Recent developments in the regulation of monoamine oxidase form and function: Is the current model restricting our understanding of the breadth of contribution of monoamine oxidase to brain [dys]function? Curr Top Med Chem 12:2163–2176.

29. Tipton KF, Davey GP, McDonald AG (2011) Kinetic behavior and reversible inhibition of monoamine oxidases–enzymes that many want dead. Int Rev Neurobiol 100:43–64. doi: 10.1016/B978-0-12-386467-3.00003-0.

30. Mattson MP (2007) Calcium and neurodegeneration. Aging Cell 6:337–350.

31. Zündorf G, Reiser G (2011) Calcium dysregulation and homeostasis of neural calcium in the molecular mechanisms of neurodegenerative diseases provide multiple targets for neuroprotection. Antioxid Redox Signal 14:1275–1288. doi: 10.1089/ars.2010.3359.

32. Calì T, Ottolini D, Brini M (2012) Mitochondrial Ca(2+) and neurodegeneration. Cell Calcium 52:73–85. doi: 10.1016/j.ceca.2012.04.015.

33. Ginhoux F, Greter M, Leboeuf M, Nandi S, See P, Gokhan S, Mehler MF, Conway SJ, Ng LG, Stanley ER, Samokhvalov IM, Merad M (2010) Fate mapping analysis reveals that adult microglia derive from primitive macrophages. Science 330:841–845. doi: 10.1126/science.1194637.

34. Thomas DD (2015) Breathing new life into nitric oxide signaling: A brief overview of the interplay between oxygen and nitric oxide. Redox Biol 5:225–233. doi: 10.1016/j.redox.2015.05.002.

35. Baxter PS, Hardingham GE (2016) Adaptive regulation of the brain's antioxidant defences by neurons and astrocytes. Free Radic Biol Med 100:147–152. doi: 10.1016/j.freeradbiomed.2016.06.027.

36. Ren X, Zou L, Zhang X, Branco V, Wang J, Carvalho C, Holmgren A, Lu J (2017) Redox signaling mediated by thioredoxin and glutathione systems in the central nervous system. Antioxid Redox Signal 27:989–1010. doi: 10.1089/ars.2016.6925.

37. Chance B, Williams GR, Holmes WF, Higgins J (1955) Respiratory enzymes in oxidative phosphorylation. V. A mechanism for oxidative phosphorylation. J Biol Chem 217:439–451.

38. Lenaz G, Genova ML (2009) Mobility and function of coenzyme Q (ubiquinone) in the mitochondrial respiratory chain. Biochim Biophys Acta 1787:563–573. doi: 10.1016/j.bbabio.2009.02.019.

39. Enriquez JA, Lenaz G (2014) Coenzyme q and the respiratory chain: Coenzyme q pool and mitochondrial supercomplexes. Mol Syndromol 5:119–140. doi: 10.1159/000363364.

40. Kühlbrandt W (2015) Structure and function of mitochondrial membrane protein complexes. BMC Biol 13:89. doi: 10.1186/s12915-015-0201-x.

41. Logan A, Pell VR, Shaffer KJ, Evans C, Stanley NJ, Robb EL, Prime TA, Chouchani ET, Cochemé HM, Fearnley IM, Vidoni S, James AM, Porteous CM, Partridge L, Krieg T, Smith RA, Murphy MP (2016) Assessing the mitochondrial membrane potential in cells and in vivo using targeted click chemistry and mass spectrometry. Cell Metab 23:379–385. doi: 10.1016/j.cmet.2015.11.014.

42. Otsu K, Sato K, Ikeda Y, Imai H, Nakagawa Y, Ohba Y, Fujii J (2005) An abortive apoptotic pathway induced by singlet oxygen is due to the suppression of caspase activation. Biochem J 389:197–206.

43. Suto D, Sato K, Ohba Y, Yoshimura T, Fujii J (2005) Suppression of the pro-apoptotic function of cytochrome c by singlet oxygen via a haem redox state-independent mechanism. Biochem J 392:399–406.

44. Brown GC, Borutaite V (2008) Regulation of apoptosis by the redox state of cytochrome c. Biochim Biophys Acta 1777:877–881. doi: 10.1016/j.bbabio.2008.03.024.

45. Marí M, Morales A, Colell A, García-Ruiz C, Fernández-Checa JC (2009) Mitochondrial glutathione, a key survival antioxidant. Antioxid Redox Signal 11:2685–2700. doi: 10.1089/ARS.2009.2695.

46. Marí M, Morales A, Colell A, García-Ruiz C, Kaplowitz N, Fernández-Checa JC (2013) Mitochondrial glutathione: Features, regulation and role in disease. Biochim Biophys Acta 1830:3317–3328. doi: 10.1016/j.bbagen.2012.10.018.

47. Ribas V, García-Ruiz C, Fernández-Checa JC (2014) Glutathione and mitochondria. Front Pharmacol 5:151. doi: 10.3389/fphar.2014.00151.

48. Calabrese G, Morgan B, Riemer J (2017) Mitochondrial glutathione: Regulation and functions. Antioxid Redox Signal 27:1162–1177. doi: 10.1089/ars.2017.7121.

49. Handy DE, Lubos E, Yang Y, Galbraith JD, Kelly N, Zhang YY, Leopold JA, Loscalzo J (2009) Glutathione peroxidase-1 regulates mitochondrial function to modulate redox-dependent cellular responses. J Biol Chem 284:11913–11921. doi: 10.1074/jbc.M900392200.

50. Bresciani G, Da Cruz IB, González-Gallego J (2015) Manganese superoxide dismutase and oxidative stress modulation. Adv Clin Chem 68:87–130. doi: 10.1016/bs.acc.2014.11.001.

51. Miriyala S, Spasojevic I, Tovmasyan A, Salvemini D, Vujaskovic Z, St Clair D, Batinic-Haberle I (2012) Manganese superoxide dismutase, MnSOD and its mimics. Biochim Biophys Acta 1822:794–814. doi: 10.1016/j.bbadis.2011.12.002.

52. Kalyanaraman B (2013) Teaching the basics of redox biology to medical and graduate students: Oxidants, antioxidants and disease mechanisms. Redox Biol 1:244–257. doi: 10.1016/j.redox.2013.01.014.

53. Batinic-Haberle I, Tovmasyan A, Spasojevic I (2015) An educational overview of the chemistry, biochemistry and therapeutic aspects of Mn porphyrins-from superoxide dismutation to H2O2-driven pathways. Redox Biol 5:43–65. doi: 10.1016/j.redox.2015.01.017.

54. Hattori F, Murayama N, Noshita T, Oikawa S (2003) Mitochondrial peroxiredoxin-3 protects hippocampal neurons from excitotoxic injury in vivo. J Neurochem 86:860–868.

55. Cox AG, Winterbourn CC, Hampton MB (2009) Mitochondrial peroxiredoxin involvement in antioxidant defence and redox signalling. Biochem J 425:313–325. doi: 10.1042/BJ20091541.

56. Drechsel DA, Patel M (2010) Respiration-dependent H2O2 removal in brain mitochondria via the thioredoxin/peroxiredoxin system. J Biol Chem 285:27850–27858. doi: 10.1074/jbc.M110.101196.

57. Starkov AA, Andreyev AY, Zhang SF, Starkova NN, Korneeva M, Syromyatnikov M, Popov VN (2014) Scavenging of H2O2 by mouse brain mitochondria. J Bioenerg Biomembr 46:471–477. doi: 10.1007/s10863-014-9581-9.

58. Schulz JB, Gerlach M, Gille G, Kuhn W, Müngersdorf M, Riederer P, Südmeyer M, Ludolph A (2011) Basic science in Parkinson's disease: Its impact on clinical practice. J Neurol 258:S299–S306. doi: 10.1007/s00415-011-6040-y.

59. Krüger R, Klucken J, Weiss D, Tönges L, Kolber P, Unterecker S, Lorrain M, Baas H, Müller T, Riederer P (2017) Classification of advanced stages of Parkinson's disease: Translation into stratified treatments. J Neural Transm (Vienna) 124:1015–1027. doi: 10.1007/s00702-017-1707-x.

60. Müller T, Gerlach M, Youdim MB, Riederer P (2012) Psychiatric, nonmotor aspects of Parkinson's disease. Handb Clin Neurol 106:477–490. doi: 10.1016/B978-0-444-52002-9.00028-0.

61. Elbaz A, Tranchant C (2007) Epidemiologic studies of environmental exposures in Parkinson's disease. J Neurol Sci 262:37–44.

62. Sian-Hülsmann J, Mandel S, Youdim MB, Riederer P (2011) The relevance of iron in the pathogenesis of Parkinson's disease. J Neurochem 118:939–957. doi: 10.1111/j.1471-4159.2010.07132.x.

63. Caudle WM, Guillot TS, Lazo CR, Miller GW (2012) Industrial toxicants and Parkinson's disease. Neurotoxicology 33:178–188. doi: 10.1016/j.neuro.2012.01.010.

64. Goldman SM (2014) Environmental toxins and Parkinson's disease. Annu Rev Pharmacol Toxicol 54:141–164. doi: 10.1146/annurev-pharmtox-011613-135937.

65. Deng H, Wang P, Jankovic J (2018) The genetics of Parkinson disease. Ageing Res Rev 42:72–85. doi: 10.1016/j.arr.2017.12.007.

66. Kuzuhara S, Mori H, Izumiyama N, Yoshimura M, Ihara Y (1988) Lewy bodies are ubiquitinated. A light and electron microscopic immunocytochemical study. Acta Neuropathol 75:345–353.

67. Ii K, Ito H, Tanaka K, Hirano A (1997) Immunocytochemical co-localization of the proteasome in ubiquitinated structures in neurodegenerative diseases and the elderly. J Neuropathol Exp Neurol 56:125–131.

68. Goedert M (2001) Alpha-synuclein and neurodegenerative diseases. Nat Rev Neurosci 2:492–501.

69. Auluck PK, Chan HY, Trojanowski JQ, Lee VM, Bonini NM (2002) Chaperone suppression of alpha-synuclein toxicity in a Drosophila model for Parkinson's disease. Science 295:865–868.

70. Schulz JB, Lindenau J, Seyfried J, Dichgans J (2000) Glutathione, oxidative stress and neurodegeneration. Eur J Biochem 267:4904–4911.

71. Maguire-Zeiss KA, Short DW, Federoff HJ (2005) Synuclein, dopamine and oxidative stress: Co-conspirators in Parkinson's disease? Brain Res Mol Brain Res 134:18–23.

72. Norris EH, Giasson BI, Hodara R, Xu S, Trojanowski JQ, Ischiropoulos H, Lee VM (2005) Reversible inhibition of alpha-synuclein fibrillization by dopaminochrome-mediated conformational alterations. J Biol Chem 280:21212–21219.

73. Parker WD Jr, Boyson SJ, Parks JK (1989) Abnormalities of the electron transport chain in idiopathic Parkinson's disease. Ann Neurol 26:719–723.

74. Schapira AH, Cooper JM, Dexter D, Clark JB, Jenner P, Marsden CD (1990) Mitochondrial complex I deficiency in Parkinson's disease. J Neurochem 54:823–827.

75. Parker WD Jr, Parks JK, Swerdlow RH (2008) Complex I deficiency in Parkinson's disease frontal cortex. Brain Res 1189:215–218.

76. Thomas RR, Keeney PM, Bennett JP (2012) Impaired complex-I mitochondrial biogenesis in Parkinson disease frontal cortex. J Parkinsons Dis 2:67–76. doi: 10.3233/JPD-2012-11074.

77. Janetzky B, Hauck S, Youdim MB, Riederer P, Jellinger K, Pantucek F, Zöchling R, Boissl KW, Reichmann H (1994) Unaltered aconitase activity, but decreased complex I activity in substantia nigra pars compacta of patients with Parkinson's disease. Neurosci Lett 169:126–128.

78. Penn AM, Roberts T, Hodder J, Allen PS, Zhu G, Martin WR (1995) Generalized mitochondrial dysfunction in Parkinson's disease detected by magnetic resonance spectroscopy of muscle. Neurology 45:2097–2099.

79. Beal MF (2001) Experimental models of Parkinson's disease. Nat Rev Neurosci 2:325–334.

80. Schober A (2004) Classic toxin-induced animal models of Parkinson's disease: 6-OHDA and MPTP. Cell Tissue Res 318:215–224.

81. Büeler H (2009) Impaired mitochondrial dynamics and function in the pathogenesis of Parkinson's disease. Exp Neurol 218:235–246. doi: 10.1016/j.expneurol.2009.03.006.

82. Blesa J, Przedborski S (2014) Parkinson's disease: Animal models and dopaminergic cell vulnerability. Front Neuroanat 8:155. doi: 10.3389/fnana.2014.00155.

83. Zhang L, Shimoji M, Thomas B, Moore DJ, Yu SW, Marupudi NI, Torp R, Torgner IA, Ottersen OP, Dawson TM, Dawson VL (2005) Mitochondrial localization of the Parkinson's disease related protein DJ-1: Implications for pathogenesis. Hum Mol Genet 14:2063–2073.

84. Miller DW, Ahmad R, Hague S, Baptista MJ, Canet-Aviles R, McLendon C, Carter DM, Zhu PP, Stadler J, Chandran J, Klinefelter GR, Blackstone C, Cookson MR (2003) L166P mutant DJ-1, causative for recessive Parkinson's disease, is degraded through the ubiquitin-proteasome system. J Biol Chem 278:36588–36595.

85. Bonifati V, Rizzu P, van Baren MJ, Schaap O, Breedveld GJ, Krieger E, Dekker MC, Squitieri F, Ibanez P, Joosse M, van Dongen JW, Vanacore N, van Swieten JC, Brice A, Meco G, van Duijn CM, Oostra BA, Heutink P (2003) Mutations in the DJ-1 gene associated with autosomal recessive early-onset parkinsonism. Science 299:256–259.

86. Gautier CA, Kitada T, Shen J (2008) Loss of PINK1 causes mitochondrial functional defects and increased sensitivity to oxidative stress. Proc Natl Acad Sci U S A 105:11364–11369. doi: 10.1073/pnas.0802076105.

87. Wood-Kaczmar A, Gandhi S, Yao Z, Abramov AY, Miljan EA, Keen G, Stanyer L, Hargreaves I, Klupsch K, Deas E, Downward J, Mansfield L, Jat P, Taylor J, Heales S, Duchen MR, Latchman D, Tabrizi SJ, Wood NW (2008) PINK1 is necessary for long term survival and mitochondrial function in human dopaminergic neurons. PLoS One 3:e2455. doi: 10.1371/journal.pone.0002455.

88. Gegg ME, Cooper JM, Schapira AH, Taanman JW (2009) Silencing of PINK1 expression affects mitochondrial DNA and oxidative phosphorylation in dopaminergic cells. PLoS One 4:e4756. doi: 10.1371/journal.pone.0004756.

89. Marongiu R, Spencer B, Crews L, Adame A, Patrick C, Trejo M, Dallapiccola B, Valente EM, Masliah E (2009) Mutant Pink1 induces mitochondrial dysfunction in a neuronal cell model of Parkinson's disease by disturbing calcium flux. J Neurochem 108:1561–1574. doi: 10.1111/j.1471-4159.2009.05932.x.

90. Papa S, Sardanelli AM, Capitanio N, Piccoli C (2009) Mitochondrial respiratory dysfunction and mutations in

mitochondrial DNA in PINK1 familial parkinsonism. J Bioenerg Biomembr 41:509–516. doi: 10.1007/s10863-009-9252-4.

91. Gandhi S, Wood-Kaczmar A, Yao Z, Plun-Favreau H, Deas E, Klupsch K, Downward J, Latchman DS, Tabrizi SJ, Wood NW, Duchen MR, Abramov AY (2009) PINK1-associated Parkinson's disease is caused by neuronal vulnerability to calcium-induced cell death. Mol Cell 33:627–638. doi: 10.1016/j.molcel.2009.02.013.

92. Müftüoglu M, Elibol B, Dalmizrak O, Ercan A, Kulaksiz G, Ogüs H, Dalkara T, Ozer N (2004) Mitochondrial complex I and IV activities in leukocytes from patients with parkin mutations. Mov Disord 19:544–548.

93. Mortiboys H, Thomas KJ, Koopman WJ, Klaffke S, Abou-Sleiman P, Olpin S, Wood NW, Willems PH, Smeitink JA, Cookson MR, Bandmann O (2008) Mitochondrial function and morphology are impaired in parkin-mutant fibroblasts. Ann Neurol 64:555–565. doi: 10.1002/ana.21492.

94. Deng H, Dodson MW, Huang H, Guo M (2008) The Parkinson's disease genes pink1 and parkin promote mitochondrial fission and/or inhibit fusion in Drosophila. Proc Natl Acad Sci U S A 105:14503–14508. doi: 10.1073/pnas.0803998105.

95. Rojas-Charry L, Cookson MR, Niño A, Arboleda H, Arboleda G (2014) Downregulation of Pink1 influences mitochondrial fusion-fission machinery and sensitizes to neurotoxins in dopaminergic cells. Neurotoxicology 44:140–148. doi: 10.1016/j.neuro.2014.04.007.

96. Winklhofer KF, Haass C (2010) Mitochondrial dysfunction in Parkinson's disease. Biochim Biophys Acta 1802:29–44. doi: 10.1016/j.bbadis.2009.08.013.

97. Pilsl A, Winklhofer KF (2012) Parkin, PINK1 and mitochondrial integrity: Emerging concepts of mitochondrial dysfunction in Parkinson's disease. Acta Neuropathol 123:173–188. doi: 10.1007/s00401-011-0902-3.

98. Scott L, Dawson VL, Dawson TM (2017) Trumping neurodegeneration: Targeting common pathways regulated by autosomal recessive Parkinson's disease genes. Exp Neurol 298:191–201. doi: 10.1016/j.expneurol.2017.04.008.

99. McWilliams TG, Muqit MM (2017) PINK1 and Parkin: Emerging themes in mitochondrial homeostasis. Curr Opin Cell Biol 45:83–91. doi: 10.1016/j.ceb.2017.03.013.

100. Harper JW, Ordureau A, Heo JM (2018) Building and decoding ubiquitin chains for mitophagy. Nat Rev Mol Cell Biol 19:93–108. doi: 10.1038/nrm.2017.129.

101. Spillantini MG, Schmidt ML, Lee VM, Trojanowski JQ, Jakes R, Goedert M (1997) Alpha-synuclein in Lewy bodies. Nature 388:839–840.

102. Parihar MS, Parihar A, Fujita M, Hashimoto M, Ghafourifar P (2008) Mitochondrial association of alpha-synuclein causes oxidative stress. Cell Mol Life Sci 65:1272–1284. doi: 10.1007/s00018-008-7589-1.

103. Devi L, Raghavendran V, Prabhu BM, Avadhani NG, Anandatheerthavarada HK (2008) Mitochondrial import and accumulation of alpha-synuclein impair complex I in human dopaminergic neuronal cultures and Parkinson disease brain. J Biol Chem 283:9089–9100. doi: 10.1074/jbc.M710012200.

104. Ellis CE, Murphy EJ, Mitchell DC, Golovko MY, Scaglia F, Barceló-Coblijn GC, Nussbaum RL (2005) Mitochondrial lipid abnormality and electron transport chain impairment in mice lacking alpha-synuclein. Mol Cell Biol 25:10190–10201.

105. Shoshan-Barmatz V, Nahon-Crystal E, Shteinfer-Kuzmine A, Gupta R (2018) VDAC1, mitochondrial dysfunction, and Alzheimer's disease. Pharmacol Res 131:87–101. doi: 10.1016/j.phrs.2018.03.010.

106. Iqbal K, Alonso AC, Gong CX, Khatoon S, Singh TJ, Grundke-Iqbal I (1994) Mechanism of neurofibrillary degeneration in Alzheimer's disease. Mol Neurobiol 9:119–123.

107. Ittner LM, Götz J (2011) Amyloid-β and tau-a toxic pas de deux in Alzheimer's disease. Nat Rev Neurosci 12:65–72. doi: 10.1038/nrn2967.

108. Mietelska-Porowska A, Wasik U, Goras M, Filipek A, Niewiadomska G (2014) Tau protein modifications and interactions: Their role in function and dysfunction. Int J Mol Sci 15:4671–4713. doi: 10.3390/ijms15034671.

109. Silva DF, Selfridge JE, Lu J, Lezi E, Cardoso SM, Swerdlow RH (2012) Mitochondrial abnormalities in Alzheimer's disease: Possible targets for therapeutic intervention. Adv Pharmacol 64:83–126. doi: 10.1016/B978-0-12-394816-8.00003-9.

110. Ow SY, Dunstan DE (2014) A brief overview of amyloids and Alzheimer's disease. Protein Sci 23:1315–1331. doi: 10.1002/pro.2524.

111. Hahr JY (2015) Physiology of the Alzheimer's disease. Med Hypotheses 85:944–946. doi: 10.1016/j.mehy.2015.09.005.

112. Andrew RJ, Kellett KA, Thinakaran G, Hooper NM (2016) A Greek tragedy: The growing complexity of Alzheimer amyloid precursor protein proteolysis. J Biol Chem 291:19235–19244. doi: 10.1074/jbc.R116.746032.

113. Zhu X, Lee HG, Casadesus G, Avila J, Drew K, Perry G, Smith MA (2005) Oxidative imbalance in Alzheimer's disease. Mol Neurobiol 31:205–217.

114. Jeong S (2017) Molecular and cellular basis of neurodegeneration in Alzheimer's disease. Mol Cells 40:613–620. doi: 10.14348/molcells.2017.0096.

115. Butterfield DA, Boyd-Kimball D (2018) Oxidative stress, amyloid-β peptide, and altered key molecular pathways in the pathogenesis and progression of Alzheimer's disease. J Alzheimers Dis 62:1345–1367. doi: 10.3233/JAD-170543.

116. Li K, Wei Q, Liu FF, Hu F, Xie AJ, Zhu LQ, Liu D (2018) Synaptic dysfunction in Alzheimer's disease: Aβ, Tau, and epigenetic alterations. Mol Neurobiol 55:3021–3032. doi: 10.1007/s12035-017-0533-3.

117. Martins RN, Villemagne V, Sohrabi HR, Chatterjee P, Shah TM, Verdile G, Fraser P, Taddei K, Gupta VB, Rainey-Smith SR, Hone E, Pedrini S, Lim WL, Martins I, Frost S, Gupta S, O'Bryant S, Rembach A, Ames D, Ellis K, Fuller SJ, Brown B, Gardener SL, Fernando B, Bharadwaj P, Burnham S, Laws SM, Barron AM, Goozee K, Wahjoepramono EJ, Asih PR, Doecke JD, Salvado O, Bush A, Rowe CC, Gandy SE, Masters CL. (2018) Alzheimer's disease: A journey from amyloid peptides and oxidative stress, to biomarker technologies and disease prevention strategies-gains from AIBL and DIAN cohort studies. J Alzheimers Dis 62:965–992. doi: 10.3233/JAD-171145.

118. Wojsiat J, Zoltowska KM, Laskowska-Kaszub K, Wojda U (2018) Oxidant/antioxidant imbalance in Alzheimer's disease: Therapeutic and diagnostic prospects. Oxid Med Cell Longev 2018:6435861. doi: 10.1155/2018/6435861.

119. Sultana R, Poon HF, Cai J, Pierce WM, Merchant M, Klein JB, Markesbery WR, Butterfield DA (2006) Identification of nitrated proteins in Alzheimer's disease brain using a redox proteomics approach. Neurobiol Dis 22:76–87.

120. Manczak M, Reddy PH (2012) Abnormal interaction of VDAC1 with amyloid beta and phosphorylated tau causes mitochondrial dysfunction in Alzheimer's disease. Hum Mol Genet 21:5131–5146. doi: 10.1093/hmg/dds360.

121. Reddy PH (2013) Is the mitochondrial outermembrane protein VDAC1 therapeutic target for Alzheimer's disease? Biochim Biophys Acta 1832:67–75. doi: 10.1016/j.bbadis.2012.09.003.

122. Sorbi S, Bird ED, Blass JP (1983) Decreased pyruvate dehydrogenase complex activity in Huntington and Alzheimer brain. Ann Neurol 13:72–78.

123. Sheu KF, Kim YT, Blass JP, Weksler ME (1985) An immunochemical study of the pyruvate dehydrogenase deficit in Alzheimer's disease brain. Ann Neurol 17:444–449.

124. Sheu KF, Cooper AJ, Koike K, Koike M, Lindsay JG, Blass JP (1994) Abnormality of the alpha-ketoglutarate dehydrogenase complex in fibroblasts from familial Alzheimer's disease. Ann Neurol 35:312–318.

125. Kish SJ, Bergeron C, Rajput A, Dozic S, Mastrogiacomo F, Chang LJ, Wilson JM, DiStefano LM, Nobrega JN (1992) Brain cytochrome oxidase in Alzheimer's disease. J Neurochem 59:776–779.

126. Mastrogiacomo F, Bergeron C, Kish SJ (1993) Brain alpha-ketoglutarate dehydrogenase complex activity in Alzheimer's disease. J Neurochem 61:2007–2014.

127. Kish SJ (1997) Brain energy metabolizing enzymes in Alzheimer's disease: Alpha-ketoglutarate dehydrogenase complex and cytochrome oxidase. Ann N Y Acad Sci 826:218–228.

128. Kish SJ, Mastrogiacomo F, Guttman M, Furukawa Y, Taanman JW, Dozić S, Pandolfo M, Lamarche J, DiStefano L, Chang LJ (1999) Decreased brain protein levels of cytochrome oxidase subunits in Alzheimer's disease and in hereditary spinocerebellar ataxia disorders: A nonspecific change?. J Neurochem 72:700–707.

129. Gibson GE, Sheu KF, Blass JP (1998) Abnormalities of mitochondrial enzymes in Alzheimer disease. J Neural Transm (Vienna) 105:855–870.

130. Parker WD Jr, Parks JK (1995) Cytochrome c oxidase in Alzheimer's disease brain: Purification and characterization. Neurology 45:482–486.

131. Hirai K, Aliev G, Nunomura A, Fujioka H, Russell RL, Atwood CS, Johnson AB, Kress Y, Vinters HV, Tabaton M, Shimohama S, Cash AD, Siedlak SL, Harris PL, Jones PK, Petersen RB, Perry G, Smith MA (2001) Mitochondrial abnormalities in Alzheimer's disease. J Neurosci 21:3017–3023.

132. Mecocci P, MacGarvey U, Beal MF (1994) Oxidative damage to mitochondrial DNA is increased in Alzheimer's disease. Ann Neurol 36:747–751.

133. Wang J, Xiong S, Xie C, Markesbery WR, Lovell MA (2005) Increased oxidative damage in nuclear and mitochondrial DNA in Alzheimer's disease. J Neurochem 93:953–962.

134. Devi L, Prabhu BM, Galati DF, Avadhani NG, Anandatheerthavarada HK (2006) Accumulation of amyloid precursor protein in the mitochondrial import channels of human Alzheimer's disease brain is associated with mitochondrial dysfunction. J Neurosci 26:9057–9068.

135. Anandatheerthavarada HK, Biswas G, Robin MA, Avadhani NG (2003) Mitochondrial targeting and a novel transmembrane arrest of Alzheimer's amyloid precursor protein impairs mitochondrial function in neuronal cells. J Cell Biol 161:41–54.

136. Takuma K, Yao J, Huang J, Xu H, Chen X, Luddy J, Trillat AC, Stern DM, Arancio O, Yan SS (2005) ABAD enhances Abeta-induced cell stress via mitochondrial dysfunction. FASEB J 19:597–598.

137. Mecocci P, Boccardi V, Cecchetti R, Bastiani P, Scamosci M, Ruggiero C, Baroni M (2018) A long journey into aging, brain aging, and Alzheimer's disease following the oxidative stress tracks. J Alzheimers Dis 62:1319–1335. doi: 10.3233/JAD-170732.

138. Onyango IG (2018) Modulation of mitochondrial bioenergetics as a therapeutic strategy in Alzheimer's disease. Neural Regen Res 13:19–25. doi: 10.4103/1673-5374.224362.

139. Schoonen WG, Wanamarta AH, van der Klei-van Moorsel JM, Jakobs C, Joenje H (1990) Hyperoxia-induced clonogenic killing of HeLa cells associated with respiratory failure and selective inactivation of Krebs cycle enzymes. Mutat Res 237:173–181.

140. Schoonen WG, Wanamarta AH, van der Klei-van Moorsel JM, Jakobs C, Joenje H (1991) Characterization of oxygen-resistant Chinese hamster ovary cells. III. Relative resistance of succinate and alpha-ketoglutarate dehydrogenases to hyperoxic inactivation. Free Radic Biol Med 10:111–118.

141. Damiani E, Donati A, Girardis M (2018) Oxygen in the critically ill. friend or foe? Curr Opin Anaesthesiol 31:129–135. doi: 10.1097/ACO.0000000000000559.

142. Kayton A, Timoney P, Vargo L, Perez JA (2018) A review of oxygen physiology and appropriate management of oxygen levels in premature neonates. Adv Neonatal Care 18:98–104. doi: 10.1097/ANC.0000000000000434.

143. Ciarlone GE, Dean JB (2016) Normobaric hyperoxia stimulates superoxide and nitric oxide production in the caudal solitary complex of rat brain slices. Am J Physiol Cell Physiol 311:C1014–C1026. doi: 10.1152/ajpcell.00160.2016.

144. Terraneo L, Paroni R, Bianciardi P, Giallongo T, Carelli S, Gorio A, Samaja M (2017) Brain adaptation to hypoxia and hyperoxia in mice. Redox Biol 11:12–20. doi: 10.1016/j.redox.2016.10.018.

145. Wang X, Su B, Fujioka H, Zhu X (2008) Dynamin-like protein 1 reduction underlies mitochondrial morphology and distribution abnormalities in fibroblasts from sporadic Alzheimer's disease patients. Am J Pathol 173:470–482. doi: 10.2353/ajpath.2008.071208.

146. Singh S, Sharma S (2017) Dynamin-related protein-1 as potential therapeutic target in various diseases. Inflammopharmacology 25:383–392. doi: 10.1007/s10787-017-0347-y.

147. Rakusan K, Tomanek RJ (1986) Distribution of mitochondria in normal and hypertrophic myocytes from the rat heart. J Mol Cell Cardiol 18:299–305.

148. Wang X, Su B, Siedlak SL, Moreira PI, Fujioka H, Wang Y, Casadesus G, Zhu X (2008) Amyloid-beta overproduction causes abnormal mitochondrial dynamics via differential modulation of mitochondrial fission/fusion proteins. Proc Natl Acad Sci U S A 105:19318–19323. doi: 10.1073/pnas.0804871105.

149. Pereira C, Santos MS, Oliveira C (1998) Mitochondrial function impairment induced by amyloid beta-peptide on PC12 cells. Neuroreport 9:1749–1755.

150. Cardoso SM, Santos S, Swerdlow RH, Oliveira CR (2001) Functional mitochondria are required for amyloid beta-mediated neurotoxicity. FASEB J 15:1439–1441.

151. Kaminsky YG, Kosenko EA (2008) Effects of amyloid-beta peptides on hydrogen peroxide-metabolizing enzymes in rat brain in vivo. Free Radic Res 42:564–573. doi: 10.1080/10715760802159057.

152. Resende R, Moreira PI, Proença T, Deshpande A, Busciglio J, Pereira C, Oliveira CR (2008) Brain oxidative stress in a triple-transgenic mouse model of Alzheimer disease. Free Radic Biol Med 44:2051–2057. doi: 10.1016/j.freeradbiomed.2008.03.012.

153. Moreira PI, Carvalho C, Zhu X, Smith MA, Perry G (2010) Mitochondrial dysfunction is a trigger of Alzheimer's disease pathophysiology. Biochim Biophys Acta 1802:2–10. doi: 10.1016/j.bbadis.2009.10.006.

154. Munoz-Sanjuan I, Bates GP (2011) The importance of integrating basic and clinical research toward the development of new therapies for Huntington disease. J Clin Invest 121:476–483. doi: 10.1172/JCI45364.

155. Weydt P, La Spada AR (2006) Targeting protein aggregation in neurodegeneration–lessons from polyglutamine disorders. Expert Opin Ther Targets 10:505–513.

156. Truant R, Atwal RS, Desmond C, Munsie L, Tran T (2008) Huntington's disease: Revisiting the aggregation hypothesis in polyglutamine neurodegenerative diseases. FEBS J 275:4252–4262. doi: 10.1111/j.1742-4658.2008.06561.x.

157. Williams AJ, Paulson HL (2008) Polyglutamine neurodegeneration: Protein misfolding revisited. Trends Neurosci 31:521–528. doi: 10.1016/j.tins.2008.07.004.

158. Weydt P, Pineda VV, Torrence AE, Libby RT, Satterfield TF, Lazarowski ER, Gilbert ML, Morton GJ, Bammler TK, Strand AD, Cui L, Beyer RP, Easley CN, Smith AC, Krainc D, Luquet S, Sweet IR, Schwartz MW, La Spada AR (2006) Thermoregulatory and metabolic defects in Huntington's disease transgenic mice implicate PGC-1alpha in Huntington's disease neurodegeneration. Cell Metab 4:349–362.

159. Orr AL, Li S, Wang CE, Li H, Wang J, Rong J, Xu X, Mastroberardino PG, Greenamyre JT, Li XJ (2008) N-terminal mutant huntingtin associates with mitochondria and impairs mitochondrial trafficking. J Neurosci 28:2783–2792. doi: 10.1523/JNEUROSCI.0106-08.2008.

160. Costa V, Scorrano L (2012) Shaping the role of mitochondria in the pathogenesis of Huntington's disease. EMBO J 31:1853–1864. doi: 10.1038/emboj.2012.65.

161. Browne SE, Bowling AC, MacGarvey U, Baik MJ, Berger SC, Muqit MM, Bird ED, Beal MF (1997) Oxidative damage and metabolic dysfunction in Huntington's disease: Selective vulnerability of the basal ganglia. Ann Neurol 41:646–653.

162. Polidori MC, Mecocci P, Browne SE, Senin U, Beal MF (1999) Oxidative damage to mitochondrial DNA in Huntington's disease parietal cortex. Neurosci Lett 272:53–56.

163. Alam ZI, Halliwell B, Jenner P (2000) No evidence for increased oxidative damage to lipids, proteins, or DNA in Huntington's disease. J Neurochem 75:840–846.

164. Giacomello M, Hudec R, Lopreiato R (2011) Huntington's disease, calcium, and mitochondria. Biofactors 37:206–218. doi: 10.1002/biof.162.

165. Panov AV, Gutekunst CA, Leavitt BR, Hayden MR, Burke JR, Strittmatter WJ, Greenamyre JT (2002) Early mitochondrial calcium defects in Huntington's disease are a direct effect of polyglutamines. Nat Neurosci 5:731–736.

166. Choo YS, Johnson GV, MacDonald M, Detloff PJ, Lesort M (2004) Mutant huntingtin directly increases susceptibility of mitochondria to the calcium-induced permeability transition and cytochrome c release. Hum Mol Genet 13:1407–1420.

167. Mahad D, Ziabreva I, Lassmann H, Turnbull D (2008) Mitochondrial defects in acute multiple sclerosis lesions. Brain 131:1722–1735. doi: 10.1093/brain/awn105.

168. Mao P, Reddy PH (2010) Is multiple sclerosis a mitochondrial disease? Biochim Biophys Acta 1802:66–79. doi: 10.1016/j.bbadis.2009.07.002.

169. Streck EL, Gonçalves CL, Furlanetto CB, Scaini G, Dal-Pizzol F, Quevedo J (2014) Mitochondria and the central nervous system: Searching for a pathophysiological basis of psychiatric disorders. Rev Bras Psiquiatr 36:156–167.

170. Rajasekaran A, Venkatasubramanian G, Berk M, Debnath M (2015) Mitochondrial dysfunction in schizophrenia: Pathways, mechanisms and implications. Neurosci Biobehav Rev 48:10–21. doi: 10.1016/j.neubiorev.2014.11.005.

171. El-Hattab AW, Scaglia F (2016) Mitochondrial cytopathies. Cell Calcium 60:199–206. doi: 10.1016/j.ceca.2016.03.003.

172. Scaini G, Rezin GT, Carvalho AF, Streck EL, Berk M, Quevedo J (2016) Mitochondrial dysfunction in bipolar disorder: Evidence, pathophysiology and translational implications. Neurosci Biobehav Rev 68:694–713. doi: 10.1016/j.neubiorev.2016.06.040.

173. Kato T (2017) Neurobiological basis of bipolar disorder: Mitochondrial dysfunction hypothesis and beyond. Schizophr Res 187:62–66. doi: 10.1016/j.schres.2016.10.037.

174. McFarland R, Taylor RW, Turnbull DM (2010) A neurological perspective on mitochondrial disease. Lancet Neurol 9:829–840. doi: 10.1016/S1474-4422(10)70116-2.

175. Clay HB, Sillivan S, Konradi C (2011) Mitochondrial dysfunction and pathology in bipolar disorder and schizophrenia. Int J Dev Neurosci 29:311–324. doi: 10.1016/j.ijdevneu.2010.08.007.

176. Birtić S, Dussort P, Pierre FX, Bily AC, Roller M (2015) Carnosic acid. Phytochemistry 115:9–19. doi: 10.1016/j.phytochem.2014.12.026.

177. Satoh T, Kosaka K, Itoh K, Kobayashi A, Yamamoto M, Shimojo Y, Kitajima C, Cui J, Kamins J, Okamoto S, Izumi M, Shirasawa T, Lipton SA (2008) Carnosic acid, a catechol-type electrophilic compound, protects neurons both in vitro and in vivo through activation of the Keap1/Nrf2 pathway via S-alkylation of targeted cysteines on Keap1. J Neurochem 104:1116–1131.

178. Rezaie T, McKercher SR, Kosaka K, Seki M, Wheeler L, Viswanath V, Chun T, Joshi R, Valencia M, Sasaki S, Tozawa T, Satoh T, Lipton SA (2012) Protective effect of carnosic acid, a pro-electrophilic compound, in models of oxidative stress and light-induced retinal degeneration. Invest Ophthalmol Vis Sci 53:7847–7854. doi: 10.1167/iovs.12-10793.

179. Satoh T, McKercher SR, Lipton SA (2013) Nrf2/ARE-mediated antioxidant actions of pro-electrophilic drugs. Free Radic Biol Med 65:645–657. doi: 10.1016/j.freeradbiomed.2013.07.022.

180. Zhang DD, Lo SC, Cross JV, Templeton DJ, Hannink M (2004) Keap1 is a redox-regulated substrate adaptor protein for a Cul3-dependent ubiquitin ligase complex. Mol Cell Biol 24:10941–10953. doi: 10.1128/MCB.24.24.10941-10953.2004.

181. Eggler AL, Liu G, Pezzuto JM, van Breemen RB, Mesecar AD (2005) Modifying specific cysteines of the electrophile-sensing human Keap1 protein is insufficient to disrupt binding to the Nrf2 domain Neh2. Proc Natl Acad Sci U S A 102:10070–10075. doi: 10.1073/pnas.0502402102.

182. Hong F, Freeman ML, Liebler DC (2005) Identification of sensor cysteines in human Keap1 modified by the cancer chemopreventive agent sulforaphane. Chem Res Toxicol 18:1917–1926.

183. Talalay P (2000) Chemoprotection against cancer by induction of phase 2 enzymes. Biofactors 12:5–11.

184. Itoh K, Tong KI, Yamamoto M (2004) Molecular mechanism activating Nrf2-Keap1 pathway in regulation of adaptive response to electrophiles. Free Radic Biol Med 36:1208–1213.

185. Erkan N, Ayranci G, Ayranci E (2008) Antioxidant activities of rosemary (Rosmarinus Officinalis L.) extract, blackseed (Nigella sativa L.) essential oil, carnosic acid, rosmarinic acid and sesamol. Food Chem 110:76–82. doi: 10.1016/j.foodchem.2008.01.058.

186. Pérez-Fons L, Garzón MT, Micol V (2010) Relationship between the antioxidant capacity and effect of rosemary (Rosmarinus officinalis L.). polyphenols on membrane phospholipid order. J Agric Food Chem 58:161–171. doi: 10.1021/jf9026487.

187. de Oliveira MR (2016) The dietary components carnosic acid and carnosol as neuroprotective agents: A mechanistic view. Mol Neurobiol 53:6155–6168. doi: 10.1007/s12035-015-9519-1.

188. Bahri S, Jameleddine S, Shlyonsky V (2016) Relevance of carnosic acid to the treatment of several health disorders: Molecular targets and mechanisms. Biomed Pharmacother 84:569–582. doi: 10.1016/j.biopha.2016.09.067.

189. Albalawi A, Alhasani RHA, Biswas L, Reilly J, Shu X (2017) Protective effect of carnosic acid against acrylamide-induced toxicity in RPE cells. Food Chem Toxicol 108:543–553. doi: 10.1016/j.fct.2017.01.026.

190. de Oliveira MR, de Souza ICC, Fürstenau CR (2018) Carnosic acid induces anti-inflammatory effects in paraquat-treated SH-SY5Y cells through a mechanism involving a crosstalk between the Nrf2/HO-1 axis and NF-κB. Mol Neurobiol 55:890–897. doi: 10.1007/s12035-017-0389-6.

191. González-Vallinas M, Reglero G, Ramírez de Molina A (2015) Rosemary (Rosmarinus officinalis L.) extract as a potential complementary agent in anticancer therapy. Nutr Cancer 67:1221–1229. doi: 10.1080/01635581.2015.1082110.

192. Doolaege EH, Raes K, De Vos F, Verhé R, De Smet S (2011) Absorption, distribution and elimination of carnosic acid, a natural antioxidant from rosmarinus officinalis, in rats. Plant Foods Hum Nutr 66:196–202. doi: 10.1007/s11130-011-0233-5.

193. Romo Vaquero M, García Villalba R, Larrosa M, Yáñez-Gascón MJ, Fromentin E, Flanagan J, Roller M, Tomás-Barberán FA, Espín JC, García-Conesa MT (2013) Bioavailability of the major bioactive diterpenoids in a rosemary extract: Metabolic profile in the intestine, liver, plasma, and brain of Zucker rats. Mol Nutr Food Res 57:1834–1846. doi: 10.1002/mnfr.201300052.

194. Miller DM, Singh IN, Wang JA, Hall ED (2015) Nrf2-ARE activator carnosic acid decreases mitochondrial dysfunction, oxidative damage and neuronal cytoskeletal degradation following traumatic brain injury in mice. Exp Neurol 264:103–110. doi: 10.1016/j.expneurol.2014.11.008.

195. Kang K, Tarchick MJ, Yu X, Beight C, Bu P, Yu M (2016) Carnosic acid slows photoreceptor degeneration in the Pde6b(rd10) mouse model of retinitis pigmentosa. Sci Rep 6:22632. doi: 10.1038/srep22632.

196. Ozarowski M, Mikolajczak PL, Bogacz A, Gryszczynska A, Kujawska M, Jodynis-Liebert J, Piasecka A, Napieczynska H, Szulc M, Kujawski R, Bartkowiak-Wieczorek J, Cichocka J, Bobkiewicz-Kozlowska T, Czerny B, Mrozikiewicz PM (2013) Rosmarinus officinalis L. leaf extract improves memory impairment and affects acetylcholinesterase and butyrylcholinesterase activities in rat brain. Fitoterapia 91:261–271. doi: 10.1016/j.fitote.2013.09.012.

197. Zhang D, Lee B, Nutter A, Song P, Dolatabadi N, Parker J, Sanz-Blasco S, Newmeyer T, Ambasudhan R, McKercher SR, Masliah E, Lipton SA (2015) Protection from cyanide-induced brain injury by the Nrf2 transcriptional activator carnosic acid. J Neurochem 133:898–908. doi: 10.1111/jnc.13074.

198. Vaka SRK, Murthy SN, Repka MA, Nagy T (2011) Upregulation of endogenous neurotrophin levels in the brain by intranasal administration of carnosic acid. J Pharm Sci 100:3139–3145. doi: 10.1002/jps.22528.

199. Vaka SR, Shivakumar HN, Repka MA, Murthy SN (2013) Formulation and evaluation of carnosic acid nanoparticulate system for upregulation of neurotrophins in the brain upon intranasal administration. J Drug Target 21:44–53. doi: 10.3109/1061186X.2012.725405.

200. Farr SA, Niehoff ML, Ceddia MA, Herrlinger KA, Lewis BJ, Feng S, Welleford A, Butterfield DA, Morley JE (2016) Effect of botanical extracts containing carnosic acid or rosmarinic acid on learning and memory in SAMP8 mice. Physiol Behav 165:328–338. doi: 10.1016/j.physbeh.2016.08.013.

201. Meng P, Yoshida H, Tanji K, Matsumiya T, Xing F, Hayakari R, Wang L, Tsuruga K, Tanaka H, Mimura J, Kosaka K, Itoh K, Takahashi I, Kawaguchi S, Imaizumi T (2015) Carnosic acid attenuates apoptosis induced by amyloid-β 1-42 or 1-43 in SH-SY5Y human neuroblastoma cells. Neurosci Res 94:1–9. doi: 10.1016/j.neures.2014.12.003.

202. de Oliveira MR, Ferreira GC, Schuck PF, Dal Bosco SM (2015) Role for the PI3K/Akt/Nrf2 signaling pathway in the protective effects of carnosic acid against methylglyoxal-induced neurotoxicity in SH-SY5Y neuroblastoma cells. Chem Biol Interact 242:396–406. doi: 10.1016/j.cbi.2015.11.003.

203. de Oliveira MR, Ferreira GC, Schuck PF (2016) Protective effect of carnosic acid against paraquat-induced redox impairment and mitochondrial dysfunction in SH-SY5Y cells: Role for PI3K/Akt/Nrf2 pathway. Toxicol In Vitro 32:41–54. doi: 10.1016/j.tiv.2015.12.005.

204. de Oliveira MR, Peres A, Ferreira GC, Schuck PF, Bosco SM (2016) Carnosic acid affords mitochondrial protection in chlorpyrifos-treated Sh-Sy5y Cells. Neurotox Res 30:367–379. doi: 10.1007/s12640-016-9620-x.

205. de Oliveira MR, Peres A, Ferreira GC, Schuck PF, Gama CS, Bosco SMD (2017) Carnosic acid protects mitochondria of human neuroblastoma SH-SY5Y cells exposed to paraquat through activation of the Nrf2/HO-1 axis. Mol Neurobiol 54:5961–5972. doi: 10.1007/s12035-016-0100-3.

206. de Oliveira MR, Da Costa Ferreira G, Peres A, Bosco SMD (2018) Carnosic acid suppresses the H2O2-induced mitochondria-related bioenergetics disturbances and redox impairment in SH-SY5Y cells: Role for Nrf2. Mol Neurobiol 55:968–979. doi: 10.1007/s12035-016-0372-7.

207. Miller DM, Singh IN, Wang JA, Hall ED (2013) Administration of the Nrf2-ARE activators sulforaphane and carnosic acid attenuates 4-hydroxy-2-nonenal-induced mitochondrial dysfunction ex vivo. Free Radic Biol Med 57:1–9. doi: 10.1016/j.freeradbiomed.2012.12.011.

208. Green DR, Galluzzi L, Kroemer G (2014) Metabolic control of cell death. Science 345:1250256. doi: 10.1126/science.1250256.

209. Lin CY, Chen JH, Fu RH, Tsai CW (2014) Induction of Pi form of glutathione S-transferase by carnosic acid is mediated through PI3K/Akt/NF-κB pathway and protects against neurotoxicity. Chem Res Toxicol 27:1958–1966. doi: 10.1021/tx5003063.

210. Wang L, Chen Y, Sternberg P, Cai J (2008) Essential roles of the PI3 kinase/Akt pathway in regulating Nrf2-dependent antioxidant functions in the RPE. Invest Ophthalmol Vis Sci 49:1671–1678. doi: 10.1167/iovs.07-1099.

211. Zhang J, Wang X, Vikash V, Ye Q, Wu D, Liu Y, Dong W (2016) ROS and ROS-mediated cellular signaling. Oxid Med Cell Longev 2016:4350965. doi: 10.1155/2016/4350965.

212. Huang XS, Chen HP, Yu HH, Yan YF, Liao ZP, Huang QR (2014) Nrf2-dependent upregulation of antioxidative enzymes: A novel pathway for hypoxic preconditioning-mediated delayed cardioprotection. Mol Cell Biochem 385:33–41. doi: 10.1007/s11010-013-1812-6.

213. Yan YF, Yang WJ, Xu Q, Chen HP, Huang XS, Qiu LY, Liao ZP, Huang QR (2015) DJ-1 upregulates anti-oxidant enzymes and attenuates hypoxia/re-oxygenation-induced oxidative stress by activation of the nuclear factor erythroid 2-like 2 signaling pathway. Mol Med Rep 12:4734–4742. doi: 10.3892/mmr.2015.3947.

214. Gozzelino R, Jeney V, Soares MP (2010) Mechanisms of cell protection by heme oxygenase-1. Annu Rev Pharmacol Toxicol 50:323–354. doi: 10.1146/annurev.pharmtox.010909.105600.

215. Lin CY, Tsai CW, Tsai CW (2016) Carnosic acid protects SH-SY5Y cells against 6-hydroxydopamine-induced cell death through upregulation of parkin pathway. Neuropharmacology 110:109–117. doi: 10.1016/j.neuropharm.2016.04.017.

216. Tanaka K, Suzuki T, Hattori N, Mizuno Y (2004) Ubiquitin, proteasome and parkin. Biochim Biophys Acta 1695:235–247.

217. Um JW, Im E, Lee HJ, Min B, Yoo L, Yoo J, Lübbert H, Stichel-Gunkel C, Cho HS, Yoon JB, Chung KC (2010) Parkin directly modulates 26S proteasome activity. J Neurosci 30:11805–11814. doi: 10.1523/JNEUROSCI.2862-09.2010.

218. Dennissen FJ, Kholod N, van Leeuwen FW (2012) The ubiquitin proteasome system in neurodegenerative diseases: Culprit, accomplice or victim? Prog Neurobiol 96:190–207. doi: 10.1016/j.pneurobio.2012.01.003.

219. Deger JM, Gerson JE, Kayed R (2015) The interrelationship of proteasome impairment and oligomeric intermediates in neurodegeneration. Aging Cell 14:715–724. doi: 10.1111/acel.12359.

220. Liu J, Su H, Qu QM (2016) Carnosic acid prevents beta-amyloid-induced injury in human neuroblastoma SH-SY5Y cells via the induction of autophagy. Neurochem Res 41:2311–2323. doi: 10.1007/s11064-016-1945-6.

221. Vaishnav RA, Singh IN, Miller DM, Hall ED (2010) Lipid peroxidation-derived reactive aldehydes directly and differentially impair spinal cord and brain mitochondrial function. J Neurotrauma 27:1311–1320. doi: 10.1089/neu.2009.1172.

222. Zarkovic K (2003) 4-hydroxynonenal and neurodegenerative diseases. Mol Aspects Med 24:293–303.

223. Fernández-Checa JC, Kaplowitz N, García-Ruiz C, Colell A, Miranda M, Marí M, Ardite E, Morales A (1997) GSH transport in mitochondria: Defense against TNF-induced oxidative stress and alcohol-induced defect. Am J Physiol 273:G7–17.

25 Mitochondrial Dysfunction and Its Impact on Oxidative Capacity of Muscle

Teet Seene

Institute of Sport Sciences and Physiotherapy, University of Tartu, Ujula 4, 51008 Tartu, Estonia

CONTENTS

1 INTRODUCTION

Striated muscle tissue is subdivided according to oxidative capacity of fibers: it includes cells with high oxidative capacity such as the heart muscle, cells with higher oxidative capacity (type I and IIA fibers of skeletal muscle) and cells with low oxidative capacity (type IIB/X fibers of skeletal muscle). According to muscle fibers which have high or higher oxidative capacity, intracellular phosphotransfer systems constitute a major mechanism linking the mitochondria and ATPases within specific structure and intracellular energetic units (Seppet et al. 2001; Saks et al. 2004). Mitochondria are located between the contractile structures in muscle fibers with high oxidative capacity (Vendelin et al. 2005). The efficacy of metabolic signalling significantly depends on morpho-functional relationship of the interaction between mitochondria and sarcomeres (Seppet et al. 2005). Hypoxia disturbs the connections between mitochondria and sarcomeres as sarcomeric structures disintegrate the muscle fiber structure and cause destruction and cell death (Seppet et al. 2005). Apoptosis is one of the reasons of the intensive protein degradation and loss of muscle nuclei associated with atrophy of fibers (Dirks and Leeuwenburgh 2005). Disruption of desmin damnifies the linking of mitochondria to Z-disc and skeletal muscle demonstrates impaired oxidative phosphorylation (Saks et al. 2001). The 5′ adenosine monophosphate-activated protein kinase (AMPK) activates skeletal muscle fibers during acute muscle activity (Aschenbach et al. 2004). Exhaustive muscle activity leads to the depletion of the energy potential, muscular fatigue and destruction of muscle fibers (Abbiss and Laursen 2005).

As children's muscle mass is much less rigid in comparison with adults, they generate lower absolute power during intensive muscle activity. Children's muscles are better equipped for oxidative rather than glycolytic pathways during increased functional activity and have lower capacity to activate type II (fast-twitch) fibers (Ratel et al. 2006). Oxidative capacity of skeletal muscle increases with aerobic type of muscle activity, and an age-associated decline in oxidative capacity is related to the reduction in fitness (Russ and Kent-Braun 2004). Aerobic type of muscle activity positively influences structural changes to capillarity (Harris 2005). Type IIB/IIX muscle fibers exhibit increased ADP concentrations in response to increased workload, which conforms the respiratory control in skeletal muscles (Saks et al. 2001). The decline of muscle mass in elderly people is caused by type II fiber atrophy and loss in the number of fast-twitch (FT) muscle fibers. Increased variability in fiber size, accumulation of ungrouped, scattered and angulated fibers expansion of extracellular space are characteristic to muscle atrophy (Kim et al. 2008; Buford et al. 2010; Seene and Kaasik 2012). Loss of fiber number, decreased production of anabolic hormones testosterone, growth hormone, insulin-like growth factor 1 (IGF-1) an increase in the release of catabolic agents are principal causes of sarcopenia. Interleukin-6 also amplifies the rate of muscle wasting (Roubenoff 2003; Goldspink, Harridge 2004). Aging skeletal muscle becomes less powerful, fat is redistributed from the depot to muscle (Toth and Tchernof 2000) intensive collagen synthesis and posttranslational changes in its structure reduce the elasticity of ligaments (Kjær et al. 2006; Heinemeier et al. 2009).

Mitochondrial dysfunction increases with age and chronic diseases, by loss of efficiency in the electron transport chain, reductions in the synthesis of adenosine-5'-triphosphate (ATP), inadequate number of mitochondria and inability to provide substrates to mitochondria.

2 SKELETAL MUSCLE OXIDATIVE CAPACITY AND MITOCHONDRIAL FUNCTION

The maximal oxygen consumption (VO2 max) per cross-sectional area (CSA) of muscle fibers in different vertebrates varies about hundred-fold range (van Wessel et al. 2010). Muscle fibers with higher oxidative capacity (type I and IIA) have small CSA compared to fibers with low oxidative capacity (type IIB/X). There exists relationship between VO2 max, succinate dehydrogenase (SDH) (Bekedam et al. 2003), oxoglutarate dehydrogenase activities (Blomstrand et al. 1997) and consequently to the number of "energy factories" (mitochondria) in different types of muscle fibers (Reichmann et al. 1991). In striated muscle tissue oxidative capacity of cells varies several folds, so even between FT fibers (type IIA and IIB) the difference is more than two-folds (Seene et al. 2007). Mitochondria contain about 12–16% of type 1 fiber, 10–12% of type II A and 2–5% of IIB/X fibers in volume. Muscle fibers with a large CSA have low SDH activity and vice versa (Kayar and Banchero 1987; Rivero et al. 1999). Type I and IIA muscle fibers have a relatively high oxidative capacity, and small fiber CSA compared to type IIB/IIX fibers. Muscle fiber CSA is linked to its oxidative capacity. Why muscle fibers with higher oxidative capacity remain relatively small compared to muscle fibers with low oxidative capacity is still unknown. One explanation is oxygen diffusion distance which is smaller in type I fibers (Fig. 1) and about quarter better oxygen supply. It is particularly important for long lasting muscle activity that at start these fibers recruited low intensity muscular activity (Seene et al. 2009; Seene and Kaasik 2013a).

Physiological function of muscle fiber type is an outcome of MyHC (myosin heavy chain) isoform expressed within fibers. Some fibers, (hybrid fibers), express a combination of two or more MyHC isoforms (Stephenson 2001; Caiozzo et al. 2003). Laboratory animal experiments have shown that the relative proportions of hybrid fibers vary significantly from muscle to muscle (Caiozzo et al. 2003). In human skeletal muscle, hybrid fiber types represent a significant population of fibers, but the stability of this fiber phenotype is currently unclear. The ATP synthesis is depend on the efficiency of mitochondria to convert energy of metabolites to reduced nicotinamide adenine dinucleotide (NADH) and transfer electrons from NADH to the electron transport chain (Nicholls 2010; Rich and Marechal 2010; Divakaruni and Brand 2011). This ability is different in muscle fibers with different oxidative capacity. A s a result of the electron transport process the reactive oxygen species (ROS) are produced as a by-product of oxidative phosphorylation in ST muscle fibers and FT IIA fibers. Sources of ROS and the related reactive nitrogen species (RNS) are mitochondria. Above mentioned free radicals can damage cellular lipids, proteins, DNA (Spector and Yorek 1985; Richter, Park and Ames 1998; Stadtman 2002). Dismutase enzymes and other antioxidants control excess of ROS/RNS. (Duchen and Szabadkai 2010; Spiteller 2010). Proton leaks, excess of oxygen consumption and ROS production induces considerable damage to lipids such as cardiolipin in inner mitochondrial membrane (Chicco and Sparagna 2007; Spiteller 2010). Oxidative damage of phospholipids in the mitochondrial inner membrane (Fig. 2) may cause partial loss of the electrochemical gradient. Cardiolipin provides stability to the cytochrome complexes in the mitochondrial inner membrane (Chicco and Sparagna 2007). Antioxidant defense is mediated by glutathione peroxidase, catalase, superoxide dismutase (Fridovich 1995; Chandra Jagetia et al. 2003) and are somehow different in muscles with different oxidative capacity. In cardiac and ST skeletal muscle (muscle with high and higher oxidative capacity respectively) the permeability of the outer mitochondrial membrane (Fig. 2) to adenine nucleotides is low and

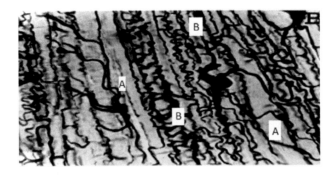

FIGURE 1 Capillary network in skeletal muscle. Oxygen diffusion distance: A -Large B -Small. Magnification x 200.

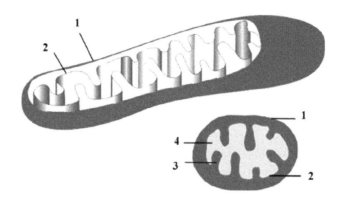

FIGURE 2 Structure of mitochondria. 1-Outer membrane 2-Inner membrane 3-Cristae 4-Matrix.

controlled by a cytoplasmic protein that is sensitive to trypsin and chymotrypsin. This protein may participate in feedback signal transduction by a mechanism of vectorial-ligand conduction. This protein factor is not expressed in FT muscle in which cellular mechanism regulation of respiration is different from that of ST muscles (Kuznetsov et al. 1996).

3 MITOCHONDRIAL DYSFUNCTION AND FATIGUE

Loss of energy is linked to loss of mitochondrial function and represents the diminished production of ATP in cells and it may be the reason of occurrence of fatigue and exertion (Booth et al. 2012). As a result of aging and chronic diseases, oxidative damage of mitochondrial membranes impairs cell function (Wei and Lee 2002; Huang and Manton 2004). Aging population with chronic fatigue syndrome gets oxidative damage of DNA and lipids, as a result of excess of oxidative stress (Logan and Wong 2001; Manuel et al. 2001). Oxidized blood markers and membrane lipids, are their indicators (Fulle et al. 2000; Richards et al. 2005). Elevated levels of peroxynitrite caused by excessive nitric oxide may result in lipid peroxidation, loss of mitochondrial function and change in cytokine levels - an indicative of the positive feedback on nitric oxide production (Pall 2000). ROS are also involved in tissue damage (Pansarasa et al. 2002; Kang et al. 2009). ROS include superoxide anion, hydrogen peroxide and hydroxyl radical and these can cause cell injuries, including lipid peroxidation, enzyme inactivation, changes in intracellular redox state and DNA damage (Urso and Clarkson 2003; Halliwell and Gutteridge 2007). Cells possess enzymes as defense mechanisms to reduce the risk of oxidative injury, these include superoxide dismutase, glutathione peroxidase and catalase for superoxide radicals and organic hydrogen peroxides, respectively (Duntas 2005; Yaegaki et al. 2008). There is evidence that an increase in ROS

production occurs during muscular activities and the resulting oxidative damage arises in muscle, liver, blood and other tissues (Venditti and Di Meo 1997, Itoh et al. 1998). Exhaustive muscular activity has been associated with enhancement of oxygen consumption in skeletal muscles (El-Sayed et al. 2005; Malek and Offert 2009; Santalia et al. 2009). This leads to an increase in lipid peroxidation and inhibition of key mitochondrial enzymes, such as citrate synthase and malate dehydrogenase (Urso and Clarkson 2003; Margonis et al. 2007). Mitochondrial dysfunction can lead to mitochondrial diseases which are higher in cells where mitochondria use more energy, such as cells of striated muscle tissue, than most other cells in our body. As 90% of energy we need for life is produced in mitochondria, main symptoms of mitochondrial diseases are muscle weakness and loss of its coordination. Characteristic symptoms of mitochondrial myopathies are muscle weakness and exercise intolerance or exertional fatigue. In this situation interaction between myofibrillar and mitochondrial compartments are disturbed and as a result of that muscle fatigue (Seene et al. 2017b).

4 CHANGES IN EXHAUSTED FAST-TWITCH AND SLOW-TWITCH SKELETAL MUSCLES

Exhaustive muscular activity is accompanied by a decreased synthesis rate of muscle proteins, particularly myofibrillar proteins and increased protein (such as Myosin and Actin) degradation rate in skeletal muscle (Seene et al. 2004; Seene et al. 2005, Seene et al. 2008). The destruction of myofibrils is shown after exhaustive exercise in skeletal muscles, mainly in FT oxidative-glycolytic muscle fibers (OG) and ST oxidative muscle fibers (O) (Seene et al. 2008). Thin myofibrils, as a result of destruction and splitting of myofibrils, are typical in case of exercise caused myopathy and glucocorticoid caused myopathy. (Fig. 3)

Fig. 4 is showing in the cross-sections of myofibrils disappearance of myosin and actin filaments.

In both cases of myopathies myosin isoforms relative content is changing (MyHC and Myosin Light Chain or MyLC) isoforms and these changes are in good agreement with changes in muscle contractile properties. Relative content of myosin isoforms is different in fiber types and ST and FT muscles (Fig. 5). The relative content of MyHC IIb isoform decreases and that of IIa isoform increases. These changes in MyHC isoforms show that contractile properties of ST and FT muscles change in different ways in accordance with muscle oxidative capacity (Seene et al. 2007; Seene et al. 2008). Changes in MyLC isoforms during exhaustive muscle

activity are much smaller in comparison with subsequent changes in MyHC isoforms. The most significant changes in MyLC isoforms during exhaustive muscular activity appeared in FT muscles (Seene et al. 2008). Regeneration of MyHC IIb and MyLC 3f isoforms, which have high affinity to each other in FT muscle fibers, after tissue damage, proceeds at different speeds (Alev et al. 2009). MyLC 3f isoform regenerates about twice faster than that of MyHC IIb isoform in FT muscle fibers with low oxidative capacity (Alev 2009). It has been shown that MyLC 1 isoform can negatively affect myoblast proliferation by facilitating myoblast withdrawal from cell cycle and differentiation (Zhang et al. 2009).

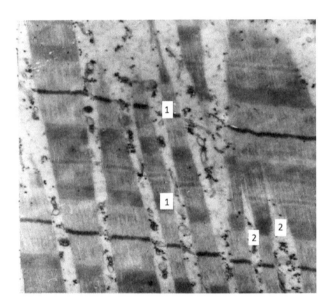

FIGURE 3   Destruction of myofibrils in exercise caused myopathic muscle. 1-Thin myofibrils 2-Splitting of myofibrls Magnification x 20500.

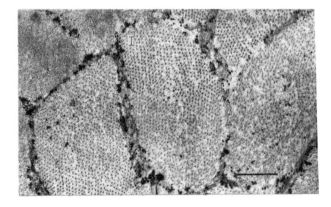

FIGURE 4   Destruction of myosin and actin filaments in myofibrils of glucocorticoid caused myopathic muscle. As a result of destruction in cross-section of myofibrils some myosin and actin filaments absent. Magnification x 44000.

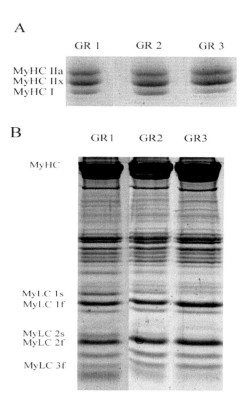

FIGURE 5 Example of changes in myosin heavy (MyHC) (A) and light chain (MyLC) (B) isoforms relative content in skeletal muscle.

5 ROLE OF CYTOKINES IN EXHAUSTED SKELETAL MUSCLE

Contracting muscle fibres release cytokines, which in turn create several different effects in other organs, including brain. Subsequently these different mechanisms create sensations of fatigue and exhaustion in the mind of the physically exhausted subjects. Exhaustive muscular activity induces an anti- inflammatory effect in skeletal muscle, especially in FT muscle fibers and a pro- inflammatory effect in adipose tissues (Neto et al. 2009). This effect contributes to increased lipolysis to provide energy for the muscle. Muscle damage during exhaustive muscular activity increases energy and protein needs (Lowery and Forsythe 2006). Basal metabolic rate increases by 32% after skeletal muscle trauma (Long et al. 1979). The acquisition of new muscle mass is energy costly process as 2300–3500 kcal surpluses is required to build each pound of new muscle tissue (Williams 2004). Higher fractional synthesis rate of MyHC in skeletal muscle increases protein intakes about two times (Brodsky et al. 2004). The DNA and protein content ratio in FT muscles decreases during exhaustive muscular activity showing signs of myopathy as a result of muscular overload (Seene et al. 2004).

Myopathy in exercise is by slow turnover of MyHC in FT muscle fibers, depressed neuromuscular and depressed α-motoneuron excitability (Seene et al. 2008). The decreased synthesis and increased degradation rate of contractile proteins, that was observed in chronically exhausted muscles, is in good agreement with the increased occurrence of destructive processes in FT fibers (Seene et al. 2004; Seene et al. 2008). Contrary to the decreased turnover rate of contractile proteins, chronically exhausted, over trained athletes show a persistent high synthesis rate or concentration of heat shock proteins (HSP) during exhaustive training which might show the increased stress tolerance of affected cells and induction of cellular repair process (Salo et al. 1991).

During migratory flights, which can last 50–100 h, muscle damage occurs mainly in young birds. Experienced migrants may avoid damage behaviourally, or have efficient biochemical and physiological defense against muscle injury (Guglielmo et al. 2001). Damaged muscle tissue releases cytokines, which act in the hypothalamus to re-set the regulatory mechanisms that, among other things shuts down functions that might promote further damage. Muscle fiber phenotype maintenance and transition depends on motoneuron-specific impulse patterns, neuromuscular activity and mechanical load. Depending on the type, intensity, duration of changes in any of these factors, muscle fibers adjust their phenotype to meet the altered functional demands (Pette and Staron 2001). Cytokines play an important role in the immune reaction, metabolic and cellular signal transduction; they are also capable of increasing HSP synthesis (Liu and Steinacker 2001). HSP may act as a cytokine in reaction to exhaustive muscular activity, stimulate tumor necrosis factor-alpha (TNF-α), interleukin β (ILB), IL-8 in monocytes, activate CD 14 - dependent and Ca2+ dependent pathways (Steinacker and Liu 2002). In contrast, mild and regular aerobic kind of activity does not lead to damage and promotes muscular adaptation (Seene et al. 2007; Carcia-Pallares et al. 2009). Signalling pathways and secondary messenger factors are involved in transcriptional, translational and posttranslational processes.

6   EXHAUSTION AND STRUCTURAL CHANGES IN MITOCHONDRIAL AND MYOFIBRILLAR COMPARTMENTS

The interaction between two muscle cell's compartments, contractile apparatus and "energy factories" of mitochondria depends on the oxidative capacity of muscle fibers. Type I and IIA fibers contain more mitochondria than IIB fibers (Seene et al. 2017a). There are large sized mitochondria tightly packed with cristae, and their small forms contain relatively few cristae (Seene et al. 2009; Seene and Kaasik 2013a). The intensive development of the mitochondrial apparatus in the recovery period after regular muscular activity vividly reflects the adaptive processes to muscle activity, which is intended to supply the increased energy requirements of muscle cells with higher oxidative capacity (Seene et al. 2007). Peripheral sarcoplasm, both type I and type IIA muscle fibers contain short canals of the granular sarcoplasmic reticulum as well as polyribosomes and several golgi complexes near the nucleus.

Exhaustive muscle activity induces significant destructive changes, most of all in myofibrils of type I and type IIA muscle fibers, including damage of myosin and actin filaments and the disturbance of the regularity of the Z-line in sarcomeres (Seene et al. 1999). The destruction of myofibrils and mitochondria (Fig. 6) is characteristic of both, fiber types, I and IIA, but it is still more typical of type I muscle fibers. In the A- disc myosin, filaments are absent and the destruction of these myofilaments may cover the whole sarcomere. Recovery of muscle structures after their destruction takes longer time in elderly population in comparison to young adults (Seene and Kaasik 2013b). In both age groups, muscle fibers, with higher oxidative capacity (type I and IIA), regenerate faster than type IIB/IIX fibers (Seene and Kaasik 2013b). Actin filaments are less damaged than myosin filaments (Seene et al. 2009). These structural changes are in accordance with biochemical ones (Magaudda et al. 2004; Seene et al. 2004). Relatively small structural destruction take place in exhausted type IIX/IIB muscle fibers as these fibers are less recruited. A characteristic

change in type IIX/IIB fibers is the focal destruction of myofibrils. The number of mitochondria in type IIX/IIB fibers decreases. Mitochondria in type IIB fibers are located in small groups near nuclei and between myofibrils on the level of Z-line but not in each sarcomere (Seene et al. 1999). The effectiveness of metabolic signalling strongly depends on structural-functional relationships of the interaction between mitochondria and sarcomeres (Seppet et al. 2005). Hypoxia disturbs connections between mitochondria and sarcomeres as sarcomeric components disintegrate the muscle cell structure, cause cell injury and death (Seppet et al. 2005). The apoptosis is responsible for the initiation of protein degradation and loss of muscle nuclei associated with local atrophy, disruption of desmin impairs the linking of mitochondria to Z-disc and skeletal muscle exhibits impaired oxidative phosphorylation (Saks et al. 2001; Dirks and Leeuwenburgh, 2005). The 5′adenosine monophosphate-activated protein kinase (AMPK) becomes activated in skeletal muscle during muscular activity and maintain muscle energy homeostasis (Aschenbach et al. 2004; Nader 2006). The structural changes of neuromuscular junctions in muscle tissue during regular muscular activity depend on type of muscle fiber and age (Seene et al. 2017). In the beginning of regular muscular activity in the aerobic zone of metabolism, a lot of neuromuscular junction terminals become branching (Seene et al. 2017). Type IIA muscle fibers cover a large postsynaptic area as the well-developed synaptic apparatus provides intensive renewal of the structures of the muscle fiber (Seene and Umnova 1992). The axon terminals of type I muscle fibers are relatively small, round or oval shape and closely located. The surface of the neighbouring neuromuscular contacts is smooth and sarcoplasm near the terminals of the type I muscle fiber contains greater number of mitochondria, which contain a lot of cristae (Seene and Umnova 1992). The axon terminals of type IIB fibers are elliptical and their synaptic vesicles are more generously provided with acetylcholine and other trophic factors. At the same time, the postsynaptic folds of the neighbouring synapses are linked with each other. In comparison with type IIA muscle fibers, the postsynaptic folds of type IIB fibers are longer and more regular and they cover a much larger area of the sarcoplasm (Seene et al. 2017). In type IIB fibers, the contact area is the largest between the ending and the surface of the muscle fiber.

Exhausted muscle and Myopathic muscle axon terminals branch off often and this leaves impression that their range is quite large. There are also regions with widened synaptic slot which contain membranous structures, vesicles and terminals contain few mitochondria (Seene et al. 1999). In these regions nerve-muscle transfer is disrupted. There are some glycogen granules, few mitochondria and many lysosomes in the terminals of neuromuscular synapses and in the postsynaptic area (Seene and Umnova 1992). Coated vesicles disappear in

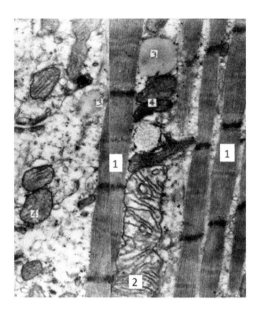

FIGURE 6 Destruction of cell structures in exercise caused myopathic muscle. 1-Thin myofibrils 2-Swelling of mitochondria 3-Lipid droplets 4-Mitochondria Magnification x27000.

the sarcoplasm of the postsynaptic area of type IIA muscle fibers (Seene et al. 1999). Disappearance of coated vesicles is associated with the reduced re-synthesis of acetylcholine in nerve endings, and as these vesicles also carry the proteins of choline receptors onto the postsynaptic membrane, the result becomes a problem with transfer of nerve impulses (Seene et al. 1999). Disturbed connection with the rough sarcoplasmic reticulum causes problems of the regulation of muscle fiber protein metabolism (Dauber et al. 1999). If subsynaptic folds open into T-tubules, they participate in the formation of inter-myofibrillar triads (Dauber and Meister 1986). T-tubules in the sole plate form an extensive network, which together with the sarcoplasmic reticulum can form triads, the position of which makes them unusable for triggering muscle contraction (Dauber et al. 2000).

7 CONCLUSIONS

The ATP synthesis in diferent muscle fibers depends first of all from their oxidative capacity as the efficiency of mitochondria to convert energy of metabolites to reduced NADH and transfer electrons from NADH to the electron transport chain is related with oxidative capacity of fibers. Loss of muscle energy is related to the loss of mitochondrial function and diminished production of ATP in muscle fibers. So, loss of mitochondrial function may be the reason of development of fatigue and exertion. Aging, chronic diseases, oxidative damage of mitochondrial membranes impairs muscle function. As a result of excess of oxidative stress, aging population with chronic fatigue syndrome is the target of oxidative damage of DNA and lipids. The decline of muscle mass in elderly people is caused by type II fiber atrophy and loss in the number of fast-twitch muscle fibers. The efficiency of interaction between contractile apparatus and mitochondria depends on the oxidative capacity of muscle fibers. Type I and IIA fibers, with higher oxidative capacity contain more mitochondria than IIB fibers and fibers with low oxidative capacity. In muscle fibers with higher oxidative capacity intracellular phosphotransfer systems constitute a major mechanism linking the mitochondria and ATPase within specific structures-intracellular energetic units. Mitochondria are located between the contractile structures in muscle fibers with high oxidative capacity. The efficacy of metabolic signalling significantly depends on morpho-functional relationships of the interaction between mitochondria and sarcomeres. Hypoxia disturb the connections between mitochondria and sarcomeres as sarcomeric structures disintegrate the muscle fiber structure and cause destruction and cell death . Apoptosis is one of the reasons of the intensive protein degradation, loss of muscle nuclei associated with atrophy of fiber. Exhaustive muscle activity is leading to the depletion of the energy potential, muscular fatigue and destruction of muscle fibers.

Aerobic type of muscle activity positively influences structural changes to capillarity. Type IIB/IIX muscle fibers exhibit increased ADP concentrations in response to increased workload, which conforms to the respiratory control in skeletal muscles. The intensive development of the mitochondrial apparatus as a result of regular moderate muscle activity is covered the increased energy requirements of muscle fibers, mainly fibers with higher oxidative capacity as a result of efficacy of ATP synthesis.

Acknowledgments

This study was supported by the Estonian Research Council, Research project number TKKSB 1787.

REFERENCES

Abbiss CR, Laursen PB. 2005 Models to explain fatigue during prolonged endurance cycling. *Sports Med* 35: 865–898.

Alev K. 2009 *Myosin light and heavy chain isoforms: endurance training: skeletal muscle plasticity, oxidative capacity and regeneration capability, MyLC and MyHC isoforms turnover rate and relative content.* Saarbrücken: VDM Verlag Dr. Müller Aktiengesellschaft & Co.KG.

Alev K, Kaasik P, Pehme A, et al. 2009 Physiological role of myosin light and heavy chain isoforms in fast- and slow-twitch muscles: effect of exercise. *Biol Sport* 26: 215–234.

Aschenbach WG, Sakamoto K, Goodyear LJ. 2004 5'Adenosine monophosphate-activated protein kinase, metabolism and exercise. *Sports Med* 34: 91–103.

Bekedam MA, van Beek-Harmsen BJ, Boonstra A, et al. 2003 Maximum rate of oxygen consumption related to succinate dehydrogenase activity in skeletal muscle fbres of chronic heart failure patients and controls. *Clin Physiol Funct Imaging* 23: 337–343.

Blomstrand E, Rådegran G, Saltin B. 1997 Maximum rate of oxygen uptake by human skeletal muscle in relation to maximal activities of enzymes in the Krebs cycle. *J Physiol* 501: 455–460.

Booth NE, Myhill S, McLaren-Howard J. 2012 Mitochondrial dysfunction and the pathophysiology of Myalgic Encephalomyelitis/Chronic Fatigue Syndrome (ME/CFS). *Int J Clin Esp Med* 5: 208–220.

Brodsky IG, Suzara D, Hornberger TA, et al. 2004 Isoenergetic dietary protein restriction decreases myosin heavy chain IIx fraction and myosin heavy chain production in humans. *J Nutr* 234: 328–334.

Buford TW, Anton SD, Judge AR, et al. 2010 Models of accelerated sarcopenia: critical pieces for solving the puzzle of age-related muscle atrophy. *Ageing Res Rev* 9: 369–383.

Caiozzo VJ, Baker MJ, Huang K, et al. 2003 Single-fiber myosin heavy chain polymorphism: how many patterns and what proportions? *Am J Physiol* 285: 570–580.

Carcia-Pallares J, Sanches-Medina L, Carrasco I, et al. 2009 Endurance and neuromuscular changes in world-class level kayakers durjng a periodized training cycle. *Eur J Appl Physiol* 2106: 629–638.

Chandra Jagetia G, Rajanikant GK, Rao SK, et al. 2003 Alteration in the glutathione, glutathione peroxidase, superoxide dismutase and lipid peroxidation by ascorbic acid in the skin of mice exposed to fractionated gamma radiation. *Clin Chim Acta* 332: 111–121.

Chicco AJ, Sparagna GC. 2007 Role of cardiolipin alterations in mitochondrial dysfunction and disease. *Am J Physiol Cell Physiol* 292: C33–C44.

Dauber W, Meister A. 1986 Ultrastructure of junctional folds of motor end plates in extensor digitorum longus muscle of mice. *J Ultrastructure Mol Structure Res* 97: 158–164.

Dauber W, Voight T, Härtel X, et al. 2000 The T-tubular network and its triads in the sole plate sarcoplasm of the motor end-plate of mammals. *J Muscle Res Cell Motility* 21: 443–449.

Dauber W, Voight T, Heini A. 1999 Junctions between subsynaptic folds and rough sarcoplasmic reticulum of muscle fibers. *J Muscle Res Cell Motility* 20: 697–701.

Dirks AJ, Leeuwenburgh C. 2005 The role of apoptosis in age-related skeletal muscle atrophy. *Sports Med* 35: 473–483.

Divakaruni AS, Brand MD. 2011 The regulation and physiology of mitochondrial proton leak. *Physiology (Bethesda)* 26: 192–205.

Duchen MR, Szabadkai G. 2010 Roles of mitochondria in human disease. *Essays Biochem* 47: 115–137.

Duntas LH. 2005 Oxidants, antioxidants in physical exercise and relation to thvroid function. *Horm Metab Res* 37: 572–576.

El-Sayed MS, Ali N, El-Sayed Ali Z. 2005 Haemorheology in exercise and training. *Sports Med* 35: 649–670.

Fridovich I. 1995 Superoxide radical and superoxide dismutases. *Annu Rev Biochem* 64: 97–112.

Fulle S, Mecocci P, Fano G, et al. 2000 Specific oxidative alterations in vastus lateralis muscle of patients with the diagnosis of chronic fatigue syndrome. *Free Radic Biol Med* 229: 1252–1259.

Goldspink G, Harridge SDR. 2004 Growth factors and muscle ageing. *Exper Gerontol* 39: 1433–1438.

Guglielmo CG, Piersma T, Williams TD. 2001 A sport-physiological perspective on bird migration: evidence for flight – induced muscle damage. *J Experl Biol* 204: 2683–2690.

Halliwell B, Gutteridge JMC. 2007 *Free radical biology and medicine.* Oxford: Clarendon Press.

Harris BA. 2005 The influence of endurance and resistance exercise on muscle capillarization in the elderly: a review. *Acta Physiol Scand* 185: 89–97.

Heinemeier KM, Olesen JL, Haddad F, et al. 2009 Effect of unloading followed by reloading on expression of collagen and related growth factors in rat tendon and muscle. *J Appl Physiol* 106: 178–186.

Huang H, Manton KG. 2004 The role of oxidative damage in mitochondria during aging: a review. *Front Biosci* 9: 1100–1117.

Itoh H, Ohkuwa T, Yamamoto T, et al. 1998 Effects of endurance physical training on hydroxyl radical generation in rat tissues. *Life Sci* 63: 1921–1929.

Kang C, O'Moore KM, Dickman JR, et al. 2009 Exercise activation of muscle peroxisome proliferator-activated receptor-gamma coactivator-1alpha signaling is redox sensitive. *Free Radic Biol Med* 47: 1394–1400.

Kayar SR, Banchero N. 1987 Volume density and distribution of mitochondria in myocardial growth and hypertrophy. *Respir Physiol* 70: 275–286.

Kim JH, Kwak HB, Leeuwenburgh C, et al. 2008 Lifelong exercise and mild (8%) caloric restriction attenuate age-induced alterations in plantaris muscle morphology, oxidative stress and IGF-1 in the fischer-344 rat. *Exper Gerontol* 43: 317–329.

Kjær M, Magnusson MP, Krogsgaard M, et al. 2006 Extracellular matrix adaptation of tendon and skeletal muscle to exercise. *J Anat* 208: 445–450.

Kuznetsov AV, Tiivel T, Sikk P, et al. 1996 Striking differences between the kinetics of regulation of respiration by ADP in slow-twitch and fast-twitch muscles in vivo. *Eur J Biochem* 241: 909–915.

Liu Y, Steinacker JM. 2001 Changes in skeletal muscle heat shock proteins: pathological signifrcance. *Front Biosci* 6: D12–D25.

Logan AC, Wong C. 2001 Chronic fatigue syndrome: oxidative stress and dietary modifications. *Altern Med Rev* 6: 450–459.

Long CL, Schaffel N, Geiger, JW., et al. 1979 Metabolic response to injury and illness: estimation of energy and protein needs from indirect calorimetry and nitrogen balance. *JPEN* 3: 452–456.

Lowery L, Forsythe C. 2006 Protein and overtraining: Potential application for free-living athletes. *J Int Soc Sports Nutr* 3: 42–50.

Magaudda L, Di Mauro D, Trimarchi F, et al. 2004 Effects of physical exercise on skeletal muscle fiber: ultrastructural and molecular aspects. *Basic Appl Myol* 14: 17–21.

Malek MH, Offert IM. 2009 Global deletion of thrombospondin-l increases cardiac and skeletal muscle capillary and exercise capacity in mice. *Exp Physiol* 94: 749–760.

Manuel Y, Keenoy B, Moorkens G, et al. 2001 Antioxidant status and lipoprotein peroxidation in chronic fatigue syndrome. *Life Sci* 68: 2037–2049.

Margonis K, Fatouros IG, Jamurtas AZ, et al. 2007 Oxidative stress biomarkers responses to physical overtraining: implications for diagnosis. *Free Radic Biol Med* 43: 901–910.

Nader GA. 2006 Concurrent strength and endurance training: from molecules to man. *Med Sci Sports Exerc* 38: 1965–1970.

Neto JCR, Lira FS, Oyama LM., et al. 2009 Exhaustive exercise causes an anti-inflammatory effect in skeletal musle and a pro-inflammatory effect in adipose tissue in rats. *Eur J Appl Physiol* 106: 697–704.

Nicholls DG. 2010 Mitochondrial ion circuits. *Essays Biochem* 47: 25–35.

Nicolson GL, Settineri R. 2011 Lipid replacement therapy: a functional food approach with new formulations for reducing cellular oxidative damage, cancer-associated fatigue and the adverse effects of cancer therapy. *Funct Foods Health Dis* 1: 135–160.

Pall ML. 2000 Elevated, sustained peroxynitrite levels as the cause of chronic fatigue syndrome. *Med Hypotheses* 54: 115–125.

Pansarasa O, D'Antona G, Gualea MR, et al. 2002 "Oxidative stress": effects of mild endurance training and testosterone treatment on rat gastrocnemius muscle. *Eur J Appl Physiol* 47: 550–555.

Pette D, Staron RS. 2001 Transitions of muscle fiber phenotypic profiles. *Histochem Cell Biol* 115: 359–372.

Ratel S, Duché P, Williams CA. 2006 Muscle fatigue during high-intensity exercise in children. *Sports Med* 36: 1031–1065.

Reichmann H, Wasl R, Simoneau JA, et al. 1991 Enzyme-activities of fatty acid oxidation and the respiratory chain in chronically stimulated fast-twitch muscle of the rabbit. *Plugers Arch* 418: 572–574.

Rich PR, Marechal A. 2010 The mitochondrial respiratory chain. *Essays Biochem* 47: 1–23.

Richards RS, Roberts TK, McGregor NR, et al. 2005 Blood parameters indicative of oxidative stress are associated with symptom expression in chronic fatigue syndrome. *Redox Rep* 5: 35–41.

Richter C, Park JW, Ames BN. 1998 Normal oxidative damage to mitochondrial and nuclear DNA is extensive. *Proc Nat Acad Sci U S A* 85: 6465–6466.

Rivero JL, Talmadge RJ, Edgerton VR. 1999 Interrelationships of myof brillar ATPase activity and metabolic properties of myosin heavy chain based fbre types in rat skeletal muscle. *Histochem Cell Biol* 111: 277–287.

Roubenoff R. 2003 Catabolism of aging: is it an inflammatory process? *Curr Opin Clin Nutr* 6: 295–299.

Russ DW, Kent-Braun JA. 2004 Is skeletal muscle oxidative capacity decreased in old age? *Sports Med* 34: 221–229.

Saks VA, Kaambre T, Sikk P, et al. 2001 Intracellular energetics units in red muscle cells. *Biochem J* 356: 643–657.

Saks VA, Kuznetsov AV, Vendelin M, et al. 2004 Functional coupling as a basic mechanism of feedback regulation of cardiac energy metabolism. *Mol Cellular Biochem* 256: 185–199.

Salo DC, Donovan CM, Davies KJA. 1991 HSP70 and other possible heat shock or oxidative stress proteins are induced in skeletal muscle, heart, and liver during exercise. *Free Radic Bio Med* 11: 239–246.

Santalia A, Naranjo J, Terrados N. 2009 Muscle efficiency improves over time in world-class cyclists. *Med Sci Sports Exerc* 41: 1096–1101.

Seene T, Alev K, Kaasik P, et al. 2005 Endurance training: volume-dependent adaptational changes in myosin. *Int J Sports Med* 26: 815–821.

Seene T, Alev K, Kaasik P, et al. 2007 Changes in fast-twitch muscle oxidative capacity and myosin isoforms modulation during endurance training. *J Sports Med Phys Fitness* 47: 124–132.

Seene T, Kaasik P. 2012 Muscle weakness in the elderly: role of sarcopenia, dynapenia, and possibilities for rehabilitation. *Eur Rev Aging Phys Act* 9: 109–117.

Seene T, Kaasik P. 2013a Biological characteristics of structural and functional remodelling in skeletal muscle: effectof exercise. *Adv Stud Biol* 5: 251–278.

Seene T, Kaasik P. 2013b Muscle damage and regeneration: response to exercise training. *Health* 5: 136–145.

Seene T, Kaasik P, Alev K, et al. 2004 Composition and turn-over of contractile proteins in volume-overtrained skeletal muscle. *Int J Sports Med* 25: 438–445.

Seene T, Kaasik P, Seppet E. 2017a Changes in myofibrillar and mitochondrial compartments during increased activity: dependance from oxidative capacity of muscle. *Health* 9: 779–798.

Seene T, Kaasik P, Seppet E. 2017b Crosstalk between mitochondria and myofibrils in adult and aging striated muscle tissue: effect of increased functional activity. *Asian J Res Med Pharm Sci* 1: 1–13.

Seene T, Kaasik P, Umnova M. 2009 Structural rearrangements in contractile apparatus and resulting skeletal muscle remodelling: effect of exercise training. *J Sports Med Phys Fitness* 49: 410–423.

Seene T, Umnova M. 1992 Relations between the changes in the turnover rate of contractile proteins, activation of satellite cells and ultra-structural response of neuromuscular junctions in the fast-oxidative-glucolytic muscle fibres in endurance trained rats. *Basic Appl Myol* 2: 39–46.

Seene T, Umnova M, Kaasik P. 1999 The exercise myopathy. In *Overload, performance incompetence and regeneration in sport*, Eds. M Lehmann et al., 119–130. New York, Boston, Dordrecht, London, Moscow: Kluwer Academic/Plenum Publishers.

Seene T, Umnova M, Kaasik P, et al. 2008 Overtraining syndrome in athietic population. In *Skeletal muscle damage and repair*, Eds. PM Tiidus, 173–184, 305–307. Champaign, IL: Human Kinetics.

Seene T, Umnova M, Kaasik P. 2017 Morphological peculiarities of neuromuscular junctions among diferent fiiber types: effect of exercise. *Eur J Transl Myol* 27: 139–146.

Seppet EK, Eimre M, Anmann T, et al. 2005 Intracellular energetic units in healthy and diseased hearts. *Exp Clin Cardiol* 10: 173–183.

Seppet EK, Käämbre T, Sikk P, et al. 2001 Functional complexes of mitochondria with Ca, MgATPases of myofibrils and sarcoplasmic reticulum in muscle cells. *Biochim Biophys Acta* 1504: 379–395.

Spector AA, Yorek MA. 1985 Membrane lipid composition and cellular function. *J Lipid Res* 26: 1015–1035.

Spiteller G. 2010 Is lipid peroxidation of polyunsaturated acids the only source of free radicals that induce aging and age-related diseases? *Rejuvenation Res* 13: 91–103.

Stadtman E. 2002 Introduction to serial reviews on oxidatively modified proteins in aging and disease. *Free Radic Biol Med* 32: 789.

Steinacker JM, Liu Y. 2002 Stress proteins and applied exercise physiology. In Exercise and stress response: the role of stress proteins, Eds. M Locke, EG Noble, 197–216. Boca Raton, FL: CRC Press.

Stephenson GM. 2001 Hybrid skeletal muscle fibres: a rare or common phenomenon? *Clin Exp Pharmacol Physiol* 28: 692–702.

Toth MJ, Tchernof A. 2000 Lipid metabolism in the elderly. *Europ J Clin Nutr* 54: S121–S125.

Urso ML, Clarkson PM. 2003 Oxidative stress, exercise, and antioxidant supplementation. *Toxicology* 189: 41–54.

van Wessel T, de Haan A, van der Laarse WJ, et al. 2010 The muscle fiber type-fiber size paradox: hypertrophy or oxidative metabolism? *Eur J Appl Physiol* 110: 665–694.

Vendelin M, Béraud N, Guerrero K, et al. 2005 Mitochondrial regular arrangement in muscle cells: a "crystal-like" pattern. *Am J Physiol Cell Physiol* 288: C757–C767.

Venditti P, Di Meo S. 1997 Effects of training on antioxidant capacity, tissue damage and endurance of adult male rats. *Int J Sports Med* 18: 497–502.

Wei YH, Lee HC. 2002 Oxidative stress, mitochondrial DNA mutation, and impairment of antioxidant enzymes in aging. *Exp Biol Med (Maywood)* 227: 671–682.

Williams M. 2004 *Nutrition for health fitness and sport*. Boston, MA: McGrow-Hill Publishing Company.

Yaegaki M, Umeda T, Takahashi I, et al. 2008 Measuring neutrophil functions might be a good predictive marker of overtraining in athletes. *Luminescence* 23: 281–286.

Zhang SZ, Xu Y, Xie HQ, et al. 2009 The possible role of myosin light chain in myoblast proliferation. *Biol Res* 42: 121–132.

26 The Effects of Resveratrol on the Brain Mitochondria

Marcos Roberto de Oliveira

Grupo de Estudos em Neuroquímica e Neurobiologia de Moléculas Bioativas, Universidade Federal de Mato Grosso (UFMT), Av. Fernando Corrêa da Costa, 2367, Cuiaba, MT, Brazil

CONTENTS

1 INTRODUCTION

There is strong evidence indicating a role for natural compounds in mediating neuroprotection in animals. The mechanism of action by which these agents exert benefits in the brain involves antioxidant and anti-apoptotic abilities, as well as anti-inflammatory actions [1–4]. Moreover, natural compounds can suppress tumor growth in central tissues by triggering apoptosis in tumor cells [3]. Other mechanisms associated with the neuroprotective action of natural compounds include modulation of autophagy, causing the removal of damaged molecules and organelles in brain cells [5–7]. This is just a summary of the molecular mechanisms underlying the benefits caused by natural compounds in the mammalian brain. Several studies have been conducted in order to explore exactly how the natural compounds ameliorate central functions, such as cognition and locomotion in experimental models and humans affected by brain disorders [8,9].

In that context, the phytoalexin resveratrol (RES; also called 3,4',5-trihydroxystilbene; $C_{14}H_{12}O_3$) has attracted special attention regarding the capacity to induce neuroprotection. RES is obtained mainly from grapes and is a component of red wine, whose moderate consumption by humans causes protective effects in the brain, heart, and liver, among other organs [10–14]. The protection caused by RES in the mammalian organs has been widely demonstrated in *in vivo* experimental models [15–17]. RES exhibits an oral absorption of about 75% occurring by transepithelial diffusion [18]. Bioavailability studies in humans have demonstrated that, after the ingestion of 25 mg RES (similar to the moderate ingestion of red wine) in a single oral dose, unmetabolized RES might be found at concentrations lower than 10 ng/mL in the plasma at 0.5–2 h after its ingestion [19,20]. Nonetheless, the concentration of RES more total metabolite was detected as being around 400–500 ng/mL in the plasma [19,20]. Thus, the bioavailability of RES after oral intake may be considered low. The methylated RES analogs present improved bioavailability, as demonstrated experimentally [21–24]. Glucuronide and sulfated metabolites of RES have been detected in the human urine after the ingestion of RES [20]. Furthermore, RES may be metabolized by the intestinal microbiota [25–27]. Indeed, it has been postulated that some of the benefits induced by RES in humans may be associated with the metabolism of RES by the gut microbiota, as reviewed by Bird et al. [28].

The effects of RES on the mitochondrial function and dynamics in mammalian cells attracted the attention of several research groups [29–31]. These organelles play a central role in the maintenance of redox and bioenergetics homeostasis, as well as are involved in the cell fate due to the function mitochondria present in the triggering of the intrinsic apoptotic pathway [32–35]. Mitochondrial dysfunction has been observed in several human pathologies, such as neurodegeneration, cardiovascular diseases, metabolic disorders, and cancer, as well as during intoxication (drugs, pollutants, other chemicals) [36–39]. Therefore, pharmacological strategies leading to attenuation of mitochondrial dysfunction are of clinical interest [40,41]. In this regard, natural compounds may be an interesting choice aiming to protect mitochondria, as well as these molecules may serve as models to generate new drugs that would cause more potent effects on the organelles, ensuring better cytoprotection [42–45]. The effects of RES on brain mitochondria are of particular interest, because mitochondrial impairment may lead to neuronal loss by the activation of the mitochondria-related apoptotic signaling pathway [31]. However, the majority of the brain areas are not able to generate new neurons in the adulthood [46]. The adult neurogenesis is restricted to the hippocampus and striatum [47–51]. Thus, neuronal loss affects brain functions aggressively in adults.

2 IN VITRO EXPERIMENTAL MODELS

The effects of RES on mitochondrial function and dynamics have been studied in several experimental models focusing on different parameters (Table 1). The utilization of purified mitochondria allows the researchers investigating the whether and how RES would bind to mitochondrial proteins. Moreover, the researchers are able to perform kinetic studies using purified mitochondria. In cultured cells, the mitochondria-related signaling pathways may be investigated by using different strategies (pharmacological inhibitors, small interfering RNA, among others). Furthermore, gene expression and alterations in the levels of specific proteins may be checked in cultured cells. Importantly, mitochondria-related apoptotic (i.e., the intrinsic apoptotic pathway) indexes have been successfully evaluated in cultured cells. Thus, the mechanism of action by which RES would interfere in mitochondrial function and dynamics, as well as in cell fate, can be examined.

3 ISOLATED MITOCHONDRIA

Zheng and Ramirez (2000) have shown that RES inhibited the F0F1-ATPase enzyme (the so called complex V or ATP synthase/ATPase) in mitochondria isolated from rat brain with IC_{50} of 13 or 18.5 μM, depending on the chemical utilized to solubilize the organelles (digitonin or CHAPS, respectively) [52]. The authors also demonstrated that the effects elicited by RES on the activity of

TABLE 1

Main findings regarding the effects of resveratrol (RES) on mitochondria in *in vitro* experimental models.

Main findings	References
– RES inhibited the F0F1-ATPase enzyme in mitochondria isolated from the rat brain	[52]
– RES decreased the levels of lipid peroxide in mitochondria isolated from the rat brain and submitted to a challenge with Fe^{2+}/ADP	[55]
– RES increased the production of H_2O_2 by the organelles in the presence of antimycin A	
– RES prevented the loss of MMP and the decrease in the synthesis of ATP in N2a cells exposed to Aβ$_{25-35}$ peptide	[58]
– RES modulated SIRT1, PGC-1α, and MFN2 and restored the levels of ATP in the mutant hSOD1-G93A-bearing motor neuron-like cell experimental model of ALS	[59]
– RES activated AMPK and upregulated PGC-1α in SH-SY5Y cells	[61]
– RES increased the immunocontents of the mitochondrial complexes I and III	
– RES upregulated Nrf2 and the expression of HO-1 and NQO1 enzymes	
– inhibition of AMPK suppressed the RES-induced effects	
– RES suppressed the effects of high glucose on the levels of complex I and PGC-1α in SH-SY5Y cells	[65]
– inhibition of AMPK blocked the effects elicited by RES	
– RES restored the activity of the complexes I and V and the synthesis of ATP in NPC isolated from the dentate gyrus of adult Ts65Dn mice	[66]
– inhibition of AMPK abrogated the RES-induced effects regarding mitochondrial function	
– RES upregulated AMPK, PGC-1α, NRF-1, and T-FAM	
– RES increased the levels of mitochondrial DNA and of the complexes I and V	
– RES promoted AMPK activation and upregulation of PGC-1α and NRF-1, leading to an increase in the levels of ATP in rat primary visual cortical neurons	[67]
– RES prevented the rotenone-induced decrease in the levels of ATP by triggering mitochondrial biogenesis in PC12 cells	[68]
– RES upregulated PGC-1α, Drp1, Fis1, OPA1, and MFN2	
– RES increased the amounts of mitochondrial DNA and the mitochondrial mass	
– RES abrogated the loss of MMP in primary cortical and striatal neurons obtained from the embryos of YAC128 mice	[73]
– RES did not activate SIRT1 in the primary cultures	
– RES prevented the decrease in the activity of complex IV and loss of MMP in primary cultures of astrocytes obtained from rat and submitted to an azide treatment	[76]

(Continued)

TABLE 1 (Cont.)

Main findings	References
– RES upregulated Mn-SOD and decreased the mitochondrial production of O_2^-	
– ZnPP IX abrogated the effects of RES on mitochondria	
– RES induced anti-apoptotic effects in SH-SY5Y cells exposed to dopamine: prevented loss of MMP, activation of caspase-3, and cleavage of PARP	[85]
– RES restored MMP and decreased the levels of Bax and the release of cytochrome c from the mitochondria in primary culture of CECs isolated from mouse and submitted to a treatment with oxLDL	[86]
– RES decreased the activity of the caspases-9, -3, and -6	
– RES suppressed apoptosis in SH-SY5Y cells by blocking the activation of caspase-3 and the cleavage of PARP in SH-SY5Y cells exposed to rotenone	[87]
– the inhibition of AMPK affected the activity of SIRT1 and the levels of LC3-II	
– blockade of autophagy suppressed the RES-induced mitochondrial protection	
– RES triggered autophagy by activating the AMPK/SIRT1 signaling pathway in SH-SY5Y cells	
– RES caused mitochondrial protection and blocked the mitochondria-dependent apoptosis in SH-SY5Y cells exposed to $PrP_{106-126}$	[89]
– RES downregulated the levels of BAX in the mitochondria and the release of cytochrome c from the organelles	
– blockade of autophagy suppressed the anti-apoptotic effects elicited by RES	
– inhibition of autophagy or inhibition of HO-1 abrogated the mitochondria-related anti-apoptotic effects caused by RES in SH-SY5Y cells exposed to rotenone	[90]
– RES activated autophagy by an HO-1-dependent manner	
– RES blocked the release of cytochrome c from the mitochondria and the consequent activation of caspase-3 in RGC-5 cells deprived of serum	[91]
– RES upregulated the levels of SIRT1, NRF1, and T-FAM	
– inhibition of SIRT1 suppressed the anti-apoptotic effects elicited by RES	
– RES blocked the activation of caspase-3 and the cleavage of PARP in SH-SY5Y cells exposed to sodium fluoride	[92]
– the levels of the acetylated form of p53 were reduced by RES	
– RES probably activated SIRT1, as suggested by the authors	
– the authors did not find evidence of the involvement of mitochondrial biogenesis in mediating the anti-apoptotic effects elicited by RES in ARPE-19 cells exposed to acrolein	[93]
– RES triggered the intrinsic apoptotic pathway (that is dependent on the mitochondria) in the human Y79 retinoblastoma and SK-N-AS neuroblastoma tumor cell lines	[94]
– RES induced loss of MMP, release of cytochrome c from the mitochondria, and activation of the caspases-9 and -3	
– RES downregulated the levels of Bcl-2 and Bcl-xL in the rat B103 neuroblastoma cells, promoting apoptosis by a mitochondria-dependent manner	[96]
– RES activated the caspases-9 and -3	
– RES caused an increase in the fragmentation of DNA	
– RES caused an increase in the levels of Bax in the mitochondria, leading to cytochrome c release from the organelles in the U251 glioma cell line	[97]
– RES activated caspase-3 and caused DNA fragmentation, inducing apoptotis by a mitochondria-related manner	

the F0F1-ATPase enzyme can be specific, since oligomycin, a potent inhibitor of the enzyme, suppressed the RES-induced inhibition. As a consequence, RES also reduced the synthesis of ATP by submitochondrial particles isolated from the rat brain (IC_{50} = 27.7 μM). Interestingly, the combination of RES with other compounds (quercetin and kaempferol) present in red wine significantly increased the inhibitory effect on the F0F1-ATPase enzyme caused by the single compounds alone. Similar data were previously published by Zini et al. (1998), showing that RES can inhibit both F0F1-ATPase and complex III enzyme activities [53]. In this context, Gledhill et al. (2007) have demonstrated that RES would be able to inhibit the rotatory mechanism of the bovine heart mitochondria F0F1-ATPase by direct binding to the enzyme,

decreasing the production of ATP [54]. Even though the exact mechanism by which RES inhibited the F0F1-ATPase enzyme was not shown in those works (with exception to the work by Gledhill et al.), data obtained by the authors may explain, at least in part, the steps by which RES affects the levels of ATP and AMP in mammalian cells. These nucleotides are signaling molecules with a very important role in the signaling pathways associated with mitochondrial biogenesis; for example Moreira et al. (2013) showed that a pre-treatment with RES at 25 μM caused an antioxidant effect in brain mitochondria exposed to a pro-oxidant challenge with the Fe^{2+}/ADP system [55]. RES decreased the formation of lipid peroxides in the organelles isolated from both male and female brain. RES at 10 μM did not protect brain mitochondria in that experimental model. Nonetheless, it was observed that RES amplified the production of H_2O_2 by brain mitochondria obtained from male rats when the organelles were incubated with antimycin A (an inhibitor of the mitochondrial complex III activity). RES (10–25 μM) also reduced the state 3 respiration without affecting the respiratory control ratio (RCR). Taken together, these data may indicate that RES would be able to decrease the O_2 consumption by the mitochondria during state 4. Interestingly, RES did not alter the maximum MMP in the absence of ADP. On the other hand, 25 μM RES partially reduced the mitochondrial depolarization induced by ADP. Moreover, RES decreased the consumption of O_2 in submitochondrial particles incubated with NADH, a substrate of the mitochondrial complex I. A similar effect was seen in mitochondria isolated from the brains of male and female rats. The effect of RES on the activity of complex I is very likely to be specific, since RES did not alter respiration in mitochondria incubated with succinate, a substrate of the complex II. Therefore, data obtained by the authors are in agreement with other works demonstrating the ability of RES in modulating the activity of complex I in brain mitochondria [56]. It is important to mention that the effects of RES on the activity of complex I depends on the concentration utilized and on the source of the mitochondria, since RES at 1–5 μM increased the activity of complex I in other experimental models using liver cells [57]. Therefore, the mechanism by which RES modulates the activity of mitochondrial enzymes is complex and deserves further investigation in order to better understand the impacts RES would present on mitochondrial function in the different mammalian tissues.

4 CULTURED CELLS

4.1 MITOCHONDRIAL FUNCTION EFFECTS

Manczak et al. (2010) have studied the effects of RES in an *in vitro* experimental model of Alzheimer's disease [58]. The authors have found that a pretreatment (6 h) with RES (5 μM) prevented the decrease in the ATP levels and the disruption in the MMP in N2a cells exposed to amyloid-β_{25-35} ($A\beta_{25-35}$) peptide. Interestingly, RES did not modulate the activity of the complex IV in the Aβ-treated cells. Additionally, RES did not stimulate mitochondrial biogenesis in that experimental model. Therefore, the data regarding the effect of RES on the levels of ATP should be associated with other type of signaling mediated by the polyphenol in the $A\beta_{25-35}$-treated cells.

Wang et al. (2011) have found that RES (10 μM for 8–48 h) modulated SIRT1 in the mutant hSOD1-G93A-bearing motor neuron-like cell experimental model of amyotrophic lateral sclerosis (ALS), restoring the levels of ATP in the cells [59]. RES also can upregulate MFN2 and PGC-1α expression in hSOD1-G93A cells. Nonetheless, it was not checked the link between the increase in the ATP levels and the regulation of the expression of

proteins involved in mitochondrial dynamics. As a comparison, similar data were observed by Mancuso et al. (2014) in an *in vivo* experimental model of ALS with the SOD1-G93A ALS mice [60]. RES (approximately 160 mg/kg.day^{-1} added to the diet for different periods) upregulated both immunocontent and activity of SIRT1 in spinal motoneurons of the ALS mice. RES also normalized the levels of the components of the OXPHOS system and of the active form of adenosine monophosphate-activated kinase (AMPK). Contrasting with the *in vitro* data, RES did not change the levels of mitofusin 2 (MFN2) in the *in vivo* experimental model of ALS in mice. Thus, it is very likely that the effects may vary according to the RES concentrations the brain cells will receive. It is important to investigate whether nanotechnology-based strategies would be more effective in triggering mitochondrial protection regarding the use of RES as a neuroprotective agent *in vivo*.

The RES-induced upregulation of the PGC-1α signaling pathway has been seen in other reports. Cao et al. (2014) have reported that RES (100 μM for 24 h) promoted activation of AMPK and upregulation of PGC-1α in the dopaminergic SH-SY5Y cell line [61]. Moreover, RES increased the immunocontents of the mitochondrial complexes I and III in that experimental model. RES also enhanced the expression of the transcription factor nuclear factor erythroid 2-related factor 2 (Nrf2) and the cytoprotective enzymes heme oxygenase-1 (HO-1) and NAD(P)H dehydrogenase [quinone] 1 (NQO1). These effects were abrogated by Compound C, a specific inhibitor of AMPK activation. The link between Nrf2 activation and the promotion of the expression of mitochondrial complexes were not examined by the authors. However, there is evidence pointing to Nrf2 as a regulator of the mitochondrial function [62,63]. Nrf2 is a master regulator of the redox biology in mammalian cells [64]. Furthermore, Nrf2 modulates the expression of phase II detoxification enzymes [64]. Thus, the activation of this transcription factors may be involved in cytoprotection in different ways, from maintaining the redox balance to ameliorating the mitochondria-related bioenergetics status in mammalian cells. Nonetheless, whether there is a role for Nrf2 in the RES-induced effects on mitochondrial function and dynamics, it remains to be elucidated.

In the work by Peng et al. (2016), it was demonstrated that a cotreatment with RES (10 μM for 24 h) prevented the high glucose-induced downregulation in complex I immunocontent in SH-SY5Y cells [65]. Moreover, RES suppressed the reduction in the levels of PGC-1α in the high glucose group. The authors have shown that the enzyme AMPK would be associated with the benefits elicited by RES, since Compound C blocked the RES-induced effects. Even though the authors have not examined whether the activity of mitochondrial complexes were altered by RES, it is very likely that RES, by restoring the amounts of complex I, restored mitochondrial function by upregulating complex I activity.

Nonetheless, this speculation deserves further efforts in order to elucidate the impact of RES on mitochondrial function (i.e., TCA enzymes activity, OXPHOS activity, synthesis of ATP, MMP, among others metabolic pathways occurring in the organelles) in cells challenged with high glucose levels.

Similarly, Valenti et al. (2016) have found that RES (10 μM for 24 h) rescued neural progenitor cells (NPC) isolated from the dentate gyrus of adult Ts65Dn mice (an animal model of Down syndrome because carry a partial triplication of mouse chromosome 16, which is analogous to the human portion 21q21–21q22.3) [66]. The authors have reported that RES restored mitochondrial function (enzyme activity of the complexes I and V; ATP synthesis) by a mechanism dependent on AMPK, since Compound C suppressed the effects induced by RES. RES also upregulated sirtuin 1 (SIRT1) activity (without altering SIRT1 levels) and the immunocontents of AMPK, peroxisome proliferator-activated receptor-γ coactivator-1α (PGC-1α), nuclear respiratory factor-1 (NRF-1), mitochondrial transcription factor A (TFAM). The activation of such proteins leads to mitochondrial biogenesis, i.e., the synthesis of new mitochondria [41]. Accordingly, the authors observed that RES treatment enhanced the levels of mitochondrial DNA, as well as the amounts of the complexes I and V in Ts65Dn NPC cells. Thus, RES would be an interesting alternative regarding the treatment of mitochondrial impairment in Down syndrome by modulating mitochondrial biogenesis. In a similar fashion, it was published by another research group that RES (20 μM for 1 h) promoted AMPK activation and upregulation of PGC-1α and NRF-1, causing an increase in the levels of ATP in rat primary visual cortical neurons [67]. In the same work, the authors have found evidence of mitochondrial biogenesis in the visual cortex of rats treated with RES (20 mg/kg.day^{-1} for 1 week) and subjected to monocular deprivation. RES also activated the AMPK/PGC-1α/ NRF-1 axis *in vivo*, leading to an increase in the number of mitochondria and in the levels of ATP in neurons of the visual cortex. In accordance with this, Peng et al. (2016) have reported that a pretreatment (for 24 h) with RES (60 μM) prevented the rotenone-induced decrease in the levels of ATP in PC12 cells by a mechanism associated with the triggering of mitochondrial biogenesis [68]. RES upregulated PGC-1α and the levels of dynamin-related protein 1 (Drp1), mitochondrial fission 1 (Fis1), optic atrophy 1 (OPA1), and MFN2, consequently enhancing the amount of mitochondrial DNA and mitochondrial mass. An effect of the inhibition of mitochondrial biogenesis on cell survival and mitochondrial function was not tested by the authors. However, data obtained by those research groups confirm that RES can promote mitochondrial biogenesis in brain cells. Moreover, strong evidence points to a role for RES in the maintenance of mitochondrial dynamics, i.e., mitochondrial fusion and fission, in the mechanism of

mitochondrial function protection, since the proteins Drp1, Fis1, OPA1, and MFN2 modulate mitochondrial fusion and fission in mammalian cells. OPA1 and MFN2 coordinate mitochondrial fusion, whereas and Drp1 and Fis1 control mitochondrial fission [69–72].

Naia et al. (2017) have reported that RES (1–5 μM for 96 h) rescued mitochondria regarding loss of MMP in primary cortical and striatal neurons obtained from the embryos of YAC128 mice (which are an experimental model of Huntington's disease – HD) [73]. Interestingly, RES did not modulate the acetylation of H3 protein, indicating that RES did not alter SIRT1 activity in that experimental model (SIRT1 is a deacetylase and H3 is a well-known target of SIRT1). The authors also tested whether RES would be able to modulate genes related with mitochondrial biogenesis in an *in vivo* experimental model using YAC 128 mice. RES (1 mg/kg.day^{-1} for 28 days, subcutaneous route) decreased the levels of acetylated H3 in both cerebral cortex and striatum of the wild type animals and only in the striatum of the YAC128 animals. The authors next examined whether RES would modulate the expression of genes related with mitochondrial function in the cerebral cortex of mice. It was observed that RES upregulated the expression of MT-ND5 (mitochondrially encoded NADH: ubiquinone oxidoreductase core subunit 5) and MT-CO1 (mitochondrially encoded cytochrome c oxidase I) genes in the cerebral cortex of YAC128 mice. Interestingly, RES can downregulate the expression of those genes in the cerebral cortex of wild type mice. Other analyses were performed by the authors and reinforced the hypothesis that RES induced mitochondrial biogenesis, ameliorating mitochondrial function in the neurons of cerebral areas affected in HD. The biochemical alterations promoted by RES were accompanied by benefits in the animal behavior in that experimental model. It was previously shown that mutant huntingtin interacts with mitochondrial membranes, causing loss of MMP, bioenergetics decline, and increased production of ROS by the organelles [74,75]. Thus, RES would be an interesting choice to be used as an alternative drug in the treatment of HD. However, safety of this treatment should be better studied in further research.

In the work by Bellaver et al. (2016), it was demonstrated that RES exhibits mitochondria-related protective effects by a mechanism associated with the HO-1 enzyme in primary cultures of astrocytes obtained from rat [76]. RES (100 μM) pretreatment (for 1 h) prevented the loss of MMP and the decrease in the activity of complex IV induced by azide (a mitochondrial toxin). The benefits elicited by RES on the mitochondrial function were suppressed by zinc protoporphyrin IX (ZnPP IX), a specific inhibitor of the HO-1 enzyme. The authors also found that RES increased the activity of Mn-SOD and decreased the production of O_2^- by the mitochondria in an HO-1-dependent fashion. Actually, a role for HO-1 in mediating mitochondrial protection has been seen in several works exploring the effects of natural compounds in mammalian cells [77–79]. However, the exact mechanism by which this enzyme prevents or restores mitochondrial function remains to be completely understood. At least in part, it may be associated with the production of cytoprotective agents by the HO-1/biliverdin reductase (BVR) axis [80]. HO-1 generates carbon monoxide (CO), free iron, and biliverdin during the degradation of heme [81]. Biliverdin is converted into bilirubin, a potent antioxidant and cytoprotective molecule, by BVR [80]. It would be necessary to examine whether bilirubin and/or CO would be really involved in the maintenance of mitochondrial homeostasis in RES-treated cells, since other mechanisms downstream the activation of HO-1 may be associated with the benefits induced by RES on the organelles. It is important to mention that the authors have detected a decrease in the production of nitric oxide (NO) in the RES-treated astrocytes exposed to azide in that experimental model. NO$^{\cdot}$ may interfere with the binding of O$_2$ to the complex IV site, enhancing the production of O$_2^{-}$ [82]. Additionally, the combination of O$_2^{-}$ with NO$^{\cdot}$ can lead to the formation of peroxynitrite (ONOO^{-}), which causes protein nitration by reacting with tyrosine residues in proteins [83]. Nitration of proteins may affect the function of these molecules by either gain or loss of function [84]. In neurons and glial cells, it can trigger cell death by different routes [84].

4.2 MITOCHONDRIA-RELATED ANTI-APOPTOTIC EFFECTS

Mitochondria are central players in the intrinsic apoptotic pathway triggering, as reviewed by Green et al. (2014) [35]. Thus, several research groups have investigated whether RES would cause neuroprotection by a mitochondria-related way. In the work by Lee et al. (2007), RES (5 µM) pretreatment (for 1 h) was found to protect SH-SY5Y cells against the dopamine-induced mitochondria-related cell death [85]. RES prevented loss of MMP and activation of caspase-3, consequently decreasing the cleavage of poly [ADP-ribose] polymerase (PARP). The mechanism of action by which RES suppressed the loss of MMP in the dopamine-treated cells was not checked by the authors; however, it might be mediated by the modulation of Bcl-2 by RES, as demonstrated by the authors. Similar effects were seen by Chang et al. (2011) in a primary culture of cerebrovascular endothelial cells (CECs) isolated from mouse [86]. The authors have found that RES (10 µM) during the treatment with oxidized low-density lipoprotein (oxLDL) attenuated the mitochondria-related apoptosis in CECs by restoring MMP and downregulating the levels of mitochondrial Bax and the release of cytochrome c from the organelles. Consequently, RES decreased the activity of the pro-apoptotic caspases-9, -3, and -6. However, the mechanism involved in the anti-apoptotic effects caused by RES in the CECs exposed to oxLDL was not evaluated in that work.

Wu et al. (2011) have shown that RES (50 µM) pretreatment (for 24 h) inhibited the activation of caspase-3 and the cleavage of PARP in SH-SY5Y cells exposed to rotenone (a mitochondrial toxin) [87]. The inhibition of the AMPK protein significantly reduced the acetylation of histone 3 (H3, a target of SIRT1) and the levels of LC3-II in RES-treated SH-SY5Y cells. Similar effects were seen in the levels of LC3-II when the cells were treated with siRNA against SIRT1, indicating a role for the deacetylase in the activation of autophagy in the cells exposed to RES. Also, blockade of the autophagy-related signaling suppressed the RES-induced mitochondria-related cytoprotection. The utilization of siRNA targeting Beclin 1, for example, attenuated the effects of RES on the release of cytochrome c to the cytosol, as well as on the levels of cleaved PARP. Thus, RES triggered autophagy by an AMPK/SIRT1 axis-dependent manner in SH-SY5Y cells, causing neuroprotection. Beclin 1 is a protein associated with the regulation of autophagy by controlling the formation of the autophagosome [5]. The inhibition of autophagy abrogates the clearance of proteins that would be prone for aggregation, such as α-synuclein [5]. Moreover, autophagy removes dysfunctional mitochondria from the cells (also called mitophagy) [88]. In this regard, Jeong et al. (2012) reported that RES pretreatment at 2 µM for 12 h promoted mitochondrial protection and mitochondria-related anti-apoptotic effects by a mechanism associated with the triggering of autophagy in SH-SY5Y cells exposed to the amino acid residues 106–126 of the prion protein (PrP$_{106-126}$) [89]. The anti-apoptotic effects induced by RES on the levels of BAX in the mitochondria and of cytochrome c in the cytosol were abrogated by the administration of 3-methyladenine (3-MA; an inhibitor of autophagy) to the cells. Furthermore, RES attenuated the loss of MMP by the same mechanism in the PrP$_{106-126}$-treated cells. Thus, RES activated autophagy promoting cytoprotection by a mitochondria-associated manner. In this same line, Lin et al. (2014) have reported that the inhibition of autophagy by bafilomycin A1 (BAF) abrogated the RES-induced inhibition of caspase-3 in SH-SY5Y cells exposed to rotenone [90]. The authors have utilized RES at 20 µM in a pretreatment model for 24 h prior to the challenge with rotenone. Interestingly, the autophagy-related cytoprotective effects elicited by RES involved the activation of HO-1, since the inhibition of this enzyme by ZnPP IX (a specific inhibitor of HO-1) suppressed the RES-induced autophagy and cytoprotection. Whether there is a link between the induction of HO-1 by RES and the mitochondrial protection mediated by this polyphenol, it remains to be determined.

A role for the RES-induced mitochondrial biogenesis has been seen during the cytoprotection the polyphenol caused in the retinal ganglion RGC-5 cell line deprived of serum [91]. RES (40 µM) increased the levels of mitochondrial DNA and the number of mitochondria by upregulating the levels of SIRT1, NRF1, and TFAM in different periods of serum deprivation. RES efficiently suppressed the serum deprivation-triggered release of cytochrome c and activation of caspase-3 in the RGC-5 cells, inhibiting the mitochondria-mediated apoptosis. Nicotinamide, a specific inhibitor of SIRT1, abrogated the anti-apoptotic effects elicited by RES in RGC-5 cells. Therefore, it is plausible that the induction of mitochondrial biogenesis induced by RES attenuated the pro-apoptotic effects caused by serum deprivation, at least in part, by restoring mitochondrial integrity and function.

Tu et al. (2018) have found that a pretreatment (2 h) with RES (20 μM) partially blocked the activation of caspase-3 and the cleavage of PARP in SH-SY5Y cells exposed to sodium fluoride, a toxicant [92]. The authors did not analyze specific mitochondria-related parameters regarding the involvement of the organelles in the triggering of cell death in that experimental model. However, it was suggested that the mechanism of action of RES would be associated with the deacetylase SIRT1, since the levels of the acetylated form of p53 (a target of SIRT1) were reduced by RES. Nonetheless, it was not observed alterations in the levels of mitochondrial DNA and of the expression of TFAM in acrolein-treated ARPE-19 cells during the cytoprotection caused by RES (10–20 μM), as reported by Sheu et al. (2013) [93]. Thus, mitochondrial biogenesis may be not involved in a generalized form in the anti-apoptotic effects elicited by RES in brain cells.

Therefore, strong evidence points to SIRT1 as a central player in the mitochondria-related anti-apoptotic effects elicited by RES in brain cells. SIRT1 exhibits a role in the autophagy process and in the promotion of mitochondrial biogenesis [69]. A simultaneous dual role for SIRT1 in these phenomena may not be discarded, since the mammalian cells present subpopulations of mitochondria playing different roles in the same time. Nonetheless, it remains to be examined whether it would present a benefit to brain cells triggering mitophagy of damaged mitochondria simultaneously with the synthesis of new mitochondria.

4.3 MITOCHONDRIA-RELATED PRO-APOPTOTIC EFFECTS

There is interest in discovering whether a drug may trigger cell death specially in the case of tumor cells. There are several natural compounds that can induce apoptosis (or other types of cell death) in tumor cells. However, the complete mechanism underlying the pro-apoptotic effects elicited by these molecules are still on debate. In this context, Sareen et al. (2006) have tested the effects of RES (50–100 μM) for different times (24–96 h) on the human Y79 retinoblastoma and SK-N-AS neuroblastoma cell lines [94]. RES significantly increased the rates of the mitochondria-related cell death by inducing loss of MMP, release of cytochrome c from the organelles, and activation of the caspases-9 and -3. In that context, van Ginkel et al. (2007) have reported that RES

(2–50 mg/kg.day^{-1} for 5 weeks, oral route) reduced the tumor growth in NGP and SK-N-AS xenograft models (an experimental design using neuroblastoma cells to induce tumor in animals) [95]. In the same work, the authors demonstrated that RES at 50–200 μM induced cell death by a mitochondria-related manner in NGP and SH-SY5Y cells. The SK-N-AS cell line seems to be more resistant to RES, because only RES at 100–200 μM was able to induce apoptosis in those cells. Importantly, it was examined the RES bioavailability in that study. It was found a RES concentration ranging from 2 to 10 μM in the serum of mice 30 min after RES ingestion. The authors also found that RES did not accumulate in the tumor after the 5 weeks of treatment. Thus, RES induced a chemotherapeutic effect in cancer cell lines by affecting mitochondrial signaling. The concentrations of RES utilized in those works are high when considering the intake of RES by humans. However, the study of nanotechnology-related strategies would be very useful in order to improve the delivery of RES to cancer cells *in vivo*. Further research is welcome in this research area in all the pharmacological aspects related with RES.

Rahman et al. (2012) also found evidence of RES in triggering the mitochondria-related apoptotic pathway [96]. RES (5–20 μM for 24–48 h) decreased the viability of the rat B103 neuroblastoma cells and downregulated the levels of B-cell lymphoma 2 (Bcl-2) and B-cell lymphoma extra-large (Bcl-xL), which are anti-apoptotic proteins. RES treatment also caused the activation of the pro-apoptotic enzymes caspase-9 and caspase-3 and the fragmentation of nuclear DNA. It was not examined by the authors whether RES would be able to modulate the levels of the pro-apoptotic proteins responsible for the release of cytochrome c from the mitochondria, such as Bcl-2-associated X protein (Bax). Also, the mitochondrial and cytosolic levels of cytochrome c were not determined in that work. Thus, it would be needed further analyses in order to confirm a role for mitochondria in triggering apoptosis in B103 cells. Apoptosis is not an event that depends only on the mitochondrial action. Other signaling pathways may be involved in the activation of caspases and in the triggering of cell death [35].

The mitochondria-related apoptotic role of RES was also tested in glioma cells. Jiang et al. (2005) have reported that RES (100 μM for 48 h) increased the levels of Bax in the mitochondria, causing an increase in the release of cytochrome c to the cytosol [97]. Consequently, RES treatment enhanced the activity of caspase-3 and caused DNA fragmentation in the U251 glioma cell line. Similar cytotoxic effects were seen in the U87 cell line when exposed to RES. As previously discussed [30], the concentrations of RES used in the works involving the pro-apoptotic effects of this polyphenol in tumor cells are considered high when considering the amounts of RES ingested by humans [30]. Therefore, strategies should be studied in order to ensure the delivery of RES to tumor cells without causing intoxication during chemotherapy, for example.

4.4 In Vivo Experimental Models

In *in vivo* experimental models, it is analyzed the effects of a given compound in a more complex way, since it is necessary to consider the bioavailability of the molecule in the different species it is studied (Table 2). Moreover, the metabolic rates of a rat are different from those observed in humans, for example. In spite of this, several works have demonstrated that RES induces mitochondria-related effects *in vivo*. Therefore, the RES that is administrated by a specific route (oral, intraperitoneal, other) reaches a given tissue and triggers mitochondria-related signals. However, it is necessary to consider RES metabolism by the gut microbiota and the modifications RES may undergone after reacting with detoxification enzymes mainly in the liver. Thus, it is plausible considering that molecules generated after RES metabolism may promote the mitochondria-associated effects *in vivo*. In addition, experimental models involving animals have been crucial to better understand whether and how a natural compound promotes cytoprotection or toxicity in the mammalian organism.

4.5 Mitochondrial Function Effects

Yousuf et al. (2009) evidenced that RES (10^{-7} g/kg, 15 before and 2 h after induction of cerebral damage, subcutaneous route) restored mitochondrial function in the transient rat middle cerebral artery occlusion (MCAO) model of ischemia in the brain [98]. RES restored the activity of the mitochondrial complexes I, II, and III, as well as the levels of ATP in the ischemic hippocampus. Furthermore, RES attenuated cytochrome c release in the ischemic group. RES significantly reduced the infarction size in the rat brain, ameliorating rat locomotion (as assessed through different behavioral tasks). It was not investigated the mechanism underlying such effects in that experimental model. Nonetheless, as will be discussed here, RES can trigger a panoply of signaling pathways involved with the mitochondrial homeostasis in the mammalian brain.

Yu and Yang (2010) have demonstrated that RES (20 mg/kg.day^{-1}) activated the AMPK/PGC-1α/NRF-1 signaling pathway in the visual cortex of rats subjected to monocular deprivation, leading to an increase in the number of mitochondria and in the amounts of ATP in that central area [67]. An acute treatment with RES at a lower dose (0.3 mg/kg 30 min after induction of depression by lipopolysaccharide – LPS, i.p. single injection) also prevented the bioenergetics deficits observed in the hippocampus of mice during inflammation [99]. RES suppressed the loss of MMP and the decline in the levels of ATP in the hippocampus of the LPS-treated mice.

The mechanism of action by which RES maintained the mitochondrial homeostasis in the mice hippocampus during inflammation was not examined by the authors. However, the data clearly shown that the mitochondria-related benefits resulting from the administration of RES depends on the animal species and on the circumstances the organelles face in the cells.

Cao et al. (2014) have tested the mitochondria-related effects of RES (100 mg/kg.day^{-1}) on the offspring of rats subjected to prenatal stress during the days 14–20 of pregnancy [61]. The authors have found that RES restored the levels of activated AMPK in the hippocampus of the offspring (both male and female rats) of stressed rats. Furthermore, RES increased the levels of mitochondrial DNA and enhanced the immunocontents of PGC-1α and of the complexes I, III, IV, and V in the hippocampus of male rats. Similar effects were observed in the hippocampus of female rats (increased levels of mitochondrial DNA and of the complexes I and IV, upregulation of PGC-1α). RES also upregulated the levels of Nrf2 and of enzymes modulated by this transcription factor in both male and female offspring, such as NQO1, HO-1, and the catalytic (GCLC) and regulatory (GCLM) subunits of the γ-glutamate-cysteine ligase (γ-GCL) enzyme. In that context, RES also elicited antioxidant effects by reducing the levels of protein carbonyl and increasing the GSH/GSSG ratio in the offspring hippocampus. RES also modulated neurotrophic factors and promoted benefits regarding rat cognition (as assessed in the T maze and Morris water maze tests). Data obtained by the authors are very interesting, since it was clearly demonstrated that RES promoted several benefits in the offspring of rats stressed during pregnancy, mimicking an event that is very common in the human life. It would be particularly interesting continuing that research in other experimental models and performing analyses in also in humans in order to explore this aspect of RES, applying this knowledge to improve life quality.

On the other hand, Palomera-Avalos et al. (2017) have published that RES (added to the diet at 1 g/kg w/w, resulting in daily doses about 160 mg/kg to the animals) ingestion caused benefits in the senescence-accelerated prone mouse (SAMP8) subjected to high fat diet (which causes metabolic stress by affecting mitochondrial function) [100]. RES intake did not alter the immunocontents of SIRT1 and PGC-1α in the hippocampus of SAMP8; however, RES upregulated the levels of p-AMPK (phosphorylated form), as well as the amounts of the mitochondrial complexes I, II, III, IV, and V. The levels of OPA1 and MFN2 were also enhanced by RES in the hippocampus of SAMP8, suggesting a role for mitochondrial dynamics modulation in the mechanism of action induced by RES. Actually, the authors have found larger mitochondria in the hippocampus of the SAMP8, which is an evidence of mitochondrial fusion. Importantly, RES ameliorated mice cognition in that experimental model.

TABLE 2

Main findings regarding the effects of resveratrol (RES) on mitochondria in *in vivo* experimental models.

Main findings	References
– RES upregulated both protein level and activity of SIRT1 in the spinal motoneurons of SOD1-G93A ALS mice – RES restored the levels of the components of the OXPHOS and of the active form of AMPK	[60]
– RES induced mitochondrial biogenesis in the visual cortex of rats subjected to monocular deprivation – RES activated the AMPK/PGC-1α/NRF-1 axis, causing an increase in the number of mitochondria and in the levels of ATP in neurons of the visual cortex	[67]
– RES decreased the levels of acetylated H3 in both cerebral cortex and striatum of the wild type animals and only in the striatum of the YAC128 animals, indicating SIRT1 activation – RES enhanced the expression of the MT-ND5 and MT-CO1 genes in the cerebral cortex of YAC128 mice	[73]
– RES restored mitochondrial function in the transient rat MCAO model of ischemia in the hippocampus by acting on the activity of the complexes I, II, and III – RES blocked the release of cytochrome c from the mitochondria in of the MCAO group	[98]
– RES activated the AMPK/PGC-1α/NRF-1 signaling pathway in the visual cortex of rats subjected to monocular deprivation, causing an increase in the number of mitochondria and in the levels of ATP	[67]
– RES prevented the bioenergetics deficits observed in the hippocampus of mice during inflammation induced by LPS – RES blocked the loss of MMP and the ATP decline	[99]
– RES restored the levels of activated AMPK in the hippocampus of the offspring of stressed rats – RES increased the levels of mitochondrial DNA – RES enhanced the immunocontents of PGC-1α and of the complexes I, III, IV, and V in the hippocampus of male rats (similar effects were seen in the hippocampus of female rats) – RES caused the activation of Nrf2, leading to an increase in the expression of NQO1, HO-1, GCLC, and GCLM in both male and female offspring – RES induced antioxidant effects in the offspring hippocampus by reducing the levels of protein carbonyl and increasing the GSH/GSSG ratio	[61]
– RES upregulated the levels of p-AMPK (phosphorylated form), as well as the amounts of the mitochondrial complexes I, II, III, IV, and V in the hippocampus of SAMP8 – RES enhanced the levels of OPA and MFN2 in the SAMP8 hippocampus – RES caused an increase in the number of larger mitochondria in the SAMP8 hippocampus	[100]
– RES restored the activity of complex I and reduced the levels of nitrated proteins and of 4-hydroxynonenal in the organelles in the cerebral cortex of rats treated with Li-pilocarpine	[101]
– RES enhanced the levels and the activity of Mn-SOD in the rat brain – RES upregulated the activity of citrate synthase	[103]
– RES increased the activity of Mn-SOD in the hippocampus of OVX female rats exposed to D-galactose – RES prevented the loss of cristae in the hippocampal mitochondria	[104]
– RES upregulated the Bcl-2/Bax ratio and attenuated the loss of MMP in hippocampal neurons of rats subjected to MCAO	[108]
– RES protected rat brain in an experimental model of early brain injury after subarachnoid hemorrhage by downregulating TXNIP and ASK1 proteins – RES increased the levels of Trx2 – RES caused mitochondria-related anti-apoptotic effects by decreasing the amounts of Bax and blocking the release of cytochrome from the organelles	[109]

Whether mitochondrial fusion is involved with the attenuation induced by RES in the high fat diet-elicited cognitive decline, it remains to be examined. Additionally, it was not investigated whether RES modulated the OXPHOS activity. Only the immunocontents of the OXPHOS components were analyzed in that work. It would be interesting testing if inhibition of mitochondrial fusion would be involved in the neuroprotective effects induced by RES in future works.

Recently, Folbergrová et al. (2018) have found that RES (25 mg/kg, i.p., in both acute and chronic protocols, as detailed in Table 2) protected mitochondria in the cerebral cortex of young rats subjected to epileptogenesis induced by Li-pilocarpine [101]. RES restored the activity of complex I and reduced the levels of nitrated proteins in the organelles. Besides, RES decreased the amounts of 4-hydroxynonenal in cortical mitochondria obtained from Li-pilocarpine-treated rats. The mechanism of action of RES in protecting mitochondria was not addressed by the authors in that work. Therefore, it is necessary to find how RES rescued mitochondrial function due to the pharmacological importance of the data obtained by the authors. Several patients present a drug resistance regarding the treatment of epilepsy [102]. Thus, it is necessary to obtain new drugs that would be able to attenuate the impact of epilepsy mainly in young individuals. This would increase the life quality of the subjects suffering from epilepsy in a very important moment of their life, in which learning capability, as well as other cognitive functions, needs to be preserved.

4.6   Mitochondria-Related Redox Effects

Robb et al. (2008) have examined the redox effects of RES in several tissues of C57 BL6 mice [103]. The animals were treated with RES at 200 mg/kg.day^{-1} for 5 weeks through different delivery methods. However, only in one method (RES-containing high fat diet) it was possible to be a redox effect of RES on brain mitochondria. RES enhanced the immunocontent and the activity of the mitochondria-located enzyme Mn-SOD in the rat brain. Interestingly, RES also upregulated the activity of citrate synthase in the brain tissue of the animals subjected to the same diet. The mechanism of action by which RES modulated those enzymes was not addressed by the authors in that work. However, it serves to demonstrate the ability RES possesses to cross the blood-brain barrier, causing mitochondria-associated redox effects that may be beneficial to the quality and function of the organelles, as well as being related with neuroprotection.

Mn-SOD upregulation in the brain was also observed in the work by Zhao et al. (2012) [104]. The authors have described increased activity of Mn-SOD, as well as of other antioxidant enzymes, in the hippocampus of ovariectomized (OVX) female rats exposed to D-galactose and treated with RES (20–80 mg/kg.day^{-1} for 12 weeks, intragastric route). RES was also effective in preventing the loss of cristae in the mitochondria of the hippocampus of the OVX rats exposed to D-galactose. It is noteworthy that RES at 40–80 mg/kg day^{-1} doses exerted very similar effects regarding mitochondrial parameters when compared to estradiol replacement therapy. Therefore, RES would be an interesting alternative treatment during the postmenopausal period. Nonetheless, studies are necessary regarding both RES bioavailability and toxicity in humans before using this polyphenol as a treatment. It would be particularly important since women following menopause exhibit increased rates of Alzheimer's disease than men at the same age [105,106]. On the other hand, Gueguen et al. (2015) have demonstrated that RES is able to bind to the complex I of brain mitochondria, promoting redox impairment in the organelles of aged mice [107]. The authors have demonstrated, by using different strategies, that the polyphenol modulated mitochondrial function according to its concentration. Additionally, it was shown that RES enhances the antioxidant defenses in brain tissue, but it was not sufficient to attenuate oxidative stress in that experimental model.

4.7   Mitochondria-Related Anti-Apoptotic Effects

Wang et al. (2014) have described a mitochondria-related anti-apoptotic role for RES in an experimental model of cerebral ischemia (induced by middle cerebral artery occlusion – MCAO) in rats [108]. RES (30 mg/kg.day^{-1} for 7 days prior the induction of cerebral ischemia, i.p.) reduced the levels of neurological deficits and infarct size by a mechanism involving an upregulation of the Bcl-2/Bax ratio and by attenuating the loss of MMP in hippocampal neurons. Other apoptosis-related parameters were not explored by the authors, but it is evident a role for RES in protecting mitochondria.

Liang et al. (2018) recently demonstrated that RES (5 mg/kg i.p. single dose) protected rat brain in an experimental model of early brain injury after subarachnoid hemorrhage by downregulating thioredoxin-interacting protein (TXNIP) and apoptosis signal-regulating kinase-1 (ASK1) proteins [109]. RES also increased the immunocontent of thioredoxin 2 (Trx2), a mitochondria-located cytoprotective protein. RES blocked the mitochondria-related apoptotic pathway by reducing the amounts of Bax and by blocking the release of cytochrome c to the cytosol and the consequent activation of caspase-3. TXNIP mediates cell death by inhibiting the interaction between Trx2 and ASK1, favoring the triggering of cell death by promoting the release of cytochrome c from the mitochondria [110]. Thus, RES attenuated the activation of the mitochondria-related pro-apoptotic pathway involving TXNIP/Trx2/ASK1 proteins in that experimental model. Similar data were shown in the work by Kairisalo et al. (2011), in which the authors found that RES at 50 µM activated the X-linked inhibitor of apoptosis (XIAP)/nuclear factor-κB (NF-κB)/ Trx2 signaling pathway in neuron-like PC6.3 cells [111]. RES also can upregulate the expression of Mn-SOD, attenuating the production of ROS in that experimental model. Mudò et al. (2012) also observed similar effects in the mouse SN4741 dopaminergic cells treated with RES at 10 µM for 6 h [112]. The authors described upregulation in the levels of SIRT1, PGC-1α Trx2, and Mn-SOD in the RES-treated group. The levels of XIAP did not change in that experimental model. On the other hand, Jang and Surh (2003) have found different data in PC12 cells exposed to amyloid-β_{25-35} (Aβ_{25-35}) and RES (10–25 µM) simultaneously for 36 h [113]. RES inhibited the transcription factor NF-κB in the Aβ_{25-35}-treated cells, consequently restoring MMP and decreasing the mitochondria-related pro-apoptotic signal.

Therefore, RES modulates the mitochondria-related anti-apoptotic signals by a complex manner depending on the cell type and also on the concentration and duration of the treatment. There are several points of regulation that should be checked in the future, as discussed above, in order to use RES as a neuroprotective agent in human diseases. It is specially recommended performing assays aiming to examine, for example, whether RES would be able to induce toxicity, mainly neurotoxicity.

5 CONCLUSION

As discussed here, RES interacts with mitochondria directly (as observed in assays using mitochondria isolated from the brain) and indirectly (as seen in protocols involving cultured cells and experimental animals). RES directly binds to mitochondrial proteins, modulating the activity of these components of the OXPHOS. In cultured cells and in experimental animals, RES activates signaling pathways that lead to alterations in both mitochondrial function and dynamics. In several studies, RES exhibited the ability to promote neuroprotection by a mitochondria-related manner. On the other hand, RES triggered apoptosis in tumor cells by the intrinsic apoptotic pathway, which depends on mitochondrial signals. Therefore, RES is a potent protective agent considering mitochondria and future studies should investigate whether this polyphenol would exert such effects in humans. The potential toxicity induced by RES should also be evaluated to avoid intoxications.

REFERENCES

1. Schaffer S, Asseburg H, Kuntz S, Muller WE, Eckert GP (2012) Effects of polyphenols on brain ageing and Alzheimer's disease: Focus on mitochondria. *Mol Neurobiol* 46:161–178. doi: 10.1007/s12035-012-8282-9

2. Virmani A, Pinto L, Binienda Z, Ali S (2013) Food, nutrigenomics, and neurodegeneration–neuroprotection by what you eat! *Mol Neurobiol* 48:353–362. doi: 10.1007/s12035-013-8498-3

3. de Oliveira MR (2016) The dietary components carnosic acid and carnosol as neuroprotective agents: A mechanistic view. *Mol Neurobiol* 53:6155–6168. doi: 10.1007/s12035-015-9519-1

4. de Oliveira MR (2018) Carnosic acid as a promising agent in protecting mitochondria of brain cells. *Mol Neurobiol* in press. doi: 10.1007/s12035-017-0842-6

5. Nakka VP, Prakash-Babu P, Vemuganti R (2016) Crosstalk between endoplasmic reticulum stress, oxidative stress, and autophagy: Potential therapeutic targets for acute CNS injuries. *Mol Neurobiol* 53:532–544. doi: 10.1007/s12035-014-9029-6

6. Lin CY, Tsai CW (2017) Carnosic acid attenuates 6-hydroxydopamine-induced neurotoxicity in SH-SY5Y cells by inducing autophagy through an enhanced interaction of Parkin and Beclin1. *Mol Neurobiol* 54:2813–2822. doi: 10.1007/s12035-016-9873-7

7. Fu RH, Huang LC, Lin CY, Tsai CW (2018) Modulation of ARTS and XIAP by Parkin is associated with carnosic acid protects SH-SY5Y cells against 6-hydroxydopamine-induced apoptosis. *Mol Neurobiol* 55:1786–1794. doi: 10.1007/s12035-017-0443-4

8. Nishimoto S, Koike S, Inoue N, Suzuki T, Ogasawara Y (2017) Activation of Nrf2 attenuates carbonyl stress induced by methylglyoxal in human neuroblastoma cells: Increase in GSH levels is a critical event for the detoxification mechanism. *Biochem Biophys Res Commun* 483:874–879. doi: 10.1016/j.bbrc.2017.01.024

9. Bagli E, Goussia A, Moschos MM, Agnantis N, Kitsos G (2016) Natural compounds and neuroprotection: Mechanisms of action and novel delivery systems. *In Vivo* 30:535–547.

10. Lippi G, Franchini M, Favaloro EJ, Targher G (2010) Moderate red wine consumption and cardiovascular disease risk: Beyond the "French paradox". *Semin Thromb Hemost* 36:59–70. doi: 10.1055/s-0030-1248725

11. Silva P, Fernandes E, Carvalho F (2015) Dual effect of red wine on liver redox status: A concise and mechanistic review. *Arch Toxicol* 89:1681–1693. doi: 10.1007/s00204-015-1538-1

12. Sung MM, Byrne NJ, Robertson IM, Kim TT, Samokhvalov V, Levasseur J, Soltys CL, Fung D, Tyreman N, Denou E, Jones KE, Seubert JM, Schertzer JD, Dyck JR (2017) Resveratrol improves exercise performance and skeletal muscle oxidative capacity in heart failure. *Am J Physiol Heart Circ Physiol* 312:H842–H853. doi: 10.1152/ajpheart.00455.2016

13. Li YR, Li S, Lin CC (2018) Effect of resveratrol and pterostilbene on aging and longevity. *Biofactors* 44:69–82. doi: 10.1002/biof.1400

14. Rauf A, Imran M, Butt MS, Nadeem M, Peters DG, Mubarak MS (2018) Resveratrol as an anti-cancer agent: A review. *Crit Rev Food Sci Nutr* 58:1428–1447. doi: 10.1080/10408398.2016.1263597

15. Sebai H, Gadacha W, Sani M, Aouani E, Ghanem-Boughanmi N, Ben-Attia M (2009) Protective effect of resveratrol against lipopolysaccharide-induced oxidative stress in rat brain. *Brain Inj* 23:1089–1094. doi: 10.3109/02699050903379370

16. Singh N, Agrawal M, Doré S (2013) Neuroprotective properties and mechanisms of resveratrol in in vitro and in vivo experimental cerebral stroke models. *ACS Chem Neurosci* 4:1151–1162. doi: 10.1021/cn400094w

17. Su Q, Pu H, Hu C (2016) Neuroprotection by combination of resveratrol and enriched environment against ischemic brain injury in rats. *Neurol Res* 38:60–68. doi: 10.1080/01616412.2015.1133027

18. Kaldas MI, Walle UK, Walle T (2003) Resveratrol transport and metabolism by human intestinal CaCo-2 cells. *J Pharm Pharmacol* 55:307–312.

19. Goldberg DM, Yan J, Soleas GJ (2003) Absorption of three wine-related polyphenols in three different matrices by healthy subjects. *Clin Biochem* 36:79–87.

20. Walle T, Hsieh F, DeLegge MH, Oatis JE Jr, Walle UK (2004) High absorption but very low bioavailability of oral resveratrol in humans. *Drug Metab Dispos* 32:1377–1382. doi: 10.1124/dmd.104.000885

21. Lin HS, Yue BD, Ho PC (2009) Determination of pterostilbene in rat plasma by a simple HPLC-UV method and its application in pre-clinical pharmacokinetic study. *Biomed Chromatogr* 23:1308–1315. doi: 10.1002/bmc.1254

22. Lin HS, Ho PC (2009) A rapid HPLC method for the quantification of 3,5,4′-trimethoxy-trans-stilbene (TMS) in rat plasma and its application in pharmacokinetic study. *J Pharm Biomed Anal* 49:387–392. doi: 10.1016/j.jpba.2008.10.042

23. Cai H, Sale S, Britton RG, Brown K, Steward WP, Gescher AJ (2011) Pharmacokinetics in mice and metabolism in murine and human liver fractions of the putative cancer chemopreventive agents 3′,4′,5′,5,7-pentamethoxyflavone and tricin (4′,5,7-trihydroxy-3′,5′-dimethoxyflavone). *Cancer*

Chemother Pharmacol 67:255–263. doi: 10.1007/s00280-010-1313-1

24. Walle T, Wen X, Walle UK (2007) Improving metabolic stability of cancer chemoprotective polyphenols. *Expert Opin Drug Metab Toxicol* 3:379–388.

25. Walle T (2011) Bioavailability of resveratrol. *Ann N Y Acad Sci* 1215:9–15. doi: 10.1111/j.1749-6632.2010.05842.x

26. Stevens JF, Maier CS (2016) The chemistry of gut microbial metabolism of polyphenols. *Phytochem Rev* 15:425–444. doi: 10.1007/s11101-016-9459-z

27. Pan MH, Wu JC, Ho CT, Lai CS (2018) Antiobesity molecular mechanisms of action: Resveratrol and pterostilbene. *Biofactors* 44:50–60. doi: 10.1002/biof.1409

28. Bird JK, Raederstorff D, Weber P, Steinert RE (2017) Cardiovascular and antiobesity effects of resveratrol mediated through the gut microbiota. *Adv Nutr* 8:839–849. doi: 10.3945/an.117.016568

29. Kulkarni SS, Cantó C (2015) The molecular targets of resveratrol. *Biochim Biophys Acta* 1852:1114–1123. doi: 10.1016/j.bbadis.2014.10.005

30. de Oliveira MR, Nabavi SF, Manayi A, Daglia M, Hajheydari Z, Nabavi SM (2016) Resveratrol and the mitochondria: From triggering the intrinsic apoptotic pathway to inducing mitochondrial biogenesis, a mechanistic view. *Biochim Biophys Acta* 1860:727–745. doi: 10.1016/j.bbagen.2016.01.017

31. Jardim FR, de Rossi FT, Nascimento MX, Da Silva Barros RG, Borges PA, Prescilio IC, de Oliveira MR (2018) Resveratrol and brain mitochondria: A review. *Mol Neurobiol* 55:2085–2101. doi: 10.1007/s12035-017-0448-z

32. Osellame LD, Blacker TS, Duchen MR (2012) Cellular and molecular mechanisms of mitochondrial function. *Best Pract Res Clin Endocrinol Metab* 26:711–723. doi: 10.1016/j.beem.2012.05.003

33. Sinha K, Das J, Pal PB, Sil PC (2013) Oxidative stress: The mitochondria-dependent and mitochondria-independent pathways of apoptosis. *Arch Toxicol* 87:1157–1180. doi: 10.1007/s00204-013-1034-4

34. Friedman JR, Nunnari J (2014) Mitochondrial form and function. *Nature* 505:335–343. doi: 10.1038/nature12985

35. Green DR, Galluzzi L, Kroemer G (2014) Metabolic control of cell death. *Science* 345. doi: 10.1126/science.1250256

36. Naoi M, Maruyama W, Shamoto-Nagai M, Yi H, Akao Y, Tanaka M (2005) Oxidative stress in mitochondria: Decision to survival and death of neurons in neurodegenerative disorders. *Mol Neurobiol* 31:81–93.

37. de Oliveira MR (2015) Vitamin A and retinoids as mitochondrial toxicants. *Oxid Med Cell Longev* 2015:140267. doi: 10.1155/2015/140267

38. de Oliveira MR (2016) Fluoxetine and the mitochondria: A review of the toxicological aspects. *Toxicol Lett* 258:185–191. doi: 10.1016/j.toxlet.2016.07.001

39. de Oliveira MR, Jardim FR (2016) Cocaine and mitochondria-related signaling in the brain: A mechanistic view and future directions. *Neurochem Int* 92:58–66. doi: 10.1016/j.neuint.2015.12.006

40. Picard M, Wallace DC, Burelle Y (2016) The rise of mitochondria in medicine. *Mitochondrion* 30:105–116. doi: 10.1016/j.mito.2016.07.003

41. Whitaker RM, Corum D, Beeson CC, Schnellmann RG (2016) Mitochondrial biogenesis as a pharmacological target: A new approach to acute and chronic diseases. *Annu Rev Pharmacol Toxicol* 56:229–249. doi: 10.1146/annurev-pharmtox-010715-103155

42. Gruber J, Fong S, Chen CB, Yoong S, Pastorin G, Schaffer S, Cheah I, Halliwell B (2013) Mitochondria-targeted antioxidants and metabolic modulators as pharmacological interventions to slow ageing. *Biotechnol Adv* 31:563–592. doi: 10.1016/j.biotechadv.2012.09.005

43. de Oliveira MR, Nabavi SM, Braidy N, Setzer WN, Ahmed T, Nabavi SF (2016) Quercetin and the mitochondria: A mechanistic view. *Biotechnol Adv* 34:532–549. doi: 10.1016/j.biotechadv.2015.12.014

44. de Oliveira MR, Jardim FR, Setzer WN, Nabavi SM, Nabavi SF (2016) Curcumin, mitochondrial biogenesis, and mitophagy: Exploring recent data and indicating future needs. *Biotechnol Adv* 34:813–826. doi: 10.1016/j.biotechadv.2016.04.004

45. de Oliveira MR, Nabavi SF, Nabavi SM, Jardim FR (2017) Omega-3 polyunsaturated fatty acids and mitochondria, back to the future. *Trends Food Sci Technol* 67:76–92. doi: 10.1016/j.tifs.2017.06.019

46. Luo C, Ikegaya Y, Koyama R (2016) Microglia and neurogenesis in the epileptic dentate gyrus. *Neurogenesis (Austin)* 3:e1235525.

47. Knoth R, Singec I, Ditter M, Pantazis G, Capetian P, Meyer RP, Horvat V, Volk B, Kempermann G (2010) Murine features of neurogenesis in the human hippocampus across the lifespan from 0 to 100 years. *PLoS One* 5: e8809. doi: 10.1371/journal.pone.0008809

48. Boldrini M, Santiago AN, Hen R, Dwork AJ, Rosoklija GB, Tamir H, Arango V, John Mann J (2013) Hippocampal granule neuron number and dentate gyrus volume in antidepressant-treated and untreated major depression. *Neuropsychopharmacology* 38:1068–1077. doi: 10.1038/npp.2013.5

49. Ho NF, Hooker JM, Sahay A, Holt DJ, Roffman JL (2013) In vivo imaging of adult human hippocampal neurogenesis: Progress, pitfalls and promise. *Mol Psychiatry* 18:404–416. doi: 10.1038/mp.2013.8

50. Ernst A, Alkass K, Bernard S, Salehpour M, Perl S, Tisdale J, Possnert G, Druid H, Frisén J (2014) Neurogenesis in the striatum of the adult human brain. *Cell* 156:1072–1083. doi: 10.1016/j.cell.2014.01.044

51. Bergmann O, Spalding KL, Frisén J (2015) Adult neurogenesis in humans. *Cold Spring Harb Perspect Biol* 7: a018994. doi: 10.1101/cshperspect.a018994

52. Zheng J, Ramirez VD (2000) Inhibition of mitochondrial proton F0F1-ATPase/ATP synthase by polyphenolic phytochemicals. *Br J Pharmacol* 130:1115–1123. doi: 10.1038/sj.bjp.0703397

53. Zini R, Morin C, Bertelli A, Bertelli AA, Tillement JP (1999) Effects of resveratrol on the rat brain respiratory chain. *Drugs Exp Clin Res* 25:87–97.

54. Gledhill JR, Montgomery MG, Leslie AG, Walker JE (2007) Mechanism of inhibition of bovine F1-ATPase by resveratrol and related polyphenols. *Proc Natl Acad Sci U S A* 104:13632–13637.

55. Moreira AC, Silva AM, Santos MS, Sardão VA (2013) Resveratrol affects differently rat liver and brain mitochondrial bioenergetics and oxidative stress in vitro: Investigation of the role of gender. *Food Chem Toxicol* 53:18–26. doi: 10.1016/j.fct.2012.11.031

56. Fišar Z, Hroudová J, Singh N, Kopřivová A, Macečková D (2016) Effect of simvastatin, coenzyme Q10, resveratrol, acetylcysteine and acetylcarnitine on mitochondrial respiration. *Folia Biol (Praha)* 62:53–66.

57. Desquiret-Dumas V, Gueguen N, Leman G, Baron S, Nivet-Antoine V, Chupin S, Chevrollier A, Vessières E, Ayer A, Ferré M, Bonneau D, Henrion D, Reynier P,

Procaccio V (2013) Resveratrol induces a mitochondrial complex I-dependent increase in NADH oxidation responsible for sirtuin activation in liver cells. *J Biol Chem* 288:36662–36675. doi: 10.1074/jbc.M113.466490

58. Manczak M, Mao P, Calkins MJ, Cornea A, Reddy AP, Murphy MP, Szeto HH, Park B, Reddy PH (2010) Mitochondria-targeted antioxidants protect against amyloid-beta toxicity in Alzheimer's disease neurons. *J Alzheimers Dis* 20 (Suppl 2):S609–S631. doi: 10.3233/JAD-2010-100564

59. Wang J, Zhang Y, Tang L, Zhang N, Fan D (2011) Protective effects of resveratrol through the up-regulation of SIRT1 expression in the mutant hSOD1-G93A-bearing motor neuron-like cell culture model of amyotrophic lateral sclerosis. *Neurosci Lett* 503:250–255. doi: 10.1016/j.neulet.2011.08.047

60. Mancuso R, Del Valle J, Modol L, Martinez A, Granado-Serrano AB, Ramirez-Núñez O, Pallás M, Portero-Otin M, Osta R, Navarro X (2014) Resveratrol improves motoneuron function and extends survival in SOD1(G93A) ALS mice. *Neurotherapeutics* 11:419–432. doi: 10.1007/s13311-013-0253-y

61. Cao K, Zheng A, Xu J, Li H, Liu J, Peng Y, Long J, Zou X, Li Y, Chen C, Liu J, Feng Z (2014) AMPK activation prevents prenatal stress-induced cognitive impairment: Modulation of mitochondrial content and oxidative stress. *Free Radic Biol Med* 75:156–166. doi: 10.1016/j.freeradbiomed.2014.07.029

62. Dinkova-Kostova AT, Abramov AY (2015) The emerging role of Nrf2 in mitochondrial function. *Free Radic Biol Med* 88:179–188. doi: 10.1016/j.freeradbiomed.2015.04.036

63. Esteras N, Dinkova-Kostova AT, Abramov AY (2016) Nrf2 activation in the treatment of neurodegenerative diseases: A focus on its role in mitochondrial bioenergetics and function. *Biol Chem* 397:383–400. doi: 10.1515/hsz-2015-0295

64. Keum YS (2012) Regulation of Nrf2-mediated phase II detoxification and anti-oxidant genes. *Biomol Ther (Seoul)* 20:144–151. doi: 10.4062/biomolther.2012.20.2.144

65. Peng Y, Liu J, Shi L, Tang Y, Gao D, Long J, Liu J (2016) Mitochondrial dysfunction precedes depression of AMPK/AKT signaling in insulin resistance induced by high glucose in primary cortical neurons. *J Neurochem* 137:701–713. doi: 10.1111/jnc.13563

66. Valenti D, de Bari L, de Rasmo D, Signorile A, Henrion-Caude A, Contestabile A, Vacca RA (2016) The polyphenols resveratrol and epigallocatechin-3-gallate restore the severe impairment of mitochondria in hippocampal progenitor cells from a Down syndrome mouse model. *Biochim Biophys Acta* 1862:1093–1104. doi: 10.1016/j.bbadis.2016.03.003

67. Yu L, Yang SJ (2010) AMP-activated protein kinase mediates activity-dependent regulation of peroxisome proliferator-activated receptor gamma coactivator-1alpha and nuclear respiratory factor 1 expression in rat visual cortical neurons. *Neuroscience* 169:23–38. doi: 10.1016/j.neuroscience.2010.04.063

68. Peng K, Tao Y, Zhang J, Wang J, Ye F, Dan G, Zhao Y, Cai Y, Zhao J, Wu Q, Zou Z, Cao J, Sai Y (2016) Resveratrol regulates mitochondrial biogenesis and fission/fusion to attenuate rotenone-induced neurotoxicity. *Oxid Med Cell Longev* 2016:6705621. doi: 10.1155/2016/6705621

69. Twig G, Hyde B, Shirihai OS (2008) Mitochondrial fusion, fission and autophagy as a quality control axis: The bioenergetic view. *Biochim Biophys Acta* 1777:1092–1097. doi: 10.1016/j.bbabio.2008.05.001

70. Lee H, Yoon Y (2016) Mitochondrial fission and fusion. *Biochem Soc Trans* 44:1725–1735. doi: 10.1042/BST20160129

71. Westermann B (2012) Bioenergetic role of mitochondrial fusion and fission. *Biochim Biophys Acta* 1817:1833–1838. doi: 10.1016/j.bbabio.2012.02.033

72. Otera H, Ishihara N, Mihara K (2013) New insights into the function and regulation of mitochondrial fission. *Biochim Biophys Acta* 1833:1256–1268. doi: 10.1016/j.bbamcr.2013.02.002

73. Naia L, Rosenstock TR, Oliveira AM, Oliveira-Sousa SI, Caldeira GL, Carmo C, Laço MN, Hayden MR, Oliveira CR, Rego AC (2017) Comparative mitochondrial-based protective effects of resveratrol and nicotinamide in huntington's disease models. *Mol Neurobiol* 54:5385–5399. doi: 10.1007/s12035-016-0048-3

74. Panov AV, Gutekunst CA, Leavitt BR, Hayden MR, Burke JR, Strittmatter WJ, Greenamyre JT (2002) Early mitochondrial calcium defects in huntington's disease are a direct effect of polyglutamines. *Nat Neurosci* 5:731–736.

75. Yano H, Baranov SV, Baranova OV, Kim J, Pan Y, Yablonska S, Carlisle DL, Ferrante RJ, Kim AH, Friedlander RM (2014) Inhibition of mitochondrial protein import by mutant huntingtin. *Nat Neurosci* 17:822–831. doi: 10.1038/nn.3721

76. Bellaver B, Bobermin LD, Souza DG, Rodrigues MD, de Assis AM, Wajner M, Gonçalves CA, Souza DO, Quincozes-Santos A (2016) Signaling mechanisms underlying the glioprotective effects of resveratrol against mitochondrial dysfunction. *Biochim Biophys Acta* 1862:1827–1838. doi: 10.1016/j.bbadis.2016.06.018

77. de Oliveira MR, Peres A, Ferreira GC, Schuck PF, Gama CS, Bosco SMD (2017) Carnosic acid protects mitochondria of human neuroblastoma SH-SY5Y cells exposed to paraquat through activation of the Nrf2/HO-1axis. *Mol Neurobiol* 54:5961–5972. doi: 10.1007/s12035-016-0100-3

78. de Oliveira MR, Da Costa Ferreira G, Brasil FB, Peres A (2018) Pinocembrin suppresses H2O2-induced mitochondrial dysfunction by a mechanism dependent on the Nrf2/HO-1 Axis in SH-SY5Y Cells. *Mol Neurobiol* 55:989–1003. doi: 10.1007/s12035-016-0380-7

79. Wang Y, Miao Y, Mir AZ, Cheng L, Wang L, Zhao L, Cui Q, Zhao W, Wang H (2016) Inhibition of beta-amyloid-induced neurotoxicity by pinocembrin through Nrf2/HO-1 pathway in SH-SY5Y cells. *J Neurol Sci* 368:223–230. doi: 10.1016/j.jns.2016.07.010

80. O'Brien L, Hosick PA, John K, Stec DE, Hinds TD Jr (2015) Biliverdin reductase isozymes in metabolism. *Trends Endocrinol Metab* 26:212–220. doi: 10.1016/j.tem.2015.02.001

81. Wegiel B, Nemeth Z, Correa-Costa M, Bulmer AC, Otterbein LE (2014) Heme oxygenase-1: A metabolic nike. *Antioxid Redox Signal* 20:1709–1722. doi: 10.1089/ars.2013.5667

82. Sarti P, Forte E, Mastronicola D, Giuffrè A, Arese M (2012) Cytochrome c oxidase and nitric oxide in action: Molecular mechanisms and pathophysiological implications. *Biochim Biophys Acta* 1817:610–619. doi: 10.1016/j.bbabio.2011.09.002

83. Moncada S, Bolaños JP (2006) Nitric oxide, cell bioenergetics and neurodegeneration. *J Neurochem* 97:1676–1689.

84. Calcerrada P, Peluffo G, Radi R (2011) Nitric oxide-derived oxidants with a focus on peroxynitrite: Molecular targets, cellular responses and therapeutic implications. *Curr Pharm Des* 17:3905–3932.

85. Lee MK, Kang SJ, Poncz M, Song KJ, Park KS (2007) Resveratrol protects SH-SY5Y neuroblastoma cells from apoptosis induced by dopamine. *Exp Mol Med* 39:376–384.

86. Chang HC, Chen TG, Tai YT, Chen TL, Chiu WT, Chen RM (2011) Resveratrol attenuates oxidized LDL-evoked Lox-1 signaling and consequently protects against apoptotic insults to cerebrovascular endothelial cells. *J Cereb Blood Flow Metab* 31:842–854. doi: 10.1038/jcbfm.2010.180

87. Wu Y, Li X, Zhu JX, Xie W, Le W, Fan Z, Jankovic J, Pan T (2011) Resveratrol-activated AMPK/SIRT1/autophagy in cellular models of Parkinson's disease. *Neurosignals* 19:163–174. doi: 10.1159/000328516

88. Youle RJ, Narendra DP (2011) Mechanisms of mitophagy. *Nat Rev Mol Cell Biol* 12:9–14. doi: 10.1038/nrm3028

89. Jeong JK, Moon MH, Bae BC, Lee YJ, Seol JW, Kang HS, Kim JS, Kang SJ, Park SY (2012) Autophagy induced by resveratrol prevents human prion protein-mediated neurotoxicity. *Neurosci Res* 73:99–105. doi: 10.1016/j.neures.2012.03.005

90. Lin TK, Chen SD, Chuang YC, Lin HY, Huang CR, Chuang JH, Wang PW, Huang ST, Tiao MM, Chen JB, Liou CW (2014) Resveratrol partially prevents rotenone-induced neurotoxicity in dopaminergic SH-SY5Y cells through induction of heme oxygenase-1 dependent autophagy. *Int J Mol Sci* 15:1625–1646. doi: 10.3390/ijms15011625

91. Chen S, Fan Q, Li A, Liao D, Ge J, Laties AM, Zhang X (2013) Dynamic mobilization of PGC-1α mediates mitochondrial biogenesis for the protection of RGC-5 cells by resveratrol during serum deprivation. *Apoptosis* 18:786–799. doi: 10.1007/s10495-013-0837-3

92. Tu W, Zhang Q, Liu Y, Han L, Wang Q, Chen P, Zhang S, Wang A, Zhou X (2018) Fluoride induces apoptosis via inhibiting SIRT1 activity to activate mitochondrial p53 pathway in human neuroblastoma SH-SY5Y cells. *Toxicol Appl Pharmacol* 347:60–69. doi: 10.1016/j.taap.2018.03.030

93. Sheu SJ, Liu NC, Ou CC, Bee YS, Chen SC, Lin HC, Chan JY (2013) Resveratrol stimulates mitochondrial bioenergetics to protect retinal pigment epithelial cells from oxidative damage. *Invest Ophthalmol Vis Sci* 54:6426–6438. doi: 10.1167/iovs.13-12024

94. Sareen D, van Ginkel PR, Takach JC, Mohiuddin A, Darjatmoko SR, Albert DM, Polans AS (2006) Mitochondria as the primary target of resveratrol-induced apoptosis in human retinoblastoma cells. *Invest Ophthalmol Vis Sci* 47:3708–3716.

95. van Ginkel PR, Sareen D, Subramanian L, Walker Q, Darjatmoko SR, Lindstrom MJ, Kulkarni A, Albert DM, Polans AS (2007) Resveratrol inhibits tumor growth of human neuroblastoma and mediates apoptosis by directly targeting mitochondria. *Clin Cancer Res* 13:5162–5169.

96. Rahman MA, Kim NH, Kim SH, Oh SM, Huh SO (2012) Antiproliferative and cytotoxic effects of resveratrol in mitochondria-mediated apoptosis in rat b103 neuroblastoma cells. *Korean J Physiol Pharmacol* 16:321–326. doi: 10.4196/kjpp.2012.16.5.321

97. Jiang H, Zhang L, Kuo J, Kuo K, Gautam SC, Groc L, Rodriguez AI, Koubi D, Hunter TJ, Corcoran GB, Seidman MD, Levine RA (2005) Resveratrol-induced

apoptotic death in human U251 glioma cells. *Mol Cancer Ther* 4:554–561.

98. Yousuf S, Atif F, Ahmad M, Hoda N, Ishrat T, Khan B, Islam F (2009) Resveratrol exerts its neuroprotective effect by modulating mitochondrial dysfunctions and associated cell death during cerebral ischemia. *Brain Res* 1250:242–253. doi: 10.1016/j.brainres.2008.10.068

99. Chen WJ, Du JK, Hu X, Yu Q, Li DX, Wang CN, Zhu XY, Liu YJ (2017) Protective effects of resveratrol on mitochondrial function in the hippocampus improves inflammation-induced depressive-like behavior. *Physiol Behav* 182:54–61. doi: 10.1016/j.physbeh.2017.09.024

100. Palomera-Avalos V, Griñán-Ferré C, Puigoriol-Ilamola D, Camins A, Sanfeliu C, Canudas AM, Pallàs M (2017) Resveratrol protects SAMP8 brain under metabolic stress: Focus on mitochondrial function and Wnt pathway. *Mol Neurobiol* 54:1661–1676. doi: 10.1007/s12035-016-9770-0

101. Folbergrová J, Ješina P, Kubová H, Otáhal J (2018) Effect of resveratrol on oxidative stress and mitochondrial dysfunction in immature brain during epileptogenesis. *Mol Neurobiol* in press. doi: 10.1007/s12035-018-0924-0

102. de Zélicourt M, de Toffol B, Vespignani H, Laurendeau C, Lévy-Bachelot L, Murat C, Fagnani F (2014) Management of focal epilepsy in adults treated with polytherapy in France: The direct cost of drug resistance (ESPERA study). *Seizure* 23:349–356. doi: 10.1016/j.seizure.2014.01.016

103. Robb EL, Winkelmolen L, Visanji N, Brotchie J, Stuart JA (2008) Dietary resveratrol administration increases MnSOD expression and activity in mouse brain. *Biochem Biophys Res Commun* 372:254–259. doi: 10.1016/j.bbrc.2008.05.028

104. Zhao H, Niu Q, Li X, Liu T, Xu Y, Han H, Wang W, Fan N, Tian Q, Zhang H, Wang Z (2012) Long-term resveratrol consumption protects ovariectomized rats chronically treated with D-galactose from developing memory decline without effects on the uterus. *Brain Res* 1467:67–80. doi: 10.1016/j.brainres.2012.05.040

105. Halbreich U, Lumley LA, Palter S, Manning C, Gengo F, Joe SH (1995) Possible acceleration of age effects on cognition following menopause. *J Psychiatr Res* 29:153–163.

106. Morrison JH, Hof PR (1997) Life and death of neurons in the aging brain. *Science* 278:412–419.

107. Gueguen N, Desquiret-Dumas V, Leman G, Chupin S, Baron S, Nivet-Antoine V, Vessières E, Ayer A, Henrion D, Lenaers G, Reynier P, Procaccio V (2015) Resveratrol directly binds to mitochondrial complex I and increases oxidative stress in brain mitochondria of aged mice. *PLoS One* 10:e0144290. doi: 10.1371/journal.pone.0144290

108. Wang R, Liu YY, Liu XY, Jia SW, Zhao J, Cui D, Wang L (2014) Resveratrol protects neurons and the myocardium by reducing oxidative stress and ameliorating mitochondria damage in a cerebral ischemia rat model. *Cell Physiol Biochem* 34:854–864. doi: 10.1159/000366304

109. Liang Y, Che X, Zhao Q, Darwazeh R, Zhang H, Jiang D, Zhao J, Xiang X, Qin W, Liu L, He Z (2018) Thioredoxin-interacting protein mediates mitochondrion-dependent apoptosis in early brain injury after subarachnoid hemorrhage. *Mol Cell Biochem* in press. doi: 10.1007/s11010-018-3381-1

110. Shalev A (2014) Minireview: Thioredoxin-interacting protein: Regulation and function in the pancreatic β-cell. *Mol Endocrinol* 28:1211–1220. doi: 10.1210/me.2014-1095

111. Kairisalo M, Bonomo A, Hyrskyluoto A, Mudò G, Belluardo N, Korhonen L, Lindholm D (2011)

Resveratrol reduces oxidative stress and cell death and increases mitochondrial antioxidants and XIAP in PC6.3-cells. *Neurosci Lett* 488:263–266. doi: 10.1016/j.neulet.2010.11.042

112. Mudò G, Mäkelä J, Di Liberto V, Tselykh TV, Olivieri M, Piepponen P, Eriksson O, Mälkiä A, Bonomo A, Kairisalo M, Aguirre JA, Korhonen L, Belluardo N, Lindholm D (2012) Transgenic expression and activation of PGC-1α protect dopaminergic neurons in the MPTP mouse model of Parkinson's disease. *Cell Mol Life Sci* 69:1153–1165. doi: 10.1007/s00018-011-0850-z

113. Jang JH, Surh YJ (2003) Protective effect of resveratrol on beta-amyloid-induced oxidative PC12 cell death. *Free Radic Biol Med* 34:1100–1110.

27 Pathophysiology and Biomarkers of Mitochondrial Dysfunction Induced by Heavy Exercise

Sergej M. Ostojic

Faculty of Sport and Physical Education, University of Novi Sad, Novi Sad, Serbia
University of Belgrade School of Medicine, Belgrade, Serbia

CONTENTS

1 BACKGROUND

Exercise-induced mitochondrial dysfunction (EIMD) is a complex and an intriguing phenomenon that is suggested to mediate many negative consequences of heavy exercise in health and disease (Filler et al. 2014; Ostojic 2016; de Gregorio et al. 2018). Although recognized ~ 60 years ago as an adverse effect of exhaustive exercise upon the fine structure of heart mitochondria in dogs exposed to forced swimming (Laguens et al. 1966), EIMD still remains a subclinical entity that is difficult to predict, prevent and monitor. Heavy (and repetitive) exercise appears to overstress the organelle in various tissues (e.g., skeletal muscle, myocardium, brain, liver, blood cells, and adipose tissue), with damaged mitochondria hardly recover (if at all) and demonstrate poor viability and irregular life cycle (Figure 1). In this chapter, different pathophysiological features of EIMD have been described, along with factors that mediate EIMD in human and animal models, and possible biomarkers of EIMD in medicine.

2 PATHOPHYSIOLOGY OF EIMD

Mitochondria are highly responsive cellular structures, and perhaps the first organelles to react to different exogenous stimuli, including exercise. Exhaustive exercise represents one of the best vehicles to provoke mitochondrial overload, with several pioneering studies reported massive and severe functional and structural changes to mitochondria after heavy exercise in different animal models (Laguens et al. 1966; Laguens and Gómez-Dumm 1967). Abnormal mitochondria and signs of muscle fiber necrosis have also been found in experienced marathon runners (Hikida et al. 1983). Gohil et al. (1984) were the first to coin a term 'exercise-induced decrease in mitochondrial activity' when describing compromised oxidative capacity of brown adipose tissue mitochondria found in rats after exhaustive exercise. An increase in mitochondrial size, a disruption of the cristae along with fusion of neighboring organelles appearing as long and slender structures, or bulky forms with a subplasmalemmal or perinuclear localization, comprise typical changes seen via electron microscope, with mitochondrial damage more marked in subjects who exercised more heavily (Gohil et al. 1984). A mechanism of EIMD has not been addressed in those seminal studies, yet it appeared that many morphological alterations remain permanent, specifically for myocardial mitochondria. Several hypotheses popped up during the following years in the EIMD causation study, with mitochondrial dysfunction from extreme exercise suggested to damages biomolecules, harms muscle function, decreases age longevity and can causes premature aging (Poulsen et al. 1996; St Clair Gibson et al. 2000).

Exhaustive exercise can alter mitochondrial function beyond repair through many routes, including turbulent biochemical perturbations, heat stress and/or mechanical

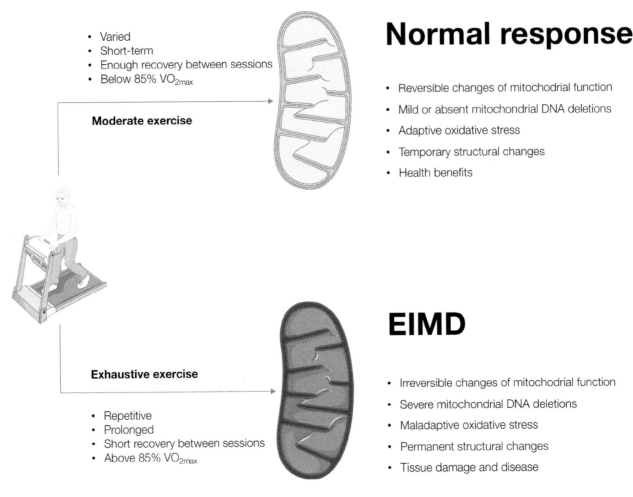

- Varied
- Short-term
- Enough recovery between sessions
- Below 85% VO$_{2max}$

Moderate exercise

Normal response

- Reversible changes of mitochodrial function
- Mild or absent mitochondrial DNA deletions
- Adaptive oxidative stress
- Temporary structural changes
- Health benefits

Exhaustive exercise

- Repetitive
- Prolonged
- Short recovery between sessions
- Above 85% VO$_{2max}$

EIMD

- Irreversible changes of mitochodrial function
- Severe mitochondrial DNA deletions
- Maladaptive oxidative stress
- Permanent structural changes
- Tissue damage and disease

FIGURE 1   A paradigm of exercise-induced mitochondrial dysfunction (EIMD).

overload induced by exercise, with pathways possibly interconnected in a cascade of reactions (for review see Ostojic 2016). Biochemical attribution of exhaustive EIMD perhaps start with overproduction of reactive oxidative species (ROS) and reactive nitrogen species (RNS), harmful compounds mass-produced by oxygen utilization in the skeletal muscle during strenuous exercise (Powers et al. 2011). Subsequently, ROS and RNS could induce hard-to-repair deletions of mitochondrial DNA (mtDNA) that diminish mitochondrial biogenesis, down-regulate gene expression and result in the degeneration and necrosis of the organelle after heavy exercise (Huang et al. 2009), or inhibit function of regulatory enzymes in mitochondrial bioenergetics (Feng et al. 2011). Mice subjected to running to exhaustion experienced augmented ROS production in skeletal muscle mitochondria, accompanied by increased activity of cyclooxygenase (a key regulatory enzyme controlling inflammation), citrate synthase (a member of the Krebs cycle enzymes family), and several pro-inflammatory markers (e.g., tumor necrosis factor alpha, interleukin 1-beta, monocyte chemoattractant protein-1) (Lee et al.

2015). Heavy exercise elevated the mitochondrial ROS generation rate by up to 90% in rats assigned to incremental treadmill running-to-exhaustion, with oxidative stress go along with decreased mitochondrial respiration rate and drop in mitochondrial adenosine triphosphate (ATP) synthase activity in the myocardium (Li et al. 2016). Japanese group detected a large-scale deletion (7052 bp) of mtDNA with mitochondrial ultrastructural changes in the soleus muscle of rats forced to run until reaching complete exhaustion (Sakai et al. 1999), and the authors suggested that repetitive exercise (without proper recovery between sessions) could start a vicious circle of recurring deletions and irreparable morphological changes of the organelle. Aguiar et al. (2008) reported that intense physical exercise induces mitochondrial dysfunction in the brain cortex of mice, characterized by elevated thiobarbituric acid reactive species levels and decreased cytochrome C oxidase activity, followed by low cortical brain-derived neurotropic factor, a growth factor that supports the survival of existing neurons, and encourage the growth and differentiation of new neurons and synapses. Other critical responses to exhaustive exercise that might be linked to oxidative stress in the organelle include a decrease in expression of peroxisome proliferator-activated receptor gamma co-activator 1-alpha (PPARGC1-α), a master regulator of mitochondrial biogenesis (Feng et al. 2011), a drop in activities of several oxidative phosphorylation-regulating enzymes (e.g., aconitase, pyruvate dehydrogenase, alpha-oxoglutarate dehydrogenases, alpha-glycerophosphate dehydrogenases) (Rasmussen et al. 2001; Larsen et al. 2016), and an increase in poly(adenosine diphosphate-ribose) polymerase (PARP), a member of family of proteins involved in genomic stability and programmed cell death (Huang et al. 2009). In addition, a recent study (Zhang et al. 2017) revealed the regulating role of mitochondrial aldehyde dehydrogenase 2 (ALDH2), an enzyme that metabolizes acetaldehyde and toxic aldehydes, in mitochondrial dysfunction due to exhaustive exercise. The authors found that transgenic mice expressing ALDH2 in skeletal muscle, and challenged with exhaustive exercise to induce skeletal muscle injury, have displayed a mild or absent mitochondrial dysfunction as compared to wild-type mice. ALDH2 may reverse skeletal muscle EIMD by regulating mitochondria dynamic remodeling and enhancing the quality of mitochondria.

Heavy exercise-induced hyperthermia might be another component that sets in motion (or contribute) to mitochondrial dysfunction. It appears that heat shock proteins of 70 kD (HSP70s) are overexpressed in chronically exercised skeletal muscle (González et al. 2000), with HSP70s could jeopardize mitochondrial signaling, chaperoning and macromolecular integrity (Mosser et al. 2000). In line with this, a 20% decrease in mitochondrial coupling (as evaluated via adenosine diphosphate-to-oxygen ratio) was observed at 43°C as compared to 37°C in rat mixed skeletal muscle mitochondria (Willis and Jackman 1994), suggesting that temperature may impact the economy of mitochondrial oxygen utilization during heavy aerobic exercise. In addition, mechanical stress during strenuous exercise could physically affect mitochondrial permeability transition pore, thus precluding normal calcium-proton flow through inner membrane that results in mitochondrial swelling and injury (Tuan et al. 2008), or causes mitochondrial aggregation (St Clair Gibson et al. 1998). Whatever the initial trigger is, mitochondria appear to develop an irreversible damage if not allowed to fully recover from exhaustive exercise and repair from arduous biochemical, thermal and/or mechanical overload. For instance, a multi-year heavy endurance running (up to 6537 km/year) induced a permanent mitochondrial dysfunction in a 28-year-old male elite athlete, as characterized by abnormal mitochondria with dense matrices and coarse, abnormal cristae, grouped in large subsacrolemmal and sarcomere aggregates in vastus lateralis muscle (St Clair Gibson et al. 1998). This was accompanied by a long-term progressive decline in running performance, associated with an increasing inability to tolerate exhaustive exercise, with abnormalities were permanent and present in the second muscle

biopsy taken 4 months after initial discovery. Other studies confirmed that strenuous exercise can lead to apoptotic cell death of muscle fibers, with dysfunctional mitochondria could play a significant role in apoptotic signaling (Koçtürk et al. 2008).

3 FACTORS THAT AFFECT EIMD

Mitochondrial dysfunction triggered by exercise appears to be impacted by different traits such as exercise load, age, and gender being often acknowledged as main factors (see below). In addition, the organelles from specific tissues (such as myocardial mitochondria) appear to have a higher susceptibility to exercise-induced injury (Huang et al. 2009; Li et al. 2016; Oláh et al. 2015; Ping et al. 2015), and accompanying pathology might also be a possible modifier for EIMD (Hambrecht et al. 1997; Menshikova et al. 2005; Abel and Doenst 2011).

3.1 Exercise Dosage

While acute high-intensity exercise usually causes short-term, benign and reversible changes in mitochondrial structure and function (Tonkonogi and Sahlin 2002), EIMD appears to be provoked by repetitive heavy-intensity or prolonged exhaustive exercise (Gollnick et al. 1990; Chen and Gollnick 1994; St Clair Gibson et al. 2000). However, no precise exercise volume or specific threshold has been identified as EIMD-inducing so far. Typically, EIMD is caused by a single session of prolonged exhaustive exercise (e.g., supramaximal plantar flexion at 120% of maximal aerobic power) or several exhaustive sessions (e.g., five 1-min cycling bouts at 90 rpm to exhaustion), with chronic model comprehend consecutive sessions of heavy exercise during weeks to months to years (e.g., multi-year endurance training ~ 100 km/week). Endurance running, cycling or swimming to exhaustion are the most prevalent modes of exercise, with exercise intensity often set at ≥ 85% of maximal oxygen uptake (VO_{2max}) to exercise until exhaustion. Experimental models for EIMD are routinely employed in exercise-naïve subjects, while chronic EIMD is usually seen in well-trained population or athletes. It appears that mitochondrial dysfunction found in trained participants subjected to chronic exhaustive exercise is more pronounced and often irrevocable as compared to untrained population, and includes severe deletions of mtDNA and apoptosis-induced DNA strand breaks in cardiac myocytes a high percentage of abnormal swollen mitochondria in the skeletal muscle (and compromised leucocyte mitochondrial permeability (St Clair Gibson et al. 1998; Hsu et al. 2002; Huang et al. 2009), thus suggesting a dose-response or cumulative effect of EIMD. A recent preliminary study reports possible mitochondrial dysfunction after resistance exercise with heavy exercise significantly upregulated total-ULK1, a central autophagy marker (Cardinale et al. 2017).

3.2 AGE

Many studies suggest age as a predisposing factor for exercise-induced mitochondrial dysfunction. When exposed to exhaustive exercise, old subjects seem to be more vulnerable to mitochondrial dysfunction, with heavy exercise perhaps expand age-related changes in mitochondrial structure and function (Conley et al. 2007). Lee et al. (2015) have shown that strenuous exercise in old mice resulted in the decreasing of both fusion (mitofusin-2) and fission (dynamin-related-1) proteins that may contribute to alteration of mitochondrial morphology and function, and also negatively affected mtDNA-to-nDNA ratio, total and nuclear PPARGC1-α, and mitochondrial transcription factor A, as compared to younger counterparts who exercised, or to old sedentary control group. Age also impacted exercise-induced mitochondrial ATP synthesis during supramaximal plantar flexion (120% of maximal aerobic power) in humans (Layec et al. 2015), with an abnormal elevation in exercise-induced skeletal muscle metabolic demand in the mitochondria of old versus young activity-matched subjects.

3.3 GENDER

Gender-related differences for EIMD appears to be scarcely evaluated so far, with only one interventional animal study recruited female subjects, while a single human study recruited both healthy men and women for exhaustive EIMD intervention. Terblanche et al. (2001) evaluated the effects of exhaustive exercise on mitochondrial capacities to oxidize pyruvate, 2-oxoglutarate, palmitoylcarnitine, succinate, and ferrocytochrome c in various tissues of female rats. Activities of specific mitochondrial pathways were reduced in the heart, liver and brown adipose tissue after exhaustive exercise, suggesting a greater susceptibility of female population to EIMD. A U.S.-France study recruited 40 men and women who were exposed to high-intensity exercise while monitoring exercise-induced mitochondrial ATP supply (Layec et al. 2015). Although mitochondrial dysfunction has been found, the authors omitted to provide gender comparison. Nevertheless, a possible sex-related difference in response to exhaustive exercise are plausible with possible sex-specific energy utilization in the mitochondria involves sarcolemmal K(ATP) channel (Kent-Braun et al. 2002; Chicco et al. 2007).

4 CLINICAL BIOMARKERS OF EIMD

Dysfunctional mitochondria can be a source of many damage-associated molecular patterns (DMAPs), including mtDNA, cytochrome C, and ROS (Li et al. 2016), with many DMAPs might be detectable in the circulation and challenged as possible biomarkers of EIMD. Leucocyte mitochondrial transmembrane potential (MTP, as evaluated with flow cytometry) could become an applicable and sensible marker for monitoring EIMD. Hsu et al. (2002) found a significant MTP decline in peripheral blood

polymorphonuclear leucocytes, monocytes and lymphocytes in trained male runners who performed exhaustive aerobic exercise at intensities above 60% VO_{2max}. The authors demonstrated that heavy exercise has accumulative effects on the leucocyte mitochondrial functional status, and alterations appeared to be exercise intensity-dependent. Blood cell mtDNA quantification (mtDNA-to-nDNA ratio) could also be explored as a possible marker of mitochondrial dysfunction induced by heavy exercise. Although not directly related to exhaustive exercise, Côté et al. (2007) linked drop in mtDNA-to-nDNA ratio with mitochondrial dysfunction and poor recovery in patients with multiple organ failure. Using circulating mtDNA as another non-invasive sensitive marker for EIMD should also be considered, since the recent study has shown lower mtDNA in plasma of clinical patients with mitochondrial dysfunction-related disorders with levels positively associated with disease stage (Xia et al. 2014). Future studies should evaluate other biomarkers of mitochondrial function, including metabolomics, circulating serum markers (e.g. fibroblast growth factor 21, growth differentiation factor 15, glutamate dehydrogenase, microRNAs), exercise physiology, and both structural and functional imaging (Steele et al. 2017), to assess mitochondrial viability during exercise.

5 OPEN QUESTIONS

Not all studies found impaired mitochondrial function by high intensity exercise (Tonkonogi et al. 1999), suggesting many open frontiers for this perplexing phenomenon (Figure 2). An EIMD protocol used (e.g., acute vs. chronic exercise model), a sample recruited (e.g., trained vs. untrained subjects), different experimental frameworks (e.g., humans vs. animal model vs. isolated mitochondria), a criterion used for an exhaustion point determination (e.g., subjectively reported response vs. objectively measured outcomes), a follow-up period (e.g., 3 min post-exercise vs. 4-month follow-up) are just few issues that might blur a comparison of different EIMD studies, and limit understanding of etiological pathways and clinical outcomes of EIMD. Besides pathophysiology and diagnosis, future research needs to be directed towards management of EIMD (also prevention), with evaluation of effectiveness and safety of novel mitochondria-targeted agents (Ostojic 2017).

6 CONCLUSION

Mitochondria appears to have a finite capacity for adaptation to exhaustive exercise, with a maladaptive accumulation of mitochondrial damage (including large deletions of mtDNA, dysregulation of controlling proteins, and pro-apoptotic signaling) appears to lead to a pathological cascade and tissue damage. EIMD seems to be provoked by prolonged sessions of heavy repetitive exhaustive exercise, accompanied by inadequate

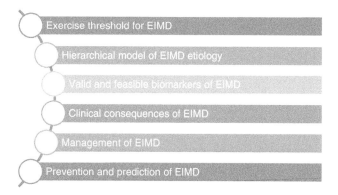

FIGURE 2 Open questions for exercise-induced mitochondrial dysfunction (EIMD) research.

recovery between sessions, with EIMD specifically tackles sensible tissues in predisposed individuals (e.g., elderly, athletes) thus inducing adverse health consequences. There is no consensus when normal adaptive response to exercise turns into irreversible mitochondrial damage, yet regular monitoring of mitochondria-specific biomarkers may help to identify the turning point.

ACKNOWLEDGMENTS

This work was supported by the Serbian Ministry of Education, Science and Technological Development (175037), the Provincial Secretariat for Science and Technological Development (114-451-710), the Faculty of Sport and Physical Education, Novi Sad, Serbia (2017-FSPE), and the Centre for Health, Exercise and Sport Sciences, Belgrade.

REFERENCES

Abel ED, Doenst T. Mitochondrial adaptations to physiological vs. pathological cardiac hypertrophy. Cardiovasc Res 2011;90:234–242.

Aguiar AS Jr, Tuon T, Pinho CA, et al. Intense exercise induces mitochondrial dysfunction in mice brain. Neurochem Res 2008;33:51–58.

Cardinale DA, Lilja M, Mandić M, et al. Resistance training with co-ingestion of anti-inflammatory drugs attenuates mitochondrial function. Front Physiol 2017;8:1074.

Chen J, Gollnick PD. Effect of exercise on hexokinase distribution and mitochondrial respiration in skeletal muscle. Pflugers Arch 1994;427:257–263.

Chicco AJ, Johnson MS, Armstrong CJ, et al. Sex-specific and exercise-acquired cardioprotection is abolished by sarcolemmal KATP channel blockade in the rat heart. Am J Physiol Heart Circ Physiol 2007;292:H2432–H2437.

Conley KE, Jubrias SA, Amara CE, et al. Mitochondrial dysfunction: Impact on exercise performance and cellular aging. Exerc Sport Sci Rev 2007;35:43–49.

Côté HC, Day AG, Heyland DK. Longitudinal increases in mitochondrial DNA levels in blood cells are associated with survival in critically ill patients. Crit Care 2007;11:R88.

de Gregorio C, Di Nunzio D, Di Bella G. Athlete's heart and left heart disease. Adv Exp Med Biol 2018;1067:313–325.

Feng Z, Bai L, Yan J, et al. Mitochondrial dynamic remodeling in strenuous exercise-induced muscle and mitochondrial dysfunction: Regulatory effects of hydroxytyrosol. Free Radic Biol Med 2011;50:1437–1446.

Filler K, Lyon D, Bennett J, et al. Association of mitochondrial dysfunction and fatigue: A review of the literature. BBA Clin 2014;1:12–23.

Gohil K, Henderson S, Terblanche SE, et al. Effects of training and exhaustive exercise on the mitochondrial oxidative capacity of brown adipose tissue. Biosci Rep 1984;4:987–993.

Gollnick PD, Bertocci LA, Kelso TB, et al. The effect of high-intensity exercise on the respiratory capacity of skeletal muscle. Pflugers Arch 1990;415:407–413.

González B, Hernando R, Manso R. Stress proteins of 70 kDa in chronically exercised skeletal muscle. Pflugers Arch 2000;440:42–49.

Hambrecht R, Fiehn E, Yu J, et al. Effects of endurance training on mitochondrial ultrastructure and fiber type distribution in skeletal muscle of patients with stable chronic heart failure. J Am Coll Cardiol 1997;29:1067–1073.

Hikida RS, Staron RS, Hagerman FC, et al. Muscle fiber necrosis associated with human marathon runners. J Neurol Sci 1983;59:185–203.

Hsu TG, Hsu KM, Kong CW, et al. Leukocyte mitochondria alterations after aerobic exercise in trained human subjects. Med Sci Sports Exerc 2002;34:438–442.

Huang CC, Lin TJ, Chen CC, et al. Endurance training accelerates exhaustive exercise-induced mitochondrial DNA deletion and apoptosis of left ventricle myocardium in rats. Eur J Appl Physiol 2009;107:697–706.

Kent-Braun JA, Ng AV, Doyle JW, et al. Human skeletal muscle responses vary with age and gender during fatigue due to incremental isometric exercise. J Appl Physiol 2002;93:1813–1823.

Koçtürk S, Kayatekin BM, Resmi H, et al. The apoptotic response to strenuous exercise of the gastrocnemius and solues muscle fibers in rats. Eur J Appl Physiol 2008;102:515–524.

Laguens RP, Gómez-Dumm CL. Fine structure of myocardial mitochondria in rats after exercise for one-half to two hours. Circ Res 1967;21:271–279.

Laguens RP, Lozada BB, Gómez Dumm CL, et al. Effect of acute and exhaustive exercise upon the fine structure of heart mitochondria. Experientia 1966;22:244–246.

Larsen FJ, Schiffer FJ, Ørtenblad TA, et al. High-intensity sprint training inhibits mitochondrial respiration through aconitase inactivation. FASEB J 2016;30:417–427.

Layec G, Trinity JD, Hart CR, et al. Impact of age on exercise-induced ATP supply during supramaximal plantar flexion in humans. Am J Physiol Regul Integr Comp Physiol 2015;309:R378–R388.

Lee S, Kim M, Lim W, et al. Strenuous exercise induces mitochondrial damage in skeletal muscle of old mice. Biochem Biophys Res Commun 2015;461:354–360.

Li H, Miao W, Ma J, et al. Acute exercise-induced mitochondrial stress triggers an inflammatory response in the myocardium via NLRP3 inflammasome activation with mitophagy. Oxid Med Cell Longev 2016;2016:1987149.

Menshikova EV, Ritov VB, Toledo FG, et al. Effects of weight loss and physical activity on skeletal muscle mitochondrial function in obesity. Am J Physiol Endocrinol Metab 2005;288: E818–E825.

Mosser DD, Caron AW, Bourget L, et al. The chaperone function of hsp70 is required for protection against stress-induced apoptosis. Mol Cell Biol 2000;20:7146–7159.

Oláh A, Németh BT, Mátyás C, et al. Cardiac effects of acute exhaustive exercise in a rat model. Int J Cardiol 2015;182:258–266.

Ostojic SM. Exercise-induced mitochondrial dysfunction: A myth or reality? Clin Sci 2016;130:1407–1416.

Ostojic SM. Mitochondria-targeted nutraceuticals in sports medicine: A new perspective. Res Sports Med 2017;25:91–100.

Ping Z, Zhang LF, Cui YJ, et al. The protective effects of salidroside from exhaustive exercise-induced heart injury by enhancing the PGC-1α-NRF1/NRF2 pathway and mitochondrial respiratory function in rats. Oxid Med Cell Longev 2015;2015:876825.

Poulsen HE, Loft S, Vistisen K. Extreme exercise and oxidative DNA modification. J Sports Sci 1996;14:343–346.

Powers SK, Nelson WB, Hudson MB. Exercise-induced oxidative stress in humans: Cause and consequences. Free Radic Biol Med 2011;51:942–950.

Rasmussen UF, Krustrup P, Bangsbo J, et al. The effect of high-intensity exhaustive exercise studied in isolated mitochondria from human skeletal muscle. Pflugers Arch 2001;443:180–187.

Sakai Y, Iwamura Y, Hayashi J, et al. Acute exercise causes mitochondrial DNA deletion in rat skeletal muscle. Muscle Nerve 1999;22:258–261.

St Clair Gibson A, Lambert MI, Collins M, et al. Chornic exercise activity and the fatigued athlete myopathic syndrome (FAMS). Int Sports Med J 2000;1:1–7.

St Clair Gibson A, Lambert MI, Weston AR, et al. Exercise-induced mitochondrial dysfunction in an elite athlete. Clin J Sport Med 1998;8:52–55.

Steele HE, Horvath R, Lyon JJ, et al. Monitoring clinical progression with mitochondrial disease biomarkers. Brain 2017;140:2530–2540.

Terblanche SE, Gohil K, Packer L, et al. The effects of endurance training and exhaustive exercise on mitochondrial enzymes in tissues of the rat (Rattus Norvegicus). Comp Biochem Physiol A Mol Integr Physiol 2001;128:889–896.

Tonkonogi M, Sahlin K. Physical exercise and mitochondrial function in human skeletal muscle. Exerc Sport Sci Rev 2002;30:129–137.

Tonkonogi M, Walsh B, Tiivel T, et al. Mitochondrial function in human skeletal muscle is not impaired by high intensity exercise. Pflugers Arch 1999;437:562–568.

Tuan TC, Hsu TG, Fong MC, et al. Deleterious effects of short-term, high-intensity exercise on immune function: Evidence from leucocyte mitochondrial alterations and apoptosis. Br J Sports Med 2008;42:11–15.

Willis WT, Jackman MR. Mitochondrial function during heavy exercise. Med Sci Sports Exerc 1994;26:1347–1353.

Xia P, Wang HJ, Geng TT, et al. Mitochondrial DNA levels in blood and tissue samples from breast cancer patients of different stages. Asian Pac J Cancer Prev 2014;15:1339–1344.

Zhang Q, Zheng J, Qiu J, et al. ALDH2 restores exhaustive exercise-induced mitochondrial dysfunction in skeletal muscle. Biochem Biophys Res Commun 2017;485:753–760.

28 Mitochondrial Pathologies and Their Neuromuscular Manifestations

Carlos Ortez and Andrés Nascimento

Neuromuscular pathology unit, Paediatric Neurology Department, Hospital Sant Joan de Déu (1), Center for Biomedical Research on Rare Diseases (CIBERER), Instituto de Salud Carlos III and Institute of Pediatric Research Sant Joan de Déu, Barcelona, Spain

CONTENTS

1 INTRODUCTION

The clinical presentation of mitochondrial diseases (MiDs) is extremely heterogeneous, and combined with the large number of genetic causes (mitochondrial and nuclear DNA) and environmental factors, makes their diagnosis challenging. Mitochondria are highly dynamic organelles that are the major contributor of ATP, the cellular unit of energy, via oxidative phosphorylation (OXPHOS)[1].

The disorders may develop at any age, with isolated or multiple systems involvement, and in any pattern of inheritance. However, post-mitotic tissues that are highly dependent on oxidative metabolism, such as neurons, muscle, and cardiac cells, seem to be preferentially vulnerable to energy imbalance[2,3].

Since the first reports of human disease due to defects in mitochondrial DNA (mtDNA) in 1988[4,5], the number of disease-associated mtDNA mutations has expanded rapidly with identification of classic mitochondrial syndromes such as MELAS (mitochondrial encephalomyopathy, lactic acidosis, and stroke-like episodes), MERRF (myoclonic epilepsy with ragged red fibers), NARP (neurogenic muscle weakness, ataxia, and retinitis pigmentosa), Kearns-Sayre syndrome and maternally inherited Leigh syndrome[2,4,5].

The phenotypic spectra, associated with known pathogenic variants, are increasing whereas in the last few years next-generation sequencing is rapidly identifying new disease-causing nuclear genetic mutations[6].

Myopathy, ocular myopathy, fatigability, peripheral neuropathy, seizures, ataxia, stroke-like episodes, psychomotor retardation, migraine-like headaches, encephalopathy, cognitive regression and movement disorders are neurological and neuromuscular manifestations commonly found in these diseases[7]. Some of the neuromuscular manifestations can be isolated but often evolve into a multisystem disease. Here we present an overview of the main neuromuscular manifestations to facilitate their recognition and to guide the diagnosis process of a MiDs in children.

Despite the wide classification forms existing for the mitochondrial diseases, the present review will focus on a practical and clinical approach on the bases of a neuromuscular "guide sign" associated with the common classification based on the molecular defect and inheritance pattern. For that purpose, we have classified the neuromuscular manifestations into two groups: One with muscular involvement as the principal feature, and in the other hand, the group with peripheral nerve defects.

2   NEUROMUSCULAR MANIFESTATIONS AND DIAGNOSIS: KEY MESSAGES

Unlike adult patients, in pediatric population the clinical manifestation varies within the age and the development process of the central and peripheral system. Generally, the most serious clinical phenotypes are present during the first months of life with multisystem manifestations associated to unspecific neuromuscular signs such as weakness and hypotonia, due to central, peripheral or mixed causes, usually difficult to differentiate; it is for this reason that the biochemical biomarkers and the recognition of the phenotypes have become very important in the diagnosis process.

The most evocative signs of MiDs remain the muscular signs, particularly the ocular myopathy and exercise intolerance. Progressive external ophthalmoplegia (PEO) is characterized by progressive weakness or paresis of the extra-ocular eye muscles leading to bilateral gaze limited in all directions, usually without diplopia, and associated with ptosis[2,7]. Ptosis may be present as isolated sign. Myopathy begins usually in the teens or during adulthood and can occasionally be congenital[2]. The weakness is slowly progressive, symmetric, and more often proximal than distal. The severity of the myopathy varies from mild to severe forms that can lead to respiratory failure[1,2].

Exercise intolerance usually appears progressively with cramps, myalgia, myoglobinuria or fatigue and may occasionally be combined with nausea, shortness of breath and dizziness or may occur even after normal daily activities. Each sign can be either isolated or associated with other neurological symptoms and/or systemic manifestations, such as growth retardation, diabetes, deafness, optic atrophy, pigmentary retinitis, cataracts, and cardiomyopathy[2,7].

Diverse complementary studies and biomarkers are especially helpful in the diagnosis process. It is known that the Creatine kinase (CK) and lactate may be high in serum at rest, and the association of both is suggestive of MiDs. In pediatric patients the increase of CK is not a frequent feature of the MiDs, and when high levels are found the differential diagnosis with muscular dystrophies its open. If the CK level are increased but in an intermittent form, other metabolic myopathies such as McArdle disease, Myoadenylate deaminase deficiencymyoadenilato deamanase, electron transfer flavoprotein, Glycogenosis, Beta-oxidations, or lipid metabolism disorders should be included in the differential diagnosis[2,8] (Fig. 1). The EMG may be normal, neurogenic or myopathic. In some cases it is possible to identify alterations at the endplate thus conditioning to establish the differential diagnosis with myasthenic syndromes.

Defects on the respiratory chain (RC) are usually identified by muscle spectrophotometry. Ragged red fibers (RRF) and/or cytochrome oxidase negative (COX-) fibers could be shown through muscle histology, sometimes with SDH-positive fibers or lipid accumulation in type I fibers. These typical histology features are less frequently found in pediatric than adult patients. Abnormal mitochondrial morphology and the presence of paracrystalline inclusions can be observed at electron microscopy[2,7].

The double genetic origin of the mitochondrial proteins explains the fact that neuromuscular manifestations associated to MiDs can be caused by mutations of either mitochondrial DNA or nuclear DNA, the main clinical and genetic characteristics are summarized in Tables 1 and 2.

3   MITOCHONDRIAL DISORDERS RELATED TO MTDNA MUTATIONS

- **MiDs with muscular involvement related to mtDNA mutations (sporadic or maternally inherited)**

There are multiples copies of circular mtDNA in one mitochondrion, and some of these copies may carry nucleotide variants, i.e., base changes, not present in all mtDNA molecules (mtDNA polymorphisms or mtDNA mutations). Homoplasmy is characterized by the identity of all copies of the mitochondrial genome. Patients harbouring pathogenic mtDNA defects frequently have a mixture of mutated and wild-type mtDNA, and this is called heteroplasmy; the percentage of mutated mtDNA can vary among different organs within the same individual. Cells are able to tolerate high percentage levels of mutated mtDNA. Thus, the phenotypic expression of a pathogenic mtDNA mutation can be evident only if the number of mutant mtDNAs exceeds a certain threshold, which varies among tissues (30 to 80% of mutated mtDNA), depending upon their constitutive and acute requirements for OXPHOS[2,7]. Mutations in mtDNA that impair mitochondrial protein synthesis include mtDNA rearrangements (deletions or duplications), mutations in tRNA genes, and mutations in protein-coding genes[2,7].

- **Single deletions of mtDNA**

Single large deletions of mtDNA, ranging in size from 1.1 to 10 kb, are usually due to a sporadic mutational event occurring during embryogenesis and thereby affecting a variable number of tissues. In most cases, these large deletions are not transmitted to the offspring, although a few cases of mother-to-child transmission have been reported. Large single mtDNA deletions, but also large-scale tandem mtDNA duplications are involved in a clinical spectrum including Kearns-Sayre syndrome, PEO or isolated ptosis[2,9].

Kearns-Sayre syndrome (KSS) is a multisystem disorder characterized by childhood-onset PEO and pigmentary retinopathy (<20 years of age), associated with at least one of the following signs: cardiac conduction block, hyperproteinorachia or cerebellar ataxia[2,7,10]. Other manifestations are frequent including limb weakness, hearing loss, dementia, diabetes, hypoparathyroidism and short stature (growth hormone deficiency). Also cerebral folate deficiency have being associated in most of the patients, so folinic acid supplementation its recommended in early stages of the disease[11,12].

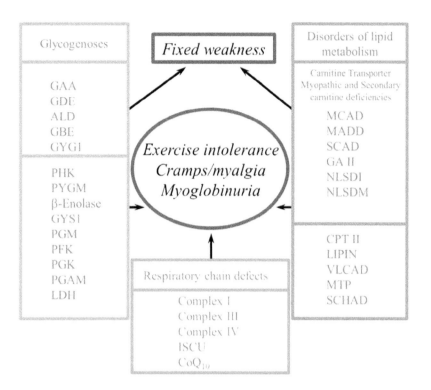

FIGURE 1 Etiology of metabolic myopathies. There are two major clinical syndromes in metabolic myopathies. Deficient enzymes are indicated by abbreviations or gene symbols as follows: GAA acid maltase (acid alpha-glucosidase) (GSD II), GDE glycogen debranching enzyme (GSD III), ALD aldolase (GSD IX), GBE glycogen branching enzyme (GSD IV), GYG1 glycogenin (GSD 0), PHK phosphorylase kinase (GSD VIII), PYGM myophosphorylase (GSD V), GYS1 glycogen synthetase, PGM phosphogluco-mutase (GD XIV), PFK phosphofructokinase (GSD VII), PGK phosphoglycerate kinase (GSD IX), PGAM phosphoglycerate mutase (GSD X), LDH lactate dehydrogenase (GSD XI), MTF mitochondrial trifunctional enzyme, MCAD medium-chain acyl-CoA dehydrogenase, MADD multiple acyl-CoA dehydrogenase, SCAD short-chain acyl-CoA dehydrogenase, GAII glutaric acid-uria type II, NLSDI neutral lipid storage diseases with ichthyosis (Chanarin-Dorfman disease), NLSDM neutral storage disease with myopathy, CPT II carnitine palmitoyltransferase II, VLCAD very long-chain acyl-CoA dehydrogenase, SCHAD short-chain 3-hydroxyacyl-CoA dehydrogenase, ISCU non-heme iron-sulfur (Fe-S) protein, CoQ10 coenzyme Q10. Figure adapted from Barca E, et al. Metabolic Myoglobinuria. Curr Neurol Neurosci Rep 15:69; 2015[8].

Muscle biopsy shows RRF, SDH+ and COX- fibers with defect of RC complexes containing mtDNA-encoded subunits. Nearly 90% of individuals with KSS have a large-scale mtDNA deletion that is usually present in muscle, but undetectable in blood cells, necessitating muscle biopsy[2,7].

PEO is a milder phenotype, characterized by ptosis, ophthalmoplegia, oropharyngeal weakness, and a variably proximal limb weakness (Figure 2). In PEO, mtDNA deletions are confined to skeletal muscle. Some individuals with PEO (< 20%) have a pathogenic point mutation of mtDNA, such as the m.3243A>G mutation or less frequently other rare mtDNA mutations. In a recent study, rare pathogenic mtDNA mutations were identified in 55 out of 743 patients with suspected MiDs, and PEO and/or ptosis were present in 42% of the cases (23/55)[2,13].

• **Mitochondrial DNA point mutations**

Mitochondrial DNA point mutations are most frequently heteroplasmic, often leading to childhod-onset, whereas the homoplasmic state is generally associated with adult-onset. There are mtDNA point mutations in tRNA genes: (tRNA-Leu, tRNA-Pro, tRNA-Phe, tRNA-Met, tRNA-Ser, tRNA-Asp, tRNA-Gln, tRNA-Ala, tRNA-Tyr, mtRNA-Lys, mtRNA-Trp) mostly described with adult-onset myopathy and exercise intolerance with or without PEO[13]; and in genes coding for RC subunit (*MTCO3, MTCO2, MTCYB, MTND1, MTND2, MTND4 ...*) associated with childhood-onset myopathy, exercise intolerance, myoglobinuria, or MELAS-like disorder[1,2,7,13].

• **Mutations in mitocondrial tRNAs**

A study on the prevalence of myopathy in 50 patients with MELAS and the A3243G mutation in tRNA-Leu(UUR) found that clinical myopathy was present in 50% of patients and manifested most commonly in the fifth decade, but varied considerably in severity and distribution, including mild to moderate limb weakness, ptosis, and ophthalmoplegia[14].

TABLE 1

Neurological and neuromuscular manifestations associated with mtDNA mutations.

Predominant signs	Onset	Phenotype	Common mutations and/ or responsible genes	Other mutations and/ or responsible genes	% in patients with rare mutations*
Muscle					
PEO	Child to teens	KSS	Large deletion		42%
	Child to adult	PEO or PEO "plus"	Large deletion, m.3243A>G	mt tRNA	
	Adult	Ptosis	Large deletion, m.3243A>G	mt tRNA	
Myopathy	Adult	Myopathy and exercise intolerance ±PEO	mt tRNA	+	25%
	Child	Myopathy _ exercise intolerance, myoglobinuria, MELAS-like disorder	*MTCYB, MTCO1, MTCO2, MTCO3*	mt tRNA	
	Child to adult	Isolated exercise intolerance	*MTCYB*	MTNDs	
	Infant	MNGIE-like	tRNATrp, tRNAVal, large deletion	-	
	First months	Benign infantile mitochondrial myopathy with reversible COX deficiency	m.14674T>C (tRNAGlu)	-	
Peripheral nervous system					
Peripheral neuropathy	Child to adult	NARP, sensory neuropathy	m.8993T>C (*MTATP6*)	*MTATP6*	14.5%
		Episodic weakness with motor neuropathy	*MTATP6, MTATP8*	-	
	Child to teens	dHMN	m.9182T>C (*MTATP6*)	-	
LMNd	Young adult	ALS-like	m.6020del5ins (*MTCO1*)	-	-

ALS-like: amyotrophic lateral sclerosis-like; dHMN: distal hereditary motor neuropathy; KSS: Kearns-Sayre syndrome; LHON: Leber hereditary optic neuropathy; LMNd: lower motor neuron disorders; MELAS: mitochondrial encephalomyopathy, lactic acidosis, and stroke-like episodes; MERRF: myoclonic epilepsy with ragged red fibers; MNGIE: myopathy and PEO; neuropathy; gastrointestinal; encephalopathy; MTNDs: mtDNA-encoded of subunits of complex I; NARP: neurogenic muscle weakness, ataxia, and retinitis pigmentosa; OA: optic atrophy; PEO: progressive external ophtalmoplegia; SPG: spastic paraplegia hereditary. a Data according Bannwarth et al., 2013[13]. Table Adapted of Chaussenot et al. Revue neurologique 170:323–338; 2014[2].

The association of PEO with neurogastrointestinal syndrome, MNGIE-like phenotype, has been observed with mutations m.5532G>A in mtRNATrp[15] and m.1630A>G in mtRNAVal[16]. Benign infantile mitochondrial myopathy with reversible COX deficiency, due to the m.14674T>C tRNAGlu mutation is a specific phenotype beginning between the first week and 2.5 months of life[17]. The severe initial weakness, predominant in proximal muscles, leads to respiratory failure, dysphagia, PEO and hypotonia associated with sensory axonal neuropathy, seizures, encephalopathy and hepatomegaly. Serum lactate and CK are elevated. Muscle histology shows RRF, SDH+, and COX-fibers. Spontaneous improvement begins commonly between 5 and 20 months leading to a normal state after 2 to 3 years, with a residual mild weakness in some patients.

Two unrelated patients, both harboring the 7472-insertion mutation in tRNASer (UCN), had very different clinical presentations: one had a multisystem disorder akin to myoclonus epilepsy and ragged-red fibers (MERRF), whereas the other had a pure myopathy. The only difference at the molecular level was that the myopathic patient also had a novel homoplasmic A7472C transition in the

same gene, which further lengthens the expanded homopolymeric C run caused by the insertion[18]. These findings raise the interesting possibility that concurrent mtDNA mutations may drastically influence the phenotype.

An unusual myopathic presentation, with congenital distal arthrogryposis in two siblings, was associated with the T3271C mutation in tRNALeu (UUR), which typically causes MELAS[19]. Although the proposita did have a stroke at age 15, a younger brother had infantile arthrogryposis but no stroke (there was no maternal family history of stroke).

• **Mutations in protein-coding genes**

The importance of mutations in mtDNA COX genes in causing exercise intolerance and myoglobinuria was confirmed by three reports on four patients. Three of them were young adults with life-long exercise intolerance, exercise-related myalgia, and at least one episode of myoglobinuria; all three were sporadic cases harboring presumably somatic, heteroplasmic mutations confined to skeletal muscle and affecting COX I (G6708A)[20], COX II (T7989C)[21], and COX III (T9789C)[22]. The fourth patient, who harbored a

TABLE 2

Muscular manifestations associated with mutations in nuclear genes.

Inheritance	Main phenotype	Onset	Specific features	Clinical characteristics	Genes involved
Recessive	Myopathies with mtDNA instability	Congenital or infant	mtDNA depletion	Fatal myopathy	TK2
			mtDNA depletion	With Leigh-like syndrome and methylmalonic aciduria	SUCLA2
			mtDNA depletion	With multisystemic (gastro-intestinal and renal)	RRM2B
		Infant	Multiple mtDNA deletions	With cataracts and deafness	GFER
		Child	mtDNA depletion	Severe myopathy	TK2
		Juvenile o adult	mtDNA depletion and/or multiple mtDNA deletions	Moderate myopathy	TK2, DGUOK
	Myopathies	Child	CoQ10 deficiency	Myopathy or encephalomyopathy with myoglobinuria	CABCB1
			Complex I assembly defect	With exercise intolerance	ACAD9
			Complex II deficiency	With exercise intolerance	ISCU, FDX2
			Movement disorders	With extrapyramidal movement disorders	MICU1
		Infant to teens	Lactic acidosis and sideroblastic anemia	MLASA	PUS1, YARS2
	PEO ± myopathy	Child to teens	Multiple mtDNA deletions	PEO or PEO+	RRM2B, MGME1
		Adult	Multiple mtDNA deletions	Isolated ptosis	POLG
			Multiple mtDNA deletions	PEO or PEO+	POLG, DGUOK, MPV17, TK2
			Multiple mtDNA deletions	SANDO	POLG, MPV17
		Child to adult	Multiple mtDNA deletions or mtDNA depletion	MNGIE	TYMP, RRM2B
				MNGIE-like without leucoencephalopathy	POLG
Dominant	PEO or PEO+	Adult	Multiple mtDNA deletions	PEO or PEO+	TWINKLE, RRM2B, ANT1, POLG, POLG2, DNA2
		Child to adult	Multiple mtDNA deletions	PEO and optic atrophy ± axonal neuropathy	OPA1, MFN2
	SANDO	Adult	Multiple mtDNA deletions	SANDO	TWINKLE, POLG

CoQ10: coenzyme Q10; MNGIE: myopathy and PEO, neuropathy, gastrointestinal, encephalopathy; MLASA: myopathy with lactic acidosis and sideroblastic anemia; mtDNA: mitochondrial DNA; PEO: progressive external ophthalmoplegia; SANDO: sensory ataxic neuropathy, dysarthria and ophthalmoparesis. Table Adapted of Chaussenot et al. Revue neurologique 170:323–338; 2014[2].

heteroplasmic mutation (G7970T) in COX II, had a severe encephalomyopathy with short stature, bilateral cataracts, sensorineural hear-ing loss, myopathy, ataxia, cardiac arrhythmia, and depression[22]. Mutations in ND1 and in ND4 have been associated with pure myopathy[9]. A mutation in ND2 (G4810A) joins the list of mutations causing severe exercise intolerance (initially diagnosed as chronic fatigue syndrome) followed by mild ptosis and PEO[23].

Many phenotypes have been described with either childhood-onset isolated myopathy wherein improvement of the severe initial weakness is associated with reduction of mutation load and a disappearance of COX- fibers, or myopathy "plus" wherein the myopathy is associated with at least one of the following signs: PEO, myalgia, recurrent myoglobinuria or exercise intolerance. MTCYB mutations have been reported mainly in childhood-onset myopathy with

myoglobinuria, exercise intolerance and isolated complex III deficiency[23,24]. MTCO3 mutations were described in an early-onset myopathy "plus" (between 4 and 20 years) with exercise intolerance, myalgia, myoglobinuria after prolonged exercise associated with recurrent encephalopathy, lactic acidemia and isolated complex IV deficiency[25].

4 MITOCHONDRIAL DISORDERS WITH MUSCULAR MANIFESTATIONS DUE TO MUTATIONS IN NUCLEAR DNA (NDNA)

Most inherited mitochondrial diseases are due to mutations in nuclear genes encoding more than 1500 proteins with a variety of housekeeping or specialized functions, which are targeted to mitochondria. Mutations in nDNA directly

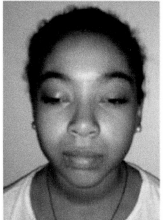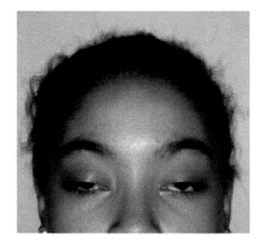

FIGURE 2   16 years old girl diagnosis of PEO and single mtDNA deletion. a) asymtomtic patient at 5 years of age. b and c) bilateral ptosis and external oftalmoplegia at 14 years old.

affecting the mitochondrial respiratory chain fall into three major groups[1,7]:

   I. Mutations in genes that encode subunits of respiratory chain complexes: these are known for complexes I and II, and coenzyme Q10 (CoQ10).
   II. Mutations in genes encoding ancillary proteins that are needed for the correct assembly and function of respiratory chain complexes.
   III. Defects in intergenomic signaling, that is, mutations in nuclear genes that control the abundance and quality of mtDNA and result in multiple deletions of mtDNA, mtDNA depletion, or both conditions at once.

From a clinical point of view, we will describe this group of disorder based on the inheritance pattern and the association with progressive external ophthalmoplegia as "guide sign":

4.1   AUTOSOMAL RECESSIVE MUSCULAR DISORDERS

- **Mitochondrial myopathies due to defects in intergenomic signaling (with or with out PEO)**

These are Mendelian disorders in which mutations in nuclear genes cause qualitative (multiple deletions) or quantitative (depletion) alterations of mtDNA[1,2]. The Thymidine kinase 2 (*TK2*) gene encodes one of the enzyme of mitochondrial expression involved in the deoxyribonucleotide triphosphates metabolism necessary for replication of mtDNA[2,26]. Myopathies due to *TK2* mutation should be consider the paradigm of mitochondrial disorder with isolated muscular manifestations in pediatric patients.

Fatal infantile myopathy with severe mtDNA depletion is a condition due to *TK2* mutations with a well-characterized phenotype, consisting of a severe and rapidly progressive

generalized muscular weakness and respiratory failure during the first few years of life. The first report was performed in 2001 by Saada and coworkers[26], in four affected children with severe myopathy, elevated creatine kinase, multiple defects of mitochondrial respiratory chain activity and reduction of mtDNA copy number in muscle tissue. Since then more than 50 patients have been reported and other neurological signs have been associated[26-30].

Proximal weakness and hypotonia appear mainly between birth and 24 months and evolve into respiratory failure, the main cause of death. Other features include facial diplegia, PEO, exercise intolerance, and severe feeding difficulties. The marked increase of CK in serum, unusual in mitochondrial myopathies, is a useful diagnostic clue. Most patients have lactic acidosis, a myopathic pattern on EMG, numerous RRF and COX fibers on muscle biopsy and generalized deficiency of RC with reduction of mtDNA copy number. Recently our group has identified a new mitochondrial biomarker, the growth and differentiation factor-15 (GDF15) useful to identify patients with TK2 mutations in witch not significant increased of CK or serum lactate were observed[31,32].

Recently late-onset or moderate myopathy with mtDNA depletion or multiple mtDNA deletions have also been observed. Reports of childhood or adult myopathic cases revealed that *TK2* mutations can have a wide rang of onset with a slow progression and a worsening in adulthood associated with mtDNA multiple deletions and depletion[28,32-34]. Pediatric cases with clinical onset after 12–24 months of age usually develop sub-acute myopathy leading to longer survival. They are able to walk during some years but generally lose this capacity during their adolescence. Most of the patients showed progressive weakness in proximal (upper > lower limbs) and axial muscles. Some cases showed bulbar and facial muscle involvement, whereas respiratory insufficiency will be evident in adulthood (Figure 3).

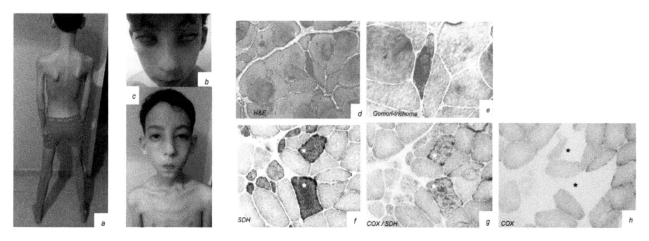

FIGURE 3 patient 15 years with TK2 myopathy chilhood form: a) severe generalized muscle weakness and wasting and scapular winging; b & c) Facial muscles weakness and mild ptosis; d) muscle biopsy at the age of 14 years shows an increased variability in muscle fiber diameter with internal nuclei in some of it; e) Ragged red fibres with mildly increased connective tissue; f, g & h) *showed a mosaic pattern with* ragged blue and *COX negative* fibers.

TK2 activity measured in fibroblast cell lines showed no correlation with the severity of the clinical phenotype. A correlation between disease severity and degree of mtDNA depletion suggest that lower residual mtDNA at diagnosis is associated with a more progressive clinical course. Unfortunately, because mtDNA depletion varies in different stages of the disease and between contiguous areas of the same muscle biopsy, severity of mtDNA depletion cannot be use to predict outcome in individual TK2 deficient patients[35,36] and no genotype-phenotype correlation was identified.

A collaborative international study (in Press; Garone et al.), confirm the phenotypic heterogeneity and propose four major clinical forms: (1) infantile myopathy with onset in the first year of life and rapid progression to early death; (2) childhood-onset myopathy or SMA-like proximal limb weakness progressing to loss of ambulation in few years after disease onset; (3) adult myopathy with subclinical or mild myopathy at disease onset in late childhood or adolescence and slow progression to inability to walk, respiratory failure, or both in adulthood; and (4) adult autosomal recessive PEO. Molecular genetics defects range from severe mtDNA depletion in the infantile/childhood myopathy to a coexisting mtDNA multiple deletions and mtDNA copy number reduction in the adult myopathy to isolated mtDNA multiple deletions in arPEO (Figure 4).

TK2 deficiency is a MiDs that may be under diagnosed due to its diverse spectrum of clinical presentations. Recently, molecular bypass treatment with deoxypyrimidine monophosphates has been demonstrated to prolong the life span and to ameliorate clinical, molecular and biochemical defects in the Tk2 H126N knockin mouse model[37]. The increasing cellular availability of the deficient deoxyribonucleoside triphosphates precursor by direct administration of the deoxyribonucleosides or inhibition of its catabolism is a potential treatment for mtDNA depletion syndrome caused by defects in dNTP metabolism[38].

Few patients have being treated with oral oligonucleosides/oligonucleotides, into the compassionate drug use program, showing improvement of clinical manifestations (personal observations). The potential translational application of this treatment underscores the need for a better understanding of the human disease, natural history and biomarkers in advance of potential clinical trials.

Mutation in *DGUOK* gene (classically associated with mitochondrial DNA depletion syndrome - hepatocerebral type) were identified in one case of juvenile-onset mitochondrial myopathy with mtDNA depletion[39] and late-onset myopathy without PEO associated with multiple mtDNA deletions[40]. Two other genes may also be involved in mtDNA depletion syndrome with early-onset myopathy but frequently associated to multisystem manifestations: *SUCLA2* with methylmalonic aciduria and Leigh-like syndrome (basal ganglial involvement, failure to thrive, epilepsy, psychomotor retardation, hearing loss and abnormal movements)[41,42] and RRM2B with a multisystem disorder including seizures, gastrointestinal dysmotility, hearing loss and renal tubulopathy[43-45].

Early-onset myopathy with congenital cataracts due to *GFER* mutations has been described with presence of multiple mtDNA deletions and combined RC defects in muscle. Motor developmental delay, diffuse muscular hypotrophy and hearing loss were observed. Lactate and CK were high in serum[46].

• **Autosomal recessive PEO (arPEO) and multiple mtDNA deletions**

POLG mutations are the main cause of arPEO leading to a large spectrum of neuromuscular symptoms. arPEO have usually an adult-onset (from 20 to 62 years old). PEO is frequently associated with myopathy and/or sensory-motor neuropathy with sensitive ataxia[47,48], while isolated ptosis

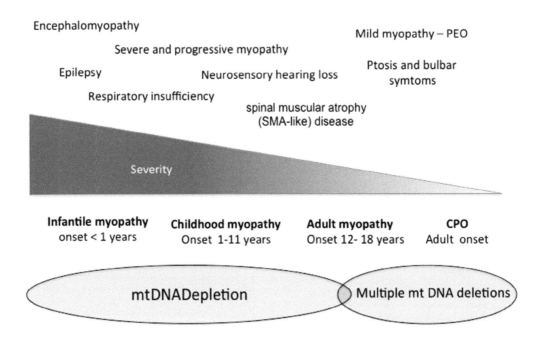

FIGURE 4 Clinical spectrum of TK2 myopathy.

has also been reported[49] . The sensory neuronopathy of SANDO phenotype (Sensory Ataxic Neuropathy, Dysarthria and Ophthalmoparesis) leads to profound sensory ataxia and commonly precedes onset of PEO by years to decades[50]. Other manifestations observed with "PEO+" phenotype appear over years or decades like psychiatric illness, dementia, extrapyramidal movement disorders, cerebellar ataxia, migraine, sensorineural deafness, and gastrointestinal dysmotility[51-53]. Unlike multisystemic MiDs, signs like retinopathy, diabetes, and cardiomyopathy are not common in POLG-related disorders. On the basis of these additional signs, different acronyms have been associated to recessive forms due to POLG1 mutations: Mitochondrial Recessive Ataxia Syndrome (MIRAS), Ataxia Neuropathy Spectrum (ANS), childhood Myo-Cerebro- Hepatopathy Spectrum (MCHS), Spino-Cerebellar Ataxia with Epilepsy (SCAE), and Myoclonic Epilepsy, Myopathy, and Sensory Ataxia (MEMSA)[50].

RRM2B mutations are also a common cause of arPEO, associated with an earlier age of onset (from birth to 20 years old), a more severe multisystemic disorder with muscular alteration (widespread COX- fibers) compared to the frequent dominant inherited forms[48,54-56] Myopathy and bulbar dysfunction are predominant with PEO. Sensorineural hearing loss and gastrointestinal problems, including irritable bowel symptoms and low weight, are additional discriminatory features seen in RRM2B-related MiDs[56]. In contrast, CNS involvement is present less frequently (except cerebellar ataxia) compared to other arPEO[55].

Recently, other genes have been reported in arPEO with multiple mtDNA deletions. In DGUOK-related MiDs, adult- onset PEO (20 to 69 years) was described with mild

myopathy, dysphagia, deafness, high serum CK level and COX- fibers[40]. In the same way, MPV17 mutations were reported in a case of adult-onset axonal sensorimotor neuropathy with PEO, distal myopathy, Parkinsonism and multisystem disorder including diabetes, hearing loss, gastrointestinal dysmotility and fatty liver[57]. Muscle biopsy showed 10% of COX- fibers and normal RC activities. *TK2* mutations have been described with late adult-onset PEO associated with slowly progressive proximal myopathy, bulbar dysphagia and dysarthria[58]. Finally, MGME1 has been identified as a new gene involved in mtDNA instability with depletion and multiple deletions. Before 20 years of age, patients developed PEO and myopathy with proximal and face weakness, leading progressively to respiratory failure and emaciation[59]. Mental retardation, cardiopathy and gastrointestinal signs were observed. Muscle analysis showed COX- fibers, and RC defect.

MNGIE (myopathy and PEO; neuropathy; gastrointestinal; encephalopathy) is a multisystemic disorder beginning usually in the first three decades of life (5 months to 53 years). This disease is characterized by complete PEO, myopathy with proximal weakness and exercise intolerance, gastrointestinal dysmotility (diarrhea, vomiting, abdominal pain, episodic intestinal pseudo-obstruction, gastroparesis), cachexia, sensory neuropathy (axonal or demyelinating), and diffuse leukoencephalopathy[60]. The course is progressive with additional features like hearing loss, retinal degeneration, optic atrophy and cognitive impairment[61]. Brain MRI shows widespread and symmetric white matter involvement, sparing corpus callosum without a lactate peak. Lactic acidosis is frequently observed. Muscle studies identified neurogenic and mitochondrial changes (many COX - fibers), multiple mtDNA deletions and mtDNA depletion. *TYMP* mutations are the main cause of the disease and are responsible for a severe deficiency in thymidine phosphorylase activity. Recessive mutations in *RRM2B* have been reported in a 30-year-old female with adult-onset MNGIE and mitochondrial DNA depletion[62]. MNGIE-like phenotype with absence of leukoencephalopathy was observed with *POLG* mutations[63,64].

5 MYOPATHIES WITHOUT PEO

• Mutations in respiratory chain subunits

Myopathy with Coenzyme Q10 (CoQ10) deficiencies includes pure myopathy or encephalomyopathy with recurrent myoglobinuria. CoQ10 deficiencies may be either primary due to mutations in ubiquinone biosynthetic genes, such as *CACB1*[65], or secondary when the responsible gene is not directly related to this pathway, such as *ETFDH*[66].

Although the first patients with muscle CoQ10 deficiency were reported in 1989[67], interest in this entity was recently rekindled by the descriptions of different clinical phenotypes[68]. A "pure" mitochondrial myopathy due to CoQ10 deficiency was described by Lalani et al [69]. in an 11-year-old boy with recent subacute onset of exercise intolerance and proximal weakness. Muscle biopsy showed RRF and lipid storage and CoQ10 concentration in muscle was 46% of normal. Oral CoQ10 supplementation (300mg daily) resulted in "sustained clinical improvement" and normalization of both serum creatine kinase and lactate levels.

• Mutations in genes encoding ancillary proteins

Mostly of the Mendelian disorders due to defects in complexes III, IV and V are due to mutations in genes encoding proteins needed for the proper assembly or function of the complexes. Ogilvie and coworkers[70] identified pathogenic mutations in the first assembly gene for complex I (B17.2L) associated with cavitating leukoencephalopathy.

Myopathy and exercise intolerance with defect of complex I assembly was described with *ACAD9*, which is, to date the first nuclear gene involved in isolated muscular phenotype with complex I deficiency[71]. Exercise intolerance appears in childhood combined with hyperlactatemia, strong nauseas and mental slowing after exercise, and improvement by high doses of riboflavin.

None of the known mutations in *COX10* gene causes isolated myopathy, although muscle is almost invariably involved in these infantile and generalized disorders. Targeted disruption of this gene in skeletal muscle of mice, however, produced a COX-deficient myopathy[72].

• Other myopathies

Myopathy and exercise intolerance with complex II deficiency, or combined RC defect, were associated with two different genes encoding components of the Fe-S cluster biogenesis machinery: *ISCU*[73] and *FDX2*[74]. The onset is in childhood. Myopathy is slowly progressive with exercise intolerance after minor exertion, marked rhabdomyolysis with myoglobinuria after exercise, hyperlactatemia and some times mild hypertrophic cardiomyopathy. Muscle pathology shows SDH- and COX- fibers, with increased iron staining in SDH- fibers and mitochondrial iron-rich inclusions.

Myopathy with lactic acidosis and sideroblastic anemia (MLASA) may be caused by two genes affecting protein translation, *YARS2* associated with an earlier age of onset (infancy–childhood)[75] than the one generally reported for *PUS1* (childhood–adolescence)[76]. Severity is variable even within affected families, with additional features like delayed milestones, mental retardation, ptosis, short stature, dysphagia, and respiratory failure. Muscle biopsy shows RC defect and histological mitochondrial changes only in *YARS2* involvement[77].

Recently, childhood-onset myopathy with extrapyramidal movement disorders (chorea, tremor, dystonia, orofacial dyskinesia) involving *MICU1* mutations has been reported[78]. Other features are occasional: ataxia, microcephaly, PEO, optic atrophy, and axonal neuropathy. Serum CK is high and brain MRI shows increased signal in globus pallidus.

5.1 AUTOSOMAL DOMINANT MUSCULAR DISORDERS WITH PEO

Autosomal dominant (Ad) PEO is an invariable adult-onset phenotype with multiple mtDNA deletions, also associated with a generalized myopathy[79] (Table 2). Some affected individuals, so-called PEO+, have variable degrees of sensorineural hearing loss, axonal neuropathy, ataxia, depression, Parkinsonism, hypogonadism and cataracts[63]. The adPEO is genetically heterogeneous. In the cohort of PEO patients with multiple mtDNA deletions reported by Ronchi et al.[80], *POLG* mutations (19%), mainly recessive, were followed by *TWINKLE* mutations (18%), mainly dominant, then *ANT1* (6.7%) and *OPA1* (2.2%), with no mutations detected in RRM2B, despite a previous study reporting mutations in this gene in 13% of PEO[54]. Serum CK and lactate may be increased. Muscle biopsy commonly shows RRF and COX-fibers and may present decreased RC complex activities.

A subset of adPEO is caused by mutations in *ANT1* gene with a mild phenotype beginning between 20 and 35 years, including PEO, exercise intolerance and muscle weakness[81-83]. Some signs are occasional such as dysphagia, dysphonia, respiratory, cataracts, hearing loss, dementia, and bipolar affective disorder.

POLG-related adPEO or "adPEO+" are less frequent than arPEO and are usually characterized by a generalized myopathy, with an age of onset between 16 and 40 years, often associated with hearing loss, sensory axonal neuropathy, ataxia, depression, Parkinsonism, bulbar syndrome, facial diplegia, and cataracts[79,84,85]. An unusual adult-onset presentation with a distal myopathy predominant in upper limbs and cachexia was described by Pitceathly et al. in two patients, who also developed PEO, bulbar syndrome and cataracts[86].

Patients with a dominant mutation in *RRM2B* develop PEO and proximal muscle weakness in mainly the fifth decade of life (age range 15–70 years)[56]. The presence of significant bulbar weakness, hearing loss and gastrointestinal symptoms should guide clinicians towards *RRM2B* genetic analysis.

TWINKLE is the most frequent gene implicated in adPEO[87]. PEO occurs mostly before 40 years (from 17 to 73 years). Patients develop frequently proximal myopathy with respiratory failure, sensory ataxic neuropathy, bulbar weakness, corresponding to SANDO and also Parkinsonism[88].

To date, only a few cases of adPEO have been reported with a mutation in *POLG2* gene[89,90]. PEO occurs in infancy and adulthood (<40 years of age), usually associated with myopathy. Other signs may be present as dysarthria, dysphagia, cerebellar ataxia, neuropathy, cardiac conduction defect, and impaired glucose tolerance. Muscle biopsy showed no abnormalities or few COX- fibers. Familial history may be absent probably due to variable penetrance[91,92].

Mutations in *DNA2* have been identified by Ronchi et al [80] . in three families, with sporadic or familial progressive proximal myopathy, associated with mild PEO and respiratory features (exertional dyspnea and episodic obstructive dyspnea). The age of onset varied from childhood to the fourth decade. Muscle histology showed only a few COX- fibers.

OPA1 and *MFN2*, two genes implicated in the mitochondrial fusion pathway, have been reported with some cases of Dominant Optic Atrophy "Plus" including PEO. In 45% of patients with Optic Atrophy "Plus" phenotype related to *OPA1*, myopathic signs, including mainly PEO were observed during the 3rd decade[93]. Muscle biopsy shows COX- fibers and multiple mtDNA deletions. Rouzier et al. described a dominant MFN2 mutation in one large family with dominant optic atrophy, axonal neuropathy and multiple mtDNA deletions, in which some individuals presented PEO, mitochondrial myopathy and ataxia[94].

6   PERIPHERAL NERVOUS SYSTEM MANIFESTATIONS IN MIDS

Peripheral nerves in MiDs may be a direct consequence of the genetic defect (primary mitochondrial PNPs) or may be secondary to manifestations of the underlying MID, which are well known as risk factors for secondary PNPs, such as diabetes, renal insufficiency, hypothyroidism, hyperthyroidism, or hypoparathyroidism (secondary mitochondrial PNPs)[95].

Polyneuropathy is observed with relative frequency in MiDs, but rarely as isolated or dominant feature, except in some phenotypes including particularly NARP syndrome, CMT2A due to MFN2 mutations, CMT2K and

CMT4A due to GDAP1 mutations. Here we will describe the main clinical features of these disorders.

6.1   MITOCHONDRIAL DYNAMICS AND AXONAL TRANSPORT

Early microscopic observations led to the idea that mitochondria were small, individual organelles randomly distributed within the cytoplasm. However, many teams have shown that mitochondria are highly structured, forming a complex network of interconnected tubules[96,97]. The mitochondrial network constitutes a dynamic system, constantly adapting to cellular requirements by changing its shape and position through processes of fission and fusion[98]. Indeed, the bioenergetic status of the cell is closely related to the structure of the mitochondrial network[99]. Insights acquired over the past decade show that mitochondrial dynamics plays a crucial role in several interdependent cellular processes such as bioenergetics, neuronal axonal transport, calcium homeostasis and apoptosis. In neurons, mitochondria are transported along the cytoskeleton of the axon, from the cell body to the periphery (anterograde transport) and from the periphery to the cell body (retrograde transport). Motor proteins of the kinesin (KIF) superfamily interact with mitochondria through the outer membrane proteins Miro1 and Miro2 and participate in anterograde transport. In contrast, cytoplasmic dynein motors mediate retrograde transport[100,-101]. Another aspect of mitochondrial dynamics concerns fusion and fission mechanisms.

Two mitochondria can fuse through mergers of successive outer and inner mitochondrial membranes (mitochondrial fusion). The reverse phenomenon is the mitochondrial fission. Relatively few proteins directly involved in fusion and fission processes have been identified so far. Three GTPase proteins are known to regulate the fusion process: mitofusin 1 (MFN1), mitofusin 2 (MFN2), and optic atrophy protein 1 (OPA1). MFN1 and MFN2 are mitochondrial transmembrane proteins localized in the outer membrane and appear to play similar roles in mitochondrial fusion. OPA1 has been shown to be the mediator of inner membrane fusion[102,103]. Mitochondrial fission mainly involves the dynamin-related protein 1 (Drp1), also referred to as the dynamin-like protein 1 (DLP1), a member of the conserved dynamin-related large GTPase superfamily. This cytosolic protein is recruited on the external surface of the mitochondrial outer membrane by Fis1, which acts as a receptor to Drp1. Drp1 can also oligomerize and assemble into rings or spirals surrounding the mitochondrial outer membrane[104]. GTP hydrolysis leads to the formation of a constriction initiating mitochondrial fission. It has been shown that GDAP1 is a regulator of mitochondrial fission[104,105] and that GDAP1-induced fission is dependent on fission factors Fis1 and Drp1[106]. Mitochondrial fusion and fission not only determine mitochondrial morphology and size, but also regulate mitochondrial distribution.

6.2 Peripheral Neuropathies Associated to mtDNA Mutations

The peripheral nervous system (PNS) is frequently involved with variable degrees of severity in some of the classical recognized phenotypes associated to mtDNA mutation or rearrengments as: MERRF[107,108], MELAS[109-112], Leigh, MILS (Maternally Inherited Leigh Syndrome)[113,114], PEO, and KSS. The typical muscular or central nervous system involvement in these disorders are predominantly associated to sensory or sensori-motor axonal polyneuropathy, but demyelinating type has also been described in few cases[109-114].

• **Neurogenic weakness, Ataxia, Retinitis Pigmentosa (NARP)**

NARP syndrome is due to the transversion m.8993T>G and, less commonly, to the transition m.8993T>C in the MT-ATP6 gene, encoding subunit 6 of ATP synthase[115]. The T8993C mutation is generally considered to be less severe than the T8993G mutation. The clinical presentation of patients with heteroplasmic NARP mutations is dominated by proximal neurogenic muscle weakness with sensory neuropathy, ataxia due to cerebellar atrophy, epilepsy, learning difficulties and pigmentary retinopathy. Associated myopathy is rare and muscle biopsy is usually normal. Electromyography (EMG) and nerve conduction studies may demonstrate peripheral neuropathy (which may be a sensory or sensori-motor axonal polyneuropathy)[116,117]. Onset of symptoms is often in early childhood; individuals can be relatively stable for many years, but may suffer episodic deterioration. When the heteroplasmy level is higher than 80% patients frequently have a more early and severe impairment which is presented as Leigh syndrome.

• **CMT2 related to MT-ATP6 mutation.**

It is still unclear why some patients harboring homoplasmic levels of MT-ATP6 mutation develop CMT2 phenotype, whereas other reported patients with similar mutant loads develop MILS. mtDNA mutation m.9185T_C in MT-ATP6 gene at homoplasmic level has been found in patients with a typical CMT2 phenotype, responsible for severe motor-predominant axonal neuropathy, associated with learning difficulties, hearing loss, and retinal degeneration, with normal muscle biopsy[117].

6.3 Peripheral Neuropathies Due to nDNA Mutations

Mitochondrial neuropathies can be caused by mutations of nuclear DNA affecting genes involved in mtDNA maintenance, or to mitochondrial bioenergetics or dynamics.

6.3.1 Mitochondrial Bioenergetics

• **Demyelinating polyneuropathies in SURF1-related disorders.**

SURF1 mutations represent a major cause of autosomal recessive Leigh syndrome associated with complex 4 (cytochrome oxidase) deficiencies. In patients with SURF1-related Leigh syndrome (usually with infantile-onset), a clinically evident PNP could be present in up to 40% of cases[118], even if the most prominent and life-threatening symptoms are related to CNS dysfunction. In these patients, nerve conduction velocities are decreased and nerve biopsy is suggestive of a myelinopathy (loss of large myelinated fibers, presence of thinly myelinated fibers). In some patients, however, the CNS dysfunction can be very limited or appear later than the PNP[118]. The absence of disabling CNS dysfunction allows a much longer life span, two of the three patients reported to date being older than 40 years of age. These patients had initially been classified as autosomal recessive forms of demyelinating CMT (CMT4).

• **X-linked CMT with PDK3 mutation.**

In a family affected by an X-linked dominant form of CMT (CMTX6) a missense mutation c.473G>A (p.R158H) has recently been identified in the pyruvate dehydrogenase kinase isoenzyme 3 (PDK3) gene[119]. PDK3 is a nuclear-coded protein regulating the pyruvate dehydrogenase complex in the mitochondrial matrix, which is a key enzyme linking glycolysis to the energy-producing Krebs cycle and lipogenic pathways. Combined effects of R158H PDK3 are likely to lock PDC in a predominantly phosphorylated inactive state leading to impaired ATP production and or lactate accumulation. The clinical phenotype is milder than that seen in PDC deficiency, and is limited to the peripheral nerves. Reduced PDC activity has previously been detected in fibroblast lines from several unrelated and genetically undefined CMT patients[120]. The cellular energy depletion is likely the underlying cause of the peripheral neuropathy, although lactate accumulation may also contribute.

• **X-linked CMT with AIF mutation.**

Cowchock syndrome (CMTX4) is a slowly progressive X-linked recessive disorder manifesting by axonal sensori-motor neuropathy, deafness, and cognitive impairment. Exome sequencing of affected individuals from a family identified a missense change c.1478A>T (p.Glu493Val) in AIFM1, the gene encoding apoptosis-inducing factor (AIF) mitochondrion-associated AIF, is a phylogenetically conserved mitochondrial flavoprotein, and one of the key caspase-independent death effectors. In this family, only males displayed the phenotype, consistent with an X-linked recessive mode of inheritance. Cranial MRI in two subjects showed multiple punctate T2 hyperintensities in the supratentorial white matter. Authors investigated the consequences of this mutation in cultured skin fibroblasts and skeletal muscle from an affected individual. They showed that the mutation alters the redox properties of AIF protein and results in increased cell death via apoptosis, without affecting the activity of the respiratory chain complexes[121].

- **MNGIE (Mitochondrial Neuro-Gastro-Intestinal Ence-phalopathy).**

MNGIE is a multisystem mitochondrial disorder, which has an autosomal recessive mode of inheritance. MNGIE is due to mutations in the *TYMP* gene encoding thymidine phosphorylase, which promotes the phosphorylation of thymidine to thymine and deoxyribose-1-phosphate[122]. The clinical presentation is dominated by digestive symptoms with severe gastrointestinal dysmotility resulting in progressive cachexia. Clinical manifestations include borboryghms, postprandial emesis, early satiety, dysphagia, gastroesophageal reflux disease, episodic abdominal distensive pain, ascites, diarrhea, and usually begin before the age of 20[123]. Additional manifestations can include progressive external ophthalmoplegia, pigmentary retinopathy, dysphonia, myopathy and diffuse leukoencephalopathy (present in all patients). PNS involvement in MNGIE is constant but with varying degrees of severity and usually presents as a sensorimotor demyelinating PNP, responsible for distal weakness and atrophy predominantly affecting the lower limbs[123]. MNGIE is suspected in patients with elevated thymidine and decreased thymidine phosphorylase activity in blood. The diagnosis is confirmed by the identification of mutations in both alleles of the TYMP gene.

- **Navajo Neuro-hepatopathy.**

A very particular autosomal recessive infantile-onset PNP was reported in 1976 in Native Americans from the Navajo population in the southwestern United States[124]. This sensory-motor PNP is manifested by distal weakness, areflexia, and a loss of sensation that results in corneal ulcerations and acral mutilation. Later on, patients develop neurogenic arthropathics. Nerve conduction velocities are reduced and nerve biopsy shows abnormalities of both types of nerve fibers (severe reduction of myelinated fibers associated with marked abnormalities of unmyelinated fibers). Affected patients also have leukoencephalopathy at MRI and liver dysfunction, which can progress to liver failure and death[125]. This syndromic PNP is associated with a founder mutation (p.R50Q) in the MPV17 gene[126]. Recessive mutations in this latter gene were previously found in patients with hepato-cerebral mtDNA depletion syndrome[127].

6.3.2 Mitochondrial Dynamics (Fusion and Fission)

The alteration of mitochondrial fusion and fission processes significantly affects neuronal function because of the dependence of motoneurons on a high metabolic rate[128]. Some proteins involved in mitochondrial dynamics are responsible for hereditary neuropathies of Charcot-Marie-Tooth type when mutated, such as MFN2 (CMT2A2) and GDAP1 (CMT4A/2K).

- **Neuropathies associated with MFN2 mutations.**

MFN2 is a mitochondrial transmembrane GTPase located in the mitochondrial outer membrane. It contributes to the maintenance of the mitochondrial network by promoting mitochondrial membrane[129]. Moreover, MFN2 interferes with mitochondrial axonal transport by interacting with the adaptor Miro/Milton complex, by which kinesin attaches mitochondria to microtubules[130]. As MFN2 plays a role in calcium crosstalk between endoplasmic reticulum and mito-chondria, and MFN2 mutants induce a higher cytoplasmic calcium concentration, it is possible that such cytosolic calcium increase might have an indirect effect on axonal transport by calcium binding to Miro[131]. In addition to its contribution to mitochondrial network dynamics, MFN2 promotes mitochondrial energization by regulating oxidative phosphorylation (OXPHOS). The expression of the MFN2 gene has been described as key pathogenic factor of metabolic abnormalities associated with type 2 diabetes, obesity, and weigh loss. Studies perfomed on skin fibroblasts have shown that the mitochondrial energetic metabolism was greatly altered in patients with MFN2 mutations with a significant coupling defect leading to reduced OXPHOS efficacy (reduction of ATP/O).

A transgenic mouse model has been generated by expressing in neurons a mutated form of human MFN2 (mutation p.R94Q)[132]. These mice show the main clinical symptoms encountered in patients with CMT2A. Oxygraphic and enzymatic measurements performed on isolated mitochondria from the brain tissue of these mice have shown that the R94Q mutation induces a combined defect of complexes II and V linked to the opening of mKATP channel, which could participate in the pathophysiology of the disease[94]. Rouzier et al. found abnormalities of mtDNA maintenance associated with a novel MFN2 missense muta-tion (c.629A>T, p.D210V) in patients suffered from optic atrophy beginning in early childhood, associated with axonal neuropathy and mitochondrial myopathy in adult life[133]. Axonal transport defect may play an important role in MFN2-related CMT2A2. This hypothesis was reinforced by a study showing a marked disruption of mitochondrial transport in cultured sensory neurons expressing disease-mutated forms of MFN2[134]. In addition, the accumulation of mitochondria in the distal part of sural nerve axons has been observed in CMT2A2 patients[135]. CMT2A2 is the most prevalent type of axonal sensori-motor dominant forms of CMT with a frequency of up to 20% among all CMT2 patients[136,137]. In some cases the inheritance was recessive[138] and semi-dominant inheritance has also been described[139]. Typical clinical symptoms of CMT2A are progressive distal limb muscle weakness with atrophy, distal sensory loss, and stepping gait which can lead to wheelchair dependency. *MFN2* mutations have been associated with clinical and MRI findings suggesting central nervous system involvement, wath included progressive cebellar atrophy[140,141]. Moreover, mutations of the MFN2 gene have been incriminated in hereditary motor and sensory neuropathy type VI (HMSN VI, MIM 601152), a form of CMT associated with optic atrophy[142]; this emphasizes the importance of mitochondrial network dynamics in optic atrophies, as well as peripheral neuropathies. Figure 5 describes spastic paraplejia, progressive optic and cerebellar atrophy in a patient with mutation in *MFN2* gene.

• **Neuropathies associated with GDAP1 mutations.**

Although not having GTPase-domain-like pro-fusion or pro-fission protein, the mitochondrial outer membrane protein GDAP1 was shown to influence mitochondrial dynamics. Overexpression in COS cells induced fragmentation of mito-chondria without inducing apoptosis and down regulation by siRNA enhanced the formation of tubular networks. When expressed in cells, recessive mutations of GDAP1 show reduced fission activity, whereas dominant mutations interfere with mitochondrial fusion and lead to elevated production of reactive oxygen species[143]. An interaction between GDAP1 and tubulins was recently demonstrated[136]. In the presence of GDAP1 mutations, this interaction between mitochondria and the microtubule cytoskeleton would be altered, which might affect mitochondrial axonal transport and movement within the cell and may explain the pathophysiology of the GDAP1-related Charcot-Marie-Tooth disease. Interestingly, we found that *GDAP1* mutations in fibroblasts from CMT2K patients carrying the C240Y mutation

are associated with mitochondrial complex I defect[137]. The association between complex I deficiency and *GDAP1* mutations suggests that GDAP1 may play a protective role against oxidative stress. This hypothesis is based on the fact that GDAP1 has structural domains related to the cytosolic GSTs; however, no experimental GST activity has been demonstrated so far. GSTs play a role in the detoxification of xenobiotics and endogenous toxicants. Recently, Noack et al. have shown that *GDAP1* overexpression protected against oxidative stress caused by depletion of the intracellular antioxidant glutathione (GHS) and against effectors of GHS depletion that affect the mitochondrial membrane integrity-[144]. Mutations in *GDAP1* are associated either with autosomal recessive forms of CMT (CMT4A) or with autosomal dominant axonal (CMT2K; see mitodyn.org). The dominant forms, which are less severe than the recessive forms, are associated with later onset (after the second decade) and a slower course of the disease. In these cases, patients may be able to walk even after the fourth decade[145]. The recessive forms of *GDAP1*-related CMT are severe with onset during the first decade, ussualy the clinical symtoms started at 18 month of life. The rapid progression of these disorders leads to functional disability, associated with additional clinical features such as a hoarse voice due to vocal cord paresis, diaphragmatic paralysis and facial weakness.

6.3.3 *Involvement of Mitochondrial Axonal Transport in Other CMT2*

• **Heat Shock Proteins (HSP)**

HSPB1 and HSPB8 are small Heat Shock Proteins that act as ATP-independent chaperones in protein folding, but

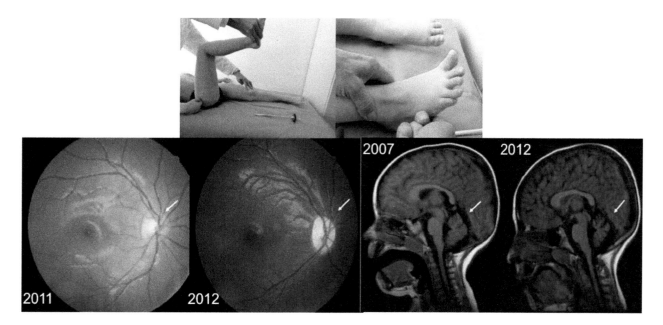

FIGURE 5 Spasticity, babinski sign (spastic paraplejia), progressive optic and cerebellar atrophy in patient with mutation (c.749G>A) in MFN2 gene.

are also implicated in architecture of the cytoskeleton. Previous studies have shown that disruption of assembly and aggregation of neurofilaments are the results of *HSPB1* mutations in CMT2F[146,147] and *HSPB8* in CMT2L[148]. A report showed that the CMT2-causing mutant HSPB1S135F decreased the abundance of acetylated a-tubulin (a guidance cue used by motor proteins to move their cargoes) and impaired mitochondrial anterograde and retrograde movement in cultured dorsal root ganglion sensory neurons isolated from transgenic mice[149], suggesting the pathogenic role of alpha-tubulin deacetylation in mutant HSPB1-induced neuropathies. Interestingly, increasing alpha-tubulin acetylation by pharmacological inhibition of histone deacetylase 6 (HDAC6) corrected the axonal transport defects caused by *HSPB1* mutations and rescued the CMT phenotype of symptomatic mutant HSPB1 mice, offering perspectives for using *HDAC6* inhibitors as a therapeutic strategy for hereditary axonopathies[149]. Compared to *HSPB1*, the role of HSPB8 is not as well characterized, but the CMT2-causing K141E mutation in *HSPB8* induces similar functional and structural deficiency.

- **Kinesin family member 1B mutation (KIF1B)**

Indirectly, mitochondria dysfunction could be involved by mutations in genes encoding molecular motors of anterograde and retrograde transport. A missense mutation in *KIF1B-beta*, a monomeric motor for anterograde transport of mitochondria, has been identified in affected members of a Japanese family presenting an autosomal dominant CMT2A1 resulting in a p.Gln98Leu substitution[150].

- **Dynein heavy chain 1 (DYNC1H1)**

Mutation in *DYNC1H1* gene Has been implicated in varius neurodegenerative disorders with including hereditary sensory – motor neuropathy and spastic hereditary paraplejia. Using exome sequencing of affected individuals in a large CMT2 family permitted to identify the c.917A>G mutation in DYNC1H1[151]. DYNC1H1 encodes the dynein heavy chain 1 which forms the core of the cytoplasmic dynein, a large multisubunit motor protein complex having a key role in retrograde mitochondrial axonal transport along cytoskeletal microtubules in neurons. Those results are supported by previous animal studies that have implicated disruption of DYNC1H1 in neuropathic disease[152,153].

6.4 Secondary Mitochondrial PNPs

Secondary mitochondrial PNPs may be due to diabetes, renal insufficiency, thyroid dysfunction, or hypoparathyroidism in syndromic as well as non-syndromic MIDs. If such risk factors are part of the phenotype it is often difficult to differentiate if PNP is only due to the underlying mutation, secondary due to the risk factor for PNP inherent to the MID, or both. MIDs, in which diabetes has been reported, include MELAS, MERRF, KSS, Leigh syndrome, and Mendelian PEO. MIDs, which go along with renal insufficiency, include MELAS, KSS, Leigh syndrome, and AHS. A further example of a secondary PNP in a syndromic MID is Wolfram syndrome, which presents as diabetes mellitus and optic atrophy. Additional features may include diabetes insipidus, renal abnormalities, psychiatric abnormalities, deafness, ataxia, and PNP[13,154]. Whether MIDs are prone to develop malignancy and consecutively paraneoplastic PNP is unknown.

7 CONCLUSION

Defects in the mitochondrial respiratory chain impair energy production and frequently involve skeletal muscle and peripheral nerves, but rarely as an isolated or dominant feature, except in some phenotypes including: PEO, NARP, and TK2 myopathy.

Several neuromuscular manifestations have been associated into the clinical and genetic heterogeneity of MiDs. The multisystem nature of these disorders may not be evident at onset but may become apparent only during the course of the disease.

In pediatric patients the clinical manifestation varies within the age and the development process of peripheral and central nervous system. The most serious clinical phenotypes are present during the first months of life with multisystem manifestations, however in older population the clinical features become more specific allowing a better phenotype recognition.

We want to highlight the importance of searching for the neuromuscular "clue clinical signs" associated to inheritance pattern as key elements to guide the complex diagnosis process and genetic studies in MiDs. The PEO and axonal neuropathy are the most useful sings to help the physician in the challenge of diagnosing these diseases.

REFERENCES

1. DiMauro S. Mitochondrial myopathies. Curr Opin Rheumatol 18:636–41; 2006.
2. Chaussenot A, Paquis-Flucklinger V. An overview of neurological and neuromuscular signs in mitochondrial diseases. Rev Neurol (Paris) 170:323–338; 2014.
3. Schapira AH. Mitochondrial disease. Lancet 368:70–82; 2006.
4. Holt I, Harding A, Morgan-Hugues JA. Deletion of muscle mitochondrial DNA in patients with mitochondrial myopathies. Nature 331:717–19; 1988.
5. Wallace D, Singh G, Lott MT, Schurr TG, Lezza AMS, Elsas LJ. Mitochondrial DNA mutation associated with Leber's hereditary optic neuropathy. Science 242:1427–30; 1988.
6. Pitceathly RD, McFarland R. Mitochondrial myopathies in adults and children: management and therapy development. Curr Opin Neurol 27:576–82; 2014.

7. DiMauro S, Schon EA. Mitochondrial disorders in the nervous system. Annu Rev Neurosci 31:91–123; 2008.

8. Barca E, Emmanuele V, DiMauro SB. Metabolic Myoglobinuria. Curr Neurol Neurosci Rep 15:69; 2015.

9. Hays AP, Oskoui M, Tanji K, et al. Mitochondrial neurology II: myopathies and peripheral neuropathies. In: DiMauro S, Hirano M, Schon EA, editors. Mitochondrial medicine. London: Informa Healthcare; 2006. pp. 45–74.

10. Kearns TP, Sayre GP. Retinitis pigmentosa, external ophthalmophegia, and complete heart block: unusual syndrome with histologic study in one of two cases. AMA Arch Ophthalmol 60:280–89; 1958.

11. Pineda M, Ormazabal A, López-Gallardo E, et al. Cerebral folate deficiency and leukoencephalopathy caused by a mitochondrial DNA deletion. Ann Neurol 59:394–98; 2006

12. Quijada-Fraile P, O'Callaghan M, Martín-Hernández E, et al. Follow-up of folinic acid supplementation for patients with cerebral folate deficiency and Kearns-Sayre syndrome. Orphanet J Rare Dis 24;9:217; 2014.

13. Bannwarth S, Procaccio V, Lebre AS, et al. Prevalence of rare mitochondrial DNA mutations in mitochondrial disorders. J Med Genet 50:704–14; 2013.

14. Ka¨Rrpa¨ M, Herva R, Moslemi A-R, et al. Spectrum of myopathic findings in 50 patients with the 3243A>G mutation in mitochondrial DNA. Brain 128:1861–69; 2005.

15. Maniura-Weber K, Taylor RW, Johnson MA, et al. A novel point mutation in the mitochondrial tRNA(Trp) gene produces a neurogastrointestinal syndrome. Eur J Hum Genet 12:509–12; 2004.

16. Horvath R, Bender A, Abicht A, et al. Heteroplasmic mutation in the anticodon- stem of mitochondrial tRNA (Val) causing MNGIE-like gastrointestinal dysmotility and cachexia. Neurol 256:810–15; 2009.

17. Horvath R, Kemp JP, Tuppen HA, et al. Molecular basis of infantile reversible cytochrome c oxidase deficiency myopathy. Brain 132:3165–74; 2009.

18. Pulkes T, Liolitsa D, Eunson LH, et al. New phenotypic diversity associated with the mitochondrial tRNASer(UCN) gene mutation. Neuromusc Disord 15:364–71; 2005.

19. McPherson E, Zabel C. Mitochondrial mutation in a child with distal arthro- gryposis. Am J Med Genet 140A:184–85; 2006.

20. Kollberg G, Moslemi A-R, Lindberg C, et al. Mitochondrial myopathy and rhabdomyolysis associated with a novel nonsense mutation in the gene encoding cytochrome c oxidase subunit I. J Neuropath Exp Neurol 64:123–28; 2005.

21. McFarland R, Taylor RW, Chinnery PF, et al. A novel sporadic mutation in cytochrome c oxidase subunit II as a cause of rhabdomyolysis. Neuromusc Disord 14:162–66; 2004.

22. Horvath R, Schoser BGH, Muller-Hocker J, et al. Mutations in mtDNA- encoded cytochrome c oxidase subunit genes causing isolated myopathy or severe encephalomyopathy. Neuromusc Disord 15:851–57; 2005.

23. Pulkes T, Liolitsa D, Wills AJ, et al. Nonsense mutations in mitochondrial DNA associated with myalgia and exercise intolerance. Neurology 64:1091–92; 2005.

24. Andreu AL, Hanna MG, Reichmann H, et al. Exercise intolerance due to mutations in the cytochrome b gene of mitochondrial DNA. New Eng J Med 341:1037–44; 1999.

25. Horvath R, Scharfe C, Hoeltzenbein M, et al. Childhood-onset mitochondrial myopathy and lactic acidosis caused by a stop mutation in the mitochondrial cytochrome c oxidase III gene. J Med Genet 39:812–16; 2002.

26. Saada A, Shaag A, Mandel H, Nevo Y, Eriksson S, Elpeleg O. Mutant mitochondrial thymidine kinase in mitochondrial DNA depletion myopathy. Nat Genet 29:342–44; 2001.

27. Mancuso M, Filosto M, Bonilla E, et al. Mitochondrial myopathy of childhood associated with mitochondrial DNA depletion and a homozygous mutation (T77M) in the TK2 gene. Arch Neurol 60:1007–09; 2003.

28. Oskoui M, Davidzon G, Pascual J, et al. Clinical spectrum of mitochondrial DNA depletion due to mutations in the thymidine kinase 2 gene. Arch Neurol 63:1122–26; 2006.

29. Go¨ Tz A, Isohanni P, Pihko H, et al. Thymidine kinase 2 defects can cause multi-tissue mtDNA depletion syndrome. Brain 131:2841–50; 2008.

30. Martí R, Nascimento A, Colomer J, et al. Hearing loss in a patient with the myopathic form of mitochondrial DNA depletion syndrome and a novel mutation in the TK2 gene. Pediatr Res 68:151–54; 2010.

31. Kalko SG, Paco S, Jou C, et al. Transcriptomic profiling of TK2 deficient human skeletal muscle suggests a role for the p53 signalling pathway and identifies growth and differentiation factor-15 as a potential novel biomarker for mitochondrial myopathies. BMC Genomic 15:91; 2014.

32. Montero R, Yubero D, Villarroya J, et al. GDF-15 is elevated in children with mitochondrial diseases and is induced by mitochondrial dysfunction. PLoS One 11;11: e0148709; 2016.

33. Behin A, Jardel C, Claeys KG, et al. Adult cases of mitochondrial DNA depletion due to TK2 defect: an expanding spectrum. Neurology 78:644–48; 2012.

34. Cámara Y, Carreño-Gago L, Martín MA, et al. Severe TK2 enzyme activity deficiency in patients with mild forms of myopathy. Neurology 84:2286–88; 2015.

35. Vilà, MR, Segovia-Silvestre, T, Gamez, J, et al. Reversion of mtDNA depletion in a patient with TK2 deficiency. Neurology 60:1203–05; 2003.

36. Vilà, MR, Villarroya, J, Garcia-Arumi, E, et al. Selective muscle fiber loss and molecular compensation in mitochondrial myopathy due to TK2 deficiency. J Neurol Sci 267:137–41; 2008.

37. Garone C, Garcia-Diaz B, Emmanuele V, et al. Deoxypyrimidine monophosphate bypass therapy for thymidine kinase 2 deficiency. EMBO Mol Med 6:1016–27; 2014.

38. Cámara Y, González-Vioque E, Scarpelli M, et al. Administration of deoxyribonucleosides or inhibition of their catabolism as a pharmacological approach for mitochondrial DNA depletion syndrome. Hum Mol Genet. 23:2459–67; 2014.

39. Buchaklian AH, Helbling D, Ware SM, Dimmock DP. Recessive deoxyguanosine kinase deficiency causes juvenile-onset mitochondrial myopathy. Mol Genet Metab 107:92–94; 2012.

40. Ronchi D, Garone C, Bordoni A, et al. Next generation sequencing reveals DGUOK mutations in adult patients with mitochondrial DNA multiple deletions. Brain 135:3404–15; 2012.

41. Elpeleg O, Miller C, Hershkovitz E, et al. Deficiency of the ADP-forming succinyl-CoA synthase activity is associated with encephalomyopathy and mitochondrial DNA depletion. Am J Hum Genet 76:1081–86; 2005.

42. Ostergaard E, Christensen E, Kristensen E, et al. Deficiency of the alpha subunit of succinate-coenzyme A ligase causes fatal infantile lactic acidosis with mitochondrial DNA depletion. Am J Hum Genet 81:383–87; 2007.

43. Bourdon A, Minai L, Serre V, et al. Mutation of RRM2B, encoding p53-controlled ribonucleotide reductase (p53R2), causes severe mitochondrial DNA depletion. Nat Genet 39:776–80; 2007.

44. Kollberg G, Darin N, Benan K, et al. A novel homozygous RRM2B missense mutation in association with severe mtDNA depletion. Neuromuscul Disord 19:147–50; 2009.

45. Spinazzola A, Invernizzi F, Carrara F, et al. Clinical and molecular features of mitochondrial DNA depletion syndromes. J Inherit Metab Dis 32:143–58; 2009.

46. Di Fonzo A, Ronchi D, Lodi T, et al. The mitochondrial disulfide relay system protein GFER is mutated in autosomal recessive myopathy with cataract and combined respiratory chain deficiency. Am J Hum Genet 84:594–04; 2009.

47. Di Fonzo A, Bordoni A, Crimi M, et al. POLG mutations in sporadic mitochondrial disorders with multiple mtDNA deletions. Hum Mutat 22:498–99; 2003.

48. Horvath R, Hudson G, Ferrari G, et al. Phenotypic spectrum associated with mutations of the mitochondrial polymerase g gene. Brain 129:1674–84; 2006.

49. Luoma PT, Luo N, Loscher WN, et al. Functional defects due to spacer-region mutations of human mitochondrial DNA polymerase in a family with an ataxia-myopathy syndrome. Hum Mol Genet 14:1907–20; 2005.

50. Van Goethem G, Martin JJ, Dermaut B, et al. Recessive POLG mutations presenting with sensory and ataxic neuropathy in compound heterozygote patients with progressive external ophthalmoplegia. Neuromuscul Disord 13:133–42; 2003.

51. Luoma P, Melberg A, Rinne JO, et al. Parkinsonism, premature menopause, and mitochondrial DNA polymerase-gamma mutations: clinical and molecular genetic study. Lancet 364:875–82; 2004.

52. Davidzon G, Greene P, Mancuso M, et al. Early-onset familial parkinsonism due to POLG mutations. Ann Neurol 59:859–62; 2006.

53. Rouzier C, Chaussenot A, Serre V, et al. Quantitative multiplex PCR of short fluorescent fragments for the detection of large intragenic POLG rearrangements in a large French cohort. Eur J Hum Genet 22:542–50; 2014.

54. Fratter C, Raman P, Alston CL, et al. RRM2B mutations are frequent in familial PEO with multiple mtDNA deletions. Neurology 76:2032–34; 2011.

55. Pitceathly RD, Fassone E, Taanman JW, et al. Kearns-Sayre syndrome caused by defective R1/p53R2 assembly. J Med Genet 48:610–17; 2011.

56. Pitceathly RD, Smith C, Fratter C, et al. Adults with RRM2B-related mitochondrial disease have distinct clinical and molecular characteristics. Brain 135:3392–403; 2012.

57. Garone C, Rubio JC, Calvo SE, et al. MPV17 mutations causing adult-onset multisystemic disorder with multiple mitochondrial DNA deletions. Arch Neurol 69:1648–51; 2012.

58. Tyynismaa H, Sun R, Ahola-Erkkila S, et al. Thymidine kinase 2 mutations in autosomal recessive progressive external ophthalmoplegia with multiple mitochondrial DNA deletions. Hum Mol Genet 21:66–75; 2012.

59. Kornblum C, Nicholls TJ, Haack TB, et al. Loss-of-function mutations in MGME1 impair mtDNA replication and cause multisystemic mitochondrial disease. Nat Genet 45:214–19; 2013.

60. Hirano M, Nishigaki Y, Martı R. MNGIE: a disease of two genomes. Neurologist 10:8–17; 2004.

61. Garone C, Tadesse S, Hirano M. Clinical and genetic spectrum of mitochondrial neurogastrointestinal encephalomyopathy. Brain 134:3326–32; 2011.

62. Shaibani A, Shchelochkov OA, Zhang S, et al. Mitochondrial neurogastrointestinal encephalopathy due to mutations in RRM2B. Arch Neurol 66:1028–32; 2009.

63. Van Goethem G, Schwartz M, Lofgren A, Dermaut B, Van Broeckhoven C, Vissing J. Novel POLG mutations in progressive external ophthalmoplegia mimicking mitochondrial neurogastrointestinal encephalomyopathy. Eur J Hum Genet 11:547–49; 2003.

64. Giordano C, Powell H, Leopizzi M, et al. Fatal congenital myopathy and gastrointestinal pseudo-obstruction due to POLG1 mutations. Neurology 72:1103–05; 2009.

65. Aure´ K, Benoist JF, Ogier de Baulny H, Romero NB, Rigal O, Lombe`S A. Progression despite replacement of a myopathic form of coenzyme Q10 defect. Neurology 63:727–29; 2004.

66. Quinzii CM, Lo´pez LC, Naini A, DiMauro S, Hirano M. Human CoQ10 deficiencies. Biofactors 32:113–18; 2008.

67. Ogasahara S, Engel AG, Frens D, Mack D. Muscle coenzyme Q deficiency in familial mitochondrial encephalomyopathy. Proc Nat Acad Sci U S A 86:2379–82; 1989

68. Hirano M, Kaufmann P, De Vivo DC, Tanji K. Mitochondrial neurology. I: encephalopathies. In: DiMauro S, Hirano M, Schon EA, editors. Mitochon- drial medicine. London: Informa Healthcare; 2006. pp. 27–44.

69. Lalani S, Vladutiu GD, Plunkett K, et al. Isolated mitochondrial myopathy associated with muscle coenzyme Q10 deficiency. Arch Neurol 62:317–20; 2005.

70. Ogilvie I, Kennaway NG, Shoubridge EA. A molecular chaperone for mito – Chondrial complex I assembly is mutated in a progressive encephalopathy. J Clin Invest 115:2784–92; 2005.

71. Gerards M, van Den Bosch BJ, Danhauser K, et al. Riboflavin-responsive oxidative phosphorylation complex I deficiency caused by defective ACAD9: new function for an old gene. Brain 134:210–219; 2011.

72. Diaz F, Thomas CK, Garcia S, et al. Mice lacking COX10 in skeletal muscle recapitulate the phenotype of progressive myopathies associated with cytochrome c oxidase deficiency. Hum Mol Genet 14:2737–48; 2005.

73. Mochel F, Knight MA, Tong WH, et al. Splice mutation in the iron-sulfur cluster scaffold protein ISCU causes myopathy with exercise intolerance. Am J Hum Genet 82:652–60; 2008.

74. Spiegel R, Saada A, Halvardson J, et al. Deleterious mutation in FDX1L gene is associated with a novel mitochondrial muscle myopathy. Eur J Hum Genet Eur J Hum Genet. 22:902–06; 2014.

75. Riley LG, Cooper S, Hickey P, et al. Mutation of the mitochondrial tyrosyl-tRNA synthetase gene, YARS2, causes myopathy, lactic acidosis, and sideroblastic anemia: MLASA syndrome. Am J Hum Genet 87:52–59; 2010.

76. Casas K, Fischel-Ghodsian N. Mitochondrial myopathy and sideroblastic anemia. Am J Med Genet 125:201–04; 2004.

77. Riley LG, Menezes MJ, Rudinger-Thirion J, et al. Phenotypic variability and identification of novel YARS2 mutations in YARS2 mitochondrial myopathy, lactic acidosis and sideroblastic anaemia. Orphanet J Rare Dis 8:193; 2013.

78. Logan CV, Szabadkai G, Sharpe JA, et al. UK10K Consortium. Loss-of-function mutations in MICU1 cause a brain and muscle disorder linked to primary alterations in mitochondrial calcium signaling. Nat Genet 46:188–93; 2014.

79. Van Goethem G, Dermaut B, Lofgren A, Martin JJ, Van Broeckhoven C. Mutation of POLG is associated with progressive external ophthalmoplegia characterized by mtDNA deletions. Nat Genet 28:211–12; 2001.

80. Ronchi D, Di Fonzo A, Lin W, et al. Mutations in DNA2 link progressive myopathy to mitochondrial DNA instability. Am J Hum Genet 92:293–300; 2013.

81. Kaukonen J, Juselius JK, Tiranti V, et al. Role of adenine nucleotide translocator 1 in mtDNA maintenance. Science 289:782–85; 2000.

82. Deschauer M, Hudson G, Muller T, Taylor RW, Chinnery PF, Zierz S. A novel ANT1 gene mutation with probable germline mosaicism in autosomal dominant progressive external ophthalmoplegia. Neuromuscul Disord 15:311–15; 2005.

83. Kawamata H, Tiranti V, Magrane J, Chinopoulos C, Manfredi G. adPEO mutations in ANT1 impair ADP-ATP translocation in muscle mitochondria. Hum Mol Genet 20:2964–74; 2011.

84. Chalmers RM, Brockington M, Howard RS, Lecky BRF, Morgan-Hughes JA, Harding AE. Mitochondrial encephalopathy with multiple mitochondrial DNA deletions: a report of two families and two sporadic cases with unusual clinical and neuropathological features. J Neurol Sci 143:41–45; 1996.

85. Hudson G, Chinnery PF. Mitochondrial DNA polymerase- gamma and human disease. Hum Mol Genet 15: R244–52; 2006.

86. Pitceathly RD, Tomlinson SE, Hargreaves I, et al. Distal myopathy with cachexia: an unrecognised phenotype caused by dominantly-inherited mitochondrial polymerase g mutations. J Neurol Neurosurg Psychiatry 84:107–10; 2013.

87. Van Hove JL, Cunningham V, Rice C, et al. Finding twinkle in the eyes of a 71-year-old lady: a case report and review of the genotypic and phenotypic spectrum of TWINKLE-related dominant disease. Am J Med Genet A 149A:861–67; 2009.

88. Kiferle L, Orsucci D, Mancuso M, et al. Twinkle mutation in an Italian family with external progressive ophthalmoplegia and Parkinsonism: a case report and an update on the state of art. Neurosci Lett 556:1–4; 2013.

89. Longley MJ, Clark S, Yu Wai Man C, et al. Mutant POLG2 disrupts DNA polymerase subunits and causes progressive external ophthalmoplegia. Am J Hum Genet 78:1026–34; 2006.

90. Walter MC, Czermin B, Muller-Ziermann S, et al. Late-onset ptosis and myopathy in a patient with a heterozygous insertion in POLG2. J Neurol 257:1517–23; 2010.

91. Young MJ, Longley MJ, Li FY, Kasiviswanathan R, Wong LJ, Copeland WC. Biochemical analysis of human POLG2 variants associated with mitochondrial disease. Hum Mol Genet 20:3052–66; 2011.

92. Craig K, Young MJ, Blakely EL, et al. A p.R369G POLG2 mutation associated with adPEO and multiple mtDNA deletions causes decreased affinity between polymerase g subunits. Mitochondrion 12:313–19; 2012.

93. Hudson G, Amati-Bonneau P, Blakely EL, et al. Mutation of OPA1 causes dominant optic atrophy with external ophthalmoplegia, ataxia, deafness and multiple mitochondrial DNA deletions: a novel disorder of mtDNA maintenance. Brain 131:329–37; 2008.

94. Rouzier C, Bannwarth S, Chaussenot A, et al. The MFN2 gene is responsible for mitochondrial DNA instability and optic atrophy 'plus' phenotype. Brain 135:23–34; 2012.

95. Cassereau J, Codron P, Funalot B. Inherited peripheral neuropathies due to mitochondrial disorders. revue neurologique 17:366–74; 2014.

96. Chan DC. Mitochondrial fusion and fission in mammals. Annu Rev Cell Dev Biol 22:79–99; 2006

97. Rojo M, Legros F, Chateau D, Lombes A. Membrane topology and mitochondrial targeting of mitofusins, ubiquitous mammalian homologs of the transmembrane GTPase Fzo. J Cell Sci 115:1663–74; 2002.

98. Detmer SA, Chan DC. Functions and dysfunctions of mitochondrial dynamics. Nat Rev Mol Cell Biol 8:870–79; 2007.

99. Benard G, Bellance N, James D, et al. Mitochondrial bioenergetics and structural network organization. J Cell Sci 120:838–48; 2007.

100. Caviston JP, Holzbaur EL. Microtubule motors at the intersection of trafficking and transport. Trends Cell Biol 16:530–37; 2006.

101. Hirokawa N, Takemura R. Molecular motors in neuronal development, intracellular transport and diseases. Curr Opin Neurobiol 14:564–73; 2004.

102. Cipolat S, Martins de Brito O, Dal Zilio B, Scorrano L. OPA1 requires mitofusin 1 to promote mitochondrial fusion. Proc Natl Acad Sci U S A 101:15927–32; 2004.

103. Olichon A, Baricault L, Gas N, et al. Loss of OPA1 perturbates the mitochondrial inner membrane structure and integrity, leading to cytochrome c release and apoptosis. J Biol Chem 278:7743–46; 2003.

104. Ingerman E, Perkins EM, Marino M, et al. Dnm1 forms spirals that are structurally tailored to fit mitochondria. J Cell Biol 170:1021–27; 2005.

105. Niemann A, Ruegg M, La Padula V, Schenone A, Suter U. Ganglioside-induced differentiation associated protein 1 is a regulator of the mitochondrial network: new implications for Charcot-Marie-Tooth disease. J Cell Biol 170:1067–78; 2005.

106. Niemann A, Wagner KM, Ruegg M, Suter U. GDAP1 mutations differ in their effects on mitocondrial dynamics and apoptosis depending on the mode of inheritance. Neurobiol Dis 36:509–20; 2009.

107. Shoffner JM, Lott MT, Lezza AM, Seibel P, Ballinger SW, Wallace DC. Myoclonic epilepsy and ragged-red fiber disease (MERRF) is associated with a mitochondrial DNA tRNA(Lys) mutation. Cell 61:931–37; 1990.

108. Fukuhara N, Tokiguchi S, Shirakawa K, Tsubaki T. Myoclonus epilepsy associated with ragged-red fibres (mitochondrial abnormalities): disease entity or a syndrome? Light-and electron-microscopic studies of two cases and review of literature. J Neurol Sci 47:117–33; 1980.

109. Goto Y, Nonaka I, Horai S. A mutation in the tRNA (Leu)(UUR) gene associated with the MELAS subgroup of mitochondrial encephalomyopathies. Nature 348:651–53; 1990.

110. Kobayashi Y, Momoi MY, Tominaga K, et al. A point mutation in the mitocondrial tRNA(Leu)(UUR) gene in MELAS (mitochondrial myopathy, encephalopathy, lactic acidosis and stroke-like episodes). Biochem Biophys Res Commun 173:816–22; 1990.

111. Pavlakis SG, Phillips PC, DiMauro S, De Vivo DC, Rowland LP. Mitochondrial myopathy, encephalopathy, lactic acidosis, and strokelike episodes: a distinctive clinical syndrome. Ann Neurol 16:481–88; 1984.

112. Rusanen H, Majamaa K, Tolonen U, Remes AM, Myllyla R, Hassinen IE. Demyelinating polyneuropathy in a patient with the tRNA(Leu)(UUR) mutation at base pair

3243 of the mitochondrial DNA. Neurology 45:1188–92; 1995.

113. Antonelli FM, Shanske S, Macaya A, DeVivo DC, DiMauro S. The mutation at nt 8993 of mitochondrial DNA is a common cause of Leigh's syndrome. Ann Neurol 34:827–34; 1993.

114. Leigh D. Subacute necrotizing encephalomyelopathy in an infant. J Neurol Neurosurg Psychiatry 14:216–21; 1951.

115. Pitceathly RD, Murphy SM, Cottenie E, et al. Genetic dysfunction of MT-ATP6 causes axonal Charcot-Marie-Tooth disease. Neurology 79:1145–54; 2012.

116. Holt IJ, Harding AE, Petty RK, Morgan-Hughes JA. A new mitochondrial disease associated with mitochondrial DNA heteroplasmy. Am J Hum Genet 46:428–33; 1990.

117. Gelfand JM, Duncan JL, Racine CA, et al. Heterogeneous patterns of tissue injury in NARP syndrome. J Neurol 258:440–48; 2011.

118. Echaniz-Laguna A, Ghezzi D, Chassagne M, et al. SURF1 deficiency causes demyelinating Charcot-Marie-Tooth disease. Neurology 81;17:1523–30; 2013

119. Kennerson ML, Yiu EM, Chuang DT, et al. A new locus for X-linked dominant Charcot-Marie- Tooth disease (CMTX6) is caused by mutations in the pyruvate dehydrogenase kinase isoenzyme 3 (PDK3) gene. Hum Mol Genet 22:1404–16; 2013.

120. Williams LL. Pyruvate oxidation in Charcot-Marie-Tooth disease. Neurology 29:1492–98; 1979.

121. Rinaldi C, Grunseich C, Sevrioukova IF, et al. Cowchock síndrome is associated with a mutation in apoptosis-inducing factor. Am J Hum Genet 91:1095–102; 2012.

122. Nishino I, Spinazzola A, Hirano M. Thymidine phosphorylase gene mutations in MNGIE, a human mitochondrial disorder. Science 283:689–92; 1999.

123. Hirano M, Silvestri G, Blake DM, et al. Mitochondrial neurogastrointestinal encephalomyopathy (MNGIE): clinical, biochemical, and genetic features of an autosomal recessive mitocondrial disorder. Neurology 44:721–27; 1994.

124. Appenzeller O, Kornfeld M, Snyder R. Acromutilating, paralyzing neuropathy with corneal ulceration in Navajo children. Arch Neurol 33:733–38; 1976.

125. Holve S, Hu D, Shub M, Tyson RW, Sokol RJ. Liver disease in Navajo neuropathy. J Pediatr 135:482–93; 1999.

126. Karadimas CL, Vu TH, Holve SA, et al. Navajo neurohepatopathy is caused by a mutation in the MPV17 gene. Am J Hum Genet 79:544–48; 2006

127. Spinazzola A, Viscomi C, Fernandez-Vizarra E, et al. MPV17 encodes an inner mitochondrial membrane protein and is mutated in infantile hepatic mitochondrial DNA depletion. Nat Genet 38:570–75; 2006.

128. Ishihara N, Eura Y, Mihara K. Mitofusin 1 and 2 play distinct roles in mitochondrial fusion reactions via GTPase activity. J Cell Sci 117:6535–46; 2004.

129. Misko A, Jiang S, Wegorzewska I, Milbrandt J, Baloh RH. Mitofusin 2 is necessary for transport of axonal mitochondria and interacts with the Miro/Milton complex. J Neurosci 2010;30:4232–40; 2010.

130. de Brito OM, Scorrano L. Mitofusin 2 tethers endoplasmic reticulum to mitochondria. Nature 456:605–10:2008.

131. Cartoni R, Arnaud E, Medard JJ, et al. Expression of mitofusin 2(R94Q) in a transgenic mouse leads to Charcot-Marie-Tooth neuropathy type 2A. Brain 133:1460–69; 2010.

132. Guillet V, Gueguen N, Cartoni R, et al. Bioenergetic defect associated with mKATP channel opening in a mouse model carrying a mitofusin 2 mutation. FASEB J 25:1618–27; 2011.

133. Baloh RH, Schmidt RE, Pestronk A, Milbrandt J. Altered axonal mitochondrial transport in the pathogenesis of Charcot-Marie-Tooth disease from mitofusin 2 mutations. J Neurosci 27:422–30; 2007.

134. Vallat JM, Ouvrier RA, Pollard JD, et al. Histopathological findings in hereditary motor and sensory neuropathy of axonal type with onset in early childhood associated with mitofusin 2 mutations. J Neuropathol Exp Neurol 67:1097–102; 2008.

135. Verhoeven K, Claeys KG, Zuchner S, et al. MFN2 mutation distribution and genotype/phenotype correlation in Charcot-Marie-Tooth type 2. Brain 129:2093–102; 2006.

138. Zuchner S, Mersiyanova IV, Muglia M, et al. Mutations in the mitocondrial GTPase mitofusin 2 cause Charcot-Marie-Tooth neuropathy type 2A. Nat Genet 36:449–51; 2004.

139. Polke JM, Laura M, Pareyson D, et al. Recessive axonal Charcot-Marie-Tooth disease due to compound heterozygous mitofusin 2 mutations. Neurology 77:168–73; 2011.

140. Nicholson GA, Magdelaine C, Zhu D, et al. Severe early-onset axonal neuropathy with homozygous and compound heterozygous MFN2 mutations. Neurology 70:1678–81; 2008.

141. Chung KW, Kim SB, Park KD, et al. Early onset severe and late-onset mild Charcot-Marie- Tooth disease with mitofusin 2 (MFN2) mutations. Brain 129:2103–18; 2006.

142. Vucic S, Kennerson M, Zhu D, Miedema E, Kok C, Nicholson GA. CMT with pyramidal features. Charcot-Marie-Tooth. Neurology 60:696–99; 2003.

143. Zuchner S, De Jonghe P, Jordanova A, et al. Axonal neuropathy with optic atrophy is caused by mutations in mitofusin 2. Ann Neurol 59:276–81; 2006.

136. Estela A, Pla-Martin D, Sanchez-Piris M, Sesaki H, Palau F. Charcot-Marie-Tooth-related gene GDAP1 complements cell cycle delay at G2/M phase in Saccharomyces cerevisiae fis1 gene-defective cells. J Biol Chem 286:36777–86; 2011.

137. Cassereau J, Chevrollier A, Gueguen N, et al. Mitochondrial complex I deficiency in GDAP1-related autosomal dominant Charcot- Marie-Tooth disease (CMT2K). Neurogenetics 10:145–50; 2009.

144. Noack R, Frede S, Albrecht P, et al. Charcot-Marie-Tooth disease CMT4A: GDAP1 increases cellular glutathione and the mitocondrial membrane potential. Hum Mol Genet 21:150–62; 2012.

145. Cassereau J, Chevrollier A, Gueguen N, et al. Mitochondrial dysfunction and pathophysiology of Charcot-Marie-Tooth disease involving GDAP1 mutations. Exp Neurol 227:31–41; 2011.

146. Ackerley S, James PA, Kalli A, French S, Davies KE, Talbot K. A mutation in the small heat-shock protein HSPB1 leadingto distal hereditary motor neuronopathy disrupts neurofilament assembly and the axonal transport of specific cellular cargoes. Hum Mol Genet 15:347–54; 2006.

147. Evgrafov OV, Mersiyanova I, Irobi J, et al. Mutant small heat-shock protein 27 causes axonal Charcot-Marie-Tooth disease and distal hereditary motor neuropathy. Nat Genet 36:602–06; 2004.

148. Irobi J, Van Impe K, Seeman P, et al. Hot-spot residue in small heat-shock protein 22 causes distal motor neuropathy. Nat Genet 36:597–601; 2004.

149. d'Ydewalle C, Krishnan J, Chiheb DM, et al. HDAC6 inhibitors reverse axonal loss in a mouse model of mutant HSPB1-induced Charcot- Marie-Tooth disease. Nat Med 17:968–74; 2011.

150. Zhao C, Takita J, Tanaka Y, et al. Charcot-Marie-Tooth disease type 2A caused by mutation in a microtubule motor KIF1Bbeta. Cell 105:587–97; 2001.

151. Weedon MN, Hastings R, Caswell R, et al. Exome sequencing identifies a DYNC1H1 mutation in a large pedigree with dominant axonal Charcot-Marie-Tooth disease. Am J Hum Genet 89:308–12; 2011.

152. Chen XJ, Levedakou EN, Millen KJ, Wollmann RL, Soliven B, Popko B. Proprioceptive sensory neuropathy in mice with a mutation in the cytoplasmic Dynein heavy chain 1 gene. J Neurosci 2007;27:14515–24; 2007.

153. Courchesne SL, Pazyra-Murphy MF, Lee DJ, Segal RA. Neuromuscular junction defects in mice with mutation of dynein heavy chain 1. PloS One 6:e16753; 2011.

154. Gómez-Zaera M, Strom TM, Rodríguez B, Estivill X, Meitinger T, Nunes V. Presence of a major WFS1 mutation in Spanish Wolfram syndrome pedigrees. Mol Genet Metab 72:72–81; 2001.

29 Mitochondrial Stress and Cellular Senescence

Irene L. Tan and Michael C. Velarde

(Institute of Biology, College of Science, University of the Philippines Diliman, Quezon City, Philippines)

CONTENTS

1 INTRODUCTION

Normal human fibroblasts have a limited number of proliferation cycles in culture. This finite replicative life span of diploid cells is termed cellular senescence and was first described by Hayflick and Moorhead more than five decades ago (Hayflick and Moorhead 1961). The maximum number of cell divisions for a given cell is also termed as the Hayflick limit. Senescent cells enter a state of permanent cell cycle arrest and they fail to complete DNA synthesis (Di Leonardo et al. 1994; Herbig et al. 2004; Serrano et al. 1997), as evidenced by reduced incorporation of 3H-thymidine (3H-dT) and the thymidine analog, bromodeoxyuridine (BrdU) (Duque and Rakic 2011).

Senescent cells have low expression of proliferation markers, such as proliferating cell nuclear antigen (PCNA) and Ki-67 antigen (Ki-67) (Bologna-Molina et al. 2013). Genes that encode for pro-proliferative proteins, such as replication-dependent histones, c-fos, cyclin A, and cyclin B, are also repressed in the senescence state (Narita et al. 2003; Seshadri and Campisi 1990; Stein et al. 1991). In contrast, cell cycle inhibitors, such as the cyclin-dependent kinase inhibitors (CDKIs) p21 (also CDKN1A or p21Cip1) and p16 (also CDKN2A or p16INK4a) are upregulated in senescent cells. These CDKIs are regulated by the retinoblastoma (RB) and p53 tumor-suppressor pathways.

Despite the permanent state of cell cycle arrest, senescent cells still remain metabolically active (James et al. 2016). They often acquire resistance to apoptosis (Crecenzi, Palumbo, and Brady 2003; Marcotte, Lacelle, and Wang 2004). Hence, it is not surprising that these cells persist in culture and *in vivo* (Sanders et al. 2013). Indeed, the number of senescent cells tends to increase with age (López-Otín et al. 2013). The mechanisms by which senescent cells resist apoptosis are not fully understood. One hypothesis is through down regulation of caspase-3, an important mediator of apoptosis (Marcotte, Lacelle, and Wang 2004).

Structurally, senescent cells differ from proliferating, quiescent, or differentiated cells by their relatively large cell size and flattened morphology (Marquez and Velarde 2017). They are characterized by increased vesicle formation (Sadaie et al. 2015). Senescent cells also have enlarged nuclei and nucleoli (Bemiller and Lee 1978; Mitsui and Schneider 1976; Yoon et al. 2016). Senescent cells also have ultrastructural changes, including age-dependent accumulation of misfolded proteins in endoplasmic reticulum (Kaufman 2002; Nuss et al. 2008), enlarged mitochondria (Goldstein and Korczack 1981), increased number of large lysosomes (Brandes et al. 1972), scattered cytoplasmic granules, and structural alterations of Golgi apparatus (Cho et al. 2011). In addition, senescent cells also have increased abundance of cytoskeletal components, such as microfilaments, microtubules, and intermediate filaments (Wang and Gundersen 1984).

Senescent cells form senescence-associated DNA-damage foci or DNA segments with chromatin alterations reinforcing senescence (DNA-SCARS) which contain DNA damage response (DDR) proteins that form distinct persistent structures (Di Micco et al. 2006; Rodier et al. 2011; Takai, Smogorzewska, and de Lange 2003).When present at telomeres, these persistent foci are termed telomere dysfunction–induced foci (TIF) (Herbig et al. 2004). Persistent DNA damage foci are usually found in aging mice, primate tissues, and tissues subjected to genotoxic insults (Herbig et al. 2006; Rodier et al. 2009).

2 MITOCHONDRIAL DYSFUNCTION ASSOCIATED SENESCENCE (MiDAS)

Mitochondrial stress induces cellular senescence, also referred to as mitochondrial dysfunction-associated senescence (MiDAS) (Wiley et al. 2016). Mitochondrial stress due to oxidative damage decreases the total number of functional mitochondria or impair function of the mitochondrial electron transport chain (ETC), resulting in decreased ATP production and reduced mitochondrial function (Nicolson 2014). Inhibition of mitochondrial electron transport complex I by rotenone causes cellular aging in normal human fibroblasts and primary mouse cells (Miwa et al. 2014; Moiseeva et al. 2009). Inhibition of complex II activity by down-regulating expression of iron-sulfur subunit also promotes premature senescence (Yoon et al. 2003). Likewise, exposure to the complex III inhibitor antimycin A and the complex V inhibitor oligomycin A are linked to cellular senescence, as observed by up-regulation of p16, p21, and p27- CDK inhibitors (CDKIs) (Stöckl et al. 2006). Knockdown of the Rieske iron sulfur protein (RISP), which transfers electrons from ubiquinol to cytochrome c1 in complex III of the ETC, also triggers the senescence phenotype (Moiseeva et al. 2009).

3 MiDAS VERSUS OTHER FORMS OF SENESCENCE

Other than mitochondrial stress, a myriad of stimuli can also induce cellular senescence, including telomere shortening, DNA damage, chromatin perturbations, tumor suppressor activation, oncogenic stimulation, cytotoxic chemical exposure, and oxidative stress (Campisi 2013; Hernandez-Segura, Nehme, and Demaria 2018; Loaiza and Demaria 2016; Marquez and Velarde 2017). Replicative senescence (RS) occurs in cells with dysfunctional telomeres (Victorelli and Passos 2017). Stress-induced senescence (SIS) is caused by stressors or mechanisms other than telomere shortening (DeMagalhães and Passos 2018; Herbig et al. 2004; Ramirez et al. 2001). This may include DNA-damaging agents (e.g., UV, ionizing radiation), chemotherapeutic drugs, and oxidative stress (Hernandez-Segura, Nehme, and Demaria 2018; Toussaint, Medrano, and von Zglinicki 2000). Oncogene-induced senescence (OIS) results from the hyperactivation of mitogenic oncogenes or the inactivation of tumor suppressor genes (Campisi and d'Adda Di Fagagna 2007; Gorgoulis and Halazonetis 2010). Mitogenic oncogenes include RAS (Serrano et al. 1997), RAF (Zhu et al. 1998), MEK (Lin et al. 1998) and BRAF (Michaloglou et al. 2005). Paracrine senescence (PS) occurs when molecules secreted by a primary senescent cell induces senescence in a surrounding non-senescent cell (Acosta et al. 2013). This is also referred to as "bystander senescence" (Hubackova et al. 2012) or senescence-induced senescence (Nelson et al. 2012).

3.1 COMMON FEATURES OF MiDAS WITH OTHER TYPES OF SENESCENCE

All senescent cells, including those induced through MiDAS, generally exhibit common features such as arrested growth, increased levels of cell cycle inhibitors, SA-β-Gal staining, flattened and enlarged morphology, HMGB1 secretion, and lamin B1 decline (Chandeck and Mooi 2010; Serrano et al. 1997; Toussaint, Medrano, and von Zglinicki 2000; Wiley et al. 2016) (Figure 1).

3.1.1 Growth Arrest

One common hallmark of cellular senescence is irreversible growth arrest. Senescent cells enter a state of permanent cell proliferation arrest, usually arrested at the G1or G2 phase of the cell cycle (Di Leonardo et al. 1994; Serrano et al. 1997). Several types of mitochondrial perturbations, such as decreased SIRT3 and SIRT5 expression, ETC inhibition (e.g., rotenone, antimycin A), mitochondrial DNA depletion, and reduced mitochondrial chaperone HSPA can induce senescence growth arrest in human fibroblasts (Wiley et al. 2016). In addition, inhibition of ATP synthase (Stöckl et al. 2006), increased adenosine monophosphate (AMP) levels (Zwerschke et al. 2003), knockdown of malic enzymes (ME1and ME2) (Jiang et al. 2013); and knockdown of

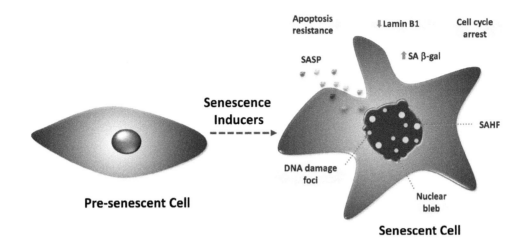

FIGURE 1   Characteristics of senescent cells. Senescent cells exhibit one or more of the following distinct manifestations: permanent growth arrest, apoptosis resistance, increased nuclear size, enlarged cell size, pronounced nuclear blebs, increased DNA damage foci, marked senescence associated heterochromatin foci (SAHF), increased activity of SA β-gal, decreased lamin B1 expression, and the senescence-associated secretory phenotype (SASP).

RISP (Moiseeva et al. 2009) cause cells to undergo cell cycle arrest. Growth arrest is also observed in RS (Hayflick and Moorhead 1961), SIS, OIS (Gorgoulis and Halazonetis 2010), and PS (Acosta et al. 2013; Hubackova et al. 2012; Nelson et al. 2012).

3.1.2 Upregulation of Cyclin-Dependent Kinase Inhibitors (CDKI).

Senescent cells show increased expression of cell cycle inhibitors such as p21 and p16 (Alcorta et al. 1996). Various forms of mitochondrial perturbations trigger high expression of p21 (Jiang et al. 2013; Moiseeva et al. 2009) and p16 (Lee et al. 2012; Wiley et al. 2016). Similarly, RS (Campisi and d'Adda Di Fagagna 2007; Itahana et al. 2003), SIS (Toussaint, Medrano, and von Zglinicki 2000), PS (Acosta et al. 2013; Hubackova et al. 2012), and OIS (Gorgoulis and Halazonetis 2010; Serrano et al. 1997) are characterized by increased expression of these cell cycle inhibitors. p21 is a downstream target of p53 (Elbendary et al. 1996). It binds to cyclin and CDK complexes to inhibit cell cycle progression (Besson, Dowdy, and Roberts 2008). p16 also inhibits CDK activities, particularly CDK 4/6, to prevent RB phosphorylation, leading to G1 cell cycle arrest (Rayess, Wang, and Srivatsan 2012). Inactivation of the p53 pathway reverses growth arrest in senescent cells induced by MiDAS (Wiley et al. 2016) and other senescence pathways, such as RS (Beauséjour et al. 2003) and OIS (Courtois-Cox, Jones, and Cichowski 2008). However, p16/RB inactivation cannot completely reverse the growth arrest phenotype in senescent cells, as it provides a more stringent and dominant, second barrier to cell proliferation (Beauséjour et al. 2003).

3.1.3 Senescence-Associated β-Galactosidase (SA-β-gal) Activity

The endogenous lysosomal β-D-galactosidase, coded by lysosomal-β-galactosidase gene (GLB1), is a lysosomal enzyme optimally active at pH 4.0–4.5 and is overexpressed in senescent cells at a suboptimal pH of 6.0 (Kurz et al. 2000; Lee et al. 2006). Senescent cells induced through MiDAS have increased senescence-associated β-galactosidase (SA-β-gal) activity (Jiang et al. 2013; Lee et al. 2012, 2007; Moiseeva et al. 2009; Park et al. 2010; Stöckl et al. 2006; Wiley et al. 2016; Yoon et al. 2003; Zwerschke et al. 2003), similar to senescent cells induced by DNA-damaging agents such as doxorubicin (Lee et al. 2006), OIS (Bartkova et al. 2006; Gorgoulis and Halazonetis 2010), RS (Itahana et al. 2003; Lee et al. 2006), and PS (Acosta et al. 2013; Hubackova et al. 2012). Increased amount of the enzyme during senescence is associated with increased lysosomal content or activity (Lee et al. 2006).

3.1.4 Flattened and Enlarged Cell Morphology

Senescent cells induced by inhibiting the mitochondrial electron transport chain (rotenone) in mouse embryonic fibroblasts exhibit the characteristic enlarged and flattened cell morphology along with increased nuclear size compared to proliferating cells (Figure 2). This senescent morphology is also observed in cells with reduced mitochondrial fission protein hFis1 and in cells subjected to other types of mitochondrial stresses (Brack et al. 2007; Lee et al. 2012, 2007; Moiseeva et al. 2009; Zwerschke et al. 2003). Senescent cells in culture assume a flattened and enlarged morphology (Muñoz-Espín and Serrano 2014). This senescent phenotype is also shared by RS,

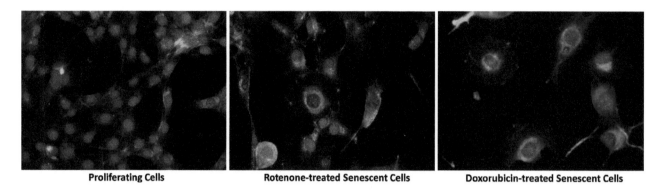

Proliferating Cells   **Rotenone-treated Senescent Cells**   **Doxorubicin-treated Senescent Cells**

FIGURE 2   Morphology of proliferating cells compared to senescent cells. Fluorescence microscopy images of DAPI and phalloidin staining of mouse embryonic fibroblast cells (NIH 3T3) treated with (A) vehicle (control), (B) 1 μM of the mitochondrial ETC complex I inhibitor rotenone, and (C) 300 nM of the DNA topoisomerase inhibitor doxorubicin for 1 day to induce senescence. Fluorescent photomicrographs were taken at 400x magnification. Abbreviation: DAPI, 4', 6-diamidino-2-phenylindole.

SIS (Toussaint, Medrano, and von Zglinicki 2000), OIS (Serrano et al. 1997), and PIS (Acosta et al. 2013).

3.1.5 HMGB1 Secretion

High mobility group box (HMGB1) proteins regulate gene expression by bending DNA to allow transcription factors to bind to promoter regions (Davalos et al. 2013). However, during cellular damage, the HMGB1 protein is released from the nucleus in a p53-dependent manner and activates nuclear factor-κB (NF-κB) and other proinflammatory pathways to signal tissue damage (Davalos et al. 2013; Nacarelli, Liu, and Zhang 2017). MiDAS is associated with secretion of HMGB1 protein similar to the senescent phenotype induced by SIS (Wiley et al. 2016). Cells induced to senesce by exposure to X-ray radiation, overexpression of Ras oncogene, upregulation of the p16INK4a tumor suppressor, or repetitive culture also show HMGB1 secretion (Davalos et al. 2013).

3.1.6 Lamin B1 Loss

Lamin B1 is a protein which comprises part of the nuclear lamina and helps in maintaining proper nuclear structure and function (Camps, Erdos, and Ried 2015). Senescent cells induced through MiDAS have lower expression of lamin B1 relative to dividing and undifferentiated cells (Wiley et al. 2016). Lamin B1 loss is also observed in RS and OIS (Freund et al. 2012). Cells induced to senesce by DNA-damaging agents, such as UV and X-ray irradiation, also exhibit decline in lamin B1 (Freund et al. 2012; Wang et al. 2017). Loss of lamin B1 in senescent cells is linked to the formation of SAHF (Sadaie et al. 2013) and nuclear blebs (Shimi et al. 2011).

Senescence-associated heterochromatin foci (SAHF) are protein-DNA complexes which contain heterochromatin-associated proteins, including heterochromatin protein 1 (HP1) and trimethylated histone 3 at lysine 9 (H3K9me3),

while excluding the euchromatin-associated markers acetylated histone 3 at lysine 9 (H3K9ac) and methylated histone 3 at lysine 4 (H3K4me) (Narita et al. 2003). SAHF is also associated with co-recruitment of heterochromatin proteins and the RB tumor suppressor to E2F-responsive promoters for the stable repression of E2F target genes (Aird et al. 2013; Narita et al. 2003). SAHF formation causes extensive chromatin reorganization in the nucleus that may result in silencing of proliferation-promoting genes (Aird and Zhang 2013). These chromatin alterations form punctate DNA foci called SAHF (Corpet and Stucki 2014; Narita et al. 2003).

Nuclear blebs are hemispherical bulges on the surface of the nucleus formed during cellular senescence (Capell and Collins 2006). Depletion of B-type lamins causes fibers in the lamin meshwork to separate and create blebs (Funkhouser et al. 2013; Shimi et al. 2011). Activation of either the p53 or pRB tumor suppressor pathway is sufficient to reduce nuclear Lamin B1, independent of p38MAPK, NF-κB, ATM, and ROS signaling pathways (Freund et al. 2012).

3.2   Unique Features of MiDAS

Although senescent cells share several features, their phenotypes may also vary depending on the type of senescence trigger (Hernandez-Segura, Nehme, and Demaria 2018). This variability in phenotype is a reflection of the heterogeneous gene expression profiles of different senescence programs (Hernandez-Segura et al. 2017). MiDAS differs from other types of senescence in their secretory profile, NF-kB activity, AMP-activated protein kinase (AMPK) activity, and NAD+/NADH levels (Figure 3).

3.2.1 Secretory Phenotype

The key characteristic of MIDAS is its secretory phenotype (Hernandez-Segura, Nehme, and Demaria 2018). Various types of mitochondrial perturbations can result in a distinct senescent secretory profile referred to as the MiDAS

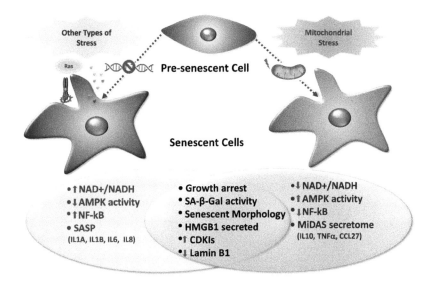

FIGURE 3 Comparison of mitochondrial dysfunction associated-senescence (MiDAS) and other types of senescence. MiDAS and other types of senescence share similarities such as permanent cell cycle arrest, prominent SA β-gal activity, decreased lamin B1, increased levels of CDKI expression, and HMGB1secreted. However, they differ in their secretory profile, AMPK activity, NAD+/NADH ratio, and NF-kB activity. The MiDAS secretome also lacks an IL1/NF-kB-dependent arm.

secretome, which is different from the canonical secretory phenotype caused by SIS or OIS (Wiley et al. 2016). The senescence-associated secretory phenotype (SASP), which is also referred to as senescence messaging secretome (SMS) (Kuilman and Peeper 2009), is a hallmark of many senescent cells (Coppé et al. 2008; Rodier et al. 2009). SASP include various types of cytokines, chemokines, growth factors, and proteases that are responsible for either the harmful or beneficial effects of senescent cells (Coppé et al. 2010; Freund et al. 2010; Tchkonia et al. 2013). The SASP includes several families of soluble and insoluble factors that may act in a paracrine manner, activating cell-surface receptors, and corresponding cell signaling pathways that may lead to multiple pathologies, including cancer. Generally, SASP secretion can be grouped into these categories: soluble signaling factors, secreted proteases, and secreted insoluble proteins/extracellular matrix (ECM) components (Coppé et al. 2010).SASP factors include IL-6, IL-8, monocyte chemoattractant proteins (MCPs), macrophage inflammatory proteins (MIPs), and granulocyte/macrophage colony–stimulating factor (GM-CSF) (Adams 2009; Coppé et al. 2010; Davalos et al. 2010).

The SASP facilitates diverse biological activities, including hampered proliferative ability (Acosta et al. 2013), tissue remodeling (Lujambio et al. 2013; Xue et al. 2007), and wound healing (Lujambio et al. 2013; Xue et al. 2007). Many SASP components directly or indirectly promote inflammation through upregulation of inflammation-associated genes (Shelton et al. 1999). Growth-regulated oncogenes (GROs) promote cell proliferation (Yang et al. 2006), while vascular endothelial growth factor (VEGF) stimulates formation of blood vessels (Coppé et al. 2006). Some SASP components can

induce epithelial-to-mesenchymal transition (Laberge et al. 2012). Others can alter stem cell proliferation or differentiation (Brack et al. 2007; Krtolica et al. 2011). The SASP is also hypothesized to play a role in aging phenotypes and age-related pathologies (Campisi 2013; Coppé et al. 2010; Velarde, Demaria, and Campisi 2013).

RS, OIS, SIS, and PS all produce the SASP (Acosta et al. 2013; Hubackova et al. 2012; Kwon, Corbett, and Louis 2015). However, MiDAS lack secretion of several cytokines belonging to the pro-inflammatory IL1-arm (e.g., IL1a, IL1b, IL6, and IL8); instead, they secrete other factors such as IL10, TNFα, and CCL27 (Wiley et al. 2016). This subset of major pro-inflammatory SASP factors lacking in MiDAS is supposed to be more common and consistent compared with other SASP components (Chien et al. 2011; Coppé et al. 2008; Herranz et al. 2015; Kuilman et al. 2008; Laberge et al. 2015).

NF-κB is a master regulator of the SASP (Chien et al. 2011). It induces expression of the pro-inflammatory IL1-arm of the SASP in RS, OIS, SIS, and PS (Hubackova et al. 2012; Watanabe et al. 2017). Upon activation, the p65 subunit of NFκB translocates into the nucleus and induces expression of the SASP genes (Hubackova et al. 2012). In MiDAS, p53 inhibits NF-kB activity, and consequentially reduces the IL1 arm of the SASP (Wiley et al. 2016).

3.2.2 Signaling Pathways of Senescence

The p53 and p16 pathways are common features of senescent cells and are important for the establishment of the growth arrest phenotype in these cells (Figure 4). These molecular pathways converge on the activation of

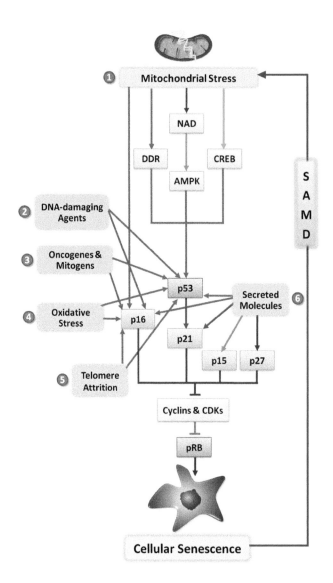

FIGURE 4   Pathways of cellular senescence. Various triggers or stressors (1–6) converge on the activation of CDKIs and RB leading to senescence. p16, p15, p21, and p27 inhibit cyclin and CDK complexes (maroon red arrows), preventing the phosphorylation of the tumor suppressor RB. Mitochondrial stress (1) activates the DDR, AMPK, and CREB, which directly activates p53 (light blue arrow) and its downstream target p21. Mitochondrial stress, which results in MiDAS, also upregulates p16 (pink arrow). DNA-damaging agents (2), oncogenes and mitogens (3), oxidative stress (4), and telomere attrition (5) also activate the p53 and p16 pathways. Secreted molecules (6), which may bring about paracrine senescence, may activate not only p16 or p53 but also upregulate other CDKIs such as p15, p21, or p27 through the SMAD complex. Cellular senescence has been associated with the accumulation of dysfunctional mitochondria or SAMD (bright red arrow).

cyclin-CDK inhibitors and the tumor suppressor RB, culminating in proliferative arrest (Muñoz-Espín and Serrano 2014). The hypophosphorylated form of RB sustains the repression on E2F proliferation-associated genes responsible for the G1 to S transition of the cell cycle (Choi and Anders 2014). However, signaling pathways upstream of these cell cycle regulators vary across the different types of senescence. DNA damaging agents (e.g. ionizing radiation, UV rays, drugs) and telomere loss trigger a DNA-damage response (DDR), which activates p53 and its downstream target p21 in SIS and RS (Moiseeva et al. 2009; Muñoz-Espín and Serrano

2014). Oxidative stress (e.g., ROS) in SIS activates p16 and p53 through the MKK3/MAPKK3 and MKK6/MAPKK6 and through the downstream effector p38 (Sun et al. 2007). In OIS, oncogenes or loss of tumor suppressors activate p53 through the DDR (Moiseeva et al. 2009) or ARF pathway (Efeyan and Serrano 2007). OIS may also de-repress the *CDKN2A* gene to induce p16 expression (Gil and Peters 2006). The *CDKN2A* locus codes for the cell cycle inhibitors p16 and the p53 activator ARF (Sharpless and Sherr 2015). ARF protein or p19 inhibits the p53-degrading activity of MDM2 (Sherr 1998), stabilizing p53 levels. In PS, some components of the SASP can trigger senescence by invoking the DDR pathway (Hubackova et al. 2012; Nelson et al. 2012). TGFβ activation in PS increases nuclear localization of the SMAD complex, which triggers upregulation of cell cycle inhibitors (e.g., p21, p27 and p15) (Acosta et al. 2013; Hubackova et al. 2012; Muñoz-Espín and Serrano 2014).

3.2.3 Senescence Rescue

Inactivation of the p53 and p16 pathways may help prevent or reverse the growth arrest phenotype in senescent cells. However, senescence rescue may vary depending on the nature of senescence induction. For example, growth arrest in OIS cannot be bypassed by expression of the human telomerase reverse transcriptase subunit, unlike that in RS (Bodnar et al. 1998). In MiDAS, pyruvate can prevent the permanent growth arrest in these cells and also rescue the IL-1 arm of the SASP (Wiley et al. 2016).

4 ESTABLISHMENT OF MiDAS

The mechanism for the establishment of MiDAS is still unclear. However, several signaling pathways have been proposed. This includes imbalanced bioenergetics, increased mitochondrial ROS generation, impaired calcium homeostasis, increased mitochondrial fusion, and decreased mitophagy (Kurz et al. 2004; Ziegler, Wiley, and Velarde 2015) (Figure 5).

4.1 IMBALANCED BIOENERGETICS

During oxidative phosphorylation, electrons are passed along the respiratory chain complexes (I, II, III and IV) embedded in the inner mitochondrial membrane. The energy released by this electron transfer is harnessed to pump protons across the membrane, creating an electrochemical gradient that is utilized by ATP synthase (complex V) to produce ATP (Saraste 1999). Hence, impaired maintenance of the electrochemical transmembrane potential of the inner mitochondrial membrane or changes in ETC efficiency leads to poor ATP production and imbalanced bioenergetics (Nicolson 2014; Zwerschke et al.

2003). Reduced ATP levels lead to increased cellular AMP, which can then activate AMP-activated protein kinase (AMPK) and promote AMPK-induced cell cycle arrest (Jones et al. 2005; Mihaylova and Shaw 2011; Moiseeva et al. 2009; Zwerschke et al. 2003). Indeed, AMPK activation has been associated with induction of the senescence phenotype (Humbert et al. 2010; Lee et al. 2012; Mandal et al. 2010; Peyton et al. 2012; Wang et al. 2002; Wang et al. 2003; Wiley et al. 2016; Zwerschke et al. 2003). Activation of AMPK increases p53 phosphorylation and p53, p21, and p27 protein expression, while decreasing RB protein phosphorylation (Peyton et al. 2012). Activated AMPK also downregulates cyclins A, B1, and E (Mandal et al. 2010; Peyton et al. 2012). Moreover, increased AMPK activity in senescent cells reduces cytoplasmic expression of the RNA-binding protein HuR, preventing HuR from stabilizing the expression of mRNAs involved in cell cycle, including cyclin A and cyclin B1 (Wang et al. 2003). Knockdown of genes important in mitochondrial metabolism (e.g. RISP, ME) also turns on AMPK-p53 senescent pathway (Jiang et al. 2013; Moiseeva et al. 2009).

Low ATP production may be associated with low intracellular concentration of the reduced form of nicotinamide adenine dinucleotide (NADH), NAD+, a cofactor required for glycolysis and mitochondrial electron transport chain. Decreased levels of intracellular NAD+ can activate AMPK and promote senescence in a p53-dependent manner (Wiley et al. 2016). In addition, NAD+ is also important for many enzymatic reactions such as DNA repair and protein acetylation (Genova and Lenaz 2014; Ziegler, Wiley, and Velarde 2015). NAD+ associates with poly-ADP ribose polymerase (PARP), which senses DNA damage and facilitates DNA repair (Choi and Mostoslavsky 2014; Fang and Bohr 2017). PARP requires NAD+ to produce poly-ADP ribose needed to recruit DNA repair proteins at the site of genetic lesion (Fang and Bohr 2017). NAD+ is also needed by sirtuin 1 (SIRT1), a deacetylase that regulates genetic stability, cellular metabolism, and longevity (Imai and Guarente 2016; Longo and Kennedy 2006). SIRT1 delays cellular senescence through activation of ERK and S6K1 (Huang et al. 2008) or by silencing p53 through deacetylation

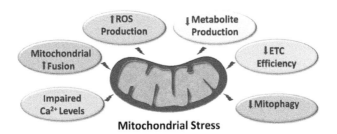

FIGURE 5 **The mechanisms for the establishment of MiDAS.** Perturbations of mitochondrial structure and function include imbalanced bioenergetics, increased mitochondrial ROS generation, impaired calcium homeostasis, increased mitochondrial fusion, and decreased mitophagy.

(Langley et al. 2002). However, SIRT1-overexpression in primary mouse embryo fibroblasts (MEFs) antagonizes promyelocytic leukemia protein (PML)-induced acetylation of p53 and rescues the cells from premature cellular senescence (Langley et al. 2002). Consequently, perturbations in the PARPs and sirtuins contributes to the onset of cellular senescence (Efimova et al. 2010).

NAD+ may also be regulated by controlling malate dehydrogenase activity (MDH). MDH1 is a component of the malate-aspartate shuttle that catalyzes the reduction of oxaloacetate to malate by converting NADH to NAD+ (Lee et al. 2012). Depleting MDH1 in young HDFs and IMR90 human fibroblast reduces NAD+/NADH ratio, inhibits SIRT1, and induces cellular senescence (Lee et al. 2012). Moreover, old human dermal fibroblasts (HDFs) have reduced MDH1 activity and cytosolic NAD+ levels.

4.2 Increased Mitochondrial ROS Generation

Perturbations in complex I activity is proposed to cause premature senescence in part by increased ROS production in mitochondria (Passos et al. 2007; Saretzki, Murphy, and von Zglinicki 2003). Approximately 90% of endogenous ROS is generated through the mitochondrial ETC (Pole, Dimri, and Dimri 2016). Highly reactive intermediates, including superoxide anion radical, singlet oxygen, and hydrogen peroxide (H_2O_2) are produced as oxygen molecules (O_2) undergo a series of reduction steps during oxidative phosphorylation (Genestra 2007; Murphy 2009). ROS can attack molecules such as proteins, carbohydrates, lipids, and nucleic acids in biological systems (Yoshikawa and Naito 2002). On the other hand, antioxidants can attenuate oxidative damage in mitochondria due to ROS (Kelso et al. 2001). Hence, the imbalance between ROS production and antioxidant activity results in oxidative stress in the cell (Davies 1999; Yoshikawa and Naito 2002). Increased levels of mitochondrial-derived ROS can further decrease ETC efficiency, generating additional ROS through positive feedback loop to cause oxidative damage (Balaban, Nemoto, and Finkel 2005).

ROS can accelerate telomere loss and induce DNA damage to invoke a DDR which leads to RS (Chen et al. 1995; Zglinicki 2002; Zglinicki, Pilger, and Sitte 2000). Telomeres are more susceptible to oxidative stress-induced DNA strand breaks than other parts of the genome (Oikawa and Kawanishi 1999; Petersen, Saretzki, and von Zglinicki 1998). These telomeric DNA breaks are also not readily repaired by the DNA repair machinery (Opresko et al. 2005). Dysfunctional telomeres trigger a canonical DDR (D'Adda Di Fagagna et al. 2003; Herbig et al. 2004; Takai, Smogorzewska, and de Lange 2003). The DDR allows cells to detect damaged DNA, especially double-strand breaks (DSBs), and to respond by arresting cell cycle progression and repairing the damage. However, telomere attrition elicits DDR without effective repair, leading to cell cycle arrest (D'Adda Di Fagagna et al. 2003;

Fumagalli et al. 2013; Herbig et al. 2004; Takai, Smogorzewska, and de Lange 2003). In turn, loss of telomeric DNA and telomere-bound factors leads to a DDR capable of inducing senescence (D'Adda Di Fagagna et al. 2003; van Steensel, Smogorzewska, and de Lange 1998). Moreover, downstream effectors of the DDR also creates a positive feedback loop of generating more ROS, which stabilizes the senescent growth arrest phenotype (Macip et al. 2003, 2002; Passos et al. 2010).

The contribution of mitochondrial ROS to cellular senescence is supported by studies on mitochondrial antioxidants. Mitochondria-specific antioxidant mitoQ [10-(6′-ubiquinonyl) decyltriphenylphosphonium bromide] inhibits telomere shortening and extends lifespan in fibroblasts exposed to mild oxidative stress (Saretzki, Murphy, and von Zglinicki 2003). Deficiency in the mitochondrial antioxidant enzyme superoxide dismutase 2 (SOD2) leads to senescence in mouse and human keratinocytes (Velarde et al. 2015, 2012). SOD2 deficiency hampers mitochondrial complex II activity and causes nuclear DNA damage, consequently leading to cellular senescence (Velarde et al. 2012). However, while SOD2 is important in limiting mitochondrial ROS-mediated senescence, excessive amounts of antioxidant proteins (e.g. SOD2 and catalase) do not necessarily inhibit the senescence phenotype in hyperoxia-induced senescent cells (Klimova et al. 2009), suggesting that endogenous levels of antioxidant proteins is sufficient to limit oxidative stress.

The mitochondrial ROS-generating enzyme NADPH oxidase 4 (Nox4) is also implicated in cellular senescence. High expression and activity of Nox4 are observed in senescent lung fibroblasts (Sanders et al. 2015). Nox4 gene is associated with the euchromatin marker acetylated H4 lysine 16 (H4K16Ac) and inversely associated with the heterochromatin marker trimethylated histone 4 at lysine 20 (H4K20Me3) (Sanders et al. 2015). ROS generated from Nox4 is a mediator of OIS (Weyemi et al. 2012).

While ROS induces genetic damage (Parrinello et al. 2003), accelerates telomere attrition (Zglinicki 2002), and acts as signaling molecules to induce cellular senescence (Passos et al. 2010), the nature and concentration of ROS determines the extent of whether the ROS generated is beneficial or harmful to the cell. Exposure to low concentrations of H_2O_2 may be beneficial because of its mitogenic effect on cells, while increasing levels of ROS cause oxidative stress, which triggers cells to enter a permanent cell-cycle arrest (Davies 1999). Hence, the contribution of oxidative stress to the establishment of cellular senescence may be more complex and may be dependent on context.

4.3 Impaired Calcium Homeostasis

Mitochondria sequester calcium ions to regulate oxidative phosphorylation, to induce the mitochondrial permeability transition (MPT) related to apoptosis, and to maintain mitochondrial and cellular calcium homeostasis (Gunter et al. 2004). Increased Ca^{2+} uptake depolarizes the mitochondrial membrane, causing a temporary decline in ATP production (Nguyen and Jafri 2005). Increased calcium loading in mitochondria also elevates cytosolic NADH levels and reduces sirtuin activity (Marcu et al. 2014), which can cause premature cell cycle arrest (Bolinches-Amoros et al. 2014; Ziegler, Wiley, and Velarde 2015). Conversely, depletion of the calcium channels inositol 1,4,5-trisphosphate receptor, type 2 (ITPR2) and mitochondrial calcium uniporter (MCU) allows escape from OIS (Wiel et al. 2014). In response to stress, mitochondria can also release stored calcium ions into the cytosol and trigger a retrograde response signaling to the nucleus and activate specific nuclear transcription factors (Butow and Avadhani 2004). For example, high intracellular calcium concentration results in phosphorylation of cAMP-responsive element-binding protein (CREB), increasing p53 recruitment to promoters, elevating p21 expression, and inhibiting cell proliferation (Arnould et al. 2002).

4.4 INCREASED MITOCHONDRIAL FUSION

Mitochondrial dynamics is fundamental for the maintenance of cellular homeostasis (Twig and Shirihai 2011). Mitochondrial dynamics refers to the ability of mitochondria to change their size and shape depending on the balance between the contrasting processes of fusion and fission. The morphology and functional status of mitochondria is affected by these dynamic processes in that an imbalance in fission and fusion events leads to fragmentation and elongation, respectively (Chan 2006). Dynamin 1-like (DNM1L or DRP1) and fission 1 (FIS1) are the molecular modulators of mitochondrial fission, whereas optic atrophy 1 (OPA1) and mitofusin 1 and 2 (Mfn1 and Mfn2) regulate mitochondrial fusion (Yue and Yao 2016).

Establishment of senescence is linked to a preponderance of fusion over fission events, resulting in abnormally enlarged mitochondria (Lee et al. 2007; Park et al. 2010; Yoon et al. 2006). Inhibiting fission by downregulating FIS1 protein results in mitochondrial elongation and senescence (Yoon et al. 2006). Cells with short hairpin RNA (shRNA) knockdowns of FIS1 protein, which recruits the pro-fission DNM1L or DRP1, harbor elongated mitochondria and exhibited the senescent phenotype (Lee et al. 2007). Reducing the expression levels of the membrane-associated ring finger C3HC4 5 (MARCH5), a mitochondrial E3 ubiquitin ligase which blocks DRP1 promotes mitochondrial elongation and senescence (Park et al. 2010). Sustained mitochondrial elongation decreases mitochondrial membrane potential, increases ROS generation, promotes DNA damage, and induces senescence (Lee et al. 2007; Yoon et al. 2006). Conversely, increasing mitochondrial fission prevents senescence caused by mitochondrial elongation (Lee et al. 2007). However, the reason why mitochondrial elongation occurs in senescent cells or how mitochondrial fusion contributes to the permanent cell growth arrest phenotype is not yet fully understood.

4.5 DECREASED MITOPHAGY

Mitophagy refers to the selective removal of mitochondria by autophagic degradation (Korolchuk et al. 2017). At the molecular level, mitophagy is controlled by Parkin, an E3 ubiquitin ligase, and PTEN-induced putative kinase protein 1 (PINK1) (Youle and Narendra 2011). Reduced mitochondrial membrane potential triggers the recruitment of PINK1 to the mitochondrial outer membrane and the subsequent phosphorylation of Parkin from the cytosol (Springer and Kahle 2011). Phosphorylated Parkin (Ser65) promotes the ubiquitination of Mfn1 and Mfn2 curtailing mitochondrial fusion. Consequently, damaged mitochondria are incorporated into autophagosomes, which eventually merge with lysosomes, resulting in their degradation (Narendra et al. 2008; Youle and Narendra 2011). PINK1 and Parkin, therefore, identify and label damaged mitochondria for autophagic degradation (Springer and Kahle 2011). Reduction of mitophagy is observed in senescent cells both *in vitro* and *in vivo* (Dalle Pezze et al. 2014; García-Prat et al. 2016). Cytoplasmic p53 accumulation interacts with Parkin, preventing its translocation to damaged mitochondria, thereby suppressing mitophagy (Ahmad et al. 2015). Reduced expression of PINK1 also impairs mitophagy and suppresses mitochondrial fission (Bueno et al. 2015; Dalle Pezze et al. 2014). However, it remains unclear whether reduced mitophagy plays a major role in the establishment of cellular senescence.

5 SENESCENCE-ASSOCIATED MITOCHONDRIAL DYSFUNCTION (SAMD)

Mitochondria play a role in the establishment of cellular senescence (Correia-Melo et al. 2016). Transcriptome analyses reveal that several senescence phenotypes, including the SASP, are dependent on mitochondrial function – genes that are usually downregulated in senescence become upregulated upon depletion of mitochondria, while genes upregulated in senescence become downregulated in mitochondrial-depleted senescent cells (Correia-Melo et al. 2016). Senescent cells induced by RS (Passos et al. 2007), OIS (Moiseeva et al. 2009), and SIS also have increased mitochondrial mass (Passos et al. 2010; Tai et al. 2017). Moreover, when cells are depleted of mitochondria through Parkin-mediated mitochondrial clearance, general reduction in senescence-associated phenotypes such as cell size, SA β-gal activity, heterochromatin foci, ROS production, and p21 and p16 expression are observed (Correia-Melo et al. 2016). Hence, mitochondrial function is thought to be important in cellular senescence.

Mitochondrial stress is a general feature of senescent cells (Allen et al. 1999; Hutter et al. 2004; Moiseeva et al. 2009; Passos et al. 2007; Zwerschke et al. 2003). This stress response, also referred as SAMD (Figure 5), is characterized by decreased efficiency in oxidative phosphorylation and increased production of mitochondrial-derived ROS (Korolchuk et al. 2017). Accumulation of dysfunctional mitochondria occurs in fibroblasts, senescent epithelial cells, enterocytes, and neurons (Jurk et al. 2012, 2014; Ohtani and Hara 2013; Wang et al. 2009). However, how mitochondria become dysfunctional in senescence still remains to be elucidated. One hypothesis is that mitophagy, which is responsible for mitochondrial turnover, is reduced in senescent cells and results in a SAMD phenotype, as decreased mitophagy prevents clearance of dysfunctional mitochondria (Dalle Pezze et al. 2014).

The SAMD, SASP, and cellular aging do not necessarily conform to a linear cause-effect relationship (Korolchuk et al. 2017). It is unclear whether mitochondrial stress is initially a cause or a consequence of cellular senescence (Correia-Melo and Passos 2015; Korolchuk et al. 2017; Passos et al. 2006). There is evidence that multiple positive feedback loops exist among these components of the senescent phenotype: between SASP factors and DDR (Acosta et al. 2008; Kuilman et al. 2008), between SAMD and DDR (Passos et al. 2010), and between SASP and SAMD (Korolchuk et al. 2017). Indeed, this perspective highlights the complexity and flexibility of the cellular senescence process. Whether mitochondrial stress serves as a trigger or a feature of senescence or both, remains to be elucidated. Hence, the resolution of this dilemma between mitochondrial stress and cellular senescence will be an impetus for lively research in the years ahead.

6   CONCLUSION

Mitochondrial stress is described here as an inducer of cellular senescence. Cellular senescence, which is primarily characterized by permanent cell cycle arrest just like mitochondrial stress, has been associated with aging and age-related disorders. While elevated levels of mitochondrial-derived ROS is a major contributor to senescence, other stresses such as bioenergetics imbalance, disturbances in mitochondrial homeostasis, impaired levels of mitochondrial metabolites, and perturbations in the ETC have also been linked to cellular senescence. However, the role of mitochondrial stress in cellular senescence might be more complex than originally thought, as conditions such as MiDAS and SAMD have been described. Hence, a better understanding of the pathways that link the various forms of mitochondrial stressors with cellular senescence need to be further elucidated.

ACKNOWLEDGMENTS

Some of the photomicrographs were generated from a project funded by the Department of Science and Technology, Philippine Council for Health Research and Development (DOST-PCHRD FPI60028, MCV).

REFERENCES

Acosta, Juan Carlos, Ana Banito, Torsten Wuestefeld, Athena Georgilis, Peggy Janich, Jennifer P. Morton, Dimitris Athineos, et al. 2013. "A Complex Secretory Program Orchestrated by the Inflammasome Controls Paracrine Senescence." *Nature Cell Biology* 15 (8): 978–90. doi: 10.1038/ncb2784.

Acosta, Juan Carlos, Ana O'Loghlen, Ana Banito, Maria V Guijarro, Arnaud Augert, Selina Raguz, Marzia Fumagalli, et al. 2008. "Chemokine Signaling via the CXCR2 Receptor Reinforces Senescence." *Cell* 133 (6): 1006–18. doi: 10.1016/j.cell.2008.03.038.

Adams, Peter D. 2009. "Healing and Hurting: Molecular Mechanisms, Functions, and Pathologies of Cellular Senescence." *Molecular Cell* 36 (1): 2–14. doi:10.1016/j.molcel.2009.09.021.

Ahmad, Tanveer, Isaac K. Sundar, Chad A. Lerner, Janice Gerloff, Ana M. Tormos, Hongwei Yao, and Irfan Rahman. 2015. "Impaired Mitophagy Leads to Cigarette Smoke Stress-Induced Cellular Senescence: Implications for Chronic Obstructive Pulmonary Disease." *FASEB Journal* 29 (7): 2912–29. doi:10.1096/fj.14-268276.

Aird, Katherine M., Gao Zhang, Hua Li, Zhigang Tu, Benjamin G. Bitler, Azat Garipov, Hong Wu, et al. 2013. "Suppression of Nucleotide Metabolism Underlies the Establishment and Maintenance of Oncogene-Induced Senescence." *Cell Reports* 3 (4): 1252–65. doi: 10.1016/j.celrep.2013.03.004.

Aird, Katherine M, and Rugang Zhang. 2013. "Detection of Senescence-Associated Heterochromatin Foci (SAHF)." *Methods in Molecular Biology* 965: 185–96. doi: 10.1007/978-1-62703-239-1_12.

Alcorta, David A, Yue Xiong, Dawn Phelps, Greg Hannon, David Beach, and J Carl Barrett. 1996. "Involvement of the Cyclin-Dependent Kinase Inhibitor P16 (INK4a) in Replicative Senescence of Normal Human Fibroblasts." *Biochemistry* 93: 13742–47.

Allen, Robert C., Maria Tresini, Bart P. Keogh, David L. Doggett, Vincent J. Cristofalo. 1999. "Differences in Electron Transport Potential, Antioxidant Defenses, and Oxidant Generation in Young and Senescent Fetal Lung Fibroblasts (WI-38)." *Journal of Cellular Physiology* 180 (1): 114–22. doi: 10.1002/(SICI)1097-4652(199907)180:1<114::AID-JCP13>3.0.CO;2-0.

Arnould, Thierry T, Sébastien SVankoningsloo, Patricia Renard, Andrée AHoubion, Noelle NNinane, Catherine CDemazy, Jose Remacle, and Martine Raes. 2002. "CREB Activation Induced by Mitochondrial Dysfunction is a New Signaling Pathway that Impairs Cell Proliferation." *The EMBO Journal* 21 (1–2): 53–63.

Balaban, Robert S., Shino Nemoto, and Toren Finkel. 2005. "Mitochondria, Oxidants, and Aging." *Cell* 120 (4): 483–95. doi:10.1016/J.CELL.2005.02.001.

Bartkova, Jirina, Nousin Rezaei, Michalis Liontos, Panagiotis Karakaidos, Dimitris Kletsas, Natalia Issaeva, Leandros Vassilios F. Vassiliou, et al. 2006. "Oncogene-Induced Senescence Is Part of the Tumorigenesis Barrier Imposed by DNA Damage Checkpoints." *Nature* 444 (7119): 633–7. doi: 10.1038/nature05268.

Beauséjour, Christian M., Ana Krtolica, Francesco Galimi, Masashi Narita, Scott W. Lowe, Paul Yaswen, and Judith Campisi. 2003. "Reversal of Human Cellular Senescence: Roles of the P53 and P16 Pathways." *EMBO Journal* 22 (16): 4212–2. doi:10.1093/emboj/cdg417.

Bemiller, Paraskevi M., and Lieh Hau Lee. 1978. "Nucleolar Changes in Senescing WI-38 Cells." *Mechanisms of Ageing and Development* 8 (6): 417–27.

Besson, Arnaud, Steven F. Dowdy, and James M. Roberts. 2008. "CDK Inhibitors: Cell Cycle Regulators and Beyond." *Developmental Cell* 14 (2): 159–69. doi:10.1016/j.devcel.2008.01.013.

Bodnar, Andrea G., Michel Ouellette, Maria Frolkis, Shawn E. Holt, Choy-Pik Chiu, Gregg B. Morin, Calvin B. Harley, et al. 1998. "Extension of Life-Span by Introduction of Telomerase into Normal Human Cells." *Science* 279 (5349): 349–52.

Bolinches-Amoros, Arantxa, Bela Molla, David Pla-Martin, Francesc Palau, and Pilar Gonzalez-Cabo. 2014. "Mitochondrial Dysfunction Induced by Frataxin Deficiency is Associated with Cellular Senescence and Abnormal Calcium Metabolism." *Frontiers in Cellular Neuroscience* 8: 124. doi:10.3389/fncel.2014.00124.

Bologna-Molina, Ronell, Adalberto Mosqueda-Taylor, Nelly Molina-Frechero, Ana-Dolores Mori-Estevez, and Guillermo Sánchez-Acuña. 2013. "Comparison of the Value of PCNA and Ki-67 as Markers of Cell Proliferation in Ameloblastic Tumors." *Medicina Oral, Patologia Oral y Cirugia Bucal* 18 (2): 174–9. doi:10.4317/MEDORAL.18573.

Brack, Andrew S., Michael J. Conboy, Sudeep Roy, Mark Lee, Calvin J. Kuo, Charles Keller, and Thomas A. Rando. 2007. "Increased Wnt Signaling During Aging Alters Muscle Stem Cell Fate." *Science* 317: 807–10. doi:10.1126/science.1144090.

Brandes, David, Donald G. Murphy, Elsa B. Anton, and Sandra Barnard. 1972. "Ultrastructural and Cytochemical Changes in Cultured Human Lung Cells." *Journal of Ultrasructure Research* 39 (5–6): 465–83. doi:10.1016/S0022-5320(72)90114-1.

Bueno, Marta, Yen-chun Lai, Yair Romero, Judith Brands, and Claudette Stcroix. 2015. "PINK1 Deficiency Impairs Mitochondrial Homeostasis Promoting Lung Fibrosis." *The Journal of Clinical Investigation* 125 (2): 521–38. doi:10.1172/JCI74942.).

Butow, Ronald A., and Narayan G. Avadhani. 2004. "Mitochondrial Signaling : The Retrograde Response." *Molecular Cell* 14: 1–15.

Campisi, Judith. 2013. "Aging, Cellular Senescence, and Cancer." *Annual Review of Physiology* 75: 685–705. doi:10.1146/annurev-physiol-030212-183653.

Campisi, Judith, and Fabrizio d'Adda Di Fagagna. 2007. "Cellular Senescence: When Bad Things Happen to Good Cells." *Nature Reviews Molecular Cell Biology* 8: 729–40. doi:10.1038/nrm2233.

Camps, Jordi, Michael R. Erdos, and Thomas Ried. 2015. "The Role of Lamin Bl for the Maintenance of Nuclear Structure and Function." *Nucleus* 6 (1): 8–14. doi:10.1080/19491034.2014.1003510.

Capell, Brian C., and Francis S. Collins. 2006. "Human Laminopathies: Nuclei Gone Genetically Awry." *Nature Reviews Genetics* 7 (12): 940–52. doi:10.1038/nrg1906.

Chan, David C. 2006. "Mitochondrial Fusion and Fission in Mammals." *Annual Review of Cell and Developmental Biology* 22 (1): 79–99. doi:10.1146/annurev.cellbio.22.010305.104638.

Chandeck, Charlotte, and Wolter J. Mooi. 2010. "Oncogene-Induced Cellular Senescence." *Advances in Anatomic Pathology* 17 (1): 42–8. doi:10.1097/PAP.0b013e3181c66f4e.

Chen, Qin, Ann Fischer, Joshua D. Reagan, Liang-Jun Yan, Bruce N. Ames, and Barker Hall. 1995. "Oxidative DNA Damage and Senescence of Human Diploid Fibroblast Cells (8-Oxoguanine/Protein Oxidation/Oxygen Tension/a-Phenyl-t-Butyl Nitrone/Replicative Life Span)." *Cell Biology* 92: 4337–41.

Chien, Yuchen, Claudio Scuoppo, Xiaowo Wang, Xueping Fang, Brian Balgley, Jessica E Bolden, Prem Premsrirut, et al. 2011. "Control of the Senescence-Associated Secretory Phenotype by NF-KB Promotes Senescence and Enhances Chemosensitivity." *Genes & Development* 25 (20): 2125–36. doi: 10.1101/gad.17276711.

Cho, Joon-Ho, Deepak Kumar Saini, Ajith Karunarathne, Vani Kalyanaraman, and Narendra Gautam. 2011. "Alteration of Golgi Structure in Senescent Cells and Its Regulation by a G Protein γ Subunit." *Cellular Signalling* 23 (5): 785–93. doi:10.1016/j.cellsig.2011.01.001.

Choi, Jee-Eun, and Raul Mostoslavsky. 2014. "Sirtuins, Metabolism, and DNA Repair." *Current Opinion in Genetics & Development* 26: 24–32. doi:10.1016/j.gde.2014.05.005.

Choi, Yoon J., and Lars Anders. 2014. "Signaling through Cyclin D-Dependent Kinases." *Oncogene* 33 (15): 1890–903. doi:10.1038/onc.2013.137.

Coppé, Jean-Philippe, Pierre-Yves Desprez, Ana Krtolica, and Judith Campisi. 2010. "The Senescence-Associated Secretory Phenotype: The Dark Side of Tumor Suppression." *Annual Review of Pathology* 5: 99–118. doi:10.1146/annurev-pathol-121808-102144.

Coppé, Jean-Philippe, Katalin Kauser, Judith Campisi, and Christian M. Beauséjour. 2006. "Secretion of Vascular Endothelial Growth Factor by Primary Human Fibroblasts at Senescence." *Journal of Biological Chemistry* 281 (40): 29568–74. doi:10.1074/jbc.M603307200.

Coppé, Jean-Philippe, Christopher K Patil, Francis Rodier, Yu Sun, Denise P Muñoz, Joshua Goldstein, Peter S Nelson, et al. 2008. "Senescence-Associated Secretory Phenotypes Reveal Cell-Nonautonomous Functions of Oncogenic RAS and the P53 Tumor Suppressor." *PLoS Biology* 6 (12): 2853–68. doi:10.1371/journal.pbio.0060301.

Corpet, Armelle, and Manuel Stucki. 2014. "Chromatin Maintenance and Dynamics in Senescence: A Spotlight on SAHF Formation and the Epigenome of Senescent Cells." *Chromosoma* 123 (5): 423–36. doi:10.1007/s00412-014-0469-6.

Correia-Melo, Clara, Francisco D.M. Marques, Rhys Anderson, Graeme Hewitt, Rachael Hewitt, John Cole, Bernadette M. Carroll, et al. 2016. "Mitochondria Are Required for Pro-Ageing Features of the Senescent Phenotype." *The EMBO Journal* 35 (7): 724–42. doi: 10.15252/embj.201592862.

Correia-Melo, Clara, and João F. Passos. 2015. "Mitochondria: Are They Causal Players in Cellular Senescence?" *Biochimica et Biophysica Acta – Bioenergetics* 1847: 1373–9. doi:10.1016/j.bbabio.2015.05.017.

Courtois-Cox, Stéphanie, Stephen L. Jones, and Karen Cichowski. 2008. "Many Roads Lead to Oncogene-Induced Senescence." *Oncogene* 27: 2801–9. doi:10.1038/sj.onc.1210950.

Crecenzi, Elvira, Giuseppe Palumbo, and Hugh J. M. Brady. 2003. "Bcl-2 Activates a Programme of Premature Senescence in Human Carcinoma Cells." *Biochemical Journal* 375 (2): 263–274. doi:10.1042/bj20030868.

D'Adda Di Fagagna, Fabrizio, Philip M. Reaper, Lorena Clay-Farrace, Heike Fiegler, Philippa Carr, Thomas Von Zglinicki, Gabriele Saretzki, et al. 2003. "A DNA Damage Checkpoint Response in Telomere-Initiated Senescence." *Nature* 426 (6963): 194–98. doi:10.1038/nature02118.

Davalos, Albert R., Jean Philippe Coppe, Judith Campisi, and Pierre Yves Desprez. 2010. "Senescent Cells as a Source of

Inflammatory Factors for Tumor Progression." *Cancer and Metastasis Reviews* 29 (2): 273–83. doi:10.1007/s10555-010-9220-9.

Davalos, Albert R., Misako Kawahara, Gautam K. Malhotra, Nicholas Schaum, Jiahao Huang, Urvi Ved, Christian M. Beausejour, et al. 2013. "P53-Dependent Release of Alarmin HMGB1 Is a Central Mediator of Senescent Phenotypes." *Journal of Cell Biology* 201 (4): 613–29. doi:10.1083/jcb.201206006.

Davies, Kelvin. 1999. "The Broad Spectrum of Responses to Oxidants in Proliferating Cells: A New Paradigm for Oxidative Stress." *IUBMB Life* 48 (1): 41–7. doi:10.1080/713803463.

DeMagalhães, João Pedro, and João F. Passos. 2018. "Stress, Cell Senescence and Organismal Ageing." *Mechanisms of Ageing and Development* 170: 2–9. doi:10.1016/j.mad.2017.07.001.

Di Leonardo, Aldo, Steven P Linke, Kris Clarkin, and Geoffrey M Wahl. 1994. "DNA Damage Triggers a Prolonged P53- Dependent G1 Arrest and a Long-Term Induction of Cip1 in Normal Human Fibroblasts." *Genes & Development* 8: 2540–51.

Di Micco, Raffaella, Marzia Fumagalli, Angelo Cicalese, Sara Piccinin, Patrizia Gasparini, Chiara Luise, Catherine Schurra, et al. 2006. "Oncogene-Induced Senescence is a DNA Damage Response Triggered by DNA Hyper-Replication." *Nature* 444 (7119): 638–42. doi:10.1038/nature05327.

Duque, Alvaro, and Pasko Rakic. 2011. "Different Effects of Bromodeoxyuridine and [3H]Thymidine Incorporation into DNA on Cell Proliferation, Position, and Fate." *Journal of Neuroscience* 31 (42): 15205–17. doi:10.1523/JNEUROSCI.3092-11.2011.

Efeyan, Alejo, and Manuel Serrano. 2007. "P53: Guardian of the Genome and Policeman of the Oncogenes." *Cell Cycle* 6 (9): 1006–10. doi:10.4161/cc.6.9.4211.

Efimova, Elena V., Helena J. Mauceri, Daniel W. Golden, Edwardine Labay, Vytautas P. Bindokas, Thomas E. Darga, Chaitali Chakraborty, et al. 2010. "Poly (ADP-Ribose) Polymerase Inhibitor Induces Accelerated Senescence in Irradiated Breast Cancer Cells and Tumors." *Cancer Research* 70 (15): 6277–82. doi: 10.1158/0008-5472.CAN-09-4224.

Elbendary, Alaa A., Frank F.D.Cirisano, A.rthur C.Evans, P. enelopeL.Davis, J.Dirk .Iglehart, JJeffrey . R.Marks, and Andrew Berchuck. 1996. "Relationship between P21 Expression and Mutation of the P53 Tumor Suppressor Gene in Normal and Malignant Ovarian Epithelial Cells." *Clinical Cancer Research* 2 (9): 1571–5.

Evangelou, Konstantinos, Nikolaos Lougiakis, Sophia V. Rizou, Athanassios Kotsinas, Dimitris Kletsas, Daniel Muñoz-Espín, Nikolaos G. Kastrinakis, Nicole Pouli, Panagiotis Marakos, Paul Townsend, Manuel Serrano, Jiri Bartek, and Vassilis G. Gorgoulis. 2017. "Robust, Universal Biomarker Assay to Detect Senescent Cells in Biological Specimens." *Aging Cell* 16 (1): 192–97. doi:10.1111/acel.12545.

Fang, Evandro F, and Vilhelm A. Bohr. 2017. "NAD : The Convergence of DNA Repair and Mitophagy." *Autophagy* 13 (2): 442–3. doi:10.1080/15548627.2016.1257467.

Freund, Adam, Remi-Martin Laberge, Marco Demaria, and Judith Campisi. 2012. "Lamin B1 Loss Is a Senescence-Associated Biomarker." *Molecular Biology of the Cell* 23 (11): 2066–75. doi:10.1091/mbc.E11-10-0884.

Freund, Adam, Arturo V. Orjalo, Pierre-Yves Desprez, and Judith Campisi. 2010. "Inflammatory Networks during

Cellular Senescence: Causes and Consequences." *Trends in Molecular Medicine* 16 (5): 238–46. doi:10.1016/j.molmed.2010.03.003.

Fumagalli, Marzia, Francesca Rossiello, Michela Clerici, Sara Barozzi, Davide Cittaro, Jessica M. Kaplunov, Gabriele Bucci, et al. 2013. "Telomeric DNA Damage Is Irreparable and Causes Persistent DNA Damage Response Activation." *Nature Cell Biology* 14 (4): 355–65. doi:10.1038/ncb2466.

Funkhouser, Chloe M., Rastko Sknepnek, Takeshi Shimi, Anne E. Goldman, Robert D. Goldman, and Monica Olvera de la Cruz. 2013. "Mechanical Model of Blebbing in Nuclear Lamin Meshworks." *Proceedings of the National Academy of Sciences* 110 (9): 3248–53. doi:10.1073/pnas.1300215110.

García-Prat, Laura, Marta Martínez-Vicente, Eusebio Perdiguero, Laura Ortet, Javier Rodríguez-Ubreva, Elena Rebollo, Vanessa Ruiz-Bonilla, et al. 2016. "Autophagy Maintains Stemness by Preventing Senescence." *Nature* 529 (7584): 37–42. doi: 10.1038/nature16187.

Genestra, Marcelo. 2007. "Oxyl Radicals, Redox-Sensitive Signalling Cascades and Antioxidants." *Cellular Signalling* 19 (9): 1807–19. doi:10.1016/j.cellsig.2007.04.009.

Genova, Maria Luisa, and Giorgio Lenaz. 2014. "Functional Role of Mitochondrial Respiratory Supercomplexes." *Biochimica et Biophysica Acta – Bioenergetics.* doi:10.1016/j.bbabio.2013.11.002.

Gil, Jesús, and Gordon Peters. 2006. "Regulation of the INK4b–ARF–INK4a Tumour Suppressor Locus: All for One or One for All." *Nature Reviews Molecular Cell Biology* 7 (9): 667–77. doi:10.1038/nrm1987.

Goldstein, Samuel, and Louise B. Korczack. 1981. "Status of Mitochondria in Living Human Fibroblasts during Growth and Senescence in Vitro: Use of the Laser Dye Rhodamine 123." *The Journal of Cell Biology* 91: 392–98. doi:10.1083/JCB.91.2.392.

Gorgoulis, Vassilis G., and Thanos D. Halazonetis. 2010. "Oncogene-Induced Senescence: The Bright and Dark Side of the Response." *Current Opinion in Cell Biology* 22 (6): 816–27. doi:10.1016/J.CEB.2010.07.013.

Gunter, Thomas E., David I. Yule, Karlene K. Gunter, Roman A. Eliseev, and Jason D. Salter. 2004. "Calcium and Mitochondria." *FEBS Letters* 567 (1): 96–102. doi:10.1016/j.febslet.2004.03.071.

Herbig, U., Wendy A. Jobling, Benjamin P.C. Chen, David J. Chen, and John M. Sedivy. 2004. "Telomere Shortening Triggers Senescence of Human Cells through a Pathway Involving ATM, P53, and P21 CIP1, but Not P16 INK4a Essential for t-Loop Formation, and Overexpression of a Dominant-Negative Variant Results in Telomere Un- Capping and Loss of The." *Molecular Cell* 14: 501–13.

Herbig, Utz, Mark Ferreira, Laura Condel, Dee Carey, and John M. Sedivy. 2006. "Cellular Senescence in Aging Primates." *Science* 311 (5765): 1257. doi:10.1126/science.1122446.

Hernandez-Segura, Alejandra, Tristan V. de Jong, Simon Melov, Victor Guryev, Judith Campisi, and Marco Demaria. 2017. "Unmasking Transcriptional Heterogeneity in Senescent Cells." *Current Biology* 27 (17): 2652–60.e4. doi:10.1016/j.cub.2017.07.033.

Hernandez-Segura, Alejandra, Jamil Nehme, and Marco Demaria. 2018. "Hallmarks of Cellular Senescence." *Trends in Cell Biology* 28 (6): 436–53. doi:10.1016/j.tcb.2018.02.001.

Herranz, Nicolás, Suchira Gallage, Massimiliano Mellone, Torsten Wuestefeld, Sabrina Klotz, Christopher J. Hanley, Selina Raguz, et al. 2015. "MTOR Regulates MAPKAPK2

Translation to Control the Senescence-Associated Secretory Phenotype." *Nature Cell Biology* 17 (9): 1205–17. doi:10.1038/ncb3225.

Huang, Jing, Qini Gan, Limin Han, Jian Li, Hai Zhang, Ying Sun, Zongyu Zhang, and Tanjun Tong. 2008. "SIRT1 Overexpression Antagonizes Cellular Senescence with Activated ERK/S6k1 Signaling in Human Diploid Fibroblasts." *PLoS One* 3 (3): 1–9. doi:10.1371/journal.pone.0001710.

Hubackova, Sona, Katerina Krejcikova, Jiri Bartek, and Zdenek Hodny. 2012. "IL1-and TGFβ-Nox4 Signaling, Oxidative Stress and DNA Damage Response Are Shared Features of Replicative, Oncogene-Induced, and Drug-Induced Paracrine 'Bystander Senescence.'." *Aging* 4 (12): 932–51. doi:10.18632/aging.100520.

Humbert, Nicolas, Naveenan Navaratnam, Arnaud Augert, Marco Da Costa, Sébastien Martien, Jing Wang, Dolores Martinez, et al. 2010. "Regulation of Ploidy and Senescence by the AMPK-Related Kinase NUAK1." *The EMBO Journal* 29 (2): 376–86. doi: 10.1038/emboj.2009.342.

Hutter, Eveline, Kathrin Renner, Gerald Pfister, Petra Stöckl, Pidder Jansen-Dürr, and Erich Gnaiger. 2004. "Senescence-Associated Changes in Respiration and Oxidative Phosphorylation in Primary Human Fibroblasts." *The Biochemical Journal* 380 (June): 919–28. doi:10.1042/BJ20040095.

Imai, Shin-ichiro, and Leonard Guarente. 2016. "It Takes Two to Tango : NAD + and Sirtuins in Aging/Longevity Control." *Aging and Mechanisms of Disease* 2 (1607): 1–6. doi:10.1038/npjamd.2016.17.

Itahana, Koji, Ying Zou, Yoko Itahana, Jose-Luis Martinez, Christian Beausejour, Jacqueline J.L. Jacobs, Maarten Van Lohuizen, et al. 2003. "Control of the Replicative Life Span of Human Fibroblasts by P16 and the Polycomb Protein Bmi-1." *Molecular and Cellular Biology* 23 (1): 389–401.

James, Emma L., James A.E. Lane, Ryan D. Michalek, Edward D. Karoly, and E. Kenneth Parkinson. 2016. "Replicatively Senescent Human Fibroblasts Reveal a Distinct Intracellular Metabolic Profile with Alterations in NAD+ and Nicotinamide Metabolism." *Scientific Reports* 6 (38489): 1–15. doi:10.1038/srep38489.

Jiang, Peng, Wenjing Du, Anthony Mancuso, Kathryn E. Wellen, and Xiaolu Yang. 2013. "Reciprocal Regulation of P53 and Malic Enzymes Modulates Metabolism and Senescence." *Nature* 493: 689–93. doi:10.1038/nature11776.

Jones, Russell G., David R. Plas, Sara Kubek, Monica Buzzai, James Mu, Yang Xu, Morris J. Birnbaum, and Craig B Thompson. 2005. "AMP-Activated Protein Kinase Induces a P53-Dependent Metabolic Checkpoint." *Molecular Cell* 18: 283–93. doi:10.1016/j.molcel.2005.03.027.

Jurk, Diana, Chunfang Wang, Satomi Miwa, Mandy Maddick, Viktor Korolchuk, Avgi Tsolou, Efstathios S. Gonos, et al. 2012. "Postmitotic Neurons Develop a P21-Dependent Senescence-like Phenotype Driven by a DNA Damage Response." *Aging Cell* 11 (6): 996–1004. doi: 10.1111/j.1474-9726.2012.00870.x.

Jurk, Diana, Caroline Wilson, João F. Passos, Fiona Oakley, Clara Correia-Melo, Laura Greaves, Gabriele Saretzki, et al. 2014. "Chronic Inflammation Induces Telomere Dysfunction and Accelerates Ageing in Mice." *Nature Communications* 5: 1–14. doi:10.1038/ncomms5172.

Kaufman, Randal J. 2002. "Orchestrating the Unfolded Protein Response in Health and Disease." *The Journal of Clinical Investigation* 110 (10): 1389–98. doi:10.1172/JCI200216886.The.

Kelso, Geoffrey F, Carolyn M. Porteous, Carolyn V. Coulter, Gillian Hughes, William K. Porteous, Elizabeth C. Ledgerwood, Robin A.J. Smith, and Michael P. Murphy. 2001. "Selective Targeting of a Redox-Active Ubiquinone to Mitochondria within Cells: Antioxidant and Antiapoptotic Properties." *Journal of Biological Chemistry* 276 (7): 4588–96. doi:10.1074/jbc.M009093200.

Klimova, Tatyana A., Eric L. Bell, Emelyn H. Shroff, Frank D. Weinberg, Colleen M. Snyder, Goberdan P. Dimri, Paul T. Schumacker, G.R. Scott Budinger, and Navdeep S. Chandel. 2009. "Hyperoxia-Induced Premature Senescence Requires P53 and PRb, but Not Mitochondrial Matrix ROS." *The FASEB Journal* 23 (3): 783–94. doi:10.1096/fj.08-114256.

Korolchuk, Viktor I., Satomi Miwa, Bernadette Carroll, and Thomas Von Zglinicki. 2017. "Mitochondria in Cell Senescence: Is Mitophagy the Weakest Link?" *EBioMedicine* 21: 7–13. doi:10.1016/j.ebiom.2017.03.020.

Krtolica, Ana, Nick Larocque, Olga Genbacev, Dusko Ilic, Jean-Philippe Coppe, Christopher K. Patil, Tamara Zdravkovic, et al. 2011. "GROα Regulates Human Embryonic Stem Cell Self-Renewal or Adoption of a Neuronal Fate." *Differentiation* 81 (4): 222–32. doi:10.1016/j.diff.2011.01.001.

Kuilman, Thomas, Chrysiis Michaloglou, Liesbeth C. W. Vredeveld, Sirith Douma, Remco van Doorn, Christophe J. Desmet, Lucien A. Aarden, et al. 2008. "Oncogene-Induced Senescence Relayed by an Interleukin-Dependent Inflammatory Network." *Cell* 133 (6): 1019–31. doi:10.1016/j.cell.2008.03.039.

Kuilman, Thomas, and Daniel S. Peeper. 2009. "Senescence-Messaging Secretome: SMS-Ing Cellular Stress." *Nature Reviews Cancer* 9 (2): 81–94. doi:10.1038/nrc2560.

Kurz, David J., Stephanie Decary, Ying Hong, and Jorge D. Erusalimsky. 2000. "Senescence-Associated (Beta)-Galactosidase Reflects an Increase in Lysosomal Mass during Replicative Ageing of Human Endothelial Cells." *Journal of Cell Science* 113: 3613–22.

Kurz, David J., Stephanie Decary, Ying Hong, Elisabeth Trivier, Alexander Akhmedov, and Jorge D. Erusalimsky. 2004. "Chronic Oxidative Stress Compromises Telomere Integrity and Accelerates the Onset of Senescence in Human Endothelial Cells." *Journal of Cell Science* 117 (May): 2417–26. doi:10.1242/jcs.01097.

Kwon, Guim, John A. Corbett, and St Louis. 2015. "Interleukin-LP-Induced Nitric Oxide Synthase Expression by Rat Pancreatic P-Cells : Evidence for the Involvement of Nuclear Factor KB in the Signaling Mechanism *." *Endocrinology* 136 (11): 4790–95.

Laberge, Remi Martin, Pierre Awad, Judith Campisi, and Pierre Yves Desprez. 2012. "Epithelial-Mesenchymal Transition Induced by Senescent Fibroblasts." *Cancer Microenvironment.* doi:10.1007/s12307-011-0069-4.

Laberge, Remi-Martin, Yu Sun, Arturo V. Orjalo, Christopher K Patil, Adam Freund, Lili Zhou, Samuel C. Curran, et al. 2015. "MTOR Regulates the Pro-Tumorigenic Senescence-Associated Secretory Phenotype by Promoting IL1A Translation." *Nature Cell Biology* 17 (8): 1049–61. doi: 10.1038/ncb3195.

Langley, Emma, Mark Pearson, Mario Faretta, Uta Maria Bauer, Roy A. Frye, Saverio Minucci, Pier Giuseppe Pelicci, et al. 2002. "Human SIR2 Deacetylates P53 and Antagonizes PML/P53-Induced Cellular Senescence." *EMBO Journal* 21 (10): 2383–96. doi:10.1093/emboj/21.10.2383.

Lee, Bo Yun, Jung A. Han, Jun Sub Im, Amelia Morrone, Kimberly Johung, Edward C. Goodwin, Wim J. Kleijer, et al.

2006. "Senescence-Associated β-Galactosidase Is Lysosomal β-Galactosidase." *Aging Cell* 5 (2): 187–95. doi:10.1111/j.1474-9726.2006.00199.x.

Lee, Seon Yong Jeong, Won Chung Lim, Sujeong Kim, Yong Yea Park, Xuejun Sun, Richard J. Youle, et al. 2007. "Mitochondrial Fission and Fusion Mediators, HFis1 and OPA1, Modulate Cellular Senescence." *Journal of Biological Chemistry* 282 (31): 22977–83. doi:10.1074/jbc.M700679200.

Lee, Seung Min, So Hee Dho, Sung Kyu Ju, Jin Soo Maeng, Jeong Yoon Kim, and Ki Sun Kwon. 2012. "Cytosolic Malate Dehydrogenase Regulates Senescence in Human Fibroblasts." *Biogerontology* 13 (5): 525–36. doi:10.1007/s10522-012-9397-0.

Hayflick, Leonard, and Paul S. Moorhead. 1961. "The Serial Cultivation of Human Diploid Cell Strains." *Experimental Cell Research* 25 (3): 585–21. doi:10.1016/0014-4827(61)90192-6.

Lin, Athena W., Marta Barradas, James C. Stone, Linda Van Aelst, Manuel Serrano, and Scott W. Lowe. 1998. "Premature Senescence Involving P53 and P16 Is Activated in Response to Constitutive MEK/MAPK Mitogenic Signaling." *Genes and Development* 12 (19): 3008–19. doi:10.1101/gad.12.19.3008.

Loaiza, Natalia, and Marco Demaria. 2016. "Cellular Senescence and Tumor Promotion: Is Aging the Key?" *Biochimica et Biophysica Acta.* doi:10.1016/j.bbcan.2016.01.007.

Longo, Valter D., and Brian K. Kennedy. 2006. "Sirtuins in Aging and Age-Related Disease." *Cell* 126 (2): 257–68. doi:10.1016/j.cell.2006.07.002.

López-Otín, Carlos, Maria A. Blasco, Linda Partridge, Manuel Serrano, and Guido Kroemer. 2013. "The Hallmarks of Aging." *Cell* 153 (6): 1194–217. doi:10.1016/J.CELL.2013.05.039.

Lujambio, Amaia, Leila Akkari, Janelle Simon, Danielle Grace, Darjus F. Tschaharganeh, Jessica E. Bolden, Zhen Zhao, et al. 2013. "Non-Cell-Autonomous Tumor Suppression by P53." *Cell* 153 (2): 449–60. doi: 10.1016/j.cell.2013.03.020.

Macip, Salvador, Makoto Igarashi, Petra Berggren, Jian Yu, Sam W. Lee, and Stuart A. Aaronson. 2003. "Influence of Induced Reactive Oxygen Species in P53-Mediated Cell Fate Decisions." *Molecular and Cellular Biology* 23 (23): 8576–85. doi:10.1128/MCB.23.23.8576-8585.2003.

Macip, Salvador, Makoto Igarashi, Li Fang, Angus Chen, Zhen Qiang Pan, Sam W. Lee, and Stuart A. Aaronson. 2002. "Inhibition of P21-Mediated ROS Accumulation Can Rescue P21-Induced Senescence." *EMBO Journal* 21 (9): 2180–88. doi:10.1093/emboj/21.9.2180.

Mandal, Sudip, William A. Freije, Preeta Guptan, and Utpal Banerjee. 2010. "Metabolic Control of G1-S Transition: Cyclin E Degradation by P53-Induced Activation of the Ubiquitin-Proteasome System." *Journal of Cell Biology* 188 (4): 473–479. doi:10.1083/jcb.200912024.

Marcotte, Richard, Chantale Lacelle, and Eugenia Wang. 2004. "Senescent Fibroblasts Resist Apoptosis by Downregulating Caspase-3." *Mechanisms of Ageing and Development* 125: 777–83. doi:10.1016/j.mad.2004.07.007.

Marcu, Raluca, Brian M. Wiczer, Christopher K. Neeley, and Brian J. Hawkins. 2014. "Mitochondrial Matrix Ca2+ Accumulation Regulates Cytosolic NAD+/NADH Metabolism, Protein Acetylation, and Sirtuin Expression." *Molecular and Cellular Biology* 34 (15): 2890–2902. doi:10.1128/MCB.00068-14.

Marquez, Cielo Mae D., and Michael C. Velarde. 2017. "Senescent-Cells-as-Drivers-of-Age-Related-Diseases.Pdf." in *Aging: Exploring a Complex Phenomenon*, ed S.I. Ahmad, 305–34. Boca Rota, FL: CRC Press/Taylor & Francis.

Michaloglou, Chrysiis, Liesbeth C. W. Vredeveld, Maria S. Soengas, Christophe Denoyelle, Thomas Kuilman, Chantal M.A.M. Van Der Horst, Donné M. Majoor, et al. 2005. "BRAFE600-Associated Senescence-like Cell Cycle Arrest of Human Naevi." *Nature* 436 (7051): 720–24. doi:10.1038/nature03890.

Mihaylova, Maria M., and Reuben J. Shaw. 2011. "The AMPK Signalling Pathway Coordinates Cell Growth, Autophagy and Metabolism." *Nature Cell Biology.* doi:10.1038/ncb2329.

Mitsui, Youji, and Edward L. Schneider. 1976. "Increased Nuclear Sizes in Senescent Human Diploid Fibroblast Cultures." *Experimental Cell Research* 100 (1): 147–52. doi:10.1016/0014-4827(76)90336-0.

Miwa, Satomi, Howsun Jow, Karen Baty, Amy Johnson, Rafal Czapiewski, Gabriele Saretzki, Achim Treumann, et al. 2014. "Low Abundance of the Matrix Arm of Complex I in Mitochondria Predicts Longevity in Mice." *Nature Communications* 5 (3837): 1–12. doi:10.1038/ncomms4837.

Moiseeva, Olga, Véronique V. Bourdeau, Antoine A. Roux, Xavier X. Deschenes-Simard, and Gerardo G. Ferbeyre. 2009. "Mitochondrial Dysfunction Contributes to Oncogene-Induced Senescence." *Molecular and Cellular Biology* 29 (16): 4495–507. doi:10.1128/MCB.01868-08.

Muñoz-Espín, Daniel, and Manuel Serrano. 2014. "Cellular Senescence: From Physiology to Pathology." *Nature Reviews Molecular Cell Biology* 15 (7): 482–96. doi:10.1038/nrm3823.

Murphy, Michael P. 2009. "How Mitochondria Produce Reactive Oxygen Species." *Biochemical Journal* 417 (1): 1–13. doi:10.1042/BJ20081386.

Nacarelli, Timothy, Pingyu Liu, and Rugang Zhang. 2017. "Epigenetic Basis of Cellular Senescence and Its Implications in Aging." *Genes* 8: 12. doi:10.3390/genes8120343.

Narendra, Derek, Atsushi Tanaka, Der-fen Suen, and Richard J. Youle. 2008. "Parkin Is Recruited Selectively to Impaired Mitochondria and Promotes Their Autophagy." *Journal of Cell Biology* 183 (5): 795–803. doi:10.1083/jcb.200809125.

Narita, Masashi, Sabrina Nûnez, Edith Heard, Masako Narita, Athena W. Lin, Stephen A. Hearn, David L. Spector, et al. 2003. "Rb-Mediated Heterochromatin Formation and Silencing of E2F Target Genes during Cellular Senescence." *Cell* 113 (6): 703–16.

Nelson, Glyn, James Wordsworth, Chunfang Wang, Diana Jurk, Conor Lawless, Carmen Martin-Ruiz, and Thomas von Zglinicki. 2012. "A Senescent Cell Bystander Effect: Senescence-Induced Senescence." *Aging Cell* 11 (2): 345–49. doi:10.1111/j.1474-9726.2012.00795.x.

Nguyen, My-Hanh T., and M. Saleet Jafri. 2005. "Mitochondrial Calcium Signaling and Energy Metabolism." *Annals of the New York Academy of Sciences* 1047 (1): 127–37. doi:10.1196/annals.1341.012.

Nicolson, Garth L. 2014. "Mitochondrial Dysfunction and Chronic Disease: Treatment With Natural Supplements." *Integrative Medicine* 13 (4): 35–43.

Nuss, Jonathan E., Kashyap B. Choksi, James H. DeFord, and John Papaconstantinou. 2008. "Decreased Enzyme Activities of Chaperones PDI and BiP in Aged Mouse Livers." *Biochemical and Biophysical Research Communications* 365 (January): 355–61. doi:10.1016/j.bbrc.2007.10.194.

Ohtani, Naoko, and Eiji Hara. 2013. "Roles and Mechanisms of Cellular Senescence in Regulation of Tissue Homeostasis." *Cancer Science* 104 (5): 525–30. doi:10.1111/cas.12118.

Oikawa, Shinji, and Shosuke Kawanishi. 1999. "Site-Specific DNA Damage at GGG Sequence by Oxidative Stress May Accelerate Telomere Shortening." *FEBS Letters* 453: 365–68.

Opresko, Patricia L., Jinshui Fan, Shamika Danzy, and David M Wilson III. 2005. "Oxidative Damage in Telomeric DNA Disrupts Recognition by TRF1 and TRF2." *Nucleic Acids Research* 33 (4): 1230–39. doi:10.1093/nar/gki273.

Park, Yong-Yea, Seungmin Lee, Mariusz Karbowski, Albert Neutzner, Richard J. Youle, and Hyeseong Cho. 2010. "Loss of MARCH5 Mitochondrial E3 Ubiquitin Ligase Induces Cellular Senescence through Dynamin-Related Protein 1 and Mitofusin 1." *Journal of Cell Science* 123 (4): 619–26. doi:10.1242/jcs.061481.

Parrinello, Simona, Enrique Samper, Ana Krtolica, Joshua Goldstein, Simon Melov, and Judith Campisi. 2003. "Oxygen Sensitivity Severely Limits the Replicative Lifespan of Murine Fibroblasts." *Nature Cell Biology* 5 (8): 741–47. doi:10.1038/ncb1024.

Passos, João F., Glyn Nelson, Chunfang Wang, Torsten Richter, Cedric Simillion, Carole J. Proctor, Satomi Miwa, et al. 2010. "Feedback between P21 and Reactive Oxygen Production Is Necessary for Cell Senescence." *Molecular Systems Biology* 6 (347): 1–14. doi: 10.1038/msb.2010.5.

Passos, João F., Gabriele Saretzki, Shaheda Ahmed, Glyn Nelson, Torsten Richter, Heiko Peters, Ilka Wappler, et al. 2007. "Mitochondrial Dysfunction Accounts for the Stochastic Heterogeneity in Telomere-Dependent Senescence." *PLoS Biology* 5 (5): 1140–51. doi: 10.1371/journal.pbio.0050110.

Passos, João F., Thomas von Zglinicki, and Gabriele Saretzki. 2006. "Mitochondrial Dysfunction and Cell Senescence: Cause or Consequence?" *Rejuvenation Research* 9 (1): 64–8. doi:10.1089/rej.2006.9.64.

Petersen, Simone, Gabriele Saretzki, and Thomas von Zglinicki. 1998. "Preferential Accumulation of Single-Stranded Regions in Telomeres of Human Fibroblasts." *Experimental Cell Research* 239 (1): 152–60. doi:10.1006/EXCR.1997.3893.

Peyton, Kelly J., Xiao-ming Liu, Yajie Yu, Benjamin Yates, and William Durante. 2012. "Activation of AMP-Activated Protein Kinase Inhibits the Proliferation of Human Endothelial Cells." *The Journal of Pharmacology and Experimental Therapeutics* 342 (3): 827–34. doi:10.1124/jpet.112.194712.

Dalle Pezze, Piero, Glyn Nelson, Elsje G. Otten, Viktor I. Korolchuk, Thomas B.L. Kirkwood, Thomas von Zglinicki, and Daryl P. Shanley. 2014. "Dynamic Modelling of Pathways to Cellular Senescence Reveals Strategies for Targeted Interventions." Edited by Andrey Rzhetsky. *PLoS Computational Biology* 10 (8): 1–20. doi:10.1371/journal.pcbi.1003728.

Pole, Akshaj, Manjari Dimri, and Goberdhan P. Dimri. 2016. "Oxidative Stress, Cellular Senescence and Ageing Akshaj Pole APA Citation Oxidative Stress, Cellular Senescence and Ageing." *AIMS Molecular Science* 3 (3): 300–24. doi:10.3934/molsci.2016.3.300.

Ramirez, Ruben D., Carmela P. Morales, Brittney-Shea Herbert, Jeffrey M. Rohde, Christina Passons, Jerry W. Shay, and Woodring E.Wright. 2001. "Putative Telomere-Independent Mechanisms of Replicative Aging Reflect Inadequate Growth Conditions." *Genes & Development* 15 (4): 398–403. doi:10.1101/gad.859201.

Rayess, Hani, Marilene B. Wang, and Eri S. Srivatsan. 2012. "Cellular Senescence and Tumor Suppressor Gene P16." *International Journal of Cancer* 130 (8): 1715–25. doi:10.1002/ijc.27316.

Rodier, Francis, Denise P. Munoz, Robert Teachenor, Victoria Chu, Oanh Le, Dipa Bhaumik, Jean-Philippe Coppe, Eric Campeau, Christian M. Beauséjour, Sahn-Ho Kim, Albert R. Davalos, and Judith Campisi. 2011. "DNA-SCARS: Distinct Nuclear Structures That Sustain Damage-Induced Senescence Growth Arrest and Inflammatory Cytokine Secretion." *Journal of Cell Science* 124 (1): 68–81. doi: 10.1242/jcs.071340.

Rodier, Francis, Jean Philippe Coppé, Christopher K. Patil, Wieteke A. M. Hoeijmakers, Denise P. Muñoz, Saba R. Raza, Adam Freund, et al. 2009. "Persistent DNA Damage Signalling Triggers Senescence-Associated Inflammatory Cytokine Secretion." *Nature Cell Biology* 11 (8): 973–979. doi:10.1038/ncb1909.

Sadaie, Mahito, Christian Dillon, Masashi Narita, Andrew R. J. Young, Claire J. Cairney, Lauren S. Godwin, Christopher J. Torrance, 2015. "Cell-Based Screen for Altered Nuclear Phenotypes Reveals Senescence Progression in Polyploid Cells after Aurora Kinase B Inhibition." *Molecular Biology of the Cell* 26 (17): 2971–85. doi:10.1091/mbc.E15-01-0003.

Sadaie, Mahito, Rafik Salama, Thomas Carroll, Kosuke Tomimatsu, Tamir Chandra, Andrew R. J. Young, Masako Narita, et al. 2013. "Redistribution of the Lamin B1 Genomic Binding Profile Affects Rearrangement of Heterochromatic Domains and SAHF Formation during Senescence." *Genes and Development* 27 (16): 1800–8. doi: 10.1101/gad.217281.113.

Sanders, Yan Y., Hui Liu, Gang Liu, and Victor J. Thannickal. 2015. "Epigenetic Mechanisms Regulate NADPH Oxidase-4 Expression in Cellular Senescence." *Free Radical Biology and Medicine* 79: 197–205. doi:10.1016/J.FREERADBIOMED.2014.12.008.

Sanders, Yan Y., Hui Liu, Xiangyu Zhang, Louise Hecker, Karen Bernard, Leena Desai, Gang Liu, et al.. 2013. "Histone Modifications in Senescence-Associated Resistance to Apoptosis by Oxidative Stress." *Redox Biology* 1 (1): 8–16. doi:10.1016/J.REDOX.2012.11.004.

Saraste, Matti. 1999. "Oxidative Phosphorylation at the Fin de Sie`cle." *Science* 283 (5407): 1488–93.

Saretzki, Gabriele, Michael P. Murphy, and Thomas von Zglinicki. 2003. "MitoQ Counteracts Telomere Shortening and Elongates Lifespan of Fibroblasts under Mild Oxidative Stress." *Aging Cell* 2 (2): 141–3. doi:10.1046/j.1474-9728.2003.00040.x.

Serrano, Manuel, Athena W. Lin, Mila E. McCurrach, David Beach, and Scott W. Lowe. 1997. "Oncogenic Ras Provokes Premature Cell Senescence Associated with Accumulation of P53 and P16INK4a." *Cell* 88 (5): 593–602. doi:10.1016/S0092-8674(00)81902-9.

Seshadri, Tara, and Judith Campisi. 1990. "Repression of C-Fos Transcription and an Altered Genetic Program in Senescent Human Fibroblasts." *Science* 247 (4939): 205–9.

Sharpless, Norman E., and Charles J. Sherr. 2015. "Forging a Signature of in Vivo Senescence." *Nature Reviews Cancer* 15 (7): 397–408. doi:10.1038/nrc3960.

Shelton, Dawne N., Edwin Chang, Peter S.Whittier, Donghee Choi, and Walter D. Funk. 1999. "Microarray Analysis of Replicative Senescence." *Current Biology* 9 (17): 939–45.

Sherr, Charles J. 1998. "Tumor Surveillance via the ARF-P53 Pathway." *Genes & Development* 12 (19): 2984–91. doi:10.1101/gad.12.19.2984.

Shimi, Takeshi, Veronika Butin-Israeli, Stephen A. Adam, Robert B. Hamanaka, Anne E. Goldman, Catherine A. Lucas, Dale K. Shumaker, et al. 2011. "The Role of Nuclear Lamin B1 in Cell Proliferation and Senescence." *Genes & Development* 25: 2579–93. doi:10.1101/gad.179515.111.

Springer, Wolfdieter, and Philipp J. Kahle. 2011. "Regulation of PINK1-Parkin-Mediated Mitophagy." *Autophagy* 7 (3): 266–78. doi:10.4161/auto.7.3.14348.

Stein, Gretchen H., Linda. F. Drullinger, Ryan S. Robetorye, Olivia M. Pereira-Smith, and James R. Smith. 1991. "Senescent Cells Fail to Express Cdc2, CycA, and CycB in Response to Mitogen Stimulation." *Proceedings of the National Academy of Sciences* 88 (24): 11012–16.

Stöckl, Petra, Eveline Hütter, Werner Zwerschke, and Pidder Jansen-Dürr. 2006. "Sustained Inhibition of Oxidative Phosphorylation Impairs Cell Proliferation and Induces Premature Senescence in Human Fibroblasts." *Experimental Gerontology* 41 (7): 674–82. doi:10.1016/J.EXGER.2006.04.009.

Sun, Peiqing, Naoto Yoshizuka, Liguo New, Bettina A. Moser, Yilei Li, Rong Liao, Changchuan Xie, et al. 2007. "PRAK Is Essential for Ras -Induced Senescence and Tumor Suppression." *Cell* 1: 295–308. doi:10.1016/j.cell.2006.11.050.

Tai, Haoran, Zhe Wang, Hui Gong, Xiaojuan Han, Jiao Zhou, Xiaobo Wang, Xiawei Wei, et al. 2017. "Autophagy Impairment with Lysosomal and Mitochondrial Dysfunction Is an Important Characteristic of Oxidative Stress-Induced Senescence." *Autophagy* 13 (1): 99–113. doi: 10.1080/15548627.2016.1247143.

Takai, Hiroyuki, Agata Smogorzewska, and Titia de Lange. 2003. "DNA Damage Foci at Dysfunctional Telomeres." *Current Biology* 13 (17): 1549–56. doi:10.1016/S0960-9822(03)00542-6.

Tchkonia, Tamara, Yi Zhu, Jan Van Deursen, Judith Campisi, and James L. Kirkland. 2013. "Cellular Senescence and the Senescent Secretory Phenotype: Therapeutic Opportunities." *The Journal of Clinical Investigation* 123: 3. doi:10.1172/JCI64098.

Toussaint, Olivier O., Estela E. Medrano, and Thomas von Zglinicki. 2000. "Cellular and Molecular Mechanisms of Stress-Induced Premature Senescence (SIPS) of Human Diploid Fibroblasts and Melanocytes." *Experimental Gerontology* 35: 927–45.

Twig, Gilad, and Orian S. Shirihai. 2011. "The Interplay Between Mitochondrial Dynamics and Mitophagy." *Antioxidants & Redox Signaling* 14 (10): 1939–51. doi:10.1089/ars.2010.3779.

van Steensel, B., A. Smogorzewska, and T. de Lange. 1998. "TRF2 Protects Human Telomeres from End-to-End Fusions." *Cell* 92 (3): 401–13.

Velarde, Michael C., Marco Demaria, and Judith Campisi. 2013. "Senescent Cells and Their Secretory Phenotype as Targets for Cancer Therapy Cancer and Aging." *Interdisciplinary Topics in Gerontology* 38: 17–27. doi:10.1159/000343572.

Velarde, Michael C., Marco Demaria, Simon Melov, and Judith Campisi. 2015. "Pleiotropic Age-Dependent Effects of Mitochondrial Dysfunction on Epidermal Stem Cells." *Proceedings of the National Academy of Sciences* 112 (33): 10407–12. doi:10.1073/pnas.1505675112.

Velarde, Michael C., James M. Flynn, Nicholas U. Day, Simon Melov, and Judith Campisi. 2012. "Mitochondrial Oxidative Stress Caused by Sod2 Deficiency Promotes Cellular Senescence and Aging Phenotypes in the Skin." *Aging* 4 (1): 3–12. doi:10.18632/aging.100423.

Victorelli, Stella, and João F. Passos. 2017. "Telomeres and Cell Senescence - Size Matters Not." *EBioMedicine* The Authors. doi:10.1016/j.ebiom.2017.03.027.

Wang, Audrey Shimei, Peh Fern Ong, Alexandre Choj, and Carlos Clavel. 2017. "Loss of Lamin B1 Is a Biomarker to Quantify Cellular Senescence in Photoaged Skin." *Scientific Reports* 7 (15678): 1–8. doi:10.1038/s41598-017-15901-9.

Wang, Chunfang, Diana Jurk, Mandy Maddick, Glyn Nelson, Carmen Martin-Ruiz, and Thomas Von Zglinicki. 2009. "DNA Damage Response and Cellular Senescence in Tissues of Aging Mice." *Aging Cell* 8 (3): 311–23. doi:10.1111/j.1474-9726.2009.00481.x.

Wang, Eugenia, and Doris Gundersen. 1984. "Increased Organization of Cytoskeleton Accompanying the Aging of Human Fibroblasts in Vitro." *Experimental Cell Research* 154 (1): 191–202. doi:10.1016/0014-4827(84)90679-7.

Wang, Weiping, Joan X. Chen, Rong Liao, Qingdong Deng, Jennifer J. Zhou, Shuang Huang, and Peiqing Sun. 2002. "Sequential Activation of the MEK – Extracellular Signal-Regulated Kinase and MKK3/6-P38 Mitogen-Activated Protein Kinase Pathways Mediates Oncogenic Ras-Induced Premature Senescence." *Molecular and Cellular Biology* 22 (10): 3389–403. doi:10.1128/MCB.22.10.3389.

Wang, Wengong, Xiaoling Yang, Isabel López de Silanes, David Carling, and Myriam Gorospe. 2003. "Increased AMP:ATP Ratio and AMP-Activated Protein Kinase Activity during Cellular Senescence Linked to Reduced HuR Function." *The Journal of Biological Chemistry* 278 (29): 27016–23. doi:10.1074/jbc.M300318200.

Watanabe, Sugiko, Shimpei Kawamoto, Naoko Ohtani, and Eiji Hara. 2017. "Impact of Senescence-Associated Secretory Phenotype and Its Potential as a Therapeutic Target for Senescence-Associated Diseases." *Cancer Science* 108 (4): 563–69. doi:10.1111/cas.13184.

Weyemi, Urbain, Odile Lagente-Chevallier, Myriem Boufraqech, Fanny Prenois, Françoise Courtin, Bernard Caillou, Monique Talbot, Michle Dardalhon, Abir Al Ghuzlan, Jean-Michel Bidart, Martin Schlumberger, and Corinne Dupuy. 2012. "ROS-Generating NADPH Oxidase NOX4 Is a Critical Mediator in Oncogenic H-Ras-Induced DNA Damage and Subsequent Senescence." *Oncogene* 31 (9): 1117–29. doi: 10.1038/onc.2011.327.

Wiel, Clotilde, Hélène Lallet-Daher, Delphine Gitenay, Baptiste Gras, Benjamin Le Calvé, Arnaud Augert, Mylène Ferrand, et al. 2014. "Endoplasmic Reticulum Calcium Release through ITPR2 Channels Leads to Mitochondrial Calcium Accumulation and Senescence." *Nature Communications* 5: 3792. doi:10.1038/ncomms4792.

Wiley, Christopher D., Michael C. Velarde, Pacome Lecot, Su Liu, Ethan A. Sarnoski, Adam Freund, Kotaro Shirakawa, et al. 2016. "Mitochondrial Dysfunction Induces Senescence with a Distinct Secretory Phenotype." *Cell Metabolism* 23 (2): 303–14. doi: 10.1016/j.cmet.2015.11.011.

Xue, Wen, Lars Zender, Cornelius Miething, Ross A. Dickins, Eva Hernando, Valery Krizhanovsky, Carlos Cordon-Cardo, et al. 2007. "Senescence and Tumour Clearance Is Triggered by P53 Restoration in Murine Liver Carcinomas." *Nature* 445 (7128): 656–60. doi:10.1038/nature05529.

Yang, Gong, Daniel G. Rosen, Zhihong Zhang, Robert C. Bast Jr., Gordon B. Mills, Justin A. Colacino, Imelda Mercado-Uribe, and Jinsong Liu. 2006. "The Chemokine Growth-Regulated Oncogene 1 (Gro-1) Links RAS Signaling to the Senescence of Stromal Fibroblasts and Ovarian Tumorigenesis." *Proceedings of the National Academy of Sciences* 103 (44): 16472–77. doi:10.1073/pnas.0605752103.

Yoon, Kyoung B., Kyeong R. Park, Soo Y. Kim, and Sun-Young Han. 2016. "Induction of Nuclear Enlargement and Senescence by Sirtuin Inhibitors in Glioblastoma

Cells." *Immune Network* 16 (3): 183–8. doi:10.4110/in.2016.16.3.183.

Yoon, Young Sil, Hae Ok Byun, Hyeseong Cho, Bu Kyoung Kim, and Gyesoon Yoon. 2003. "Complex II Defect via Down-Regulation of Iron-Sulfur Subunit Induces Mitochondrial Dysfunction and Cell Cycle Delay in Iron Chelation-Induced Senescence-Associated Growth Arrest." *Journal of Biological Chemistry* 278 (51): 51577–86. doi:10.1074/jbc.M308489200.

Yoon, Young Sil, Dong Sun Yoon, In Kyoung Lim, Soo Han Yoon, Hae Young Chung, Manuel Rojo, Florence Malka, et al. 2006. "Formation of Elongated Giant Mitochondria in DFO-Induced Cellular Senescence: Involvement of Enhanced Fusion Process through Modulation of Fis1." *Journal of Cellular Physiology* 209 (2): 468–80. doi:10.1002/jcp.20753.

Yoshikawa, Toshikazu, and Yuji Naito. 2002. "What Is Oxidative Stress?" *Japan Medical Association* 124 (11): 1549–53. doi:10.1016/S0026-0495(00)80077-3.

Youle, Richard J., and Derek P. Narendra. 2011. "Mechanisms of Mitophagy." *Nature Reviews Molecular Cell Biology* 12 (1): 9–14. doi:10.1038/nrm3028.

Yue, Li, and Hongwei Yao. 2016. "Mitochondrial Dysfunction in Inflammatory Responses and Cellular Senescence: Pathogenesis and Pharmacological Targets for Chronic Lung Diseases." *British Journal of Pharmacology* 2305–18. doi:10.1111/bph.13518.

Zglinicki, Thomas von. 2002. "Oxidative Stress Shortens Telomeres." *Trends in Biochemical Sciences* 27 (7): 339–44.

Zglinicki, Thomas von, Rita Pilger, and Nicolle Sitte. 2000. "Accumulation of Single-Strand Breaks Is the Major Cause of Telomere Shortening in Human Fibroblasts." *Free Radical Biology and Medicine* 28 (1): 64–74. doi:10.1016/S0891-5849(99)00207-5.

Zhu, Jiyue, Douglas Woods, Martin McMahon, and J. Michael Bishop. 1998. "Senescence of Human Fibroblasts Induced by Oncogenic Raf." *Genes and Development* 12 (19): 2997–3007. doi:10.1101/gad.12.19.2997.

Ziegler, Dorian V., Christopher D. Wiley, and Michael C. Velarde. 2015. "Mitochondrial Effectors of Cellular Senescence: Beyond the Free Radical Theory of Aging." *Aging Cell* 14 (1): 1–7. doi:10.1111/acel.12287.

Zwerschke, Werner, Sybille Mazurek, Petra S. Ockl, Eveline Utter, Erich Eigenbrodt, and Pidder Jansen-Urr. 2003. "Metabolic Analysis of Senescent Human Fibroblasts Reveals a Role for AMP in Cellular Senescence." *Biochemical Journal* 376: 403–11.

30 Role of Mitochondrial Dysfunction in Human Obesity

David Albuquerque
Research Center for Anthropology and Health (CIAS), University of Coimbra, Department of Life Sciences, Coimbra, Portugal
Centre for Biomedical Research (CBMR), University of Algarve, Faro, Portugal

Sara Carmo-Silva
Center for Neurosciences and Cell Biology (CNC), University of Coimbra, Coimbra, Portugal

Daniel Álvarez-Vaca
Hospital Universitario Doctor Peset, Valencia, Spain

Célia Aveleira
Center for Neurosciences and Cell Biology (CNC), University of Coimbra, Coimbra, Portugal

Clévio Nóbrega
Centre for Biomedical Research (CBMR), University of Algarve, Faro, Portugal
Center for Neurosciences and Cell Biology (CNC), University of Coimbra, Coimbra, Portugal
Department of Biomedical Science and Medicine (DCBM), University of Algarve, Portugal
Algarve Biomedical Center (ABC), University of Algarve and University Hospital of Algarve, Portugal.

CONTENTS

1 INTRODUCTION

Obesity is considered a major public health concern that has reached epidemic worldwide. If current trends continue, by 2030 an estimated 38% of the world's adult will be overweight and 20% will be obese (Hruby and Hu, 2015). While growth trends in obesity in developed countries continue to increase drastically, including in children, the prevalence of obesity in developing countries continues also to trend upwards toward developed countries levels (Hruby and Hu, 2015). Despite the increase in preventive campaigns, obesity is still growing in prevalence worldwide, and the causes of weight gain are extremely complex. Obesity represents a major risk factor for several other diseases as, some types of cancer, musculosketal disorders, and causes several changes in the body, such as inflammation, oxidative stress, apoptosis, and mitochondrial dysfunction (Albuquerque et al., 2015).

This epidemic has been first attributed to a radical change in modern human lifestyle. These changes, including in our form of transportation, communication, entertainment and eating behaviors have been associated with significantly reduced demands for physical activity and increase in calories intake. Obesity has been considered for a long time as the result of a "social globalization," where physical activity decreased and food intake increased drastically, especially regarding the consumption of "junk" food, a product of the industrialization of societies (Albuquerque et al., 2017). Over time, a range of innovative studies emerged describing several factors that can be related with the increased prevalence of obesity including: societal, economics, cultural, genetics, metabolomics, gut microbiota, etc. (Figure 1) (Henstridge and Bozaoglu, 2017). All these studies generate a huge amount of results with a high potential for future preventive and treatment approaches in obesity, and overall to better understand the complexity of this etiology. Some of these evidences emerged and suggest that whole-body energy metabolism is regulated and coordinated by circuits within the central nervous system (Timper and Brüning, 2017). In addition, mitochondrial dysfunction has been implicated in various human metabolic disorders (Bhatti et al., 2017), including obesity (Bournat and Brown, 2010), thus, the function of mitochondria in hypothalamic circuits has become in the last two decades an important focus of research.

Obesity has a very complex etiology due to its heterogeneity. In this book chapter, we pretend to provide a general point of view about the notion that some clues can explain that weigh gain may be linked at the cellular level, more particularly, in the mitochondria. We will discuss about the possible implication of mitochondria in the development of an obese phenotype. However, it is important to keep in mind that several other factors (e.g., socioeconomic, cultural, genetic, epigenetic, microbiota, etc.) are implicated, and it is the set of all of them that contribute to this condition.

2   MITOCHONDRIA

Mitochondria are fundamental and highly dynamic organelles present in eukaryotic cells surrounded by two membranes and have their own genome. Mitochondria have emerged as playing a critical role in energy generation, control of intermediate metabolism, homeostasis of intracellular calcium and regulation of apoptosis events (Duchen, 2004). Mitochondria are the parts of the cells that turn food (hydrocarbons, fats and proteins) that we eat and oxygen into adenosine triphosphate (ATP), the primary energy source of the cellular function. However, when mitochondria are not functioning properly, food and oxygen cannot be used properly affecting the entire organism (Duchen, 2004). Furthermore, mitochondria are involved in the production and elimination of reactive oxygen species (ROS) (Nickel et al., 2014).

This small spherical organelle is present in each cell with thousand copies and holds its own genome. The human mitochondrial genome possesses a maternal

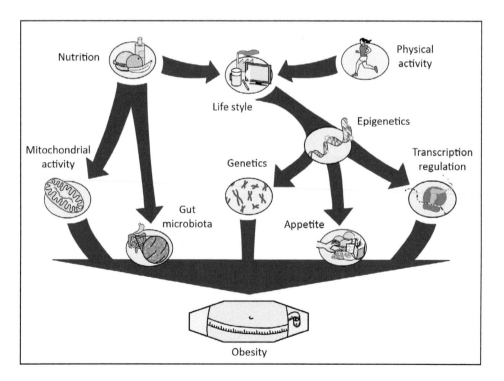

FIGURE 1   Obesity is a complex and heterogeneous condition resulting from a combination of multiple factors, including genetic and lifestyle influences, which interact with other players in a diverse and complex manner.

inheritance and have been used in several kinds of studies including, maternal family ties, Eva mitochondrial, etc. (Vigilant et al., 1991; Howell et al. 2000). Several mutations have been also identified associated with diverse forms of human diseases (Taylor and Turnbull, 2005).

2.1 Mitochondria in the Hypothalamus as an Energy Balance Regulator

Obesity, among other aspects, results when energy intake exceeds that expended. The balance between energy intake and energy expenditure is regulated by an intricate system, managed by the central nervous system (Myers and Olson, 2012). This metabolic homeostasis is maintained mainly through neurons located at the hypothalamus. This brain region integrates peripheral signals such as hormones and nutrients and orchestrates a neuronal response to regulate appetite and energy expenditure, thus body weight (Schwartz et al., 2013).

In a particular region of the hypothalamus, we can find the arcuate nucleus (ARC). Within this region, there are two distinct functionally antagonist types of neurons population that regulate food intake and energy expenditure (the orexigenic (hunger signaling) neuropeptine Y (NPY) and agouti-related peptide (AgRP)-expressing (NPY/AgRP) neurons, and the anorexigenic (satiety signaling) pro-opiomelanocortin (POMC)-expressing POMC neurons) (Timper and Brüning, 2017). Both of these neurons play an important role of the control circuits in ARC that regulate food intake and energy homeostasis. NPY neurons are activated by ghrelin release from the stomach, and inhibited by insulin and leptin after nutrient intake (Schwartz et al., 2013). Leptin deficiency in the fasting state or in ob/ob mice markedly elevates NPY mRNA. Leptin administration decreases the ARC NPY mRNA levels in ob/ob mice and fasted rodents (Elias et al., 1999). POMC is the precursor of alpha-melanocyte stimulating hormone (α-MSH), which decreases feeding by acting through its receptors melanocortin 3 receptor (*MC3R*) and *MC4R*, in response to insulin and leptin action (Schwartz et al., 2013). Obesity can impair this intricate neuronal network and deregulate food intake and energy expenditure processes.

We know that excessive intake of saturated fats can promote damage to the hypothalamus and its functions, namely through inflammation (Velloso and Schwartz, 2011). Some authors showed that even one day of high fat diet is enough to promote the rise of pro-inflammatory markers within the hypothalamus (Thaler et al., 2012; Waise et al., 2015). Other showed the promotion of a chronic inflammatory process within the hypothalamus of rodents, upon a high fat diet regimen (Zhang et al., 2008; Milanski et al., 2009). From these inflammatory processes that occur upon high fat diet feeding and obesity, derive phenomena such as ER stress and autophagy impairment, which can ultimately culminate in mitochondrial dysfunction

within the hypothalamus (Cavadas et al., 2016). Evidence show that hypothalamic mitochondrial dysfunction might occur at a later stage of obesity, further supporting this phenomenon as a result from all the other damage to the cellular functioning in the hypothalamic neurons (Carraro et al., 2018).

Mitochondria play a pivotal role in energy metabolism, functioning as the powerhouse of the cell by producing energy in form of ATP from substrates that can derive from diet, such as carbohydrates and lipids (de Mello et al., 2018). Neurons have especially high energetic needs, thus rely on the mitochondrial oxidative phosphorylation (OXPHOS) to meet these requirements (Bélanger et al., 2011). Besides energy generation, mitochondria is also necessary for ROS production, that determines the function of neurons, especially those involved in the regulation of appetite and energy expenditure within the hypothalamus, NPY/AgRP and POMC neurons (Jin and Diano, 2018). Furthermore, mitochondria dynamics can deeply change upon metabolic stress such as an increase on fat intake, that will compromise the mitochondria performance within the neurons of the hypothalamus (Ni et al., 2015; Wai and Langer, 2016).

Metabolic imbalance such as nutrient excess, can promote mitochondrial dysfunction within the hypothalamus. This mitochondrial dysfunction is not necessarily the inability to generate ATP, but irregularities in other mitochondria functions, such as calcium signaling, ROS production, apoptosis, changes in mitochondria size and morphology and even mutations in mitochondrial DNA (de Mello et al., 2018).

All of these alterations that occur in obesity within the neurons of the hypothalamus, will eventually lead to an exacerbation of the metabolic damage of weight excess, further deregulating food intake signaling pathways and compromising the regulation of the hypothalamus to the periphery (Carmo-Silva and Cavadas, 2017).

2.2 Mitochondria Deregulation on Hypothalamic Neuronal Network

Mitochondria can sense and adapt to the external stimuli such as nutrients in order to serve the energetic demands of neurons. This dynamic plays a determinant role in the response of AgRP/NPY and POMC neurons activity to different energetic environments (Figure 2) (Diano et al., 2011; Schneeberger et al., 2013).

Studies show that fasting or ghrelin administration can promote the activation of AgRP/NPY neurons, as a physiological response to promote feeding. This is associated with an increase in mitochondria density and a decrease in mitochondria size, suggesting fission processes (Coppola et al., 2007; Andrews et al., 2008). In a context of obesity, the orexigenic neuronal population AgRP/NPY appears to be more resistant to mitochondrial damage. In obese mice, the mitochondria within these

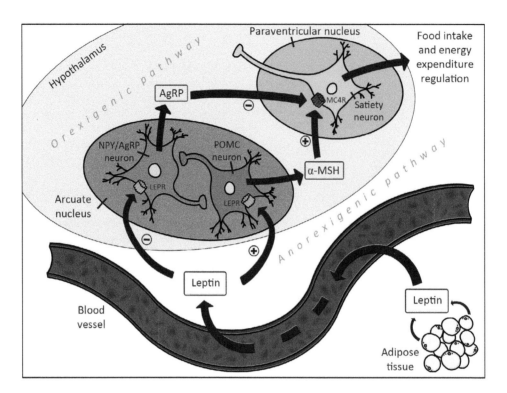

FIGURE 2 The central nervous system plays an important role in regulating food intake (the hypothalamic leptin/melanocortin pathway acting as the central regulator), receiving several signals from peripheral tissues for the regulation of energy expenditure. The melanocortin-4 receptor (MC4R) is highly expressed in the paraventricular nucleus of the hypothalamus, having a key role in the control of appetite. Leptin released from adipose tissues binds to leptin receptor (LEPR). On the one hand, proopiomelanocortin (POMC) neurons in the arcuate nucleus are activated by leptin and produce alpha melanocyte stimulating hormone (α-MSH), which then activates the MC4R in the paraventricular nucleus, triggering an anorexigenic response. On the other hand, neurons expressing neuropeptide Y (NPY) and agouti-related protein (AGRP) are inhibited by leptin, but in its absence, NPY and AGRP would act as potent inhibitors of MC4R signaling, triggering an orexigenic response.

neurons increase in size through fusion processes which can promote an increase in AgRP/NPY firing and activity (Dietrich et al., 2013). This can promote an increase in food intake and further exacerbate weight gain. Interestingly, the deletion of Mfn2 (Mitofusin 2), thus preventing mitochondria fusion, decreased AgRP/NPY firing and was enough to protect mice from diet induced obesity (Dietrich et al., 2013).

POMC neurons, responsible for signaling satiety have higher propensity for cell death in obesity, which might be linked to mitochondrial dysfunction (Jin and Diano, 2018). The higher caloric intake in obesity can decrease levels of PGC-1α and Mfn2, leading to reduced mitochondrial function and predisposition for apoptosis (Pintus et al., 2012). The impaired mitochondrial network within POMC neurons causes alterations in calcium signaling and decreased excitability of this neuronal population, which can further impair the food intake regulation in obese individuals (Paeger et al., 2017). Furthermore, mitochondria-ER interaction, mediated by Mfn2, is compromised in POMC neurons of obese mice, when compared to diet-resistant mice (Schneeberger

et al., 2013; Long et al., 2014). Deletion of Mfn2 in POMC neurons is sufficient to promote obesity (Schneeberger et al., 2013), interestingly, this deletion in AgRP/NPY neurons can protect mice from obesity (Dietrich et al., 2013). This change observed upon Mfn2 deletion in POMC neurons of the hypothalamus, is associated with impaired complex I activity and impaired OXPHOS (Schneeberger et al., 2013; Cunarro et al., 2018). Interestingly, recent evidence show that partial impairment of mitochondrial OXPHOS in POMC neurons can restore their firing in obesity and improve the obesity-associated peripheral changes such insulin resistance (Timper et al., 2018).

ROS is an important signaling molecule that indicates positive or negative energy states at the level of AgRP/NPY and POMC neurons. Suppression of ROS diminished POMC cell activation and elevated the activity of AgRP/NPY neurons and feeding, whereas ROS activated POMC neurons and reduced feeding. In *ob/ob* mice, injection of ROS promoted hunger, and increased levels of hypothalamic ROS which seem to activate POMC neurons and reduced feeding (Drougard et al., 2015). Mitochondrial

ROS production is negatively regulated by UCP2 (uncoupling protein 2), which highly expressed in the hypothalamus. In diet induced obese mice, UCP2 can affect POMC activation by reducing ROS levels which are necessary for this neuronal population activity (Diano et al., 2011). In opposition, hypothalamic AMPK activation can promote UCP2 activation, leading to the increase in the number of mitochondria and limiting ROS generation, that culminates in the activation of AgRP/NPY neurons (Coppola et al., 2007; Andrews et al., 2008; Jin and Diano, 2018).

These evidences show that mitochondrial dynamics can determine the transitions from fasting-feeding, acting differently on the two subsets of neuronal populations within the hypothalamus that regulate food intake drive; but studies also connected mitochondrial activity with *Bmal1* gene regulation (Jacobi et al., 2015). *Bmal1* gene is part of a set of clock genes that regulate the intrinsic biological clock, the circadian rhythm, and are responsible for the switching in feeding and fasting behaviour of all living organisms. The regulation of the circadian rhythm starts at the SCN (suprachiasmatic nucleus of the hypothalamus) and its deeply affected by nutrient availability (e.g., obesity) (Eckel-Mahan and Sassone-Corsi, 2013). Some authors suggest this link between mitochondria function, oxidative stress and circadian rhythm and its impact on metabolism (Wilking et al., 2013), however, this has yet to be studied in the context of hypothalamic neurons.

These observations, show just how intricate this neuronal system is and how can be completely deregulated through nutrient excess and consequent mitochondrial dysfunction. As in other pathologies and even other brain regions, also for hypothalamic physiology, it appears that mitochondrial dysfunction and oxidative stress can function either way, being deleterious or beneficial, just in the right amount.

3 THE MELANOCORTIN PATHWAY/ENERGY REGULATION

As mentioned earlier, obesity was considered for long time the result of an imbalance between energy expenditure and energy intake (Hill et al., 2012). This equilibrium is essential to the viability of the organism. Furthermore, persistent unbalancing between energy expenditure/energy intake in either direction has fundamental physiologic repercussions, as the development of obesity (Hill et al., 2012). This regulation of body weight in this manner is depending upon the continued assessment of both acute and chronic nutritional state. This information is made through several circulating humoral factors that communicate with the brain, adjusting appropriately the energy intake/expenditure equilibrium to maintain body weight within a normal range (Galgani and Ravussin, 2008). Acute signals such as ghrelin released in response to changes in the

immediate metabolic environment which contribute to short-term appetite indices of hunger and satiety, whereas longer-term regulation is governed by signals positively correlated with adiposity reflecting the overall metabolic state of the individual (Guyenet and Schwartz, 2012). It is precisely the integration of these signals which modulates an intricate network of neurons that adjust the physiological and behavioral responses. One of the most studied systems is the melanocortin pathway, which plays a key role to energy homeostasis in humans (Garfield et al., 2009). While the identification of genetic variants associated with obesity poses a major challenge, significant highlights in obesity gene research highlights the discovery of several mutations which result in human with a severe obese phenotype, namely leptin (*LEP*), leptin receptor (*LEPR*), melanocortin 4 receptor (*MC4R*), pro-opiomelanocortin (*POMC*), and so on (Huvenne and Dubern, 2014). Monogenic forms of human obesity resulted in the deficiency of critical molecules and disruption of the melanocortin system in energy homeostasis, unravelling the molecular circuitry that regulates body weight. So, monogenic forms of human obesity appear to be linked to mutations in this system. For example, mutations in *MC4R* gene lead to obesity by MC4R deficiency what causes hyperphagia with impaired satiety (Hinney et al. 2013). Other genes mutations have been identified and linked to some heritable obesity syndromes, such as mutations on *BDNF* gene causing WAGRO syndrome (Han et al., 2008), or on *ALMS1* gene causing Alström syndrome (Collin et al., 2002), and so on (Huvenne and Dubern, 2014). All these data offer a new perspective of the genetic cause of at least some human obesity syndromes.

4 INFLAMMATION AND OBESITY

Inflammation could be understood as a set of events triggered to maintain tissue and organ homeostasis in response to a harmful stimulus. It results from a coordinated action of several cells and mediators, delivering a short and acute inflammatory response and/or a long and chronic inflammation state (Ellulu et al., 2017). Adipose tissue is an active endocrine organ responsible for the synthesis and secretion of several hormones. Obesity and the metabolic syndrome are characterized by a chronic low-grade inflammation state, as it is not associated to an infection or autoimmunity, neither to tissue injury (Saltiel and Olefsky, 2017). However, this chronic inflammation state is characterized by an elevation of inflammatory markers and cytokines, and by the presence of macrophages infiltrated into the white adipose tissue (Johnson et al., 2012). This low-grade inflammation is observed in the adipose tissue and it spreads to a systemic inflammation, which definitively contributes to the metabolic syndrome associated features. Behind the activation of the inflammatory state seems to be the increase of lipids accumulation in adipocytes, which in turn induce the production of pro-inflammatory cyto-kines, as well as an increase in the number of macrophages (Monteiro and Azevedo, 2010).

5 OXIDATIVE STRESS AND OBESITY

The mitochondria harbor a complex enzymatic system responsible for the production of energy from the oxidation of sugars, fats and proteins. The mitochondria ATP energy production results also in the production of reactive oxygen species (ROS). The ROS homeostasis in cells results from a balance between ROS production and antioxidant reactions. An imbalance in this equilibrium towards ROS production will cause damage to cells, leading to oxidative stress. Importantly, a reduction in antioxidant enzymes levels or activity will also lead to oxidative stress (Bhat et al., 2015). The consequences of oxidative stress include the oxidation of biomolecules and consequent loss of function damaging cells and tissues.

In the context of obesity, oxidative stress can trigger or potentiate an inflammatory response (Roberts and Sindhu, 2009). As referred, obesity results in a simple view from an imbalance between energy intake and energy expenditure. Frequently, the surplus in the energy intake results from an increased amount of ingested food. This excess of food leads to an excess of biomolecules available for the production of energy in mitochondria. This increase energy production is also accompanied by an increase in the production of ROS, contributing to oxidative stress (Matsuda and Shimomura, 2013). The increase in ROS production is particularly observed in adipocytes, which in turn will contribute to an increase in the production of inflammatory adipokines (Aroor and DeMarco, 2014). Moreover, the accumulation of visceral fat in obesity increase the levels of systemic oxidative stress, which might contribute to the acceleration of oxidative stress in other organs affected by obesity and related morbidities (Fujita et al., 2006). On the other hand, inflammation events will also potentiate more ROS production and thus contributing to further oxidative stress (Bondia-Pons et al., 2012). Finally, this excess of ROS production and inflammation might underlie mitochondrial dysfunction.

6 MITOCHONDRIAL DYSFUNCTION IN OBESITY

The classical definition of mitochondrial dysfunction refers to the incapacity of these organelles to produce and maintain enough ATP levels needed to the energy production in cells (Brand and Nicholls, 2011). However, currently the concept of mitochondrial dysfunction is much broader including for example alterations in the size and morphology of the mitochondria, alterations in ROS production or even alterations in apoptosis events, among other (Lahera et al., 2017). In the context of obesity, there are multiple evidence and studies linking

a dysfunction in mitochondria as a consequence of the metabolic dysfunction, but also as factor exacerbating the effects of that dysfunction.

7 MITOCHONDRIA BIOGENESIS AND DYNAMICS IN OBESITY

Mitochondrial biogenesis can be defined as the growth and division of pre-existing mitochondria, whereas mitochondria dynamics refers to the balance between mitochondrial fusion and fission processes. Several studies reported alterations in both biogenesis and dynamics of mitochondria in the context of obesity or obesity-related complications, such type 2 diabetes. Moreover, these reported alterations were observed in different tissues such as adipose tissue or liver, in animal models and human subjects.

The adipose tissue is an important contributor to energy balance and metabolic homeostasis, and thus several studies looked for the mitochondria biogenesis and dynamics in adipocytes. For example, the mitochondrial biogenesis was found to be strongly suppressed in mouse models of obesity and diabetes (Rong et al., 2007). On the other hand, it was also shown that the knockdown of important mitochondria dynamics regulators *in vivo* lead to the development of obesity and related comorbidities such as hepatic steatosis (Quirós et al., 2012). In human studies, data suggested that mitochondrial biogenesis is downregulated in acquired obesity (Heinonen et al., 2015), although others reported no alterations in the mitochondrial content in adipocyte tissue between obese and non-obese subjects (Yin et al., 2014).

In the skeletal muscle, the mitochondrial size was reduced in obese and insulin resistant Zucker rats (Bach et al., 2003). Other study also reported smaller and shorter mitochondria and increased mitochondrial fission machinery in the skeletal muscle of genetically obese and diet-induced mice (Jheng et al., 2012). On the other hand, it was detected an increase in mitochondrial fission proteins in the skeletal muscle of ob/ob mice (Jheng et al., 2012), and in high-fat diet (HFD) induced mice (Liu et al., 2014). In human subjects it was described a reduced mitochondrial mass, which was related with mitochondrial dysfunction in the skeletal muscle of obese and type 2 diabetic subjects (Zorzano et al., 2009).

Also, in the liver, signs of mitochondrial dysfunction were related with obesity, for example by an enhancement of fission processes in the liver of genetically obese mice (Holmström et al., 2012) and also diet-induced obese mice (Putti et al., 2015). The liver plays an important role in glucose homeostasis and development of metabolic alterations. In this line, it was found that deletion of an essential protein involved in mitochondrial functions led to several metabolic alterations, including glucose intolerance and enhanced hepatic gluconeogenesis (Sebastián et al., 2012).

Altogether these selected studies clearly highlight that obesity and related metabolic abnormalities produce alterations in mitochondria biogenesis and dynamics leading to its dysfunction. On the other hand, the deletion of important proteins involved in these mitochondrial processes leads also to the development of metabolic dysfunctions.

7.1 APOPTOSIS AND OBESITY

Apoptosis refers to a process of programmed cell death, triggered by two main pathways: the extrinsic pathway which is receptor-mediated and the intrinsic pathway which is triggered by mitochondria. The extrinsic pathway is triggered by an external signal and mediated by the cellular membrane receptors belonging to the tumor necrosis factor (TNF) gene family (Kiraz et al., 2016). On the other hand, in the intrinsic pathway apoptosis which is triggered by the mitochondria with the activation of the pro-apoptotic Bax protein, in turn trigger downstream signals leading to apoptosis (Kiraz et al., 2016). Apoptosis is an important regulator of tissue and physiological homeostasis, occurring normally for example in the development and aging processes, or as a defense mechanism when a cell is damaged by disease or a toxic stimulus (Elmore, 2007).

As it was already referred, the increase in caloric intake promotes the dysfunction of mitochondria, thus increasing the cell susceptibility to apoptosis (Hernández-Aguilera et al., 2013). On the other hand, caloric restriction has a positive effect in mitochondrial homeostasis, for example reducing the ROS production, or by promoting the expressing of genes that stimulate mitochondrial biogenesis and fusion (Pintus et al., 2012). In the context of obesity, it was shown that apoptotic proteins are upregulated and apoptosis was found to be highly increased in the adipose tissue (Herold et al., 2013).

8 INHERITANCE OF OBESITY VIA MITOCHONDRIA

In the literature, we can find some studies that found an association between body mass index (BMI) in mothers and offspring over three generations (Murrin et al., 2012). Furthermore, some of these studies showed evidences that obesity can be transmitted from mothers to offspring through dysfunctional mitochondria (Wu et al., 2015; Ridler, 2016). It is not surprising when we know that mitochondria are maternally inherited, thus, any defect on these organelles can be a potential mechanism for transmission of metabolic dysfunction. As exposed in previous sections, mitochondria play a key role in metabolic regulation and mitochondria dysfunction has been clearly identified not only associated with obesity but also with insulin resistance, cardiovascular disease, and so on (Bournat and Brown, 2010; Bhatti et al., 2017).

Some evidences indicate that maternal obesity can damage mitochondria in the oocyte, due to increased inflammation and oxidative stress (Igosheva et al., 2010; Boots et al., 2016; Boudoures et al., 2016), but the mechanisms underlying this damage remain to be elucidated. Mitochondria are maternally inherited, thus defects in these organelles present a potential mechanism for transmission of metabolic dysfunction. In a study conducted by Saben and colleagues (Saben et al., 2016), female fed a high fat and high-sugar diet developed obesity and the metabolic syndrome. After three generations, they observed that female offspring born to obese mothers had abnormal enlarged mitochondria and with aberrant cristae morphology in the muscle, when compared with mitochondria in control offspring of normal-fed mothers. Importantly, was the observation of aberrant mitochondria morphology and protein expression in the oocytes of first- and second-generation offspring of obese mothers. Another experiment in mouse conducted by Boudoures and colleagues (Boudoures et al., 2017), suggests that the oocyte itself, independently of the uterine environment, can have effects on embryo quality and offspring health. These observational data suggest that mitochondria dysfunction is transmitted via germ line. The inheritance of this dysfunction and its potential persistence through several generations can explain this persistence and the metabolic syndrome through our populations. Further investigations on this area are needed in an attempt to elucidate if this phenomenon is done by nuclear or epigenetics effect on mitochondrial DNA.

9 CONCLUSION

All the presented evidence shows just how intricate the pathophysiology of obesity is, depending on the central nervous system (hypothalamus) and periphery communication, as well as impact of other factors such as genetics, age, gender, and environment. Mitochondria dysfunction poses as a consequence of nutrient excess occurring in obesity, compromising cellular function and viability, threatening the normal physiology of whole body. However, mitochondria dysfunction per se, occurring in normal aging or even upon genetic events, can trigger obesity and other metabolic alterations. Furthermore, evidence also show that mitochondrial dysfunction and oxidative stress, are necessary for cell physiology to cope with stress and regulate the normal cellular processes. Thus, therapeutic approaches for obesity targeting mitochondria will always have to consider the specific characteristic of the individual that might determine mitochondria activity and fitness.

ACKNOWLEDGMENTS

David Albuquerque has a Post-Doc fellowship (SFRH/BPD/109043/2015) from the Portuguese Foundation for Science and Technology (FCT). Clévio Nóbrega's laboratory is funded by the French Muscular Dystrophy Association (AFM-Téléthon), the Ataxia UK, and by the FCT.

REFERENCES

Albuquerque D, Nóbrega C, Manco L. et al. The contribution of genetics and environment to obesity. Br Med Bull 2017; 123:159–73.

Albuquerque D, Stice E, Rodríguez-López R. et al. Current review of genetics of human obesity: From molecular mechanisms to an evolutionary perspective. Mol Genet Genomics 2015; 290:1191–221.

Andrews ZB, Liu Z-W, Walllingford N. et al. UCP2 mediates ghrelin's action on NPY/AgRP neurons by lowering free radicals. Nature 2008; 454:846–51.

Aroor AR, DeMarco VG. Oxidative stress and obesity: The chicken or the egg?. Diabetes 2014; 63:2216–18.

Bach D, Pich S, Soriano FX. et al. Mitofusin-2 determines mitochondrial network architecture and mitochondrial metabolism. J Biol Chem 2003; 278:17190–97.

Bélanger M, Allaman I, Magistretti PJ. Brain energy metabolism: Focus on astrocyte-neuron metabolic cooperation. Cell Metab 2011; 14:724–38.

Bhat AH, Dar KB, Anees S. et al. Oxidative stress, mitochondrial dysfunction and neurodegenerative diseases; a mechanistic insight. Biomed Pharmacother 2015; 74:101–10.

Bhatti JS, Bhatti GK, Reddy PH. Mitochondrial dysfunction and oxidative stress in metabolic disorders — A step towards mitochondria based therapeutic strategies. Biochim Biophys Acta - Mol Basis Dis 2017; 1863:1066–77.

Bondia-Pons I, Ryan L, Martinez JA. Oxidative stress and inflammation interactions in human obesity. J Physiol Biochem 2012; 68:701–11.

Boots CE, Boudoures AL, Zhang W. et al. Obesity-induced oocyte mitochondrial defects are partially prevented and rescued by supplementation with co-enzyme Q10 in a mouse model. Hum Reprod 2016; 31:2090–97.

Boudoures AL, Chi M, Thompson A. et al. The effects of voluntary exercise on oocyte quality in a diet-induced obese murine model. Reproduction 2016; 151:261–70.

Boudoures AL, Saben J, Drury A. et al. Obesity-exposed oocytes accumulate and transmit damaged mitochondria due to an inability to activate mitophagy. Dev Biol 2017; 426:126–38.

Bournat JC, Brown CW. Mitochondrial dysfunction in obesity. Curr Opin Endocrinol Diabetes Obes 2010; 17:446–452.

Brand MD, Nicholls DG. Assessing mitochondrial dysfunction in cells. Biochem J 2011; 435:297–312.

Carmo-Silva S, Cavadas C. Hypothalamic dysfunction in obesity and metabolic disorders. Adv Neurobiol 2017; 19:73–116.

Carraro RS, Souza GF, Solon C. et al. Hypothalamic mitochondrial abnormalities occur downstream of inflammation in diet-induced obesity. Mol Cell Endocrinol 2018; 460:238–45.

Cavadas C, Aveleira CA, Souza GFP. et al. The pathophysiology of defective proteostasis in the hypothalamus - from obesity to ageing. Nat Rev Endocrinol 2016; 12:723–33.

Collin GB, Marshall JD, Ikeda A. et al. Mutations in ALMS1 cause obesity, type 2 diabetes and neurosensory degeneration in Alström syndrome. Nat Genet 2002; 31:74–78.

Coppola A, Liu Z-W, Andrews ZB. et al. A central thermogenic-like mechanism in feeding regulation: An interplay between arcuate nucleus T3 and UCP2. Cell Metab 2007; 5:21–33.

Cunarro J, Casado S, Lugilde J. et al. Hypothalamic mitochondrial dysfunction as a target in obesity and metabolic disease. Front Endocrinol (Lausanne) 2018; 9:283.

de Mello AH, Costa AB, Engel JDG. et al. Mitochondrial dysfunction in obesity. Life Sci 2018; 192:26–32.

Diano S, Liu Z-W, Jeong JK. et al. Peroxisome proliferation-associated control of reactive oxygen species sets melanocortin tone and feeding in diet-induced obesity. Nat Med 2011; 17:1121–27.

Dietrich MO, Liu Z-W, Horvath TL. Mitochondrial dynamics controlled by mitofusins regulate Agrp neuronal activity and diet-induced obesity. Cell 2013; 155:188–99.

Drougard A, Fournel A, Valet P. et al. Impact of hypothalamic reactive oxygen species in the regulation of energy metabolism and food intake. Front Neurosci 2015; 9:56.

Duchen MR. Roles of mitochondria in health and disease. Diabetes 2004; 53(Suppl 1):S96–102.

Eckel-Mahan K, Sassone-Corsi P. Metabolism and the circadian clock converge. Physiol Rev 2013; 93:107–35.

Elias CF, Aschkenasi C, Lee C. et al. Leptin differentially regulates NPY and POMC neurons projecting to the lateral hypothalamic area. Neuron 1999; 23:775–86.

Ellulu MS, Patimah I, Khaza'ai H. et al. Obesity and inflammation: The linking mechanism and the complications. Arch Med Sci 2017; 4:851–63.

Elmore S. Apoptosis: A review of programmed cell death. Toxicol Pathol 2007; 35:495–516.

Fujita K, Nishizawa H, Funahashi T. et al. Systemic oxidative stress is associated with visceral fat accumulation and the metabolic syndrome. Circ J 2006; 70:1437–42.

Galgani J, Ravussin E. Energy metabolism, fuel selection and body weight regulation. Int J Obes (Lond) 2008; 32(Suppl 7):S109–19.

Garfield AS, Lam DD, Marston OJ. et al. Role of central melanocortin pathways in energy homeostasis. Trends Endocrinol Metab 2009; 20:203–15.

Guyenet SJ, Schwartz MW. Clinical review: Regulation of food intake, energy balance, and body fat mass: Implications for the pathogenesis and treatment of obesity. J Clin Endocrinol Metab 2012; 97:745–55.

Han JC, Liu Q-R, Jones M. et al. Brain-derived neurotrophic factor and obesity in the WAGR syndrome. N Engl J Med 2008; 359:918–27.

Heinonen S, Buzkova J, Muniandy M. et al. Impaired mitochondrial biogenesis in adipose tissue in acquired obesity. Diabetes 2015; 64:3135–45.

Henstridge D, Bozaoglu K. The omics of obesity. in Adiposity – Omics and molecular understanding (ed.) Gordeladze J. London: InTechOpen Publisher, 2017, pp. 575–483.

Hernández-Aguilera A, Rull A, Rodríguez-Gallego E. et al. Mitochondrial dysfunction: A basic mechanism in inflammation-related non-communicable diseases and therapeutic opportunities. Mediators Inflamm 2013; 2013:135698.

Herold C, Rennekampff HO, Engeli S. Apoptotic pathways in adipose tissue. Apoptosis 2013; 18:911–16.

Hill JO, Wyatt HR, Peters JC. Energy balance and obesity. Circulation 2012; 126:126–32.

Hinney A, Volckmar A-L, KN. Melanocortin-4 receptor in energy homeostasis and obesity pathogenesis. Prog Mol Biol Transl Sci 2013; 114:147–91.

Holmström MH, Iglesias-Gutierrez E, Zierath JR. et al. Tissue-specific control of mitochondrial respiration in obesity-related insulin resistance and diabetes. Am J Physiol Metab 2012; 302:E731–E739.

Howell N, Chinnery PF, Ghosh SS. et al. Transmission of the human mitochondrial genome. Hum Reprod 2000; 15 (Suppl 2):235–45.

Hruby A, Hu FB. The epidemiology of obesity: A big picture. Pharmacoeconomics 2015; 33:673–89.

Huvenne H, Dubern B. Monogenic forms of obesity. in Molecular mechanisms underpinning the development of obesity (eds.) Nóbrega C, Ródriguez-López R. Cham: Springer International Publishing, 2014, pp. 9–21.

Igosheva N, Abramov AY, Poston L. et al. Maternal diet-induced obesity alters mitochondrial activity and redox status in mouse oocytes and Zygotes. PLoS One 2010; 5:e10074.

Jacobi D, Liu S, Burkewitz K. et al. Hepatic Bmal1 Regulates Rhythmic Mitochondrial Dynamics and Promotes Metabolic Fitness. Cell Metab 2015; 22:709–20.

Jheng H-F, Tsai P-J, Guo S-M. et al. Mitochondrial fission contributes to mitochondrial dysfunction and insulin resistance in skeletal muscle. Mol Cell Biol 2012; 32:309–19.

Jin S, Diano S. Mitochondrial dynamics and hypothalamic regulation of metabolism. Endocrinology 2018; 159:3596–604.

Johnson AR, Justin Milner J, Makowski L. The inflammation highway: Metabolism accelerates inflammatory traffic in obesity. Immunol Rev 2012; 249:218–38.

Kiraz Y, Adan A, Kartal Yandim M. et al. Major apoptotic mechanisms and genes involved in apoptosis. Tumor Biol 2016; 37:8471–86.

Lahera V, de Las Heras N, López-Farré A. et al Role of mitochondrial dysfunction in hypertension and obesity. Curr Hypertens 2017; Rep 19:11.

Liu R, Jin P, Liqun Yu. et al. Impaired mitochondrial dynamics and bioenergetics in diabetic skeletal muscle. PLoS One 2014; 9:e92810.

Long L, Toda C, Jeong JK. et al. PPARγ ablation sensitizes proopiomelanocortin neurons to leptin during high-fat feeding. J Clin Invest 2014; 124:4017–4027.

Matsuda M, Shimomura I. Increased oxidative stress in obesity: Implications for metabolic syndrome, diabetes, hypertension, dyslipidemia, atherosclerosis, and cancer. Obes Res Clin Pract 2013; 7:e330–e341.

Milanski M, Degasperi G, Coope A. et al. Saturated fatty acids produce an inflammatory response predominantly through the activation of TLR4 signaling in hypothalamus: Implications for the pathogenesis of obesity. J Neurosci 2009; 29:359–70.

Monteiro R, Azevedo I. Chronic inflammation in obesity and the metabolic syndrome. Mediators Inflamm 2010; 2010: pii289645.

Murrin CM, Kelly GE, Tremblay RE. et al. Body mass index and height over three generations: Evidence from the Lifeways cross-generational cohort study. BMC Public Health 2012; 12:81.

Myers MG, Olson DP. Central nervous system control of metabolism. Nature 2012; 491:357–63.

Ni H-M, Williams JA, Ding W-X. Mitochondrial dynamics and mitochondrial quality control. Redox Biol 2015; 4:6–13.

Nickel A, Kohlhaas M, Maack C. Mitochondrial reactive oxygen species production and elimination. J Mol Cell Cardiol 2014; 73:26–33.

Paeger L, Pippow A, Hess S. et al. Energy imbalance alters Ca2+ handling and excitability of POMC neurons. Elife 2017; 6: e25641.

Pintus F, Floris G, Rufini A. Nutrient availability links mitochondria, apoptosis, and obesity. Aging (Albany NY) 2012; 4:734–41.

Putti R, Sica R, Migliaccio V. et al. Diet impact on mitochondrial bioenergetics and dynamics. Front Physiol 2015; 6:109.

Quirós PM, Ramsay AJ, Sala D. et al. Loss of mitochondrial protease OMA1 alters processing of the GTPase OPA1

and causes obesity and defective thermogenesis in mice. EMBO J 2012; 31:2117–33.

Ridler C. Inheritance via mitochondria. Nat Rev Endocrinol 2016; 12:497–497.

Roberts CK, Sindhu KK. Oxidative stress and metabolic syndrome. Life Sci 2009; 84:705–12.

Rong JX, Qiu Y, Hansen MK. et al. Adipose mitochondrial biogenesis is suppressed in db/db and high-fat diet-fed mice and improved by rosiglitazone. Diabetes 2007; 56:1751–60.

Saben JL, Boudoures AL, Asghar Z. et al. Maternal metabolic syndrome programs mitochondrial dysfunction via germline changes across three generations. Cell Rep 2016; 16:1–8.

Saltiel AR, Olefsky JM. Inflammatory mechanisms linking obesity and metabolic disease. J Clin Invest 2017; 127:1–4.

Schneeberger M, Dietrich MO, Sebastián D. et al. Mitofusin 2 in POMC neurons connects ER stress with leptin resistance and energy imbalance. Cell 2013; 155:172–187.

Schwartz MW, Seeley RJ, Tschöp MH. et al. Cooperation between brain and islet in glucose homeostasis and diabetes. Nature 2013; 503:59–66.

Sebastián D, Hernández-Alvarez MI, Segalés J. et al. Mitofusin 2 (Mfn2) links mitochondrial and endoplasmic reticulum function with insulin signaling and is essential for normal glucose homeostasis. Proc Natl Acad Sci U S A 2012; 109:5523–28.

Taylor RW, Turnbull DM. Mitochondrial DNA mutations in human disease. Nat Rev Genet 2005; 6:389–402.

Thaler JP, Yi C-X, Schur EA. et al. Obesity is associated with hypothalamic injury in rodents and humans. J Clin Invest 2012; 122:153–62.

Timper K, Brüning JC. Hypothalamic circuits regulating appetite and energy homeostasis: Pathways to obesity. Dis Model Mech 2017; 10:679–89.

Timper K, Paeger L, Sánchez-Lasheras C. et al. Mild impairment of mitochondrial OXPHOS promotes fatty acid utilization in POMC neurons and improves glucose homeostasis in obesity. Cell Rep 2018; 25:383–397.e10.

Velloso LA, Schwartz MW. Altered hypothalamic function in diet-induced obesity. Int J Obes 2011; 35:1455–65.

Vigilant L, Stoneking M, Harpending H. et al. African populations and the evolution of human mitochondrial DNA. Science 1991; 253:1503–07.

Wai T, Langer T. Mitochondrial dynamics and metabolic regulation. Trends Endocrinol Metab 2016; 27:105–17.

Waise TMZ, Toshinai K, Naznin F. et al. One-day high-fat diet induces inflammation in the nodose ganglion and hypothalamus of mice. Biochem Biophys Res Commun 2015; 464:1157–62.

Wilking M, Ndiaye M, Mukhtar H. et al. Circadian rhythm connections to oxidative stress: Implications for human health. Antioxid Redox Signal 2013; 19:192–208.

Wu LL, Russell DL, Wong SL. et al. Mitochondrial dysfunction in oocytes of obese mothers: Transmission to offspring and reversal by pharmacological endoplasmic reticulum stress inhibitors. Development 2015; 142:681–91.

Yin X, Lanza IR, Swain JM. et al. Adipocyte mitochondrial function is reduced in human obesity independent of fat cell size. J Clin Endocrinol Metab 2014; 99:E209–E216.

Zhang X, Zhang G, Zhang H. et al. Hypothalamic IKKbeta/NF-kappaB and ER stress link overnutrition to energy imbalance and obesity. Cell 2008; 135:61–73.

Zorzano A, Liesa M, Palacín M. Role of mitochondrial dynamics proteins in the pathophysiology of obesity and type 2 diabetes. Int J Biochem Cell Biol 2009; 41:1846–54.

31 Roles of Melatonin in Maintaining Mitochondrial Welfare
Focus on Renal Cells

Feres José Mocayar Marón

Área de Química Biológica. Departamento de Morfofisiología. Facultad de Ciencias Médicas. Universidad Nacional de Cuyo. Mendoza, Argentina

Emiliano Diez

Instituto de Medicina y Biología Experimental de Cuyo (IMBECU); Consejo Nacional de Investigaciones Científicas y Tecnológicas (CONICET), Mendoza, Argentina
Instituto de Fisiología, Departamento de Morfofisiología, Facultad de Ciencias Médicas, Universidad Nacional de Cuyo, Mendoza, Argentina

Russel J. Reiter

Department of Cellular and Structural Biology, University of Texas Health Science at San Antonio, San Antonio, TX, USA

Walter Manucha

Instituto de Medicina y Biología Experimental de Cuyo (IMBECU); Consejo Nacional de Investigaciones Científicas y Tecnológicas (CONICET), Mendoza, Argentina
Área de Farmacología. Departamento de Patología. Facultad de Ciencias Médicas, Universidad Nacional de Cuyo, Mendoza, Argentina.

CONTENTS

1 INTRODUCTION

In 1958 melatonin was discovered, and although initially it was conceived only produced by the pineal gland, currently the synthesis in multiple organs, cells, and subcellular domains is now recognized. Original works in medicine extensively describe their biological actions on sleep and wake cycle's regulation. Subsequently, some authors describing effects on the blood pressure. In this context, it is relevant to highlight that most melatonin actions are due to their specific receptors. However, other actions are due to its role as an antioxidant, anti-inflammatory, and immunomodulatory, particularly in the protection of nuclear and mitochondrial DNA. Latest knowledge about melatonin highpoints many protective functions linked to the mitochondria for cellular viability.

Mitochondrial dysfunction is closely related to loss of melatonin regulation. Indeed, melatonin acts on mitochondria regulation to maintain a healthy renal as well as cardiovascular system. Therefore, it is interesting to deepen our knowledge about the most innovative aspects of melatonin for mitochondrial welfare, with particular emphasis on the renal cells.

2 BEYOND THE PINEAL GLAND: MELATONIN SYNTHESIS AND METABOLISM WITHIN THE MITOCHONDRIA

Melatonin (*N*-acetyl-5-methoxytryptamine) is a highly conserved molecule that is found not only in vertebrates but also in multiple phylogenetically-diverse species; its origin can be traced back an estimated 2.5 to 3.2 billion years ago[1]. In addition to its presence in vertebrates, melatonin has also been identified in invertebrates, bacteria, algae, plants, and fungi. Lerner et al. isolated and identified melatonin 60 years ago[2]. It was called melatonin since it lightens frog skin by contracting melanophores (i.e., melanophore-contracting hormone; Greek: μέλας = black - τόνος = tension, in the sense of contraction)[3] and because it is a chemical derivative of serotonin[4].

Melatonin synthesis and secretion were initially described in the pineal gland; its synthesis is regulated by the environmental light/dark cycle via the suprachiasmatic nucleus. The pineal gland is part of the photoneuroendocrine system, and acts as a transducer, converting afferent neurally-coded photic information into melatonin, the chemical expression of darkness[5]. The precursors of melatonin are the aromatic amino acid tryptophan and serotonin. The pineal parenchyma captures tryptophan; then it is hydroxylated and decarboxylated to serotonin. Serotonin is then N-acetylated by the enzyme aralkylamine N-acetyltransferase (AANAT), and finally, the product is converted into melatonin by the N-Acetylserotonin-O-methyltransferase (ASMT), also known as hydroxyindole-O-methyltransferase (HIOMT)[6]. Melatonin is not stored within the gland but quickly diffuses into the bloodstream and cerebrospinal fluid[7].

Plasma melatonin displays a bi-exponential decay with a first distribution half-life of 2 minutes and a second metabolic half-life of 20 min[8]. Circulating melatonin is taken up by hepatocytes where hepatic P450 monooxygenases such as CYP1A1, CYP1A2, and CYP1B1 catalyze it 6-hydroxylation[8]. The liver clears more than 90% of the indolamine[9]. The product is then conjugated to sulfate, released from liver cells and excreted in the urine as 6-sulfatoxymelatonin[10]. Other sites of melatonin catabolism may be the lung, kidney, small intestine, and brain[11]. Also, all cells avidly take up melatonin[12].

The concentrations of melatonin in the blood follow a circadian pattern with a low level during daytime and high levels at night[13]. During the day, light impacts the retina and blunts via a polysynaptic neural circuit where it inhibits melatonin production[14]. The mechanism causing the rhythmicity in melatonin production and secretion varies among different species[15]. In humans, rhythmicity is acquired between the second and third month of postdelivery life; in utero, the mother provides melatonin to the foetus[16,17]. In rodents, the regulation of the biosynthetic pathway is mediated by transcriptional modifications in the AANAT gene[18]; while in humans, this control is achieved by post-transcriptional/translational modifications[19].

When pineal gland is removed, barely measurable melatonin quantities are measured in plasma and urine of rats, and the blood levels of the agent no longer exhibit the classic circadian oscillation[20]. Since small amounts of melatonin are present after pinealectomy, the existence of extrapineal melatonin sources has been proposed and proven. The presence and effects of melatonin

in the cell were initially thought to be exclusively related to its uptake from the blood. Notwithstanding it has become evident, that perhaps all cells have the molecular machinery to synthesize the indoleamine[14,21,22].

Melatonin is detected in almost all biological fluids including cerebrospinal fluid, saliva, bile, ovarian follicular fluid, anterior chamber of the eye, synovial fluid, amniotic fluid, and breast milk[23,24]. The expression of mRNAs encoding two critical enzymes of the melatonin biosynthesis was found in several tissues using the reverse transcription-polymerase chain reaction (RT-PCR)[25]. Among these tissues are thymus, spleen, heart, skeletal muscle, liver, stomach, gut, salivary glands, thyroid gland, placenta, testes, ovaries, kidney, lens cochlea, skin, blood vessels, pancreas, cerebral cortex, and striatum[23,26,27]. Notably, the gastrointestinal tract emerges as a major extrapineal melatonin-producing site since its melatonin concentration surpasses blood levels by 10–100 times[28], and it is much larger than the pineal gland[29].

Of interest in this report is the analysis of the melatonin intracellular distribution. Melatonin levels within the cell membrane, cytosol, nucleus, and mitochondria do not oscillate with a 24-hour cycle, although there are slight variations in their concentrations during the light-dark cycle. The highest concentration of melatonin was found in the cell membrane, followed by the mitochondria, nucleus, and cytosol. Interestingly, melatonin content in all subcellular organelles was higher than those found in plasma, and it was not diminished by pinealectomy[14]. The results suggest that individual cells produce melatonin[30].

Based on the endosymbiotic theory, mitochondria evolved from the engulfment of the bacteria phylum α-proteobacteria[31]. Rhodospirillum rubrum, a member of the α-proteobacteria family, might be phylogenetically the oldest organism which can synthesize melatonin[32]. It has been hypothesized that in most organisms, melatonin biosynthesis occurs in mitochondria and then transferred from mitochondria and chloroplasts (where melatonin is also produced) to other compartments of the cell[33]. In the primitive eukaryotes, mitochondria and chloroplasts are presumably the exclusive sites of melatonin synthesis. Through evolution, the genes for melatonin biosynthesis contained in the bacteria were slowly integrated into the nuclear genome[22] and mitochondria have retained the innate capacity to produce melatonin[34]. The indolamine synthesis in the mitochondria may be function in endocrine, paracrine and autocrine or "auto-mitocrine" capacity[35] (Figure 1).

Pinealocytes contain more mitochondria than the ones found in neurons, although pineal cells do not have the highest metabolic rate[36]. During darkness, there is an increase in the size of the mitochondria in pinealocytes, as well as more fusion events[37]. These changes correlate with a higher melatonin synthesis by the gland. The morphologies of pinealocyte mitochondria show apparent dynamic alterations related to their fission, fusion and mitophagic processes during a 24-hour period. If surgical removal of the superior cervical ganglia is performed, the pineal gland is denervated and therefore loses its ability to synthesize melatonin. A decrease in the mitochondrial mass is also observed in the denervated pineal gland[38].

In most cells, mitochondria emerge as the primary site of melatonin production[21]. The enzymes involved in its biosynthesis, as well as its local production, have been found in the mitochondria in oocytes[39], pinealocytes[40,41], plant cells[42], embryonic cells (two-cell, eight-cell, and blastocyst stages)[43] and in nervous system cells[35]. He and colleagues demonstrated that mitochondria are the primary sites for melatonin synthesis in oocytes and that they produce a large quantity of melatonin during their maturation. They showed that cultured mitochondria in the presence of serotonin produce more melatonin than non-serotonin enriched medium. AANAT was detected in mitochondria of oocytes by immuno-electron microscopy[39]. Suofu et al. provided evidence that AANAT and ASMT are present in brain mitochondria as well as the synthesis and release of melatonin from the mitochondrial matrix. They also found that brain AANAT is not regulated in a circadian rhythm. Also, the chaperone 14-3-3, which inhibits AANAT degradation and increases the affinity for its substrate is also expressed in mitochondria[35]. The mechanism by which these enzymes are regulated in mitochondria remains unknown. While Suofu et al. identified that AANAT exclusively localized in the matrix of mitochondria, Yang et al. reported that a significant portion of AANAT localizes close to the mitochondria out-membrane[43] (Figure 1).

Mitochondria are not the only a site of melatonin production but also of its degradation. The mitochondrial P450 complex participates together with the microsomal system in the hepatic degradation of the indolamine in rat liver. More specifically CYP2E1 and CYP3A seem to be involved[44]. A recent study *in vitro* demonstrates that CYP1B1 is localized in the mitochondria of cancer cells; This cytochrome metabolizes melatonin to form N-acetyl-serotonin, prompting mitochondria-dependent apoptosis[45]. In the photosynthetic bacterium, *Rhodospirillum rubrum*, mitochondrial cytochrome C metabolizes melatonin to N-acetyl-N-formyl-5-methoxykynuramine (AFMK) through the catalytic iron atom[46] (Figure 1).

3   MELATONIN MECHANISMS OF ACTION AT THE MITOCHONDRIAL LEVEL

In addition to its chronobiological actions, melatonin intervenes in numerous physiological and pathophysiological processes[47]. A recent meta-analysis data by Posadzki et al. show that the indoleamine exerts its effects by the following processes: anti-oxidative, anti-inflammatory, anti-apoptotic, anti-nociceptive, anti-hypertensive, cytoprotective (neuroprotective, cardioprotective, nephroprotective), anti-proliferative, oncostatic, and by enhancing mitochondrial function, and protecting nuclear and mitochondrial DNA and regulating homeostasis[48].

Due to its amphipathic nature, melatonin crosses membranes readily, which then exerts effects independent of its membrane receptors in all cells[49]. However, the presence of several targets could explain melatonin´s pleiotropism. Among others targets are: melatonin G-protein-coupled receptors in the plasma membrane (e.g., MT1 and MT2)[50]; binding to intracellular proteins, such as calmodulin, sirtuins, quinone reductase, tubulin, and calreticulin; and orphan nuclear receptors (e.g., RZR/RORα: retinoid-related orphan nuclear hormone receptor and GPR50: X-linked melatonin-related orphan receptor)[51]. Melatonin´s antioxidant proprieties are the result of direct free radical scavenging and indirect pathways which involves membrane or mitochondrial receptors[52,53]. It was recently reported that the transmission of microRNAs by exosomes might contribute to melatonin´s actions[54].

Although its amphipathic nature allows a widespread distribution, melatonin may have several means by which it penetrates lipid membranes[55]. Melatonin is remarkably high concentrated intracellularly, particularly within the mitochondria. The indolamine might use transport proteins to promote its movement across membranes[56]. Glucose transporter type 1 (GLUT-1), a member of the solute carrier family 2 glucose transporters family (SLC2/GLUT), was correlated with the intracellular transport of melatonin into various cell lines[57]. Considering that GLUT-1 is also expressed in the mitochondrial membrane[58], it can be assumed that the transporter act as a gatekeeper for the indolamine. Recently, it was reported that melatonin and its sulfation metabolites are the substrates for the oligopeptide transporter (PEPT) 1/2 and organic anion transporter (OAT) 3, respectively[59]. This evidence may explain why melatonin can be accumulated in mitochondria against a concentration gradient (Figure 1).

The effects of melatonin on mitochondrial function have been widely described[30,60–62]. While receptors are typically located in either the cytosol, the nucleus or the plasma membrane, they are also detected in the mitochondria[63,64]. Recently, the presence of melatoninergic receptors in the mitochondrial membrane has been reported. Wang and colleagues found significant levels of the MT1 receptor in mitochondria from wild-type mice brains[65]. MT1 receptors seem to be located in the outer mitochondrial membrane of neurons[66]. The activation of the mitochondrial MT1/G protein system inhibits cytochrome c release and blocks adenylate cyclase activity[35]. A recent investigation also reveals the expression of MT1 and MT2 on the mitochondrial membranes of gastric endothelial cells; MT1 was expressed in the mitochondrial membrane while MT2 was also shown in the nucleus[67] (Figure 1).

Melatonin and its metabolites perform a marked role as direct free radical scavengers and as a regulator of mitochondrial bioenergetic function[68]. Melatonin prevents oxidative stress/nitrosative stress-induced mitochondrial dysfunction[69]. The indolamine can directly scavenge two •OH radicals[70]. After scavenging reactive oxygen species (ROS), melatonin is converted into several principal metabolites, cyclic 3-hydroxymelatonin (3-OHM), AFMK and N-acetyl-5-methoxykynuramine (AMK) (Figure 1), all of which have antioxidant and neuroprotective properties[71]. Besides its capacity to directly scavenge oxidative species, melatonin has several indirect actions, e.g., it can increase the superoxide dismutase (SOD) and catalase activities and induces the expression and activity of glutathione peroxidase and glutathione reductase, adjusting the redox cycle of glutathione[72-74]. Melatonin enhances the transfer of electrons between the mitochondrial respiratory complexes and ATP synthesis by reducing electron leakage and free radical formation[75,76]. Another "mito-protective" and antioxidant mechanism of melatonin is the inhibition of the ubiquitination of the redox-sensitive transcription factor nuclear factor erythroid 2-related factor (Nrf2), which plays a significant role in the defense against oxidative stress, via the induction of antioxidant enzymes[77,78].

A critical function of melatonin is to reduce mitochondrial oxidation and inflammation. Numerous reports have shown that mitochondria are an essential target for melatonin's protection[79]. The mechanisms implicated include the preservation of function complex I and III, the inhibition of mitochondrial permeability transition pore opening, and reducing cytochrome c release[80]. The mitochondrial changes in a mouse model of global inflammation were modulated by melatonin treatment, which produced beneficial effects on mitochondrial morphology and dynamics by mitofusin-2 (Mtf2) and the intrinsic apoptotic cascade modulation[81]. The Mtf2 protein is involved in mitochondrial reduction; nonetheless, infliximab -a tumor necrosis factor alpha (TNF-α) inhibitor-, normalizes Mtf2 levels. Because of these findings, Carraro et al. propose that inflammation is the critical mechanism activated in the progress of mitochondrial abnormalities[82].

With regard to autophagy, the indolamine appears to have different effects depending on cell type. Some researchers report that melatonin inhibits autophagy in cells and organisms which are exposed to various stressors. Melatonin actions might be dual, inducing or inhibiting autophagy, based on cellular requirements and oxidative stress levels[83]. Mitophagy, a particular form of autophagy, by which the elimination of whole mitochondria is produced[84], it appears to be also regulated by melatonin. Melatonin may enhance mitophagy by increasing the expression of mitophagy proteins (Beclin-1 and microtubule-associated protein 1A/1B-light chain 3β -LC-3β-)[85] and (Parkin and PTEN-induced kinase 1 -PINK-1-)[86], by activation of Parkin translocation via inhibition of mammalian sterile 20-like kinase 1 (Mst1)[87], and by enhancing adenosine 5′-monophosphate-activated protein kinase (AMPK), among others[88]. However, some recent articles suggest a negative role of melatonin concerning mitophagy[89,90].

4   RENAL MITOCHONDRIAL PROTECTION BY MELATONIN

There is a clear correlation between the global increase in cardiovascular disease (CVD) which is linked to a higher frequency of chronic kidney disease (CKD). Even though cardiovascular events increase in patients with CKD, there is a lack of specific treatments for these related diseases. However, there are promising experimental and clinical data suggesting that the modulation of inflammation and oxidative stress could mitigate the progression of kidney disease[91]. Related to this, endogenous melatonin secretion is impaired during CKD[92]. Moreover, the impairment is related to the degree of CKD and worsens with the progression of CKD[93]. Similarly, low melatonin levels have been demonstrated to correlate with the degree of inflammation in patients; this is consistent with the melatonin effects on the regulation of energy metabolism[94], and its anti-inflammatory actions[95].

Melatonin functions as an antioxidant and anti-inflammatory molecule in every cell, including renal cells. However, the definition of melatonin´s specific action in renal tissue is quite recent. Melatonin exerts an antioxidant effect in the kidneys of Zucker diabetic fatty rat, a model of type 2 diabetes. The mechanism suggested includes the decrease of NADPH oxidase and the increase of glutathione peroxidase activities[96]. Uncoupling ROS generation and elimination from mitochondria lead to inflammation, cell death, tissue damage, and disease progression (Figure 2).

Mitochondrial function is also a crucial point of interaction for circadian alterations in CKD associated with renin-angiotensin system (RAS) up-regulation[97,98]. RAS tonically modulates melatonin synthesis by inducing the tryptophan hydroxylase activity via angiotensin II receptors type 1 (AT-1)[99]. Accordingly, a relationship between angiotensin II and melatonin has been suggested in the modulation of circadian rhythms (Figure 2). Since melatonin improves abnormalities in CKD, the beneficial effects of RAS blockade may be enhanced with melatonin therapy[100]. Consistent with this, we found that the kidney in a CKD model exhibited an AT-1 downregulation linked to the melatonin-cytoprotective effect. Specifically, melatonin restored vitamin D receptors/heat shock protein 70 (VDR/Hsp70) and decreased AT-1 expression, oxidative stress, fibrosis, and apoptosis, as well as conserved mitochondrial ultrastructure by prevented mitochondrial edema with dilated cristae and high NADPH oxidase activity (Figure 2). Because of these findings, we postulate a possible feedback (or reciprocal regulation) between AT-1 and melatonin (Figure 2). Mitochondrial dysfunction by an over-activation of the RAS linked to the NADPH oxidase activity is a consequence shared by a multiplicity of pathologies that impact renal function such as obesity, diabetes, metabolic syndrome, impaired glucose tolerance, and

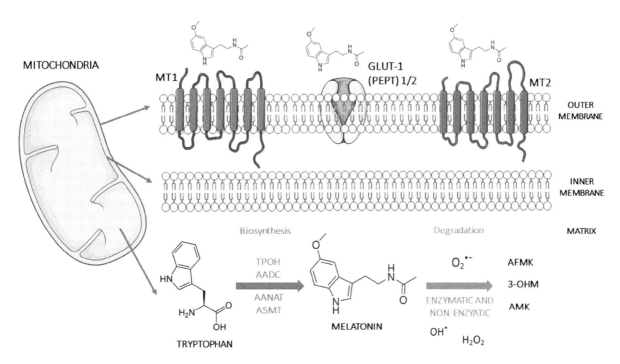

FIGURE 1 "Mitocrine" effect of melatonin. MT-1 and MT-2 G-protein-coupled receptors have been found on the outer mitochondrial membrane. Although its amphipathic nature, GLUT-1, (PEPT) 1/2 and (OAT) 3 allow melatonin transport within the mitochondria against a concentration gradient. The enzymes involved in melatonin biosynthesis and degradation are found in the mitochondrial matrix. Tryptophan-5-Hydroxylase, (TPOH); Aromatic Amino Acid Decarboxylase, (AADC); aralkylamine *N*-acetyltransferase, (AANAT); *N*-*A*cetylserotonin-*O*-methyltransferase, (ASMT); hydroxyl radical, (•OH); hydrogen peroxide, (H2O2); superoxide anion, (O−₂); cyclic 3-hydroxymelatonin, (3-OHM), *N*-acetyl-*N*-formyl-5-methoxykynuramine, (AFMK); *N*-acetyl-5-methoxykynuramine, (AMK); melatonin receptor type 1, (MT-1); Melatonin receptor type 2 (MT-2); Glucose transporter type 1, (GLUT-1); oligopeptide transporter 1/2 (PEPT 1/2); organic anion transporter 3, (OAT 3). This figure was created using Servier Medical Art templates, which are licensed under a Creative Commons Attribution 3.0 Unported License; https://smart.servier.com.

dyslipidemias. In cardiorenal patients, decreased levels of vitamin D have been reportedly linked to higher RAS activity[101]. VDR activators antagonize RAS effects, as well as modulate anti-inflammatory and antifibrotic actions. The high AT-1 expression and NADPH activity are reversed in mitochondrial fractions after treatment with VDR agonists[102]. CKD is linked to inflammation with a substantial energetic cost. Recent reports suggest that vitamin D can act as a therapeutic agent for inflammation of chronic disease by modulating cellular bioenergetics[103].

Impaired release and reduced bioavailability of nitric oxide (NO) are associated with inflammatory pathologies. Moreover, a reduced NO release induces Hsp70 with effects against oxidative stress, inflammation, and apoptosis during CVD[104]. In this context, melatonin improves signal transduction by Akt phosphorylation induction and NO levels in a cardiovascular mice model[105]. Melatonin increases the heat shock protein response (HSP) as part of its antioxidant protective effects[106]. Notably, melatonin decreases renal injury by modulation of NO, oxidative markers, and Hsp70 expression[107] (Figure 2). Thus, clarifying the signaling pathways and shows the role of Hsp70 is relevant to the

application of new treatments. Fujimoto et al. evaluated cytokine production linked to Hsp70 protein levels[108]. We recently reviewed the opposing actions of extracellular versus intracellular Hsp70 on NF-κB pathway activation[109].

The Hsp70 induction by melatonin also supports an additional mechanism linked to translocase in the outer mitochondrial membrane 70 (Tom70). Hsp70 interaction with Tom70 is essential for the initiation of the mitochondrial import processes[110] (Figure 2). Tom70 is required for the recognition, unfolding, and translocation of amino acids into the mitochondria. This protein is a crucial member of the mitochondrial outer and inner membrane transport systems. For example, specifically in diabetes, Tom70 and the inner mitochondrial membrane 44 protein (Timm44) are significantly decreased, and the binding of nuclear-encoded mitochondrial transcription factor A (TFAM) with Tom70, Timm44, and mitochondrial DNA (mtDNA) is impaired. Hence, hyperglycemia affects the ability of TFAM to access the mitochondria[111]. Similarly, Wang et al. demonstrated that Timm44 alters mitochondrial fusion and fission dynamics and protects from type 2 diabetes[112]. Also, melatonin protection of the mitochondrial function is associated with suppression of TNFα expression, which is responsible for dynamin-related protein 1-mediated mitochondrial fission (Drp1)[113] (Figure 2).

It is widely known that hypertension, obesity, and diabetes likely contribute to the increasing incidence of CKD worldwide and it requires novel strategies for treatment. The kidney has been recognized as a critical organ in the development of inflammation during such diseases. Several studies have demonstrated that treatment with antioxidants improves hypertensive conditions. Thus, an exciting finding of melatonin and renal protection during hypertension was observed by Nava et al., who showed that melatonin reduces hypertension in spontaneously hypertensive rats (SHR) with a decrease in the renal inflammation. Moreover, they suggested that the antihypertensive effects of melatonin are related to a decreased activation of NF-κB[114]. These contributions support new studies to evaluate whether melatonin would modify the course of CKD. Melatonin reduces oxidative stress, renal inflammation, proteinuria, and progression of renal damage in rats with renal mass reduction[115] (Figure 2).

In close association with the previously mentioned relationship with NO and renal protection, more recently it was shown in a model of hypertension that melatonin exerts a renoprotective effect associated with NO bioavailability. The low NO levels occur during CKD, where NO depletion increases the kidney damage. However, melatonin treatment produced renal protection due to a reduction of oxidative stress and restored NO production in the kidney[116]. Also,

melatonin attenuated urine protein excretion, serum creatinine and reduced oxidative stress in renal tissues. The kidney improvements were correlated with the increases in the inducible nitric oxide synthase (iNOS) and intercellular adhesion molecule 1 (ICAM-1) expression[117]. According, melatonin associated with exendin-4 was more efficient in reducing renal function in cardiorenal syndrome. Notably, the combined treatment produced, at the renal level, a reduction of protein expression of inflammation (TNF-α/NF-κB/MMP-9/iNOS/RANTES), oxidative stress (NOX-1/NOX-2/NOX-4/oxidized protein), apoptosis (cleaved caspase-3/cleaved PARP/Bax), DNA-damaged markers (γ-H2AX) and fibrosis (p-mad3/TFG-β)[118].

Moreover, melatonin regulates renal transcriptome and prevents prenatal L-NAME-induced fetal programming of hypertension. Interesting, genes belonging to the RAS family, including the arachidonic acid metabolism pathways, are involved in hypertension-induced by maternal NO depletion. However, maternal melatonin reprogrammed the RAS and arachidonic acid pathway. The early intervention with melatonin could improve the redox imbalance (NO and ROS) during pregnancy and reprogram the future burden of hypertension[119]. The melatonin prevention of the kidney injury was confirmed in a high salt diet-induced hypertension model. The positive effect is not mediated by blood pressure but by a direct antioxidative effect.

During CKD there is impaired night-time melatonin secretion that correlates negatively with intrarenal RAS

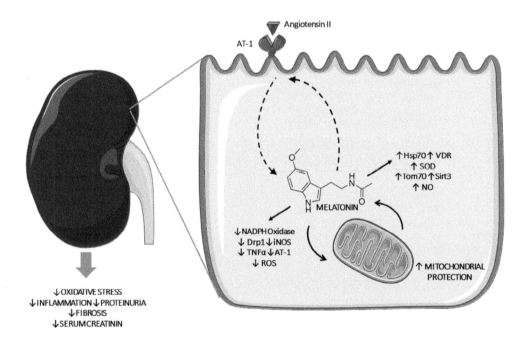

FIGURE 2　Graphical overview of renal mitochondrial protection by melatonin. AT-1, angiotensin II receptors type 1; Hsp70, heat shock protein 70; VDR, vitamin D receptors; SOD; Tom70, translocase in the outer mitochondrial membrane 70; Sirt3, sirtuin 3; NO, nitric oxide; Drp1, dynamin-related protein 1; iNOS, nitric oxide synthase; TNFα, tumor necrosis factor alpha; ROS, reactive oxygen species. This figure was created using Servier Medical Art templates, which are licensed under a Creative Commons Attribution 3.0 Unported License; https://smart.servier.com.

activity. Ishigaki et al. showed that melatonin ameliorates intrarenal RAS in a 5/6 nephrectomy rat model; this generally causes increases in intrarenal angiotensinogen, AT-1, and angiotensin II, accompanied by elevated blood pressure, oxidative stress, and interstitial fibrosis in the remnant kidneys[120]. With special attention to mitochondrial function, 5/6 nephrectomy shifted mitochondrial dynamics to fusion, which was evidenced by increased optic atrophy 1 and mitofusin 1 (Mfn1) and decreased fission 1 and dynamin-related protein 1 expression. Related to this those drugs usually used in the treatment of hypertension or cardiovascular disease, such as AT-1 receptor blockers, have shown auspicious "preventive" intracellular anti-inflammatory/antioxidant activity in addition to their original pharmacological actions.

The kidney intracellular protection was associated with an elevated number of mitochondria and upregulation of the prosurvival genes such as nicotinamide phosphoribosyltransferase (Nampt) and sirtuin 3 (Sirt3)[121]. Sirt3 is a member of the family that is localized mostly to the mitochondria and protects against inflammation and oxidative stress-related diseases, including hypertension. Melatonin elevated Sirt3 stimulated SOD activity and suppressed mitochondrial oxidative stress through AMPK[122] (Figure 2). Melatonin's high concentrations and multiple actions as an anti-inflammatory/antioxidant provide substantial protection to mitochondria which are exposed to injury[22].

Some of the reported findings are inconsistent regarding the effects of melatonin and the reduction of blood pressure. However, there is total agreement on antioxidant and anti-inflammatory effects that both, directly and indirectly, improve blood pressure. There are multiples studies involving kidney injury with hypertension consequences. Melatonin reduces oxidative stress, chronic inflammation, and apoptosis[123]. The benefits include the reduction of fibrosis, proteinuria, and impairment of parenchymal cells. The anti-inflammatory actions of melatonin induce beneficial effects at numerous pathophysiological levels related to CKD linked to hypertension. It is hoped that these findings will stimulate clinical trials to determine the possible efficacy of melatonin as a promising treatment for this growing pathological condition.

5 CONCLUSION

Melatonin previously thought of as a hormone exclusively produced in the pineal gland is present in the vast majority of cells and tissues. Mitochondria have emerged as a biosynthetic organelle, degrading and at the same time, consuming melatonin, generating a "mitocrine" system. This organelle has biosynthetic enzymes, degrading enzymes, receptors, and transporters that constitute an intracellular signaling system for melatonin. Alterations in the mitochondrial signaling pathways associated with the loss of melatonin-mediated regulation induce in renal cells

a functional decoupling with the development of renal and finally cardiovascular tissue pathology. Although it is an endogenous compound, melatonin would be useful as an exogenous pharmacological tool for many diseases; additional clinical trials are needed to demonstrate its translational power.

REFERENCES

1 Tan, D.-X., et al. The changing biological roles of melatonin during evolution: from an antioxidant to signals of darkness, sexual selection and fitness. *Biological Reviews.* **85**, 607–623 (2010). doi:10.1111/j.1469-185X.2009.00118.x

2 Lerner, A. B., Case, J. D., Takahashi, Y., Lee, T. H. & Mori, W. Isolation of melatonin, the pineal gland factor that lightens melanocytes[1]. *J Am Chem Soc.* **80**, 2587–2587 (1958).

3 Claustrat, B. & Leston, J. Melatonin: physiological effects in humans. *Neurochirurgie.* **61**, 77–84 (2015).

4 Sánchez, S., et al. The pineal gland: functional connection between melatonin and immune system in birds. *Biogenic Amines.* **18**, 147–176 (2004).

5 Reiter, R. J. Melatonin: the chemical expression of darkness. *Mol Cell Endocrinol.* **79**, C153–1158 (1991).

6 Chattoraj, A., Liu, T., Zhang, L. S., Huang, Z. & Borjigin, J. Melatonin formation in mammals: in vivo perspectives. *Rev Endocr Metab Dis.* **10**, 237–243 (2009).

7 Reiter, R. J., Tan, D. X., Kim, S. J. & Cruz, M. H. C. Delivery of pineal melatonin to the brain and SCN: role of canaliculi, cerebrospinal fluid, tanycytes and Virchow-Robin perivascular spaces. *Brain Struct Funct.* **219**, 1873–1887 (2014).

8 Claustrat, B., Brun, J. & Chazot, G. The basic physiology and pathophysiology of melatonin. *Sleep Med Rev.* **9**, 11–24 (2005).

9 Skene, D. J., et al. Contribution of CYP1A2 in the hepatic metabolism of melatonin: studies with isolated microsomal preparations and liver slices. *J Pineal Res.* **31**, 333–342 (2001).

10 Ma, X., et al. Urinary metabolites and antioxidant products of exogenous melatonin in the mouse. *J Pineal Res.* **40**, 343–349 (2006).

11 Tian, X., et al. Sulfation of melatonin: enzymatic characterization, differences of organs, species and genders, and bioactivity variation. *Biochem Pharmacol.* **94**, 282–296 (2015).

12 Reiter, R. J., et al. Melatonin, a full service anti-cancer agent: inhibition of initiation, progression and metastasis. *Int J Mol Sci.* **18**, 843 (2017).

13 Arendt, J. Melatonin and human rhythms. *Chronobiol Int.* **23**, 21–37 (2006).

14 Venegas, C., et al. Extrapineal melatonin: analysis of its subcellular distribution and daily fluctuations: brain and liver extrapineal melatonin distribution. *J Pineal Res.* **52**, 217–227 (2012).

15 Falcón, J., et al. Structural and functional evolution of the pineal melatonin system in vertebrates. *Ann NY Acad Sci.* **1163**, 101–111 (2009).

16 Ackermann, K. & Stehle, J. H. Melatonin synthesis in the human pineal gland: advantages, implications, and difficulties. *Chronobiol Int.* **23**, 369–379 (2006).

17 Reiter, R. J., Tan, D. X., Korkmaz, A. & Rosales-Corral, S. A. Melatonin and stable circadian rhythms optimize

maternal, placental and fetal physiology. *Hum Reprod Update.* **20**, 293–307 (2014).

18 Schomerus, C. Mechanisms regulating melatonin synthesis in the mammalian pineal organ. *Ann N Y Acad Sci.* **1057**, 372–383 (2005).

19 Ackermann, K., *et al.* Characterization of human melatonin synthesis using autoptic pineal tissue. *Endocrinology.* **147**, 3235–3242 (2006).

20 Ozaki, Y. & Lynch, H. J. Presence of melatonin in plasma and urine of pinealectomized rats[1]. *Endocrinology.* **99**, 641–644 (1976).

21 Tan, D.-X., *et al.* Mitochondria and chloroplasts as the original sites of melatonin synthesis: a hypothesis related to melatonin's primary function and evolution in eukaryotes. *J Pineal Res.* **54**, 127–138 (2012).

22 Reiter, R. J., *et al.* Melatonin as a mitochondria-targeted antioxidant: one of evolution's best ideas. *Cell Mol Life Sci.* **74**, 3863–3881 (2017).

23 Acuña-Castroviejo, D., *et al.* Extrapineal melatonin: sources, regulation, and potential functions. *Cell Mol Life Sci.* **71**, 2997–3025 (2014).

24 Reiter, R. J., Tan, D.-X. & Fuentes-Broto, L. Melatonin: a multitasking molecule. *Prog Brain Res.* **181**, 127–151 (2012).

25 Stefulj, J., *et al.* Gene expression of the key enzymes of melatonin synthesis in extrapineal tissues of the rat. *J Pineal Res.* **30**, 243–247 (2001).

26 Söderquist, F., Hellström, P. M. & Cunningham, J. L. Human gastroenteropancreatic expression of melatonin and its receptors MT1 and MT2. *PLoS ONE.* **10**, e0120195 (2015).

27 Shimozuma, M., *et al.* Expression and cellular localizaion of melatonin-synthesizing enzymes in rat and human salivary glands. *Histochem Cell Biol.* **135**, 389–396 (2011).

28 Bubenik, G. A. Gastrointestinal melatonin: localization, function, and clinical relevance. *Digest Dis Sci.* **47**, 2336–2348 (2002).

29 Bubenik, G. A. & Brown, G. M. Pinealectomy reduces melatonin levels in the serum but not in the gastrointestinal tract of rats. *Neurosignals.* **6**, 40–44 (1997).

30 Reiter, R. J., *et al.* Mitochondria: central organelles for melatonin⬚s antioxidant and anti-aging actions. *Molecules.* **23**, 509 (2018).

31 Gray, M. W. Mitochondrial evolution. *Cold Spring Harb Perspect Biol.* **4**, 1–16 (2012).

32 Manchester, L. C., Poeggeler, B., Alvares, F. L., Ogden, G. B. & Reiter, R. J. Melatonin immunoreactivity in the photosynthetic prokaryote Rhodospirillum rubrum: implications for an ancient antioxidant system. *Cell Mol Biol Res.* **41**, 391–395 (1995).

33 Manchester, L. C., *et al.* Melatonin: an ancient molecule that makes oxygen metabolically tolerable. *J Pineal Res.* **59**, 403–419 (2015).

34 Tan, D.-X., *et al.* Fundamental issues related to the origin of melatonin and melatonin isomers during evolution: relation to their biological functions. *Int J Mol Sci.* **15**, 15858–15890 (2014).

35 Suofu, Y., *et al.* Dual role of mitochondria in producing melatonin and driving GPCR signaling to block cytochrome c release. *Proc Natl Acad Sci USA.* **114**, E7997–E8006 (2017).

36 Tan, D.-X., Manchester, L. C., Qin, L. & Reiter, R. J. Melatonin: a mitochondrial targeting molecule involving mitochondrial protection and dynamics. *Int J Mol Sci.* **17**, 2124 (2016).

37 Karasek, M., *et al.* Comparison of the rat pinealocyte ultrastructure with melatonin concentrations during daytime and at night. *J Pineal Res.* **9**, 251–257 (1990).

38 Karasek, M., Zielinska, A., Marek, K. & Swietoslawski, J. Effect of superior cervical ganglionectomy on the ultrastructure of pinealocytes in the Djungarian hamster (Phodopus sungorus): quantitative study. *Neuro Endocrinol Lett.* **23**, 443–446 (2002).

39 He, C., *et al.* Mitochondria synthesize melatonin to ameliorate its function and improve mice oocyte's quality under in vitro conditions. *Int J Mol Sci.* **17**, 939 (2016).

40 Kerényi, N. A., Sótonyi, P. & Somogyi, E. Localizing acethyl-serotonin transferase by electron microscopy. *Histochem.* **46**, 77–80 (1975).

41 Kerényi, N. A., Balogh, I., Somogyi, E. & Sótonyi, P. Cytochemical investigation of acetyl-serotonin-transferase activity in the pineal gland. *Cell Mol Biol Incl Cyto Enzymol.* **25**, 259–262 (1979).

42 Wang, L., *et al.* Plant mitochondria synthesize melatonin and enhance the tolerance of plants to drought stress. *J Pineal Res.* **63**, e12429 (2017).

43 Yang, M., *et al.* Aanat knockdown and melatonin supplementation in embryo development: involvement of mitochondrial function and DNA methylation. *Antioxid Redox Signal.* (2018). doi:10.1089/ars.2018.7555

44 Semak, I., Korik, E., Antonova, M., Wortsman, J. & Slominski, A. Metabolism of melatonin by cytochrome P450s in rat liver mitochondria and microsomes. *J Pineal Res.* **45**, 515–523 (2008).

45 Yu, Z., *et al.* Mitochondrial cytochrome P450 (CYP) 1B1 is responsible for melatonin-induced apoptosis in neural cancer cells. *J Pineal Res.* **65**, e12478 (2018).

46 Tan, D.-X., Manchester, L. C., Esteban-Zubero, E., Zhou, Z. & Reiter, R. J. Melatonin as a potent and inducible endogenous antioxidant: synthesis and metabolism. *Molecules.* **20**, 18886–18906 (2015).

47 Hardeland, R., *et al.* Melatonin—A pleiotropic, orchestrating regulator molecule. *Prog Neurobiol.* **93**, 350–384 (2011).

48 Posadzki, P. P., *et al.* Melatonin and health: an umbrella review of health outcomes and biological mechanisms of action. *BMC Med.* **16**, 18 (2018).

49 Pandi-Perumal, S. R., *et al.* Melatonin: nature's most versatile biological signal? *FEBS J.* **273**, 2813–2838 (2006).

50 Cecon, E., Oishi, A. & Jockers, R. Melatonin receptors: molecular pharmacology and signalling in the context of system bias: melatonin receptor system bias. *Br J Pharmacol.* **175**, 3263–3280 (2018).

51 Emet, M., *et al.* A review of melatonin, its receptors and drugs. *Eurasian J Med.* **48**, 135–141 (2016).

52 Reiter, R. J., *et al.* Melatonin as an antioxidant: under promises but over delivers. *J Pineal Res.* **61**, 253–278 (2016).

53 Reiter, R. J., *et al.* Melatonin as an antioxidant: biochemical mechanisms and pathophysiological implications in humans. *Acta Biochimica Polonica* **50**, 1129–1146 (2003).

54 Hardeland, R. Extended signaling by melatonin. *Cell Cellular Lif Sci J.* **3**(2), 000123 (2018).

55 Hardeland, R. Melatonin — More than just a pineal hormone. *Biomed J Sci Tech Res.* **1**, 1–4 (2017).

56 Mayo, J. C., Sainz, R. M., González-Menéndez, P., Hevia, D. & Cernuda-Cernuda, R. Melatonin transport into mitochondria. *Cell Mol Life Sci.* **74**, 3927–3940 (2017).

57 Hevia, D., et al. Melatonin uptake through glucose transporters: a new target for melatonin inhibition of cancer. *J Pineal Res.* **58**, 234–250 (2015).

58 Kc, S., Cárcamo, J. M. & Golde, D. W. Vitamin C enters mitochondria via facilitative glucose transporter 1 (Glut1) and confers mitochondrial protection against oxidative injury. *FASEB J.* **19**, 1657–1667 (2005).

59 Huo, X., et al. Human transporters, PEPT1/2, facilitate melatonin transportation into mitochondria of cancer cells: an implication of the therapeutic potential. *J Pineal Res.* **62**, e12390 (2017).

60 Srinivasan, V., Spence, D. W., Pandi-Perumal, S. R., Brown, G. M. & Cardinali, D. P. Melatonin in mitochondrial dysfunction and related disorders. *Int J Alzheimers Dis.* **2011**, 1–16 (2011).

61 Hardeland, R. Melatonin and the electron transport chain. *Cell Mol Life Sci.* **74**, 3883–3896 (2017).

62 Jou, M.-J., et al. Melatonin protects against common deletion of mitochondrial DNA-augmented mitochondrial oxidative stress and apoptosis. *J Pineal Res.* **43**, 389–403 (2007).

63 Hardeland*, R. Mitochondrial hormone receptors – an emerging field of signaling in the cell's powerhouse. *BJSTR.* **1**, 001–004 (2017).

64 Lahuna, O. & Jockers, R. Signalisation mitochondriale des récepteurs couplés aux protéines G. *Biologie Aujourd'hui.* **212**, 21–26 (2018).

65 Wang, X., et al. The melatonin MT1 receptor axis modulates mutant Huntingtin-mediated toxicity. *J Neurosci.* **31**, 14496–14507 (2011).

66 Suofu, Y., Carlisle, D. L., Vilardaga, J.-P. & Friedlander, R. M. Reply to Ahluwalia et al.: contributions of melatonin receptors are tissue-dependent. *Proc Nat Aca Sci.* **115**, E1944–E1944 (2018).

67 Ahluwalia, A., Brzozowska, I. M., Hoa, N., Jones, M. K. & Tarnawski, A. S. Melatonin signaling in mitochondria extends beyond neurons and neuroprotection: implications for angiogenesis and cardio/gastroprotection. *Proc Natl Acad Sci USA.* **115**, E1942–E1943 (2018).

68 Galano, A. & Reiter, R. J. Melatonin and its metabolites vs oxidative stress: from individual actions to collective protection. *J Pineal Res.* **65**, e12514 (2018).

69 León, J., Acuña-Castroviejo, D., Escames, G., Tan, D.-X. & Reiter, R. J. Melatonin mitigates mitochondrial malfunction. *J Pineal Res.* **38**, 1–9 (2005).

70 Reiter, R., Paredes, S., Korkmaz, A., Jou, M.-J. & Tan, D.-X. Melatonin combats molecular terrorism at the mitochondrial level. *Interdiscip Toxicol.* **1**, 137–149 (2008).

71 Galano, A., Tan, D. X. & Reiter, R. J. On the free radical scavenging activities of melatonin's metabolites, AFMK and AMK. *J Pineal Res.* **54**, 245–257 (2013).

72 Lopez, L. C., Acuna-Castroviejo, D., Pino, A. del, Tejada, M. & Escames, G. Mitochondrial disorders therapy: the utility of melatonin. *Open Biol J.* **3**, 53–65 (2010).

73 Reiter, R. J., et al. A review of the evidence supporting melatonin's role as an antioxidant. *J Pineal Res.* **18**, 1–11 (1995).

74 Reiter, R. J., Tan, D.-X. & Allegra, M. Melatonin: reducing molecular pathology and dysfunction due to free radicals and associated reactants. *Neuro Endocrinol Lett.* **23**(Suppl 1), 3–8 (2002).

75 Acuña-Castroviejo, D., et al. Melatonin, mitochondria, and cellular bioenergetics. *J Pineal Res.* **30**, 65–74 (2001).

76 Escames, G., et al. The role of mitochondria in brain aging and the effects of melatonin. *Curr Neuropharmacol.* **8**, 182–193 (2010).

77 Santofimia-Castaño, P., et al. Melatonin induces the expression of Nrf2-regulated antioxidant enzymes via PKC and Ca2+ influx activation in mouse pancreatic acinar cells. *Free Radic Bio Med.* **87**, 226–236 (2015).

78 Ali, T., Rehman, S. U., Shah, F. A. & Kim, M. O. Acute dose of melatonin via Nrf2 dependently prevents acute ethanol-induced neurotoxicity in the developing rodent brain. *J Neuroinflammation.* **15**, 119 (2018).

79 Paradies, G., Petrosillo, G., Paradies, V., Reiter, R. J. & Ruggiero, F. M. Melatonin, cardiolipin and mitochondrial bioenergetics in health and disease. *J Pineal Res.* **48**, 297–310 (2010).

80 Petrosillo, G., et al. Protective effect of melatonin against mitochondrial dysfunction associated with cardiac ischemia- reperfusion: role of cardiolipin. *FASEB J.* **20**, 269–276 (2006).

81 Stacchiotti, A., et al. Mitochondrial and metabolic dysfunction in renal convoluted tubules of obese mice: protective role of melatonin. *PLoS ONE.* **9**, e111141 (2014).

82 Carraro, R. S., et al. Hypothalamic mitochondrial abnormalities occur downstream of inflammation in diet-induced obesity. *Mol Cell Endocrinol.* **460**, 238–245 (2018).

83 Coto-Montes, A., et al. Role of melatonin in the regulation of autophagy and mitophagy: a review. *Mol Cell Endocrinol.* **361**, 12–23 (2012).

84 Pickles, S., Vigié, P. & Youle, R. J. Mitophagy and quality control mechanisms in mitochondrial maintenance. *Curr Biol.* **28**, R170–R185 (2018).

85 Pan, P., Zhang, H., Su, L., Wang, X. & Liu, D. Melatonin balance the autophagy and apoptosis by regulating UCP2 in the LPS-Induced cardiomyopathy. *Molecules.* **23**, 675 (2018).

86 Cao, S., et al. Melatonin-mediated mitophagy protects against early brain injury after subarachnoid hemorrhage through inhibition of NLRP3 inflammasome activation. *Sci Rep.* **7**, 2417 (2017).

87 Wang, S., et al. Melatonin activates Parkin translocation and rescues the impaired mitophagy activity of diabetic cardiomyopathy through Mst1 inhibition. *J Cell Mol Med.* **22**, 5132–5144 (2018).

88 Kang, J.-W., Hong, J.-M. & Lee, S.-M. Melatonin enhances mitophagy and mitochondrial biogenesis in rats with carbon tetrachloride-induced liver fibrosis. *J Pineal Res.* **60**, 383–393 (2016).

89 Zhou, H., et al. Melatonin protects cardiac microvasculature against ischemia/reperfusion injury via suppression of mitochondrial fission-VDAC1-HK2-mPTP-mitophagy axis. *J Pineal Res.* **63**, e12413 (2017).

90 Zhou, H., et al. Melatonin suppresses platelet activation and function against cardiac ischemia/reperfusion injury via PPARⵡ/FUNDC1/mitophagy pathways. *J Pineal Res.* **63**, e12438 (2017).

91 Whaley-Connell, A. & Sowers, J. R. Obesity and kidney disease: from population to basic science and the search for new therapeutic targets. *Kidney Int.* **92**, 313–323 (2017).

92 Ishigaki, S., et al. Impaired endogenous nighttime melatonin secretion relates to intrarenal renin-angiotensin system activation and renal damage in patients with chronic kidney disease. *Clin Exp Nephrol.* **20**, 878–884 (2016).

93 Russcher, M., *et al.* The role of melatonin treatment in chronic kidney disease. *Front Biosci (Landmark Ed)*. **17**, 2644–2656 (2012).

94 Szewczyk-Golec, K., Woźniak, A. & Reiter, R. J. Inter-relationships of the chronobiotic, melatonin, with leptin and adiponectin: implications for obesity. *J Pineal Res*. **59**, 277–291 (2015).

95 Nabavi, S. M., *et al.* Anti-inflammatory effects of melatonin: a mechanistic review. *Crit Rev Food Sci Nutr.* 1–62 (2018). doi:10.1080/10408398.2018.1487927

96 Winiarska, K., *et al.* NADPH oxidase inhibitor, apocynin, improves renal glutathione status in Zucker diabetic fatty rats: a comparison with melatonin. *Chem Biol Interact.* **218**, 12–19 (2014).

97 Baltatu, O. C., Amaral, F. G., Campos, L. A. & Cipolla-Neto, J. Melatonin, mitochondria and hypertension. *Cell Mol Life Sci.* **74**, 3955–3964 (2017).

98 Ohashi, N., Isobe, S., Ishigaki, S. & Yasuda, H. Circadian rhythm of blood pressure and the renin-angiotensin system in the kidney. *Hypertens Res.* **40**, 413–422 (2017).

99 Baltatu, O., *et al.* Locally synthesized angiotensin modulates pineal melatonin generation. *J Neurochem.* **80**, 328–334 (2002).

100 Campos, L. A., *et al.* The angiotensin-melatonin axis. *Int J Hypertens.* **2013**, 1–7 (2013).

101 Vaidya, A., Sun, B., Larson, C., Forman, J. P. & Williams, J. S. Vitamin D3 therapy corrects the tissue sensitivity to angiotensin ii akin to the action of a converting enzyme inhibitor in obese hypertensives: an interventional study. *J Clin Endocrinol Metab.* **97**, 2456–2465 (2012).

102 García, I. M., *et al.* Role of mitochondria in paricalcitol-mediated cytoprotection during obstructive nephropathy. *Am J Physiol Renal Physiol.* **302**, F1595–11605 (2012).

103 Calton, E. K., Keane, K. N. & Soares, M. J. The potential regulatory role of vitamin D in the bioenergetics of inflammation. *Curr Opin Clin Nutr Metab Care.* **18**, 367–373 (2015).

104 Molina, M. N., Ferder, L. & Manucha, W. Emerging role of nitric oxide and heat shock proteins in insulin resistance. *Curr Hypertens Rep.* **18**, 1 (2016).

105 Sartori, C., *et al.* Melatonin improves glucose homeostasis and endothelial vascular function in high-fat diet-fed insulin-resistant mice. *Endocrinology.* **150**, 5311–5317 (2009).

106 Xu, W., *et al.* Melatonin enhances thermotolerance by promoting cellular protein protection in tomato plants. *J Pineal Res.* **61**, 457–469 (2016).

107 Motawi, T. K., Ahmed, S. A., A Hamed, M., El-Maraghy, S. A. & M Aziz, W. Melatonin and/or rowatinex attenuate streptozotocin-induced diabetic renal injury in rats. *J Biomed Res.* (2017). doi:10.7555/JBR.31.20160028

108 Fujimoto, E., Imai, A., Utsuyama, M. & Sato, K. Effects of in vitro heat shock on immune cells in diet-induced obese mice. *J Therm Biol.* **69**, 124–131 (2017).

109 Mazzei, L., Docherty, N. G. & Manucha, W. Mediators and mechanisms of heat shock protein 70 based

110 Fan, A. C. Y. & Young, J. C. Function of cytosolic chaperones in Tom70-mediated mitochondrial import. *Protein Pept Lett.* **18**, 122–131 (2011).

111 Santos, J. M. & Kowluru, R. A. Impaired transport of mitochondrial transcription factor A (TFAM) and the metabolic memory phenomenon associated with the progression of diabetic retinopathy. *Diabetes Metab Res Rev.* **29**, 204–213 (2013).

112 Wang, Y., *et al.* Translocase of inner mitochondrial membrane 44 alters the mitochondrial fusion and fission dynamics and protects from type 2 diabetes. *Metab Clin Exp.* **64**, 677–688 (2015).

113 Ding, M., *et al.* Dynamin-related protein 1-mediated mitochondrial fission contributes to post-traumatic cardiac dysfunction in rats and the protective effect of melatonin. *J Pineal Res.* **64**, e12447 (2018).

114 Nava, M., Quiroz, Y., Vaziri, N. & Rodriguez-Iturbe, B. Melatonin reduces renal interstitial inflammation and improves hypertension in spontaneously hypertensive rats. *Am J Physiol Renal Physiol.* **284**, F447–4454 (2003).

115 Quiroz, Y., Ferrebuz, A., Romero, F., Vaziri, N. D. & Rodriguez-Iturbe, B. Melatonin ameliorates oxidative stress, inflammation, proteinuria, and progression of renal damage in rats with renal mass reduction. *Am J Physiol Renal Physiol.* **294**, F336–3344 (2008).

116 Cheng, M.-C., Wu, T.-H., Huang, L.-T. & Tain, Y.-L. Renoprotective effects of melatonin in young spontaneously hypertensive rats with L-NAME. *Pediatrics & Neonatology.* **55**, 189–195 (2014).

117 Qiao, Y.-F., *et al.* Melatonin attenuates hypertension-induced renal injury partially through inhibiting oxidative stress in rats. *Mol Med Rep.* **13**, 21–26 (2016).

118 Chen, K.-H., *et al.* Combined therapy with melatonin and exendin-4 effectively attenuated the deterioration of renal function in rat cardiorenal syndrome. *Am J Transl Res.* **9**, 214–229 (2017).

119 Tain, Y.-L., Lee, C.-T., Chan, J. Y. H. & Hsu, C.-N. Maternal melatonin or N-acetylcysteine therapy regulates hydrogen sulfide-generating pathway and renal transcriptome to prevent prenatal NG-Nitro-L-arginine-methyl ester (L-NAME)-induced fetal programming of hypertension in adult male offspring. *Am J Obstet Gynecol.* **215**, 636.e1–636.e72 (2016).

120 Ishigaki, S., *et al.* Melatonin ameliorates intrarenal renin-angiotensin system in a 5/6 nephrectomy rat model. *Clin Exp Nephrol.* **22**, 539–549 (2018).

121 Benigni, A., *et al.* Disruption of the Ang II type 1 receptor promotes longevity in mice. *J Clin Invest.* **119**, 524–530 (2009).

122 Chen, Y., *et al.* Melatonin protects hepatocytes against bile acid-induced mitochondrial oxidative stress via the AMPK-SIRT3-SOD2 pathway. *Free Radic Res.* **49**, 1275–1284 (2015).

123 Hrenak, J., *et al.* Melatonin and renal protection: novel perspectives from animal experiments and human studies (review). *Curr Pharm Des.* **21**, 936–949 (2015).

cytoprotection in obstructive nephropathy. *Cell Stress Chaperon.* **20**, 893–906 (2015).

32 Mitochondrial Dysfunction Affecting the Peripheral Nervous System in Diabetic Neuropathy and Avenues for Therapy

Jennifer Jossart and Taylor N. Dennis
Department of Biochemistry, University of California, Riverside, Riverside, CA 92010, USA

J. Jefferson P. Perry
Department of Biochemistry, University of California, Riverside, Riverside, CA 92010, USA
School of Biotechnology, Amrita University, Kollam, Kerala 690525, India

CONTENTS

1 DIABETIC NEUROPATHY

Peripheral neuropathy is highly prevalent within the general population, and it is estimated to occur in 8% of individuals overall, and this rises to 15% in individuals over 40 (1). Diabetes is the major cause of these neuropathy cases, and this troubling due to the rapidly growing incidence of type 2 diabetes mellitus (T2DM) in populations across the globe; it has been estimated that the number of individuals with diabetes is expected to double between 2000 and 2030, to a total of 366 million people worldwide (2). One of the most common complications of diabetes is neuropathy, which is expected to affect at least 50% of diabetic individuals during their lifetime (3–6). Neuropathy is also the major cause of mortality, often involving cardiovascular disease or other occurrences such as end-stage renal failure; in addition to being the main cause of morbidity in diabetes which includes 50–75% of all non-traumatic amputations cases (7,8) that accounts for an estimated 80,000 cases in the United States alone.

Diabetic neuropathy has come to define a number of neuropathic syndromes caused by diabetes mellitus that includes focal, multifocal, and symmetrical neuropathies, which may involve small or large fibers, or both.

Neuropathy and its symptoms typically develop progressively over time and they can be distinct among individuals, being based on the nerves that are afflicted. The development of diabetic autonomic neuropathy in an individual can be of significant concern, as it can range from a subclinical impairment of cardiovascular reflexes to severe cardiovascular, gastrointestinal, or genitourinary dysfunction. In particular, diabetic autonomic neuropathy can cause silent myocardial infarction, and it is known to shorten the lifespan on patients with this condition and results in death of 25–50% of these patients over a 5 to 10 year time course (9,10). However, the most prevalent condition is diabetic peripheral neuropathy (DPN, also referred to as distal symmetrical polyneuropathy, DSPN) that is estimated to occur in 75% of cases of neuropathy (4,11). This form of neuropathy can potentially affect all peripheral nerves, which includes both sensory and motor neurons, and to a lesser extent the autonomic nervous system and a diabetic individual can have one or a combination of these systems affected.

Key risk factors for the prevalence of neuropathy are hyperglycemia, hypertension, hyperlipidemia, age, height, cigarette smoking, and the duration that the individual has had diabetes (12,13). Interestingly, the control of hypoglycemia may have a considerably more protective effect in patients with type 1 diabetes mellitus (T1DM), as tight glucose control was observed to decrease incidence of neuropathy by 60–70% in individuals with T1DM (14,15). Also, The Diabetes Control and Complications Trial in 1995 noted that the general incidence of neuropathy was at 2% yearly, and this level decreases to 0.56% with intensive insulin treatment of individuals with T1DM (15). However, glycemic control in patients with T2DM produced only a marginal effect, reducing incidence of neuropathy by 5–7% (16,17), and such individuals with good glucose control were noted to be still at risk of developing DPN (18). This suggests that there are potential differences in pathological mechanisms behind DPN in the two types (19), and where modifiable cardiovascular risk factors, such serum triglyceride levels, hypertension and body mass index, may play important roles in the development of DPN in persons with T2DM (12,13).

There are diagnoses of a neuropathy condition that causes nerve pain, and which is driven by treatment with insulin and oral hypoglycemic agents. This treatment-induced neuropathy in diabetes (TIND), originally identified as "insulin neuritis," is an acute neuropathy for patients who have had a prolonged period of hyperglycemia and that can have additional symptoms, such as peripheral nerve damage and decreased microvascular health resulting in retinopathy or nephropathy (20). TIND can affect people with T1DM or T2DM, and may be an under-diagnosed condition with one study suggesting approximately 10% of individuals under observation had TIND (21). Treatment-induced neuropathy has been clinically defined as the acute onset of neuropathic pain and/or autonomic dysfunction occurring within 8 weeks of a significant improvement in glycemic control (21). TIND is different from DPN in that autonomic symptoms and pain are noted to be more extensive in TIND, and opioid treatment for the pain produces less of a response than in DPN. The mechanisms behind TIND are not well-defined, but rapid glucose control via insulin could be leading to deleterious blood vessel changes resulting in hypoxia of and therefore damage to small neuronal fibers (22,23).

2 DIABETIC PERIPHERAL NEUROPATHY

DPN has been defined as a "length-dependent, symmetrical sensorimotor polyneuropathy that is attributable to metabolic and micro-vessel alterations, resulting from chronic hyperglycemia exposure (diabetes) and cardiovascular risk covariates" (6). The sensory symptoms spread in a 'stocking-glove' distribution, starting in the toes and moving over time to reach the knees, and then moving to the upper limbs. A variety of symptoms may be present that includes prickling and tingling sensations (paresthesias), electric shocks, burning feeling, or pain insensitivity. Hyperalgesia, an increased sensitivity to pain, or allodynia, a painful response to otherwise

painless stimuli, are other notable DPN symptoms that are mediated through neuropathy of the small fibers. Of individuals with DPN, 20–30% will suffer from neuropathic pain (5,6,19,24), and this is typically observed to be worse at night. Notably however, the stronger painful symptoms do not appear to directly correlate with severity of damage to axons (4).

An individual affected by DPN can also have a substantial reduction in their quality of life, aside from the risk of neuropathic pain. Disabilities stemming from diabetes are largely driven by DPN, due to gait disturbance and fall-related injuries, which manifests due to motor neuron dysfunction that occurs in later stages of the disease. Sensory loss can lead to un-noticed foot ulceration, and the lifetime risk for developing a foot ulcer in DPN is estimated at 15%, and there is a significant chance that infection, gangrene, and impaired wound healing could lead to amputation. Unsurprisingly, DPN contributes a significant 27% of diabetic health care costs (25,26), which amounts to $25 billion being spent per annum in the United States for treatment of diabetic complications. For an individual with diabetes, the presence of DPN doubles healthcare costs, and these costs quadruple with neuropathic pain (27).

The pathology of the disease is still being defined, but it appears to be related to metabolic dysfunction. In addition to hyperglycemia, increased levels of oxidative stress, polyol pathway activation, advanced glycation end products (AGEs), inflammation, and mitochondrial dysfunction have all been implicated (3,13). Other important aspects to the disease are defects within the microvasculature, as endothelial cell proliferation and hypertrophy with endoneurial microangiopathy (disease of the small capillaries in the layer connective tissue that surrounds the myelin sheath) are observed in nerve biopsies from DPN individuals (28,29). Sensory and motor nerves in DPN individuals have shown a reduced nerve conduction velocity, a decreased epidermal nerve fiber density (30–32) and defective axon regeneration that inhibits tissue re-innervation (30,33). These phenotypes are attributed to the dying-back of the distal axons occurring from loss of both the myelinated and unmyelinated fibers, where the neurodegeneration occurs in a fiber-length-dependent pattern, being most acute in the longest axons (34). In DPN, the Schwann cells that function to provide support and protection to the peripheral nervous system via myelination, are also noted to have ultrastructural abnormalities that may also contribute to the distal degeneration observed in peripheral neurons (35).

3 MITOCHONDRIAL DYSFUNCTION IN DPN

Peripheral nerve fiber endings in the skin are highly plastic and they maintain extensive fields of innervation. Unlike the central nervous system, that only has an extremely limited regenerative capacity, innervation occurs through axonal regeneration and collateral sprouting that provides outgrowths from the shafts of existing neurons (36–38). These regenerative and sprouting processes involving growth-cone motility require a high expenditure of cellular energy, at approximately 50% of the ATP reserves (39), and they are set in motion through actin treadmilling in the axon (40,41). This high demand for ATP is met through copious amounts of mitochondria, where electron tomography studies on the axoplasm, the cytoplasm of the peripheral neuron, have shown condensed and abundant mitochondria that are predominantly in the paranode–node–paranode region of a myelinated nerve fiber (42). Unmyelinated axons in the peripheral nervous system have even higher energy requirements, consuming 2.5 to 10 times the energy required for an action potential, as compared to a myelinated neuron (43). Thus, it is perhaps not surprising that mitochondrial dysfunction has been implicated as a causative factor in DPN, as well as other distal axonopathy diseases that include chemotherapy-induced peripheral neuropathy (CIPN), HIV-associated distal-symmetric neuropathy (HIV-DSP), Charcot-Marie-Tooth disease type 2 (CMT2), and Friedreich ataxia (44–46).

There is now a growing body of evidence that strongly supports the concept that abnormal mitochondrial function is a primary cause of the distal degeneration of sensory neurons in DPN (44,45,47,48). Initially, the mitochondria of Schwann cells, the cells of the peripheral nervous system that produce the axonal myelin sheath, were observed to have subtle ultrastructural changes that included enlargement of their mitochondria and effacement of the mitochondrial cristae (35). Such Schwann cell abnormalities are regarded as being significant in diabetic neuropathy (49–51), and proteomic studies on cultured primary Schwann cells treated with high glucose revealed an increase in the expression of proteins connected with mitochondrial dysfunction (52). Sensory neurons are observed to have a condensed mitochondrial network, especially in the unmyelinated neurons that require the extremely high ATP production rates (42,43). Studies on the perikarya, the cell body of a neuron, and the afferent projections of sensory neurons of diabetic humans and of animal models of diabetes, had determined that ultrastructure of these mitochondria appears relatively normal (35,53,54). However, results from several mouse models of type 1 diabetes, including non-obese diabetic (NOD) mice, the spontaneously genetically diabetic Akita mouse, and the STZ-treated mice and STZ-treated NOD/severe combined immunodeficiency (SCID) mice all noted far greater changes to their neuronal mitochondrial ultrastructure (55,56). In these mice, it was observed that the dorsal root and the sympathetic ganglia accumulated small, hyperchromatic and dense mitochondria, and these alterations are potentially due to changes in mitochondrial biogenesis and/or trafficking

(56–59). Notably, studies on STZ-diabetic rats also revealed that the mitochondrial inner membrane potential is depolarized in mitochondria that are present in either the adult sensory neuron perikarya (60–62) or the axons (63,64). Treating these rats with low doses of insulin, which although did no significantly affect the levels of hyperglycemia, could prevent this depolarization (60,61). Also, insulin as well as other neurotrophic growth factors could alter this depolarization in cultured neurons, through a phosphoinositide 3-kinase dependent pathway (65). These observations of mitochondrial depolarization in the neurons of STZ-diabetic rats (63,64) are also supported by similar observations of diabetic mitochondrial dysfunction in both skeletal muscle and cardiac tissue (47,66,67).

Nutrient excess is thought to be driving the depolarization effect, and could occur as a direct response to glucose levels. Extracellular glucose is driven across the plasma membrane of neuron by GLUT3 in an insulin independent manner resulting in high levels inside the cell under hyperglycemic conditions (68,69). High levels of glucose, and also fatty acids, can cause an excess of NADH and $FADH_2$ electron donors, as generated from β-oxidation and the glycolytic citric acid cycle. High levels of NADH and $FADH_2$ disturb the inner mitochondrial membrane potential, and high glucose concentrations down-regulates the AMP-activated protein kinase (AMPK)/peroxisome proliferator-activated receptor γ coactivator-1α (PGC- 1α) signaling axis (47,63,70,71). The AMPK/PGC-1α signaling axis is a fundamental energy sensing metabolic pathway controlling mitochondrial function and biogenesis (70–75), where AMPK is required for optimal mitochondrial function under the stress of high ATP demands. This signaling axis is also known to be critical for axonal plasticity and growth-cone motility (39,76–78), and it has a key role in the expression of oxidative phosphorylation proteins. Along these lines, studies in skeletal muscle, liver and cardiac tissues have shown that the activity of AMPK and PGC-1α signaling axis declines under diabetic conditions (66,79–82).

Gene array and proteomic studies have been undertaken to characterize the effect of altered gene expression on mitochondrial biology (52,79,81,83). Skeletal muscle from T2DM patients in gene array studies diminished expression of oxidative phosphorylation genes (81). SILAC-based quantitative proteomics analysis determined that the levels of the mitochondrial Complexes I–V and proteins involved the TCA cycle and combating oxidative stress were largely down-regulated (47,59). This is similar to results from proteomic studies on cardiac tissue of Akita mice that showed decreased transcripts for PGC-1α and the oxidative phosphorylation genes (79). This disturbance of mitochondrial oxidative phosphorylation results in a decrease in ATP production, and results from animal models and *in vitro* cultures indicates that the antioxidant response is also greatly diminished, resulting in high levels of reactive oxygen species that further decrease the production of ATP (84). This constraint of ATP production is likely to be exceedingly challenging to distal unmyelinated axons present in the epidermis, where ATP requirements can greatly fluctuate as compared to the perikarya of the neuron. This lack of sufficient ATP perturbs both sprouting and plasticity, leading to a gradual pruning of the axonal network and thus shrinking of the sensory innervation fields, with the end result of loss of innervation and function of the nerve fibers.

4   INITIAL ATTEMPTS AT DPN THERAPY

Current treatments include strong glycemic control that can reduce the risk of DPN and the rate of progression of DPN in patients with T1DM (15). Unfortunately, only palliative treatment is available for individuals with T2DM who instead decrease their risk of developing DPN through lifestyle changes that are predominantly aimed at combating obesity, including diet and exercise. Several enzymes and cellular pathways have been therapeutically targeted with the aim of developing a DPN-based therapy, including aldose reductase inhibitors, protein kinase C-beta inhibitors, agents acting on the AGE pathway, compounds targeting the hexosamine pathway, or reducing reactive oxygen species production. However, within the United States, no clinical trial for a treatment DPN has been met with approval by the FDA (3,85). Thus there is a clear and urgent unmet medical need for developing a therapy that can substantially limit or reverse the progression of DPN. There was general apprehension about developing new therapies to combat the disease due to the lack of a molecular-based understanding of its manifestation, but this view may be changing due to recent results highlighting mitochondrial constraint in ATP production that may be behind DPN. Aldose reductase treatments also showed a potential benefit for DPN (86,87), but a Cochrane review of 32 randomized controlled trials showed no significant difference between these treatments and placebo (88). Another therapy with potential is the antioxidant α-lipoic acid, which was noted to have a robust therapeutic effect in rodent models of DPN (reviewed in (89)) and it has been approved for the clinical trial in Europe (90). However, α-lipoic acid was not approved for DPN treatment in the United States, as it did not show a change in end point after therapy, and therefore new approaches that include combination therapies of α-lipoic acid and ameliorating the inflammatory response are now being tested to hopefully provide more therapeutic benefit in humans (91,92).

5   A NEW HOPE FOR DPN THERAPY THROUGH TARGETING THE MUSCARINIC PATHWAY

A potential novel therapeutic avenue has opened up for DPN treatment, due to the discovery of significant new insights into the molecular mechanisms of neurite outgrowth. It has been revealed that adult sensory neurons are under a cholinergic constraint, which limits outgrowth, and that functions through muscarinic receptors (MRs) (93). Mice lacking the muscarinic 1 receptor (M_1R) were observed to have an enhanced neurite outgrowth, and strikingly, such deficient mice that were made diabetic through streptozotocin (STZ) treatment were protected from sensory neuropathy (93). In rodent models, targeting M_1R through the selective antagonists of pirenzepine, VU0255035, or the snake venom toxin MT7 prevented or reversed the peripheral neuropathy indices that were measured (93). This included nerve conduction slowing, a common end-point measurement in clinical trials, which was counteracted, in addition to thermal hypoalgesia and the loss of intra-epidermal nerve fibers that is a marker for DPN, were prevented by antimuscarinic drugs. Encouragingly, studies with these antimuscarinic drugs on the animal models did not show any significant side effects, including echocardiogram studies that revealed no alterations in cardiac function or structure. These therapies also did not affect the overall diabetic state, and thus are unlikely to be reversing DPN pathology by affecting pancreatic function and insulin levels. Research on both T1DM and T2DM rodent models to define the mechanisms of action of these M_1R antagonists, revealed that the nerve protective and repair promoting properties where associated with deactivation of AMPK (93). The central role of mitochondrial function is further supported by results from studies on ciliary neurotrophic factor and C-terminal inhibitors of heat shock protein 90, known to promote mitochondrial function, revealing an improved neuropathy state in diabetic rodents (94,95). Mitochondrial dysfunction may not be exclusive to the nervous system in diabetes however, as a lower AMPK activity and subsequent decreased expression of mitochondrial complex proteins have been observed in kidney mesangial cells in diabetic nephropathy (96), suggesting added potential benefits of antimuscarinic therapy.

This M_1R constraint mechanism could extend beyond DPN, as pirenzepine and MT7 where both observed to prevent chemotherapeutic or HIV models of peripheral neuropathy. Moreover, release of muscarinic receptor driven inhibition of mitochondrial activity could have an even wider applicability than neuropathies closely related to DPN, and instead could potentially benefit for other conditions that have a lessened energy capacity under stress. For example, screening for novel therapeutic compounds that promote remyelination has also highlighted the therapeutic potential of roles of antimuscarinics as myelination enhancers in multiple sclerosis (97). Excitingly, there are antimuscarinic drugs that have already been used in the clinic, with greater than 20 years of clinical application in Europe for various conditions, as well as the use of topical pirenzepine for myopia in children, and thus the safety profiles have been well characterized (98). This could allow for a rapid progression into clinical trials for antimuscarinics to treat DPN and other forms of peripheral neuropathy. Such clinical trials could include studies on early indices of DPN, as antimuscarinics may be most effective during the early stages of distal dying-back of neurons, rather than at later stages of complete loss of fibers.

6 CONCLUSION

Mitochondrial dysfunction appears to be a central component to diabetic neuropathy, with nutrient excess in the diabetic condition causing depolarization of the mitochondrial membrane and thereby a loss of ATP production. This lack of sufficient ATP may stress neurons in the peripheral nervous system, which require relatively high levels of energy to promote and support nerve outgrowth, and thus loss of ATP likely causes dying-back of neurons and loss of nerve fibers. The antimuscarinic drugs give strong support for this model that is centered on the role that mitochondria play in DPN, as it appears that the M_1R receptor provides a critical constraint on neurite outgrowth through limiting mitochondrial function by perturbing AMPK activity. Markedly, animal models of DPN when treated with M_1R antagonists can display reversal of DPN indices, suggesting first-in-class treatment for DPN. Key questions remain, such as what are all the components of the M_1R to AMPK signaling pathway, and can these other parts also be targeting for DPN therapy. This may be the case, as a recent study on the novel compound, J147, which is a derivative of curcumin and a neuroprotective drug candidate for Alzheimer's disease, can also ameliorate multiple DPN indices (99). It appears that J147 targets the mitochondrial α-F_1-ATP synthase, which leads to an increase in intracellular calcium levels, and thereby calcium/calmodulin-dependent protein kinase kinase β (CAMKK2)-dependent activation of the AMPK/mTOR pathway (100). Thus, there may be other proteins and complexes suitable for targeting, in order to develop therapies to treat DPN.

ACKNOWLEDGMENTS

We would like to thank Dr. Anna Travesa Centrich for critical reading of the manuscript. The Perry lab is supported by UC CRCC grant CRN-18-524,906 and UCOP grant LFR-17-476,732.

CONFLICT OF INTEREST

J. J. P. Perry declares that he is a stakeholder in WinSanTor Inc.

REFERENCES

1. Gregg EW, Sorlie P, Paulose-Ram R, Gu Q, Eberhardt MS, Wolz M, Burt V, Curtin L, Engelgau M, Geiss L, national h, nutrition examination s. Prevalence of lower-extremity disease in the US adult population ≥40 years of age with and without diabetes: 1999–2000 national health and nutrition examination survey. Diabetes Care. 2004;27 (7):1591–1597. PubMed PMID: 15220233.
2. Hossain P, Kawar B, El Nahas M. Obesity and diabetes in the developing world–a growing challenge. N Engl J Med. 2007;356(3):213–15. doi: 10.1056/NEJMp068177. PubMed PMID: 17229948.
3. Singh R, Kishore L, Kaur N. Diabetic peripheral neuropathy: current perspective and future directions. Pharmacol Res. 2014;80:21–35. doi: 10.1016/j.phrs.2013.12.005. PubMed PMID: 24373831.
4. Tesfaye S, Selvarajah D. Advances in the epidemiology, pathogenesis and management of diabetic peripheral neuropathy. Diabetes Metab Res Rev. 2012;28(Suppl 1):8–14. doi: 10.1002/dmrr.2239. PubMed PMID: 22271716.
5. Tesfaye S, Vileikyte L, Rayman G, Sindrup SH, Perkins BA, Baconja M, Vinik AI, Boulton AJ, Toronto Expert Panel on Diabetic N. Painful diabetic peripheral neuropathy: consensus recommendations on diagnosis, assessment and management. Diabetes Metab Res Rev. 2011;27 (7):629–38. doi: 10.1002/dmrr.1225. PubMed PMID: 21695762.
6. Tesfaye S, Boulton AJ, Dyck PJ, Freeman R, Horowitz M, Kempler P, Lauria G, Malik RA, Spallone V, Vinik A, Bernardi L, Valensi P, Toronto Diabetic Neuropathy Expert G. Diabetic neuropathies: update on definitions, diagnostic criteria, estimation of severity, and treatments. Diabetes Care. 2010;33(10):2285–93. doi: 10.2337/dc10-1303. PubMed PMID: 20876709; PMCID: PMC2945176.
7. Trautner C, Haastert B, Spraul M, Giani G, Berger M. Unchanged incidence of lower-limb amputations in a German City, 1990–1998. Diabetes Care. 2001;24(5):855–9. PubMed PMID: 11347743.
8. Almaraz MC, Gonzalez-Romero S, Bravo M, Caballero FF, Palomo MJ, Vallejo R, Esteva I, Calleja F, Soriguer F. Incidence of lower limb amputations in individuals with and without diabetes mellitus in Andalusia (Spain) from 1998 to 2006. Diabetes Res Clin Pract. 2012;95(3):399–405. doi: 10.1016/j.diabres.2011.10.035. PubMed PMID: 22133651.
9. Levitt NS, Stansberry KB, Wynchank S, Vinik AI. The natural progression of autonomic neuropathy and autonomic function tests in a cohort of people with IDDM. Diabetes Care. 1996;19(7):751–4. PubMed PMID: 8799632.
10. Rathmann W, Ziegler D, Jahnke M, Haastert B, Gries FA. Mortality in diabetic patients with cardiovascular autonomic neuropathy. Diabet Med. 1993;10(9):820–4. PubMed PMID: 8281726.
11. Bansal V, Kalita J, Misra UK. Diabetic neuropathy. Postgrad Med J. 2006;82(964):95–100. doi: 10.1136/pgmj.2005.036137. PubMed PMID: 16461471; PMCID: PMC2596705.
12. Tesfaye S, Chaturvedi N, Eaton SE, Ward JD, Manes C, Ionescu-Tirgoviste C, Witte DR, Fuller JH, Group EPCS. Vascular risk factors and diabetic neuropathy. N Engl J Med. 2005;352(4):341–50. doi: 10.1056/NEJMoa032782. PubMed PMID: 15673800.
13. Callaghan B, Feldman E. The metabolic syndrome and neuropathy: therapeutic challenges and opportunities. Ann Neurol. 2013;74(3):397–403. doi: 10.1002/ana.23986. PubMed PMID: 23929529; PMCID: PMC3881591.
14. Linn T, Ortac K, Laube H, Federlin K. Intensive therapy in adult insulin-dependent diabetes mellitus is associated

with improved insulin sensitivity and reserve: a random-ized, controlled, prospective study over 5 years in newly diagnosed patients. Metabolism. 1996;45(12):1508–13. PubMed PMID: 8969284.

15. Diabetes C, Complications Trial Research G, Nathan DM, Genuth S, Lachin J, Cleary P, Crofford O, Davis M, Rand L, Siebert C. The effect of intensive treatment of diabetes on the development and progression of long-term complications in insulin-dependent diabetes mellitus. N Engl J Med. 1993;329(14):977–86. doi: 10.1056/NEJM199309303291401. PubMed PMID: 8366922.

16. Duckworth W, Abraira C, Moritz T, Reda D, Emanuele N, Reaven PD, Zieve FJ, Marks J, Davis SN, Hayward R, Warren SR, Goldman S, McCarren M, Vitek ME, Henderson WG, Huang GD. Investigators V. Glucose control and vascular complications in veterans with type 2 diabetes. N Engl J Med. 2009;360(2):129–39. doi: 10.1056/NEJMoa0808431. PubMed PMID: 19092145.

17. Ismail-Beigi F, Craven T, Banerji MA, Basile J, Calles J, Cohen RM, Cuddihy R, Cushman WC, Genuth S, Grimm RH, Jr., Hamilton BP, Hoogwerf B, Karl D, Katz L, Krikorian A, O'Connor P, Pop-Busui R, Schubart U, Simmons D, Taylor H, Thomas A, Weiss D, Hramiak I, group At. Effect of intensive treatment of hyperglycaemia on microvascular outcomes in type 2 diabetes: an analysis of the ACCORD randomised trial. Lancet. 2010;376 (9739):419–430. doi: 10.1016/S0140-6736(10)60576-4. PubMed PMID: 20594588; PMCID: PMC4123233.

18. Tesfaye S, Stevens LK, Stephenson JM, Fuller JH, Plater M, Ionescu-Tirgoviste C, Nuber A, Pozza G, Ward JD. Prevalence of diabetic peripheral neuropathy and its relation to glycaemic control and potential risk factors: the EURODIAB IDDM Complications Study. Diabetologia. 1996;39(11):1377–84. PubMed PMID: 8933008.

19. Callaghan BC, Hur J, Feldman EL. Diabetic neuropathy: one disease or two? Curr Opin Neurol. 2012;25(5):536–41. doi: 10.1097/WCO.0b013e328357a797. PubMed PMID: 22892951; PMCID: PMC4239661.

20. Knopp M, Srikantha M, Rajabally YA. Insulin neuritis and diabetic cachectic neuropathy: a review. Curr Diabetes Rev. 2013;9(3):267–4. PubMed PMID: 23506377.

21. Gibbons CH, Freeman R. Treatment-induced diabetic neuropathy: a reversible painful autonomic neuropathy. Ann Neurol. 2010;67(4):534–41. doi: 10.1002/ana.21952. PubMed PMID: 20437589; PMCID: PMC3057039.

22. Tran C, Philippe J, Ochsner F, Kuntzer T, Truffert A. Acute painful diabetic neuropathy: an uncommon, remittent type of acute distal small fibre neuropathy. Swiss Med Wkly. 2015;145:w14131. doi: 10.4414/smw.2015.14131. PubMed PMID: 25941879.

23. Tesfaye S, Malik R, Harris N, Jakubowski JJ, Mody C, Rennie IG, Ward JD. Arterio-venous shunting and proliferating new vessels in acute painful neuropathy of rapid glycaemic control (insulin neuritis). Diabetologia. 1996;39 (3):329–35. PubMed PMID: 8721779.

24. Quattrini C, Tesfaye S. Understanding the impact of painful diabetic neuropathy. Diabetes Metab Res Rev. 2003;19 (Suppl 1):S2—8. doi: 10.1002/dmrr.360. PubMed PMID: 12577252.

25. American Diabetes A. Economic costs of diabetes in the U.S. in 2012. Diabetes Care. 2013;36(4):1033–46. doi: 10.2337/dc12-2625. PubMed PMID: 23468086; PMCID: PMC3609540.

26. Gordois A, Scuffham P, Shearer A, Oglesby A, Tobian JA. The health care costs of diabetic peripheral neuropathy in

the US. Diabetes Care. 2003;26(6):1790–5. PubMed PMID: 12766111.

27. Sadosky A, Mardekian J, Parsons B, Hopps M, Bienen EJ, Markman J. Healthcare utilization and costs in diabetes relative to the clinical spectrum of painful diabetic peripheral neuropathy. J Diabetes Complications. 2015;29 (2):212–17. doi: 10.1016/j.jdiacomp.2014.10.013. PubMed PMID: 25498300.

28. Malik RA, Tesfaye S, Thompson SD, Veves A, Sharma AK, Boulton AJ, Ward JD. Endoneurial localisation of microvascular damage in human diabetic neuropathy. Diabetologia. 1993;36(5):454–9. PubMed PMID: 8314451.

29. Malik RA, Tesfaye S, Newrick PG, Walker D, Rajbhandari SM, Siddique I, Sharma AK, Boulton AJ, King RH, Thomas PK, Ward JD. Sural nerve pathology in diabetic patients with minimal but progressive neuropathy. Diabetologia. 2005;48(3):578–85. doi: 10.1007/s00125-004-1663-5. PubMed PMID: 15729579.

30. Ebenezer GJ, O'Donnell R, Hauer P, Cimino NP, McArthur JC, Polydefkis M. Impaired neurovascular repair in subjects with diabetes following experimental intracutaneous axotomy. Brain. 2011;134(Pt 6):1853–63. doi: 10.1093/brain/awr086. PubMed PMID: 21616974; PMCID: PMC3140859.

31. Kennedy WR, Wendelschafer-Crabb G, Johnson T. Quantitation of epidermal nerves in diabetic neuropathy. Neurology. 1996;47(4):1042–8. PubMed PMID: 8857742.

32. Quattrini C, Tavakoli M, Jeziorska M, Kallinikos P, Tesfaye S, Finnigan J, Marshall A, Boulton AJ, Efron N, Malik RA. Surrogate markers of small fiber damage in human diabetic neuropathy. Diabetes. 2007;56(8):2148–54. doi: 10.2337/db07-0285. PubMed PMID: 17513704.

33. Polydefkis M, Hauer P, Sheth S, Sirdofsky M, Griffin JW, McArthur JC. The time course of epidermal nerve fibre regeneration: studies in normal controls and in people with diabetes, with and without neuropathy. Brain. 2004;127(Pt 7):1606–15. doi: 10.1093/brain/awh175. PubMed PMID: 15128618.

34. Said G. Diabetic neuropathy–A review. Nat Clin Pract Neurol. 2007;3(6):331–40. doi: 10.1038/ncpneuro0504. PubMed PMID: 17549059.

35. Kalichman MW, Powell HC, Mizisin AP. Reactive, degenerative, and proliferative Schwann cell responses in experimental galactose and human diabetic neuropathy. Acta Neuropathol. 1998;95(1):47–56. PubMed PMID: 9452821.

36. Diamond J, Coughlin M, Macintyre L, Holmes M, Visheau B. Evidence that endogenous beta nerve growth factor is responsible for the collateral sprouting, but not the regeneration, of nociceptive axons in adult rats. Proc Natl Acad Sci U S A. 1987;84(18):6596–600. PubMed PMID: 3306683; PMCID: PMC299126.

37. Diamond J, Foerster A, Holmes M, Coughlin M. Sensory nerves in adult rats regenerate and restore sensory function to the skin independently of endogenous NGF. J Neurosci. 1992;12(4):1467–76. PubMed PMID: 1313494.

38. Diamond J, Holmes M, Coughlin M. Endogenous NGF and nerve impulses regulate the collateral sprouting of sensory axons in the skin of the adult rat. J Neurosci. 1992;12 (4):1454–66. PubMed PMID: 1556603.

39. Bernstein BW, Bamburg JR. Actin-ATP hydrolysis is a major energy drain for neurons. J Neurosci. 2003;23(1):1–6. PubMed PMID: 12514193.

40. Bray D. Growth cone formation and navigation: axonal growth. Curr Opin Cell Biol. 1989;1(1):87–90. PubMed PMID: 2698213.

41. Bray D, White JG. Cortical flow in animal cells. Science. 1988;239(4842):883–8. PubMed PMID: 3277283.

42. Perkins GA, Ellisman MH. Mitochondrial configurations in peripheral nerve suggest differential ATP production. J Struct Biol. 2011;173(1):117–27. doi: 10.1016/j.jsb.2010.06.017. PubMed PMID: 20600951; PMCID: PMC3078762.

43. Wang SS, Shultz JR, Burish MJ, Harrison KH, Hof PR, Towns LC, Wagers MW, Wyatt KD. Functional trade-offs in white matter axonal scaling. J Neurosci. 2008;28 (15):4047–56. doi: 10.1523/JNEUROSCI.5559-05.2008. PubMed PMID: 18400904; PMCID: PMC2779774.

44. Bennett GJ, Doyle T, Salvemini D. Mitotoxicity in distal symmetrical sensory peripheral neuropathies. Nat Rev Neurol. 2014;10(6):326–336. doi: 10.1038/nrneurol.2014.77. PubMed PMID: 24840972; PMCID: PMC4870000.

45. Cashman CR, Hoke A. Mechanisms of distal axonal degeneration in peripheral neuropathies. Neurosci Lett. 2015;596:33–50. doi: 10.1016/j.neulet.2015.01.048. PubMed PMID: 25617478; PMCID: PMC4428955.

46. Chowdhury SK, Smith DR, Fernyhough P. The role of aberrant mitochondrial bioenergetics in diabetic neuropathy. Neurobiol Dis. 2013;51:56–65. doi: 10.1016/j.nbd.2012.03.016. PubMed PMID: 22446165.

47. Chowdhury SK, Dobrowsky RT, Fernyhough P. Nutrient excess and altered mitochondrial proteome and function contribute to neurodegeneration in diabetes. Mitochondrion. 2011;11(6):845–54. doi: 10.1016/j.mito.2011.06.007. PubMed PMID: 21742060; PMCID: PMC3375692.

48. Casanova-Molla J, Morales M, Garrabou G, Sola-Valls N, Soriano A, Calvo M, Grau JM, Valls-Sole J. Mitochondrial loss indicates early axonal damage in small fiber neuropathies. J Peripher Nerv Syst. 2012;17(2):147–57. doi: 10.1111/j.1529-8027.2012.00396.x. PubMed PMID: 22734900.

49. Dobrowsky RT, Rouen S, Yu C. Altered neurotrophism in diabetic neuropathy: spelunking the caves of peripheral nerve. J Pharmacol Exp Ther. 2005;313(2):485–91. doi: 10.1124/jpet.104.079921. PubMed PMID: 15608075.

50. Eckersley L. Role of the Schwann cell in diabetic neuropathy. Int Rev Neurobiol. 2002;50:293–321. PubMed PMID: 12198814.

51. Kennedy JM, Zochodne DW. Impaired peripheral nerve regeneration in diabetes mellitus. J Peripher Nerv Syst. 2005;10(2):144–57. doi: 10.1111/j.1085-9489.2005.0010205.x. PubMed PMID: 15958126.

52. Zhang L, Yu C, Vasquez FE, Galeva N, Onyango I, Swerdlow RH, Dobrowsky RT. Hyperglycemia alters the schwann cell mitochondrial proteome and decreases coupled respiration in the absence of superoxide production. J Proteome Res. 2010;9(1):458–71. doi: 10.1021/pr900818g. PubMed PMID: 19905032; PMCID: PMC2801777.

53. Schmidt RE, Dorsey D, Parvin CA, Beaudet LN, Plurad SB, Roth KA. Dystrophic axonal swellings develop as a function of age and diabetes in human dorsal root ganglia. J Neuropathol Exp Neurol. 1997;56(9):1028–43. PubMed PMID: 9291944.

54. Kamiya H, Zhang W, Sima AA. Degeneration of the Golgi and neuronal loss in dorsal root ganglia in diabetic BioBreeding/Worcester rats. Diabetologia. 2006;49 (11):2763–74. doi: 10.1007/s00125-006-0379-0. PubMed PMID: 17047923.

55. Schmidt RE, Dorsey DA, Beaudet LN, Frederick KE, Parvin CA, Plurad SB, Levisetti MG. Non-obese diabetic mice rapidly develop dramatic sympathetic neuritic dystrophy: a new experimental model of diabetic autonomic neuropathy. Am J Pathol. 2003;163(5):2077–91. doi: 10.1016/S0002-9440(10)63565-1. PubMed PMID: 14578206; PMCID: PMC1892436.

56. Schmidt RE, Green KG, Snipes LL, Feng D. Neuritic dystrophy and neuronopathy in Akita (Ins2(Akita)) diabetic mouse sympathetic ganglia. Exp Neurol. 2009;216(1):207–18. doi: 10.1016/j.expneurol.2008.11.019. PubMed PMID: 19111542; PMCID: PMC2672346.

57. Fernyhough P, Calcutt NA. Abnormal calcium homeostasis in peripheral neuropathies. Cell Calcium. 2010;47 (2):130–9. doi: 10.1016/j.ceca.2009.11.008. PubMed PMID: 20034667; PMCID: PMC2834846.

58. Schmidt RE, Parvin CA, Green KG. Synaptic ultrastructural alterations anticipate the development of neuroaxonal dystrophy in sympathetic ganglia of aged and diabetic mice. J Neuropathol Exp Neurol. 2008;67(12):1166–86. doi: 10.1097/NEN.0b013e318190d6db. PubMed PMID: 19018240; PMCID: PMC2665250.

59. Vincent AM, Edwards JL, McLean LL, Hong Y, Cerri F, Lopez I, Quattrini A, Feldman EI. Mitochondrial biogenesis and fission in axons in cell culture and animal models of diabetic neuropathy. Acta Neuropathol. 2010;120 (4):477–89. doi: 10.1007/s00401-010-0697-7. PubMed PMID: 20473509; PMCID: PMC4254759.

60. Huang TJ, Price SA, Chilton L, Calcutt NA, Tomlinson DR, Verkhratsky A, Fernyhough P. Insulin prevents depolarization of the mitochondrial inner membrane in sensory neurons of type 1 diabetic rats in the presence of sustained hyperglycemia. Diabetes. 2003;52(8):2129–36. PubMed PMID: 12882932.

61. Huang TJ, Sayers NM, Verkhratsky A, Fernyhough P. Neurotrophin-3 prevents mitochondrial dysfunction in sensory neurons of streptozotocin-diabetic rats. Exp Neurol. 2005;194(1):279–83. doi: 10.1016/j.expneurol.2005.03.001. PubMed PMID: 15899264.

62. Srinivasan S, Stevens M, Wiley JW. Diabetic peripheral neuropathy: evidence for apoptosis and associated mitochondrial dysfunction. Diabetes. 2000;49(11):1932–1938. PubMed PMID: 11078462.

63. Akude E, Zherebitskaya E, Chowdhury SK, Smith DR, Dobrowsky RT, Fernyhough P. Diminished superoxide generation is associated with respiratory chain dysfunction and changes in the mitochondrial proteome of sensory neurons from diabetic rats. Diabetes. 2011;60(1):288–97. doi: 10.2337/db10-0818. PubMed PMID: 20876714; PMCID: PMC3012184.

64. Chowdhury SK, Zherebitskaya E, Smith DR, Akude E, Chattopadhyay S, Jolivalt CG, Calcutt NA, Fernyhough P. Mitochondrial respiratory chain dysfunction in dorsal root ganglia of streptozotocin-induced diabetic rats and its correction by insulin treatment. Diabetes. 2010;59 (4):1082–91. doi: 10.2337/db09-1299. PubMed PMID: 20103706; PMCID: PMC2844817.

65. Huang TJ, Verkhratsky A, Fernyhough P. Insulin enhances mitochondrial inner membrane potential and increases ATP levels through phosphoinositide 3-kinase in adult sensory neurons. Mol Cell Neurosci. 2005;28 (1):42–54. doi: 10.1016/j.mcn.2004.08.009. PubMed PMID: 15607940.

66. Szendroedi J, Phielix E, Roden M. The role of mitochondria in insulin resistance and type 2 diabetes mellitus. Nat Rev Endocrinol. 2011;8(2):92–103. doi: 10.1038/nrendo.2011.138. PubMed PMID: 21912398.

67. Bugger H, Abel ED. Mitochondria in the diabetic heart. Cardiovasc Res. 2010;88(2):229–40. doi: 10.1093/cvr/cvq239. PubMed PMID: 20639213; PMCID: PMC2952534.

68. Simpson IA, Carruthers A, Vannucci SJ. Supply and demand in cerebral energy metabolism: the role of nutrient transporters. J Cereb Blood Flow Metab. 2007;27(11):1766–91. doi: 10.1038/sj.jcbfm.9600521. PubMed PMID: 17579656; PMCID: PMC2094104.

69. Tomlinson DR, Gardiner NJ. Glucose neurotoxicity. Nat Rev Neurosci. 2008;9(1):36–45. doi: 10.1038/nrn2294. PubMed PMID: 18094705.

70. Hardie DG. AMPK: a key regulator of energy balance in the single cell and the whole organism. Int J Obes (Lond). 2008;32(Suppl 4:S7–12. doi: 10.1038/ijo.2008.116. PubMed PMID: 18719601.

71. Canto C, Auwerx J. PGC-1alpha, SIRT1 and AMPK, an energy sensing network that controls energy expenditure. Curr Opin Lipidol. 2009;20(2):98–105. doi: 10.1097/MOL.0b013e328328d0a4. PubMed PMID: 19276888; PMCID: PMC3627054.

72. Dominy JE, Jr., Lee Y, Gerhart-Hines Z, Puigserver P. Nutrient-dependent regulation of PGC-1alpha's acetylation state and metabolic function through the enzymatic activities of Sirt1/GCN5. Biochim Biophys Acta. 2010;1804(8):1676–83. doi: 10.1016/j.bbapap.2009.11.023. PubMed PMID: 20005308; PMCID: PMC2886158.

73. Feige JN, Auwerx J. Transcriptional coregulators in the control of energy homeostasis. Trends Cell Biol. 2007;17(6):292–301. doi: 10.1016/j.tcb.2007.04.001. PubMed PMID: 17475497.

74. Puigserver P. Tissue-specific regulation of metabolic pathways through the transcriptional coactivator PGC1-alpha. Int J Obes (Lond). 2005;29(Supp l):S5-9. doi: 10.1038/sj.ijo.0802905. PubMed PMID: 15711583.

75. Rodgers JT, Lerin C, Gerhart-Hines Z, Puigserver P. Metabolic adaptations through the PGC-1 alpha and SIRT1 pathways. FEBS Lett. 2008;582(1):46–53. doi: 10.1016/j.febslet.2007.11.034. PubMed PMID: 18036349; PMCID: PMC2275806.

76. Dasgupta B, Milbrandt J. Resveratrol stimulates AMP kinase activity in neurons. Proc Natl Acad Sci U S A. 2007;104(17):7217–22. doi: 10.1073/pnas.0610068104. PubMed PMID: 17438283; PMCID: PMC1855377.

77. Samuel MA, Voinescu PE, Lilley BN, de Cabo R, Foretz M, Viollet B, Pawlyk B, Sandberg MA, Vavvas DG, Sanes JR. LKB1 and AMPK regulate synaptic remodeling in old age. Nat Neurosci. 2014;17(9):1190–7. doi: 10.1038/nn.3772. PubMed PMID: 25086610; PMCID: PMC5369022.

78. Tao K, Matsuki N, Koyama R. AMP-activated protein kinase mediates activity-dependent axon branching by recruiting mitochondria to axon. Dev Neurobiol. 2014;74(6):557–73. doi: 10.1002/dneu.22149. PubMed PMID: 24218086.

79. Bugger H, Chen D, Riehle C, Soto J, Theobald HA, Hu XX, Ganesan B, Weimer BC, Abel ED. Tissue-specific remodeling of the mitochondrial proteome in type 1 diabetic akita mice. Diabetes. 2009;58(9):1986–97. doi: 10.2337/db09-0259. PubMed PMID: 19542201; PMCID: PMC2731527.

80. Mootha VK, Lindgren CM, Eriksson KF, Subramanian A, Sihag S, Lehar J, Puigserver P, Carlsson E, Ridderstrale M, Laurila E, Houstis N, Daly MJ, Patterson N, Mesirov JP, Golub TR, Tamayo P, Spiegelman B, Lander ES, Hirschhorn JN, Altshuler D, Groop LC. PGC-1alpha-responsive genes involved in oxidative phosphorylation are coordinately downregulated in human diabetes. Nat Genet. 2003;34(3):267–73. doi: 10.1038/ng1180. PubMed PMID: 12808457.

81. Patti ME, Butte AJ, Crunkhorn S, Cusi K, Berria R, Kashyap S, Miyazaki Y, Kohane I, Costello M, Saccone R, Landaker EJ, Goldfine AB, Mun E, DeFronzo R, Finlayson J, Kahn CR, Mandarino LJ. Coordinated reduction of genes of oxidative metabolism in humans with insulin resistance and diabetes: potential role of PGC1 and NRF1. Proc Natl Acad Sci U S A. 2003;100(14):8466–71. doi: 10.1073/pnas.1032913100. PubMed PMID: 12832613; PMCID: PMC166252.

82. Richardson DK, Kashyap S, Bajaj M, Cusi K, Mandarino SJ, Finlayson J, DeFronzo RA, Jenkinson CP, Mandarino LJ. Lipid infusion decreases the expression of nuclear encoded mitochondrial genes and increases the expression of extracellular matrix genes in human skeletal muscle. J Biol Chem. 2005;280(11):10290–7. doi: 10.1074/jbc.M408985200. PubMed PMID: 15598661.

83. Pande M, Hur J, Hong Y, Backus C, Hayes JM, Oh SS, Kretzler M, Feldman EL. Transcriptional profiling of diabetic neuropathy in the BKS db/db mouse: a model of type 2 diabetes. Diabetes. 2011;60(7):1981–9. doi: 10.2337/db10-1541. PubMed PMID: 21617178; PMCID: PMC3121428.

84. Fernyhough P. Mitochondrial dysfunction in diabetic neuropathy: a series of unfortunate metabolic events. Curr Diab Rep. 2015;15(11):89. doi: 10.1007/s11892-015-0671-9. PubMed PMID: 26370700.

85. Ziegler D, Low PA, Litchy WJ, Boulton AJ, Vinik AI, Freeman R, Samigullin R, Tritschler H, Munzel U, Maus J, Schutte K, Dyck PJ. Efficacy and safety of antioxidant treatment with alpha-lipoic acid over 4 years in diabetic polyneuropathy: the NATHAN 1 trial. Diabetes Care. 2011;34(9):2054–60. doi: 10.2337/dc11-0503. PubMed PMID: 21775755; PMCID: PMC3161301.

86. Hotta N, Akanuma Y, Kawamori R, Matsuoka K, Oka Y, Shichiri M, Toyota T, Nakashima M, Yoshimura I, Sakamoto N, Shigeta Y. Long-term clinical effects of epalrestat, an aldose reductase inhibitor, on diabetic peripheral neuropathy: the 3-year, multicenter, comparative Aldose Reductase Inhibitor-Diabetes Complications Trial. Diabetes Care. 2006;29(7):1538–44. doi: 10.2337/dc05-2370. PubMed PMID: 16801576.

87. Bril V, Buchanan RA. Long-term effects of ranirestat (AS-3201) on peripheral nerve function in patients with diabetic sensorimotor polyneuropathy. Diabetes Care. 2006;29(1):68–72. PubMed PMID: 16373898.

88. Chalk C, Benstead TJ, Moore F. Aldose reductase inhibitors for the treatment of diabetic polyneuropathy. Cochrane Database Syst Rev. 2007;(4):CD004572. doi: 10.1002/14651858.CD004572.pub2. PubMed PMID: 17943821.

89. Cameron NE, Cotter MA. Effects of antioxidants on nerve and vascular dysfunction in experimental diabetes. Diabetes Res Clin Pract. 1999;45(2–3):137–46. PubMed PMID: 10588366.

90. Papanas N, Ziegler D. Efficacy of alpha-lipoic acid in diabetic neuropathy. Expert Opin Pharmacother. 2014;15(18):2721–31. doi: 10.1517/14656566.2014.972935. PubMed PMID: 25381809.

91. Yorek MS, Obrosov A, Shevalye H, Coppey LJ, Kardon RH, Yorek MA. Early vs. late intervention of high fat/low

dose streptozotocin treated C57Bl/6J mice with enalapril, alpha-lipoic acid, menhaden oil or their combination: effect on diabetic neuropathy related endpoints. Neuropharmacology. 2017;116:122–31. doi: 10.1016/j.neuropharm.2016.12.022. PubMed PMID: 28025096; PMCID: PMC5385152.

92. Chen L, Li B, Chen B, Shao Y, Luo Q, Shi X, Chen Y. Thymoquinone alleviates the experimental diabetic peripheral neuropathy by modulation of inflammation. Sci Rep. 2016;6:31656. doi: 10.1038/srep31656. PubMed PMID: 27545310; PMCID: PMC4992870.

93. Calcutt NA, Smith DR, Frizzi K, Sabbir MG, Chowdhury SK, Mixcoatl-Zecuatl T, Saleh A, Muttalib N, Van der Ploeg R, Ochoa J, Gopaul A, Tessler L, Wess J, Jolivalt CG, Fernyhough P. Selective antagonism of muscarinic receptors is neuroprotective in peripheral neuropathy. J Clin Invest. 2017;127(2):608–22. doi: 10.1172/JCI88321. PubMed PMID: 28094765; PMCID: PMC5272197 WinSanTor Inc., which has licensed intellectual property from the University of Manitoba and UCSD. This intellectual property includes data presented in the current manuscript.

94. Urban MJ, Pan P, Farmer KL, Zhao H, Blagg BS, Dobrowsky RT. Modulating molecular chaperones improves sensory fiber recovery and mitochondrial function in diabetic peripheral neuropathy. Exp Neurol. 2012;235(1):388–96. doi: 10.1016/j.expneurol.2012.03.005. PubMed PMID: 22465570; PMCID: PMC3336191.

95. Saleh A, Roy Chowdhury SK, Smith DR, Balakrishnan S, Tessler L, Martens C, Morrow D, Schartner E, Frizzi KE, Calcutt Na, Fernyhough P. Ciliary neurotrophic factor activates NF-kappaB to enhance mitochondrial bioenergetics and prevent neuropathy in sensory neurons

of streptozotocin-induced diabetic rodents. Neuropharmacology. 2013;65:65–73. doi: 10.1016/j.neuropharm.2012.09.015. PubMed PMID: 23022047; PMCID: PMC3521091.

96. Sharma K. Mitochondrial hormesis and diabetic complications. Diabetes. 2015;64(3):663–72. doi: 10.2337/db14-0874. PubMed PMID: 25713188; PMCID: PMC4338592.

97. Mei F, Fancy SPJ, Shen YA, Niu J, Zhao C, Presley B, Miao E, Lee S, Mayoral SR, Redmond SA, Etxeberria A, Xiao L, Franklin RJM, Green A, Hauser SL, Chan JR. Micropillar arrays as a high-throughput screening platform for therapeutics in multiple sclerosis. Nat Med. 2014;20(8):954–60. doi: 10.1038/nm.3618. PubMed PMID: 24997607; PMCID: PMC4830134.

98. Siatkowski RM, Cotter SA, Crockett RS, Miller JM, Novack GD, Zadnik K. Group USPS. Two-year multicenter, randomized, double-masked, placebo-controlled, parallel safety and efficacy study of 2% pirenzepine ophthalmic gel in children with myopia. J AAPOS. 2008;12(4):332–9. doi: 10.1016/j.jaapos.2007.10.014. PubMed PMID: 18359651.

99. Daugherty DJ, Marquez A, Calcutt NA, Schubert D. A novel curcumin derivative for the treatment of diabetic neuropathy. Neuropharmacology. 2018;129:26–35. doi: 10.1016/j.neuropharm.2017.11.007. PubMed PMID: 29122628; PMCID: PMC5841546.

100. Goldberg J, Currais A, Prior M, Fischer W, Chiruta C, Ratliff E, Daugherty D, Dargusch R, Finley K, Esparza-Molto PB, Cuezva JM, Maher P, Petrascheck M, Schubert D. The mitochondrial ATP synthase is a shared drug target for aging and dementia. Aging Cell. 2018;17(2). doi: 10.1111/acel.12715. PubMed PMID: 29316249; PMCID: PMC5847861.

33 Mitochondrial Dysfunction and Oxidative Stress in the Pathogenesis of Metabolic Syndrome

Qiaozhu Su and Hao Wang
Institute for Global Food Security, School of Biological Sciences, Queen's University Belfast, Belfast, UK, BT9 5AG

CONTENTS

1   INTRODUCTION

Mitochondria are the double membrane, cytoplasmic organelles that contain their self-replicating genome. In humans, the 16,569 base pairs of mitochondrial DNA (mtDNA) encode for 37 genes[1], which are inherited solely from the mother and packaged in high-ordered nucleoprotein structures called nucleoids[2]. Although nucleoids are distributed throughout the mitochondrial matrix, they are often located in the proximity of the cristae where they accommodate many copies of the respiratory chain components, or oxidative phosphorylation (OXPHOS) complexes (I–IV). Together with adenosine triphosphate (ATP) synthase (complex V), they form the machinery for ATP generation. A small intermembrane space exists between the outer and inner mitochondrial membranes. Outer mitochondrial membrane and intermembrane space are relatively more permeable than the

inner mitochondrial membrane which has much more restricted permeability and contains enzymes involved in the process of electron transport chain and ATP generation. The inner membrane surrounds the mitochondrial matrix, wherein the electrons produced by the citric acid cycle (TCA) are taken in by electron transport chain for the production of ATP. An electrochemical gradient generated across the inner membrane drives the process of OXPHOS[3].

Mitochondria play a significant role in the cells by metabolizing nutrients and producing the "energy currency" ATP; also is responsible for various processes such as energy metabolism, generation of reactive oxygen species (ROS) and calcium homeostasis, cell survival and death[4,5]. About 90% of the body's cellular energy is produced by mitochondria in the form of ATP via TCA cycle and the electron transport chain (ETC). Mitochondrial ETC is composed of five multi-subunit enzyme complexes I, II,

III, IV and V located in the inner mitochondrial membrane[6]. The electrons donated by coenzymes, NADH and FADH2 during TCA cycle and fatty acid β-oxidation are accepted and transferred to components of ETC at complex I (NADH ubiquinone reductase) or complex II, Succinate dehydrogenase, and then consecutively to complex III, Ubiquinol-cytochrome c reductase, to complex IV, Cytochrome c oxidase, and finally to oxygen through complex V, ATP synthase. The transfer of electrons along the electron transport chain is coupled with the transport of protons across the inner membrane, establishing the electrochemical gradient that generates ATP[7].

Mitochondria continuously function to metabolize nutrients and generate ATP. However, the flow of electrons through the ETC is an imperfect process in which 0.4 to 4% of oxygen consumed by mitochondria is incompletely reduced and leads to the production of ROS, which are a family of free radicals that includes superoxide anions ($O2^{\bullet-}$), hydroxyl($\bullet OH$), peroxyl radicals and other non-radicals capable of generating free radicals[5,8,9]. Free radicals are fundamental to cellular physiological process and are continuously maintained at an appropriate level in the body. However, excessive generation of $O2^{\bullet-}$ can interact with various cellular components and generate "secondary" ROS[10,11]. It has been observed that $\bullet OH$ interact, and react with the nitrogenous bases, purine/pyrimidine and the deoxyribose backbone of DNA[5]. Overproduction of ROS further reacts with mitochondrial proteins/enzymes, membranes, lipids, and nucleic acids, especially thiamine and guanosine, which leads to the interruption of ATP generation and other essential functions in mitochondria[9,10]. Other reactive species, such as nitric oxide (NO) and reactive nitrogen species (RNS), are also produced by ETC, which also reacts with cellular proteins and glutathione through nitration.

Most eukaryote cells possess numerous defense systems, involving both enzymatic and non-enzymatic mechanisms, to alleviate the oxidative damage, induced by ROS which either directly diminish their generation of these or scavenge them, mediated by various antioxidants. Enzymatic defense system regulates the action of various antioxidant enzymes, such as superoxide dismutase (SOD), catalase (CAT), glutathione reductase (GR) and glutathione peroxidase (GPx), while the non-enzymatic defenses are mediated via a large variety of antioxidant compounds to protect cells against oxidative stress, including vitamin E and C, glutathione (GSH), various carotenoids and flavonoids. Under physiological conditions, the overproduction of ROS is restricted in mitochondria via the enzymatic and non-enzymatic defense systems which protect cellular organelle from oxidative damage induced by ROS. However, when the antioxidant defenses are overwhelmed, the overproduction of ROS leads to oxidative damage to the proteins, DNA, and lipids in mitochondria,[12] which result in impaired function of the enzymes involved in the respiratory chain and ultimately compromise mitochondrial function, reduce mitochondrial biogenesis and OXPHOS as well as a broad range of pathologic conditions, such as metabolic syndrome, aging and neurodegenerative disorders[11,13–15].

2 THE INDEPENDENCE AND CROSSTALK BETWEEN NUCLEAR DNA AND MITOCHONDRIAL DNA

2.1 MITOCHONDRIAL DNA

Each eukaryotic cell contains several hundred copies of mitochondrion, which are unique organelles under dual genome regulation. Although the majority of DNA is enclosed within the nucleus (nDNA), mitochondria also contain their own and separate DNA, the mitochondrial (mtDNA), which is the small circular chromosome located in mitochondria. The human mtDNA is a double-stranded, circular molecule of 16, 569 base pairs and contains 37 genes coding for two rRNAs, 22 tRNAs and 13 subunits of the ETC complexes. The human mtDNA encodes the peptide, humanin, known to have important pro-survival and metabolic regulatory functions. Mitochondrial-encoded subunits of the oxidative phosphorylation system assemble with nuclear-encoded subunits into enzymatic complexes. Recent findings show that mitochondrial translation is linked to other mitochondrial functions, as well as to cellular processes. In turn, translation in mitochondria controls cellular proliferation, and mitochondrial ribosomal subunits contribute to the cytoplasmic stress response. Thus, translation in mitochondria is apparently integrated into cellular processes[16].

2.2 MITOKINE

Although mitochondria possess their own genome for encoding 13 proteins, most of the ~1,200 mitochondrial proteins originate from the nuclear genome. Therefore, the nucleus and mitochondria have to constantly communicate with each other to adjust their activities in order to ensure cellular homeostasis and adaptation to mitochondrial stress. This communication is defined as mitonuclear communication. For instance, a mtDNA-encoded peptide, the mitochondrial open reading frame of the 12S rRNA-c (MOTS-c), could translocate into the nucleus upon stress induction. In the nucleus, MOTS-c collaborates with other transcription factors to regulate the transcription of stress response genes. This example clearly demonstrates that a mitochondria-derived peptide can function in the nucleus to regulate nuclear transcription in response to stress. It is very likely that MOTS-c may bear some characteristics of a "mitokine," factor that mediates mitohormesis to influence cell survival and organismal health[17]. It has been reported that mtDNA mutations, abnormal mtDNA copy number and mitochondrial tRNA (mt-tRNA) mutations are linked to a range of human diseases, including Alzheimer's disease, premature aging and cancer[18].

2.3 INTERACTION BETWEEN nRNA AND mtRNA

Mitochondria are not self-supporting entities but rely heavily for their functions on imported nuclear gene products. mtDNA replication is a continuous process which is independent of cell division. nDNA encodes all the proteins responsible for replication, as well as other proteins necessary for the transcription and copy number control of this multicopy genome[19]. It also requires constant supply of deoxyribonucleotide triphosphates (dNTPs) and interaction with other mitochondrial machinery for mixing and unifying the mitochondrial compartment[20]. nDNA also encodes respiratory chain subunits[19]. Perturbation of mtDNA maintenance causes defects in nuclear genes involved in different aspects of mtDNA replication, such as mtDNA depletion and/or multiple mtDNA deletions, resulting in variable disorders in the affected tissues[20]. Numerous nuclear mutations on genes expressing mitochondrial protein have been reported in association with very tissue specific clinical disease, suggesting that there are unidentified factors determining the tissue specificity in mitochondrial translation. Most of these gene defects result in histological alterations and multiple respiratory chain defects in the affected organs. In addition, the clinical symptom is usually early-onset, severe, and often life-threatening, implicating the essential role of mitochondrial translation in organ/tissue development[21].

3 REGULATION OF MITOCHONDRIAL BIOGENESIS AND DYNAMICS

Mitochondrial biogenesis maintains the number and size of mitochondria which involves regulation from both nuclear and mitochondrial genomes. Nuclear genome encoded regulatory factors are essential for mitochondrial biogenesis and function. These factors act exclusively within the mitochondria to control of mitochondrial transcription, translation, and other functions whereas others govern the expression of nuclear genes required for mitochondrial biogenesis and metabolism. Physiologic stimuli, including physical exercise, dietary restrictions, temperature, and muscle myogenesis, are involved in mitochondrial biogenesis via mediation of transcription factors.

3.1 PGC-1 FAMILY IN MITOCHONDRIAL BIOGENESIS

The peroxisome proliferator-activated receptor γ coactivator-1 (PGC-1) family of transcriptional coactivators, consisting of PGC-1α, PGC-1β and PRC, plays a crucial role in a regulatory network governing the transcriptional control of mitochondrial biogenesis and respiratory function[22,23]. PGC-1α acts as a critical mediator of mitochondrial biogenesis induced by cues that signal high-energy needs, and governs the process of mitochondrial biogenesis by interacting with a various transcription factors/proteins such as nuclear respiratory factors (NRF-1 and NRF-2), mitochondrial transcription factor A (Tfam), uncoupling proteins (UCP2), PPARs, thyroid hormone,

glucocorticoid, estrogen and estrogen related receptors (ERR) α and γ[24-26]. Among them, NRF-1, NRF-2, and Tfam regulate the transcription of the main mitochondrial enzymes and mtDNA synthesis[27]. On the other hand, PGC-1β is to the maintenance of basal mitochondrial function[28].

3.2 mTOR, AMPK AND SIRTUINS (SIRTS)

In addition, three major nutritional sensors in cell metabolism, including mTOR, AMPK and Sirtuins (Sirts), are also involved in the regulation of mitochondrial biogenesis and dynamics by regulating mitochondrial fusion, fission, and turnover through mitophagy and autophagy. In energy deprived state, AMPK and Sirt1 regulate PGC-1α through phosphorylation and deacetylation, respectively[29]. Both basic research and clinical studies revealed the alterations in the morphology and number of mitochondria in heart, skeletal muscles and liver tissues in pathogenic conditions[4,30-34]. A tight-tuning interaction between the activity of these nutritional sensors, mitochondrial biogenesis rate and dynamics, affects aging and age-related diseases such as metabolic disease[35].

mTOR is a serine/threonine kinase that is activated in high calorie intake and induce anabolic activities. mTOR signalling is associated with insulin resistance, adipogenesis, angiogenesis, inflammation and cancer development[36]. The signalling pathway triggered by mTOR is counterbalanced by AMPK and Sirts signalling which is activated when energy is insufficient, such as low food intake, starvation or calorie restriction. AMPK stimulates cellular catabolism of carbohydrates, proteins and lipids, and inhibits many anabolic enzymes which regulate glycogen and lipid biosynthesis[37]. Sirts are a protein family that regulates the activity of enzymes by deacetylation. Sirt3, a Sirt family member located in the mitochondria with deacetylase activity, is involved in the catabolism of lipids and regulates the activity of ETC complexes II, III and IV via deacetylation[38], modulation of the activity of Mn-SOD[39], regulation of autophagy via deacetylation of FOXO3[40] and modulation of mitochondrial dynamics by activating OPA1[41]. Of note, an exceptional case in mammalian cells is that Sirt4 and Sirt5 show ADP-ribosyl-transferase activity instead of deacetylase activity[42]. Sirts act as metabolic sensors by detecting fluctuations in the $NAD^+/NADH$ ratio, such as in the case of reduced nutrient availability. In addition, these proteins are involved in various key cellular processes, from cell cycle control to metabolism and antioxidant protection[42]. The impact of Sirts on mitochondrial physiology and cell metabolism has been shown to be associated with their effect on longevity. Regulation of mitochondrial biogenesis by Sirts also contributes to the health benefits of calorie restriction[43-45]. For instance, Sirt3 is induced in skeletal muscle after calorie restriction and its levels decrease under high-fat diet, indicating calorie-dependent regulation[46]. All these nutrient sensor, mTOR, AMPK and Sirts, regulate the activity of PGC1a. Sirt1 not only activates transcriptional activity of PGC1a by deacetylation[47], but also induces PGC1a expression which is evidenced by the reduced levels of PGC1a in mice lacking Sirt1[48]. AMPK activates PGC1a by phosphorylation and induces its own transcription via a positive feedback loop[49] or by inducing its deacetylation through Sirt1 activation[50].

3.3 INTERACTION BETWEEN SESTRINS, mTORC1 AND AMPK

The mTORC1 and AMPK also regulate sestrins to module stress gene expression and the maintenance of autophagy, mitochondrial turnover and ageing-related diseases[51]. Sestrins are highly conserved protein family encoded by genes known as Sens that are upregulated after environmental stress, hypoxia and DNA damage[52]. Sestrins repress mTORC1 activity by activating AMPK[52]. Inactivation of Sens in invertebrates produces diverse metabolic pathologies resembling accelerated aging and metabolic stress, oxidative damage, lipid accumulation, mitochondrial dysfunction, and muscle degeneration[53]. Depletion of Sens2 and Sens3 genes in mice establishes the important role of these proteins in the regulation of lipid metabolism, obesity-associated mitochondrial disorders and the suppression of aging[54,55]. For instance, Sesn2 KO mice show higher glucose intolerance, reduced insulin sensitivity and development of hepatic steatosis[55]. Furthermore, Sens3 was demonstrated to be associated with the activation of mTORC2 and AKT phosphorylation and enhanced insulin sensitivity and glucose metabolism[56]. In humans, Protein levels of Sestrin1 and 3 are decreased in aged muscle[57]. Serum from frail elderly people shows significant reductions of Sestrin1 and Sestrin2, indicating the importance of these factors in the maintenance of muscle capacity during aging[58]. Since Sestrins are regulated by many stress factors, such as ROS, these proteins have been emerging as interesting regulators of mitochondrial maintenance and ageing-related diseases[59].

3.4 MITOCHONDRIA TURNOVER

AMPK and mTORC1 (mTOR complex 1) control the removal of damaged mitochondria through the regulation of autophagy and mitophagy[60,61]. AMPK promotes autophagy by phosphorylating ULK-1 and ULK-2 in their activation residues[62] whereas mTORC1 inhibits autophagy by phosphorylating the inhibitory sites of unc-51-like kinase 1/2 (ULK-1 and ULK-2), initial mediators of this process. In mitochondrial removal, Forkhead box proteins (FoxO) proteins, a family of transcription factors involved in the regulation of cell growth, proliferation, differentiation and longevity, play an important role. FoxO proteins induce several stress response genes to rescue cells and organs from different stress condition[63]. Further studies show that the health beneficial effects of calorie restriction, polyphenols or

exercise on mitochondrial dysfunction during ageing are also mediated by FoxO proteins and regulated by metabolic regulators such as mTOR, AMPK and Sirts signalling molecules[64].

Furthermore, the shape of mitochondria, biogenesis ratio, turnover and dynamics, fusion/fission processes are also dynamically changing depending on the bioenergetic state of the cell[65]. Mitochondrial fusion is associated with active mitochondrial function whereas fission is associated with mitochondria damage and their removal by mitophagy and autophagy[66]. Mitofusins 1 and 2 (Mfn1 and Mfn2), the large GTPases, and autosomal dominant optic atrophy-1 (OPA1), are the main proteins involved in mitochondrial fusion. Mfn1 and Mfn2 are located at the surface of the outer membrane of mitochondria and act in the fusion process of these membranes whereas OPA1 is an intermembrane protein anchored to the inner membrane and acts in the fusion of inner membranes[67,68]. Mfn2 is also involved in tethering mitochondria with endoplasmic reticulum (ER) which regulates the communication between mitochondria and ER, particularly in nutrient-surplus induced ER stress in metabolic syndrome[69-71]. In mitochondrial fission, the dynamin related protein 1 (DRP1) and the fission protein 1 (FIS1) are the main proteins involved in this process. DRP1 is a GTPase located mainly in the cytosol with a fraction localized at punctate structures on the outer mitochondria membrane[72]. FIS1 is a protein integrated in the outer mitochondrial membrane that recruits DRP1[73,74]. Another regulatory protein, prohibitin, is a complex of two proteins, PHB-1 and PHB-2, which are ubiquitous and form a ring-like complex at the inner membrane of the mitochondria[75] where prohibitin stabilizes OPA1 and regulates mitochondrial fusion and cristae morphogenesis. However, the exact molecular mechanism of this complex in lipid overload-induced mitochondrial dysfunction and oxidative stress response remains unclear at the present time[76,77].

4 PPARα SIGNALING IN MITOCHONDRIAL FATTY ACID β-OXIDATION

4.1 THE PPAR FAMILY

Peroxisome proliferator-activated receptors (PPARs) are a family of nuclear receptors, which consists of three members, PPARα, PPARγ and PPARβ/δ. Among them, PPARα is expressed highly in tissues with high fatty acid oxidation (FAO) rates, such as liver, heart, skeletal muscle, brown adipose tissue, and kidney, although it is also universally expressed in other tissues, including the intestine, vascular endothelium, smooth muscle and immune cells[78]. PPARα is an essential nutritional sensor, which regulates transcription of genes involved in fatty acid (FA) catabolism, lipogenesis and ketone body synthesis, in response to feeding and starvation[79]. These include genes regulating mitochondrial β-oxidation, FA transport and hepatic glucose production[80]. Moreover,

PPARα also displays anti-inflammatory properties by inhibiting pro-inflammatory and acute phase response (APR) signalling pathways, as evidenced by rodent models of systemic inflammation, atherosclerosis and non-alcoholic steatohepatitis (NASH)[81,82].

4.2 PPARα Regulates Fatty Acid β–Oxidation

FA derivatives from lipolysis, lipogenesis or FA catabolism are ligands of PPARa. Free fatty acids, including long-chain polyunsaturated FAs (LCPUFAs) and n-3 LCPUFAs, bind and activate the PPAR signaling to regulate the transcription of a cluster of genes involved in lipid and lipoprotein metabolism, FA β-oxidation in tissues with high oxidative rates, such as heart, liver, and muscle. Substrates of acyl-CoA oxidase 1 (ACOX1), the first rate-limiting peroxisomal β-oxidation enzyme, may also serve as PPARα agonists. Hydrolysis of hepatic intracellular triglyceride also yields lipid ligands for PPARa[83]. A range of synthetic PPARα agonists, including gemfibrozil, fenofibrate and ciprofibrate, have been synthesized and used in the treatment of primary hypertriglyceridemia or complexed dyslipidaemia[84] in clinical.

Regulation of FA β-oxidation is a major physiological function of PPARα. FAs are transported into cells by membrane-bound fatty acid transport proteins (FATPs)[85]. FATP1 is a direct PPARα target gene, which catalyses esterification of long-chain FAs (LCFAs) and very long-chain FAs (VCFA) into acyl-CoA derivatives in a ATP-dependent manor[86,87]. FAT/CD36 is another plasma membrane FA transporter which is positively regulated by PPARα signalling[88]. In rodents and primates, FA transport across the mitochondrial membrane is regulated by Carnitine palmitoyltransferase I (CPT-I) and Carnitine palmitoyltransferase II (CPT-II), two protein enzymes localized in the outer and inner mitochondrial membrane, respectively[89–91]. Moreover, PPARα regulates the crucial reaction of mitochondrial β-oxidation by directly controlling expression of enzymes involved in FA β-oxidation catabolic steps, including MCAD, LCAD, and VLCAD[92,93]. In the liver, PPARα activation, in combination with PPARβ/δ agonist, improves hepatic steatosis, inflammation and liver fibrosis in animal model of nonalcoholic fatty liver disease (NAFLD)[94]. PPARs further play essential roles during placental, embryonic fetal development, and in the physiological processes of oxidative stress, inflammation, and neurodegeneration[95–97].

5 MITOCHONDRIAL UNFOLDED PROTEIN RESPONSE (UPR^MT)

5.1 Mitochondrial Unfolded Protein Response (UPR^MT)

Mitochondria regulate energy (ATP) production through oxidative phosphorylation, maintaining redox homeostasis, modulating calcium levels, and controlling cell death through apoptosis. These physiological activities take place in the mitochondrial matrix and involve multiple regulations. Perturbation of these regulations induces various mitochondrial stresses. For instance, abnormal mtDNA maintenance and translation induces protein stress, mtDNA replication and transcription stress as well as mtDNA depletion associated stress[98] which ultimately affect the assembly of the oxidative phosphorylation complexes. About 85% of the mitochondrial structural subunits of the respiratory chain complexes and ATP synthase, as well as all of the required factors for assembly, are synthesized by cytoplasmic ribosomes. Multiple regulations are taken to ensure the protein quality control that precedes the integration of the newly synthesized proteins into the inner membrane, including co- and post-translation or after importing into mitochondria[99]. In contrast, the 13 mtDNA-encoding proteins are co-translationally integrated directly into the inner membrane. The mitochondrial inner membrane is one of the most protein-rich membranes in cells. Thus, quality control of the incorporated proteins is essential for maintaining metabolic homeostasis of mitochondria. However, the protein-folding environment in mitochondria is challenged by organelle architecture, presence of ROS and the difficulties associated with assembly of the electron transport chain, which require both the cytoplasmic and mitochondrial gene expression machineries coordinate tightly and coupled with mitochondrial membrane dynamics to fulfil this complex regulation. It has been reported that up to 75% of *de novo* mitochondrial translation products are turned over by quality control pathways in human fibrosis cell line[100]. Disrupting the coordination of these processes disturbs protein homeostasis and results in protein stress, which will subsequently activate an adaptive response, the mitochondrial protein unfolded response (UPR^mt), and organelle membrane stress. In addition, accumulation of newly synthesized proteins by mitochondrial ribosomes can dissipate the mitochondrial membrane potential and trigger remodelling of mitochondrial morphology, resulting in mitochondrial fragmentation[101].

5.2 Mitochondrial Retrograde Signalling in UPR^MT

Mitochondrial stress further activates the retrograde signals to regulate the stress responses and the quality control signalling. Mitochondrial retrograde signaling is a pathway of communication from mitochondria to the nucleus under physiological and pathophysiological conditions. Activation of UPR^mt releases peptides from mitochondria which direct the signals to nucleus[102]. This action triggers nuclear genetic programmes for mitochondrial maintenance and enhances quality control by upregulating expression of genes that carry a conserved amino acid response element (AARE) in their gene promotor regions[103]. The AARE is the binding element of the activating transcription factors (ATFs) that have been linked to the unfolded protein response in the ER (UPR[ER])[104] and the UPR[mt105]. UPR[mt] was first described in cultured mammalian cells by pioneering experiments from the Hoogenraad laboratory which overexpressed a nuclear-encoded unstable mutant mitochondrial matrix protein[105], and was then further characterized in the nematode Caenorhabditis elegans (C. elegans)[106]. These studies revealed that expression of mitochondria-specific molecular chaperone, the heat shock protein 60 (HSP60), and the mitochondrial protease ClpP were induced upon UPR[mt] in culture cells[105,106]. In C. elegans, the activating transcription factor associated with stress 1 (ATFS-1), an orthologue of the mammalian ATF proteins, is constitutively imported into mitochondria and degraded. In response to mitochondrial dysfunction and reduced protein import to mitochondria, a nuclear localization signal in the carboxyl terminus of ATFS-1 protein redirects ATFS-1 to the nucleus to activate a protective transcriptional response, resulting in upregulated expression of mitochondrial heat shock proteins (e.g., HSP60)[107,108]. In this worm, deficiency of the mitochondrial respiratory chain complex IV in neurons induces upregulation of HSP expression in the intestine, which was proposed as an effect of humoral spreading of the stress response via a secreted "mitokine"[109]. The AARE element is conserved between worm and mammals, suggesting that the response may be highly conserved. However, despite the conservative response of ATF-AARE induction upon exposure to mitochondrial insults between lower organisms and mammals, the downstream targets in mammalian postmitotic tissues[103] are quite different from those in cultured cells or in C. elegans. In mammals, there is little evidence of the existence of unfolded mitochondrial proteins. Furthermore, in tissues from patients with mitochondrial diseases, there is only a subtle induction of the UPR[mt]-associated HSPs and the protease ClpP which is often accompanied by the induction of a mitochondrial biogenesis programme[110–112]. Instead, in mammals, defects in mtDNA maintenance and translation enhance AARE transcripts which enhance translation of proteins involved in the regulation of lipid and glucose metabolism, as well as the anabolic one-carbon cycle[98,113,114]. Significant nuclear transcriptional stress response has been observed in patients, mice and cell lines with mtDNA maintenance and translation defects[98,103,112,113,115]. The downstream targets are regulated both in a cell-autonomous manner and in a non-cell-autonomous manner via paracrine or endocrine effects. Because this response integrates transcriptional components of UPR[mt], metabolites and redox signalling, it has been defined as the mitochondrial integrated stress response (ISR[mt]).

6 MITOCHONDRIA-ASSOCIATED MEMBRANES STRUCTURE IN MITOCHONDRIAL AND ER SIGNALING TRANSDUCTION

6.1 THE STRUCTURE OF MITOCHONDRIA-ASSOCIATED MEMBRANES (MAMS)

Mitochondria and ER are two membrane-bound subcellular organelles in eukaryotic cells which perform unique intracellular tasks via highly dynamic communication[116–118]. ER is an active organelle involved in protein folding and modifying secretory and membrane proteins. Direct ER-mitochondria communication is facilitated by the physical interaction of their membranes in the dedicated structural domains known as mitochondria-associated membranes (MAMs), which shelter various proteins with different functions and play a crucial role in cellular functions, such as facilitating calcium (Ca^{2+}) transfer, lipid transport, cell survival signaling transduction between these two organelles and also act as platforms for cellular signaling transduction[119,120]. Recent studies further revealed the pivotal roles of MAMs in nutrient and hormonal signaling, implicating the regulatory role of MAMs in metabolic homeostasis and metabolic diseases. The ER-mitochondria interaction was first reported in 1959[121]. However, the MAM fractions were not isolated until 1990[122] by subcellular fractionation using Percoll density gradients[122,123] and were further visualized in living cells using fluorescent proteins at the end of the 90s[124]. Ever since their discovery, the importance of these contact sites in organelle cross-talk has been evolved by the improvement of imaging and analytic techniques, such as electron microscopy and electron tomography[125–128]. ER-mitochondria contact sites permit reciprocal regulation of function in both organelles, thereby impacting various cellular activities, including energy metabolism, Ca^{2+} handling[124], lipid homeostasis[122] as well as regulation of cell survival and death[125–127,129].

6.2 THE MAMS IN CELLULAR METABOLISM

The ER-mitochondria interfaces are enriched in protein enzymes that participate in the synthesis and transport of phospholipids and glycosphingolipids[122,130], such as phosphatidylserine (PS) synthase 1 and 2 that are involved in phospholipid (PL) synthesis. Newly synthesized PS is preferentially transferred from ER to mitochondria[123], where it is used for the synthesis of PE[127]. In mammals, PE is also transferred back to the ER for its conversion to PC via PE methylation by the hepatic enzyme PE-N-methyltransferase[131]. Other lipid biosynthetic enzymes such as glycerol 3-phosphate acyltransferase, acylcoA synthase 4, diacylglycerol acyltransferase 2 and the microsomal triacylglycerol transfer protein are also localized at MAMs, which are considered structural components of the MAMs, contributing both to the stability and dynamics of this structure[127,132]. Another well-known protein complex identified at MAM contact site is the Ca^{2+} channeling complex made of the inositol 1,4,5-trisphosphate receptor (IP3R) at the ER, the cytosolic chaperone glucose-regulated-protein 75 (GRP75) and the the mitochondrial porin voltage-dependent anion channel (VDAC) at the outer mitochondrial membrane (OMM) for mediating the transfer of Ca^{2+} from ER to mitochondria[133]. Once Ca^{2+} ions have passed through the OMM, they enter into the matrix through a mitochondrial Ca^{2+} uniporter (MCU). Recently, another mitochondrial matrix protein, cyclophilin D (CYPD), was found to interact with and regulate this Ca^{2+} channeling complex. CYPD belongs to the family of the peptidyl-prolyl cis-trans isomerases and regulates the opening of the mitochondrial permeability transition pore in stress conditions, particularly during myocardial ischemia-reperfusion injury[134]. At the MAM interface, CYPD forms a high molecular complex with the IP3R-Grp75-VDAC complex in both cardiomyocytes and hepatocytes and regulated Ca^{2+} exchange between these two organelles[135–137].

Proteins involved in mitochondrial dynamics are also part of MAMs. Mitofusins, Mfn-1 and Mfn-2, are two related GTPases that localize to OMM, where they participate in mitochondrial fusion. In fact, Mfn-2 is also present on the ER, where it forms heterotypic or homotypic interactions with mitochondrial Mfn-1 or Mfn-2, respectively, thus forming ER-mitochondria bridges[69] and regulating Ca^{2+} homeostasis. Depletion of Mfn-2 in embryonic fibroblasts was found to increase the distance between ER and mitochondria[69], and cells lacking Mfn-2 alter composition of MAMs, resulting in a loosening connection between ER and mitochondria[69]. The mitochondrial ubiquitin ligase MITOL[138] has also been reported to regulate Mfn-2-dependent ER-mitochondria tethering at the ER-mitochondria interface[138]. However, this well-accepted concept has been challenged recently as both increased ER-mitochondria interactions and increased Ca^{2+} transfer were observed in Mfn2 knockout (KO) cells[139,140]. More research is therefore required to clarify the precise role of Mfn-2 at MAMs.

Phosphofurin acidic cluster-sorting proteins 2 (PACS2) is a multifunctional homeostatic regulator that is present at the MAMs and plays a role in sorting proteins on the ER and mitochondria[141] as well as the trafficking of ion channels between secretory pathway compartments[142]. Depletion of PACS2 augments organelle distance and favours cell apoptosis by promoting caspase-dependent cleavage of Bap31[141]. In contrast, overexpression of PACS2 was shown to increase ER-mitochondria interaction and Ca^{2+} exchange in liver[143], suggesting that the presence of PACS2 is sufficient to control MAM integrity and function. PACS2 further regulates the activity of calnexin (CNX), an ER chaperone compartmentalized

in MAM which couples with sarco/endoplasmic reticulum Ca^{2+}-ATPase (SERCA) to modulate local Ca^{2+}. More than 80% of CXN localize to the ER, mainly in MAMs, under physiological conditions. Depletion of Pacs2 causes a relocation of CXN from ER to the plasma membrane, therefore affecting mitochondrial and ER Ca^{2+} homeostasis[144]. In addition, the ER-resident sigma 1 receptor (Sig-1R), a novel "ligand-operated" chaperone[145], the 78 kDa glucose-regulated protein (GRP78), the promyelocytic leukaemia (PML) tumour suppressor[146] and Bap31[147] are also associated with MAMs and involved in the regulations of Ca^{2+} signalling, apoptotic signalling, cell survival, and autophagy at the MAM interface[148].

7 THE MAMS IN MITOCHONDRIAL AND ER STRESS

ER plays a key role in protein folding. Endogenous or exogenous insults compromising this ER function, leads to the accumulation of misfolded proteins, inducing ER stress and triggering activation of an adaptive response, the unfolded protein response in the ER (UPR^{ER}). Activation of UPR^{ER} restores ER homeostasis by enhancing expression of stress response genes, such as ER chaperones, inhibiting global protein synthesis, promoting degradation of misfolded proteins, and increasing total ER volume[149]. This response also triggers changes in cellular metabolism to provide metabolic support for cellular adaptation. It has been shown that in early stages of ER stress, there is a microtubule-dependent redistribution of the ER and the mitochondrial network toward the perinuclear area of the cell and an increase on MAMs in this region[150] which is accompanied with an increase in mitochondrial Ca^{2+} uptake and mitochondrial ATP production[150]. Disruption of MAMs or blockage of Ca^{2+} transfer increases cell death in response to ER stress, suggesting that this reorganization favours cell survival[149,150]. Moreover, the MAM-resident chaperone, Sig-1R, also plays an important role in regulating ER-mitochondria contacts during ER stress. Sig-1R promotes the efficiency of Ca^{2+} transfer to mitochondria and ATP production to favour cell survival during conditions of ER stress[151,152]. Sig-1R stabilizes an ER-stress sensor, IRE-1α, at the MAMs, which prolongs activation of IRE-1α to support cell survival[153] in the early stages of ER stress. Mfn-2 is another protein associated with the MAM structure and involved in ER stress response. Induction of ER stress increases Mfn-2 levels[154] whereas Mfn-2 deficiency promotes ER stress[154,155], highlighting the importance of MAM function in the ER stress response. Furthermore, upon ER stress, Ca^{2+} is released from the ER which augments the production of mitochondrial ROS. Accumulation of ROS within ER and mitochondria disturbs fundamental organelle functions. Prolonged ER stress is known to potentially elicit inflammatory responses via UPR pathways. ROS generated through inflammation or mitochondrial dysfunction could further accelerate ER malfunction. Ample evidences suggest mitochondrial dysfunction and ER stress as major contributing factors in the onset and development of metabolic disease, including obesity, diabetes, cardiovascular disease. Although the pathogenesis of diseases may be diverse, ER stress and/or mitochondrial oxidative stress associated metabolic inflammation could be the fundamental mechanism that initiates and increases the severity of the disease[156].

8   ER-MITOCHONDRIAL INTERFACES IN METABOLIC SYNDROME

Metabolic syndrome is a cluster of metabolic disorders, including hypertension, hyperglycemia, obesity, hyperlipidemia, NAFLD and cardiovascular diseases (CAD)[157–160]. Mitochondrial dysfunction and oxidative stress have been found to be involved in the process of these chronic metabolic diseases[161,162]. In metabolic syndrome, mitochondria become less efficient, produce less ATP but high ROS[163]. High fat diets have been reported to shift the cellular redox environment to a more oxidized state in mitochondria in adipose tissue, skeletal muscle and liver in both rodents and humans which decreases the redox-buffering capacity. Attenuation of mitochondrial H_2O_2 emission, either by treating rats with a cell-permeable small peptide antioxidant, SS31, or by overexpressing a human catalase specifically targeting to mitochondria in mouse muscle, completely preserves insulin sensitivity despite of a high fat diet[160]. These experiments indicate that the pathophysiological alterations in metabolic syndrome is associated with impaired mitochondrial functions, such as reduced oxidative capacity and antioxidant defense by the upregulated mitochondrial H_2O_2 emission, suppressed OXPHOS and decreased ATP production[13,14,164–166]. Emerging evidences have demonstrated that mitochondrial dysfunction plays key roles in the pathophysiology of obesity and its associated complications, such as type 2 diabetes and cardiovascular disease. This is supported by the close association between mitochondrial dysfunction with reduced insulin sensitivity, increased secretion of inflammatory cytokines[167] and impaired β-cell function in the pancreas[168]. In mouse models, hepatic mitochondrial dysfunction is linked to the development of NAFLD and NASH induced by lipogenic diets[169,170]. Mitochondrial dysfunction has also been linked with a reduced ability of skeletal muscle to respond to insulin[171].

As discussed above, mitochondria and ER are two organelles intimately connected and physically linked via MAMs which make them functionally interdependent. Mitochondria dysfunction is expected to affect the ER homeostasis and vice versa. Depletion of Mfn-2, a protein residing in the MAM structures, resulted in ER stress and reduced insulin sensitivity in the skeletal muscle and liver[155]. Further animal study showed that high-fat-fed mice displayed symptoms of both mitochondrial dysfunction and ER stress in the liver and skeletal muscle[172]. The fact that obese mice are usually accompanied with increased ER-mitochondria contacts and simultaneous mitochondrial Ca^{2+} overload as well as mitochondrial dysfunction provides more direct evidence to support the intimate association between mitochondria and ER in the onset and development of metabolic disorders[143]. Of note, there is also an exceptional case being reported: one study showed that reduced ER-mitochondria contacts were recorded in the POMC neurons of a high fat diet induced obese mice despite there was ER stress[173], suggesting the roles of MAMs may be tissue- or context-specific.

Cardiovascular diseases are common complication of obesity, insulin resistance and diabetes. An ex-vivo experiment using cultured rat neonatal cardiomyocytes showed that that insulin can induce Ca^{2+} releasing from the ER[174], which was subsequent uptaken into mitochondria[175]. Ca^{2+} transfer from ER to mitochondria is essential for insulin signalling transduction in cardiomyocytes[175]. This evidence strongly suggests that ER-mitochondria interactions play important roles in the regulation of energy metabolism and the pathophysiology of obesity-related metabolic diseases. Further deciphering the intracellular mechanisms involved in the pathogenesis of metabolic syndrome may shed lights on reducing the deleterious consequences of obesity and its associated metabolic disease.

9   CONCLUSION

Mitochondria are the cytoplasmic organelles responsible for cell energy production, survival and cell death. Mitochondrial abnormalities, including impaired mitochondrial dynamics, defects in mitochondrial biogenesis, mitochondrial dysfunction, and oxidative stress contribute to the onset of metabolic syndrome, aging progress, and cancer development. Maintaining mitochondrial dynamics (fission-fusion balance) and mitochondrial homeostasis are essential for the prevention and treatment of metabolic syndromes. However, there are gaps on our knowledge between metabolic syndromes and mitochondrial structural/functional changes. Therefore, further research is urgently needed to advance our understanding on the underlying molecular mechanisms.

REFERENCES

1. Anderson S, Bankier AT, Barrell BG, Debruijn MHL, Coulson AR, Drouin J, Eperon IC, Nierlich DP, Roe BA, Sanger F, Schreier PH, Smith AJH, Staden R and Young IG. Sequence and organization of the human mitochondrial genome. *Nature.* 1981;290:457–65.
2. Chen XJ and Butow RA. The organization and inheritance of the mitochondrial genome. *Nat Rev Genet.* 2005;6:815–25.

3. Sherratt HS. Mitochondria: Structure and function. *Rev Neurol (Paris)*. 1991;147:417–30.

4. Kim JA, Wei Y and Sowers JR. Role of mitochondrial dysfunction in insulin resistance. *Circ Res*. 2008;102:401–14.

5. Halliwell B and Gutteridge JM. Free radicals in biology and medicine. Oxford University Press; 2015.

6. Dallner G and Sindelar PJ. Regulation of ubiquinone metabolism. *Free Radical Bio Med*. 2000;29:285–94.

7. Venditti P, Di Stefano L and Di Meo S. Mitochondrial metabolism of reactive oxygen species. *Mitochondrion*. 2013;13:71–82.

8. Halliwell B. Reactive species and antioxidants. Redox biology is a fundamental theme of aerobic life. *Plant Physiol*. 2006;141:312–22.

9. Murphy MP. How mitochondria produce reactive oxygen species. *Biochem J*. 2009;417:1–13.

10. Droge W. Free radicals in the physiological control of cell function. *Physiol Rev*. 2002;82:47–95.

11. Valko M, Rhodes CJ, Moncol J, Izakovic M and Mazur M. Free radicals, metals and antioxidants in oxidative stress-induced cancer. *Chem-Biol Interact*. 2006;160:1–40.

12. Beckman KB and Ames BN. Endogenous oxidative damage of mtDNA. *Mutat Res-Fund Mol M*. 1999;424:51–8.

13. Finkel T and Holbrook NJ. Oxidants, oxidative stress and the biology of ageing. *Nature*. 2000;408:239–47.

14. Barnham KJ, Masters CL and Bush AI. Neurodegenerative diseases and oxidative stress. *Nat Rev Drug Discov*. 2004;3:205–14.

15. Thanan R, Oikawa S, Hiraku Y, Ohnishi S, Ma N, Pinlaor S, Yongvanit P, Kawanishi S and Murata M. Oxidative stress and its significant roles in neurodegenerative diseases and cancer. *Int J Mol Sci*. 2014;16:193–217.

16. Richter-Dennerlein R, Dennerlein S and Rehling P. Integrating mitochondrial translation into the cellular context. *Nat Rev Mol Cell Bio*. 2015;16:586–92.

17. Yong CQY and Tang BL. A mitochondrial encoded messenger at the nucleus. *Cells*. 2018;7:E105.

18. Ding L and Liu Y. Borrowing nuclear DNA helicases to protect mitochondrial DNA. *Int J Mol Sci*. 2015;16:10870–87.

19. Rusecka J, Kaliszewska M, Bartnik E and Tonska K. Nuclear genes involved in mitochondrial diseases caused by instability of mitochondrial DNA. *J Appl Genet*. 2018;59:43–57.

20. Almannai M, El-Hattab AW and Scaglia F. Mitochondrial DNA replication: Clinical syndromes. *Essays Biochem*. 2018;62:297–308.

21. Boczonadi V, Ricci G and Horvath R. Mitochondrial DNA transcription and translation: Clinical syndromes. *Essays Biochem*. 2018;62:321–40.

22. Scarpulla RC, Vega RB and Kelly DP. Transcriptional integration of mitochondrial biogenesis. *Trends Endocrinol Metab*. 2012;23:459–66.

23. Scarpulla RC. Metabolic control of mitochondrial biogenesis through the PGC-1 family regulatory network. *Biochim Biophys Acta*. 2011;1813:1269–78.

24. Ventura-Clapier R, Garnier A and Veksler V. Transcriptional control of mitochondrial biogenesis: The central role of PGC-1alpha. *Cardiovasc Res*. 2008;79:208–17.

25. Jornayvaz FR and Shulman GI. Regulation of mitochondrial biogenesis. *Essays Biochem*. 2010;47:69–84.

26. Hock MB and Kralli A. Transcriptional control of mitochondrial biogenesis and function. *Annu Rev Physiol*. 2009;71:177–203.

27. Virbasius JV and Scarpulla RC. Activation of the human mitochondrial transcription factor A gene by nuclear respiratory factors: A potential regulatory link between nuclear and mitochondrial gene expression in organelle biogenesis. *Proc Natl Acad Sci U S A*. 1994;91:1309–13.

28. Villena JA. New insights into PGC-1 coactivators: Redefining their role in the regulation of mitochondrial function and beyond. *FEBS J*. 2015;282:647–72.

29. Reznick RM and Shulman GI. The role of AMP-activated protein kinase in mitochondrial biogenesis. *J Physiol*. 2006;574:33–39.

30. Ren J, Pulakat L, Whaley-Connell A and Sowers JR. Mitochondrial biogenesis in the metabolic syndrome and cardiovascular disease. *J Mol Med (Berl)*. 2010;88:993–1001.

31. Nisoli E, Clementi E, Carruba MO and Moncada S. Defective mitochondrial biogenesis: A hallmark of the high cardiovascular risk in the metabolic syndrome? *Circ Res*. 2007;100:795–806.

32. Bugger H and Abel ED. Molecular mechanisms for myocardial mitochondrial dysfunction in the metabolic syndrome. *Clin Sci (Lond)*. 2008;114:195–210.

33. Ritov VB, Menshikova EV, He J, Ferrell RE, Goodpaster BH and Kelley DE. Deficiency of subsarcolemmal mitochondria in obesity and type 2 diabetes. *Diabetes*. 2005;54:8–14.

34. Lowell BB and Shulman GI. Mitochondrial dysfunction and type 2 diabetes. *Science*. 2005;307:384–87.

35. Lopez-Lluch G, Hernandez-Camacho JD, Fernandez-Ayala DJM and Navas P. Mitochondrial dysfunction in metabolism and ageing: Shared mechanisms and outcomes?. *Biogerontology*. 2018;19:461–80.

36. Laplante M and Sabatini DM. mTOR signaling at a glance. *J Cell Sci*. 2009;122:3589–94.

37. Hardie DG, Ross FA and Hawley SA. AMPK: A nutrient and energy sensor that maintains energy homeostasis. *Nat Rev Mol Cell Bio*. 2012;13:251–62.

38. Kendrick AA, Choudhury M, Rahman SM, McCurdy CE, Friederich M, Van Hove JLK, Watson PA, Birdsey N, Bao JJ, Gius D, Sack MN, Jing EX, Kahn CR, Friedman JE and Jonscher KR. Fatty liver is associated with reduced SIRT3 activity and mitochondrial protein hyperacetylation. *Biochem J*. 2011;433:505–14.

39. Qiu XL, Brown K, Hirschey MD, Verdin E and Chen D. Calorie restriction reduces oxidative stress by SIRT3-mediated SOD2 activation. *Cell Metab*. 2010;12:662–67.

40. Kume S, Uzu T, Horiike K, Chin-Kanasaki M, Isshiki K, Araki S, Sugimoto T, Haneda M, Kashiwagi A and Koya D. Calorie restriction enhances cell adaptation to hypoxia through Sirt1-dependent mitochondrial autophagy in mouse aged kidney. *J Clin Invest*. 2010;120:1043–55.

41. Samant SA, Zhang HJ, Hong ZG, Pillai VB, Sundaresan NR, Wolfgeher D, Archer SL, Chan DC and Gupta MP. SIRT3 Deacetylates and Activates OPA1 To Regulate Mitochondrial Dynamics during Stress. *Mol Cell Biol*. 2014;34:807–19.

42. Calvo SS-C, Navas P and López-Lluch G. Sirtuin-Dependent Metabolic Control and Its Role in the Aging Process. Bioenergetics. InTech; 2012.

43. Guarente L. Sirtuins, aging, and metabolism. *Cold Spring Harb Sym*. 2011;76:81–90.

44. Guarente L. Calorie restriction and sirtuins revisited. *Gene Dev*. 2013;27:2072–85.

45. Imai S and Guarente L. Ten years of NAD-dependent SIR2 family deacetylases: Implications for metabolic diseases. *Trends Pharmacol Sci.* 2010;31:212–20.

46. Palacios OM, Carmona JJ, Michan S, Chen KY, Manabe Y, Ward JL, 3rd, Goodyear LJ and Tong Q. Diet and exercise signals regulate SIRT3 and activate AMPK and PGC-1alpha in skeletal muscle. *Aging (Albany NY).* 2009;1:771–83.

47. Rodgers JT, Lerin C, Haas W, Gygi SP, Spiegelman BM and Puigserver P. Nutrient control of glucose homeostasis through a complex of PGC-1alpha and SIRT1. *Nature.* 2005;434:113–18.

48. Gerhart-Hines Z, Rodgers JT, Bare O, Lerin C, Kim SH, Mostoslavsky R, Alt FW, Wu Z and Puigserver P. Metabolic control of muscle mitochondrial function and fatty acid oxidation through SIRT1/PGC-1alpha. *EMBO J.* 2007;26:1913–23.

49. Jäger S, Handschin C, Pierre JS- and Spiegelman BM. AMP-activated protein kinase (AMPK) action in skeletal muscle via direct phosphorylation of PGC-1α. *Proc Natl Acad Sci U S A.* 2007;104:12017–22.

50. Canto C, Jiang LQ, Deshmukh AS, Mataki C, Coste A, Lagouge M, Zierath JR and Auwerx J. Interdependence of AMPK and SIRT1 for Metabolic Adaptation to Fasting and Exercise in Skeletal Muscle. *Cell Metab.* 2010;11:213–19.

51. Ishihara M, Urushido M, Hamada K, Matsumoto T, Shimamura Y, Ogata K, Inoue K, Taniguchi Y, Horino T, Fujieda M, Fujimoto S and Terada Y. Sestrin-2 and BNIP3 regulate autophagy and mitophagy in renal tubular cells in acute kidney injury. *Am J Physiol Renal Physiol.* 2013;305:F495–509.

52. Budanov AV and Karin M. p53 target genes sestrin1 and sestrin2 connect genotoxic stress and mTOR signaling. *Cell.* 2008;134:451–60.

53. Lee JH, Budanov AV and Karin M. Sestrins orchestrate cellular metabolism to attenuate aging. *Cell Metab.* 2013;18:792–801.

54. Bae SH, Sung SH, Oh SY, Lim JM, Lee SK, Park YN, Lee HE, Kang D and Rhee SG. Sestrins activate Nrf2 by promoting p62-dependent autophagic degradation of keap1 and prevent oxidative liver damage. *Cell Metab.* 2013;17:73–84.

55. Lee JH, Budanov AV, Talukdar S, Park EJ, Park HL, Park HW, Bandyopadhyay G, Li N, Aghajan M, Jang I, Wolfe AM, Perkins GA, Ellisman MH, Bier E, Scadeng M, Foretz M, Viollet B, Olefsky J and Karin M. Maintenance of metabolic homeostasis by Sestrin2 and Sestrin3. *Cell Metab.* 2012;16:311–21.

56. Tao R, Xiong X, Liangpunsakul S and Dong XC. Sestrin 3 protein enhances hepatic insulin sensitivity by direct activation of the mTORC2-Akt signaling. *Diabetes.* 2015;64:1211–23.

57. Zeng N, D'Souza RF, Mitchell CJ and Cameron-Smith D. Sestrins are differentially expressed with age in the skeletal muscle of men: A cross-sectional analysis. *Exp Gerontol.* 2018;110:23–34.

58. Rai N, Venugopalan G, Pradhan R, Ambastha A, Upadhyay AD, Dwivedi S, Dey AB and Dey S. Exploration of novel anti-oxidant protein sestrin in frailty syndrome in elderly. *Aging Dis.* 2018;9:220–27.

59. Wang M, Xu Y, Liu J, Ye J, Yuan W, Jiang H, Wang Z, Jiang H and Wan J. Recent insights into the biological functions of sestrins in health and disease. *Cell Physiol Biochem.* 2017;43:1731–41.

60. Chan EY. mTORC1 phosphorylates the ULK1-mAtg13-FIP200 autophagy regulatory complex. *Sci Signal.* 2009;2: pe51.

61. Kim J, Kundu M, Viollet B and Guan KL. AMPK and mTOR regulate autophagy through direct phosphorylation of Ulk1. *Nat Cell Biol.* 2011;13:132–41.

62. Alers S, Loffler AS, Wesselborg S and Stork B. Role of AMPK-mTOR-Ulk1/2 in the regulation of autophagy: Cross talk, shortcuts, and feedbacks. *Mol Cell Biol.* 2012;32:2–11.

63. Goto T and Takano M. Transcriptional Role of FOXO1 in Drug Resistance through Antioxidant Defense Systems. *Adv Exp Med Biol.* 2009;665:171–79.

64. Lopez-Lluch G and Navas P. Calorie restriction as an intervention in ageing. *J Physiol-London.* 2016;594: 2043–60.

65. Liesa M, Palacin M and Zorzano A. Mitochondrial dynamics in mammalian health and disease. *Physiol Rev.* 2009;89:799–845.

66. Lopez-Lluch G, Irusta PM, Navas P and de Cabo R. Mitochondrial biogenesis and healthy aging. *Exp Gerontol.* 2008;43:813–19.

67. Chen HC and Chan DC. Emerging functions of mammalian mitochondrial fusion and fission. *Hum Mol Genet.* 2005;14:R283–9.

68. Olichon A, Emorine LJ, Descoins E, Pelloquin L, Brichese L, Gas N, Guillou E, Delettre C, Valette A, Hamel CP, Ducommun B, Lenaers G and Belenguer P. The human dynamin-related protein OPA1 is anchored to the mitochondrial inner membrane facing the inter-membrane space. *FEBS Lett.* 2002;523:171–76.

69. de Brito OM and Scorrano L. Mitofusin 2 tethers endoplasmic reticulum to mitochondria. *Nature.* 2008;456: 605–10.

70. de Brito OM and Scorrano L. Mitofusin-2 regulates mitochondrial and endoplasmic reticulum morphology and tethering: The role of Ras. *Mitochondrion.* 2009;9:222–26.

71. Naon D, Zaninello M, Giacomello M, Varanita T, Grespi F, Lakshminaranayan S, Serafini A, Semenzato M, Herkenne S, Hernandez-Alvarez MI, Zorzano A, De Stefani D, Dorn GW, 2nd and Scorrano L. Critical reappraisal confirms that Mitofusin 2 is an endoplasmic reticulum-mitochondria tether. *Proc Natl Acad Sci U S A.* 2016;113: 11249–54.

72. Smirnova E, Griparic L, Shurland DL and van der Bliek AM. Dynamin-related protein Drp1 is required for mitochondrial division in mammalian cells. *Mol Biol Cell.* 2001;12:2245–56.

73. Santel A and Frank S. Shaping mitochondria: The complex posttranslational regulation of the mitochondrial fission protein DRP1. *IUBMB Life.* 2008;60:448–55.

74. Yoon Y, Krueger EW, Oswald BJ and McNiven MA. The mitochondrial protein hFis1 regulates mitochondrial fission in mammalian cells through an interaction with the dynamin-like protein DLP1. *Mol Cell Biol.* 2003; 23:5409–20.

75. Artal-Sanz M and Tavernarakis N. Prohibitin couples diapause signalling to mitochondrial metabolism during ageing in C. elegans. *Nature.* 2009;461:793–97.

76. Lourenco AB, Munoz-Jimenez C, Venegas-Caleron M and Artal-Sanz M. Analysis of the effect of the mitochondrial prohibitin complex, a context-dependent modulator of longevity, on the C. elegans metabolome. *Biochim Biophys Acta.* 2015;1847:1457–68.

77. Theiss AL, Vijay-Kumar M, Obertone TS, Jones DP, Hansen JM, Gewirtz AT, Merlin D and Sitaraman SV. Prohibitin is a novel regulator of antioxidant response that attenuates colonic inflammation in mice. *Gastroenterology.* 2009;137:199–208, 208 e1–6.

78. Lefebvre P, Chinetti G, Fruchart JC and Staels B. Sorting out the roles of PPAR alpha in energy metabolism and vascular homeostasis. *J Clin Invest.* 2006;116:571–80.

79. Hashimoto T, Cook WS, Qi C, Yeldandi AV, Reddy JK and Rao MS. Defect in peroxisome proliferator-activated receptor alpha-inducible fatty acid oxidation determines the severity of hepatic steatosis in response to fasting. *J Biol Chem.* 2000;275:28918–28.

80. Xu J, Xiao G, Trujillo C, Chang V, Blanco L, Joseph SB, Bassilian S, Saad MF, Tontonoz P, Lee WN and Kurland IJ. Peroxisome proliferator-activated receptor alpha (PPARalpha) influences substrate utilization for hepatic glucose production. *J Biol Chem.* 2002;277: 50237–44.

81. Gervois P, Kleemann R, Pilon A, Percevault F, Koenig W, Staels B and Kooistra T. Global suppression of IL-6-induced acute phase response gene expression after chronic in vivo treatment with the peroxisome proliferator-activated receptor-alpha activator fenofibrate. *J Biol Chem.* 2004;279:16154–60.

82. Ip E, Farrell GC, Robertson G, Hall P, Kirsch R and Leclercq I. Central role of PPARalpha-dependent hepatic lipid turnover in dietary steatohepatitis in mice. *Hepatology.* 2003;38:123–32.

83. Sapiro JM, Mashek MT, Greenberg AS and Mashek DG. Hepatic triacylglycerol hydrolysis regulates peroxisome proliferator-activated receptor alpha activity. *J Lipid Res.* 2009;50:1621–29.

84. Staels B, Maes M and Zambon A. Fibrates and future PPARalpha agonists in the treatment of cardiovascular disease. *Nat Clin Pract Cardiovasc Med.* 2008;5:542–53.

85. Schaffer JE and Lodish HF. Expression cloning and characterization of a novel adipocyte long-chain fatty-acid transport protein. *Cell.* 1994;79:427–36.

86. Martin G, Schoonjans K, Lefebvre AM, Staels B and Auwerx J. Coordinate regulation of the expression of the fatty acid transport protein and acyl-CoA synthetase genes by PPARalpha and PPARgamma activators. *J Biol Chem.* 1997;272:28210–17.

87. Frohnert BI, Hui TY and Bernlohr DA. Identification of a functional peroxisome proliferator-responsive element in the murine fatty acid transport protein gene. *J Biol Chem.* 1999;274:3970–77.

88. Motojima K, Passilly P, Peters JM, Gonzalez FJ and Latruffe N. Expression of putative fatty acid transporter genes are regulated by peroxisome proliferator-activated receptor alpha and gamma activators in a tissue- and inducer-specific manner. *J Biol Chem.* 1998; 273:16710–14.

89. Louet JF, Chatelain F, Decaux JF, Park EA, Kohl C, Pineau T, Girard J and Pegorier JP. Long-chain fatty acids regulate liver carnitine palmitoyltransferase I gene (L-CPT I) expression through a peroxisome-proliferator-activated receptor alpha (PPARalpha)-independent pathway. *Biochem J.* 2001;354:189–97.

90. Mascaro C, Acosta E, Ortiz JA, Marrero PF, Hegardt FG and Haro D. Control of human muscle-type carnitine palmitoyltransferase I gene transcription by peroxisome proliferator-activated receptor. *J Biol Chem.* 1998; 273:8560–63.

91. Barrero MJ, Camarero N, Marrero PF and Diego H. Control of human carnitine palmitoyltransferase II gene transcription by peroxisome proliferator-activated receptor through a partially conserved peroxisome proliferator-responsive element. *Biochem J.* 2003;369: 721–29.

92. Gulick T, Cresci S, Caira T, Moore DD and Kelly DP. The peroxisome proliferator-activated receptor regulates mitochondrial fatty acid oxidative enzyme gene expression. *Proc Natl Acad Sci U S A.* 1994;91:11012–16.

93. Aoyama T, Peters JM, Iritani N, Nakajima T, Furihata K. Hashimoto T and Gonzalez FJ. Altered constitutive expression of fatty acid-metabolizing enzymes in mice lacking the peroxisome proliferator-activated receptor alpha (PPARalpha). *J Biol Chem.* 1998;273:5678–84.

94. Pawlak M, Lefebvre P and Staels B. Molecular mechanism of PPARalpha action and its impact on lipid metabolism, inflammation and fibrosis in non-alcoholic fatty liver disease. *J Hepatol.* 2015;62:720–33.

95. Bordet R, Ouk T, Petrault O, Gele P, Gautier S, Laprais M, Deplanque D, Duriez P, Staels B, Fruchart JC and Bastide M. PPAR: A new pharmacological target for neuroprotection in stroke and neurodegenerative diseases. *Biochem Soc Trans.* 2006;34:1341–46.

96. Sung B, Park S, Yu BP and Chung HY. Modulation of PPAR in aging, inflammation, and calorie restriction. *J Gerontol A Biol Sci Med Sci.* 2004;59:997–1006.

97. Jawerbaum A and Capobianco E. Review: Effects of PPAR activation in the placenta and the fetus: Implications in maternal diabetes. *Placenta.* 2011;32(Suppl 2): S212–7.

98. Nikkanen J, Forsstrom S, Euro L, Paetau I, Kohnz RA, Wang LY, Chilov D, Viinamaki J, Roivainen A, Marjamaki P, Liljenback H, Ahola S, Buzkova J, Terzioglu M, Khan NA, Pirnes-Karhu S, Paetau A, Lonnqvist T, Sajantila A, Isohanni P, Tyynismaa H, Nomura DK, Battersby BJ, Velagapudi V, Carroll CJ and Suomalainen A. Mitochondrial DNA replication defects disturb cellular dNTP pools and remodel one-carbon metabolism. *Cell Metab.* 2016;23:635–48.

99. Itakura E, Zavodszky E, Shao S, Wohlever ML, Keenan RJ and Hegde RS. Ubiquilins chaperone and triage mitochondrial membrane proteins for degradation. *Mol Cell.* 2016;63:21–33.

100. Galper JB. Mitochondrial protein synthesis in HeLa cells. *J Cell Biol.* 1974;60:755–63.

101. Richter U, Lahtinen T, Marttinen P, Suomi F and Battersby BJ. Quality control of mitochondrial protein synthesis is required for membrane integrity and cell fitness. *J Cell Biol.* 2015;211:373–89.

102. Hill S, Sataranatarajan K and Remmen HV. Role of signaling molecules in mitochondrial stress response. *Front Genet.* 2018;9:225.

103. Tyynismaa H, Carroll CJ, Raimundo N, Ahola-Erkkila S, Wenz T, Ruhanen H, Guse K, Hemminki A, Peltola-Mjosund KE, Tulkki V, Oresic M, Moraes CT, Pietilainen K, Hovatta I and Suomalainen A. Mitochondrial myopathy induces a starvation-like response. *Hum Mol Genet.* 2010;19:3948–58.

104. Yoshida H, Haze K, Yanagi H, Yura T and Mori K. Identification of the cis-acting endoplasmic reticulum stress response element responsible for transcriptional induction of mammalian glucose-regulated proteins. Involvement of basic leucine zipper transcription factors. *J Biol Chem.* 1998;273:33741–49.

105. Zhao Q, Wang J, Levichkin IV, Stasinopoulos S, Ryan MT and Hoogenraad NJ. A mitochondrial specific stress response in mammalian cells. *EMBO J.* 2002;21: 4411–19.

106. Haynes CM, Petrova K, Benedetti C, Yang Y and Ron D. ClpP mediates activation of a mitochondrial unfolded protein response in C-elegans. *Dev Cell.* 2007;13:467–80.

107. Nargund AM, Pellegrino MW, Fiorese CJ, Baker BM and Haynes CM. Mitochondrial import efficiency of ATFS-1 regulates mitochondrial UPR activation. *Science.* 2012;337:587–90.

108. Haynes CM, Fiorese CJ and Lin YF. Evaluating and responding to mitochondrial dysfunction: The mitochondrial unfolded-protein response and beyond. *Trends Cell Biol.* 2013;23:311–18.

109. Durieux J, Wolff S and Dillin A. The cell-non-autonomous nature of electron transport chain-mediated longevity. *Cell.* 2011;144:79–91.

110. Khan NA, Auranen M, Paetau I, Pirinen E, Euro L, Forsstrom S, Pasila L, Velagapudi V, Carroll CJ, Auwerx J and Suomalainen A. Effective treatment of mitochondrial myopathy by nicotinamide riboside, a vitamin B3. *EMBO Mol Med.* 2014;6:721–31.

111. Cerutti R, Pirinen E, Lamperti C, Marchet S, Sauve AA, Li W, Leoni V, Schon EA, Dantzer F, Auwerx J, Viscomi C and Zeviani M. NAD(+)-dependent activation of Sirt1 corrects the phenotype in a mouse model of mitochondrial disease. *Cell Metab.* 2014;19:1042–49.

112. Dogan SA, Pujol C, Maiti P, Kukat A, Wang S, Hermans S, Senft K, Wibom R, Rugarli EI and Trifunovic A. Tissue-specific loss of DARS2 activates stress responses independently of respiratory chain deficiency in the heart. *Cell Metab.* 2014;19:458–69.

113. Bao XR, Ong SE, Goldberger O, Peng J, Sharma R, Thompson DA, Vafai SB, Cox AG, Marutani E, Ichinose F, Goessling W, Regev A, Carr SA, Clish CB and Mootha VK. Mitochondrial dysfunction remodels one-carbon metabolism in human cells. *Elife.* 2016;5:e10575.

114. Ost M, Keipert S, van Schothorst EM, Donner V, van der Stelt I, Kipp AP, Petzke KJ, Jove M, Pamplona R, Portero-Otin M, Keijer J and Klaus S. Muscle mitohormesis promotes cellular survival via serine/glycine pathway flux. *FASEB J.* 2015;29:1314–28.

115. Richter U, Lahtinen T, Marttinen P, Myohanen M, Greco D, Cannino G, Jacobs HT, Lietzen N, Nyman TA and Battersby BJ. A mitochondrial ribosomal and RNA decay pathway blocks cell proliferation. *Curr Biol.* 2013;23: 535–41.

116. Porter KR, Claude A and Fullam EF. A study of tissue culture cells by electron microscopy: Methods and preliminary observations. *J Exp Med.* 1945;81:233–46.

117. Benda C. About the spermatogenesis of vertebrates and higher invertebrates, part II. The histogenesis of sperm. *Arch Anat Physiol.* 1898;73:393–98.

118. Hogeboom GH, Claude A and Hotch-Kiss RD. The distribution of cytochrome oxidase and succinoxidase in the cytoplasm of the mammalian liver cell. *J Biol Chem.* 1946;165:615–29.

119. Giorgi C, Missiroli S, Patergnani S, Duszynski J, Wieckowski MR and Pinton P. Mitochondria-associated membranes: Composition, molecular mechanisms, and physiopathological implications. *Antioxid Redox Signal.* 2015;22:995–1019.

120. Lopez-Crisosto C, Bravo-Sagua R, Rodriguez-Pena M, Mera C, Castro PF, Quest AF, Rothermel BA, Cifuentes M and Lavandero S. ER-to-mitochondria miscommunication and metabolic diseases. *Biochim Biophys Acta.* 2015;1852:2096–105.

121. Copeland DE and Dalton AJ. An association between mitochondria and the endoplasmic reticulum in cells of the pseudobranch gland of a teleost. *J Biophys Biochem Cytol.* 1959;5:393–96.

122. Vance JE. Phospholipid synthesis in a membrane fraction associated with mitochondria. *J Biol Chem.* 1990;265:7248–56.

123. Vance JE. Newly made phosphatidylserine and phosphatidylethanolamine are preferentially translocated between rat liver mitochondria and endoplasmic reticulum. *J Biol Chem.* 1991;266:89–97.

124. Rizzuto R, Pinton P, Carrington W, Fay FS, Fogarty KE, Lifshitz LM, Tuft RA and Pozzan T. Close contacts with the endoplasmic reticulum as determinants of mitochondrial Ca2+ responses. *Science.* 1998;280:1763–66.

125. Bravo-Sagua R, Torrealba N, Paredes F, Morales PE, Pennanen C, Lopez-Crisosto C, Troncoso R, Criollo A, Chiong M, Hill JA, Simmen T, Quest AF and Lavandero S. Organelle communication: Signaling crossroads between homeostasis and disease. *Int J Biochem Cell Biol.* 2014;50:55–59.

126. Marchi S, Patergnani S and Pinton P. The endoplasmic reticulum-mitochondria connection: One touch, multiple functions. *Biochim Biophys Acta.* 2014;1837:461–69.

127. Vance JE. MAM (mitochondria-associated membranes) in mammalian cells: Lipids and beyond. *Biochim Biophys Acta.* 2014;1841:595–609.

128. Shore GC and Tata JR. Two fractions of rough endoplasmic reticulum from rat liver. I. Recovery of rapidly sedimenting endoplasmic reticulum in association with mitochondria. *J Cell Biol.* 1977;72:714–25.

129. Hayashi T, Rizzuto R, Hajnoczky G and Su TP. MAM: More than just a housekeeper. *Trends Cell Biol.* 2009;19:81–88.

130. Ardail D, Popa I, Bodennec J, Louisot P, Schmitt D and Portoukalian J. The mitochondria-associated endoplasmic-reticulum subcompartment (MAM fraction) of rat liver contains highly active sphingolipid-specific glycosyltransferases. *Biochem J.* 2003;371:1013–19.

131. Cui Z, Vance J, Chen M, Voelker D and Vance D. Cloning and expression of a novel phosphatidylethanolamine N-methyltransferase. A specific biochemical and cytological marker for a unique membrane fraction in rat liver. *J Biol Chem.* 1993;268:16655–63.

132. Stone SJ and Vance JE. Phosphatidylserine synthase-1 and -2 are localized to mitochondria-associated membranes. *J Biol Chem.* 2000;275:34534–40.

133. Szabadkai G, Bianchi K, Varnai P, De Stefani D, Wieckowski MR, Cavagna D, Nagy AI, Balla T and Rizzuto R. Chaperone-mediated coupling of endoplasmic reticulum and mitochondrial Ca2+ channels. *J Cell Biol.* 2006;175:901–11.

134. Alam MR, Baetz D and Ovize M. Cyclophilin D and myocardial ischemia-reperfusion injury: A fresh perspective. *J Mol Cell Cardiol.* 2015;78:80–89.

135. Paillard M, Tubbs E, Thiebaut PA, Gomez L, Fauconnier J, Da Silva CC, Teixeira G, Mewton N, Belaidi E, Durand A, Abrial M, Lacampagne A, Rieusset J and Ovize M. Depressing mitochondria-reticulum interactions protects cardiomyocytes from lethal hypoxia-reoxygenation injury. *Circulation.* 2013;128:1555–65.

136. Tubbs E, Theurey P, Vial G, Bendridi N, Bravard A, Chauvin MA, Ji-Cao J, Zoulim F, Bartosch B, Ovize M, Vidal H and Rieusset J. Mitochondria-associated endoplasmic reticulum membrane (MAM) integrity is required for insulin signaling and is implicated in hepatic insulin resistance. *Diabetes*. 2014;63:3279–94.

137. Rieusset J, Fauconnier J, Paillard M, Belaidi E, Tubbs E, Chauvin MA, Durand A, Bravard A, Teixeira G, Bartosch B, Michelet M, Theurey P, Vial G, Demion M, Blond E, Zoulim F, Gomez L, Vidal H, Lacampagne A and Ovize M. Disruption of calcium transfer from ER to mitochondria links alterations of mitochondria-associated ER membrane integrity to hepatic insulin resistance. *Diabetologia*. 2016;59:614–23.

138. Sugiura A, Nagashima S, Tokuyama T, Amo T, Matsuki Y, Ishido S, Kudo Y, McBride HM, Fukuda T, Matsushita N, Inatome R and Yanagi S. MITOL regulates endoplasmic reticulum-mitochondria contacts via Mitofusin2. *Mol Cell*. 2013;51:20–34.

139. Cosson P, Marchetti A, Ravazzola M and Orci L. Mitofusin-2 independent juxtaposition of endoplasmic reticulum and mitochondria: An ultrastructural study. *PLoS One*. 2012;7:e46293.

140. Filadi R, Greotti E, Turacchio G, Luini A, Pozzan T and Pizzo P. Mitofusin 2 ablation increases endoplasmic reticulum-mitochondria coupling. *Proc Natl Acad Sci U S A*. 2015;112:E2174–81.

141. Simmen T, Aslan JE, Blagoveshchenskaya AD, Thomas L, Wan L, Xiang Y, Feliciangeli SF, Hung CH, Crump CM and Thomas G. PACS-2 controls endoplasmic reticulum-mitochondria communication and Bid-mediated apoptosis. *EMBO J*. 2005;24:717–29.

142. Kottgen M, Benzing T, Simmen T, Tauber R, Buchholz B, Feliciangeli S, Huber TB, Schermer B, Kramer-Zucker A, Hopker K, Simmen KC, Tschucke CC, Sandford R, Kim E, Thomas G and Walz G. Trafficking of TRPP2 by PACS proteins represents a novel mechanism of ion channel regulation. *EMBO J*. 2005;24:705–16.

143. Arruda AP, Pers BM, Parlakgul G, Guney E, Inouye K and Hotamisligil GS. Chronic enrichment of hepatic endoplasmic reticulum-mitochondria contact leads to mitochondrial dysfunction in obesity. *Nat Med*. 2014;20:1427–35.

144. Rizzuto R, De Stefani D, Raffaello A and Mammucari C. Mitochondria as sensors and regulators of calcium signalling. *Nat Rev Mol Cell Bio*. 2012;13:566–78.

145. Hayashi T and Su TP. Sigma-1 receptor chaperones at the ER-mitochondrion interface regulate Ca(2+) signaling and cell survival. *Cell*. 2007;131:596–610.

146. Giorgi C, Ito K, Lin HK, Santangelo C, Wieckowski MR, Lebiedzinska M, Bononi A, Bonora M, Duszynski J, Bernardi R, Rizzuto R, Tacchetti C, Pinton P and Pandolfi PP. PML regulates apoptosis at endoplasmic reticulum by modulating calcium release. *Science*. 2010;330:1247–51.

147. Iwasawa R, Mahul-Mellier AL, Datler C, Pazarentzos E and Grimm S. Fis1 and Bap31 bridge the mitochondria-ER interface to establish a platform for apoptosis induction. *EMBO J*. 2011;30:556–68.

148. Missiroli S, Bonora M, Patergnani S, Poletti F, Perrone M, Gafa R, Magri E, Raimondi A, Lanza G, Tacchetti C, Kroemer G, Pandolfi PP, Pinton P and Giorgi C. PML at Mitochondria-Associated Membranes Is Critical for the Repression of Autophagy and Cancer Development. *Cell Rep*. 2016;16:2415–27.

149. Bravo R, Gutierrez T, Paredes F, Gatica D, Rodriguez AE, Pedrozo Z, Chiong M, Parra V, Quest AF, Rothermel BA and Lavandero S. Endoplasmic reticulum: ER stress regulates mitochondrial bioenergetics. *Int J Biochem Cell Biol*. 2012;44:16–20.

150. Bravo R, Vicencio JM, Parra V, Troncoso R, Munoz JP, Bui M, Quiroga C, Rodriguez AE, Verdejo HE, Ferreira J, Iglewski M, Chiong M, Simmen T, Zorzano A, Hill JA, Rothermel BA, Szabadkai G and Lavandero S. Increased ER-mitochondrial coupling promotes mitochondrial respiration and bioenergetics during early phases of ER stress. *J Cell Sci*. 2011;124:2143–52.

151. Bravo-Sagua R, Rodriguez AE, Kuzmicic J, Gutierrez T, Lopez-Crisosto C, Quiroga C, Diaz-Elizondo J, Chiong M, Gillette TG, Rothermel BA and Lavandero S. Cell death and survival through the endoplasmic reticulum-mitochondrial axis. *Curr Mol Med*. 2013;13:317–29.

152. Shioda N, Ishikawa K, Tagashira H, Ishizuka T, Yawo H and Fukunaga K. Expression of a truncated form of the endoplasmic reticulum chaperone protein, sigma1 receptor, promotes mitochondrial energy depletion and apoptosis. *J Biol Chem*. 2012;287:23318–31.

153. Mori T, Hayashi T, Hayashi E and Su TP. Sigma-1 receptor chaperone at the ER-mitochondrion interface mediates the mitochondrion-ER-nucleus signaling for cellular survival. *PLoS One*. 2013;8:e76941.

154. Ngoh GA, Papanicolaou KN and Walsh K. Loss of mitofusin 2 promotes endoplasmic reticulum stress. *J Biol Chem*. 2012;287:20321–32.

155. Sebastian D, Hernandez-Alvarez MI, Segales J, Sorianello E, Munoz JP, Sala D, Waget A, Liesa M, Paz JC, Gopalacharyulu P, Oresic M, Pich S, Burcelin R, Palacin M and Zorzano A. Mitofusin 2 (Mfn2) links mitochondrial and endoplasmic reticulum function with insulin signaling and is essential for normal glucose homeostasis. *Proc Natl Acad Sci U S A*. 2012;109:5523–28.

156. Chaudhari N, Talwar P, Parimisetty A, Lefebvre d'Hellencourt C and Ravanan P. A molecular web: Endoplasmic reticulum stress, inflammation, and oxidative stress. *Front Cell Neurosci*. 2014;8:213.

157. Schwarz PEH, Reimann M, Li J, Bergmann A, Licinio J, Wong ML and Bornstein S. The metabolic syndrome - A global challenge for prevention. *Horm Metab Res*. 2007;39:777–80.

158. Grundy SM. Metabolic syndrome pandemic. *Arterioscl Throm Vas*. 2008;28:629–36.

159. Danaei G, Lu Y, Singh G, Stevens G, Cowan M, Farzadfar F, Lin J, Finucane M, Rao M and Khang Y. Cardiovascular disease, chronic kidney disease, and diabetes mortality burden of cardiometabolic risk factors from 1980 to 2010: A comparative risk assessment. *Lancet*. 2014;2:634–47.

160. Pan WH, Yeh WT and Weng LC. Epidemiology of metabolic syndrome in Asia. *Asia Pac J Clin Nutr*. 2008;17:37–42.

161. Christian P and Su Q. MicroRNA regulation of mitochondrial and ER stress signaling pathways: Implications for lipoprotein metabolism in metabolic syndrome. *Am J Physiol Endocrinol Metab*. 2014;307:E729–37.

162. Su Q, Tsai J, Xu E, Qiu W, Bereczki E, Santha M and Adeli K. Apolipoprotein B100 acts as a molecular link between lipid-induced endoplasmic reticulum stress and hepatic insulin resistance. *Hepatology*. 2009;50:77–84.

163. Lopez-Lluch G. Mitochondrial activity and dynamics changes regarding metabolism in ageing and obesity. *Mech Ageing Dev*. 2017;162:108–21.

164. Tsutsui H, Kinugawa S and Matsushima S. Oxidative stress and heart failure. *Am J Physiol Heart Circ Physiol.* 2011;301:H2181–90.

165. Roberts CK and Sindhu KK. Oxidative stress and metabolic syndrome. *Life Sci.* 2009;84:705–12.

166. Dickinson BC and Chang CJ. Chemistry and biology of reactive oxygen species in signaling or stress responses. *Nat Chem Biol.* 2011;7:504–11.

167. Wang CH, Wang CC, Huang HC and Wei YH. Mitochondrial dysfunction leads to impairment of insulin sensitivity and adiponectin secretion in adipocytes. *FEBS J.* 2013;280:1039–50.

168. Supale S, Thorel F, Merkwirth C, Gjinovci A, Herrera PL, Scorrano L, Meda P, Langer T and Maechler P. Loss of prohibitin induces mitochondrial damages altering beta-cell function and survival and is responsible for gradual diabetes development. *Diabetes.* 2013;62:3488–99.

169. Handa P, Maliken BD, Nelson JE, Morgan-Stevenson V, Messner DJ, Dhillon BK, Klintworth HM, Beauchamp M, Yeh MM, Elfers CT, Roth CL and Kowdley KV. Reduced adiponectin signaling due to weight gain results in nonalcoholic steatohepatitis through impaired mitochondrial biogenesis. *Hepatology.* 2014;60:133–45.

170. Su Q, Baker C, Christian P, Naples M, Tong X, Zhang K, Santha M and Adeli K. Hepatic mitochondrial and ER stress induced by defective PPARalpha signaling in the pathogenesis of hepatic steatosis. *Am J Physiol Endocrinol Metab.* 2014;306:E1264–73.

171. Hoeks J and Schrauwen P. Muscle mitochondria and insulin resistance: A human perspective. *Trends Endocrinol Metab.* 2012;23:444–50.

172. Yuzefovych LV, Musiyenko SI, Wilson GL and Rachek LI. Mitochondrial DNA damage and dysfunction, and oxidative stress are associated with endoplasmic reticulum stress, protein degradation and apoptosis in high fat diet-induced insulin resistance mice. *PLoS One.* 2013;8: e54059.

173. Schneeberger M, Dietrich MO, Sebastian D, Imbernon M, Castano C, Garcia A, Esteban Y, Gonzalez-Franquesa A, Rodriguez IC, Bortolozzi A, Garcia-Roves PM, Gomis R, Nogueiras R, Horvath TL, Zorzano A and Claret M. Mitofusin 2 in POMC neurons connects ER stress with leptin resistance and energy imbalance. *Cell.* 2013;155:172–87.

174. Contreras-Ferrat AE, Toro B, Bravo R, Parra V, Vasquez C, Ibarra C, Mears D, Chiong M, Jaimovich E, Klip A and Lavandero S. An inositol 1,4,5-triphosphate (IP3)-IP3 receptor pathway is required for insulin-stimulated glucose transporter 4 translocation and glucose uptake in cardiomyocytes. *Endocrinology.* 2010;151:4665–77.

175. Gutierrez T, Parra V, Troncoso R, Pennanen C, Contreras-Ferrat A, Vasquez-Trincado C, Morales PE, Lopez-Crisosto C, Sotomayor-Flores C, Chiong M, Rothermel BA and Lavandero S. Alteration in mitochondrial Ca(2+) uptake disrupts insulin signaling in hypertrophic cardiomyocytes. *Cell Commun Signal.* 2014;12:68.

Section IV

Immunity and Toxicity

34 Regulation of Antiviral Immunity by Mitochondrial Dynamics

Mohsin Khan, Hasan Imam, and Saiful Anam Mir

Division of Infectious Diseases, School of Medicine, University of California, San Diego, La Jolla CA 92093, USA.

CONTENTS

1 INTRODUCTION

Discovered over a century ago, mitochondria have become an emerging and intense subject of investigation. Altman first concluded that the mitochondria are the elementary organisms living within the cells and called them "bioblast" (mentioned in Akira et al., 2006; Barral et al., 2009; Ernster and Schatz, 1981; Khan et al., 2015; O'Rourke, 2010). Soon after this discovery, hundreds of reports, originating from a cumulative efforts of century-long research, have provided substantial evidence suggesting that the mitochondria are the central part of cellular metabolism (O'Rourke, 2010). However, recent advancement in the field and continuous investigation have proved that these cellular "power houses" are not only the main players of metabolism, but also the junction where multiple signaling pathways converge (McBride et al., 2006). Of note, mitochondria are the signaling hub for innate immune signaling, which is one of those multiple events that converge at mitochondrial platform (Chan, 2006; Ernster and Schatz, 1981). Innate immune signaling is triggered by the pathogen associated molecular patterns (PAMPs) that are recognized by pathogen recognition receptors (PRRs) to facilitate the downstream signaling that produces effector molecules (such as interferons in case of viral infection) (Akira et al., 2006; Barral et al., 2009; Castanier et al., 2010).

Mitochondria associated antiviral signaling protein (MAVS) and stimulator of interferon genes (STING) are the central adapter that facilitate the viral RNA or DNA mediated induction/synthesis of interferon within the infected cells (Barber, 2015; McBride et al., 2006; Seth et al., 2005). Interestingly, MAVS is primarily localized to the mitochondria and many reports have established that the mitochondrial dynamics potentially modulates the MAVS mediated interferon signaling (Castanier et al., 2010; Seth et al., 2005). On the other hand, substantial evidence have established that STING can be localized to MAMs (mitochondria associated membranes) (Bockler and Westermann, 2014; Ma and Damania, 2016). Therefore, mitochondria are the center where signaling, induced by DNA and RNA viruses, converges (Akira et al., 2006; Bockler and Westermann, 2014; Castanier et al., 2010; Horner et al., 2011; Khan et al., 2015; Loo and Gale, 2011). Viral infection can substantially modulate cell physiology and completely take over the host cell signaling events (Anand and Tikoo, 2013; Khan et al., 2015). It is evident that such alterations often affect cellular homeostasis (Khan et al., 2015). Mitochondria are quite sensitive to such physiological alterations and once any stress signal is triggered, it causes substantial damage to the mitochondria leading to significant changes in overall functioning of mitochondrial machinery (Chan, 2006, 2012). Interestingly, viruses have devised strategies to subvert the events of mitochondrial dynamics, causing damage or altering depolarization, which can potentially deregulate the cellular antiviral defense network (West et al., 2011). Beyond the role of mitochondria in metabolism and apoptosis, recent reports have clearly implicated their importance in innate immune signaling (Khan et al., 2015; West et al., 2011). Multiple investigations have revealed that mitochondrial dynamics is a critical regulator of antiviral signaling in the host cell (Castanier et al., 2010; Khan et al., 2015; Kim et al., 2014). Therefore, the purpose of this chapter is to give a comprehensive account of recent developments pertaining to the essential roles of mitochondrial dynamics in regulating antiviral signaling.

2 KEY EVENTS OF MITOCHONDRIAL MORPHODYNAMICS

Mitochondrial dynamics and mitophagy are interlinked processes, which are required to maintain mitochondrial homeostasis (Ashrafi and Schwarz 2013; Chan, 2006, 2012). Morphological and molecular evidence show that a single mitochondrion undergoes several rounds of fission and fusion (Ashrafi and Schwarz, 2013; Chan, 2012). Altogether, fusion, fission and mitophagy enable the segregation of impaired mitochondria and deliver it to the lysosomes for degradation by mitophagy (Chen and Chan, 2009; Khan et al., 2015). Impaired mitochondria are first sensed by cellular machinery and then eliminated by mitophagy (Khan et al., 2015). Mitochondrial dynamics is usually coordinated by dynamin-related proteins and all the key proteins of this machinery are listed in Table 1. These key proteins undergo several post-translational modifications, which regulate their function (Chan, 2006, 2012; Khan et al., 2015). Before we discuss the direct implication of mitochondrial dynamics in antiviral response, it is imperative to first understand the molecular events associated with mitochondrial dynamics. These key events are depicted in the Figure 1 and discussed later in this chapter.

2.1 MITOCHONDRIAL FUSION MACHINERY AND MECHANISM

During this process, two mitochondria fuse together to exchange membrane components and matrix proteins (Twig and Shirihai, 2011). Fusion often leads to balance

TABLE 1

Key proteins of mitochondrial dynamics machinery and a brief summary of their posttranslational modifications.

Protein	Enzymatic activity and function	Posttranslational modification and its role
Drp1	GTPase/Fission	• S616 phosphorylation by CDK1, Erk1/2 and PKC leads to enhance fission activity. • S616 phosphorylation by CDK5 leads to lower fission activity. • S637 Phosphorylation by PKA leads to lower fission activity.
Fis1	None/Fission	• MITOL mediated ubiquitination and proteasomal degradation leads to reduce fission.
Opa1	GTPase/Fusion	• Acetylation of K926/931by stress lowers GTPase activity. • De- acetylation K926/931 by SIRT3 leads to enhancement of GTPase activity
Mfn1	GTPase/Fusion	• Parkin and MARCH3 mediated ubiquitination and proteasomal degradation leads to reduce fusion.
Mfn2	GTPase/Fusion	• Parkin mediated ubiquitination and proteasomal degradation leads to reduce fusion. • PINK1 mediated Phosphorylation of T111, S442 helps parkin recruitment.
Parkin	Kinase/Mitophagy	• PINK1 mediated S65 phosphorylation leads to enhance parkin activity and support mitophagy. • Parkin's auto-ubiquitination leads to proteasomal degradation.
PINK1	E3 ubiquitin ligase/Mitophagy	• PINK1 mediated auto-phosphorylation of S228, T257, S402 leads to self-activation and support mitophagy.

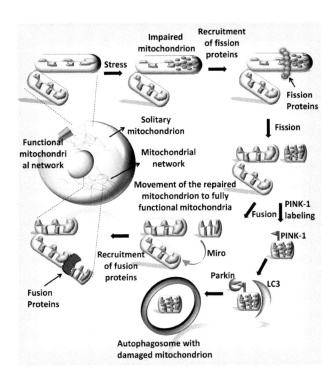

Figure 1 Events of mitochondrial dynamics. Mitochondrial pool consists healthy, differentially polarized and damaged mitochondria. A fraction of mitochondria in this pool is often reversibly damaged and can be repaired by maintaining the membrane potential. However highly polarized and irreversibly damaged mitochondria from this pool are eliminated selectively from the cell by mitophagy. If a single mitochondrion is partially impaired or depolarized, it can be rescued by undergoing fusion with fully functional mitochondrion and exchange the matrix and membrane surface in such a way that the damage signal is uniformly diluted and newly fused mitochondria can revert back to the cellular function. On the other hand, it is also possible that a damaged part of impaired mitochondria is removed by fission and subsequently degraded by mitophagy. In short, the impaired and segregated daughter mitochondria are either restored by undergoing repair and incorporate back to the functional pool of mitochondria or, if the damage is irreversible, completely removed by mitophagy. For mechanistic details, please refer to the text section.

the damage response in which a partially damaged mitochondrion fuses with a fully healthy mitochondrion and this fusion leads to dilute the damaged content accumulated in the impaired mitochondrion (Twig et al., 2008; Twig and Shirihai, 2011). In eukaryotic cells, a single mitochondrion can exist in pre-fusion and post-fusion phases. In pre-fusion stage, the mitochondria are solitary while post-fusion phase represent complex network of mitochondria (Twig et al., 2008). The first fusion regulatory protein, fuzzy onion (Fzo) was identified in *Drosophila*. Fzo deleted strain showed accumulation of fragmented mitochondria, which proved that Fzo can control mitochondrial fusion. In Fzo mutant flies, mitochondria aggregate and fail to fuse, which lead to the accumulation of fragmented mitochondria in spermatid (Hales and Fuller, 1997; Hermann et al., 1998). In mammals, the molecules that govern fusion machinery are Mitofusin 1/2 (Mfn1 and Mfn2) and Optic atrophy 1 protein (OPA1) (Khan et al., 2015). Elimination of any of these proteins leads to embryonic lethality and mitochondrial dysfunction (Chen et al., 2003). Mfn1 and Mfn2 are homologs of Fzo which have similar but

distinct roles in mitochondrial morphology control (Chen et al., 2003). These are ubiquitously expressed and membrane anchored dynamin family proteins that are uniformly localized to the outer mitochondrial membrane (OMM).

Mitochondrial fusion occurs via fusion of the outer and inner mitochondrial membrane (IMM) which results in the union of two separate mitochondrion and simultaneous mixing of their inner contents (Santel and Fuller, 2001). Mfn1 and Mfn2 coordinate OMM fusion while the IMM fusion is mediated by Opa1. Mutation in any of these proteins can cause abnormal mitochondrial morphology and remarkable reduction in mitochondrial fusion (Chen and Chan, 2009; Chen et al., 2005). Mfn1 and Mfn2 depleted cells showed poor cell growth, decreased cellular respiration associated with the heterogeneity of mitochondrial membrane potential (Chan, 2012). OPA1; a dynamin-related GTPase is localized in the inner mitochondrial membrane (IMM) of many mammalian cells and orchestrate IMM fusion (Frezza et al., 2006). OPA1 protein plays multiple important roles in mitochondrial cristae architecture, bioenergetics and apoptosis (Frezza et al., 2006; Head et al., 2009). The presence of Mfns on both the fusing mitochondria is the prerequisite for outer membrane fusion. Oligomerization of the Mfn1/2 on adjacent mitochondria brings the opposing membranes within close proximity and by membrane fusion, it eventually promotes fusion. Homotypic and heterotypic physical interaction between Mfn1/2 result in Mfn1 homotypic oligomers, Mfn2 homotypic oligomers and Mfn1–Mfn2 heterotypic oligomers (Chen et al., 2003).

Post-translational modifications and proteolysis of OPA1 and Mfns are well reported (Khan et al., 2015). Because of differential RNA splicing, OPA1 exists as distinct isoforms (Baker et al., 2014; Ehses et al., 2009; Head et al., 2009). The longer isoform of OPA1 contains membrane anchor. Although, the short isoform lacks membrane anchor it can interact with membrane proteins. Both long and short OPA1 isoforms can impel IMM fusion, however under stress condition only the long isoform promotes mitochondrial fusion. SIRT1-mediated deacetylation can regulate the activity of OPA1 (Samant et al., 2014). Reduction in mitochondrial membrane potential triggers the proteolysis of OPA1 by OMA1 (Overlapping with the M-AAA Protease 1 Homolog) (Baker et al., 2014; Ehses et al., 2009). Under normal cellular condition OMA1 is constitutively active. Various stress stimulation can increase its catalytic activity several fold. Study by Baker et al, identified a stress sensor N-terminal domain of OMA1, which is responsible for its stress-induced activation (Baker et al., 2014).

2.2 Mitochondrial Fission Machinery and Mechanism

In eukaryotic cells, mitochondrial fission (or fragmentation) is a multistep process. It is categorized by the division of one mitochondrion into two daughter mitochondria. Recruitment of the dynamin related protein 1 (Drp1) plays a crucial role to initiate mitochondrial division (Frank et al., 2001). Drp1 has membrane remodeling property and it can localize to mitochondria and peroxisomes. Upon activation, it drives membrane scission in a GTP-dependent manner (Kraus and Ryan, 2017). Inhibition of Drp1 showed drastic elongation of both mitochondria and peroxisomes (Gandre-Babbe and van der Bliek, 2008; Koch et al., 2005; Lee et al., 2004). When mitochondrial fission occurs, Drp1 is recruited to the outer mitochondrial membrane (OMM) to form a ring like structure which causes the constriction of the OMM (Kraus and Ryan, 2017). Membrane scission starts through GTP hydrolysis, which marks a potential future site of mitochondrial scission (Kraus and Ryan, 2017). Fission has also been implicated in sorting the mitochondria with mutant mtDNA copies (Shitara et al., 2000). It is presumed that spontaneous depolarization during solitary stage can results into impairment of mitochondria (Twig et al., 2008;

Twig and Shirihai, 2011) and sometimes, asymmetric fission can induce remarkable alteration in the mitochondrial membrane potential ($\Delta\Psi_m$) (Twig and Shirihai, 2011).

In mammalian cells, the central player of mitochondrial fission, Drp1 is primarily localized in the cytosol and the accessory proteins, MiD (mitochondrial dynamics protein) and Mff (mitochondrial fission factor) mediates its recruitments to the mitochondria by acting as ligands for Drp1 (Chang and Blackstone, 2007; James et al., 2003; Loson et al., 2013). Although it's not clear about the exact mechanism of mitochondrial fission but recent evidence indicates that endoplasmic reticulum (ER) tubules are the key players in this event (Friedman et al., 2011). ER tubules can wrap around and constrict the mitochondria to mark them for subsequent scission by Drp1 (Friedman et al., 2011). Since ER-mitochondrial contact sites are enriched with Drp1, it is assumed that ER-mitochondria connection directs mitochondrial fragmentation (Friedman et al., 2011; Lee and Yoon, 2014). In addition, the cells depleted for Drp1 and Mff still showed mitochondrial constriction at ER contact sites which indicates that ER-mediated flagging of mitochondrial fission sites occurs before the mitochondrial recruitment of Drp1 (Friedman et al., 2011). ER-mediated constriction of mitochondrial tubules requires mechanical force and it seems to be generated by actin assembly via interaction with ER-associated INF2 (Inverted Formin 2) at the ER-mitochondrial contact site (Korobova et al., 2013). Fission1 (Fis1) protein was the first molecule identified as Drp1 receptor in yeast, which has a C-tail protein anchored on OMM and displays uniform distribution (Cerveny and Jensen, 2003; Gomes and Scorrano, 2008; Lee et al., 2004). Another C-tail anchored protein on OMM is Mff, which is a receptor of Drp1 and interacts transiently through its N-terminal cytoplasmic domain. It colocalizes mainly with Drp1 foci on the OMM (Gandre-Babbe and van der Bliek, 2008; Loson et al., 2013). Mff knockdown results in mitochondrial elongation but its overexpression stimulates mitochondrial recruitment of Drp1 and mitochondrial fragmentation (Gandre-Babbe and van der Bliek, 2008; Otera et al., 2010). In the absence of Fis1 and Mff, the MiDs can mediate mitochondrial fission (Loson et al., 2013). Silencing any of these genes results in increased mitochondrial length and interconnectivity, that indicates that these proteins can positively regulate mitochondrial fission. Interestingly, overexpression of MiDs also results in elongated mitochondria, which is associated with increased phosphorylation of Drp1 at S637 site and for this reason Drp1 function is negatively regulated (Chang and Blackstone, 2007). When cells were mutated for Drp1, they become resistant to mitochondrial fragmentation induced by depolarization but they also exhibit substantial mitochondrial fragmentation when treated with apoptotic stimuli such as actinomycin D and etoposide. These observations prove that mitochondrial fission can also occur through Drp1 independent mechanisms. In support of this, it was observed that pore-forming toxin listeriolysin O, secreted by *Listeria monocytogens*, can induce mitochondrial fragmentation which is dependent on actin cytoskeleton but independent of traditional fission protein Drp1 (Stavru et al., 2013).

Mitochondrial fission directed by various physiological cues can be regulated by several post-translational modifications of Drp1. In normal cells, Drp1 is mostly present in the cytosol, where a minute fraction of this protein is associated with the mitochondria. Enhanced fission does not require Drp1 upregulation but it requires Drp1 mitochondrial recruitment, spiral assembly and GTPase activity that directly influences mitochondrial fission (Kraus and Ryan, 2017). Cdk1/cyclin B phosphorylates Drp1 at S616 residue during mitosis, resulting in the upregulation of mitochondrial recruitment of Drp1 and fission activity (Kashatus et al., 2011; Taguchi et al., 2007). Aurora A kinase also promotes mitochondrial fission during mitosis by phosphorylation of small Ras-like GTPase, RalA leading to its relocation to mitochondria, followed by the recruitment of its effector RalBP1 (Kashatus et al., 2011). CDK1-mediated Drp1 S616 phosphorylation and its recruitment to the mitochondria is then facilitated by RalBP1. Yu et al. showed that stimulation of liver derived cells with high glucose induces Ca^+ – mediated MAPK signaling, which then lead to ERK1/2-dependent Drp1 S616 phosphorylation and increased mitochondrial fission (Yu et al., 2011). However, if the Drp1 is phosphorylated at S637 residue, it can reduce fission activity. CAMP-dependent protein kinase or protein kinase A (PKA) phosphorylates Drp1 on S637 (Dickey and Strack, 2011; Merrill et al., 2011). PKA mediated phosphorylation of Drp1 is affected during hypoxia because of proteasomal degradation of mitochondrial A-kinase anchoring protein 121 (Akap1) that anchors mitochondria to PKA (Dickey and Strack, 2011; Merrill et al., 2011). Drp1 is dephosphorylated at S637 residue by calcium-dependent phosphatase, calcineurin and facilitates mitochondrial recruitment of Drp1 followed by mitochondrial fragmentation. Phosphatase, PP2A/Bβ 2 can also dephosphorylate S637 of Drp1 (Dickey and Strack, 2011). On the other hand, phosphorylation of Drp1 at S637 residue can also result in enhanced fission. Phosphorylation of S637 by CaMKIα and dephosphorylation of S637 by calcineurin mysteriously show similar phenotypes (Han et al., 2008). These reports established the importance of upstream kinases rather than the phosphorylation site as a vital factor, which determines Drp1 activity. Further studies are required to explain how phosphorylation and dephosphorylation of the same site of a particular protein can have similar phenotypes. Drp1 activity can also be regulated by ubiquitination, sumoylation and s-nitrosylation (Braschi et al., 2009; Figueroa-Romero et al., 2009; Wasiak et al., 2007; Yonashiro et al., 2006; Zunino et al., 2007). Additionally, Drp1 degradation via proteasomal pathways can be mediated by parkin and mitochondria ubiquitin ligase (MITOL) (Yonashiro et al., 2006).

2.3 Mitophagy Machinery and Mechanism

Autophagy is a self-eating process, in which the damaged or excessed cell organelles and misfolded protein aggregates are actively engulfed by phagophore followed by fusion with lysosomes for degradation (Feng et al., 2014). Although autophagy is nonselective process, recent evidence showed that selective cargo or damaged organelles could be eliminated by selective autophagy (Bauckman et al., 2015). Selective autophagy of mitochondria is termed as mitophagy (Ashrafi and Schwarz, 2013). Parkin and PINK1 proteins are among the major key players involved in mitophagy but recent reports suggest that parkin independent mitophagy also exists (Allen et al., 2013; Ashrafi and Schwarz, 2013). It is worth mentioning that multiple mechanisms have been reported to explain mitophagy. However, the most widely recognized mechanism is the stabilization of PINK1 via mitochondrial membrane potential and the events associated with this mechanism are depicted in Figure 2.

PINK1 mediated parkin recruitment to the mitochondria has been explained by several studies. On depolarized mitochondria, PINK1 auto-phosphorylation is necessary for parkin recruitment and then E3 ligase activity of parkin is stimulated through direct phosphorylation by PINK1 at S65 residue. S65 residue of ubiquitin is also phosphorylated by PINK1 that consequently unlocks parkin's autoinhibition at the active cysteine site (Kane et al., 2014; Ordureau et al., 2014). Parkin translocation to damaged mitochondria is inhibited by ubiquitin mutant S65A. Small molecules that imitate the S65 phospho-ubiquitin have been proposed as vital therapeutic tool against the Parkinson's disease (Kazlauskaite et al., 2014). A direct physical interaction between parkin and PINK1 has been suggested by several studies (Narendra et al., 2010). During parkin recruitment to mitochondria, other substrates of PINK1 kinase are also involved. In PINK1 deficient flies, mitochondrial morphology can be rescued by parkin overexpression but more severe phenotype was observed in parkin deficient flies, which indicates PINK1 independent functions of parkin. PINK1 can also phosphorylate Mfn2 to promote parkin recruitment on depolarized mitochondria (Chen and Dorn, 2013; Park et al., 2006).

For subsequent autophagic removal of damaged mitochondria, it is essential for parkin to ubiquitinate and induce OMM protein degradation such as Mfn1, Mfn2, and TOM complex. Parkin dependent mitophagy is impaired by the inhibition of proteasomal activity. Proteasomal machinery is directly recruited for mitochondrial degradation (Chan et al., 2011). K-63 linked polyubiquitin conjugated proteins binds to p62 adaptor protein and plays a role in recruiting mitochondria to autophagosomes (Chan et al., 2011). However, mitochondrial elimination has also been reported in absence of p62 clustering, Histone deacetylase 6 (HDAC6) is recruited to mitochondria in parkin dependent manner

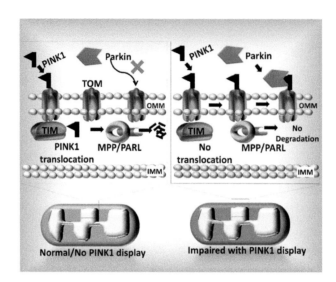

Figure 2   Mechanism of mitophagy. In the process of marking and segregation of defective mitochondria, mitochondrial membrane potential ($\Delta\Psi_m$) plays a vital role. Parkin dependent mitophagy involves selective labeling of impaired mitochondria by PTEN-induced putative kinase 1 (PINK1) followed by parkin recruitment to the damaged mitochondria and subsequent mitophagy. PINK1 uses a mitochondrial targeting sequence (MTS) to be imported into the healthy mitochondria via the translocase of outer mitochondrial membrane (TOM) and translocase of inner mitochondrial membrane (TIM). PINK1 is processed by the mitochondrial processing protease (MPP) at the IMM to excise the MTS, and then cleaved by PARL (presenilin-associated rhomboid-like protease) to produce a 52 kDa protein, which is then degraded quickly by proteasome. However, the loss of membrane potential ($\Delta\Psi_m$) in the damaged mitochondria inhibits the import of PINK1 to the IMM and rescue PINK1 from degradation. Rescued PINK1 is then displayed on the surface of damaged mitochondria and mark it for parkin recruitment to initiate mitophagy.

and it binds with ubiquitinated mitochondrial proteins (Lee et al., 2010). To promote mitophagy, PINK1 and parkin can directly interact with Beclin-1 and PI3K (phosphatidylinositol-4,5-bisphosphate 3-kinase) complex. Recruitment of Ambra 1 (activating molecule in Beclin-1 regulated autophagy 1) is also facilitated by parkin (Strappazzon et al., 2014).

Autosomal recessive genetic disorder Parkinsonism is most prevalently caused by mutations in both parkin and PINK1. Usually it is associated with accumulation of damaged mitochondria leading to neurodegeneration in Parkinson's disease (PD) (Detmer and Chan, 2007). PINK1/parkin pathway plays an important role in mitochondrial quality control by facilitating segregation of impaired mitochondria by preventing their fusion with healthy functional mitochondria (Narendra et al., 2008). This separation is done in two ways: by restricting the cytosolic motility of damaged mitochondria and by degrading the fusion proteins such as Mfns. Parkin can recognize and degrade Miro after being phosphorylated by PINK1. The kinesin motor is released from the mitochondria and hinders their motility after Miro degradation (Sarraf et al., 2013; Wang et al., 2011). All these observations make it clear that mitophagy and mitochondrial dynamics are interlinked and controlled in a synchronized way.

PINK1-parkin-independent pathways of mitophagy are supported by multiple studies. There is no evidence

for dramatic loss of mitochondria after parkin deficiency and parkin deficient mice are normal unless they are exposed to any stress stimuli (Kubli et al., 2013). In primary human fibroblasts isolated from PD patients with parkin mutation, iron chelation induced mitophagy is independent of PINK1 stabilization and parkin activation (Allen et al., 2013). Ambra-1 can directly interact with LC3 via its LC3 interacting region (LIR) to induce parkin independent mitophagy (Strappazzon et al., 2014). In mammalian cells, loss of AMPK (5′ AMP-activated protein kinase) or ULK1 (Unc-51 Like Autophagy Activating Kinase) leads to accumulation of p62 and reduced mitophagy, which indicates that ULK1 is necessary for cellular and mitochondrial homeostasis during starvation (Kim et al., 2011). The FUNDC1 (FUN14 domain-containing protein) is phosphorylated by the ULK1 after translocation to mitochondria and can directly bind to LC3 to regulate mitophagy (Wu et al., 2014). FUNDC1 can also cause hypoxia induced mitophagy. Cardiolipin externalization to the OMM can serve as elimination signal of damaged mitochondria in neurons (Chu et al., 2014, 2013). The phospholipid cardiolipin of IMM interacts with LC3 to facilitate mitochondrial degradation, when it is externalized to the OMM (Chu et al., 2014, 2013; Montessuit et al., 2010). Whether parkin-dependent and -independent pathways function simultaneously is not known yet.

3 MITOCHONDRIAL DYNAMICS AND VIRAL PATHOGENESIS

Viral infection subverts mitochondrial signaling pathways to establish productive infection. Regulation of mitochondrial dynamics and mitophagy upon viral infection is one such example of virus-mitochondria crosstalk (Khan et al., 2015). The functional implications of mitochondrial dynamics during viral infection suggest that targeting mitochondrial dynamics is an efficient viral invasion strategy. As summarized in Figure 3, it is also evident that multiple viruses have evolved with a strategy to modulate mitochondrial dynamics in various ways. In this section, we have summarized the overall effects of selected viral infection on mitochondrial dynamics and briefed the consequences of these effects on viral pathogenesis (Table 2).

3.1 HUMAN IMMUNODEFICIENCY VIRUS (HIV)

Human immunodeficiency virus (HIV) is composed of positive sense, single-stranded RNA. Entry of HIV into the central nervous system (CNS) can cause HIV-associated neurocognitive disorders (HAND) in around one-third of the individuals (Zayyad and Spudich, 2015). Transactivator of transcription (Tat) protein is associated with HIV mediated neurotoxicity via an unknown mechanism. Tat is a 101-amino-acid protein actively secreted from HIV infected macrophages, astrocytes and microglia. Neurons can rapidly internalize Tat protein and promote trimming of neurites, cell death and mitochondrial dysfunction (Bagashev and Sawaya, 2013). Tat exposed neurons showed morphologic and functional changes in mitochondria and loss of mitochondrial membrane potential. Tat transgenic mice or neurons exposed to HIV showed neuronal loss and also exhibited smaller mitochondria compared to the controls (Guitart-Mampel et al., 2017; Rozzi et al., 2017). Rozzi et al. showed Tat-mediated impairment of mitochondrial dynamics in neurons (Rozzi et al., 2018). This impairment

can contribute to cell death (Rozzi et al., 2018). They also explained Tat mediated alterations in mitochondrial membrane potential shortly after exposure and changes in mitochondrial size and subcellular localization in a calcineurin-dependent manner (Rozzi et al., 2018). Other reports explained that gp120 is associated with neuropathological mechanisms of axonal degeneration by altering mitochondrial dynamics. HIV encephalitis (HIVE) patients as well as gp120 transgenic mice showed enlarged mitochondria and altered mitochondrial protein levels in the neurons. In vitro experiments confirmed gp120 mediated alteration of mitochondrial dynamics in neurons (Teodorof-Diedrich and Spector, 2018). In summary, HIV infection can modulate mitochondrial dynamics in Tat and gp120 dependent manner.

3.2 HEPATITIS B VIRUS (HBV)

Hepatitis B Virus (HBV) is a DNA virus, which, during its replication, makes an RNA intermediate termed pre-genomic RNA, which is then encapsidated and converts to HBV DNA via reverse transcription (Seeger and Mason, 2000). HBV encoded regulatory protein X (HBx) can translocate to mitochondria and by its physical interaction with human voltage-dependent anion channel (HVDAC3), it causes mitochondrial dysfunction (Rahmani et al., 2000; Waris et al., 2001). HBx alters the mitochondrial transmembrane potential and affects the level of calcium and ROS, which cause mitochondria damage (Li et al., 2007; Zhong et al., 2017). By these events, HBx is known to activate latent transcription factors such as STAT-3, NF-kB, and NFAT (Waris et al., 2001). In addition, HBx is also associated with HBV mediated early autophagy induction that helps viral DNA replication (Sir et al., 2010; Tang et al., 2009). First, Kim et al. explained that HBV can induce Drp1-dependent mitochondrial fission and Parkin-dependent mitophagy (Kim et al., 2013a). Cells expressing full-length HBV genome or only HBx show disruption of mitochondrial dynamics but

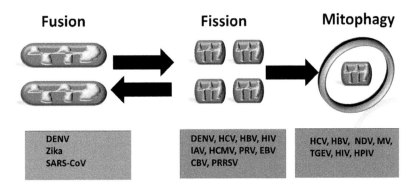

Figure 3 Mitochondrial dynamics and viral infection. In the context of the events associated with mitochondrial dynamics, multiple viruses usurp various mechanisms to subvert these pathways. As described in the text, each virus has unique mechanism to modulate mitochondrial dynamics that ultimately is required for the establishment of successful infection.

TABLE 2

Summary of the involvement of mitochondrial dynamics in multiple virus pathogenesis.

Virus	Key viral protein	Effect
HIV	*Tat* *Gp120*	Induction of PINK1, Parkin and Drp-1 and incomplete mitophagy lead to activation of microglia via induction of mitochondrial damage.
HBV	HBx	Induction of mitochondrial fission and mitophagy help to establish persistent viral replication and inhibition of apoptosis. HBV induced parkin helps virus to suppress interferon production via LUBAC.
HCV	E1, E2 and Core	Induction of mitochondrial fission and mitophagy support persistent viral replication and inhibition of apoptosis and antiviral response.
SARS-CoV	ORF-9b	Mitochondrial elongation to regulate MAVS signaling
Pseudorabies Virus	Glycoprotein B (GB)	Mitochondrial fission and altered Miro activity to regulate intracellular calcium.
Influenza A Virus	PB1-F2	Induction of mitochondrial fission and regulation of NLRP3 inflammasome activity and MAVS signaling.
HCMV	vMIA	Mitochondrial fission to regulate apoptosis via Bax.
EBV	LMP2A	Mitochondrial fission to regulate apoptosis.
Measles virus	Unknown	Mitophagy and regulation of innate immune signaling.
Newcastle disease virus	Unknown	Induction of mitophagy to supports viral replication.
Dengue Virus	NS4B	Mitochondrial elongation and alleviation of RIG-I mediated interferon response to favor replication.
Classical Swine Fever Virus	Unknown	Mitochondrial Fission and complete mitophagy to support viral replication and inhibition of apoptosis.
Human Parainfluenza Virus	Matrix protein	Induction of mitophagy for the regulation of innate immune signaling.
Coxsackievirus B	Unknown	Mitochondrial fission and mitophagy which help induction of viral egress via extracellular vesicles.
PRRSV	Unknown	Induction of mitochondrial fission and mitophagy lead to attenuation of apoptosis.
TGEV	Nucleocapsid	Mitophagy leads to attenuation of oxidative stress.

HBx-mutated (HBV-ΔX) genome had no effect, which clearly indicates that HBx is sufficient to promote mitochondrial damage and altering mitochondrial dynamics (Kim et al., 2013a). This HBx mediated disruption of mitochondrial dynamics is necessary to maintain persistent infection by inhibition of apoptosis of HBV infected cells. HBV infection can cause apoptosis in Parkin-depleted cells by caspase-3 activation, cytochrome C release and poly (ADP-ribose) polymerase (PARP) cleavage, which altogether suggest that alteration/maintenance/modulation of mitochondrial dynamics is required to maintain cell viability of HBV infected cells (Kim et al., 2013a).

3.3 HEPATITIS C VIRUS (HCV)

Hepatitis C virus (HCV) is positive-sense single stranded RNA virus belongs to *Flaviviridae* family (El-Serag, 2012). The 9.5 kb RNA genome is translated into ~3000 amino acid precursor polyprotein. This polyprotein is then processed by viral and host cell's proteases and produces three structural and seven non-structural HCV proteins (Bartenschlager et al., 2011; Coller et al., 2012; Gastaminza et al., 2008). Majority of HCV proteins are associated with ER and are capable to induce ER stress and autophagy, which play a vital role in HCV replication and RNA translation (Gong et al., 2001; Ivanov et

al., 2011). Interestingly, HCV proteins can colocalize with the OMM and MAM (Horner et al., 2011). HCV core and NS5a proteins perturb the complex I activity and promote mitochondrial Ca^{2+} uptake, mitochondrial permeability transition and ROS production (Amako et al., 2011; Ivanov et al., 2011; Ke and Chen, 2011; Kim et al., 2014; Panteva et al., 2003). To facilitate viral invasion and enabling virus to escape from innate immune response, HCV NS3/4a protease cleaves MAM-associated MAVS (Gokhale et al., 2014; Horner et al., 2011). HCV infection promotes perinuclear clustering of mitochondria and Parkin-mediated elimination of damaged mitochondria via mitophagy (Kim et al., 2013b, 2014). HCV infection stimulates Parkin and PINK1 gene expression at transcriptional level. Mitophagy is functionally associated with HCV replication as silencing of Parkin and PINK1 downregulates HCV replication (Kim et al., 2013b, 2014). HCV infection can also modulate mitochondrial dynamics by stimulating Drp1 and Mff gene expression and promoting Drp1 recruitment to the mitochondria by stimulating S616 phosphorylation of Drp1. These events lead to mitochondrial fission followed by mitophagy. Interference of HCV mediated mitochondrial fission by Drp1 or Mff inhibition leads to the accumulation of swollen mitochondria that resisted mitophagic degradation and also suppressed viral secretion and cellular glycolysis. A concomitant reduction in cellular ATP levels and increased IFN synthesis was also observed in Drp1 silenced cells infected with HCV. Inhibition of HCV mediated mitochondrial fission and mitophagy triggered apoptosis by increasing cytochrome C release, caspase 3 activity and cleavage of poly (ADP-ribose) polymerase (Kim et al., 2013b, 2014). In summary, HCV hijacks mitochondrial dynamics pathways that results in the establishment of chronic infection and helps infected cells to escape from apoptotic death.

3.4 SEVERE ACUTE RESPIRATORY SYNDROME-CORONAVIRUS (SARS-CoV)

Severe acute respiratory syndrome-coronavirus (SARS-CoV) is a novel coronavirus recently emerged as a significantly fatal pathogen. How SARS-CoV cripples host cell's antiviral response is not fully characterized yet. SARS-CoV encoded protein, ORF-9b (open reading frame-9b) localizes to OMM and downregulates Drp1 (Shi et al., 2014). It was also observed that this depletion in Drp1 level is independent of autophagy but dependent of proteasomal machinery. ORF-9b induces ubiquitination of Drp1 and its subsequent degradation by proteasome. In HEK293 cells, expression of ORF-9b caused ~ 70% reduction in Drp1 levels. Reduction in Drp1 level inhibits mitochondrial fragmentation and infected cells predominantly show elongated population of mitochondria. ORF-9b also targets the MAVS signalosome by seizing PCBP2 (poly C-binding protein 2) and the E3 ligase ITCH (Itchy E3 Ubiquitin Protein Ligase) to promote MAVS degradation. In summary, SARS-CoV ORF-9b interrupts with mitochondrial dynamics (Shi et al., 2014) to evade the host cell innate immunity.

3.5 PSEUDORABIES VIRUS (PRV)

Pseudorabies virus (PRV) is also known to disrupt mitochondrial dynamics and cause calcium imbalance in infected cells. PRV encoded proteins interfere with mitochondrial function via direct translocation to mitochondria (Kramer and Enquist, 2012). These events induced by PRV are required to block apoptotic pathways. Glycoprotein B (gB) of PRV induces mitochondrial fission during PRV infection and results in electrical coupling of neurons, which subsequently increases intracellular calcium (Kramer and Enquist, 2012). Modulation in calcium regulates the mitochondrial motility. Imbalance in calcium regulates the activity of calcium sensitive cellular protein Miro, which then reduces recruitment of kinesin-1 to mitochondria. Kinesin 1 recruitment is required for mitochondrial motility. PRV mediated disruption of mitochondrial dynamics is required for efficient growth and spreading of the virus, which demonstrate that altered mitochondrial transport increases alphaherpesvirus infection and pathogenesis (Kramer and Enquist, 2012).

3.6 INFLUENZA A VIRUS (IAV)

Influenza A virus (IAV) is a major concern for morbidity and economic loss. Previous study explains the promotion of ULK1 phosphorylation and induction of mitophagy during IAV infection. This event is governed by NOD2 (Nucleotide Binding Oligomerization Domain Containing) and RIPK2 (Receptor interacting protein kinase 2) that respond to IAV infection. In Ripk2$^{-/-}$ cells, defective mitophagy was found and this defect leads to trigger inflammasome activation. Accumulation of damaged mitochondria is responsible for higher inflammation and therefore indicates that mitochondrial quality control is required to reduce inflammasome activation and IL-18 production during IAV infection. In case of IAV infection, mitophagy is Parkin independent and mediated by S555 phosphorylation of ULK1. Studies however indicate that NOD2-RIPK2 signaling mediated protection against IAV is mitophagy dependent. IAV protein PB1-F2 binds to MAVS and affects interferon synthesis. PB1-F2 leads to the reduction of mitochondrial membrane potential via translocation to mitochondrial inner membrane space through TOM40 channel. PB1-F2 also regulates innate immune response, triggers mitochondrial fragmentation and activation of NLRP3 inflammasome. C-terminal domain of PB1-F2 is required for regulation of mitochondrial function (Lupfer et al., 2013; Varga et al., 2012; Yoshizumi et al., 2014).

3.7 PORCINE REPRODUCTIVE AND RESPIRATORY SYNDROME VIRUS (PRRSV)

Porcine reproductive and respiratory syndrome virus (PRRSV) infection results in severe mortality and morbidity in piglets and also, reproductive and respiratory distress in adult pigs, mainly through induction of mitochondrial dysfunction and elevation of ROS (Li et al., 2016; Yan et al., 2015). Electron microscopic analysis of PRRSV infected cultured Marc145 cells revealed fragmentation of mitochondria with swelling and distortion of mitochondrial cristae within the cells (Li et al., 2016). These observations indicated that PRRSV induces mitochondrial fission in infected cells. PRRSV infection resulted in enhanced phosphorylation of Drp1 at Ser616, promoting its translocation from cytosol to mitochondrial membrane- one of the key events for mitochondrial fission (Li et al., 2016). PRRSV causes diminution of mitochondrial DNA (mtDNA) copy number and also increases expression of parkin and PINK at both mRNA and protein level indicating induction of parkin-dependent mitophagy. Depletion of Drp1 caused a decline in mtDNA copy number whereas silencing parkin had an opposite effect (Li et al., 2016). Furthermore, depletion of both Drp1 and parkin activity results in significant activation of caspase 3 in PRRSV infected cells compared to the uninfected cells demonstrating that PRRSV induced mitochondrial fission and autophagy induction in turn attenuates apoptosis and thereby facilitate persistent viral infection within the host cell (Li et al., 2016).

3.8 TRANSMISSIBLE GASTROENTERITIS VIRUS (TGEV)

Transmissible gastroenteritis virus (TGEV) causes watery diarrhea, vomiting and dehydration in young piglets up to two weeks age (Zhu et al., 2016). Infection of TGEV in IPEC-J2 cells decreases mitochondrial membrane potential, total mitochondrial mass with swollen mitochondria and sometimes cristae (Zhu et al., 2016). Presence of autophagosome-like or mitophagosome-like double membrane structures surrounding mitochondria indicates induction of mitophagy to eliminate damaged mitochondria (Zhu et al., 2016). SQSTM1 degradation is the marker for autophagy mediated degradation pathway. TGEV infection induces a complete autophagic response as evident by increased expression of BECN1 and also increased ratio of LC3-II/LC3-I implying initiation of autophagic pathway followed by degradation of SQSTM1 (Zhu et al., 2016). Presence of nucleocapsid protein of TGEV (N) after infection on the mitochondria and subsequent decrease in mitochondrial mass indicate that viral N protein may contribute to mitochondrial dysfunction and induces mitophagy (Zhu et al., 2016). Inhibition of ATG5, which is involved in autophagic vesicle formation during viral infection, leads to reduced viral progeny yield through reduced apoptosis in IPEC-J2 cells. Molecular analysis further reveals that redox-sensitive chaperone protein DJ-1 plays an important role in reducing oxidative stress and inducing mitophagy response during TGEV infection in intestinal epithelial cells (Zhu et al., 2016).

3.9 HUMAN CYTOMEGALOVIRUS (HCMV)

Human cytomegalovirus (HCMV) is a β-herpesvirus, causes infection for lifetime with endogenous reaction episodes and severe infection found to be associated with immunosuppression (Sinzger and Jahn, 1996). HCMV infected fibroblasts show disrupted reticular mitochondrial network and the mitochondria become punctate and dispersed (Sinzger and Jahn, 1996). The expression of immediate-early gene product, vMIA (viral mitochondrion-localized inhibitor of apoptosis) is responsible for these changes (Arnoult et al., 2004; Goldmacher, 2002; Goldmacher et al., 1999). In short, mitochondrial dynamics is altered by vMIA during HCMV infection (McCormick et al., 2003). In all HCMV strains, vMIA is conserved and it has two domains. The first domain (between amino acids 5 and 34) contains mitochondrial signal sequence that targets vMIA to the mitochondria. The second domain (between amino acids 118 and 147) displays anti-apoptotic properties (Goldmacher, 2002; Goldmacher et al., 1999). A recombinant fusion protein consisting of these two domains derived from vMIA is

fully functional. Inhibition of apoptosis through vMIA depends on its interaction with the pro-apoptotic Bax and Bak proteins. In general, Bax helps to induce mitochondrial fusion by changing its sub-mitochondrial distribution and activating the assembly of Mfn2 (Arnoult et al., 2004; Goldmacher, 2002; Hayajneh et al., 2001; McCormick et al., 2003).

3.10 EPSTEIN-BARR VIRUS (EBV)

Epstein-Barr virus (EBV) belongs to gamma herpesvirus family. As an oncogenic virus, it is associated with various epithelial and lymphoid malignancies (Young and Rickinson, 2004). In nasopharyngeal carcinoma, latent membrane protein 2A (LMP2A) of EBV can induce epithelial–mesenchymal transition and increase the invasive ability (Pal et al., 2014). Study reveals that during EBV associated epithelial cancers, mitochondrial dynamics also plays crucial role. In gastric and breast cancer cells, LMP2A causes mitochondrial fragmentation via induction of Drp1 (Pal et al., 2014). This event is associated with enhanced cell migration. LMP2A mediated Notch pathway is also responsible for enhanced fission while inhibitors of this pathway decreases Drp1 expression (Pal et al., 2014). Since several viruses including human papilloma virus, adenovirus and simian virus 40 can activate the Notch pathway upon infection, signaling events associated with Notch and mitochondrial dynamics are the critical pathway necessary for viral function and maintenance (Pal et al., 2014).

3.11 MEASLES VIRUS (MV)

Measles virus (MV) is a paramyxovirus of the genus *Morbillivirus*. It causes Measles, an infection of the respiratory system, immune system and skin (Xia et al., 2014b). In NSCLC cells, defect in autophagy decreased viral titers and cell death which suggests that autophagy is required for productive MV replication (Xia et al., 2014a, 2014b). Published report also shows that Edmonston strain of Measles virus (MV-Edm) induces p62 mediated mitophagy in non-small cell lung cancer cells (NSCLC). MV-Edm induced mitophagy can cause reduction of MAVS protein and dampens down the innate immune response. In MV-Edm infected cells, mitophagy is inhibited by p62 silencing which results in the preservation of mitochondrial mass. An interesting study by Xia et al. indicates that MV seizes RIG-I/MAVS signaling via mitophagy to mitigate the innate immune response (Xia et al., 2014a). MV-Edm triggered mitophagy challenges the apoptosis of infected NSCLC cells by preventing cytochrome C release which leads to increased viral replication. In summary, p62 mediated mitophagy induced by MV-Edm is required for suppression of antiviral response and promote viral replication (Xia et al., 2014a, 2014b).

3.12 Newcastle Disease Virus (NDV)

Newcastle disease virus (NDV) is a member of Paramyxoviridae family. It is a causative agent of highly contagious and fatal disease for poultry and other avian species (Brown and Bevins, 2017). La Sota strain of NDV is reported to induce autophagy and autophagic flux in NSCLC cells, which altogether promote NDV replication by blocking caspase- dependent apoptosis (Meng et al., 2014). In order to control cytochrome c release, NDV induces p62-mediated mitophagy, which blocks intrinsic pro-apoptotic signaling. In the presence of autophagy inhibitor such as 3-MA (3-methyladenine), NDV treated NSCLC cells showed increased oncolysis (Meng et al., 2014). This study uncovers a novel way for NDV mediated mitophagy subversion to favor viral replication via suppression of apoptosis (Meng et al., 2014).

3.13 Dengue Virus (DENV)

Dengue virus (DENV) is a member of *Flaviviridae* family with a positive-sense RNA genome encoding a viral polyprotein. In case of DENV infection, reports are in little contradiction. First Yu et al, observed that DENV induces mitochondrial fission by degrading Mfns (Yu et al., 2015). In DENV infected cells, mitochondrial fusion is impaired by DENV protease NS2B3 mediated cleavage of two mitofusins Mfn1 and Mfn2, which are necessary for mitochondrial fusion (Yu et al., 2015). Mfn1 and Mfn2 are required for efficient antiviral signaling and maintenance of mitochondrial membrane potential (MMP) to attenuate DENV-mediated cell death (Yu et al., 2015). By subverting interferon production and enabling MMP disruption, DENV protease mediated Mfn1 and Mfn2 cleavage can suppress mitochondrial fusion and also deteriorates DENV–induced cytopathic effects (Yu et al., 2015). Imbalance in mitochondrial dynamics happens because of DENV infection through inhibition of Drp1 mediated mitochondrial fission. However, the reports by Barbier et al. and Chatel-Chaix et al, showed that during DENV infection mitochondrial length and respiration is increased (Barbier et al., 2017; Chatel-Chaix et al., 2016). In these reports, DENV was found to enhance mitochondrial elongation via Drp1 downregulation and reduced S616 phosphorylation. It was also concluded that the mitochondrial elongation is responsible for better DENV replication and alleviation of antiviral response. Association of DENV NS3 and NS4b proteins with subcellular fraction of mitochondria is also reported (Chatel-Chaix et al., 2016). During DENV infection NS4b can induce mitochondrial elongation which is physically linked with virus-induced convoluted membranes. DENV infection mediated antiviral immunity is alleviated due to mitochondrial elongation (Chatel-Chaix et al., 2016). In summary, the role of mitochondrial dynamics in DENV infection remains controversial and needs further investigation.

3.14 Classical Swine Fever Virus (CSFV)

Classical swine fever virus (CSFV) is an enveloped RNA virus belongs to *Flaviviridae* family, which is associated with hepatitis C and dengue virus (Gou et al., 2017). In CSFV infected cells, oxidative stress is induced and mitochondrial transmembrane potential is disturbed. CSFV infection induces mitochondrial fission and mitophagy to inhibit host cell apoptosis (Gou et al., 2017). Expression of PINK and Parkin and their mitochondrial translocations are increased during CSFV infection. In CSFV infected cells, Mfn2 was ubiquitinated and degraded via PINK1 and Parkin pathways. During infection of CSFV, Drp1 translocation into mitochondria induces mitochondrial fission that results in mitochondrial fragmentation (Gou et al., 2017). Silencing of Drp1 and parkin during CSFV infection upregulate apoptotic signals, preserves mitochondrial proteins and reduces viral replication (Gou et al., 2017). All these events clearly indicate that mitochondrial fission and mitophagy induced by CSFV are required to increase viral persistence and cell survival.

3.15 Human Parainfluenza Virus 3 (HPIV3)

Human Parainfluenza Virus 3 (HPIV3) is a member of the *Paramyxoviridae* family with a non-segmented negative-strand RNA genome (Ding et al., 2017). HPIV3 is the causative agent of severe respiratory tract diseases in infants and young children. Although it was reported that HPIV3 infection can induce autophagy but the mechanism is still unclear. Ding et al. showed that HPIV3 mediated matrix protein (M) can translocate to mitochondria and interact with mitochondrial Tu translation elongation factor (TUFM) to induce mitophagy, which ultimately results in type I interferon (IFN) response inhibition (Ding et al., 2017). For autophagosome formation during mitophagy, M interacts with LC3 protein which is Parkin-PINK1 pathway independent. This study indicates that a viral protein can induce mitophagy by joining mitochondria and autophagosomes (Ding et al., 2017).

3.16 Coxsackievirus B (CVB)

Coxsackievirus B (CVB) is a member of the genus Enterovirus and causative agent of numerous life threatening systemic inflammatory diseases. CVB can trigger the release of virus containing infectious extracellular microvesicles (EMVs), originated from autophagosomes and these EMVs have LC3 (Sin et al., 2017). Sin et al. demonstrated that CVB can subvert mitophagy machinery to support the dissemination of virus in released EMVs (Sin et al., 2017). CVB infection activates Drp1 mediated mitochondrial fragmentation. Mitochondrion-containing autophagosomes are released from cells instead of degradation by lysosomes. Mitophagy is induced when CVB is localized to mitochondria and subsequently released from the cell in an autophagosome-bound mitochondrion-virus complex. Suppression of mitophagy markedly reduces the production of virus (Sin et al., 2017).

4 MITOCHONDRIAL DYNAMICS GOVERNS INNATE ANTIVIRAL RESPONSE

The antiviral signaling plays a crucial role in the mammalian immune response modulation. Upon infection, viral DNA or RNA is sensed by cytosolic PRRs (Zevini et al., 2017). Cytosolic sensing of viral nucleic acid triggers a complex signaling cascade, in which MAVS and STING play a central role. These two adapter proteins mediate the innate immune signaling through the induction of interferon and inflammatory cytokines. Interestingly, both the pathways of MAVS and STING signaling are regulated by mitochondria and therefore these cellular power houses have emerged as a central hub that play a critical role in interferon production (Zevini et al., 2017). In this section, we will discuss how mitochondrial dynamics is intricately linked to the antiviral response mediated by MAVS and STING.

4.1 DNA Sensing Pathway

Sensing of cytosolic DNA is mainly governed by cGAS (Cyclic GMP-AMP synthase)-STING pathways (Sun et al., 2013). Please note that this pathway has been reported to be activated by mitochondrial or cellular DNA during stress (Dhanwani et al., 2018). However in this section we will mainly focus on the signaling response induced by viral DNA. cGAS contains two DNA binding domains. Under normal conditions, cGAS exists in an autoinhibited state. Binding with cytosolic DNA induces conformational changes in such a way that synthesis of cyclic GMP-AMP (cGAMP) complex (Civril et al., 2013) starts. cGAMP acts as a secondary messenger and can efficiently bind with the STING. cGAMP-STING interaction induces conformational changes in STING, which further facilitates the traffics of STING from ER to Golgi (Civril et al., 2013). During this process, STING recruits TBK1 (TANK binding kinase 1), which in turn activates IRF3 via phosphorylation. Activated IRF3 gets dimerized and translocate into the nucleus to facilitate the production of interferon (Civril et al., 2013; Sun et al., 2013).

4.2 RNA Sensing Pathway

In higher organisms, viral RNA is sensed by TLRs (Toll-like receptors), RIG-I (Retinoic acid inducible gene I) and RLRs (RIG-I like receptors) (Yoneyama et al., 2004). RIG-I and RLRs contain RNA helicase domain. RIG-I contains a C-terminal regulatory domain that can sense the 5′ triphosphate of RNA (Yoneyama et al., 2004). Multiple RLRs detect distinct classes of RNA

viruses (Zevini et al., 2017). Binding of viral RNA with RIG-I and RLRs induces conformational changes and facilitate their interaction with MAVS (Seth et al., 2005). MAVS then activates cytosolic kinase TBK1 followed by IRF3 activation. Activated IRF3 translocates into the nucleus and activate type I interferon production and also other inflammatory molecules (Seth et al., 2005; Zevini et al., 2017).

4.3 Crosstalk between MAVS-STING Signaling

In the above section, it is mentioned that the MAVS and STING signaling are specific for the viral RNA and DNA respectively (Zevini et al., 2017). Although these two signaling events, in principle are predominantly RNA or DNA specific, crosstalk between these two has been reported in case of multiple viruses. For instance, STING is an interacting partner of RIG-I and MAVS (Zevini et al., 2017). In fact viral infection stabilizes this interaction. MAVS is predominantly localized to the mitochondria while STING is associated with ER (Zevini et al., 2017). In general mitochondria-ER contacts sites exist in the cytosol and these sites are known as MAM (mitochondria associated membrane) (Horner et al., 2011). It is known that STING-MAVS interaction occurs at MAMs and regulates interferon signaling. Elimination of STING inhibits the interferon signaling in Japanese encephalitis virus (JEV) infected cells (Ding et al., 2018; Nazmi et al., 2012). Notably JEV is an RNA virus and role of STING in JEV induced interferon response strongly supports the existence of MAVS-STING crosstalk. It is worth mentioning that the MAVS-STING crosstalk is important for multiple reasons. First, it represents a common step where DNA and RNA sensing events converges, and, second, it highlights the importance of MAMs, which determines the viral invasion strategies.

4.4 Mitochondrial Dynamics Regulates Antiviral Signaling

As mentioned in the above sections, multiple studies have shown the involvement of mitochondrial dynamics in the regulation of antiviral response (Castanier et al., 2010). For instance, Mfn2 was found to inhibit RLS signaling by interacting with MAVS (Yasukawa et al., 2009). It was observed that ectopic expression of Mfn2 cripples IRF3 activation whereas genetic elimination of Mfn2 was associated with enhanced MAVS signaling (Yasukawa et al., 2009). Although Mfn1 and Mfn2, both are the key players of mitochondrial dynamics, Yasukawa et al. observed that Mfn1 manipulation exerts no effect on RLR signaling (Yasukawa et al., 2009). However, Castanier et al, report did claim that both Mfn1 and Mfn2 are able to regulate RLR signaling (Castanier et al., 2010). Precisely, this report claimed that elongated mitochondria are more efficient in mediating RLR signaling events. Cells depleted for Mfn1 or Mfn2, which exhibits predominantly fragmented mitochondria, are weak responders of RLR agonists while cells depleted for Drp1, which are enriched in elongated mitochondrial population, produce type I interferon more efficiently (Castanier et al., 2010). The involvement of mitochondrial dynamics in antiviral signaling is not only confined to MAVS. It has also been shown that it regulates STING mediated signaling. It was observed that during viral infection, elongated mitochondria promote ER-mitochondria contacts facilitating the MAVS-STING interaction to augment RLR signaling (Kwon et al., 2017). In recent reports, it was observed that the parkin and PINK1 are able to mitigate STING induced inflammation in mice and loss of STING in these mice completely rescues the inflammatory response (Sliter et al., 2018). On the other hand some other investigations have concluded that parkin is a negative regulator of RLR signaling. First, Khan et al, observed that HBV infection upregulates parkin expression and induces parkin translocation to mitochondria (Khan et al., 2016). HBx, the regulatory protein of HBV is a key player in these events. Once translocated to mitochondria, parkin interacts with the MAVS signaling complex and makes HBV infected cells refractory for interferon production (Khan et al., 2016). Interestingly, mitochondrial parkin acts as a recruiter for LUBAC (Linear ubiquitin assembly complex). LUBAC is a multimeric protein complex that has ability to catalyse linear ubiquitin chains (also known as M-Ub chains) (Khan et al., 2016). These M1-Ub chains have been shown to regulate interferon signaling (Belgnaoui et al., 2012). In this investigation, Khan et al also observed that parkin is an interacting partner of LUBAC. Once translocated to the mitochondria, parkin can efficiently induce the accumulation of M-Ub chain via LUBAC recruitment (Khan et al., 2016). These M1-Ub chains can non-covalently associate with the MAVS complex and disrupts the formation of MAVS signalasome. Recent advancements in the field of selective mitophagy and detailed characterization, parkin, PINK1 and LUBAC have provided a solid platform to further investigate and identify new therapeutic targets for viral infection. For example, inhibition of mitophagy in HBV or HCV infected cells induces apoptosis (Kim et al., 2013a, 2013b, 2014). Moreover, LUBAC has been observed to regulate MAVS and inflammatory signaling in HBV- and HCV-infected cells (Belgnaoui et al., 2012; Chen et al., 2015; Khan et al., 2016). Therefore, now it is conceivable that LUBAC, parkin and mitophagy are linked with each other. It is also possible that LUBAC may be involved in the pathogenesis of multiple viruses to ensure cell survival and reduced IFN synthesis (Chen et al., 2015; Khan et al., 2016). Altogether, these key observations strongly point out that M1-Ub chain specific proteases, USPs or deubiquitinases are promising tools to be studied in details as it holds a great promise to design novel therapeutic targets for viral infection.

4.5 CONCLUSION

A number of viral invasion mechanisms that dampen host cell defense network have been reported. Although the function of mitochondria in regulating interferon production requires further investigation, recent studies have clearly established that mitochondria are the hub of antiviral signaling. Mitochondrial dynamics and mitophagy appear to modulate interferon production in multiple ways. On the other hand, there are diverse mechanisms, proposed to explain how different viruses subvert mitochondrial dynamics and mitophagy. Mitochondria and especially MAMs are the sites where MAVS and STING signaling converge. Therefore, future investigation should be focused on further elucidation of how MAMs' organization is modulated upon viral infection. Deeper understanding and future studies in this direction can potentially be translated into useful drug targets and develop potent antivirals.

REFERENCES

Akira, S., Uematsu, S., and Takeuchi, O. (2006). Pathogen recognition and innate immunity. Cell *124*, 783–801.

Allen, G.F., Toth, R., James, J., and Ganley, I.G. (2013). Loss of iron triggers PINK1/Parkin-independent mitophagy. EMBO Rep *14*, 1127–1135.

Amako, Y., Syed, G.H., and Siddiqui, A. (2011). Protein kinase D negatively regulates hepatitis C virus secretion through phosphorylation of oxysterol-binding protein and ceramide transfer protein. J Biol Chem *286*, 11265–11274.

Anand, S.K., and Tikoo, S.K. (2013). Viruses as modulators of mitochondrial functions. Adv Virol *2013*, 738794.

Arnoult, D., Bartle, L.M., Skaletskaya, A., Poncet, D., Zamzami, N., Park, P.U., Sharpe, J., Youle, R.J., and Goldmacher, V.S. (2004). Cytomegalovirus cell death suppressor vMIA blocks Bax- but not Bak-mediated apoptosis by binding and sequestering Bax at mitochondria. Proc Natl Acad Sci U S A *101*, 7988–7993.

Ashrafi, G., and Schwarz, T.L. (2013). The pathways of mitophagy for quality control and clearance of mitochondria. Cell Death Differ *20*, 31–42.

Bagashev, A., and Sawaya, B.E. (2013). Roles and functions of HIV-1 Tat protein in the CNS: an overview. Virol J *10*, 358.

Baker, M.J., Lampe, P.A., Stojanovski, D., Korwitz, A., Anand, R., Tatsuta, T., and Langer, T. (2014). Stress-induced OMA1 activation and autocatalytic turnover regulate OPA1-dependent mitochondrial dynamics. EMBO J *33*, 578–593.

Barber, G.N. (2015). STING: infection, inflammation and cancer. Nat Rev Immunol *15*, 760–770.

Barbier, V., Lang, D., Valois, S., Rothman, A.L., and Medin, C.L. (2017). Dengue virus induces mitochondrial elongation through impairment of Drp1-triggered mitochondrial fission. Virology *500*, 149–160.

Barral, P.M., Sarkar, D., Su, Z.Z., Barber, G.N., DeSalle, R., Racaniello, V.R., and Fisher, P.B. (2009). Functions of the cytoplasmic RNA sensors RIG-I and MDA-5: key regulators of innate immunity. Pharmacol Ther *124*, 219–234.

Bartenschlager, R., Penin, F., Lohmann, V., and Andre, P. (2011). Assembly of infectious hepatitis C virus particles. Trends Microbiol *19*, 95–103.

Bauckman, K.A., Owusu-Boaitey, N., and Mysorekar, I.U. (2015). Selective autophagy: xenophagy. Methods 75, 120–127.

Belgnaoui, S.M., Paz, S., Samuel, S., Goulet, M.L., Sun, Q., Kikkert, M., Iwai, K., Dikic, I., Hiscott, J., and Lin, R. (2012). Linear ubiquitination of NEMO negatively regulates the interferon antiviral response through disruption of the MAVS-TRAF3 complex. Cell Host Microbe 12, 211–222.

Bockler, S., and Westermann, B. (2014). ER-mitochondria contacts as sites of mitophagosome formation. Autophagy 10, 1346–1347.

Braschi, E., Zunino, R., and McBride, H.M. (2009). MAPL is a new mitochondrial SUMO E3 ligase that regulates mitochondrial fission. EMBO Rep 10, 748–754.

Brown, V.R., and Bevins, S.N. (2017). A review of virulent Newcastle disease viruses in the United States and the role of wild birds in viral persistence and spread. Vet Res 48, 68.

Castanier, C., Garcin, D., Vazquez, A., and Arnoult, D. (2010). Mitochondrial dynamics regulate the RIG-I-like receptor antiviral pathway. EMBO Rep 11, 133–138.

Cerveny, K.L., and Jensen, R.E. (2003). The WD-repeats of Net2p interact with Dnm1p and Fis1p to regulate division of mitochondria. Mol Biol Cell 14, 4126–4139.

Chan, D.C. (2006). Mitochondria: dynamic organelles in disease, aging, and development. Cell 125, 1241–1252.

Chan, D.C. (2012). Fusion and fission: interlinked processes critical for mitochondrial health. Annu Rev Genet 46, 265–287.

Chan, N.C., Salazar, A.M., Pham, A.H., Sweredoski, M.J., Kolawa, N.J., Graham, R.L., Hess, S., and Chan, D.C. (2011). Broad activation of the ubiquitin-proteasome system by Parkin is critical for mitophagy. Hum Mol Genet 20, 1726–1737.

Chang, C.R., and Blackstone, C. (2007). Cyclic AMP-dependent protein kinase phosphorylation of Drp1 regulates its GTPase activity and mitochondrial morphology. J Biol Chem 282, 21583–21587.

Chatel-Chaix, L., Cortese, M., Romero-Brey, I., Bender, S., Neufeldt, C.J., Fischl, W., Scaturro, P., Schieber, N., Schwab, Y., Fischer, B., et al. (2016). Dengue virus perturbs mitochondrial morphodynamics to dampen innate immune responses. Cell Host Microbe 20, 342–356.

Chen, H., and Chan, D.C. (2009). Mitochondrial dynamics–fusion, fission, movement, and mitophagy–in neurodegenerative diseases. Hum Mol Genet 18, R169–176.

Chen, H., Chomyn, A., and Chan, D.C. (2005). Disruption of fusion results in mitochondrial heterogeneity and dysfunction. J Biol Chem 280, 26185–26192.

Chen, H., Detmer, S.A., Ewald, A.J., Griffin, E.E., Fraser, S.E., and Chan, D.C. (2003). Mitofusins Mfn1 and Mfn2 coordinately regulate mitochondrial fusion and are essential for embryonic development. J Cell Biol 160, 189–200.

Chen, Y., and Dorn, G.W., 2nd. (2013). PINK1-phosphorylated mitofusin 2 is a Parkin receptor for culling damaged mitochondria. Science 340, 471–475.

Chen, Y., He, L., Peng, Y., Shi, X., Chen, J., Zhong, J., Chen, X., Cheng, G., and Deng, H. (2015). The hepatitis C virus protein NS3 suppresses TNF-alpha-stimulated activation of NF-kappaB by targeting LUBAC. Sci Signal 8, ra118.

Chu, C.T., Bayir, H., and Kagan, V.E. (2014). LC3 binds externalized cardiolipin on injured mitochondria to signal mitophagy in neurons: implications for Parkinson disease. Autophagy 10, 376–378.

Chu, C.T., Ji, J., Dagda, R.K., Jiang, J.F., Tyurina, Y.Y., Kapralov, A.A., Tyurin, V.A., Yanamala, N., Shrivastava, I.H.,

Mohammadyani, D., et al. (2013). Cardiolipin externalization to the outer mitochondrial membrane acts as an elimination signal for mitophagy in neuronal cells. Nat Cell Biol 15, 1197–1205.

Civril, F., Deimling, T., de Oliveira Mann, C.C., Ablasser, A., Moldt, M., Witte, G., Hornung, V., and Hopfner, K.P. (2013). Structural mechanism of cytosolic DNA sensing by cGAS. Nature 498, 332–337.

Coller, K.E., Heaton, N.S., Berger, K.L., Cooper, J.D., Saunders, J.L., and Randall, G. (2012). Molecular determinants and dynamics of hepatitis C virus secretion. PLoS Pathogens 8, e1002466.

Detmer, S.A., and Chan, D.C. (2007). Functions and dysfunctions of mitochondrial dynamics. Nat Rev Mol Cell Biol 8, 870–879.

Dhanwani, R., Takahashi, M., and Sharma, S. (2018). Cytosolic sensing of immuno-stimulatory DNA, the enemy within. Curr Opin Immunol 50, 82–87.

Dickey, A.S., and Strack, S. (2011). PKA/AKAP1 and PP2A/Bbeta2 regulate neuronal morphogenesis via Drp1 phosphorylation and mitochondrial bioenergetics. J Neurosci 31, 15716–15726.

Ding, B., Zhang, L., Li, Z., Zhong, Y., Tang, Q., Qin, Y., and Chen, M. (2017). The matrix protein of human parainfluenza virus type 3 induces mitophagy that suppresses interferon responses. Cell Host Microbe 21, 538–547 e534.

Ding, Q., Gaska, J.M., Douam, F., Wei, L., Kim, D., Balev, M., Heller, B., and Ploss, A. (2018). Species-specific disruption of STING-dependent antiviral cellular defenses by the Zika virus NS2B3 protease. Proc Natl Acad Sci U S A 115, E6310–E6318.

Ehses, S., Raschke, I., Mancuso, G., Bernacchia, A., Geimer, S., Tondera, D., Martinou, J.C., Westermann, B., Rugarli, E.I., and Langer, T. (2009). Regulation of OPA1 processing and mitochondrial fusion by m-AAA protease isoenzymes and OMA1. J Cell Biol 187, 1023–1036.

El-Serag, H.B. (2012). Epidemiology of viral hepatitis and hepatocellular carcinoma. Gastroenterology 142, 1264–1273 e1261.

Ernster, L., and Schatz, G. (1981). Mitochondria: a historical review. J Cell Biol 91, 227s–255s.

Feng, Y., He, D., Yao, Z., and Klionsky, D.J. (2014). The machinery of macroautophagy. Cell Res 24, 24–41.

Figueroa-Romero, C., Iniguez-Lluhi, J.A., Stadler, J., Chang, C.R., Arnoult, D., Keller, P.J., Hong, Y., Blackstone, C., and Feldman, E.L. (2009). SUMOylation of the mitochondrial fission protein Drp1 occurs at multiple nonconsensus sites within the B domain and is linked to its activity cycle. FASEB J 23, 3917–3927.

Frank, S., Gaume, B., Bergmann-Leitner, E.S., Leitner, W.W., Robert, E.G., Catez, F., Smith, C.L., and Youle, R.J. (2001). The role of dynamin-related protein 1, a mediator of mitochondrial fission, in apoptosis. Dev Cell 1, 515–525.

Frezza, C., Cipolat, S., Martins de Brito, O., Micaroni, M., Beznoussenko, G.V., Rudka, T., Bartoli, D., Polishuck, R.S., Danial, N.N., De Strooper, B., et al. (2006). OPA1 controls apoptotic cristae remodeling independently from mitochondrial fusion. Cell 126, 177–189.

Friedman, J.R., Lackner, L.L., West, M., DiBenedetto, J.R., Nunnari, J., and Voeltz, G.K. (2011). ER tubules mark sites of mitochondrial division. Science 334, 358–362.

Gandre-Babbe, S., and van der Bliek, A.M. (2008). The novel tail-anchored membrane protein Mff controls mitochondrial and peroxisomal fission in mammalian cells. Mol Biol Cell 19, 2402–2412.

Gastaminza, P., Cheng, G., Wieland, S., Zhong, J., Liao, W., and Chisari, F.V. (2008). Cellular determinants of hepatitis C virus assembly, maturation, degradation, and secretion. J Virol *82*, 2120–2129.

Gokhale, N.S., Vazquez, C., and Horner, S.M. (2014). Hepatitis C virus. Strategies to evade antiviral responses. Future Virol *9*, 1061–1075.

Goldmacher, V.S. (2002). vMIA, a viral inhibitor of apoptosis targeting mitochondria. Biochimie *84*, 177–185.

Goldmacher, V.S., Bartle, L.M., Skaletskaya, A., Dionne, C.A., Kedersha, N.L., Vater, C.A., Han, J.W., Lutz, R.J., Watanabe, S., Cahir McFarland, E.D., *et al.* (1999). A cytomegalovirus-encoded mitochondria-localized inhibitor of apoptosis structurally unrelated to Bcl-2. Proc Natl Acad Sci U S A *96*, 12536–12541.

Gomes, L.C., and Scorrano, L. (2008). High levels of Fis1, a pro-fission mitochondrial protein, trigger autophagy. Biochim Biophys Acta *1777*, 860–866.

Gong, G., Waris, G., Tanveer, R., and Siddiqui, A. (2001). Human hepatitis C virus NS5A protein alters intracellular calcium levels, induces oxidative stress, and activates STAT-3 and NF-kappa B. Proc Natl Acad Sci U S A *98*, 9599–9604.

Gou, H., Zhao, M., Xu, H., Yuan, J., He, W., Zhu, M., Ding, H., Yi, L., and Chen, J. (2017). CSFV induced mitochondrial fission and mitophagy to inhibit apoptosis. Oncotarget *8*, 39382–39400.

Guitart-Mampel, M., Hernandez, A.S., Moren, C., Catalan-Garcia, M., Tobias, E., Gonzalez-Casacuberta, I., Juarez-Flores, D.L., Gatell, J.M., Cardellach, F., Milisenda, J.C., *et al.* (2017). Imbalance in mitochondrial dynamics and apoptosis in pregnancies among HIV-infected women on HAART with obstetric complications. J Antimicrob Chemother *72*, 2578–2586.

Hales, K.G., and Fuller, M.T. (1997). Developmentally regulated mitochondrial fusion mediated by a conserved, novel, predicted GTPase. Cell *90*, 121–129.

Han, X.J., Lu, Y.F., Li, S.A., Kaitsuka, T., Sato, Y., Tomizawa, K., Nairn, A.C., Takei, K., Matsui, H., and Matsushita, M. (2008). CaM kinase I alpha-induced phosphorylation of Drp1 regulates mitochondrial morphology. J Cell Biol *182*, 573–585.

Hayajneh, W.A., Colberg-Poley, A.M., Skaletskaya, A., Bartle, L.M., Lesperance, M.M., Contopoulos-Ioannidis, D.G., Kedersha, N.L., and Goldmacher, V.S. (2001). The sequence and antiapoptotic functional domains of the human cytomegalovirus UL37 exon 1 immediate early protein are conserved in multiple primary strains. Virology *279*, 233–240.

Head, B., Griparic, L., Amiri, M., Gandre-Babbe, S., and van der Bliek, A.M. (2009). Inducible proteolytic inactivation of OPA1 mediated by the OMA1 protease in mammalian cells. J Cell Biol *187*, 959–966.

Hermann, G.J., Thatcher, J.W., Mills, J.P., Hales, K.G., Fuller, M.T., Nunnari, J., and Shaw, J.M. (1998). Mitochondrial fusion in yeast requires the transmembrane GTPase Fzo1p. J Cell Biol *143*, 359–373.

Horner, S.M., Liu, H.M., Park, H.S., Briley, J., and Gale, M., Jr. (2011). Mitochondrial-associated endoplasmic reticulum membranes (MAM) form innate immune synapses and are targeted by hepatitis C virus. Proc Natl Acad Sci U S A *108*, 14590–14595.

Ivanov, A.V., Smirnova, O.A., Ivanova, O.N., Masalova, O.V., Kochetkov, S.N., and Isaguliants, M.G. (2011). Hepatitis C virus proteins activate NRF2/ARE pathway by distinct ROS-dependent and independent mechanisms in HUH7 cells. PLoS One *6*, e24957.

James, D.I., Parone, P.A., Mattenberger, Y., and Martinou, J.C. (2003). hFis1, a novel component of the mammalian mitochondrial fission machinery. J Biol Chem *278*, 36373–36379.

Kane, L.A., Lazarou, M., Fogel, A.I., Li, Y., Yamano, K., Sarraf, S.A., Banerjee, S., and Youle, R.J. (2014). PINK1 phosphorylates ubiquitin to activate Parkin E3 ubiquitin ligase activity. J Cell Biol *205*, 143–153.

Kashatus, D.F., Lim, K.H., Brady, D.C., Pershing, N.L., Cox, A.D., and Counter, C.M. (2011). RALA and RALBP1 regulate mitochondrial fission at mitosis. Nat Cell Biol *13*, 1108–1115.

Kazlauskaite, A., Kondapalli, C., Gourlay, R., Campbell, D.G., Ritorto, M.S., Hofmann, K., Alessi, D.R., Knebel, A., Trost, M., and Muqit, M.M. (2014). Parkin is activated by PINK1-dependent phosphorylation of ubiquitin at Ser65. Biochem J *460*, 127–139.

Ke, P.Y., and Chen, S.S. (2011). Activation of the unfolded protein response and autophagy after hepatitis C virus infection suppresses innate antiviral immunity in vitro. J Clin Invest *121*, 37–56.

Khan, M., Syed, G.H., Kim, S.J., and Siddiqui, A. (2015). Mitochondrial dynamics and viral infections: a close nexus. Biochim Biophys Acta *1853*, 2822–2833.

Khan, M., Syed, G.H., Kim, S.J., and Siddiqui, A. (2016). Hepatitis B virus-induced Parkin-dependent recruitment of Linear Ubiquitin Assembly Complex (LUBAC) to mitochondria and attenuation of innate immunity. PLoS Pathogens *12*, e1005693.

Kim, J., Kundu, M., Viollet, B., and Guan, K.L. (2011). AMPK and mTOR regulate autophagy through direct phosphorylation of Ulk1. Nat Cell Biol *13*, 132–141.

Kim, S.J., Khan, M., Quan, J., Till, A., Subramani, S., and Siddiqui, A. (2013a). Hepatitis B virus disrupts mitochondrial dynamics: induces fission and mitophagy to attenuate apoptosis. PLoS Pathogens *9*, e1003722.

Kim, S.J., Syed, G.H., Khan, M., Chiu, W.W., Sohail, M.A., Gish, R.G., and Siddiqui, A. (2014). Hepatitis C virus triggers mitochondrial fission and attenuates apoptosis to promote viral persistence. Proc Natl Acad Sci U S A *111*, 6413–6418.

Kim, S.J., Syed, G.H., and Siddiqui, A. (2013b). Hepatitis C virus induces the mitochondrial translocation of Parkin and subsequent mitophagy. PLoS Pathogens *9*, e1003285.

Koch, A., Yoon, Y., Bonekamp, N.A., McNiven, M.A., and Schrader, M. (2005). A role for Fis1 in both mitochondrial and peroxisomal fission in mammalian cells. Mol Biol Cell *16*, 5077–5086.

Korobova, F., Ramabhadran, V., and Higgs, H.N. (2013). An actin-dependent step in mitochondrial fission mediated by the ER-associated formin INF2. Science *339*, 464–467.

Kramer, T., and Enquist, L.W. (2012). Alphaherpesvirus infection disrupts mitochondrial transport in neurons. Cell Host Microbe *11*, 504–514.

Kraus, F., and Ryan, M.T. (2017). The constriction and scission machineries involved in mitochondrial fission. J Cell Sci *130*, 2953–2960.

Kubli, D.A., Zhang, X., Lee, Y., Hanna, R.A., Quinsay, M.N., Nguyen, C.K., Jimenez, R., Petrosyan, S., Murphy, A.N., and Gustafsson, A.B. (2013). Parkin protein deficiency exacerbates cardiac injury and reduces survival following myocardial infarction. J Biol Chem *288*, 915–926.

Kwon, D., Park, E., and Kang, S.J. (2017). Stimulator of IFN genes-mediated DNA-sensing pathway is suppressed by

NLRP3 agonists and regulated by mitofusin 1 and TBC1D15, mitochondrial dynamics mediators. FASEB J *31*, 4866–4878.

Lee, H., and Yoon, Y. (2014). Mitochondrial fission: regulation and ER connection. Mol Cells *37*, 89–94.

Lee, J.Y., Nagano, Y., Taylor, J.P., Lim, K.L., and Yao, T.P. (2010). Disease-causing mutations in Parkin impair mitochondrial ubiquitination, aggregation, and HDAC6-dependent mitophagy. J Cell Biol *189*, 671–679.

Lee, Y.J., Jeong, S.Y., Karbowski, M., Smith, C.L., and Youle, R.J. (2004). Roles of the mammalian mitochondrial fission and fusion mediators Fis1, Drp1, and Opa1 in apoptosis. Mol Biol Cell *15*, 5001–5011.

Li, B., Gao, B., Ye, L., Han, X., Wang, W., Kong, L., Fang, X., Zeng, Y., Zheng, H., Li, S., *et al.* (2007). Hepatitis B virus X protein (HBx) activates ATF6 and IRE1-XBP1 pathways of unfolded protein response. Virus Res *124*, 44–49.

Li, S., Wang, J., Zhou, A., Khan, F.A., Hu, L., and Zhang, S. (2016). Porcine reproductive and respiratory syndrome virus triggers mitochondrial fission and mitophagy to attenuate apoptosis. Oncotarget *7*, 56002–56012.

Loo, Y.M., and Gale, M., Jr. (2011). Immune signaling by RIG-I-like receptors. Immunity *34*, 680–692.

Loson, O.C., Song, Z., Chen, H., and Chan, D.C. (2013). Fis1, Mff, MiD49, and MiD51 mediate Drp1 recruitment in mitochondrial fission. Mol Biol Cell *24*, 659–667.

Lupfer, C., Thomas, P.G., Anand, P.K., Vogel, P., Milasta, S., Martinez, J., Huang, G., Green, M., Kundu, M., Chi, H., *et al.* (2013). Receptor interacting protein kinase 2-mediated mitophagy regulates inflammasome activation during virus infection. Nat Immunol *14*, 480–488.

Ma, Z., and Damania, B. (2016). The cGAS-STING defense pathway and its counteraction by viruses. Cell Host Microbe *19*, 150–158.

McBride, H.M., Neuspiel, M., and Wasiak, S. (2006). Mitochondria: more than just a powerhouse. Curr Biol *16*, R551–560.

McCormick, A.L., Smith, V.L., Chow, D., and Mocarski, E.S. (2003). Disruption of mitochondrial networks by the human cytomegalovirus UL37 gene product viral mitochondrion-localized inhibitor of apoptosis. J Virol *77*, 631–641.

Meng, G., Xia, M., Wang, D., Chen, A., Wang, Y., Wang, H., Yu, D., and Wei, J. (2014). Mitophagy promotes replication of oncolytic newcastle disease virus by blocking intrinsic apoptosis in lung cancer cells. Oncotarget *5*, 6365–6374.

Merrill, R.A., Dagda, R.K., Dickey, A.S., Cribbs, J.T., Green, S. H., Usachev, Y.M., and Strack, S. (2011). Mechanism of neuroprotective mitochondrial remodeling by PKA/ AKAP1. PLoS Biol *9*, e1000612.

Montessuit, S., Somasekharan, S.P., Terrones, O., Lucken-Ardjomande, S., Herzig, S., Schwarzenbacher, R., Manstein, D.J., Bossy-Wetzel, E., Basanez, G., Meda, P., *et al.* (2010). Membrane remodeling induced by the dynamin-related protein Drp1 stimulates Bax oligomerization. Cell *142*, 889–901.

Narendra, D., Tanaka, A., Suen, D.F., and Youle, R.J. (2008). Parkin is recruited selectively to impaired mitochondria and promotes their autophagy. J Cell Biol *183*, 795–803.

Narendra, D.P., Jin, S.M., Tanaka, A., Suen, D.F., Gautier, C. A., Shen, J., Cookson, M.R., and Youle, R.J. (2010). PINK1 is selectively stabilized on impaired mitochondria to activate Parkin. PLoS Biol *8*, e1000298.

Nazmi, A., Mukhopadhyay, R., Dutta, K., and Basu, A. (2012). STING mediates neuronal innate immune response following Japanese encephalitis virus infection. Sci Rep *2*, 347.

O'Rourke, B. (2010). From bioblasts to mitochondria: ever expanding roles of mitochondria in cell physiology. Front Physiol *1*, 7.

Ordureau, A., Sarraf, S.A., Duda, D.M., Heo, J.M., Jedrychowski, M.P., Sviderskiy, V.O., Olszewski, J.L., Koerber, J.T., Xie, T., Beausoleil, S.A., *et al.* (2014). Quantitative proteomics reveal a feedforward mechanism for mitochondrial Parkin translocation and ubiquitin chain synthesis. Mol Cell *56*, 360—375.

Otera, H., Wang, C., Cleland, M.M., Setoguchi, K., Yokota, S., Youle, R.J., and Mihara, K. (2010). Mff is an essential factor for mitochondrial recruitment of Drp1 during mitochondrial fission in mammalian cells. J Cell Biol *191*, 1141–1158.

Pal, A.D., Basak, N.P., Banerjee, A.S., and Banerjee, S. (2014). Epstein-Barr virus latent membrane protein-2A alters mitochondrial dynamics promoting cellular migration mediated by Notch signaling pathway. Carcinogenesis *35*, 1592–1601.

Panteva, M., Korkaya, H., and Jameel, S. (2003). Hepatitis viruses and the MAPK pathway: is this a survival strategy? Virus Res *92*, 131–140.

Park, J., Lee, S.B., Lee, S., Kim, Y., Song, S., Kim, S., Bae, E., Kim, J., Shong, M., Kim, J.M., *et al.* (2006). Mitochondrial dysfunction in Drosophila PINK1 mutants is complemented by Parkin. Nature *441*, 1157–1161.

Rahmani, Z., Huh, K.W., Lasher, R., and Siddiqui, A. (2000). Hepatitis B virus X protein colocalizes to mitochondria with a human voltage-dependent anion channel, HVDAC3, and alters its transmembrane potential. J Virol *74*, 2840–2846.

Rozzi, S.J., Avdoshina, V., Fields, J.A., and Mocchetti, I. (2018). Human immunodeficiency virus Tat impairs mitochondrial fission in neurons. Cell Death Discov *4*, 8.

Rozzi, S.J., Avdoshina, V., Fields, J.A., Trejo, M., Ton, H.T., Ahern, G.P., and Mocchetti, I. (2017). Human immunodeficiency virus promotes mitochondrial toxicity. Neurotox Res *32*, 723–733.

Samant, S.A., Zhang, H.J., Hong, Z., Pillai, V.B., Sundaresan, N.R., Wolfgeher, D., Archer, S.L., Chan, D.C., and Gupta, M.P. (2014). SIRT3 deacetylates and activates OPA1 to regulate mitochondrial dynamics during stress. Mol Cell Biol *34*, 807–819.

Santel, A., and Fuller, M.T. (2001). Control of mitochondrial morphology by a human mitofusin. J Cell Sci *114*, 867–874.

Sarraf, S.A., Raman, M., Guarani-Pereira, V., Sowa, M.E., Huttlin, E.L., Gygi, S.P., and Harper, J.W. (2013). Landscape of the PARKIN-dependent ubiquitylome in response to mitochondrial depolarization. Nature *496*, 372–376.

Seeger, C., and Mason, W.S. (2000). Hepatitis B virus biology. Microbiol Mol Biol Rev *64*, 51–68.

Seth, R.B., Sun, L., Ea, C.K., and Chen, Z.J. (2005). Identification and characterization of MAVS, a mitochondrial antiviral signaling protein that activates NF-kappaB and IRF 3. Cell *122*, 669–682.

Shi, C.S., Qi, H.Y., Boularan, C., Huang, N.N., Abu-Asab, M., Shelhamer, J.H., and Kehrl, J.H. (2014). SARS-Coronavirus open reading frame-9b suppresses innate immunity by targeting mitochondria and the MAVS/TRAF3/ TRAF6 signalosome. J Immunol *193*, 3080–3089.

Shitara, H., Kaneda, H., Sato, A., Inoue, K., Ogura, A., Yonekawa, H., and Hayashi, J.I. (2000). Selective and continuous elimination of mitochondria microinjected into mouse eggs from spermatids, but not from liver cells, occurs throughout embryogenesis. Genetics *156*, 1277–1284.

Sin, J., McIntyre, L., Stotland, A., Feuer, R., and Gottlieb, R.A. (2017). Coxsackievirus B escapes the infected cell in ejected mitophagosomes. J Virol *91*, e01347–1417.

Sinzger, C., and Jahn, G. (1996). Human cytomegalovirus cell tropism and pathogenesis. Intervirology *39*, 302–319.

Sir, D., Tian, Y., Chen, W.L., Ann, D.K., Yen, T.S., and Ou, J.H. (2010). The early autophagic pathway is activated by hepatitis B virus and required for viral DNA replication. Proc Natl Acad Sci U S A *107*, 4383–4388.

Sliter, D.A., Martinez, J., Hao, L., Chen, X., Sun, N., Fischer, T. D., Burman, J.L., Li, Y., Zhang, Z., Narendra, D.P., *et al.* (2018). Parkin and PINK1 mitigate STING-induced inflammation. Nature *561*, 258–262.

Stavru, F., Palmer, A.E., Wang, C., Youle, R.J., and Cossart, P. (2013). Atypical mitochondrial fission upon bacterial infection. Proc Natl Acad Sci U S A *110*, 16003–16008.

Strappazzon, F., Nazio, F., Corrado, M., Cianfanelli, V., Romagnoli, A., Fimia, G.M., Campello, S., Nardacci, R., Piacentini, M., Campanella, M., *et al.* (2014). AMBRA1 is able to induce mitophagy via LC3 binding, regardless of PARKIN and p62/SQSTM1. Cell Death Differ *22*, 419–432.

Sun, L., Wu, J., Du, F., Chen, X., and Chen, Z.J. (2013). Cyclic GMP-AMP synthase is a cytosolic DNA sensor that activates the type I interferon pathway. Science *339*, 786–791.

Taguchi, N., Ishihara, N., Jofuku, A., Oka, T., and Mihara, K. (2007). Mitotic phosphorylation of dynamin-related GTPase Drp1 participates in mitochondrial fission. J Biol Chem *282*, 11521–11529.

Tang, H., Da, L., Mao, Y., Li, Y., Li, D., Xu, Z., Li, F., Wang, Y., Tiollais, P., Li, T., *et al.* (2009). Hepatitis B virus X protein sensitizes cells to starvation-induced autophagy via up-regulation of beclin 1 expression. Hepatology *49*, 60–71.

Teodorof-Diedrich, C., and Spector, S.A. (2018). Human immunodeficiency virus type-1 gp120 and tat induce mitochondrial fragmentation and incomplete mitophagy in human neurons. J Virol *92*, e00993–18.

Twig, G., Hyde, B., and Shirihai, O.S. (2008). Mitochondrial fusion, fission and autophagy as a quality control axis: the bioenergetic view. Biochim Biophys Acta *1777*, 1092–1097.

Twig, G., and Shirihai, O.S. (2011). The interplay between mitochondrial dynamics and mitophagy. Antioxid Redox Signal *14*, 1939–1951.

Varga, Z.T., Grant, A., Manicassamy, B., and Palese, P. (2012). Influenza virus protein PB1-F2 inhibits the induction of type I interferon by binding to MAVS and decreasing mitochondrial membrane potential. J Virol *86*, 8359–8366.

Wang, X., Winter, D., Ashrafi, G., Schlehe, J., Wong, Y.L., Selkoe, D., Rice, S., Steen, J., LaVoie, M.J., and Schwarz, T.L. (2011). PINK1 and Parkin target Miro for phosphorylation and degradation to arrest mitochondrial motility. Cell *147*, 893–906.

Waris, G., Huh, K.W., and Siddiqui, A. (2001). Mitochondrially associated hepatitis B virus X protein constitutively activates transcription factors STAT-3 and NF-kappa B via oxidative stress. Mol Cell Biol *21*, 7721–7730.

Wasiak, S., Zunino, R., and McBride, H.M. (2007). Bax/Bak promote sumoylation of DRP1 and its stable association with mitochondria during apoptotic cell death. J Cell Biol *177*, 439–450.

West, A.P., Shadel, G.S., and Ghosh, S. (2011). Mitochondria in innate immune responses. Nat Rev Immunol *11*, 389–402.

Wu, W., Tian, W., Hu, Z., Chen, G., Huang, L., Li, W., Zhang, X., Xue, P., Zhou, C., Liu, L., *et al.* (2014). ULK1

translocates to mitochondria and phosphorylates FUNDC1 to regulate mitophagy. EMBO Rep *15*, 566–575.

Xia, M., Gonzalez, P., Li, C., Meng, G., Jiang, A., Wang, H., Gao, Q., Debatin, K.M., Beltinger, C., and Wei, J. (2014a). Mitophagy enhances oncolytic measles virus replication by mitigating DDX58/RIG-I-like receptor signaling. J Virol *88*, 5152–5164.

Xia, M., Meng, G., Jiang, A., Chen, A., Dahlhaus, M., Gonzalez, P., Beltinger, C., and Wei, J. (2014b). Mitophagy switches cell death from apoptosis to necrosis in NSCLC cells treated with oncolytic measles virus. Oncotarget *5*, 3907–3918.

Yan, Y., Xin, A., Liu, Q., Huang, H., Shao, Z., Zang, Y., Chen, L., Sun, Y., and Gao, H. (2015). Induction of ROS generation and NF-kappaB activation in MARC-145 cells by a novel porcine reproductive and respiratory syndrome virus in Southwest of China isolate. BMC Vet Res *11*, 232.

Yasukawa, K., Oshiumi, H., Takeda, M., Ishihara, N., Yanagi, Y., Seya, T., Kawabata, S., and Koshiba, T. (2009). Mitofusin 2 inhibits mitochondrial antiviral signaling. Sci Signal *2*, ra47.

Yonashiro, R., Ishido, S., Kyo, S., Fukuda, T., Goto, E., Matsuki, Y., Ohmura-Hoshino, M., Sada, K., Hotta, H., Yamamura, H., *et al.* (2006). A novel mitochondrial ubiquitin ligase plays a critical role in mitochondrial dynamics. EMBO J *25*, 3618–3626.

Yoneyama, M., Kikuchi, M., Natsukawa, T., Shinobu, N., Imaizumi, T., Miyagishi, M., Taira, K., Akira, S., and Fujita, T. (2004). The RNA helicase RIG-I has an essential function in double-stranded RNA-induced innate antiviral responses. Nat Immunol *5*, 730–737.

Yoshizumi, T., Ichinohe, T., Sasaki, O., Otera, H., Kawabata, S., Mihara, K., and Koshiba, T. (2014). Influenza A virus protein PB1-F2 translocates into mitochondria via Tom40 channels and impairs innate immunity. Nat Commun *5*, 4713.

Young, L.S., and Rickinson, A.B. (2004). Epstein-Barr virus: 40 years on. Nat Rev Cancer *4*, 757–768.

Yu, C.Y., Liang, J.J., Li, J.K., Lee, Y.L., Chang, B.L., Su, C.I., Huang, W.J., Lai, M.M., and Lin, Y.L. (2015). Dengue virus impairs mitochondrial fusion by cleaving mitofusins. PLoS Pathogens *11*, e1005350.

Yu, T., Jhun, B.S., and Yoon, Y. (2011). High-glucose stimulation increases reactive oxygen species production through the calcium and mitogen-activated protein kinase-mediated activation of mitochondrial fission. Antioxid Redox Signal *14*, 425–437.

Zayyad, Z., and Spudich, S. (2015). Neuropathogenesis of HIV: from initial neuroinvasion to HIV-Associated Neurocognitive Disorder (HAND). Curr HIV/AIDS Rep *12*, 16–24.

Zevini, A., Olagnier, D., and Hiscott, J. (2017). Crosstalk between cytoplasmic RIG-I and STING sensing pathways. Trends Immunol *38*, 194–205.

Zhong, L., Shu, W., Dai, W., Gao, B., and Xiong, S. (2017). Reactive oxygen species-mediated c-Jun NH2-terminal kinase activation contributes to hepatitis B virus X protein-induced autophagy via regulation of the beclin-1/Bcl-2 interaction. J Virol *91*, e00001–17.

Zhu, L., Mou, C., Yang, X., Lin, J., and Yang, Q. (2016). Mitophagy in TGEV infection counteracts oxidative stress and apoptosis. Oncotarget *7*, 27122–27141.

Zunino, R., Schauss, A., Rippstein, P., Andrade-Navarro, M., and McBride, H.M. (2007). The SUMO protease SENP5 is required to maintain mitochondrial morphology and function. J Cell Sci *120*, 1178–1188.

35 Mitochondrial Dysfunction, Immune Systems, Their Diseases, and Possible Treatments

Elise Jacquin
INSERM UMR1231 Lipides, Nutrition, Cancer, Université Bourgogne Franche-Comté, Dijon, France

Eric Hervouet
Université Bourgogne Franche-Comté, INSERM, EFS BFC, UMR1098, Interactions Hôte-Greffon-Tumeur/Ingénierie Cellulaire et Génique, Besançon, France
DImaCell Platform, Université, Bourgogne Franche-Comté, Besançon, France
EPIGENEXP Platform, Université, Bourgogne Franche-Comté, Besançon, France

Michaël, Boyer-Guittaut
Université Bourgogne Franche-Comté, INSERM, EFS BFC, UMR1098, Interactions Hôte-Greffon-Tumeur/Ingénierie Cellulaire et Génique, Besançon, France
DImaCell Platform, Université, Bourgogne Franche-Comté, Besançon, France

CONTENTS

Abbreviations

AMPK	AMP-activated protein kinase
AD	Alzheimer's disease
CAR T-cells	chimeric antigen receptor T-cells
cGAS	cyclic GMP-AMP synthase
cGAMP	cyclic GMP-AMP
CL	cardiolipin
DAMPs	damage-associated molecular patterns
DC	dendritic cell
ETC	Electron Transport Chain
HIF1α	hypoxia-inducible factor-1 α
HUVEC	human umbilical vascular endothelial cells
IFN	interferon
IRF	interferon regulatory factors
LPS	lipopolysaccharides
MAVS	mitochondrial antiviral signaling
MDVs	mitochondria-derived vesicles
Mfn1/2	Mitofusin1/2
mitAP	mitochondrial antigen presentation
MMP	mitochondrial membrane potential
mtDNA	mitochondrial DNA
mTOR	mammalian Target of Rapamycin
mtROS	mitochondrial ROS
MPTP	1-méthyl-4-phényl-1,2,3,6-tétrahydropyridine
NAC	N-acetylcysteine
NAD$^+$	nicotinamide adenine dinucleotide
NLR	NOD-like receptors
NLRP3	NOD-, LRR- and pyrin domain-containing protein 3
OXPHOS	oxidative phosphorylation
PAMPs	pathogen-associated molecular patterns
PD	Parkinson's disease
PGC-1α	PPARγ co-activator 1 α
PRR	pattern recognition receptor
RIG-I	retinoic acid-inducible gene I
RLR	retinoic acid-inducible gene I-like receptor
ROS	reactive oxygen species
STING	stimulator of interferon genes
TCA	tricarboxylic acid cycle
TFAM	transcription factor A, mitochondrial
TILs	Tumor infiltrating lymphocytes
TLR9	Toll-like receptor 9.

1 INTRODUCTION

Mitochondria biogenesis and metabolism have been extensively studied for decades in the context of pathologies because of their ability to produce energy rich adenosine triphosphate (ATP) through oxidative phosphorylation (OXPHOS), to generate signalling intermediates or toxic molecules, such as reactive oxygen species (ROS), and to liberate factors involved in programmed cell death and apoptosis (for a review, see[1]). All these studies led to the conclusion that mitochondria are positioned at the crossroad of cell survival, proliferation and cell death.

Mitochondria are dynamic double-membrane organelles which possess their own genome encoding 13 proteins of the electron transport chain (ETC), 2 ribosomal proteins and 22 transfer RNAs. The mitochondria network is highly dynamic and undergoes fusion and fission and regulating its metabolic activities. Due to its essential metabolic roles, mitochondrial dysfunction has been linked to the development of numerous diseases such as cancer, neurodegenerative diseases, myopathies, cardiovascular diseases, or autoimmune diseases. More recent data have described mitochondria as main actors of the regulation of the immune system and the activation of inflammation, which could explain its importance in the etiology of these different diseases; but the question remains whether the role of mitochondria in these diseases is indeed linked to the regulation of the immune system.

Inflammation corresponds to the response of the organisms to infection or tissue damage leading to the recruitment of immune system effectors from the blood circulation to the site of infection or injury in order to eliminate the danger and/or stress. But, if inflammation is not properly controlled following infection clearance or damage repair, chronic inflammation can lead to loss of tissue homeostasis and even tissue degradation (for a review, see[2]). The process of inflammation is closely linked to the activation of the innate immune system which is the first line of defense against foreign pathogens (bacteria, virus or harmful chemical or physical agents) or stress. Cellular innate immunity is linked to the activity of pattern recognition receptors (PRRs) which recognize pathogen-derived molecules (*e.g.* proteins, LPS or nucleic acids) called pathogen-associated molecular patterns (PAMPs). Amongst this family of receptors, we can cite Toll-like receptors (TLRs) and retinoic acid-inducible gene and I-like receptors (RLRs).[3] Activation of these receptors following the binding or their ligands will induce intracellular signalling cascades leading to the activation of the NFκB transcription factor and interferon regulatory factors (IRFs) and induction of their target genes, mostly pro-inflammatory cytokines necessary to resolve the infection.[4]

2 PHYSIOLOGICAL ROLE FOR MITOCHONDRIAL METABOLISM IN IMMUNE CELL ACTIVATION

Mitochondria can act as a platform for molecules involved in signalling pathways regulating as an example the viral or bacterial infection. Following infection, cytosolic receptors (Toll-like receptors, TLR, retinoic-acid-inducible gene-I, RIG-I, or Melanoma Differentiation-Associated protein 5, MDA5) detect PAMPs and in particular, viral RNA or bacterial DNA acting as antigens. The activated cytosolic receptors then form a complex with their adaptor, MAVS (mitochondrial antiviral signalling protein), which is located at the outer membrane of the mitochondria.[5] Once activated, this complex then induces the transcription of type I interferon (IFNα/β) through the nuclear translocation of NFκB and IRFs which subsequently activate innate antigen response. Interestingly, oxidative stress and mitochondrial ROS (mtROS) production can activate MAVS and downstream NFκB and IRFs signalling pathways, therefore contributing to innate immunity, even in the absence of viral infection.[6]

Mitochondrial dynamics (fission and fusion events) have been shown to play important roles in immune cell responses to infection. Indeed, activation of MAVS following viral infection promotes mitochondrial elongation through Mitofusin-1 (Mfn1). This change in mitochondrial dynamics in turn favors interaction between activated MAVS and the endoplasmic reticulum protein STING (stimulator of interferon genes), thus inducing type I IFN and pro-inflammatory cytokine production.[7] Deletion of Drp1 (a protein involved in mitochondria fission) and aberrant mitochondria elongation in macrophages has been shown to lead to NRLP3 activation, caspase 1 cleavage as well as IL-1β secretion and aberrant inflammation.[8] Furthermore, Mitofusin-2 (Mfn2), involved in mitochondria fusion, has been shown to interact with NRLP3 and drive inflammasome formation and secretion of mature IL-1β following viral infection.[9] Mitochondria metabolism is also regulated during immune cell stimulation. Activated M1 macrophages, as well as dendritic cells (DCs) present a decrease of oxidative phosphorylation in mitochondria together with an increase in glycolysis (for a review, see[10]). Upon activation with LPS, macrophages switch their metabolism from OXPHOS to glycolysis leading to the accumulation of tricarboxylic acid cycle (TCA) intermediates, such as succinate. This metabolite promotes the expression and secretion of the pro-inflammatory cytokine IL-1β by stabilizing the transcription factor HIF1α, thus enhancing the pro-inflammatory functions of M1 macrophages.[11]

Unlike DCs and macrophages, antigen-specific activation of T-cells along with co-stimulatory signals results in a concomitant increase in aerobic glycolysis and OXPHOS. Enhanced glycolysis provides the metabolites required for cell proliferation, mitochondrial metabolism and mtROS production which are essential and sufficient to support T-cell activation and IL-2 production.[12] T-cell activation then induces mitochondrial biogenesis leading to increased mitochondrial mass, area and mtDNA amount, events which are coherent with the important increase in cell size and proliferation.[13] Mitochondrial dynamics have also been shown to control the fate of T-cells after activation. Indeed, the reduced OXPHOS and high glycolysis observed in effector T-cells have been shown to result from the rounded and fragmented morphology of mitochondria whereas fusion events reprogram metabolism towards OXPHOS and FAO (Fatty acid oxidation) during the formation of memory T-cells which display better survival in stress conditions.[14] A recent study indicated that early CD28 co-stimulation of naive T-cells is essential to prime mitochondria remodeling and later generation of memory T-cells.[15] Mitochondrial morphology has also been linked to CD4 T-cell effector functions. Conditional deletion of the mitochondrial transcription factor TFAM in CD4 T cell, induces severe mitochondrial dysfunctions with impaired cristae organization and ETC function favoring the differentiation of CD4 T-cells towards the pro-inflammatory Th1 subset and exacerbating inflammatory responses.[16] Contrary to effector CD4 T-cells which depend on glycolysis, regulatory T-cells (Treg) rely on lipid oxidation and display higher mitochondrial potentially highlighting a high sensitivity to mitochondrial dysfunction.[17]

3 CONSEQUENCES OF MITOCHONDRIAL DAMAGE IN IMMUNE CELLS

As described above, the activation of innate immunity require the activation of PRRs in the context of viral infection. However, these receptors can also be activated without any pathogen infection. Indeed, an alteration of mitochondria metabolism and integrity can lead to the production, externalization or release of different mitochondrial factors (mtROS, mtDNA, ATP, cardiolipin, succinate, N-formyl peptides, TFAM) in the cytosol, at the cell surface or in the extracellular space which would act as mitochondrial damage-associated molecular patterns (mtDAMPs), also known as alarmins, and trigger several immune system pathways and inflammation (Figure 1).[18]

3.1 INFLAMMASOME

Inflammasomes are multi-protein complexes which can be activated by pathogen-derived PAMPs or released DAMPs following organelle disruption or cell death. mtROS production by mitochondria, translocation of the inner membrane lipid cardiolipin (CL) to the outer membrane of mitochondria due to mitochondrial membrane depolarization as well as a release of mtDNA have been described to recruit and activate the NOD-, LRR- and pyrin domain-containing protein 3 (NRLP3) inflammasome leading to the activation of Caspase 1 and the cleavage of the pro-inflammatory cytokines, pro-IL-1β and pro-IL-18 giving their secreted mature forms (Figure 1).[19]

3.2 TLR9

The PPR TLR9 recognizes hypomethylated CpG motifs from bacterial DNA. Upon stimulation, TLR9 translocates

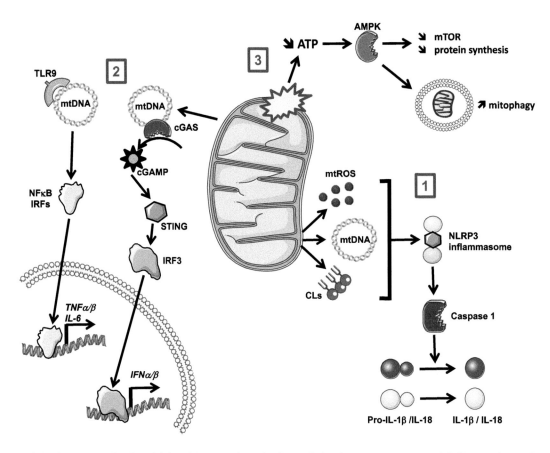

FIGURE 1 Links between mitochondrial DAMPs and activation of the immune system and inflammation. (1) Release of mtDAMPs (mtROS, mtDNA, CLs) from dysfunctional mitochondria leads to the activation of the NRLP3 inflammasome, the activation of Caspase 1 and the processing of pro-IL-1b/IL-18 into mature pro-inflammatory IL-1b and IL-18 cytokines. (2) mtDNA is also able to activate the TLR9/NFkB/IRFs pathway and the production of TNFa/b and IL-6 cytokines but also to enhance the cGAS/cGAMP/STING signaling cascade to lead to the production of Type I interferon IFNa/b. (3) Damage to mitochondria induces a decrease of the ATP/AMP ratio which activates the AMPK kinase, the inhibition of mTOR and the induction of mitophagy.

AMPK: AMP-activated protein kinase; cGAS: cyclic GMP-AMP synthase; cGAMP: cyclic GMP-AMP; CL: cardiolipin; DAMPs: damage-associated molecular patterns; IFN: interferon; IRF: interferon regulatory factors; mtDNA: mitochondrial DNA; mTOR: mammalian Target of Rapamycin; mtROS: mitochondrial ROS; NLRP3: NOD-, LRR- and pyrin domain-containing protein 3; STING: stimulator of interferon genes; TLR9: Toll-like receptor 9.

to endosomes where it activates NFκB signaling pathway and the production of the pro-inflammatory cytokines, TNFα and IL-6. The endosymbiotic origin of mitochondria, which are thought to be derived from a former α-probacterium and phagocytosed by cells to form a new symbiotic entity,[20] has led to postulate that mtDNA, which displays frequent unmethylated CpG repeats, can be recognized by TLR9 receptor like bacterial DNA. Indeed, several studies have shown that mtDNA release in the cytosol after mitochondrial damage or even circulating mtDNA can trigger inflammatory response through TLR9 activation (for a review, see[21]). A series of studies have demonstrated that mtDNA can also activate the cGAS (cyclic GMP-AMP synthase)-STING (stimulator of interferon genes) pathway of cytosolic DNA sensing.[22] mtDNA released from endogenous mitochondria can bind and

activates cGAS, which produces cGAMP (cyclic GMP-AMP). This cyclic dinucleotide then acts as a second messenger to activate STING, leading to the activation of the downstream transcription factor IRF3 (interferon regulatory factor 3) and the transcription of its target genes such as Type I interferon (IFNα/β).[23] Type I IFN binding to their receptor IFNAR can then activate the transcription of IFN-stimulated genes (interferon-α/β receptor) which are essential in the context of viral infection (Figure 1).

3.3 CARDIOLIPIN

mitochondrial-specific phospholipids, cardiolipins (CLs), can also be found in the extracellular compartment, following the fragmentation of dying cells and the release of mitochondria fragments. There, CLs can act as "eat me" signals and stimulate mitochondrion engulfment by phago-cytes; CD36 receptor further activate the immune system (for a review, see[24]). Once antigen-presenting cells (APCs), such as dendritic cells and macrophages, are activated by CLs, they can activate T- and B-cells to induce adaptive immune response and the production of anti-CL anti-bodies which can lead to the development of an auto-immune disease called anti-phospholipid syndrome (APS). A recent work has shown that a viral infection inducing mitochondria damage and externalization of CLs, com-bined with a defect in mitophagy, can lead to an auto-immune response and the development of autoimmune disease.[25]

3.4 N-FORMYL PEPTIDES

Mitochondria, which are supposedly derived from pro-teobacteria, share some similarities with these organisms, one being the fact that mitochondria protein translation starts with N-formyl-methionine leading to the synthesis of N-formyl-peptides (NFP). These peptides, if released in the extracellular space can be recognized by neutro-phils, monocytes, and DCs, thanks to their G-protein-coupled formyl-peptide receptor 1 (FPR1) which can lead to calcium mobilization, cell migration and innate immune response activation.[26]

3.5 ATP AND TCA INTERMEDIATES AND THE REGULATION OF IMMUNE SYSTEM

During the last one decade, it has been extensively described that mitochondria products, such as ATP as well as intermediates of the TCA cycle, such as succinate or cit-rate, are able to regulate inflammation, innate immunity as well as adaptive immunity. For example, mitochondrial damage results in a decrease in ATP production but to an increase in AMP levels leading to the activation of the energy sensor, the AMPK kinase which, in response, modulates several signaling pathways. AMPK especially inhibits mTOR and protein synthesis but also activates autophagy/mitophagy to regulate the elimination of dam-aged mitochondria (Figure 1).[27] Regarding TCA inter-mediates, it has been shown that a break in the Krebs cycle at the succinate dehydrogenase level, the complex II of the ETC, leads to an accumulation of succinate which is known to present a pro-inflammatory activity and limit the production of anti-inflammatory cytokines such as IL-10 (for a review, see[28]).

Below we will now describe how the alteration of mitochondrial activity and biogenesis can regulate the innate and adaptive immune system and lead to the development of different pathologies from autoimmune diseases to neurodegeneration. We will then discuss whether mitochondria function might be considered as a future strategy to treat these numerous pathologies.

4 MITOCHONDRIAL DYSFUNCTION, IMMUNE SYSTEM AND PATHOLOGIES

As discussed above, mitochondria are located at the crossroad of life and death and participate in many cel-lular signaling pathways. Therefore, maintaining their homeostasis is crucial and any alteration of their activity or of their metabolism would lead to drastic cell changes and disruption of many physiological functions. Alter-ation of mitochondrial activity may have numerous causes but the activation of the immune system is often linked to the release of DAMPs, one of them being mtDNA from cells containing dysfunctional mitochon-dria. It is also well admitted that the ETC is responsible for most of the production of intracellular ROS and that these molecules, which levels are directly correlated to mitochondria metabolism, play a major role as signaling intermediates in the regulation of mitochondria-driven inflammation. Immune system is indeed particularly sen-sitive to changes in mitochondrial activity leading to inflammation, dysregulation of innate and adaptive immunity and, later on, to the development of immune-linked pathologies.

4.1 MITOCHONDRIAL DYSFUNCTION, IMMUNE SYSTEM AND NEURODEGENERATION/AGING:

Neurodegenerative diseases have become a major health problem linked to the increase of life expectancy world-wide. The main problem of these diseases is that patients have often reached advanced, non-reversible stages when diagnosed. The incoming challenges for researchers are therefore to characterize the different causes initiating and driving these pathologies in order to select new tar-gets to improve early diagnosis and treatments.

Most of these pathologies, such as Alzheimer's disease (AD) or Parkinson's disease (PD), are often linked to the intra- or extracellular accumulation of protein aggre-gates leading to neuronal death and early symptoms. But, the initial causes of these diseases are still contro-versial and are not clearly understood. The main reason for this lack of information is that these triggers are probably happening years before the detection of the first symptoms. Genetic causes have been hypothesized in some cases but other factors are clearly involved. During the last decades, metabolism and, especially mitochondria have been described to be involved in neuronal death. These two hypotheses may also be linked since we described before that mtDAMPs are fac-tors which can activate the innate immune system and inflammation. In the brain, mitochondria dysfunction would lead to the release of mtDAMPs and therefore to the activation of neuroinflammation. This process is defined as the activation of macrophages cells present in the microglia and astrocytes but also as an increase in cytokine levels sometimes associated with the disruption of the brain-blood barrier leading to a recruitment of peripheral immune cells such as lymphocytes.

In Alzheimer's disease, the first fact proposing a role for inflammation in the development of the disease came from epidemiological studies suggesting that the long-term use of non-steroidal anti-inflammatory drugs could lead to a decreased risk of developing AD.[29] Microglia and activated astrocytes are cell populations which have been closely associated to the presence of amyloid plaques in AD and the expression of IL-1 by microglia cells is increased sixfold in AD and correlates with plaques.[30] Other cytokines produced by macrophages or other immune cells, such as IL-6, TNFα, IL-8 or TGFβ, have also been described to be increased in AD.[31] Even if inflammation played a role during AD, it has to be demonstrated that this activation was linked to mitochondrial dysfunction and mtDAMPs released in the CNS. In several postmortem analysis, it has been described that, in AD brains, the number of intact mitochondria was decreased, that mitophagy was increased and that mutations and deletions of mtDNA were also increased.[32] Mitochondrial dysfunction could then lead to the liberation of mtDAMPs, as described above, and the activation of neuroinflammation. For example, it has been shown that the injection of whole mitochondria lysates into the brain of rodents increased amyloid precursor protein (APP) as well as Aβ$_{1-42}$. Similar injections into the hippocampus of rodents also led to the activation of pro-inflammatory phenotypes illustrated by an increase in *TNFα* mRNA levels.[33] These data therefore suggest that mitochondrial dysfunction and the release of mtDAMPs may lead to the activation of neuroinflammation and favor the progression of AD symptoms.

In Parkinson's disease, early evidence of the involvement of neuroinflammation came from data showing activated microglia in the *substantia nigra* detected by imagery studies (PET) analyzing benzodiazepine sites, a marker of activated microglial cells.[34] Postmortem studies also showed increased levels of TNFα, IL-1β and IFNγ in PD brains.[35] Like AD, PD risk is diminished following long-term use of non-steroidal anti-inflammatory drugs. Indeed, aspirin use is correlated to decreased PD risk but the risk is even further decreased with non-aspirin-linked drugs or for women who used aspirin in a chronic manner compared to men.[36] Studies have shown that neuroinflammation can be linked to PD but peripheral inflammation may play a role too. Indeed, it has been shown that CD8+/CD4+ ratio was increased in the blood of PD patients as well as the ratio of IFNγ+/IL-4+ T-cells.[37] This perturbation may then affect the brain since the brain-blood barrier has been described to be altered in PD patients. Other studies also showed that CD8+ and CD4+ T-cell reactivity is increased in the *substantia nigra* of PD patients and that MPTP (1-methyl 4-phenyl 1,2,3,6-tetrahydro pyridine), a compound used to mimic PD in rodent models, led to an increase in

T-cell infiltration in the *substantia nigra*.[38] It was also shown that dopaminergic cell death was reduced when MPTP was used in a T-cell immunodeficient model (*Rag1*$^{-/-}$ and *Tcrb*$^{-/-}$ mice) lacking mature T-cells compared to wild type mice, illustrating the potential role of T cells in this pathology. It is noteworthy that this decrease was still observed when CD4+ T-cells, but not CD8+ T-cells, were reinjected suggesting that dopaminergic cell death in this PD model is mainly linked to CD4+ T-cells. PD initiation/evolution has been extensively linked to mitochondrial dysfunction even if the initial molecular steps of the disease are still controversial. Nevertheless, ETC-targeting drugs, such as MPTP, rotenone or paraquat which inhibit complex I activity, can trigger PD symptoms or pathways since they are commonly used to mimic PD in rodent models. It has also been shown that mtDNA mutations or deletions were observed in PD patients confirming that mitochondrial defects are important for the development of the pathology. It is therefore now admitted that mitochondrial dysfunction as well as neuroinflammation play an important role during PD etiology but there is no consensus whether one is a consequence or a cause of the other one or if they are even related. Recent data seem to support this hypothesis since they described that mitochondria play a role in antigen presentation *via* MHC class I via the intervention of PD-linked proteins. Indeed, in 2016, Matheoud and collaborators described that Mitochondrial Antigen Presentation (MitAP) on MHC class I at the cell surface did not require autophagy but used a different pathway involving the formation of mitochondrial-derived vesicles (MDVs) requiring the intervention of Rab9 and Sortin nexin 9 (Snx9). Interestingly, Parkin and PINK1, two effectors of mitophagy involved in the etiology of PD, inhibited the recruitment of Rab9 and Snx9 to the mitochondria and therefore repressed the formation of MDVs and MitAP in both macrophages and dendritic cells.[39] These data established a potential link between the development of PD, mitochondrial dysfunction and the activation of autoimmune pathways. An even more recent study published in *Nature* in 2018 established a clear link between Parkin, PINK1 and STING-mediated inflammation and its consequence on the development of PD.[40] Indeed, in this study, it was shown that deletion of *Parkin* or *Pink1* gene led to a strong inflammatory phenotype characterized by high levels of IL-6, IFNβ1, IL-12, IL-13, CCL2 and CCL4 in the serum of these knock-out mice after exhaustive exercise (acute mitochondrial stress). The same phenotype was observed in the *Parkin*$^{-/-}$/*mutator* mice which accumulated mtDNA mutations (expression of a proofreading defective mtDNA polymerase, *PolG*, inducing a chronic mitochondrial stress). The proinflammatory phenotype linked to exercise or mtDNA mutation accumulation was completely reversed when STING, a regulator of the IFN type I pathway in response to cytosolic DNA, was deleted. The deletion of

STING also diminished the loss of dopaminergic neurons in the *substantia nigra* of *Parkin⁻/⁻/mutator* aging mice, suggesting that inflammation played a role in the development of PD. This phenotype was also observed in humans carrying a mono-or biallelic mutation in *PRKN* since they presented elevated pro-inflammatory cytokine levels (IL-6, IL-1β, CCL2 and CCL4) compared to control serums. These data therefore support a role of PINK1-Parkin-mediated mitochondrion degradation in the down-regulation of neuroinflammation and subsequent neuronal cell death and demonstrated that mtDNA is the key inflammatory trigger in absence of *Parkin*. Therefore, Parkin and PINK1 may prevent the liberation of cytosolic and circulating mtDNA by degrading dysfunctional mitochondria by mitophagy and consequently inhibit neuroinflammation and neurodegeneration. These data are supported by another study which described that dopaminergic neurons of the *substantia nigra* present higher basal mitophagy levels than those observed in other dopamine-producing neurons.[41]

4.2 MITOCHONDRIAL DYSFUNCTION, IMMUNE SYSTEM AND CANCER:

A growing body of evidence indicates that mitochondrial dysfunction influences cancer development by modulating tumor cell properties but also by affecting immune cells of the tumor microenvironment. This later effect can be indirect since mitochondrial dysfunctions occurring in tumor cells can indirectly impact immune cells. Cancer treatment often leads to tumor cell stress and/or death leading to the release of DAMPs in the tumor microenvironment. Immune cells such as DCs, macrophages or natural killer cells can recognize these DAMPs through PRRs expressed at their surface which triggers their activation. In response, activated immune cells can directly kill tumor cells or stimulate adaptive immune cells with cytotoxic activity thus limiting tumor progression.[42] As described in a previous section, recognition of mtDNA can also trigger downstream signaling pathways leading to the production of pro-inflammatory cytokines (for review see[21]), thus creating an inflammatory tumor microenvironment which can favor tumor progression (for a review, see[43]).

While it is not clear whether changes in mitochondrial dynamics are a cause or a consequence of the process, it is well known that tumor cells rewire their energy metabolism towards glycolysis and increase their glucose uptake. This phenomenon, known as the Warburg effect, is thought to provide tumor cells with survival and proliferative advantages.[44] This switch in tumor cell metabolism also impact immune cells from the tumor microenvironment. For instance, tumor-derived lactic acid generating by glycolysis triggers the polarization of tumor associated macrophages (TAM). Indeed, lactic acid acts as a signalling molecule which induces a M2-phenotype in TAMS in an HIF1-α -dependent manner, characterized by immunosuppressive and tumor-promoting functions.[45]

The elevated consumption of glucose by tumor cells creates a competition with tumor infiltrating lymphocytes (TILs) for glucose availability. The restriction in glucose imposed by tumor cells has been shown to alter mTOR signaling and glycolytic metabolism in T cells and to impair their production of IFN-γ, a cytokine with antitumor properties.[46] This suggests that tumor-imposed restriction in glucose in the tumor microenvironment alters TIL metabolism and effector functions, thus promoting tumor progression.

Mitochondrial dysfunction appears to intrinsically play critical roles in the ability of TILs to exert antitumor functions. In an elegant study, Scharping et al.[47] showed that chronic activation and tumor microenvironment-specific signals led to a loss of mitochondrial mass and functions in CD8+ TILs. While mitochondrial defects in CD8+ TILS were associated with upregulated levels of the co-inhibitory molecules PD-1, known as a major regulator of T cell exhaustion, PD-1 blockade was insufficient to restore mitochondrial functions in TILs. Instead, mitochondrial dysfunction in tumor-specific CD8+ TILs was rather due to a loss of PPARγ co-activator 1α (PGC1α), a regulator of mitochondrial biogenesis, driven by chronic AKT signalling.[47] In a follow-up study, the same research group demonstrated the importance of the costimulatory molecule 4-1BB/CD137, highly expressed on exhausted T cells, in the mitochondrial activity and metabolic functions of CD8+ TILs. The authors showed that 4-1BB stimulation resulted in p38-MAPK-dependent upregulation of PGC1α and increased mitochondrial functions allowing for metabolic sufficiency of CD8+ TILS.[48] Accordingly, 4-1BB stimulation was shown to increase mitochondrial mass and function, probably due to OPA-1 (optic atrophy-1)-dependent mitochondrial fusion mechanisms.[49] Thus, the loss of 4-1BB co-stimulation in immunosuppressive tumor microenvironment might be, at least partly, responsible for mitochondrial dysfunctions in TILs leading to impaired antitumor capacities.

A recent study focusing on human clear renal cell carcinoma (ccRCC) showed that CD8+ TILs isolated from ccRCC patients display an exhausted phenotype as well as impaired activation and proliferation capacities. Phenotypic analyses of these CD8+ TILs revealed metabolic defects including mitochondrial fragmentation associated with increased mtROS production. Even though the cytotoxic activity of these CD8+ TILs was not assessed in the study, neutralization of mtROS partially restored CD8+ TIL activation capacity, supporting the critical role of mitochondrial function in T-cell antitumor properties.[50]

4.3 Mitochondrial Dysfunction, Immune System and Auto-Immune Disease

Oxidative stress has been described to play a major role in auto-immune rheumatic diseases such as rheumatoid arthritis (RA) (for a review, see[51]). Oxidative stress appears when an excess of ROS (reactive oxygen species) and RNS (reactive nitrogen species) cannot be eliminated by the antioxidant response linked to superoxide dismutase, catalase, glutathione peroxidase and peroxiredoxin activities. It is known that the major producer of ROS species is the ETC and that dysfunctional mitochondria would to an excess of ROS. Regarding immune system regulation, ROS are necessary for antiviral responses but, when produced in excess, they lead to the oxidation of proteins or lipids creating neo-antigens. These neo-antigens being recognized as "non-self" antigens, this generates autoimmune responses and production of auto-antibodies. Therefore, it has been proposed that the detection of these auto-antibodies could be used in the future for early diagnosis.

As shown above, activation of the adaptive immune system and production of auto-antibodies is a major cause of auto-immune disease development. One of the auto-antibodies linked to auto-immune diseases is the anti-cardiolipin (aCL) directed against this mitochondrial phospholipid released in the extracellular compartment from dying cells presenting fragmented mitochondria. A clinical study performed in 2016 has linked the production of anti-phospholipid (aPL) antibodies with systemic lupus erythematosus (SLE).[52] Among the 317 SLE patients included in the cohort, 117 (37%) presented a positive aPL profile (detection of anti-cardiolipin and/or anti- β2Glycoprotein I antibodies associated with a positive lupus anticoagulant test). These patients were more likely to develop deep venous thrombosis, pulmonary embolism, cardiac valvular disease, cognitive dysfunction and antiphospholipid syndrome (APS). This study revealed that the SLE patients with aPL antibodies had a higher risk to develop APS than controls. Furthermore, mitochondrial dysfunction plays a central role in the phenotype of SLE patients. Indeed, in T lymphocytes of SLE patients, mitochondrial dysfunction and increased ROS production have been shown to trigger MAVS activation and Type I IFN production, associated with active disease(6).

4.4 Mitochondrial Dysfunction, Immune System and Inflammatory Bowel Disease (IBD)

Inflammatory bowel disease (IBD) is a family of diseases comprising different pathologies such as ulcerative colitis (UC) or Crohn's disease (CD). Their main feature is an activation of the immune system leading to chronic inflammation targeting the intestine. In a recent publication, Boyapati and collaborators emitted the idea that mtDNA would be released during the course of the disease and would contribute to the activation of the pro-inflammatory immune response.[53] To confirm their hypothesis, they measured the levels of circulating mtDNA in plasma samples collected from a cohort of 67 UC, 30 CD, and 40 controls as well as in plasma of a mouse model of colitis induced with DSS (dextran sulfate-sodium). They also used mass spectrometry to detect a second mitochondrial DAMP, mitochondria-derived formyl peptides, and EM to detect mitochondria damage and TRL9 expression, the target of mtDNA, in human intestinal IBD mucosa. Their results indeed confirmed that plasma circulating mtDNA levels were increased in both UC and CD compared to healthy controls but they also detected elevated fecal mtDNA in patients. In the colitis-induced mouse model, they also detected higher circulating mtDNA. At last, they observed damaged mitochondria in inflamed UC mucosa and elevated TLR9 levels in immune inflammatory cells in UC and CD compared to controls. These data confirmed that mtDNA is released during IBD and could therefore serve as a circulating biomarker or a therapeutic target to regulate the activation of the TLR9 pro-inflammatory pathway.

4.5 Mitochondrial Dysfunction, Immune System, and Pre-Eclampsia

Pre-eclampsia, which affects about 8 million pregnancies worldwide every year, has been linked to a dysfunction of endothelial cell proliferation and vascular development.[54] The causes of pre-eclampsia are still largely unknown but this pathology has been linked to inflammation and the release of pro-inflammatory cytokines leading to endothelial dysfunction. A recent study has shown that circulating mtDNA levels as well as TLR9 expression are increased in patients presenting pre-eclampsia and that mtDNA and ROS may play a role in this process.[55] It is hypothesized that mtDNA released in the maternal blood can interact with TLR9 in dendritic cells, activate the NFkB pathway, and induce the production of the pro-inflammatory cytokine IL-6.[56] A symptom of pre-eclampsia is characterized by hypertension which has also been described to be induced by inflammation or oxidative stress. Indeed, macrophages and T-cells can accumulate in the vascular system of *in vivo* models of hypertension and oxidative stress is known to induce endothelial cell dysfunction. An *in vitro* study has shown that incubation of endothelial cells HUVECs with plasma from women who suffered from pre-eclampsia led to an induction of mtROS production and an increase in *TLR9* mRNA in HUVECs. This increase was abolished by an antioxidant, demonstrating the link between mitochondrial dysfunction, mtROS production and TLR9 pathway.[57] Another study analyzed the circulating immune cell profiles of women suffering from pre-eclampsia and demonstrated an increase in the number of immune cells, and especially lymphocytes and monocytes, in these patients.[58] Recently, pre-eclampsia has been linked to anti-phospholipids (aPL) antibodies, some of them being anti-cardiolipin (aCL), and therefore APS.[59] For this study, the authors collected samples from 148 women who delivered early (≤ 36 weeks) and 148 controls and these samples were tested for aPL detection. The results showed that 11.5% of patients displayed aPL compared to only 1.4% of controls demonstrating that women with early delivery are more likely to present APS.

4.6 MITOCHONDRIAL DYSFUNCTION, IMMUNE SYSTEM AND GULF WAR ILLNESS (GWI)

Gulf War Illness (GWI) is a chronic disease affecting a high percentage of veterans of the first Gulf war of 1991. Symptoms are multiple such as fatigue, depression or memory deficits but the causes are still unknown.[60] One hypothesis is that these veterans have been subjected to environmental and chemical toxins. Even if the triggers are not known, mitochondrial dysfunction is one well documented hypothesis. Indeed, GWI-linked chemicals (pesticides, pollutants from burning oil fields ...) have been described to increase ROS production and oxidative stress which alter mitochondrial activity and therefore further increase ROS production and oxidative stress. For example, it has been shown that administration of the antioxidant coenzyme Q10 can improve mitochondrial activity and ameliorate some GWI symptoms.[61] More interestingly, studies also documented changes in the circulating cytokine profiles of these veterans with an increase in pro-inflammatory cytokines (IFNγ, TNFα, and IL-6) correlated with an increase in NKκB signalling. So, the hypothesis would be that chemicals inhaled during the war would have induced an alteration of mitochondrial activity followed by a dysregulation of the immune system and a chronic inflammation.[62]

4.7 MITOCHONDRIAL DYSFUNCTION, IMMUNE SYSTEM, AND TYPE II DIABETES MELLITUS

Type II diabetes mellitus (T2DM) has been characterized as an autoimmune disease since uncontrolled immune responses lead to pancreatic β cell dysfunction and insulin resistance with pro-inflammatory IL-1β inducing pancreatic cell apoptosis.[63] This chronic inflammation has been supposed to be linked to mitochondrial dysfunction, ROS production and NRLP3 inflammasome activation. A study has also analyzed the mononuclear cells of 27 T2DM patients and controls and the data obtained showed that diabetic mitochondria were smaller and more spherical and that mitochondrial mass was lower in T2DM patients compared to controls.[64] Similar results were obtained in another study comparing 10 samples of lymphocytes from T2DM to 10 controls. This work

showed that mitochondrial membrane potential (MMP) and ROS were increased in diabetic lymphocytes compared to controls.[65] This link between mitochondrial dysfunction and immune system deregulation has also been established in gestational diabetes mellitus (GDM), which has been associated with an increased risk of developing T2DM. Indeed, in GDM, it was observed an increase in mitochondria size, ROS production but a decrease of MMP in monocytes/macrophages of GDM and this observation was linked to increased IL-1β, TNFα and IL-6 levels in the serum of GDM pregnancies and decreased HDAC2 activity.[66] At last, a study described a link between altered mitochondrial metabolism and the activation of pro-inflammatory cytokine secretion. Indeed, in T2DM, it is known that incomplete β-oxidation of fatty acids in mitochondria leads to the accumulation of byproducts (medium- and long-chain acylcarnitines) in the serum of diabetic patients.[67] The use of L-C14 carnitine, a byproduct of incomplete FAO, induced PRR, JNK, and ERK signaling pathways leading to the expression and secretion of pro-inflammatory cytokines suggesting a link in T2DM between alteration of mitochondrial metabolism and chronic inflammation.

5 MITOPHAGY, IMMUNE SYSTEM, AND THE PREVENTION OF PATHOLOGIES

Autophagy is a cellular catabolic process, described for the first time in the 1960s by Christian de Duve.[68] This field attracted some important interest since the 1990s with the characterization of this mechanism in yeast and the discovery of the homolog proteins in humans leading to the attribution of the Nobel Prize to Yoshinori Ohsumi in 2016 for its contribution to the development of the autophagy field.[69] Autophagy is an intracellular degradation process leading to the formation of an intracellular double-membrane vesicle, called autophagosome, which engulfs part of the cytoplasm (soluble proteins, protein aggregates or even organelles) and then fuses with the lysosome to induce the degradation of this vesicle and its content. This process is highly regulated by more than 40 ATG (AuTophaGy-related) proteins and can be induced by numerous stress conditions such as starvation, hypoxia, oxidative stress, chemical drugs ? For years, this process has been described as non-selective and the complement of the proteasome since it can degrade aggregates or organelles while proteasome only proteolyze soluble proteins. But, during the last decade, a selective function of autophagy has emerged. Indeed, several teams have described specific adapters (P62/SQSTM1, NBR1, OPTN, NDP52, NIX, BNIP3/3L) which interact with specific cargo (the aggregate or organelle to degrade) and a member of the ATG8 family (LC3 or GABARAP family) to induce the specific recruitment of the cargo into the autophagosome leading to its degradation. This selective process has been named according to the cargo degraded: aggrephagy, xeno-phagy, ribophagy, pexophagy, nucleophagy, glycophagy, mitophagy.

As described before, mitochondria play an essential role in the regulation of the immune system and mito-chondrial dysfunction seems to alter this regulation and play a role in the origin or development of different dis-eases. Mitophagy, as a selective autophagy targeting the degradation of damaged mitochondria could therefore regulate the immune system and inflammation.

5.1 MITOPHAGY AND ACTIVATION OF IMMUNE SYSTEM CELLS

Natural killer (NK) cells belong to the innate immune response but possess features of adaptive immune cells such as clonal expansion, contraction and the generation of long-lived memory cells. O'Sullivan and collaborators showed that during viral infection, NK accumulated damaged mitochondria and that ROS-linked mitophagy induction was necessary during the contraction phase to allow for their survival.[70] Moreover, the use of mTOR inhibitors or AMPK inducers enhanced BNIP3- and BNIP3L-induced mitophagy and the generation of NK memory cells, demonstrating the importance of mito-phagy in the establishment of NK cell memory. Macro-phages are essential players of both immune and adaptive immunity and their activation is essential to fully activate the immune system via the presentation of antigens to T-cells. They can also activate the adaptive immune system by secreting pro-inflammatory cytokines such as TNFα. It has been shown that the induction of macrophages by TNFα, but not by IL-1β or IFN-γ, induced mitophagy, confirmed by the observation of a decrease in mitochondrial proteins by mass spectrom-etry, and the proteolysis and presentation of self-mitochondrial antigens (MitAg) at the cell surface by MHC class I.[71] In 2016, Corrado and collaborators have shown a link between autophagy and apoptosis in the regulation of immune tolerance.[72] Indeed, they described that, following induction of T-cell receptor signaling, the activation of protein kinase A inhibited autophagy-mediated degradation of damaged mitochondria leading to an increase in cytochrome c release and the subse-quent induction of apoptosis. Therefore, to induce AICD (activation-induced cell death) through T-cell receptor signaling, autophagy should be inhibited allow-ing for an accumulation of damaged mitochondria and apoptosis.

5.2 MITOPHAGY DYSFUNCTION AND THE DEVELOPMENT OF IMMUNE SYSTEM-LINKED DISEASES

Numerous studies have clearly established a link between altered autophagy and the development of diseases but few recent projects have clearly established whether overall autophagy inhibition or specific mitophagy inhib-ition would be the cause of these diseases. As described earlier, multiple inflammatory stimuli can lead to the activation of the NRLP3 inflammasome and the matur-ation of pro-IL-1α/β and pro-IL-18 in active pro-inflammatory cytokines. But, it has also been demon-strated that these cytokines can lead to the hyperactiva-tion of pro-inflammatory responses and the development of pathologies if secreted in an uncontrolled manner (for a review, see[73]). It has been shown, for example, that an inhibition of autophagy in macrophages or dendritic cells would lead to the accumulation of damaged mito-chondria and increased ROS production driving NRLP3 inflammasome activation and hypersecretion of the IL-1 family cytokines.[74] A more recent study also demon-strated that macrophages from Parkin-/- mice presented increased IL-1β secretion in response to NRLP3 acti-vators which were linked to alteration of mitochondria integrity, PARKIN-dependent ubiquitination of mito-chondria and P62/SQSTM1-induced mitophagy.[75] Alter-ation of autophagy/mitophagy and the accumulation of dysfunctional mitochondria has also been linked to the secretion of Type I IFN. Indeed, it has been shown that Atg7-/- embryonic fibroblasts presented an hypersecre-tion of IFNα/β following dsDNA induction and that this effect was due to dysfunctional mitochondria and ROS production and the activation of the RIG-I-like receptor (RLR) signalling pathway.[76] Release of mtDNA from damaged mitochondria has also been linked to the secre-tion of type I IFN and activation of the immune system. This effect is linked to the activation of the cGAS/STING/IRF3-dependent DNA sensing pathway.[23] There-fore, it is clear that inhibition of autophagy, or the more selective mitophagy, would lead to accumulation of dam-aged mitochondria as well as ROS production and release of mtDNA which in turn will alter the cytokine secretion profile and the immune response.

In 2016, researchers have designed mice with a specific deletion of the Atg7 gene in the myeloid lin-eage (Atg7 cKO) to address the link between inhibition of autophagy and the apparition of type 2 diabetes (T2D).[77] Atg7 cKO mice develop diabetes together with inflammatory cytokine expression and inflammasome activation in adipose tissue and an increase of IL-1β secretion from macrophages induced by LPS. It was also observed an increase in ROS production in macrophages from Atg7 cKO mice suggesting the accumulation of dys-functional mitochondria. These results suggest that autophagy/mitophagy is necessary to degrade damaged mitochondria and prevent ROS production as well as inflammasome activation during type 2 diabetes. Proteins linked to mitophagy have also been linked to the regula-tion of inflammation and its link with cancer progres-sion. Indeed, a recent study has shown that PINK1 and PARK2, two proteins linked to mitophagy and PD, sup-pressed pancreatic tumor initiation via the control of mitochondria iron-linked immunometabolism. In mouse models, depletion of these two proteins increased Kras-driven tumor initiation. PINK1-PARK2 induced the specific degradation of SLC25A37 and SLC25A28, two mitochondrial iron transporters, leading to iron accumulation, HIF1α-linked induction of glycolysis and AIM2 (absent in melanoma 2)-dependent inflammasome activation in tumor cells. AIM2 also induced HMGB1 release from cancer cells and expression of PD-L1. In patients with pancreatic cancers, low PARK2 expression but high SLC25A37 and AIM2 expression were linked to poor prognosis. Therefore, alteration of mitochondrial iron metabolism correlated to increased inflammation may contribute to pancreatic cancer development and progression.[78]

Chronic inflammation has been described in numerous cases as a trigger to induce more aggressive pathologies. It is for example the case in colitis-associated cancer (CAC) in which inflammasome activity is altered leading to continuous inflammation. A new small molecule, Andrographolide (Andro), has been described to inhibit NRLP3-inflammasome activity in macrophages and the release of pro-inflammatory cytokine IL-1β, and to attenuate colitis and tumor growth.[79] More interestingly, Andro has been shown to induce mitophagy in macrophages and therefore to inhibit the accumulation of damaged mitochondria presenting mitochondrial membrane potential collapse. This effect was confirmed to be mitophagy-specific since *Becn1* knockdown as well as autophagy inhibitors inhibited this effect. This result demonstrated that Andro inhibited NRLP3-inflammasome activity by degrading damaged mitochondria and then leading to the attenuation of CAC.

Rheumatoid arthritis (RA) has been linked to the invasion of apoptosis-resistant fibroblast-like synoviocytes (FLSs) in bone and cartilage. Kim and collaborators studied the link between IL-17 and Th17, producers of IL-17, and FLS resistance to apoptosis and the link with mitochondrial dysfunction, autophagy and apopstosis.[80] Their data showed that Th17 and IL-17 induced mitochondria dysfunction and then autophagy in FLS of RA patients. This induction of autophagy would lead to the degradation of damaged mitochondria and then inhibition of apoptosis leading to FLS resistance to apoptosis and their proliferation in joints.

As described before, the development of T2DM has been linked to increased uncontrolled inflammation as well as metabolic modifications and especially, mitochondrial dysfunction. A recent study investigated whether levels of mitophagy would be different in metabolically healthy obese (MHO), metabolically non-healthy obese (MHNO) and metabolically abnormal diabetic obese (MADO) patients, data which could explain the apparition of mitochondrial dysfunction as well as inflammation.[81] In this study, they indeed observed a decrease in mitophagy and TEM analyses revealed an increase in damaged mitochondria in the MADO group compared to the data obtained from the MHO or MHNO group (20 patients each). Therefore, a constant level of mitophagy may be a factor of good prognosis for MHO patients and prevent the development of T2DM in these patients.

6 THERAPEUTICS TARGETING THE IMMUNE SYSTEM THROUGH MITOCHONDRIA ACTIVITY

6.1 TARGETING MITOCHONDRIAL METABOLISM TO REGULATE THE IMMUNE SYSTEM AND INFLAMMATION

As described above, mitochondria metabolism evolves during immune cell differentiation or activation. Therefore, activating or inhibiting mitochondrial functions would allow for the reprogramming of some immune cells. One example would be the reprogramming of M1 pro-inflammatory macrophages towards M2 anti-inflammatory macrophages in case of inflammatory auto-immune diseases. One example was described in the treatment of T2DM. In this pathology, one of the current treatments is the use of Metformin, an inhibitor of complex I of the ETC, which decreases ROS production and activates AMPK due to cellular ATP decrease. More interestingly, a recent study showed that Metformin would also change the cytokine profile of macrophages induced by LPS. Indeed, treatment with Metformin, as well as Rotenone, another complex I inhibitor, led to the inhibition of pro-inflammatory IL-1β secretion while increasing the production of the anti-inflammatory IL-10 cytokine therefore reprogramming macrophages towards an anti-inflammatory phenotype (Figure 2A.1).[82] This second effect of Metformin could therefore be useful to diminish inflammation during T2DM.

Other studies have shown that increasing skeletal muscle mitochondrial metabolism and functions may prevent or treat T2DM.[83] For example, many studies described the positive effect of exercise on the improvement of mitochondrial function and consequently T2DM in humans. This effect would be linked to the expression of genes involved in mitochondria biogenesis, PGC-1α, as well as fusion, Mfn1/2 and Opa1. Numerous clinical trials did, or are currently, targeting the way to restore mitochondrial function *via* exercise in T2DM (for examples, see *ClinicalTrials.gov*, Identifier: NCT00382473 or NCT02977442) (Figure 2A.2). Others described that calorie restriction but also sleep, meaning circadian rhythms, and cold exposure might regulate mitochondrial function and therefore intervene in T2DM. One study for example described that a calorie restriction of 25% of energy level requirements in humans led to increased PGC-1α and TFAM expression levels and mtDNA content in skeletal muscle.[84]

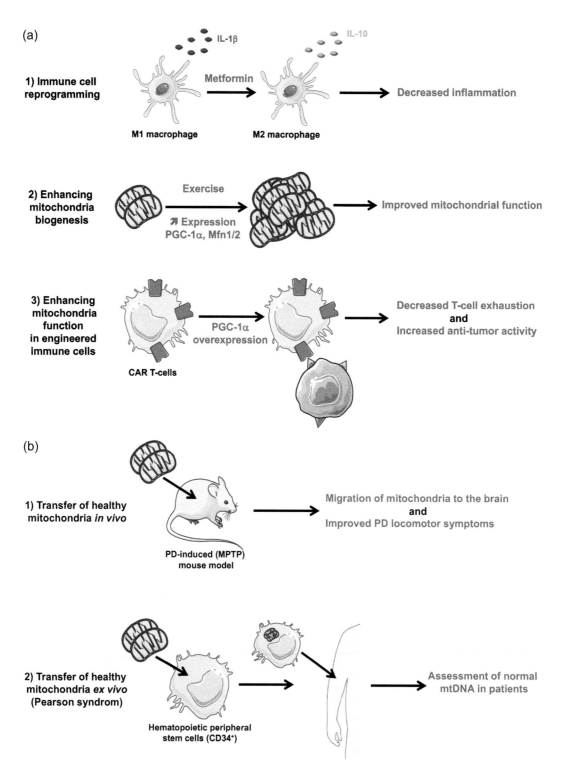

FIGURE 2 New mitochondria-derived therapeutics developed to regulate immune system, inflammation and disease progression. CAR T-cells: chimeric antigen receptor T-cells; Mfn1/2: Mitofusin1/2; MPTP: 1-méthyl-4-phényl-1,2,3,6-tétrahydropyridine; PD: Parkinson's disease; PGC-1a: PPARg co-activator 1a.

In a recent study, Escribano-Lopez and collaborators have tested the effect of the mitochondrial antioxidant MitoQ[85] on the regulation of oxidative stress and leuko-cyte-endothelium interaction in 98 T2DM and 71 control patients. They observed that mitochondrial ROS were higher in cells from T2DM patients and that ROS levels were decreased with MitoQ. TNFα and NFkB-p65 levels were also higher in T2DM but also decreased upon the addition of MitoQ. These data suggested that MitoQ had anti-oxidant and anti-inflammatory effects in T2DM patients through the inhibition of ROS and TNFα production.

6.2 Inducing Mitochondria Biogenesis to Counteract Mitochondrial Dysfunction

Inherited or sporadic mitochondrial diseases linked to mutations or deletions in mitochondrial DNA lead to the impairment of the ETC and decreased energy production. Mitochondrial diseases are considered as rare diseases but the symptoms develop early and there are currently no cures. One way to treat these young patients would be to activate mitochondria biogenesis in order to increase the overall number of mitochondria in the body. Previous studies have shown that Nicotinamide Riboside (NR), a functional B3 vitamin and a NAD^+ natural precursor, might be able to activate mitochondrial biogenesis, *via* the PGC-1α factor, and therefore increase energy production. It has, for example, been shown in a mouse model of mitochondrial myopathy that administration of NR led to the delay of disease progression and that this effect was dependent on mitochondrial biogenesis in skeletal muscle and adipose tissue.[86] One clinical trial posted in 2018 (NCT03432871) will target the potential of this compound to increase mitochondrial biogenesis (respiratory chain enzyme analysis and mtDNA quantification) in patients with Progressive External Ophthalmoplegia (PEO), a condition linked to a single deletion of mtDNA. Interestingly, this compound has also be shown to modulate NRLP3 inflammasome and to reduce inflammation in a model of T2DM in rodents.[87] In this study, the authors showed that administration of NR rescued mitochondrial integrity in the liver of obese and diabetic mice, inhibited NRLP3 and Caspase 1 activation and decreased pro-inflammatory cytokines levels (TNFα and IL-1) in hepatic cells.

6.3 Links between Cancer Immunotherapies and Mitochondrial Dysfunction

During the last one decade, immunotherapies emerged as a new line of treatment of liquid as well as solid tumors. Amongst these protocols, we can cite antibodies against cancer-related membrane receptors (e.g., anti-HER2) or growth factors (e.g., anti-VEGF), antibodies against checkpoint inhibitors (e.g., anti-PDL1 or PD1) or the development of T-cells expressing chimeric antigen receptors (CAR T-cells). But, several studies also demonstrated that the efficiency of these new treatments can be linked to the metabolic activity of the targeted cells. It has been shown, for example, that expression of PD-1 on CD8 T-cells, a marker of T-cell exhaustion, inhibited glycolysis and mitochondria metabolic pathways and down-regulated PGC-1α expression leading to

diminished mitochondria biogenesis.[88] They also showed that overexpressing PGC-1α in T-cells transferred to mice carrying tumors counteracted metabolic alterations with increased mitochondria number as well as enhanced anti-tumor T-cell activity which led to the idea that improving mitochondria metabolism might decrease CD8 T-cell exhaustion and could be used in combination with anti-checkpoint immunotherapies (Figure 2A.3).

The efficiency of treatments using adoptive T-cell transfer in case of metastatic cancers has also been shown to be linked to the metabolic status of the transferred T-cells. Indeed, highly glycolytic T-cells showed shorter life expectancy and lower anti-tumor activity in the patient than the highly respiratory ones which presented longer persistence *in vivo*.[89] Therefore, sorting T-cells according to their metabolic activity before transfer could improve treatment efficiency in term of anti-tumor activity as well as duration. Furthermore, Buck and collaborators recently showed that pharmacological enforcement of mitochondrial fusion improved the anti-tumor functions of CD8 T-cells upon adoptive transfer.[14] These data suggest a potential interest in targeting mitochondria in the context of cellular immunotherapy of cancer. In line with this, 4-1BB co-stimulation appears like an interesting strategy to enhance CD8 T cell-mediated antitumor responses through mitochondrial function improvement in the context of both T cell adoptive transfer and PD-1 blockade.[48]

A recent study also showed that the design of CAR-T cells might also regulate the T-cell metabolic pathways and their function and anti-tumor activity when transferred to the patient. Indeed, the authors observed that the choice of the co-induction domains expressed at the surface of the CAR-T cell modified different metabolic pathways.[90] The expression of 4-1BB enhanced CD8 memory T-cells persistence presenting increased OXPHOS and mitochondria biogenesis while the expression of CD28 reprogrammed T-cells towards a glycolytic phenotype. These data will have to be taken into account when designing future CAR-T cells to ensure their survival, persistence and anti-tumor activity in patients.

6.4 A New Line of Treatment: The In Vitro or In Vivo Transfer of Healthy Mitochondria

A recent study has shown that transferring healthy functional mitochondria to treat mitochondria-derived diseases such as PD could be a future line of treatment.[91] In this study, they purified mitochondria from human hepatocytes and demonstrated that they could enter neuronal cells *in vitro* and could be administered *in vivo* in an MPTP-induced PD mouse model. The authors showed that purified mitochondria could increase the survival rate of SH5Y-SY cells treated with an inhibitor of complex I, MPP+, but also an increase in respiratory complex I activity and ATP production together with a decrease in ROS. *In vivo*, injected mitochondria could migrate in the animal and reach all organs including the brain even after a systemic injection. They also showed improvement in PD locomotor symptoms in a MPTP-induced mouse PD model demonstrating that this new therapeutic protocol might of interest in the future for the treatment of diseases linked to dysfunctional mitochondria, such as auto-immune diseases (Figure 2B.1).

These data open avenues for the development of new treatments using the transfer of healthy mitochondria into target cells *in vitro* or *in vivo* in order to increase their function, activity and survival. This new protocol would be particularly interesting when using methods requiring the transfer of immune cells. This has already been tested in a protocol considering the transfer of healthy mitochondria to treat Non-Inherited Mitochondrial disorders, such as the Pearson Syndrome. This syndrome is linked to a deletion of mtDNA leading to a deficit in the function of the ETC. The symptoms are refractory sideroblastic anemia and vacuolization of bone marrow precursors. In this new line of treatment, autologous hematopoietic peripheral stem cells (CD34$^+$) were enriched with normal and healthy mitochondria obtained from healthy donor blood cells before their reinjection into the patient (*Clinicaltrials.gov*, NCT03384420). The aim of this study was to assess the level of normal mitochondrial DNA present in the patient CD34$^+$ cells after enrichment with normal mitochondria (Figure 2B.2).

Another therapeutic approach to reduce mutations in the mtDNA and restore ETC activity in mitochondrial diseases would be to use MitoTALENs.[92] In a recent study, two MitoTALENs have been designed to target two common point mutations linked to mitochondrial diseases (myoclonic epilepsy with ragged red fibers, MERRF and MELAS/Leigh syndrome). Using cybrid cell models, they showed that the transfection of MitoTALENs efficiently eliminated the deficient mitochondria carrying the pathogenic mtDNA in these cells but not the healthy ones leading to recovered oxidative phosphorylation activity as well as respiratory capacity. Regarding the use of the CRISPR/Cas9 technology to edit the mitochondrial genome, a lot of controversy remains since it is not clear whether an efficient endogenous nucleic acid import system into mammalian mitochondria exists. But, a study in 2015 described that the expression of a mitochondria-addressed Cas9 enzyme (mitoCas9) together with sgRNAs targeting the mitochondrial genes *Cox1* and *3* led to the specific cleavage of the targeted sequences in mtDNA proving the feasibility of the concept.[93] Nevertheless, this strategy will have to be improved before consideration for mtDNA editing in order to correct mitochondrial mutations linked to mitochondria-linked genetic diseases.

6.5 MITOCHONDRIAL DAMPs USED AS MARKERS OF DIAGNOSIS OR TREATMENT EFFICIENCY

As shown above, alteration of mitochondrial structure would lead to the release of several factors, classified as mtDAMPs and the activation the immune system. Therefore, if these molecules can be detected in the vascular system and specifically linked to pathologies, they could be used as diagnostic markers or treatment efficiency markers. One of them, mtDNA, is particularly interesting since it can be found in blood and can be easily detected thanks to the recent development of sensitive molecular biology techniques.

For example, a recent study quantified the levels of free mtDNA in the cerebrospinal fluid (CSF) of patients with multiple sclerosis.[94] The authors showed that mtDNA concentration was increased in the cerebrospinal fluid of patients compared to controls and that there was an inverse correlation between the diagnosis of the disease and the concentration of mtDNA suggesting that mtDNA could be involved in the early inflammatory steps of the disease and therefore used as an early diagnostic marker. In AD and PD, mtDNA has been proposed to serve as a biomarker. Reduced levels of circulating mtDNA have been shown in the CSF of patients (cohort of 282 patients) presenting a genetically increased AD risk but no symptoms and clinically symptomatic patients compared to controls or patients with frontotemporal lobar degeneration.[95] However, these data have to be analyzed carefully, since it has been previously shown that another AD biomarker, Aβ, is increased in the brain but reduced in the CSF. Another study showed that patients with vitiligo (56 patients and 46 controls) presented a higher copy number of circulating mtDNA compared to controls and that degradation products of mtDNA (8-hydroxy-2-deoxyguanosine; 8-OHdG) were increased in these patients supposing a role of oxidative DNA damage in the development of vitiligo.[96]

The quantification of mtDAMPs, and more particularly mtDNA, to characterize the risk of developing different diseases or to follow the efficiency of a treatment is a new growing area as shown by the high number of clinical trials currently opened (see *Clinicaltrials.gov*). Levels of mtDNA are for example quantified as biomarker in Chronic Lymphocytic Leukemia (NCT03595969), Irritable Bowel Syndrome (IBS) (NCT01028898), Sepsis (NCT02019992), Intra-amniotic infection (NCT03306719) or blood mtDNA damage in AD (NCT03340571). The quantification of mtDNA can also be analyzed during the course of treatment in order to determine whether its levels or mutations could be linked to a risk of relapse. This has, for example, been done in patients with Acute Myeloid Leukemia (NCT01138319). Indeed, in this clinical trial, the goal was to determine whether mutations of mtDNA might be linked to the expansion of Acute Myeloid Leukemia and predict the outcome of the disease, remission or relapse.

7 CONCLUSIONS

In this chapter, we discussed how mitochondria function regulates immune system activity as well as inflammation. But, more importantly, we reviewed the current literature to show how mitochondria dysfunction might alter the immune system to lead to the development of various pathologies from autoimmune diseases to cancer and neurodegenerative diseases. The deregulation of the immune system is mainly linked to the release of mtDAMPs in the cytosol of immune cells, in the extracellular space or in the vascular system. These mitochondrial factors then lead to the activation of an acute or chronic pro-inflammatory phenotype which has been linked to several diseases. The question remains whether mitochondrial dysfunction, and the consequent activation of inflammation, is a trigger or a consequence of the initiation of these diseases. It seems clear that for some neurodegenerative diseases, such as PD, this event is an early step of the disease but for others, such as the immunomodulation of cancers, mitochondrial dysfunction and inflammation might only favor tumor progression. More interestingly, the mitochondria-linked factors and signaling pathways regulating immune cells and activating inflammation are fairly described and therefore may now be considered as future biomarkers or targets for the development of future therapeutics. Pre-clinical trials, as well as clinical trials, are already conducted to correlate mtDAMPs levels with disease progression or stages. Other trials are studying different ways to inhibit mitochondrial dysfunction, activate mitochondria biogenesis or transfer healthy mitochondria form healthy donors into patients to ameliorate mitochondria function and prevent inflammation. These data would then be essential to validate the rational of targeting mitochondria to treat inflammation-linked diseases. But, the future might be to associate a diagnosis based on circulating mtDAMPs levels associated to a cytokine profile and then adapt a specific mitochondria-targeted therapeutic for each patient, a protocol defined as Theranostics and proposed by many as the future of pathology treatment.

ACKNOWLEDGMENTS

EJ is supported by funding from the "Fondation pour la Recherche Médicale" (ARF20170938687). MBG and EH are supported by funding from institutional grants from INSERM, EFS and Univ. Bourgogne Franche-Comté and by the "Ligue Contre le Cancer" (001AC.2015) and "Région Bourgogne Franche-Comté" (2014C-15449).

REFERENCES

1. Mills, E. L., Kelly, B. & O'Neill, L. A. J. Mitochondria are the powerhouses of immunity. *Nat. Immunol.* **18**, 488–498 (2017).
2. Dela Cruz, C. S. & Kang, M.-J. Mitochondrial dysfunction and damage associated molecular patterns (DAMPs) in chronic inflammatory diseases. *Mitochondrion* **41**, 37–44 (2018).
3. Janeway, C. A. & Medzhitov, R. Innate immune recognition. *Annu. Rev. Immunol.* **20**, 197–216 (2002).
4. Iwasaki, A. & Medzhitov, R. Control of adaptive immunity by the innate immune system. *Nat. Immunol.* **16**, 343–353 (2015).
5. Zevini, A., Olagnier, D. & Hiscott, J. Crosstalk between cytoplasmic RIG-I and STING sensing pathways. *Trends Immunol.* **38**, 194–205 (2017).
6. Buskiewicz, I. A. et al. Reactive oxygen species induce virus-independent MAVS oligomerization in systemic lupus erythematosus. *Sci. Signal.* **9**, ra115 (2016).
7. Castanier, C., Garcin, D., Vazquez, A. & Arnoult, D. Mitochondrial dynamics regulate the RIG-I-like receptor antiviral pathway. *EMBO Rep* **11**, 133–138 (2010).
8. Park, S. et al. Defective mitochondrial fission augments NLRP3 inflammasome activation. *Sci. Rep.* **5**, 15489 (2015).
9. Ichinohe, T., Yamazaki, T., Koshiba, T. & Yanagi, Y. Mitochondrial protein mitofusin 2 is required for NLRP3 inflammasome activation after RNA virus infection. *Proc. Natl. Acad. Sci. U. S. A.* **110**, 17963–17968 (2013).
10. Kelly, B. & O'Neill, L. A. J. Metabolic reprogramming in macrophages and dendritic cells in innate immunity. *Cell. Res.* **25**, 771–784 (2015).
11. Tannahill, G. M. et al. Succinate is an inflammatory signal that induces IL-1β through HIF-1α. *Nature* **496**, 238–242 (2013).
12. Sena, L. A. et al. Mitochondria are required for antigen-specific T cell activation through reactive oxygen species signaling. *Immunity* **38**, 225–236 (2013).
13. Ron-Harel, N. et al. Mitochondrial biogenesis and proteome remodeling promote one-carbon metabolism for T cell activation. *Cell Metab.* **24**, 104–117 (2016).
14. Buck, M. D. et al. Mitochondrial dynamics controls T cell fate through metabolic programming. *Cell* **166**, 63–76 (2016).
15. Klein Geltink, R. I. et al. Mitochondrial priming by CD28. *Cell* **171**, 385-397.e11 (2017).
16. Baixauli, F. et al. Mitochondrial respiration controls lysosomal function during inflammatory T cell responses. *Cell. Metab.* **22**, 485–498 (2015).
17. Michalek, R. D. et al. Cutting edge: Distinct glycolytic and lipid oxidative metabolic programs are essential for effector and regulatory CD4+ T cell subsets. *J. Immunol. Baltim. Md* **186**, 3299–3303 (2011).
18. Chen, G. Y. & Nuñez, G. Sterile inflammation: Sensing and reacting to damage. *Nat. Rev. Immunol.* **10**, 826–837 (2010).
19. Elliott, E. I. & Sutterwala, F. S. Initiation and perpetuation of NLRP3 inflammasome activation and assembly. *Immunol. Rev.* **265**, 35–52 (2015).
20. Gray, M. W., Burger, G. & Lang, B. F. Mitochondrial evolution. *Science* **283**, 1476–1481 (1999).
21. Boyapati, R. K., Tamborska, A., Dorward, D. A. & Ho, G.-T. Advances in the understanding of mitochondrial DNA as a pathogenic factor in inflammatory diseases. *F1000Research* **6**, 169 (2017).
22. Barber, G. N. STING-dependent cytosolic DNA sensing pathways. *Trends Immunol.* **35**, 88–93 (2014).
23. West, A. P. et al. Mitochondrial DNA stress primes the antiviral innate immune response. *Nature* **520**, 553–557 (2015).
24. Maguire, J. J. et al. Known unknowns of cardiolipin signaling: The best is yet to come. *Biochim. Biophys. Acta* **1862**, 8–24 (2017).

25. Broder, A., Chan, J. J. & Putterman, C. Dendritic cells: An important link between antiphospholipid antibodies, endothelial dysfunction, and atherosclerosis in autoimmune and non-autoimmune diseases. *Clin. Immunol. Orlando Fla* **146**, 197–206 (2013).

26. Dahlgren, C., Gabl, M., Holdfeldt, A., Winther, M. & Forsman, H. Basic characteristics of the neutrophil receptors that recognize formylated peptides, a danger-associated molecular pattern generated by bacteria and mitochondria. *Biochem. Pharmacol.* **114**, 22–39 (2016).

27. Tan, T., Zimmermann, M. & Reichert, A. S. Controlling quality and amount of mitochondria by mitophagy: Insights into the role of ubiquitination and deubiquitination. *Biol. Chem.* **397**, 637–647 (2016).

28. Mills, E. & O'Neill, L. A. J. Succinate: A metabolic signal in inflammation. *Trends Cell Biol* **24**, 313–320 (2014).

29. In T' Veld, B. A. *et al.* Nonsteroidal antiinflammatory drugs and the risk of Alzheimer's disease. *N. Engl. J. Med.* **345**, 1515–1521 (2001).

30. Griffin, W. S. *et al.* Brain interleukin 1 and S-100 immunoreactivity are elevated in Down syndrome and Alzheimer disease. *Proc. Natl. Acad. Sci. U. S. A.* **86**, 7611–7615 (1989).

31. Akiyama, H. *et al.* Inflammation and Alzheimer's disease. *Neurobiol. Aging* **21**, 383–421 (2000).

32. Swerdlow, R. H. Mitochondria and cell bioenergetics: Increasingly recognized components and a possible etiologic cause of Alzheimer's disease. *Antioxid. Redox Signal.* **16**, 1434–1455 (2012).

33. Wilkins, H. M. *et al.* Extracellular mitochondria and mitochondrial components act as damage-associated molecular pattern molecules in the mouse brain. *J. Neuroimmune Pharmacol. Off. J. Soc. NeuroImmune Pharmacol.* **11**, 622–628 (2016).

34. Gerhard, A. *et al.* In vivo imaging of microglial activation with 11C.(R)-PK11195 PET in idiopathic Parkinson's disease. *Neurobiol. Dis.* **21**, 404–412 (2006).

35. Hunot, S. *et al.* FcepsilonRII/CD23 is expressed in Parkinson's disease and induces, in vitro, production of nitric oxide and tumor necrosis factor-alpha in glial cells. *J. Neurosci. Off. J. Soc. Neurosci.* **19**, 3440–3447 (1999).

36. Wahner, A. D., Bronstein, J. M., Bordelon, Y. M. & Ritz, B. Nonsteroidal anti-inflammatory drugs may protect against Parkinson disease. *Neurology* **69**, 1836–1842 (2007).

37. Baba, Y., Kuroiwa, A., Uitti, R. J., Wszolek, Z. K. & Yamada, T. Alterations of T-lymphocyte populations in Parkinson disease. *Parkinsonism Relat. Disord.* **11**, 493–498 (2005).

38. Brochard, V. *et al.* Infiltration of CD4+ lymphocytes into the brain contributes to neurodegeneration in a mouse model of Parkinson disease. *J. Clin. Invest.* **119**, 182–192 (2009).

39. Matheoud, D. *et al.* Parkinson's Disease-Related Proteins PINK1 and Parkin Repress Mitochondrial Antigen Presentation. *Cell* **166**, 314–327 (2016).

40. Sliter, D. A. *et al.* Parkin and PINK1 mitigate STING-induced inflammation. *Nature* **561**, 258–262 (2018).

41. Guzman, J. N. *et al.* Systemic isradipine treatment diminishes calcium-dependent mitochondrial oxidant stress. *J. Clin. Invest.* **128**, 2266–2280 (2018).

42. Fang, H. *et al.* TLR4 is essential for dendritic cell activation and anti-tumor T-cell response enhancement by DAMPs released from chemically stressed cancer cells. *Cell. Mol. Immunol.* **11**, 150–159 (2014).

43. Hernandez, C., Huebener, P. & Schwabe, R. F. Damage-associated molecular patterns in cancer: A double-edged sword. *Oncogene* **35**, 5931–5941 (2016).

44. Warburg, O., Wind, F. & Negelein, E. The metabolism of tumors in the body. *J. Gen. Physiol.* **8**, 519–530 (1927).

45. Colegio, O. R. *et al.* Functional polarization of tumour-associated macrophages by tumour-derived lactic acid. *Nature* **513**, 559–563 (2014).

46. Chang, C.-H. *et al.* Metabolic Competition in the Tumor Microenvironment Is a Driver of Cancer Progression. *Cell* **162**, 1229–1241 (2015).

47. Scharping, N. E. *et al.* The tumor microenvironment represses T cell mitochondrial biogenesis to drive intratumoral T cell metabolic insufficiency and dysfunction. *Immunity* **45**, 374–388 (2016).

48. Menk, A. V. *et al.* 4-1BB costimulation induces T cell mitochondrial function and biogenesis enabling cancer immunotherapeutic responses. *J. Exp. Med.* **215**, 1091–1100 (2018).

49. Teijeira, A. *et al.* Mitochondrial morphological and functional reprogramming following CD137 (4-1BB) costimulation. *Cancer Immunol. Res.* **6**, 798–811 (2018).

50. Siska, P. J. *et al.* Mitochondrial dysregulation and glycolytic insufficiency functionally impair CD8 T cells infiltrating human renal cell carcinoma. *JCI Insight* **2**, 93411 (2017).

51. Smallwood, M. J. *et al.* Oxidative stress in autoimmune rheumatic diseases. *Free Radic. Biol. Med.* **125**, 3–14 (2018). doi: 10.1016/j.freeradbiomed.2018.05.086.

52. Taraborelli, M. *et al.* The role of clinically significant antiphospholipid antibodies in systemic lupus erythematosus. *Reumatismo* **68**, 137–143 (2016).

53. Boyapati, R. K. *et al.* Mitochondrial DNA is a pro-inflammatory damage-associated molecular pattern released during active IBD. *Inflamm. Bowel Dis.* **24**, 2113–2122 (2018). doi: 10.1093/ibd/izy095.

54. McCarthy, C. M. & Kenny, L. C. Immunostimulatory role of mitochondrial DAMPs: Alarming for pre-eclampsia? *Am. J. Reprod. Immunol* **76**, 341–347 (2016).

55. Qiu, C., Hevner, K., Enquobahrie, D. A. & Williams, M. A. A case-control study of maternal blood mitochondrial DNA copy number and preeclampsia risk. *Int. J. Mol. Epidemiol. Genet.* **3**, 237–244 (2012).

56. Scharfe-Nugent, A. *et al.* TLR9 provokes inflammation in response to fetal DNA: Mechanism for fetal loss in preterm birth and preeclampsia. *J. Immunol. Baltim. Md* **188**, 5706–5712 (2012).

57. McCarthy, C. & Kenny, L. C. Therapeutically targeting mitochondrial redox signalling alleviates endothelial dysfunction in preeclampsia. *Sci. Rep.* **6**, 32683 (2016).

58. Brien, M.-E. *et al.* Distinct inflammatory profile in preeclampsia and postpartum preeclampsia reveal unique mechanisms. *Biol. Reprod.* **100**, 187–194 (2018). doi: 10.1093/biolre/ioy164.

59. Gibbins, K. J., Tebo, A. E., Nielsen, S. K. & Branch, D. W. Antiphospholipid antibodies in women with severe preeclampsia and placental insufficiency: A case-control study. *Lupus* **27**, 1903–1910 (2018). doi: 10.1177/0961203318787035.

60. White, R. F. *et al.* Recent research on Gulf War illness and other health problems in veterans of the 1991 Gulf War: Effects of toxicant exposures during deployment. *Cortex J. Devoted Study Nerv. Syst. Behav.* **74**, 449–475 (2016).

61. Golomb, B. A. *et al.* Coenzyme Q10 benefits symptoms in Gulf War veterans: Results of a randomized double-blind study. *Neural Comput* **26**, 2594–2651 (2014).

62. Broderick, G. *et al.* Altered immune pathway activity under exercise challenge in Gulf War Illness: An exploratory analysis. *Brain. Behav. Immun.* **28**, 159–169 (2013).

63. Donath, M. Y. & Shoelson, S. E. Type 2 diabetes as an inflammatory disease. *Nat. Rev. Immunol.* **11**, 98–107 (2011).

64. Widlansky, M. E. *et al.* Altered mitochondrial membrane potential, mass, and morphology in the mononuclear cells of humans with type 2 diabetes. *Transl. Res. J. Lab. Clin. Med.* **156**, 15–25 (2010).

65. Anthonsen, S., Larsen, J., Pedersen, P. L., Dalgaard, L. T. & Kvetny, J. Basal and T₃-induced ROS production in lymphocyte mitochondria is increased in type 2 diabetic patients. *Horm. Metab. Res. Horm. Stoffwechselforschung Horm. Metab.* **45**, 261–266 (2013).

66. Qu, X. *et al.* Association of downregulated HDAC 2 with the impaired mitochondrial function and cytokine secretion in the monocytes/macrophages from gestational diabetes mellitus patients. *Cell Biol. Int.* **40**, 642–651 (2016).

67. Rutkowsky, J. M. *et al.* Acylcarnitines activate proinflammatory signaling pathways. *Am. J. Physiol. Endocrinol. Metab.* **306**, E1378–1387 (2014).

68. Deter, R. L. & De Duve, C. Influence of glucagon, an inducer of cellular autophagy, on some physical properties of rat liver lysosomes. *J. Cell Biol.* **33**, 437–449 (1967).

69. Hornyak, T. Profile: Yoshinori Ohsumi: The rise and rise of a biology superstar. *Nature* **543**, S19 (2017).

70. O'Sullivan, T. E., Johnson, L. R., Kang, H. H. & Sun, J. C. BNIP3- and BNIP3L-mediated mitophagy promotes the generation of natural killer cell memory. *Immunity* **43**, 331–342 (2015).

71. Bell, C. *et al.* Quantitative proteomics reveals the induction of mitophagy in tumor necrosis factor-α-activated (TNFα) macrophages. *Mol. Cell. Proteomics* **12**, 2394–2407 (2013).

72. Corrado, M. *et al.* Macroautophagy inhibition maintains fragmented mitochondria to foster T cell receptor-dependent apoptosis. *EMBO J* **35**, 1793–1809 (2016).

73. Afonina, I. S., Müller, C., Martin, S. J. & Beyaert, R. Proteolytic processing of interleukin-1 family cytokines: Variations on a common theme. *Immunity* **42**, 991–1004 (2015).

74. Nakahira, K. *et al.* Autophagy proteins regulate innate immune responses by inhibiting the release of mitochondrial DNA mediated by the NALP3 inflammasome. *Nat. Immunol.* **12**, 222–230 (2011).

75. Zhong, Z. *et al.* NF-κB restricts inflammasome activation via elimination of damaged mitochondria. *Cell* **164**, 896–910 (2016).

76. Tal, M. C. *et al.* Absence of autophagy results in reactive oxygen species-dependent amplification of RLR signaling. *Proc. Natl. Acad. Sci. U. S. A.* **106**, 2770–2775 (2009).

77. Lee, H.-Y. *et al.* Autophagy deficiency in myeloid cells increases susceptibility to obesity-induced diabetes and experimental colitis. *Autophagy* **12**, 1390–1403 (2016).

78. Li, C. *et al.* PINK1 and PARK2 suppress pancreatic tumorigenesis through control of mitochondrial iron-mediated immunometabolism. *Dev. Cell* **46**, 441–455.e8 (2018).

79. Guo, W. *et al.* Small molecule-driven mitophagy-mediated NLRP3 inflammasome inhibition is responsible for the prevention of colitis-associated cancer. *Autophagy* **10**, 972–985 (2014).

80. Kim, E. K. *et al.* IL-17-mediated mitochondrial dysfunction impairs apoptosis in rheumatoid arthritis synovial fibroblasts through activation of autophagy. *Cell Death Dis.* **8**, e2565 (2017).

81. Bhansali, S., Bhansali, A. & Dhawan, V. Favourable metabolic profile sustains mitophagy and prevents metabolic abnormalities in metabolically healthy obese individuals. *Diabetol. Metab. Syndr.* **9**, 99 (2017).

82. Kelly, B., Tannahill, G. M., Murphy, M. P. & O'Neill, L. A. J. Metformin inhibits the production of reactive oxygen species from NADH:Ubiquinone oxidoreductase to limit induction of interleukin-1β (IL-1β) and boosts interleukin-10 (IL-10) in Lipopolysaccharide (LPS)-activated macrophages. *J. Biol. Chem.* **290**, 20348–20359 (2015).

83. Hesselink, M. K. C., Schrauwen-Hinderling, V. & Schrauwen, P. Skeletal muscle mitochondria as a target to prevent or treat type 2 diabetes mellitus. *Nat. Rev. Endocrinol.* **12**, 633–645 (2016).

84. Civitarese, A. E. *et al.* Calorie restriction increases muscle mitochondrial biogenesis in healthy humans. *PLoS Med.* **4**, e76 (2007).

85. Escribano-Lopez, I. *et al.* The mitochondria-targeted antioxidant MitoQ modulates oxidative stress, inflammation and leukocyte-endothelium interactions in leukocytes isolated from type 2 diabetic patients. *Redox Biol* **10**, 200–205 (2016).

86. Khan, N. A. *et al.* Effective treatment of mitochondrial myopathy by nicotinamide riboside, a vitamin B3. *EMBO Mol. Med.* **6**, 721–731 (2014).

87. Lee, H. J., Hong, Y.-S., Jun, W. & Yang, S. J. Nicotinamide riboside ameliorates hepatic metaflammation by modulating NLRP3 inflammasome in a rodent model of Type 2 Diabetes. *J. Med. Food* **18**, 1207–1213 (2015).

88. Bengsch, B. *et al.* Bioenergetic insufficiencies due to metabolic alterations regulated by the inhibitory receptor PD-1 are an early driver of CD8(+) T Cell exhaustion. *Immunity* **45**, 358–373 (2016).

89. Sukumar, M. *et al.* Mitochondrial membrane potential identifies cells with enhanced stemness for cellular therapy. *Cell Metab* **23**, 63–76 (2016).

90. Kawalekar, O. U. *et al.* Distinct signaling of coreceptors regulates specific metabolism pathways and impacts memory development in CAR T Cells. *Immunity* **44**, 380–390 (2016).

91. Shi, X., Zhao, M., Fu, C. & Fu, A. Intravenous administration of mitochondria for treating experimental Parkinson's disease. *Mitochondrion* **34**, 91–100 (2017).

92. Hashimoto, M. *et al.* MitoTALEN: A general approach to reduce mutant mtDNA loads and restore oxidative phosphorylation function in mitochondrial diseases. *Mol. Ther. J. Am. Soc. Gene Ther.* **23**, 1592–1599 (2015).

93. Jo, A. *et al.* Efficient mitochondrial genome editing by CRISPR/Cas9. *BioMed Res. Int.* **2015**, 305716 (2015).

94. Varhaug, K. N. *et al.* Increased levels of cell-free mitochondrial DNA in the cerebrospinal fluid of patients with multiple sclerosis. *Mitochondrion* **34**, 32–35 (2017).

95. Podlesniy, P. *et al.* Low cerebrospinal fluid concentration of mitochondrial DNA in preclinical Alzheimer disease. *Ann. Neurol.* **74**, 655–668 (2013).

96. Vaseghi, H., Houshmand, M. & Jadali, Z. Increased levels of mitochondrial DNA copy number in patients with vitiligo. *Clin. Exp. Dermatol.* **42**, 749–754 (2017).

36 Organophosphorus Compound-Induced Mitochondrial Disruption

Alan J. Hargreaves

Interdisciplinary Biomedical Research Centre, School of Science and Technology, Nottingham Trent University, Nottingham NG11 8NS, UK

CONTENTS

1 INTRODUCTION

Despite widespread concern about the risks to human health posed by organophosphate toxicity, OPs are used extensively worldwide in agriculture (e.g., as herbicides and insecticides) and in a range of other applications (e.g., as flame retardants and lubricants) (Abou-Donia and Lapadula, 1990; Chambers, 1992; Krieger, 2001; Abou-Donia, 2003; Gupta, 2006; Hargreaves, 2012; Costa, 2018).

As shown in Figure 1, the generalized chemical structure of an OP insecticide or lubricant consists of a pentavalent phosphorous atom linked by a double bond to an atom of oxygen or sulphur, in the case of organophosphates and organophosphorothioates, respectively. The phosphorous atom is also covalently linked to 3 other groups, two of which are usually referred to as "R" groups while the third is designated the "leaving group" (LG). These substituent groups can take a variety of forms, giving a significant degree of structural diversity, as illustrated in the specific examples shown in Figure 2.

Occupational, accidental or deliberate exposure to OPs can result in both acute and delayed or chronic forms of neurotoxicity in humans, which are discussed later. For example, there are several epidemiological and experimental studies to support the view that some OP pesticides (e.g., chlorpyrifos) can induce developmental neurotoxicity (Grandjean and Landrigan, 2006; Slotkin and Seidler, 2008; Flaskos and Sachana, 2010; Hargreaves, 2012).

Furthermore, diazinon, which is used as an insecticide for crop protection and as a veterinary medicine in sheep dips, was originally considered to be moderately toxic based on LD_{50} assays in rats (Olsen, 1998). However, due to the adverse effects on pesticide handlers, farm workers and other non-target species, this compound is now either banned or restricted in use in many countries (EPA, 2000, 2006, 2017; APVMA, 2007; COT, 2007; DPI, 2007; EC, 2007).

The *ortho* isomer (TOCP) of the OP lubricant tricresyl phosphate (TCP) has been linked to numerous outbreaks of a neurodegenerative conditions termed organophosphate induced delayed neuropathy (OPIDN), including Ginger Jake poisoning in the years of prohibition in the United States (Bishop and Stewart, 1930; Zeligs, 1938; Abou-Donia and Lapadula, 1990; Lotti, 1992; Abou-Donia, 2003). Concern over the risk of contracting OPIDN from handling materials containing TCP led to the imposition of a limit of the amounts of TOCP in aviation fluids in order to minimise the risk of contracting OPIDN (Harris et al., 1997). However, there are other OPs in aviation fluids, such as triphenyl phosphate (Winder, 2005, 2006). The influence of these and of non-OP components on toxicity is not known, and there is still concern that OP leakage into aircraft cabins may be linked to air cabin sickness (CAQPCCA, 2002; CAA, 2004; Winder, 2005, 2006). In addition, human exposure to certain OPs used as pesticides has been

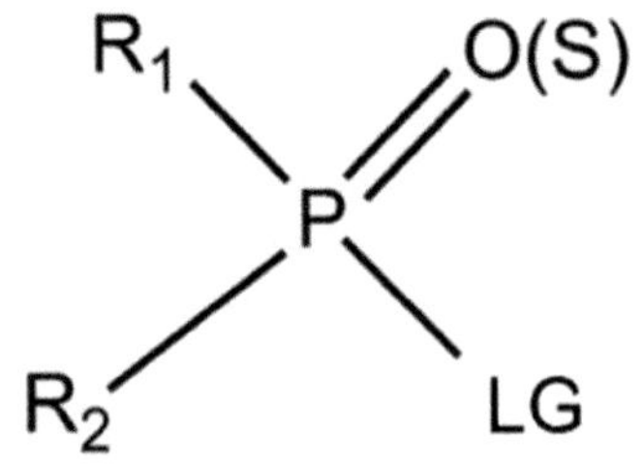

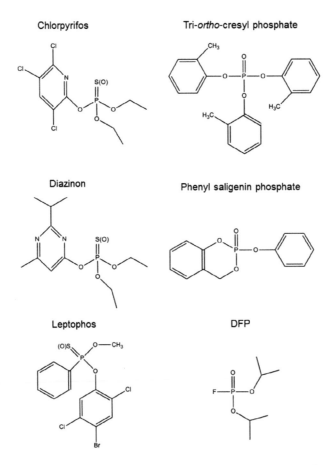

Figure 1 Generalised structure of typical organophosphorous compounds. Many OPs have a pentavalent phosphorous atom linked by a double bond to an oxygen or sulphur in organophosphates or organophosphorothioates, respectively. R_1 and R_2 can be ethoxy or methoxy groups but other substituents are possible. The group labelled 'LG' is the leaving group, which is typically an ester linked aliphatic, homocyclic or heterocyclic arrangement and in some compounds (e.g. chlorpyrifos) contains halogen groups.

Figure 2 Structures of typical OPs. Shown are the chemical structures of some typical OPs discussed in this chapter. In order to become potent inhibitors of AChE, the organophosphorothioates (e.g. diazinon, chlorpyrifos and leptophos) require bioactivation to their 'oxon' derivatives (e.g. chlorpyrifos oxon) by specific microsomal cytochrome P450s, as a result of which the sulphur atom is replaced by an oxygen atom.

associated with the induction of OPIDN following recovery from acute toxicity. This has been observed in cases of accidental and deliberate exposure to OPs such as chlorpyrifos and leptophos, whereas diazinon is not believed to present a significant risk of OPIDN (Chow et al., 1986; Richardson et al., 1993; de Blaquiere et al., 2000; Krieger, 2001; Gupta, 2006).

Human exposure to OPs has been linked to several types of neurotoxicity, including acute and delayed effects (Abou-Donia and Lapadula, 1990; Chambers, 1992; Krieger, 2001;

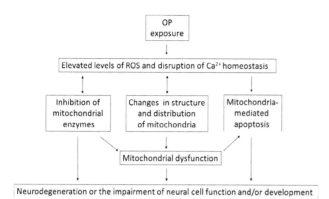

Figure 3 Schematic representation of the role of mitochondrial dysfunction in OP-induced toxicity. OP exposure leads to disruption of Ca^{2+} homeostasis and elevated levels of ROS. These changes have a variety of molecular effects including the disruption of the function and distribution of mitochondria. These together with the induction of mitochondria-mediated apoptosis contribute to the neurodegenerative effects (e.g. in OPIDN or OP-induced brain lesions) and the impairment of neural function (e.g. cognitive and behavioural effects in adults) and developmental neurotoxicity (e.g. impairment of neural differentiation).

Abou-Donia, 2003; Gupta, 2006; Hargreaves, 2012; Costa, 2018). These are discussed briefly in the next section.

2 HUMAN NEUROTOXICITY INDUCED BY OPs

2.1 ACUTE TOXICITY

OP insecticides exert their insecticidal effect primarily through the inhibition of acetylcholinesterase (AChE) in insects but, depending on the level of exposure, can also disrupt cholinergic neurotransmission in non-target species including humans (Chambers, 1992; Krieger, 2001; Gupta, 2006; Hargreaves, 2012). Organophosphorothioates are weak inhibitors of AChE but become much more potent inhibitors when converted into organophosphates by specific cytochrome P450s (Tang et al., 2001). AChE inhibition initially involves covalent binding of the OP to the serine residue in the catalytic triad of AChE, with the concomitant loss of the leaving group. It becomes irreversibly inhibited following the loss of an R group, a process referred to as aging (Worek et al., 2002). If unchecked the resultant accumulation of acetylcholine (ACh) results in hyperstimulation of muscarinic and nicitonic ACh receptors (AChRs). The symptoms, which progressively worsen without treatment, depend on the level of hyperstimulation of each receptor type. These include bradycardia, bronchoconstriction, abdominal cramps, hypersalivation, hypertension, tachycardia, fibrillation, fasciculation, striated muscle necrosis, tremor, poor movement coordination, seizures, respiratory failure, coma, and death (Hargreaves, 2012).

Patients may be placed on ventilators and pharmacological intervention typically involves the use of the muscarinic acetylcholine receptor antagonist atropine and AChE reactivators such as oximes (e.g., pralidoxime). However, the latter are not effective if aging has already occurred, in which case recovery is by the synthesis of new AChE, which may take several days (Thiermann et al., 1999; Eddleston et al., 2008).

After recovery from acute exposure to OP pesticides (e.g., chlorpyrifos and leptophos), there have been recorded cases of delayed neuropathy (El Sebae et al., 1977; Abou-Donia and Graham, 1979; Richardson et al., 1993). However, not all OPs are strong AChE inhibitors; for example, TOCP is a weak AChE inhibitor but can induce delayed neuropathy in non-target organisms (Aldridge, 1954; Gupta, 2006).

2.2 INTERMEDIATE SYNDROME AND ORGANOPHOSPHATE-INDUCED DELAYED NEUROTOXICITY

Several clinical case studies have been published showing that, within 2–3 days of exposure to OPs, some individuals may develop a myopathic condition known as intermediate syndrome, the symptoms of which include muscle weakness in the proximal limb, neck and diaphragm. The pathophysiology and molecular basis of this condition are poorly understood but are thought to be associated with lesions at the neuromuscular junctions (Civil Aviation Authority (CAA), 2004; Karalliedde et al., 2006). Patients then either suffer no further complications or they proceed to develop OPIDN, the clinical signs of which are not detected until at least 2–3 weeks after a single exposure to OPIDN-inducing OPs. Studies of a major outbreak in the last century showed that approximately 50,000 people suffered from OPIDN due to ingestion of Ginger Jake tonic adulterated with tricresyl phosphate, exhibiting symptoms to varying degrees of severity depending on the amount and duration of exposure (Bishop and Stewart, 1930; Zeligs, 1938). Possible symptoms range, in increasing order and severity, from cramp and burning or stinging sensations in the calf muscles, ankles and feet, numbness and paraesthesia in the lower limbs, leading to limb weakness which can include the hands and arms, foot drag when walking, deterioration of balance and limb reflexes, with flaccid paralysis, bladder and bowel problems in the more severe cases (Bishop and Stewart, 1930; Zeligs, 1938; Abou-Donia and Lapadula, 1990; Hargreaves, 2012).

2.3 MOLECULAR BASIS OF OPIDN

Research into the biochemical basis of OPIDN has shown that certain molecular events occur within 24 hours of exposure, thus preceding the onset of clinical signs. The first of these molecular lesions is the inhibition and aging of neuropathy target esterase (NTE) which, like AChE is a member of the serine esterase family; this is the main molecular initiating event in OPIDN (Johnson, 1974, 1982,

1990; Bal-Price et al., 2015). However, it may not be the serine esterase activity *per se*, which is responsible for triggering OPIDN, as NTE is now known to exhibit several other biochemical properties such as phospholipase and ion channel activities whose disruption may contribute to the development of the neuropathy (Glynn, 1999a, 1999b, 2000, 2005). A second major biochemical lesion detected in animal models of OPIDN and in cellular models of OP-induced neurite disruption are increased phosphorylation and increased degradation of cytoskeletal proteins (Suwita et al., 1986; Zhao et al., 2004).

Such changes have been observed within 24 hours of exposure *in vivo* and in cellular models of OP induced neurite outgrowth and retraction (Hargreaves et al., 2006; Sindi et al., 2016). Other key molecular events downstream from inhibition and aging of NTE include cytoskeletal disruption, oxidative stress and mitochondrial dysfunction (Hargreaves, 2012; Karami-Mohajeri and Abdollahi, 2013; Bal-Price et al., 2015), the latter of which is of particular interest to the current review and is discussed in more detail later.

2.4 OTHER FORMS OF CHRONIC OP-INDUCED NEUROTOXICITY

Neurobehavioral effects can possibly be the result of developmental neurotoxicity, which is well studied for organophosphorothioate insecticides. For example, pre- or post-natal exposure of rodents to chlorpyrifos and diazinon can result in changes to a range of neurodevelopmentally important genes, proteins and associated morphological changes, with consequent effects on the pattern of brain development and neurobehavioural and cognitive deficits (Slotkin et al., 2005; Ricceri et al., 2006; Slotkin et al., 2006, 2007, 2008; Slotkin and Seidler, 2008, 2012; Timofeeva et al., 2008). Both OPs and their oxon metabolites are also able to inhibit neural differentiation in primary culture reaggregates and in several cell lines of both neuronal and glial origin (Monnet-Tschudi et al., 2000; Qiao et al., 2001; Guizzetti et al., 2005; Sachana et al., 2008; Sidiropoulou et al., 2009a).

Occupational exposure of pesticide handlers has also been linked to the development of behavioural changes and cognitive impairment, as determined by epidemiological studies (Rosenstock et al., 1991; Jett et al., 2001; Abdel Rasoul et al., 2008; Ross et al., 2010). It has also been suggested, albeit controversially, that a condition known as air cabin sickness is caused by exposure to traces of OP from contaminated aircraft cabin air (Winder, 2006). However, the molecular basis for these forms of OP-induced chronic toxicity is not known.

Thus, not only do OPs exert toxicity act via AChE inhibition and subsequent cholinergic crisis, they can also induce a variety of delayed neurotoxic effects including OPIDN, developmental neurotoxicity and other forms of chronic delayed neurotoxicity involving a range of other types of molecular lesions and symptoms. It is also believed that OPs can form covalent adducts with a range of protein targets other than the serine esterases AChE and NTE (Grigoryan and Lockridge, 2009; Lockridge and Schopfer, 2010). Of particular interest here, is the ability of OPs to induce changes in mitochondria, since mitochondrial disruption leading to dysfunction is proposed to be a key molecular event in the adverse outcome pathway for OPIDN (Bal-Price et al., 2015) and in other human neurodegenerative conditions that involve elevated levels of reactive oxygen species (ROS) (Lin and Beal, 2006).

3 THE ROLE OF MITOCHONDRIAL DISRUPTION IN OP TOXICITY

3.1 MITOCHONDRIAL MORPHOLOGY, DISTRIBUTION AND TRANSPORT

Middlemore-Risher et al. (2011), in a study on primary cultures of post-natal day zero rat cortical neurons, demonstrated that exposure to chlorpyrifos (1–20 µM) and chlorpyrifos oxon (5 nM – 20 µM) for 1–24 hours resulted in the elongation and decreased numbers of mitochondria, indicative OP induced mitochondrial fusion. The rate of transport of MC along axons was also reduced as a result of OP exposure. The fact that changes were observed in the absence of significant inhibition of AChE or disruption of mitochondrial membrane potential and were not blocked by AChR antagonists, suggests that they involved the interaction of OPs with novel non-cholinergic targets involved in mitochondrial transport and fusion. The lack of effect of exposure on ROS levels suggested that these changes were not caused by oxidative stress (Middlemore-Risher et al., 2011).

The precise molecular basis of these effects is not known. However, Gearhart et al. (2007) having previously shown that several sub-acute doses of chlorpyrifos inhibited axonal transport in rat sciatic nerve (Refs) hypothesised that this might occur through direct effects on the activity of the MT-associated motor protein kinesin, which regulates the anterograde axonal transport of mitochondria and membrane vesicles along MTs (Yang et al., 2016). To test this hypothesis, they performed in vitro experiments on MT proteins purified from bovine brain in kinesin motility assays on sterile coverslips. When bovine brain microtubules were pre-incubated with chlorpyrifos (0–10 µM), chlorpyrifos oxon (0–10 µM) or DFP (0–0.59 nM) there was no significant effect on kinesin-mediated fluorescently-labelled microtubule movements. However, when bovine brain kinesin was preincubated with the OPs, a concentration-dependent increase in detachment of MTs from kinesin-coated coverslips was observed, with no effect on the rate of movement of those MTs remaining attached, suggesting that these OPs disrupt kinesin-mediated transport along microtubules by directly affecting the ability of kinesin to interact with MTs (Gearhart et al., 2007). A study by Grigoryan et al. (2009), in which mass spectrometry was used to identify non-esterase proteins in nerve tissue that can form covalent protein-bound lysine adducts with OPs, supports the notion that this might involve the formation of covalent adducts with kinesin and possibly other MT proteins such as tubulin. Thus, the direct targeting of kinesin-dependent intracellular transport could account for some of the long-term sub-acute neurodegenerative effects of organophosphates on the peripheral and central nervous system.

Changes in mitochondrial structure following acute exposure to OPs have also been observed at the ultrastructural level by electron microscopy (Satar et al., 2004). For this study methamidophos alone or with counter OP poisoning treatments (atropine and pralidoxime), were administered to male Wistar rats. Methamidophos was given by gavage at the median LD_{50} concentration and cholinergic signs started to appear within 5 minutes, at which point the OP-only treated rats were sacrificed for analysis. Treatments were administered either at the first signs of cholinergic symptoms (5 min) or when all cholinergic signs were exhibited (8 min) and rats sacrificed after signs had subsided. In the livers of rats exposed to OP without atropine and pralidoxime treatment, there was evidence of ultrastructural changes. Interestingly, there was evidence of vacuolation in the mitochondrial matrix and accumulation of lipid droplets and glycogen granules, consistent with disruption of mitochondria and metabolism. However, quantitative data with respect to these changes were not presented. No such changes were observed in liver samples from either of the pharmacological treatments or in the vehicle control, suggesting that they were induced rapidly by acute exposure to this OP and that the effects were reversible following therapy with atropine and pralidoxime. However, since the treated animals were kept alive until all symptoms disappeared, the possibility that the effects of initial OP administration on mitochondria were time related rather than treatment related cannot be ruled out completely. However, ultrastructural changes to rat liver mitochondria, typically swelling and blebbing, have also been observed following 12 weeks chronic exposure) to the OP pesticide dichlorvos (Binkumar et al., 2010) suggesting that sub-acute effects on mitochondrial structure can be cumulative.

In another study, the same dosing regimen of dichlorvos induced mitochondrial swelling, loss of cristae and chromatin condensation in rat brain neurons (Wani et al., 2011). These changes may be associated with mitochondria-induced apoptosis (see later) and were attenuated in animals co-treated with the mitochondria-specific antioxidant MitoQ, suggesting that they were directly linked to OP-induced oxidative stress in mitochondria. In another study by Xin et al. (2011), in which OPIDN was induced by TOCP in hens, structural changes were observed in mitochondria including vacuolation and fission, which became more severe as OPIDN developed, suggesting an early involvement of mitochondrial dysfunction in the pathogenesis of this condition.

Similar ultrastructural changes to mitochondria, together with increased mitochondrial lipid peroxidation, have been observed in fallopian tubes of rats following 4 weeks sub-chronic administration of methyl parathion (Guney et al., 2007), further suggesting that mitochondria are affected by OP exposure in multiple tissues.

3.2 MITOCHONDRIAL HEALTH AND ENZYME ACTIVITIES

Lukaszewicz-Hussain and Moniuszko-Jakoniuk (2004) studied the effects of the OP pesticide chlorphenvinphos administered to rats intragastrically by tube as a single dose of 2%, 10%, or 50% of LD_{50} on the activities of antioxidant enzymes and the levels of reduced glutathione and hydrogen peroxide (H_2O_2) in liver tissue of male Wistar rats. Tissues were removed and analysed at 1, 24, and 48 hours post administration. At 2% LD_{50}, transient initial reductions were observed in the activity of glutathione peroxidase, superoxide dismutase and isocitrate dehydrogenase and in the concentration of reduced glutathione, with all values recovering to levels at or near "control" levels at 48 hours. The concentration of mitochondrial H_2O_2 increased initially then reduced to a level closer to but still significantly higher than the control after 48 hours. The disruptive effects of OP were more sustained at the two higher concentrations, suggesting a dose dependent effect. These findings suggest the that antioxidant defences in liver mitochondria are depleted even at very low exposure levels in association with increased levels of ROS. However, in this study there was only a single control value for each parameter and the time point was not made clear.

In the study of by Binkumar et al. (2010) of chronic dichlorvos exposure in rats, prior to the observation of ultrastructural changes in rat liver mitochondria, reductions were observed in the activities of complex I (NADH dehydrogenase), complex II (succinate dehydrogenase), complex IV (cytochrome C oxidase) and in the rate of oxygen consumption by purified mitochondria. Reductions were also observed in the synthesis and levels of ATP in mitochondria, whereas there was an increase in the rates of ATP hydrolysis in the liver of OP-treated rats compared to untreated control livers. The liver mitochondria of OP-treated rats also exhibited increased Ca^{2+} uptake and elevated levels of ROS, which possibly led to the higher levels of alanine aminotransferase (ALT), aspartate transaminase (AST) and alkaline phosphatase (ALP) that were detected in the serum.

Venkatesh et al. (2009) carried out a study of the effects of acute exposure to monocrotophos (at 80% LD_{50}) on skeletal muscle mitochondria in female Wistar rats. Within hours of exposure, and in parallel with symptoms of cholinergic crisis (e.g., chewing and tremor), ATP synthase activity was found to be reduced in muscle mitochondria, whereas mitochondrial respiration was unaffected. The uptake of Ca^{2+} by skeletal muscle mitochondria of OP-treated animals was reduced to 40% of that observed in the non-OP-treated control muscle mitochondria. Under the same conditions elevated levels of NO were detected in the plasma and in muscle mitochondria of OP-treated animals. Further analyses indicated decreased levels of reduced -SH groups on the ATP synthase subunits, an effect that was blocked by pre-treatment with the nitric oxide synthase inhibitor L-NG nitroarginine methyl ester prior to monocrotophos exposure, suggesting a role for NO in this effect. It was suggested that impairment of mitochondrial calcium uptake could result in higher cytosolic calcium levels. However, further work is needed to determine whether these effects may contribute to muscle weakness in acute organophosphate exposure. Indeed, in a similar study by Raghupathy et al. (2010) it was found that high levels of cholinesterase inhibition (30–60%) rather than reduced levels of ATP was the main contributor to acute monocrotophos-induced severe muscle weakness in the rat. Nevertheless this effect is not only observed on acute exposure to monocrotophos, as a study on rats exposed to an acute oral dose of the OP pesticide malathion for 24 hours also showed significant reductions in the activities of complexes I, IV, and V but no significant effect on complex II in muscle mitochondria (Karami-Mohajeri and Abdollahi, 2013), suggesting that in addition to acetylcholinesterase inhibition, there are certain major effects on mitochondrial proteins involved in ATP synthesis. Yen et al. (2004), studied acute mevinphos poisoning in rats after bilateral microinjection of 10 nmol into the rostral ventrolateral medulla (RVLM), a brain region involved in the basal and reflex control of cardiac function. They found reduced mitochondrial complex I and IV enzyme activities, but no effect on complexes II and III, in RVLM but not in heart muscle. This led to cardiovascular complications associated with cardiotoxicity of OPs.

It is of course of further interest to look at the potential involvement of mitochondrial dysfunction in delayed neuropathy, for example at later time intervals after exposure to OPIDN-inducing OPs and/or post administration of anti-cholinergic treatments following initial acute intoxication by exposure to an OP insecticide, in order to determine whether the effects on mitochondrial enzymes persisted in delayed neuropathic conditions. In this regard, acute exposure of hens to chlorpyrifos with subsequent atropine/pralidoxime therapy to inhibit hyperstimulation of cholinergic neurones and promote AChE reactivation respectively, led to symptoms of delayed neuropathy as adjudged by the development of mild, moderate or severe ataxia, events that precede the paralysis typical of OPIDN (Salama et al., 2014). The extent of NTE inhibition, which was only measured in brain, after 3 days exposure was much lower (ca. 50% of the control value) than that observed for an OPIDN inducing dose of TOCP (ca. 95% of the control value), suggesting that full-blown OPIDN may not necessarily be involved in this case. It could be that chlorpyrifos-induced inhibition of NTE would have been greater in spinal cord or sciatic

nerve than that detected in brain, but this was not tested. Nevertheless, the data showed that there was a significant reduction in complex I activity in the chlorpyrifos treated hens showing symptoms of ataxia after 12–16 days exposure, suggesting that mitochondrial disruption was involved in this delayed neuropathic effect of chlorpyrifos. The observed reduction in complex I activity was, however, similar to that for TOCP-treated hens in a positive control group for OPIDN.

In an earlier study, where TOCP-induced OPIDN in hens administered oral doses of TOCP was the main focus, 12–16 days' exposure to the OPIDN-inducing lubricant resulted not only in the ultrastructural changes in mitochondria discussed above but also in reduced activity of the mitochondrial inner membrane enzyme succinate dehydrogenase well in advance the detection of clinical signs of OPIDN in hens (Xin et al., 2011). These changes were particularly prevalent in mitochondria isolated from spinal cord tissue within 21 days of exposure to OPIDN-inducing doses, indicating that (as in OPIDN) nerves with very long fibre tracts are the most sensitive to OP-induced neurodegeneration in OPIDN. However, under the experimental conditions used, there was a slight but not statistically significant reduction in the mitochondrial permeability transition, suggesting that this phenomenon was not a major factor in the development of OPIDN following exposure to single oral doses of TOCP (185–750 mg/kg).

3.3 Mitochondria-Mediated Apoptosis

The mechanism of neuronal cell death in OPIDN was studied in hens induced to suffer OPIDN from a single oral dose of TOCP (750 mg/kg) (Zou et al., 2013). Apoptotic neurons were detectable within 5 days of exposure and gradually increased in number with time, as indicated by TUNEL assays, although their numbers were not quantified. Increased release of cytochrome c to the cytoplasm was detected by immunohistochemical staining and Western blot analysis, peaking at 5 days, indicating involvement of the mitochondrial apoptotic pathway. Taken together with the fact that reduced levels of anti-apoptotic proteins, pro-caspases 3 and 9, and elevated levels of pro-apoptotic proteins Bax and PARP-1 were also observed and showed a similar pattern of change, this suggests that the induction of OPIDN by TOCP administration in hens is associated with elevated levels of neuronal cell apoptosis in hen spinal cords by the activation of the mitochondrial apoptotic pathway, consistent with the pattern of mitochondrial ultrastructural changes described earlier.

The mechanism of OP-induced delayed brain injury was studied in rats given an intraperitoneal dose of DFP (9 mg/kg) and subsequent treatment including atropine and pyridostigmine bromide to reduce mortality and peripheral signs (Kim et al., 1999). Necrotic brain injury was detected within 1 hour of exposure, whereas apoptotic cells appeared in the thalamus and amygdala regions after 12 hours, as determined by TUNEL staining. The fact that apoptosis but not necrotic changes were attenuated on pre-treatment with L NG-nitro arginine suggest that apoptosis was induced by oxidative stress involving the NO pathway.

Kaur et al. (2007) demonstrated that chronic low-level exposure to dichlorvos resulted in inhibition of mitochondrial complex I and cytochrome oxidase in rat brain, resulting in the generation of reactive oxygen species (ROS). Enhanced ROS production led to disruption of cellular antioxidant defence systems and release of cytochrome c from mitochondria to the cytosol, resulting in apoptotic cell death. In the chronic dichlorvos exposure study of Wani et al. (2011) mitochondria-mediated neuronal apoptosis was observed in rat brain following the ultrastructural changes discussed earlier. The fact that neuronal apoptosis was attenuated when rats were pre-treated with the mitochondria specific antioxidant MitoQ, suggests that oxidative damage to mitochondria was involved and that MitoQ is a potential protective agent against dichlorvos-induced neuronal apoptosis.

Numerous cell culture studies have shown that some OPs are capable of inducing cell death via mitochondria mediated apoptosis. For example, in cultures of retinoic acid treated Ntera2 cells diazinon (10 µM) was found to induce neuronal apoptosis in about 15% of cells after 24 hours exposure, as determined by caspase activation and nuclear fragmentation (Aluigi, Guida and Falugi, 2010). However, as differentiation markers were not characterised in this study it is not clear if the cells were mitotic (non-differentiated) or differentiated or whether specific neural cell types were affected. The short incubation time would suggest that the former is more likely to be the case, meaning that the cells would largely be exhibiting a tumour phenotype at this stage. On the other hand, a study by Rush et al. (2010) using primary cultures from rat brain cortex containing a mixture of neuronal and glial cells found significant cytotoxicity with diazinon (10 µM) and chlorpyrifos (100 µM) but not their oxon metabolites. Diazinon was found to induce mitochondria-mediated apoptosis of neuronal cells, as it had no effect on the viability of pure glial cultures whereas chlorpyrifos-induced cell death was concluded to involve mainly necrosis.

However, this pattern of toxicity is not observed in other cell types, since low µM levels of phenyl saligenin phosphate, a structural congener of the neuropathic metabolite of TOCP (Jortner and Ehrich, 1987), induced apoptosis in cultured mitotic and differentiating H9c2 cardiomyocyte-like cells, whereas concentrations of below 100 µM diazinon, chlorpyrifos and their oxon metabolites had no cytotoxic effect up to 48 hours exposure (Felemban et al., 2015). However, chlorpyrifos has been shown to be capable of inducing apoptosis in cultured human HeLa cervical carcinoma cells and HEK293 embryonic kidney cells but not at concentrations below 90 µM (Li et al., 2015). Work on cultured

cells may or may not be a reliable indicator of likely *in vivo* effects and would benefit from validation in an *in vivo* model.

In this respect, in a study of developmental exposure of rats to chlorpyrifos and diazinon administered at postnatal days 1–4, found that such toxicity was associated with the upregulation of various groups of genes including those encoding proteins involved in the regulation of the cell cycle and apoptosis. This was observed both in animals and in cultured PC12 cells, irrespective of the degree of AChE inhibition (Slotkin and Seidler, 2012). This suggests that these OPs may exert developmental effects on these gene families by acting directly on neuronal cells *in vivo* and that neuronotypic cells in culture are valid models for molecular studies of developmental neurotoxicity. Differences were observed between mitotic (undifferentiated) and differentiating PC12 cultures in the profile of gene changes on exposure to chlorpyrifos (30 µM) and were proposed to reflect variations in vulnerability to OP exposure at different developmental stage effects (Slotkin and Seidler, 2012). The OP pesticide monocrotophos (1–10 µM) was also found to induce mitochondria mediated apoptosis in mitotic PC12 cells within 6 hours of exposure (Kashyap et al., 2010, 2011).

Another study in which mitotic and differentiating human SH-SY5Y cells were exposed to chlorpyrifos (25–100 µM) also reported mitochondria-mediated apoptosis and proposed that this effect was mediated by dynamin-related protein 1, which was required for the translocation of p53 to mitochondria under chlorpyrifos-induced oxidative stress (Park et al., 2015). Thus, there is clear evidence from both *in vivo* and cell culture studies that numerous OPs can induce mitochondria-mediated apoptosis, and that the latter may be one of the key events involved in OP-induced delayed neurotoxicity. However, the ability of OPs to induce apoptosis in different cell types and tissues is variable possibly due to differences in uptake and in the ability of cells to metabolize OPs, and in some cases the use of very high concentrations begs the question as to whether the levels entering specific cells and tissues in such studies are physiologically relevant

4 CONCLUSION

There is a significant body of evidence to support the view that mitochondrial dysfunction is a common lesion in OP toxicity in a range of tissue types and that this may be related to oxidative stress in mitochondria. Studies on OPIDN suggest that it may be a key molecular event in this condition (Bal-Price et al., 2015). Further work required to uncover the mitochondrial mechanism(s) underlying this and other types of OP-induced delayed neurotoxicity would be worthwhile.

Numerous studies have linked OP toxicity to the induction of cell death, but the mechanisms involved are not fully understood and may vary depending on the nature of the OP, the individual affected, the dose and duration of exposure. Indeed, in most studies the cellular or tissue concentration of OP was not determined.

As summarized in Figure 3, mitochondrial disruption can be the result of direct effects of OPs on mitochondrial proteins or on cytoskeletal networks involved in the transport along axons and throughout the cells. As this phenomenon may give rise to and be exacerbated by elevated levels of ROS and intracellular Ca^{2+}, further work is needed to determine exactly how this contributes to the various neurotoxicity pathways induced following OP exposure and to develop pharmacological approaches targeting mitochondria to determine whether protective strategies can be established.

REFERENCES

Abdel Rasoul, G.M., Abou Salem, M.E., Mechael, A.A., Hendy, O.M., Rohlman, D.M. and Ismail, A.A. (2008) Effects of occupational pesticide exposure on children applying pesticides. Neurotoxicology 29, 833–888.

Abou-Donia, M.B. (2003) Organophosphorus ester induced chronic neurotoxicity. Archives of Environmental Health 58, 484–497.

Abou-Donia, M.B. and Graham, D.G. (1979) Delayed neurotoxicity of sub chronic oral administration of leptophos to hens: Recovery during four months after exposure. Journal of Toxicology and Environmental Health 5, 1133–1147.

Abou-Donia, M.B. and Lapadula, D.M. (1990) Mechanisms of organophosphorus ester induced delayed neurotoxicity: Type I and II. Annual Reviews of Pharmacology and Toxicology 30, 405–440.

Aldridge, W.N. (1954) Tricresyl phosphates and cholinesterase. Biochemical Journal 56, 185–189.

Aluigi, M.G., Guida, C. and Falugi, C. (2010) Apoptosis as a specific biomarker of diazinon toxicity in NTera2-D1 cells. Chemico-Biological Interactions 187, 299–303.

Australian Pesticides and Veterinary Medicines Authority (APVMA) APVMA suspends the use of diazinon for sheep dipping and jetting. *Media Release*, May 2007.

Bal-Price, A., Crofton, K.M., Shafer, T., Sachana, M., Behl, M., Forsby, A., Hargreaves, A.J., Landesmann, B., Lein, P.J., Louisse, J., Monnet-Tschudi, F., Paini, A., Rolaki, A., Schrattenholz, A., Suñol, C., van Thriel, C., Whelan, M., and Fritsche, E. (2015) Workshop report: Adverse Outcome Pathways (AOP) relevant to neurotoxicity. Critical Reviews in Toxicology 45, 83–91.

Binkumar, B.K., Bal, A., Kandamilla, R., Sunkaria, A. and Gill, K. (2010) Mitochondrial energy metabolism impairment and liver dysfunction following chronic exposure to dichlorvos. Toxicology 270 (2–3), 77–84.

Bishop, E.L. and Stewart, H.C. (1930) Incidence of partial paralysis. American Journal of Public Health 20, 1307–1312.

Chambers, J.E. (1992) Organophosphate compounds an overview. In: Chambers JE, Levi P eds. Organophosphates: Chemistry, fate and effects. San Diego, CA: Academic Press, 3–17.

Civil Aviation Authority (CAA) (2004) Cabin air quality. CAA paper 2004: 04.

Committee on Air Quality in Passenger Cabins of Commercial Aircraft (CAQPCCA) (2002) Board on environmental studies and toxicology, national research council, national academy of sciences. The airliner cabin environment and health of passengers and crew. Washington, DC: National Academic Press.

Committee on Toxicology (COT) of Chemicals in Food, Consumer Products and the Environment (2007) Organophosphates. Department of Health.

Costa, L.G. (2018) Organphosphorus compounds at 80: Some old and new issues. Toxicological Sciences 162 (1), 24–35.

Department of Primary Industries (DPI) (2007) Sheep body lice: Control and eradication. Agriculture Notes of the DPI, State of Victoria, 2007: AG1110.

Eddleston, M., Buckley, N.A., Eyer, P., Dawson, A.H. (2008) Management of acute organophosphorus pesticide poisoning. The Lancet 371, 597–607.

El Sebae, A.H., Soliman, S.A., Abo Elamayem, M. and Ahmed, N.S. (1977) Neurotoxicity of organophosphorus insecticides leptophos and EPN. Journal of Environmental Science and Health 12 (4), 269–288.

Environmental Protection Agency (EPA) USA (2000 Annual report. Office of Pesticide Programs.

EPA (2006) Reregistration eligibility decision for diazinon. www3.epa.gov/pesticides/chem_search/reg_actions/reregis tration/red_PC-057801_31-Jul-06.pdf

EPA (2017) Biological evaluation chapters for diazinon ESA assessment. www.epa.gov/endangered-species/biological-evaluation-chapters-diazinon-esa-assessment

European Commission (EC) (2007) Decision of 6 June 2007concerning the non-inclusion of diazinon in annex I to the council directive 91/414/EEC and the withdrawal of authorisations for plant protection agents containing that substance. Official Journal of the European Union 140, 9–10.

Felemban, S.G., Garner, A.C., Smida F.A., Boocock, D.J., Hargreaves, A.J. and Dickenson, J.M. (2015) Phenyl saligenin phosphate induced caspase on mitotic and differentiated rat H9c2 cardiomyoblasts. Chemical Research in Toxicology 28, 2179–2191.

Flaskos, J. and Sachana, M. (2010) Developmental neurotoxicity of anticholinesterase pesticides. In: Satoh, T. and Gupta, R.C. eds. Anticholinesterase pesticides. Metabolism, neurotoxicity and epidemiology. Hoboken, NJ: Wiley, 203–224.

Gearhart, D.A., Sickles, D.W., Buccafusco, J.J., Prendergast, M.A. and Terry, Jr., A.V (2007) Chlorpyrifos, chlorpyrifos-oxon, and diisopropylfluorophosphate inhibit kinesin-dependent microtubule motility. Toxicology and Applied Pharmacology 218, 20–29.

Glynn, P. (1999a) Neuropathy target esterase. Biochemical Journal 344, 625–631.

Glynn, P. (1999b) Molecular cloning of neuropathy target esterase. Chemico-Biological Interactions 119–120, 513–517.

Glynn, P. (2000) Neural development and neuropathy target esterase: Two faces of neuropathy target esterase. Progress in Neurobiology 61, 61–74.

Glynn, P. (2005) Neuropathy target esterase and phospholipid deacylation. Biochimica et Biophysica Acta 1736, 87–93.

Grandjean, P., and Landrigan, P.J. (2006) Developmental neurotoxicity of industrial chemicals. The Lancet 368, 2167–2178.

Grigoryan, H., Li, B., Xue, W., Grigoryan, M., Schopfer, L.M. and Lockridge, O. (2009) Mass spectral characterization of organophosphate-labelled lysine in peptides. Analytical Biochemistry 394, 92–100.

Grigoryan, H. and Lockridge, O. (2009) Nanoimages show disruption of tubulin polymerisation by chlorpyrifos oxon: Implications for neurotoxicity. Toxicology and Applied Pharmacology 240, 143–148.

Guizzetti, M., Pathak, S., Giordano, G. and Costa, L.G. (2005) Effects of organophosphorus insecticides and their metabolites on astroglial cell proliferation. Toxicology 215 (3), 182–190.

Guney, M., Oral, B., Demirin, H., Take, G., Giray, S.G., Altuntas, I. and Mungani, T. (2007) Fallopian damage induced by organophosphate insecticide methyl parathion, and protective effect of vitamins E and C on ultrastructural changes in rats. Toxicology and Industrial Health 2007 (23), 429–438.

Gupta R.C. ed. (2006) Toxicology of organophosphate and carbamate compounds. San Diego, CA: Elsevier.

Hargreaves, A.J. (2012) Neurodegeneration induced by organophosphorous compounds. Advances in Experimental Medicine and Biology 724, 189–204.

Hargreaves, A.J., Fowler, M.J., Sachana, M., Flaskos, J., Bountouri, M., Coutts, I.C., Glynn, P., Harris, W., and McLean, W.G. (2006) Inhibition of neurite outgrowth in differentiating mouse N2a neuroblastoma cells by phenyl saligenin phosphate: Effects on neurofilament heavy chain phosphorylation, MAP kinase (ERK 1/2) activation and neuropathy target esterase activity. Biochemical Pharmacology 71, 1240–1247.

Harris, M.O., McLure, P., Chessin, R.L., Corcoran, J.J. (1997) Toxicological profile for hydraulic fluids. USA: Agency for Toxic Substances and Disease Registry (ATDSR).

Jett, D.A., Navoa, R.V., Beckles, R.A., McLemore, G.L. (2001) Cognitive function and cholinergic neurochemistry in weanling rats exposed to chlorpyrifos. Toxicology and Applied Pharmacology 174, 89–98.

Johnson, M.K. (1974) The primary biochemical lesion leading to the delayed neurotoxic effects of some organophosphorus esters. Journal of Neurochemistry 23, 785–789.

Johnson, M.K. (1982) The target for initiation of delayed neurotoxicity by organophosphorus esters: Biochemical studies and toxicological applications. In: Hodgson, E., Bend, J.R. and Philpot, R.M. eds. Reviews in biochemical toxicology. New York: Elsevier, vol. 4, 141–212.

Johnson, M.K. (1990) Organophosphates and delayed neuropathy—is NTE alive and well? Toxicology and Applied Pharmacology 102, 385–389.

Jortner, B.S. and Ehrich, M. (1987) Neuropathological effects of phenyl saligenin phosphate in chickens. Neurotoxicology 8 (2), 303–314.

Kamanyire, R. and Karalliedde, L. (2004) Organophosphate toxicity and occupational exposure. Occupational Medicine 54, 69–75.

Karalliedde, L., Baker, D., Marrs, T.C. (2006) Organophosphate-induced intermediate syndrome: Aetiology and relationships with myopathy. Toxicological Reviews 25, 1–14.

Karami-Mohajeri, S., and Abdollahi, M. (2013) Mitochondrial dysfunction and organophosphorus compounds. Toxicology and Applied Pharmacology 270, 39–44.

Kashyap, M.P., Singh, A.K., Kumar, V., Tripathi, V.K., Srivastava, R.K., Agrawal, M., Khanna, V.K., Yadav, S., Jain, S.K. and Pant, A.B. (2011) Monocrotophos induced apoptosis in PC12 Cells: Role of xenobiotic metabolizing cytochrome P450s. PLoS ONE 6 (3), e17757.

Kashyap, M.P., Singh, A.K., Siddiqui, M.A., Kumar, V., Tripathi, V.K., Khanna, V.K., Yadav, S., Jain, S.K. and Pant, A.B. (2010) Caspase cascade regulated mitochondria mediated apoptosis in monocrotophos exposed PC12 cells. Chemical Research in Toxicology 23 (11), 1663–1672.

Kaur, P., Radotra, B., Minz, R.W., Gill, D.P. (2007) Impaired mitochondrial energy metabolism and neuronal apoptotic cell death after chronic dichlorvos (OP) exposure in rat brain. Neurotoxicology 28 (6), 1208–1219.

Kim, Y.B., Hur, G.H., Shin, S., Sok, D.E., Kang, J.K. and Lee, Y.S. (1999) Organophosphate-induced brain injuries: Delayed apoptosis mediated by nitric oxide. Environmental Toxicology and Pharmacology 7, 147–152.

Krieger, R.I. ed. (2001) Handbook of pesticide toxicology: Principles, 2nd edition. San Diego, CA: Academic Press.

Li, D., Huang, Q., Lu, M., Zhang, L., Yang, Z., Zong, M. and Tao, L. (2015) The organophosphate insecticide chlorpyrifos confers its genotoxic effects by inducing DNA damage and cell apoptosis. Chemosphere 135, 387–393.

Lin, M.T. and Beal, M.F. (2006) Mitochondrial dysfunction and oxidative stress in neurodegenerative diseases. Nature 443, 787–795.

Lockridge, O. and Schopfer, L.M. (2010) Review of tyrosine and lysine as new motifs for organophosphate binding to proteins that have no active site serine. Chemico-Biological Interactions 187, 344–388.

Lotti, M. (1992) The pathogenesis of organophosphate polyneuropathy. Critical Reviews in Toxicology 21, 465–487.

Lukaszewicz-Hussain, A. and Moniuszko-Jakoniuk, J. (2004) Chlorphenvinphos, an organophosphate insecticide, affects liver mitochondria antioxidative enzymes, glutathione and hydrogen peroxide concentration. Polish Journal of Environmental Studies 13 (4), 397–401.

Middlemore-Risher, M.L., Adam, B.L., Lambert, N.A. and Terry, Jr., A.V. (2011) Effects of chlorpyrifos and chlorpyrifos-oxon on the dynamics and movement of mitochondria in rat cortical neurons. The Journal of Pharmacology and Experimental Therapeutics 339 (2), 341–349.

Monnet-Tschudi, F., Zurich, M.G., Schilter, B., Costa, L.G. and Honegger, P. (2000) Maturation-dependent effects of chlorpyrifos and parathion and their oxygen analogs on acetylcholinesterase and neuronal and glial markers in aggregating brain cultures. Toxicology and Applied Pharmacology 165 (3), 175–183.

Olsen, K.R. ed. (1998) Poisoning & drug overdose. Stamford, CT: Appleton and Lange.

Park, J.H., Ko, J., Hwang, J., Koh, H.C. (2015) Dynamin-related protein 1 mediates mitochondria-dependent apoptosis in chlorpyrifos-treated SH-SY5Y cells. Neurotoxicology 51, 145–157.

Qiao, D., Seidler, F.J. and Slotkin, T.A. (2001) Developmental toxicity of chlorpyrifos modelled in vitro: Comparative effects of metabolites and other cholinesterase inhibitors on DNA synthesis in PC12 and C6 cells. Environmental Health Perspectives 109 (9), 909–913.

Raghupathy, V., Poornima, S., Sivaguru, A., Ramachandran, A., Zachariah, A. and Oommen, A. (2010) Monocrotophos toxicity and bioenergetics of muscle weakness in the rat. Toxicology 277, 6–10.

Ricceri, L., Venrosi, A., Capone, F., Cometa, M.F., Lorenzini, P., Fortuna, F. and Calamandrei, G. (2006) Developmental neurotoxicity of organophosphorous pesticides: Fetal and neonatal exposure to chlorpyrifos alters sex-specific behaviors in mice at adulthood. Toxicological Sciences 93 (1), 105–113.

Rosenstock, L., Keifer, M., Daniell, W.E., McConnell, R. and Claypoole, K. (1991) Chronic central nervous system effects of acute organophosphate pesticide intoxication. The Lancet 338, 223–227.

Ross, S.J.M., Brewin, C.R., Curran, H.V., Furlong, C.E., Abraham-Smith, K.M. and Harrison, V. (2010) Neuropsychological and psychiatric functioning in sheep farmers exposed to low levels of organophosphate pesticides. Neurotoxicology and Teratology 32, 452–459.

Rush, T., Liu, X.Q., Hjelmhaug, J. and Lobner, D. (2010) Mechanisms of chlorpyrifos and diazinon induced neurotoxicity in cortical culture. Neuroscience 166, 899–906.

Sachana, M., Flaskos, J. and Hargreaves, A.J. (2008) Effects of chlorpyrifos and chlorpyrifos-methyl on the outgrowth of axon-like processes, tubulin and GAP-43 in N2a cell. Toxicology Mechanisms and Methods 105 (6), 405–441.

Salama, M., El-Morsy, D., El-Gamal, M., Shabka, O. and Mohammed, W.M.Y. (2014) Mitochondrial complex I inhibition as a possible mechanism of chlorpyrifos induced neurotoxicity. Annals of Neurosciences 21 (3), 85–89.

Satar, S., Satar, D., Tap, O., Koseoglu, Z. and Kaya, M. (2004) Ultrastructural changes in rat liver treated with pralidoxime following acute organophosphate poisoning. Mount Sinai Journal of Medicine 71 (6), 405–410.

Sidiropoulou, E., Sachana, M., Flaskos, J., Harris, W., Hargreaves, A.J. and Woldehiwet, Z. (2009a) Diazinon oxon affects the differentiation of mouse N2a neuroblastoma cells. Archive of Toxicology 83 (4), 373–380.

Sidiropoulou, E., Sachana, M., Flaskos, J., Harris, W., Hargreaves, A.J. and Woldehiwet, Z. (2009b) Diazinon oxon affects the differentiation of rat C 6 glioma cells. Toxicology In Vitro 23 (8), 1548–1552.

Sindi, R.A., Harris, W., Arnott, G., Flaskos, J., Lloyd Mills, C. and Hargreaves, A.J. (2016) Chlorpyrifos- and chlorpyrifos oxon-induced neurite retraction in pre-differentiated N2a cells is associated with transient hyperphosphorylation of neurofilament heavy chain and ERK 1/2. Toxicology and Applied Pharmacology 308, 20–31.

'Slotkin T.A., Oliver, C.A. and Seidler, F.J. (2005) Critical periods for the role of oxidative stress in developmental neurotoxicity of chlorpyrifos and terbutaline alone or in combination. Developmental Brain Research 157, 172–180.

Slotkin, T.A., Ryde, I.T., Levin, E.D., and Seidler, F.J. (2008) Developmental neurotoxicity of low dose diazinon exposure of neonatal rats: Effects on serotonin systems in adolescence and adulthood. Brain Research Bulletin 75, 640–647.

Slotkin, T.A. and Seidler, F.J. (2008) Developmental neurotoxicants target differentiation into the serotonin phenotype: Chlorpyrifos, diazinon, dieldrin and divalent nickel. Toxicology and Applied Pharmacology 233, 211–219.

Slotkin, T.A. and Seidler, F.J. (2012) Developmental neurotoxicity of organophosphates targets cell cycle and apoptosis, revealed by transcriptional profiles in vivo and in vitro. Neurotoxicology and Teratology 34 (2), 232–241.

Slotkin, T.S., MacKillop, E.A., Ryde, I.T., Tate, C.A. and Seidler, F.J. (2007) Screening of developmental toxicity using PC12 cells. Comparisons of organophosphates with a carbamate, an organochlorine and divalent nickel. Environmental Health Perspectives 115, 93–101.

Slotkin, T.S., Tate, C.A., Ryde, I.T., Levin, E.D. and Seidler, F.J. (2006) Organophosphate insecticides target the serotonergic system in developing rat brain regions: Disparate effects of diazinon and parathion at doses spanning the threshold for cholinesterase inhibition. Environmental Health Perspectives 114, 1542–1546.

Suwita, E., Lapadula, D.M. and Abou-Donia, M.B. (1986) Calcium and calmodulin enhanced in vitro phosphorylation of hen brain cold stable microtubules and spinal cord neurofilament proteins following a single oral dose of tri-o-cresyl phosphate. Proceedings of the National Academy of Sciences USA 83, 6174–6178.

Tang, J., Cao, Y., Rose, R.L., Brimfield, A.A., Dai, D., Goldstein, J.A. and Hodgson, E. (2001) Metabolism of chlorpyrifos by human cytochrome P450 isoforms and human, mouse and rat liver microsomes. Drug Metabolism and Disposition 29, 1201–1204.

Thiermann, H., Szinicz, L., Eyer, F., Worek, F., Eyer, P., Felgenhauer, N. and Zilker, T. (1999) Modern strategies in therapy of organophosphate poisoning. Toxicology Letters 1107, 233–239.

Timofeeva, O.A., Roegge, C.S., Seidler, F.J., Slotkin, T.A. and Levin, E.D. (2008) Persistent cognitive alterations in rats after early postnatal exposure to low doses of the organophosphate pesticide, diazinon. Neurotoxicology and Teratology 30 (1), 38–45.

Venkatesh, S., Ramachandran, A., Zachariah, A. and Oommen, A. (2009) Mitochondrial ATP synthase inhibition and nitric oxide are involved in muscle weakness that occurs in acute exposure of rats to monocrotophos. Toxicology Mechanisms and Methods 19 (3), 239–245.

Wani, W.Y, Gudup, S., Sunkaria, A., Bal, A., Singh, P.P., Kandimalla, R.J.L., Sharma, D.R. and Gill, K.D. (2011) Protective efficacy of mitochondrial targeted antioxidant MitoQ against dichlorvos induced oxidative stress and cell death in rat brain. Neuropharmacology 61, 1193–1201.

Winder, C., ed. (2005) Contaminated air protection. Proceedings of the Air Safety and Cabin Air Quality International Aero Industry Conference. British Airline Pilots Association (BALPA) and the University of New South Wales, Australia.

Winder, C. (2006) Air monitoring studies for air cabin contamination. Current Topics in Toxicology 3, 33–48.

Worek, F., Reiter, G., Eyer, P. and Szinicz, L. (2002) Reactivation kinetics of acetylcholinesterase from different species inhibited by highly toxic organophosphates. Archive of Toxicology 76, 523–529.

Xin, X., Zeng, T., Dou, D.D., Sheng Zhao, S. Du, J.Y., Pei, J.J., Xie, K.Q. and Zhao, X.L. (2011) Changes of mitochondrial ultrastructures and function in central nervous tissue of hens treated with tri-ortho-cresyl phosphate (TOCP). Human and Experimental Toxicology 30 (8), 1062–1072.

Yang, R., Bentley, M., Huang, C.F. and Banker, G. (2016) Analyzing kinesin motor domain translocation in cultured hippocampal neurons. Methods in Cell Biology 131, 217–232.

Yen, D.H.T., Chan, J.Y.H., Tseng, H.P., Huang, C.I., Lee, C.H., Chan, S.H.H., and Chang, A.Y.W. (2004) Depression of mitochondrial respiratory enzyme activity in rostral ventrolateral medulla during acute mevinphos intoxication in the rat. Shock 21 (4), 358–363.

Zhao, X.L., Zhu, Z.P., Zhang, T.L., Zhang, C.L., Yu, H.L. and Xe, K.Q. (2004) Tri-ortho-cresyl phosphate (TOCP) decreases the levels of cytoskeletal proteins in hen sciatic nerve. Toxicology Letters 152, 139–147.

Zou, C., Kou, R., Gao, Y., Xie, K.Q. and Song, F. (2013) Activation of mitochondria-mediated apoptotic pathway in tri-ortho-cresyl phosphate-induced delayed neuropathy. Neurochemistry International 62 (7), 965–972.

37 Mitochondrial Dysfunction in Friedreich Ataxia

Rosella Abeti, Annalisa Baccaro, and Paola Giunti

Ataxia Centre, Department of Clinical and Movement Neurosciences, University College London, Queen Square Institute of Neurology, London WC1N 3BG, United Kingdom

CONTENTS

Abbreviations

FRDA	Friedreich's Ataxia
FXN	frataxin
DRG	dorsal root ganglia
LOFA	late onset Friedreich ataxia
SCA	spinocerebellar ataxia
MRI	magnetic resonance imaging
MPP	mitochondrial processing peptidase
ISC	iron sulphur cluster
Fe-S	iron sulphur
ETC	electron transport chain
NADH	nicotinamide adenine dinucleotide
ROS	reactive oxygen species
CGNs	cerebellar granule neurons
dPUFAs	deuterium at bis-allylic sites polyunsaturated fatty acids
GSH	reduced glutathione
IMM	inner mitochondrial membrane
Nrf2	nuclear factor (erythroid-derived 2)-like 2
Omav	omaveloxone
Ca^{2+}	calcium
SERCA	sarcolemmal ATPase
RyR2	ryanodine receptors
MCU	mitochondrial calcium uniporter 1.

1 PATHOGENESIS OF FRIEDREICH'S ATAXIA

Friedreich's Ataxia (FRDA) is the commonest hereditary form of ataxia affecting the Western European population (Pandolfo, 2008). At the end of the 19th century, Nikolaus Friedreich described its pathology for the first time (Friedreich, 1863). More than one hundred fifty years after the discovery, we know more about the genetic origin and the pathological features. However, the pharmacological intervention for this disorder remains elusive (Schulz and Pandolfo, 2013).

FRDA is an autosomal recessive neurodegenerative disorder caused by an intronic GAA repeat expansion within the *FXN* gene, and the 96% of the patients are homozygous, while the remaining 4% are compound heterozygous carrying the GAA repeat mutation on one allele and point mutations on the other one (Campuzzano et al., 1996; Gelea et al., 2016). The pathology affects the Caucasian population independent of gender, and rarely affects Southeast Asian and African individuals (Delatycki and Corben, 2013).

Expanded GAA repeats may vary in length during disease progression by a process called somatic instability, which usually starts during the embryonic stage and continues along life with an elongation of the repeat expansion (De Biase et al., 2007; Long et al., 2017). Usually GAA repeat expansions have been characterized in terms of the overall repeat size, and inversely correlated with the age at onset. However, this correlation is subjected to high variability due to genetic modifiers inducing somatic instability. Therefore Al-Mahdawi and co-workers, studied DNA samples from 238 FRDA patients and 7 carriers using long-rate PCR and the restriction enzyme MboII. This study showed in few cases the presence of large interruptions in GAA expansion, while small once were occurring with an higher frequency. (Al-Mahdawi et al., 2018). Small interruptions may be related with FRDA disease phenotype, similarly to other genetic disorders, such as spinocerebellar ataxia 1 (SCA1) in which small interruptions, within the CAG repeat expansion, were relevant modifiers to the phenotype. (Menon et al., 2013).

The *FXN* gene is located in the chromosome 9q13 and the mutation impedes its transcription (Grabczyk and Usdin, 2000), reducing the level of downstream frataxin (FXN) protein in the cells. FXN is a mitochondrial protein located in the inner mitochondrial membrane and causing iron biogenesis deficiency. The range of GAA repeat expansion can span between 70 and 1700, however, the average patients have been shown to have around 600 and 900 repeats (Pandolfo, 2001).

Typically, the first symptoms of FRDA appear at young age during the first two decades of life. However, there are also late onset and very late onset cases. The time of onset depends on the length of the shorter GAA expanded allele; the longer are the expansions the earlier is the onset of the symptoms (Filla et al., 1996; Mateo et al., 2004). The diagnosis is confirmed using genetic techniques that reveal the length of the repeat expansion.

2 CLINICAL FEATURES

FRDA is characterized by progressive neurodegeneration of dorsal root ganglia (DRG) and cerebellar atrophy. Patients also develop axonal neuropathy and, at later stage of the disease, loss of the myelination (Hughes et al., 1968; Koeppen et al., 2009). The clinical features include progressive gait and limb ataxia, dysarthria, muscle weakness, lack of balance and coordination, peripheral sensory neuropathy, skeletal abnormalities such as pes cavus or equinovarus deformity of the feet and scoliosis, loss of joint position, or vibration sense (Harding, 1981; Delatycki and Corben, 2013; Parkinson et al., 2013). Cognitive function may also be affected (Mantovan et al., 2006), although concrete thinking is present; some patients lack of verbal fluency, motor, and reaction times (Corben et al., 2006; Nieto et al., 2012). FRDA is a multi-systemic disorder; therefore, patients develop non-neurological signs, such as hypertrophic cardiomyopathy, diabetes and urological problems.

Among the above manifestations, cardiomyopathy affects around 85% of patients and may result in a premature death caused through heart failure (Casazza and Morpugno, 1996; Koeppen et al., 2015). Heart disease is commonly asymptomatic; however, some patients can present general symptoms of heart failure or palpitation (Payne et al., 2011). The commonest damage is hypertrophic cardiomyopathy, which can involve either the left ventricles or interventricular septum. At molecular level, it was found a clear iron accumulation (Sanchez-Casis et al., 1977) and mitochondrial dysfunction linked to calcium dyshomeostasis (Abeti et al., 2018b).

Another clinical symptom of FRDA is diabetes mellitus that occurs in 5%–40% of the cases (McCormick et al., 2017). The mechanism by which diabetes occurs is not well understood but may be associated to an insulin resistance of peripheral tissue and a decrease of insulin secretion from β-pancreatic cells (Finocchiaro et al., 1988; Cnop et al., 2012; Cnop et al., 2013; Ran et al., 2017), probably related to a mitochondrial dysfunction.

3 THE FRATAXIN PROTEIN AND THE MITOCHONDRIAL DYSFUNCTION

The genetic mutation leads to a progressive decrease of the mitochondrial protein FXN, which resides in the inner mitochondrial membrane. FXN is a small essential protein of 210 amino acids whose structure in the C-terminal region is conserved in all organisms from bacteria to human (Adinolfi et al., 2002).

After transduction, the protein translocates into the mitochondrial membrane for two proteolytic steps. Mitochondrial processing peptidase (MPP) first cleaves the 23kDa precursor to intermediate between Gly41 and Leu42 residues. Then MPP by a second proteolytic cleavage between Lys80 and Ser81 converts the intermediate of 19 kDa to mature FXN which is the predominant form of 14.2 kDa (Kouticova et al., 1998; Schmucker et al., 2008). FXN is ubiquitously expressed, but the highest levels are found in tissues with high-energy demand, such as the nervous system, heart and liver.

Mitochondrial dysfunction in FRDA is strictly linked to frataxin functional role in the iron biogenesis (Stemmler et al., 2010). FXN deficiency leads to impaired ISCs formation which in turn affect the ISC-containing proteins (including complex I, II, and III of the mitochondrial electron transport chain and aconitase) (Pastore and Puccio, 2013).

Importantly the FXN protein controls an important step of iron mitochondrial uptake. This happens between the iron transported by mitoferrin 1 and mitoferrin 2 and the inorganic sulphide to form a transient Fe-S on the ISCU scaffold protein (Ferecatu et al., 2014). Amongst the possible physiological functions of the FXN, it was found that it acts as a ferritin-like scavenger to control the iron availability (Adinolfi et al., 2002). In addition, a high affinity was found between the FXN and the ferrochetalase suggesting an involvement of the protein into the heme group synthases (Lesuisse et al., 2003).

Puccio and colleagues found that cardiomyopathy in FRDA is associated to the ISCs formation through lack of FXN (Puccio et al., 2001). The first investigation hypothesized that frataxin was the iron donor during the ISCs biogenesis (Yoon and Cowan, 2003) but later Adinolfi and colleagues found that the role of FXN is also associated to the iron transport (Adinolfi et al., 2009). Additionally, different studies showed that FXN is linked to the electron transport chain (ETC) complex within the mitochondria, in particular to complex I, II, and III. Although, few proteins might be inhibited by the reduced level of ISCs, Abeti and colleagues were able to identify that Complex I was prevalently defective in cellular models of FRDA (Abeti et al., 2016, 2018b). The reason for this given is the fact that Complex I requires more ISCs than other complexes. Experiments conducted on the NADH redox state and pool revealed that Complex I was inhibited and that there was a lack of substrate (Abeti et al., 2018b).

Mitochondria are the sites where the cell produces energy and generates reactive oxygen species (ROS) (Nickel et al., 2014). Increased level of free radicals in cells seriously damage proteins, lipids and nucleic acids. The pathogenic mechanism triggered by the reduced production of FXN leads to the generation of oxidative stress, mitochondrial energy imbalance and an increase

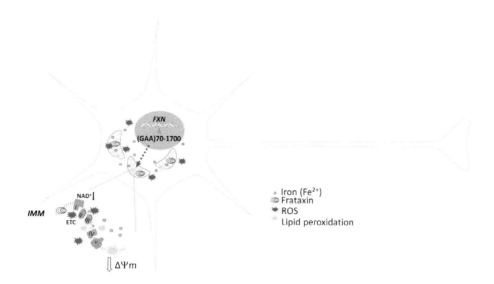

Figure 1 Description of the predicted molecular pathophysiology in FRDA neurons. Genetic mutation within the *FXN* gene impedes its transcription causing a decrease of the mitochondrial protein, FXN. Localized onto the inner mitochondrial membrane, FXN, serves iron biogenesis. FXN deficiency results in iron accumulation, oxidative stress (including lipid oxidation) and mitochondrial membrane potential ($\Delta\Psi_m$) depolarization, inducingmitochondrial dysfunction.

in lipid peroxidation, as shown in cerebellar granule neurons (CGNs), and mouse fibroblasts – see Figure 1 (Abeti et al., 2015, 2016).

Lipid peroxidation has been proven to be one of the causes of inducing neuronal death in FRDA cell models, as by pre-treating the cells with deuterium at bis-allylic sites polyunsaturated fatty acids (dPUFAs), the phenotype was rescued (Abeti et al., 2016). Another marker of pathophysiology was the deemed level of reduced glutathione (GSH) (Abeti et al., 2016, 2018a).

Amongst the therapeutic strategies applied to FRDA, over the years, antioxidants have been shown to have a positive outcome (Cooper and Schapira, 2007; Cooper et al., 2008). However, due to lack of natural history, these studies did not proceed to final conclusion.

In human fibroblasts from patients a marked sensitivity was found to pro-oxidant agents (Paupe et al., 2009; Shan et al., 2013), which was due to the nuclear factor (erythroid-derived 2)-like 2 (Nrf2) failure to translocate from cytosol to nucleus (Paupe et al., 2009). Recently, we demonstrated that Omaveloxolone (Omav) (which specifically triggers the Nrf2 pathway) had a beneficial effect on oxidative stress and mitochondrial dysfunction in FRDA models (Creelan et al., 2017; Abeti et al., 2018a). The turnover of Nrf2 is regulated through ubiquitination where Omav operates; pre-treating the cells with this compound the oxidative stress was reduced and the cell survival was promoted (Abeti et al., 2016, 2018a). Importantly, Nrf2 is also involved in mitochondrial biogenesis; indeed, Hayashi and colleagues found a reduced mtDNA copy number in knockout Nrf2 mice (Hayashi et al., 2017).

The studies on oxidative stress and mitochondrial dysfunction were important to characterize the physiological function of FXN, which could also be involved indirectly in calcium (Ca^{2+}) homeostasis. A novel study highlighted that in FRDA-like cells the Ca^{2+} level in the stores was altered (Abeti et al., 2018b). By investigating the Ca^{2+} homeostasis the authors found that both FRDA-like neurons and cardiomyocytes have a decreased ER/SR Ca^{2+} content. Neurons were unable to go back to resting state after plasma membrane depolarization, suggesting a probable defect on the sarcolemmal ATPase (SERCA), while cardiomyocytes revealed that the reduced Ca^{2+} content was due to an over activity of ryanodine receptors (RyR2). Moreover, it was found that mitochondrial also inhibited Ca^{2+} uptake, similarly to other FRDA-like models (Bolinches-Amorós et al., 2014) and that Vitamin E, a lipophilic antioxidant, restored the signal and prevented apoptosis induced by Hypoxia Reperfusion (Abeti et al., 2018b).

4 CONCLUSION

From clinical features to molecular abnormalities, we summarized the current knowledge of FRDA. The genetic mutation characterised by a GAA repeat expansion is subjected to genomic instability which correlates to the disease phenotype. We have identified that small interruptions act as genetic modifiers. Meanwhile, the progressive deficiency of FXN leads to mitochondrial energy imbalance causing alterations within specific tissues. We have demonstrated that Complex I inhibition, lipid peroxidation and aberrant Ca^{2+} signalling play an important role in the pathogenesis of FRDA. Although we have made substantial progresses in understanding the role of the mitochondrial protein, FXN, further investigations are essential to develop novel therapeutic strategies.

REFERENCES

Abeti R, Baccaro A, Esteras N, Giunti P. Novel Nrf2-inducer prevents mitochondrial defects and oxidative stress in Friedreich's ataxia models. *Front Cell Neurosci*, 2018a; 12:188.

Abeti R, Brown AF, Maiolino M, Patel S, Giunti P. Calcium deregulation: novel insights to understand Friedreich's ataxia pathophysiology. *Front Cell Neurosci*, 2018b; 12:264.

Abeti R, Parkinson MH, Hargreaves IP, Angelova PR, Sandi C, Pook MA, Abramov AY. Mitocondrial energy imbalance and lipid peroxidation cause cell death in Friedreich's ataxia. *Cell Death Dis*, 2016; 7:e2237.

Abeti R, Uzun E, Renganathan I, Honda T, Pook MA, Giunti P. Targeting lipid and mitochondiral imbalance in Friedreich's ataxia. *Pharmacol Res*, 2015; 99:344–350.

Adinolfi S, Ianuzzi C, Prischi F, Pastore C, Iannetti S, Martin S, Bonomi F, Pastore A. Bactrial frataxin CyaY is the gate keeper of iron sulfur formation catalysed by IscS. *Nat Struct Mol Biol*, 2009; 16:390–396.

Adinolfi S, Trifuoggi M, Politou AS, Martin S, Pastore A. A structural approach to understanding the iron-binding properties of phylogenetically different frataxins. *Hum Mol Genet*, 2002; 11:1865–1877.

Al-Mahdawi S, Ging H, Bayot A, Cavalcanti F, La Cognata V, Cavallaro S, Giunti P, Pook MP. Large interruptions of GAA repeat expansion mutations in Friedreich ataxia are very rare. *Front Cell Neurosci*, 2018; 12:443. doi: 10.3389/fncel.2018.00443.

Bolinches-Amorós A, Mollá B, Pla-Martín D, Palau F,González-Cabo P. Mitochondrial dysfunction induced by frataxin deficiency is associated with cellular senescence and abnormal calcium metabolism. *Front Cell Neurosci*, 2014; 8:124. doi: 10.3389/fncel.2014.00124.

Campuzzano V, Montermini L, Dolores Molt M, Pianese L, Cossée M, Cavalcanti F, Monros E, Rodius F, Duclos F, Monticelli A, Zara F, Cañizares J, Koutnikova H, Bidichandani SI, Gellera C, Brice A, Trouillas P, De Michele G, Filla A, De Frutos R, Palau F, Patel PI, Di Donato S, Mandel JL, Cocozza S, Koenig M, Pandolfo M. Friedreich's ataxia: autosomal recessive disease caused by an intronic GAA triplet repeat expansion. *Science*, 1996; 271:143–1427.

Casazza F, Morpugno M. The varying evolution of Friedreich's ataxia cardiomyophaty. Am J Cardiol, 1996; 77:895–898.

Cnop M, Igoillo-Esteve M, Rai M, Begu A, Serroukh Y, Depondt C, Musuaya AE, Marhfour I, Ladrière L, Moles Lopez X, Lefkaditis D, Moore F, Brion JP, Cooper JM, Schapira AH, Clark A, Koeppen AH, Marchetti P, Pandolfo M, Eizirik DL, Féry F. Central role and mechanisms of ⊠-cell dysfunction and death in friedreich ataxia-associated diabetes. *Ann Neurol*, 2012; 72(6):971-982. doi: 10.1002/ana.23698.

Cnop M, Mulder H, Igoillo-Esteve M. Diabetes in Friedreich ataxia. *J Neurochem*, 2013; 126(Suppl 1): 94-102. Review.

Cooper JM, Korlipara LV, Hart PE, Bradley JL, Schapira AH. Coenzyme Q10 and vitamin E deficiency in Friedreich's ataxia: predictor of efficacy of vitamin E and coenzyme Q10 therapy. *Eur J Neurol*, 2008; 15(12):1371–1379.

Cooper JM, Schapira AH. Friedreich's ataxia: coenzyme Q10 and vitamin E therapy. *Mitochondrion*, 2007; 7:S127–135.

Corben LA, Georgiou-Karistianis N, Fahey MC, Storey E, Churchyard A, Horne M, Bradshaw J, Delatycki MB. Towards an understanding of cognitive function in Friedreich ataxia. *Brain Res Bull*, 2006; 70(3):197–202.

Creelan BC, Gabrilovich DI, Gray JE, Williams CC, Tanvetyanon T, Haura EB, Weber JS, Gibney GT, Markowitz J, Proksch JW, Reisman SA, McKee MD, Chin MP, Meyer CJ, Antonia SJ. Safety, pharmacokinetics, and pharmacodynamics of oral omaveloxolone (RTA 408), a synthetic triterpenoid, in a first-in-human trial of patients with advanced solid tumors. *Onco Targets Ther*, 2017; 10:4239–4250.

De Biase I, Rasmussen A, Monticelli A, Al-Mahdawi S, Pook MA, Cocozza S, Bidichandani SI. Somatic instability of the expanded GAA triplet-repeat sequence in Friedreich ataxia progresses throughout life. *Genomics*, 2007; 90(1): 1–5.

Delatycki MB, Corben LA. Clinical features of Friedreich ataxia. *J Chil Neurol*, 2013; 27(9):1133–1137.

Ferecatu I, Gonçalves S, Golinelli-Cohen MP, Clémancey M, Martelli A, Riquier S, Guittet E, Latour JM, Puccio H, Drapier JC, Lescop E, Bouton C. The diabetes drug target MitoNEET governs a novel trafficking pathway to rebuild an Fe-S cluster into cytosolic aconitase/iron regulatory protein 1. *J Biol Chem*, 2014; 289(41):28070–28086.

Filla A, De Michele G, Cavalcanti F, Pianese L, Monticelli A, Campanella G, Cocozza S. The relationship between trinucleotide (GAA) repeat length and clinical features in Friedreich ataxia. *Am J Hum Genet*, 1996; 59:554–560.

Finocchiaro G, Baio G, Micossi P, Pozza G, Donato S. Glucose metabolism alterations in Friedreich's ataxia. *Mov Disord*, 1988; 15:1255–1258.

Galea CA, Huq A, Lockhart PJ, Tai G, Corben LA, Yiu EM, Gurrin LC, Lynch DR, Gelbard S, Durr A, Pousset F, Parkinson M, Labrum R, Giunti P, Perlman SL, Delatycki MB, Evans-Galea MV Compound heterozygous FXN mutations and clinical outcome in Friedreich ataxia. *Ann Neurol*, 2016; 79(3):485-95.

Grabczyk E, Usdin K. The GAA-TCC triplet repeat expanded in Friedreich's ataxia impedes transcription elongation by T7 RNA polymerase in a length and supercoil dependent manner. *Nucl Acids Res*, 2000; 94:7452–7457.

Harding AE. Friedreich's ataxia: a clinical and genetic study of 90 families with an analysis of early diagnostic criteria and intrafamilial clustering of clinical features. *Brain*, 1981; 104:586–620.

Hayashi G, Jasoliya M, Sahdeo S, Saccà F, Pane C, Filla A, Marsili A, Puorro G, Lanzillo R, Brescia Morra V, Cortopassi GA. Dimethyl fumarate mediates Nrf2-dependent mitochondrial biogenesis in mice and humans. *Hum Mol Genet*, 2017; 26(15):2864–2873.

Hughes JT, Brownell B, Hewer RL. The peripheral sensory pathway in Friedreich's ataxia: an examination by light and electron microscopy of the posterior nerve roots, posterior root ganglia, and peripheral sensory nerves in cases of Friedreich's ataxia. *Brain*, 1968; 91(4):803–818.

Koeppen AH, Morral JA, Davis AN, Qian J, Petrocine SV, Knutson MD, Gibson WM, Cusack MJ, Li D. The dorsal root ganglion in Friedreich's ataxia. *Acta Neuropathol*, 2009; 118(6): 763–776.

Koeppen AH, Ramirez RL, Becker AB, Bjork ST, Levi S, Santambrogio P, Parsons PJ, Kruger PC, Yang KX, Feustel PJ, Mazurkiewicz JE. The pathogenesis of cardiomyopathy in Friedreich ataxia. *PLoS ONE*, 2015; 10 (3):e0116396.

Kouticova H, Campunzano V, Koening M. Maturation of wild-type and mutated frataxin by the mitochondrial processing peptidase. *Hum Mol Genet*, 1998; 7:1485–1489.

Lesuisse E, Santos R, Matzanke BF, Knight SA, Camadro JM, Dancis A. Iron use for haeme synthesis is under control of the yeast frataxin homologue (Yfh1). *Hum Mol Genet*, 2003; 12:879–889.

Long A, Napierala SJ, Polak U, Hauser L, Koeppen AH, Lynch DR, Napierala M. Somatic instability of the expanded GAA repeats in Friedreich's ataxia. *PLoS ONE*, 2017; 12:e0189990. doi: 10.1371/journal.pone.0189990.

Mantovan MC, Martinuzzi A, Squarzanti F, Bolla A, Silvestri I, Liessi G, Macchi C, Ruzza G, Trevisan CP, Angelini C. Exploring mental status in Friedreich's ataxia: a combined neuropsychological, behavioral and neuroimaging study. *Eur J Neurol*, 2006; 13(8):827–835.

Mateo I, Llorca J, Volpini V, Corral J, Berciano J, Combarros O. Expanded GAA repeats and clinical variation in Friedreich's ataxia. *Acta Neurol Scand*, 2004; 109(1):75–78.

McCormick A, Farmer J, Perlman S, Delatycki M, Wilmot G, Matthews K, Yoon G, Hoyle C, Subramony SH, Zesiewicz T, Lynch DR, McCormack SE. Impact of diabetes in the Friedreich ataxia clinical outcome measures study. *Ann Clin Transl Neurol*, 2017; 4(9):622–631.

Menon RP, Nethisinghe S, Faggiano S, Vannocci T, Rezaei H, Pemble S, Sweeney MG, Wood NW, Davis MB, Pastore A, Giunti P. The role of interruptions in polyQ in the pathology of SCA1. *PLoS ONE*, 2013; 9:e1003648.

Nickel A, Kohlhaas M, Maack C. Mitochondrial reactive oxygen species production and elimination. *J Mol Cell Cardiol*, 2014; 73:26-33.

Friedreich N. Ueber degenerative atrophie der spinalen hinterstränge. *Arch Pathol Anat Phys Klin Med*, 1863; 26 (3–4): 391–419

Nieto A, De Nóbrega E, Montón F, Hess S, Barroso J. Cognition in Friedreich ataxia. *Cerebellum*, 2012; 11(4):834–844.

Pandolfo M. Molecular basis of Friedreich ataxia. *Mov Disord*, 2001; 16:815–821.

Pandolfo M. Friedreich ataxia. *Arch Neurol*, 2008; 65(10):1296–1303.

Parkinson MH, Boesch S, Nachbauer W, Mariotti C, Giunti P. Clinical features of Friedreich's ataxia: classical and atypical phenotypes. *J Neurochem*, 2013; 126(1):103–117.

Pastore A, Puccio H. Frataxin: a protein in search for a function. *J Neurochem*, 2013; 126(1):43–52.

Paupe V, Dassa EP, Gonçalves S, Auchère F, Lönn M, Holmgren A, Rustin P. Impaired nuclear Nrf2 translocation undermines the oxidative stress response in Friedreich ataxia. *PLoS ONE*, 2009; 4(1):e4253.

Payne RM, Pride PM, Babbey CM. Cardiomyopathy of Friedreich's ataxia: Use of mouse models to understand human disease and guide therapeutic development. *Pediatric Cardiology*, 2011; 32(3):366–37.

Puccio H, Simon D, Coss M, Criqui-Filipe P, Tiziano F, Melki J, Hindelang C, Matyas R, Rustin P, Koeing M. Mouse models for Friedreich ataxia exhibit cardiomyopathy, sensory nerve defect and Fe-S enzyme deficiency followed by intramitochondrial iron deposits. *Nat Genet*, 2001; 27:181–186.

Ran S, Abeti R and Giunti P. *Chapter 17.* Friedreich's ataxia and diabetes. Diabetes Associated with Single Gene Defects. Front Diabetes. Basel, Karger, 2017; vol 25, pp 1–10.

Sanchez-Casis G, Cote M, Barbeau A. Pathology of the heart in Friedreich's ataxia: review of the literature and report of one case. *Can J Neurol Sci*, 1977; 3:349–354.

Schmucker S, Argentini M, Carelle-Calmes N, Martelli A, Puccio H. The in vivo mitochondrial two-step maturation of human frataxin. *Hum Mol Genet*, 2008; 17 (22):3521-31.

Schulz JB, Pandolfo M. 150 years of Friedreich ataxia: from its discovery to therapy. *J Neurochem*, 2013; 126(1):1–3.

Shan Y, Schoenfeld RA, Hayashi G, Napoli E, Akiyama T, Iodi Carstens M, Carstens EE, Pook MA, Cortopassi GA. Frataxin deficiency leads to defects in expression of antioxidants and Nrf2 expression in dorsal root ganglia of the Friedreich's ataxia YG8R mouse model. *Antioxid Redox Signal*, 2013; 19(13): 1481–1493.

Stemmler TL, Lesuisse E, Pain D, Dancis A. Frataxin and mitochondrial FeS cluster biogenesis. *J Biol Chem*, 2010; 285 (35):26737–43.

Yoon T, Cowan JA. Iron-sulfur cluster biosynthesis. Characterization of frataxin as an iron donor for assembly of [2Fe-2S] clusters in ISU-type proteins. *J Am Chem Soc*, 2003; 125:6078–6084.

Index

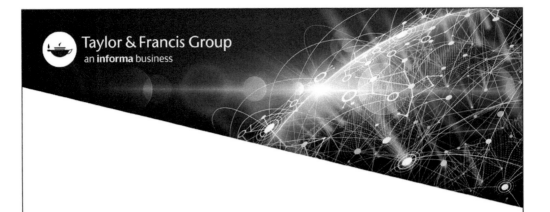